Fifth Hutton Symposium
The Origin of Granites and Related Rocks

Proceedings of a Symposium held in Toyohashi, Japan
2–6 September 2003

Edited by

S. Ishihara
Geological Survey of Japan
AIST-Central 7, 1-1-1 Higashi
Tsukuba 305-8567
Japan

W.E. Stephens
School of Geography & Geosciences
University of St Andrews
Irvine Building, North Street
St Andrews KY16 9AL, Scotland
UK

S.L. Harley
Grant Institute of Earth Science
School of Earth Sciences
University of Edinburgh
King's Buildings, West Mains Road
Edinburgh EH9 3JW, Scotland
UK

M. Arima
Geological Institute
Yokohama National University
Hodogaya-ku, Yokohama 240-8501
Japan

T. Nakajima
Geological Survey of Japan
AIST-Central 7, 1-1-1 Higashi
Tsukuba 305-8567
Japan

AF698978

WITHDRAWN

Copublished in volume format by arrangement with, and
with the permission of, The Royal Society of Edinburgh

THE
GEOLOGICAL
SOCIETY
OF AMERICA

Special Paper 389

3300 Penrose Place, P.O. Box 9140 ▪ Boulder, Colorado 80301-9140 USA

2005

Copyright © 2005, The Royal Society of Edinburgh

COPYRIGHT: It is the policy of the Royal Society of Edinburgh not to charge any royalty for the production of a single copy of any one article made for private study or research. Specific permission will not be required for photocopying multiple copies of copyrighted materials to be used for bona fide educational purposes, provided this is done by a member of the staff of the university, school, or other comparable institution, for distribution without profit to student members of that institution and provided the copies are made from the original publication. Requests for copying or reprinting of any article for any other purpose should be sent to the Royal Society of Edinburgh.

The papers in this volume were originally published together as Transactions of the Royal Society of Edinburgh: Earth Sciences, Volume 95, Parts 1 and 2 (ISBN 0 902198 34 3). Copies of that issue are available from the Royal Society of Edinburgh's distribution agents.

Copyright is not claimed on any material prepared wholly by government employees within the scope of their employment.

Copublished as a limited edition in volume format in the United States by The Geological Society of America, Inc., 3300 Penrose Place, P.O. Box 9140, Boulder, Colorado 80301-9140, USA

www.geosociety.org

Printed in Dorchester, UK

Library of Congress Cataloging-in-Publication Data

Hutton Symposium on the Origin of Granites and Related Rocks (5th : 2003 : Toyohashi-shi, Japan)
Fifth Hutton Symposium ; the origin of granites and related rocks ; proceedings of a symposium held in Toyohashi, Japan, 2–6 September 2003 / edited by S. Ishihara ... [et al.].
p. cm. -- (Special paper ; 389)
Includes bibliographical references and index.
ISBN 0-8137-2389-2 (pbk.)
1. Granite--Congresses. I. Title: Origins of granites and related rocks. II Ishihara, Shunso. III Title. IV. Special papers (Geological Society of America) ; 389.

QE462.G7H88 2003
552'.3—dc22

2005040012

Cover: Kurama stone of Kyoto City, Sanyo Belt. The Kurama stone consists of weathered boulders that belong to late Cretaceous magnetite-series quartz diorite to granodiorite at north of Kyoto City. The granitoids are unusual because of the adakitic composition occurring in the Inner Zone of southwest Japan, as well as their very high content of pyrrhotite (<2%). The pyrrhotite has $\delta^{34}S$ −4%, indicating that sedimentary wall rocks were incorporated into the granitic magmas. This sulfide decomposes quickly, giving rise to different colors on the boulder surface every year. This characteristic was loved by noble people in Kyoto, who used the boulders in their Japanese gardens as tools and for viewing during tea ceremony (Shunso Ishihara).

10 9 8 7 6 5 4 3 2 1

FIFTH HUTTON SYMPOSIUM
THE ORIGIN OF GRANITES AND RELATED ROCKS

Proceedings of a Symposium held in
Toyohashi, Japan
2–6 September 2003

Edited by S. Ishihara, W. E. Stephens, S. L. Harley,
M. Arima and T. Nakajima

ISBN 0 902198 34 3

Printed in the United Kingdom by Henry Ling Limited, at the Dorset Press, Dorchester, DT1 1HD

CONTENTS

W. E. STEPHENS and S. L. HARLEY
Preface v

S. ISHIHARA, M. ARIMA and T. NAKAJIMA
On the Fifth Hutton Symposium vii

B. W. CHAPPELL
Towards a unified model for granite genesis 1

J. L. VIGNERESSE
A new paradigm for granite generation 11

S. ISHIHARA
The redox state of granitoids relative to tectonic setting and earth history: the magnetite-ilmenite series 30 years later 23

M. BROWN
The mechanism of melt extraction from lower continental crust of orogens 35

S. N. OLSEN, B. D. MARSH and L. P. BAUMGARTNER
Modelling mid-crustal migmatite terrains as feeder zones for granite plutons: the competing dynamics of melt transfer by bulk versus porous flow 49

A. WHITTINGTON, P. RICHET, H. BEHRENS, F. HOLTZ and B. SCAILLET
Experimental temperature-X(H_2O)-viscosity relationship for leucogranites and comparison with synthetic silicic liquids 59

P. I. NABELEK and M. LIU
Petrologic and thermal constraints on the origin of leucogranites in collisional orogens 73

S. L. HARLEY and P. THOMPSON
The influence of cordierite on melting and mineral-melt equilibria in ultra-high-temperature metamorphism 87

S. ERDMANN, D. B. CLARKE and M. A. MACDONALD
Origin of chemically zoned and unzoned cordierites from the South Mountain and Musquodoboit Batholiths, Nova Scotia 99

T. KAWAKAMI
Tourmaline and boron as indicators of the presence, segregation and extraction of melt in pelitic migmatites: examples from the Ryoke metamorphic belt, SW Japan 111

B. W. CHAPPELL, A. J. R. WHITE, I. S. WILLIAMS and D. WYBORN
Low- and high-temperature granites 125

V. JANOUŠEK, F. FINGER, M. ROBERTS, J. FRÝDA, C. PIN and D. DOLEJŠ
Deciphering the petrogenesis of deeply buried granites: whole-rock geochemical constraints on the origin of largely undepleted felsic granulites from the Moldanubian Zone of the Bohemian Massif 141

T. SHIMURA, M. OWADA, Y. OSANI, M. KOMATSU and H. KAGAMI
Variety and genesis of the pyroxene-bearing S- and I-type granitoids from the Hidaka Metamorphic Belt, Hokkaido, northern Japan 161

Y-J. JWA
Possible source rocks of Mesozoic granites in South Korea: implications for crustal evolution in NE Asia 181

C. J. BRYANT, B. W. CHAPPELL, V. C. BENNETT and M. T. MCCULLOCH
Lithium isotopic compositions of the New England Batholith: correlations with inferred source rock compositions 199

U. POLLER, D. P. GLADKOCHUB, T. V. DONSKAYA, A. M. MAZUKABZOV, E. V. SKLYAROV and W. TODT
Timing of Early Proterozoic magmatism along the Southern margin of the Siberian Craton (Kitoy area) 215

J-S. QIU, D-Z. WANG, B. I. A. McINNES, S-Y. JIANG, R-C. WANG and S. KANISAWA
Two subgroups of A-type granites in the coastal area of Zhejiang and Fujian Provinces, SE China: age and geochemical constraints on their petrogenesis 227

C-H. CHEN, W. LIN, C-Y. LAN and C-Y. LEE
Geochemical, Sr and Nd isotopic characteristics and tectonic implications for three stages of igneous rock in the Late Yanshanian (Cretaceous) orogeny, SE China 237

T. NAKAJIMA, H. KAMIYAMA, I. S. WILLIAMS and K. TANI
Mafic rocks from the Ryoke Belt, southwest Japan: implications for Cretaceous Ryoke/San-yo granitic magma genesis 249

Y. SHIMIZU, S. ARAI, T. MORISHITA, H. YURIMOTO and F. GERVILLA
Petrochemical characteristics of felsic veins in mantle xenoliths from Tallante (SE Spain): an insight into activity of silicic melt within the mantle wedge 265

B. E. HARPER, C. F. MILLER, G. C. KOTEAS, N. L. CATES, R. A. WIEBE, D. S. LAZZARESCHI and J. W. CRIBB
Granites, dynamic magma chamber processes and pluton construction: Aztec Wash pluton, Eldorado Mountains, Nevada, USA 277

K. BENN
Late Archaean Kenogamissi complex, Abitibi Subprovince, Ontario: doming, folding and deformation-assisted melt remobilisation during syntectonic batholith emplacement 297

S. W. RICHARDS and W. J. COLLINS
Growth of wedge-shaped plutons at the base of active half-grabens 309

K. SATO, S. V. KOVALENKO, N. P. ROMANOVSKY, M. NEDACHI, N. V. BERDNIKOV and T. ISHIHARA
Crustal control on the redox state of gramitoid magmas: tectonic implications from the granitoid and metallogenic provinces in the circum-Japan Sea Region 319

C. J. R. HART, J. L. MAIR, R. J. GOLDFARB and D. I. GROVES
Source and redox controls on metallogenic variations in intrusion-related ore systems, Tombstone-Tungsten Belt, Yukon Territory, Canada 339

R. V. METCALF
Volcanic-plutonic links, plutons as magma chambers and crust-mantle interaction: a lithospheric scale view of magma systems 357

G. A. R. GUALDA, D. L. COOK, R. CHOPRA, L. QIN, A. T. ANDERSON JR and M. RIVERS
Fragmentation, nucleation and migration of crystals and bubbles in the Bishop Tuff rhyolitic magma 375

AUTHOR INDEX 391

Transactions of the Royal Society of Edinburgh: Earth Sciences, **95**, v–vi, 2004

Preface

W. E. Stephens and S. L. Harley

Much has been written in recent years about James Hutton and his *Theory of the Earth*, published between 1785 and 1795. The Hutton Symposia, the Proceedings of which are also published by the Royal Society of Edinburgh in *Transactions*, specifically celebrate his contributions to understanding the origin of granites, one of the cornerstones of his theory. The recent biography of Hutton (Repcheck 2003) is a most welcome analysis of the historical context of his work. It shows how his ideas can be traced directly to Newton by his application of the laws of science to understanding nature without recourse to unexplainable factors, but also how these ideas were revolutionary in rejecting the established religious constraints on the age of the Earth and the duration of surface and internal processes. Jack Repcheck argues that had Hutton been a better communicator of his own theory, then his name might now rank in public esteem among rare scientific geniuses such as Newton, Darwin and Einstein.

The present Hutton volume, the fifth in the series, testifies that after more than two centuries of research the 'problem' of the origin of granites is still very much with us, and that the considerable advances in understanding over that time serve to confirm that a singular 'origin' is no longer scientifically tenable, notwithstanding the 'Huttonian' title our volume. The volume contains abundant evidence for the complexity of the natural processes involved in generating granite magma and controlling its ascent and final emplacement, as well as in influencing its crystallisation and hydrothermal activity. These combine in so many different ways that we now recognise the enormous diversity of granites as different manifestations of thermal perturbation in the lithosphere. This would please James Hutton, for magmatism was central to his theory as the primary mechanism of transporting heat to the near-surface, where it acted to consolidate sediments (although we now know that burial is sufficient for that purpose). He rigorously applied scientific method by making and testing predictions and, in the case of granites, he anticipated the main features of a contact between granite and country rocks and knew just where he should find such a contact to inspect in the field (Glen Tilt). Indeed, he was an enormous enthusiast for examining rocks *in situ* and transmitted that enthusiasm to those accompanying him in the field, including his friend and pioneer chemist, Joseph Black, and other leading intellectuals of that special time known as the Scottish Enlightenment.

This volume is a compilation of invited papers presented at the Fifth Hutton Symposium on the Origin of Granites and Related Rocks held at Toyohashi, Japan in early September 2003. In the following 'Introduction', Shunso Ishihara and his colleagues describe the very successful conference and accompanying field excursions. All participants are enormously grateful to the organisers for arranging an excellent scientific programme and, for the many participants from beyond Asia, a wonderful cultural experience. Papers in this volume address all the major aspects of granite studies, some of which have a strong flavour of Far Eastern granite geology, reflecting the location of the meeting.

Opening the volume are three papers that take rather different approaches to generic models of granite magmatism. Chappell proposes a unified model of crustal granites based on the interaction between source rock composition and varying degrees of partial melting, fractional crystallisation and restite separation. Vigneresse reviews the major processes of melting, melt segregation, ascent and emplacement and links them in a new paradigm to explain the diversity of granite types. Ishihara surveys the occurrences of reduced and oxidised (ilmenite and magnetite series) granites in both time and space, and shows how they can be linked to global tectonic processes and temporal changes in redox conditions (a theme developed on a regional scale later in the volume).

The relationships between melt production, fluid activity and melt migration and segregation are central to modelling of the deformation and thermal evolutions of orogenic belts. These issues are explored from a variety of directions in a series of contributions on the anatectic regime as monitored through experiment and natural rock observation. Brown presents an overview of current approaches to modelling how melt migrates and is extracted from the deep crust, focusing in particular on the concept of ductile fracture propagation. Olsen *et al.* also consider the physical links between zones of melt production and melt transfer, concentrating on competing mechanisms for melt transfer. The efficiencies of these physical processes are strongly influenced by melt composition, especially water contents, which impact strongly on viscosities. This aspect is addressed by Whittington *et al.* through experiments on leucogranites and related modelling of the T–H_2O–viscosity relations of crustal melts. Nabelek & Liu also consider leucogranites, concluding from thermal modelling that these melts are generated in collisional orogenesis only when shear heating along crustal-scale shear zones occurs, providing an effective mechanism for raising temperatures into the dehydration-melting regime. The mineral-melt systematics of this regime, and even higher-temperature deep crustal metamorphism, are addressed by Harley & Thompson. These authors illustrate the role that hydrous and CO_2-bearing cordierite plays in limited the amounts of melt produced through biotite dehydration melting, and combine cordierite-melt volatile partitioning data with mineral equilibria in model pelite systems to evaluate fluid activities in high- and ultrahigh-temperature migmatites. Cordierite is also recognised as an important mineral involved in melting and melt crystallisation by Erdmann *et al.*, who establish chemical and textural criteria for discriminating cordierites formed at different stages of the magmatic evolution of granitoids. Kawakami adopts an alternative approach to establishing the progress of melting in pelites and politic migmatites, using tourmaline as an index mineral for the onset of melting and identifying melt segregation, loss and wall-rock reaction using boron as a chemical tracer.

Source rock controls on magma composition continue to be a major focus of research. Chappell *et al.* develop their model for low- and high-temperature granites in relation to source compositions and thermal conditions. Janousek *et al.* find evidence for old upper crustal granites taken to deep crustal depths for the sources of some granulites of the Bohemian Massif. The occurrence of pyroxene bearing tonalities of both I- and S-type affinity in the Hikada Belt of Hokkaido is attributed by Shimura *et al.* to granulite source rocks of different bulk composition. A detailed radiogenic isotope study of Mesozoic granites in South Korea by Jwa provides strong evidence for regional variation in source rock composition and for vertical stratification of the crust in both age and composition. Bryant *et al.* examine the variations in Li isotopes with other petrogenetic indicators in the New England Batholith of Australia and show that this relatively novel isotope system can provide useful new constraints on source rocks and crustal processes.

The relationships between geochronology, granite geochemistry and tectonothermal events are providing new insights into crustal and mantle processes on the lithospheric scale. Poller *et al.* use precise U–Pb geochronology to show that Archaean crustal material is reworked in the Proterozoic (with a minor mantle influence) along the edge of the Siberian craton. The tectonic significance of A-types is still relatively poorly understood and Qiu *et al.* describe different subgroups in the Cretaceous of Fujian Province (SE China) which they attribute to distinct differentiation processes controlled by varying fluorine contents. Geochemistry is used by Chen *et al.* to document the changes from syn-orogenic to post-orogenic crustal granites, which they relate to early lithospheric thickening by basaltic underplating followed by delamination. Geochemistry is also used by Nakajima *et al.* to interpret the relationships between mafic rocks and granites in the Ryoke Belt of Southwest Japan as indicating a major episode of crustal growth during the Cretaceous. Shimizu *et al.* studied felsic veins preserved in peridotite xenoliths at Tallante, Spain and recognised adakitic signatures in melts generated by partial melting of a detached slab beneath the Betic area during the Tertiary.

Insights into the dynamics of magma chamber processes were gained from the study of petrological and compositional variations in a tilted section of the Aztec Wash pluton, Nevada by Harper *et al.*, whilst Richards & Collins address the age-old 'room problem' by using combined field and geophysical data to show that plutons from the Bega batholith of the Lachlan orogen (Australia) are wedge-shaped bodies emplaced in active extensional environments. Deep-level interactions between magmas, melts and host rocks in a complex dynamic environment are described by Benn for the late Archaean Kenogamissi complex (Ontario) and evidence is presented for the magmatic differentiation of batholiths by remobilising and extracting of fractionated liquids.

Granite terranes are sometimes economically important for their associated metalliferous deposits and the significance of redox conditions is highlighted by two contributions to this volume. Sato *et al.* show how the redox state of granites is correlated with the nature of the associated country rocks, which in turn determine the associated mineralisation in the circum-Japan Sea region. Hart *et al.* find strong evidence to attribute the contrasting metallogeny associated with three distinct plutonic suites in the northern Canadian cordillera to different source rocks and redox controls.

The volume is completed by two studies of volcanic terranes. Compositional variations in Tertiary volcanics of the northern Colorado River region (USA) are related by Metcalf to contemporaneous plutonic rocks through magma chamber processes, and Gualda *et al.* present a study of the crystals and bubbles in pumice samples from the large volume Bishop Tuff rhyolite, proposing a scenario of rising bubbles and sinking crystals that may have considerable relevance for the study of granite plutons and batholiths.

Although James Hutton would be unfamiliar with many of the techniques used in these studies, he would recognise in most contributions the need for very long periods of time for processes to operate. Before Hutton, such time was not available, and it is this part of his enormous scientific legacy that has enabled us to discover so much about the origin of granites.

Reference

Repcheck, J. 2003. *The Man Who Found Time: James Hutton and the Discovery of the Earth's Antiquity*, 247 pp. Croydon, UK: Simon & Schuster Ltd.

On the Fifth Hutton Symposium

Shunso Ishihara, Makoto Arima and Takashi Nakajima

The Fifth Hutton Symposium on the Origin of Granites and Related Rocks was held from 2–6 September 2003, at Aichi University, Toyohashi, central Japan, an area underlain by Cretaceous Ryoke granites and metamorphic rocks. It was the first of the series to be held in Asia and the first held in an active island arc setting, past meetings having been held in Edinburgh, Scotland (1987), Canberra, Australia (1991), College Park, Maryland, USA (1995) and Clermont-Ferrand, France (1999). The majority of participants were, as was to be expected, from Asia: 94 from Japan, 13 from Korea, 11 from China, nine from India and five from Taiwan, although the participation from China was adversely affected by the outbreak of the SARS virus. Other participants were primarily from Australia (17), the USA (15), Canada (six), Germany (six), France (five), and the UK (four). The total number of attendees was 205, of whom 50 were students. It is hoped that the Symposium will have given encouragement and stimulation to the study of granites, particularly to these young geologists.

To demonstrate the complexity of Japanese geology, several short field excursions of three to five days were organised for delegates. Four of these took place before the Symposium and a further one after it. Some 51 participants, of whom about 90% were non-Japanese, joined these excursions. In addition, there were four one-day mid-Symposium field trips, which were enjoyed by 110 participants in total.

Two hundred and ten abstracts were originally submitted, and the authors of 116 of these applied to present orally, as oral presentation has become a pre-requisite for governmental support in some Asian countries. In an attempt to accommodate as many of these as possible, a time limit of 20 minutes was assigned for each presentation, resulting in 64 oral presentations in total over the week of the Symposium. This was around double the number at previous Symposia (there were 35 oral presentations in Maryland and 32 in Clermont-Ferrand).

The Opening Session of the Fifth Hutton Symposium comprised three invited speakers. First, Shunso Ishihara introduced his 30-year study of granitoids and emphasised that the redox state of granitoids reflects that of both the source material and the tectonic setting. He further addressed the secular variation of the redox state of granitoids throughout Earth's history and found the first oxidised magma occurred at 3·1 Ga. He suggested that such variation was related to the evolution of the Earth's hydrosphere.

Secondly, Jon Blundy presented his latest experimental studies of the lower crustal differentiation of hydrous basalts and discussed the generation of silicic calc-alkaline magmas in the lower crust. He emphasised the important role of water in the ascent and emplacement of granitic magmas. He argued that crystalline resorption processes during the ascent of calc-alkaline magmas, in which water-rich magma is superheated, leads to partial or complete digestion of entrained crystalline phases.

In the third talk of this opening session, Bruce Chappell emphasised that the dependence of both the physical and chemical properties of magma on source compositions will provide a unifying basis for understanding the origins of granites. He defined high- and low-temperature granites and suggested that high-temperature granites formed from complete melts where zircon was soluble and that fractional crystallisation was an important process in their evolution. By contrast, in low-temperature granites, zircon was saturated and the magma could never have been completely molten. These rocks evolved by the removal of entrained crystals of restite.

Following this Opening Session, ten thematic Sessions on differing topics acted as foci of the meeting. These were:

A1: UHT and UHP Metamorphism, Melting Processes under Extreme Conditions and Crustal Evolution;
A2: Anatexis and Melt Segregation in Continental Lower Crust;
B1: Processes in Granitic Magma Chambers;
B2: Links between Volcanic and Plutonic Processes;
C1: Juvenile Granitoids in a Subduction Setting;
C2: Granitoids with Intra-crustal Reworking Processes;
C3: Tectonic Controls on the Generation and Intrusion of Granitic Magmas;
C4: New Insights into the Asian Granites;
D1: Behaviour of Metals and Volatiles during Magmatic Processes; and
D2: Granitoid-series and Metallogenesis.

Contributions to Session A1 (6 orals, 7 posters) highlighted the growing body of mineralogical and petrological evidence that UHT metamorphism is common in the lower crust. Furthermore, it is clear that the pressures of melting and peritectic mineralogy exert a strong control over melt diversity, whereas trace components significantly enhance stability of the same mineral phases under extreme conditions. The origin of the thermal perturbation necessary for UHT metamorphism remains elusive, but mantle heat is implicated.

Session A2 (7 orals, 15 posters) addressed more directly the issues of crystal melting and melt extraction from the lower crust, as well as the influence of source composition and magmatic processes as recorded by refractory accessory phases. Source composition may be investigated via migmatites and granulites, particularly the questions of segregation and initial transport of melt out of the system. Source composition, melting reactions and associated processes control source fertility, whereas rapid overstepping of reactions may be significant in some circumstances. Variation in melt composition may be controlled at source by both protolith composition and degree of melting, whereas in the shallow crust magmatic processes such as crystal fractionation may be responsible. Shear hearing may be significant in achieving anatectic temperatures in crustal-scale shear zone systems.

Session B1 (6 orals, 20 posters) focused on mafic magma replenishment of silicic magma chambers, the effects of assimilation with fractional crystallisation at the floors of magma chambers, and the role of degassing in the development of the final textures of granites. In addition, the new use of the CSD (crystal size distribution) method for understanding the development of textures in both volcanic and plutonic rocks, and petrographic and chemical criteria for understanding the origin of cordierite grains in granite were introduced. The largest number of posters focused on mafic magma replenishment into silicic magma chambers, producing sheets and enclaves within the granite and even remobilising some existing granite. These studies will continue to provide new insights.

Session B2 (6 orals, 7 posters) focused on the relationship between volcanic and plutonic processes, especially in terms of the temporal and spatial relationships between these processes. Missing links between them were also an important subject for this session. Issues discussed included the comparison of mafic enclaves in both volcanic and plutonic rocks, the petrographical characteristics of solidifying granite, the precise time-space analysis of extremely young exposed volcano-plutons and the lithosphere-scale view of magma systems. In order to understand the link between volcanic and plutonic processes, questions were addressed such as how shallow level magma chambers are deformed when a caldera collapses, and the similarities and differences between magma pumping systems in volcanoes and plutons.

Session C1 (7 orals, 14 posters) discussed juvenile granitoids, which represent granitic magma derived either directly from the mantle, or through melting of mantle-derived basaltic crust. This session addressed several important issues, including (1) the similarities and differences between Archaean and recent slab melt products (TTGs and adakites, respectively) and (2) the subduction zone factory, involving extensive interactions between oceanic lithosphere, continental crust, mantle lithosphere and deep asthenospheric mantle, and possibly core. The role of subduction for both crustal generation and mantle evolution was discussed. The systematics of Li isotopes in the granites of the New England area, Australia, was an example of their application to granite petrogenesis. In addition, the petrogenesis of granitic rocks in the oceanic environment, including plagiogranites from ophiolite complexes and intra-oceanic arcs, was discussed. Although major advances in understanding subduction zone processes have been made in the last 10 to 20 years, the relative roles of partial melting versus the more commonly accepted dehydration of subducted slabs and sediments needs further study. Fruitful avenues of research include melt inclusion studies from high-Mg andesite, adakites and Nb-enriched basalts, as well as continuing studies of mantle wedge xenoliths. Studies detailing the diversity of magmatism in modern and ancient arc environments would also prove fruitful.

Session C2 (7 orals, 17 posters) discussed non-juvenile granites formed in varous tectonic settings, including subduction-related ones. Post-collisional granites were discussed in relation to their generation in environments where the roll-back of the subducting lithosphere took place due to a change in the convergence rate following collision with the continental lithosphere. Most of the papers in this session dealt with the geochemistry or geochronology of granites in the context of their tectonic settings. The role of gabbroids in the formation of batholithic granites was discussed with reference to Southwest Japan. The $\delta^{18}O$ values of zircon in A-type granites of eastern China are uniformly low (3·5–5·1‰), suggestive of partial melting of crustal materials which suffered from high-temperature alteration. Most of the geochemical studies discussed the relative proportions of the contributions of oceanic crust, mantle and continental crustal materials. Systematic zircon U/Pb geochronological studies and their tectonic interpretation were presented by several authors. The Kaymaz granite of western Turkey exhibits such high radioactivity that a possible radioactive hazard of this granite was highlighted.

Session C3 (6 orals, 6 posters) discussed the origin of the tabular shape of granite plutons common to different parts of the world and attributed it to brittle/ductile boundaries in the Earth's crust and crustal magma traps. Case studies of plutonic intrusions in various tectonic settings, including transcurrent shear zones, extensional, compressional and island-arc settings, were presented. Contributors in this session were increasingly aware of the importance of the rheology of magmas, and used multidisciplinary approaches that combined field studies and numerical modelling to understand the mechanism of ascent and emplacement of granite plutons. New techniques for non-destructive fabric analysis and phase topology, and sophisticated statistical analysis of granite fabrics were introduced. Two contributions involved AMS techniques to elucidate the detailed structural setting of island arc granites.

Session C4 (5 orals, 14 posters) presented geochemical studies of granitoids of the Central Asia Fold Belt. Abundant juvenile granitoids were shown to be present, implying the amalgamation of primitive island arcs in that region. A-type granites along the eastern margin of the Asian continent are indeed 'orogenic granites', which are the latest phase of the orogenic activity. Mesozoic granitoids and ore deposits of the Russian Far East were summarised for the first time in terms of fluid and melt inclusion studies, and those of the Circum-Japan Sea region were updated by modern petrological and isotopic investigations.

Session D1 (5 orals, 4 posters) discussed the halogen contents of hydrous rock-forming minerals such as mica and apatite, as well as the unmixing of melts consisting of Cu-Cl-salt-hydrate. One paper presented as a poster described the behaviour of metals in magmatic processes related to mineralisation, and other posters discussed hydrothermal alteration. An oral paper presented a case study on one of the oldest copper vein deposits in India.

Session D2 (5 orals, 8 posters) focused on petrography (two papers), redox states on a regional scale (two papers), and the petrochemical characteristics of granitic rocks associated with Au mineralisation. Posters reported on the origin of oxidised and reduced granitoids, genetic implications and metallogenesis in Korea and Australia, exploration of gold deposits in China, geochemistry of granitoids of Siberia, transport of granitic melts, and the hydrothermal alteration of skarn.

From the 63 oral presentations, 30 papers were submitted for peer review, and 27 of these are now published in this Proceedings volume.

The Fifth Hutton Symposium was organised by a Committee comprising Shunso Ishihara (President, Geological Survey of Japan), Takashi Nakajima (Secretary General, Geological Survey of Japan), Makoto Arima (Yokohama National University, Programme Committee Chair), Yoshikuni Hiroi (Chiba University), Akira Imai (Kyushu University), Masaki Takahashi (Nihon University), Toshio Kutsukake (Aichi University, Symposium-site Chair), Toshiaki Shimura (Niigata University, Excursion Chair) and Yoshinobu Kawano (Saga University, Information Coordinator).

The Organising Committee is grateful for the patronage of Aichi University, The Geological Survey of Japan/AIST, The Daiko Foundation, The Inoue Foundation for Science, The Iwatani Naoji Foundation, The Tokyo Geographical Society, The Toyohashi Association for Kanko and Convention and Toyohashi City Council. Also co-sponsoring the meeting were The Geological Society of Japan, The Geochemical Society of Japan, The Japanese Association of Mineralogists, Petrologists and Economic Geologists, The National Committee for Mineralogy, Science Council of Japan, The Society of Resource Geology and The Volcanological Society of Japan.

On behalf of the Fifth Hutton Symposium Organising Committee, we wish to acknowledge the patronage and sponsorship of the above organisations, and also personal help provided by Bruce Chappell, Mike Brown and Ed Stephens. Without their assistance, the Fifth Hutton Symposium could not have been organised successfully.

The venue for the Sixth Hutton Symposium was decided, by voting, on the final day in Toyohashi and it is to be held in South Africa. We look forward to meeting all the delegates there again in 2007.

Transactions of the Royal Society of Edinburgh: Earth Sciences, **95**, 1–10, 2004

Towards a unified model for granite genesis

B. W. Chappell

ABSTRACT: Most granites result from partial melting within the crust. Granite melts produced at the lowest temperatures of partial melting mainly comprise close to equal amounts of the haplogranite components *Qz*, *Ab* and *Or*, with H_2O. Many felsic granites were formed by partial melting under such conditions and are low-temperature types, with crystals of zircon and other restite minerals present in the initial magma. Such magmas evolve in composition, at least initially, through fractionation of that restite. If one of the four haplogranite components either becomes depleted or too low in amount to contribute further to the melt, then melting may proceed to higher temperatures without a contribution from that component. Melting will advance to significantly higher temperatures if there is a critical deficiency in one or more components and a high-temperature granite magma forms, in which zircon is completely soluble. Such magmas are extracted from the source in a completely molten state and may evolve by fractional crystallisation. They are monzonitic, tonalitic or A-type, depending on whether the critical deficiency occurred in the *Qz*, *Or* or H_2O component. If the *Ab* component is critically deficient, as in pelitic rocks, the rocks may be infertile for granite production. The control that source rock compositions exert on both the physical and chemical properties of granite magmas provides a unifying element in granite genesis.

KEY WORDS: Crystal fractionation, fractional crystallisation, haplogranite, high-temperature granite, low-temperature granite, partial melt, restite fractionation, tonalite.

Since the experimental studies of Tuttle & Bowen (1958), it has generally been accepted that granites are the products of magmatic processes taking place within the Earth's crust. These rocks all have in common that they formed as a result of processes involving reactions (melting or crystallisation) between crystals and silicate melt at the lowest magmatic temperatures appropriate to the bulk chemical composition of a particular system. Therefore, it would be surprising if we did not ultimately find that the genesis of granites can be represented in a more unified way than has been the case up to the present time. The present paper suggests a way in which that might happen. It is illustrated with examples of various Australian granites, mainly from the mid-Palaeozoic Lachlan Fold Belt (LFB), but also from the mainly late-Palaeozoic New England Fold Belt. While the present author's experience of granites has principally been in those two areas, the development of the ideas in this paper has also drawn on personal observations of rocks in other places. Most important among those have been the North American Cordillera, and the Caledonian and Hercynian fold belts of Europe. Granites of the LFB are diverse and they encompass a wide range of rock types, including approximately equal amounts of I- and S-type granites, and a small amount of distinctive A-type granites. But there are important varieties of granites which are not present in the LFB, and therefore, must be examined elsewhere. These include the Cordilleran-like tonalites and, of course, their older compositional analogues, the Archaean tonalite–trondjemite–granodiorite associations (TTGs).

1. Introduction

The Hutton Symposia, the first of which was held in 1987, celebrate the life and work of James Hutton, in particular his observations on granites. In 1785, he observed that dykes of granite in Glen Tilt, Scotland, are intrusive into the adjacent schists, and he made similar observations in Galloway and Arran during the following two years. This showed that granites are plutonic rocks, and not deposits from a primeval ocean as the Neptunists had proposed. However, it was some 50 years before Hutton's views on granites and more generally on the evolution of the Earth (Hutton 1788) became widely accepted (see Repcheck 2003). From that time, an intense debate developed over the origin of these plutonic rocks, i.e. whether granites have an origin that is igneous (magmatic) or metasomatic (produced by granitisation). This debate was only resolved satisfactorily by the publication of the now classic memoir of Tuttle & Bowen in 1958, which undoubtedly has been the most significant single contribution to our understanding of the origin of granites and related rocks.

2. Tuttle & Bowen and haplogranite compositions

Publication of the Tuttle & Bowen (1958) memoir resolved arguments about the origins of granites firmly in favour of the magmatic view. These authors showed that: (1) hydrous silicate melts which occur at the lowest temperatures contain close to equal proportions of the *Qz*, *Ab* and *Or* components; and (2) the most felsic granites and rhyolites containing more than 80% $Qz+Ab+Or$, the haplogranites, have compositions which are very close to those lowest-temperature melts (Fig. 1). Thus, the study of Tuttle & Bowen (1958) showed that haplogranites were once H_2O-bearing melts, or cumulative rocks from such melts (see section 2.3 below), with compositions determined during equilibrium between quartz and feldspars at magmatic temperatures. The fact that felsic granites and related volcanic rocks have that distinctive composition was the defining observation that resolved the Granite Controversy of the mid-twentieth century in favour of the magmatic view. The ideas of granitisation foundered when confronted with the restricted ranges of major element abundances observed in felsic granites, and this marked the beginning of modern, post-granitisation, studies of granites.

The experiments of Tuttle & Bowen (1958) were carried out under H_2O-saturated conditions, and it is likely that most granite melts do not form under those conditions (e.g. Clemens 1984), although they may evolve to H_2O-saturation as a result

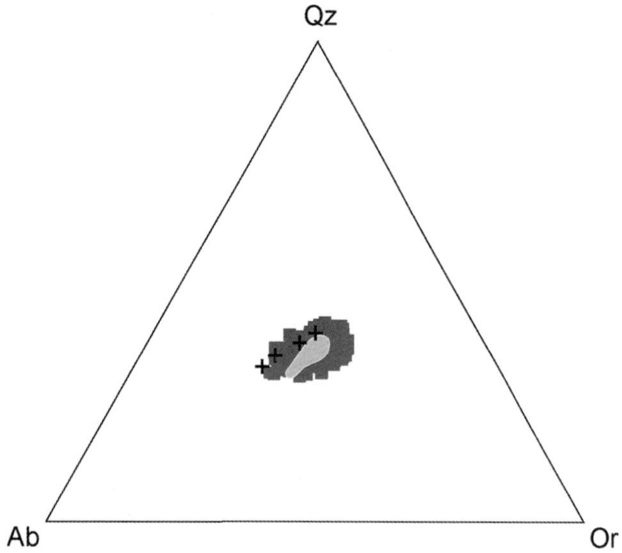

Figure 1 The *Qz-Ab-Or* face of the tetrahedron for the system *Qz-Ab-Or*-H_2O studied by Tuttle & Bowen (1958), with positions of the ternary temperature minima at 50, 100, 200 and 300 MPa shown as crosses in positions of increasing *Ab* content. The central part of the contoured field of 511 analysed haplogranites drawn by the above authors is also shown as a lightly shaded area. The darker surrounding area contains the plotted positions of all 445 analysed granites of the Lachlan Fold Belt which contain more than 90% *Qz+Ab+Or* on a normative corundum-free basis.

of the removal of anhydrous phases during fractional crystallisation. Experiments subsequent to Tuttle & Bowen (1958) have shown that the position of the minimum melt compositions in the haplogranite system do not alter greatly under H_2O-undersaturated conditions, moving towards the right in Figure 1 as the a_{H_2O} decreases (Johannes & Holtz 1990). The present paper will refer to melts produced under all of those conditions as Quasi Tuttle & Bowen (QTB) melts.

Bowen had been the chief proponent of a magmatic origin for granites. He took the view that granites result from fractional crystallisation of basaltic magma, perhaps somewhat modified by the solution of granite and other salic material (Bowen 1949). He did not regard granite magmas as likely to be primary, but noted (Bowen 1947) that studies showing that granite is a late-crystallising residuum of fractional crystallisation imply that it would, likewise, be an early product of 'selective fusion' of appropriate material, which we now generally call partial melting. However, the unequivocal establishment of the magmatic nature of granites by Bowen and his colleagues in fact rapidly lead to the almost universal recognition that granites are not only magmatic, but also that, generally, those magmas are primary. Such concepts were incorporated by Read (1948) in his Granite Series, which evolved out of his earlier view that granites are non-magmatic. The Granite Series proposed that all granites are related to each other in their process of formation and are the result initially of the partial melting of metasedimentary rocks. These are what we now call the S-type granites (Chappell & White 2001). It is significant that Read did not recognise the existence of related volcanic rocks, which are now recognised as an important component of such magmatism.

The restriction in compositions of felsic granites is illustrated by the data from the granites of the LFB also represented in Figure 1. These comprise 445 analysed samples from that region containing more than 90% of the three normative components *Qz*, *Ab* and *Or*, from a total of 2220 analyses of granites and other rocks with which they are compositionally gradational. Most haplogranites are peraluminous, those of S-type origin more strongly so, and therefore, the analyses have been selected on a normative corundum-free basis, so as not to discriminate against the S-type granites. The samples which are plotted in Figure 1 include 228 granites which have been classified as I-type, 174 as S-type and 22 as A-type, with 21 unassigned. These data again show the relatively restricted range of haplogranite compositions, in terms of the *Qz*, *Ab* and *Or* components. The point could be made that the compositions represented in Figure 1 are from only one region, but the result would have been very similar if felsic granites from any region were plotted. The LFB simply provides a large number of modern analyses of such rocks to serve as the basis for a more general discussion. Also, the LFB compositions are from a variety of source rocks. They include the products of partial melting of older igneous rocks and sedimentary rocks, in which the partial melts underwent little or no compositional change. Some resulted from fractional crystallisation of less felsic melts, both I- and S-type. We will now examine the processes which constrain those compositions.

2.1. Felsic granites as products of fractional crystallisation or fractional melting

The observations of Tuttle & Bowen (1958) showed that the haplogranites must have attained their compositions within the Earth's crust, where quartz and feldspars are stable. It has been noted above that this could involve an origin by either fractional crystallisation or fractional melting. Granites which formed in both ways undoubtedly occur, but the present author adopts the view that most of the more felsic granites are products of partial melting of the crust, with compositions which may have been modified to a degree by fractional crystallisation. Such an origin, rather than a complete process of fractional crystallisation from initial mafic melts, is favoured in the general case for the following reasons.

2.1.1. Direct observations of rocks of the deep crust. Studies of metamorphic rocks show that the deep crust can reach temperatures high enough for quartzofeldspathic rocks to partly melt under QTB conditions. Also, we can make direct observations of partial melting of the deep crust (e.g. in migmatites), although the leucosome compositions may (e.g. Kalsbeek *et al.* 2001) or may not (e.g. Sawyer 1996) closely match those of granites.

2.1.2. Occurrence of S-type granites. S-type granites are found in many fold belts. These have compositions which show that they were derived largely or entirely from sedimentary source rocks. They clearly are not the products of fractional crystallisation of basalt, even with extensive contamination by sedimentary materials. In eastern Australia, S-type granites occur in each of the three fold belts of early-, mid- and late-Palaeozoic age. They comprise a little more than half of the 63 000 km^2 of granites in the mid-Palaeozoic LFB, and related volcanic rocks are also very important in that region. Elsewhere, they occur widely in Palaeozoic fold belts, such as the Caledonian and Hercynian belts of Europe extending into North America. There is argument about the proportion of basaltic material that might have been present in the source materials for such granites [e.g. the contrasting views of Collins (1998) and Chappell *et al.* (2000)], but even if there was a component of fractionated basalt among the source components, there would be general agreement that the source comprised dominantly solid metasedimentary materials which were partially melted. The occurrence of large bodies of granite formed from S-type magmas tells us that partial melting of the crust on a large scale to produce granite batholiths is possible.

2.1.3. Relative abundance of felsic and mafic rocks. Granites and rhyolites dominate over gabbros and basalts in most plutonic/volcanic rock provinces. There are granite terranes in which the granitic rocks are accompanied by abundant mafic rocks; for example the western half of the Peninsular Ranges Batholith of southern and Baja California (Silver & Chappell 1988). But there are many more areas in which mafic rocks are completely or almost completely absent. Such a comment applies fairly generally to the S-type granite occurrences listed above. In the LFB, a large area containing both S- and I-type granites, the amount of gabbro is very small. Among the 843 named plutonic units on the map of that area (Chappell *et al.* 1991), 26 are gabbro, all of which are relatively small with the largest being 14·4 km^2 in area, with a combined area of 71 km^2. That is slightly more than 0.1% of the total area of plutonic rocks. There is a small additional area of gabbroic rock associated with the high-temperature granites of the Boggy Plain Supersuite, but they are gradational in the field into granites, and are cumulate rocks from a less mafic melt and do not represent primary mafic compositions (Wyborn *et al.* 2001; Chappell & Wyborn 2004). The classic use of fractional crystallisation as a mechanism to account for the production of a series of granitic rocks was that of Nockolds (1940) for the Garabal Hill–Glen Fyne complex (32 km^2) of Scotland. However, granitic rocks in that complex are much more voluminous than their mafic associates. The ratio of areas of granodiorite to pyroxene-mica diorite, Nockold's presumed parent, at Garabal Hill–Glen Fyne, is approximately 100:1. Although such gabbroic rocks are more common among the plutonic rock suites of Scotland than they are in many other granite terranes, a dominance of felsic rocks is typical. While the mafic rocks of that region were probably derived from material that contributed to the partial melting of the crust giving the more felsic rocks, the amount of material that fractionated mafic melts contributed to the mass of the felsic granites was, at most, very small.

2.1.4. Lack of fractionation of trace elements. Trace elements are not strongly fractionated in most felsic granites and their abundances are consistent with the rocks forming from melts which were either primary or only slightly fractionated from primary partial melt compositions. Many felsic granites with major element QTB compositions do not have the elevated Rb contents and lowered Sr contents which would indicate feldspar fractionation, although some do, of course. The rare-earth elements (REE) are sensitive indicators of fractional crystallisation of haplogranite melts. Monazite is a ubiquitous mineral in S-type granites and its removal during fractional crystallisation removes Th and the light REE very rapidly (Chappell 1999), but many felsic S-type granites do not show this feature and cannot have undergone fractional crystallisation to any extent. Those S-type granites which have undergone monazite fractionation are compositionally quite distinctive, with dramatic decreases in the abundances of Th and the light REE down to about one-tenth of the unfractionated levels, but they comprise a very small fraction of all S-type granites. Again, Y and the heavy REE are observed to increase very rapidly by factors of two or more in felsic I-type melts during fractional crystallisation (Chappell 1999), but many more felsic I-type granites show no sign of such enrichments. And when fractionated I- and S-type granites are seen, their compositions can often be traced to less-fractionated granites as the primary composition, not to mafic rocks.

The linear trends on variation diagrams which characterise the low-temperature granite suites often extend to haplogranite compositions and elements whose abundances are sensitive to fractional crystallisation, such as Rb, Sr and Y in I-type suites, are generally not displaced from that trend. For example, this is the case for Sr in the Glenbog Suite of the LFB, which ranges in SiO_2 content from 66·2% to 73·8%. Three analysed rocks from the most felsic unit of the Moruya Suite, the Bodalla Monzogranite, have Sr contents of 255, 123 and 119 ppm, and SiO_2 values of 71·9%, 74·9% and 74·7%. The first of these lies very close to the linear trend for the Moruya suite that extends from 60·1% SiO_2. The displacement of the other two samples from that trend is the result of fractional crystallisation of feldspars, and fractionation is further supported by Y values of 14, 25 and 23 ppm for those three rocks. The first sample from Bodalla represents the composition of the primary unfractionated partial melt. Restite crystals separated from that melt to produce the compositional variation seen in the more mafic units of the Moruya Suite on the one hand, with the much less abundant fractionated granites on the other. Variation diagrams for Sr in the Glenbog and Moruya Suites were provided by Chappell (1996a).

Granites with fractionated trace element abundances generally show petrographic evidence for late-stage hydrothermal activity; for example, in many 'tin granites'. This is not unexpected since H_2O abundances must increase in melts undergoing fractional crystallisation. However, most felsic granites are relatively unaltered, although some are not, which would be consistent with lower H_2O contents related to an origin by partial melting rather than fractional crystallisation.

2.1.5. Evidence for low temperatures. Many felsic granites which are cogenetic with less felsic granites do not lie at the end of a liquid line of descent, and therefore, the melts themselves must be regarded a primary component of that suite of rocks. These are the low-temperature granite suites of Chappell *et al.* (1998, 2004) in which the melt phase of the magma involved in their production existed at low magmatic temperatures and had a felsic composition. World-wide, these low-temperature granites appear to dominate over high-temperature granites.

2.2. Felsic rocks in volcanic arcs

Smith *et al.* (2003) described widespread felsic volcanism in the intra-oceanic Kermadec arc of the SW Pacific. They ascribed the origin of those rocks to crustal anatexis since such processes can account for features of those rocks which are difficult to explain by fractional crystallisation. The volcanic rocks have SiO_2 contents in the range from 64% to 71% and K_2O contents up to 2·2%, and very low Rb contents from 7 to 43 ppm, in the analyses quoted in that paper. Such compositions are appropriate for products of partial melting of mafic source rocks in primitive crust. Such an origin is consistent with a more widespread origin of felsic magmas by partial melting within the crust rather than fractional crystallisation of basaltic parents.

2.3. Cumulative haplogranites

Tuttle & Bowen (1958) showed that haplogranites have the compositions of hydrous melts occurring in equilibrium with quartz and feldspars. It is generally assumed that haplogranites formed by the solidification of such a melt through crystallisation as a closed system (equilibrium crystallisation). However, the occurrence of sequences of felsic granites with progressive changes in minor element abundances which represent evolved melt compositions (e.g. Chappell & White 1998; Chappell 1999) implies that cumulative haplogranites must also exist, even though they are not often recognised. Such rocks would be produced by the separation of dominantly quartz and feldspars which precipitated under eutectic conditions from a QTB melt. Such rocks would contain approximately the same relative proportions of *Qz*, *Ab* and *Or* as rocks formed by the equilibrium crystallisation of that same melt, with the precise proportions dependent on the *An* and H_2O contents of the melt. Therefore, these cumulative rocks would plot in the same distinctive fashion on the *Qz-Ab-Or* diagram. Thus, such figures cannot be used as a simple criterion for a haplogranite having formed by the equilibrium crystallisation of a melt. While the relative amounts of the *Qz*, *Ab* and *Or* components need not be affected by the presence of accumulated crystals, the abundances of Ca and some trace elements would be. Plagioclase crystals in equilibrium with a felsic melt will be relatively *An*-rich, and accessory minerals such as monazite may be present. Rocks formed by the accumulation of such minerals will show related chemical features. Cumulative rocks are rare at the present levels of exposure of the LFB, although they must occur at depth in some places. They are more common in some other granite terranes and they are the subject of a separate communication (Chappell & Wyborn 2004).

3. Causes of variations within granite suites

There is far less agreement about details of the origins of the less felsic granites, although they are clearly magmatic. Many processes have been proposed to account for the details of formation of such granites and the generation of compositional variation within granite suites. All of these processes probably operate to some extent, and can be firmly established in particular examples. The question is: On what scale and how frequently did a particular process operate? Mechanisms which have been determined in one or a few cases should not necessarily be regarded as general processes. Neither should those processes established in small areas be regarded as necessarily applying to granite bodies of batholithic dimensions. In approximately the sequence in which these different processes could operate to produce compositional variation within granite suites, they are:

(1) variation inherited from heterogeneous source rocks;
(2) varying degrees of partial melting;
(3) magma mixing and/or mingling;
(4) restite fractionation (crystal fractionation of entrained crystals);
(5) fractional crystallisation (crystal fractionation of precipitated crystals);
(6) assimilation or contamination; and
(7) hydrothermal alteration.

These processes all effect granite compositions at some scale. Each could operate alone, or else simultaneously [e.g. (5) and (6) in the AFC process], or sequentially [e.g. (4) followed by (5), or (5) followed by (7)].

Apart from (2), these different processes were examined in some detail at a previous Hutton Symposium (Chappell 1996b), and the reader is referred to that discussion. The two forms of crystal fractionation [(4) and (5)], and more briefly, the varying degrees of partial melting (2) will be considered here as part of the 'unified model'. Magma mixing and/or mingling is not assigned a significant role in that model and the reasons for that are summarised below (see section 3.2).

3.1. Crystal fractionation

Crystal fractionation, either of entrained (4) or of precipitated (5) crystals dominates in producing variation within granite plutons, and in the suites to which such compositionally variable plutons belong. Varying degrees of partial melting of the source rocks (2) probably contribute to, and may dominate, variations within those low-K or 'tonalitic' suites in which most of the variation occurs between the plutons of a suite, with relatively little variation within plutons (see section 6.5 below).

The process of restite crystal fractionation leads to changes in composition through the removal of entrained crystals residual from partial melting of the source (Chappell *et al.* 1987) and is characterised by 'linear' trends on variation diagrams. It is characteristic of the low-temperature granites, i.e. of many I-types and all S-types, and may be followed by low-temperature fractional crystallisation of the separated melt in both types.

Fractional crystallisation is the classic process in which the precipitation and removal of crystals from a largely or completely molten magma leads to a series of more evolved compositions. This occurs among the high-temperature granite magmas and sometimes in the late stages of evolution of the low-temperature granites. It is characterised by curved trends on element variation diagrams. It is an important process in the concentration of metals and H_2O at the magmatic stage, and is more likely to lead to mineralisation (White & Chappell 2004).

3.2. Magma mixing and mingling

Magma mixing and/or mingling is a mechanism that attracts much support, but is regarded here as a process that may be locally important but has no significant role in producing larger scale variations. An important feature of many granite suites used as one argument for such a process are the 'linear' compositional variations which are a distinctive feature of many low-temperature granite suites. The fact that such variations are not seen among the high-temperature granite suites would seem to rule out this process for those rocks. When the variations within the low-temperature suites are examined in detail, a similar conclusion must be drawn, at least for the granites of southeastern Australia. In that region, correlations are observed between the relative abundances of elements at the two extremes of compositions in rock suites. For example, the sequence of increasing abundances of Sr in the more felsic granites of individual suites of the Bega Batholith is the same as for the more mafic granites of those suites. This precludes an origin by mixing or mingling of previously unrelated materials (Chappell 1996a) and is consistent with the suites evolving by the 'unmixing' of felsic partial melt from mafic restite, which would be consistent with that correlation between the compositions of the two 'end-members'. This conclusion is supported by more limited isotopic data which show which there is no systematic variation in isotopic properties as chemical compositions change within those rock suites (Chappell & McCulloch 1990), although there are distinct isotopic differences between the suites.

It could perhaps be argued that the granites of southeastern Australia are unusual in this respect and that large-scale mixing and/or mingling could be important in low-temperature granites in other places. While these granites may well have some aspects which are not commonly found elsewhere, they do have diverse compositions which indicate derivation from a variety of source rocks. The granites display features which are commonly taken as evidence of mixing or mingling processes among plutonic rocks, such as linear patterns of chemical variation, the development of local hybrid rocks and the presence of mafic enclaves. They also show patterns of isotopic variation on a regional scale that could be taken as evidence for mixing processes (Keay *et al.* 1997). Despite these features, the granites of that region, as we have seen, have compositional characteristics which preclude magma mixing or mingling as significant contributors to development of the large-scale variations. In view of the mechanical difficulties of mixing or mingling two components on a large scale, the view is taken that this is likely to be the case elsewhere. These arguments do not exclude the possibility that mixing of different components may have contributed to the compositions of the source rocks of various suites, and the 'unified model' includes that possibility (see section 8 below).

4. Origin of low- and high-temperature granites

There are two contrasting patterns of Zr variation in granite suites of southeastern Australia, and this led to recognition of the low- and high-temperature granite groups by Chappell *et al.* (1998, 2004). Zircon was saturated throughout the evolution of low-temperature granite suites, which results in Zr abundances plotting on linear trends on Harker diagrams, generally decreasing with increasing SiO_2. In contrast, among the high-temperature granite suites, zircon was not saturated in the magma from which the more mafic rocks (less than ~68% SiO_2) formed, at which compositions Zr abundances increase with increasing SiO_2. In contrast, zircon was saturated at more felsic compositions, and consequently, Zr falls in abundance as the rocks become more felsic. We must now ask why granites have formed at varying temperatures, particularly when it is considered that this is not a function of the composition of an individual granite? For example, tonalites of the Cobargo Suite of the LFB apparently formed at temperatures ~800°C and are 'low-temperature', whereas Cordilleran-like tonalites formed at temperatures greater than 1000°C and are 'high-temperature'. Nevertheless, both types can have very similar major element compositions (see Chappell *et al.* 2004, table 2), and hence, similar temperatures for the start of melting (solidus), and when the H_2O contents of the melts are comparable, of crystallisation (liquidus). The two types cannot be discriminated using major element compositions.

The answer, as with so many things related to granites, lies in the source rock compositions. Granites will be low-temperature in nature if there are sufficient low-temperature melt components in the source region to form an extractable magma at low magmatic temperatures. If one of the low-temperature melt components is low in abundance, then the source is said to be critically deficient in that haplogranite component. In that case, melting must continue to higher temperatures if granite magma is to be produced, and a high-temperature granite then results. The source may have become critically deficient in a haplogranite component (relatively infertile) as a result of igneous, sedimentary or metasomatic processes which occurred prior to the partial melting event. Or it may occur if a component is removed from the solid source materials by partial melting before magma can be extracted.

5. Partial melting at low magmatic temperatures

5.1. Low-temperature haplogranites

To produce a low-temperature I- or S-type haplogranite by partial melting, the source rocks must contain sufficient *Qz*, *Ab* and *Or*, with H_2O, to form either an extractable haplogranite melt, or else an extractable magma containing restite crystals from which a haplogranite melt can later fractionate by removal of those crystals. Such a source composition cannot be too far removed from the central part of the *Qz-Ab-Or* diagram (Fig. 2) and sufficient H_2O must be present. Such source rocks would be broadly quartzofeldspathic, and the H_2O would generally be present in hydrous minerals, of which biotite would generally be the most important immediately prior to melting (e.g. see Whitney 1988). Granites formed in this way are reasonably abundant in the LFB, and examples are the Maffra (I-type) and Happy Jacks (S-type) granites for which analyses were listed by Chappell (1999).

Of course, some haplogranites are, as Bowen (1949) considered, the low-temperature products of fractional crystallisation of higher temperature melts. In the LFB, there are excellent examples of such fractionated haplogranites among the felsic rocks of the high-temperature Boggy Plain Supersuite (Wyborn *et al.* 2001), although their source melts were probably intermediate in composition, rather than basaltic, because the mafic rocks with which they are associated had a cumulate origin (Wyborn 1983; Chappell & Wyborn 2004) and are not primary magmatic compositions.

5.2. Restite-bearing low-temperature granites

To produce an extractable magma containing crystals of restite, the amount of melt must exceed a critical melt fraction (van der Molen & Paterson 1979) or second percolation threshold (Vigneresse *et al.* 1996) so that the restite framework breaks down. To do that, the source rocks must contain enough of each of *Qz*, *Ab* and *Or* to form the required amount of melt. As before, the bulk source composition cannot be too far removed from the central part of the *Qz-Ab-Or* diagram, with H_2O also present. This is indicated by the positions of the less felsic low-temperature Moruya, Moonbi and Bullenbalong Suites in Figure 2. Partial melting again takes place at low temperatures, and in this case, the restite does not completely disengage from the melt, if at all. At least initially, the magma probably moves *en masse* as a mixture of crystals and melt that may later fractionate restite from melt to produce a range of rock compositions. A discussion of magma mobility under these conditions has been provided in section 1 of Sawyer (1996).

Under conditions where magma entrains crystals of restite, the compositions of the more mafic granites will reflect those of their source rocks and crystals of older zircon would always be present, although ages may sometimes not be sufficiently older than the magmatic event for those older ages to be recognised. Low-temperature granites of this type are the dominant plutonic rocks of the LFB, and their associated volcanic rocks are also important. Among the I-type granites, they are exemplified by the Moruya Suite (Griffin *et al.* 1978) and other suites of the large Bega Batholith (8940 km^2), for which the most mafic rocks are tonalites and rare quartz diorites containing a little less than 60% SiO_2. In more potassic systems, the most mafic rocks are monzonites and monzodiorites with similar SiO_2 contents, such as those of the Moonbi Suite of the younger New England Fold Belt (Chappell 1966, 1996a).

Partial melting at low temperatures produces a wide range of S-type granites, depending on the composition of the source rocks. At temperatures sufficient to produce biotite breakdown, magmas such as those of the Bullenbalong Supersuite of the LFB are formed (White & Chappell 1988). The most mafic S-type granites of that supersuite contain ~65% SiO_2 and more than 20% biotite, and have compositions, both chemical and mineralogical, close to those of their source materials. The originally mafic magmas may evolve by restite fractionation to produce a wide range of rock (including volcanic) compositions.

The concept of restite fractionation has been controversial ever since it was first proposed by White & Chappell (1977); for example, see Wall *et al.* (1987) and Collins (1998). However, the presence of inherited zircon in many granites implies that they formed at low magmatic temperatures (Chappell *et al.* 1998, 2004). Suites of these low-temperature granites must have evolved by restite fractionation, at least at more mafic compositions. In some cases, fractional crystallisation operated at more felsic compositions; for example, as in the Koetong Suite of the LFB (Chappell & White 1998), for which the initial magma probably formed at slightly higher temperatures (see section 5.3 below).

5.3. Slightly higher temperatures of partial melting

If there is a small deficiency in the source rocks of one of the components of a minimum-temperature melt, then melting may continue to slightly higher temperatures, at a melt composition further from the component for which the source rocks have become depleted, illustrated schematically by the relative positions of the Jindabyne, Inlet and Koetong Suites in Figure 2. Such melts will also contain less SiO_2, and also more CaO, MgO and FeO, for example. In these early stages of depletion of a component in the source rocks, restite will still be present in the magma, including older zircon. Evolution of the magma may take place by restite fractionation and by fractional crystallisation after the restite has been removed. These are still low-temperature granites in the sense of Chappell *et al.* (1998), although it may be desirable to place them in a mid-temperature group in the future, when their overall characteristics are better defined (Chappell *et al.* 2004).

The Jindabyne Suite of the LFB (Hine *et al.* 1978) comprises nine plutons of I-type granodiorite to tonalite with SiO_2 contents in the range from 56% to 68%. Unlike the low-temperature I-type Moruya Suite, the rocks of the Jindabyne Suite do not extend to SiO_2 contents greater than 70% and the P_2O_5 values do not decrease with increasing SiO_2, as is shown in figure 6 of White & Chappell (1977) and by Chappell *et al.* (2004). Hine *et al.* (1978) ascribed these differences to a higher temperature of partial melting to produce what they called a non-minimum temperature melt. However, the Jindabyne Suite is not high-temperature because it contains calcic cores of plagioclase crystals, interpreted as restite (Chappell & White 2004), the compositional variation was dominantly the result of restite fractionation, and it contains inherited zircon crystals (I. S. Williams pers. comm.). The Inlet Monzonite pluton of the New England Fold Belt (Chappell *et al.* 1998) is a small pluton (12·5 km^2) that is a potassic analogue of the Jindabyne Suite. Formation of the non-minimum temperature Jindabyne and Inlet Suites is ascribed to the disappearance of *Or* and *Qz* components, respectively, from the residual mineral assemblage during partial melting,

The mineralised S-type Koetong Suite of LFB (Chappell & White 1998) formed after melting to higher temperatures, presumably because of a deficiency in *Ab* as a result of the source being slightly more pelitic. The more felsic rocks of that suite evolved by fractional crystallisation, which was an essential precursor to mineralisation. This suggests that a slightly higher temperature of partial melting may be necessary for the concentrating of Sn, W and U, for example, by fractional crystallisation.

6. Production of the high-temperature granites

We have seen that there are four essential components of a source rock, *Qz*, *Ab*, *Or* and H_2O, for the production of a haplogranite melt by partial melting. It follows that there are four end-member cases for which source rocks would not form an extractable low-temperature magma because they are or become critically deficient in one of those four components. These circumstances are illustrated in Figure 2, and each of these cases will now be discussed.

6.1. Source rocks critically deficient in *Qz*

If the source rocks are critically deficient in the *Qz* component, then melting will continue to much higher temperatures by dissolving more of the *Ab* and *Or* components of the source rocks, as well as more CaO, MgO and FeO, for example, leading to more mafic compositions. The melt compositions will be monzonitic or monzodioritic (Fig. 2), and the magmas, being less viscous and hence free of restite, will generally evolve by fractional crystallisation. Old zircon will not be found in these high-temperature granites, except as xenocrysts in felsic fractionated melt. Both the source rocks and the granites are broadly monzonitic. In the LFB, such rocks are represented by the members of the Boggy Plain Supersuite of Wyborn *et al.* (1987).

6.2. Source rocks critically deficient in *Ab*

Source rocks which are critically deficient in *Ab* are mature detrital sedimentary rocks since, when plagioclase is destroyed during weathering, part or most of the *Ab* component is lost. The abundances of the three other haplogranite components would generally be increased by weathering with *Qz*, and *Or* plus H_2O derived from high-temperature products of clay minerals, available in these rocks. Hence, the abundance of *Ab* appears to be the critical factor in this case. If feldspars are absent or low in amount, the low Na contents will make these rocks infertile in terms of granite production, although *Ab* may be present in the restite mineral assemblage as a component of plagioclase crystals, which may be more Ca-rich if some melt has been produced. Since S-type granites that do not contain inherited zircon are not known at the present time, it is possible that high-temperature S-type granites do not exist, so that the corresponding part of Figure 2 is represented by infertile pelites and high-K leucosomes. Most of the exposed Ordovician sedimentary rocks of the LFB, which form the general matrix to the exposed granites of that region, are in this compositional category, and those rocks would generally have been infertile during production of the granites. If more feldspar were present, either in a feldspathic greywacke (Chappell *et al.* 2000) or as igneous material (Gray 1984; Collins 1998), then these rocks may be fertile during partial melting.

6.3. Source rocks critically deficient in *Or*

When *Or*, in part supplied by the breakdown of mica, becomes depleted during the partial melting event, the composition of the melt moves away from the composition of that component with increasing temperatures (Fig. 2). This is by far the most common case of high-temperature granite magmatism, applying as it does to the Cordilleran-like tonalites, the TTGs, and perhaps, to some plutonic rocks of the island arcs. Both the source rocks and the derived granites are relatively low in K. Among southeastern Australian granites, the tonalites and granodiorites of the Clarence River Supersuite of the New England Fold Belt (Bryant *et al.* 1997) provide an excellent example of these rocks. The TTGs are an important component of the Archaean plutonic rocks of Western Australia, which include the rocks to which that acronym was first applied by Jahn *et al.* (1981).

The Clarence River rocks are compositionally very similar to the dominantly tonalitic rocks of the Peninsular Ranges batholith, particularly the western part. Silver & Chappell (1988) described the rocks of that batholith as calcic, and pointed to the contrast with the more potassic central and eastern parts of the Sierra Nevada batholith to the north. Variability among the granitic rocks of the Cordillera is further illustrated by the Coastal Batholith of Peru (Pitcher *et al.* 1987), which includes substantial amounts of tonalite, but overall, is more potassic than the Peninsular Ranges batholith and is calc-alkaline in character rather than calcic. The differences in K contents of Cordilleran granites directly reflect

Q

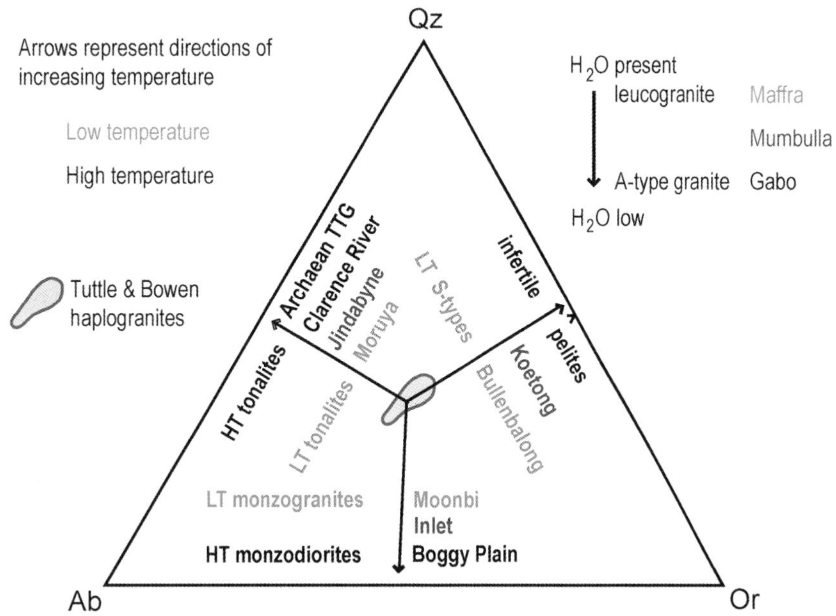

Figure 2 A schematic representation of the relationship between some granite types, and the relative proportions of the four haplogranite components *Qz*, *Ab*, *Or* and H_2O in the source rocks. The shaded area in the centre of the figure is the haplogranite field of Tuttle & Bowen (1958), also shown in Figure 1.

corresponding differences in their source rocks. In all cases, however, the *Or* component in the source rocks became depleted during partial melting, although at a later stage of melting in the more potassic source rocks. This resulted in the formation of plutonic and volcanic rocks which are dominantly, at least, high-temperature in nature. The different K contents of the source rocks and the derived melts was significant for the later high-level evolution of these rocks (see section 6.5 below).

6.4. Source rocks in which the availability of H_2O is restricted

If the source rocks contain sufficient *Qz*, *Ab* and *Or*, but the amount of H_2O available to enter the melt is restricted, at least at lower temperatures, then melting may continue to higher temperatures to produce a high-temperature or A-type haplogranite melt. Such a melt will initially be very felsic and contain similar proportions of *Qz*, *Ab* and *Or* to the low-temperature haplogranites, but will be enriched in certain trace elements, e.g. Zr, Ga, Y and the REE, because these are likely to be more soluble in the melt at higher temperatures and/or lower water contents. Melting may continue to higher temperatures, at which time the A-type melts become slightly less felsic. The LFB contains some excellent examples of A-type granites, although they comprise only 0.6% of the total area of granite exposures. They were initially described in the LFB by Collins *et al.* (1982), and have been discussed further by King *et al.* (1997, 2001). The major element compositions of the A-type granites, being close to the QTB composition, are distinctive but not diagnostic. In cases where these rocks can be distinguished, it is on the basis of their trace element compositions, such as high abundances of Zr ($\sim$500 ppm in the Gabo Island and Danswell Creek Suites of the LFB), which show that they formed at higher temperatures than other felsic granites. Those A-type granites of the LFB which are interpreted to represent primary unfractionated compositions have calculated zircon saturation temperatures (Watson & Harrison 1983) close to 900 °C. The temperatures of formation of the A-type granites are generally lower than those of other high-temperature granites. Also, the criterion used for recognising those other high-temperature granites, i.e. the absence of zircon in more mafic members of rock suites, is not applicable to such felsic compositions. Fractionated haplogranites which are known to be A-type from their close association with unfractionated A-type granites have calculated zircon saturation temperatures down to about 750 °C and are indistinguishable from fractionated low-temperature haplogranites on a purely compositional basis; for example, the Dunskeig A-type granite of King *et al.* (2001).

6.5. Variations within tonalitic and higher-K suites

The Marulan Batholith of the LFB (Carr *et al.* 1992) illustrates an important feature of many high-temperature tonalite suites, i.e. in contrast to the higher-K suites, the amount of variation within individual plutons is restricted, and the within-suite variation is largely provided by variations between different plutons. Large and compositionally uniform plutons of tonalite and granodiorite are a feature of the Peninsular Ranges batholith (Silver & Chappell 1988), and L. T. Silver (pers. comm.) proposes that the variation within cogenetic granites resulted from varying degrees of partial melting of the source rocks, rather than fractional crystallisation (see section 3 above).

In contrast to the low-K plutons, those of the high-temperature suites with higher K contents are almost invariably zoned. Wyborn *et al.* (2001) have pointed out that the higher amounts of K mean that the melts would contain more low-melting components. Therefore, if cumulate rocks are forming (e.g. at the walls of a pluton), those components would have too great a bulk to be completely retained between the cumulus minerals, which means that some melt would be displaced back into the main body of melt, modifying its composition. During the crystallisation of tonalite melts, that low-temperature interstitial melt has less bulk and would be retained between the cumulus minerals, and hence, captured within the solidification zone, so that the remaining melt does not change in composition. It may be because the higher-K magmas have a greater propensity to fractionate, rather than any geochemical association between various elements, that results in those magmas being more commonly associated with significant mineralisation.

Atherton & Sanderson (1987) have ascribed compositional variations within some units of the Coastal Batholith of Peru to differentiation (fractional crystallisation) at or near the present level of exposure. The operation of that process is supported by the presence of some concentrically and gradationally zoned plutons in which fractionation occurred *in situ*, while in other cases, separate pulses of magma came from below. Those authors point out that, under those circumstances, the composition of the original magma is not absolutely clear. However, the presence of more K-rich fractionated rocks would imply that it was more potassic than was the case for the tonalitic rocks of the Peninsular Ranges batholith. These observations suggest that the parental magmas of the Coastal Batholith might, at least in some cases, have been sufficiently potassic to undergo fractional crystallisation along the lines proposed by Wyborn *et al.* (2001). In that case, some of the tonalites of the Coastal Batholith could be cumulate rocks, or *cumulate tonalites*, in contrast with the *primary tonalites* of the Peninsular Ranges batholith. Both of these types contrast with the low-temperature *restitic tonalites*, such as those of the Moruya Suite (see section 5.2 above).

7. Granite compositions and the tectonic environment of formation

There has long been an aspiration to link granite compositions and the tectonic conditions of their formation, perhaps most notably by Pearce *et al.* (1984), and more recently, for example, by Barbarin (1999). In contrast, the present paper ascribes the compositional characteristics, at least of the more common types of granites, to the compositional features of their source rocks. Such a compositional nexus does not exclude possible tectonic correlations. However, it must be realised that, to the extent that compositions can be used as tectonic indicators, it is likely to be specifically the source rocks and their conditions of formation which are being assessed, as Chappell & Stephens (1988) pointed out.

Among the more abundant types of granites, it is probably the dominant tonalites and low-K granodiorites of much of the Cordilleran environment which have the closest relationship to a particular tectonic environment. Those rocks have, as their source rocks before them had, compositional features which reflect their primitive nature as relatively new components of the crust. The I-type granites of the LFB include a few rocks with very similar compositions, such as the mafic members of the Moruya Suite (Griffin *et al.* 1978), but they did not necessarily form in an analogous tectonic environment. Rather, they reflect the compositions of their source rocks, which may have previously been produced in a Cordilleran-like environment. Patterns of chemical variation show that the Moruya magmas were all saturated in zircon, that they are low-temperature varieties and that the suite includes significant amounts of felsic granodiorite, all distinct differences from most Cordilleran granites. This illustrates the difficulty of simply using bulk compositions of granites as indicators of tectonic conditions without considering their broader rock associations and petrogenesis. More broadly, Barbarin (1999) stated that most I-type granites of the LFB belong to his 'amphibole-rich calc-alkaline granite' or ACG group. He also assigned the main granite components of the western margins of the American continents to that group, but they are distinctly different from the I-type granites of the LFB, which are mostly low-temperature, on the whole distinctly more felsic and potassic, and in places, associated very closely in the field with abundant S-type granites.

8. A unified model of granite genesis

A unified model can be summarised as follows:

1. Plutonic and volcanic rocks which are associated in space and time are generally cogenetic, and may be comagmatic. Compositions of the two may be the same or they may be complementary.
2. Most granites and related volcanic rocks were formed from processes which were initiated by heating and partial melting of the crust.
3. To a significant degree, both the chemical and physical properties of magmas which formed in a particular case were largely predetermined by the composition of their source rocks. Each source rock responds to heating in a way that is a physicochemical expression of its composition. This applies most precisely if the source rocks behave as a closed system, but it is still relevant in systems which are open for some components (e.g. H_2O), or to any mixed sources.
4. In turn, those magmatic properties largely determine the evolution of the magma.
5. Although granites show a great diversity in their compositions and in details of their evolution, those features mostly relate back to the compositions of their source rocks, which are the unifying element in granite genesis.
6. Source rock compositions are the 'genetic code' for granites which largely predetermine the evolution of a granite magma. However, features inherited from the sources may be partly modified by environmental factors, such as country rock compositions and structures, and the stress regime.
7. Tectonic conditions do not directly determine granite compositions, but may affect them indirectly by an association with distinctive source rock compositions and by providing heat.
8. In detail, individual samples of granite evolve in a many ways, but it is important to distinguish between processes which occur commonly and on a large scale, and those that do not. Crystal fractionation, either of entrained or previously precipitated crystals, appears to be the most common process for compositional evolution. For some conditions, it is likely that diverse melt compositions result from varying degrees of partial melting of source rocks with similar compositions.

9. Acknowledgements

The author is grateful to the organisers of the Fifth Hutton Symposium on the Origin of Granites and Related Rocks for their invitation to present this paper at the opening session of that conference in Toyohashi, Japan. The helpful comments of many persons, particularly Professors Penny King and Allan White, and Dr Carlos Rapela and a second reviewer, are gratefully acknowledged.

10. References

Atherton, M. P. & Sanderson, L. M. 1987. The chemical variation and evolution of the super-units of the segmented Coastal Batholith. *In* Pitcher, W. S., Atherton, M. P., Cobbing, E. J. & Beckinsale, R. D. (eds) *Magmatism at a Plate Edge: the Peruvian Andes*, 208–27. Glasgow: Blackie.

Barbarin, B. 1999. A review of the relationships between granitoid types, their origins and their geodynamic environment. *Lithos* **46**, 605–26.

Bowen, N. L. 1947. Magmas. *Bulletin of the Geological Society of America* **58**, 263–80.

Bowen, N. L. 1949. The granite problem and the method of multiple prejudices. *In* Gilluly, J. (ed.) *Origin of Granite. The Geological Society of America Memoir* **28**, 79–90.

Bryant, C. J., Arculus, R. J. & Chappell, B. W. 1997. Clarence River Supersuite: 250 Ma Cordilleran tonalitic I-type intrusions in eastern Australia. *Journal of Petrology* **38**, 975–1001.

Carr, P. F., Chappell, B. W. & Jones, B. G. 1992. Geochemical contrasts between intersecting Devonian and Carboniferous magmatic belts in the eastern Lachlan Fold Belt. [Abstract.] *Transactions of the Royal Society of Edinburgh: Earth Sciences* **83**, 487.

Chappell, B. W. 1966. *Petrogenesis of the granites at Moonbi, New South Wales*. Unpublished Ph.D. thesis. Canberra: Australian National University.

Chappell, B. W. 1996a. Magma mixing and the production of compositional variation within granite suites: evidence from the granites of southeastern Australia. *Journal of Petrology* **37**, 449–70.

Chappell, B. W. 1996b. Compositional variation within granite suites of the Lachlan Fold Belt: its causes and implications for the physical state of granite magma. *Transactions of the Royal Society of Edinburgh: Earth Sciences* **87**, 159–70.

Chappell, B. W. 1999. Aluminium saturation in I- and S-type granites and the characterization of fractionated haplogranites. *Lithos* **46**, 535–51.

Chappell, B. W., White, A. J. R. & Wyborn, D. 1987. The importance of residual source material (restite) in granite petrogenesis. *Journal of Petrology* **28**, 1111–38.

Chappell, B. W., English, P. M., King, P. L., White, A. J. R. & Wyborn, D. 1991. Granites and related rocks of the Lachlan Fold Belt (1:1 250 000 scale map). Bureau of Mineral Resources, Geology and Geophysics, Australia.

Chappell, B. W., Bryant, C. J., Wyborn, D., White, A. J. R. & Williams, I. S. 1998. High- and low-temperature I-type granites. *Resource Geology* **48**, 225–36.

Chappell, B. W., White, A. J. R., Williams, I. S., Wyborn, D. & Wyborn, L. A. I. 2000. Lachlan Fold Belt granites revisited: high- and low-temperature granites and their implications. *Australian Journal of Earth Sciences* **47**, 123–38.

Chappell, B. W., White, A. J. R., Williams, I. S. & Wyborn, D. 2004. Low- and high-temperature granites. *Transactions of the Royal Society of Edinburgh: Earth Sciences* **95**, 125–140.

Chappell, B. W. & McCulloch, M. T. 1990. Possible mixed source rocks in the Bega Batholith: constraints provided by combined chemical and isotopic studies. *In Seventh International Conference on Geochronology, Cosmochronology and Isotope Geology. Geological Society of Australia, Abstracts* **27**, 17.

Chappell, B. W. & Stephens, W. E. 1988. Origin of infracrustal (I-type) granite magmas. *Transactions of the Royal Society of Edinburgh: Earth Sciences* **79**, 71–86.

Chappell, B. W. & White, A. J. R. 1998. Development of P-rich granites by sequential restite fractionation and fractional crystallisation: the Koetong Suite in the Lachlan Fold Belt. *Acta Universitatis Carolinae Geologica* **42**, 23–7.

Chappell, B. W. & White, A. J. R. 2001. Two contrasting granite types: 25 years later. *Australian Journal of Earth Sciences* **48**, 489–99.

Chappell, B. W. & Wyborn, D. 2004. Cumulate and cumulative granitic rocks. *Resource Geology* **54**, 227–40.

Clemens, J. D. 1984. Water contents of silicic to intermediate magmas. *Lithos* **17**, 273–87.

Collins, W. J. 1998. Evaluation of petrogenetic models for Lachlan Fold Belt granitoids: implications for crustal architecture and tectonic models. *Australian Journal of Earth Sciences* **45**, 483–500.

Collins, W. J., Beams, S. D., White, A. J. R. & Chappell, B. W. 1982. Nature and origin of A-type granites with particular reference to south-eastern Australia. *Contributions to Mineralogy and Petrology* **80**, 189–200.

Gray, C. M. 1984. An isotopic mixing model for the origin of granitic rocks in southeastern Australia. *Earth and Planetary Science Letters* **70**, 47–60.

Griffin, T. J., White, A. J. R. & Chappell, B. W. 1978. The Moruya Batholith and geochemical contrasts between the Moruya and Jindabyne suites. *Journal of the Geological Society of Australia* **25**, 235–47.

Hine, R., Williams, I. S., Chappell, B. W. & White, A. J. R. 1978. Contrasts between I- and S-type granitoids of the Kosciusko Batholith. *Journal of the Geological Society of Australia* **25**, 219–34.

Hutton, J. 1788. Theory of the Earth; or an investigation of the laws observable in the composition, dissolution, and restoration of land upon the globe. *Transactions of the Royal Society of Edinburgh* **1**, 209–304.

Jahn, B. M., Glikson, A. Y., Peucat, J. J. & Hickman, A. H. 1981. REE geochemistry and isotopic data of Archaean silicic volcanics and granitoids from the Pilbara Block, western Australia: implications for the early crustal evolution. *Geochimica et Cosmochimica Acta* **45**, 1633–52.

Johannes, W. & Holtz, F. 1990. Formation and composition of H_2O-undersaturated granitic melts. *In* Ashworth, J. R. & Brown, M. (eds) *High-temperature Metamorphism and Crustal Anatexis*, 87–104. London: Unwin Hyman.

Kalsbeek, F., Jepsen, H. F. & Nutman, A. P. 2001. From source migmatites to plutons: tracking the origin of ca. 435 Ma S-type granites in the East Greenland Caledonian orogen. *Lithos* 57, 1–21.

Keay, S., Collins, W. J. & McCulloch, M. T. 1997. A three-component Sr-Nd isotopic mixing model for granitoid genesis, Lachlan fold belt, eastern Australia. *Geology* **25**, 307–10.

King, P. L., White, A. J. R., Chappell, B. W. & Allen, C. M. 1997. Characterization and origin of the aluminous A-type granites of the Lachlan Fold Belt, southeastern Australia. *Journal of Petrology* **38**, 371–91.

King, P. L., Chappell, B. W., Allen, C. M. & White, A. J. R. 2001. Are A-type granites the high temperature felsic granites? Evidence from fractionated granites of the Wangrah Suite. *Australian Journal of Earth Sciences* **48**, 501–14.

Nockolds, S. R. 1940. The Garabal Hill-Glen Fyne igneous complex. *Quarterly Journal of the Geological Society of London* **96**, 451–511.

Pearce, J. A., Harris, M. B. W. & Tindle, A. G. 1984. Trace element discrimination diagrams for the tectonic interpretation of granitic rocks. *Journal of Petrology* **25**, 956–83.

Pitcher, W. S., Atherton, M. P., Cobbing, E. J. & Beckinsale, R. D. 1987. *Magmatism at a Plate Edge: the Peruvian Andes*. Glasgow: Blackie.

Read, H. H. 1948. Granites and granites. *The Geological Society of America Memoir* **28**, 1–19.

Repcheck, J. 2003. *The Man Who Found Time: James Hutton and the Discovery of the Earth's Antiquity*. New York, NY: Perseus Publishing.

Sawyer, E. W. 1996. Melt segregation and magma flow in migmatites: implications for the generation of granitic magmas. *Transactions of the Royal Society of Edinburgh: Earth Sciences* **87**, 85–94.

Silver, L. T. & Chappell, B. W. 1988. The Peninsular Ranges batholith: an insight into the evolution of the Cordilleran batholiths of southwestern North America. *Transactions of the Royal Society of Edinburgh: Earth Sciences* **79**, 105–21.

Smith, I. E. M., Worthington, J., Stewart, R. B., Price, R. C. & Gamble, J. A. 2003. Felsic volcanism in the Kermadec arc, SW Pacific: crustal recycling in an oceanic setting. *In* Larter, R. D. & Leat, P. T. (eds) *Intra-oceanic Subduction Systems: Tectonic and Magmatic Processes. Geological Society, London, Special Publication* **219**, 99–118.

Tuttle, O. F. & Bowen, N. L. 1958. Origin of granite in the light of experimental studies in the system $NaAlSi_3O_8$-$KAlSi_3O_8$-SiO_2-H_2O. *Geological Society of America Memoir* **74**.

van der Molen, I. & Paterson, M. S. 1979. Experimental deformation of partially melted granite. *Contributions to Mineralogy and Petrology* **70**, 299–318.

Vigneresse, J. L., Barbey, P. & Cuney, M. 1996. Rheological transitions during partial melting and crystallization with application to felsic magma segregation and transfer. *Journal of Petrology* **37**, 1579–600.

Wall, V. J., Clemens, J. D. & Clarke, D. B. 1987. Models for granitoid evolution and source compositions. *Journal of Geology* **95**, 731–49.

Watson, E. B. & Harrison, T. M. 1983. Zircon saturation revisited: temperature and composition effects in a variety of crustal magma types. *Earth and Planetary Science Letters* **64**, 295–304.

White, A. J. R. & Chappell, B. W. 1977. Ultrametamorphism and granitoid genesis. *Tectonophysics* **43**, 7–22.

White, A. J. R. & Chappell, B. W. 1988. Some supracrustal (S-type) granites of the Lachlan Fold Belt. *Transactions of the Royal Society of Edinburgh: Earth Sciences* **79**, 169–81.

White, A. J. R. & Chappell, B. W. 2004. Petrographic discrimination of low- and high-temperature I-type granites. *Resource Geology* **54**, 215–26.

Whitney, J. 1988. The origin of grantite: the role and source of water in the evolution of granitic magmas. *Geological Society of America Bulletin* **100**, 1886–97.

Wyborn, D. 1983. *Fractionation processes in the Boggy Plain zoned pluton*. Unpublished Ph.D. thesis. Canberra: Australian National University.

Wyborn, D., Turner, B. S. & Chappell, B. W. 1987. The Boggy Plain Supersuite: a distinctive belt of I-type igneous rocks of potential economic significance in the Lachlan Fold Belt. *Australian Journal of Earth Sciences* **34**, 21–43.

Wyborn, D., Chappell, B. W. & James, M. 2001. Examples of convective fractionation in high temperature granites from the Lachlan Fold Belt. *Australian Journal of Earth Sciences* **48**, 531–41.

B. W. CHAPPELL, ARC National Key Centre for Geochemical Evolution and Metallogeny of Continents, Department of Earth and Planetary Sciences, Macquarie University, Sydney, NSW 2109, Australia.

MS received 2 December 2003. Accepted for publication 9 September 2004.

Transactions of the Royal Society of Edinburgh: Earth Sciences, **95**, 11–22, 2004

A new paradigm for granite generation

Jean Louis Vigneresse

ABSTRACT: Ideas about granite generation have evolved considerably during the past two decades. The present paper lists the ideas which were accepted and later modified concerning the processes acting during the four stages of granite generation: melting, melt segregation and ascent, and emplacement. The active role of the mantle constitutes a fifth stage.

Fluid-assisted melting, deduced from metamorphic observations, was used to explain granite and granulite formation. Water seepage into meta-sedimentary rocks can produce granitic melt by decreasing melting temperature. CO_2 released by the mantle helps to transform rocks into granulites. However, dehydration melting is now considered to be the origin of most granitic melts, as confirmed by experimental melting. Hydrous minerals are involved, beginning with muscovites, followed by biotite at higher temperatures. At even deeper conditions, hornblende dehydration melting leads to calc-alkaline magmas.

Melt segregation was first attributed to compaction and gravity forces caused by the density contrast between melt and its matrix. This was found insufficient for magma segregation in the continental crust because magmas were transposed from mantle conditions (decompression melting) to crustal conditions (dehydration melting). Rheology of two-phase materials requires that melt segregation is discontinuous in time, occurring in successive bursts. Analogue and numerical models confirm the discontinuous melt segregation. Compaction and shear localisation interact non-linearly, so that melt segregates into tiny conduits. Melt segregation occurs at a low degree of melting.

Global diapiric ascent and fractional crystallisation in large convective batholiths have also been shown to be inadequate and at least partly erroneous. Diapiric ascent cannot overcome the crustal brittle-ductile transition. Fracture-induced ascent influences the neutral buoyancy level at which ascent should stop but does not. Non-random orientation of magma feeders within the ambient stress field indicates that deformation controls magma ascent.

Detailed gravity and structural analyses indicate that granite plutons are built from several magma injections, each of small size and with evolving chemical composition. Detailed mapping of the contact between successive magma batches documents either continuous feeding, leading to normal petrographic zoning, or over periods separated in time, commonly leading to reverse zoning. The local deformation field controls magma emplacement and influences the shape of plutons.

A typical source for granite magmas involves three components from the mantle, lower and intermediate crusts. The role of the mantle in driving and controlling essential crustal processes appears necessary in providing stress and heat, as well as specific episodes of time for granite generation. These mechanisms constitute a new paradigm for granite generation.

KEY WORDS: Ascent, emplacement, granite emplacement, mantle, melting, segregation.

The Earth is a thermal machine that progressively loses its energy with time. The heat generated by radiogenic elements in the core, mantle and crust is progressively lost through the convective motion of matter that drives plate tectonics (Bercovici 2003). It generates basaltic magma that adds material to the crust at Mid Ocean Ridges. In the continental crust, energy is lost through material transport during granitic intrusions and through deformation that manifests itself as seismic activity. Granitic magma formation is also a major way to transfer elements, such as Rb and Ba, but also Th and U, which are incompatible in mantle minerals, to the crust. It leads to a chemical differentiation of the continental crust with time.

Specific studies conducted on granitic bodies have documented internal structures, revealing the role of deformation. Joint structural and geophysical approaches identified patterns in granitic plutons, such as root zones from where magma was derived from the lower crust. At the same time, experimental studies on magma formation led to renewed ideas about crustal-derived magmas. They resulted from two different paths. On the one hand, new analytical techniques allowed more precise measurements involving a smaller sample volume. On the other, new concepts also modified scientists' interpretations of field observations (Castro *et al.* 1999). The renewed data sets led researchers to consider four stages of granite generation: melting, melt segregation, ascent and emplacement (Petford *et al.* 1997). Each stage has its own time and length scales, with an associated physical process. The four stages are embedded within a global context, which is mantle controlled.

The present paper suggests a new paradigm for granite generation. It is based on the author's own experience, gained through studies conducted on the several stages of granite generation. It also involves data and ideas which needed to be collected and put together in a synthetic form. It is organised in five major sections, according to the four stages and the global context. Definitions of the terms used throughout the paper relate to felsic igneous bodies. A granitic pluton is considered as a unit for granitic intrusion. Batholiths, such as the Sierra Nevada Batholith, are consequently built from several plutons. A single unit is built from several batches of magma with specific mineral facies or chemical compositions.

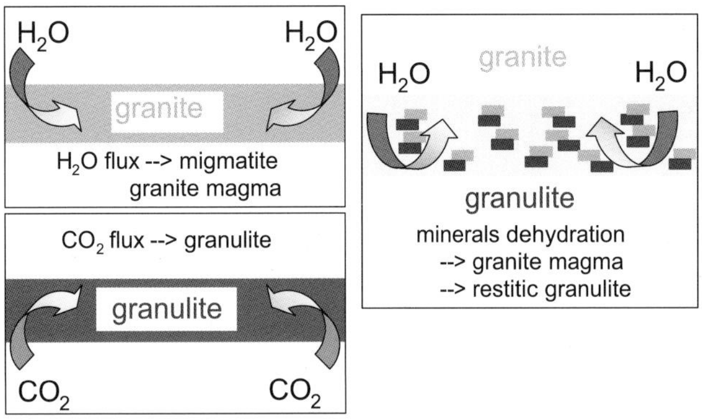

Figure 1 Melting as viewed in the old (left) and new (right) paradigms: The old paradigm considered a generalised metamorphism, resulting in granitic melt generation by a flux of water and granulite production by a flux of CO_2. In the new paradigm, water liberated from hydrous minerals during dehydration melting is immediately reincorporated to granitic melts, leading to residual granulites.

1. Melting

1.1. The old paradigm

Melting is the initial stage of magma formation. It depends on the available heat from either a thickened crust and/or from the underlying mantle. Older models of granitic magma production, especially for crustal-derived magmas, were inspired from a generalised concept of metamorphism (Fig. 1). Crustal rocks are progressively buried, increasing the ambient temperature. A flux of H_2O enhances melting by decreasing the liquidus temperature (Wyllie 1977). The parallel model of a CO_2 flux originated from the mantle results in granulite formation in the lower crust (Touret 1977).

Pioneering experiments (Tuttle & Bowen 1958) attributed granitic magma to H_2O-saturated or weakly under-saturated melting. However, experiments could not document water exceeding 2% in weight in granitic magmas (Maaløe & Wyllie 1975), weakening the idea of a water flux control. Andesitic magma, which forms a large part of subduction-related magmatism, could not derive from mantle peridotite because of the limited amount of water. It must derive from intermediate liquids formed from subducted sediments and mantle peridotite (Wyllie *et al.* 1976). In the continental crust, granitic magmas could be generated under regional metamorphism, provided that heat was there, resulting in H_2O-under-saturated granitic melt and migmatites (Wyllie *et al.* 1976). Hence, the model requires elevated temperature. A granitic magma with 2% water needs a temperature from 800 to 1000°C, under a pressure of 500–1000 MPa. Under such conditions, the continental crust can produce water-under-saturated liquids of granitic composition and granodioritic magma at higher temperatures (Wyllie 1977).

1.2. The new paradigm

In contradiction to a water-saturated model of melting, observations document that the lower crust has no free water, no hydrous minerals and too low a permeability (Yardley & Valley 1997). During experimental melting, dehydration breakdown of hydrous minerals (micas and amphiboles) results in a melt with granitic composition (Thompson 1982; Clemens 1990). The reactions liberate water that is immediately reincorporated to produce magma (Fig. 1).

Muscovite is the first mineral to melt above 700°C and melting is completed below 800°C in low-pressure conditions. Increasing pressure shifts the reaction to higher temperatures (Patiño Douce & Beard 1995). In both conditions, melt productivity remains weak, around 10–15%. The reaction also involves plagioclase and quartz, both in similar proportions and equivalent to the muscovite content. Biotite dehydration occurs between 820 and 950°C, with a weak pressure effect. The reaction involves biotite, as well as plagioclase and quartz in similar proportions, but each about three-quarters of the amount of biotite (Harris *et al.* 1995). The maximum amount of melt is about three-quarters of the initial volume of biotite in the source. Melts are peraluminous in composition with a K/Na ratio commonly greater than one (Whitney 1990).

Progressive dehydration melting of amphiboles starts at about 1000°C and 800 MPa (Rapp & Watson 1995). Reaction temperature slightly increases with pressure (Wyllie 1977). In vapour-absent conditions, hornblende and plagioclase react to form clinopyroxene and garnet plus melt. The melt is tonalitic in composition (Percival 1983) with K/Na commonly lower than one. At higher temperatures, amphiboles continue to react with plagioclase and clinopyroxene, producing more melt, up to 45%. At intermediate pressure, but lower than 1·0 GPa, the reaction consumes only hornblende and quartz, producing plagioclase and pyroxenes (Harlov 2002). In this case, the amount of hornblende should be greater than quartz (Patiño Douce & Beard 1995).

Mafic magmas derived from the mantle are necessary to generate the high temperature for hornblende dehydration melting. They also participate in granite formation in the continental crust. Mafic enclaves are ubiquitous in subduction-related granitic magmas, documenting several types of magma mixing (Didier & Barbarin 1991). Mixing takes place either deep in the crust, during the ascent or even after emplacement by a continuous feeding of mafic dykes (Collins *et al.* 2000; Weinberg *et al.* 2001). Three sources of magma have been suggested, involving the mantle, the intermediate crust and the lower crust (Collins 1996; Clarke *et al.* 2000). Balancing between the three sources may reproduce all types of granitic magmas. Trace element chemistry and isotopic signatures document the large mass transfer that has occurred between mafic and felsic magmas (Holden *et al.* 1991).

The dehydration reactions also imply the production of a restitic material of granulitic composition (Thompson 1982). A global estimate is that one volume of granitic magma and three volumes of granulitic restite result from the melting of four volumes of amphibolitic compositions (Wedepohl *et al.* 1991). Thus, dehydration melting has the double advantage of explaining granitic magma formation and crustal differentiation without the intervention of any external fluid flux.

2. Segregation

Two major points have long hindered comprehension of melt segregation. One was the requirement of a large volume of melt before segregation could take place. The second was the very incomplete understanding of the rheology of partially molten rocks (PMRs).

2.1. The old paradigm

Melting starts at mineral triple or quadruple junctions, where the activation energy is lowest (Jurewicz & Watson 1985). Once melt films are connected throughout the system, segregation starts, leading to the loss of cohesion of the source for about 20–30% melt. The commonly accepted model of melt segregation required a huge volume (10–20 km thick) of crust with highly variable melt fraction and a force to concentrate melts. Gravity-driven compaction is the preferred model (Fig. 2) of such force (Miller *et al.* 1988), in contrast to extensional fracturing (Shaw 1980).

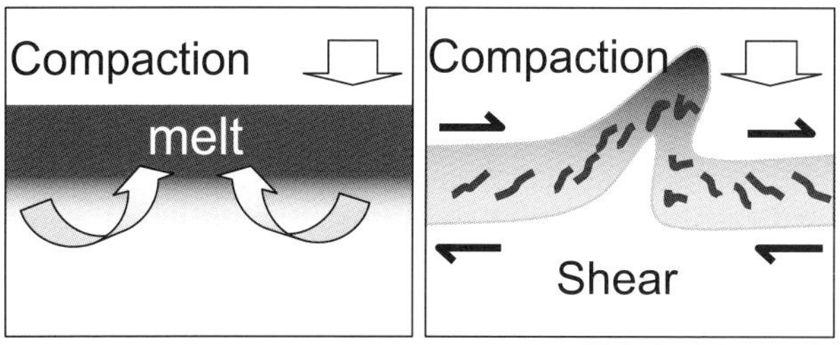

Figure 2 Melt segregation: Within the old paradigm (left), a layer (in grey) with large melt fraction segregates by compaction (arrow). Melt can escape from its source region when overcoming a critical percentage. In the new paradigm (right), compaction and shear instabilities develop and interact, resulting in small melt packets (in grey) periodically escaping the system. The figure is adapted from experiments (Barraud *et al.* 2001).

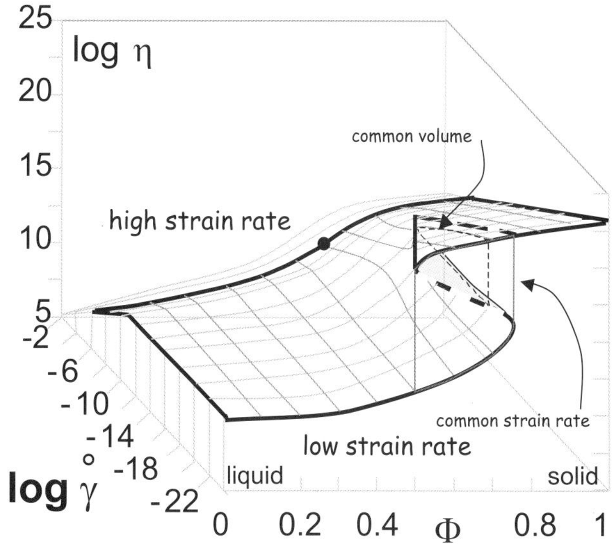

Figure 3 Rheology of partially molten rocks: The three-dimensional diagram plots viscosity (η) as a function of strain rate (γ°) and a percentage of the solid fraction (φ). For high strain rate (e.g. tectonic strains, the viscosity contrast is the lowest and the transition is continuous between the two end-members. In contrast, at low strain rates, the curve adopts a cusp shape and three values of viscosity exist for a given value of solid fraction. Two are metastable, and the intermediate value does not exist in nature. Different paths exist, depending on whether strain develops under constant strain rate, constant volume or constant stress.

The binary description of PMR rheology implied a critical value of melt percentage at which the system shifts from a solid-like to a liquid-like rheology. A rheological critical melt percentage (RCMP) marks the transition between the two behaviours (Arzi 1978). The binary description was later transposed to the deformation of rock samples bearing a small amount of melt (Van der Molen & Paterson 1979). In such a binary model, melting is symmetric to crystallisation. The RCMP value has been fixed at 25–30% melt, corresponding to the range observed in migmatites (Arzi 1978; Miller *et al.* 1988).

Numerical models attributed magma segregation to compaction (McKenzie 1984; Brown *et al.* 1995). They were directly transposed from mantle conditions to the continental crust. Nevertheless, they resulted in compaction lengths of the order of 100 m (McKenzie 1984), which do not match field observations on migmatites.

2.2. The new paradigm

Detailed observations on melt segregation in anatectic rocks (Gleason *et al.* 1999; Sawyer 2000; Harlov 2002) or in crystallising magmas (Bouchez *et al.* 1992; Sawyer 2001) document the difficulty of finding adequate rocks in the temperature range corresponding to that of melting. Sawyer (2000) examined rocks in the granulitic and amphibolitic facies, as well as meta-pelites during incipient melting. Local replacement of mineral by melt (Sawyer 2000; Harlov 2002) and later connection of melt along grain boundaries document anisotropic segregation. New optical techniques able to investigate the nature of grain boundaries (Dimanov *et al.* 1999; Kohlstedt *et al.* 2000) document the presence of amorphous material at grain boundaries. They suggest that a melt film can be traced even for very low melt fractions (Wirth & Frantz 2000).

Those observations suggest new interpretations on the rheology of PMR. Obviously, a binary model is false, and melting is not symmetric to crystallisation (Burg & Vigneresse 2002). Hence, latent heat makes the liquidus temperature different from that of the solidus. In the new model, three stages are considered (Fig. 3): two end-members and a transitional stage bracketed by two thresholds (Vigneresse *et al.* 1996). The later are the liquid percolation threshold (LPT) and the melt escape threshold (MET) for melting. Identically, a rigidity percolation threshold (RPT) and particle locking threshold (PLT) exist for crystallisation. The individual threshold values are not dual to each other during melting and crystallisation. In the transitional stage, non-linear feedback loops develop which have a positive effect during melting, but a negative one during crystallisation (Burg & Vigneresse 2002). Instabilities result from the heterogeneous distribution of melt with temperature and place. The transitional stage greatly depends on strain rate. Whereas the melt presents a constant viscosity, the matrix follows a power law rheology. Decreasing strain rate exacerbates the viscosity contrast, which lead to two types of reaction (Fig. 3). At a high strain rate, such as during tectonics, the viscosity contrast between the two phases is the lowest. The transition is progressive between the two end-members and an effective viscosity can be assigned, whatever the melt content. Conversely, at lower strain rates, the contrast between the two phases is the greatest. Three situations are observed, depending on the melt content. At low and high melt content, the system adopts the rheology of the respective end-member. In between the thresholds, a complex situation develops during which the two phases, i.e. matrix and melt, coexist. Jumps continuously occur from the value of one phase to that of the other, precluding assignation of a bulk viscosity value to the system. Strain preferentially partitions in the melt (Vigneresse & Tikoff 1999), which segregates discontinuously, yielding successive bursts of melt. The effects differ on whether they develop at constant strain rate, constant stress or constant amount of melt (Fig. 3) (Vigneresse & Burg 2004).

Analogue models of melt segregation (Barraud *et al.* 2001; Bons *et al.* 2001) demonstrate segregation of discrete magma packets (Fig. 2). No quantitative interpretation is presently given for the discontinuous melt motion in those experiments. Numerical models using classical descriptions (Rabinowicz *et al.* 2001; Ricard *et al.* 2001) or using cellular automata (Vigneresse & Burg 2000) also document instabilities. They all consider low melt concentration located at grain boundaries (5–10% melt), which results in an equivalent porous flow. For granitic melts which segregate in crustal conditions, compaction is not efficient enough and horizontal shear must be added (Rabinowicz & Vigneresse 2004). A competition starts between instabilities initiated by compaction and by shear. They cooperate and lead to a compaction length that reduces to decimetres in scale, analogous to what is observed in migmatites (Rabinowicz & Vigneresse 2004). Melt extraction is discontinuous and successive cycles can develop during which a limited percentage (5%) of melt is expelled. The heat

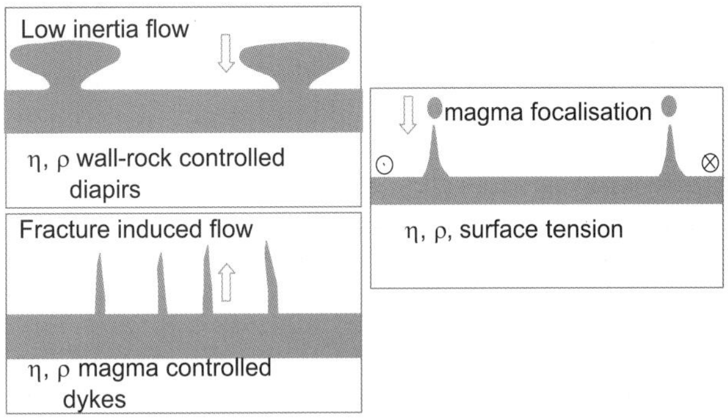

Figure 4 Magma ascent toward the upper crust: In the old paradigm (left), the ascent develops through low inertia flow (diapirs) and fracture-induced flow (dykes). The viscosity and density values of the wall rock and of the magma, respectively, control the ascent in the two processes. In the new paradigm, rheological instabilities develop under compaction and shear, leading to individual bursts of melt and magma focalisation.

available for additional melting controls the cycle frequency. However, progressive melt removal also stiffens the crust, limiting the total number of cycles. The total amount of melt that can be extracted remains below 25% in volume (Rabinowicz & Vigneresse 2004).

3. Ascent

3.1. The old paradigm

Granitic magmas ascend from the lower crust where they form by melting to the middle crust where they build large plutons. In the old paradigm, there were two basic models (Fig. 4) of magma ascent: inertia flow or fracture propagation (Vigneresse & Clemens 2000). In both cases, a nearly liquid magma ascends to coalesce and form large plutons. Both models assume a neutral buoyancy level at which the density contrast between magma and its host rocks vanishes, stopping the ascent.

In diapiric ascent or Rayleigh-Taylor instability, a lighter and less viscous layer forms a dome, followed by the rise of a characteristic inverted-teardrop-shaped body. The viscosity contrast between magma and its host rocks drives the instability. Combined analogue and numerical experiments refined inertia flow models. A Stokes flow used either a rigid (Cruden 1988) or a soft sphere (Schmelling *et al.* 1988). Rayleigh-Taylor instabilities assumed either viscosity contrast (Weinberg 1996) or not (Weijemars 1986). Models in which both the ascending body and its host are viscous (Mahon *et al.* 1988) also examined the thermal effects. Most recent models consider a viscous intrusion into a ductile and/or brittle crust (Barnichon *et al.* 1999). Those models restrict diapirism to the lower plastic deforming crust. Former models had already demonstrated, though sometimes in a cryptic form, that diapirism is not compatible with a brittle upper crust (Mahon *et al.* 1988; Weinberg & Podladchickov 1994). In those models, a less dense and less viscous sphere, typically 5 km in radius, starts from a depth of about 30–40 km, and slows down its ascent at mid-crustal depth before stopping at a depth of between 14 and 21 km (see Vigneresse & Clemens 2000, fig. 1).

Fracture propagation, or dyking, represents an alternative for magma ascent. The magma viscosity drives the forces at the tip of a fracture, leading to the upward propagation of the fracture. The internal magma pressure must balance the elastic response of the rock. The first models (Weertman 1971) used an elastic crack with an internal pressure gradient caused by magma (Pollard & Muller 1976) or viscous liquid fill (Lister & Kerr 1991; Rubin 1997).

At present, magma ascent as a whole body (i.e. diapirism) is commonly considered to be unrealistic. Even salt diapirism is driven by tectonics, as evidenced by precise measurements of extrusion rate (Talbot *et al.* 2000). For magma, dyking is actually the preferred model. Nevertheless, it faces numerous problems relating the diameter of the feeding pipes to magma viscosity (Petford *et al.* 1993; Clemens *et al.* 1997). Small zones observed from gravity data inversion and interpreted as feeders for the pluton (Améglio *et al.* 1997) argue for dyking. Nevertheless, gravity data precision limits the detection of these feeders to those about 300 m in width at about 5 km in depth. Larger feeders should be detectable from gravity measurements, which obviously is not the case.

A neutral buoyancy level is not supported by field observations. A single pluton shows various types of magma, which all present with a varying composition and, hence, density. The apparent occurrence of magmas with different densities at the same level contradicts the existence of a neutral buoyancy level. Gravity measurements performed over granitic intrusion document negative or positive anomalies between the massif and its surroundings. The anomalies indicate that successive magma pulses stopped at a similar level, whatever their density (Vigneresse & Clemens 2000).

3.2. The new paradigm

In the new paradigm, deformation is intimately linked to granitic magma ascent (Fig. 4), providing the required forces to overcome the strength barrier at the ductile/brittle transition (Vigneresse 1995b). Nevertheless, magma ascent is not related to a specific type of deformation such as extensional tectonics. Numerous examples document granites associated with strike slip conditions or compressional thrusts (Schmidt *et al.* 1990; Kalakay *et al.* 2001). The best way to extract a melt from its matrix is by compression and shear (Vigneresse & Burg 2000). However, by itself, the ascent must be in a locally extensional near-field setting because of the lesser magnitude of the tectonic stresses magma must overcome (Vigneresse 1995a).

Complex relationships exist between granitic intrusion and faulting (Paterson & Schmidt 1999), excluding a genetic link between intrusion and faulting. Intrusions are randomly distributed with respect to faults, which precludes that faults could channel magma during the ascent (Paterson & Schmidt 1999). From gravity data inversion, the root zones of a pluton are not connected to adjacent faults, leading to a similar conclusion (Vigneresse 1995a). However, the shape of the walls in a pluton are controlled by adjacent faulting as with the Mortagne Massif, France (Guineberteau *et al.* 1987).

Differences in ascent are found in analogue models depending on whether they develop within a homogeneous crust or not. Laccoliths form during intrusion into a homogeneous crust (Roman-Berdiel *et al.* 1995), while wedge-shaped massifs develop when intruding a heterogeneous crust (Roman-Berdiel 1999). When the deformation pattern is complex (transtension or transpression), intrinsic non-linear relationships form loops between magma production and tectonics, driving mechanisms to complex situations (Saint Blanquat *et al.* 1998).

4. Emplacement

4.1. The old paradigm

In the old paradigm, the ascent of a large volume of magma as a whole requires a large space to be created that would be later filled with magma. The so-called 'room problem' has been a

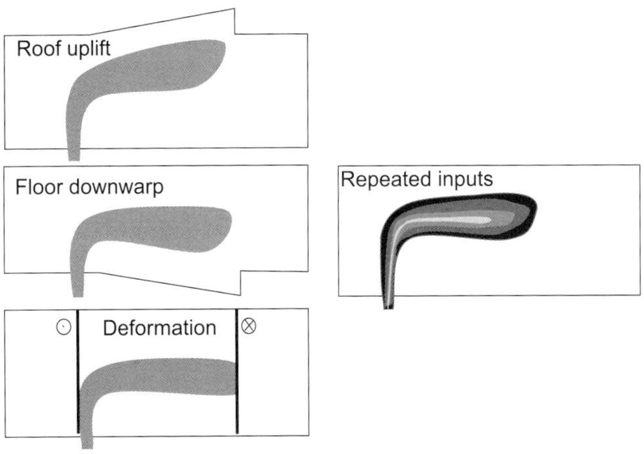

Figure 5 Granitic magma emplacement in the intermediate crust: In the old paradigm (left), the so called room problem is solved by roof lifting (top), floor downwarping (centre) or deformation-assisted opening (bottom). In the new paradigm, repeated inputs of magma leave time for the surrounding rocks to relax and to accommodate magma emplacement.

continuous source of disputes during the past few decades (Read 1957; Paterson & Vernon 1995; Hutton 1997). Several mechanisms have been suggested which all contribute to create space for intrusion to take place. Wall assimilation has been advanced on the basis of isotopic and trace elements (Taylor 1980; De Paolo 1981). Large blocks of surrounding rocks are commonly observed which have fallen into the magma and present various degrees of assimilation. Room accommodation could also be realised by either uplifting (Fig. 5) the overlying rocks (Kerr & Pollard 1998) or depressing the floor (Fig. 5) to compensate magma ascent (Cruden 1998). Tectonic-aided fractures (Fig. 5) can also facilitate magma emplacement by inducing pull-apart-like structures which are rapidly filled by magma (Hutton 1982; Tikoff & Teyssier 1992).

In all those emplacement mechanisms, granitic magma acts passively. Nevertheless, it remains difficult to conceive how magma in a quasi-liquid state can develop enough strength to sustain those stresses induced by surrounding crustal rocks.

4.2. The new paradigm

A significant advance was initiated when structures were identified within granitic plutons (see review in Bouchez *et al.* 1997). Documenting internal structures, together with the determination of the three-dimensional (3D) shape of the massif by gravity data inversion, revealed that granite plutons recorded strain history (Guillet *et al.* 1985). Magma registers the very late strain increment (Benn 1994; Paterson *et al.* 1998). Present structural analyses commonly involve the anisotropy of magnetic susceptibility (AMS). The bulk shape of a granitic massif is obtained after mapping gravity anomalies, i.e. density changes caused by facies variations, over a pluton. Full 3D inversion is recommended (Vigneresse 1990). The shape of a massif at depth can later be correlated with its internal structures. Zones where the depth of the floor becomes deeper, provided they also correlate with steeply plunging magmatic lineations, are interpreted to be reflecting magma feeders (Vigneresse & Bouchez 1997).

Field observations, especially structures at the contact with wall rock, reveal that a forceful intrusion, with a major vertical movement of the magma, is often unlikely in granite plutons. In many cases, intrusions contain horizontal foliation and lineations close to, but orthogonal to, the walls of the granitic body. No flattening is observed within the granitic body when approaching the contact. By contrast, the wall rocks may show flattening approaching the granite (Brun *et al.* 1990), showing that magma inflated, pushing the host rocks during a large-scale horizontal movement of magma.

Granitic plutons have been classed by their shape into tabular- and wedge-shaped bodies (Améglio & Vigneresse 1999). A majority of batholiths belongs to the first category, whereas the second includes special cases often associated with tectonically controlled emplacement (Améglio *et al.* 1997). Tabular intrusions show a limited thickness (5 ± 2 km in average), with their floors gently dipping towards a magma feeder, marked by a pronounced deepening. Within the massif, magma flows and regularly fills its room within the upper crust. Other examples describe magma as being intrusive within concordant structures and sub-parallel to host rocks anisotropy, for which the brittle-ductile transition is the best example (Vigneresse 1995a). In the second case, walls steeply plunge inward and the floor is deep (about 10 km at present) and flat, without any evidence of a localised deeper root zone. They fit models of tectonically assisted emplacement, the walls being often associated with shear zones.

The regional stress field developed by magma intrusion under partial crystallisation (Parsons *et al.* 1992; Vigneresse *et al.* 1999) explains the geometry of some granitic intrusion emplaced at high levels in the crust (Hogan *et al.* 1998). Magma intrudes along planes perpendicular to the minor stress component (σ_3), which are usually vertical planes in extensional or strike-slip environments. They represent planes of easy opening (Ramsay & Huber 1983). During crystallisation, the framework of touching crystals rapidly reaches 50% concentration which guarantees the rigidity of the body (Guyon *et al.* 1990). Above, the magma is able to sustain and to transmit normal stresses. It locally alters the nearby stress field (Parsons *et al.* 1992; Vigneresse *et al.* 1999), increasing the magnitude of the two minor stress components (σ_3 and σ_2). Two minor stress components may switch together when they have similar amplitudes. Step by step, opening along the initial vertical plane slows down and the stress pattern switches to another configuration (Vigneresse *et al.* 1999).

In a purely extensional stress field, the horizontal stress components are about one-quarter of the vertical load. Successive switches between the horizontal components may lead to the formation of vertical and cylindrical plutons (Fig. 6). The latter builds up from successive vertical sheets oriented along planes which rotate according to successive stress exchanges. This is the case for many 'anorogenic' granites, such as those in Namibia (Bowden & Turner 1974). In the case of transtension, switch occurs from strike-slip to extension, whereas the horizontal stress components switch from strike-slip condition to compression in cases of transpression. Elongated thin vertical sheets of magma result in transtensional situations, as observed in Southern Chile (Wilson & Grocott 1999). Batholiths built during transpression are wider, oriented along the shear direction (Fig. 6). They may develop intrinsic shear zones within the granitic magma; for instance, as observed in the Sierra Nevada (Saint Blanquat *et al.* 1998). When being emplaced during compression, massifs adopt a thin plate horizontal shape (Fig. 6).

Petrographic zoning in granitic plutons is a function of the rate at which successive batches of magma are emplaced (Hecht *et al.* 1997; Hecht & Vigneresse 1999). In the case of rapidly delivered magma batches, the earlier magma has no time to solidify. The later pulses of magma are emplaced into the core, pushing aside the former batch. If those batches consist of magma fractionated at depth, the resulting zonation of the pluton is normal, the latter material intrusive towards the core being more felsic. In contrast, when the rate of emplacement is slow, magma has time to cool down, reaching about 50% crystals or more. It becomes rigid before a new

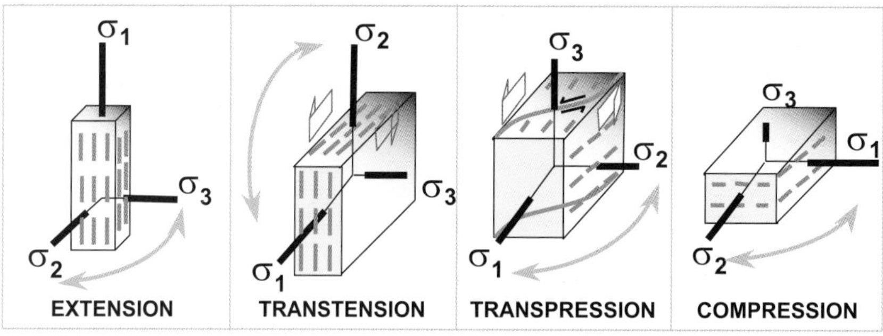

Figure 6 Shape of granitic plutons depending on the switches between the stress components ($\sigma_1 \geq \sigma_2 \geq \sigma_3$) occurring in extensional, transtension, transpression and compressional environments (see text).

magma batch comes in. The later batch cannot develop enough force to overcome the yield strength of the now solidified magma. Thus, it is forced to intrude beside the earlier intrusion. A reverse zoning results, with the older magma in the core, and the younger, more evolved magma at the rims on both sides.

A granitic pluton is built from a sequence of successive intrusions (Fig. 5) with variously evolved magma (Vigneresse & Bouchez 1997; Petford *et al.* 2000). It has been verified by numerous examples on the scale of the large batholiths which form the Sierra Nevada (McNulty *et al.* 2000) and on a smaller scale in southern Spain (Aranguren *et al.* 1997). At an even smaller scale, in the Himalayan granites, magma heterogeneity is documented by the large variation in isotopic ratios (Deniel *et al.* 1987) and by the variable degree of melting shown by the distribution of rare-earth elements in zircons and monazites (Brouand *et al.* 1990).

In conclusion, granite magmatism takes place because host rocks allow emplacement within the crust. Discontinuous intrusions of magma leave time for the crust to relax the stress induced by magma intrusion, avoiding the room problem. The competition between the rate of magma emplacement and that of its cooling to reach rigidity determines the type of zoning in a granitic pluton. Both emplacement and zoning are dynamic.

5. The necessary mantle implication

5.1. The old paradigm

Granite magmas are, by essence, of crustal origin, as indicated by isotopic data. Therefore, initial isotopic strontium ratio (I_{Sr}) constitutes a good indicator of crustal contamination (Rollinson 1993). Hence, the mantle presents a relatively low and uniform $^{87}Sr{:}^{86}Sr$ ratio that slowly increases linearly with time. In contrast, the ratio linearly evolves more rapidly from the time crustal rocks were formed and it starts from a value already higher compared to the mantle. The Sm:Nd ratio of the mantle is higher than that of the crust, implying that their isotopic ratios are in the same trend. However, the ratio for the crust and siliceous igneous rocks remains relatively uniform. The initial $^{143}Nd{:}^{144}Nd$ ratio of the mantle linearly varies with age from the Bulk Earth composition. A coefficient, ε_{Nd}, represents the relative deviation of this ratio from the chondritic ratio. Starting with a null value of ε_{Nd} at the formation of the Earth, mantle rocks display a linear increase with time, whereas crustal rocks show negative ε_{Nd} values (Allègre & Ben Othman 1980).

The early finding of granitic magmas with high values of I_{Sr} and negative ε_{Nd} values indicated that a large amount of crust was necessary to explain granite genesis (Allègre & Ben Othman 1980). It also suggested that granites were mostly crustally derived.

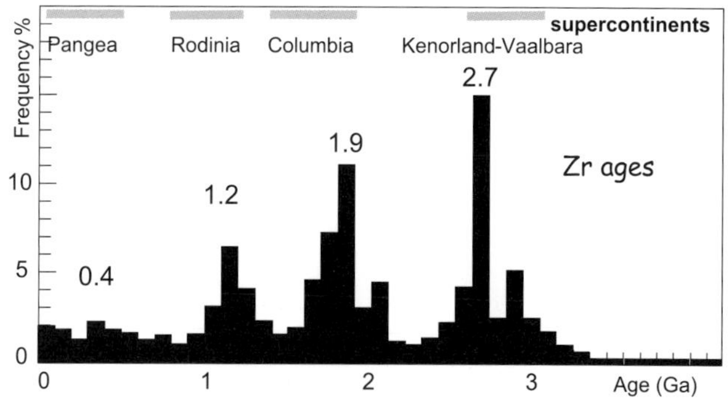

Figure 7 Frequency diagram of zircon ages magmatism (redrawn from Condie 1998). The various peaks show the exacerbated magma production during amalgamation of supercontinents. It indirectly shows the role of mantle in producing granites by the increased amount of heat and stress generated during those periods.

5.2. The new paradigm

In the new paradigm, mantle acts as a material, heat and stress provider. Granites are obviously linked to plate tectonics (Pitcher 1993). They preferentially develop at plate margins, and show affinity with plate convergence and, to a lesser extent, with plate formation in the case of alkaline magmatism. Thus, it is not surprising to observe peaks in a frequency diagram of zircon ages of magmatic events (Condie 1998). These peaks reflect the abundant magma production linked to supercontinent cycles (Fig. 7) and document the role of the mantle during granite generation.

The importance of the mantle as a chemical reservoir is illustrated by the ubiquitous presence of mafic dykes and enclaves in a granitic magma (Didier & Barbarin 1991; Collins *et al.* 2000). Isotopic studies confirm the role of the mantle component with granitic rocks presenting positive ε_{Nd}. The diagram (Fig. 8) compiled for Hercynian and Caledonian granites from Europe and Australia, plus some granites from China (Jahn *et al.* 2000), demonstrates a balance of mantle and crustal components. Models of contamination indicate that most of data points can be explained by 25–66% of the crustal component, the volume of the mantle component thus consisting of three-quarters to one-third of the total material.

The mantle is also the heat engine that controls crustal melting. Mantle heat adds to the crustal heat produced by the radioactive decay of elements (U, Th and K). Common values of continental surface heat flow range from 40 to $120\,mW\,m^{-2}$, depending on crustal thickness and age of the geothermal province (Morgan 1984). The average value is about $60\,mW\,m^{-2}$. A common accepted balance is 10–$15\,mW\,m^{-2}$ from the mantle, $5\,mW\,m^{-2}$ from the lower crust, which is notably depleted in both U and Th, the remainder from the uppermost crust. Increasing the thickness

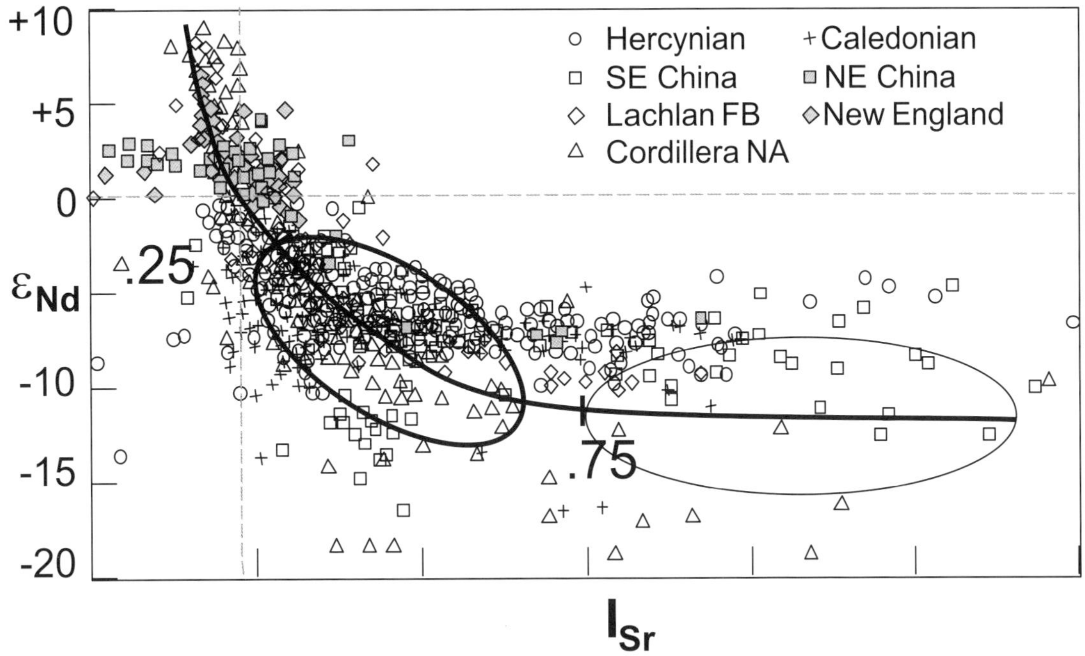

Figure 8 Isotopic ratios (I_{Sr} and ε_{Nd}) for Hercynian and Caledonian granites of Europe and south western Australia, north western Cordillera and from China (data from Jahn *et al.* 2000). Superimposed on the data sets are the field of the depleted mantle and upper crust. The field of the crustal component is shifted toward more negative ε_{Nd} depending on the age of the protolith. Contamination by an older crustal reservoir (f. i. 1·0 Ga) shifts the values towards negative ε_{Nd} (-10 units) and higher I_{Sr}. An average trend of mixing mantle and crustal components is indicated with their respective amount (0·25, 0·50 and 0·75). Most of the granitic magmas demonstrate the important contribution of the mantle (0·50–0·66).

of the heat-producing elements increases the surface heat flow, hence the geothermal gradient, which is 15–17°C/km^{-1} in standard conditions (Thompson 1999). In consequence, the temperature hardly reaches more than 525–600°C at 35 km or at the Moho. It should be compared to the minimum temperature of 820°C for biotite dehydration and 1000 °C for hornblende dehydration reactions (Patiño Douce & Beard 1995). Steady-state geotherms are not high enough to cause melting, attesting to the need for an additional heat source by either increasing crustal thickness or adding heat flux from the mantle. Doubling the average surface heat flow requires doubling of the crustal thickness or quadrupling the mantle heat flow.

Actually, high surface heat flow values (>100 mW m^{-2}) are measured in regions (e.g. the Alps, Himalayas and Andes) of thickened crust (Sclater *et al.* 1981). However, crustal thickening is not a mechanism likely to generate huge amounts of melt in a short time because increasing crustal thickness by thrusting takes time (Thompson 1999). Crustal erosion competes with heat diffusion and the crust returns to its normal thickness in about 100 Ma (England & Thompson 1984). No more than 20% melt is produced in about 70 Ma after crustal thickness doubling in cases of a standard geotherm and high (30 mW m^{-2}) mantle heat flow (Thompson 1999). In the case of a hot geotherm (40 mW m^{-2}), corresponding to tonalitic melts, melting starts 10 Ma after crustal thickening. A maximum (50%) of melt is produced 60 Ma later (Thompson 1999). Conversely, models in which the crust is thinned to half its initial thickness cannot achieve melting temperatures using a standard geotherm (Loosveld 1989) and melting is confined into the mantle. Only hot mode delamination, which would bring hotter mantle material to the Moho depth, can lead to high enough temperatures in the crust (Thompson 1999).

Other mechanisms for crustal melting require either advective or diffusive heat. Advective heat transport brings hot material from the mantle and rapidly heats up the crust. The time scale to induce melting is shorter by about one or two orders of magnitude compared to crustal thickening. However, latent heat effects restrict the melt volume. Six volumes of basalt are required to generate one volume of granitic magma (Bergantz 1995). The time for diffusive heat to progress varies 0·03 L^2 with distance (L) in kilometres and using a thermal diffusivity of 10^{-6} m^2 s^{-1} (Crank 1975). When the heat source is at the base of the crust, it takes 3 Ma for heat to induce crustal melting at 22 km in depth, corresponding to about 600 MPa. When the heat source is at the base of the lithosphere, corresponding to heat released from the asthenosphere, the time delay may reach about 300 Ma. For instance, it corresponds to the formation of the so-called 'anorogenic' granites which lead to the 1·6–1·3 Ga formation of the rapakivi granites in Fennoscandia (Rämö & Haapala 1995). This situation requires that the continental plate remains essentially stationary over the heat source for this long. It can develop only in the case of a supercontinent cycle (Condie 1998), during which continental fragments amalgamate and remain stationary above a converging, and thus descending, mantle convective cell.

The mantle is also indirectly linked to granite generation as a stress provider to the continental crust. Because plate tectonics are partly driven by mantle convection, it generates permanent stresses within the plates (Zoback 1992). The stress pattern is characteristic of the continental crust. Stiffness of the thick and brittle upper crust allows large stress amplitude to develop. At the brittle/ductile transition, 100 MPa is necessary to induce extensional fractures in the continental crust and values of 600 MPa are common in cases of compressional environments (Vigneresse *et al.* 1999). Such high values contrast with stress that develops in a mantle environment. Stresses originating from the mantle control magma segregation, ascent and emplacement of granitic magmas in the crust. The high stress level in the continental crust precludes the transposition to crustal conditions of mechanisms acting into the mantle.

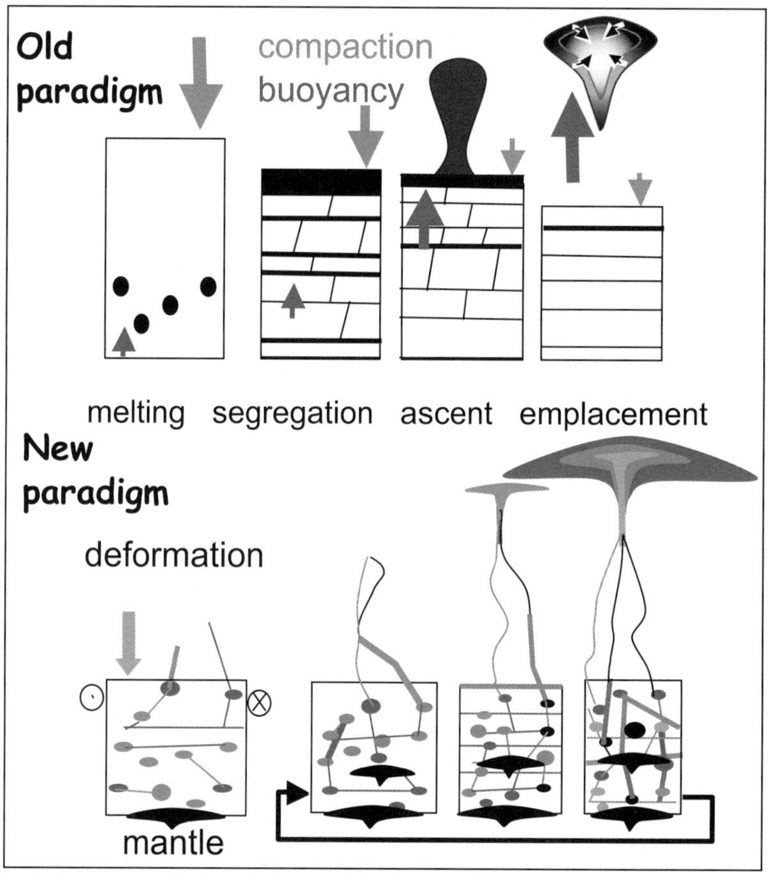

Figure 9 MASH+AFC model (modified from Hildreth & Moorbath 1988) and the suggested new paradigm. In the old paradigm (top), four stages are identified: melting, segregation, ascent and emplacement. They develop successively, and each of them has its own characteristic length and time scale. In the new paradigm (bottom), the four phases are discontinuous, but they may develop at any time during the whole process. In addition, the external settings, represented by the mantle, act as a provider of heat and stress and as a chemical reservoir.

6. Global model

6.1. The old paradigm: MASH and AFC models

More accurate isotopic and trace element analytical methods allowed modelling of element variation during assimilation and fractional crystallisation or (AFC; De Paolo 1981; Powell 1984). The models were initiated with the equations describing isotope and trace elements behaviour (Shaw 1970) during fractionation in an open system or one periodically recharged with magma (O'Hara 1977). The most widely used AFC formalism describes the evolution of isotopes, and major and trace elements contained in a magma chamber affected by fractional crystallisation and wall rock assimilation at the same time (De Paolo 1981). Mixing with another magma may also be involved (Bergantz 2000). The necessary parameters are the distribution coefficients and the balance between the rate of assimilation to the rate of fractional crystallisation. When the ratio approaches infinite values, a simple binary model is approached. The variations of the distribution coefficients reflect the respective characters of compatible and incompatible elements within the given phase. Modifications to AFC have been introduced allowing magma replenishment (O'Hara & Matthews 1981), or decoupling between fractional crystallisation and assimilation (Cribb & Barton 1996). Energy-constrained assimilation and fractional crystallisation with replenishment has recently been suggested (Spera & Bohrson 2001). However, the basic assumption of a global ascent and emplacement of felsic magma with a volume similar to a pluton is still the basis of all AFC models.

The AFC models also require a coherent set of partition coefficients for felsic magmas (Bea *et al.* 1994). Their determination for major elements is ruled by Henry's laws, assuming equilibrium between melt and mineral formation (Rollinson 1993). This is valid for major elements contained within the major mineral phases. In contrast, incompatible elements (Th, Zr or REE) are incorporated as a separate phase in accessory minerals. The initial assumptions are consequently violated since elements are equally partitioned in the major minerals, but preferentially in accessory minerals. Thus, REE can hardly be used for modelling the genesis of granites (Bea 1996). It is necessary to check for which mineral assemblage a given set of partition coefficients is valid. For instance, the often-referred compilation of partition coefficients by Rollinson (1993) is only valid for basalts and basaltic andesites (Roberts & Clemens 1995). Determining a full set of partition coefficients for a granitic magma is a very difficult task and values commonly vary by several orders of magnitude for felsic melts. This has drastic effects on the fractional crystallisation curves.

Assimilation and fractional crystallisation (AFC) modelling is used on either trace element or isotopic data, but rarely on both at the same time (Roberts & Clemens 1995). When restricted to a set of elements, AFC modelling can always come out with a solution by modifying the respective rate of fractionation/assimilation or the degree of partial melting.

A global model for igneous intrusions is commonly designated as melting, assimilation, storage and homogenisation (MASH; Hildreth & Moorbath 1988). It applies to entire magma generation (Fig. 9a). Mantle and crustal magmas mix close to the mantle–crust boundary, establishing the characteristic chemical signature of magmas at the melting stage. The age and consequent chemical composition of the lower crustal component controls the isotopic variations, forming the assimilation stage. The depth at which mixing occurs controls the storage stage. Finally, AFC processes may substantially modify the composition of the ascending magmas, resulting in the homogenisation stage (Hildreth & Moorbath 1988). The MASH model is the first holistic approach to magma generation in convergent plate settings. Its major points consist of: (1) the importance of the crustally derived magma, with less involvement of mantle component; and (2) the role played by the mixing of magmas deep in the crust, in contrast to earlier ideas which implied an upper crustal magma chamber.

In conclusion, the current generally preferred model of granite generation assumes four separate stages which make up a global process (Fig. 9). A large amount of melt is required before segregation, and therefore, the bulk volume of buoyant magma drives magma ascent. A magma chamber is created by the input of magma that later differentiates and evolves chemically. The MASH model is basically static.

6.2. Toward a new paradigm

The new suggested paradigm is dynamic. Magma segregates by developing alternate periods of melt segregation with periods of matrix relaxation (Rabinowicz & Vigneresse 2004). Magma ascent also proceeds from a similar mechanism, discontinuous in space and time. Building a pluton consists of aggregating discrete pulses of magma, each having a unique chemical history and evolution (Bergantz 2000; Petford *et al.* 2000). Local mixing and chemical re-equilibration may take place, induced by the disturbance caused by the arrival of a new batch of magma into a more or less partially crystallised pool of magma. The discrete succession of magma batches interacts with the deformation field, locally altering the stress pattern. The intrusive body continuously modifies its internal shape and chemistry (Fig. 9), resulting in a petrographic zoning.

The discontinuous character of magma flow has important consequences for granite generation itself. It severely reduces the implications of results obtained from numerical models based on bulk magma chamber evolution. Most thermomechanical models of convective motion, crystallisation, cooling and chemical evolution caused by crystal-mush interactions assume a stationary process or a closed chemical system (see reviews in Cashman & Bergantz 1991; Ryan 1994). These models may remain valid, but on a much smaller scale. They could apply successively to the evolution of a body equivalent to a pluton in size, assuming that each pulse evolves individually. However, complex interactions and instabilities develop inherently with the decreasing time lag between magma batches.

The role of the mantle appears essential to all crustal processes. The mantle also certainly has an indirect role to play, since it brings the initial heat source or induces a stress pattern to the crust. It may also act as a counteractive parameter when its action switches off the controlling parameters.

7. Implications

The implications of the new paradigm are twofold. First, the changing context of several inputs of magma to build a granitic pluton certainly implies revisiting all models of chemical differentiation and fractionation. All schemes which incorporate a bulk *en masse* movement of magma should be reconsidered. Secondly, the successive inputs of magma, with varying chemical composition or temperature, have consequences on the way that the magma globally evolves. This should lead to a reconsideration of the very late stage of emplacement, which is often closely related to ore formation.

8. Acknowledgements

This paper is the result of a long time spent on granite magmas genesis partly funded by U Mining Companies and CREGU. The paper benefited from active support from CREGU. I appreciated collaborating with many people in previous papers (L. Améglio, J. L. Bouchez, J. P. Burg, J. Clemens, J. Dehls, P. Guillet, B. Guineberteau, L. Hecht, N. Petford, M. Rabinowicz, B. Tikoff and certainly many others). They are especially thanked for providing data and ideas. P. Lagrange offered his skill and judicious advice for the illustrations. Two anonymous referees greatly helped in editing the manuscript.

9. References

Allègre, C. J. & Ben Othman, D. 1980. Nd-Sr isotopic relationship in granitoid rocks and continental crust development: a chemical approach to orogenesis. *Nature* **286**, 335–42.

Améglio, L. & Vigneresse, J. L. 1999. Geophysical imaging of the shape of granitic intrusions at depth: a review. *In* Castro, A., Fernandez, C. & Vigneresse, J. L. (eds) *Understanding Granites: Integrating New and Classical Techniques. Geological Society, London, Special Publication* **168**, 39–54. London: Geological Society.

Améglio, L., Vigneresse, J. L. & Bouchez, J. L. 1997. An assessment of combined fabrics and gravity data in granites. *In* Bouchez, J. L., Hutton, D. & Stephens, W. E. (eds) *Granite From Melt Segregation to Emplacement Fabrics*, 199–214. Dordrecht: Kluwer Academic Publishers.

Aranguren, A., Larrea, F. J., Carracedo, M., Cuevas, J. & Tubia, J. M. 1997. The Los Pedroches batholith (southern Spain): polyphase interplay between shear zones in transtension and setting of granites. *In* Bouchez, J. L., Hutton, D. H. W. & Stephens, W. E. (eds) *Granite: From Segregation of Melt to Emplacement Fabrics*, 215–29. Dordrecht: Kluwer Academic Publishers.

Arzi, A. A. 1978. Critical phenomena in the rheology of partially melted rocks. *Tectonophysics* **44**, 173–84.

Barnichon, J. D., Hanevith, H. D., Hoffer, B., Charlier, R., Jongmans, D. & Duchesne, J. C. 1999. The deformation of the Egersund-Ogna anorthosite massif, south Norway: finite element modelling of diapirism. *Tectonophysics* **303**, 109–30.

Barraud, J., Gardien, V., Allemand, P. & Grandjean, P. 2001. Analog modelling of melt segregation and migration during deformation. *Physics and Chemistry of the Earth* **A26**, 317–23.

Bea, F. 1996. Residence of REE, Y, Th and U in granites and crustal protoliths. Implications for the chemistry of crustal melts. *Journal of Petrology* **37**, 521–52.

Bea, F., Pereira, M. D. & Stroh, A. 1994. Mineral/leucosome trace-element partitioning in a peraluminous migmatite (a laser ablation-ICP-MS study). *Chemical Geology* **117**, 291–312.

Benn, K. 1994. Overprinting of magmatic fabrics in a granite by small strains: numerical modelling. *Tectonophysics* **213**, 153–62.

Bercovici, D. 2003. The generation of plate tectonics from mantle convection. *Earth and Planetary Science Letters* **205**, 107–21.

Bergantz, G. W. 1995. Changing techniques and paradigms for the evaluation of magmatic processes. *Journal of Geophysical Research* **B100**, 17603–13.

Bergantz, G. W. 2000. On the dynamics of magma mixing by reintrusion: implications for pluton assembly processes. *Journal of Structural Geology* **22**, 1297–309.

Bons, P. D., Dougherty-Page, J. & Elburg, M. A. 2001. Stepwise accumulation and ascent of magmas. *Journal of Metamorphic Geology* **19**, 627–33.

Bouchez, J. L., Delas, C., Gleizes, G., Nédelec, A. & Cuney, M. 1992. Submagmatic microfractures in granites. *Geology* **20**, 35–8.

Bouchez, J. L., Hutton, D. & Stephens, W. E. 1997. *Granite: From Melt Segregation to Emplacement Fabrics.* Dordrecht: Kluwer Academic Publishers.

Bowden, P. & Turner, D. C. 1974. Peralkaline and associated ring-complexes in the Nigeria-Niger province, West Africa. *In* Sorenson, H. (ed.) *The Alkaline Rocks.* New York, NY: Wiley.

Brouand, M., Banzet, G. & Barbey, P. 1990. Zircon behavior during crustal anatexis – evidence from the Tibetan Slab migmatites. *Journal of Volcanic and Geothermal Research* **44**, 143–61.

Brown, M., Averkin, Y. A., McLellan, E. L. & Sawyer, E. 1995. Melt segregation in migmatites. *Journal of Geophysical Research* **B100**, 15655–79.

Brun, J. P., Gapais, D., Cogné, J. P., Ledru, P. & Vigneresse, J. L. 1990. The Flamanville granite (NW France): an unequivocal example of an expanding pluton. *Geological Journal* **25**, 271–86.

Burg, J. P. & Vigneresse, J. L. 2002. Non-linear feedback loops in the rheology of cooling-crystallising felsic magma and heating-melting felsic rock. *In* De Meer, S., Drury, M. R., De Bresser, J. H. P. & Pennock, G. M. (eds) *Deformation Mechanisms, Rheology and Tectonics: Current Status and Future Perspectives. Geological Society, London, Special Publications* **200**, 275–92. London: Geological Society.

Cashman, K. V. & Bergantz, G. W. 1991. Magmatic processes. *Reviews of Geophysics* **29**, 500–12.

Castro, A., Fernandez, C. & Vigneresse, J. L. 1999. *Understanding Granites: Integrating New and Classical Techniques. Geological Society, London, Special Publication* **168**. London: Geological Society.

Clarke, D. B., Fallon, R. & Heaman, L. M. 2000. Interaction among upper crustal, lower crustal and mantle materials in the Port Mouton pluton, Meguma Lithotectonic Zone, southwest Nova Scotia. *Canadian Journal of Earth Sciences* **37**, 579–600.

Collins, W. J. 1996. Lachlan Fold Belt granitoids: products of three-component mixing. *Transactions of the Royal Society of Edinburgh: Earth Sciences* **87**, 171–81.

Clemens, J. D. 1990. The granulite–granite connexion. *In* Vielzeuf, D. & Vidal, P. (eds) *Granulite and Crustal Evolution*, 25–36. *NATO ASI Series* **C311**. Dordrecht: Kluwer Academic Publishers.

Clemens, J. D., Petford, N. & Mawer, C. K. 1997. Ascent mechanisms of granitic magmas: causes and consequences. *In* Holness, M. B. (ed.) *Deformation-enhanced Fluid Transport in the Earth's Crust and Mantle*, 144–71. London: Chapman & Hall.

Collins, W. J., Richards, S. R., Healy, B. E. & Ellison, P. I. 2000. Origin of heterogeneous mafic enclaves by two-stage hybridisation in magma conduits (dykes) below and in granitic magma chambers. *Transactions of the Royal Society of Edinburgh: Earth Sciences* **91**, 27–45.

Condie, K. C. 1998. Episodic continental growth and supercontinents: a mantle avalanche connection? *Earth and Planetary Science Letters* **163**, 97–108.

Crank, J. 1975. *The Mathematics of Diffusion*. Oxford: Oxford University Press.

Cribb, J. W. & Barton, M. 1996. Geochemical effects of decoupled fractional crystallisation and assimilation. *Lithos* **37**, 293–307.

Cruden, A. R. 1988. Deformation around a rising diapir modeled by creeping flow past a sphere. *Tectonics* **7**, 1091–101.

Cruden, A. R. 1998. On the emplacement of tabular granites. *Journal of the Geological Society of London* **155**, 853–62.

De Paolo, D. J. 1981. Trace element and isotopic effects of combined wallrock assimilation and fractional crystallisation. *Earth and Planetary Science Letters* **53**, 189–202.

Deniel, C., Vidal, P., Fernandez, A., Le Fort, P. & Peucat, J. J. 1987. Isotopic study of the Manaslu granite (Himalaya, Nepal): inferences on the age and source of the Himalayan leucogranites. *Contributions to Mineralogy and Petrology* **96**, 78–92.

Didier, J. & Barbarin, B. 1991. *Enclaves and Granite Petrology. Developments in Petrology* **13**. Amsterdam: Elsevier.

Dimanov, A., Dresen, G., Xiao, X. & Wirth, R. 1999. Grain boundary diffusion creep of synthetic anorthite aggregates: the effect of water. *Journal of Geophysical Research* **B104**, 10483–97.

England, P. C. & Thompson, A. B. 1984. Pressure-temperature-time paths of regional metamorphism. I. Heat transfer during evolution of thickened continental crust. *Journal of Petrology* **25**, 894–928.

Gleason, G. C., Bruce, V. & Green, H. W. 1999. Experimental investigation of melt topology in partially molten quartzo feldspathic aggregates under hydrostatic and non hydrostatic stress. *Journal of Metamorphic Geology* **17**, 705–22.

Guillet, P., Bouchez, J. L. & Vigneresse, J. L. 1985. Le complexe granitique de Plouaret (Bretagne) Mise en évidence structurale et gravimétrique de diapirs emboités. *Bulletin de la Société Géologique de France* **8**, 503–13.

Guineberteau, B., Bouchez, J. L. & Vigneresse, J. L. 1987. The Mortagne granite pluton (France) emplaced by pull apart along a shear zone: structural and gravimetric arguments and regional implication. *Geological Society of American Bulletin* **99**, 763–70.

Guyon, E., Roux, S., Hansen, A., Bideau, D., Troadec, J. P. & Crapo, H. 1990. Non-local and non-linear problems in the mechanics of disordered systems: application to granular media and rigidity problems. *Reports on Progress in Physics* **53**, 373–419.

Harlov, D. E. 2002. Review of petrographic and mineralogical evidence for fluid induced dehydration of amphibolite facies to granulite facies rocks. *Zeitschrift für Geologische Wissenschaften* **30**, 13–36.

Harris, N., Ayres, M. & Massey, J. 1995. Geochemistry of granitic melts produced during the incongruent melting of muscovit: implications for the extraction of Himalayan leucogranite magmas. *Journal of Geophysical Research* **B100**, 15767–77.

Hecht, L. & Vigneresse, J. L. 1999. A multidisciplinary approach combining geochemical, gravity and structural data: implications for pluton emplacement and zonation. *In* Castro, A., Fernandez, C. & Vigneresse, J. L. (eds) *Understanding Granites: Integrating New and Classical techniques. Geological Society, London, Special Publication* **168**, 95–110. London: Geological Society.

Hecht, L., Vigneresse, J. L. & Morteani, J. 1997. Constraints on the origin of zonation of the granite complexes in the Fichtelgebirge (Germany and Czech Republik): evidence from a gravity and geochemical study. *Geologische Rundschau* **87**, S93–109.

Hildreth, E. W. & Moorbath, S. 1988. Crustal contributions to arc magmatism in the Andes of Central Chile. *Contributions to Mineralogy and Petrology* **76**, 177–95.

Hogan, J. P., Price, J. P. & Gilbert, M. C. 1998. Magma traps and driving pressure: consequences for pluton shape and emplacement in an extensional regime. *Journal of Structural Geology* **20**, 1155–68.

Holden, P., Halliday, A. N., Stephens, W. E. & Henney, P. J. 1991. Chemical and isotopic evidences for major mass transfer between mafic enclaves and felsic magma. *Chemical Geology* **92**, 135–52.

Hutton, D. H. W. 1982. A tectonic model for the emplacement of the main Donegal granite, NW Ireland. *Journal of the Geological Society, London* **139**, 615–31.

Hutton, D. H. W. 1997. Syntectonic granites and the principle of effective stress: a general solution to the space problem? *In* Bouchez, J. L., Hutton, D. H. W. & Stephens, W. E. (eds) *Granite: From Segregation of Melt to Emplacement Fabrics*, 189–97. Dordrecht: Kluwer Academic Publishers.

Jahn, B. M., Wu, F. & Chen, B. 2000. Granitoids of the Central Asian Orogenic Belt and continental growth in the Phanerozoic. *Transactions of the Royal Society of Edinburgh: Earth Sciences* **91**, 181–93.

Jurewicz, S. R. & Watson, E. B. 1985. Distribution of partial melt in a felsic system: the importance of surface energy. *Contributions to Mineralogy and Petrology* **85**, 25–9.

Kalakay, T. J., John, B. E. & Lageson, D. R. 2001. Fault-controlled pluton emplacement in the Sevier fold-and-thrust belt of southwest Montana, USA. *Journal of Structural Geology* **23**, 1151–65.

Kerr, A. D. & Pollard, D. D. 1998. Toward more realistic formulations for the analysis of laccoliths. *Journal of Structural Geology* **20**, 1783–93.

Kohlstedt, D. L., Bai, Q, Wand, Z. C. & Mei, S. 2000. Rheology of partially molten rocks. *In* Bagdassarov, N., Laporte, D. & Thompson, A. B. (eds) *Physics and Chemistry of Partially Molten Rocks*, 3–28. Dordrecht: Kluwer Academic Publishers.

Lister, J. R. & Kerr, R. C. 1991. Fluid-mechanical models of crack propagation and their application to magma transport in dykes. *Journal of Geophysical Research* **B96**, 10049–77.

Loosveld, R. J. H. 1989. The synchronism of crustal thickening and low pressure facies metamorphism in the Mount Isa Inlier, Australia. 2. Fast convective thinning of mantle lithosphere during crustal thickening. *Tectonophysics* **165**, 191–218.

Maaløe, S. & Wyllie, P. J. 1975. Water content of a granite magma deduced from the sequence of crystallization determined experimentally with water-undersaturated conditions. *Contributions to Mineralogy and Petrology* **52**, 175–91.

Mahon, K. I., Harrison, T. M. & Drew, D. A. 1988. Ascent of a granitoid diapir in a temperature varying medium. *Journal of Geophysical Research* **B93**, 1174–88.

McKenzie, D. 1984. The generation and compaction of partially molten rocks. *Journal of Petrology* **25**, 713–65.

McNulty, B. A., Tobisch, O. T., Cruden, A. R. & Gilder, S. 2000. Multistage emplacement of the Mount Givens pluton, central Sierra Nevada batholith, California. *Geological Society of America Bulletin* **112**, 119–35.

Miller, C. F., Watson, E. B. & Harrison, T. M. 1988. Perspectives on the source, segregation and transport of granitoid magmas. *Transactions of the Royal Society of Edinburgh: Earth Sciences* **79**, 135–56.

Morgan, P. 1984. The thermal structure and thermal evolution of the continental lithosphere. *Physics and Chemistry of the Earth* **15**, 107–93.

O'Hara, M. J. & Mathews, R. E. 1981. Geochemical evolution in an advancing, periodically replenished, periodically tapped, continuously fractionated magma chamber. *Journal of the Geological Society, London* **138**, 237–77.

O'Hara, M. J. 1977. Geochemical evolution during fractional crystallization of a periodically refilled magma chamber. *Nature* **266**, 503–7.

Parsons, T., Sleep, N. H. & Thompson, G. A. 1992. Host rock rheology controls on the emplacement of tabular intrusions: Implications for underplating of extending crust. *Tectonics* **11**, 1348–56.

Paterson, S. R., Fowler, T. K., Schmidt, K. L., Yoshinobu, A. S., Yuan, E. S. & Miller, R. B. 1998. Interpreting magmatic fabric patterns in plutons. *Lithos* **44**, 53–82.

Paterson, S. R. & Vernon, R. H. 1995. Bursting the bubble of ballooning plutons – A return to nested diapirs emplaced by multiple processes. *Geological Society of America Bulletin* **107** (11), 1356–80.

Paterson, S. R. & Schmidt, K. L. 1999. Is there a close spatial relationship between faults and plutons? *Journal of Structural Geology* **21**, 1131–42.

Patiño Douce, A. E. & Beard, J. S. 1995. Dehydration melting of biotite gneiss and quartz amphibolite from 3 to 15 kbar. *Journal of Petrology* **96**, 707–38.

Percival, J. A. 1983. High grade metamorphism in the Chapleau-Foleyet area, Ontario. *American Mineralogist* **68**, 667–86.

Petford, N., Clemens, J. D. & Vigneresse, J. L. 1997. Application of information theory to the formation of granitic rocks. *In* Bouchez, J. L., Hutton, D. & Stephens, W. E. (eds) *Granite From Melt Segregation to Emplacement Fabrics*, 3–10. Dordrecht: Kluwer Academic Publishers.

Petford, N., Cruden, A. R., McCaffrey, K. J. W. & Vigneresse, J. L. 2000. Granite magma formation, transport and emplacement in the Earth's crust. *Nature* **408**, 669–73.

Petford, N., Kerr, R. C. & Lister, J. R. 1993. Dike transport of granitoid magmas. *Geology* **21**, 845–8.

Pitcher, W. S. 1993. *The Nature and Origin of Granite*. London: Chapman & Hall.

Rämö, O. T. & Haapala, I. 1995. One hundred years of rapakivi granite. *Mineralogy and Petrology* **52**, 129–85.

Pollard, D. D. & Muller, O. H. 1976. The effect of gradients in regional stress and magma pressure on the form of sheet intrusions in cross section. *Journal of Geophysical Research* **B81**, 975–84.

Powell, R. 1984. Inversion of the assimilation and fractional crystallization (AFC) equations: characterization of contaminants from isotope and trace element relationships in volcanic suites. *Journal of the Geological Society, London* **141**, 447–52.

Rabinowicz, M. & Vigneresse, J. L. 2004. Melt segregation under compaction and shear channelling: Application to granitic magma segregation in a continental crust. *Journal of Geophysical Research* **109** (B4), article number B04407.

Rabinowicz, M., Genthon, P., Ceuleneer, G. & Hillairet, M. 2001. Compaction in a mantle mush with high melt concentrations and the generation of magma chambers. *Earth and Planetary Science Letters* **188**, 313–28.

Ramsay, J. & Huber, M. 1983. *The Techniques of Modern Structural Geology. Vol. 1: Strain Analysis*. London: Academic Press.

Rapp, R. P. & Watson, E. B. 1995. Dehydration melting of metabasalt at 8–32 kbar: implications for continental growth and crust-mantle recycling. *Journal of Petrology* **36**, 891–931.

Read, H. H. 1957. *The Granite Controversy*. London: Thomas Murby and Co. London.

Ricard, Y., Bercovici, D. & Schubert, G. 2001. A two-phase model for compaction and damage. 2. Applications to compaction, deformation and the role of interfacial surface tension. *Journal of Geophysical Research* **B106**, 8907–24.

Roberts, M. P. & Clemens, J. D. 1995. Feasability of AFC models for the petrogenesis of calc-alkaline magma series. *Contributions to Mineralogy and Petrology* **121**, 139–47.

Rollinson, H. R. 1993. *Using Geochemical Data: Evaluation, Presentation, Interpretation*. London: Longman Scientific and Technical.

Roman-Berdiel, T. 1999. Geometry of granite emplacement in the upper crust: contributions of analogue modelling. *In* Castro, A., Fernandez, C. & Vigneresse, J. L. (eds) *Understanding Granites: Integrating New and Classical Techniques. Geological Society, London, Special Publication* **168**, 77–94. London: Geological Society.

Roman-Berdiel, T., Gapais, D. & Brun, J. P. 1995. Analogue models of laccolith formation. *Journal of Structural Geology* **17**, 1337–46.

Rubin, A. M. 1997. Dike ascent in partially melted rocks. *Journal of Geophysical Research* **B103**, 20901–19.

Ryan, M. P. 1994. *Magmatic Systems*. Orlando, FL: Academic Press.

Saint Blanquat, M. de, Tikoff, B., Teyssier, C. & Vigneresse, J. L. 1998. Transpressional kinematics and magmatic arcs. *In* Holdsworth, R. E., Strachan, R. A. & Dewey, J. F. (eds) *Continental Transpressional and Transtensional Tectonics. Geological Society, London, Special Publication* **135**, 327–40. London: Geological Society.

Sawyer, E. W. 2000. Grain scale and outcrop-scale distribution and movement of melt in a crystallizing granite. *Transactions of the Royal Society of Edinburgh: Earth Sciences* **91**, 73–85.

Sawyer, E. W. 2001. Melt migration in the continental crust: distribution and movement of melt in anatectic rocks. *Journal of Metamorphic Geology* **19**, 291–309.

Schmelling, H., Cruden, A. R. & Marquart, G. 1988. Finite deformation in and around a fluid sphere moving through a viscous medium: implications for diapiric ascent. *Tectonophysics* **149**, 17–34.

Schmidt, C. J., Smedes, H. W. & O'Neill, J. M. 1990. Syncompressioal emplacement of the Boulder and Tobacco Root batholiths (Montana-USA) by pull apart along old fault zones. *Geological Journal* **25**, 305–18.

Sclater, J. G., Parsons, B. & Jaupart, C. 1981. Oceans and continents: similarities and differences in the mechanisms of heat loss. *Journal of Geophysical Research* **B86**, 6966–83.

Shaw, D. M. 1970. Trace element fractionation during anatexis. *Geochimica et Cosmochimica Acta* **34**, 237–43.

Shaw, H. R. 1980. Fracture mechanisms of magma transport from the mantle to surface. *In* Hargraves, R. B. (ed.) *Physics of Magmatic Processes*. Princeton, NJ: Princeton University Press.

Spera, F. J. & Bohrson, W. A. 2001. Energy-constrained open-system magmatic processes II: Applications of Energy Constrained Assimilation-Fractional crystallisation (EC-AFC) model to magmatic systems. *Journal of Petrology* **42**, 1919–41.

Talbot, C. J., Medvedev, S., Alavi, M., Sharivar, H. & Heidari, E. 2000. Salt extrusion at Kuh-e-Jahani, Iran, from June 1994 to November 1997. *In* Vendeville, B., Mart, Y. & Vigneresse, J. L. (eds) *Salt Shale and Igneous Diapirs In and Around Europe. Geological Society, London, Special Publication* **174**, 93–110. London: Geological Society.

Taylor, H. P., Jr. 1980. The effects of assimilation of country rocks by magmas on $^{18}O/^{16}O$ and $^{87}Sr/^{86}Sr$ systematics in igneous rocks. *Earth and Planetary Science Letters* **47**, 243–54.

Thompson, A. B. 1982. Dehydration melting of pelitic rocks and the generation of H_2O-undersaturated granitic liquids. *American Journal of Science* **282**, 1567–95.

Thompson, A. B. 1999. Some time-space relationships for crustal melting and granitic intrusion at various depths. *In* Castro, A., Fernandez, C. & Vigneresse, J. L. (eds) *Understanding Granites: Integrating New and Classical Techniques. Geological Society, London, Special Publication* **168**, 7–25. London: Geological Society.

Tikoff, B. & Teyssier, C. 1992. Crustal-scale, en-echelon 'P-shear' tensional bridges: a possible solution to the batholithic room problem. *Geology* **20**, 927–30.

Touret, J. 1977. The significance of fluid inclusions in metamorphic rocks. *In* Fraser, D. G. (ed.) *Thermodynamics in Geology*, 203–25. *NATO ASI Series* **C30**. Dordrecht: Kluwer Academic Publishers.

Tuttle, O. F. & Bowen, N. L. 1958. *Origin of Granite in the Light of Experimental Studies in the System $NaAlSi_3O_8$-$KAlSi_3O_8$-SiO_2-H_2O. Geological Society of America Memoir* **74**. Washington: Geological Society of America.

Van der Molen, I. & Paterson, M. S. 1979. Experimental deformation of partially-melted granite. *Contributions to Mineralogy and Petrology* **70**, 299–318.

Vigneresse, J. L. 1990. Use and misuse of geophysical data to determine the shape at depth of granitic intrusions. *Geological Journal* **25**, 249–60.

Vigneresse, J. L. 1995a. Control of granite emplacement by regional deformation. *Tectonophysics* **249**, 173–86.

Vigneresse, J. L. 1995b. Crustal regime of deformation and ascent of granitic magma. *Tectonophysics* **249**, 187–202.

Vigneresse, J. L., Barbey, P. & Cuney, M. 1996. Rheological transitions during partial melting and crystallisation with application to felsic magma segregation and transfer. *Journal of Petrology* **37**, 1579–600.

Vigneresse, J. L., Tikoff, B. & Améglio, L. 1999. Modification of the regional stress field by magma intrusion and formation of tabular granitic plutons. *Tectonophysics* **302**, 203–24.

Vigneresse, J. L. & Bouchez, J. L. 1997. Successive granitic magma batches during pluton emplacement: the case of Cabeza de Araya (Spain). *Journal of Petrology* **38**, 1767–76.

Vigneresse, J. L. & Burg, J. P. 2000. Continuous versus discontinuous melt segregation in migmatites: insights from a cellular automaton model. *Terra Nova* **12**, 188–92.

Vigneresse, J. L. & Burg, J. P. 2004. Some insights on the rheology of partially molten rocks. *In* Grocott, J., McCaffrey, K., Taylor, G. & Tikoff, B. (eds) *Vertical Coupling and Decoupling in the Lithosphere. Geological Society, London, Special Publication* **227**, 327–36. London: Geological Society.

Vigneresse, J. L. & Clemens, J. D. 2000. Granitic magma ascent and emplacement: neither diapirism nor neutral buoyancy. *In* Vendeville, B., Mart, Y. & Vigneresse, J. L. (eds) *Salt, Shale and Igneous Diapirs In and Around Europe. Geological Society, London, Special Publication* **174**, 1–19. London: Geological Society.

Vigneresse, J. L. & Tikoff, B. 1999. Strain partitioning during partial melting and crystallising felsic magmas. *Tectonophysics* **312**, 117–32.

Wedepohl, K. H., Heinrichs, H. & Bridgewater, D. 1991. Chemical characteristics and genesis of the quartz-feldspathic rocks in the Archean crust of Greenland. *Contributions to Mineralogy and Petrology* **107**, 163–79.

Weertman, J. 1971. Theory of water-filled crevasses in glaciers applied to vertical magma transport beneath ocean ridges. *Journal of Geophysical Research* **B76**, 1171–83.

Weijermars, R. 1986. Flow behaviour and physical chemistry of bouncing putties and related polymers in view of tectonic laboratory applications. *Tectonophysics* **124**, 325–58.

Weinberg, R. F. 1996. The ascent mechanism of felsic magmas: news and views. *Transactions of the Royal Society of Edinburgh: Earth Sciences* **87**, 95–103.

Weinberg, R. F., Sial, A. N. & Pessoa, R. R. 2001. Magma flow within the Tavares pluton, northeast Brazil: Compositional and thermal convection. *Geological Society of America Bulletin* **113**, 508–20.

Weinberg, R. F. & Podladchikov, Y. Y. 1994. Diapiric ascent of magmas through power law crust and mantle. *Journal of Geophysical Research* **B99**, 9543–59.

Whitney, J. A. 1990. Origin and evolution of silicic magmas. *Reviews in Economic Geology* **4**, 183–201.

Wilson, J. & Grocott, J. 1999. The emplacement of the granitic Las Tazas complex, northern Chile: the relationship between local and regional strain. *Journal of Structural Geology* **21**, 1513–23.

Wirth, R. & Franz, L. 2000. Thin amorphous intergranular layers at mineral interfaces in xenoliths: the early stage of melting. *In* Bagdassarov, N., Laporte, D. & Thompson, A. B. (eds) *Physics and Chemistry of Partially Molten Rocks*, 229–68. Dordrecht: Kluwer Academic Publishers.

Wyllie, P. J. 1977. From crucibles through subduction to batholiths. *In* Saxena, S. K. & Battacharji, S. (eds) *Energetics of Geological Processes*, 389–433. Berlin: Springer-Verlag.

Wyllie, P. J., Huang, W. L., Stern, C. R. & Maaløe, S. 1976. Granitic magmas: possible and impossible sources, water contents and crystallization sequences. *Canadian Journal of Earth Sciences* **13**, 1007–19.

Yardley, B. W. D. & Valley, J. W. 1997. The petrologic case for a dry lower crust. *Journal of Geophysical Research* **B102**, 12173–85.

Zoback, M. L. 1992. First- and second-order patterns of stress in the lithosphere: the World Stress Map Project. *Journal of Geophysical Research* **B97**, 11703–28.

JEAN LOUIS VIGNERESSE
CREGU, UMR CNRS 7566 G2R, BP 23, F-54501 Vandoeuvre Cedex, France.
e-mail: jean-louis.vigneresse@g2r.uhp-nancy.fr

MS received 29 September 2003. Accepted for publication 9 May 2004.

Transactions of the Royal Society of Edinburgh: Earth Sciences, **95**, 23–33, 2004

The redox state of granitoids relative to tectonic setting and earth history: The magnetite–ilmenite series 30 years later

Shunso Ishihara

ABSTRACT: The redox state variation of orogenic granitoids along convergent plate margins is examined in the Phanerozoic Circum-Pacific Belt and in some Cryptozoic terranes. The Phanerozoic granitoids of the NW and NE Pacific Rims can be divided into reduced ilmenite series occurring in the accretionary terranes with compressional tectonic setting, and oxidised magnetite series intruding crystalline basements under extensional to intermediate regional stress regime. The ilmenite-series granitoids have negative but the magnetite series have positive $\delta^{34}S$ values, which show a positive correlation with magnetic susceptibility of the granitoids. The negative $\delta^{34}S$ sulphur originated in biogenic sulphur from accreted pelitic sediments and positive $\delta^{34}S$ values show that sulphate sulphur migrated from seawater through subduction processes. The whole rock $\delta^{18}O$ values are higher than 8 permil in the ilmenite series, but lower than 8 permil in the magnetite series, and as a whole show negative correlation with the magnetic susceptibility of the granitoids. The higher $\delta^{18}O$ values reflect those of accreted sediments, whilst the lower $\delta^{18}O$ values represent magmatic values of an oxidised mafic protolith at depth.

The predominance of ilmenite-series granitoids of the NW Pacific rim can be explained by well-developed accretionary terranes in which mafic magmas from depth mingled with felsic magmas from the accretionary complex to form granodioritic magmas, whilst that of magnetite-series granitoids is postulated to be oxidised igneous sources for the magma generation and an extensional and/or intermediate tectonic setting for the magma ascent. The absence of the accretionary wedges by tectonic erosion and/or no fore-arc sedimentation also helped to form magnetite-series granitoids. Potassic granitoids are generally of oxidised type. A-type granites in late orogenic environments also belong to the magnetite series. Adakitic high-Sr/Y granitoids are oxidised in the Mesozoic–Cenozoic but are reduced in the Archaean TTG, reflecting the redox state of the then-current sea-floor environment. The oldest magnetite-series granite so far known is the 3105 Ma-old biotite granite of the Nelspruit batholith, South Africa.

KEY WORDS: Accretionary wedge, Archaean, ilmenite series, magnetite series, O-isotope, Phanerozoic, S-isotope.

Orogenic magmatism along convergent plate boundaries is the most important process producing granitoids and creating continental crust. The granitoids were simply classified into magnetite-series and ilmenite-series (Ishihara 1977), based upon the presence and absence of the rock-forming mineral magnetite, reflecting the oxygen fugacity of the granitic magmas. Free oxygen is most available on the earth's surface; therefore the redox state of granitoids is related not only to igneous processes in the lithosphere but also to biogenic evolution in the hydrosphere and atmosphere of the Earth. The formation of the magnetite-series granitoids required an oxidised mafic source, whilst the ilmenite series could have been formed by assimilation of organic carbon from the accreted sediments in lower-middle crust processes (Ishihara & Matsuhisa 1999) or through subduction processes (Takagi 2004) during the Phanerozoic.

These orogenic granitoids are important sources for metallic mineral resources, as their redox state primarily controls the types of metals concentrated in given ore deposits (Ishihara 1977; Blevin & Chappell 1992). In this paper the genesis of these granitoids is considered in terms of the magnetite/ilmenite-series classification for the Phanerozoic Japanese islands and Circum-Pacific region (Ishihara 1998), as well as some Cryptozoic terranes. The Japanese islands are composed of two basic basement units: the Hida meta-igneous crystalline terrane, which is a part of the Sino–Korean Paraplatform, and accretionary terranes of Palaeozoic to Tertiary ages (Fig. 1).

1. History of the discovery

Molybdenum (Mo) and tungsten (W) are known to be most closely associated with granites in space and time. A quantitative expression of the ore deposits in the 1960s (i.e., production and remaining ore reserves) in the major Mo and W mineralised region of Southwest Japan, indicated that many molybdenites recorded in W deposits are just mineral occurrences and were absent in many of tungsten deposits in the Sanyo Province; thus an obvious paired zoning, Mo in the north and W in the south, was delineated (Fig. 2).

Moreover, 98% of the Mo ores and 45% of the W ores are hosted in the granites, i.e. they are endogranitic, implying that the host granites must be different. Granitoids of the Sanyo W Province were studied in 1970, and their modal analyses indicated almost no opaque minerals in those from the Sanyo W Province, in contrast to abundant opaques in the Sanin Mo province. Wet chemical analyses of major elements indicated similar contents for most of the components, but a clear difference in the ferric/ferrous ratio (Ishihara 1971). It was obvious from these data that the difference between the

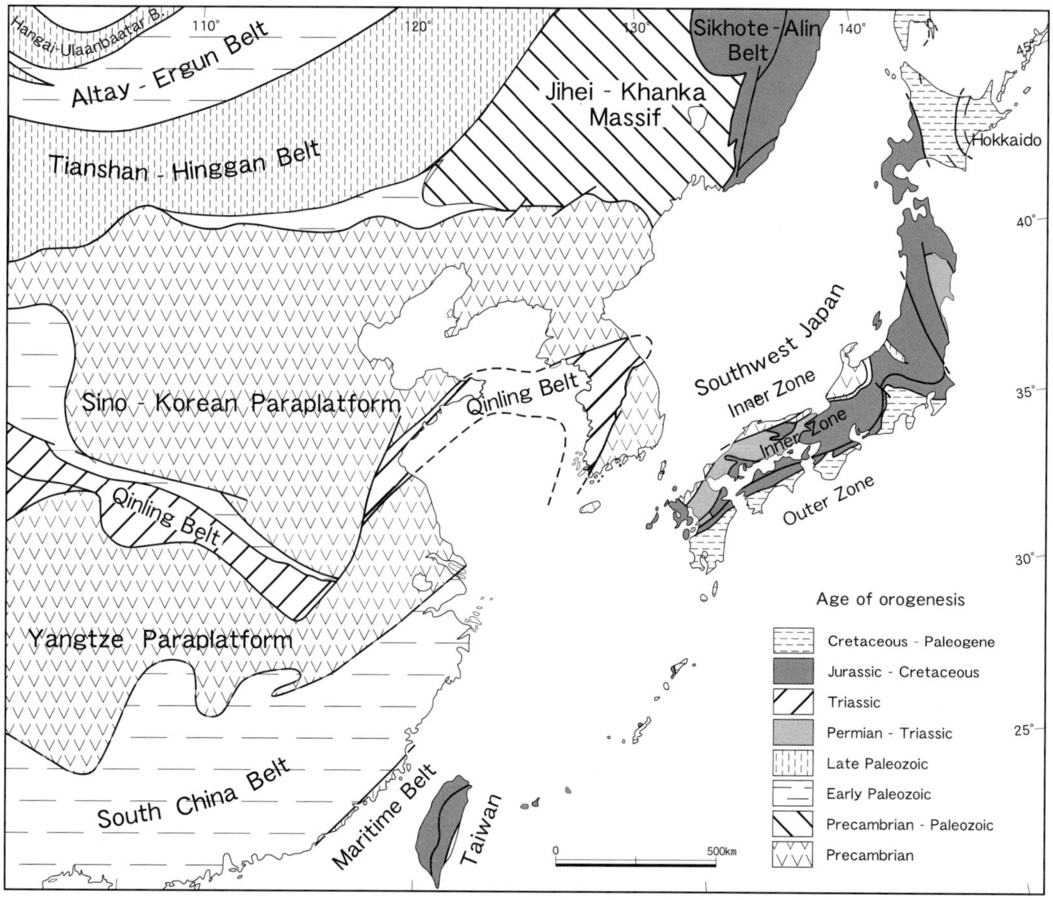

Figure 1 Straigraphic division of East Asia (from Teraoka & Okumura 2003). Typical accretionary terranes of Jurassic to Palaeogene age occur in the Sikhote Alin, Japan and Taiwan. The Palaeozoic Altay-Ergun, Tianshan-Hinggan and South China Belts are mainly composed of accretionary terranes.

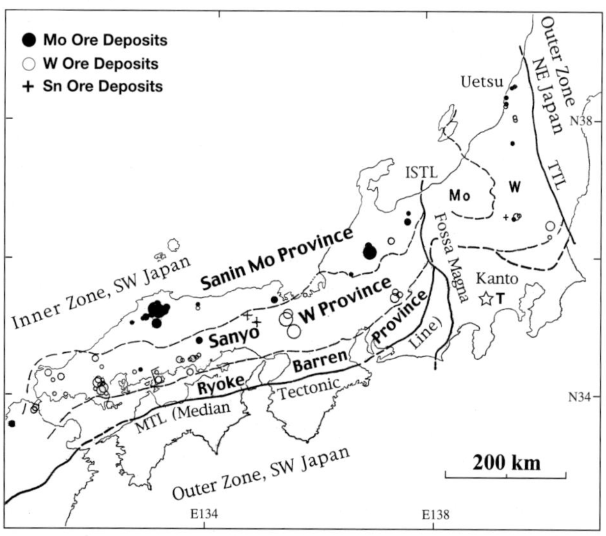

Figure 2 Late Cretaceous–Palaeogene Mo, W and barren metallogenic provinces in the Inner Zone of SW Japan (Ishihara, 1971), where magnetite/ilmenite-series granitoids were first identified. Open star with T, Tanzawa tonalite pluton.

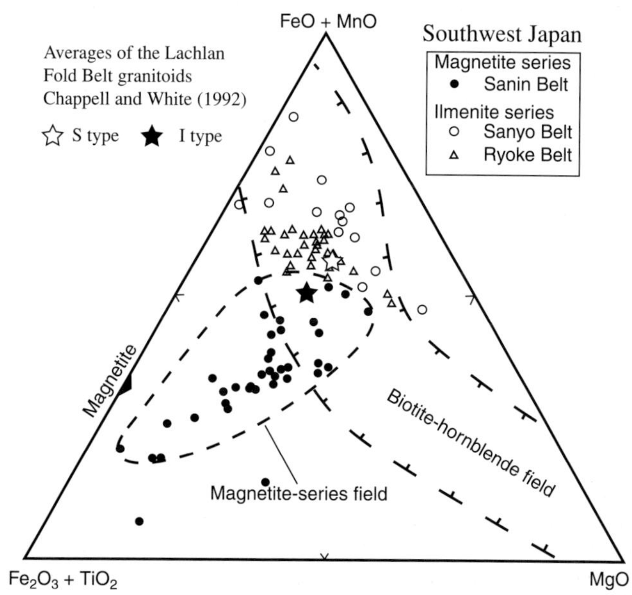

Figure 3 Bulk femic compositions of late Cretaceous–Palaeogene granitoids of the Inner Zone of SW Japan. Revised from Ishihara (1971). Note that the granitoids of Sanin Mo province plot towards the magnetite-haematite side line, while the averaged S and I types of the Lachlan Fold Belt (Chappell & White 1992) plot in the ilmenite-series field.

granitoids in the two provinces is the content of magnetite. This varying content, easily detected by magnetic susceptibility, was confirmed in the following years by using the Bison apparatus on the powdered samples (Kanaya and Ishihara 1973; Ishihara 1979).

The granitoids studied are mostly hornblende-biotite granodiorite or biotite monzogranite. Bulk analyses of the magnetite-free Sanyo and Ryoke granitoids plot in the biotite composition field of Heinrich (1946) in the FeO+MnO–Fe_2O_3+TiO_2–MgO diagram (Fig. 3). Average compositions of S and I type granitoids of the Lachlan Fold Belt (Chappell & White 1992) also plot in the same field, indicating the general

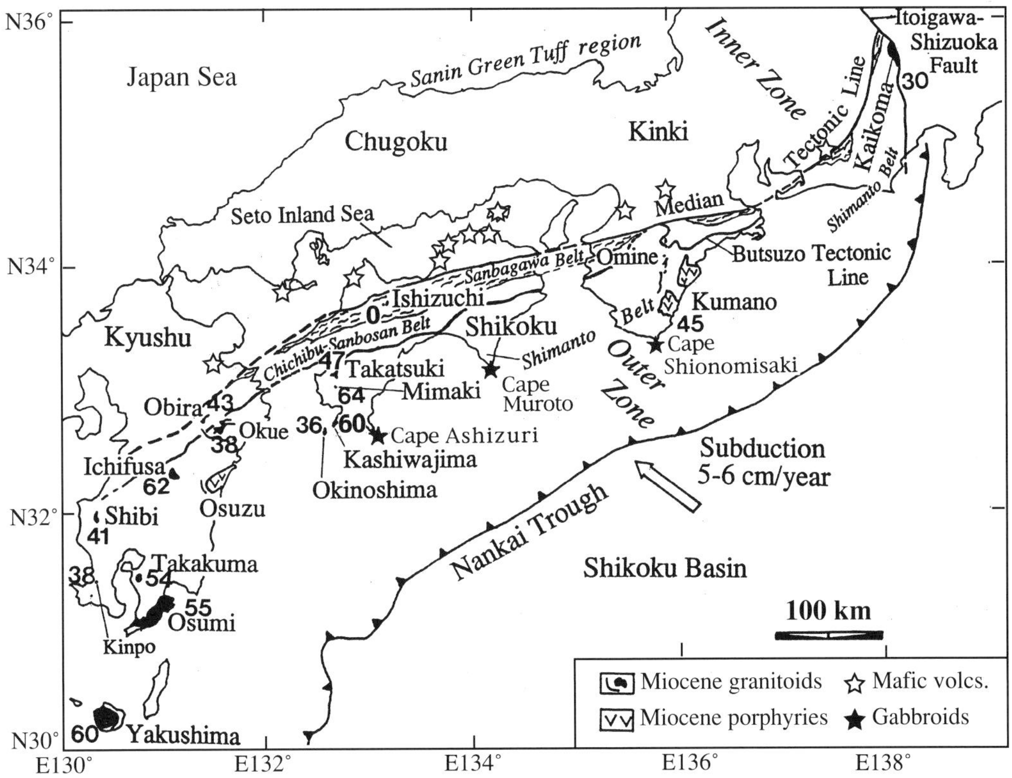

Figure 4 Distribution of Miocene granitoids and gabbroids in the Outer Zone of SW Japan. Numerals indicate percentage of accretionary wedge component in the granitoids, as calculated from the whole rock $\delta^{18}O$ values of Ishihara and Matsuhisa (1999). Mafic volcanics (basalts and high-Mg andesites) and gabbroids occur only at both sides of the ilmenite-series granitoids belt, implying that the mafic magmas were mingled with the wedge-source magmas to form the granitoids. The magnetite-series counterparts occur along the Sanin Green Tuff Belt along the Japan-Sea coast.

reduced nature of the Lachlan granitoids, especially the S-types.

The granitoids of the Sanin Province, in contrast, plot towards the magnetite–haematite side, implying that they contain haematitised magnetite, a feature confirmed by microscopic observation (Tsusue & Ishihara 1974). We now have good quality aeromagnetic data, which are useful in discriminating the two series of granitoids in regions where little or no overlapping of later magnetite-series igneous activity is known.

2. Granitoids in orogenic belts

Along the convergent plate boundary of the NW Pacific Rim, Phanerozoic granitic magmatism occurred in two contrasting basements: (1) the accretionary sedimentary unit and (2) the old crystalline unit. The former comprises sedimentary terrains formed by accretionary wedges accreted just prior to granitic intrusion. Their ages vary from the late Palaeozoic to Tertiary, culminating in the Jurassic (Isozaki 1998), and represented in the islands of Japan, Far East Russia, and the NE Pacific Rim. The sedimentary rocks are regionally un-metamorphosed or metamorphosed, and involved in the magma generation of ilmenite-series granitoids. These terranes have been intruded by granitoids from Triassic to Tertiary in age.

The old crystalline units are represented by the Sino–Korean Paraplatform, which is composed of Precambrian gneisses and granitoids seen in North China and the Korean peninsula (Fig. 1). This basement was intruded by Palaeozoic to Cretaceous granitoids in which the peak magmatism is Jurassic (Yanshanian and Daebo cycles) in the continental region. In Japan, similar granitoids occur in the Hida terrain (formerly called Funatsu Granite). Miocene magmatism of the Green Tuff region, Japan, is much younger, but may be the same in the basement setting, intruding mainly Cretaceous crystalline rocks.

In the NE Pacific rim, the major accretionary terranes are also Jurassic in age, intruded by the Jurassic–Cretaceous Coast Plutonic complex in Canada, and by Cretaceous and Tertiary plutonic rocks in the inland areas (Nokleberg *et al.* 2000; Hart *et al.* 2004). In the coastal Andes, on the other hand, the basement setting appears to be of crystalline type, because the Mesozoic to Cenozoic granitoids often intrude Precambrian crystalline rocks and late Palaeozoic granitoids, as well as into Jurassic and younger coeval volcanic rocks (Ishihara *et al.* 1984; Jaillard *et al.* 2000).

3. Granitoids with accretionary components – ilmenite series

The ilmenite-series granitoids are most widely distributed in the islands of Japan as well as in East Asia, in contrast to the predominance of magnetite-series granitoids in the East Pacific Rim (Ishihara 1977). The granitoids occur in major parts of the Inner Zone of SW Japan, i.e. late Cretaceous Sanyo and Ryoke Belts, and in the Outer Zone of the late Cretaceous Abukuma Belt, Tertiary Hidaka Belt and Miocene SW Japan, the last of which is the best preserved and discussed in the following paragraphs.

In the Sikhote Alin Belt (Fig. 1), Mesozoic granitoids are largely of ilmenite series associated with Sn and F mineralisations (Sato *et al.* 2002). In the South China Belt, Jurassic granitoids are mainly ilmenite series associated with W-mineralisations whereas magnetite series increase toward the Cretaceous granitoids (Ishihara & Wang 1999).

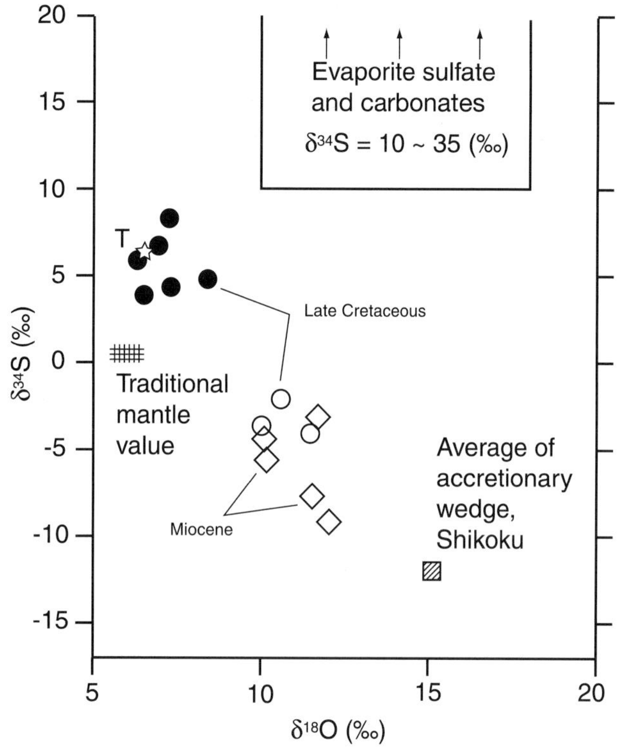

Figure 5 Whole rock $\delta^{34}S$ versus $\delta^{18}O$ values of the late Cretaceous to Miocene granitoids in SW Japan. (star with T) Tanzawa tonalite; (solid symbols) magnetite series; (open symbols) ilmenite series. Note that the SW Japan granitoids have no relationship with the sulphate–sulphur of evaporite and carbonates, which is common in the Cino-Korean Paraplatform (Ishihara *et al.* 2002a), but with that of the biogenic sulphur from the accretionary complexes.

3.1. Outer zone granitoids of SW Japan

The Miocene ilmenite-series granitoids occur mostly in the Cretaceous–Tertiary accretionary complex as isolated plutons ranging from 426 km^2 (Osumi body) to small stocks (Fig. 4). The average composition of the whole zone would be approximately monzogranite–granodiorite. These granites have high sulphur contents, locally up to 3550 p.p.m. (Ishihara *et al.* 1999), mostly occurring as pyrrhotite, but also as rare arsenopyrite and chalcopyrite. Their $\delta^{34}S$ values vary from -3 to -13 permil (Fig. 5), in contrast to the positive $\delta^{34}S$ values of the magnetite series (Sasaki & Ishihara 1979). Thus, the isotopic values are positively correlated with magnetic susceptibility (Ishihara & Sasaki 1989). The negative S isotopes are a characteristic product of biogenic sulphur in the euxenic environment on ocean floors where anaerobic bacteria live. An average $\delta^{34}S$ value of the Cretaceous–Tertiary accretionary wedges is $-13{\cdot}7$ permil. Thus, the Miocene granitoids have received sulphur from the intruded accretionary wedges.

In the largest Osumi pluton, the $\delta^{34}S$ values are fairly constant ($-8{\cdot}3$ to $-7{\cdot}0$‰) as compared to the large variation in the S content (70 to 1360 p.p.m., Ishihara *et al.* 1999), indicating that biogenic sulphur from the accretionary sediments was incorporated in the granitic magmas from the beginning of the magma generation and was well homogenised in the granitic body.

The Miocene granitoids have high $\delta^{18}O$ values of 9–13 permil (Fig. 5). Thus, they are negatively correlated with the $\delta^{34}S$ values. ^{18}O is concentrated by very low temperature fractionation, such as precipitation of cherts and illite on the ocean floor. The $\delta^{18}O$ values of the Outer Zone granitoids are much higher than that of typical Miocene low-K tonalite at Tanzawa in the southern Fossa Magna (T in Fig. 5), both occurring near the volcanic fronts. Both isotopic systems suggest that significant sedimentary components were involved in the genesis of the ilmenite-series granitoids.

The Tertiary granitoids are seen, together with coeval gabbroids, in the Hidaka accretionary wedges in the axial zone of Hokkaido, whilst the Miocene granitoids occur sporadically in the Cretaceous–Palaeogene Shimanto accretionary wedge of the Outer Zone of SW Japan (Fig. 4). Miocene gabbroids

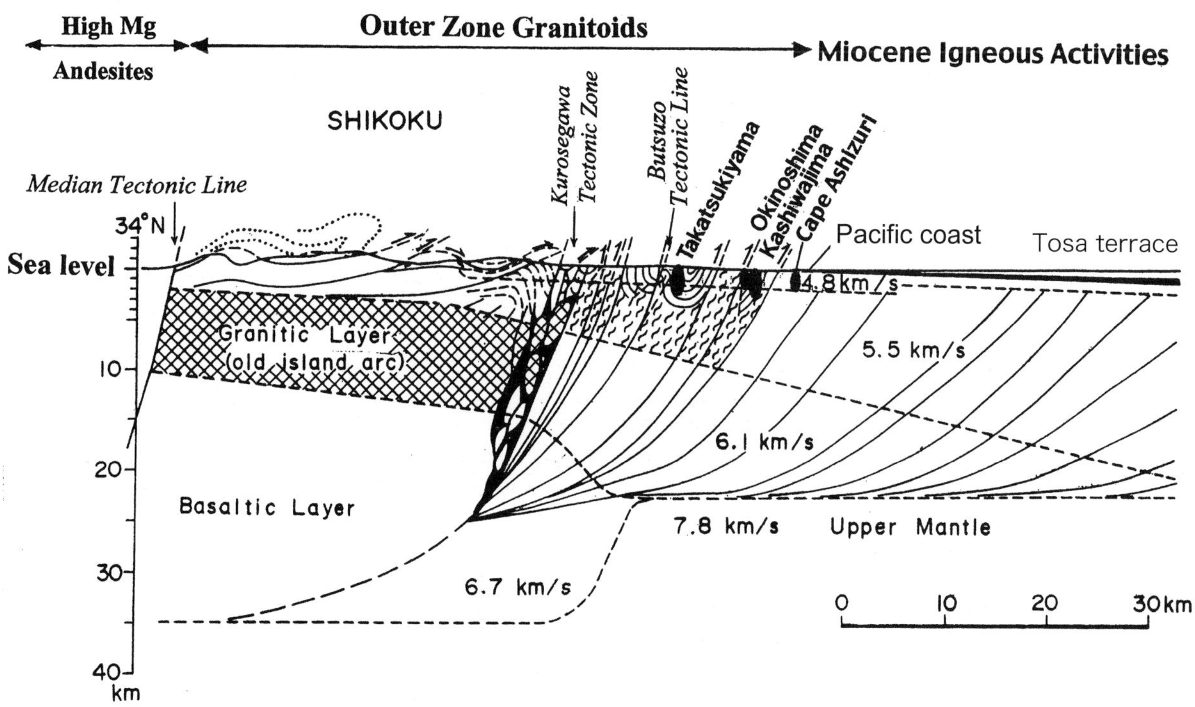

Figure 6 Seismic profile and assumed geologic section of central Shikoku Island. The 6·1 km/s layer to the north of the Kurosegawa Tectonic Zone is considered by Hada *et al.* (1982) to be an old island arc for the presence of granitoids in the Tectonic Zone. Note that no granitic layer is present underneath the southern Shimanto Belt where many of the Miocene plutons are emplaced. Modified from Hada *et al.* (1982).

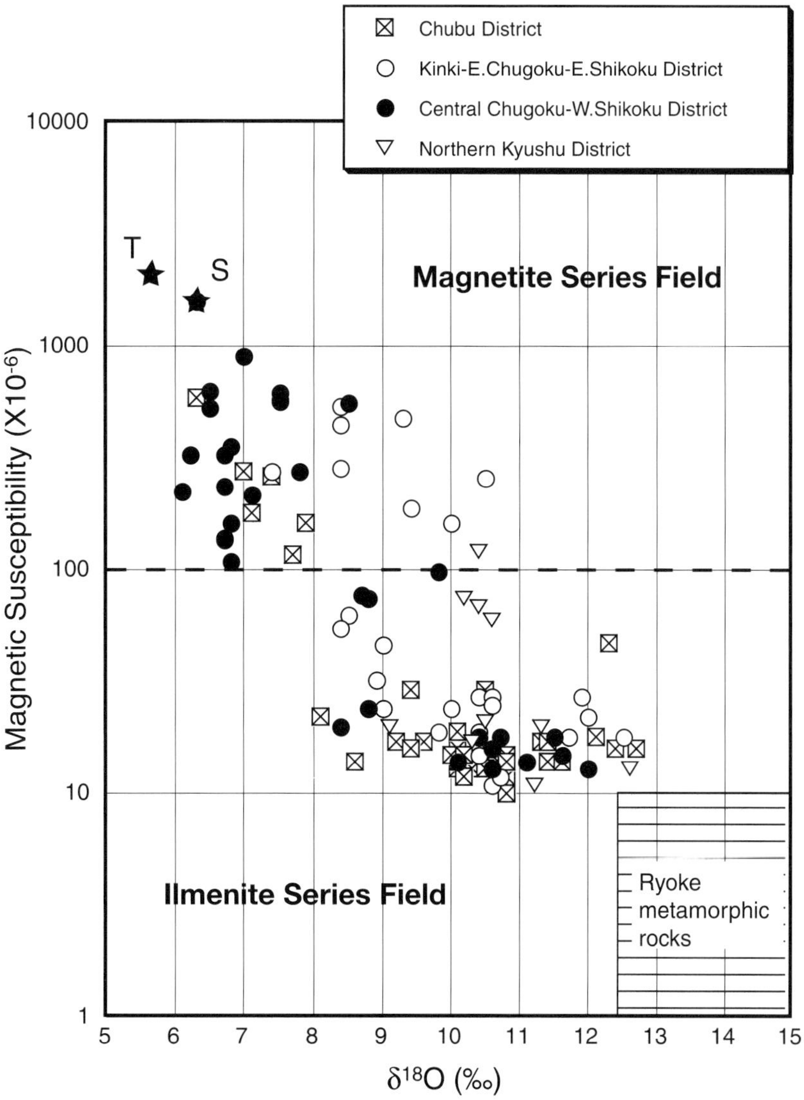

Figure 7 Magnetic susceptibility versus $\delta^{18}O$ values of the late Cretaceous–Palaeogene granitoids, SW Japan. The granitoids plot between the gabbroids of the Miocene Tanzawa body (star with T) and those of the Palaeogene body in Shimane Prefecture (star with S), and Jurassic accretionary wedges of the Ryoke metamorphic rocks. The original data from Ishihara and Matsuhisa (2002). Several granitoids which have possibly interacted with meteoric water in the Sanyo Belt are excluded.

occur at three cape areas: Cape Ashizuri, Cape Muroto and Cape Shionomisaki. To the north of the Median Tectonic Line (Fig. 4), high Mg andesites occur sporadically. These mantle-derived mafic rocks do not occur with the granitoid zone, but mingling of the mafic and felsic magmas is observed (e.g., Okueyama, Takatsukiyama and Cape Ashizuri). These observations indicate that the gabbroids and andesites must have provided heat and materials to the lower part of the accretionary complex from the upper mantle, and their mingling and mixing resulted in the generation of the ilmenite-series granodiorite–granitic magmas.

Seismic data for central Shikoku (Fig. 6) indicate that a basaltic layer of 6·7 km/s does not exist below the accretionary wedges of the Shimanto Supergroup; the latter is comprised of 6·1–4·8 km/s layers, probably the metamorphic and non-metamorphic accretionary complex, directly overlaying a 7·8 km/s unit, probably upper mantle peridotite. A 6·1 km/s layer to the north of the Kurosegawa Tectonic Zone is considered to represent an old island arc crust, given the nature of granitoids occurring in the tectonic zone (Hada *et al.* 1982). Severe crustal shortening has been observed in the accretionary wedges along the fore-arc side. This geologic setting was formed by southeastward migration of SW Japan, related to repeated openings of the marginal basin, the last of which formed the Japan Sea in the late Miocene. Miocene gabbroids, with the very low initial $^{87}Sr/^{86}Sr$ ratio of 0·7036 (Shibata and Ishihara 1979), may have originated in ultra-mafic to mafic materials in the upper mantle or underplated lower crust amphibolites (Kawate and Arima 1998). Any felsic magmas generated in this zone must have originated in the accretionary wedges, in the absence of any other felsic protolith below the Shimanto Supergroup.

Sulphur is a minor but oxygen is a major component in these granitoids. Assuming the granitoids were formed by mixing of the two end-members, namely the arc front gabbro-basalt and the accretionary wedge sediments, the observed $\delta^{18}O$ values can be converted to mass fraction (wt.%) of the accretionary sediments by simple mass-balance calculation (Ishihara & Matsuhisa 2002), as shown in Figure 4. The maximum is 67%, with mostly 50–30% of the sediments input in those granitoids intruding the accretionary terrane of Shikoku and Kyushu. A value of less than zero was obtained at the Ishizuchi-yama body occurring in the Sanbagawa meta-igneous metamorphic belt (Fig. 4).

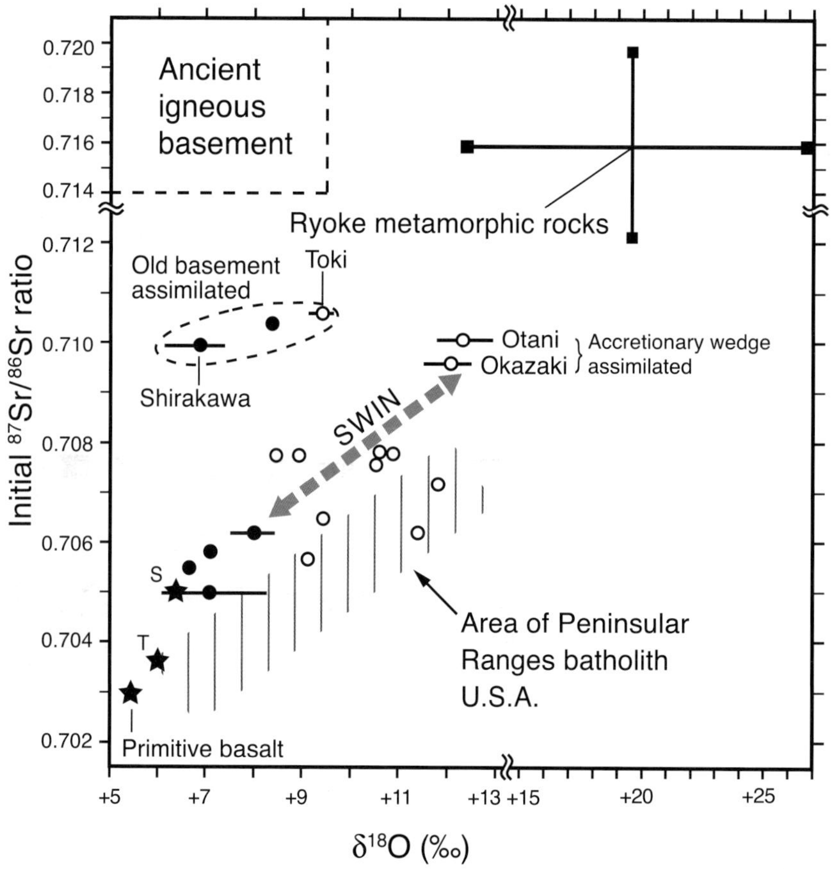

Figure 8 Initial $^{87}Sr/^{86}Sr$ ratio versus $\delta^{18}O$ ratios of the late Cretaceous–Palaeogene batholith of the Inner Zone of SW Japan (SWIN). Revised from Ishihara & Matsuhisa (2002). The original data for the Peninsular Range batholith are taken from Taylor (1986). (solid circle) magnetite series; (open circle) ilmenite series; (star with S) Palaeocene gabbro in Shimane Prefecture; (star with T) Miocene gabbro in Tanzawa; (primitive basalt) Quaternary tholeiitic basalt of Hachijo-jima (Matsuhisa *et al.* 1973).

3.2. Inner zone batholith of SW Japan

An older cycle of Cretaceous–Palaeogene granitoids occurs in the Inner Zone of SW Japan. The Inner Zone granitoids are mostly ilmenite series on the fore-arc side, and magnetite series on the back-arc side, similar to the asymmetric zoning of the Miocene granitoids. They are mostly granodiorite and monzogranite of calc-alkaline series. Alkaline rocks are absent, but adakitic magnetite-series quartz diorite-granodiorite stocks occur very locally in the early stage of the granitic activities at the northern margin of the W province (Kiji *et al.* 2000).

Whole rock $\delta^{18}O$ values of these granitoids are generally high in the ilmenite-series granitoids to the south, but low in the magnetite-series granitoids to the north. The ilmenite/magnetite-series boundary is located at around 8 permil (Ishihara & Matsuhisa 2002). The whole rock values are negatively correlated with magnetic susceptibility of the granitoids. All the data fall in an area between primitive gabbroids and the Ryoke metamorphic rocks (Fig. 7).

Gabbroids are rather dominant in the southernmost Ryoke Belt, where mingling of mafic and granitic magmas is often observed (Yokoyama 1984; Ishihara 2003). The Inner Zone granitoids were formed by mixing of mafic magmas, originated at depth, with felsic magmas derived from the Jurassic accretionary complex, and some older continental blocks. A mass-balance calculation on the $\delta^{18}O$ values between the gabbroids and the Ryoke metamorphic rocks indicates up to 32% of the metamorphic rocks mixed into the granitic magmas (Ishihara and Matsuhisa 2002). In this regard, the Ryoke metamorphic and granitic rocks presently exposed may represent a deeper facies of the Miocene Outer Zone granitoids of SW Japan.

The Inner Zone granitoids are different from the Miocene Outer Zone granitoids, based upon sporadic occurrences of high initial $^{87}Sr/^{86}Sr$ rocks, particularly around the Hida massif. The high initial $^{87}Sr/^{86}Sr$ rocks are observed in both ilmenite and magnetite series (Fig. 8), and are independent of the reducing agents. The observation indicates the presence of old, high Rb blocks of continental fragments, beneath the upper Palaeozoic to Jurassic accretionary wedges. The figure also indicates that the Inner Zone granitoids are more mature than the primitive arcs of the southern Fossa Magna as represented by the Tanzawa pluton (T, Fig. 8) and the Peninsular Range Batholith, reflecting the absence of old continental fragments in the latter regions.

In conclusion, the major part of the SW Japan granitoids, i.e., the ilmenite-series granitoids, were formed from the melting of accretionary wedges under the influence of heat and mafic and adakitic magmas from depth. Due to the opening of marginal basins, compressional tectonic setting occurred repeatedly from the late Cretaceous onwards, and mafic ascending magmas along the frontal zone were forced to interact with felsic magmas generated from the accretionary wedges. Along the back-arc side, on the other hand, intrinsically oxidised gabbroids and granitoids may have been intruded without crustal assimilation and solidified as the magnetite-series in the extensional tectonic setting, related to the opening of the marginal basin, and to the lack of organic carbon in the basement.

4. Granitoids in crystalline basements – magnetite series

The Quaternary volcanoes of the islands of Japan, composed of I-type magnetite-series rocks, occur along the volcanic front which cuts through various crystalline basements consisting of Mesozoic metamorphic, granitic and sedimentary rocks, overlain by the Neogene volcano-sedimentary rocks of the Green Tuff Belt. Miocene plutonism occurred in the same basement within the back-arc side of NE Japan and along the Japan Sea coast of SW Japan. It is possible that the crystalline basement provided no reducing agents such as organic carbon for magma generation.

4.1. Granitoids related to marginal basins

The Miocene plutonic rocks are of oxidised type, except where granitoids locally interacted with wall-rock sediments (e.g., Tokuwa granodiorite, Shimizu 1986; Chichibu mine stock, Ishihara *et al.* 1987). The magmatism appears to have occurred along rift zones formed by the opening of the Japan Sea and sheared fractures in the Fossa Magna (cf. Fig. 2). Therefore, the tectonic setting is considered generally extensional-intermediate, which allowed oxidised mafic to intermediate magmas to ascend without interaction with upper crustal rocks.

Palaeogene magnetite-series granitoids are also present in the Mo province of the Inner Zone of SW Japan (Fig. 2). They occur essentially in the pre-Jurassic metamorphic and granitic basement of the Hida Belt, which corresponds to the eastern edge of the Sino–Korean Platform (Fig. 1), and also in the Sangun metamorphic rocks. The Hida metamorphic rocks locally contain graphite-rich gneiss layers (e.g., Sennotani, Genda, Amo mines), but are largely meta-igneous in composition. Their tectonic setting is unclear but may be extensional, due to the opening of the proto-Japan Sea. Late Cretaceous granitoids of the Gyeongsang basin, South Korea, intruding Cretaceous non-marine sediments and overlying volcanic rocks at the southeastern edge of the Sino–Korean Platform, are mostly of magnetite series (Jin *et al.* 2001). A-type granites at Namsan, in the Gyeongsang Basin (Koh *et al.* 1996), which is late orogenic A-type and equivalent to the post-orogenic PA type (Hong *et al.* 1996), are strongly oxidised magnetite series.

4.2. Early Cretaceous Kitakami granitoids

Early Cretaceous granitoids occur along a narrow fault-bounded zone of the Kitakami Belt (Fig. 9). Compositionally calcic, they comprise some gabbroids, much quartz diorite, tonalite and granodiorite with some granite. High-Sr/Y adakitic tonalites and granodiorites are abundant in Zone II (Tsuchiya & Kanisawa 1994), but K-alkaline rocks also occur in the western part. These plutonic rocks are of oxidised type, having high values of magnetic susceptibility; highest in the K-alkaline rocks, high in the high Sr/Y granitoids, and lowest in the low Sr/Y granitoids (Fig. 10).

Country rocks for the early Cretaceous granitoids are continental margin sediments with some older metamorphic and granitic blocks in the southwestern Kitakami Mountains, and the Jurassic-Cretaceous accretionary wedges in the northeastern Kitakami Mountains. Yet the majority of the granitoids are of magnetite series with only local reduction due to assimilation of the basement sediments (Ishihara *et al.* 1985). The granitoids are associated with coeval volcanic rocks along the eastern coastal zone of the mountains where the granitoids are most magnetic. These granitoids are similar to those of the Peninsular Range Batholith, where coastal volcano-plutonic complexes are more magnetic than those inland (Gastil *et al.* 1990), but different from the granitoids of northern Chile, where the magnetic susceptibility of the I-type magnetite series increases towards the continent (Ishihara *et al.* 1984).

Due to the mixed occurrence of gabbroid and granitoids of K-alkalic, high-Sr/Y and low-Sr/Y series, the Kitakami

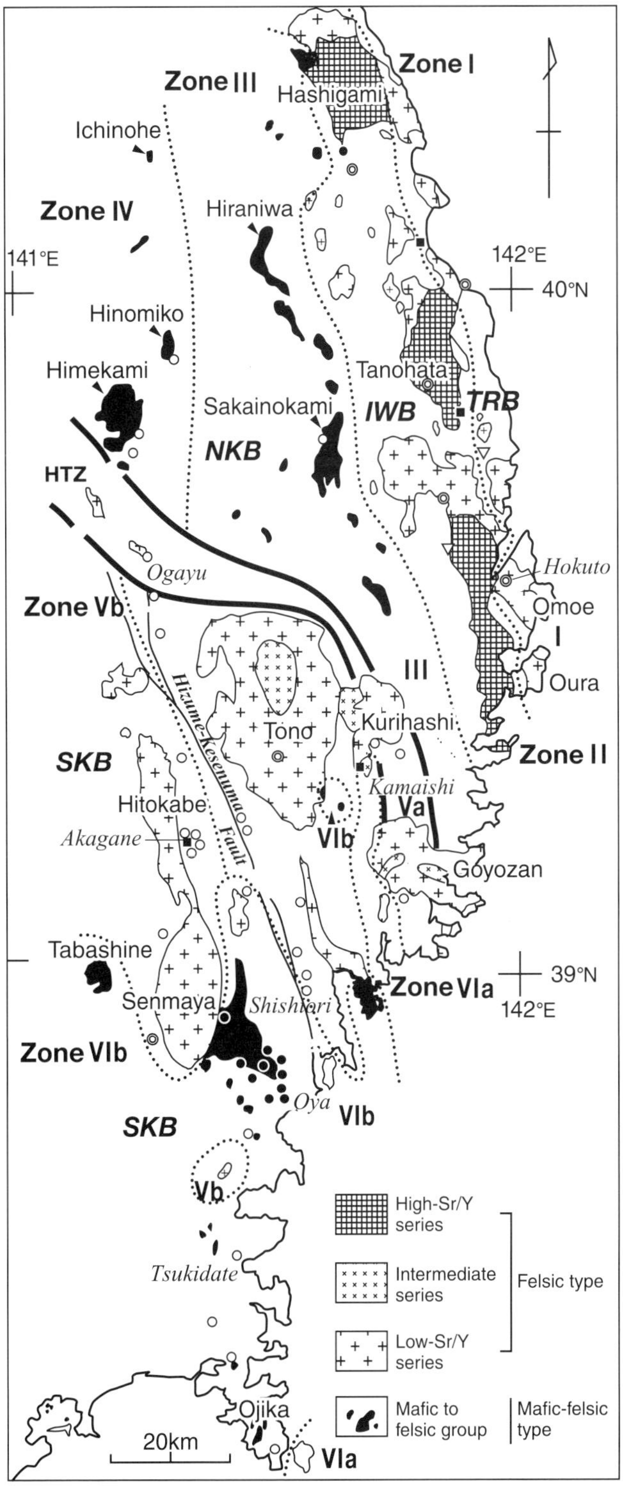

Figure 9 Distribution of the Cretaceous granitoid series and the studied ore deposits in the Kitakami Mountains (simplified from Ishihara & Muakami 2004). Adakitic rocks occur typically in Zone II and partly in Zone Va. Zone IV plutons are alkaline, whilst Zone VIb plutons are K-subalkaline. ◎ W and Mo deposits; ■ Skarn-type Fe and Cu-Pb-Zn deposits; ● high sulphide-type Au deposits; ○ low sulphide-type Au deposits; ▽ volcanogenic deposits. (HTZ) Hayachine Tectonic Zone; (SKB) Southern Kitakami belt; (NKB) Northern Kitakami belt; (IWB) Iwaizumi belt; (TRB) Taro belt.

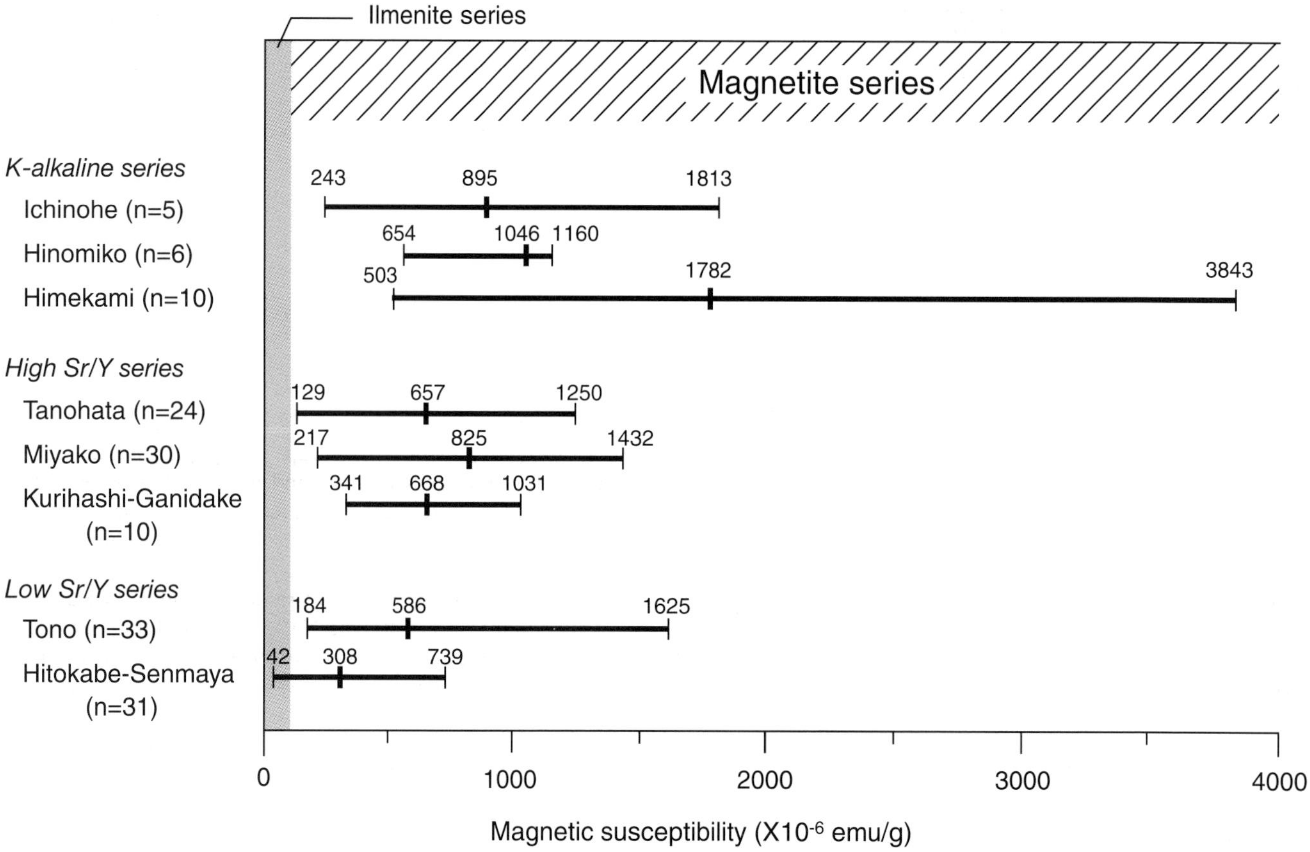

Figure 10 Range and average of magnetic susceptibility of the K-alkaline, high-Sr/Y and low Sr/Y granitoids in the Kitakami Belt, Japan. Original data from Geological Survey of Japan (1974).

plutonic rocks are considered to be a mixture of various magmas originating in the subducted slab, the upper mantle, and the lower crust (Tsuchiya & Kanisawa 1994; Ujiie-Mikoshiba *et al.* 2004; Ishihara & Murakami 2004). N–S and NNW-trending faults cut the accretionary complexes. The Kitakami block can be considered to have an intermediate status in the tectonic setting when the magmas intruded. Reducing agents from the accretionary wedges reacted only locally with the magmas of the marginal facies.

Similar magmatic arcs are predominant along the East Pacific Rim. Quartz diorite to granodiorite of the Coast Plutonic Complex, Canada, belong mostly to the magnetite series (J. Roddick & L. Carmel, personal com. 2001). Quartz diorite to granite of the Sierra Nevada batholith is dominantly magnetite series (Bateman *et al.* 1991). The coastal plutonic rocks in Peru (Ishihara *et al.* 2000) and northern Chile (Ishihara *et al.* 1984) are also mainly quartz diorite to granodiorite in composition. They generally belong to the magnetite series, interpreted to reflect the absence of the accretionary wedges due to tectonic erosion or little fore-arc sedimentation occurring from the Jurassic onward.

5. The redox state of granitoids in early Earth history

Phanerozoic adakitic plutons in the Kitakami Belt (Tsuchiya & Kanisawa 1994), and small copper-related porphyries in the Philippines (Sajona & Maury 1998), Erdenet, Mongolia (Morozumi 2003), Tongshankou, China (Wang *et al.* 2004) and northern Chile (Oyarzun *et al.* 2001) are oxidised, magnetite series. Archaean TTG suites are mostly high-Sr/Y adakitic series, and are often considered products of slab-melting, reflecting hotter oceanic crust in the Archaean (Martin 1986; Drummond & Defant 1990). The TTG suites in the Kaapvaal Craton of South Africa are all identified as ilmenite or intermediate series based on magnetic susceptibility (Ishihara *et al.* 2000, 2001), reflecting the generally reduced nature of the early (>3·2 Ga) Archaean hydrosphere and

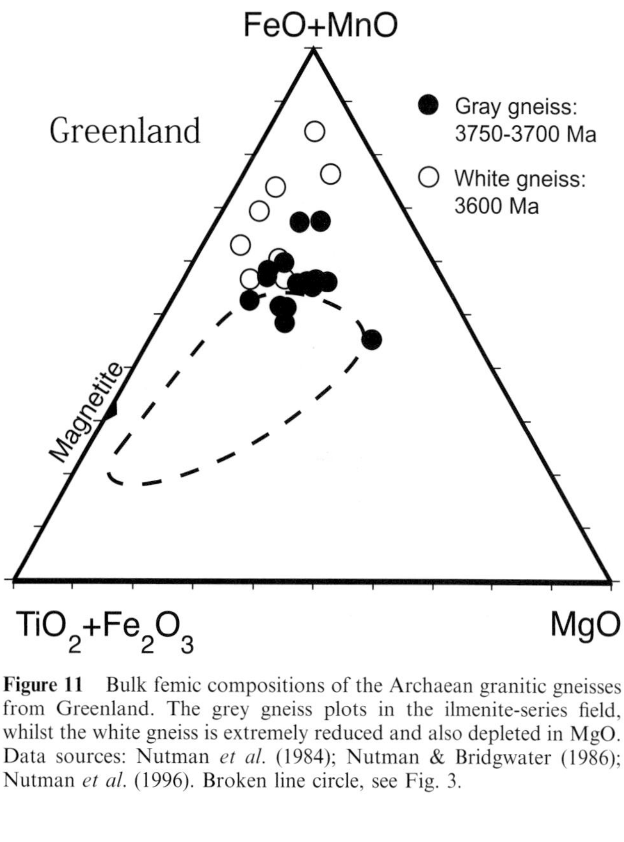

Figure 11 Bulk femic compositions of the Archaean granitic gneisses from Greenland. The grey gneiss plots in the ilmenite-series field, whilst the white gneiss is extremely reduced and also depleted in MgO. Data sources: Nutman *et al.* (1984); Nutman & Bridgwater (1986); Nutman *et al.* (1996). Broken line circle, see Fig. 3.

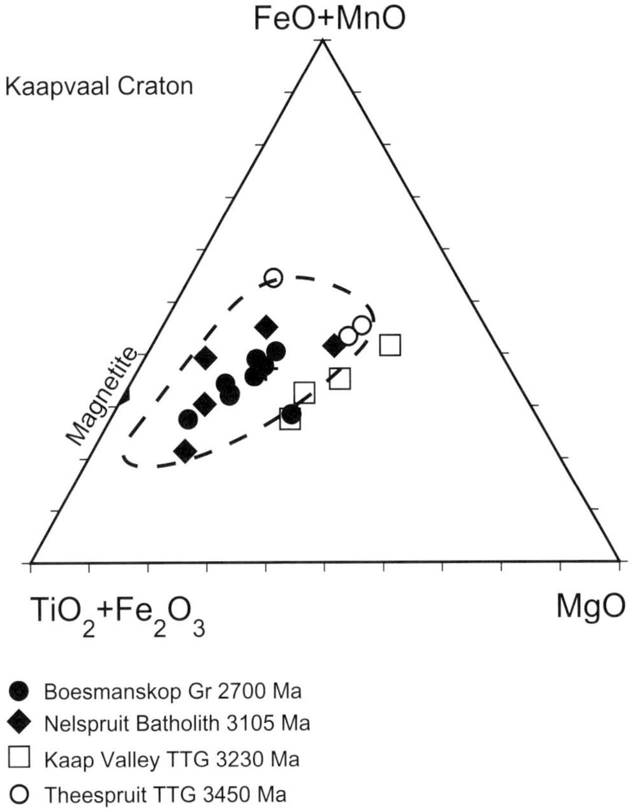

Figure 12 Bulk femic compositions of the Archaean granitoids from the Barberton Mountains, South Africa. Data sources: Viljoen & Viljoen (1969); Hunter (1974). The broken circle shows the representative area of the magnetite-series granitoids of the Sanin Province, Japan.

lithosphere (Huston & Logan 2004), and different source materials for the TTG magmatism (Smithies 2000; Smithies *et al.* 2003).

Some of the oldest Archaean granitoids are gneisses of the Central Dome of W. Greenland (Nutman & Bridewater 1986). Gray granitic gneiss (3750–3700 Ma) plot in the very reduced area of the FeO+MnO–Fe_2O_3+TiO_2–MgO diagram (Fig. 11), whilst the younger white granitic gneisses (3600 Ma) are slightly oxidised and MgO rich, but still largely reduced.

In South Africa, TTG suites occur closely associated with greenstones. The oldest TTG suites of the Barberton region (older than 3450 Ma) such as Rooihoogkte, Theespruit and other bodies, as well as the younger 3230 Ma Nelshoogte body, are reduced, ilmenite series (Ishihara *et al.* 2002b). The younger 3230 Ma Kaap Valley TTG pluton, and TTG and calc-alkaline granitoids of the Johannesburg Dome pluton (3340–2950 Ma) correspond to the ilmenite to intermediate series based on measurement of the magnetic susceptibility (Ishihara *et al.* 2002b).

Chemically, the Theespruit TTG plot in the least oxidised field (Fig. 12), whilst the Kaap Valley pluton falls in the ilmenite to intermediate series and is characterised by a high MgO content. The TTGs originated in reduced oceanic crust, but the Kaap Valley magmas may have had a high mantle component in the source materials.

The oldest known oxidised granite in South Africa is the biotite granite of the 3105 Ma Nelspruit batholith belonging to the late-tectonic cycle (Robb *et al.* 1983). Here, 86% of the 284 samples taken from the 100–230 km batholith give magnetic susceptibility values indicative of the magnetite series (Ishihara *et al.* 2002b). Euhedral magnetite is common and the coexisting biotite has a high Mg/Fe ratio (Ishihara *et al.* 2004). This granite and the post-tectonic K-alkaline granite are clearly the most oxidised in the FeO+MnO–Fe_2O_3+TiO_2–MgO diagram (Fig. 12). The Nelspruit granite is considered to have been generated from older Archaean continental crust (Anhaeusser & Robb 1980), implying that an oxidised crust existed before 3105 Ma ago in the Kaapvaal Craton.

Oxidised granitoids become common in the late Archaean and Proterozoic, especially after 2700 Ma. Good examples are those of the Eastern Goldfield Province, eastern Yilgarn Craton (Smithies & Witt 1997; Smithies & Champion 1999). Here the granitoids plot in the oxidised granitoid field in the FeO+MnO–Fe_2O_3+TiO_2–MgO diagram (Fig. 13). The felsic alkaline granites are even more oxidised than the typical

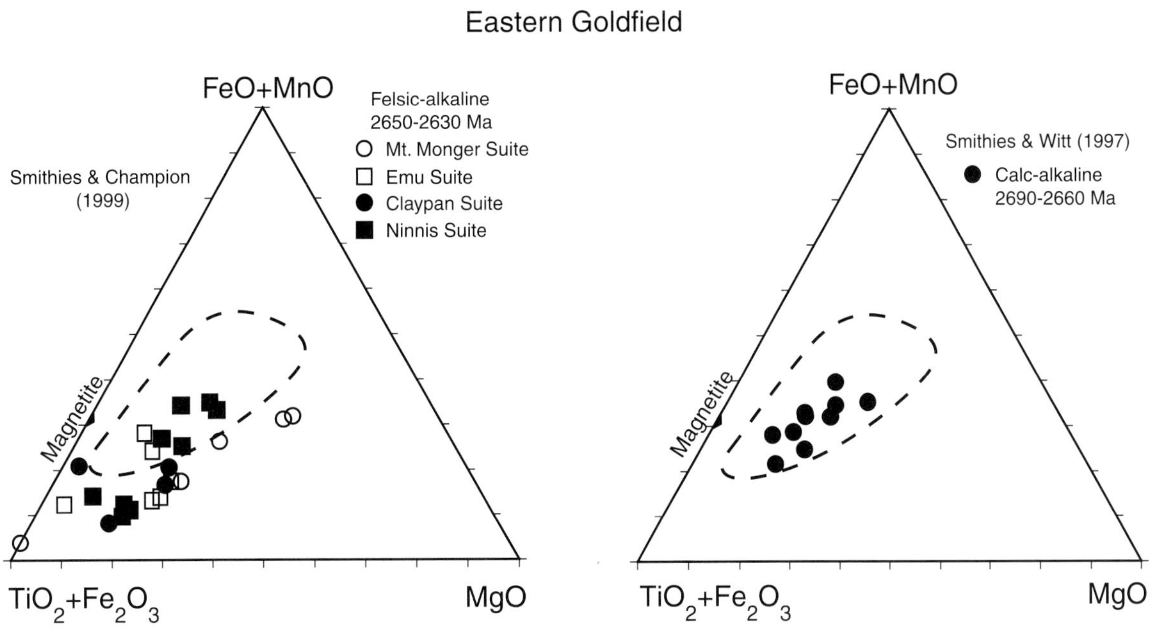

Figure 13 Bulk femic compositions of the Archaean granitoids, Eastern Goldfield, Yilgarn Craton, Western Australia. The felsic alkaline granitoids are extremely oxidised. Data sources: Smithies & Witt (1997); Smithies & Champion (1999). The broken circle shows the representative area of the magnetite-series granitoids of the Sanin Province, Japan.

magnetite-series, suggesting the presence of haematitised magnetite in the alkaline granites. The earth lithosphere appears to have become oxidised by this time, reflecting the increase of oxygen on the Earth's surface.

6. Concluding remarks

The orogenic belts of NW and NE Pacific Rims are characterised by granitic magmatism intruded into accretionary sediments and older cratons. Ilmenite-series granitoids were formed by mafic magmas from the upper mantle mixed with the accretionary sediments within the continental crust, as best shown by the Miocene and late Cretaceous granitoids of SW Japan. The mafic magmas may have been of oxidised type, but their original nature has been completely erased by reduced felsic magmas generated within the accretionary prism.

Miocene granitoids of the Green Tuff region are generally of magnetite series, because they were originally oxidised magmas and had little chance to mix with basement rocks, due to the development of vertical faulting, rather than horizontal compression, during the ascent and emplacement of the granitic magmas. The Cretaceous granitoids of the Kitakami Mountains, part of which occur within the Jurassic–Cretaceous accretionary complexes, can be explained by the same vertical-faulting tectonic setting. The magnetite-series occurring in the Sino–Korean Paraplatform had little chance of mixing with reducing agents in the source region, because carbon was concentrated in limited horizons as graphite gneisses.

The redox state of the granitoids varies in magma types. K-alkaline granitoids are generally oxidised. Anorogenic A-types are low in both fH_2O and fO_2 by definition (Loiselle & Wones 1979). Those having A-type chemistry in orogenic belts (the PA type of Hong *et al.* 1996), are strongly oxidised. The so-called adakites are oxidised in the Phanerozoic orogenic belt, but are reduced in the Greenland and Archaean Kaapvaal cratons, reflecting the reduced state of atmosphere, hydrosphere and lithosphere at that time. The first oxidised-type granite (*senso stricto*) appeared in the 3·1 Ga Nelspruit batholith in the Barberton region, South Africa.

7. Acknowledgements

The author wishes to acknowledge Dr. David Champion for his critical reviewing of the manuscript, which greatly improved the final version of this contribution.

8. References

Anhaeusser, C. R. & Robb, L. J. 1980. Magmatic cycles and the evolution of the Archaean granitic crust in the eastern Transvaal and Swaziland. *Special Publication of the Geological Society of Australia* **7**, 457–67.

Bateman, P. C., Dodge, F. C. W. & Kistler, R. W. 1991. Magnetic susceptibility and relation to initial $^{87}Sr/^{86}Sr$ for granitoids of the central Sierra Nevada, California. *Journal of Geophysical Research* **96**, 19555–68.

Blevin, P. L. & Chappell, B.W. 1992. The role of magma sources, oxidation states and fractionation in determining the granite metallogeny of eastern Australia. *Transactions of the Royal Society of Edinburgh: Earth Sciences* **83**, 305–16.

Chappell, B. W. & White, A. J. R. 1992. I- and S-type granites in the Lachlan Fold Belt. *Transactions of the Royal Society of Edinburgh: Earth Sciences* **83**, 1–26.

Drummond, M. S. & Defant, M. J. 1990. A model for trondhjemite-tonalite-dacite genesis and crustal growth via slab melting: Archean to modern comparisons. *Journal of Geophysical Research* **95B**, 21503–21.

Gastil, G., Diamond, J., Knaack, C., Walawender, M., Marshall, M., Boyles, C. & Chadwick, B. 1990. The problems of the magnetite/ilmenite boundary in southern and Baja California, California. *In The Geology of North America. Geological Society of America, Bulletin* **174**, 19–32.

Geological Survey of Japan 1974. *Cretaceous granitic rocks in the Kitakami Mountains – Petrography and zonal arrangement. Report of the Geological Survey of Japan,* **251**, 139 p. [In Japanese with English abstract.]

Hada, M., Suzuki, T., Okano, K. & Kimura, S. 1982. Crustal section based on the geological and geophysical features in the Outer Zone of southwest Japan. *Memoirs of the Geological Society of Japan* **21**, 197–211.

Hart, C. J. R., Goldfarb, R. J., Lewis, L. L. & Mair, J. L. 2004. The northern Cordilleran mid-Cretaceous plutonic province: Ilmenite/magnetite-series granitoids and intrusion-related mineralization. *Resource Geology* **54**, 253–80.

Heinrich, E. W. 1946. Studies in the mica group; the biotite-phlogopite series. *American Journal of Science* **244**, 836–48.

Hong, D., Wang, S., Han, B. & Jin, M. 1996. Post-orogenic alkaline granites from China and comparisons with anorogenic alkaline granites elsewhere. *Journal of Southeast Asian Earth Sciences* **13**, 13–27.

Hunter, D. R. 1974. Crustal development in the Kaapvaal Craton, I. The Archean. *Precambrian Research* **1**, 259–94.

Huston, D. L. & Logan, G. A. 2004. Barite, BIFs and bugs: evidence for the evolution of the Earth's early hydrosphere. *Earth and Planetary Science Letters* **220**, 41–55.

Ishihara, S. 1971. Modal and chemical composition of the granitic rocks related to the major molybdenum and tungsten deposits in the Inner Zone of Southwest Japan. *Journal of the Geological Society of Japan* **77**, 441–52.

Ishihara, S. 1977. The magnetite-series and ilmenite-series granitic rocks. *Mining Geology* **27**, 293–305.

Ishihara, S. 1979. Lateral variation of magnetic susceptibility of the Japanese granitoids. *Journal of the Geological Society of Japan* **85**, 509–23.

Ishihara, S. 1998. Granitoid series and mineralization in the Circum-Pacific Phanerozoic granitic belts. *Resource Geology* **48**, 219–24.

Ishihara, S. 2003. Chemical contrast of the Late Cretaceous granitoids of the Sanyo and Ryoke Belts, Southwest Japan: Okayama-Kagawa transect. *Bulletin of the Geological Survey of Japan* **54**, 95–116.

Ishihara, S., Ulriksen, C. E., Sato, K., Terashima, S., Sato, T. & Endo, Y. 1984. Plutonic rocks of North-Central Chile. *Bulletin of the Geological Survey of Japan* **35**, 503–36.

Ishihara, S., Matsuhisa, Y., Sasaki, A. & Terashima, S. 1985. Wall rock assimilation by magnetite-series granitoids at the Miyako pluton, Kitakami, northeastern Japan. *Journal of the Geological Society of Japan* **91**, 679–90.

Ishihara, S., Terashima, S. & Tsukimura, K. 1987. Spatial distribution of magnetic susceptibility and ore elements, and cause of local reduction on magnetite-series granitoids and related ore deposit at Chichibu, central Japan. *Mining Geology* **37**, 15–28.

Ishihara, S., Yamamoto, M. & Sasaki, A. 1999. Sulfur and carbon contents and $\delta^{34}S$ ratio of Miocene ilmenite-series granitoids: Osumi and Shibi-san plutons, Kyushu, SW Japan. *Bulletin of the Geological Survey of Japan* **50**, 671–682.

Ishihara, S., Hashimoto, M. & Machida, M. 2000. Magnetite/ilmenite-series classification and magnetic susceptibility of the Mesozoic–Cenozoic batholiths in Peru. *Resource Geology* **50**, 123–9.

Ishihara, S., Anhaeusser, C. R. & Robb, L. J. 2001. Granitoid-series evaluation of the Archean Johannesburg Dome granitoids, South Africa. *Bulletin of the Geological Survey of Japan* **53**, 1–9.

Ishihara, S., Jin, M. S. & Kajiwara, Y. 2002a. Source diversity of ore sulfur from Mesozoic–Cenozoic mineral deposits in the Korean peninsula region. *Resource Geology* **50**, 203–13.

Ishihara, S., Robb, L. J., Anhaeusser, C. R. & Imai, A. 2002b. Granitoid series in terms of magnetic susceptibility: a case study from the Barberton region, South Africa. *Gondwana Research* **5**, 581–9.

Ishihara, S., Robb, L. J., Anhaeusser, C. R. & Imai, A. 2004. On the oldest oxidized granitoids from the Archean Kaapvaal Craton, southern Africa. *Geological Society of America, Special Paper* (in press).

Ishihara, S. & Matsuhisa, Y. 1999. Oxygen isotopic constraints on the genesis of the Miocene Outer Zone granitoids in Japan. *Lithos* **46**, 523–34.

Ishihara, S. & Matsuhisa, Y. 2002. Oxygen isotopic constraints on the genesis of the Cretaceous–Paleogene granitoids in the Inner Zone of southwest Japan. *Bulletin of the Geological Survey of Japan* **53**, 421–38.

Ishihara, S. & Murakami, H. 2004. Magma types related to orogenic gold and Cu-Fe skarn deposits in the Kitakami Mountains, Japan. *Resource Geology* **54** (3) 281–98.

Ishihara, S. & Sasaki, A. 1989. Sulfur isotopic ratios of the magnetite-series and ilmenite-series granitoids of the Sierra Nevada batholith: a reconnaissance study. *Geology*, **12**, 788–91.

Ishihara, S. & Wang, P. A. 1999. The ilmenite-series and magnetite-series clssification of the Yanshanian granitoids of South China. *Bulletin of the Geological Survey of Japan* **50**, 661–70.

Isozaki, Y. 1998. Origin and growth of accretionary orogen of the Japanese Islands: from birth at rift zone to Miyashiro-type orogeny. *Memoirs of the Geological Society of Japan* **50**, 89–106.

Jaillard, E., Herail, G., Monfret, T., Diaz-Martines, E., Baby, P. Lavenu, A. & Dumon, J. F. 2000. Tectonic evolution of the Andes of Ecuador, Peru, Bolivia and northernmost Chile. *In* Cordani, U. G., Milano, E. J., Thomaz Filho, A. & Campos, D. A. (eds) *Tectonic Evolution of South America*, 481–560. Santiago, Chile: Servicio Nacional de Geologia y Mineria.

Jin, M. S., Lee, Y. S. & Ishihara, S. 2001. Granitoids and their magnetic susceptibility in South Korea. *Resource Geology* **51**, 189–203.

Kanaya, H. & Ishihara, S. 1973. Regional variation of magnetic susceptibility of the granitic rocks in Japan. *Journal of the Japanese Association of Mineralogists, Petrologists and Economic Geologists* **68**, 211–24.

Kawate, S. & Arima, M. 1998. Petrogenesis of the Tanzawa plutonic complex, central Japan: Exposed felsic middle crust of the Izu-Bonin-Mariana arc. *The Island Arc* **7**, 342–58.

Kiji, M., Ozawa, H. & Murata, M. 2000. Cretaceous adakitic Tanba granitoids in northern Kyoto, San'yo belt, Southwest Japan. *Journal of Mineralogical and Petrological Sciences* **29**, 136–49.

Koh, J. S., Yun, S. H. & Lee, S. W. 1996. Petrology and goechemical characteristics of A-type granites with particular reference to the Namsan Granites, Kyeongju. *Journal of the Korean Society of Petroleum Geologists* **5**, 142–60.

Loiselle. M. C. & Wones, D. R. 1979. Characteristics and origin of anorogenic granites. *Geological Society of America, Abstract with Program* **11**, 468.

Martin, H. 1986. Effect of steeper Archean geothermal gradient on geochemistry of subduction-zone magmas. *Geology* **14**, 753–56.

Matsuhisa, Y., Matsubaya, O. & Sakai, H. 1973. Oxygen isotope variations in magmatic differentiation processes of the volcanic rocks in Japan. *Contributions of Mineralogy and Petrology* **39**, 277–88.

Morozumi, H. 2003. Geochemical characteristics of granitoids of the Erdenet porphyry copper deposit, Mongolia. *Resource Geology* **53**, 311–16.

Nokleberg, W. J., Parfenov, L. M., Monger, J. W. H., Norton, I. O., Khanchuk, A. I., Stone, D. B., Scoese, C. R., Scholl, D. W. & Fujita, K. 2000. Phanerozoic tectonic evolution of the Circum-North Pacific. *US Geological Survey, Professional Paper* **1626**, 122 p.

Nutman A. P., Bridgwater, D. & Fryer, B. J. 1984. The iron-rich suite from the Amitsoq gneisses of southern West Greenland: early Archean plutonic rocks of mixed crustal and mantle origin. *Contributions to Mineralogy and Petrology* **87**, 24–34.

Nutman, A. P., McGregor, V. R., Friend, C. R. L., Bennett, V. C. & Kinny, P. D. 1996. The Itsaq gneiss complex of southern West Greenland; the world's most extensive record of early crustal evolution (3900–3600 Ma). *Precambrian Research* **78**, 1–39.

Nutman A. P. & Bridgwater, D. 1986. Early Archean Amitsoq tonalites and granites of the Isukasia area, southern West Greenland: development of the oldest-known sial. *Contributions to Mineralogy and Petrology* **94**, 137–148.

Oyarzun, R., Mrques, A., Lillo, J., Lopez, I. & Rivera, S. 2001. Giant versus small porphyry copper deposits of Cenozoic age in northern Chile: adakitic versus normal calc-alkaline magmatism. *Mineralium Deposita* **36**, 794–8.

Robb, L. J., Anhaeusser, C. R. & Van Nierop, D. A. 1983. The recognition of the Nelspruit batholith north of the Barberton greenstone belt and its significance in terms of Archaean crustal evolution. *Special Publication of the Geological Society of South Africa* **9**, 117–30.

Sajona, F. G. & Maury, R. C. 1998. Association of adakites with gold and copper mineralization in the Philippines. *Comptes Rendue de l'Academie des Sciences de Paris, Earth and Planetary Sciences* **326**, 27–34.

Sasaki, A. & Ishihara, S. 1979. Sulfur isotopic composition of the magnetite-series and ilmenite-series granitoids in Japan. *Contributions to Mineralogy and Petrology* **68**, 107–55.

Sato, K., Vrublevsky, A. A., Rodionov, S. M., Romanovsky, N. P. & Nedachi, M. 2002. Mid-Cretaceous episodic magmatism and tine mineralization in Khingan-Okhotsk volcano-plutonic belt, Far East Russia. *Resource Geology* **52**, 1–14.

Shibata, K. & Ishihara, S. 1979. Initial $^{87}Sr/^{86}Sr$ ratios of plutonic rocks from Japan. *Contributions to Mineralogy and Petrology* **70**, 381–90.

Shimizu, M. 1986. The Tokuwa batholith, central Japan – an example of occurrence of ilmenite-series and magnetite-series granitoids in a batholith. *Bulletin of the University Museum, University of Tokyo* **28**, 152 p.

Smithies, R. H. 2000. The Archean tonalite-trondhjemite-granodiorite (TTG) series is not an analogue of Cenozoic adakite. *Earth and Planetary Science Letters* **182**, 115–25.

Smithies, R. H., Champion, D. C. & Cassidy, 2003. Formation of Earth's early Archean continental crust. *Precambrian Research* **127**, 89–101.

Smithies, R. H. & Champion, D. C. 1999. Late Archean felsic alkaline igneous rocks in the Eastern Goldfields, Yilgarn Craton, Western Australia: a result of lower crustal delamination? *Journal of the Geological Society, London* **156**, 561–76.

Smithies, R. H. & Witt, W. K. 1997. Distinct basement terranes identified from granite geochemistry in late Archean granite-greenstones, Yilgarn Craton, Western Australia. *Precambrian Research* **83**, 185–201.

Takagi, T. 2004. Origin of magnetite and ilmenite-series granitic rocks in the Japan Arc. *American Journal of Science* **304**, 169–202.

Taylor, H. P., Jr. 1986. Igneous rocks: II. Isotopic case studies of Circum Pacific magmatism. *Reviews in Mineralogy* **16**, 273–317.

Teraoka, Y. & Okumura, K. 2003. Geological map of East Asia. 1:3,000,000. Geological Survey of Japan, AIST.

Tsuchiya, N, & Kanisawa, S. 1994. Early Cretaceous Sr-rich silicic magmatism by slab melting in the Kitakami Mountains, northeast Japan. *Journal of Geophysical Research* **99**, 22205–20.

Tsusue, A. & Ishihara, S. 1974. The iron-titanium oxides in the granitic rocks of Southwest Japan. *Mining Geology* **24**, 13–30.

Ujiie-Mikoshiba, M., Kanisawa, S., Matsuhisa, Y. & Togashi, S. 2004. Geochemical and isotopic characteristics of the Cretaceous Orikabe plutonic complex, Kitakami Mountains, Japan: magmatic evolution in a zone pluton and significance of a subduction-related mafic parental magma. *Contributions to Mineralogy and Petrology* **146**, 433–49.

Viljoen, M. J. & Viljoen, R. P. 1969. The geochemical evolution of the granitic rocks of the Barberton region. *Report on the Upper Mantle Project*, 189–211. Washington, DC: US Geodynamics Committee, National Academy of Sciences–National Research Council.

Wang, Q., Zhao, Z. H., Bao, Z. W., Xu, J. F., Liu, W., Li, C. F. Bai, Z. H. & Xiong, X. L. 2004. Geochemistry and petrogenesis of the Tngshankou and Yinzu adakitic intrusive rocks and the associated porphyry copper-molybdenum mineralization in Southwest Hubei, East China. *Resource Geology* **54**, 137–52.

Yokoyama, S. 1984. Geological and petrological studies of the late Mesozoic dyke swarms in the Inner Zone of Southwest Japan. *Geological Report, Hiroshima University* **24**, 1–63.

SHUNSO ISHIHARA, Geological Survey of Japan, Central 7, AIST, Tsukuba, 305–8567 Japan.
e-mail: s-ishihara@aist.go.jp

MS received 2 October 2003. Accepted for publication 9 September 2004.

Transactions of the Royal Society of Edinburgh: Earth Sciences, **95**, 35–48, 2004

The mechanism of melt extraction from lower continental crust of orogens

Michael Brown

ABSTRACT: Melt extraction is a process with a length scale that spans many orders of magnitude. Studies of residual migmatites and granulites suggest that melt has migrated from grain boundaries to networks of leucosome-filled structures to steeply inclined cylindrical or tabular granites inferred to have infilled ascent conduits. For example, in anatectic rocks from southern Brittany, France, during decompression-induced biotite-breakdown melting, melt is inferred to have been expressed from foliation-parallel structures analogous to compaction bands to dilation and shear bands, based on location of residual leucosome, and from this network of structures to ascent conduits, preserved as dykes of granite. The leucosome-filled deformation band network is elongated parallel to a sub-horizontal lineation, suggesting that mesoscale melt flow was focused primarily in the plane of the foliation along the lineation to developing dilatant transverse structures. The leucosome network connects with petrographic continuity to granite in dykes; however, the orientation of dykes discordant to fabric anisotropy suggests that their formation was controlled by stress, which indicates that the process is a fracture phenomenon. Blunt fracture tips and zigzag propagation paths indicate that the dykes represent ductile opening-mode fractures; these are postulated to have formed by coalescence of melt pockets. The structures record a transition from accumulation to draining; quantitative volume fluxes are calculated and presented for the generalised extraction process. The anatectic system may have converged to a critical state at some combination of melt fraction and melt distribution that enabled formation of ductile opening-mode fractures, but fractal distribution of inferred mesoscale melt-filled structures has not been demonstrated; this may reflect the inherent anisotropy and/or residual nature of the drained source. Melt extraction has been modelled as a self-organised critical phenomenon, but the mechanism of extraction is not described and the relationship between these models and the spatial and temporal granularity of lower continental crust is not addressed. Self-organised critical phenomena are driven systems involving 'avalanches' with a fractal frequency-size distribution; thus, the distribution of melt batch sizes might be expected to be fractal, but this has not yet been demonstrated in nature.

KEY WORDS: crustal melting, deformation bands, ductile opening-mode fractures, magma ascent, melt flow networks, melt fluxes, spatial patterning.

Melt extraction, magma ascent and pluton emplacement represent a significant and irreversible heat and mass transfer from lower to upper continental crust (e.g. Brown *et al.* 1995a). In many cases, the size, shape and spatial distribution of upper crustal plutons probably reflect aspects of melting in the lower crust, and extraction and emplacement are presumed complementary processes between which there must be feedback modulated by the ascent mechanism (e.g. Brown & Solar 1999). Superimposed on this intrinsic control are conduit-forming and space-making processes induced by regional deformation (Brown 2001a, b). However, in the extreme case where magma ascent is strongly controlled by a major steeply inclined crustal-scale shear zone, such as the Peri-Adriatic Fault System, a linear alignment of plutons in the upper crust may not necessarily reflect a linear source region in the partially molten lower crust (e.g. Rosenberg 2004).

Melting occurs at distributed multiphase grain boundaries in a source (mm^3), whereas a typical pluton represents a large volume of melt accumulated at a sink (10^3–10^4 km^3); plutons commonly crystallise within 10 s to 100 s of ky, and batholiths within several my (cf. Glazner *et al.* 2004). Assuming 10–20 vol.% melt extraction, source volumes are 5×10^3–10^5 km^3. Thus, melt extraction, from segregation to emplacement, is a process with a length scale that spans at least seven orders of magnitude, occurring on a time scale of $<10^5$–$>10^6$ y. In the exposed lower continental crust of deeply eroded orogens, where evidence of melting is preserved in residual migmatites and granulites, magma chambers of any kind related to the inferred melting event appear to be absent. Therefore, either melt must accumulate in a network that feeds drainage channels (e.g. Brown & Solar 1999), or there must be some other mechanism by which melt batches aggregate to increase in size and ascend through the crust (e.g. Bons *et al.* 2001, 2004), or the source must ascend *en masse* (Marsh 1982; Olsen *et al.* 2004).

To a first approximation, assuming no mass addition from the mantle, the volume of granite emplaced in the upper crust will be equal to the volume of melt extracted from the lower crust, and rates of melt production, extraction, ascent and emplacement are likely to have been balanced at the crustal scale. Volume fluxes [Fig. 1; Q_{Pr}, melt production (implicit in source); Q_{Ex}, melt extraction; Q_{As}, melt ascent; and Q_{Em}, melt emplacement] over distinct time intervals have been considered qualitatively by, for example, Thompson (1999) and Brown (2001a, b), although the role of deformation in enhancing rates was not considered explicitly by either author. These authors argued that relationships between pairs of these quantities can be used to deduce the relative efficiencies and magnitudes of

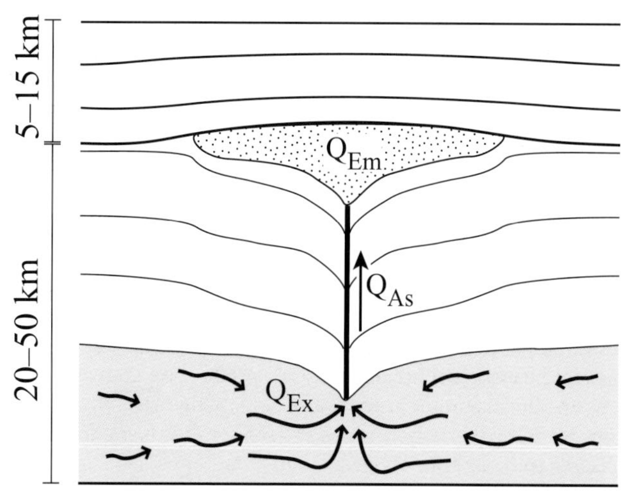

Figure 1 A simple model of melt extraction, ascent and emplacement in which the rates of melt production, extraction, ascent and emplacement are likely to have been balanced at the crustal scale. The rate of melt production (Q_{Pr}) is controlled by heat flow and is not shown explicitly in the figure. Volume fluxes for melt extraction (Q_{Ex}), melt ascent (Q_{As}) and melt emplacement (Q_{Em}) are shown schematically. The depths shown on the left-hand side of the figure are intended as a general indication of the likely range in orogenic environments. From Cruden 1998, modified by permission of the Geological Society.

melt-flow mechanisms, and although this analysis is simplistic in relation to orogens where feedback occurs between thermally induced weakening, melt-induced weakening and deformation, a simple qualitative approach nonetheless provides a starting point.

If melt production, extraction, ascent and emplacement are equal (i.e. $Q_{Pr}=Q_{Ex}$, $Q_{Ex}=Q_{As}$ and $Q_{As}=Q_{Em}$, then $Q_{Pr}=Q_{Em}$), the system may be described by percolative flow (cf. Olsen *et al.* 2004). These rates vary with time, but if the rate of melt production is faster than the rate of extraction at any time, the source will stagnate, melt fraction will increase and the volume change related to the melting reaction will increase the fluid pressure in the source. Thus, when $Q_{Pr}>Q_{Ex}$, melt-induced dilatancy potentially occurs in the source, and Q_{Ex} is increased by channelled flow. On the other hand, if melt emplacement is faster than melt extraction ($Q_{Em}>Q_{Ex}$), episodic extraction is more likely (cf. Handy *et al.* 2001) or the melt flux is rapidly exhausted. Finally, if the rate of melt production is less than the rate of melt extraction ($Q_{Pr}<Q_{Ex}$), the source will be drained slowly. Ultimately, since Q_{Pr} is controlled by the heat flow, it must exert control over the other fluxes.

To improve our knowledge of melt extraction from the continental crust, we need to: develop realistic models of how and at what rate melt segregates from residual material; establish the spatial (pervasive or channelled) and temporal (continuous or transient) characteristics of melt extraction; and pursue realistic models of magma ascent. To this end, the present author briefly summarises postulated mechanisms for magma ascent, and briefly reviews the microstructures and the mesoscale spatial patterning of former melt-bearing rocks, such as residual migmatites and granulites. Then, he uses one particular example from southern Brittany, France, related to extensional exhumation of lower crust to form a core complex, to illustrate in more detail one mechanism of melt extraction. The author presents quantitative volume fluxes related to the extraction process calculated for the general case, and in the final section, develops a general model for melt extraction from the lower continental crust of orogens.

1. Mechanisms proposed for magma ascent in orogenic belts

Dyking by brittle fracture is not a viable mechanism for extracting melt from partially molten crustal sources (e.g. Rubin 1998), although brittle fracture is a viable mechanism for second-stage ascent of melt from an upper crustal magma chamber into sub-solidus crust (e.g. Weinberg 1996). Alternatives for initial melt ascent through suprasolidus crust include mobile melt-filled hydrofractures (Dahm 2000; Bons *et al.* 2001, 2004) and various mechanisms to bridge grain-boundary flow to channelled flow in ductile crust (e.g. Brown 1994; Collins & Sawyer 1996; Brown & Solar 1998a; Sawyer *et al.* 1999; Weinberg 1999; Leitch & Weinberg 2002; Olsen *et al.* 2004).

The slow rates of migration of viscous silicic melt through pores and shear-enhanced compaction of residue probably require that melt drained in a dyke propagation event must have been previously extracted from the pores and reside an accumulation network; additionally, there must be sufficient hydraulic head to drive ascent (Brown & Solar 1998b). In general, tabular intrusions are favoured in nature (Emerman & Marrett 1990; Rubin 1995; Weinberg 1999), and initiating dykes by ductile fracture avoids the problems and assumptions inherent in brittle fracture. By ductile, the present author means fracturing associated with distributed inelastic deformation at the fracture tip (cf. Rutter 1986), but no particular stress-strain relationship or deformation mechanism is implied (cf. Eichhubl 2004).

Suprasolidus crust is an over-pressured system with supra-magmastatic fluid pressure gradients and spatial and temporal variations in permeability. Permeability has a power law dependence on melt fraction (Wark & Watson 1998), which may lead to compaction-generated flow instabilities, manifested as self-propagating high-porosity structures (Connolly & Podladchikov 1998; Rabinowicz & Vigneresse 2004). These high-porosity structures may grow by influx of melt, which may lead to fully-segregated melt flow (Richardson *et al.* 1996; Podladchikov & Connolly 2001). Thus, melt flow in high porosity structures may represent an intermediate mechanism bridging porous flow to channelled flow.

Diapirism involves ascent of melt plus residue; it is not generally considered to be a viable process for emplacement of upper crustal plutons (Emerman & Marrett 1990; Clemens & Mawer 1992; Rubin 1993), although subsequent modelling contradicts this view (Weinberg & Podladchikov 1994; Burov *et al.* 2003). Diapirism may occur in the lower crust (Barbey *et al.* 1999; Teyssier & Whitney 2002), and it is a mechanism by which (some) plutons are inferred to have been emplaced in the deep crust in continental arcs (e.g. Miller & Patterson 1999).

2. Petrological and field-based arguments and observations

2.1. Microstructural scale

Under equilibrium conditions in isotropic crust, melting begins at multiphase grain junctions which include quartz and feldspar, and a hydrate phase. However, Earth's crust is anisotropic and in a state of stress, and variations in bulk composition and grain size influence the sites at which melting begins, and orientation of hydrate phases in a metamorphic fabric may influence the spatial development of melt pockets (Sawyer 2001). Differential stress varies within any given volume of crust, and these variations give rise to pressure gradients throughout the volume of crust. Melting may begin at sites of lower or higher P, once the initial thermal overstep is close to that required to overcome the activation energy for the melting reaction. For high $a(H_2O)$ melting, where the dP/dT of reaction is negative, small increases in P may promote the start of reaction, whereas, for moderate-to-low $a(H_2O)$ hydrate-breakdown melting, where the dP/dT of reaction is positive, small decreases in P may promote the start of reaction. Furthermore, for moderate-to-low $a(H_2O)$ melting, the solid products of the reaction may have difficulty nucleating (Powell & Downes 1990). Thus, once melting is established at a particular site, it is energetically favourable for melting to continue at that site and for the solid products to grow at that site (Fig. 2) by a process of diffusional mass transfer (Powell & Downes 1990; Hand & Dirks 1992).

The grain-scale distribution and connectivity of silicic melt is of primary importance in determining the physical properties [e.g. permeability, rheology (which is sensitive to the presence and distribution of melt) and effective diffusivity (which may be increased by orders of magnitude by the presence of interconnected melt)] of melt-bearing lower continental crust. Permeability is especially important because it, melt viscosity and matrix viscosity are the main parameters which control the dynamics of melt segregation. Under conditions of mechanical equilibrium, silicic melt wets matrix minerals in crustal rocks (Laporte & Watson 1995; Laporte & Provost 2000); thus, in melt-bearing crust, melt is expected to be interconnected at very low melt fraction (e.g. Laporte *et al.* 1997; Lupulescu & Watson 1999; Wark *et al.* 2003). Also, deformation experiments on analogue materials show evidence of dynamic wetting of grain boundaries, which enhances early interconnection of melt, and inter-granular fracture and grain boundary sliding (Rosenberg & Handy 2000, 2001). Thus, melt is likely to be connected in all common crustal protoliths (Laporte *et al.* 1997).

This conclusion is supported by observations of microstructures in natural anatectic systems, which provide evidence of the former presence of melt in grain-scale pores (e.g. Brown *et al.* 1999; Brown 2001a, b). Melting and crystallisation from melt may be inferred using the following criteria: the igneous microstructure of leucosomes in migmatites and granulites (e.g. Cuney & Barbey 1982; Vernon & Collins 1988); mineral pseudomorphs after grain-boundary melt films and pockets [in contact migmatites (e.g. Holness & Clemens 1999; Marchildon & Brown 2001, 2002), in regional migmatites (e.g. Marchildon & Brown 2003) and in granulites (e.g. Sawyer 1999, 2001; Guernina & Sawyer 2003)]; magmatic rims on subsolidus cores of grains [e.g. rational faces, overgrowths of different compositions (e.g. Marchildon & Brown 2001, 2002; Sawyer 2001)]; spatial distribution of like and unlike phases [i.e. mineral phase distribution (Ashworth & McLellan 1985)]; and, annealed microfractures (Watt *et al.* 2000; Marchildon & Brown 2001, 2002).

2.2. Mesoscale structures

First, the present author evaluates whether features preserved in lower crust represent what intuition leads us to infer – *viz.* melt accumulation networks and magma ascent conduits (Brown 1994). For flow, melt must form an interconnected network and the host must be permeable and for extraction, there must be a driving force and a sink. It is important to ascertain whether the conduit topology infilled by remnant leucosome and the structures infilled by granite represent evidence of active crustal-scale melt transfer from lower to upper crust, or the preserved record of a non-productive system that failed to support crustal-scale melt flow. Active crustal-scale melt flow is inferred if the bulk rock chemical composition is residual for anatectic rocks (migmatites and granulites) in comparison with protolith compositions, and the difference is consistent with net melt loss (e.g. Sawyer *et al.* 1999; Solar & Brown 2001; Guernina & Sawyer 2003); the net difference represents a minimum estimate of melt loss attributable to melt flux through any level of observation and partial crystallisation of migrating melt. Also, melt loss is implied if peritectic phases survive without significant retrogression (Powell & Downes 1990; Brown 2002; White & Powell 2002). For the purpose of this paper, the present author assumes melt loss to be the common situation based on published examples (Sawyer *et al.* 1999; Solar & Brown 2001; Guernina & Sawyer 2003), but emphasises that net melt loss must be evaluated as part of any specific case study.

In the case of isotropic protoliths with randomly oriented pores of melt, analytical (Sleep 1988) and numerical (Simakin & Talbot 2001a) solutions, and finite element modelling (Simakin & Talbot 2001b), indicate that melt-filled veins will form parallel to σ_1 and perpendicular to σ_3 by closure of pores perpendicular to σ_1 and expansion of pores perpendicular to σ_3. Melt segregation into vein networks may lead to tabular bodies of melt of sufficient size that buoyant ascent becomes possible. Alternatively, in appropriate circumstances, bottom-to-top directionality (vertical fracture alignment) may emerge by non-linear evolution of an isotropic fracture network formed by thermal expansion (Petford & Koenders 1998). Such models may apply where lower crustal protoliths are isotropic and heating is rapid, which might be the case in the lower parts of arcs during the subduction period of the orogenic cycle, although Rubin (1993, 1995) has argued that melt-filled fractures are unlikely to propagate out of a melt-bearing source, unless the source is already pre-warmed by dyking from deeper levels.

In anisotropic protoliths, the scale of compositional layering and the strain field generally control sites of early melt concentration; foliation-parallel and transverse stromata form by mass transfer down gradients in pressure (van der Molen 1985a, b; Maaløe 1992; Brown *et al.* 1995b). With increasing melt fraction (MF), melt becomes progressively concentrated into a developing network of leucosome-filled structures (cf. Brown & Rushmer 1997), including interboudin partitions, shear surfaces, and axial surfaces and hinges of folds; the exact form of the network being controlled by the style of deformation (Brown *et al.* 1995b; Brown *et al.* 1999; Sawyer *et al.* 1999; Guernina & Sawyer 2003; Marchildon & Brown 2003). The role of anisotropy in controlling crustal-scale melt transport has been demonstrated by several studies (e.g. D'Lemos *et al.* 1992; Collins & Sawyer 1996; Brown & Solar 1998a).

Incongruent melting that involves biotite or hornblende breakdown commonly results in melt concentration around the peritectic product minerals (e.g. Powell & Downes 1990), so that nucleation and growth of peritectic phases also controls sites of early melt concentration (Hand & Dirks 1992; Johnson & Brown 2004; Fig. 2a–d). Increasing MF leads to weakening and formation of a network of deformation bands, most probably nucleated on the local concentrations of melt (e.g. as modelled by Grujic & Mancktelow 1998). The preferential occurrence of peritectic minerals in the deformation bands, in spite of a homogeneous distribution of reactant phases in the matrix, supports the inference that, in general, deformation was localised by the concentration of melt (higher porosity) at these sites. This contrasts with the study by Hand & Dirks (1992), where strain associated with pre-existing crenulation structures apparently controlled the sites of initial incongruent melting; in general, this seems to be less common.

2.2.1. Specific examples. At Mt Hay, central Australia, based on sites of leucosome entrapment, Collins & Sawyer (1996) inferred that melt migrated parallel to the steep regional

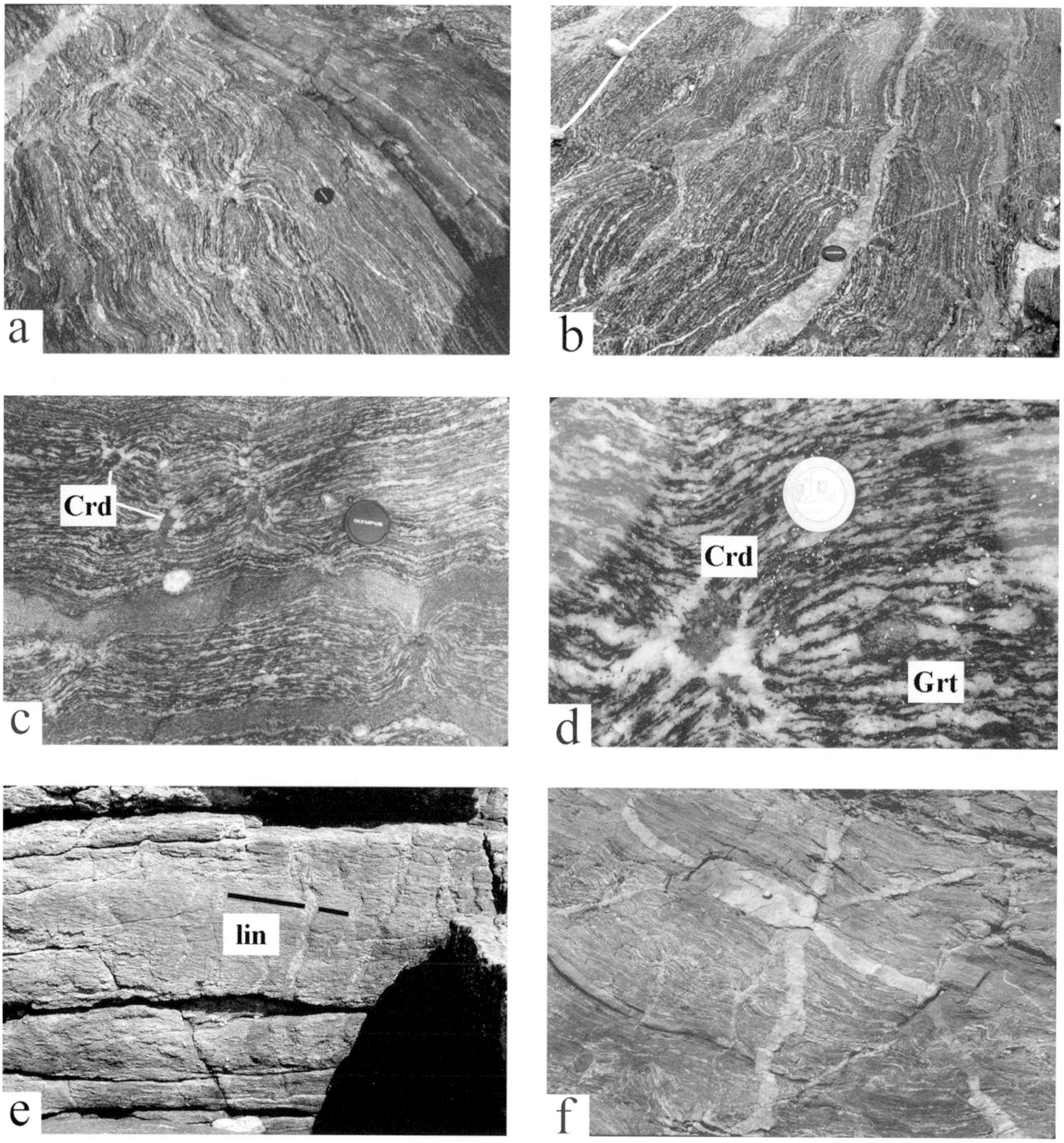

Figure 2 Features associated with melting and inferred melt-bearing structures in granulite facies anatectic migmatites from southern Brittany, France: (a) Stromatic migmatite with layer-parallel leucosome stromata and leucosome in shear bands (Port Navalo Plage, Morbihan); (b) Stromatic migmatite with layer-parallel leucosome stromata showing periodic foliation boudinage (Port Navalo Plage, Morbihan); (c) Stromatic migmatite with layer-parallel leucosome stromata showing periodic foliation boudinage; note cordierite (Crd) in interboudin partitions (Port Navalo Plage, Morbihan); (d) Close-up of the petrographic details in (c); note partially reacted garnet (Grt, in foliation below and right from coin) and cordierite in interboudin partition (below and left of coin); (e) Steeply oriented surface parallel to foliation (note subhorizontal elongation lineation) to show vertical extent of leucosome in transverse structures (Petit Mont, Arzon, Morbihan); (f) Sub-horizontal surface through stromatic migmatite to show granite dykes which intersect in a common steeply inclined cylindrical channel structure; these dykes do not clearly cross-cut each other, but rather merge where they meet. Although the dykes are discordant at outcrop scale, there is only limited evidence for cross-cutting relationships at the vein-to-grain scale, suggesting that material now in the dykes was molten at the same time as the material in the host rock included a melt phase (Petit Mont, Arzon, Morbihan).

elongation direction defined by a strong mineral lineation and coaxial folds via a network of diverse structurally controlled conduits (fold hinges and axial surfaces, interboudin partitions and shear zones) or created new channels via magma fracture, commonly parallel to foliation. Furthermore, these authors argued that porous flow of melt through the interconnected network of mesoscale structures fed larger conduits which transported melt to the upper crust. Similarly, at Wuluma Hills, central Australia, in residual granulites, Sawyer *et al.* (1999) concluded that syn-anatectic deformation enabled melt migration from grain boundaries through a network of veins to hinge zones of moderately plunging folds, and subsequently, into mostly foliation-concordant veins in shear zones which parallel the fold axial surfaces.

Based on detailed studies of granulites in the Ashuanipi subprovince, Quebec, Guernina & Sawyer (2003) suggested two alternative short-range strategies for melt migration. First, in melt-depleted layers without leucosome stromata within layers, but where layer-parallel leucosomes occur at the interfaces between layers, melt movement is inferred to have been primarily orthogonal to the layering, and melt extraction is inferred to have taken place by flow along the layer interfaces. Secondly, in melt-depleted layers with leucosome stromata within the layering as well as in the interfaces between layers, the leucosome forms networks composed of wide transverse leucosome-filled structures which are inferred to have drained individual layers and linking layer-parallel leucosomes between the layers which apparently record pathways of melt extraction out of the source. In this second case, melt movement is inferred to have been primarily within the plane of the layer towards the developing dilatant transverse structures which are now infilled by leucosome.

In the Acadian belt of western Maine, northern Appalachians, USA, in residual migmatites, the form of melt ascent conduits mimics the apparent strain ellipsoid recorded by steeply inclined host rock fabrics. Thus, steeply inclined tabular granites occur in zones of apparent flattening strain, whereas moderate-to-steeply plunging cylindrical granites occur in zones of apparent constrictional strain (Brown & Solar 1999; Solar & Brown 2001). Other studies have provided general descriptions of the relationship between mesoscale structures and ascent conduits, which vary from shear zones (e.g. D'Lemos *et al.* 1992; Brown 1994; Rosenberg 2004) to dykes (Marchildon & Brown 2003). The study by Marchildon & Brown (2003) is considered further below.

One implication derived from all of these examples is that large magma chambers do not form in lower continental crust during syn-anatectic deformation. Ultimately, melt drains from the source via discrete tabular or cylindrical conduits to form plutons. How such conduits form is controversial, but deformation-enhanced variations in pressure gradients and variations in porosity are thought to be the key.

2.3. Quantitative data about melt flow networks from southern Brittany

There have been several attempts recently to obtain quantitative data from anatectic rocks. Tanner (1999) and Bons *et al.* (2004) argued that leucosome distributions are scale invariant, but Marchildon & Brown (2003) cautioned that such a conclusion may not be supported by careful analysis of the data.

Marchildon & Brown (2003) utilised well-exposed coastal outcrops of granulite facies migmatites from southern Brittany, in the Variscides of France, to obtain descriptive and quantitative data on the one- and two-dimensional distribution of inferred melt-bearing structures, and the transitions from foliation-parallel to network to conduit flow in an attempt to understand the evolution of the architecture (connectivity and permeability) of melt-bearing systems. The structural relationships between leucosome network morphology and granite in dykes investigated by Marchildon & Brown (2003) are a late-stage feature of the evolution of the Variscan belt in southern Brittany which relate to decompression and biotite-breakdown melting producing peritectic cordierite; the earlier history is not discussed in detail in the present paper. Fugitive melt from the anatectic rocks is inferred to have accumulated as leucogranite plutons at shallower levels of the crust (Gapais *et al.* 1993; Brown & Dallmeyer 1996; Marchildon & Brown 2003).

In this example, leucosome and granite-filled structures include (Fig. 2) millimetre to centimetre scale foliation-parallel and transverse leucosome stromata (reflecting compositional layering, interboudin partitions and extensional shear surfaces) and centimetre- to metre-scale foliation-discordant steeply inclined tabular granites (representing inferred ascent conduits). Petrographic continuity (mineralogy, mode and microstructure) from structure to structure implies that the infilling material crystallised from a continuous melt-bearing network (Fig. 3). Peritectic phases are commonly located in deformation bands, particularly interboudin partitions (Fig. 2), which connotes an important contribution by diffusive mass transfer in the local concentration of melt, particularly at low melt fractions early during biotite-breakdown melting. Using the equation for dimensionless melt transport (MT) number of Olsen *et al.* (2004) to evaluate porous flow versus bulk flow for melt fractions appropriate to biotite-breakdown melting producing peritectic cordierite yields values for MT <1 which are consistent with macroscopic porous flow.

Marchildon & Brown (2003) obtained information about thickness and spacing distributions of foliation-parallel leucosome stromata from one-dimensional traverses (Figs 4 & 5). Cumulative leucosome thickness versus distance along the traverse, and log-log plots of leucosome cumulative frequency versus thickness or spacing are useful in discriminating between different possible types of thickness or spacing distributions (e.g. Gillespie *et al.* 1993, 1999; Loriga 1999). For instance, based on an analysis of synthetic datasets for different vein distribution types, Gillespie *et al.* (1999) argued that fractal distributions will yield a stepwise plot of cumulative thickness versus distance because of inherent clustering. Fractal distributions also define linear trends (with slope related to the fractal dimension, D) in log cumulative frequency versus log thickness or spacing plots. On the other hand, random vein distributions define smooth and regular cumulative thickness versus distance plots, and concave downward patterns of log cumulative frequency versus log thickness and spacing (Gillespie *et al.* 1999).

In Figure 4 the plots of cumulative leucosome thickness versus distance along the traverse are essentially linear, whereas, in Figure 5, the downward concavity of arrays on the plots of log cumulative frequency versus log thickness or spacing indicates a limited range of leucosome thicknesses [generally between 1 mm (the limit of resolution) and 10 mm (upper limit 20–30 mm)], and shows that frequency decreases rapidly and non-linearly with increasing thickness. These features are inconsistent with scale invariance. In these migmatites, the volume of leucosome (~ 50 vol.%) far exceeds that expected at the calculated *P-T* conditions (~ 25 vol.%; Johnson & Brown 2004). In general, leucosomes do not have melt compositions, but represent some combination of residual reactants and products in the deformation bands, together with any early crystallised products trapped during syn-transport crystallisation of the migrating melt as the exhumed lower crust initially cooled under suprasolidus conditions. The amount of melt present at final solidification of the leucosome is unknown, but if the minimum is that required for the leucosome to be grain-supported, the proportion of melt remaining is likely to have been small (cf. Powell & Downes 1990).

Observation of structures in mutually perpendicular two-dimensional surfaces suggests that the leucosome network is elongated in relation to a sub-horizontal lineation. Analysis of leucosome distributions using the box-counting method

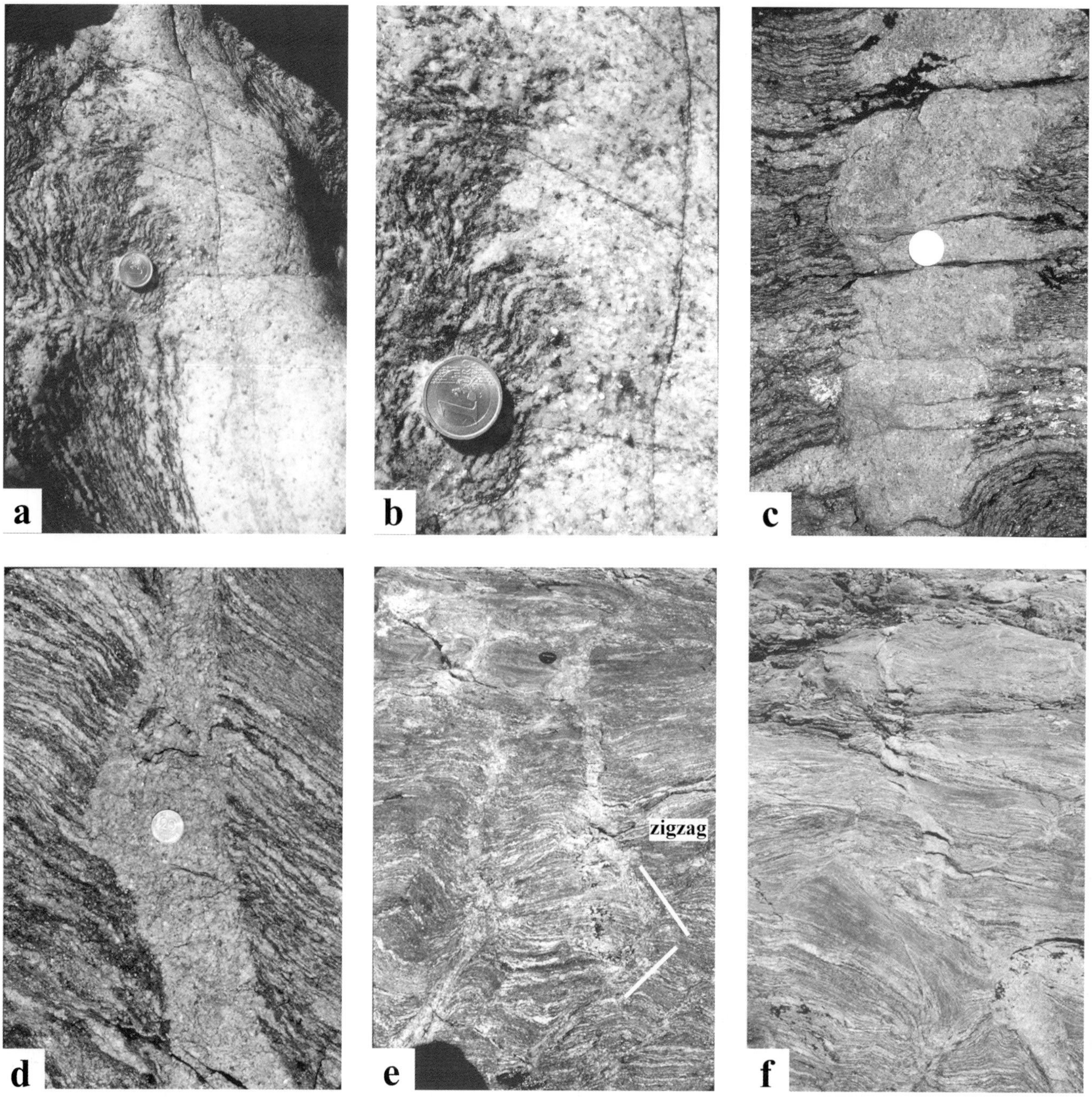

Figure 3 Leucosome in petrographic continuity (microstructure, mineralogy and mode) with granite in dykes in migmatite from southern Brittany, France. (a) Irregular interface between migmatite and granite in ductile opening-mode fracture (Port Navalo Plage, Morbihan); (b) Close-up of (a) to show petrographic continuity between leucosome and granite; (c) Highly discordant granite dyke showing petrographic continuity with the layer-parallel leucosomes it cross-cuts (Le Petit Mont, Arzon, Morbihan); (d) Highly discordant granite dyke showing petrographic continuity with the layer-parallel leucosomes it cross-cuts (Le Petit Mont, Arzon, Morbihan); (e) Thin, discordant sheet-like granite with zigzag form at tip (bottom of image; Le Petit Mont, Arzon, Morbihan); (f) Thin, discordant sheet-like granite with zigzag form (Le Petit Mont, Arzon, Morbihan).

corroborates the suggested anisotropy, but indicates that leucosome morphology is not scale invariant. Melt flow is inferred to have been primarily in the plane of the foliation and along the lineation to developing transverse structures. This network fed steeply dipping discordant planar conduits now filled by dykes of granite.

Draining of melt from deformation bands and capture of small conduits by large conduits will be favoured by the lower melt pressure in the larger conduits (cf. Sleep 1988). In southern Brittany, the formation and orientation of dykes was apparently controlled by far-field stresses associated with dextral transtension (Marchildon & Brown 2003; Cagnard *et al.* 2004). However, spacing of conduits is inferred to have been controlled by the strong intrinsic permeability anisotropy in layers with well-developed stromatic layering and the weak implied extrinsic permeability anisotropy between layers. This leads to variable focusing of melt that requires a larger collection zone for each conduit than would be the case for an isotropic protolith (cf. Eichhubl & Boles 2000).

3. Fluxes

In this section, the present author considers quantitatively fluxes associated with the processes of melt extraction from lower continental crust. Given a pluton of volume of 10^3–10^4 km^3 crystallised within a period of 10^4–10^6 y, and assuming that crystallisation times correspond to transport times,

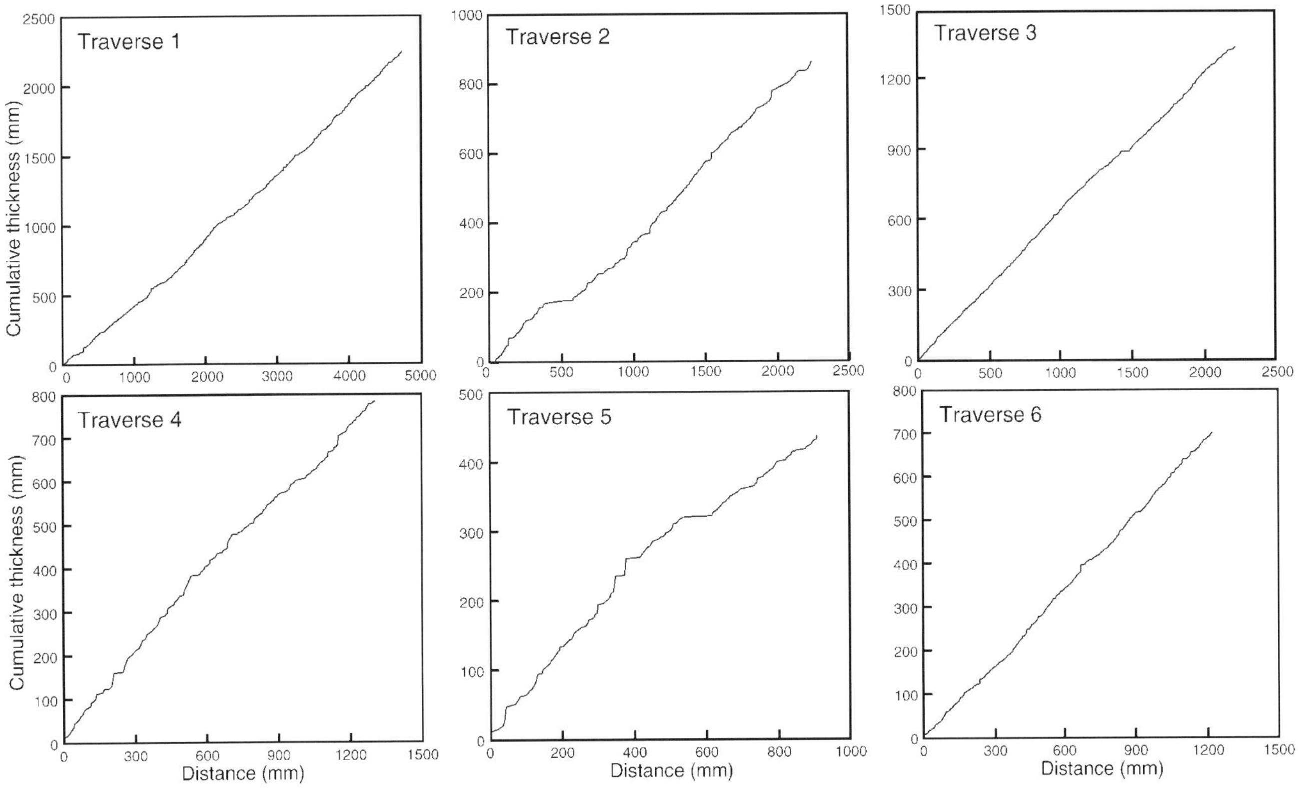

Figure 4 Information about thickness of foliation-parallel leucosome stromata from one-dimensional traverses (after Marchildon & Brown 2003) shown as plots of cumulative leucosome thickness versus distance along the traverse.

then, if a pluton is formed in a single filling event, the crystallisation time gives a maximum time for transport, and all transport estimates will be lower as a consequence. The lower limit for the crystallisation period is consistent with the minimum time of *ca.* 50 ky to extract, via a vein network, 10 vol.% of melt to the top of 10 km thickness of protolith with 0·5–1 wt.% H_2O (assuming approximately isothermal conditions of ~850°C and applied differential stress of <10 but >1 MPa, calculated based on data in Rutter & Neumann 1995). The upper limit for the crystallisation period is consistent with melt extraction from a protolith with similar H_2O content and under similar applied differential stress, but at a lower temperature of ~750°C.

Consider a source of area of 10^3–10^4 km^2 beneath which an average of 1 km^3 per km^2 of melt is extracted from 10 km depth of suprasolidus continental crust (e.g. Zen 1992). For a pluton that is 1 km thick, the melting footprint, the region of melt throughput and the pluton area are identical. This geometry allows calculation of reference melt extraction rates.

To form a pluton of 1 km thickness with a region of melt throughput to melting footprint ratio of 1 requires an integrated melt extraction flux of 1 km^3 melt per km^2 crust, which yields an average melt extraction flux of 10^{-6} to 10^{-4} km^3 melt per km^2 crust y^{-1}. This is equivalent to the situation where $Q_{Pr}=Q_{Em}$, as discussed in the introduction. These fluxes should be contrasted with the reference melt extraction velocity, which is set by the depth of melting. In order to get the last drop of melt through the source area requires transport over at least 10 km, which yields an average melt velocity of 10^{-5}–10^{-3} km melt y^{-1}. The melt velocity is faster than the melt flux because most of the crust is solid; thus, we may think of the melt velocity in units of volume of melt per area of melt per unit time and the melt flux in units of volume of melt per area of crust per unit time.

Given the distinction between melt flux and velocity, the ratio of melt footprint to the region of melt throughput is useful. Consider the extreme case of a melt-filled conduit. With reference to the average pluton volumes (above), a melt-filled conduit of area 10^{-2}–10^{-1} km^2 requires an integrated melt velocity of 10^4 to 10^6 km^3 melt km^{-2} melt, which yields an average melt velocity in the conduit of 10^{-2}–100 km melt y^{-1}. This is equivalent to a single dyke ascent model (Clemens 1998). In this limiting case, the 'crust' through which the melt flows (i.e. the conduit) is all melt, so the melt flux in the region of melt throughput will be the same. However, the melt flux through the melting footprint is the same as estimated above because the pluton is being filled from the same volume of melted crust. These estimates bracket the actual melt extraction rate.

A related issue is the transport of melt within the melting volume to the melt-filled conduit. Whether the melt-filled conduit extends to the base or to the top of the melting volume [10^3 to 10^4 km^2 by 10 km], the distance over which melt travels to the conduit is 10^1–10^2 km. This estimate yields an average melt velocity of 10^{-5}–10^{-2} km melt y^{-1}. A rough estimate of the melt extraction flux to the conduit can be made by assuming that the melt to crust volume fraction is 0·1 [1 km^3 of melt per 1 km^2 by 10 km of crust]. The average melt extraction flux to the conduit is given by the product of the melt fraction with the melt velocity to yield 10^{-6}–10^{-3} km^3 melt per km^2 crust y^{-1}.

Convergent fluxes increase approaching a sink. The melt flux just before it reaches the conduit will be greater than the average melt flux through the melting volume. If the conduit reaches to the bottom of the melting volume, all the melt generated within the melting volume will be converging on a conduit with a surface area of 1–10^{-1} km^2. The integrated melt flux converging on the conduit will be 10^3–10^5 km^3 melt per km^2 crust, which yields melt fluxes across the boundary of

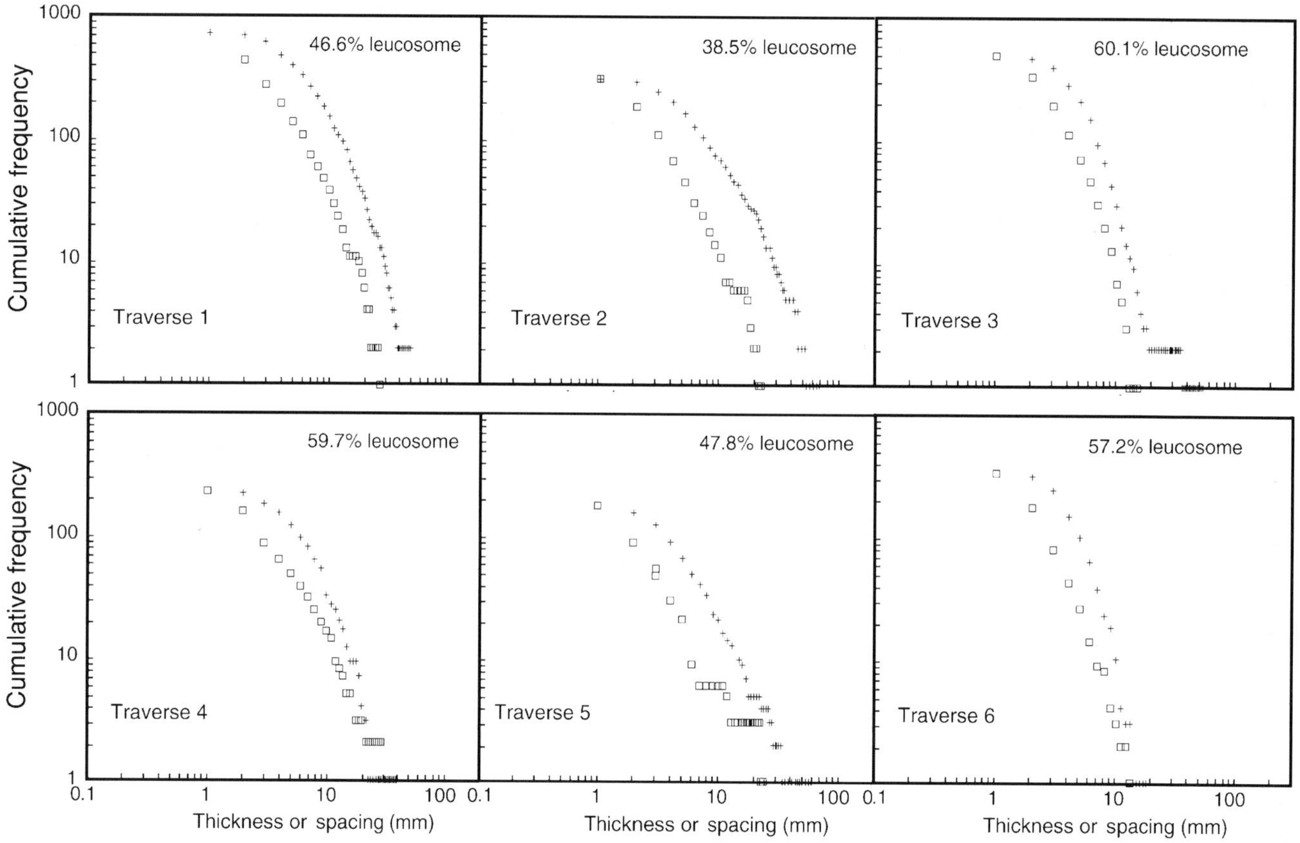

Figure 5 Log-log plots of foliation-parallel leucosome cumulative frequency versus thickness (open squares) or spacing (upright crosses) of leucosome stromata from one-dimensional traverses (after Marchildon & Brown 2003).

the conduit of 10^{-3}–10 km^3 melt per km^2 crust y^{-1}. These estimates are misleading because the melt flux falls away as distance squared from a cylindrical conduit and as distance cubed from a point sink. The mismatch between estimated transport rates in the conduit and rates at the boundary of the conduit also suggests that the MF probably increases as the conduit is approached. It requires a more rigorous analysis outside the scope of the present paper to take into account the spatial variation in MF.

The present author notes that the range of estimates for melt extraction calculated here is similar to the range of estimates derived by Harris *et al.* (2000) of 10^4–10^2 y, based on accessory phase thermometry. Furthermore, these authors calculated that melt may ascend through dykes from lower to upper crust in $<10^{-2}$ y. These rates are fast in comparison with estimates of the time scale for prograde metamorphism to the metamorphic peak, which is on the order of millions of years. This suggests a mechanism for the punctuated nature of upper crustal magmatism during orogenesis. The melt production rate in the lower crust is determined by heat flow, which means that the time scale to achieve interconnection of melt in the source determines the beginning of a melt extraction window. Once interconnection of melt has been achieved, the time scale for extraction and ascent is short in relation to the time scale associated with orogenesis, or prograde and retrograde metamorphic evolution (cf. Harris *et al.* 2000).

4. A paradigm for melt extraction from lower continental crust of orogens

Field studies have characterised the spatial and temporal granularity, and petrological studies have characterised the thermal evolution of migmatites in an attempt to understand the physical process of melt extraction from anisotropic lower crustal rocks. The sequence of mesoscale leucosome- and granite-filled structures described from lower continental crust records a transition from accumulation to draining, and melt is inferred to have been extracted by gravity-driven ascent in steeply oriented conduits. This is the outcome of multiple studies (e.g. Collins & Sawyer 1996; Brown & Solar 1999; Sawyer *et al.* 1999; Marchildon & Brown 2003), and although the specific details vary from example to example, the generic conclusion is robust. The inherent complexity of residual migmatites and granulites has led to several attempts to characterise the spatial patterning by fractal dimensions (Marchildon & Brown 2003) or to determine leucosome connectivity using probability (Petford & Koenders 1998; Weinberg 1999). However, in the present study, the author considers the dynamic evolution of melt-bearing deformation band networks and melt extraction from the source via ductile fracturing; this is a new approach that takes into account the ductile behaviour of lower crustal rocks, particularly under suprasolidus conditions.

4.1. Grain-to-vein scale

Mesoscale concentration of melt may occur in one of two ways (Powell & Downes 1990; Hand & Dirks 1992; Maaløe 1992; Brown *et al.* 1995b; Watson 2001). First, mass transfer may be by diffusional flow – chemical migration of components through a matrix by a dissolution-diffusion-crystallisation process in the matrix through which it moves (e.g. in metasedimentary protoliths with compositional anisotropy, formation of leucosome stromata with melanocratic selvedges in which the grain size is coarser than that for the same mineral in the matrix; and for hydrate dissolution under suprasolidus conditions, concentration of leucosome around a peritectic phase).

Secondly, mass transfer may be by advective flow – physical migration of melt through the matrix without chemical interactions (e.g. preferential movement of melt through structures with high melt fraction to draining conduits).

Average rates of diffusive mass transfer yield characteristic distances which may approach one metre on time scales around one million years, but there are many variables which affect rates and efficiencies of diffusion (e.g. Costa *et al.* 2003; Dohmen & Chakraborty 2003). In contrast, advective fluid flow is several orders of magnitude faster (e.g. Harris *et al.* 2000). Both processes may be inferred to have occurred in any one suprasolidus episode, based on evidence in individual outcrops of anatectic rocks, and these processes commonly act in concert. Coupling between mechanical and physicochemical aspects of rock behaviour is significant in the anatectic environment, but the complex interplay among melt fraction, melt segregation, bulk strength contrasts, mechanical properties of different minerals and the melt phase, and time means that the rheology is difficult to describe by a single flow law (e.g. Berger & Kalt 1999).

Gradients in melt fraction will affect density and buoyancy forces, and may necessitate separated flow of solids and melt rather than a bulk response. However, what happens to such a mixture of solids and melt, in which the potential range of viscosities may approach 14 orders of magnitude and each end member has a different rheology (e.g. power law with $n=3$ for the solid and Newtonian with $n=1$ for the liquid), will depend upon the level of applied stress, the differential stress, the effective stress and the strain rate, the melt fraction and changes in the proportion of melt to solid with time, and whether localisation occurs. At high strain rates, the contrast in effective viscosity is reduced, and the bulk response most likely will be coupled flow (e.g. dominated by the matrix or at high melt fraction behavior as a pseudo-fluid), whereas, at low strain rates, the rheology may be dominated by one of the phases, and phase separation (strain rate compatibility) or strain localisation (stress compatibility) may occur (Vigneresse & Burg 2000; Vigneresse 2004).

It is clear that the key to understanding melt migration is an understanding of the nature of strain localisation in melt-bearing crust, and how localisation controls melt extraction. Discordant veins of leucosome in residual migmatite and granulite have commonly been interpreted as resulting from brittle fracture induced by embrittlement as a result of high melt pore pressure (e.g. van der Molen & Paterson 1979). However, fracture toughness analysis of crystalline solids indicates that there is a change in failure from brittle to ductile modes with increasing temperature (e.g. Hirth & Tullis 1994; Hirsch & Roberts 1997; Stockhert *et al.* 1999). At middle and lower crustal depths, below the brittle-to-viscous transition zone, lower strain rates, higher temperatures and confining pressures, and the presence of a volatile or melt phase attributable to dehydration reactions or hydrate-breakdown melting reactions promote ductile processes.

For layers or elongate bands of differing rheology in extension, mean stress is higher in the weaker material, but in shortening, it is lower. Thus, in extension, the resulting difference in mean stress drives melt from layers of greater melting, which are weaker, down gradients in melt pressure to stronger layers (cf. Brown *et al.* 1995b). Also, for an individual melt-bearing elongate shear band that is stretching and thinning, mean stress is higher in the band and melt will be expelled (cf. Mancktelow 2002). Conversely, for an individual melt-bearing elongate shear band that is shortening and thickening, mean stress is lower in the band and melt is sucked into it. Also, fluid pressure enables melt to exploit planes of weakness in anisotropic protoliths, for example by dilating foliation and/or interfaces between layers of contrasting composition, in a fashion similar to that described by Barraud *et al.* (2004) from analogue experiments.

The structural control on sites of leucosome concentration is consistent with gradients in mean stress as the principal driving force for matrix compaction, melt segregation and metre-scale melt flow (Rutter 1997). Thus, melt may be pumped in cyclic fashion along shallowly oriented fabrics by a dilation–contraction mechanism, moving through foliation-parallel segments via connecting transverse structures. Also, melt may be pumped through shear systems from thinning to thickening shear bands, and as shear bands rotate into the field of finite extension, melt increasingly is expelled from the source.

The rate of volumetric strain by melt expulsion cannot exceed the rate at which the solid matrix deforms by shear-enhanced compaction, and therefore, the rate of melt expulsion is controlled by the rate of matrix compaction or the rate of volumetric strain is limited by the rate of melt expulsion. If the solid matrix is able to compact at a higher rate than the rate at which melt may be expelled, then a higher distortional strain will be the result, requiring melt-bearing shear bands to develop or more distributed flow to occur (Rutter 1997). Overall, the positive feedback relation enables interconnected networks of deformation bands to evolve to an optimum structure, controlled by the rates of strain accumulation (and melt production) in relation to rates of matrix deformation and melt expulsion.

4.2. Mesoscale-to-crustal scale

A critical point is reached at some combination of MF and distribution, probably contemporaneously with the developing ductile deformation band network reaching the percolation threshold, where conduits of a size that allows melt to flow out of the local source volume and upward through the crust are formed. For this to occur, it is likely that the MF will be in the range 0·05–0·2; such a conclusion is consistent with estimates of the MF in the crust beneath the Central Andes and the Tibetan Plateau (Schilling & Partzsch 2001).

The leucosome networks described from the southern Brittany migmatites are composed of foliation-parallel and transverse stromata (interboudin partitions and shear surfaces) which represent remnants of melt flow through a linked network of deformation bands. The present author suggests that leucosome networks in residual migmatites and granulites, such as those in southern Brittany, are analogous to the ideal deformation band networks described by Du Bernard *et al.* (2002). Ideal deformation band networks are composed of three kinds of structure (Fig. 6): shear bands, characterised by a dominant shear displacement gradient that is accompanied by porosity reduction or compaction, or porosity increase or dilation (Antonellini *et al.* 1994); compaction bands, characterised by a localised porosity reduction or compaction, and lacking macroscopic shear offset (Mollema & Antonellini 1996); and dilation bands, characterised by localised porosity increase and opening-mode with respect to their boundaries (Du Bernard *et al.* 2002).

The dynamic evolution of these networks is illustrated by foliation-parallel leucosome stromata in residual migmatites and granulites. Assuming layer-parallel extension, foliation planes (or stronger layers) initially dilate to accumulate melt, but as transverse structures form, so the foliation (or melt-rich layer) contracts as melt flows into the developing transverse structures down gradients in fluid pressure. The present author infers that foliation-parallel leucosome stromata in melt-depleted crustal rocks are structures analogous to compaction bands. As an interconnected network of deformation bands develops, the system will approach the percolation threshold,

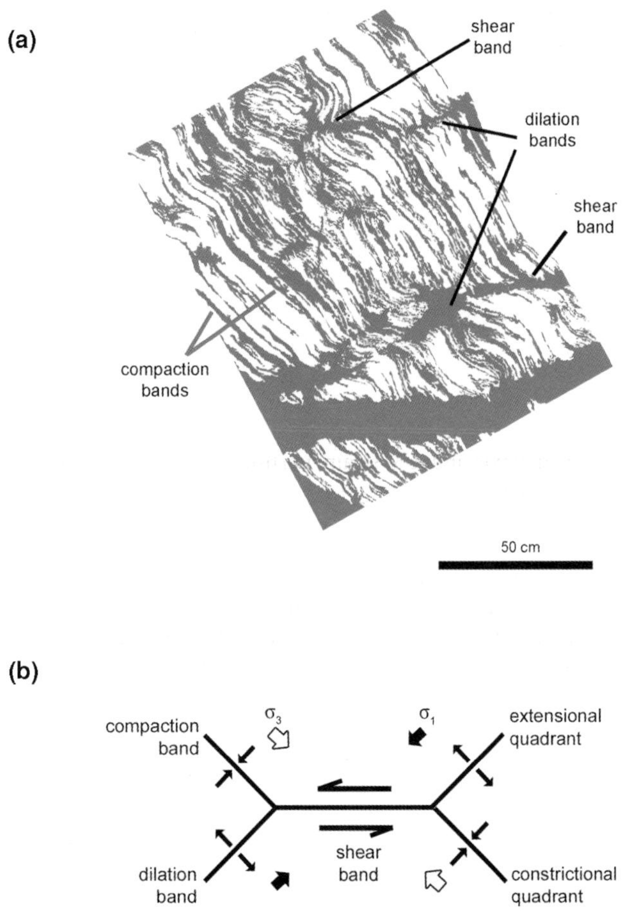

Figure 6 (a) Binary image of (sub-horizontal) surface perpendicular to foliation and parallel to lineation to show patterning of leucosomes (after Marchildon & Brown 2003; Le Petit Mont, Arzon, Morbihan). (b) Idealised diagram to show the coexistence of shear, compaction and dilation bands (σ_1 and σ_3 are greatest and least far field principal stresses, respectively; note that locally the effective stress parallel to σ_3 will be tensile because of melt pressure). Modified from Du Bernard 2002 by permission of the American Geophysical Union. © 2002 American Geophysical Union.

the permeability will change rapidly with concomitant increase in melt flow rates, and melt will drain from the system if appropriate conduits for ascent are available or conditions are appropriate for these to form. In the residue, a mix of residual and early crystallised (cumulate) phases with only minimal melt is retained as leucosome stromata in the deformation bands because shear and compactive deformation has expelled most of the melt from these sites into ductile opening-mode fractures and out of the local source volume as dykes (cf. Solar & Brown 2001; Guernina & Sawyer 2003).

The common view that melt extraction is a brittle fracturing phenomenon from the source to the upper crust is based on linear elastic fracture mechanics in which inelastic deformation in a small process zone ahead of the fracture tip, with or without chemical weakening (subcritical brittle fracture or brittle fracture, respectively), is driven by the magnified stresses at the macroscopic fracture tip (e.g. Rubin 1995; Eichhubl 2004). However, the application of brittle fracture theory to the suprasolidus environment is not intuitively obvious since it neglects thermally-activated flow processes. These thermally-activated processes overtake brittle processes with increasing depth to dominate in the crust below the brittle-to-viscous transition zone, which leads to extensive inelastic deformation, the blunting of crack tips and a change in fracture mode. Processes similar to those in creep failure of ceramics at high homologous temperature under low rates of loading, such as nucleation, growth and coalescence of pores by predominantly diffusive deformation mechanisms and side-lobe damage, are likely to be particularly relevant to failure of suprasolidus crust, leading to ductile fracture (cf. Eichhubl 2004).

As a hypothesis, the present author suggests that brittle theory is unlikely to be applicable to depths beyond 10–15 km and at elevated temperatures, and that ductile fracturing dominates in the lower crust. There are a number of geological features which characterise the nature of the dominant fracturing mechanism in the lower crust. These features include the nature of the tips of melt-filled fractures (which are blunt in ductile fracture but sharp in brittle fracture), the geometry of coalescing melt-filled fractures (which zigzag in ductile fracture but form *en echelon* arrays in brittle fracture), and the nature of the margins of the fracture, and between the host and the fracture-filling material (irregular margins indicate ductile fracture whereas sharp margins indicate brittle fracture).

In suprasolidus lower crust, the author suggests that fracture propagation takes place by development and coalescence of melt-filled pores ahead of the fracture tip, with fracture opening involving extensive inelastic deformation and diffusive mass transfer (cf. Regenauer-Lieb 1999; Eichhubl *et al.* 2001; Eichhubl & Aydin 2003; Eichhubl 2004). This coupling between fracture formation and mass transfer is significant for both opening- and shearing-mode fracturing in a suprasolidus crustal environment, where a large amount of plastic strain may accumulate before fracture. In this environment, the differential stresses are low, the effective mean stress is low, the cohesive strength is low (relative to the strength of the grains) at melt fractions greater than ~7–8 vol.%, and the rapid exhumation of the lower crust means that the loading rate is fast, which are the conditions necessary for ductile failure (P. Eichhubl, pers. comm., 2003).

In southern Brittany, dyke emplacement occurred along a preferred direction independent of anisotropy, suggesting that the orientation is controlled by stress, which with the macroscopic fracture-like discontinuity infilled by the granite, characterises the process of dyke formation as a fracture phenomenon. The petrographic continuity between leucosome in the deformation band networks and granite in dykes (Fig. 3) indicates that the ascent conduits represent either structures analogous to large-scale dilation bands formed by opening-mode failure along zones of localised porosity increase or ductile opening-mode fractures formed by pore growth and coalescence of melt pockets (e.g. Regenauer-Lieb 1999; Eichhubl *et al.* 2001; Du Bernard *et al.* 2002; Eichhubl & Aydin 2003). Zigzag propagation paths at tips of dykes (Fig. 3e, f), which commonly propagate from transverse structures, point to ductile fracture (Fig. 7) as the mode of formation of these conduits (cf. Eichhubl & Aydin 2003).

Whether each ascent conduit was the product of a single melt-draining event from its collection zone or whether there is a build-up of melt pressure before periodic activation of a draining conduit in a cyclic fashion (e.g. Handy *et al.* 2001) must be determined on a case-by-case basis (e.g. Solar & Brown 2001; Marchildon & Brown 2003). However, there is increasing evidence of common ascent conduits for different batches of melt (e.g. Brown & Solar 1999; Solar & Brown 2001; Sawyer & Bonnay 2003). Thus, studies of exhumed melt-depleted anatectic migmatite and granulite terrains suggest that melt may be lost in batches. This view is supported by detailed isotope geochemistry from some plutons (e.g. the Phillips pluton in Maine, northern Appalachians, USA) where data from representative samples are interpreted to show that the pluton was constructed from multiple batches of isotopically distinct melt (Pressley & Brown 1999). Furthermore, this

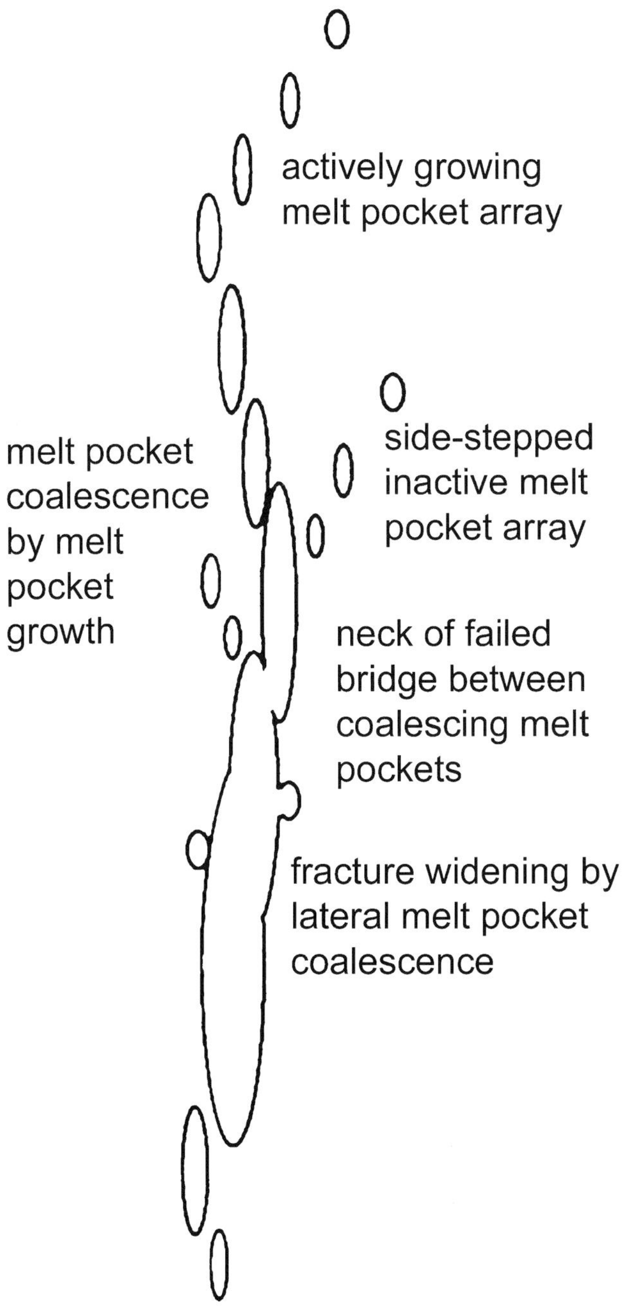

Figure 7 Sketch to illustrate the mechanism of formation of ductile opening-mode fractures by pore growth and melt pocket coalescence. Switching between left- and right-stepping direction leads to zigzag propagation paths. Modified from Eichhubets & Aydin 2003 by permission of Elsevier B.V. © 2003 Elsevier B.V.

conclusion is consistent with the occurrence of clearly discordant, but petrographically continuous dykes at the same structural level in the source as developing ductile opening-mode fractures (Figs 2 & 4); thus, at any particular structural level in the source, newly forming conduits and conduits transporting melt through that level will occur together.

4.3. Relationship to other postulated mechanisms of ascent

The model proposed in this paper combines elements of pervasive migration with dyking by ductile fracture; there is no *a priori* reason why these dykes cannot be the physical expression of mobile hydrofractures frozen in ascent. It is also possible that such a model also may represent the physical expression of successive porosity waves passing through the anatectic zone.

5. Self-organisation

Self-organisation is a fundamental property of natural systems, including orogens, driven far from equilibrium by a constant energy flux (e.g. Hodges 2000). The segregation, extraction and ascent of magma in orogens have been postulated to be self-organised, resulting from non-linear interactions and feedback relations among deformation and melting (Brown & Solar 1998a, b). Sources of non-linearity include spatial and temporal variation in melting, melt production and melt distribution, and the combination of linear and non-linear rheologies in the melting environment. Thus, deformation during anatexis induces strain partitioning at all scales, increasing the vorticity within the weaker phase, which promotes melt redistribution and bulk softening (shear thinning), and further concentrates strain in a positive feedback relationship that induces increasing non-coaxial deformation and accelerates melt displacement. As a result, the large-scale interaction between melting and deformation, and granite magma ascent and emplacement has been proposed (e.g. Brown & Dallmeyer 1996; Brown & Solar 1998a, b), although such an intimate feedback relation may not always be the case (e.g. Fitzsimons 1996). However, the studies to date have provided only qualitative descriptions of the interactions and feedback relations, and a quantitative analysis of how these non-linear effects interact in the differentiation of heterogeneous, anisotropic lower continental crust during orogenesis has not been undertaken.

One feature of self-organised phenomena is that these are driven systems which may achieve a critical state, which may lead to 'avalanches' with a fractal (power-law) frequency-size distribution (Turcotte 1999, 2001). Thus, if melt extraction reflects a critical state in melt-bearing lower crust, we might expect the distribution of melt batch sizes and/or melt flow channels to be fractal. Whereas the system as a whole may converge to a critical state that allows melt escape, fractal (scale invariant) distribution of melt batches (or leucosome-filled structures) has not yet been demonstrated satisfactorily (see Tanner 1999; Marchildon & Brown 2003; Bons *et al.* 2004). This may reflect the residual nature of the drained source in depleted anatectic migmatite and granulite terrains. In a recent example of modelling melt segregation from a homogeneous isotropic mush subject to both compactive and shear deformation, Rabinowicz & Vigneresse (2004) demonstrated that instabilities will generate parallel melt-filled veins in the σ_1–σ_2 plane, with size and spacing close to the compaction length, which is expected to be metric or sub-metric. However, the above authors do not address how this modelling relates to observations from field examples of exposed lower continental crust, where the tectonically induced metamorphic fabric clearly exerts the primary control over leucosome orientation and spacing.

6. Conclusions

Melt extraction from the lower continental crust of orogens involves the migration of melt down gradients in pressure over distances of millimetres by flow along grain boundaries to form layer or foliation-parallel leucosome stromata. With increasing melt fraction, melt moves distances of tens of centimetres along these concordant structures to developing dilatant transverse structures, such as ductile extension and shear bands, which become interconnected. These ductile deformation band networks, and similar networks of dilatant structures in examples described in the literature, represent accumulation networks in melt-bearing lower continental crust. Melt is extracted from the accumulation networks by porous flow down-gradients in pressure to ductile opening-mode fractures and shear zones, where ascent is buoyancy-driven. Extraction occurs at some critical combination of melt fraction and distribution, most probably as the developing accumulation network reaches the percolation threshold. The partially molten lower continental crust of orogens is driven far from thermal and mechanical equilibrium in relation to the adjacent subsolidus crust, so that melt extraction is via emergent structures which enable dissipation of excess energy in the system.

7. Acknowledgements

I thank Boswell A. Wing for assistance with the flux calculations, Margaret A. Baker and Barry Reno for help in the preparation of figures, and Jeanne Martin for help in the preparation of the text. This paper was improved as a consequence of review comments by Keith Benn, Jean-Louis Vigneresse and Roberto Weinberg; however, any remaining infelicities are the responsibility of the author alone.

8. References

Antonellini, M. A., Aydin, A. & Pollard, D. D. 1994. Microstructure of deformation bands in porous sandstones at Arches National Park, Utah. *Journal of Structural Geology* **16**, 941–59.

Ashworth, J. R. & McLellan, E. L. 1985. Textures. *In* Ashworth, J. R. (ed) *Migmatites*, 180–203. Glasgow: Blackie & Son Ltd.

Barbey, P., Marignac, C., Montel, J. M., Macaudière, J., Gasquet, D. & Jabbori, J. 1999. Cordierite growth textures and the conditions of genesis and emplacement of crustal granitic magmas: the Velay Granite Complex (Massif Central, France). *Journal of Petrology* **40**, 1425–41.

Barraud, J., Gardien, V., Allemand, P. & Grandjean, P. 2004. Analogue models of melt-flow networks in folding migmatites. *Journal of Structural Geology* **26**, 307–24.

Berger, A. & Kalt, A. 1999. Structures and melt fractions as indicators of rheology in cordierite-bearing migmatites of the Bayerische Wald (Variscan Belt, Germany). *Journal of Petrology* **40**, 1699–719.

Bons, P. D., Dougherty-Page, J. & Elburg, M. A. 2001. Stepwise accumulation and ascent of magmas. *Journal of Metamorphic Geology* **19**, 627–33.

Bons, P. D., Arnold, J., Elburg, M. A., Calder, J. & Soesoo, A. 2004. Melt accumulation from partially molten rocks. *Lithos*, in press.

Brown, M. 1994. The generation, segregation, ascent and emplacement of granite magma: the migmatite-to-crustally-derived granite connection in thickened orogens. *Earth-Science Reviews* **36**, 83–130.

Brown, M. 2001a. Crustal melting and granite magmatism: key issues. *Physics and Chemistry of the Earth (A)* **26**, 201–12.

Brown, M. 2001b. Orogeny, migmatites and leucogranites: a review. *Proceedings of the Indian Academy of Science* **110**, 313–36.

Brown, M. 2002. Prograde and retrograde processes in migmatites revisited. *Journal of Metamorphic Geology* **20**, 25–40.

Brown, M. Rushmer, T. & Sawyer, E. W. 1995a. Introduction to Special Section: Mechanisms and consequences of melt segregation from crustal protoliths. *Journal of Geophysical Research* **100**, 15,551–63.

Brown, M., Averkin, Y. A., McLellan, E. L. & Sawyer, E. W. 1995b. Melt segregation in migmatites. *Journal of Geophysical Research* **100**, 15,655–79.

Brown, M. A., Brown, M., Carlson, W. D. & Denison, C. 1999. Topology of syntectonic melt flow networks in the deep crust: inferences from three-dimensional images of leucosome geometry in migmatites. *American Mineralogist* **84**, 1793–818.

Brown, M. & Dallmeyer, R. D. 1996. Rapid Variscan exhumation and role of magma in core complex formation: Southern Brittany metamorphic belt, France. *Journal of Metamorphic Geology* **14**, 361–79.

Brown, M. & Rushmer, T. 1997. The role of deformation in the movement of granite melt: views from the laboratory and the field. *In* Holness, M.B. (ed.) *Deformation-enhanced Fluid Transport in the Earth's Crust and Mantle*, 111–44. *The Mineralogical Society Series* **8**. London: Chapman and Hall.

Brown, M. & Solar, G. S. 1998a. Shear zone systems and melts: feedback relations and self-organization in orogenic belts. *Journal of Structural Geology* **20**, 211–27.

Brown, M. & Solar, G. S. 1998b. Granite ascent and emplacement during contractional deformation in convergent orogens. *Journal of Structural Geology* **20**, 1365–93.

Brown, M. & Solar, G. S. 1999. The mechanism of ascent and emplacement of granite magma during transpression: a syntectonic granite paradigm. *Tectonophysics* **312**, 1–33.

Burov, E., Jaupart, C. & Guillou-Frottier, L. 2003. Ascent and emplacement of buoyant magma bodies in brittle-ductile upper crust. *Journal of Geophysical Research* **108**, B4, 2177, doi:10.1029/2002JB001904.

Cagnard, F., Gapais, D., Brun, J.-P., Gumiaux, C. & Van den Driessche, J. 2004. Late pervasive crustal-scale extension in the south Armorica Hercynian belt (Vendée, France). *Journal of Structural Geology* **26**, 435–49.

Clemens, J. D. 1998. Observations on the origins and ascent mechanisms of granitic magmas. *Journal of the Geological Society, London* **155**, 843–51.

Clemens, J. D. & Mawer, C. K. 1992. Granitic magma transport by fracture propagation. *Tectonophysics* **204**, 339–60.

Collins, W. J. & Sawyer, E. W. 1996. Pervasive magma transfer through the lower-middle crust during non-coaxial compressional deformation: an alternative to diking. *Journal of Metamorphic Geology*, **14** 565–79.

Connolly, J. A. D. & Podladchikov, Y. Y. 1998. Compaction-driven fluid flow in viscoelastic rock. *Geodinamica Acta* **11**, 55–84.

Costa, F., Chakraborty, S. & Dohmen, R. 2003. Diffusion coupling between trace and major elements and a model for calculation of magma residence times using plagioclase. *Geochimica et Cosmochimica Acta* **67**, 2189–200.

Cruden, A. R. 1998. On the emplacement of tabular granites. *Journal of the Geological Society, London* **155**, 853–62.

Cuney M. & Barbey P. 1982. Mise en evidence de phénomènes de cristallisation fractionnée dans les migmatites. *Comptes Rendus de l'Acadamie des sciences, Paris* **295**, 37–42.

Dahm, T. 2000. On the shape and velocity of fluid-filled fractures in the earth. *Geophysical Journal International* **142**, 181–92.

D'Lemos, R. S., Brown, M. & Strachan, R. A. 1992. Granite magma generation, ascent and emplacement within a transpressional orogen. *Journal of the Geological Society, London* **149**, 487 90.

Dohmen, R. & Chakraborty, S. 2003. Mechanism and kinetics of element and isotopic exchange mediated by a fluid phase. *American Mineralogist* **88**, 1251–70.

Du Bernard, X., Eichhubl, P. & Aydin, A. 2002. Dilation bands: a new form of localized failure in granular media. *Geophysical Research Letters* **29**, DOI:10.1029/2002GL015966.

Eichhubl, P. 2004. Growth of ductile opening-mode fractures in geomaterials. *In* Cosgrove, J. W. & Engelder, T. (eds) *The Initiation, Propagation, and Arrest of Joints and Other Fractures*, 11–24. *Geological Society, London, Special Publication* **231**. London: Geological Society.

Eichhubl, P., Aydin, A. & Lore, J. 2001. Opening-mode fracture in siliceous mudstone at high homologous temperature – effective of surface forces. *Geophysical Research Letters* **28**, 1299–302.

Eichhubl, P. & Aydin, A. 2003. Ductile opening-mode fracture by pore growth and coalescence during combustion alteration of siliceous mudstone. *Journal of Structural Geology* **25**, 121–34.

Eichhubl, P. & Boles, J. R. 2000. Focused fluid flow along faults in the Monterey Formation, Coastal California. *Geological Society of America Bulletin* **112**, 1667–79.

Emerman, S. H. & Marrett, R. 1990. Why dikes? *Geology* **18**, 231–3.

Fitzsimons, I. C. W. 1996. Metapelitic migmatites from Brattstrand Bluffs, East Antarctica – metamorphism, melting and exhumation of the mid crust. *Journal of Petrology* **37**, 395–414.

Gapais, D., Lagarde, J.-L., Le Corre, C., Audren, C., Jégouzo, P., Casas Sainz, A. & van den Driessche, J. 1993. La zone de cisaillement de Quiberon: témoin d'extension de la cha(ne varisque en Bretagne méridionale au Carbonifère. *Comptes Rendus de l'Acadamie des Sciences, Paris* **316**, 1123–9.

Gillespie, P. A., Howard, C. B., Walsh, J. J. & Watterson, J. 1993. Measurement and characterisation of spatial distributions of fractures. *Tectonophysics* **226**, 113–41.

Gillespie, P. A., Johnston, J. D., Loriga, M. A., McCaffrey, K. J. W., Walsh, J. J. & Watterson, J. 1999. Influence of layering on vein systematics in line samples. *In* McCaffrey, K. J. W., Lonergan, L. & Wilkinson, J. J. (eds) *Fractures, Fluid Flow and Mineralization*, 35–56. *Geological Society, London, Special Publication* **155**. London: Geological Society.

Glazner, A. F., Bartley, J. M., Coleman, D. S., Gray, W. & Taylor, R. Z. 2004. Are plutons assembled over millions of years by amalgamation from small magma chambers. *GSA Today* **14**, 4–11.

Grujic, D. & Mancktelow, N. S. 1998. Melt-bearing shear zones: analogue experiments and comparison with examples from southern Madagascar. *Journal of Structural Geology* **20**, 673–80.

Guernina, S. & Sawyer, E. W. 2003. Large-scale melt-depletion in granulite terranes: an example from the Archean Ashuanipi Subprovince of Quebec. *Journal of Metamorphic Geology* **21**, 181–201.

Hand, M. & Dirks, P. H. G. M. 1992. The influence of deformation on the formation of axial-planar leucosomes and the segregation of small melt bodies within the migmatitic Napperby Gneiss, central Australia. *Journal of Structural Geology* **14**, 591–604.

Handy, M. R., Mulch, A., Rosenau, N. & Rosenberg, C.L. 2001. The role of fault zones and melts as agents of weakening, hardening and differentiation of the continental crust: a synthesis. *In* Holdsworth, R. E., Magloughlin, J., Knipe, R. J., Strachan, R. A. & Searle, R. C. (eds) *The Nature and Tectonic Significance of Fault Zone Weakening*, 305–32. *Geological Society, London, Special Publication* **186**. London: Geological Society.

Harris, N., Vance, D. & Ayers, M. 2000. From sediment to granite: time scales of anatexis in the upper crust. *Chemical Geology* **162**, 155–67.

Hirsch, P. B. & Roberts, S. G. 1997. Modelling plastic zones and the brittle-ductile transition. *Philosophical Transactions of the Royal Society, London* **355**, 1991–2002.

Hirth, G. & Tullis, J. 1994. The brittle-plastic transition in experimentally deformed quartz aggregates. *Journal of Geophysical Research* **99**, 11,731–47.

Hodges, K. 2000. Tectonics of the Himalayas and southern Tibet from two perspectives. *Geological Society of America Bulletin* **112**, 324–50.

Holness, M. B. & Clemens, J. D. 1999. Partial melting of the Appin Quartzite driven by fracture-controlled H_2O infiltration in the aureole of the Ballachulish Igneous Complex, Scottish Highlands. *Contributions to Mineralogy and Petrology* **136**, 154–68.

Johnson, T. & Brown, M. 2004. Quantitative constraints on metamorphism in the variscides of Southern Brittany – a complementary pseudosection approach. *Journal of Petrology* **45**, 1237–59.

Laporte, D. & Provost, A. 2000. Equilibrium geometry of a fluid phase in a polycrystalline aggregate with anisotropic surface energies: dry grain boundaries. *Journal of Geophysical Research* **105**, 25937–53.

Laporte, D., Rapaille, C. & Provost, A. 1997. Wetting angles, equilibrium melt geometry, and the permeability threshold of partially molten crustal protoliths. *In* Bouchez, J. L., Hutton, D. H. W. & Stephens, W. E. (eds) *Granite: from Segregation of Melt to Emplacement Fabrics*, 31–54. Kluwer Academic Publishers.

Laporte, D. & Watson, E. B. 1995. Experimental and theoretical constraints on melt distribution in crustal sources – the effect of crystalline anisotropy on melt interconnectivity. *Chemical Geology* **124**, 161–84.

Leitch, A. M. & Weinberg, R. F. 2002. Modelling granite migration by mesoscale pervasive flow. *Earth and Planetary Science Letters* **200**, 131–46.

Loriga, M. A. 1999. Scaling systematics of vein size: an example from the Guanajuato mining district (Central Mexico). *In* McCaffrey, K. J. W., Lonergan, L. & Wilkinson, J. J. (eds) *Fractures, Fluid Flow and Mineralization*, 57–67. *Geological Society, London, Special Publication* **155**. London: Geological Society.

Lupulescu, A. & Watson, E. B. 1999. Low melt fraction connectivity of granitic and tonalitic melts in a mafic crustal rock at 800 °C and 1 GPa. *Contributions to Mineralogy and Petrology* **134**, 202–16.

Maaløe, S. 1992. Melting and diffusion processes in closed-system migmatization. *Journal of Metamorphic Geology* **10**, 503–16.

Mancktelow, N. S. 2002. Finite-element modelling of shear zone development in viscoelastic materials and its implications for localisation of partial melting. *Journal of Structural Geology* **24**, 1045–53.

Marchildon, N. & Brown, M. 2001. Melt segregation in late syntectonic anatectic migmatites: an example from the Onawa Contact Aureole, Maine, U.S.A. *Physics and Chemistry of the Earth (A)* **26**, 225–9.

Marchildon, N. & Brown, M. 2002. Grain-scale melt distribution in two contact aureole rocks: implication for controls on melt localization and deformation. *Journal of Metamorphic Geology* **20**, 381–96.

Marchildon, N. & Brown, M. 2003. Spatial distribution of melt-bearing structures in anatectic rocks from Southern Brittany: implications for melt-transfer at grain- to orogen-scale. *Tectonophysics* **364**, 215–35.

Marsh, B. D. 1982. On the mechanics of igneous diapirism, stoping and zone melting. *American Journal of Science* **282**, 808–55.

Miller, R. B. & Patterson, S. R. 1999. In defense of magmatic diapirs. *Journal of Structural Geology* **21**, 1161–73.

Mollema, P. N. & Antonellini, M. A. 1996. Compaction bands; a structural analog for anti-mode I cracks in Aeolian sandstone. *Tectonophysics* **267**, 209–28.

Olsen, S. N., Marsh, B. D. & Baumgartner, L. P. 2004. Modelling mid-crustal migmatite terrains as feeder zones for granite plutons: competing dynamics of melt transfer versus porous flow. *Transactions of the Royal Society of Edinburgh: Earth Sciences* **95**, 49–58.

Petford, N. & Koenders, M. A. 1998. Self-organization and fracture connectivity in rapidly heated continental crust. *Journal of Structural Geology* **20**, 1425–34.

Podladchikov, Yu.Yu. & Connolly, J. A. D. 2001. The transition from pervasive to segregated melt flow in ductile rock. *In EUG XI, Journal of Conference Abstracts*, 813. Cambridge: Cambridge Publications.

Powell, R. & Downes, J. 1990. Garnet porphyroblast-bearing leucosomes in metapelites: mechanisms, phase diagrams and an example from Broken Hill. *In* Ashworth, J. R. & Brown, M. (eds) *High Temperature Metamorphism and Crustal Anatexis*, 105–23. London: Unwin Hyman.

Pressley, R. A. & Brown, M. 1999. The Phillips Pluton, Maine, USA: evidence of heterogeneous crustal sources, and implications for granite ascent and emplacement mechanisms in convergent orogens. *Lithos* **46**, 335–66.

Rabinowicz, M. & Vigneresse, J.-L. 2004. Melt segregation under compaction and shear channeling: application to granitic magma segregation in a continental crust. *Journal of Geophysical Research* **109**, B04407, doi:10.1029/2002JB002372.

Regenauer-Lieb, K. 1999. Dilatant plasticity applied to Alpine collision: ductile void growth in the intraplate area beneath the Eifel volcanic field. *Journal of Geodynamics* **27**, 1–21.

Richardson, C. N., Lister, J. R. & McKenzie, D. 1996. Melt conduits in a viscous porous matrix. *Journal of Geophysical Research* **101**, 20,423–32.

Rosenberg, C. L. 2004. Shear zones in magma ascent: a model based on a review of the Tertiary magmatism in the Alps. *Tectonics* **23**, TC3002, doi: 10.1029/2003TC001526.

Rosenberg, C. L. & Handy, M. R. 2000. Syntectonic melt pathways during simple shearing of a partially molten rock analogue (Norcamphor-Benzamide). *Journal of Geophysical Research* **105**, 3135–49.

Rosenberg, C. L. & Handy, M. R. 2001. Mechanisms and orientation of melt segregation paths during pure shearing of a partially molten rock analog (norcamphor-benzamide). *Journal of Structural Geology* **23**, 1917–32.

Rubin, A. M. 1993. Dikes vs. diapirs in visco-elastic rock. *Earth and Planetary Science Letters* **117**, 653–70.

Rubin A. M. 1995. Propagation of magma-filled cracks. *Annual Review of Earth and Planetary Sciences* **23**, 287–336.

Rubin, A. M. 1998. Dike ascent in partially molten rock. *Journal of Geophysical Research* **103**, 20,901–19.

Rutter, E. H. 1986. On the nomenclature of mode of failure transitions in rocks. *Tectonophysics* **122**, 381–7.

Rutter, E. H. 1997. The influence of deformation on the extraction of crustal melts: a consideration of the role of melt-assisted granular flow. *In* Holness, M. B. (ed.) *Deformation-enhanced Fluid Transport in the Earth's Crust and Mantle*, 82–110. *Mineralogical Society Series* **8**. London: Chapman and Hall.

Rutter, E. H. & Neumann, D. H. K. 1995. Experimental deformation of partially molten Westerly granite, with implications for the extraction of granitic magmas. *Journal of Geophysical Research* **100**, 15,697–15,715.

Sawyer, E. W. 1999. Criteria for the recognition of partial melting. *Physics and Chemistry of the Earth, (A)* **24**, 269–79.

Sawyer, E. W. 2001. Melt segregation in the continental crust: Distribution and movement of melt in anatectic rocks. *Journal of Metamorphic Geology* **18**, 291–309.

Sawyer, E. W., Dombrowski, C. & Collins, W. J. 1999. Movement of melt during synchronous regional deformation and granulite-facies anatexis, an example from the Wuluma Hills, central Australia. *In* Castro, A., Fernandez, C. & Vigneresse, J.-C. (eds) *Understanding Granites: Integrating New and Classical Techniques*, 221–37. *Geological Society, London, Special Publication* **168**. London: Geological Society.
Sawyer, E. W. & Bonnay, M. 2003. Melt segregation and magma movement in the crust. *Geophysical Research Abstracts* **5**, 02458. Kattenenberg–Lindau: Copernicus Gesellschaft mbH.
Schilling, F. R. & Partzsch, G. N. 2001. Quantifying partial melt fraction in the crust beneath the central Andes and the Tibetan Plateau. *Physics and Chemistry of the Earth, A* **26**, 239–46.
Simakin, A. & Talbot, C. 2001a. Transfer of melt between microscopic pores and macroscopic veins in migmatites. *Physics and Chemistry of the Earth, A* **26**, 363–7.
Simakin, A. & Talbot, C. 2001b. Tectonic pumping of pervasive granitic melts. *Tectonophysics* **332**, 387–402.
Sleep, N. H. 1988. Tapping of melt by veins and dikes. *Journal of Geophysical Research* **93**, 10,255–72.
Solar, G. S. & Brown, M. 2001. Petrogenesis of Migmatites in Maine, USA: Possible source of peraluminous leucogranite in plutons. *Journal of Petrology* **42**, 789–823.
Stockhert, B., Brix, M. R., Kleinschrodt, R., Hurford, A. J. & Wirth, R. 1999. Thermochronometry and microstructures of quartz – a comparison with experimental flow laws and predictions on the temperature of the brittle-plastic transition. *Journal of Structural Geology* **21**, 351–69.
Tanner, D. C. 1999. The scale-invariant nature of migmatites from the Oberpfalz, NE Bavaria and its significance for melt transport. *Tectonophysics* **302**, 297–305.
Teyssier, C. & Whitney, D. L. 2002. Gneiss domes and orogeny. *Geology* **30**, 1139–42.
Thompson, A. B. 1999. Some time–space relationships for crustal melting and granitic intrusion at various depths. *In* Castro, A., Fernåndez, C. & Vigneresse, J. L. (eds) *Understanding Granites: Integrating New and Classical Techniques*, 7–25. *Geological Society, London, Special Publication* **168**. London: Geological Society.
Turcotte, D. L. 1999. Self-organized criticality. *Reports on Progress in Physics* **62**, 1377–429.
Turcotte, D. L. 2001. Self-organized criticality: Does it have anything to do with criticality and is it useful? *Nonlinear Processes in Geophysics* **8**, 193–196.
van der Molen, I. 1985a. Interlayer material transport during layer-normal shortening, I, the model. *Tectonophysics* **115**, 275–95.
van der Molen, I. 1985b. Interlayer material transport during layer-normal shortening, II, boudinage, pinch-and-swell and migmatite at Søndre Strømfjord Airport, west Greenland. *Tectonophysics* **115**, 275–95.
van der Molen, I. & Paterson, M. S. 1979. Experimental deformation of partially-melted granite. *Contributions to Mineralogy and Petrology* **70**, 299–318.
Vernon, R. H. & Collins, W. J. 1988. Igneous microstructures in migmatites. *Geology* **16**, 1126–9.
Vigneresse, J.-L. 2004. Rheology of a two-phase material with applications to partially molten rocks, plastic deformation and saturated soils. *In* Alsop, G. I., Holdsworth, R. E., McCaffrey, K. J. W. & Hand, M. (eds) *Flow Processes in Faults and Shear Zones*, 79–94. *Geological Society, London, Special Publication* **224**, London: Geological Society.
Vigneresse, J. L. & Burg, J. P. 2000. Continuous vs. discontinuous melt segregation in migmatites: Insights from a cellular automaton model. *Terra Nova* **12**, 188–92.
Vigneresse, J. L. & Burg, J. P. 2004. Strain-rate-dependent rheology of partially molten rocks. *In* Grocott, J., Tikoff, B., McCaffrey, K. J. W. & Taylor, G. (eds) *Vertical Coupling and Decoupling in the Lithosphere*, 327–36. *Geological Society, London, Special Publication* **227**. London: Geological Society.
Wark, D. A., Williams, C. A., Watson, E. B. & Price, J. D. 2003. Reassessment of pore shapes in microstructurally equilibrated rocks, with implications for permeability of the upper mantle. *Journal of Geophysical Research* **108**, doi: 10.1029/2001JB001575.
Wark, D. A. & Watson, E. B. 1998. Grain-scale permeabilities of texturally equilibrated, monomineralic rocks. *Earth and Planetary Science Letters* **164**, 591–605.
Watson, E. B 2001. Aspects of fluid/rock microstructure that might affect rock rheology: an experimentalist's historical perspective. *Geological Society of America, Abstracts with programs* **33**, 6, 50.
Watt, G. R., Oliver, N. H. S. & Griffin, B. J. 2000. Evidence for reaction-induced microfracturing in granulite facies pelitic migmatites. *Geology* **28**, 331–4.
Weinberg, R. F. 1996. The ascent mechanism of felsic magmas: news and views. *Transactions of the Royal Society of Edinburgh: Earth Sciences* **87**, 95–103.
Weinberg, R. F. 1999. Mesoscale pervasive felsic magma migration: alternatives to dyking. *Lithos* **46**, 393–410.
Weinberg, R. F. & Podladchikov, Y. Y. 1994. Diapiric ascent of magmas through power law crust and mantle. *Journal of Geophysical Research* **99**, 9543–60.
White, R. W. & Powell, R. 2002. Melt loss and the preservation of granulite facies mineral assemblages. *Journal of Metamorphic Geology* **20**, 621–32.
Zen, E. 1992. Using granite to image the thermal state of the source terrain. *Transactions of the Royal Society of Edinburgh: Earth Sciences* **83**, 107–14.

MICHAEL BROWN, Laboratory for Crustal Petrology, Department of Geology, University of Maryland, College Park, MD 20742–4211, USA.
e-mail: mbrown@geol.umd.edu

MS received 29 September 2003. Accepted for publication 18 June 2004.

Transactions of the Royal Society of Edinburgh: Earth Sciences, **95**, 49–58, 2004

Modelling mid-crustal migmatite terrains as feeder zones for granite plutons: the competing dynamics of melt transfer by bulk versus porous flow

Sakiko N. Olsen, Bruce D. Marsh and Lukas P. Baumgartner

ABSTRACT: The common association of mid-crustal migmatites with an upper-level granite pluton could indicate that the migmatites are a feeder zone for the pluton. If magma from a deeper level pervasively intrudes a high temperature metamorphic complex, most of the intruded magma would not freeze because of the prevailing temperature. The interaction between the magma and country rocks, which could include partial melting and crystallisation of the magma passing through, would modify magma to a more granitic composition, as found in the higher-level pluton.

The physical aspect of the magma transport through such a hot feeder zone is modelled by introducing a dimensionless melt transport (MT) number, which is the ratio of the rate of melt movement caused by the bulk flow of the entire mass (melt+solid) to that of porous media flow of melt only through the solid framework. The MT number is strongly dependent on the melt content of the melt-rich zone (MRZ), the diameter of the MRZ and typical particle size in the MRZ.

The ~300-Ma, diatexitic, Lauterbrunnen migmatites (LM) in the Aar massif, Swiss Alps, may be such a feeder zone for the nearby 303-Ma Gastern granite (GG). The chemical and field evidence indicates that the LM formed by an intrusion of intermediate composition magma, which interacted with country rocks to produce a magma of GG composition.

KEY WORDS: Aar massif, Diatexite, magma, dimensionless number.

It is generally argued that the magmas for most upper-crustal granite plutons have their sources in the lower crust (e.g. Clemens *et al.* 1997; Petford *et al.* 2000), and therefore, the formation of these plutons must involve magma generation deeper in the crust, transport through the crust and upper-crustal level emplacement (e.g. Brown 1994). The present authors will address the second stage, the long-range transport of magma from deep to shallow crustal levels. The two, end-member models most often advocated for this transport are transport through a dyke and diapiric ascent of a magma body. Field evidence is frequently not definitive (e.g. Petford *et al.* 2000); for example, dyke swarms of granitic compositions are uncommon, and structures attributed to diapiric ascent could be related to emplacement instead. The present authors describe a model of mid-crustal migmatites as feeder zones, in which the magma transport is by pervasive, and macroscopic, porous-media flow (see also Petford & Koenders 1998; Brown & Solar 1999; Weinberg 1999) through country rocks which have already been warmed to nearly solidus temperatures.

Mid-crustal migmatites are an enigma in and of themselves. *In situ* partial-melting is generally proposed for their genesis (e.g. Brown 1994; Sawyer 1996). P-T conditions for the migmatite formation (Fig. 1) are consistent with nearly water-saturated melting of common rocks (e.g. Wyllie & Tuttle 1964; Thompson 1999), but such melting requires both a source and a transport mechanism for a large volume of free water, neither of which have been unequivocally documented for the mid-crust. Dry melting of these rocks requires much higher temperatures. Alternatively, fluid-absent melting of common metamorphic minerals (e.g. muscovite and biotite) may provide the melt, and P-T estimates for the mid-crustal migmatites are compatible with such melting, especially muscovite dehydration melting (Fig. 1), but many of the migmatites contain neither muscovite nor a possible byproduct of muscovite-dehydration melting such as aluminosilicate. Moreover, typical geothermal gradients alone do not provide temperatures high enough for any dehydration melting, especially of biotite (Fig. 1). Regardless of the mechanism, an extra heat source seems necessary for the significant melt generation at the crustal levels of these migmatites. If the temperatures obtained from geothermometers are underestimated, for example, because of partial re-equilibration of cation exchange reactions during cooling (Frost & Chacko 1989), P-T conditions would deviate even farther from what are compatible with geothermal gradients. On the other hand, P-T estimates and resulting geothermal gradients are consistent with seismic evidence showing that only a small portion of the crust comprises melt. Hence, melting must generally occur because of local heating above the solidus (e.g. Marsh 1984). Many sources, such as mantle heat, or magma and radiogenic heat, have been proposed for this necessary extra heat (e.g. see Huerta *et al.* 1996; Jamieson *et al.* 1998; Thompson 1999).

In this model, the thermal input comes from a magma of more mafic composition which formed at a deeper level. The unavoidable dynamic expression of the presence of this magma as an active mobile body is the development of a large-scale complex of veins and dykes above. The net thermal effect of this complex is to slightly, but pervasively, preheat the overlying high-grade metamorphic complex to a thermal state conducive to the survival of small parcels of melt for long periods of time, which would make upward transfer of melt through such a complex possible. The critical thermal condition necessary for melt persistence is that the *contact* (i.e. dykelet/wall rock) temperature be at or above the solidus of the melt. This may be achieved at least in part by the

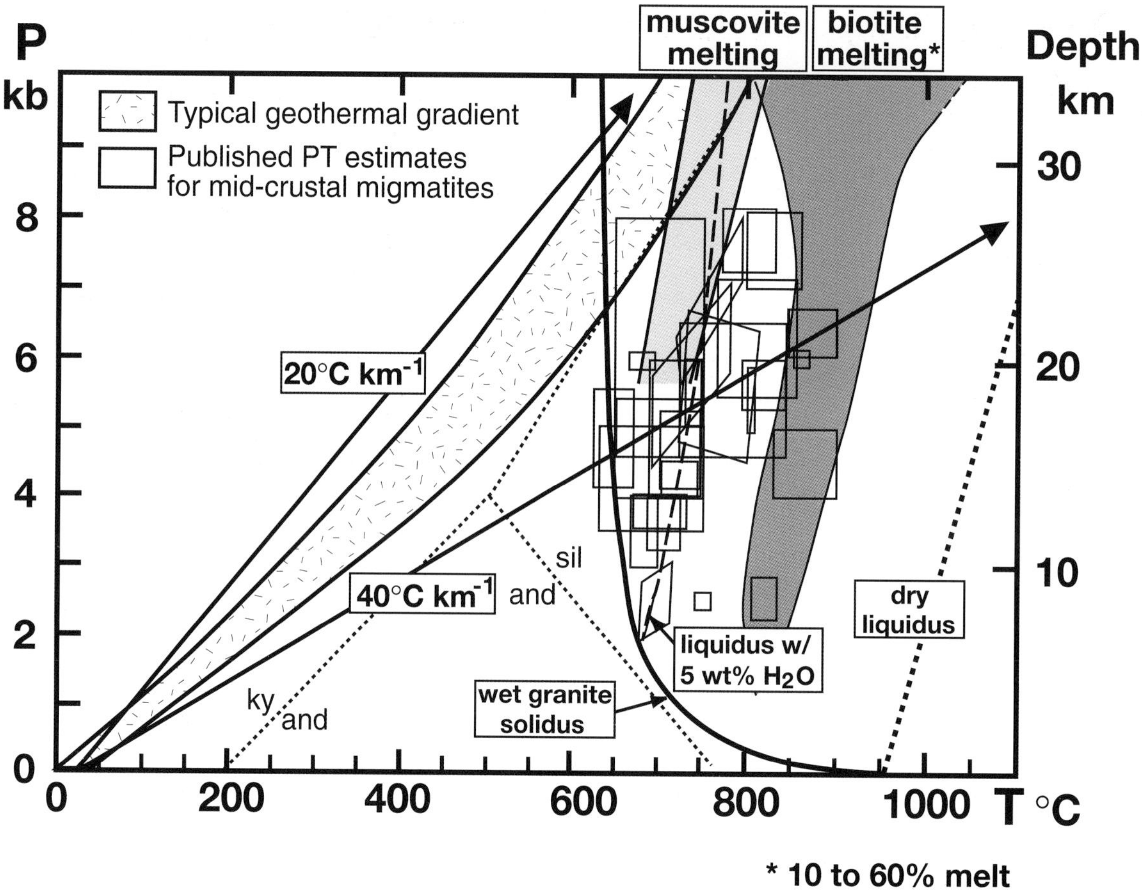

Figure 1 Geothermal gradients, experimental melting reactions and published P-T conditions for mid-crustal migmatites. Sources of data: Geothermal gradient (Brown & Mussett 1993). Experimental data: Aluminosilicates phase relations (Holdaway 1971). Granite melting curves, dry and wet (Ebadi & Johannes 1991). H_2O contents in granitic melt (Holz & Johannes 1994). Biotite fluid-absent melting reaction (Vielzeuf & Montel 1994; Montel & Vielzeuf 1997). Muscovite fluid-absent melting reaction (Patino-Douce & Harris 1998). P-T estimates for migmatites: Antarctica, Brattstrand Bluffs (Fitzsimons 1996). Antarctica, Taylor Valley (Allibone & Norris 1992). Antarctica, Victoria Land (Palmeri 1997). Argentina, Sierra Pampeana (Rapela *et al.* 2002). Australia, Cooma (Ellis & Obata 1992). Australia, Mt Stafford (Greenfield *et al.* 1996). Australia, Reynolds Range (Dirks & Hand 1991). Australia, Wuluma Hills (Sawyer *et al.* 1999). Finland, Turku area (Mengel *et al.* 2001). Finland, Rantasalmi-Sulkava area (Gupta & Johannes 1986). France, St Malo (Brown 1979). Italy, Sardinia (Cruciana *et al.* 2001). Japan, Higo district (Obata *et al.* 1994). Nambia (Jung *et al.* 1998). Spain, Iberian Massif (Viruete 1999). Spain, Trois Seigneurs, Pyranees (Wickham 1987). Sweden, Arvika (Henkes & Johannes 1981). Switzerland, Aar Massif (Olsen *et al.* 2000). Switzerland, Black Forest (Mazurek 1992). USA, Massachusetts (Thomson 2001).

generation of melt that has a solidus lower than that of its parent rock (e.g. Leitch & Weinberg 2002). An explicit, quantitative formulation of these conditions is given in Marsh (1989; e.g. see Eqn. 57).

The large scale but low amplitude of this thermal anomaly means that this complex has a long-term thermal inertia, which facilitates extensive thermal and chemical interactions during melt development on small scales within the intruded host rocks. Such interactions can form a magma of modified composition which, in turn, intrudes the upper crust as a granite pluton. In this paper, the authors present a first step in quantifying this general model of mid-crustal migmatites as feeder-zones by introducing a dimensionless number, MT, for melt transport, which is the ratio of the rate of melt movement caused by the bulk flow of the entire mass (melt + solid) to that of porous media flow of melt only through the solid framework. Finally, they apply it to a geologic example located in the Swiss Alps. In a related work, Leitch & Weinberg (2002) have examined the specific thermal problem of injecting melt segregated from an underlying partially melted zone upward through dyke-like channels into a region whose temperature is above the solidus of the injecting melt. In contrast to the present authors' treatment, they did not consider the dynamic stability of this assumed flow relative to both a Darcian flow and a regional flow, as is treated here.

1. Modelling melt transport through a feeder zone

For this feeder zone model to be valid, the melt transport must be predominantly, if not solely, by porous flow (Fig. 2a) in which the melt percolates through the interstices of a solid framework in a melt-rich zone (MRZ) to feed an overlying pluton. This is one of the two end-member transport mechanisms for the melt. The other is a bulk mass flow of the entire MRZ caused by gross buoyancy (bulk flow: Fig. 2b). The present authors evaluate the relative importance of these flows by using the MT number as defined below.

The overall rate of total melt flux, V_m, can be expressed as

$$V_m = (1 - \varphi)V_s - \left[\frac{K_D}{\mu_2}\right]\left\{\frac{\partial P}{\partial z} + \Delta\rho_2 g\right\} \qquad (1)$$

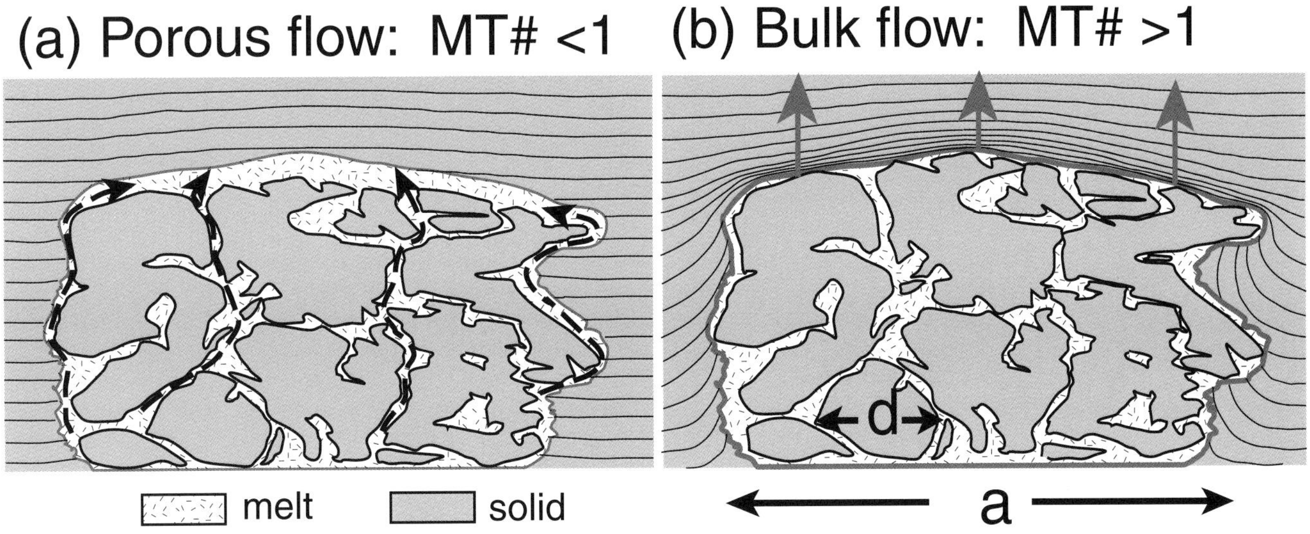

Figure 2 Models of the melt transport mechanisms.

Table 1 Definitions of the parameters in Equations 1 to 6, values used in calculations and their sources: (MRZ) melt-rich zone.

Symbol	Explanation	Values used for Figures 3 and 4 unless otherwise specified	Source
V_m	Overall velocity of melt		
V_s	Velocity of MRZ (solid+melt)		
K_D	Permeability		
P	Pressure		
z	Distance		
g	Gravitational acceleration		
e	Porosity (=1–φ)		
φ	Crystallinity		
C_1	Viscosity constant	2·5/9	Marsh (1982)
C_K	Permeability constant	100	Marsh (2002)
a	Diameter of MRZ		
d	Diameter of grains/particles		
μ_1	Viscosity of hot country rocks	10^{18} poises (=10^{17} Pa s)	Marsh (1984)
μ_2	Viscosity of melt	10^5 poises (=10^4 Pa s)	Marsh (2002): calculated using MELTS (Ghiorso & Sack 1995)
$\Delta\rho_1$	Density difference between MRZ and country rocks	Country rock=2·85 (MRZ=2·68 @e=0·5)	Miller *et al.* (1988)
$\Delta\rho_2$	Density difference between solid and melt in MRZ	Melt density=2·5	Marsh (2002)

in which the first term on the right-hand side of the equation represents bulk flow, and the remaining term represents porous flow after Darcy's law (e.g. Phillips 1991: see Table 1 for the parameters in Equations 1–6). The first term, that for bulk flow transport, is the product of the 'regional' melt content and a measure of the rate of upward motion (V_s) of the rock containing the melt. Since this upward motion is caused by large-scale buoyancy or a gravitational instability, regardless of the exact form of the flow, V_s will be proportional to Stokes' Law velocity (e.g. Leal 1992, p. 230), within the value of an unknown constant of numerical magnitude of near unity (i.e. $C_1 \sim 1$). Therefore,

$$V_s = C_1\left[\frac{\Delta\rho_1 g\, a^2}{\mu_1}\right] \qquad (2)$$

The rate of porous flow, on the other hand, is heavily dependent on the local permeability, which is estimated from the Blake–Kozeny–Carman equation (e.g. Bird *et al.* 1960),

$$K_D = \left[\frac{d^2 e^3}{(1-e)^2\, C_K}\right] \qquad (3)$$

so that the net transport is

$$V_m = (1-\varphi)C_1\left[\frac{\Delta\rho_1 g\, a^2}{\mu_1}\right] - \frac{1}{\mu_2}\left[\frac{d^2 e^3}{(1-e)^2\, C_K}\right]\left\{\frac{\partial P}{\partial z} + \Delta\rho_2 g\right\} \qquad (4)$$

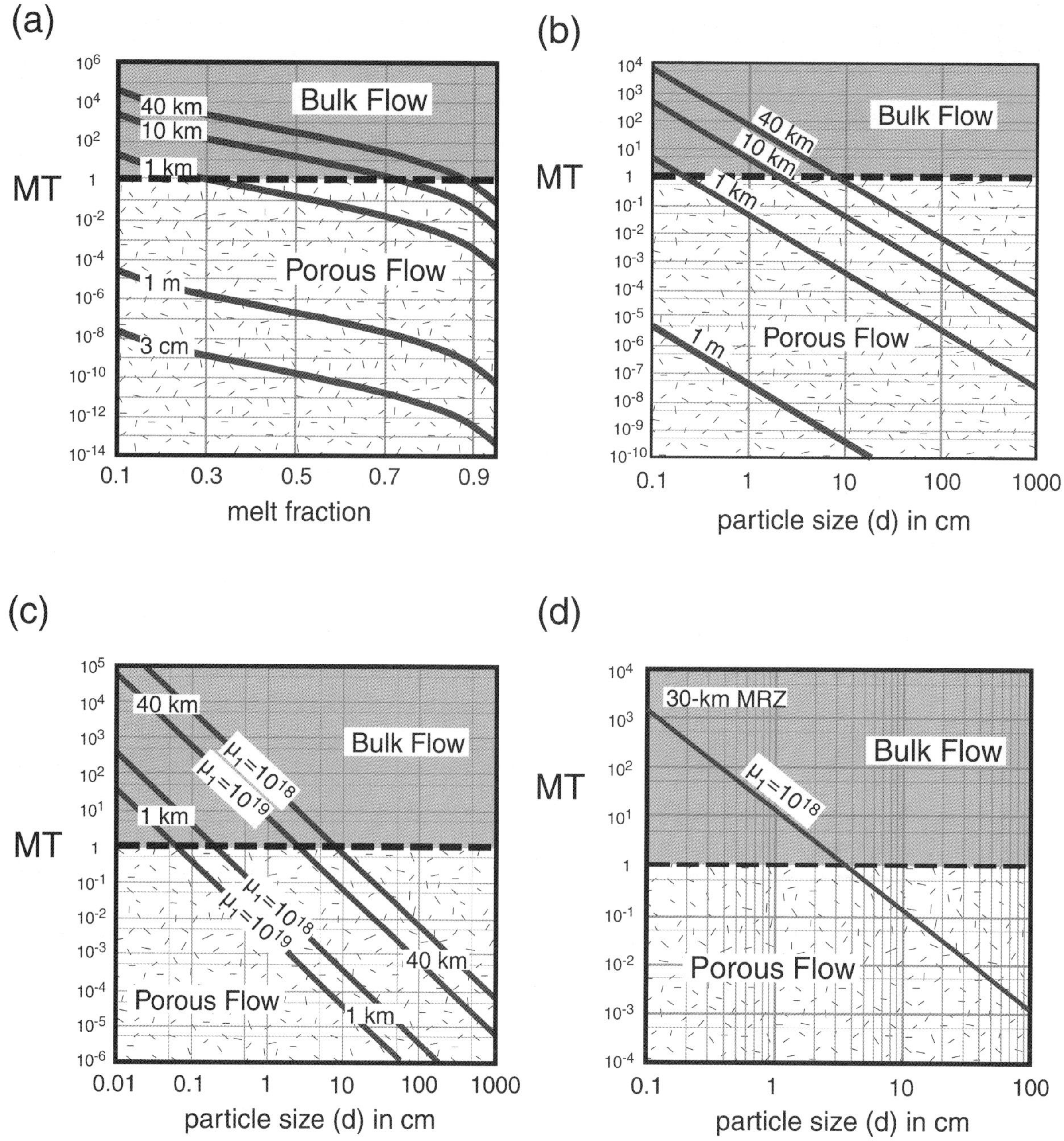

Figure 3 The results of the melt transport (MT) number calculations. For the values of the parameters used, see Table 1. (a) Effect of pluton/melt-rich zone (MRZ) size (e.g. 40 km) on MT number at particle size of 3 mm and $\mu_1 = 10^{18}$ poise. (b) Effect of MRZ/particle size on MT number at 0·4 melt fraction. (c) Effect of viscosity and MRZ size on MT number at 0·4 melt fraction. (d) Melt transport number calculation for the LM as a 30 km MRZ at 0·5 melt fraction (see text).

The intrinsic permeability is a function of the pore size. The use of particle size, rather than pore or channel size, in Equation 3 is justified because the intrinsic permeability is directly linked to the overall scale of the granularity of the solids regardless of actual size (e.g. Bird *et al.* 1960). This is because the drag exerted by the fluid against the surface of the solids controls the rate of flow.

The present authors define a melt transport (MT) number, as the ratio of the two terms in Equation 4, i.e.

$$MT = \left\{ \frac{(1-\varphi)C_1 \left[\dfrac{\Delta\rho_1 g\, a^2}{\mu_1}\right]}{\dfrac{1}{\mu_2}\left[\dfrac{d^2 e^3}{(1-e)^2\, C_K}\right]\left\{\dfrac{\partial P}{\partial z} + \Delta\rho_2 g\right\}} \right\} \qquad (5)$$

The porous flow is driven by either density (or buoyancy) differences ($\Delta\rho g$) or a local pressure gradient ($\partial P/\partial z$), or both. For now, the authors assume $\partial P/\partial z \approx 0$ for convenience, but the assumption will be justified later. Then,

$$MT = C_1 C_K \left[\frac{(1-e)}{e}\right]^2 \left[\frac{\Delta\rho_1}{\Delta\rho_2}\right]\left[\frac{\mu_2}{\mu_1}\right]\left[\frac{a}{d}\right]^2 \qquad (6)$$

The MT number is an indicator of the relative contributions of the flows by the two transport mechanisms to the overall movement of the melt. If MT >1, melt transport is dominated by bulk flow. If MT <1, melt transport is dominated by porous flow. If MT ≈ 1, both mechanisms contribute subequally to the overall flow.

Sample calculations show that the MT number is strongly dependent on melt fraction (e), the diameter (a) of the MRZ

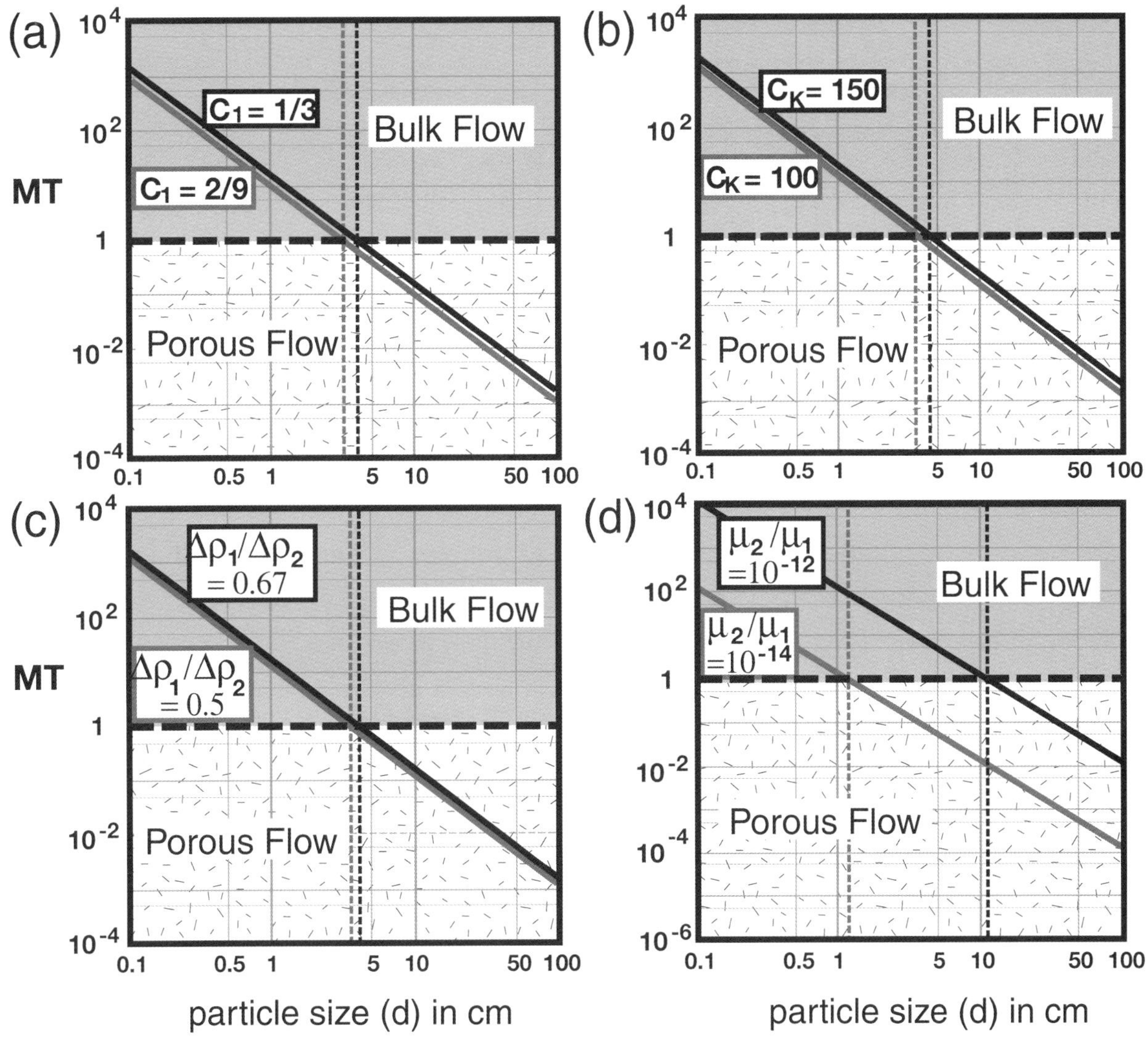

Figure 4 Further results of the melt transport (MT) number calculations showing the effects of varied values for the parameters on MT number at 0·5 melt fraction for a 30 km melt-rich zone: effects of: (a) viscosity constant; (b) permeability constant; (c) density difference; and (d) viscosity difference. For the values of the other parameters used, see Table 1.

and particle size (d), but less so on all other parameters (Figs 3 & 4; also Table 1). If the particle size is ~3 mm, the size of coarse crystals, transport by bulk flow is favoured for a large MRZ (10–40 km), except at very high melt fractions (Fig. 3a). As the size of the MRZ decreases, however, porous flow becomes more important. On the scale of typical migmatite neosomes, porous flow dominates at all melt fractions. Melt transport also depends strongly on particle size (Fig. 3b–d): as particle size increases, local drag (per unit volume of melt) is reduced, so that even for a large MRZ porous flow becomes dominant. For example, at a melt fraction of 0·4, porous flow becomes dominant in a 40 km MRZ at a particle size of ~8 cm (Fig. 3b). Melt transfer through dykes may be considered to be the extreme case of very large particle size.

Other parameters, as long as their values stay within reasonable ranges, as discussed below, do not significantly affect the critical particle size for the cross-over from bulk flow controlled by buoyancy to flow controlled by porous media. Varied values are tested in Figure 4 for a MRZ of 30 km at a melt fraction of 0·5, which are appropriate for the Alpine example discussed below (also see Fig. 3d). If C_1, the viscosity coefficient, is changed from 1/3 (typical for a gas bubble in a viscous fluid) to 2/9 (for a solid sphere in a viscous fluid, e.g. Marsh 1982), the critical particle size decreases from 3·9 to 3·2 cm, or a size change of ~15% (Fig. 4a). For granular materials, C_K, the permeability coefficient, of 150 is used in the Blake–Kozeny–Carman equation, but a coefficient of 100 fits data for the partially molten media better (Marsh 2002). The difference between these values of C_K corresponds to a difference of ~1 cm (~25% diameter change) in critical particle size (Fig. 4b). The density difference is constrained by the melt fraction because the density of MRZ is a linear sum of the densities for solid and melt. The ratio of density differences can be varied arbitrarily but the effect is insignificant (Fig. 4c). The viscosity of the country rocks (μ_1) is probably the least known of all parameters in Equation 6. The present authors constrained the viscosity to a representative value of 10^{18} poises by considering how hot crust responds to deformation (e.g. Wernicke *et al.* 2000). A lower bound comes from the viscosity of basaltic rock near its solidus, which was found by Sakuma (1953) to be $\sim 10^{15}$ poises, but holocrystalline crustal rocks must be more viscous than this. An upper bound comes from the treatment and discussion of folding of cool crustal rocks by Biot (1961) and Johnson & Fletcher (1994; see p. 30), and the flexural rigidity of the crust from glacial rebound studies (e.g. Cathles 1975), which altogether suggest a value of $\sim 10^{22}$–10^{23} poise. The viscosity of the melt (μ_2) is calculated for a dacite with SiO_2 content of 63 wt.% (Table 1). The

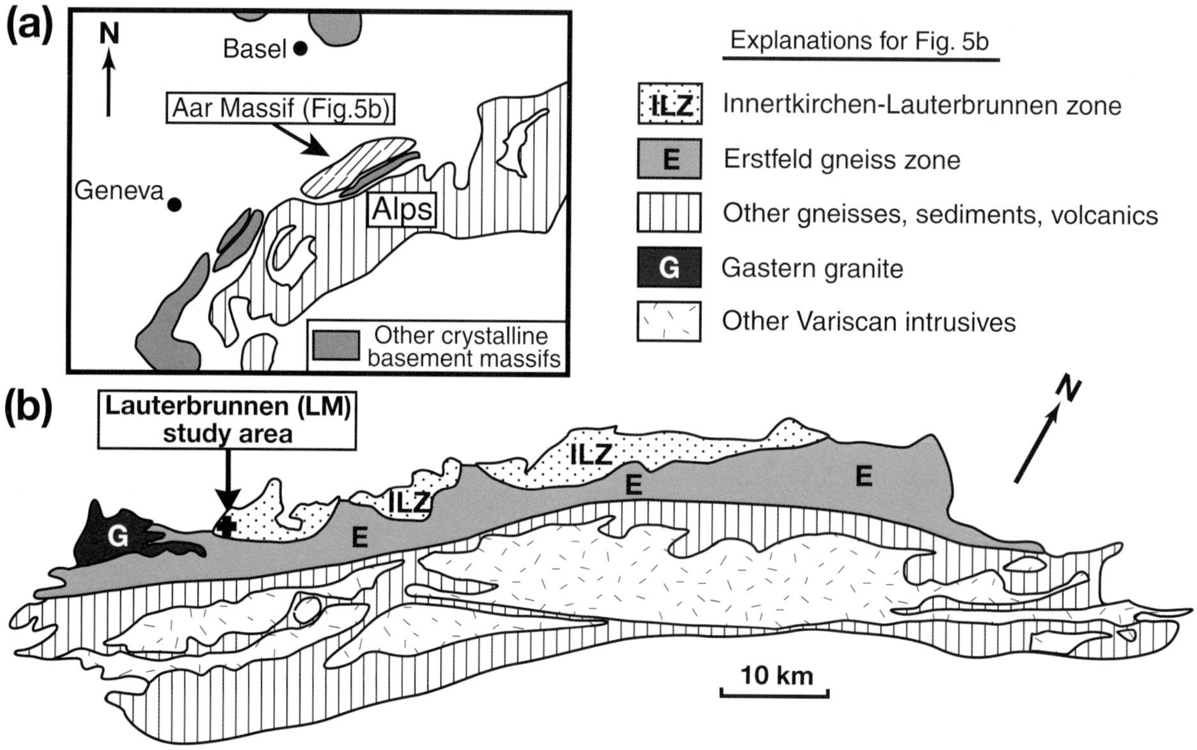

Figure 5 (a) The location of the Aar massif. (b) A simplified geologic map of the Aar massif after Schaltegger (1993) showing the study area.

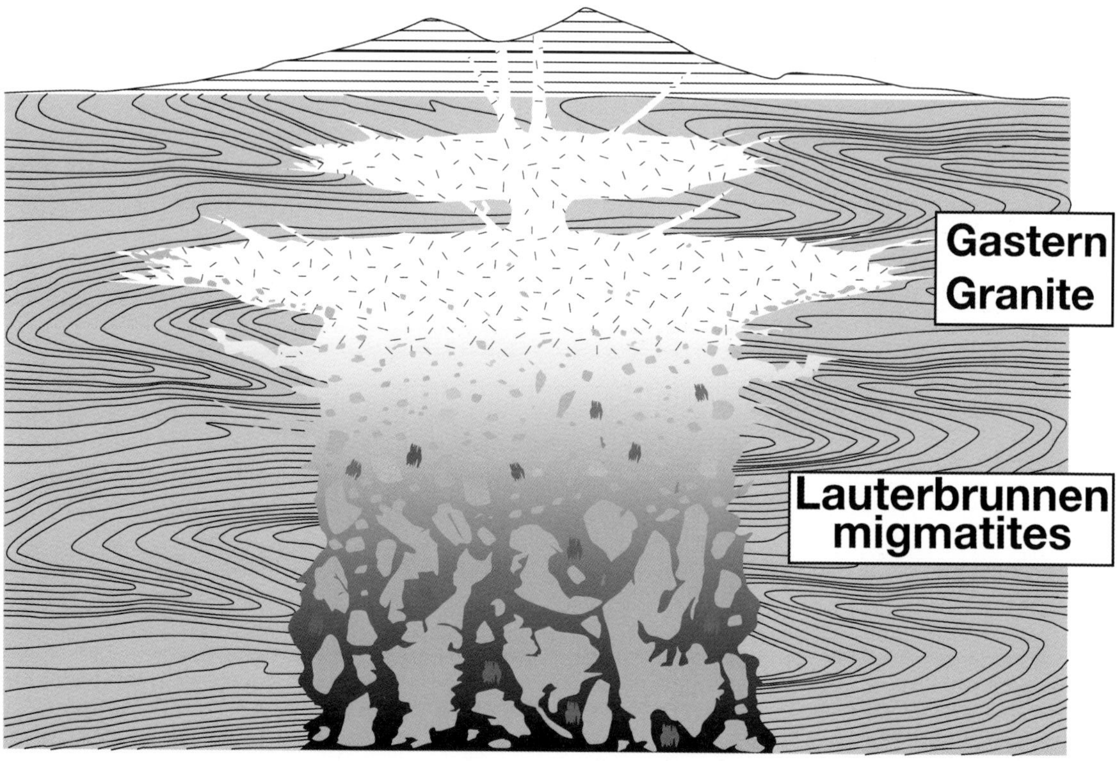

Figure 6 A schematic sketch of the Lauterbrunnen migmatites and Gastern granite showing their inferred relationship.

viscosity difference does affect the result significantly (Fig. 4d). Use of 10^{19} instead of 10^{18} for μ_1 for MRZs at a melt fraction of 0·4 reduced the critical particle size by ~68% (Fig. 3c). The use of an even larger value for μ_1 (e.g. ~10^{25} poises: Clemens *et al.* 1997) greatly favours porous flow, because the MRZ becomes virtually immobile (no bulk flow). The present authors emphasise, however, that such a large 'regional' crustal viscosity corresponding to the cool upper crust, as might be found, for example, through crustal flexure modelling, is clearly inappropriate for the locally heated mid-level crust.

Thus far, $\partial P/\partial z=0$, or no pressure gradient, has been assumed. However, a local pressure gradient, for example, one caused by regional shearing (e.g. Hollister & Crawford 1986) or a volume change consequent to melting (e.g. Brown 1994), could be an important driving force for melt movement. All of

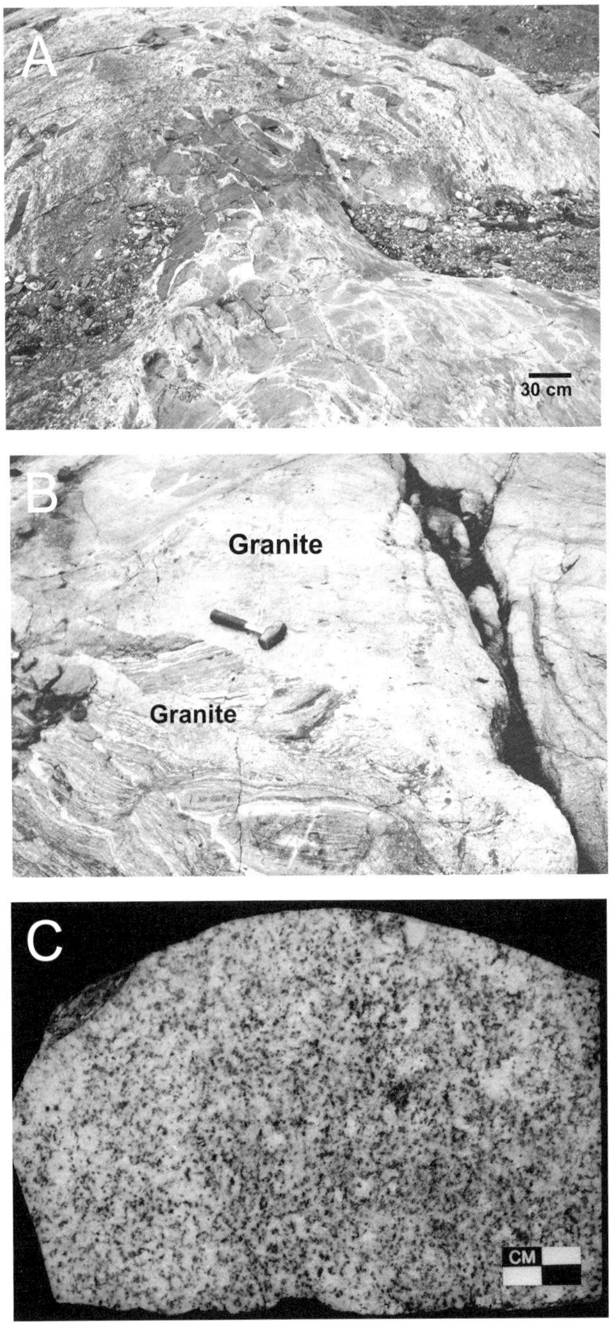

Figure 7 Field and sample photos: (a) a large, fractured enclave (lower half) and heterogeneous matrix (upper half) of the Lauterbrunnen migmatites (LM); (b) the Gastern granite (top) in intrusive contact with the country rocks considered to be an equivalent of LM (lower left); and (c) a slab of Sample LA14.

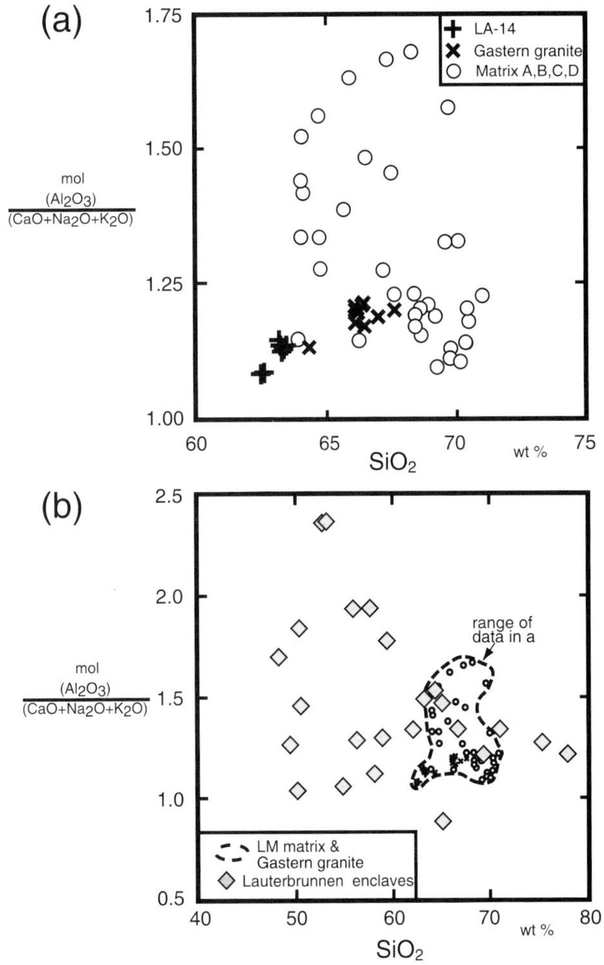

Figure 8 (a) Aluminium indices versus SiO_2 contents of the Lauterbrunnen migmatites (LM) matrix, Gastern granite and (b) LM enclaves. Matrices A, B, C and D correspond to zones A, B, C and D in Figure 9, which have somewhat different lithologies (see Rutishauser 1972).

the local pressure gradients produced by these processes enhance porous media flow. Thus, the balance is shifted further in favour of porous flow, so that the above conclusions concerning porous flow should still be generally valid: the feeder zone model is supported by the feasibility of porous media flow under geologically reasonable conditions. The effect of pressure gradient can also be gauged by varying the density difference, which exerts the same driving force as pressure gradient (Equation 4), but this does not significantly change the results (Fig. 4c).

2. Geologic example: the Lauterbrunnen migmatites

A possible geologic example of the present model for the feeder zone is the Lauterbrunnen migmatites (LM) in the Aar massif in the Central Swiss Alps (Fig. 5), which may be the feeder zone for the nearby Gastern granite (GG) (Fig. 6). The LM are typical mid-crustal migmatites. P-T conditions are constrained to from 650 to 750°C at 5–6 kb by the mineral assemblage present, which has at one time included abundant cordierite. Precise thermo-barometric determinations are not possible because of extensive greenschist facies mineral alteration during subsequent Alpine metamorphism. The U-Pb zircon ages show that the LM and GG are basically coeval ($289 \pm 9{\cdot}4$ Ma, Olsen *et al.* 2000, and 303 ± 4 Ma, Schaltegger 1993, respectively).

The LM comprise a weakly foliated, extremely heterogeneous, biotite-two-feldspar-quartz matrix containing numerous enclaves, which are a few centimetres to >30 m in size (Fig. 7a). The enclaves consist of amphibolite, calc-silicate hornfels, marble and biotite-quartz-feldspar gneiss. The LM matrix has peraluminous granitic to granodioritic compositions (Fig. 8a) and its proportion varies locally from 50% to nearly 100% of the outcrop area (Fig. 9). The enclave compositions are much more varied than those of the matrix, and many are refractory (Fig. 8b). The GG (Fig. 7b) is a homogeneous granite that cross-cuts the foliation of the country rocks, which are considered to be equivalent to the LM, but the chemical and isotopic compositions of the GG are indistinguishable from the average LM matrix (Fig. 8a). The GG composition is also reasonable as a product of the LM parental magma (see below), which has been modified by interactions with rocks similar to the LM enclaves.

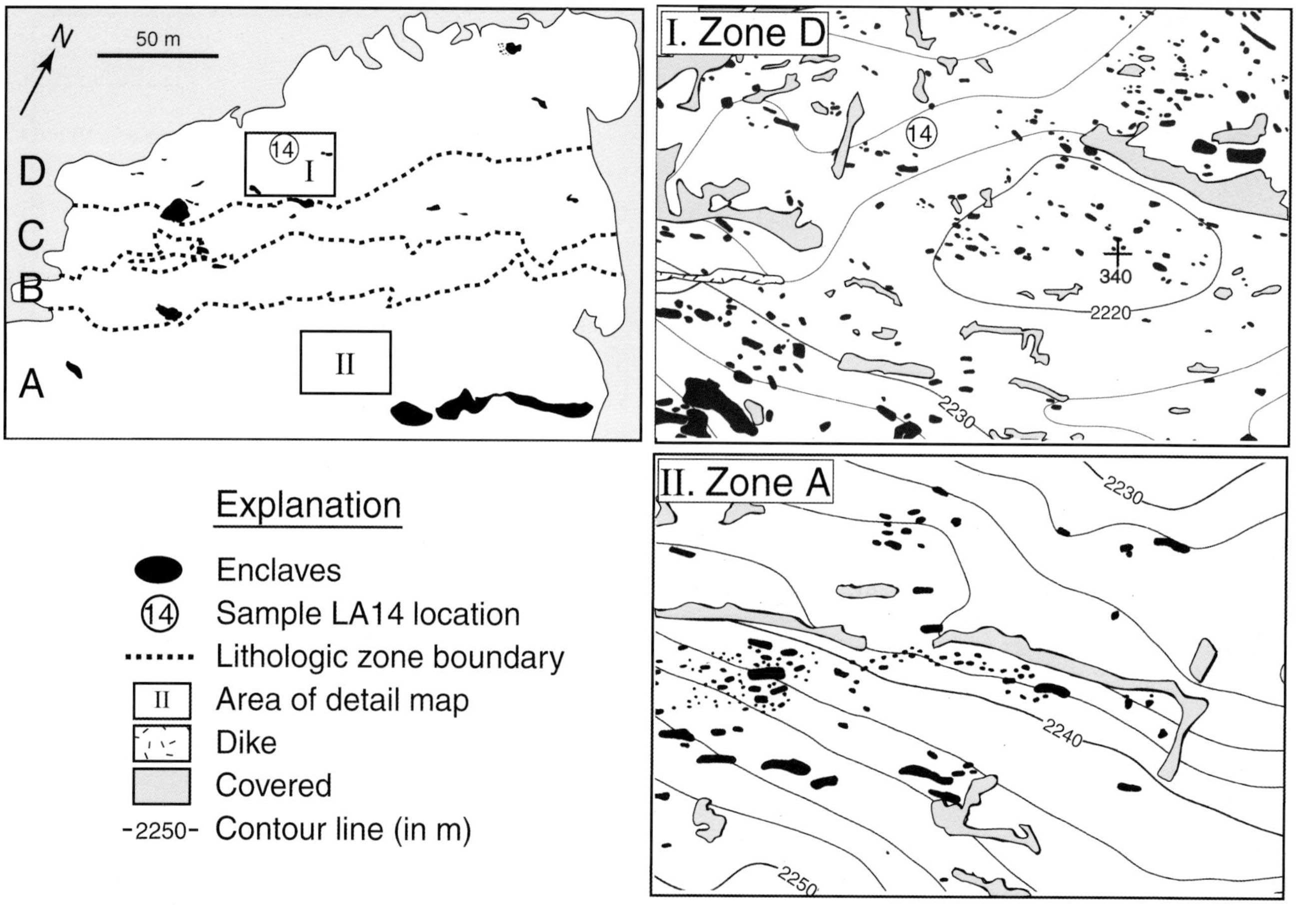

Figure 9 A simplified map of the Lauterbrunnen outcrop (upper left) after Rutishauser (1972): maps I and II show details from Rutishauser.

The evidence for a parental magma in the LM from a deep level is provided by a sample from an area of matrix that is lensoid in shape, metre-scale and uncharacteristically homogeneous in an otherwise heterogeneous matrix (Sample LA14: Fig. 7c; also Fig. 9). LA14 is interpreted to be a sample of 'melt', which is the closest in composition to the intrusive parental magma. LA14 is unique among all LM matrix samples: first, LA14 is homogeneous, unlike the typical, extremely heterogeneous, LM matrix (compare Fig. 7a and c); secondly, LA14 has abundant magmatic zircons with nearly concordant ages of $289 \pm 9{\cdot}4$ Ma, whilst other LM matrix samples have zircons with highly discordant ages (Olsen *et al.* 2000); and thirdly, the LA14 composition is the *least* silicic of all LM matrix samples (Fig. 8a). LA14 composition at 63 wt.% SiO_2 is not appropriate as an *in situ* melt at the crustal level of the LM, and hence, must reflect an external origin of the magma (intrusion). The LA14 composition is more consistent with that of a higher temperature, lower crustal melt (e.g. compare Skjerlie & Johnston 1996). LA14 is interpreted to be a sample of the intrusive magma from below, whilst the remaining heterogeneous LM matrix is interpreted as representing highly contaminated magmas which formed through physical and chemical mingling of the intrusive magma and country rocks. The abundant enclaves are preserved country rock fragments. Reaction/assimilation features commonly observed in the LM enclaves attest to the extensive interaction between the melt and the country rocks. There are no chilled margins, nor are there extensive contact metamorphic effects to indicate a large temperature gradient at the contact. A viscous, magmatic flow in the LM is inferred from a general concordancy of the strike of lithologic zones in the outcrop and in the orientation of the long axes of the enclaves (Fig. 9; Rutishauser 1972).

Magma transport in the crust during a melting event is apt to vary considerably depending on position and age within the system. One way of evaluating the competing effects of bulk transport versus porous media flow is the size of the MT number. Using this approach to understand the transport of melt (similar to LA14 in composition) through LM modelled as a MRZ of ~ 30 km in diameter at a melt fraction of 0·5, which are broadly consistent with field observations, shows that melt transport by porous flow would become dominant over that by bulk flow as effective particle size becomes larger than $\sim 3{\cdot}5$ cm (Fig. 3d). This appears to be on the order of the size of pieces which form as a larger enclave breaks up (see Fig. 7A) and therefore, is consistent with the field evidence as initiation of melt flow through an MRZ should start from the breakup of county rocks. Hence, melt is predicted to percolate through the migmatites, rather than (or in addition to) being transported by a bulk movement of the LM. This also provides the possibility of separating the melt ahead of the LM, resulting in the GG.

3. Conclusions

Mid-crustal migmatites are a common feature in high-grade metamorphic terrains. They may be the feeder zones for upper-level granite plutons in which magma from a deeper level is both the heat source and a reactive agent.

Predominantly porous-media flow could transport the melt through the hot feeder zone, which acts as a reaction vessel in which the deeper-level magma interacts with the country rocks to form an upper-level magma.

A dimensionless number analysis of the possible melt-transport processes shows that porous flow will be the dominant mechanism for the melt transport through a 30-km-diameter feeder zone at a melt fraction of 0·5, such as the LM in the Aar massif, if particle size is larger than ~3·5 cm.

The model of the LM as a feeder zone for the nearby GG, with which they have a temporal and chemical link, is supported by the evidence in the LM for a deeper-level magma. The observed reaction and assimilation, as well as the structural concordance, is consistent with solid-laden viscous flow of magma in the migmatites.

4. Acknowledgements

We thank Jürgen Abrecht and Clark Johnson for helpful discussions. Critical comments by M. Crawford, R. Weinberg and two anonymous reviewers on the earlier drafts were very helpful. Constructive comments by M. Brown and A. Whittington helped in producing a greatly improved final draft. We also thank the organisers of the Hutton V symposium for making it possible for us to present this contribution. We gratefully acknowledge the supports of NSF grants, GER-9450125 to LPB and SNO, OPP-9814332, OCE-9711575 and EAR-9725158 to BDM, and EAR-9257160 to LPB.

5. References

Allibone, A. H. & Norris, R. J. 1992. Segregation of leucogranite microplutons during syn-anatectic deformation – an example from the Taylor Valley, Antarctica. *Journal of Metamorphic Geology*, **10**, 589–600.

Biot, M. A. 1961. Theory of folding of stratified viscoelastic media and its implications in tectonics and orogenesis. *Geological Society of America Bulletin* **72**, 1595–620.

Bird, R. B., Stewart, W. E. & Lightfoot, E. N. 1960. *Transport Phenomena*. Chichester: John Wiley & Sons.

Brown, G. C. & Mussett, A. E. 1993. *The inaccessible earth*, 2nd edn. London: Chapman and Hall.

Brown, M. 1979. The petrogenesis of the St. Malo Migmatite Belt, Armorican Massif, France, with particular reference to the diatexites. *Neues Jahrbuch für Mineralogie Abhandlungen* **135**, 48–74.

Brown, M. 1994. The generation, segregation, ascent and emplacement of granite magma: the migmatite- to crustally-derived granite connection in thickened orogens. *Earth-Science Reviews* **36**, 83–130.

Brown, M. & Solar, G. S. 1999. The mechanism of ascent and emplacement of granite magma during transpression; a syntectonic granite paradigm. *Tectonophysics* **312**, 1–33.

Cathles, L. M. 1975. *The viscosity of the Earth's mantle*. Princeton, NJ: Princeton University Press.

Clemens, J. D., Petford, N. & Mawer, C. K. 1997. Ascent mechanisms of granitic magmas: causes and consequences. *In* Holness, M. B (ed.) *Deformation-enhanced fluid transport in the Earth's crust and mantle*, 145–72. London: Chapman & Hall.

Cruciani, G., Franceschelli, M., Caredda, A. M. & Carcangiu, G. 2001. Anatexis in the Hercynian basement of NE Sardinia, Italy: a case study of the migmatite of Porto Ottiolu. *Mineralogy and Petrology* **71**, 195–233.

Dirks, P. & Hand, M. 1991. Structural and metamorphic controls on the distribution of zircon in an evolving quartzofeldspathic migmatite: an example from the Reynolds Range, central Australia. *Journal of Metamorphic Geology* **9**, 191–201.

Ebadi, A. & Johannes, W. 1991. Beginning of melting and composition of first melts in the system Qz-Ab-Or-H_2O-CO_2. *Contributions to Mineralogy and Petrology* **106**, 286–95.

Ellis, D. J. & Obata, M. 1992. Migmatite and melt segregation at Cooma, New South Wales. *Transactions of the Royal Society of Edinburgh: Earth Sciences* **83**, 95–106.

Fitzsimons, I. C. W. 1996. Metapelitic migmatites from Brattstrand Bluffs, East Antarctica-metamorphism, melting and exhumation of the mid crust. *Journal of Petrology* **37**, 395–414.

Frost, B. R. & Chacko, T. 1989. The granulite uncertainty principle: limitations on thermobarometry in granulites. *Journal of Geology* **97**, 435–50.

Ghiorso, M. S. & Sack, R. O. 1995. Thermodynamic model for the interpolation and extrapolation of liquid-solid equilibria in magmatic systems at elevated temperatures and pressures. *Contributions to Mineralogy and Petrology* **90**, 107–20.

Greenfield, J. E., Clarke, G. L., Bland, M. & Clark, D. J. 1996. In-situ migmatite and hybrid diatexite at Mt Stafford, central Australia. *Journal of Metamorphic Geology* **14**, 413–26.

Gupta, L. N. & Johannes, W. 1986. Genetic model for the stromatic migmatites of the Rantasalmi-Sulkava area, Finland. *Journal of Petrology* **27**, 521–39.

Henkes, L. & Johannes, W. 1981. The petrology of a migmatite (Arvika, Varmland, Western Sweden). *Neues Jahrbuch für Mineralogie Abhandlungen* **141**, 113–33.

Holdaway, M. J. 1971. Stability of andalusite and the aluminum silicate phase diagram. *American Journal of Science* **271**, 97–131.

Hollister, L. S. & Crawford, M. L. 1986. Melt-enhanced deformation – a major tectonic process. *Geology* **14**, 558–61.

Holz, F. & Johannes, W. 1994. Maximum and minimum water contents of granitic melts: implications for chemical and physical properties of ascending magmas. *Lithos* **32**, 149–59.

Huerta, A. D., Royden, L. H. & Hodges, K. V. 1996. The interdependence of deformational and thermal processes in mountain belts. *Science* **273**, 637–9.

Jamieson, R. A., Beaumont, C., Fullsack, P. & Lee, B. 1998. Barrovian regional metamorphism: Where's the heat? *In* Treloar, P. J. & O'Brien, P. J. (eds) *What drives metamorphism and metamorphic reactions? Geological Society, London, Special Publication* **138**. London: Geological Society.

Johnson, A. M. & Fletcher, R. C. 1994. *Folding of Viscous Layers*. New York, NY: Columbia University Press.

Jung, S., Mezger, K., Masberg, P., Hoffer, E. & Hoernes, S. 1998. Petrology of intrusion-related high-grade migmatite: implications for partial melting of metasedimentary rocks and leucosome-forming process. *Journal of Metamorphic Geology* **16**, 425–45.

Leal, L. G. 1992. *Laminar flow and convective transport processes. Scaling principles and asymptotic analysis*. Boston: Butterworth-Heinemann.

Leitch, A. M. & Weinberg, R. F. 2002. Modeling granite migration by mesoscale pervasive flow. *Earth and Planetary Science Letters* **200**, 131–46.

Marsh, B. D. 1982. On the mechanics of igneous diapirism, stoping, and zone melting. *American Journal of Science* **282**, 808–55.

Marsh, B. D. 1984. Mechanics and energetics of magma formation and ascension. *In* Boyd, J. F. R. (ed.) *Explosive volcanism: inception, evolution, and hazards*, 67–83. Washington, DC: National Academy Press.

Marsh, B. D. 1989. On convective style and vigor in sheet-like magma chambers. *Journal of Petrology* **30**, 479–530.

Marsh, B. D. 2002. On bimodal differentiation by solidification front instability in basaltic magmas, part 1: basic mechanics. *Geochimica et Cosmochimica Acta* **66**, 2211–29.

Mazurek, M. 1992. Phase equilibria and oxygen isotopes in the evolution of metapelitic migmatites: a case study from the Pre-Alpine basement of Northern Switzerland. *Contributions to Mineralogy and Petrology* **109**, 494–510.

Mengel, K., Richterb, M. & Johannes, W. 2001. Leucosome-forming small-scale geochemical processes in the metapelitic migmatites of the Turku area, Finland. *Lithos* **56**, 47–73

Miller, C. F., Watson, E. B. & Harrison, T. M. 1988. Perspectives on the source, segregation and transport of granitoid magmas. *Transactions of the Royal Society of Edinburgh: Earth Sciences* **79**, 135–56.

Montel, J. M. & Vielzeuf, D. 1997. Partial melting of metagreywackes, Part II. Compositions of minerals and melts. *Contributions to Mineralogy and Petrology* **128**, 176–96.

Obata, M., Yoshimura, Y., Nagakawa, K., Odawara, S. & Osanai, Y. 1994. Crustal anatexis and melt migrations in the Higo metamorphic terrane, west-central Kyushu, Kumamoto, Japan. *Lithos* **32**, 135–47.

Olsen, S. N., Johnson, C. M., Beard, B. L. & Baumgartner, L. P. 2000. New U-Pb zircon data and constraints on the age and mode of migmatization in the Aar Massif, Central Alps. *European Journal of Mineralogy* **12**, 1245–60.

Palmeri, R. 1997. P-T paths and migmatite formation: an example from Deep Freeze Range, northern Victoria Land, Antarctica. *Lithos* **42**, 47–66.

Patino Douce, A. E. & Harris, N. 1998. Experimental constraints on Himalayan anatexis. *Journal of Petrology* **39**, 689–710.

Petford, N., Cruden, A. R, McCaffrey, K. J. W. & Vigneresse, J. L. 2000. Granite magma formation, transport and emplacement in the Earth's crust. *Nature* **408**, 669–73.

Petford, N. & Koenders, M. A. 1998. Granular flow and viscous fluctuations in low Bagnold number granitic magmas. *Journal of the Geological Society of London* **155**, 873–81.

Phillips, O. M. 1991. *Flow and reactions in permeable rocks.* Cambridge: Cambridge University Press.

Rapela, C. W., Baldo, E. G., Pankhurst, R. J. & Saavedra, J. 2002. Cordierite and leucogranite formation during emplacement of highly peraluminous magma: the El Pilón Granite Complex (Sierras Pampeanas, Argentina). *Journal of Petrology* **43**, 1003–28.

Rutishauser, H. 1972. *Die quantitative Erfassung von Migmatiten im Aufschlußbereich.* Unpublished Ph.D. thesis. University of Bern.

Sakuma, S. 1953. Elastic and viscous properties of volcanic rocks and high temperatures. Part 3. Osima lava. *Earthquake Research Institute Bulletin* **31**, 291–303.

Sawyer, E. W. 1996. Melt segregation and magma flow in migmatites: implications for the generation of granite magmas. *Transactions of the Royal Society of Edinburgh* **87**, 85–94.

Sawyer, E. W., Dombrowdki, C. & Collins, W. J. 1999. Movement of melt during synchronous regional deformation and granulite-facies anatexis, an example from the Wuluma Hills, central Australia. *In* Castro, A., Fernandez, C. & Vigneresse, I.-L. (eds) *Understanding granites; integrating new and classical techniques*, 221–37. London: Geological Society of London.

Schaltegger, U. 1993. The evolution of the polymetamorphic basement in the Central Alps unravelled by precise U-Pb zircon dating. *Contributions to Mineralogy and Petrology* **113**, 466–78.

Skjerlie, K. P. & Johnston, D. A. 1996. Vapour-absent melting from 10 to 20 kbar of crustal rocks that contain multiple hydrous phases: implications for anatexis in the deep to very deep continental crust and active continental margins. *Journal of Petrology* **37**, 661–91.

Thompson, A. B. 1999. Some time-space relationships for crustal melting and granitic intrusion at various depths: *In* Castro, A., Fernandez, C. & Vigneresse, J.-L. (eds) *Understanding granites; integrating new and classical techniques*, 7–25. London: Geological Society of London.

Thomson, J. A. 2001. A counterclockwise P-T path for anatectic pelites, south-central Massachusetts. *Contributions to Mineralogy and Petrology* **141**, 623–41.

Vielzeuf, D. & Montel, J. M. 1994. Partial melting of metagreywackes. Part I. Fluid-absent experiments and phase relationships. *Contributions to Mineralogy and Petrology* **117**, 375–93.

Viruete, J. E. 1999. Hornblende-bearing leucosome development during syn-orogenic crustal extension in the Tormes Gneiss Dome, NW Iberian Massif, Spain. *Lithos* **46**, 751–72.

Weinberg, R. F. 1999, Mesoscale pervasive felsic magma migration: alternatives to dyking. *Lithos* **46**, 393–410.

Wernicke, B., Friedrich, A. M., Niemi, N. A., Bennett, R. A. & Davis, J. L 2000. Dynamics of plate boundary fault systems from Basin and Range geodetic network (BARGEN) and geologic data. *GSA Today* **10**, 1–7.

Wickham, S. M. 1987. Crustal anatexis and granite petrogenesis during low-pressure regional metamorphism: the Trois Seigneurs Massif, Pyrenees, France. *Journal of Petrology* **28**, 127–69.

Wyllie, P. J. & Tuttle, O. F. 1964. Experimental investigation of silicate systems containing two volatile components. Part III: The effects of SO_3, P_2O_5, HCl, and Li_2O, in addition to H_2O, on the melting temperatures of albite and granite. *American Journal of Sciences* **262**, 930–9.

SAKIKO N. OLSEN and BRUCE D. MARSH, Department of Earth and Planetary Sciences, Johns Hopkins University, Baltimore, MD 21218, USA.
e-mail: olsen@jhu.edu; bmarsh@jhu.edu

LUKAS P. BAUMGARTNER, Institute for Mineralogy and Geochemistry, University of Lausanne, CH-1015 Switzerland.
e-mail: lukas.baumgartner@img.unil.ch

MS received 5 September 2003. Accepted for publication 9 May 2004.

Transactions of the Royal Society of Edinburgh: Earth Sciences, **95**, 59–71, 2004 (for 2003)

Experimental temperature–X(H_2O)–viscosity relationship for leucogranites and comparison with synthetic silicic liquids

Alan Whittington, Pascal Richet, Harald Behrens, François Holtz and Bruno Scaillet

ABSTRACT: Viscosities of liquid albite ($NaAlSi_3O_8$) and a Himalayan leucogranite were measured near the glass transition at a pressure of one atmosphere for water contents of 0, 2·8 and 3·4 wt.%. Measured viscosities range from $10^{13\cdot8}$ Pa.s at 935 K to $10^{9\cdot0}$ Pa.s at 1119 K for anhydrous granite, and from $10^{10\cdot2}$ Pa.s at 760 K to $10^{12\cdot9}$ Pa.s at 658 K for granite containing 3·4 wt.% H_2O. The leucogranite is the first naturally occurring liquid composition to be investigated over the wide range of T-X(H_2O) conditions which may be encountered in both plutonic and volcanic settings. At typical magmatic temperatures of 750°C, the viscosity of the leucogranite is $10^{11\cdot0}$ Pa.s for the anhydrous liquid, dropping to $10^{6\cdot5}$ Pa.s for a water content of 3 wt.% H_2O. For the same temperature, the viscosity of liquid $NaAlSi_3O_8$ is reduced from $10^{12\cdot2}$ to $10^{6\cdot3}$ Pa.s by the addition of 1·9 wt.% H_2O. Combined with published high-temperature viscosity data, these results confirm that water reduces the viscosity of $NaAlSi_3O_8$ liquids to a much greater degree than that of natural leucogranitic liquids. Furthermore, the viscosity of $NaAlSi_3O_8$ liquid becomes substantially non-Arrhenian at water contents as low as 1 wt.% H_2O, while that of the leucogranite appears to remain close to Arrhenian to at least 3 wt.% H_2O, and viscosity–temperature relationships for hydrous leucogranites must be nearly Arrhenian over a wide range of temperature and viscosity. Therefore, the viscosity of hydrous $NaAlSi_3O_8$ liquid does not provide a good model for natural granitic or rhyolitic liquids, especially at lower temperatures and water contents.

Qualitatively, the differences can be explained in terms of configurational entropy theory because the addition of water should lead to higher entropies of mixing in simple model compositions than in complex natural compositions. This hypothesis also explains why the water reduces magma viscosity to a larger degree at low temperatures, and is consistent with published viscosity data for hydrous liquid compositions ranging from $NaAlSi_3O_8$ and synthetic haplogranites to natural samples. Therefore, predictive models of magma viscosity need to account for compositional variations in more detail than via simple approximations of the degree of polymerisation of the melt structure.

KEY WORDS: Aluminosilicate liquid, configurational entropy, granite, rheology, rhyolite, water.

The viscosity of silicate liquids exerts strong controls over many geological processes, including rates of melt extraction, ascent and emplacement, magma mixing, and the style of volcanic eruptions (e.g. Petford 1996; Baker 1998; Dingwell 1999; Petford *et al.* 2000; Scaillet *et al.* 2000; and references therein). The shear viscosity of silicate melts depends strongly on temperature, bulk chemical composition and volatile contents, and has the potential to vary by orders of magnitude within a single geological environment. Therefore, a predictive model for magma viscosity over the full geologic range of temperature, composition and volatile content, is one major aim of experimental petrology. Additional variations in magma viscosity caused by bubbles, crystal content and strain rate are also significant, but can be accounted for by composition-independent models such as the Einstein–Roscoe equation for liquid-solid suspensions (e.g. Bottinga & Weill 1972), or the Maxwell relationship for shear relaxation time (e.g. Webb & Dingwell 1990).

There are two complementary approaches to understanding magma viscosity: one is to conduct experiments on well-characterised natural samples from a specific locality (e.g. Kushiro 1976; Scaillet *et al.* 1996; Giordano *et al.* 2000); and the other is to conduct experiments on simple analogue materials with compositions corresponding to those of minerals such as albite, anorthite and diopside (e.g. Kushiro 1978; Scarfe *et al.* 1983; Hummel & Arndt 1985). Experiments on analogue compositions have the combined aims of (1) understanding the underlying physico-chemical controls on magma viscosity, and (2) providing data which can be applied to more chemically complex natural compositions.

In view of the importance of granitic (*sensu lato*) magma in both plutonic and volcanic settings, the viscosity of hydrous $NaAlSi_3O_8$ (albitic) liquids has been extensively investigated, both at high temperature and pressure (e.g. Dingwell 1987; Persikov *et al.* 1990; Baker 1996), and at low temperatures and atmospheric pressure (Romano *et al.* 2001). These measurements have been used to model processes such as melt extraction, and whether melt ascent is more likely to proceed by dyking or diapirism (e.g. Petford *et al.* 1996; Baker 1998). Such models assume that granites and $NaAlSi_3O_8$ liquids have similar properties despite the greater complexity of natural granite compositions. However, recent high-temperature measurements on a haplogranite melt (Schulze *et al.* 1996), and in the $NaAlSi_3O_8$–SiO_2 system (Holtz *et al.* 1999) have demonstrated that the viscosities of hydrous silicic liquids within the haplogranite system are distinctly composition-dependent.

Previous studies on the effect of volatiles on magma viscosity have often been conducted using the falling sphere method, under hundreds of MPa pressure, where the volatiles are easily dissolved and the melt is not saturated with water (e.g. Kushiro 1978; Dingwell 1987; Baker & Vaillancourt 1995; Scaillet *et al.* 1996; Schulze *et al.* 1996; Holtz *et al.* 1999). Experiments under these conditions are time-consuming, cover a narrow viscosity range and require great care to produce accurate results (Holtz *et al.* 1999). In addition, they usually require extrapolation to lower temperatures for application to granite petrogenesis. Thus, as first described by Lejeune *et al.* (1994) and Richet *et al.* (1996), measurements on hydrous liquids just above the glass transition range nicely complement the observations made at high temperature. Because exsolution of volatiles is sufficiently slow that negligible water loss takes place at such high viscosities, the measurements can be performed at one bar in a precise and repeatable manner.

Recent studies of hydrous melt viscosities at low temperatures have dealt with synthetic intermediate to highly depolymerised compositions (Richet *et al.* 1996; Whittington *et al.* 2000, 2001), $XAlSi_3O_8$ (X=Li, Na, K, $Ca_{0.5}$, $Mg_{0.5}$; Romano *et al.* 2001), the haplogranite system (Dingwell *et al.* 1996) and related compositions (Dingwell *et al.* 1998a, b, 2000). Studies of naturally occurring hydrous liquid compositions have been less common. Stevenson *et al.* (1998) studied natural rhyolitic compositions containing up to 1 wt.% H_2O, Giordano *et al.* (2000) studied phonolites from Tenerife, and Giordano & Dingwell (2003) studied basalts from Etna.

The present authors report new low-temperature viscosity measurements for hydrous $NaAlSi_3O_8$ liquids and natural leucogranite liquids, with the aims of (1) enabling a more complete description of their viscosities in P-T-X(H_2O) space by combination with previous high temperature measurements, and (2) comparing the effect of water on simple (analogue) and complex (natural) silicic liquid compositions. The leucogranite is the first naturally-occurring liquid composition to have its viscosity measured over the wide range of T-X(H_2O) conditions which may be encountered in both plutonic and volcanic settings. It is an appropriate choice since, in addition to the widespread occurrence of similar granites along the Higher Himalaya, rhyolites of near-identical composition are known from Peru (Pichavant *et al.* 1988a, b).

1. Samples and experimental methods

The hydrous $NaAlSi_3O_8$ samples were synthesised from the same anhydrous starting material as used in the high-temperature study of Holtz *et al.* (1999), originally synthesised by the Schott glass company (sample AbI of Behrens *et al.* 1996). Starting material compositions are listed in Table 1. The leucogranite samples were synthesised from sample DK89 of Scaillet *et al.* (1995) from the Manaslu pluton of the Himalaya. They are almost identical in composition to sample GB4 from the nearby Gangotri pluton, which was studied under high-temperature conditions by Scaillet *et al.* (1996). Both leucogranites are strongly peraluminous (molar Al_2O_3/{$Na_2O+K_2O+CaO+MgO$}=1·12 for DK89 and 1·16 for GB4). Scaillet *et al.* (1996) measured the viscosity of GB4 under P-T conditions from 300 to 800 MPa at 800 to 1100°C and water contents of 3·98 and 6·66 wt.%, close to the source and emplacement conditions inferred by crystallisation studies (Scaillet *et al.* 1995). Although the present authors' P-T conditions are more directly relevant to volcanic processes, the results also apply to plutonic contexts because changing pressure has apparently little effect on viscosity for this composition (Scaillet *et al.* 1996).

Experimental methods are described in detail elsewhere (Richet *et al.* 1996; Whittington *et al.* 2001), and are reviewed

Table 1 Sample compositions (wt.% oxides).

Sample	DK89[a]	GB4[b]	HAB[c]	Albite[d]
SiO_2	73·61	73·89	69·19	68·74
TiO_2	0·17	0·06	0·00	0·00
Al_2O_3	15·45	15·77	18·83	19·44
FeO_t	0·89	0·75	0·00	0·00
MgO	0·19	0·14	0·00	0·00
CaO	0·84	0·58	0·00	0·00
Na_2O	3·93	4·62	11·96	11·82
K_2O	4·93	4·19	0·02	0·00
TiO_2	0·17	0·06	0·00	0·00

[a]From Scaillet *et al.* (1995). This was the sample used in the present authors' low-temperature viscosity measurements.
[b]From Scaillet *et al.* (1995). This was the sample used in the high-temperature viscosity measurements of Scaillet *et al.* (1996).
[c]From Behrens *et al.* (1996).
[d]Composition of perfectly stoichiometric albite, for comparison.

only briefly here. The dry leucogranitic glass was formed by fusion of a powdered whole-rock sample (DK89 of Scaillet *et al.* 1995) at 1600°C for about one hour, with occasional stirring until the melt was free of bubbles. The melt was quenched to glass by pouring into cylindrical graphite moulds on a copper plate. Densities were measured on several different pieces as an initial check of chemical homogeneity.

Hydration of the materials was conducted at the Institut für Mineralogie (IMH) in Hannover, Germany, using about one gram of powdered glass (for $NaAlSi_3O_8$) or powdered whole-rock sample (for the leucogranite), and variable quantities of de-ionised H_2O. These were weighed into Pt capsules and placed in internally-heated pressure vessels, oriented vertically under 2–3 kbar pressure at 1200 to 1300°C, for some days. Individual synthesis conditions are listed in Table 2. At the end of the experiment, the power supply was switched off and the samples were quenched isobarically, which avoided the exsolution of water and the formation of quench bubbles. The samples were generally crystal-free, with a few air bubbles whose volume fraction was much too small to affect the viscosity measurably. The present authors followed the method described by Behrens (1995) to measure the water contents of samples by Karl–Fischer titration. For each sample, the two ends were analysed by KFT (Table 2). Reported water contents include the addition of 0·10± 0·05 wt.% H_2O still dissolved in the melt after heating to 1300°C at atmospheric pressure. Water contents for the two ends of the samples are generally within the estimated ±0·10 wt.% accuracy of the technique; in two cases, the differences are greater, which may be because of temperature gradients within the capsule during synthesis. In these cases, the average of the two ends is a good estimate of the water content of the middle of the charge, which is the part used for viscosity measurement. This is demonstrated by sample 'Albite-2·2', where a third measurement on a piece of glass from near the middle of the capsule, adjacent to the section used for viscosity measurement, yields exactly the average of the analyses of the two ends (Table 2).

Viscosity was measured at temperatures near the glass transition at the Institut de Physique du Globe in Paris, France, as described by Neuville & Richet (1991). The general principle is the measurement of the deformation rate of a glass cylinder of known length under known uniaxial compressive stress. Temperature is measured by two thermocouples at each end of the cylinder, and thermal gradients are reduced to less than 1°C by the addition of a silver cylinder that surrounds the

Table 2 Hydration conditions, water contents and densities of hydrated glasses.

Sample	H_2O content (wt.%)			Pressure (MPa)	Temperature (°C)	Time (h)	ρ[a] (g cm^{-3})
	End A	End B	Average				
Albite A	–	–	0·036[b]	–	–	–	2·3623
Albite B	–	–	0·036[b]	–	–	–	2·3704
Albite-1·3	1·00	1·14	1·07	200	1200	65	2·3843
Albite-2·2	1·72	2·01	1·87[c]	200	1200	65	2·3752
DK89–0	–	–	–	–	–	–	2·4160
DK89–2	2·63	3·03	2·83	200	1200	48	2·3340
DK89–3	3·40	3·42	3·41	300	1200	39	2·3358

[a]Density was measured by the Archimedean method using toluene. Measurement errors are better than ± 0·0010 g cm^{-3} based on repeat measurements. For hydrous glasses, densities were measured on compacted samples.
[b]Determined by infrared spectroscopy.
[c]A third analysis from near the middle of the experimental charge, adjacent to the section used in viscosity determinations, yielded 1·87 wt.% H_2O.

Table 3 Low-temperature viscosity data for albite.*

Anhydrous A		Anhydrous B		1·07 wt.% H_2O		1·87 wt.% H_2O	
T (K)	log η (Pa s)	T (K)	log η (Pa s)	T (K)	log η (Pa s)	T (K)	log η (Pa s)
1028·1	12·11 (0·03)[3]	1004·8	12·04 (0·09)[2]	771·2	12·44 (0·04)[3]	701·1	12·32 (0·08)[1]
1048·5	11·63 (0·02)[1]	1031·6	11·41 (0·04)[3]	781·2	11·99 (0·04)[1]	711·6	12·07 (0·05)[8]
1069·5	11·16 (0·02)[2]	1057·9	10·75 (0·02)[6]	791·9	11·75 (0·02)[8]	721·6	11·70 (0·04)[3]
1090·8	10·72 (0·04)[4]	1077·3	10·40 (0·02)[1]	802·1	11·46 (0·04)[2]	730·8	11·36 (0·03)[2]
1111·7	10·33 (0·05)[8]	1110·3	9·76 (0·03)[4]	812·5	11·24 (0·06)[4]	742·2	11·14 (0·04)[6]
1132·0	9·93 (0·03)[5]	1141·4	9·25 (0·01)[5]	823·4	10·88 (0·04)[7]	751·7	10·84 (0·06)[4]
1152·5	9·57 (0·02)[6]			833·5	10·75 (0·10)[5]	762·4	10·56 (0·04)[5]
1173·7	9·26 (0·01)[7]			854·0	10·23 (0·07)[6]	777·5	10·21 (0·04)[7]

*For all samples, superscripts indicate the order in which measurements were made. Errors in parentheses are standard deviations of measurements made at three different stresses, to verify Newtonian (stress-independent) viscosity behaviour.

sample. At each temperature, three different measurements were made after increasing and decreasing stresses to check that viscosity was Newtonian. The precision of the technique is generally better than 0·04 log units. Although solubility of water in these compositions is very low at atmospheric pressure, the effect of water on the viscosity of silicate melts is so extreme under these conditions that, were significant water loss to occur, it would be immediately detectable because of a marked increase of the measured viscosity (e.g. Richet *et al.* 1996; Dingwell *et al.* 1996).

In order to estimate the possible effect of water loss by diffusion on the viscosity measurements, the present authors have calculated water diffusion profiles in two samples. Water diffusivity is almost identical in albitic and granitic melts (Behrens & Nowak 1997; Zhang & Behrens 1998), and the authors used the diffusion equation for rhyolitic glasses with water contents below 2 wt.% from Zhang & Behrens (2000) to model the profiles in albitic glasses HAB-1.3 (1·07 wt.% H_2O) and HAB-2·2 (1·87 wt.% H_2O) after the experiments. At such low water contents, the total water diffusivity is proportional to the water content. Assuming one-dimensional diffusion, the total length of the profiles is estimated to be 190 μm for HAB1–3 and 115 μm for HAB2–2. A near-surface zone of about 38 μm in HAB-1·3 and 23 μm in HAB-2·2 is relatively strongly depleted in H_2O (more than 20% compared to the initial water content). However, the fraction of the cylinder that is strongly depleted in H_2O is only 3% of the total volume in the case of run HAB-1·3, and only 2% in the case of run HAB-2·2. One advantage of the parallel-plate technique used in this study is that it measures the bulk deformation of the entire cylinder, rather than only deformation within a surface layer. This small, water-depleted volume fraction has a minor influence on the viscosity, as demonstrated by the good agreement of viscosity measurements at similar temperature at the beginning and at the end of the experiments.

2. Results and comparison with high-temperature data

2.1. $NaAlSi_3O_8$ melts

Low-temperature viscosity measurements are reported in Table 3. The data plotted in Figure 1 indicate that the two anhydrous samples do not have exactly the same composition, despite being cored from the same lump of starting glass. This once more illustrates the extreme sensitivity of glass transition temperature and viscosity of $NaAlSi_3O_8$ to minor composition differences (Richet & Bottinga 1984; Toplis *et al.* 1997). The measured difference in glass transition temperature (Tg, taken to be the temperature at which the viscosity is 10^{12} Pa.s) is 25 K. This is small in comparison to the 60 K difference between two nominally stoichiometric $NaAlSi_3O_8$ glasses found in calorimetry measurements by Richet & Bottinga (1984). The observed homogeneous deformation of both glass cylinders indicates that the bulk composition did not vary significantly within each cylinder.

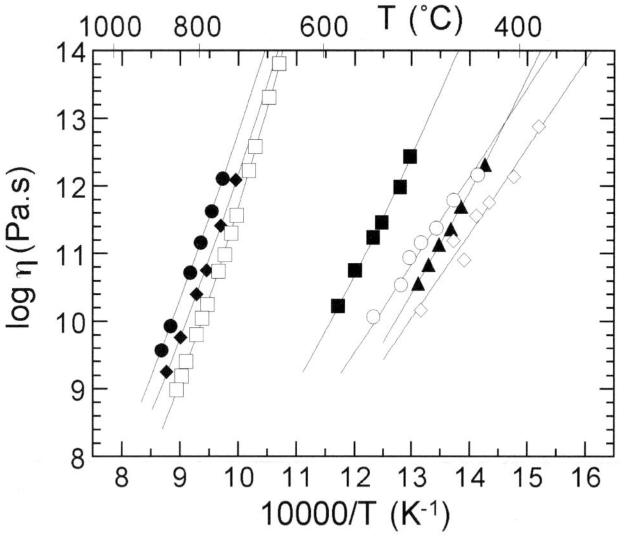

Figure 1 Low-temperature viscosity of hydrous $NaAlSi_3O_8$ and leucogranitic liquids as a function of inverse temperature. $NaAlSi_3O_8$: (filled circles) anhydrous cylinder A; (filled diamonds) anhydrous cylinder B; (filled squares) 1·07 wt.% H_2O; and (filled triangles) 1·87 wt.% H_2O. Leucogranite DK89: (open squares) anhydrous cylinders A and B; (open circles) 2·83 wt.% H_2O; and (open diamonds) 3·41 wt.% H_2O. Solid lines are TVF equations from Table 5.

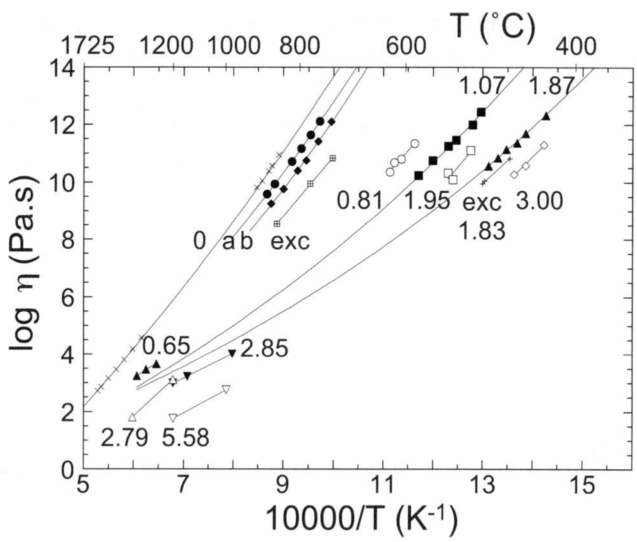

Figure 2 Comparison of low- and high-temperature data for dry and hydrous $NaAlSi_3O_8$ liquids. Numbers refer to water content in weight percentage: (x) anhydrous $NaAlSi_3O_8$ (labelled '0') from Toplis *et al.* (1997); (filled circles) anhydrous cylinder A; (filled diamonds) anhydrous cylinder B; (filled squares) 1·07 wt.% H_2O; and (black triangles) 1·87 wt.% H_2O; (all from this study; (crossed squares) anhydrous $NaAlSi_3O_8$ with excess sodium (labelled 'exc'); (open circles) 0·81 wt.% H_2O; (open squares) 1·95 wt.% H_2O; and (open diamonds) 3·00 wt.% H_2O; (all from Romano *et al.* (2001); (filled triangles) 0·65 and 2·85 wt.% H_2O (Holtz *et al.* 1999); (open triangles) 2·79; and 5·58 wt.% H_2O (Dingwell 1987). Solid lines are: TVF fits from Table 5 for samples from this study. TVF equation for anhydrous $NaAlSi_3O_8$ from Toplis *et al.* (1997). Straight-line (Arrhenian) fits for other samples with four data points or fewer.

The starting plate of $NaAlSi_3O_8$ glass is known to be slightly peralkaline overall (because of excess sodium), and the viscosity data imply that sample B is slightly more peralkaline than sample A. The slight compositional difference is also evident from the densities (Table 2). For nominal $NaAlSi_3O_8$ compositions, viscosity and density may, therefore, be a more sensitive indicator of compositional variation than analysis by electron microprobe. For comparison, the molar formula of glass AbI was measured as $Na_{1·01}Al_{0·97}Si_{3·01}O_{7·98}$ by electron microprobe (Behrens *et al.* 1996), that of the nominally albitic sample of Romano *et al.* (2001) is $Na_{1·01}Al_{0·99}Si_{3·00}O_{7·99}$, and that of peralkaline sample 'Naexc' of Romano *et al.* (2001) is $Na_{1·04}Al_{0·97}Si_{3·00}O_{7·97}$. As expected, the anhydrous viscosity curves for the present authors' two cylinders of glass AbI lie between those of the nominally albitic and peralkaline samples of Romano *et al.* (2001) (Fig. 2).

Although the viscosity difference between the two samples is about half a log unit, over the measured temperature range this is small compared to the effect of water. From anhydrous liquid to a water content of 1·07 wt.%, the glass transition temperature is reduced by 248 K, from 1032 to 784 K. Increasing the water content to 1·87 wt.% reduces Tg by a further 72 K, to 712 K. It is noticeable that the slopes of the η-1/*T* curves are less steep for the hydrous than for the anhydrous liquids (Fig. 1). This decrease in 'apparent activation energy' of viscous flow with increasing water content is a common feature of silicic liquids (e.g. Dingwell *et al.* 1996; Romano *et al.* 2001).

A satisfactory description of viscous flow can only be achieved by combining high and low temperature data (Richet 1984). This is especially true for dry or wet $NaAlSi_3O_8$ melts whose viscosity might incorrectly appear perfectly Arrhenian when investigated separately over a restricted temperature range, whether at low *T* (Toplis *et al.* 1997 and references therein) or high *T* (Baker 1996; Holtz *et al.* 1999). For hydrous melts, a difficulty might stem from the fact that high and low viscosities have generally been measured at different pressures. For example, while the present authors' measurements were performed at one bar, those of Holtz *et al.* (1999) were performed at between 3·1 and 3·6 kbars. However, the consistency of data obtained over a range of pressures (Fig. 2), and several other lines of evidence, demonstrate that viscosity changes caused by varying pressure are minor compared to those resulting from varying temperature and water content. First, the differential thermal analysis experiments by Rosenhauer *et al.* (1979) give a variation of the $NaAlSi_3O_8$ glass transition temperature of –2·5 K/kbar. Thus, the effect of 3–4 kbar pressure on lines of constant viscosity should be less than 10 K, which is minor in comparison to the compositional effects discussed above. Secondly, the viscosity of dry $NaAlSi_3O_8$ at 1400 °C varies only slightly with pressure, decreasing from $10^{4·25}$ to $10^{4·02}$ Pa.s between one bar and 4 kbar (Persikov *et al.* 1990), and from $10^{3·9}$ to $10^{3·3}$ Pa.s between 5 and 20 kbar (Kushiro 1978). Similarly, a measurement at 7·5 kbar and 1600 °C (Dingwell 1987) lies within error of available one-bar data (Toplis *et al.* 1997, and references therein). The small pressure dependence of the viscosity also holds for hydrous $NaAlSi_3O_8$ liquids. At 1200 °C, the viscosity of $NaAlSi_3O_8$ liquid with 2·8 wt.% H_2O is the same at 7·5 kbar (Dingwell 1987) and 3·1 kbar (Holtz *et al.* 1999).

Thus, the present authors conclude that the high-pressure data of Holtz *et al.* (1999) can be compared both with their new data and that of Romano *et al.* (2001), with an uncertainty of less than 0·2 log units because of the pressure difference of about 3 kbar. Whereas the present authors' two anhydrous $NaAlSi_3O_8$ samples slightly differ in composition, their viscosity values, as well as the interpolated viscosity for another $NaAlSi_3O_8$ liquid from Toplis *et al.* (1997), are all in the same narrow range. The effect of water is considerably greater than that of minor variations in composition or pressure. Although measurements at high and low temperatures on samples with identical water contents are not available, a moderate curvature of viscosity with temperature is required to reconcile the

Table 4 Calculated viscosities at high temperature.*

Sample	H_2O (wt.%)	$\log_{10}\eta$ (Pa s)	T (K)	Sample	H_2O (wt.%)	$\log_{10}\eta$ (Pa.s)	T (K)
Albite-1·3	1·07	4·00	1385	DK89–2	2·83	4·00	1297
Albite-1·3	1·07	3·50	1495	DK89–2	2·83	3·50	1367
Albite-1·3	1·07	3·00	1622	DK89–2	2·83	3·00	1443
Albite-2·2	1·87	4·00	1308	DK89–3	3·41	4·00	1269
Albite-2·2	1·87	3·50	1412	DK89–3	3·41	3·50	1340
Albite-2·2	1·87	3·00	1535	DK89–3	3·41	3·00	1418

*Viscosities were calculated using the models of Holtz *et al.* (1999) for $NaAlSi_3O_8$, and Scaillet *et al.* (1996) for the leucogranite.

Table 5 Parameters for TVF fits: $\log_{10}\eta = A + B/(T-C)$.

Sample	H_2O (wt.%)	A ($\log_{10}$ Pa. s)	B	C (K)	AAD* ($\log_{10}$ Pa. s)	T_{12}† (K)	*m*‡	Chi-square
Albite (T97)[a]	dry	− 6·39	14 900	257·5		1068	24·2	
Albite cyl. A	dry	− 4·38	1 1384	337·2	0·01	1032	24·3	0·001
Albite cyl. B	dry	− 3·66	9722	386·3	0·03	1007	25·4	0·007
Albite-1·3[b]	1·07	− 2·32	6973	297·0	0·05	784	23·1	0·03
Albite-2·2[b]	1·87	− 1·44	5713	286·8	0·03	712	22·5	0·01
LBG rhyolite[c]	dry	− 5·77	16 039	136·0		1039	20·4	
DK89–0	dry	− 3·05	7713	476·7	0·03	989	29·0	0·04
DK89–2[b]	2·83	− 5·88	12 704	9·3	0·09	720	18·1	0·05
DK89–3[b]	3·41	− 4·44	10 007	78·2	0·13	687	18·6	0·17

*AAD is the average absolute deviation between viscosities calculated using the TVF equation and the measured viscosities in Tables 3 and 6.
†T_{12} is the temperature at which the viscosity is 10^{12} Pa.s (corresponding approximately to the glass transition temperature).
‡*m* is the fragility parameter (see text for discussion).
[a]Albite (T97) is the TVF equation interpolated for perfectly stoichiometric $NaAlSi_3O_8$ liquid from Toplis *et al.* (1997).
[b]TVF equations for hydrous samples were fitted to a dataset including all experimental data listed in Tables 3 and 6, and also the calculated points at high temperature, listed in Table 4.
[c]TVF equation fitted to data for LBG rhyolite from Neuville *et al.* (1993).

two datasets. This non-linear effect can be represented with Tammann–Vogel–Fulcher (TVF) equations:

$$\log\eta = A + B/(T-C) \quad (1)$$

In order to constrain the degree of non-Arrhenian behaviour rather better over the investigated viscosity ranges, the Arrhenian equation of Holtz *et al.* (1999) was used to calculate viscosities at high temperatures for $NaAlSi_3O_8$ liquid containing 1·07 and 1·87 wt.% H_2O, within the viscosity range to which this equation applies (Table 4). A set of TVF equations was fitted, incorporating these calculated points (Table 5), and these confirm a substantial departure from Arrhenian behaviour for the viscosity of hydrous $NaAlSi_3O_8$ liquids (Fig. 2). The degree of non-Arrhenian behaviour may be quantified with the fragility parameter *m*, which may be calculated by:

$$m = \left. \frac{d\log_{10}\eta}{d\left(\dfrac{T_g}{T}\right)} \right|_{T=T_g} \quad (2)$$

(e.g. Moynihan 1995). 'Fragile' liquids have strongly non-Arrhenian viscosity-temperature relationships and have very large apparent activation energies at the glass transition (Angell 1985). Calculated fragility parameters are listed in Table 5, and range from 24·3 or 25·4 for anhydrous cylinders A and B, to 22·5 for $NaAlSi_3O_8$ liquid with 1·8 wt.% H_2O. These are comparable with values given by Romano *et al.* (2001) for $NaAlSi_3O_8$ liquids with water contents from zero (m=21·1) to about 3 wt.%. (m=24·3 and 25·6). The fragility parameter for $NaAlSi_3O_8$ liquid changes little on hydration, with an apparent slight decrease for our data, although the trends in Figure 2 illustrate increasingly non-Arrhenian behaviour with increasing water content. However, because *m* is calculated for T_g, which is at one end of the investigated viscosity range, it can be relatively insensitive to changes in the shape of the viscosity curve at high temperatures. This point is discussed further below, in reference to the leucogranites.

2.2. Leucogranite melts

Low-temperature viscosity measurements are reported in Table 6 and plotted as a function of inverse temperature in Figure 1. Viscosity measurements of hydrous leucogranites are consistent both on heating and cooling, and yield slightly shallower slopes than the hydrous $NaAlSi_3O_8$ melts. Anhydrous DK89 displays a slight but noticeable concave-up curvature, so that a better fit to the data is again achieved with a (non-Arrhenian) TVF equation.

The new data are compared in Figure 3 with high-temperature measurements from Scaillet *et al.* (1996). No high-temperature data are available for an anhydrous leucogranitic melt, but observations for a natural obsidian of similar composition indicate slight deviations from Arrhenian behaviour when considered over a wide temperature range (Neuville

Table 6 Low-temperature viscosity data for DK89.

Anhydrous		2·83 wt.% H_2O		3·41 wt.% H_2O	
T (K)	log η (Pa s)	T (K)	log η (Pa s)	T (K)	log η (Pa s)
934·5	13·81 (0·04)[b]	707·6	12·17 (0·06)[1]	658·1	12·88 (0·07)[3]
950·1	13·31 (0·04)[b]	728·4	11·79 (0·06)[2]	677·2	12·14 (0·05)[1]
971·4	12·58 (0·04)[b]	744·4	11·38 (0·12)[4]	697·6	11·76 (0·04)[2]
982·5	12·23 (0·06)[a]	759·6	11·16 (0·07)[3]	708·8	11·56 (0·22)[5]
1003·6	11·57 (0·02)[a]	771·6	10·94 (0·10)[7]	728·4	11·19 (0·18)[4]
1012·8	11·30 (0·03)[b]	780·4	10·54 (0·09)[5]	719·1	10·91 (0·07)[7]
1023·9	10·98 (0·02)[a]	811·5	10·07 (0·08)[6]	759·6	10·17 (0·05)[6]
1035·4	10·74 (0·03)[a]				
1056·6	10·25 (0·02)[a]				
1066·7	10·05 (0·03)[b]				
1077·3	9·81 (0·03)[a]				
1098·6	9·40 (0·02)[a]				
1108·0	9·19 (0·01)[b]				
1119·1	8·99 (0·01)[b]				

[a, b]For the anhydrous samples, superscripts indicate measurements on two different cylinders, DK89–0A and DK89–0B. For the hydrous samples, superscripts indicate the order in which measurements were made. Errors in parentheses are standard deviations of measurements made at three different stresses.

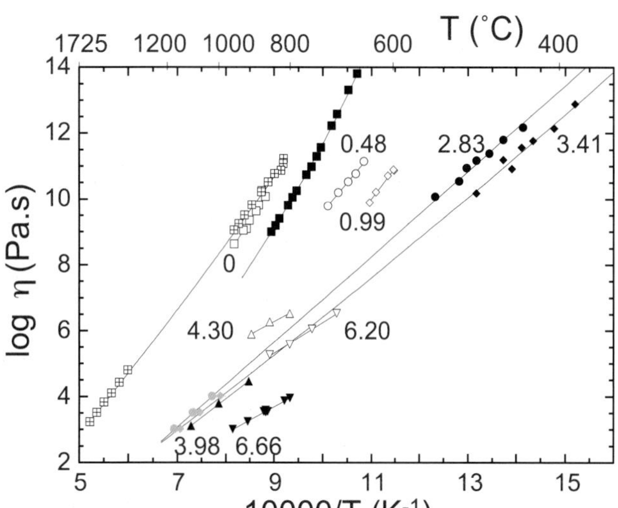

Figure 3 Comparison of low- and high-temperature viscosity data for silicic compositions: (crossed squares) anhydrous LGB rhyolite from Neuville *et al.* (1993); (filled squares) anhydrous DK89; (filled circles) 2·83 wt.% H_2O; (filled diamonds) 3·41 wt.% H_2O; all from this study; (shaded circles and diamonds) high-temperature viscosities for the same water contents; (calculated using the equation of Scaillet *et al.* (1996); (open squares) 0·04 wt.% H_2O; (open circles) 0·48 wt.% H_2O; and (open diamonds) 0·99 wt.% H_2O; all for obsidian BL6 from Stevenson *et al.* (1998); (filled triangles) Gangotri leucogranite GB4 with 3·98 and 6·66 wt.% H_2O (Scaillet *et al.* 1996); (open triangles) obsidian with 4·30 and 6·20 wt.% H_2O (Shaw 1963). Solid lines are: TVF fits from Table 5 for samples from this study; TVF equation for LGB rhyolite from Neuville *et al.* (1993); and straight-line (Arrhenian) fits for all other samples.

et al. 1993) (Fig. 3). Whilst the anhydrous liquid displays a slight curvature even over the restricted viscosity range of the low-temperature technique, the hydrous liquids behave in a near-Arrhenian fashion (compare the data for 3·41 and 3·98 wt.% H_2O samples). Again, the effect of pressure must be taken into account, but repeat measurements of a leucogranite containing 6·66 wt.% H_2O at 860 °C and pressures from 500 to 800 MPa yielded identical values within experimental uncertainty (Scaillet *et al.* 1996). In addition, increasing pressure is likely to decrease the viscosity for polymerised melts (see above). Barring pressure effects, viscosity is apparently near-Arrhenian for the hydrous leucogranites. If pressure effects were significant, high-temperature measurements at lower pressure would plot at higher viscosities than shown on Figure 3, and would require a convex-up viscosity curve. No silicate liquid composition is known to have such a curve, and it would also result in unrealistically low viscosities at lower temperatures. Thus, the present authors conclude that the pressure effect on viscosity over the range one bar to 8 kbar is negligible, and they may directly compare their new data with those of Scaillet *et al.* (1996).

Hence, viscosity–temperature relationships must be nearly Arrhenian over a wide viscosity range for hydrous leucogranitic liquids. The TVF equations calculated for the hydrous leucogranites were fitted to a dataset consisting of the low-temperature data and three high-temperature points calculated using the equation of Scaillet *et al.* (1996), which are listed in Table 4. The TVF curves demonstrate near-Arrhenian behaviour for the viscosity of hydrous leucogranite liquids (Fig. 3; Table 5). Since the anhydrous leucogranite liquid is slightly non-Arrhenian (m=29·0), adding water to natural leucogranite liquids actually decreases the 'fragility' of the liquid (m=18·1 and 18·6 for 2·83 and 3·41 wt.% H_2O, respectively). Furthermore, the calculated fragilities for the hydrous leucogranites are lower than for hydrous $NaAlSi_3O_8$, or for anhydrous rhyolite LGB (Table 5). This observation contrasts with the increasingly non-Arrhenian behaviour (and increasing fragility) induced by adding water to $NaAlSi_3O_8$ liquid noted above.

The Himalayan leucogranite investigated here is the first silicate liquid ever to display decreasing fragility with increasing water content. This result is somewhat surprising, especially since hydrous synthetic haplogranitic and andesitic melts, like $NaAlSi_3O_8$, become less Arrhenian as water content increases (Dingwell *et al.* 1996; Richet *et al.* 1996). However, the leucogranite is the first natural liquid composition to be investigated over a range of T-X(H_2O) conditions sufficient to test for variably (non-)Arrhenian behaviour. To further illustrate the magnitude of the change in behaviour of these liquids as a function of water content, the viscosity-temperature relations for $NaAlSi_3O_8$ and leucogranite liquids are plotted again in Figure 4 over a wider temperature range that includes the extrapolated viscosity intercept at theoretical infinite temperature.

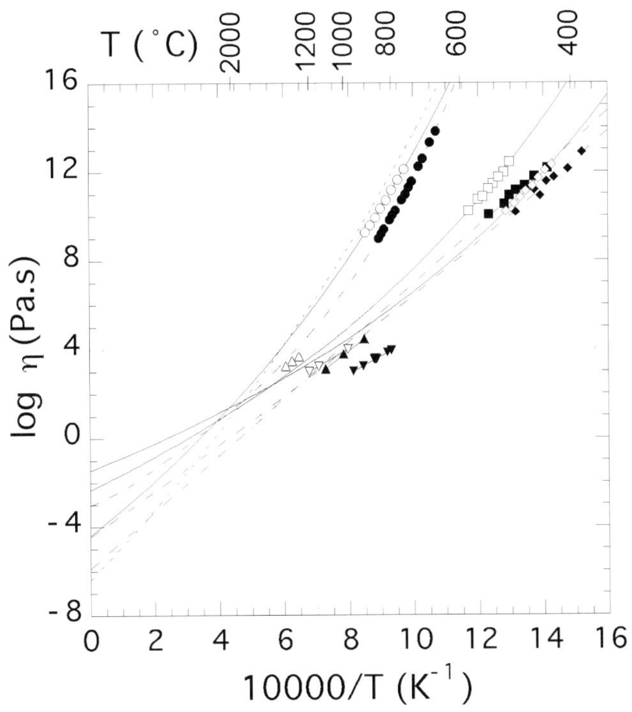

Figure 4 Viscosity-temperature relations for $NaAlSi_3O_8$ and granitic liquids: (dotted line) TVF equation for stoichiometric $NaAlSi_3O_8$, from Toplis *et al.* (1997); (open circles) anhydrous cylinder A; (open squares) 1·07 wt.% H_2O; (open diamonds) 1·87 wt.% H_2O; from this study; (open upright triangles) 0·65 wt.% H_2O; (open inverted triangles) 2·65 wt.% H_2O; from Holtz *et al.* (1999); (solid lines) TVF fits for $NaAlSi_3O_8$ (Table 5); (filled circles) anhydrous DK89; (filled squares) 2·83 wt.% H_2O; (filled diamonds) 1·87 wt.% H_2O; from this study; (filled upright triangles) 3·98 wt.% H_2O; (filled inverted triangles) 6·66 wt.% H_2O; (from Scaillet *et al.* (1996); (dashed lines) TVF fits for DK89 (Table 5).

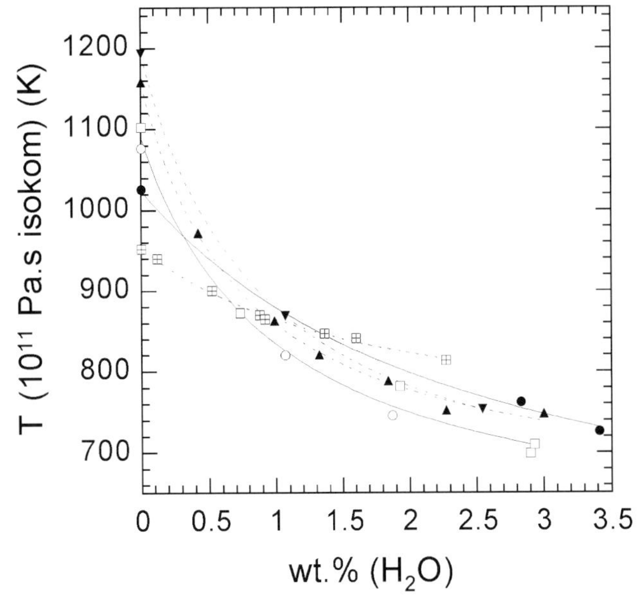

Figure 5 Temperatures of 10^{11} Pa.s isokoms for silicic liquids as a function of water content: (open circles) $NaAlSi_3O_8$ (cylinder A, this study); (open squares) $NaAlSi_3O_8$ (Romano *et al.* 2001); (filled circles) leucogranite DK89 (this study); (filled upright triangles) haplogranite HPG8 (Dingwell *et al.* 1996); (filled inverted triangles) peraluminous haplogranite (Dingwell *et al.* 1998a); (crossed squares) synthetic tephrite (Whittington *et al.* 2000).

Note that uncertainties for both high-temperature measurements by the falling sphere method, and low-temperature measurements by the parallel-plate method, are similar to or smaller than the symbol size on Figure 4. The parallel-plate viscometer reproduces the viscosity of standard glass DGG-1 to within 0·05 log units (Whittington *et al.* 2001). For falling-sphere measurements, the longer falling distance made possible by using internally heated pressure vessels rather than a piston-cylinder apparatus allows viscosity determinations to within ± 15% relative (± 0·06 log units), or better if the temperature is known to within 5 °C (see discussion by Holtz *et al.* 1999). Uncertainties in micropenetration viscometry measurements are also reported to be ± 0·06 log units (Dingwell *et al.* 1996).

When viscosity curves are plotted over a large enough temperature range, it becomes apparent that for granitic liquids the changes in the calculated fragility are mostly caused by a progressive rotation of the viscosity curves with increasing water content, and not by a dramatic change from linear to highly non-linear curves (Fig. 4). For comparison, the calculated fragility parameter for an anhydrous synthetic tephrite is 50·7. At these degrees of non-Arrhenian behaviour, the viscosity curve has a pronounced curvature even over a relatively small range of viscosity or temperature. In the next section, an explanation is sought for the contrasting behaviour of the leucogranite and $NaAlSi_3O_8$ compositions.

3. Comparison between natural and analogue silicic melt compositions

Both $NaAlSi_3O_8$ and leucogranitic melts undergo an initially rapid decrease in viscosity when small amounts of water are added, and a subsequently slower decrease with further added water, in common with all previously investigated silicic compositions (e.g. Dingwell *et al.* 1996; Romano *et al.* 2001). The effect of water on such highly polymerised liquids is greater than for partially depolymerised melts (e.g. Persikov *et al.* 1990; Giordano *et al.* 2000; Whittington *et al.* 2000). For example, the decrease in temperature of the 10^{11} Pa.s isokom (line of constant viscosity) for addition of about 2 wt.% H_2O is 340 K for $NaAlSi_3O_8$, but only 130 K for a synthetic tephrite, a relatively depolymerised liquid (Fig. 5). One measure of depolymerisation is the ratio of non-bridging oxygens (NBO) to tetrahedrally coordinated cations (T) within the liquid structure (Mysen 1988). For comparison, a theoretically fully polymerised liquid such as $NaAlSi_3O_8$ has a calculated NBO/T of 0, an $MgSiO_3$ liquid has an NBO/T of 2, and the NBO/T of the synthetic tephrite plotted in Figure 5 is 0·86.

Albitic ($NaAlSi_3O_8$) materials are commonly used as analogues for natural granitic or rhyolitic magma compositions. However, there are several important differences between their physical and chemical properties. For example, over the investigated range of T-X(H_2O) conditions, the viscosity of hydrous $NaAlSi_3O_8$ liquids is distinctly non-Arrhenian (Fig. 2), whereas hydrous leucogranites behave in a quasi-Arrhenian manner (Fig. 3). Viscosities for the two liquids at magmatic temperatures, calculated using the equations of Holtz *et al.* (1999) and Scaillet *et al.* (1996), are plotted against water content in Figure 6. The effect of small quantities of water is more important for the $NaAlSi_3O_8$ melt than for the leucogranite, although their viscosities appear to be similar at higher water contents.

The differences between the two compositions can also be investigated without extrapolation of the dataset by examining the decrease in glass transition temperature (T_g) as a function of water content (Fig. 7). There are two broad trends, with strong reductions in T_g upon hydration for $NaAlSi_3O_8$ and a synthetic haplogranite (HPG8 of Dingwell *et al.* 1996), but distinctly less strong reductions for the leucogranite, a volcanic

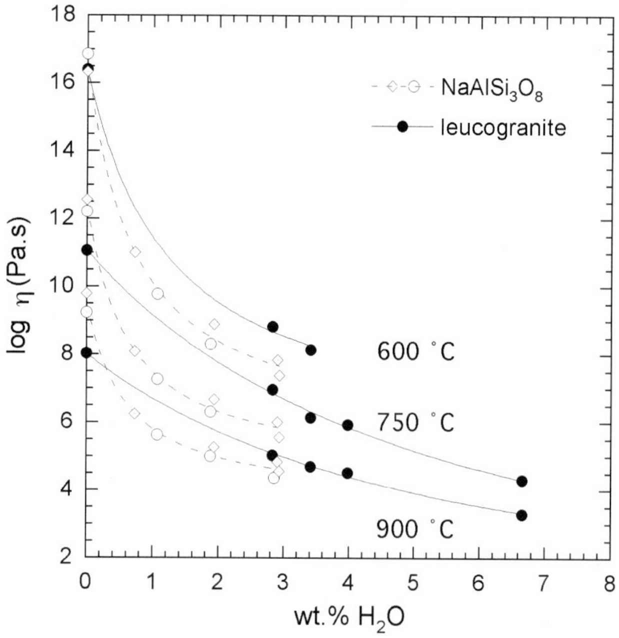

Figure 6 Effect of water on the viscosity of $NaAlSi_3O_8$ liquids and leucogranites at 900, 750 and 600°C; (filled circles) leucogranite viscosity [data for 0, 2·83 and 3·41 wt.% H_2O are calculated from the TVF equations of Table 5; data for 3·98 and 6·66 wt.% H_2O are calculated from the Arrhenian equation of Scaillet *et al.* (1996) at magmatic *T* only]; (open circles) $NaAlSi_3O_8$ viscosity [data for 0, 1·07 and 1·87 wt.% H_2O are calculated from the TVF equations of Table 5; datum for 2·85 wt.% H_2O is calculated from the Arrhenian equation of Holtz *et al.* (1999) for 900°C only]; (open diamonds) $NaAlSi_3O_8$ viscosity [data for 0, 0·73, 1·93, 2·90 and 2·93 wt.% H_2O are calculated from the TVF equations of Romano *et al.* (2001; table 4)]; Dashed lines are fitted to all the $NaAlSi_3O_8$ data, and solid lines are fitted to all the granite data. Curves are of the form $\log \eta = A + B(C + \text{wt.\%}\ H_2O)^{-6/5}$, where A, B, and C are constants, and are intended only to guide the eye.

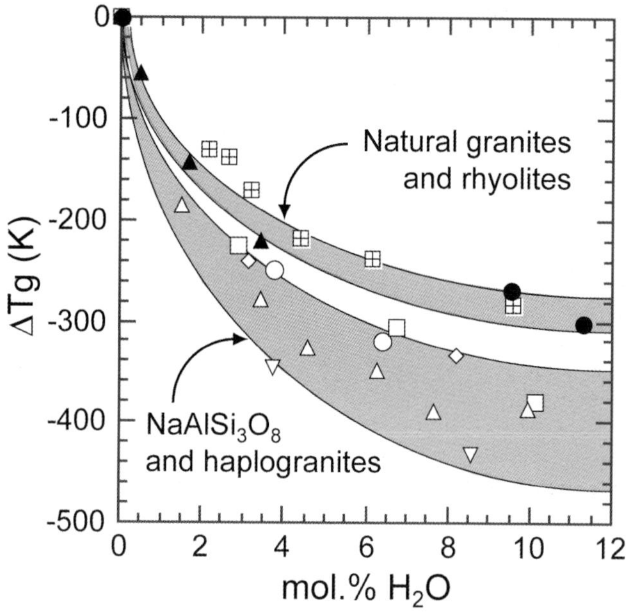

Figure 7 Change in temperature of the glass transition temperature (taken to be the 10^{12} Pa.s isokom) as a function of mol.% added water for various silicic liquids. $NaAlSi_3O_8$: (open circles) this study; and (open squares) Romano *et al.* (2001). Synthetic haplogranites: (open upright triangles) metaluminous HPG8 (Dingwell *et al.* 1996); (open inverted triangles) peraluminous HPG8Al02 (Dingwell *et al.* 1998a); (open diamonds) anorthite-bearing metaluminous HPG8An10 (Dingwell *et al.* 2000); (crossed squares) peralkaline HPG8N5 with excess sodium (Dingwell *et al.* 1998b). Natural granites and rhyolites: (filled circles) leucogranite DK89 (this study); (filled triangles) obsidian BL6 (Stevenson *et al.* 1998). Curves are sketched in to roughly define trends for model and real systems; note that HPG8N5 is the only peralkaline composition in this plot.

rhyolite (BL6 of Stevenson *et al.* 1998) and a synthetic peralkaline haplogranite (HPG8N5 of Dingwell *et al.* 1998b).

From a compositional point of view, there are four potentially important differences in major element chemistry between $NaAlSi_3O_8$ liquid and the Himalayan leucogranite. First, the ratio of aluminium to silicon is larger in $NaAlSi_3O_8$ (1:3) than in the leucogranite (~1:4 for DK89). However, the viscosity trends for albite and synthetic haplogranite HPG8 are similar, even though HPG8 has a smaller Al:Si ratio of 1:5, so the ratio of aluminium to silicon is not the controlling factor.

Secondly, the leucogranite is strongly peraluminous, whereas the $NaAlSi_3O_8$ liquid is slightly peralkaline. However, Dingwell *et al.* (1998a) demonstrated that the effect of water on the viscosity of peraluminous haplogranites was similar to or slightly greater than the effect of water on their metaluminous haplogranite composition (HPG8). In contrast, the effect of water is noticeably less for the peraluminous granite than for $NaAlSi_3O_8$ or HPG8. The synthetic peraluminous haplogranite HPG8AlO2 of Dingwell *et al.* (1998a) has almost the same molar fraction of excess alumina as the present authors' leucogranite DK89, yet it is more strongly affected by water than DK89 (Figs 5, 7). Baker & Vaillancourt (1995) measured identical viscosities for metaluminous and peraluminous granites containing 6 wt.% H_2O and 1·5 wt.% F between 800 and 1000°C, while a peralkaline granite was substantially less viscous. In a study of hydrous volcanic compositions, Stevenson *et al.* (1998) noted a similar effect of water on both calc-alkaline and peraluminous samples, although their peraluminous samples had lower viscosities for a given temperature. Hence, peraluminosity is unlikely to explain the different effects of water on $NaAlSi_3O_8$ and the leucogranite.

Thirdly, the speciation of water in these liquids may have an influence on viscosity. It is well known from infrared and nuclear magnetic resonance (NMR) spectroscopic studies that water is preferentially dissolved as hydroxyl groups (OH) at low water contents, but molecular H_2O becomes dominant at higher water contents (e.g. Stolper 1982; Eckert *et al.* 1988). *In situ* infrared measurements on hydrous haplogranite (Nowak & Behrens 1995) and sodium aluminosilicate (Shen & Keppler 1995) liquids at high *P* and *T* demonstrate that the proportion of OH/H_2O in the liquid increases with temperature. Quantification of OH and H_2O concentration in the liquids is still under discussion (e.g. Withers *et al.* 1999; Withers & Behrens 1999). However, it has been widely accepted that the water speciation in the glass phase is the same as in the liquid phase at the glass transition temperature (Dingwell & Webb 1990). Thus, the measured water speciation in quenched glass might be directly related to the 10^{12} Pa.s isokom. Withers & Behrens (1999) found almost the same OH/H_2O ratios in $NaAlSi_3O_8$ glasses (same composition as used in this study) and in rhyolitic glasses at water contents up to 6 wt.%, provided that the cooling rate of the quenched glasses was identical. This implies that differences in water speciation cannot explain the two distinct trends observed in Figure 6.

Finally, haplogranites contain only Na and K in addition to Al and Si, whereas other elements such as Ca, Mg and Fe are present in significant quantity in DK89. We can interpret these differences between natural and simple synthetic compositions in terms of configurational entropy theory (Richet 1984; Richet & Bottinga 1995). From the Adam & Gibbs (1965) theory of relaxation processes in viscous liquids, one obtains a viscosity equation of the form:

$$\log\eta = A_e + B_e / TS^{conf} \quad (3)$$

where S^{conf} is the configurational entropy, A_e is a pre-exponential constant, and B_e is the Gibbs free energy barrier opposing the cooperative rearrangements required for viscous flow. This equation has a significant advantage over the TVF equation (1), since it is based in thermodynamic theory and, in principle, has only two unknowns instead of three. In practice, the full range of thermodynamic data needed to determine S^{conf}, and to apply configurational entropy theory to hydrous leucogranites in a quantitative manner is still lacking. However, qualitative interpretations are possible, if water is treated as any other 'network modifier' oxide. In binary systems, a strong viscosity decrease is observed at low temperatures, interpreted as the result of an excess entropy of mixing. The configurational entropy is much larger at higher temperatures than at lower ones, and hence, the viscosity minima are much shallower (Neuville & Richet 1991).

The primary effect of dissolved water on viscosity results from the addition of an excess network modifier oxide to the silicate network. This greatly increases the topological contribution to the configurational entropy and is more important than the entropy of mixing. Hence, the addition of the hydrous component has a stronger effect at low water contents and low temperatures for both $NaAlSi_3O_8$ and granitic melts because entropy of mixing curves are concave-up on T-X diagrams such as Figures 5 and 7. The general pattern of an initially rapid decrease in melt viscosity with the addition of small amounts of water, and the subsequently smaller decreases with further added water, has been noted for all silicate liquid compositions which have been investigated to date. Even in the most depolymerised hydrous silicate liquid studied to date, with an NBO/T of 1·51 (synthetic foidite 'NIQ' of Whittington *et al.* 2000), the viscosity decrease accompanying increasing water content approaches linearity, but is still slightly concave-up. The effect of water on melt viscosity becomes smaller with increasing depolymerisation of the anhydrous composition (e.g. Persikov 1990; Whittington *et al.* 2000) because the additional topological contribution to the configurational entropy of an already depolymerised melt is less important than for a nominally tectosilicate composition.

Although it is convenient to look at hydrous melts as pseudo-binary mixtures of a silicate liquid and water, the leucogranite contains significant amounts of different alkalis (Na and K) as well as alkaline earths (Ca and Mg) and other elements (e.g. Fe and Ti). Under anhydrous conditions, the entropy of mixing between these elements results in lower viscosities for the leucogranite than for $NaAlSi_3O_8$, where substitutions of the form (Na<->K) and ($2Na^+$<->Ca^{2+}) cannot occur (Fig. 1). Addition of water to the system then results in a significant contribution to entropy of mixing for $NaAlSi_3O_8$ liquid, which displays a very strong viscosity decrease. In contrast, the leucogranite viscosity is less affected, since the additional entropy of mixing is relatively less important. This explanation is qualitatively consistent with the smaller viscosity reduction observed for the addition of water to haplogranite liquids with increasing anorthite content (Dingwell *et al.* 2000).

The contrasting effect of water on melt fragility (degree of non-Arrhenian behaviour) for $NaAlSi_3O_8$ and leucogranite melts suggests that the effect of water on the configurational heat capacity of melts is not linear with composition. This would be consistent with the marked nonlinear trends already described for silica-rich sodium aluminosilicates (Richet & Bottinga 1984).

In this discussion, the authors have disregarded the effect of trace elements since they comprise a very small proportion (<1 weight percentage in total) of DK89 (and of GB4). Their presence should increase the entropy of mixing in the anhydrous liquid, and so further reduce the effect of added water. In the case of many granitic liquids, particularly leucogranites, the most abundant trace elements will include alkalis such as Li and Rb, and alkaline earths such as Ba and Sr, which are implicitly included in the preceding discussion.

4. Models for the viscosity of hydrous silicic melts at magmatic temperatures

4.1. Review of literature models for molten granite viscosity

Combined with previous measurements at low viscosity, these new high-viscosity data provide strong constraints on η-T-$X(H_2O)$ relations for $NaAlSi_3O_8$ and leucogranitic liquids (Figs 2, 3). The leucogranite measurements should be applicable at least to all Himalayan leucogranites because variations in chemistry are slight between the different bodies (Le Fort *et al.* 1987). In this section, the authors compare viscosities calculated from literature models with their new data, and with the high-temperature data of Scaillet *et al.* (1996). Several different models exist, most being empirical fits to limited datasets, and therefore, with restricted predictive power, particularly in regard to compositional variation. The purpose is to test the model calculations on a natural leucogranite with well-constrained viscosity under magmatic conditions. To do this, the authors calculated viscosities using the different models for each of the T-$X(H_2O)$ conditions for which they measured viscosity in this study (i.e. for every measurement condition listed in Table 3), and also for every T-$X(H_2O)$ condition for which viscosity was measured by Scaillet *et al.* (1996). Table 6 lists the average absolute deviations (AADs) of predicted viscosities from experimental observations grouped into four different data sets: all measurements, high-temperature measurements only, low-temperature measurements only, and measurements under hydrous conditions only.

The first predictive model of hydrous melt viscosity to be developed was the empirical Arrhenian model of Shaw (1972), which gives five partial molar Arrhenian coefficients for different oxides, derived from an extensive dataset collected at high temperatures. This model developed that of Bottinga & Weill (1972) by including a coefficient for water derived from the measurements of Shaw (1963) on obsidian melts with 4·3 and 6·2 wt.% H_2O in the range 1–2 kbar and 700–900°C (Fig. 3). Shaw (1972) suggested that the model should be treated with caution at viscosities higher than 10^5 Pa.s or at water contents higher than about 6 wt.%. This model does not take into account the non-linear decrease of viscosity with water content (Fig. 6), which could not be recognised by Shaw (1972) because of the lack of experimental data available at that time.

The empirical model of Hess & Dingwell (1996) was fitted to low-temperature viscosity data for hydrous haplogranites (Dingwell *et al.* 1996) and extensive high-temperature work on $NaAlSi_3O_8$ and granite melts. The model covers a wide temperature range, but does not incorporate parameters for compositional variation and cannot be applied to anhydrous liquids. This model has been shown to successfully predict the viscosity of hydrous peraluminous haplogranites (Dingwell *et al.* 1998a), but not of peralkaline haplogranites (Dingwell *et al.* 1998b).

The models of Baker (1996, 1998) are also not applicable to anhydrous liquids. Baker (1996) measured the high-temperature viscosity of hydrous granite and $NaAlSi_3O_8$,

Table 7 Average absolute deviation (in $\log_{10}$ units) between measured viscosity of leucogranite DK89 and literature models for granite viscosity, arranged by experimental conditions

Model	Reference					
	Shaw (1972)	Baker (emp) (1998)	Baker (S^{conf}) (1996)	Hess & Dingwell (1996)	Scaillet *et al.* (1996)	Zhang *et al.* (2003)
Low *T* anhyd[a]	0·50	–	–	–	–	1·87
Low *T* hyd[b]	0·67	–	–	0·27	0·13	0·45
High *T* hyd[c]	0·08	0·51	3·12	0·26	0·02	0·21
All *T* hydrous[d]	0·41	–	–	0·27	0·08	0·34
Emplacement conditions[e]	4·54	4·96	5·09	4·75	4·40	4·60

[a]'Low *T* anhyd' refers to anhydrous measurements from this study (Table 6, column 1). Only the models of Shaw (1972) and Zhang *et al.* (2003) can be applied to anhydrous liquids.
[b]'Low *T* hyd' refers to low-temperature measurements on hydrous liquids from this study (Table 6, columns 2 and 3). The models of Baker (1996, 1998) should not be used at temperatures less than 700 and 577°, respectively.
[c]'High *T* hyd' refers to the high-temperature measurements of Scaillet *et al.* (1996), between 813 and 1100°C (n=11).
[d]'All *T* hydrous' refers to the AAD for all measurements on hydrous liquids at both high and low *T*.
[e]Emplacement conditions=calculated $\log_{10}$ viscosity (in Pa.s) for DK89 containing 6·5 wt.% H_2O at 750°C.

combined these data with those of Dingwell (1987) for hydrous $NaAlSi_3O_8$, and derived empirical equations for viscosity at high and low water contents for temperatures in excess of 700°C. These equations were updated by Baker (1998), who derived empirical equations for water contents lower and higher than $X(H_2O)$=0·25 (about 2·2 wt.%). Baker (1996) also derived an equation based on configurational entropy theory, treating the melt as a ternary system involving silicate, H_2O and OH^- end-members. These equations do not allow for compositional variation of the silicate component and they explicitly incorporate 'the accepted assumption that $NaAlSi_3O_8$ provides an analogue to granitic melts' (Baker 1996).

Most recently, Zhang *et al.* (2003) proposed an equation based on fluidity, the inverse of viscosity, as a basis for modelling the viscosity of hydrous liquids:

$$\frac{1}{\eta} = \frac{1}{\eta_1} + \frac{1}{\eta_2} x^n \qquad (4)$$

where η is viscosity and $1/\eta$ is fluidity, η_1 is the viscosity of the dry melt, x is the mole fraction of total dissolved H_2O on a 1-oxygen basis, n and η_2 are two fitting parameters, and η_2 can be interpreted to be the viscosity of the hypothetical melt consisting of pure H_2O (Zhang *et al.* 2003). The parameters n and η_2 were derived from inferred viscosity data for Mono Crater rhyolites, based on the kinetics of the water speciation reaction, for experimental conditions between 570 and 1920 K, and water contents between 6 p.p.m. and 8·2 wt.%. Zhang *et al.* (2003) then tested their model against various subsets of literature data, and reported a 2σ uncertainty of 0·36 log units in its predictions, when tested against literature data with analytical errors smaller than one log unit.

Based on high-temperature measurements of Himalayan leucogranite viscosity, Scaillet *et al.* (1996) derived an Arrhenian equation applicable only to hydrous peraluminous leucogranites. The present authors include this model here to test how reliably it may be extrapolated from the high-temperature measurement conditions to the low temperature measurements presented here.

4.2. Comparison with leucogranite data

Only the models of Shaw (1972) and Zhang *et al.* (2003) can be applied to anhydrous liquids. For the Shaw (1972) model, the average absolute deviation (AAD) between predicted anhydrous viscosities and our measured values is 0·5 log units over the investigated temperature range (935–1119 K). The Shaw (1972) model also agrees very well with the high-temperature data for hydrous leucogranites of Scaillet *et al.* (1996), but not with the new low-temperature data.

If the sample of DK89 is considered to be truly anhydrous, the Zhang *et al.* (2003) model overestimates the measured viscosity by an average of 1·9 log units. If the actual water content were as high as 0·1 wt.%, the overestimation would diminish to 0·9 log units. The likely water content of the present nominally anhydrous sample is somewhere between these two values. The Mono Crater Rhyolite, for which they derived their model parameters, is very silicic (76·6 wt.% SiO_2 compared to 73·6 wt.% for DK89) and less aluminous (12·7 wt.% Al_2O_3 compared to 15·5 wt.% for DK89), so the Zhang *et al.* (2003) model may be best suited to high-silica metaluminous rhyolites.

Turning to hydrous granites, both high- and low-temperature viscosities are well reproduced by the equations of Zhang *et al.* (2003) and Hess & Dingwell (1996), probably because they are the only ones to be based partly on low-temperature measurements. High-temperature viscosities calculated with the empirical model of Baker (1998) are within about 0·5 log units of the measurements of Scaillet *et al.* (1996), while the configurational entropy model of Baker (1996) does not reproduce the data well, probably because it was based mostly on data for $NaAlSi_3O_8$ liquid. Neither of these models was intended for use below 700°C, and application at temperatures below their intended range gives poor results. Finally, because of the quasi-Arrhenian viscosity of hydrous leucogranite melts demonstrated above, the equation of Scaillet *et al.* (1996) provides the best prediction of leucogranite viscosity at both high and low temperatures, with an average absolute deviation comparable to the TVF fits to the data.

4.3. Viscosity under magmatic conditions

The data of Table 7 emphasise that current models cannot predict viscosity over wide ranges of temperature, water content and composition simultaneously. The present authors have also included in Table 7 calculated viscosities for probable Himalayan leucogranite emplacement conditions at 750°C and containing 6·5 wt.% H_2O (Scaillet *et al.* 1995). The actual viscosity at these conditions is taken to be $10^{4·4}$ Pa.s, calculated with the equation of Scaillet *et al.* (1996), which

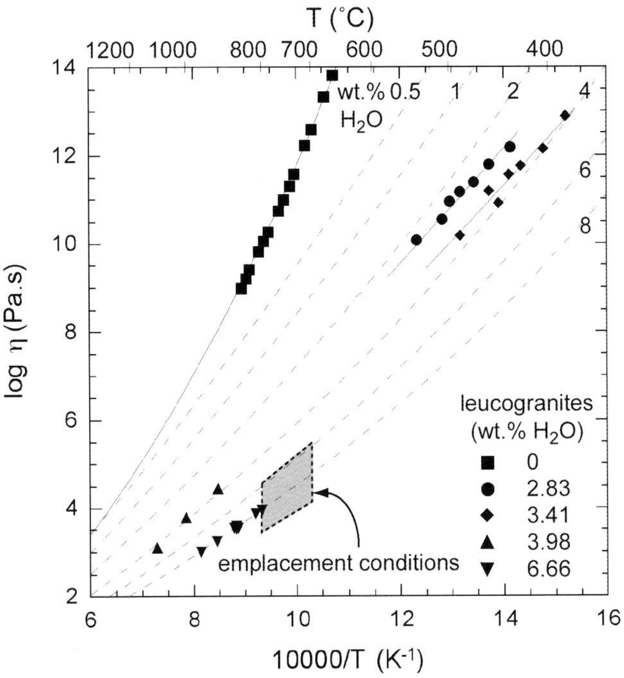

Figure 8 Viscosity-T-X(H_2O) grid for Himalayan leucogranites. Symbols for experimental data are the same as in Figure 3: (solid line) TVF equation for anhydrous DK89 (Table 5; dashed lines) predicted viscosity–temperature relationships using the equation of Hess & Dingwell (1996) for granitic liquids containing 0·5, 1, 2, 4, 6 and 8 wt.% H_2O. The shaded box outlines inferred leucogranite emplacement conditions of 750°C and 6·5 wt.% H_2O (Scaillet *et al.* 1995); (with margins of ± 50°C and ± 1 wt.% H_2O.

requires an extrapolation from their high-temperature measurements of only 50°C (Fig. 8). Despite the shortcomings of individual models when applied over a wide range of temperatures and water contents, it is reassuring that all the different models are within 0·7 log units (a factor of five) of the actual value.

The models of Shaw (1972), Hess & Dingwell (1996) and Zhang *et al.* (2003) all agree well with the high-temperature data of Scaillet *et al.* (1996), and have different advantages. The model of Shaw (1972) can be used to calculate viscosities for intermediate or basic melt compositions, although it is not suited for low water contents. Hess & Dingwell (1996) and Zhang *et al.* (2003) incorporated non-Arrhenian behaviour and the non-linearity of viscosity with water content, allowing good predictions at lower temperatures and water contents. Although neither model allows for compositional changes of the silicate phase, this is unlikely to be a problem for granites derived by low-temperature crustal anatexis, since their near-eutectic compositions will change relatively little over most of the crystallisation range (e.g. Scaillet *et al.* 1995). Finally, for the range of measurement conditions, the equation of Scaillet *et al.* (1996) reproduces all the experimental data very well and covers the likely range of magmatic conditions with little extrapolation from their high-temperature data (Fig. 8). However, this equation is composition-specific, and should not be extrapolated to both low temperatures and very high water contents simultaneously.

5. Conclusions

New measurements of the viscosity of hydrous $NaAlSi_3O_8$ liquid and a hydrous leucogranite illustrate the large viscosity changes of these liquids as a function of temperature and water content. The two compositions behave slightly differently, with a more drastic viscosity decrease for $NaAlSi_3O_8$ at low water contents. The results for both compositions are consistent with previous low-temperature measurements on both $NaAlSi_3O_8$ and natural rhyolitic liquids.

Combination of the new data with previously published data for $NaAlSi_3O_8$ (Holtz *et al.* 1999) and Himalayan leucogranites (Scaillet *et al.* 1996) allows interpolation rather than extrapolation to magmatic temperatures. The viscosity of $NaAlSi_3O_8$ liquid becomes more fragile (more non-Arrhenian) with the addition of water, while the leucogranite becomes less fragile (more Arrhenian), at least for water contents of up to about 4 wt.% H_2O. Therefore, the viscosity of hydrous $NaAlSi_3O_8$ melts cannot be used as a model for that of granitic liquids.

Such different behaviour can be qualitatively interpreted in terms of configurational entropy theory. The larger effect of water on $NaAlSi_3O_8$ liquid viscosity may be caused by a larger contribution of the entropy of mixing, resulting in a strong reduction of viscosity at low water contents, and a more modest viscosity decrease at high temperatures.

Currently, there is no fully satisfactory model for predicting the viscosity of molten granite as a function of water content, temperature and composition. The very strong viscosity decrease that occurs over the low water content range raises special difficulties in this respect. The equation of Scaillet *et al.* (1996) successfully reproduces both high- and low-temperature experimental data, and may be used with confidence in predicting the viscosity of hydrous peraluminous leucogranites, and chemically similar rhyolites, at magmatic to sub-magmatic temperatures. Over a wide range of temperature and water contents, the models of Hess & Dingwell (1996) and Zhang *et al.* (2003) are likely to be more applicable.

6. Acknowledgements

This work was supported by the EU TMR network ERBFMRX 960063 '*In situ* hydrous melts'. We thank Jasper Berndt, Susanne Ohlhorst, Frank Schulze, Max Wilke and Tony Withers for help with sample syntheses, Ali Bouhifd and Mike Toplis for valuable discussions, Becky Lange for comments on a previous version, and Don Baker and Simon Harley for constructive reviews.

7. References

Adam, G. & Gibbs, J. H. 1965. On the temperature dependence of relaxation phenomena in glass-forming liquids. *Journal of Chemical Physics* **43**, 139–46.

Angell, C. A. 1985. Strong and fragile liquids. *In*: Ngai, K. L. & Wright, G. B. (eds) *Relaxations in Complex Systems*, 3–11. Springfield, VA: U.S. Department of Commerce, National Technical Information Service.

Baker, D. R. 1996. Granitic melt viscosities: empirical and configurational entropy models for their calculation. *American Mineralogist* **81**, 126–34.

Baker, D. R. 1998. Granitic melt viscosity and dike formation. *Journal of Structural Geology* **20**, 1395–404.

Baker, D. R. & Vaillancourt, J. 1995. The low viscosity of F+H_2O-bearing granitic melts and implications for melt extraction and transport. *Earth and Planetary Science Letters* **132**, 199–211.

Behrens, H. 1995. Determination of water solubilities in high-viscosity melts: an experimental study on $NaAlSi_3O_8$and $KAlSi_3O_8$ melts. *European Journal of Mineralogy* **7**, 905–20.

Behrens, H., Romano, C., Nowak, M., Holtz, F. & Dingwell, D. B. 1996. Near-infrared spectroscopic determination of water species in glasses of the system $MAlSi_3O_8$ (M=Li, Na, K): an interlaboratory study. *Chemical Geology* **128**, 41–63.

Behrens, H. & Nowak, M. 1997. The mechanisms of water diffusion in polymerized melts. *Contributions to Mineralogy and Petrology* **126**, 377–85.

Bottinga, Y. & Weill, D. F. 1972. The viscosity of magmatic liquids: a model for calculations. *American Journal of Science* **272**, 438–75.

Dingwell, D. B. 1987. Melt viscosities in the system $NaAlSi_3O_8$–H_2O–F_2O. *In* Mysen, B. O. (ed.) *Magmatic Processes: Physicochemical Principles*, 423–33. *Geochemical Society Special Publication* **1**. University Park, PA: Geochemical Society.

Dingwell, D. B. 1999. Recent experimental progress in the physical description of silicic magma relevant to explosive volcanism. *Geological Society, London, Special Publication* **145**. London: Geological Society.

Dingwell, D. B., Romano, C. & Hess, K.-U. 1996. The effect of water on the viscosity of a haplogranitic melt under P-T-X conditions relevant to silicic volcanism. *Contributions to Mineralogy and Petrology* **124**, 19–28.

Dingwell, D. B., Hess, K.-U. & Romano, C. 1998a. Viscosity data for hydrous peraluminous granitic melts: comparison with a metaluminous model. *American Mineralogist* **83**, 236–9.

Dingwell, D. B., Hess, K.-U. & Romano, C. 1998b. Extremely fluid behaviour of hydrous peralkaline rhyolites. *Earth and Planetary Science Letters* **158**, 31–8.

Dingwell, D. B., Hess, K. U. & Romano, C. 2000. Viscosities of granitic (*sensu lato*) melts; influence of the anorthite component. *American Mineralogist* **85**, 1342–8.

Dingwell, D. B. & Webb, S. L. 1990. Relaxation in silicate melts. *European Journal of Mineralogy* **2**, 427–49.

Eckert, H., Yesinowski, J. P., Silver, L. A. & Stolper, E. M. 1988. Water in silicate glasses: quantification and structural studies by ^1H solid echo and MAS-NMR methods. *Journal of Physical Chemistry* **92**, 2055–64.

Giordano, D., Dingwell, D. B. & Romano, C. 2000. Viscosity of a Teide phonolite in the welding interval. *Journal of Volcanology and Geothermal Research* **103**, 239–45.

Giordano, D. & Dingwell, D. B. 2003. Viscosity of hydrous Etna basalt; implications for Plinian-style basaltic eruptions. *Bulletin of Volcanology* **65**, 8–14.

Hess, K.-U. & Dingwell, D. B. 1996. Viscosities of hydrous leucogranitic melts: a non-Arrhenian model. *American Mineralogist* **81**, 1297–300.

Holtz, F., Roux, J., Ohlhorst, S., Behrens, H. & Schulze, F. 1999. The effects of silica and water on the viscosity of hydrous quartzofeldspathic melts. *American Mineralogist* **84**, 27–36.

Hummel, W. & Arndt, J. 1985. Variation of viscosity with temperature and composition in the plagioclase system. *Contributions to Mineralogy and Petrology* **90**, 83–92.

Kushiro, I. 1976. Viscosities of basalt and andesite melts at high pressures. *Journal of Geophysical Research* **81**, 6351–6.

Kushiro, I. 1978. Viscosity and structural change of albite ($NaAlSi_3O_8$) melt at high pressure. *Earth and Planetary Science Letters* **41**, 87–90.

Le Fort, P., Cuney, M., Deniel, C., France-Lanord, C., Sheppard, S. M. F., Upreti, B. N. & Vidal, P. 1987. Crustal generation of the Himalayan leucogranites. *Tectonophysics* **134**, 39–57.

Lejeune, A.-M., Holtz, F., Roux, J. & Richet, P. 1994. Rheology of hydrous andesite: an experimental study at high viscosities. *EOS, Transactions of the American Geophysical Union* **44**, 724.

Moynihan, C. T. 1995. Structural relaxation and the glass transition. *Reviews in Mineralogy* **32**, 1–19.

Mysen, B. O. 1988. *Structure and properties of silicate melts*. Amsterdam: Elsevier.

Neuville, D. R., Courtial, P., Dingwell, D. B. & Richet, P. 1993. Thermodynamic and rheological properties of rhyolite and andesite melts. *Contributions to Mineralogy and Petrology* **113**, 572–81.

Neuville, D. R. & Richet, P. 1991. Viscosity and mixing in molten (Ca, Mg) pyroxenes and garnets. *Geochimica et Cosmochimica Acta* **55**, 1011–19.

Nowak, M. & Behrens, H. 1995. The speciation of water in haplogranitic melts determined by in situ near-infrared spectroscopy. *Geochimica et Cosmochimica Acta* **59**, 3445–50.

Persikov, E. S., Zharikhov, V. A., Bukhtiyarov, P. G. & Pol'skoy, S. F. 1990. The effect of volatiles on the properties of magmatic melts. *European Journal of Mineralogy* **2**, 621–42.

Petford, N. 1996. Dykes or diapirs? *Transactions of the Royal Society of Edinburgh: Earth Sciences* **87**, 105–14.

Petford, N., Cruden, A. R., McCaffrey, K. J. W. & Vigneresse, J.-L. 2000. Granite magma formation, transport and emplacement in the Earth's crust. *Nature* **408**, 669–73.

Pichavant, M., Kontak, D. J., Valencia-Herrera, J. & Clark, A. H. 1988a. The Miocene-Pliocene Macusani Volcanics, SE Peru; 2, Geochemistry and origin of a felsic peraluminous magma. *Contributions to Mineralogy and Petrology* **100**, 300–24.

Pichavant, M., Kontak, D. J., Briqueu, L., Valencia-Herrera, J. & Clark, A. H. 1988b. The Miocene–Pliocene Macusani Volcanics, SE Peru; 1, Mineralogy and magmatic evolution of a two-mica aluminosilicate-bearing ignimbrite suite. *Contributions to Mineralogy and Petrology* **100**, 325–38.

Richet, P. 1984. Viscosity and configurational entropy of silicate melts. *Geochimica et Cosmochimica Acta* **48**, 471–83.

Richet, P., Lejeune, A.-M., Holtz, F. & Roux, J. 1996. Water and the viscosity of andesite melts. *Chemical Geology* **128**, 185–97.

Richet, P. & Bottinga, Y. 1984. Glass transition and thermodynamic properties of SiO2, $NaAlSi_nO_{2n+2}$ and $KAlSi_3O_8$. *Geochimica et Cosmochimica Acta* **48**, 453–70.

Richet, R. & Bottinga, Y. 1995. Rheology and configurational entropy of silicate melts. *Reviews of Mineralogy* **32**, 67–93.

Romano, C., Hess, K-U., Mincione, V., Poe, B. & Dingwell, D. B. 2001. The viscosities of hydrous $XalSi_3O_8$ (X=Li, Na, K, $Ca_{0\cdot5}$, $Mg_{0\cdot5}$) melts. *Chemical Geology* **174**, 115–32.

Rosenhauer, M., Scarfe, C. M. & Virgo, D. 1979. Pressure dependence of the glass transition temperature in glasses of diopside, albite and sodium trisilicate composition. *Carnegie Institute of Washington Yearbook* **78**, 547–51.

Scaillet, B., Pichavant, M. & Roux, J. 1995. Experimental crystallisation of leucogranite magmas. *Journal of Petrology* **36**, 664–706.

Scaillet, B., Holtz, F., Pichavant, M. & Schmidt, M. 1996. Viscosity of Himalayan leucogranites: implications for mechanisms of granitic magma ascent. *Journal of Geophysical Research* **101**, 27691–9.

Scaillet, B., Whittington, A., Martel, C., Pichavant. M. & Holtz, F. 2000. Phase equilibrium constraints on the viscosity of silicic magmas II: implications for mafic-silicic mixing processes. *Transactions of the Royal Society of Edinburgh: Earth Sciences* **91**, 61–72.

Scarfe, C. M., Cronin, D. J., Wenwel, J. T. & Kauffman, D. A. 1983. Viscosity–temperature relationships at 1 atm in the system diopside-anorthite. *American Mineralogist* **68**, 1083–8.

Schulze, F., Behrens, H., Holtz, F., Roux, J. & Johannes, W. 1996. The influence of H_2O on the viscosity of a haplogranitic melt. *American Mineralogist* **81**, 1155–65.

Shaw, H. R. 1963. Obsidian-H_2O viscosities at 1000 and 2000 bars in the temperature range 700 to 900°C. *Journal of Geophysical Research* **68**, 6337–43.

Shaw, H. R. 1972. Viscosities of magmatic silicate liquids: an empirical method of prediction. *American Journal of Science* **272**, 870–89.

Shen, A. & Keppler, H. 1995. Infrared spectroscopy of hydrous silicate melts to 1000°C and 10 kbar – direct observation of H_2O speciation in a diamond-anvil cell. *American Mineralogist* **80**, 1335–8.

Stevenson, R. J., Bagdassarov, N. S., Dingwell, D. B. & Romano, C. 1998. The influence of trace amounts of water on the viscosity of rhyolites. *Bulletin of Volcanology* **60**, 89–97.

Stolper, E. 1982. The speciation of water in silicate melts. *Geochimica et Cosmochimica Acta* **46**, 2609–20.

Toplis, M. J., Dingwell, D. B., Hess, K-U. & Lenci, T. 1997. Viscosity, fragility, and configurational entropy of melts along the join SiO_2–$NaAlSiO_4$. *American Mineralogist* **82**, 979–90.

Webb, S. L. & Dingwell, D. B. 1990. Non-Newtonian rheology of igneous melts at high stresses and strain rates: experimental results for rhyolite, andesite, basalt, and nephelinite. *Journal of Geophysical Research* **95**, 15695–701.

Whittington, A., Richet, P. & Holtz, F. 2000. Water and the viscosity of hydrous depolymerized aluminosilicate melts. *Geochimica et Cosmochimica Acta* **64**, 3725–36.

Whittington, A., Richet, P., Linard, Y. & Holtz, F. 2001. The viscosity of hydrous phonolites and trachytes. *Chemical Geology* **174**, 209–24.

Withers, A. C., Zhang, Y. & Behrens, H. 1999. Reconciliation of experimental data on water speciation in rhyolitic glass using in situ and quenching techniques. *Earth and Planetary Science Letters* **173**, 343–9.

Withers, A. C. & Behrens, H. 1999. Temperature induced changes in the NIR spectra of hydrous albitic and rhyolitic glasses between 300 and 100 K. *Physics and Chemistry of Minerals* **27**, 119–32.

Zhang, Y., Xu, Z. & Liu, Y. 2003. Viscosity of hydrous rhyolitic melts inferred from kinetic experiments, and a new viscosity model. *American Mineralogist* **88**, 1741–52.

Zhang, Y. & Behrens, H. 1998. H_2O diffusion in silicate glasses and melts. *Mineralogical Magazine* **62A**, 1695–6.

Zhang, Y. & Behrens, H. 2000. H_2O diffusion in rhyolitic melts and glasses. *Chemical Geology* **169**, 243–62.

ALAN WHITTINGTON, Department of Geological Sciences, University of Missouri-Columbia, Columbia, MO 65211–1380, USA.

PASCAL RICHET, Laboratoire de Physique des Géomatériaux, Institut de Physique du Globe, 4 place Jussieu, 75252 Paris cedex 05, France.

HARALD BEHRENS and FRANÇOIS HOLTZ, Institut für Mineralogie, Universität Hannover, Welfengarten 1, D-30167 Hannover, Germany.

BRUNO SCAILLET, Institut des Sciences de la Terre d'Orléans, UMR 6113 CNRS-UO, 1A rue de la Férollerie, 45071 Orléans cedex 02, France.

MS received 3 September 2003. Accepted for publication 4 May 2004.

Transactions of the Royal Society of Edinburgh: Earth Sciences, **95**, 73–85, 2004

Petrologic and thermal constraints on the origin of leucogranites in collisional orogens

Peter I. Nabelek and Mian Liu

ABSTRACT: Leucogranites are typical products of collisional orogenies. They are found in orogenic terranes of different ages, including the Proterozoic Trans-Hudson orogen, as exemplified in the Black Hills, South Dakota, and the Appalachian orogen in Maine, both in the USA, and the ongoing Himalayan orogen. Characteristics of these collisional leucogranites show that they were derived from predominantly pelitic sources at the veining stages of deformation and metamorphism in upper plates of thickened crusts. Once generated, the leucogranite magmas ascended as dykes and were emplaced within shallower parts of their source sequences. In these orogenic belts, there was a strong connection between deformation, metamorphism and granite generation. However, the heat sources needed for partial melting of the source rocks remain controversial. Lack of evidence for significant intrusion of mafic magmas necessary to cause melting of upper plate source rocks suggests that leucogranite generation in collisional orogens is mainly a crustal process.

The present authors evaluate five types of thermal models which have previously been proposed for generating leucogranites in collisional orogens. The first, a thickened crust with exponentially decaying distribution of heat-producing radioactive isotopes with depth, has been shown to be insufficient for heating the upper crust to melting conditions. Four other models capable of raising the crustal temperatures sufficiently to initiate partial melting of metapelites in thickened crust include: (1) thick sequences of sedimentary rocks with high amounts of internal radioactive heat production; (2) decompression melting; (3) thinning of mantle lithosphere; and (4) shear-heating. The authors show that, for reasonable boundary conditions, shear-heating along crustal-scale shear zones is the most viable process to induce melting in upper plates of collisional orogens where pelitic source lithologies are usually located. The shear-heating model directly links partial melting to the deformation and metamorphism that typically precede leucogranite generation.

KEY WORDS: Partial melting, shear heatings, thermal models.

Leucogranites, peraluminous granites with near-eutectic composition, are common in collisional orogens, where they were produced by partial melting of deformed and metamorphosed accretionary-wedge and ocean-floor sediments. Whilst there is general agreement that leucogranites are anatectic products of pelitic crustal sources, the heat sources for their production have remained controversial. Thickened orogens do not become sufficiently hot to cause water-absent melting of pelites without additional heat input from the mantle or crustal processes, such as shear-heating or accretion of rocks with high concentrations of radioactive isotopes (e.g. Royden 1993; Thompson & Connolly 1995).

Much of the discussion concerning petrogenesis of collisional leucogranites has focused on the Tertiary High Himalaya granites, because they record recent crustal melting in an ongoing continental collision (Fig 1). Processes which have been proposed to have induced melting include fluid-fluxing of source lithologies above the Main Central Thrust, a major shear zone that separates metasedimentary sequences that have different metamorphic field gradients and crustal residence ages (Le Fort *et al.* 1987; Guillot & Le Fort 1995), melting of rapidly decompressing crustal blocks (Zeitler & Chamberlain 1991; Harris & Massey 1994), deep accretion of rocks with high concentration of radioactive isotopes (Huerta *et al.* 1998; Jamieson *et al.* 1998), and deformation-induced heating along shear zones (Zhu & Shi 1990; Harrison *et al.* 1998).

Whilst the Himalayas provide an example of recent leucogranite petrogenesis, extensive leucogranites also occur in ancient collisional orogens, including the 400–270 Ma leucogranites of Maine, USA, which were generated during the Appalachian orogeny (Tomascak *et al.* 1996b; Pressley & Brown 1999), and the leucogranites of the Black Hills, South Dakota, USA, which are the products of melting in the southern part of the Trans-Hudson orogen (Fig. 2; Redden *et al.* 1990; Nabelek *et al.* 1992a; Nabelek & Bartlett 1998). The Trans-Hudson orogeny was responsible for agglomeration of several Archaean microcontinents in the Proterozoic to form the nucleus of North America. The chemistry, mineralogy, character of occurrence and differentiation, and relationships to the host rocks of these leucogranites bear many similarities to the Himalayan leucogranites. In this paper, the present authors review these characteristics and use them to place constraints on the thermal models for generating collisional leucogranites.

1. Mineralogical and chemical characteristics of collisional leucogranites

Collisional leucogranites are characterised by their peraluminous compositions (ASI >1·1) and very low concentrations of CaO, MgO, and FeO, typically <1 wt.% for each (Fig. 3). These concentrations are lower than in the prototypical S-type granites of the Lachlan fold belt, Australia, which contain large proportions of restite (Chappell & White 1992), whereas leucogranites are generally restite-free. In leucogranites, muscovite is a characteristic mineral, along with tourmaline or biotite. Almandine-spessartine garnet and minor sillimanite can also occur. Tourmaline and biotite are often exclusive of each other. Biotite is usually found in rocks which have more than ~0·06 wt.% TiO_2, whereas tourmaline occurs

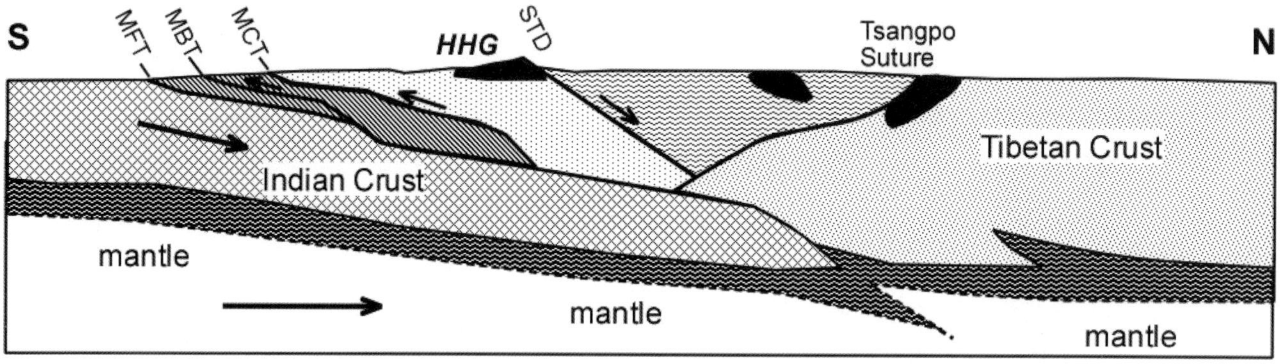

Figure 1 Cross-section across the Himalayas into southern Tibet (after Nelson *et al.* 1996). Bounding faults for major crustal blocks are shown: (MFT) Main Frontal Thrust; (MBT) Main Boundary Thrust; (MCT) Main Central Thrust; and (STD) South Tibetan Detachment. High Himalaya crystallinites are between the MCT and STD. High Himalaya leucogranites (HHG) comprise some of the highest peaks of the Himalayas. Lithologies between the MFT and the Tsangpo suture are mainly oceanic and accretionary wedge sediments which were deformed and metamorphosed to various grades during the Himalayan orogeny.

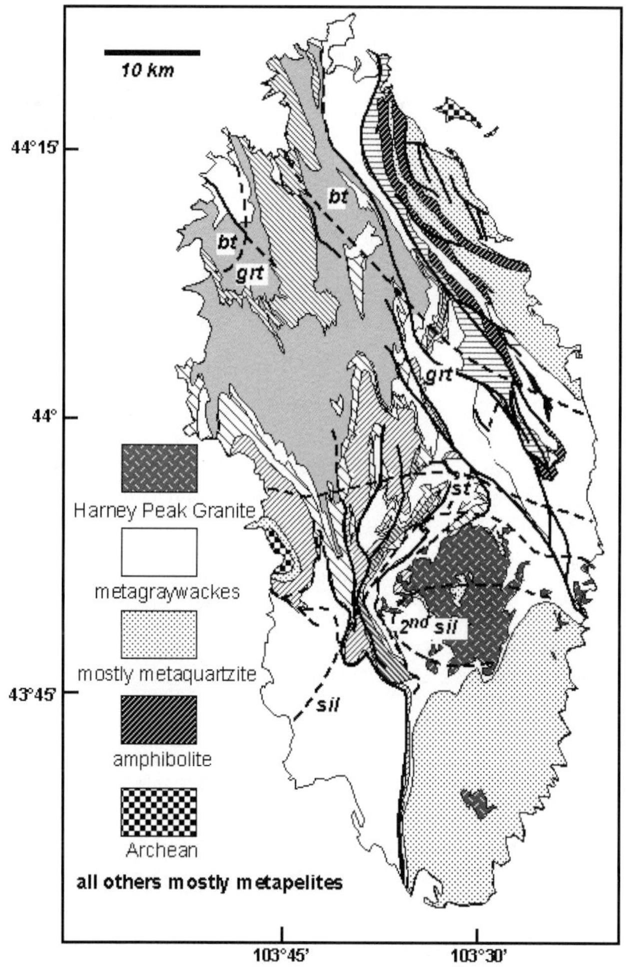

Figure 2 Geologic map of the Precambrian core of the Black Hills, South Dakota, USA. The terrane is dominated by Proterozoic metapelites to metaquartzites which were deformed into large folds and metamorphosed during the Trans-Hudson orogeny. Archaean metasedimentary rocks are found along the periphery. Intrusion of the Harney Peak granite reoriented the Trans-Hudson foliation. Metamorphic grades are shown by heavy dashed lines: (bt) biotite-chlorite; (grt) garnet-biotite-chlorite; (st) staurolite; (sil) sillimanite; and (2nd sil) second sillimanite.

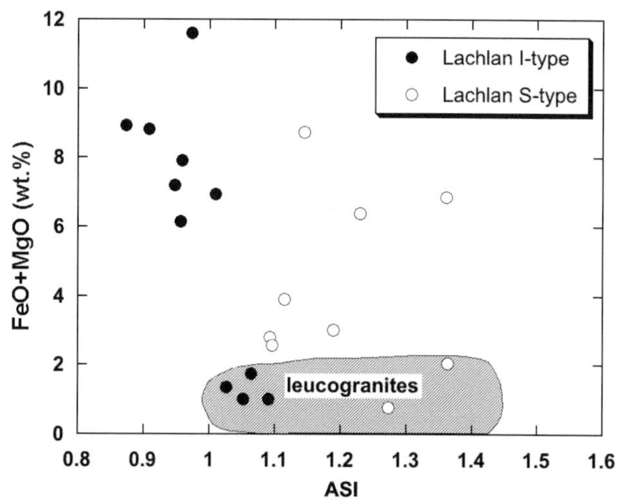

Figure 3 Comparison of leucogranite compositions from the Black Hills, High Himalayas and Maine (Nabelek *et al.* 1992a; Guillot & Le Fort 1995; Tomascak *et al.* 1996b; Pressley & Brown 1999) with compositions of representative S- and I-type granites from the Lachlan Fold Belt (Chappell & White 1992). All leucogranites from the three regions overlap and fall within the shaded field. FeO+MgO concentrations in the Lachlan suites are more elevated and variable because of a greater proportion of mafic components, restite and fractional crystallisation.

in rocks which have less TiO_2. Tourmaline and biotite-bearing granites often exist within different parts of composite plutons, such as in the Badrinath-Gangotri plutons in India and in the Harney Peak Granite and its satellite plutons in the Black Hills (Scaillet *et al.* 1990; Nabelek *et al.* 1992a). The relationship between tourmaline and biotite-bearing leucogranites has been attributed to fractional crystallisation (Scaillet *et al.* 1990; Guillot & Le Fort 1995) or production by muscovite versus biotite-dehydration melting reactions, respectively (Nabelek *et al.* 1992a). The latter alternative is based on often distinct isotopic compositions of the two leucogranite types in a given region (see below).

Leucogranite plutons generally form by intrusion of multitudes of dykes and sills into ballooning magma chambers (Duke *et al.* 1988; Searle *et al.* 1997; Brown & Solar 1998b). Magma batches are thought to migrate from source-regions into dilatant portions of shear-zone systems in crustal sequences which are undergoing contractional deformation (Brown & Solar 1998b). Shear-zone systems are integral to leucogranite magma generation, ascent and emplacement. After emplacement, individual melt sheets often locally differentiate into aplite-pegmatite couples, giving wide ranges in plagioclase/K-feldspar ratios within the sheets. This, in turn, leads to wide ranges in Ba, Sr and other trace elements that are fractionated by feldspars (Duke *et al.* 1992).

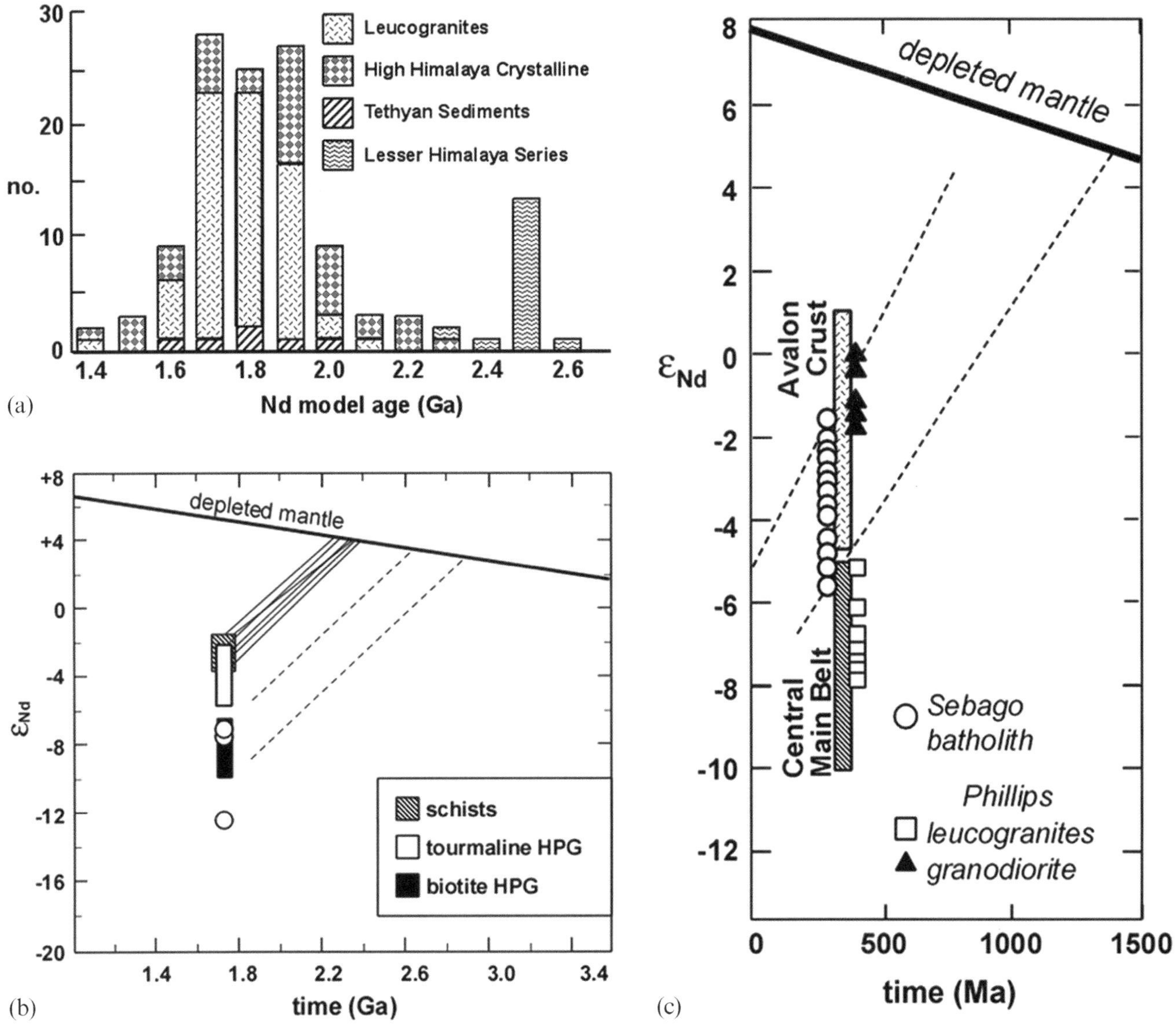

Figure 4 (a) Nd model ages for various upper crustal blocks and leucogranites in the Himalayas (after Ahmad *et al.* 2000). The range of model ages of the leucogranites is the same as that of the High Himalaya crystallinites within which they occur. (b) ε_{Nd}(1715) and selected calculated evolution trajectories of the values in tourmaline and biotite-bearing parts of the Harney Peak Granite and its satellite intrusions (after Krogstad & Walker 1996). The data indicate mixed Archaean and Proterozoic sources for the leucogranites. (c) ε_{Nd} values of the Sebago batholith and Phillips pluton in the Central Maine Belt (CMB) (after Tomascak *et al.* 1996b; Pressley & Brown 1999). The data indicate melt derivation from the host CMB rocks and Avalonian lithologies which were thrust over the Grenville basement during the Appalachian orogeny.

In addition to composite granite plutons, unzoned and large zoned pegmatites of the Li-Cs-Ta class often occupy leucogranite fields. Pegmatites occur in Maine and are prominent in the Black Hills (Norton & Redden 1990). They are the products of low-temperature, sometimes <400°C crystallisation of peraluminous melts fluxed by borate and carbonate species (Sirbescu & Nabelek 2003a, b).

2. Source rocks

Major and trace element compositions of collisional leucogranites show that metapelites, and to a lesser extent, metagraywackes are their sources (Harris & Inger 1992; Nabelek & Bartlett 1998; Solar & Brown 2001). Partial melting experiments on metapelites (Patiño-Douce & Harris 1998), and structural relationships between migmatites, melt ascent paths and plutons (Le Fort *et al.* 1987; Brown & Solar 1998b) support this link. Metapelites are more fertile melt producers than metagraywackes because metapelites contain higher proportions of muscovite. The often significant paragonite component in muscovite adds to melt production (Nabelek & Bartlett 2000).

A key feature of collisional leucogranite plutons is that they are emplaced within shallower sections of their source metasedimentary sequences. The direct connection between plutons and specific source rocks has been made using different isotopic systems. For example, the Nd model ages of the High Himalaya leucogranites are the same as those of the high-grade High Himalaya crystalline metamorphic rocks (1·4–2·2 Ga) within which they occur, but are different from the underlying Lesser Himalaya metamorphic rocks (2·3–2·6 Ga, Fig. 4a; Ahmad *et al.* 2000). Similarly, $\delta^{18}O$ values of the granites and underlying migmatitic gneisses span the same range, mostly 11–14‰, and $^{87}Sr/^{86}Sr_i$ ratios of both groups are high at >0·73

(France-Lanord & Le Fort 1988). The heterogeneity of source rocks is reflected in the isotopic heterogeneity of plutons. Within the Manaslu pluton, the two-mica and tourmaline-bearing granites have distinct $^{87}Sr/^{86}Sr_i$ ratios. The modes of the ratios of the two granite types are 0·745 and 0·759, corresponding to isotopic compositions of metagraywackes and metapelites in the High Himalaya crystallinites, respectively (Guillot & Le Fort 1995).

Analogous connections between leucogranites and isotopically heterogeneous pelitic source rocks occur in the Black Hills. The tourmaline-bearing suite has a 12·3–13·6‰ $\delta^{18}O$ range, which corresponds to isotopic compositions of the host Proterozoic metapelites and metagraywackes, whereas values of the biotite-bearing suite are lower at 10·8–12·8‰ (Nabelek *et al.* 1992b). K-feldspars from the tourmaline granites, and several simple and zoned pegmatites in the region have low values of $^{207}Pb/^{204}Pb$ for their $^{206}Pb/^{204}Pb$ ratios compared to K-feldspars in the biotite-bearing granites (Krogstad *et al.* 1993). The ratios indicate a relatively short, 100–300 m.y. crustal residence of the source for the tourmaline suite and crustal residence since the late Archaean for the source of the two-mica suite prior to melting. The two granite suites have overlapping ε_{Nd}(1715 Ma) values, −6·4 to −9·9 for ten samples of the two-mica suite, and −2·0 to −7·7 for 10 samples of the tourmaline suite (Fig. 4b; Krogstad & Walker 1996). One sample of the tourmaline suite has an unusually low average value of −12·6. The spread in ε_{Nd}(1715 Ma) values may in part reflect disequilibrium melting of the protoliths (Nabelek & Glascock 1995; Ayres & Harris 1997). Nevertheless, the high end of the range for the tourmaline suite overlaps that of the Proterozoic schists, which have model T_{DM} extraction ages ~2300 Ma, whereas the low end approaches an Archaean component. Thus, the isotopic data suggest that the source rocks for the two leucogranite suites had mixed provenance, sediments derived from Archaean microcontinents and Proterozoic ocean-floor sediments which were thrust over the Wyoming basement during the Trans-Hudson orogeny. These sedimentary sources comprise much of the Precambrian core of the Black Hills.

The relationship between leucogranites and their source rocks in Maine has been deduced principally from Nd isotopic studies of the 404 Ma Phillips pluton (Pressley & Brown 1999) and the 293 Ma Sebago batholith (Tomascak *et al.* 1996a, b) which occur within the Central Maine Belt (CMB). The Phillips pluton was emplaced within a low-strain portion of the CMB during Early Devonian deformation of metasedimentary sequences (Solar *et al.* 1998). Intrusion of the Sebago batholith and the associated pegmatites which are as young as 270 Ma appears to be related to movement along the regional Norumbega shear-zone system that is the boundary between the accreted Avalon terrane to the east and the CMB to the west. The CMB and perhaps portions of the Avalon terrane overly an older Grenville basement. The age of the Sebago batholith was used to constrain timing of the most recent deformation within the shear zones which mark the end stages of the Appalachian orogeny (Tomascak *et al.* 1996a).

Leucogranites of the Phillips pluton have ε_{Nd}(404) values which correspond to values of the CMB metasedimentary rocks and migmatites at the time of intrusion (Fig. 4c). In contrast, the Sebago batholith and small granodioritic portions of the Phillips pluton have values which suggest derivation from sedimentary rocks with Avalon terrane affinities. Thus, the sources of these leucogranites were metasedimentary sequences which were undergoing active metamorphism and deformation during magma genesis, ascent and emplacement in upper portions of the thickened Appalachian crust.

3. Metamorphism, deformation and melt production

Structural analysis in the CMB by Solar, Brown and co-workers (Brown & Solar 1998a, b; Solar *et al.* 1998) has clearly linked leucogranite magmatism to active contractional deformation and metamorphism. The links can also be made in the Black Hills and the Himalayas, where granite intrusion was, in some cases, accompanied by rapid denudation (Winslow *et al.* 1995; Davidson *et al.* 1997; Whittington & Treloar 2002). In the three terranes, regional metamorphism and deformation of the host metapelitic sequences began several tens of millions of years prior to leucogranite magmatism during early stages of continental collisions, as shown by dated monazite and garnet growth (Dahl & Frei 1998; Harris *et al.* 2000). In the Black Hills, initial garnet growth preceded leucogranite emplacement by about 35 m.y. The garnet grew at 350–400 °C as indicated by its high spessartine component and low $a(H_2O)$ in the metamorphic rocks during growth (Nabelek *et al.* 2003). Metamorphism proceeded until the onset of leucogranite generation, which marked the end stages of deformation of the host rocks. However, in all these orogens, leucogranite magmatism was clearly synkinematic, although contact aureoles were imposed on shallower, earlier-deformed parts of the overthrust sequences by growing plutons (Brown & Solar 1998b; Solar *et al.* 1998; Schneider *et al.* 1999). Previously, such contact aureoles were attributed to post-collisional generation of magmas as a result of increased heat flow after thinning of the mantle lithosphere (e.g. De Yoreo *et al.* 1991; Holm *et al.* 1997).

In the Himalayas and Maine, leucogranite plutons occur in dilatant zones in close proximity to major shear zones. The shear zones themselves typically include migmatitic rocks (Solar *et al.* 1998; Schneider *et al.* 1999; Whittington & Treloar 2002). Thus, melt migration was from areas of compressional high strain to areas of dilatancy. The largest pluton in the Black Hills, the Harney Peak Granite, also appears to have been emplaced between major faults which were undergoing active displacement on the pluton's east and west sides (Fig. 2).

4. Melt-producing reactions

The dominant collisional leucogranites are two-mica granites and muscovite-tourmaline granites. Muscovite-dehydration melting (MDM) of metapelites that also includes the breakdown of tourmaline explains the chemical characteristics of muscovite-tourmaline granites, including the necessary boron concentrations (Wilke *et al.* 2002). Muscovite-tourmaline granites typically contain very low concentrations of light rare-earth and other high field-strength elements (Fig. 5). The low concentrations result from lack of achievement of equilibrium between melts and accessory minerals, such as monazite and zircon, in residues. The disequilibrium has been ascribed to rapid removal of melts from source regions (e.g. Watt & Harley 1993). Alternatively, Nabelek & Glascock (1995) proposed that monazite may remain armoured by biotite, which does not partake in the melting reaction, thus preventing equilibration of monazite with the partial melt.

The large masses of muscovite-tourmaline granites and tourmaline-bearing migmatites in convergent orogens suggest that MDM is a characteristic melt-producing mechanism. However, two-mica granites indicate melting conditions which also include at least a partial breakdown of biotite. Evidence for biotite-dehydration melting (BDM) includes the generally higher Ti concentrations in two-mica granites and more

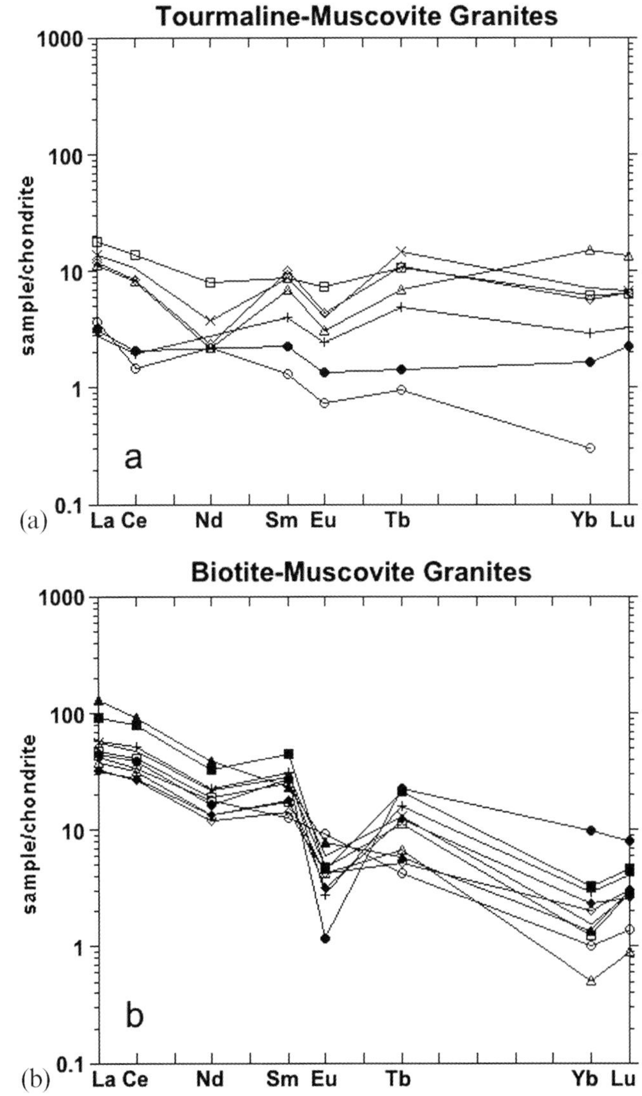

Figure 5 Rare-earth element patterns in (a) two-mica and (b) tourmaline-muscovite granites in the Black Hills. Analogous patterns occur in leucogranite suites elsewhere. REE depletion in the tourmaline-bearing suite is the result of disequilibrium melting involving monazite.

elevated light rare earth element (REE) concentrations (Fig. 5) which indicate equilibration of previously biotite-armoured monazite with melt. Titanium, which is released to the melt by the breakdown of biotite, stabilises biotite over tourmaline in the crystallising magma (Nabelek *et al.* 1992a; Scaillet *et al.* 1995; Wilke *et al.* 2002).

Earlier models for generation of the Manaslu granite in the Himalayas included water-present melting of the High Himalaya crystallinites, with water derived from the underlying Lesser Himalaya metasedimentary rocks (Le Fort *et al.* 1987; France-Lanord & Le Fort 1988). However, there is now a preponderance of evidence for generation of collisional leucogranites mostly under fluid-absent conditions at temperatures far above the water-saturated granite solidus:

(1) Harris & Inger (1992) have argued that high Rb:Sr and low Sr:Ba ratios in the Manaslu granite compared to the source rocks are inconsistent with fluid-present melting, since fluid-present melting would remove most feldspar from the residue. Feldspar removal would increase Sr concentrations in the melt above those which are observed. However, it is cautioned, that the relative ratios are also dependent on the amount of biotite in the residue, whose proportion during MDM can be sufficiently high to cause lower Rb:Sr and higher Sr:Ba ratios in partial melts than their source rocks, even when plagioclase remains in the residue (Nabelek & Bartlett 2000).

(2) Patiño-Douce & Harris (1998) have shown experimentally that fluid-present melting of Himalayan metapelites produces melts which are trondhjemitic in composition, whereas fluid-absent melting produces melts whose compositions match those of the Himalayan leucogranites. Similarly, the Black Hills leucogranites are too potassic to have been produced by water-present melting (Nabelek *et al.* 1992a) because the granite minimum in the SiO_2–$NaAlSi_3O_8$–$KAlSi_3O_8$ system lies closer to the SiO_2–$NaAlSi_3O_8$ join when $a(H_2O)_{fluid}=1$ (Holtz & Johannes 1991) than is the composition of the granites.

(3) Fractionation of oxygen isotopes among minerals in the Black Hills leucogranites show minimum equilibration temperatures of >750 °C (Nabelek *et al.* 1992b), which suggest melt production at temperatures far above the water-saturated granite solidus.

(4) Application of REE solubility models to the Black Hills and Himalaya granites yields >700 °C temperatures for both locations (Nabelek & Glascock 1995; Ayres & Harris 1997). Because the granites contain very little restite, it is unlikely that these high temperatures are anomalously high because of entrainment of monazite and zircon from source rocks.

(5) $X(CO_2)$ of primary fluid inclusions in the Black Hills leucogranites ranges from 0·55 to 0·1 (Nabelek & Ternes 1997). The range is attributed to differential degassing of CO_2 and H_2O from crystallising magma. The largest $X(CO_2)$ value can be used to estimate the concentrations of these two species in the magma at initial saturation (Holloway & Blank 1994). For the 3·5 kbar pressure of magma emplacement, the estimated concentrations are 3·5 wt.% H_2O and 1500 ppm CO_2. The concentration of water is significantly below the required ~14 wt.% at the water-saturated granite solidus at 10 kbar (Johannes & Holtz 1996).

(6) Occurrence of MDM or BDM reactions does not necessarily preclude initiation of melting under water-present conditions. However, the metasedimentary sequences within the collisional orogens described here contain large proportions of graywackes which contain substantial amounts of K-feldspar. Under water-present conditions, K-feldspar along with quartz and plagioclase should partake in melt-production at the temperature of the water-saturated granite minimum. There is little evidence for melting of this anhydrous assemblage in the collisional orogens. These melts should not be hotter than ~650 °C, they should not be peraluminous, and they should contain much less boron and titanium, characteristic elements of MDM and BDM, respectively, than the orogenic leucogranites.

(7) Fluid-present melting conditions may occur in the upper crust in shallow parts of shear zones which may be penetrated by water (e.g. Whittington *et al.* 1999). Melts produced by fluid-present melting cannot ascend far from their source because they rapidly cross the solidus as they migrate into lower-temperature regions higher in the crust. In contrast, melts formed by fluid-absent melting can remain above the liquidus until their emplacement in plutons. Water-undersaturated leucogranite magmas can reach the upper crust through dykes which are several metres wide (Scaillet *et al.* 1996), as those seen in the three described regions.

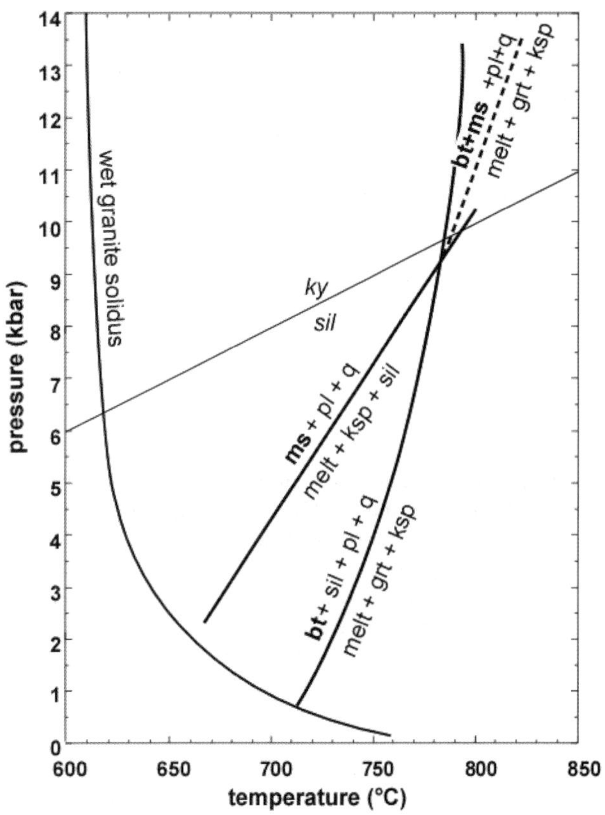

Figure 6 Plot showing positions of the muscovite-dehydration melting reaction (MDM; Patiño-Douce & Harris 1998) and the biotite-dehydration melting reaction (BDM; Le Breton & Thompson 1988). The two reactions should cross at ~10 kbar, above which both micas should be involve in the melting reaction together (dashed curve). The intersection places an upper pressure limit on the production of tourmaline-muscovite granites.

The position of the MDM and BDM reactions in the *P-T* regime in the crust places a constraint on the depths of leucogranite generation. Figure 6 shows the position of the experimentally determined MDM reaction and the initial stage of the continuous BDM reaction in metapelites (Le Breton & Thompson 1988; Patiño-Douce & Harris 1998). The reactions' slopes suggest that they should intersect at ~10 kbar. Above this pressure, muscovite and biotite should partake in the initial melting reaction together. Thus, the occurrence of muscovite-tourmaline leucogranites and migmatites places the upper limit of their source region to <10 kbar, within the upper plate of a thickened orogenic crust.

The depth of generation of two-mica leucogranites is more difficult to constrain because they can be generated either by BDM below the intersection of reactions in Figure 6, or by the combined breakdown of muscovite and biotite. Isotopic compositions of muscovite-tourmaline and two-mica portions of the Manaslu pluton suggest that they were derived from isotopically distinct metapelites and metagraywackes, respectively, in the High Himalaya crystallinites (Guillot & Le Fort 1995). Therefore, if the muscovite-tourmaline granites were generated by MDM, then the two-mica granites must have been generated by BDM at approximately the same conditions, also below 10 kbar. In the Nanga Parbat massif, Pakistan, spinel- and cordierite-bearing migmatites formed by anatexis of metapelites at ~5 kbar (Whittington *et al.* 1998), showing that biotite-breakdown conditions can also be reached at rather low pressures. Harris & Massey (1994) estimated these to have been the conditions of melting in the Himalayan migmatites.

In the Black Hills, the depth of generation of the two-mica variety of the Harney Peak Granite is difficult to determine because the appropriate Archaean metasedimentary protolith occurs both along the margins of the Precambrian terrane and probably below a dipping interface with the exposed folded Proterozoic metasedimentary sequence. However, there is no evidence for great depths of generation of the two-mica variety. Instead, its emplacement by dykes, in the same manner and within the same plutons as the muscovite-tourmaline variety, suggests magma generation within approximately the same region in the upper plate of the thickened crust as generation of the tourmaline granites.

5. Thermal constraints

The above summary shows that leucogranite generation in collisional orogens is directly related to metamorphism and deformation of the upper plate of thickened crust. Therefore, thermal models in which granite generation is simply a passive response to heat flow and heat production in the crust do not adequately reflect the dynamic nature of the granite-generation process. Successful thermal models should reproduce the metamorphic and deformation conditions of collisional orogens in addition to meeting the thermal requirements for leucogranite generation. Below, the present authors review five types of models which have been proposed to explain leucogranite generation in collisional orogens: simple crustal thickening, enhanced internal heat production in the crust, decompression melting, thinning of the mantle lithosphere and shear-heating. Although they have previously used forms of the models to address the specific problem of metamorphism, granite generation and crustal rheology in the Black Hills (Nabelek & Liu 1999; Nabelek *et al.* 2001), the models presented here have fairly simple boundary conditions which permit a more general evaluation of the potential processes which lead to metamorphism and leucogranite generation in collisional orogens.

The transient thermal evolution during crustal thickening and denudation that involves melt generation can be written as

$$\frac{\partial T}{\partial t}\left(1+\frac{L\partial f}{C_p\partial T}\right)=k\nabla^2 T-u\cdot\nabla T+\frac{1}{\rho C_p}(A_r+A_s) \quad (1)$$

(Liu & Furlong 1993). The parameter T is temperature and the term $u \cdot \nabla T$ is thermal advection associated with thickening and erosion, in which u is the velocity vector. Parameter t is time, k is thermal diffusivity ($1 \times 10^{-6}\ m^2\ s^{-1}$), ρ is density ($2900\ kg\ m^{-3}$), C_p is specific heat, L is latent heat of fusion and f is melt fraction. The rates of volumetric radioactive heating, A_r, and shear-heating, A_s, are salient parameters in the models discussed here.

The specific heat of the crustal rocks was assumed to vary with temperature, with $C_p = a + bT - cT^{-2}$. The applied coefficients, a=276, b=68·3 × 10^{-3} and c=87·5 × 10^5, are appropriate for the average mineralogy of Black Hills schists. For example, at 25°C, heat capacity is 726 J kg^{-1} and at 700°C it is 1223 J kg^{-1}. Latent heat of fusion, 1·8 × 10^5 J kg^{-1}, was invoked in the calculations when temperature reached the muscovite or biotite+muscovite dehydration-melting reactions. Melting was assumed to occur over a 25°C interval and melt fraction was 20%, as allowed for MDM of Black Hills schists (Nabelek & Bartlett 1998). Melting buffers geotherms to remain close to the schist solidus until the melting interval is exceeded.

Equation 1 was solved by the finite difference method using a modified version of the OROGEN computer program (Liu &

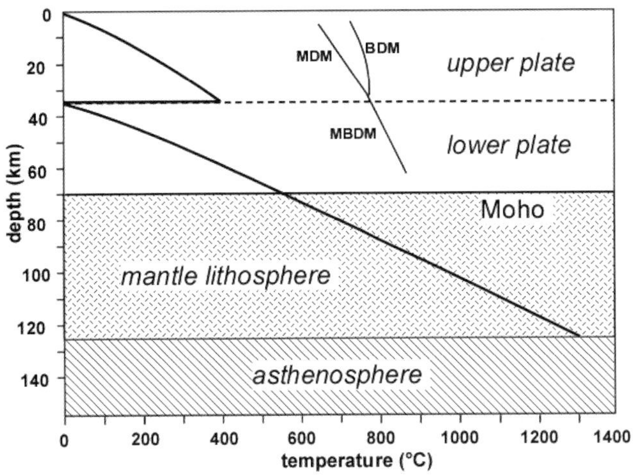

Figure 7 Initial geometry and temperature distribution for numerical simulations. The model assumes instantaneous thickening placing a 35 km upper plate over a 90 km lithosphere consisting of 35 km crust and 55 km mantle lithosphere. Muscovite-dehydration melting (MDM), biotite-dehydration melting (BDM) and combined muscovite-biotite-dehydration melting (MBDM) reactions are also shown.

Furlong 1993). The present authors assumed that the crust was thickened by stacking a 35-km sequence of oceanic sediments over a 90-km lithosphere that included a 35-km continental crust (Fig. 7). This lithologic profile is an analogue to the stack of metasedimentary sequences which lie over the Indian crust in the Himalayas, the stack of CMB sequences over the Grenville basement in Maine, and the stack of Proterozoic and Archaean sedimentary rocks over the Wyoming basement in the Black Hills.

The initial 'saw-tooth' thermal profile has an appeal to some because it reflects a rapid superposition of hot rocks over cold rocks, which could potentially produce an inverted metamorphic sequence below the thrust (e.g. Swapp & Hollister 1991). However, the details of the initial saw-tooth thermal profile for the long-term thermal evolution of orogens are unimportant, because the profile gets smoothed out by relaxation within ~10 Ma after 'instantaneous' stacking, erasing most of the initial inverted thermal profile below the thrust.

The total thickness of the lithosphere in all models was assumed to be 125 km. In the lithosphere-thinning models, the thickness was reduced to 100 and 80 km at 30 Ma. after instantaneous stacking of the crust. Temperature at the surface was fixed at 25 °C and at the base of the lithosphere at 1300 °C, maintained by convective mantle flow below. The present authors chose fixed temperature at the base of the lithosphere because both temperature and heat flux at the base of the crust are transient variables during orogeny. Models which assume a fixed heat flux at the base of the crust during the crustal heating process imply that, as the crust is heating up, there is an artificial rise in the mantle temperature to keep a constant thermal gradient across the Moho since Fourier's law requires heat flux to be proportional to the thermal gradient.

Although various rates of denudation of the thickened crust can be incorporated into the models to fit a known *P-T-t* path in a particular orogen (Nabelek & Liu 1999; Nabelek *et al.* 2001), here the authors assumed a fixed 70-km crustal thickness for all models, except for decompression melting. The assumption of the fixed crustal thickness allows us to make more general conclusions. It is also reasonable for the Himalayas, where the crust has maintained a double thickness through dynamic equilibrium between erosion and thickening, and the southern Trans-Hudson orogen, where the crust has remained >50 km thick since the collision even after ~15 km of erosion.

Internal heat generation depends on the concentration and distribution of heat-producing radioactive isotopes in the crust. The authors used 2 μW m^{-3} for volumetric internal heating at the surface (A_o), which is based on the average concentration of the major radioactive isotopes in the Black Hills schists (Nabelek & Bartlett 1998). The value is normal for crustal rocks, which generally have heat production in the range of 0·5 to 3 μW m^{-3} (Spear 1993). The present authors used two profiles for radiogenic heat production in the upper plate. The first incorporates a 'cold upper plate' (CUP) that has an exponential decrease of internal heat production with depth, $A(Z)=A_o e^{-Z/D}$, where Z is depth and D is drop-off length (Lachenbruch & Sass 1977). D was assumed to be 15 km. The second profile incorporates a 'hot upper plate' (HUP) in which internal heat production does not decrease with depth. The second profile is more appropriate for an upper plate made up mostly of pelitic metamorphic rocks. In both cases, heat production in the lower crust was assumed to decay exponentially with depth with D=15 km and A_0=2 μW m^{-3}. Higher average radiogenic heat production in the upper or lower plates, for example ≥3 μW m^{-3} (Huerta *et al.* 1998; Jamieson *et al.* 1998), would require unusual conditions which may be specific to a given orogen, but are unlikely to be generally applicable.

Geotherms and relevant *P-T-t* paths for all models shown are in time intervals of 10 Ma for the total of 50 Ma, except in cases of lithospheric thinning for which the 60 Ma geotherms are also shown. In the Black Hills and the Himalayas, leucogranite generation has occurred 30–40 Ma after initiation of collision. Therefore, geotherms for these times are most important.

5.1. Geotherms in 'cold' thickened crust

The evolution of geotherms in a thickened crust with CUP is shown in Figure 8a. The geotherms are essentially the same as those produced by others for similarly distributed internal heat production (e.g. Thompson & Connolly 1995). It is clear that simple stacking of crusts with normal internal heat generation cannot raise temperature high enough to melt the crust anywhere. Melting could only occur in rocks which are rapidly uplifted from depths of more than ~50 km, as shown by Thompson & Connolly (1995). These are not appropriate depths for source rocks of leucogranites in thickened crusts because there is no strong evidence that accretionary metapelitic rocks are injected into these depths during collisions, although this has been recently proposed for the Himalayas (Kohn & Parkinson 2002). Alternative models assumed different distributions of radioactive elements in the crust and shear-heating as means to raise the geotherms sufficiently to produce melts within the crust without the involvement of mafic magmas (Chamberlain & Sonder 1990; Royden 1993; Harrison *et al.* 1998; Huerta *et al.* 1998; Jamieson *et al.* 1998; Nabelek & Liu 1999; Nabelek *et al.* 2001).

5.2. Geotherms in 'hot' thickened crust

Stacking of HUP, with uniform 2 μW m^{-3} internal heat generation, causes elevation of geotherms, particularly in the middle and lower crust compared with the CUP case (Fig. 8b). However, only the very bottom of the crust may reach melting conditions after ~50 Ma. A greater thickness of HUP or deeper burial of lithologies with high radiogenic heat production could elevate the melting zone to shallower depths (Huerta *et al.* 1998; Jamieson *et al.* 1998), but within reasonable ranges of radioactive heat production, this model cannot sufficiently heat the upper plate to melting conditions.

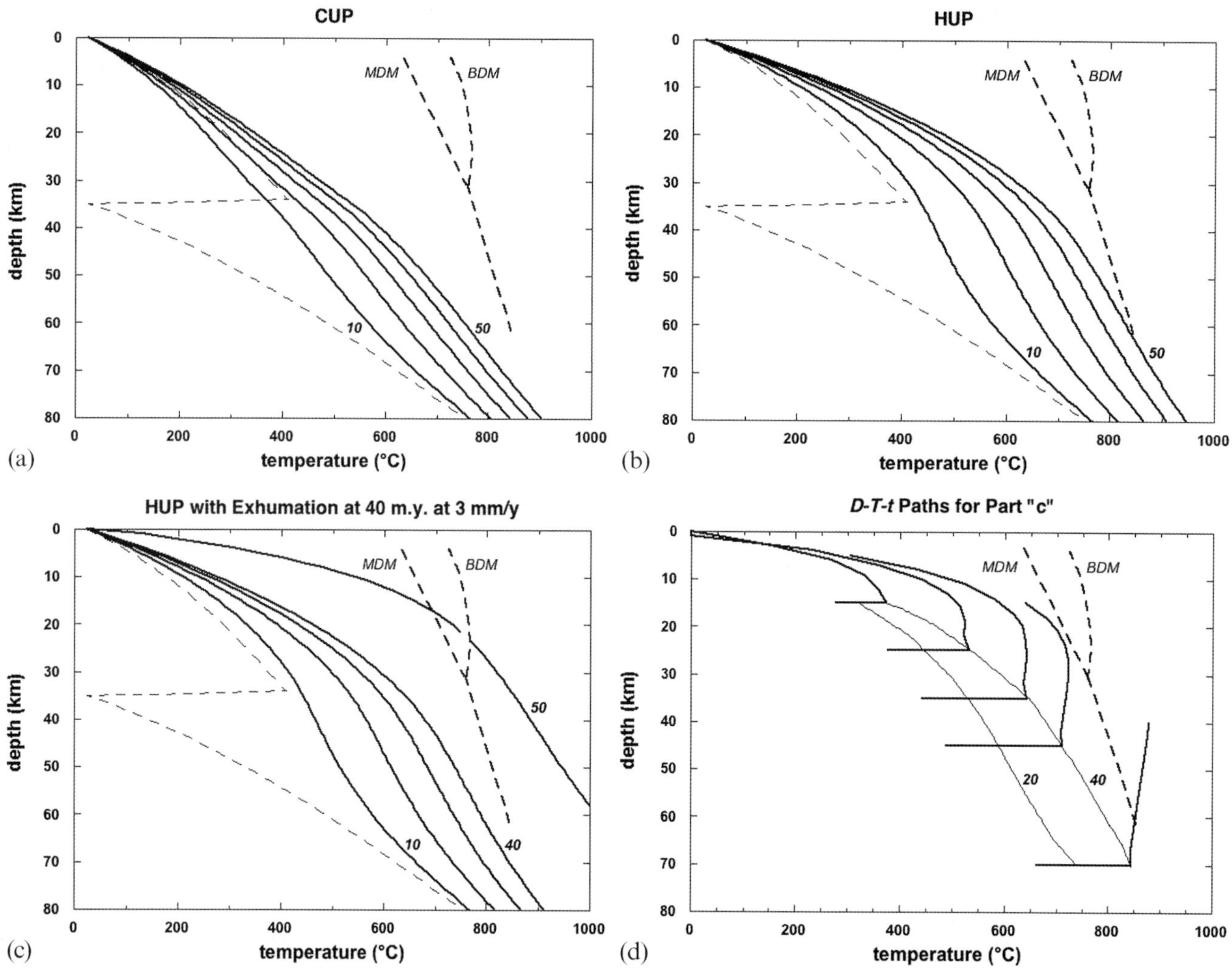

Figure 8 (a) Evolving geotherms in a stacked lithosphere with 'cold upper plate' (CUP). (b) Evolving geotherms in a stacked lithosphere with 'hot upper plate' (HUP). (c) Evolving geotherms in a stacked lithosphere with HUP in which exhumation at 3 mm y^{-1} begins 40 Ma after initial 'instantaneous' crustal thickening. (d) Exhumation paths of rocks from different depths for the model shown in (c). The 20 and 40 Ma time lines correspond to geotherms at those times. The figure shows that only rocks which originate at depths >45 km can cross the metapelite solidus during rapid exhumation.

5.3. Decompression melting

Decompression melting is a popular model for leucogranite generation in convergent orogens (e.g. Zeitler & Chamberlain 1991; Harris & Massey 1994). The model is based on petrologic evidence for rapid decompression of massifs which contain leucogranites and migmatites. A weakness of most published decompression models is that they say little about how source rocks become hot enough to eventually melt and from what depths they come during decompression. Progress in estimating the depth of burial of source rocks was made by Vance & Harris (1999), who used combined thermobarometry and dating of garnet growth to show that High Himalaya crystallinites in the Zanskar region of India reached 10 kbar and 700°C at ~5 Ma before leucogranite generation. What happed to these rocks during this time period before leucogranite generation is unclear, but they would have had to remain hot until melting has occurred.

The present authors modelled decompression melting using the HUP stack, assuming exhumation beginning at 40 Ma after the 'instantaneous' thrusting event. The exhumation was at the constant rate of 3 mm y^{-1} and lasted for 10 Ma. For decompression melting to work, the exhumation rate must be higher than the rate of thermal relaxation so that raising of geotherms to shallower levels can be sustained. The used exhumation rate is the lower estimate for exhumation of the Nanga Parbat massif in the Himalayas (Whittington 1996). In the Black Hills there is no evidence for such high exhumation rates.

Figure 8c shows that 10 Ma of rapid exhumation of hot lower-crustal rocks significantly elevates the geotherm to temperatures sufficient for melting at relatively shallow levels. However, the corresponding depth-temperature-time (*D-T-t*) paths (Fig. 8d) show that only rocks which come from depths of >45 km can cross the metapelite solidus, and only after more than ~40 Ma of thermal relaxation after initial thickening. Thus, decompression melting requires that source rocks had been buried near the bottom of the thickened crust prior to melting, much deeper than the maximum depth estimated for the High Himalaya crystallinites by Vance & Harris (1999). Although Kohn & Parkinson (2002) argued for burial of the High Himalaya crystallinites to ~100 km based on rare eclogites found in the crystallinites, it is unknown when the eclogites formed. In the Black Hills and Maine, there is no evidence for such deep burial of leucogranite protoliths. Furthermore, it is unclear why decompression paths of source rocks in the three orogens would happen to have crossed the metapelite solidus mostly below 10 kbar instead of at higher pressures during exhumation.

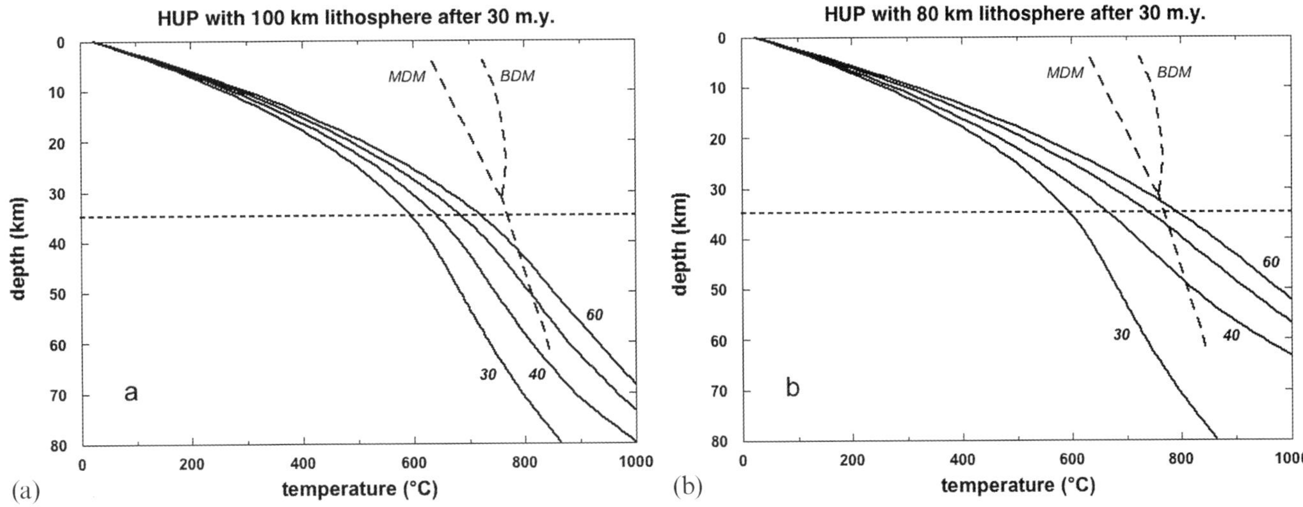

Figure 9 Evolving geotherms in a stacked lithosphere with 'hot upper plate' that has undergone 'instantaneous' thinning of the mantle lithosphere at 30 Ma after initial thickening. In part (a), the lithosphere was thinned to 100 km, and in part (b), to 80 km. Only the thinner lithosphere permits melting of the upper plate, but only after ~25 Ma of thermal relaxation after thinning.

5.4. Thinning of mantle lithosphere

Thinning of the mantle lithosphere as a means to heat up the crust has been invoked to explain the intrusion of leucogranites into deformed and metamorphosed metasedimentary rocks in several collisional orogens, including the CMB and the Trans-Hudson (e.g. De Yoreo *et al.* 1991; Holm *et al.* 1997). Because leucogranite plutons in these orogens imprinted contact aureoles on their country rocks, they were interpreted to be post-collisional. However, more recent structural evidence and thermal models suggest that the granites merely intruded upper portions of metasedimentary sequences which were still undergoing contractional deformation and melting at greater depths (Brown & Solar 1998b; Nabelek *et al.* 2001). Nevertheless, it is useful to examine the feasibility of producing leucogranites within upper-plate rocks if the mantle lithosphere is thinned after a period of doubled crustal thickness.

In the models, the present authors assumed that the lithosphere was instantly thinned 30 Ma after crustal thickening. The thinning was assumed to occur from the bottom. They modelled two cases, one with thinning to 100 km of lithosphere thickness and the other to 80 km (Fig. 9). In both cases, HUP was assumed since it favours greater elevation of geotherms. The results show that, if the lithosphere is thinned to 100 km, geotherms in the upper crust are not raised to melting conditions even when steady-state conditions are approached 30 Ma later. In the case of a 80 km lithosphere, it would take ~25 Ma after thinning to raise the geotherm to melting conditions at the upper plate-lower plate boundary. Thus, there has to be a period on the order of tens of millions of years after extreme (~45 km) lithosphere thinning before magmas can be generated from metapelites in the upper plate.

To explain 'apparent' post-collisional granites, lithosphere thinning has been assumed to begin after cessation of contractional deformation. The 25-Ma period required to heat up upper plate rocks is not supported by data from either the Black Hills or the Himalayas. In the Black Hills, metamorphic monazite within foliated metapelites grew only ~5 Ma prior to intrusion of the Harney Peak Granite (Dahl & Frei 1998; Dahl *et al.* 2002). In the Himalayas, there is about the same gap between garnet growth in the High Himalaya crystallinites, which is thought to date deformation of the rocks and leucogranite generation (Vance & Harris 1999). Five million years is insufficient for conduction to bring sufficient heat into upper parts of a thickened crust after thinning of the mantle lithosphere, even with the extreme assumption of instantaneous thinning.

The model would be helped if thinning of the lithosphere was at least partly concurrent with collision. Although the lithosphere has been thinned under northern Tibet, for example, while the Himalayan collision is still active (Molnar *et al.* 1993), the lithosphere has remained thick below the Himalayas. Below the southern Trans-Hudson orogen and the CMB, the lithosphere is thick to this day (e.g. Van Der Lee & Nolet 1997).

In subduction regions and extensional terranes, where various forms of lithosphere thinning, including slab break-off, can occur, the thinning is typically accompanied by alkalic volcanism or production of I-type lower-crustal magmas as basaltic liquids are intruded into the crust and the geotherms become elevated (e.g. Fig. 9b). There is no evidence for this style of magmatism in the Black Hills, the Himalayas or the CMB at the time of leucogranite production.

5.5. Shear-heating

Shear-heating has been invoked to explain both inverted metamorphic gradients below shear zones and leucogranite generation (Zhu & Shi 1990; England & Molnar 1993; Harrison *et al.* 1998; Nabelek *et al.* 2001). Shear-heating is attractive because it directly relates magma generation to preceding metamorphism and deformation that is evident in orogens. It is a common misconception that these shear-heating models involve simple frictional heating along fault planes. On the contrary, as applied here, shear-heating is the thermal response of rocks to strain that acts over a volume of rock that is undergoing deformation. Therefore, it can also be effective in the ductile regime within the crust. Perhaps a better term would be 'strain-heating', but the present authors retain the more commonly used term. In their models, shear-heating is assumed to operate across a broad horizontal shear zone, consistent with the occurrence of broad migmatitic shear zones in pelitic sequences which are thrust over basements during collisions.

The rate of volumetric shear-heating in strained rocks is given by

$$A_s = \tau v / dz, \qquad (2)$$

where τ is shear stress, v is thrusting velocity, and dz is the width of the shear zone (Liu & Furlong 1993). The parameter

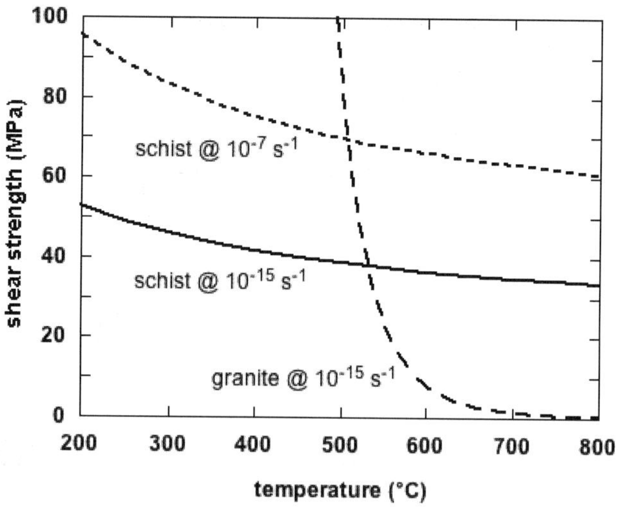

Figure 10 Shear strengths of mica schist and dry granite as a function of temperature calculated from coefficients for power-law behaviour of rheology (Kirby & Kronenberg 1987; Shea & Kronenberg 1992). Assumed strain rate for each plot is indicated in the legend.

dz accounts for partitioning of stress created by plate convergence across the width of the shear zone. The present authors assumed that the effect of shear-heating decreases in a Gaussian fashion away from centre of the shear zone. They also assumed that the width of the shear zone is 4 km. This width is approximately the spacing of faults in the Black Hills and is appropriate for scales of shear zones elsewhere. In any case, the results are relatively insensitive to *dz* values between 1 and 10 km.

Shear-heating requires that rocks retain sufficient shear strength to heat up while deforming at geologically slow rates. The authors assumed that shear stress is limited by the shear strength of rocks in the shear zone. They used 35 MPa for τ, the shear strength of a mica schist at 750°C under a strain-rate of $10^{-15}\,s^{-1}$, based on experiments of Shea & Kronenberg (1992). It is notable that the strength of a mica schist is quite insensitive to temperature, especially in comparison to a dry granite, which rapidly looses strength above 500°C (Fig. 10). Thus, the strength of a mica schist is much less sensitive to temperature increase with depth in the crust than the strength of a granite, which is commonly used to model the strength of the continental crust and effectiveness of shear-heating with depth (e.g. Leloup *et al.* 1999). Indeed, a granite rheology leads to a very weak lower crust, whereas a schist rheology leads to a quite strong lower crust even in the ductile regime (Shea & Kronenberg 1992; Nabelek *et al.* 2001). The shear strength of a mica schist has also small dependence on strain rate (Fig. 10).

In the present models, shear-heating was allowed to proceed only below the melting interval because it is presumed that strain is partitioned into the melt in partially molten rocks. Therefore, when temperature within the shear zone reaches the solidus, shear strength of the rocks should significantly drop. When shear strength drops, the temperature drops below the solidus until strength is again regained either because the rock solidifies (becomes a migmatite?) or because the melt is extracted. In this way, temperature within the shear zone in the models was kept near the solidus by cycling the strength of the rocks. The cycling reproduces a possible episodic extraction of melts from a deforming region. It is also noted that provision for heat of fusion was incorporated into the models.

Four sets of geotherms for different rates of thrusting, shear zone depths, and different heat-production profiles are shown in Figure 11. Note that the depth of the shear zone does not correspond to the upper/lower crust boundary in the examples shown. Placement of the shear zone above the main crustal boundary merely emphasises that shear-heating is effective in raising the crustal geotherms irrespective of where shear zones occur. In all cases, thrusting was assumed to occur for the whole duration of simulation. Overall, the results show that shear-heating caused by deformation associated with thrusting at the rate of 4 cm y^{-1} can easily raise temperatures in the upper plate to conditions of melting. The metapelite solidus is intersected even in a CUP when the shear zone is at ~25 km or deeper (Fig. 11a). For HUP, the shear zone can be at a much shallower depth and still generate leucogranites. Whether the shear zone is at 15 or 25 km depth, the metapelite solidus is intersected within the upper plate (Figs. 11b, c). However, it takes >10 Ma longer for melting conditions to be reached if the depth is 15 km because colder rocks need to be heated and heat generated by shear is lost faster by conduction closer to the surface.

It is notable that, irrespective of where the shear zone is in the upper plate, the geotherms cross, undoubtedly by coincidence, the metapelite solidus near the probable intersection of the MDM and BDM reactions. This allows for the occurrence of both reactions within a very narrow temperature range, such that they can occur at approximately the same time and the same depth in metapelites which contain both muscovite and biotite, and metagraywackes which may contain only biotite. Thus, tourmaline-muscovite and two-mica leucogranites can be produced within the same region of the crust.

Lowering the rate of thrusting diminishes the effectiveness of shear-heating. At 2 cm y^{-1}, the metapelite solidus is barely intersected within the middle crust if there is HUP (Fig. 11d). In the case of CUP, the metapelite solidus is not intersected. Of course, doubling of the shear strength is as effective as doubling the rate of thrusting. Shear strength would increase, for example, with higher strain rate (Fig. 10). If shear strengths of rocks were as high as empirically estimated by England & Molnar (~100 MPa; 1993) for shear zones which include inverted metamorphic gradients below them, then shear-heating would be effective in raising geotherms even at much lower rates of thrusting and in a crust with less radiogenic heat production.

6. Discussion

The model results show that heating associated with deformation in a shear zone in the upper plate can effectively raise the geotherm to initiate melting. A common objection to the shear-heating model is that, once melting begins, the contribution of shear-heating to the geotherm will drop as shear stress in the shear zone becomes negligible. This drop in temperature will reduce melt production. Moreover, because strain becomes accumulated in the partially molten zones, deformation will be reduced elsewhere in the crust (Brown & Solar 1998a). Such drop in shear stress in the crust may explain why leucogranite generation appears to have often followed deformation of metapelites during convergent orogenies. On the other hand, MDM, which can lead to, on average, ~15% melting in rocks like the Black Hills metapelites (Nabelek & Bartlett 2000), is essentially a discontinuous reaction occurring over a very narrow temperature interval (Patiño-Douce & Johnston 1991). Therefore, generation of melts by MDM is likely to be rapid and proceed to completion once melting temperatures are reached. Moreover, after initial melts are extracted, shear stress in the source region may again increase to produce additional melts from still-fertile rocks, perhaps even at a higher temperature. Thus, protracted or episodic leucogranite magmatism can occur during continental collisions (e.g. Harrison *et al.* 1998). In the Black Hills, the earliest

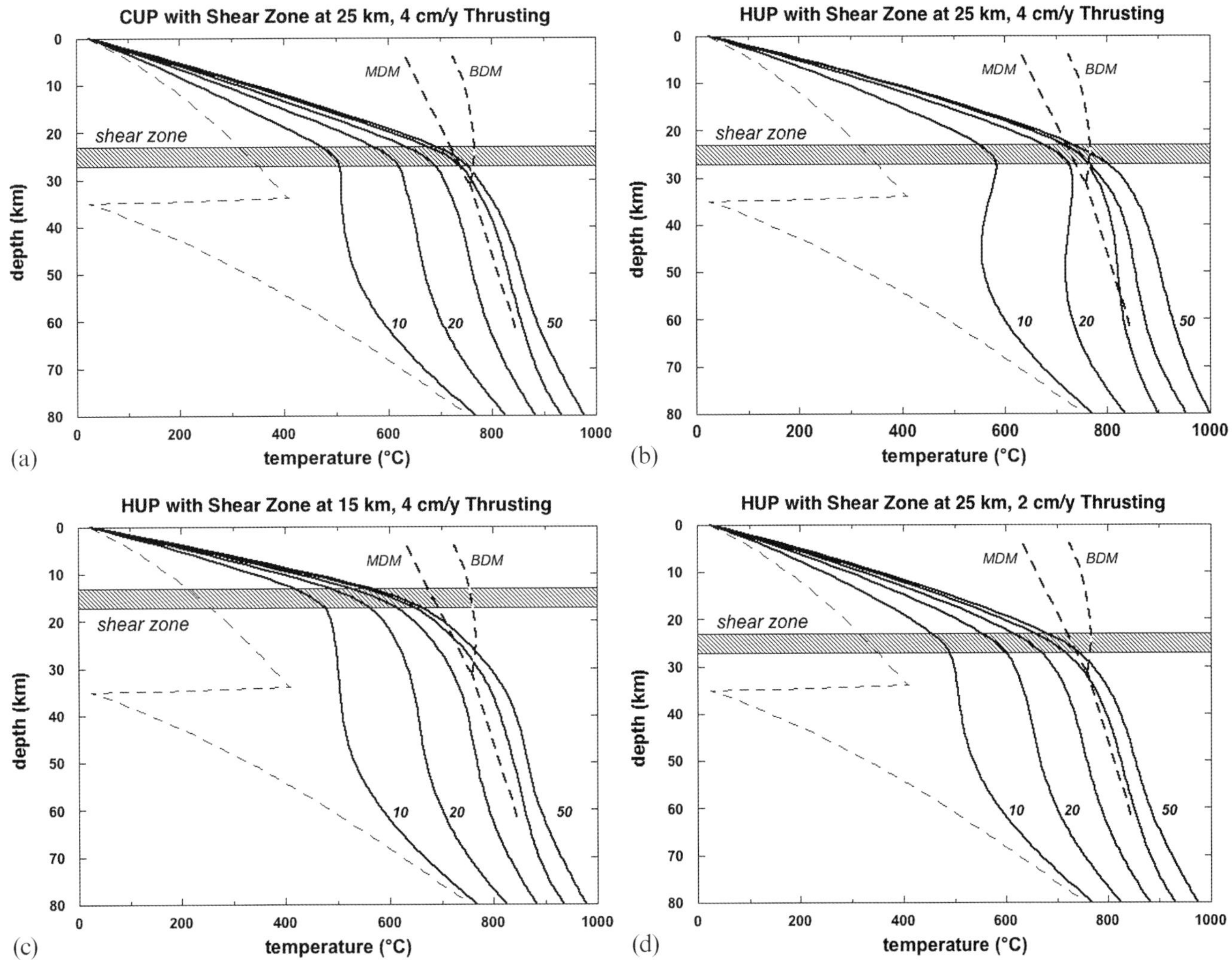

Figure 11 Evolving geotherms in thickened crust with shear-heating: (a) 25-km-deep shear zone undergoing 4 cm y^{-1} thrusting in a 'cold upper plate'' (b) 25-km-deep shear zone undergoing 4 cm y^{-1} thrusting in a 'hot upper plate' (HUP); (c) 15-km-deep shear zone undergoing 4 cm y^{-1} thrusting in a HUP; and (d) 25-km-deep shear zone undergoing 2 cm y^{-1} thrusting in a HUP.

known dates for leucogranite and pegmatite magmatism are 1720 and the latest 1702 (Redden *et al.* 1990; Krogstad & Walker 1994), indicating a ~20 Ma period of magmatism.

The evolving geotherms shown in Figure 11a–c nicely reproduce the progression from initial garnet growth in upper-crustal rocks at 400–500 °C, followed a few tens of millions of years later by leucogranite generation, and then by rapid cooling that may accompany exhumation of crustal blocks (Harris *et al.* 2000; Nabelek *et al.* 2001). Therefore, the shear-heating model ties together metamorphism, deformation and magma genesis into one dynamic process instead of a set of discrete events. Viewed in this manner, exhumation of massifs may be the result of granite generation instead of being the cause of it. As shown by Teyssier & Whitney (2002), partially molten crustal blocks may undergo buoyancy-driven exhumation that may bring deep migmatitic parts of the crust to the surface. The partially molten rocks may undergo further melting if their decompression path is rapid enough to remain nearly isothermal so that the rocks effectively move further above their solidus that has a positive dP/dT slope.

The present authors' modelling suggests that an inverted metamorphic gradient that may initially develop below a thrust of one crustal plate over another (c.f. Swapp & Hollister 1991) is unlikely to be preserved after 10 Ma, unless the thrust is rapidly exhumed and the inverted gradient 'quenched'. Only when the upper plate is hot and the thrusting fairly rapid may shear-heating maintain a thermal maximum at the shear zone for longer durations (Fig. 11b). Nevertheless, the maintained inverted thermal gradient is much smaller than that observed below the MCT in the Himalayas, for example. By the time temperatures in the upper plate reach melting conditions, the rocks below are also in excess of 700 °C, and therefore, largely dehydrated except for structural water in micas. It is highly unlikely that conditions can exist in the crust where dehydration of low-grade sediments leads to melting in an overthrusted hot plate, as in the model of Le Fort *et al.* (1987). Evidence for substantial movement on the MCT and garnet growth in the inverted sequence below it over the last 3 Ma also argues against a connection between metamorphism of the footwall sediments and generation of the High Himalaya granites (Catlos *et al.* 2001).

7. Conclusion

Shear-heating associated with deformation of the upper thickened crust provides the best explanation for the common characteristics of leucogranites and their relationships to metamorphism and structure in several collisional orogenies which have occurred throughout the Earth's history. The shear-heating model directly links partial melting of upper-crustal lithologies to the dynamic crustal processes which occur during thrusting of continental slices over the subducting continental crust. The model does not require exhumation of source rocks from the deep crust, abnormally high concentrations of heat-producing radioactive elements in the crust or extensive, very rapid thinning of the mantle lithosphere as are required by alternative models. While such conditions may be specific to a given orogen, it is unlikely that they existed in all collisional orogens where leucogranites occur. The shear-heating model emphasises the active nature of the leucogranite-generation process that is evident in all the orogens examined in this study.

8. Acknowledgements

Discussions and comments from Alan Whittington led to improvement of the paper. Constructive reviews by Gary Stevens and M. Satish-Kumar were very helpful. The study was funded in part by National Science Foundation Grant EAR-9980374 and University of Missouri Research Board Grant 01–146.

9. References

Ahmad, T., Harris, N., Bickle, M., Chapman, H., Bunbury, J. & Prince, C. 2000. Isotopic constraints on the structural relationships between the Lesser Himalayan Series and the High Himalayan Crystalline Series, Garhwal Himalaya. *Geological Society of America Bulletin* **112**, 467–77.

Ayres, M. & Harris, N. 1997. REE fractionation and Nd-isotope disequilibrium during crustal anatexis; constraints from Himalayan leucogranites. *Chemical Geology* **139**, 249–69.

Brown, M. & Solar, G. S. 1998a. Shear-zone systems and melts: feedback relations and self-organization in orogenic belts. *Journal of Structural Geology* **20**, 211–27.

Brown, M. & Solar, G. S. 1998b. Granite ascent and emplacement during contractional deformation in convergent orogens. *Journal of Structural Geology* **20**, 1365–93.

Catlos, E. J., Harrison, T. M., Kohn, M. J., Grove, M., Ryerson, F. J., Manning, C. E. & Upreti, B. N. 2001. Geochronologic and thermobarometric constraints on the evolution of the Main Central Thrust, central Nepal Himalaya. *Journal of Geophysical Research* **106**, 16,177–204.

Chamberlain, C. P. & Sonder, L. J. 1990. Heat-producing elements and the thermal and baric patterns of metamorphic belts. *Science* **250**, 763–9.

Chappell, B. W. & White, A. J. R. 1992. I- and S-type granites in the Lachlan fold belt. *Transactions of the Royal Society of Edinburgh: Earth Sciences* **83**, 1–26.

Dahl, P. S., Jercinovic, M. J. & Williams, M. L. 2002. Chemical dating of monazite in the Black Hills, South Dakota, with implications for Proterozoic thermotectonism in the eastern Wyoming Province *Geological Society of America, 2002 Annual Meeting* **34**, 68.

Dahl, P. S. & Frei, R. 1998. Step-leach Pb-Pb dating of inclusion-bearing garnet and staurolite, with implications for Early Proterozoic tectonism in the Black Hills collisional orogen, South Dakota, United States. *Geology* **26**, 111–14.

Davidson, C., Grujic, D. E., Hollister, L. S. & Schmid, S. M. 1997. Metamorphic reactions related to decompression and synkinematic intrusion of leucogranite, High Himalayan Crystallines, Bhutan. *Journal of Metamorphic Geology* **15**, 393–612.

De Yoreo, J. J., Lux, D. R. & Guidotti, C. V. 1991. Thermal modeling in low-pressure/high-temperature metamorphic belts. *Tectonophysics* **188**, 209–38.

Duke, E. F., Redden, J. A. & Papike, J. J. 1988. Calamity Peak layered granite-pegmatite complex, Black Hills, South Dakota: Part I. Structure and emplacement. *Bulletin of the Geological Society of America* **100**, 825–40.

Duke, E. F., Papike, J. J. & Laul, J. C. 1992. Geochemistry of a boron-rich peraluminous granite pluton: the Calamity Peak layered granite-pegmatite complex, Black Hills, South Dakota. *Canadian Mineralogist* **30**, 811–34.

England, P. & Molnar, P. 1993. The interpretation of inverted metamorphic isograds using simple physical calculations. *Tectonics* **12**, 145–57.

France-Lanord, C. & Le Fort, P. 1988. Crustal melting and granite genesis during the Himalayan collision orogenesis. *Transactions of the Royal Society of Edinburgh: Earth Sciences* **79**, 183–95.

Guillot, S. & Le Fort, P. 1995. Geochemical constraints on the bimodal origin of High Himalayan leucogranites. *Lithos* **35**, 221–34.

Harris, N. B. W., Vance, D. & Ayres, M. 2000. From sediment to granite; timescales of anatexis in the upper crust *Chemical Geology* **162**, 155–67.

Harris, N. & Massey, J. 1994. Decompression and anatexis of Himalayan metapelites. *Tectonics* **13**, 1537–46.

Harris, N. B. W. & Inger, S. 1992. Trace element modelling of pelite-derived granites. *Contributions to Mineralogy and Petrology* **110**, 46–56.

Harrison, T. M., Grove, M., Lovera, O. M. & Catlos, E. J. 1998. A model for the origin of Himalayan anatexis and inverted metamorphism. *Journal of Geophysical Research* **103**, 27,017–32.

Holloway, J. R. & Blank, J. G. 1994. Application of experimental results to C-O-H species in natural melts. *Reviews in Mineralogy* **30**, 187–230.

Holm, D. K., Dahl, P. S. & Lux, D. R. 1997. $^{40}Ar/^{39}Ar$ evidence for Middle Proterozoic (1300–1500 Ma) slow cooling of the southern Black Hills, South Dakota, midcontinent, North America: implications for Early Proterozoic P-T evolution and posttectonic magmatism. *Tectonics* **16**, 609–22.

Holtz, F. & Johannes, W. 1991. Genesis of peraluminous granites I. Experimental investigation of melt compositions at 3 and 5 kb and various H_2O activities. *Journal of Petrology* **32**, 935–58.

Huerta, A. D., Royden, L. H. & Hodges, K. V. 1998. The thermal structure of collisional orogens as a response to accretion, erosion, and radiogenic heat production. *Journal of Geophysical Research* **103**, 15,287–302.

Jamieson, R. A., Beaumont, C., Fullsack, P. & Lee, B. 1998. Barrovian regional metamorphism: where's the heat? *In* Treloar, P. J. & O'Brien, P. J. (eds) *What Drives Metamorphism and Metamorphic Reactions*? 23–51. London: Geological Society.

Johannes, W. & Holtz, F. 1996. *Petrogenesis and Experimental Petrology of Granitic Rocks*. Berlin: Springer-Verlag.

Kirby, S. H. & Kronenberg, A. K. 1987. Rheology of the lithosphere: selected topics. *Reviews of Geophysics* **25**, 1219–44.

Kohn, M. J. & Parkinson, C. D. 2002. Petrologic case for Eocene slab breakoff during the Indo-Asian collision. *Geology (Boulder)* **30**, 591–4.

Krogstad, E. J., Walker, R. J., Nabelek, P. I. & Russ-Nabelek, C. 1993. Lead isotopic evidence for mixed sources of Proterozoic granites and pegmatites, Black Hills, South Dakota, USA. *Geochimica et Cosmochica Acta* **57**, 4677–85.

Krogstad, E. J. & Walker, R. J. 1994. High closure temperatures of the U-Pb system in large apatites from the Tim Mountain pegmatite, Black Hills, South Dakota, USA. *Geochimica et Cosmochimica Acta* **58**, 3845–53.

Krogstad, E. J. & Walker, R. J. 1996. Evidence of heterogeneous crustal sources: the Harney Peak Granite, South Dakota, U.S.A. *Transactions of the Royal Society of Edinburgh: Earth Sciences* **87**, 331–7.

Lachenbruch, A. & Sass, J. 1977. Heat flow in the United States and the thermal regime of the crust. *In The Earth's Crust: Its Nature and Physical Properties*, 626–75. *Geophysical Monograph Series* **20**. Washington, DC: American Geophysical Union.

Le Breton, N. & Thompson, A. B. 1988. Fluid-absent (dehydration) melting of biotite in metapelites in the early stages of crustal anatexis. *Contributions to Mineralogy and Petrology* **99**, 226–37.

Le Fort, P., Cuney, M., Deniel, C., France-Lanord, C., Sheppard, S. M. F., Upreti, B. N. & Vidal, P. 1987. Crustal generation of the Himalayan leucogranites. *Tectonophysics* **134**, 39–57.

Leloup, P. H., Ricard, Y., Battaglia, J. & Lacassin, R. 1999. Shear heating in continental strike-slip shear zones; model and field examples. *Geophysical Journal International* **136**, 19–40.

Liu, M. & Furlong, K. P. 1993. Crustal shortening and Eocene extension in the southeastern Canadian Cordillera: some thermal and rheological considerations. *Tectonics* **12**, 776–86.

Molnar, P., England, P. & Martinod, J. 1993. Mantle dynamics, uplift of the Tibetan Plateau, and the Indian monsoon. *Reviews of Geophysics* **31**, 357–96.

Nabelek, P. I., Russ-Nabelek, C. & Denison, J. R. 1992a. The generation and crystallization conditions of the Proterozoic Harney Peak leucogranite, Black Hills, South Dakota, USA: petrologic and geochemical constraints. *Contributions to Mineralogy and Petrology* **110**, 173–91.

Nabelek, P. I., Russ-Nabelek, C. & Haeussler, G. T. 1992b. Stable isotope evidence for the petrogenesis and fluid evolution in the Proterozoic Harney Peak leucogranite, Black Hills, South Dakota. *Geochimica et Cosmochimica Acta* **56**, 403–17.

Nabelek, P. I., Liu, M. & Sirbescu, M. L. 2001. Thermo-rheological, shear heating model for leucogranite generation, metamorphism, and deformation during the Proterozoic Trans-Hudson Orogeny, Black Hills, South Dakota. *Tectonophysics* **342**, 371–88.

Nabelek, P. I., Wilke, M., Huff, T. A. & Wopenka, B. 2003. Methane, an important component of fluids in graphitic metapelites. *Geochimica et Cosmochimica Acta* **67** (Supplement 1), A317.

Nabelek, P. I. & Bartlett, C. D. 1998. Petrologic and geochemical links between the post-collisional Proterozoic Harney Peak leucogranite, South Dakota, USA, and its source rocks. *Lithos* **45**, 71–85.

Nabelek, P. I. & Bartlett, C. D. 2000. Fertitlity of metapelites and metagraywackes during leucogranite generation: an example from the Black Hills, U.S.A. *Transactions of the Royal Society of Edinburgh: Earth Sciences* **91**, 1–14.

Nabelek, P. I. & Glascock, M. D. 1995. REE-depleted leucogranites, Black Hills, South Dakota: a consequence of disequilibrium melting of monazite-bearing schists. *Journal of Petrology* **36**, 1055–71.

Nabelek, P. I. & Liu, M. 1999. Leucogranites in the Black Hills, South Dakota: the consequence of shear heating during continental collision. *Geology* **27**, 523–6.

Nabelek, P. I. & Ternes, K. 1997. Fluid inclusions in the Harney Peak Granite, Black Hills, South Dakota, USA: Implications for solubility and evolution of magmatic volatiles and crystallization of leucogranite magmas. *Geochimica et Cosmochimica Acta* **61**, 1447–65.

Nelson, K. D., Zhao, W., Brown, L. D., Kuo, J., Che, J., Liu, X., Klemperer, S. L., Makovsky, Y., Meissner, R., Mechie, J., Kind, R., Wenzel, F., Ni, J., Nabelek, J., Chen, L., Tan, H., Wei, W., Jones, A. G., Booker, J., Unsworth, M., Kidd, W. S. F., Hauck, M., Alsdorf, D., Ross, A., Cogan, M., Wu, C., Sandvol, E. A. & Edwards, M. 1996. Partially molten middle crust beneath southern Tibet; synthesis of Project INDEPTH results. *Science* **274**, 1684–8.

Norton, J. J. & Redden, J. A. 1990. Relations of zoned pegmatites to other pegmatites, granite, and metamorphic rocks in the southern Black Hills, South Dakota. *American Mineralogist* **75**, 631–55.

Patiño-Douce, A. E. & Harris, N. 1998. Experimental constraints on Himalayan anatexis. *Journal of Petrology* **39**, 689–710.

Patiño-Douce, A. E. & Johnston, A. D. 1991. Phase equilibria and melt productivity in the pelitic system: implications for the origin of peraluminous granitoids and aluminous granulites. *Contributions to Mineralogy and Petrology* **107**, 202–18.

Pressley, R. A. & Brown, M. 1999. The Phillips Pluton, Maine, USA: evidence of heterogeneous crustal sources, and implications for granite ascent and emplacement mechanisms in convergent orogens. *Lithos* **46**, 335–66.

Redden, J. A., Peterman, Z. E., Zartman, R. E. & Dewitt, E. 1990. U-Th-Pb zircon and monazite ages and preliminary interpretation of the tectonic development of Precambrian rocks in the Black Hills. *Geological Association of Canada, Special Paper* **37**, 229–51.

Royden, L. H. 1993. The steady state thermal structure of eroding orogenic belts and accretionary prisms. *Journal of Geophysical Research* **98**, 4487–507.

Scaillet, B., France-Lanord, C. & Le Fort, P. 1990. Badrinath-Gangotri plutons (Garhwal, India): petrological and geochemical evidence for fractionation processes in high Himalayan leucogranite. *Journal of Volcanology and Geothermal Research* **44**, 163–88.

Scaillet, B., Pichavant, M. & Roux, J. 1995. Experimental crystallization of leucogranite magmas. *Journal of Petrology* **36**, 663–705.

Scaillet, B., Holtz, F., Pichavant, M. & Schmidt, M. 1996. Viscosity of Himalayan leucogranites; implications for mechanisms of granitic magma ascent. *Journal of Geophysical Research* **101**, 27,691–9.

Schneider, D. A., Edwards, M. A., Kidd, W. S. F., Asif, K. M., Seeber, L. & Zeitler, P. K. 1999. Tectonics of Nanga Parbat, western Himalaya; synkinematic plutonism within the doubly vergent shear zones of a crustal-scale pop-up structure. *Geology* **27**, 999–1002.

Searle, M. P., Parrish, R. R., Hodges, K. V., Hurford, A., Ayres, M. W. & Whitehouse, M. J. 1997. Shisha Pangma leucogranite, South Tibetan Himalaya; field relations, geochemistry, age, origin, and emplacement. *Journal of Geology* **105**, 295–317.

Shea, W. T. & Kronenberg, A. K. 1992. Rheology and deformation of an isotropic mica schist. *Journal of Geophysical Research* **97**, 15,201–37.

Sirbescu, M.-L. C. & Nabelek, P. I. 2003a. Crystallization conditions and evolution of magmatic fluids in the Harney Peak Granite and associated pegmatites, Black Hills, South Dakota – evidence from fluid inclusions. *Geochimica et Cosmochimica Acta* **67**, 2443–65.

Sirbescu, M.-L. C. & Nabelek, P. I. 2003b. Crustal melts below 400 °C. *Geology* **31**, 685–8.

Solar, G. S., Pressley, R. A., Brown, M. & Tucker, R. D. 1998. Granite ascent in convergent orogenic belts: testing a model. *Geology* **26**, 711–4.

Solar, G. S. & Brown, M. 2001. Petrogenesis of migmatites in Maine, USA: possible source of peraluminous leucogranite in plutons? *Journal of Petrology* **42**, 789–823.

Spear, F. S. 1993. *Metamorphic Phase Equilibria and Pressure-Temperature-Time Paths*. Washington, DC: Mineralogical Society of America.

Swapp, S. M. & Hollister, L. S. 1991. Inverted metamorphism within the Tibetan slab of Bhutan; evidence for a tectonically transported heat-source. *Canadian Mineralogist* **29**, 1019–41.

Teyssier, C. & Whitney, D. L. 2002. Gneiss domes and orogeny. *Geology* **30**, 1139–42.

Thompson, A. B. & Connolly, J. A. D. 1995. Melting of the continental crust: some thermal and geological constraints on anatexis in continental collision zones and other tectonic settings. *Journal of Geophysical Research* **100**, 15,565–79.

Tomascak, P. B., Krogstad, E. J. & Walker, R. J. 1996a. U-Pb monazite geochronology of granitic rocks from Maine: implications for Late Paleozoic tectonics in the Northern Appalachians. *Journal of Geology* **104**, 185–95.

Tomascak, P. B., Krogstad, E. J. & Walker, R. J. 1996b. Nature of the crust in Maine, USA: evidence from the Sebago batholith. *Contributions to Mineralogy and Petrology* **125**, 45–59.

Van Der Lee, S. & Nolet, G. 1997. Upper mantle S velocity structure of North America. *Journal of Geophysical Research, B, Solid Earth and Planets* **102**, 22,815–38.

Vance, D. & Harris, N. 1999. Timing of prograde metamorhism in the Zanskar Himalaya. *Geology* **27**, 395–8.

Watt, G. R. & Harley, S. L. 1993. Accessory phase controls on the geochemistry of crustal melts and restites produced during water-undersaturated partial melting. *Contributions to Mineralogy and Petrology* **11**, 550–66.

Whittington, A. G. 1996. Exhumation overrated at Nanga Parbat, northern Pakistan. *Tectonophysics* **206**, 215–26.

Whittington, A., Harris, N. & Baker, J. 1998. Low-pressure crustal anatexis; the significance of spinel and cordierite from metapelitic assemblages at Nanga Parbat, northern Pakistan. *In* Treloar, P. J. & O'Brien, P. J. (eds.) *What Drives Metamorphism and Metamorphic Relations*? 183–98. London: Geological Society.

Whittington, A. G., Harris, N. B. W. & Butler, R. W. H. 1999. Contrasting anatectic styles at Nanga Parbat, northern Pakistan. *Geological Society of America, Special Paper* **328**, 129–44.

Whittington, A. G. & Treloar, P. J. 2002. Crustal anatexis and its relation to the exhumation of collisional orogenic belts, with particular reference to the Himalaya. *Mineralogical Magazine* **66**, 53–91.

Wilke, M., Nabelek, P. I. & Glascock, M. D. 2002. B and Li in metapelites from the Proterozoic Terrane in the Black Hills, South Dakota, USA: implications for the origin of leucogranitic magmas. *American Mineralogist* **87**, 491–500.

Winslow, D. M., Chamberlain, C. P. & Zeitler, P. K. 1995. Metamorphism and melting of the lithosphere due to rapid denudation, Nanga Parbat Massif Himalaya. *Journal of Geology* **103**, 395–409.

Zeitler, P. K. & Chamberlain, C. P. 1991. Petrogenetic and tectonic significance of young leucogranites from the northwestern Himalaya, Pakistan. *Tectonics* **10**, 729–41.

Zhu, Y. & Shi, Y. 1990. Shear heating and partial melting of granite: thermal structure of overthrusted terrains in the Greater Himalaya. *Chinese Journal of Geophysics* **33**, 341–51.

PETER I. NABELEK and MIAN LIU, Department of Geological Sciences, University of Missouri-Columbia, Columbia, MO 65211, USA.
e-mail: nabelekp@missouri.edu

MS received 4 November 2003. Accepted for publication 9 June 2004.

Transactions of the Royal Society of Edinburgh: Earth Sciences, **95**, 87–98, 2004

The influence of cordierite on melting and mineral-melt equilibria in ultra-high-temperature metamorphism

Simon L. Harley and P. Thompson

ABSTRACT: Experimentally constrained calibrations of the incorporation of H_2O and CO_2 into cordierite as functions of P–T-a_{H_2O}-a_{CO_2} are integrated with KFMASH grids which define mineral-melt equilibria in pelites. This is used to explore the impact of the volatile content and composition of cordierite on anatexis and melt-related processes in high-temperature (HT) and ultra-high-temperature (UHT) metamorphism. The strongly temperature-sensitive H_2O content of cordierite coexisting with dehydration melts (0·4–1·6 wt.%) causes a 10–25% relative decrease in the amount of melt produced from pelites compared with models which treat cordierite as anhydrous.

KFMASH melting grids quantified for a_{H_2O} demonstrate consistency between the measured H_2O contents in cordierite from granulite-migmatite terrains and mineral equilibria. These indicate anatexis with a_{H_2O} in the range 0·26–0·16 at 6–8 kbar and 870–930°C. The pressure-stability of cordierite+garnet with respect to orthopyroxene+sillimanite+quartz in KFMASH is strongly influenced by cordierite H_2O content, which decreases from 1·1 to 0·5 wt.% along the melting reaction Grt+Crd_H+Kfs=Opx+Sil+Qz+L. The lower-T invariant point involving biotite (8·8 kbar/900°C) that terminates this reaction has a_{H_2O} of $0{\cdot}16 \pm 0{\cdot}03$, whereas the higher-T terminating invariant point involving osumilite (7·9 kbar/940°C) occurs at a_{H_2O} $0{\cdot}08 \pm 0{\cdot}02$. Osumilite-bearing assemblages in UHT terrains imply a_{H_2O} of <0·08, and at 950–1000°C and 8–9 kbar calculated a_{H_2O} is only 0·04–0·02. Cordierites stable in osumilite-bearing assemblages or with sapphirine+quartz have maximum predicted H_2O contents of ca. 0·2 wt.%, consistent with H_2O measured in cordierites from two sapphirine-bearing UHT samples from the Napier Complex.

The addition of CO_2 to the H_2O-undersaturated (dehydration-melting) system marginally decreases the temperature of melting because of the stabilisation of cordierite, the solid product of the peritectic melting reactions. The preferential incorporation of CO_2 enhances the stability of cordierite, even at fixed a_{H_2O}, and causes the stability fields of Grt+Crd+Sil+Kfs+Qz+L and Grt+Opx+Crd+Kfs+Qz+L to expand to higher pressure, and to both higher and lower temperatures. The minimum solubility of H_2O in granitic melt is independent of the CO_2 content of cordierite, and the distribution of H_2O between melt and cordierite is similar at a given melt H_2O-content to the H_2O-only system. This enhanced stability of CO_2-bearing cordierite leads to a reduced stability range for osumilite-bearing assemblages to temperatures of ca. 950–975°C or greater. Cordierites in the Napier Complex UHT gneisses contain 0·5 and 1·05 wt.% CO_2, consistent with a role for CO_2 in stabilising cordierite with respect to osumilite in these unusual sapphirine-bearing granulites.

KEY WORDS: Fluids, granites. granulites, osumilite, sapphirine.

The H_2O-CO_2-bearing phase cordierite (Schreyer 1985) is involved in many key mineral equilibria in high-temperature (HT) and ultra-high-temperature (UHT) metamorphism (Hensen 1971, 1986; Hensen & Green 1973; Aranovich & Podlesskii 1989; Vry *et al.* 1990; Bertrand *et al.* 1991; Harley 1998a). It is also central to many of the reactions controlling melting (Vielzeuf & Holloway 1988; Patiño Douce & Johnston 1991; Carrington & Harley 1995a; Stevens *et al.* 1995; Holland *et al.* 1996; Harley 1998a; White *et al.* 2001), being a modally and texturally significant phase in regional granulite terranes which exhibit migmatisation, melt mobilisation and melt extraction (Waters 1988; Fitzsimons 1994, 1996; Harley 1994; Harley *et al.* 2002). From the viewpoint of peraluminous granite melt production in the middle crust, dehydration-melting of biotite-bearing mineral assemblages in pelites with moderate to high X_{Mg} (>0·5) is often controlled by divariant, or higher variance, cordierite-forming reactions such as the model KFMASH (K_2O-FeO-MgO-Al_2O_3-SiO_2-H_2O) reaction:

$$\text{Biotite+Sillimanite+Quartz+K-feldspar=Cordierite+Liquid} \quad \text{(R1)}$$

producing cordierite migmatites and cordierite-bearing leucogranites at 2·5–7 kbar and 750–900°C (e.g. Waters 1988; Fitzsimons 1996; Kriegsman 2001). Hence, an understanding of the consequences of cordierite as a volatile-bearing phase that can take up H_2O and CO_2 is critical to evaluation of melting processes and useful for the calculation of fluid activities potentially accompanying melt production. With respect to HT and UHT metamorphism, the stability of cordierite in both the presence and absence of melt generally defines the lower temperature (T) and pressure (P) limits of mineral assemblages such as sapphirine+quartz and orthopyroxene+sillimanite+quartz, which are critical to establishing the progress and conditions of UHT metamorphism. Therefore, an understanding of the stability limits of cordierite and their dependence on its H_2O and CO_2 content is central to any evaluation of UHT metamorphism and the role of melts and melt extraction in stabilising UHT assemblages (e.g. Harley 1998a; Kriegsman 2001; White & Powell 2002; Kelsey *et al.* 2003).

In the present contribution, the authors first overview and summarise the experimental constraints on the P–T-a_{H_2O}-a_{CO_2} dependencies on the uptake of H_2O and CO_2 in cordierite, and the distribution of H_2O between coexisting granitic melt and cordierite. They then go on to integrate this data with phase equilibrium experiments and calculations pertinent to HT/UHT melting and metamorphism in order to examine the impact of cordierite as a peritectic phase in migmatites, and the effect of CO_2 on melting and other phase relations. Finally, they also document and evaluate fluid activities calculated based on cordierite compositions from selected HT and UHT terrains.

1. Incorporation of H_2O and CO_2 into cordierite under HT/UHT conditions

The importance of cordierite as a phase with variable H_2O and CO_2 has long been recognised (Goldman *et al.* 1977; Newton & Wood 1979; Armbruster & Bloss 1982; Aines & Rossman 1984). This is expressed in its general formula (Mg, Fe)$_2$Al$_4$Si$_5$O$_{18}$.nH$_2$O.mCO$_2$, where 'n' and 'm' are the number of molecules of H_2O and CO_2 respectively, per nine cations (18 oxygen volatile-free) formula unit. It should be noted that this formula is a simplification of the chemistry of cordierite, which also generally incorporates alkalis (Li, Na, K) and Be in its structure (Goldman *et al.* 1977; Armbruster & Bloss 1982; Schreyer 1985; Kalt, 2000). However, Na and K do not appear to exert a strong influence on the H_2O and CO_2 uptake in cordierite compared with the effects of P–T and fluid composition (Thompson *et al.* 2002). The uptake of H_2O and CO_2 by cordierite under fluid-present conditions has been assessed in several experimental studies which have generally involved equilibrating (or attempting to equilibrate) natural or synthetic cordierite with H_2O, H_2O-CO_2 or CO_2 fluids under P–T conditions in the range 1–9 kbar and 500–800 °C (Mirwald *et al.* 1979; Kurepin 1985; Boberski & Schreyer 1990; LeBreton & Schreyer 1993; Carey 1995; Skippen & Gunter 1996).

Using a different approach to these earlier studies, Carrington & Harley (1996), Harley & Carrington (2001) and Thompson *et al.* (2001) equilibrated cordierite with peraluminous haplogranitic melt under both fluid-saturated and -undersaturated conditions, and in fluid-present experiments with pure H_2O and with mixed H_2O-CO_2 fluids. These experiments at P–T conditions of 3, 5 and 7 kbars and 800, 900 and 1000 °C have enabled the formulation of new expressions relating the activity of H_2O (Harley & Carrington 2001) and CO_2 (Harley *et al.* 2002) to the volatile contents of cordierite as measured by SIMS, and moreover, have defined the P–T-a_{H_2O} dependence of the distribution of H_2O between melt and cordierite, D_w (Harley & Carrington 2001). As with many of the earlier studies, and despite the spectroscopic evidence for distinct orientations for H_2O in the cordierite channels (Armbruster & Bloss 1982; Schreyer 1985; Winkler *et al.* 1994), these experiments have demonstrated that a simple, one-site model adequately describes the incorporation of both H_2O and CO_2 in P–T-a_{H_2O}-a_{CO_2} space. This one-site model defines the maximum total number of molecules of H_2O (n) and CO_2 (m) potentially incorporated into cordierite to be unity (i.e. n+m=1).

In the case of the incorporation of H_2O into cordierite, the saturation H_2O contents determined by Harley & Carrington (2001) are consistent with the majority of the previous lower-T experimental data in the cordierite-H_2O system (Mirwald *et al.* 1979; Boberski & Schreyer 1990; Mukhopadhyay & Holdaway 1994; Carey 1995; Skippen & Gunter 1996). Hence, Harley & Carrington (2001) combined their data with the previous work to obtain a P–T-a_{H_2O} relationship for cordierite that is applicable over the T range 600–1000 °C:

$$a_{H_2O}=[n/(1-n)].(\exp\{[-4203(\pm 320)/T]+11{\cdot}75(\pm 0{\cdot}33)\})/f_{H_2O(P,T)} \quad (1)$$

Setting $a_{H_2O}=1$ and solving for 'n' at specified P,T and f_{H_2O}(P,T) produces the H_2O isopleths (also known as isohydrons) presented in Figure 1. These isohydrons are within the error range of isopleths calculated from the thermodynamic data set of Holland & Powell (1998).

Of significance for melting relationships in HT and UHT metamorphism, the wt.% H_2O incorporated into melt and cordierite at specified P–T in H_2O-undersaturated conditions (i.e. Crd+L) and up to H_2O fluid saturation (i.e. Crd+L+V) are systematically correlated (Harley & Carrington 2001; Harley *et al.* 2002). When considered in terms of a_{H_2O}, a linear relationship exists between n/(1–n) in cordierite and the a_{H_2O} functions which describe H_2O uptake in melt as based on the Burnham model (Burnham 1994; Holloway & Blank 1994), but fitted using the H_2O solubility data of Holtz & Johannes (1994) and Johannes & Holtz (1996). This relationship allows modelling of the variation in D_w, the distribution of H_2O between peraluminous granitic melt and cordierite as a function of P–T and a_{H_2O}. D_w is defined as:

$$D_w=[\text{wt.\% } H_2O(\text{melt})]/[\text{wt.\% } H_2O(\text{Crd})] \quad (2)$$

The H_2O distribution curves presented in Figure 2 are calculated from the models developed by Harley & Carrington (2001) based on their experimental data. At low melt H_2O contents in the range 1–3·5 wt.% (those likely to be typical of HT/UHT melts), D_w values are generally in the range 2·5–5·0. As considered in Harley *et al.* (2002) and below, such D_w values impact on the amount of melt that can be produced in a cordierite-forming dehydration-melting reaction compared with dehydration-melting reactions in which cordierite does not participate.

The incorporation of CO_2 into cordierite and its modelling has been the subject of several investigations based on cordierite-fluid exchange experiments (e.g. Johannes & Schreyer 1981; Kurepin 1985; Le Breton & Schreyer 1993). The

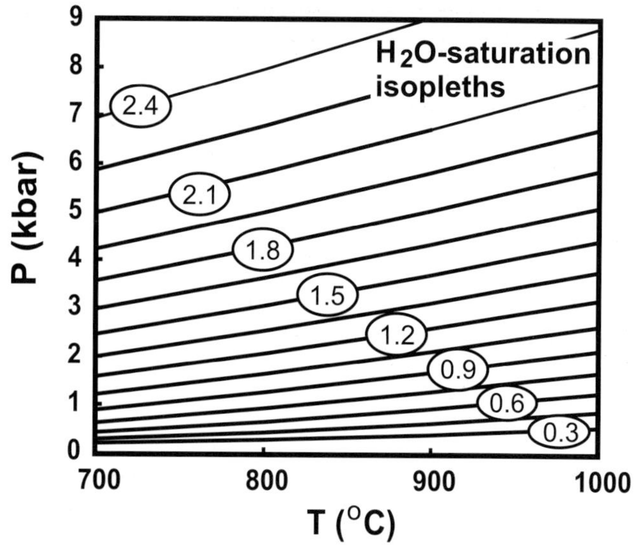

Figure 1 Pressure–temperature diagram contoured with isopleths (isohydrons) of H_2O in X_{Mg}=0·75 cordierite coexisting with pure H_2O fluid, as modelled by Harley & Carrington (2001). The isohydrons are expressed in terms of wt.% H_2O.

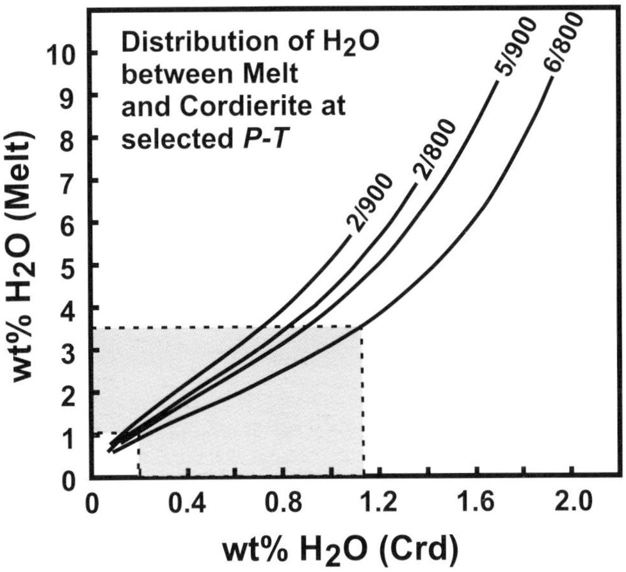

Figure 2 Modelled distribution curves of H_2O between cordierite and coexisting granitic melt, calculated from the a–X relationships for cordierite and melt presented by Harley & Carrington (2001), for selected high-temperature P–T conditions (e.g. 2/900 is 2 kbar and 900°C). Note that D_w at a selected P–T condition and cordierite H_2O content can be calculated from this diagram simply by dividing the appropriate wt.% H_2O (melt) by wt.% H_2O (Crd). The shaded field depicts the H_2O contents typical of cordierite coexisting with high-temperature/ultra-high-temperature melts.

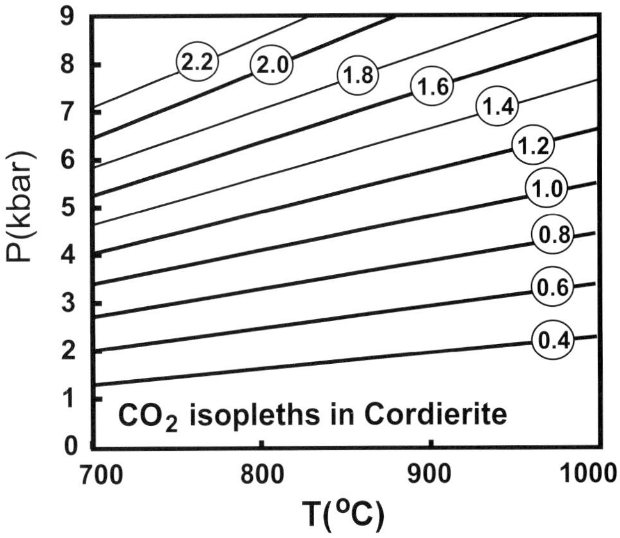

Figure 3 Pressure–temperature diagram contoured with the isopleths of CO_2 in $X_{Mg}=0{\cdot}80$ cordierite coexisting with pure CO_2 fluid, as modelled by Thompson *et al.* (2001) and Harley *et al.* (2002). The isopleths are expressed in terms of wt.% CO_2.

exchange experiments have internal inconsistencies and appear to fail to achieve equilibrium (Le Breton & Schreyer 1993), rendering previous modelling tenuous. However, Thompson *et al.* (2001) and Harley *et al.* (2002) have recently reported and summarised the results of cordierite-granitic melt experiments conducted in the presence of mixed H_2O-CO_2 fluids. Extrapolation of the linear relationships which exist between a_{CO_2} and m/(1–m) in these experiments allows calculation of the maximum quantity of CO_2 that can be taken up by cordierite at unit a_{CO_2} for each P, T condition. The P–T- a_{CO_2} dependence of the incorporation of CO_2 into cordierite P–T range 3–7 kbar and 800–1000°C is then described by the following equation:

$$a_{CO_2}=[m/(1-m)].(\exp\{[-3073(\pm 200)+(0{\cdot}3023P)]/T+13{\cdot}1(\pm 1{\cdot}2)\})/f_{CO_2(P,T)} \quad (3)$$

Setting $a_{CO_2}=1$ and solving for 'm' at specified P–T and f_{CO_2}(P,T) produces CO_2 isopleths like those presented in Figure 3. Equation 3 and the calculated isopleths, which are consistent with one of the experimental cordierites used by Johannes & Schreyer (1981), are used here in the consideration of the effects of CO_2 on HT/UHT equilibria involving cordierite.

As an independent check on the validity of the cordierite-granitic melt experiments conducted in the presence of mixed H_2O-CO_2 fluids, the H_2O contents of melt and CO_2-bearing cordierite have been measured (Thompson *et al.* 2001). The D_w calculated from these data (e.g. for a given P–T condition and melt H_2O content) are consistent with those modelled from the H_2O-undersaturated experiments of Harley & Carrington (2001). This indicates ideal or near-ideal behaviour of H_2O and CO_2 in the channel site of cordierite, and furthermore, implies that CO_2 has negligible impact on distribution of H_2O between melt and cordierite, at least at temperatures of 800°C and above. In other words, the H_2O contents of H_2O-CO_2-bearing cordierite can be used to deduce the H_2O contents of any fugitive melts formerly coexisting with cordierite under HT/UHT conditions. This applies so long as the HT/UHT melts are close to the experimental leocogranitic melts in terms of their major element chemistries, such as Na/K and SiO_2 contents (Thompson *et al.* 2002), or can be corrected to account for differences in these parameters. In the present work, the experimental cordierite-melt data will be applied to analyse mineral-melt reaction systematics in KFMASH. This involves correction of the melt H_2O model to account for the lack of Na. In practice, this correction is very small (<0·5 wt.% H_2O in melt at saturation: Holtz & Johannes 1994; Johannes & Holtz 1996), and within the error range of the experimental data and model (<0·1 wt.% H_2O in melt) for the range of low-a_{H_2O} conditions considered in the analysis presented herein.

Equations 1 and 3 allow calculation of the saturation values (molecules p.f.u. or wt.%) of both H_2O and CO_2 in cordierite at any intermediate H_2O-CO_2 fluid composition providing an appropriate fluid *a*-X relation is used for H_2O-CO_2 mixing. In the present study, as in previous work (Thompson *et al.* 2001; Harley *et al.* 2002), the authors have adopted the same fluid H_2O-CO_2 *a*-X model as applied to the experiments, i.e. the CORK model of Holland & Powell (1990). The saturation H_2O-CO_2 channel fluid contents and channel X_{CO_2} attainable in cordierite coexisting with peraluminous granitic melts produced through 'dehydration' melting in the presence of a low-a_{H_2O}, CO_2-rich fluid phase will be investigated and tested in comparison with natural rocks, following an analysis of the impact of cordierite volatile contents on melting and phase relations in HT/UHT model systems.

2. Impact of cordierite on percentage melting: H_2O distribution systematics

As noted in the introduction and demonstrated in the seminal work of Waters (1988), dehydration-melting of biotite-bearing mineral assemblages in magnesian pelites (X_{Mg} >0·5) is often controlled by cordierite-forming reactions such as the KFMASH divariant equilibrium:

$$\text{Bt+Sil+Qz+Kfs=Crd+L} \quad \text{(R1)},$$

which is related to the KFMASH univariant reaction:

$$\text{Bt+Sil+Qz+Kfs=Grt+Crd+L} \quad \text{(R2)}$$

(Fig. 5a). Reaction R2 is also recognised as a multivariant equilibrium controlling melting in more complex natural rock systems and case studies (e.g. Fitzsimons 1996; White *et al.* 2001).

Consideration of cordierite as a H_2O-bearing phase shows that its formation as a peritectic phase via reaction R1 decreases the volume percentage of melt produced from the dehydration-melting of biotite-bearing pelites. This was first recognised by Harley (1994), who defined the distribution of H_2O between melt and cordierite (now denoted as D_w=wt.% H_2O_{melt}/wt.% H_2O_{crd}), and suggested, on the basis of one reconnaissance experiment, that this distribution coefficient was $3{\cdot}3 \pm 0{\cdot}5$. Stevens *et al.* (1995) indirectly produced D_w values ($2{\cdot}1 \pm 0{\cdot}3$ to $1{\cdot}5 \pm 1{\cdot}1$ at 800–1000°C) for a series of melting experiments at 5 kbar using modal mass-balance calculations, and suggested that the proportion of melt produced would be 'severely' restricted. Direct experimental determination of cordierite and melt H_2O contents (Carrington & Harley 1996; Harley & Carrington 2001) demonstrated that dehydration-melts coexist with cordierites which contain 0·4–1·6 wt.% H_2O, depending on P–T. Modelling of D_w as a function of P,T and a_{H_2O} shows that D_w lies in the range 2·5–5·0 for dehydration melting in the P–T field 5–9 kbar and 850–1000°C (e.g. Fig. 2). Based on mass-balance calculations in KFMASH, D_w values of this magnitude cause a 15–30% relative decrease in the amount of melt produced from reaction R1 compared with the analoguous reaction assuming anhydrous cordierite or with melting reactions involving garnet, which may operate at similar P–T conditions, but in Fe-richer bulk compositions. An example of this behaviour is illustrated in Figure 4, which shows the impact of cordierite wt.% H_2O on the relative amount of melt produced when that melt contains 2 wt.% H_2O, corresponding to temperatures of 900–950°C at 5–8 kbar. In this case, the amount of melt is reduced by some 20%. The general conclusion from this is that the amount of melt produced from slightly more magnesian, cordierite-bearing pelites will be less than that produced from garnet-bearing pelite at the same P–T conditions – the cordierite-bearing lithologies will appear more refractory.

Using the same logic, for a specific bulk rock composition, the amount of melt produced at lower-P where cordierite is stable in preference to garnet in a pelite with intermediate X_{Mg} (=0·40–0·60) will be some 20% (relative) less than that predicted from analyses which ignore the impact of cordierite H_2O on the stoichiometry of the melting reactions. Moreover, since K-feldspar is consumed in proportion to the amount of melt produced (Fig. 4), some 50% (relative) less K-feldspar is consumed in the cordierite-bearing case. Hence, K-feldspar will persist to higher temperatures if it initially is present in similar modal amounts in Mg- and Fe-pelites. These considerations also apply to all calculations of pseudosections involving cordierite at HT/UHT conditions (e.g. White *et al.* 2001; Kelsey *et al.* 2003). Decreased melt production means that K-feldspar will persist to higher-T, and hence, be able to participate in osumilite-forming reactions; for example, in bulk compositions which otherwise may be modelled to lose K-feldspar, and hence, not intersect potential osumilite-producing reactions, such as those depicted in Figure 5a. Hence, consideration of cordierite as a phase that sequesters H_2O in the melting regime extends the range of KFMASH bulk compositions (in terms of A'KF) which are accessible to osumilite. Pseudosections which attempt to illustrate the mineral assemblages stable with melt in specific pelite rock compositions in HT/UHT metamorphism must recognise and incorporate the effect of H_2O in cordierite on the percentage

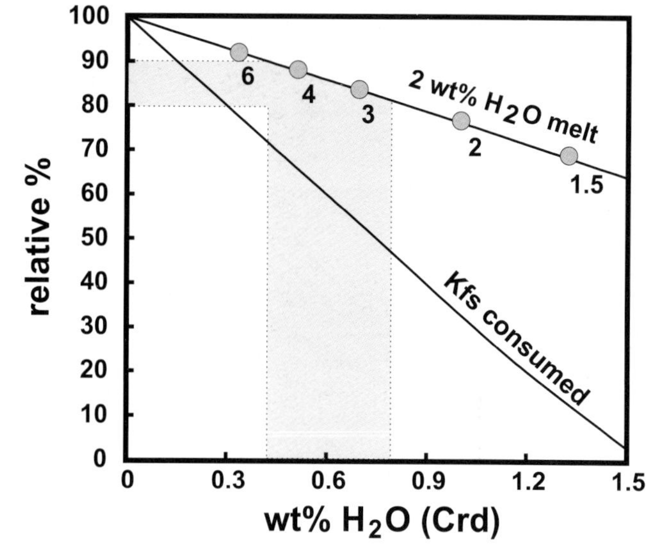

Figure 4 Relative proportion of melt produced and K-feldspar consumed (normalised to 1 for the case of anhydrous cordeierite) in a model pelite initially containing 25% biotite for the case in which the P–T conditions of melting result in a melt with 2 wt.% H_2O (e.g. 5·5 kbar/900°C to 9·5 kbar/950°C; see Fig. 5a). The relative proportions of melt produced and K-feldspar consumed in the reaction (R1) decrease as the H_2O-content of coexisting cordierite increases and D_w (numbers below filled circles) decreases [i.e. as P increases along the 2 wt.% H_2O (melt) isopleth in Fig. 5a]. The shaded field depicts the H_2O contents typical of cordierite in this case, which lead to a decrease in melt production to 80–90% of the amounts expected if the H_2O content of cordierite is ignored.

melt production and residual solid phase assemblage if they are to be realistic and informative.

3. Integration of cordierite-melt H_2O data with UHT reaction grids: predicted cordierite H_2O contents and a_{H_2O} conditions in HT/UHT metamorphism

The H_2O contents of cordierite coexisting with melt, and the a_{H_2O} conditions along KFMASH dehydration-melting reactions involved in HT/UHT metamorphism (Carrington & Harley 1995a; Harley 1998a; White *et al.* 2001; Kelsey *et al.* 2003), can be calculated based on knowledge of the variation in the minimum H_2O content of peraluminous granitic melt with P, T and D_w over the same P–T domain (Harley & Carrington 2001). Contours of minimum melt H_2O contents (Holtz & Johannes 1994; Johannes & Holtz 1996) are superimposed on the KFMASH petrogenetic grid involving the phases Grt, Crd, Bt, Sil, Opx, Qz, Kfs, L and Os developed by Carrington & Harley (1995a, b) and modified in the light of calculations by subsequent workers (Holland *et al.* 1996; White *et al.* 2001; Kelsey *et al.* 2003) in Figure 5a. Using the D_w relationships implicit in Figure 2 and described in Harley & Carrington (2001), this grid also has been contoured for the H_2O content of cordierite coexisting with these dehydration-melts. Like the melt H_2O content isopleths, but in contrast to cordierite H_2O saturation isohydrons, the cordierite H_2O isopleths relevant to dehydration melting are strongly temperature-dependent, varying from 1·2 to 1·4 wt.% H_2O at 800°C to less than 0·2 wt.% H_2O at 1000°C. Moreover, comparison of these isopleths with the assemblage fields defined by the KFMASH grid shows that cordierites coexisting with dehydration melts in the assemblage Bt+Sil+Crd+Qz+Kfs+L should generally contain >1·2 wt.% H_2O (melt H_2O >3 wt.%), whereas cordierites in the higher-T migmatite

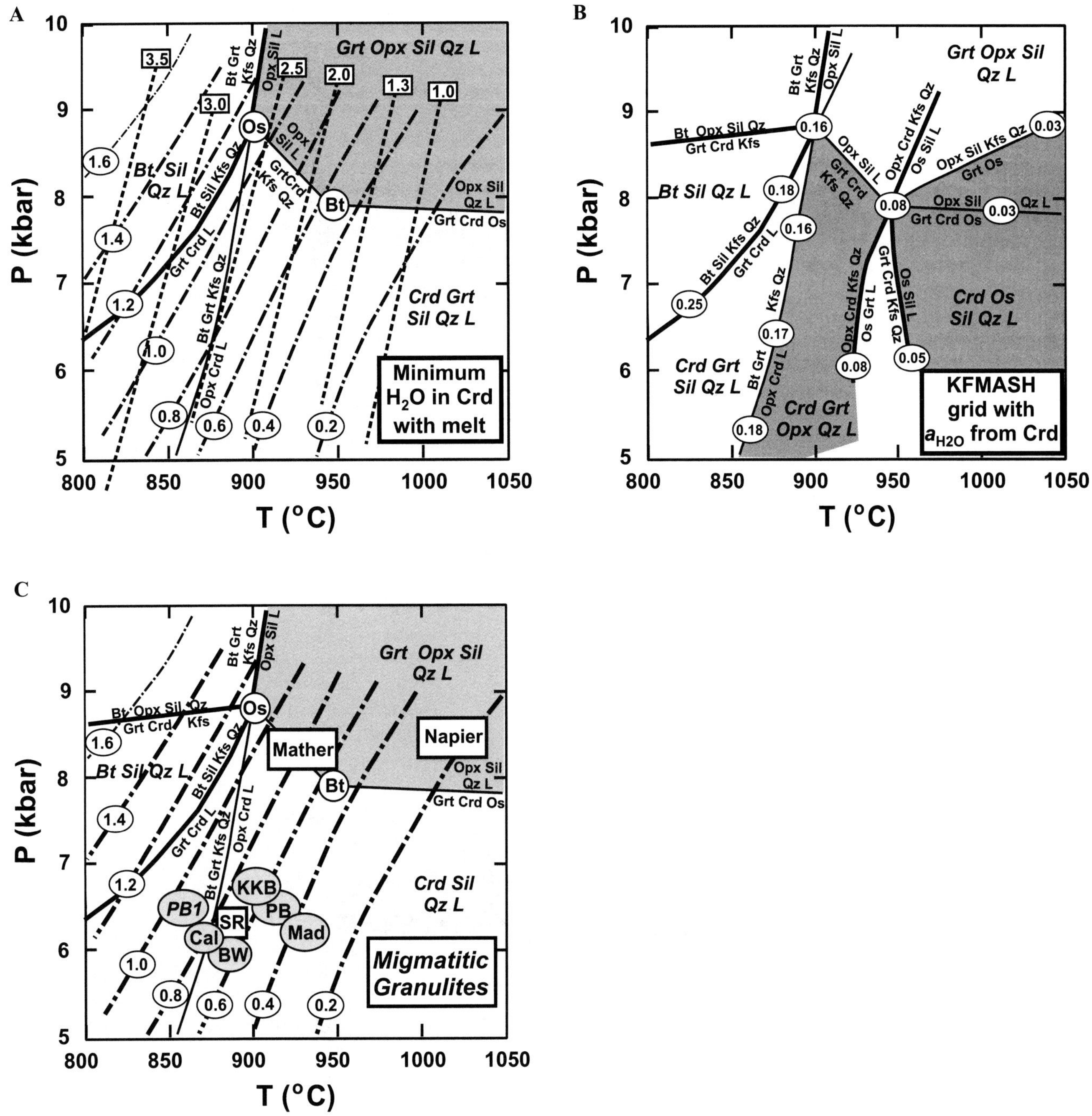

Figure 5 (A) Contoured pressure–temperature (P–T) diagram depicting cordierite-melt H_2O relationships for conditions where the melt has its minimum possible H_2O content at P and T. Contours of minimum melt wt.% H_2O (dashed lines annotated with wt.% H_2O values in rectangles) are after Johannes & Holtz (1996) and Harley & Carrington (2001). Contours of cordierite wt.% H_2O (dot-dashed lines annotated with wt.% H_2O values in ellipses) are derived from the D_w-a_{H_2O} relationships defined in Harley & Carrington (2001) and calculated using Equation 1 for a_{H_2O} compatible with those defined by the melt isopleths. The heavy solid lines are KFMASH system reactions which limit the stabilities of key cordierite-bearing assemblages (e.g. Crd+Grt+Sil+Qz+L) These reactions are based on Carrington & Harley (1995a), as modified by White *et al.* (2001), and are focused on the osumilite-absent, [Os], and biotite-absent, [Bt], invariant points (see Fig. 5b for other related KFMASH equilibria). (B) A P–T diagram depicting KFMASH system reactions based on Carrington & Harley (1995a, b), as modified by White *et al.* (2001), and focused on the osumilite-absent, [Os], and biotite-absent, [Bt], invariant points of Fig. 5a. These invariant points have associated a_{H_2O} values of 0·16 and 0·08 respectively (see text for discussion). Selected KFMASH reactions are labelled at specific P–T conditions for a_{H_2O} calculated from the modelled cordierite H_2O contents (Harley & Carrington 2001; see Fig. 5a). Selected mineral assemblage fields are illustrated by shading. (C) A P–T diagram as in Figure 5a, contoured for the H_2O contents of cordierite formed in equilibrium with KFMASH granitic melt, illustrating the P–T conditions of equilibration of the Crd+Sil migmatites (ellipses) and Crd+Opx migmatites/granulites (rectangles) listed in Table 1. The P–T conditions inferred from independent equilibria are in reasonable agreement with measured cordierite H_2O contents in most cases (Mad: Madagascar; KKB: Kerala Khondalite Belt; Cal: Calabria; BW: Bayerische Wald; SR: Strangways Range; Mather; Napier), but peak temperature estimates for Prydz Bay (*PB1*) are significantly lower than would be deduced from cordierite in this case (PB). Sources for P–T estimates are given in Table 1 along with the relevant cordierite compositional data.

assemblage Grt+Crd+Sil+Qz+Kfs+L should preserve H_2O contents in the range 1·1–0·2 wt.%, depending on T. Cordierite in the assemblage Crd+Opx+Qz+Kfs+L, which lies to higher-T than the KFMASH univariant:

$$Bt+Grt+Kfs+Qz=Opx+Crd+L \qquad (R3)$$

is constrained to contain <0·8 wt.% H_2O for all UHT P–T conditions apart from those near the [Os] invariant point at 8·8 kbar and ca. 900 °C.

The volatile content of cordierite formed through dehydration-melting is critical to the stability of cordierite+garnet with respect to assemblages such as orthopyroxene+sillimanite+quartz and sapphirine+quartz (Aranovich & Podlesskki 1989; Bertrand *et al.* 1991; Carrington 1995; Carrington & Harley 1995a; Holland *et al.* 1996; Aranovich & Berman 1996). As shown in Figure 5a, a pronounced change in melt H_2O contents in the 900–950 °C interval coupled with an increase in D_w from 2·4 to 3·0 leads to a marked decrease in cordierite H_2O from 1·1 to 0·5 wt.% along the univariant reaction that joins the two invariant points equivalent to [Os] and [Bt] in the grid of Carrington & Harley (1995a), and limits the stability of Fe-Mg garnet+cordierite in this system:

$$Grt+Crd_H+Kfs=Opx+Sil+Qz+L \qquad (R4)$$

where Crd_H denotes hydrous cordierite in equilibrium with the dehydration melt. The negative dP/dT slope of this reaction is consistent with and, at least in part, reflects this marked decrease in channel occupancy. Cordierite that persists with Opx+Sil+L at pressures above those of reaction R4 is predicted to contain 1·1–0·4 wt.% H_2O over the temperature range 900–970 °C. It is also clear from Figure 5a that cordierite coexisting with osumilite can only contain up to 0·5 wt.% H_2O and would generally contain less (<0·4 wt.%) in UHT terrains.

Using Equation 1, a_{H_2O} along the dehydration-melting reactions and at the invariant points in the KFMASH grid (Carrington & Harley 1995a; Holland *et al.* 1996) has been calculated (Fig. 5b). These a_{H_2O} values vary from 0·24 at 6·7 kbar and 825 °C to 0·16 at 8·8 kbar and 900 °C in the case of Reaction R2, and from 0·18 at 5 kbar to 0·16 at 8·8 kbar along the very steep (dP/dT=85 bar/°C) Reaction R3. The a_{H_2O} values calculated for the two key KFMASH invariant points using the present authors' cordierite and melt models are 0·16 ± 0·03 at the lower-T [Os] point (8·8 kbar/900 °C), and only 0·08 ± 0·02 at the [Bt] point (7·9 kbar/940 °C) where the osumilite+garnet assemblage becomes stable and overlaps with assemblages involving orthopyroxene+sillimanite+quartz+melt (Carrington & Harley 1995b). From this, we can infer that the osumilite-bearing assemblages equilibrated at 7–9 kbar in the Napier Complex of East Antarctica (Ellis *et al.* 1980; Grew 1982; Audibert *et al.* 1995) and other UHT terrains imply a_{H_2O} of <0·08.

The contoured KFMASH melting grids (Fig. 5a, b) are consistent with independent P–T estimates obtained from garnet+cordierite+sillimanite migmatitic granulites from the Kerala Khondalite Belt (Nandakumar & Harley 2000), southern Madagascar (Markl *et al.* 2000) and Calabria (Schenk 1984). SIMS H_2O measurements on cordierites from these terrains yield a_{H_2O} conditions in the range 0·06–0·18 (Table 1) and imply cordierite-melt equilibrium at P–T conditions in the range ca. 6–7 kbar and 860–930 °C, depending on terrain (Fig. 5c). The H_2O contents of former melts which equilibrated with these low-H_2O (0·6–1·2 wt.%) cordierites are 1·3–2·2 wt.%. Based on measured cordierite CO_2 contents, calculated a_{CO_2} in these migmatitic granulites are high (0·34–0·85) but consistent with fluid-undersaturation in all examples except Calabria. This is considered further in section 5 below.

The H_2O contents of cordierites from cordierite-sillimanite migmatites from the Bayerische Wald have been determined by informed spectroscopy (IR) techniques by Kalt (2000). The average H_2O content of cordierite based on all the measured samples in this case is 0·53 ± 0·10 wt.%, although one sample preserves cordierite with up to 0·74 wt.% H_2O. This most hydrous cordierite is consistent with equilibrium with granitic melt containing 2·5 wt.% H_2O at the upper end of the P–T conditions preferred by Kalt (2000) for the metamorphic peak in this region based on geothermobarometric calculations (BW, Fig. 5c.: ca. 6 kbar, 870 °C). The majority of samples which preserve lower H_2O contents could be interpreted to indicate melt loss at higher temperatures (up to 900 °C), but could equally reflect post-peak loss or leakage of H_2O from the cordierite channels in this case.

A similar interpretation could be put forward for cordierites in the garnet+cordierite+sillimanite migmatitic granulites of Prydz Bay, East Antarctica (Fitzsimons 1994, 1996). The H_2O contents, measured by SIMS, of these cordierites are only 0·35–0·60 wt.% and average at ca. 0·5 wt.% (Table 1; Harley *et al.* 2002). Such low H_2O contents would be consistent with cordierite-melt equilibrium at 6·5 kbar and 900–930 °C (PB, Fig. 5c), whereas geothermometry corrected for the effects of Fe-Mg exchange and phase equilibrium considerations indicate peak temperatures of 860 °C (Fitzsimons 1996). If the latter temperatures are accepted (*PB1*, Fig. 5c) then the cordierite that coexisted with melt according to Figure 5c would have contained >0·8 wt.% H_2O, implying that all of the cordierites measured from the Prydz Bay granulites have lost between 25% and 50% of their initial H_2O. Such marked H_2O loss is not required to explain the volatile contents and phase relations of most of the other examples considered here and in Harley *et al.* (2002), particularly in the cases where the cordierites contain significant CO_2 (e.g. BB4: 0·6 wt.% CO_2). Hence, although the present authors cannot discount the possibility that some H_2O leakage has occurred from cordierite, they suggest that the peak temperatures attained in the Prydz Bay granulites were higher than those deduced by Fitzsimons (1996) and in the range 900–930 °C.

Also depicted in Figure 5c are the preferred P–T conditions for three examples of cordierite-orthopyroxene HT/UHT granulites. Migmatitic garnet+orthopyroxene+cordierite granulite from Strangways Range (Arunta Complex) and garnet+orthopyroxene+cordierite ± sillimanite granulite from Mather Peninsula (Harley 1998b) both lie on the 0·7 wt.% H_2O isopleth and indicate temperatures wholly consistent with independent estimates from geothermobarometry. At UHT conditions of greater than 1000 °C at 8–9 kbar, as recorded from several localities in the Napier Complex of East Antarctica (Ellis *et al.* 1980; Sheraton *et al.* 1987; Harley & Hensen 1990; Harley & Motoyoshi 2000), a_{H_2O} values calculated using cordierite-melt equilibria (Fig. 5b; Table 1) are only 0·04–0·02. Any cordierite stable in osumilite-bearing assemblages, or with sapphirine+quartz in the presence of melt, would be expected to have H_2O contents of only 0·2 wt.%; lower H_2O cordierites would be present if melting did not occur. Two samples from the Napier Complex preserve cordierites containing 0·2 wt.% H_2O (Table 1), consistent with this expectation (Fig. 5c).

4. Melting and migmatisation in the H_2O-CO_2 system

So far, the present authors have considered only the impact of cordierite as a H_2O-bearing phase involved in melting and

Table 1 Volatiles in cordierites from selected high-temperature and ultra-high-temperature granulites/migmatites*

Sample	Cordierite							P (kbar)	T (°C)	a_{H_2O}	a_{CO_2}	Melt	
	X_{Mg} (Crd)	H_2O (wt.%)	CO_2 (wt.%)	n (pfu)	m (pfu)	m+n (pfu)	X_{CO_2}					H_2O wt.%)	D_w
Bayerische Wald (Kalt 2000)													
BW-44ii	0·60	0·740	0·051	0·255	0·007	0·262	0·027	6·0	900	0·178	0·033	2·5	3·3
BW-average	0·60	0·528	0·064	0·180	0·009	0·189	0·048	6·0	900	0·115	0·041	1·8	3·5
Crd-Sil migmatite		*0·095*	*0·018*	*0·032*	*0·003*	*0·032*	*0·013*			*0·025*	*0·012*	*0·3*	
Prydz Bay migmatitic granulites, Antarctica (Fitzsimons 1994, 1996)													
BB4 (17)	0·66	0·494	0·583	0·168	0·081	0·250	0·325	6·0	860	0·096	0·361	1·7	3·5
Grt-Crd-Sil migmatite		*0·090*	*0·118*	*0·031*	*0·016*	*0·038*	*0·045*			*0·028*	*0·122*	*0·4*	
Calabria, Serre Massif (Schenk 1984)													
8–90 (170)	0·63	0·805	1·305	0·272	0·181	0·453	0·400	6·5	875	0·162	0·853	2·2	2·7
Grt-Crd-Sil migmatite		*0·060*	*0·060*	*0·021*	*0·008*	*0·027*	*0·014*			*0·014*	*0·041*	*0·4*	
Kerala Khondalite Belt, Southern India (Nandakumar & Harley 2000)													
SHED10 (10)	0·60	0·636	0·854	0·219	0·120	0·340	0·355	6·5	900	0·129	0·557	2·0	3·1
Grt-Crd leucosome		*0·036*	*0·022*	*0·012*	*0·006*	*0·010*	*0·011*			*0·013*	*0·019*	*0·3*	
Chittikara-SLH (12)	0·60	0·633	0·762	0·218	0·108	0·326	0·330	6·5	900	0·128	0·490	2·0	3·1
Grt-Crd leucosome		*0·112*	*0·053*	*0·039*	*0·012*	*0·020*	*0·043*			*0·030*	*0·043*	*0·6*	
SE Madagascar (Markl *et al.* 2000)													
patch migmatite (8)	0·86	0·287	0·527	0·095	0·072	0·167	0·431	6·0	920	0·057	0·344	1·3	4·5
Crd leucosome		*0·050*	*0·065*	*0·017*	*0·009*	*0·022*	*0·029*			*0·016*	*0·074*	*0·3*	
Strangways Range, Arunta Complex													
9052 (5)	0·87	0·690	0·023	0·229	0·003	0·232	0·014	6·0	875	0·147	0·013	2·3	3·3
Grt-Opx-Crd patch		*0·092*	*0·005*	*0·031*	*0·001*	*0·031*	*0·004*			*0·030*	*0·004*	*0·3*	
Mather Peninsula, Rauer Islands (Harley 1998b)													
SH/88/218 (8)	0·88	0·698	0·055	0·233	0·008	0·241	0·031	8·5	920	0·091	0·021	1·8	2·6
Grt-Opx-Crd-Sil		*0·067*	*0·021*	*0·022*	*0·003*	*0·024*	*0·008*			*0·009*	*0·011*	*0·3*	
Napier Complex, Enderby Land (Harley, 1986; Sheraton *et al.* 1987)													
49752 Mt Sones (8)	0·88	0·194	0·498	0·064	0·068	0·132	0·513	8·5	1000	0·025	0·256	0·85	4·4
Grt-Spr-Sil-Crd		*0·021*	*0·026*	*0·007*	*0·004*	*0·010*	*0·018*			*0·008*	*0·024*	*0·1*	
49354 Mt Hardy (6)	0·90	0·215	1·040	0·072	0·142	0·214	0·664	8·5	1000	0·028	0·576	0·96	4·3
Spr-Qz-Crd		*0·031*	*0·021*	*0·010*	*0·003*	*0·013*	*0·026*			*0·011*	*0·025*	*0·1*	

(number) number of SIMS volatile analyses of cordierite in the sample. The *P–T* estimates used in the calculation of fluid activities are taken from the references given for each case study. Figures given in italics are the 1σ uncertainties in the quoted parameters, propagated from 1σ uncertainties in the analytical populations and SIMS calibrations, and uncertainties in the cordierite activity calibrations of Equations 4 and 8, together with ± 0·5 kbar and ± 30°C uncertainties in the *P–T* estimates used in calculation of fluid activities.

mineral reactions in HT/UHT metamorphism and anatexis. However, cordierite also incorporates CO_2 into its channels, and as noted by Thompson *et al.* (2001), the distribution of CO_2 between melt and cordierite is strongly in favour of cordierite, such that $D_C < 0{\cdot}2$ and probably $<0{\cdot}1$. The stability of cordierite as a H_2O-CO_2-bearing phase coexisting with melt and preferentially fractionating CO_2 has important consequences for the character of fluid-melt-rock interactions since the melt segregates, migrates and crystallises in granulite host rocks. This can be demonstrated by consideration of the crystallisation of two initially slightly fluid-undersaturated leucogranitic melts (e.g. 2 wt.% H_2O, 1000 ppm CO_2 at 5 kbar/ 900°C) which differ only in their X_{Mg} [i.e. Mg/(Mg+Fe)]. Cooling of the Fe-richer melt (e.g. $X_{Mg}=0{\cdot}25$) will result in the crystallisation of garnet along with quartz and feldspars at these pressures, continuously enriching the remaining melt in both H_2O and CO_2 until saturation is reached and a CO_2-rich fluid phase is generated. This fluid may be trapped as CO_2-rich but post-peak fluid inclusions in garnet, as seen in some migmatites (e.g. Santosh 1987), or it may react with reduced minerals to precipitate graphite as crystallisation proceeds (e.g. Fitzsimons & Mattey 1995). In contrast, crystallisation of a Mg-richer melt (e.g. $X_{Mg}=0{\cdot}35$) to form cordierite plus quartz and feldspars will result in a slower increase in melt H_2O (dependent on D_w) and a decrease in ppm CO_2 in the melt – the latter effect caused by the sequestering of CO_2 by cordierite (at least 5000 ppm CO_2 cf. 500–1000 ppm in the melt). The cordierite-bearing melt or leucosome is unlikely to attain fluid saturation unless external fluids are introduced or rapid decompression occurs, whereby both melt and cordierite exceed their saturation isopleths. In a closed system (i.e. melt batch crystallising in isolation) or in a system in which melt is continously in communication with cordierite, it is likely that all the available CO_2 will be taken up in the cordierite, inhihiting the formation of fluid inclusions, and indeed, of graphite, since fluid–present conditions are not attained. Conversely, if fluid inclusions (Santosh *et al.* 1993) or leucosome-hosted graphite (Fitzsimons & Mattey 1995) are observed in cordierite-bearing migmatites, it is almost certain that externally derived fluid has been introduced into the system. In principle, such fluid-present conditions should be recorded by the volatile abundances in the cordierite and the a_{H_2O} and a_{CO_2} values calculated using Equations 1 and 2. In short, it is significantly more difficult to saturate a cordierite-bearing migmatite terrain than a garnet-dominated migmatite area which lacks cordierite; for similar P–T conditions and percentage melting, higher fluid fluxes are required in the cordierite migmatite case.

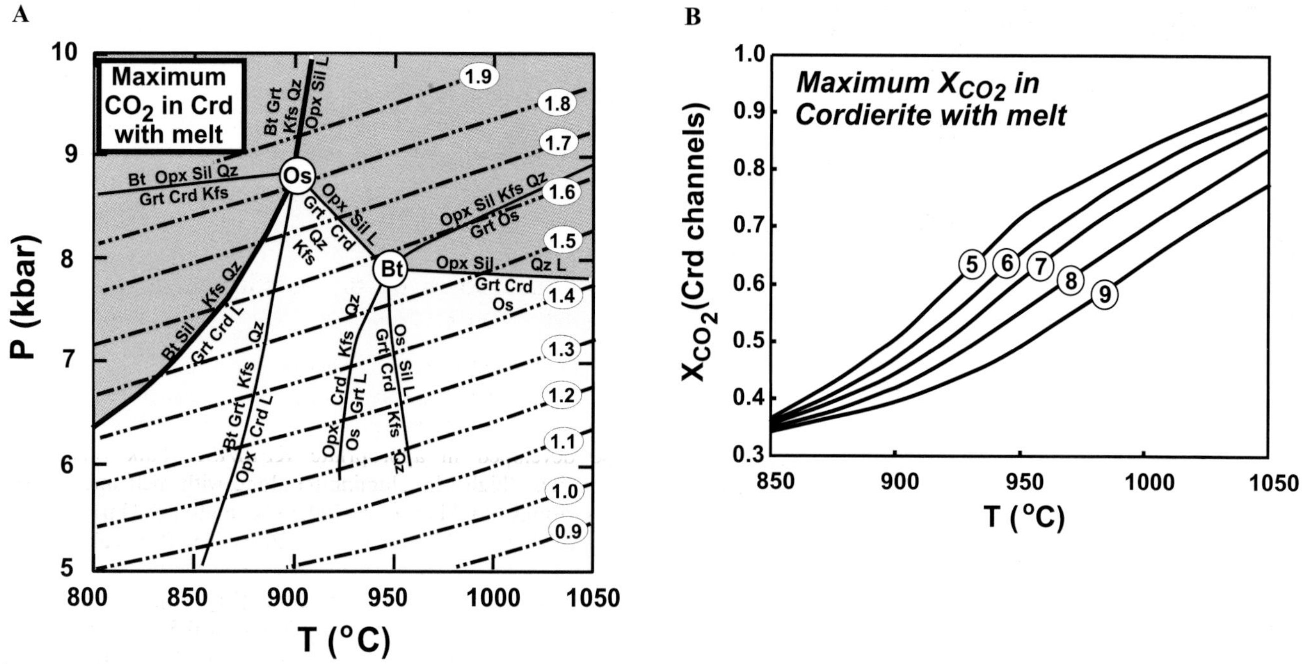

Figure 6 (A) Pressure–temperature (P–T) diagram showing isopleths of the *maximum* CO_2 contents (in wt.%) in cordierite coexisting with melts with minimum H_2O contents (i.e. melts with H_2O contents drawn in Fig. 5a). These isopleths are calculated using Equation 3 (based on Harley *et al.* 2002) in combination with Equation 1. Note that, since cordierite also cotains H_2O, which varies mainly with temperature (Fig. 5a), these CO_2 isopleths do not correspond to those presented for the pure CO_2 system in Figure 3. (B) The *maximum* X_{CO_2} composition of cordierite coexisting with melt and a free-CO_2-dominated fluid phase, plotted as a function of temperature in the melting range for various pressures (in kilobars; denoted by numbers in circles). Corderite coexisting with a low-H_2O melt at 950°C and 7 kbar may contain up to 1·4 wt.% CO_2 (Fig. 6a) and have a channel X_{CO_2} of up to 0·6 without fluid saturation being attained.

5. The effect of CO_2 in cordierite on KFMASH equilibria in HT/UHT metapelites

The contoured grid presented in Figure 5a only includes H_2O dissolved in melt and incorporated in cordierite. If CO_2 is added to this system, it is favourably incorporated into cordierite compared with melt, as noted above, and all other solid phases in the absence of graphite. Given that only a_{H_2O} is fixed, at fixed P–T, by the dehydration melting reactions of Figure 5a, it is possible for cordierites with near-constant H_2O contents but a range of CO_2 contents to be in equilibrium with melts with near-minimum H_2O contents in the KFMASH-CO_2 system. The upper limit on the CO_2 content of cordierite coexisting with melt is that calculated for fluid saturation, but cordierites with low H_2O contents and elevated CO_2 contents which are still less than those required for saturation could coexist with melt alone under fluid-absent conditions.

The maximum (saturation) CO_2 contents of cordierite coexisting with melts formed through biotite dehydration melting are presented as wt.% CO_2 contours in Figure 6a. These isopleths are steeper than and differ from those presented in Figure 3 since they are calculated for fluid saturation in the H_2O-CO_2 system in which a_{H_2O}, and hence, cordierite H_2O content is defined by Crd+L equilibrium. Hence, at lower temperatures, where both cordierite and melt have higher H_2O contents (Fig. 5a), the maximum CO_2 contents attainable in the melt environment are significantly less than those defined for the CO_2-only system of Figure 3, whereas, under HT/UHT conditions, the Crd-L-V isopleths approach those in the CO_2-only system. This is further illustrated by the cordierite channel H_2O-CO_2 composition diagram of Figure 6b, which shows that cordierites with channel X_{CO_2} of up to 0·4 may coexist with melt at 900°C, whilst, at 1000°C, the cordierite channel X_{CO_2} can be as high as 0·6–0·7. The presence of CO_2-rich, high-X_{CO_2} cordierite does not automatically imply the presence of a free carbonic fluid phase in UHT metamorphism or in HT migmatites (Vry *et al.* 1990; Fitzsimons 1994; Harley 1994).

If CO_2 is added to the KFMASH system without the addition of a separate new phase, its favourable incorporation into cordierite will enhance the stability of the mineral, even at fixed a_{H_2O}, and cause all other KFMASH univariant reactions to slide along loci which coincide with the cordierite-absent reactions emanating from the KFMASH invariant points (Carrington 1995). For the H_2O-CO_2 system, Thompson *et al.* (2001, 2002) and Harley *et al.* (2002) have shown that the minimum solubility of H_2O in granitic melt is essentially independent of the CO_2 content of the system and that D_w (Crd-L) is similar at a given melt H_2O content to the H_2O-only system. Hence, the dehydration-melting isopleth sets (Crd and L wt.% H_2O contours) of Figure 5a shift very little with the addition of CO_2 into the cordierite. This behaviour does not apply to saturation H_2O isopleths because, in that case, there is an overall decrease in total volatiles in cordierite at a given P–T as CO_2 enters its channels.

With the addition of CO_2 into cordierite, the KFMASH [Os] and [Bt] points move to higher P and T along (Crd) reactions whereas the H_2O-isopleths remain fixed. Hence, the stability fields of Grt+Crd+Sil+Kfs+Qz+L and Grt+Opx+Crd+Kfs+Qz+L expand to higher pressure, and to both higher and lower temperatures, and KFMASH-C divariant melting fields are generated from the KFMASH univariant dehydration-melting reactions (Fig. 7). Under UHT conditions (>900°C), the high-P stability of Grt+Crd+L, which is bounded up-pressure by the assemblage Opx+Sil+Qz+L, is estimated to be extended by at least 1 kbar on the basis of comparisons between the experimental data of Carrington & Harley (1995a, b) and Bertrand *et al.* (1991), and calculations by Aranovich & Podlesskii (1989). Paradoxically, the addition of CO_2 to the H_2O-undersaturated (dehydration-melting)

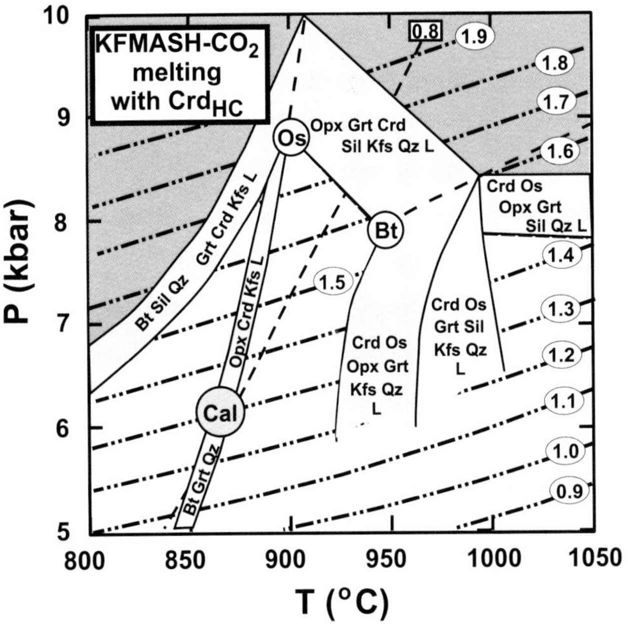

Figure 7 Pressure–temperature (P–T) diagram depicting the offset of selected KFMASH system reactions focused on the [Os] and [Bt] points of Figure 5a as a consequence of the addition of CO_2 to produce the system KFMASH-CO_2. Note that the KFMASH invariant points become lines located along (Crd) equilibria, sliding up-pressure with increasing CO_2 in cordierite. KFMASH univariant reactions become seven phase divariant fields in which cordierite can vary in H_2O and CO_2 depending on P–T (shaded fields). The maximum CO_2 contents of cordierite in these fields are dictated by the isopleths shown (dot-dashed contours in wt.%), which define CO_2 at the offset boundaries of the fields. For example, cordierite in the field Crd+Os+Opx+Grt+Kfs+Qz+L will contain 1·4 wt.% CO_2 at 7·3 kbar, 960 °C. Also shown on this diagram is the preferred P–T estimate for the Calabrian cordierite, Cal, which contains 0·8 wt.% H_2O and 1·3 wt.% CO_2 (Table 1). This cordierite lies on the saturation CO_2 isopleth, and hence, suggests equilibration in the presence of melt plus CO_2-rich fluid at ca. 6 kbar and 860 °C.

system decreases the temperature of melting, as controlled by the melting reactions which emanate down-pressure from the [Os] point, i.e. melting via equilibria such as R2 and R1 proceeds at lower temperatures in the fluid-saturated and -undersaturated H_2O-CO_2 systems compared with the H_2O-only one, not as a result of CO_2 entering the melt, but rather, reflecting the stabilisation of cordierite, the solid product of the peritectic melting reactions. By stabilising cordierite, it is possible for divariant melting in the presence of a CO_2-rich fluid to progress at temperatures some 20–40 °C lower than those required for melting via the reactions noted above (Fig. 7), provided enough H_2O is available to ensure the melt does not crystallise. In other words, the a_{H_2O} must be at least as high as that value defined by the dehydration-melting reaction at the proscribed P–T condition.

The movement of the KFMASH invariant points to higher P also have consequences for the range of volatile compositions attainable by cordierite coexisting with melt. The displaced [Os] point will occur at higher a_{H_2O} and as a result, the cordierite and melt will both have higher H_2O contents (ca. 1·2 and 3 wt.%, respectively) at the saturation condition where CO_2 is in excess but in equilibrium with both Crd and L (10 kbar, 910 °C in Fig. 7). The [Bt] point displaces to higher P and T, and therefore, will involve cordierite and melt with marginally lower H_2O contents (0·2 and 1 wt.%, respectively), but the cordierite will have high CO_2 (1·6 wt.%; Fig. 7). As a consequence of these displacements, the high-P limiting reaction involving H_2O-CO_2 cordierite (Crd_{HC}):

$$Grt+Crd_{HC}+Kfs+Qz=Opx+Sil+L \qquad (R4a)$$

occurs at higher pressures and over a broader T range (900–1000 °C) in the CO_2-bearing system compared with the KFMASH case, and at ca. 10–9 kbar (Fig. 7). At saturation (high CO_2, low a_{H_2O}, melt still present), this reaction may have a flat to moderately negative dP/dT, dictated by the change in H_2O-CO_2 partitioning between Crd and V across this T range. In the absence of relevant partitioning data, this reaction is drawn parallel to R4 in Figure 7.

6. Consequences for the stability of osumilite in melt-bearing UHT rocks

The uptake of CO_2 by cordierite and its consequent stabilisation to higher P relative to Opx+Sil+Qz has implications for the stability of osumilite and Os+Grt assemblages which may be developed in appropriate KFMASH bulk rock compositions (high-Mg, aluminous) along with melt under UHT conditions (Audibert *et al.* 1995; Carrington & Harley 1995b; Holland *et al.* 1996; White *et al.* 2001). The P–T stability field occupied by osumilite in KFMASH is limited by reactions focused on the [Bt] invariant point (Fig. 5a). Translation of this point up-pressure by a minimum of 0·5 kbar to 8·4 kbar and 990 °C, expands the stability field of Grt+Crd +Os+L towards higher pressures. However, the stability fields of Os+Sil+L and Os+Grt+L contract to higher T at pressures less than those of the [Bt] point (Fig. 7). The extent of translation estimated from Figure 7 is of the order of 40–50 °C as fluid saturation is approached (i.e. Crd_{HC}+L+V). Hence, under conditions where CO_2-rich cordierite is present, melting involving osumilite will only be seen in UHT granulites at >950–975 °C. Cordierite in the sapphirine-bearing but osumilite-absent assemblages from Mount Sones and Mount Hardy in the Napier Complex (Table 1) contain 0·5 and 1·05 wt.% CO_2, respectively, suggesting a role for CO_2 in stabilising cordierite with respect to osumilite, which does occur in other sapphirine-bearing Mg-pelites from these localities.

It is possible that the KFMASH [Bt] point, which is more completely denoted as [Bt, Spr] if sapphirine is considered as an additional UHT mineral, is rendered metastable in CO_2-saturated conditions. This will occur if the displaced (Bt) KFMASH-C reaction involving Crd, Grt, Opx, Sil, Os, Qz and L translates up P and T to intersect the group of UHT reactions which involve Spr+Qz at 1040–1050 °C (Bertrand *et al.* 1991; Audibert *et al.* 1995; Carrington & Harley 1995b; Harley 1998a). In this case, the (Bt, Crd, L) reaction of Figure 5b:

$$Opx+Sil+Kfs+Qz=Grt+Os \qquad (Bt, Crd, L)$$

is reduced to zero length, and its high-T terminating invariant point [Bt, Crd, L] will be replaced by others, such as [Bt, Sil, L] and [Bt, Grt, L], which involve Spr+Opx+Os+Qz and CO_2-rich (but H_2O-bearing) cordierite. The linking KFMASH-C reaction:

$$Os+Opx+Crd_{HC}=Spr+Qz+Kfs+L \qquad (Bt, Grt, Sil)$$

will limit cordierite stability to high P–T. Therefore, cordierite with volatile contents of H_2O <0·2 wt.% and CO_2 <1·8 wt.% could, in principle, coexist with all the other UHT phases in the absence of melt.

7. Integration with other UHT indicators: Opx–Crd relationships

Cordierite may coexist with orthopyroxene and melt over a wide range of HT/UHT conditions in rocks of appropriate

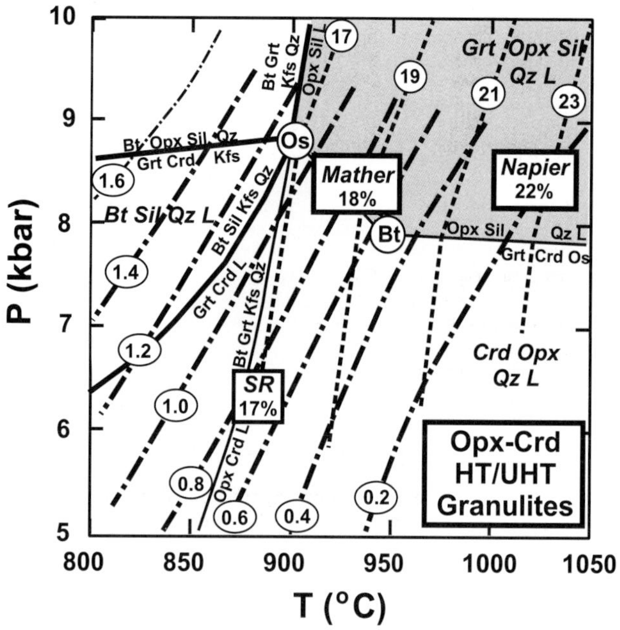

Figure 8 Pressure–temperature (P–T) diagram illustrating the P–T conditions of equilibration of the Crd+Opx migmatites/granulites listed in Table 1. This diagram is contoured for the H_2O contents of cordierite formed in equilibrium with KFMASH granitic melt (dot-dashed lines with wt.% H_2O indicated in ellipses), and for the Al_2O_3 content of orthopyroxene as modelled by Kelsey *et al.* (2003) and expressed in terms of mol% MgTs component. The agreement between the measured cordierite H_2O data and orthopyroxene MgTs contents is good in all three examples. Sources for P–T estimates are given in Table 1 along with the relevant cordierite compositional data; SR: Strangways Range.

bulk composition (Hensen & Green 1973; Carrington & Harley 1995a). At pressures less than those defined by the KFMASH [Os] and [Bt] points (Fig. 5a), relatively magnesian sub-aluminous to moderately aluminous pelites (i.e. compositions below the Grt-Crd tieline on an AFM diagram) may contain Crd+Opx+Qz+L at T>870°C. At higher pressures where the Opx+Sil +Qz+L sub-assemblage is stable, cordierite is restricted in FMAS and KFMASH to highly magnesian bulk compositions lacking garnet (e.g. Hensen 1971; Hensen & Green 1973; Bertrand *et al.* 1991; Carrington & Harley 1995a; Harley 1998a), but may be present along with garnet in slightly less magnesian bulk compositions (X_{Mg} >0·65) if the stabilising effect of CO_2 is considered. When garnet, sillimanite or sapphirine are present along with Opx+Crd, it is possible to estimate the temperatures of metamorphism using the Al_2O_3 content of the orthopyroxene (Ganguly *et al.* 1996; Aranovich & Berman 1997; Harley 1998a, Hollis & Harley 2003; Kelsey *et al.* 2003), and compare such estimates with those inferred from the H_2O contents of the cordierite. This comparison is illustrated for three examples of cordierite-bearing UHT assemblages in Figure 8.

Migmatitic Opx+Sil+Grt+Kfs ± Qz gneisses at Mather Peninsula in the Rauer Islands of East Antarctica have been described in detail by Harley (1998b). These contain magnesian garnet (X_{Mg} >0·66), aluminous orthopyroxene (X_{Mg}=0·78–0·82; Al_2O_3=8·5–10 wt.% zoning up to 12% on rims) and sillimanite in palaeosomes/mesosomes. Similarly, magnesian garnets occur in in 3–10-mm-thick leucosomes along with 2–8-mm-diameter subhedral aluminous orthopyroxene (Al_2O_3=9 wt.%) and microperthite. Locally, the leucosomes also contain cordierite (X_{Mg}=0·90; H_2O= 0·7 wt.%; CO_2=0·06 wt.%), in apparent textural equilibrium with the aluminous orthopyroxene and, by association, with the mesosome assemblage Grt+Opx+Sil. The P–T and a_{H_2O} conditions of extraction and crystallisation of the leucosomes can be assessed from the cordierite H_2O contents by reference to Figure 8. This indicates P–T conditions of ca. 8·5 kbar and 925°C, and a_{H_2O} (Table 1) of 0·09 ± 0·01. In Figure 8, a set of calculated orthopyroxene Al_2O_3 isopleths, expressed in terms of the mol% MgTs (Mg-tschermaks) component, are superimposed on the cordierite H_2O isopleths so that the P–T conditions defined above can be independently assessed. These Al_2O_3 isopleths (Kelsey *et al.* 2003), calculated for the assemblages Grt+Opx+Sil+Qz+L and Grt+Crd+Opx+Qz+L, are in reasonable agreement with those presented by Harley (1998a): over the temperature interval 910°C<T<1050°C, MgTs varies from 17 to 23 mol% in the modelled isopleth set of Kelsey *et al.* (2003), whereas MgTs ranges from 16 to 25 mol% in the Harley (1998a) isopleth diagram. It is encouraging that the MgTs isopleths, cordierite H_2O isopleths and assemblage constraints all match rather well. This suggests that, at least in the present case study, the cordierite has preserved its UHT volatile contents. This is interpreted to reflect melt loss from the Mather Peninsula migmatitic gneiss leucosomes (e.g. Kriegsman 2001; Kelsey *et al.* 2003), which are residual or in their overall chemistry and preserve anhydrous mineral assemblages only partially retrogressed to biotite-bearing ones (Harley 1998b). Melt loss early in the UHT history, and certainly below the peak conditions attained by these rocks (ca. 11–12 kbar, 1030°C; Harley 1998b), is interpreted to have enabled the preservation of the prograde melting record since the leucosomes acted as closed systems following the loss of hydrous melt components.

Grt+Crd+Opx+K-feldspar leucosomes occur as local patchy segregations in semipelitic gneisses near Phlogopite Mine in the Strangways Ranges, which comprise part of the Arunta Complex of central Australia. Cordierite H_2O contents (Table 1) are consistent with dehydration melting via Reaction R3 at 875–900°C if pressures were 6–7 kbar. The orthopyroxene in these leucosome patches contains 7·7 wt.% Al_2O_3, or 17 mol% MgTs, once again consistent with the temperatures implied by the cordierite H_2O data, and therefore, suggestive of melt loss at these conditions, followed by an absence of any fluid that could interact with and hence change the H_2O content of the preserved cordierites.

The final comparison of cordierite H_2O content with the Al_2O_3 contents of orthopyroxene in the same UHT assemblage is provided by the sapphirine+orthopyroxene+quartz granulite from Mount Hardy, East Antarctica (Table 1). This granulite occurs in the UHT sapphirine+quartz zone of the Archaean Napier Complex (Ellis *et al.* 1980; Sheraton *et al.* 1987; Harley & Hensen 1990) and is inferred to have experienced P–T conditions in the region of 8–9 kbar and greater than 1000°C (Harley & Hensen 1990), with even higher temperatures (>1120°C) possible on the basis of comparisons with a sapphirine+orthopyroxene+quartz granulite reported from Mount Riiser-Larsen nearby (Harley & Motoyoshi 2000). Cordierite not only occurs in the Mount Hardy sample as coronas on sapphirine+quartz, but also as large grains in apparent textural equilibrium with orthopyroxene rims which contain up to 10–11 wt.% Al_2O_3, or 22 mol% MgTs. Figure 8 demonstrates that the measured H_2O content of this cordierite (0·22 wt.%) is consistent with the P–T conditions inferred from the orthopyroxene, and is suggestive of the presence of some melt that may have back-reacted with the UHT assemblage to produce this cordierite. Although the CO_2 content of the cordierite is significant (1·04 wt.%), it is not high enough to saturate the cordierite at these P–T conditions, or indeed, any P–T conditions along the post-peak IBC path experienced by the Napier Complex. Hence, in this case, the present authors suggest that the cordierite formed in the absence of a free fluid phase, under low a_{H_2O} (ca. 0·03) and high a_{CO_2} (0·58) conditions mediated by the presence of a small amount of melt.

8. Conclusions

As emphasised in many previous studies (e.g. Hensen 1971; Hensen & Green 1973; Aranovich & Podlesskii 1989; Vry *et al.* 1990; Bertrand *et al.* 1991; Audibert *et al.* 1995; Carrington & Harley 1995a, b; Fitzsimons 1996; Harley 1998a; White *et al.* 2001; Harley *et al.* 2002), cordierite plays a significant role in the high-grade metamorphism and anatexis of pelites. The P–T-dependent H_2O exchange equilibrium between cordierite and granitic melt exerts a strong control on melt production from such rocks. Moreover, the variable CO_2 contents of cordierite formed coexisting with melt influences the P–T positions of melting reactions, expanding the stability fields of cordierite+melt assemblages in KFMASH-CO_2 at the expense of UHT indicator assemblages such as orthopyroxene+sillimanite+K-feldspar+quartz and osumilite+garnet.

SIMS microanalysis of natural cordierites from a selection of HT/UHT migmatitic and granulite terrains yields H_2O contents which are consistent with cordierite-melt equilibrium at temperatures >860 °C, and in most cases, in agreement with calculated KFMASH grids and independent equilibria, irrespective of the CO_2 contents of the cordierites. This gives confidence that volatiles in cordierite, analysed carefully, and with sufficient spatial resolution to evaluate and volatile zoning or leakage, can provide valuable information on the fluid and melting conditions attending HT/UHT metamorphism. With the notable exception of the Calabrian granulites, which appear to have been fluid-saturated at HT conditions and equilibrated in the presence of a CO_2-rich fluid phase at or near peak metamorphic conditions, cordierites from the migmatitic granulites studied thus far yield CO_2 and H_2O contents which produce calculated a_{CO_2} and a_{H_2O} which sum to less than unity. These imply metamorphism under fluid-absent conditions, but in the presence of melts which contain 1·3–2·5 wt.% H_2O and, in many cases, appreciable dissolved CO_2. It is clear that the volatile contents of cordierite in HT/UHT metamorphic and magmatic rocks cannot be assumed to adhere to saturation values or ignored in calculations of mineral and mineral-melt equilibria: they must be routinely measured and incorporated into the assessment and analysis of the P–T and P-T-X-fluid evolution of high-grade terrains.

9. Acknowledgements

This research was supported by NERC grants GR3/9099 and GR3/12573 to S.L.H. We thank John Craven and Richard Hinton (SIMS), and Peter Hill (electron probe) for their assistance with the analysis, Steve Elphick and Damian Carrington for advice on experimental design, and David Mattey, Jim Webster, Volker Schenk and Francois Holtz for their help in producing standards. We are indebted to Bas Hensen, Ian Fitzsimons, Damian Carrington, Ian Buick and Julie Vry for their interest in the cordierite work and discussions of its many implications over the years.

References

Aines, R. D. & Rossman, G. R. 1984. The high-temperature behavior of water and carbon dioxide in cordierite and beryl. *American Mineralogist* **69**, 319–27.

Aranovich, L. Y. & Berman, R. G. 1996. Optimized standard state and solution properties of minerals II. Comparisons, predictions, and applications. *Contributions to Mineralogy and Petrology* **126**, 25–37.

Aranovich, L. Y. & Berman, R. G. 1997. A new garnet-orthopyroxene barometer based on reversed Al_2O_3 solubility in FeO-Al_2O_3-SiO_2 orthopyroxene. *American Mineralogist* **82**, 345–53.

Aranovich, L. Y. & Podlesskii, K. K. 1989. Geothermometry of high-grade metapelites: simultaneously operating reactions. *In* Daly, J. S., Cliff, R. A. & Yardley, B. W. D. (eds) *Evolution of Metamorphic Belts. Geological Society of London, Special Publication* **43**, 45–62. London: Geological Society.

Armbruster, Th. & Bloss, F. D. 1982. Orientation and effects of channel H_2O and CO_2 in cordierite. *American Mineralogist* **67**, 284–91.

Audibert, N., Hensen, B. J. & Bertrand, P. 1995. Experimental study of phase relations involving osumulite in the system K_2O-FeO-MgO-Al_2O_3-SiO_2-H_2O at high pressure and temperature. *Journal of Metamorphic Geology* **13**, 331–44.

Bertrand, P., Ellis, D. J. & Green, D. H. 1991. The stability of sapphirine-quartz and hypersthene-sillimanite-quartz assemblages: an experimental investigation in the system FeO-MgO-Al_2O_3-SiO_2 under H_2O and CO_2. *Contributions to Mineralogy and Petrology* **108** 55–71.

Boberski, C. & Schreyer, W. 1990. Synthesis and water contents of Fe^{2+}-cordierites. *European Journal of Mineralogy* **2**, 565–84.

Burnham, C. W. 1994. Development of the Burnham model for prediction of H_2O solubility in magmas. *In* Carroll, M. R. & Holloway, J. R. (eds) *Volatiles in Magmas. Mineralogical Society of America, Reviews in Mineralogy* **30**, 123–9.

Carey, J. W. 1995. A thermodynamic formulation for hydrous cordierite. *Contributions to Mineralogy and Petrology* **119**, 155–65.

Carrington, D. P. 1995. The relative stability of garnet-cordierite and orthopyroxene-sillimanite-quartz assemblages in metapelitic granulites: experimental data. *European Journal of Mineralogy* **7**, 949–60.

Carrington, D. P. & Harley, S. L. 1995a. Partial melting and phase relations in high-grade metapelites: an experimental petrogenetic grid in the KFMASH system. *Contributions to Mineralogy and Petrology* **120**, 270–91.

Carrington, D. P. & Harley, S. L. 1995b. The stability of osumilite in metapelitic granulites. *Journal of Metamorphic Geology* **13**, 613–25.

Carrington, D. P. & Harley, S. L. 1996. Cordierite as a monitor of fluid and melt water contents in the lower crust: an experimental calibration. *Geology* **24**, 647–50.

Ellis, D. J., Sheraton, J. W., England, R. N. & Dallwitz, W. B. 1980. Osumilite-sapphirine-quartz granulites from Enderby Land, Antarctica – mineral assemblages and reactions. *Contributions to Mineralogy and Petrology* **72**, 123–43.

Fitzsimons, I. C. W. 1994. Cordierite migmatites from East Antarctica: geochemical constraints on volatile distribution during crustal anatexis. *Mineralogical Magazine* **58a**, 274–5.

Fitzsimons, I. C. W. 1996. Metapelitic migmatites from Brattstrand Bluffs, east Antarctica – metamorphism, melting and exhumation of the mid-crust. *Journal of Petrology* **37**, 395–414.

Fitzsimons, I. C. W. & Mattey, D. P. 1995. Carbon isotope constraints on volatile mixing and melt transport in granulite-facies migmatites. *Earth and Planetary Science Letters* **134**, 319–28.

Goldman, D. S., Rossman, G. R. & Dollase, W. A. 1977. Channel constituents in cordierite. *American Mineralogist* **62**, 1144–57.

Ganguly, J., Cheng, W. & Tirone, M. 1996. Thermodynamics of aluminosilicate garnet solid solution: new experimental data, an optimized model, and thermometric applications. *Contributions to Mineralogy and Petrology* **126**, 137–51.

Grew, E. S. 1982. Osumilite in the sapphirine-quartz terrane of Enderby Land, Antarctica: implications for osumilite petrogenesis in the granulite facies. *American Mineralogist* **67**, 762–87.

Harley, S. L. 1986. A sapphirine-cordierite-garnet-sillimanite granulite from Enderby Land, Antarctica: implications for FMAS petrogenetic grids in the granulite facies. *Contributions to Mineralogy and Petrology* **94**, 452–60.

Harley, S. L. 1994. Cordierite as a sensor of fluid and melt distribution in crustal metamorphism. *Mineralogical Magazine* **58a**, 374–5.

Harley, S. L. 1998a. On the occurrence and characterisation of ultra high-temperature (UHT) crustal metamorphism. *In* Treloar, P. J. & O'Brien, P. (eds) *What Controls Metamorphism and Metamorphic Reactions? Geological Society, London, Special Publication* **138**, 75–101. London: Geological Society.

Harley, S. L. 1998b. Ultrahigh temperature granulite metamorphism (1050 °C, 12 kbar) metamorphism and decompression in garnet (Mg70)-orthopyroxene-sillimanite gneisses from the Rauer Group, East Antarctica. *Journal of Metamorphic Geology* **16**, 541–62.

Harley, S. L., Thompson, P., Buick, I. S. & Hensen, B. J. 2002. Cordierite as a sensor of fluid conditions in high-grade metamorphism and crustal anatexis. *Journal of Metamorphic Geology* **20**, 71–86.

Harley, S. L. & Carrington, D. P. 2001. The distribution of H_2O between cordierite and granitic melt: improved calibration of H_2O incorporation in cordierite and its application to high-grade metamorphism and crustal anatexis. *Journal of Petrology* **42**, 1595–620.

Harley, S. L. & Hensen, B. J. 1990. Archaean and Proterozoic high-grade terranes of east Antarctica (40–80°): a case study of diversity in granulite facies metamorphism. *In* Ashworth, J. R. & Brown, M. (eds) *High-temperature Metamorphism and Crustal Anatexis. Mineralogical Society Special Publication* **2**, 320–70.

Harley, S. L. & Motoyoshi, S. L. 2000. Al zoning in orthopyroxene in a sapphirine quartzite: evidence for >1120°C UHT metamorphism in the Napier Complex, Antarctica, and implications for the entropy of sapphirine. *Contributions to Mineralogy and Petrology* **138**, 293–307.

Hensen, B. J. 1971. Theoretical phase relations involving cordierite and garnet in the system FeO-MgO-Al_2O_3-SiO_2. *Contributions to Mineralogy and Petrology* **33**, 191–214.

Hensen, B. J. 1986. Theoretical phase relations involving garnet and cordierite revisited: the influence of oxygen fugacity on the stability of sapphirine and spinel in the system Mg-Fe-Al-Si-O. *Contributions to Mineralogy and Petrology* **92**, 362–7.

Hensen, B. J. & Green, D. H. 1973. Experimental study of the stability of cordierite and garnet in pelitic compositions at high pressures and temperatures. III. Synthesis of experimental data and geological applications. *Contributions to Mineralogy and Petrology* **38**, 151–66.

Holland, T. J. B., Babu, E. V. S. S. K. & Waters, D. J. 1996. Phase relations of osumilite and dehydration melting in pelitic rocks: a simple thermodynamic model in the KFMASH system. *Contributions to Mineralogy and Petrology* **124**, 383–94.

Holland, T. J. B. & Powell, R. 1990. An internally consistent thermodynamic dataset with uncertainties and correlations: the system Na_2O-K_2O-CaO-MgO-MnO-FeO-Fe_2O_3-Al_2O_3-SiO_2-TiO_2-C-H_2-O_2. *Journal of Metamorphic Geology,* **8**, 89–124.

Holland, T. J. B. & Powell, R. 1998. An internally consistent thermodynamic dataset for phases of petrological interest. *Journal of Metamorphic Geology* **16**, 309–43.

Hollis, J. & Harley, S. L. 2003. Alumina solubility in orthopyroxene coexisting with sapphirine and quartz. *Contributions to Mineralogy and Petrology* **144**, 473–83.

Holloway, J. R. & Blank, J. G. 1994. Application of experimental results to C-O-H species in natural melts. *In* Carroll, M. R. & Holloway, J. R. (eds) *Volatiles in Magmas. Mineralogical Society of America, Reviews in Mineralogy* **30**, 187–230.

Holtz, F. & Johannes, W. 1994. Maximum and minimum water contents of granitic melts: implications for chemical and physical properties of ascending magmas. *Lithos* **32**, 149–59.

Johannes, W. & Holtz, F. 1996. *Petrogenesis and experimental petrology of granitic rocks.* Berlin: Springer-Verlag.

Johannes, W. & Schreyer, W. 1981. Experimental introduction of CO_2 and H_2O into Mg-cordierite. *American Journal of Science* **281**, 299–317.

Kalt, A. 2000. Cordierite channel volatiles as evidence for dehydration melting: an example from high-temperature metapelites of the Bayerische Wald (Variscan Belt, Germasny). *European Journal of Mineralogy* **12**, 987–98.

Kelsey, D. E., White, R. W. & Powell, R. 2003. Orthopyroxene-sillimanite-quartz assemblages: distribution, petrology, quantitative P–T-X constraints and P–T paths. *Journal of Metamorphic Geology* **21**, 439–53.

Kriegsman, L. M. 2001. Partial melting, partial melt extraction and partial back reaction in anatectic migmatites. *Lithos* **56**, 75–96.

Kurepin, V. A. 1985. H_2O and CO_2 contents of cordierite as an indicator of thermodynamical conditions of formation. *Geochemistry International* **22**, 148–56.

Le Breton, N. & Schreyer, W. 1993. Experimental CO_2 incorporation in Mg-cordierite: non-linear behaviour of the system. *European Journal of Mineralogy* **5**, 427–38.

Markl, G., Bäuerle, J. & Grujic, D. 2000. Metamorphic evolution of Pan-African granulite facies metapelites from southern Madagascar. *Precambrian Research* **102**, 47–68.

Mirwald, P. W., Maresch, W. V. & Schreyer, W. 1979. Der Wassergehalt von Mg-cordierit zwischen 500° und 800°C sowie 0·5 und 11 kbar: *Fortschritte der Mineralogie* **57**, 101–2.

Mukhopadhyay, B. & Holdaway, M. J. 1994. Cordierite-garnet-sillimantie-quartz equilibrium I. New experimental calibration in the system FeO-Al_2O_3-SiO_2-H_2O and certain P–T–X_{H_2O} relations. *Contributions to Mineralogy and Petrology* **116**, 462–72.

Nandakumar, V. & Harley, S. L. 2000. A reappraisal of the pressure-temperature path of granulites from the Kerala Khondalite Belt, southern India. *Journal of Geology* **108**, 687–703.

Newton, R. C. & Wood, B. J. 1979. Thermodynamics of water in cordierite and some petrologic consequences of cordierite as a hydrous phase. *Contributions to Mineralogy and Petrology* **68**, 391–405.

Patiño Douce, A. E. & Johnston, A. D. 1991. Phase equilibria and melt productivity in the pelitic system: implications for the origin of peraluminous granitoids and aluminous granulites. *Contributions to Mineralogy and Petrology* **107**, 202–18.

Santosh, M. 1987. Cordierite gneisses of Southern Kerala, India: petrology, fluid inclusions and implications for crustal uplift history. *Contributions to Mineralogy and Petrology* **96**, 343–56.

Santosh, M., Jackson, D. H. & Harris, N. B. W. 1993. The significance of channel and fluid inclusion CO_2 in cordierite: evidence from carbon isotopes. *Journal of Petrology* **34**, 233–58.

Schenk, V. 1984. Petrology of felsic granulites, metapelites, metabasics, ultramafics, and metacarbonates from Southern Calabria (Italy): prograde metamorphism, uplift and cooling of a former lower crust. *Journal of Petrology* **25**, 255–98.

Schreyer, W. 1985. Experimental studies on cation substitutions and fluid incorporation in cordierite. *Bulletin de Minéralogie* **108**, 273–91.

Sheraton, J. W., Tingey, R. J., Black, L. P., Offe, L. A. & Ellis, D. J. 1987. Geology of Enderby Land and Western Kemp Land, Antarctica. *Bulletin Australian Bureau of Mineral Resources* **223**, 51 pp.

Skippen, G. B. & Gunter, A. E. 1996. The thermodynamic properties of H_2O in magnesian and iron cordierite. *Contributions to Mineralogy and Petrology* **124**, 82–9.

Stevens, G., Clemens, J. D. & Droop, G. T. R. 1995. Hydrous cordierite in granulites and crustal magma production. *Geology* **23**, 925–8.

Thompson, P., Harley, S. L. & Carrington, D. P. 2001. The distribution of H_2O-CO_2 between cordierite and granitic melt under fluid-saturated conditions at 5 kbar and 900°C. *Contributions to Mineralogy and Petrology* **142**, 107–18.

Thompson, P., Harley, S. L. & Carrington, D. P. 2002. Sodium and potassium in cordierite – a potential thermometer for melts? *European Journal of Mineralogy* **14**, 459–69.

Vielzeuf, D. & Holloway, J. R. 1988. Experimental determination of the fluid-absent melting relations in the pelitic system. Consequences for crustal differentiation. *Contributions to Mineralogy and Petrology* **98**, 257–76.

Vry, J. K., Brown, P. E. & Valley, J. W. 1990. Cordierite volatile content and the role of CO_2 in high-grade metamorphism. *American Mineralogist* **75**, 71–88.

Waters, D. J. 1988. Partial melting and the formation of granulite facies assemblages in Namaqualand, South Africa. *Journal of Metamorphic Geology* **6**, 387–404.

White, R. W., Powell, R. & Holland, T. J. B. 2001. Calculation of partial melting equilibria in the system Na_2O-CaO-K_2O-FeO-MgO-Al_2O_3-SiO_2-H_2O (NCKFMASH). *Journal of Metamorphic Geology* **19**, 139–53.

White, R. W. & Powell, R. 2002. Melt loss and the preservation of granulite facies mineral assemblages. *Journal of Metamorphic Geology* **20**, 621–32.

Winkler, B., Milman, V. & Payne, M. C. 1994. Orientation, location, and total energy of hydration of channel H_2O in cordierite investigated by ab-initio total energy calculations. *American Mineralogist* **79**, 200–4.

SIMON L. HARLEY and P. THOMPSON, Grant Institute of Earth Science, School of GeoSciences, University of Edinburgh, Kings Buildings, West Mains Road, Edinburgh EH9 3JW, Scotland.
e-mail: sharley@glg.ed.ac.uk

MS received 17 October 2003. Accepted for publication 24 August 2004.

Transactions of the Royal Society of Edinburgh: Earth Sciences, **95**, 99–110, 2004

Origin of chemically zoned and unzoned cordierites from the South Mountain and Musquodoboit Batholiths, Nova Scotia

Saskia Erdmann, D. Barrie Clarke, Michael A. MacDonald

ABSTRACT: Textural relations and chemical zoning of cordierites in granites act as sensitive recorders of the conditions of their crystallisation history and underlying magma chamber processes. In this contribution, we present new data on texturally distinct and variably zoned cordierites from the late-Devonian, granitic South Mountain and Musquodoboit Batholiths, and infer the conditions of their formation. Using a combined textural (grain size, grain shape and inclusion relationships) and chemical (major element composition and compositional zoning) classification, we recognise the following six cordierite types: CG1/TT1, anhedral to subhedral macrocrysts with random inclusions and patchy normal zoning; CG2a/TT2, euhedral to subhedral macrocrysts with random inclusions and normal zoning; CG2b/TT2, euhedral to subhedral macrocrysts with random or oriented inclusions, and oscillatory zoning; CG3a/TT3, subhedral to euhedral microcrysts with no inclusions and reverse zoning; CG3b/TT4, euhedral macrocrysts with no inclusions and no zoning; and CG4/TT5, anhedral macrocrysts with random inclusions and normal zoning. The textural criteria suggest that these cordierites formed as a product of cotectic crystallisation from a melt, or as the result of a peritectic reaction involving country-rock material. The combined chemical and textural criteria suggest that: (1) normal zoning results from cotectic crystallisation during cooling, cotectic overgrowths on grains formed in a peritectic reaction with country-rock material, or cation exchange with a fluid; (2) oscillatory zoning results from cotectic crystallisation during variations in X_{Mg} of the silicate melt following magma replenishment; (3) reverse zoning results from crystallisation during pressure quenching; and (4) the unzoned cordierite results from cotectic crystallisation under fluid-rich conditions.

KEY WORDS: Chemical zoning, cordierite, granite, microstructures, xenocrysts.

The late-Devonian South Mountain and Musquodoboit Batholiths (Fig. 1), intruded into metasediments of the Cambro-Ordovician Meguma Supergroup, constitute the largest peraluminous granitoid complex of Nova Scotia, Canada (e.g. McKenzie & Clarke 1975; MacDonald & Clarke 1985). Both batholiths are connected at a depth of ~5 km (O'Reilly 1975; McKenzie 1976; Douma 1978), and the South Mountain Batholith (SMB) consists of several plutons with distinct compositions (MacDonald *et al.* 1992; MacDonald 2001). The Halifax Pluton (HP) (Fig. 1b) is one of the larger plutons in the SMB, exhibiting normal compositional zoning at the margin, and reverse zoning in the core (MacDonald & Horne 1988). The rock succession is biotite granodiorite, biotite monzogranite, leucomonzogranite and muscovite-biotite monzogranite; all units are locally intruded by aplites and pegmatites. Cordierite is present in all rock types of the pluton, including aplites and pegmatites, but is most prominent in the muscovite-biotite monzogranite and in the coarse-grained leucomonzogranite, where it constitutes up to 5 modal%. The Musquodoboit Batholith (MB), located NE of the SMB (Fig. 1a, c), mainly consists of muscovite-biotite monzogranite and minor leucomonzogranite. Cordierite is present throughout the batholith, locally with 4–10% of the mode (MacDonald & Clarke 1985; Ham 1999). The samples investigated are muscovite-biotite monzogranites and aplites from the HP and the MB, and a pegmatite dyke from the HP.

All granitoid rocks in the SMB and the MB are peraluminous (A/CNK >1), and thus, permit a primary magmatic origin for minerals such as biotite, muscovite, garnet, cordierite and andalusite, all of which occur as zoned and unzoned varieties in different units of the HP (Ding 1995) and the MB (MacDonald & Clarke 1985). However, these minerals may also have a foreign origin; they may be xenocrystic (crystals derived from country rocks which did not form as a result of a chemical reaction with the main magma), or they may have formed in a peritectic reaction, involving country-rock material and main magma ($Bt+Qtz+Plag+L1 \rightarrow Crd \pm$ other minerals$+L2$; L1=main magma, L2=hybrid magma consisting of country-rock partial melt and main magma; abbreviations after Kretz 1983). In other words: the origin of each mineral in the granites has to be questioned, and the peraluminous nature of the granitoid rocks may reflect the peraluminosity of the primary magma, as well as the peraluminosity of a random mixture of primary magma and country-rock material. Hence, only a detailed microstructural and microchemical study of the mineral assemblage may permit a correct determination of the origin of the grains in question, and moreover, may lead to a better constrained interpretation of whole-rock geochemical data (e.g. explaining variations in peraluminosity dominantly as a result of fractional crystallisation, or as a result of fractional crystallisation combined with contamination by country-rock material).

Of all minerals in the SMB and the MB, cordierite has the most striking textural and chemical variability (Maillet & Clarke 1985). Therefore, cordierite appears to be most amenable to search for grains of different origins, and to reveal key information about magma chamber processes and conditions of crystallisation at various stages of chemical evolution. In this paper, we (1) evaluate potential criteria for distinguishing between cotectic magmatic ($L \rightarrow Crd+$ other minerals),

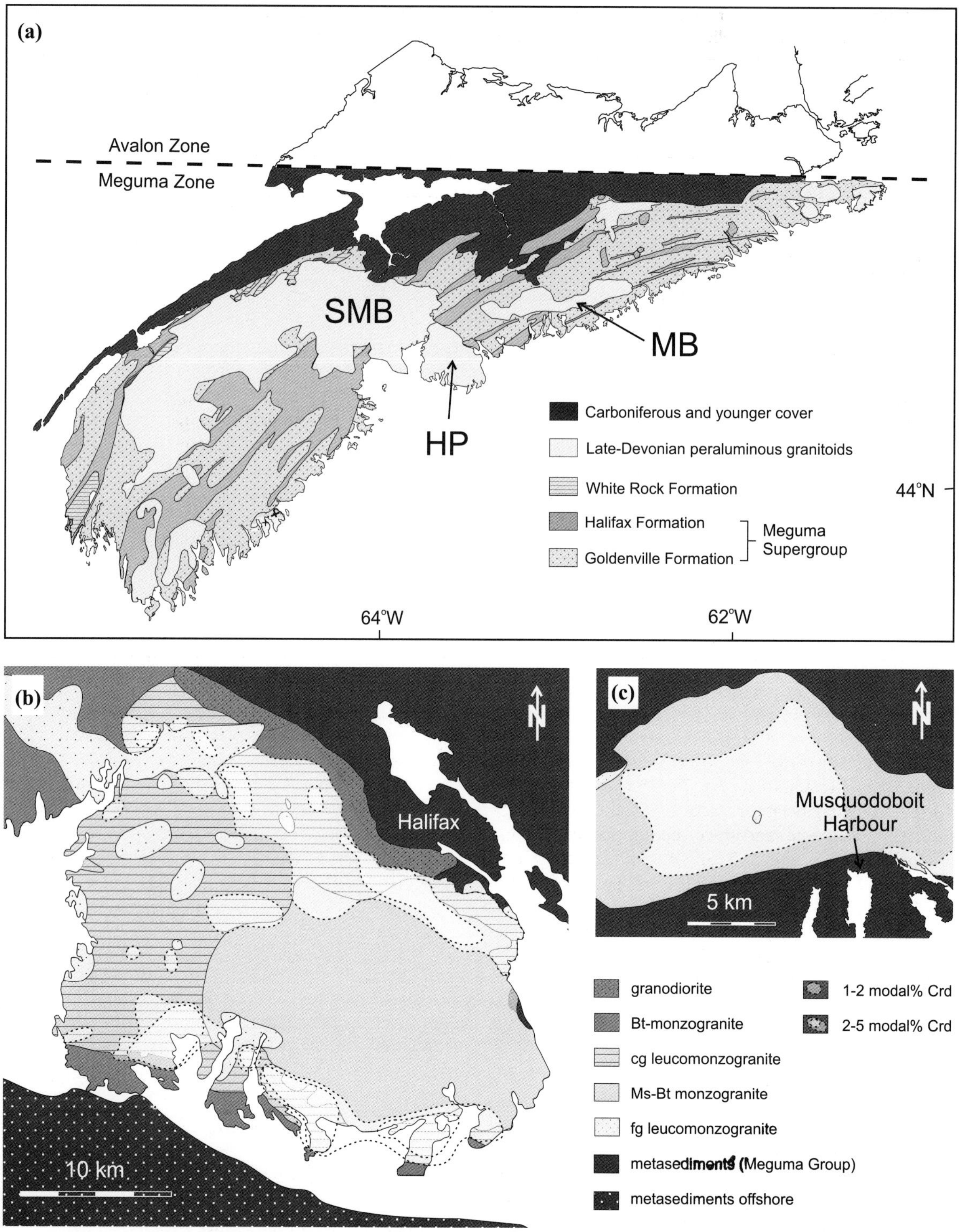

Figure 1 (a) Geological map of Nova Scotia, Canada, showing the location of the South Mountain and Musquodoboit Batholiths (SMB and MB), and the Halifax Pluton (HP) (modified after Tate 1995). (b) Detailed map of the HP. Cordierite is most abundant in zones away from the granite/metasediment contact, roughly subparallel to the margin of the pluton, and apparently independent of internal lithological contacts (modified after MacDonald 2001). (c) Detailed map of the centre of the MB. In contrast to the HP, cordierite is most abundant in one major and in several smaller zones, showing no spatial relationship to the margin of the batholith (modified after Ham 1999).

peritectic magmatic [country rock (Bt+Qtz+Plag)+magma (L1)→Crd ± other minerals+L2], and xenocrystic (derived from the country rocks) cordierite; (2) discuss the driving forces for the normal, oscillatory, and reverse zoning in the cordierites; and (3) make inferences about the origin and the growth history of the various cordierite types investigated.

1. Cordierite-bearing granitoid rocks

The monzogranites of the HP are coarse-grained. They contain quartz, plagioclase, K-feldspar, biotite, muscovite, ± cordierite, ± garnet, ± rutile, and traces of apatite, zircon, monazite, xenotime and ilmenite; X_{Mg} of the monzogranites varies

Table 1 Average whole-rock chemical compositions for monzogranites and aplites from the Halifax Pluton (HP) and the Musquodoboit Batholith (MB): (*n*) number of samples analysed by XRF; and (FeO^T) Fe total as FeO

	Monzogranites				Aplites		
	HP (*n*=26)*		MB (*n*=7)**		HP (*n*=8)***		MB (*n*=1)**
	wt.%	1σ	wt.%	1σ	wt.%	1σ	wt.%
Crd textural type:	TT1, TT2		TT1, TT2		TT3		TT5
Crd chemical group:	CG1, CG2		CG1, CG2		CG3		CG4
SiO_2	71·11	1·04	71·92	1·11	72·00	1·19	72·56
TiO_2	0·39	0·07	0·28	0·09	0·26	0·08	0·04
Al_2O_3	14·63	0·38	14·58	0·16	14·55	0·15	15·21
FeO^T	2·52	0·39	1·72	0·40	1·67	0·41	0·51
MnO	0·06	0·01	0·06	0·02	0·06	0·02	0·17
MgO	1·23	0·18	0·53	0·22	0·59	0·16	0·08
CaO	1·02	0·29	0·86	0·20	0·80	0·12	0·28
Na_2O	3·57	0·42	3·60	0·11	3·63	0·10	3·82
K_2O	4·47	0·25	4·60	0·23	4·60	0·25	4·8
P_2O_5	0·25	0·03	0·25	0·09	0·28	0·04	0·9
F	0·07	0·01	–	–	–	–	–
LOI	0·64	0·15	1·14	0·27	1·20	0·24	0·44
Total	99·89	0·39	99·68	0·64	99·80	0·61	98·81
X_{Mg}	0·33	0·04	0·23	0·07	0·26	0·01	0·14

Sources of data: *Ham *et al.* (1989, 1990); **MacDonald & Clarke (1985) and ***MacDonald (2001).

between 0·26 and 0·39, and MnO ranges from 0·04 to 0·08 wt. (Table 1; Ham *et al.* 1989, 1990). The aplites of the HP are fine-grained. They are characterized by the assemblage of quartz, K-feldspar, plagioclase, muscovite, biotite, ± cordierite, ± garnet, ± andalusite, and traces apatite, zircon, monazite, xenotime and ilmenite; X_{Mg} of the aplites varies between 0·24 and 0·27, and MnO lies between 0·04 and 0·09 wt. (MacDonald 2001). Pegmatites have a similar mineralogy as the aplites, but contain in addition tourmaline, fluorite and various disseminated sulphides (e.g. MacDonald 2001); whole-rock analysis for the pegmatites are not available.

The monzogranites of the MB are medium- to coarse-grained. They are characterised by the assemblage of quartz, K-feldspar, plagioclase, biotite, muscovite, ± cordierite, ± andalusite, ± garnet, ± rutile, and traces of apatite, zircon, monazite and ilmenite; X_{Mg} of the monzogranites ranges from 0·31 to 0·46, and MnO varies between 0·04 and 0·07 wt. (Table 1; MacDonald & Clarke 1985). The aplites of the MB are fine-grained. They comprise an assemblage of quartz, K-feldspar, plagioclase, muscovite, ± biotite, ± garnet, ± andalusite, ± tourmaline, ± cordierite, and traces of apatite, zircon, monazite, xenotime, and ilmenite; a whole-rock analysis for the aplite sample studied is not available, but several aplites of the MB have low X_{Mg} (~0·14) and are high in MnO (~0·17 wt.) (MacDonald & Clarke 1985).

2. Cordierite in the South Mountain and Musquodoboit Batholiths

2.1. Classification

Cordierite in peraluminous granites may have various origins, and two or more origins may be valid for cordierites even from a single rock (Clarke 1995). Correctly inferring the origin of any cordierite crystal depends on successfully applying a combination of the following textural and chemical criteria:

2.1.1. Grain shape. Metamorphic and xenocrystic cordierite (formed via a subsolidus or a peritectic reaction) is typically subhedral to anhedral (e.g. Phillips *et al.* 1981; Bouloton 1992; Pereira & Bea 1994; Barbey *et al.* 1999; Cenki *et al.* 2002), whereas cotectic magmatic cordierite is commonly subhedral to euhedral (e.g. Phillips *et al.* 1981; Bouloton 1992; Pereira & Bea 1994; Villaseca & Barbero 1994), but may be subhedral to anhedral in the case of grain interference during primary magmatic growth or in the case of subsequent reaction.

2.1.2. Inclusions. If cordierite contains inclusions, the sizes, shapes and compositions of the inclusions, and the inclusion patterns, may help to constrain its crystallisation history, and thus, its origin. If cordierite contains inclusions of former partial melt (now in the form of solids which crystallised from the melt), its origin is probably magmatic (either cotectic or peritectic). In contrast, inclusions which otherwise do not belong to the mineral assemblage of the igneous rock, most likely suggest a metamorphic xenocrystic or a peritectic magmatic origin. However, even cotectic magmatic cordierite can contain inclusions of 'foreign' minerals if it overgrew xenocrysts present in the melt.

2.1.3. Grain size. Dimensional compatibility of a crystal of unknown origin with other, generally accepted igneous rock-forming minerals in the sample may suggest that the mineral in question also has a magmatic origin. On the other hand dimensional incompatibility may point to, but does not necessarily require, different origins. The grain sizes of the various igneous minerals in any rock, and also the grain size of a certain mineral, can vary by orders of magnitude, and thus, grain-size criteria are not particularly reliable.

2.1.4. Mineral composition. Major-element compositions may indicate different populations of a single mineral. Metamorphic cordierites, for example, tend to have an X_{Mg} of ~0·50–0·95, whereas magmatic cordierites commonly have lower values (Deer *et al.* 1997). Other components, such as Na_2O, as well as H_2O and CO_2, are less useful for determining the origin of an unknown cordierite because their

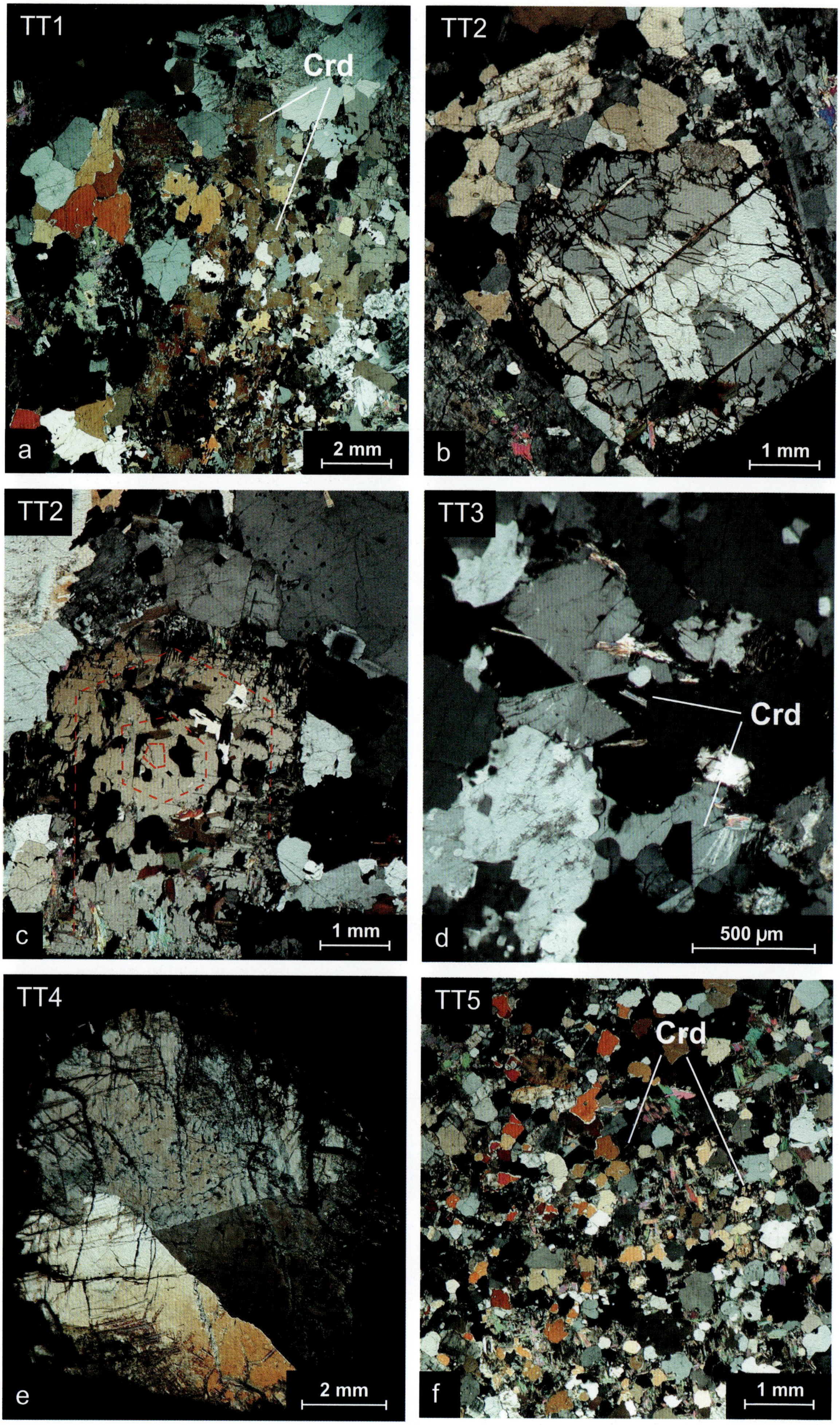
TT1
Crd
a
2 mm
TT2
b
1 mm
TT2
c
1 mm
TT3
Crd
d
500 μm
TT4
e
2 mm
TT5
Crd
f
1 mm

Table 2 Textural-chemical classification of cordierite in the South Mountain and the Musquodoboit Batholiths

Textural type:	TT1	TT2	TT3	TT4	TT5
Chemical group:	CG1	CG2	CG3	CG3	CG4
Lithology:	mng	mng	apl	peg	apl
n	1	11	5	1	3
Size	>1 mm	>1 mm	<1 mm	≫1 mm	>1 mm
Shape	ah–sh	sh–eh	sh–eh	sh–eh	ah
Inclusion type	Bt, Qtz, Acc	Bt, Qtz, Acc	–	–	Qtz
Inclusion orientation	r	r/o	–	–	r
X_{Mg}	0·45–0·64	0·37–0·50	0·20–0·32	0·24–0·25	0·35–0·36
Mn (apfu)	0·01–0·06	0·04–0·09	0·10–0·12	0·13–0·14	0·20–0·25
Zoning	nz	nz/oz	rz	uz	nz

Key: mng=monzogranite; apl=aplite; peg=pegmatite; *n*=number of crystals investigated by electron-microprobe; Acc=accesories; ah=anhedral; sh=subhedral; eh=euhedral; r=random, o=oriented; nz=normally zoned; oz=oscillatory zoned; rz=reversely zoned; uz=unzoned.

concentrations depend strongly on the crystallisation conditions (T, P_{H2O}), and the subsequent cooling history (e.g. Schreyer & Yoder 1964; Johannes & Schreyer 1981; Harley *et al.* 2002).

2.1.5. Chemical zoning. Normal and reversely zoned, as well as unzoned cordierite occurs in metamorphic and magmatic rocks (e.g. see review by Deer *et al.* 1997), and thus, those zoning patterns do not necessarily reflect a distinct origin of a grain in question. On the other hand, oscillatory zoning in cordierite is known exclusively from granites (Hiroi *et al.* 2002). Although the origin of the oscillatory zoned cordierite may, at least theoretically, be xenocrystic, a metamorphic origin seems rather improbable, except if episodic flow of a compositionally variable metamorphic fluid took place.

We have examined the textures of more than 70 cordierite grains from 15 samples, and selected 21 grains from the various rock units for the present textural and chemical study. To classify the selected grains, we used the following textural variables: size (macrocrysts >1 mm, microcrysts <1 mm); shape (euhedral, subhedral and anhedral); and inclusions (absence, presence – mineral types, assemblages – random, oriented). The chemical variables for the classification are X_{Mg}=[Mg/(Mg+Mn+Fe)] and MnO, and chemical zoning (normal, reverse, oscillatory and unzoned), in which normal zoning indicates a core to rim decrease in X_{Mg} and an increase in MnO.

2.2. Textures

Using grain size, grain shape, and inclusions, the authors recognise five main textural types of cordierite (TT1–5; Fig. 2, Table 2) in the granitic rocks of the SMB and MB. TT1 is very coarse-grained (<25 mm), anhedral to subhedral, showing randomly oriented quartz and biotite inclusions (Fig. 2a). The quartz and biotite inclusions consist of rounded single grains or monomineralic aggregates. Although lobate cordierite grain boundaries dominate, subhedral parts of the macro-crysts with few inclusions do occur. TT2 consists of coarse-grained (~3–10 mm), subhedral to euhedral, single and multiple macro-cysts, occurring as isolated grains or in clusters (Fig. 2b). Inclusions are randomly oriented and relatively rare, except in one grain, where there is a concentric pattern of euhedral to rounded biotite, and anhedral quartz inclusions, separating three growth shells of cordierite (core, intermediate and rim zone; Fig. 2c). TT3 microcrysts have a grain size of ~0·2–1·0 mm, and occur as single or multiple, subhedral to euhedral, inclusion-free grains (Fig. 2d). TT4 is coarse-grained (≤10 mm), euhedral and inclusion-free (Fig. 2e). TT5 is coarse-grained (up to 10 mm in diameter) and anhedral (Fig. 2f). This type of macro-cryst shows abundant fine-grained quartz inclusions (similar in shape and slightly smaller than the quartz of the surrounding matrix); other inclusions are lacking.

2.3. Mineral compositions and chemical zoning

Based on X_{Mg} and Mn apfu, determined by microprobe analysis, we recognise four main chemical groups of cordierite (CG1–4) in the granitic rocks of the SMB and MB (Fig. 3; Table 3). Three of these groups define a linear trend from high X_{Mg}-low Mn to low X_{Mg}-high Mn, and the fourth group lies well off this trend. Furthermore, the style of chemical zoning (normal, reverse, oscillatory and unzoned) enables us to subdivide some of these groups. CG1 has the highest X_{Mg} (0·64) and lowest Mn values (0·01–0·06 apfu). Cordierite belonging to this chemical group is characterised by patchy normal zoning, where X_{Mg} decreases towards grain boundaries and towards micro-cracks (Fig. 4a). CG2 has intermediate X_{Mg} (0·37–0·50) and Mn values (0·04–0·09 apfu), and consists of grains showing either normal or oscillatory zoning. The normally zoned grains (CG2a) have high-X_{Mg}, low-Mn cores (~0·47, 0·04 apfu), and low-X_{Mg}, high-Mn rims (~0·37, 0·07 apfu); the lowest X_{Mg} and highest Mn values occur along grain boundaries and micro-cracks (Fig. 4b). The oscillatory zoned grains (CG2b) are characterised by sharp to diffuse, asymmetric, large-scale oscillations (0·1–1·0 mm) in Fe-Mg-Mn (Figs 4c, 5a, c) with a maximum X_{Mg}-amplitude of 0·02, and a maximum Mn-amplitude of 0·001 apfu. The width of the oscillating growth layers varies between crystallographically non-equivalent faces, but resorption features are absent. Normal zoning is superimposed on the oscillatory zoning along grain boundaries and cracks. CG3 shows the lowest X_{Mg} (0·20–0·32) and relatively higher Mn values (0·10–0·14 apfu). Cordierite from this chemical group is either unzoned (CG3a) or is reversely zoned (CG3b). The reverse zoning consists of monotonic increases in X_{Mg} from core to rim (Figs 4d, 5b, d),

Figure 2 Photomicrographs (crossed-polarised light) of the five main textural cordierite types. (a) TT1: Part of an anhedral to subhedral cordierite macro-cryst with random inclusions of biotite and quartz; (b) TT2: Euhedral cordierite macro-cryst with few randomly oriented biotite inclusions; (c) TT2: Euhedral cordierite macro-cryst showing a concentric, alternating zoned texture (marked by the broken red lines); (d) TT3: Cluster of subhedral cordierite micro-crysts with no inclusions; (e) TT4: Euhedral cordierite macro-cryst with no inclusions (grain separate); (f) TT5: Large, anhedral cordierite macro-cryst with abundant random quartz inclusions.

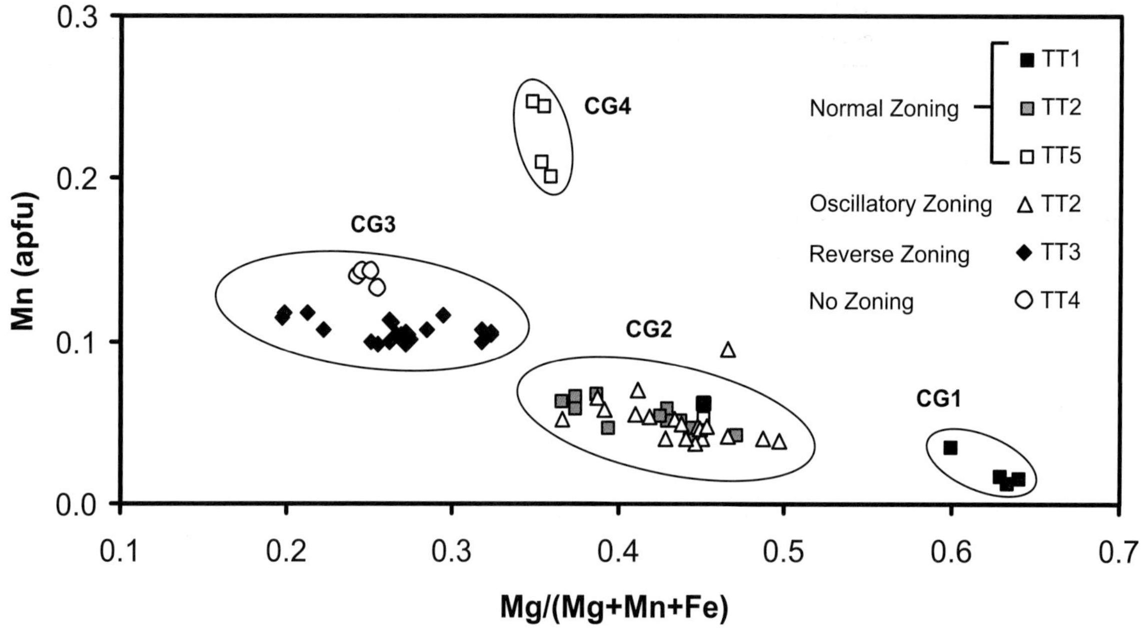

Figure 3 X_{Mg} versus Mn (apfu) plot showing the four main chemical groups. The analyses represent the 21 cordierite grains selected, including: (a) core and rim compositions for the normally and reversely zoned grains; and (b) minimum and maximum values for the oscillatory zoned grains. Although the four main chemical groups are distinct based on the present data, the major-element composition of some groups may eventually overlap for an enlarged data set.

in which the maximum compositional range within a single grain is 0·07 for X_{Mg}, and 0·003 apfu for Mn. CG4 has intermediate X_{Mg} (0·35–0·36) and the highest Mn-concentration (0·20–0·25 apfu). Cordierite belonging to this group has weak normal zoning along grain boundaries and towards micro-cracks.

2.4. Combined textural-chemical classification

We have identified five textural types and four main chemical groups of cordierite in the granitic rocks of the SMB and MB. The symbols in Figure 3 illustrate the close relationship between the textural types and chemical groups. The large, anhedral to subhedral, inclusion-ridden cordierite (TT1) is most primitive chemically (CG1). The subhedral single and multiple macrocrysts with few inclusions (TT2) have intermediate chemical compositions (CG2); normal and oscillatory zoning patterns permit a chemical subdivision of these cordierites. The large, inclusion-free cordierite (TT4) falls into the most evolved chemical group (CG3) and is unzoned. The large, normally zoned grain of cordierite with abundant random inclusions of quartz (TT5) has a composition off the chemical trend (CG4).

The high MnO composition of CG4 probably reflects a high bulk MnO of the host aplite. The more primitive composition of cordierite CG1 and CG2, compared to cordierite CG3 and CG4, can be explained by their occurrence within the monzogranites and the relatively more evolved aplites, respectively. However, the compositional difference between CG1 and CG2 cannot be explained by variation in the whole-rock geochemistry, given that the CG1 cordierite occurs in the same sample as crystals classified as CG2. The grain size of the cordierites does generally correlate with the grain size of the host rock [e.g. fine-grained cordierite (TT3) in some aplites and very coarse-grained cordierite in the pegmatite], but exceptions occur (e.g. TT5 in the aplites).

2.5. Spatial occurrence

CG1/TT1 and CG2/TT2 predominate in the monzogranites of the HP and MB. CG2/TT2 dominates in abundance; however, the relative abundance of CG2a/TT2 and CG2b/TT2 is unclear. In the samples investigated, CG2a/TT2 and CG2b/TT2 are equally abundant, but too few samples were studied to be representative. Detailed mapping (MacDonald & Horne 1988; MacDonald 2001) has revealed that cordierite (CG1/TT1 and CG2/TT2) in the HP is most abundant in zones which occur 1–5 km away from the granite/metasediment contact, roughly parallel to the margin of the pluton, and apparently oriented independently of internal lithological boundaries (Fig. 1b). In contrast, CG1/TT1 and CG2/TT2 in the MB are dominant in one major and in several smaller zones, showing no regular spatial relationship to the margin of the batholith (Fig. 1c). CG4/TT5 and CG3/TT3 occur in aplites, and CG3/TT4 in a pegmatite.

3. Other zoned minerals

Plagioclase macro-crysts from the monzogranites of the HP and MB show complex oscillatory zoning. Some have rounded, resorbed cores, and euhedral to subhedral epitactic rims, in which the dissolution surface marks a major drop in X_{Ca}; other crystals lack a core-rim texture. The composition ranges between $X_{An}=1$ and $X_{An}=36$, showing complex large- and small-scale oscillations. The large-scale oscillations are characterised by a mostly saw-tooth like pattern, consisting of reverse and normally zoned layers, <50 to 500 μm wide; superimposed are small-scale oscillations with a width of <20 μm and an An-amplitude $\leq$0·01. However, it was neither possible to correlate the zoning patterns of different plagioclase macro-crysts, nor to correlate the patterns with those of spatially related cordierite macrocrysts.

Garnet from the aplites is fine- to medium-grained, euhedral in outline, showing fine-grained quartz inclusions. The composition is Alm-Sps-dominated (X_{Alm}=0·73–0·75, X_{Sps} 0·20–0·23), in which the Sps component decreases, and the Grs and Pyp components increase from core to rim (Fig. 6). As for cordierite type CG3a/TT3 from the same

Table 3 Characteristic core-rim electron microprobe analyses of the four main chemical cordierite groups. The rather low totals for cordierite may be the result of submicroscopic decomposition. Microprobe operating conditions were 15 kV, 15 nA, 10 μm spot size, and 40-s counting times

	Cordierite								Garnet		Biotite	
Type:	CG1		CG2		CG3		CG4		Grt		Bt*	
	core	rim	core	rim	core	rim	core	rim	core	rim	core	rim
Lithology:	mng		mng		apl		apl		apl		apl	
SiO_2	46·43	47·30	47·07	46·29	45·71	46·03	46·89	47·05	36·74	36·74	33·52	33·61
TiO_2	0·01	–	0·03	0·02	–	0·01	–	–	0·07	0·02	3·27	3·27
Al_2O_3	31·42	31·76	31·47	31·33	30·61	30·61	31·39	31·13	20·58	20·64	19·12	19·33
FeO	7·96	10·25	10·42	12·16	14·71	13·74	10·52	9·95	3·88	3·88	24·02	23·81
MnO	0·18	0·67	0·57	0·68	1·22	1·20	2·21	2·71	9·67	8·56	0·39	0·41
MgO	7·73	5·06	5·11	4·18	2·20	3·02	4·04	3·81	0·83	0·97	3·66	3·68
CaO	0·01	0·01	0·03	0·00	0·03	–	0·01	–	0·29	0·40	–	–
Na_2O	0·78	1·49	1·50	1·30	1·05	0·96	1·77	1·88	–	–	0·13	0·09
K_2O	–	–	–	–	–	–	–	–	–	–	9·33	9·16
Cl	–	–	0·00	0·00	0·01	0·01	–	–	–	–	–	–
Total	94·51	96·54	96·20	95·96	95·56	95·58	96·83	96·52	100·27	99·44	93·44	93·36
Si	4·989	5·041	5·039	5·012	5·040	5·048	5·033	5·064	5·971	5·994	4·381	4·384
Ti	0·001	–	0·002	0·002	–	0·001	–	–	0·008	0·002	0·321	0·321
Al	3·979	3·990	3·972	3·999	3·979	3·956	3·971	3·949	3·942	3·969	2·945	2·972
Fe^{2+}	0·715	0·914	0·933	1·101	1·357	1·260	0·944	0·895	3·834	3·851	2·626	2·598
Fe^{3+}	–	–	–	–	–	–	–	–	0·475	0·477	–	–
Mn	0·017	0·061	0·052	0·062	0·114	0·111	0·201	0·247	1·331	1·183	0·043	0·045
Mg	1·238	0·803	0·815	0·674	0·361	0·494	0·646	0·611	0·202	0·235	0·714	0·715
Ca	0·001	0·001	0·003	0·000	0·004	–	0·002	–	0·051	0·069	–	–
Na	0·163	0·308	0·311	0·273	0·226	0·204	0·369	0·392	–	–	0·033	0·022
K	–	–	–	–	–	–	–	–	–	–	1·556	1·525
Cl	–	–	0·001	0·000	0·003	0·001	–	–	–			
Total	11·10	11·12	11·13	11·12	11·08	11·08	11·17	11·16	15·81	15·78	12·62	12·58
A	0·66	0·66	0·66	0·65	0·63	0·63	0·65	0·65	0·59	0·61	−0·46	−0·44
F	0·37	0·55	0·55	0·63	0·80	0·73	0·64	0·65	0·97	0·96	0·79	0·79
M	0·63	0·45	0·45	0·37	0·20	0·27	0·36	0·35	0·03	0·04	0·21	0·21

*Analysis given for the most common, unzoned Bt-type: mng=monzogranite; apl=aplite.

samples, the reverse zoning pattern of garnet is monotonic, but varies for the different components, possibly suggesting growth under disequilibrium conditions.

Biotite from the aplites is fine-grained and subhedral. X_{Mg} varies between 0·21 and 0·25. Most biotite crystals are unzoned, however, a few have normal zoning (Ding 1995) and some show weak reverse zoning. The origin of the various biotite zoning patterns is not yet clear.

4. Causes of zoning

Normally and reversely zoned cordierite has been reported from magmatic and metamorphic environments (e.g. see review by Deer *et al.* 1997). Normal zoning may result from epitactic overgrowths on crystals formed in a peritectic reaction (Flood & Shaw 1975), or a decrease in the degree of partial melting (Pereira & Bea 1994), whereas reverse zoning has been suggested to result from processes such as a decrease in H_2O (Birch & Gleadow 1974), an increase in temperature and the degree of partial melting (e.g Holtz & Johannes 1991; Pereira & Bea 1994), or solid-state diffusion during cooling (Ashworth & Chinner 1978). Hiroi *et al.* (2002) described, but did not explain, oscillatory-zoned cordierite in granites.

4.1. Normal zoning

Normal zoning in the cordierites investigated may be the result of magmatic growth, diffusion, or both, involving the following mechanisms:

4.1.1. Fractional crystallisation. Closed-system fractional crystallisation should result in the formation of normally zoned cordierite with a high-X_{Mg}, low-MnO core, and a low-X_{Mg}, high-MnO rim. Fractional crystallisation may account for the normal zoning in CG2a/TT2, but the evidence is not clear. However, this mechanism may have played an important role in the formation of the oscillatory zoning of CG2b/TT2 (see below).

4.1.2. Epitactic crystallisation. Epitactic overgrowths on xenocrystic or peritectic cordierite may result in normal zoning with a core of high X_{Mg}, and a rim of low X_{Mg}, with or without hiatus or resorption features. Normal zoning without hiatus or resorption features in CG1/TT1 may be the result of this mechanism. The absence of any hiatus or absorption features may support a peritectic origin of the cordierite core (compositional disequilibrium with the main magma is more likely for xenocrystic cordierite than for a peritectic cordierite), resulting from the reaction between country-rock material and main magma, whereas the rim crystallised in a cotectic reaction involving mostly the main magma, and little or no country-rock material.

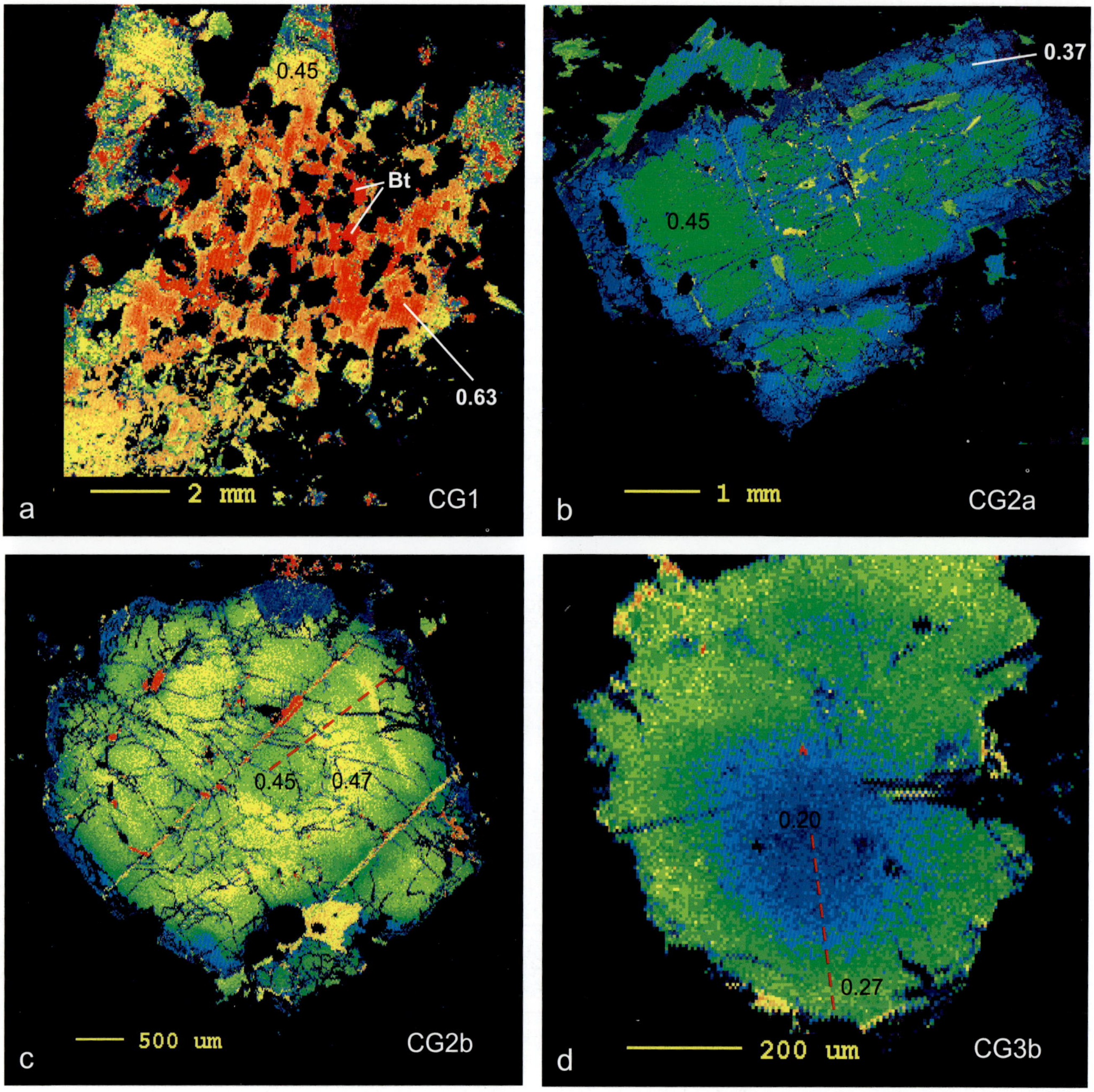

Figure 4 X-ray maps showing compositional zoning in Mg. Microprobe operating conditions were: 15 kV, 200 nA, 500 ms pixel^{-1}, 10 μm pixel^{-1}. Warm colours represent higher Mg values, but the colours are not standardised to fixed Mg-concentrations; the numbers represent X_{Mg} values of certain parts of the grains. (a) Patchy normal zoning in CG1/TT1. Note that the green and blue colours represent pinite. (b) Normally zoned CG2a/TT2, where Mg decreases along grain boundaries and micro-cracks. The dark blue colours represent pinite. (c) Oscillatory zoned CG2b/TT2 grain showing sharp to diffuse, roughly concentric growth shells. The blue colour marks pinite. (d) Reversely zoned CG3a/TT3.

4.1.3. Diffusive exchange. Diffusive exchange between cordierite and a fluid during cooling may also produce normal zoning, and will be most effective along grain boundaries and microcracks. To varying degrees all of the cordierites investigated show irregular to patchy normal zoning along grain boundaries and micro-cracks, suggesting that diffusive exchange was the dominant mechanism for the formation of normal zoning in the present sample set. Biotite, the major exchange phase, is unzoned, but relatively high diffusivities for biotite even at low temperature have probably allowed for diffusive equilibration.

4.2. Oscillatory zoning

Oscillatory-zoned cordierite may, at least theoretically, form in magmatic and metamorphic environments, in an open system, with repeated changes in T, P and/or X, or as a result of local fluctuations in the composition of a boundary layer of melt around the growing crystal (e.g. see review by Shore & Fowler 1996). The following are the most likely driving forces causing oscillatory zoning in igneous systems:

4.2.1. Advection of crystals. Magmatic convection, and therefore, advection of minerals through different T-P-X environments within the magma chamber, as first suggested by Bowen (1928), may be a mechanism for producing oscillatory zoning (the classic example is plagioclase) (e.g. Singer *et al.* 1995). Though physically possible, we cannot test the importance of this mechanism, but advection appears to be rather unlikely to produce the characteristic asymmetrical sawtooth-like oscillation patterns in the cordierites studied.

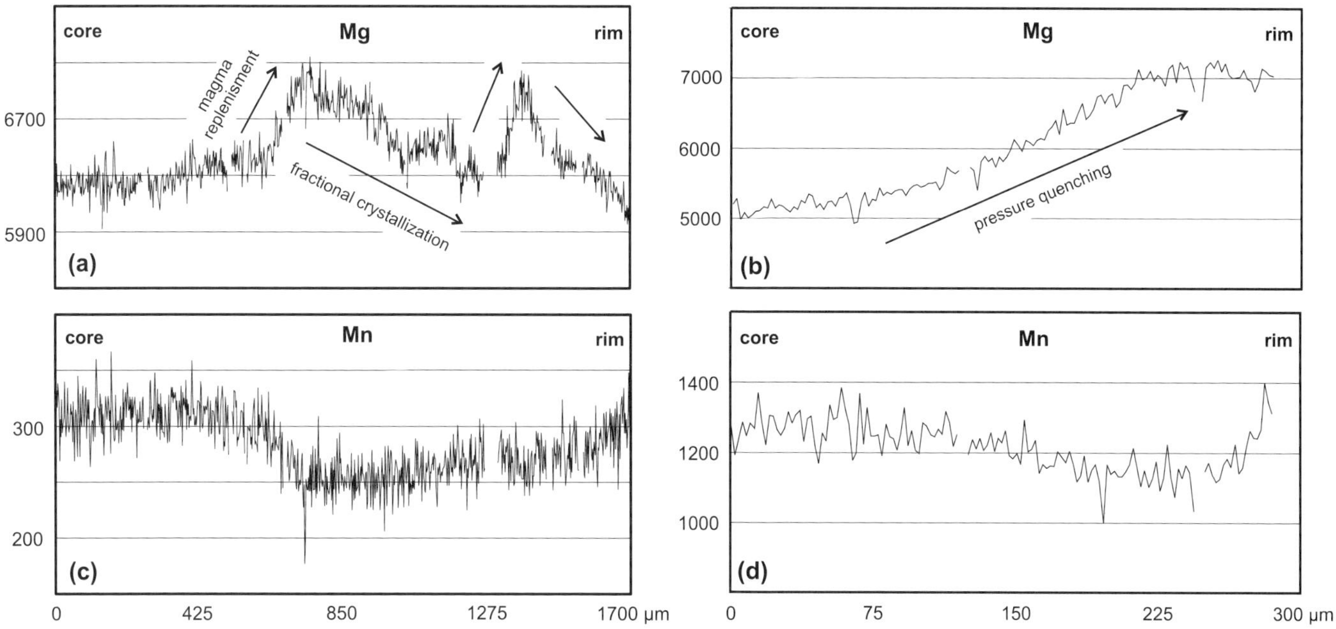

Figure 5 Plot of (a, b) Mg and (c, d) Mn concentration (qualitative analysis) versus radius for (a, c) CG2b/TT2 and (b, d) CG3a/TT3 with the suggested processes responsible. Figure 4 shows the locations of the profiles. Microprobe operating conditions were: 200 nA, 500 ms pixel^{-1}, 2 μm pixel^{-1} for CG2b/TT2 (a, c); and 200 nA, 1000 ms pixel^{-1}, 2 μm pixel^{-1} for CG3a/TT3 (b, d).

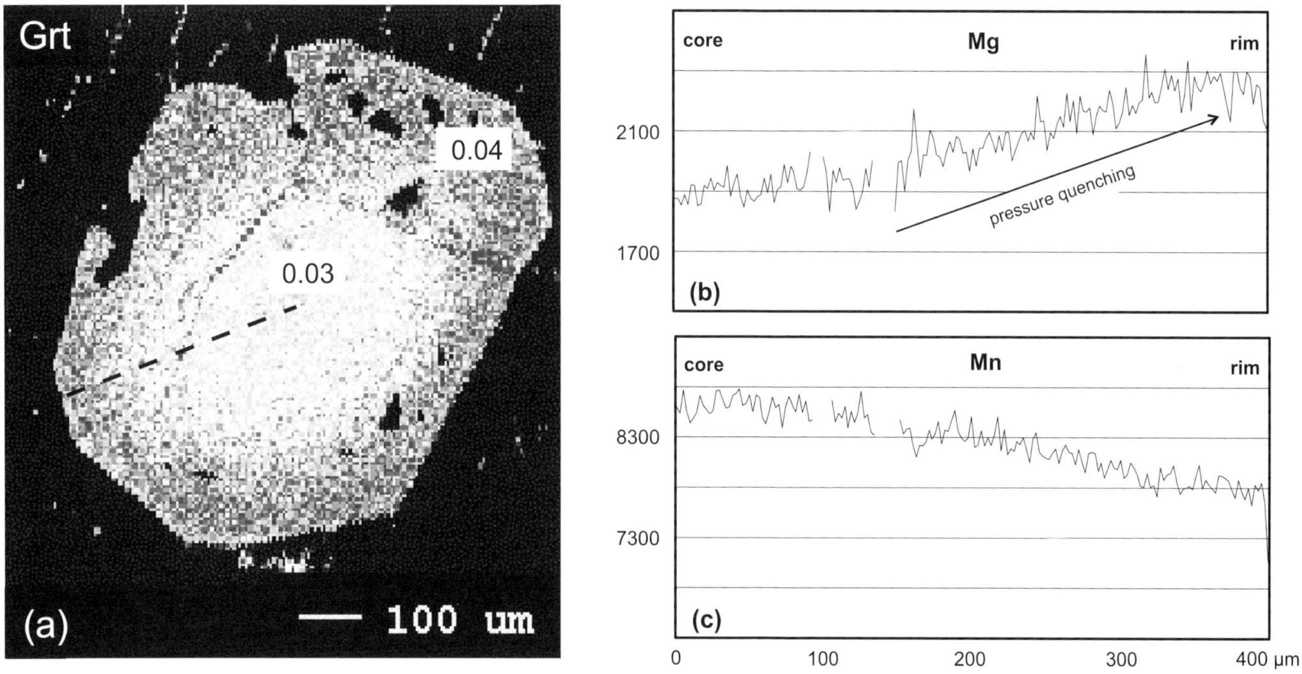

Figure 6 (a) X-ray map showing reverse zoning in Mg for garnet from an aplite. Microprobe operating conditions were: 15 kV, 200 nA, 500 ms pixel^{-1}, 10 μm pixel^{-1}. The numbers represent X_{Mg} of quantitative core and rim analysis. Plot of (b) Mg and (c) Mn concentration (qualitative analysis) versus radius for the profile marked in (a), and the suggested process responsible. Microprobe operating conditions were: 200 nA, 1000 ms pixel^{-1}, 2 μm pixel^{-1}.

4.2.2. Chemical assimilation of xenolithic material. Minerals crystallising and growing adjacent to xenoliths may record chemical changes caused by assimilation (Edwards & Russell 1996), but the formation of oscillatory zoning as a result of assimilation is unlikely in the samples studied because it would require periodically repeated chemical contamination close to the same cordierite crystal.

4.2.3. Degassing processes. Changes in fluid pressure and fluid composition may account for changes in the composition of cordierite. Based on their data set, the present authors cannot test the importance of fluid pressure and/or composition for the oscillatory zoning observed, but CG2/TT2b appears to have crystallised during an early stage of the magmatic evolution, when degassing and/or fluid flux variation were probably less important than other processes.

4.2.4. Diffusion-controlled growth. Diffusion-controlled growth may generate heterogeneous, oscillating compositional patterns (e.g. L'Heureux & Fowler 1996) if the growth rate of a crystal increases with the concentration of certain components in the surrounding melt layer, and if the partition relationship favours a higher concentration of certain species in the crystal than in the silicate melt. Although possible, we think that this mechanism is unlikely to explain the observed oscillatory zoning because the zone thicknesses in cordierite are relatively large and highly variable compared to the zone thicknesses which are usually ascribed to diffusion-controlled growth (e.g. Ginibre *et al.* 2002).

4.2.5. Magma replenishment. Magma replenishment results in increasing T, followed by compositional changes if mixing is sufficient. Repeated mass flux into, or through, the region in which the crystal growth takes place will produce periodic changes in the T-X-a_{H2O} conditions, and thus may produce oscillatory zoning. Evidence for repeated magma recharge in the HP includes the occurrence of several types of microgranitoid enclaves (Tate 1994), and the occurrence of inverse zoning of the HP itself (MacDonald & Horne 1988; MacDonald 2001). Thus, magma replenishment appears to be the most likely cause of the reversely zoned segments of the oscillatory zoning in CG2b/TT2, whereas the normally zoned segments probably result from fractional crystallisation (Fig. 5a, c).

4.3. Reverse zoning

Possible mechanisms to explain reverse zoning in the cordierites from the aplite samples investigated involve:

4.3.1. Magma recharge and increasing T. Evidence for the influx of more primitive (high X_{Mg}) and, thus, hotter magma, is absent, and unlikely in the aplites studied. Garnet-cordierite thermometry (Perchuck *et al.* 1985) suggests cordierite and garnet growth during cooling (T_{core}=780°C; T_{rim}=700°C; calculations are for P=200 MPa, garnet-cordierite core and garnet-cordierite rim compositions, respectively; Table 3). On the other hand, garnet-biotite thermometry (Ganguly & Saxena 1984) suggests growth during rising T (T_{core}=710°C; T_{rim}=760°C; calculations are for P=200 MPa, garnet-cordierite core and garnet-cordierite rim compositions, respectively; Table 3), a condition unlikely in the aplites investigated, indicating that biotite crystallised early, followed by cordierite and garnet.

4.3.2. Increasing fO_2. Rising oxygen fugacity, caused by the dissociation of exsolved magmatic H_2O, may account for gradual reverse zoning in AFM minerals by oxidising remaining Fe^{2+} in the melt to Fe^{3+}. Because Fe^{3+} becomes preferentially partitioned into Fe-oxides, this change increases the X_{Mg} of the melt, and silicates such as cordierite, garnet and biotite would incorporate more Mg towards the rim (Czmanske & Wones 1973; Lalonde & Martin 1983). Although this mechanism is consistent with reverse zoning observed in cordierite and garnet, it does not seem to apply because of the paucity of Fe-oxides in the aplite samples studied.

4.3.3. Metasomatic reaction. Reaction of cordierite and garnet with fluids of changing composition may have caused the reverse zoning observed. However, the 'freshness' of coexisting plagioclase and K-feldspar suggests that hydrothermal alteration was not important for the samples studied.

4.3.4. Pressure quenching. Under equilibrium conditions, a decrease in P may lead to the breakdown of biotite, and the formation of reversely zoned cordierite and garnet ($L+Bt+Qtz \rightarrow Grt+Crd+Kfs+H_2O$) (J. D. Clemens, pers. comm., 2004). Biotite may have crystallised prior to the rapid ascent of the aplite magma; during decompression, biotite may have partly decomposed to cordierite and garnet. With progressive decompression, the divariant reaction would result in the crystallisation of continuously more Mg-rich cordierite and garnet, producing the zoning patterns observed. Although a possible explanation, we do not see textural evidence for the consumption of biotite. Under disequilibrium conditions, on the other hand, biotite may have remained metastable, or may have undergone metastable growth, whereas cordierite and garnet may have crystallised in a cotectic reaction ($L \rightarrow Crd+Grt+$other phases). The reverse zoning may have developed if diffusivities of Fe in the melt were high compared to diffusivities of Mg towards the growing cordierite and garnet crystals.

5. Origin of the zoned and unzoned cordierites

CG1/TT1. Most likely, CG1/TT1 (Figs 2a & 4a) crystallised as a result of the reaction between xenolithic material and the main magma, as suggested by its abundant fine-grained, rounded biotite inclusions, which show no textural relationship to the magmatic biotite of the surrounding granitic matrix, and thus, may have a 'foreign' origin. A xenocrystic origin is unlikely given that cordierite from the country rocks is fine-grained. The high-X_{Mg} composition of CG1/TT1 (compared to all other cordierite types studied) probably reflects the consumption of Mg-rich biotite either in a dehydration, or a melting reaction, whereas the low-X_{Mg} margins of the cordierite grain may represent cotectic overgrowths (Fig. 4a). Superimposed on the potential magmatic normal zoning is a second normal zoning resulting from diffusive exchange along grain boundaries and micro-cracks.

CG2/TT2. The alternating, concentrically zoned texture of the unusual CG2b/TT2 grain (Fig. 2c), and the subhedral to euhedral shape, as well as the chemical oscillatory zoning of all other CG2b/TT2 grains (Figs 2b & 4c), strongly supports a cotectic origin, whereas the origin of the normally zoned CG2a/TT2 grains is less clear, and may be peritectic, cotectic or a combination of both (with epitactic, cotectic overgrowths on peritectic grains). However, both types of cordierite occur in a single rock sample and show similar compositions. A possible explanation is that some initially (maybe weakly) oscillatory zoned grains have lost their zoning by diffusion, but we do not see a range of oscillatory-zoned cordierites with high- and low-amplitude oscillations. Another possible explanation is that magma mixing events were localised and that the compositional changes were recorded only by few crystals, namely CG2b/TT2 grains. If so, the different cordierite types must have been spatially homogenised after growth, during ascent or final emplacement of the magma because evidence for compositional variation caused by magma mixing or mingling on the scale of a hand specimen is lacking. A third possible explanation may be that only CG2b/TT2 cordierite has been affected by magma recharge events, potentially in deeper parts of the magma chamber, whereas CG2a/TT2 grains may have crystallised near or at the emplacement level in zones that did not experience significant magma replenishment, but contamination by country-rock material. Although a true xenocrystic nature for CG2a/TT2 is unlikely because cordierite of the country rocks is usually fine- to medium-grained, anhedral and inclusion-rich (Maillet & Clarke 1985; MacDonald 2001), CG2a/TT2 may be the result of a reaction relationship between xenoliths and the magma, and thus, may have a 'foreign', peritectic origin [country rock (Bt+Qtz+Plag) ± L1 → Crd ± other minerals+L2], as suggested by abundant peritectic cordierite along the margin of wall rocks and xenoliths, and in the surrounding granodiorite, respectively (Fig. 7). Such a 'xenocrystic birth' of CG2a/TT2 is further suggested by the spatial distribution of cordierite in the HP in zones which roughly parallel the margin of the pluton, independent of the rock unit (Fig. 1b). Strikingly, these zones occur 1–5 km away from the metasedimentary contact, in rocks with lower percentages of xenoliths (MacDonald 2001), but perhaps

Figure 7 Photomicrograph (plane-polarised light), showing the contact between fine-grained country rocks of the Meguma Supergroup (left) and coarse-grained biotite granodiorite (right). Cordierite (now highly pinitised) formed in a peritectic reaction along the boundary between country rock and the surrounding magma. The crystals within the granodiorite show features similar to CG2a/TT2, and may indicate how a significant proportion of cordierite from the South Mountain and Musquodoboit Batholiths formed.

with higher percentages of xenocrystic/peritectic crystals than the marginal units. In the MB, however, cordierite shows no obvious relationship to the margin of the batholith, but is most abundant in several irregular zones (MacDonald & Clarke 1985), which may either represent zones or layers of a relatively cordierite-rich magma, or contaminated near-roof zones (Fig. 1c).

CG3a/TT3. The hypidiomorphic to idiomorphic grain shapes of CG3a/TT3, and comparable grain sizes to the other rock-forming minerals of the aplites, suggest former equilibrium with the melt, and thus, most likely, cotectic crystallisation from that melt. The reverse zoning of cordierite and coexisting garnet are a result of growth during pressure quenching, but the reaction mechanism is not yet clear.

CG3b/TT4. Textural relationships and the grain shape of CG3b/TT4 indicate a cotectic origin.

CG4/TT5. The anhedral grain shape of CG4/TT5 and its abundant quartz inclusions, which have similar grain shapes but slightly smaller grain sizes than the quartz grains of the surrounding granitic matrix, may be explained if cordierite consumed all phases except quartz (e.g. $Bt^- \pm Kfs^- \pm Plag^- \pm L^- + Qtz \rightarrow Crd^+ + Qtz$), but we do not see evidence for a reaction between cordierite and the other phases. Thus, the most likely explanation is that cordierite crystallised from a melt where quartz was already present. Biotite, and in some cases K-feldspar, show similar anhedral textures as cordierite, indicating that these phases may have crystallised together with cordierite, whereas the coarse grain size of cordierite and biotite (±K-feldspar) suggests the production or growth of only few nuclei.

The analysis of additional samples from the various units, and the analysis of trace elements such as Be, Cs, Li, B, as well as H_2O and CO_2, were beyond the scope of this study, but may help to further constrain the crystallisation conditions of the various cordierites investigated (e.g. Harley *et al.* 2002; Evensen & London 2003).

6. Summary and conclusions

We have considered several textural (grain size, grain shape and inclusion relations) and chemical (chemical composition and chemical zoning) criteria for classifying and determining the origin of cordierite in a given rock, and have examined in detail zoning patterns and possible explanations for normal, oscillatory and reverse zoning in cordierite from various units of two related peraluminous batholiths. They believe that grain shape and inclusion relationships are the most useful criteria with which to determine the origin of cordierite; however, chemical zoning appears (at least in our sample set) to be the most reliable criterion with which to discriminate different cordierite populations, and to detect the crystallisation and post-crystallisation history of the cordierite and the underlying magma chamber processes in detail.

Using a combination of key characteristics, we tentatively conclude that (1) type CG1/TT1 and CG2a/TT2 cordierite grew in a peritectic reaction between country rocks and melt; and (2) type CG2b/TT2, CG3a/TT3, CG3b/TT4 and CG4/TT5 cordierite crystallised in a cotectic reaction from the main magma. Because cordierite grains in granites may have multiple origins, only a detailed examination of combined textural and chemical criteria can lead to a correct determination of their origin and crystallisation history; apparent magmatic grains may be foreign, and thus, contamination of the granite by country rocks may be greater than expected. A more refined and combined consideration of all zoned and unzoned minerals from the batholiths is required to put better constraints on the suggested magma chamber processes, and their importance and timing in the evolution of the batholiths.

7. Acknowledgements

Barrie Clarke acknowledges the support of an NSERC research grant that made this study possible; Saskia Erdmann acknowledges the support of a Killam Scholarship at Dalhousie University. We thank John Clemens for generously sharing his view on the origin of the reversely zoned cordierite, Ron Vernon for his many helpful comments on an earlier draft of this manuscript, and Simon Harley and an anonymous reviewer for their many constructive contributions towards improving the manuscript. We also thank Gordon Brown for the polished thin sections, and Robert MacKay and Patricia Stoffyn for assistance at Dalhousie Regional Electron Microprobe Laboratory.

8. References

Ashworth, J. R. & Chinner, G. A. 1978. Coexisting garnet and cordierite in migmatites from the Scottish Caledonides. *Contributions to Mineralogy and Petrology* **65**, 379–94.

Barbey, P., Marignac, C., Montel, J. M., Macaudière, J., Gasquet, D. & Jabbori, J. 1999. Cordierite growth textures and the conditions of genesis and emplacement of crustal granitic magmas: the Velay granite complex (Massif Central, France). *Journal of Petrology* **40**, 1425–41.

Birch, W. D. & Gleadow, A. J. W. 1974. The genesis of garnet and cordierite in acid volcanic rocks: evidence from the Cerberean cauldron, central Victoria, Australia. *Contributions to Mineralogy and Petrology* **45**, 1–13.

Bouloton, J. 1992. Mise en évidence de cordiérite des terrains traversés dans le pluton granitique des Oulad Ouaslam (Jebilet, Maroc). *Canadian Journal of Earth Sciences* **29**, 658–68.

Bowen, N. L. 1928. *The Evolution of the Igneous Rocks*. New Jersey: Princeton University Press.

Cenki, B., Kriegsman, L. M. & Braun, I. 2002. Melt-producing and melt-consuming reactions in the Achankovil cordierite gneisses, South India. *Journal of Metamorphic Geology* **20**, 543–61.

Clarke, D. B. 1995. Cordierite in felsic igneous rocks: a synthesis. *Mineralogical Magazine* **59**, 311–25.

Czmanske, G. K & Wones, D. R. 1973. Oxidation during magmatic differentiation, Finmarka Complex, Oslo area, Norway: Part 2, the mafic silicates. *Journal of Petrology* **14**, 349–80.

Deer, W. A., Howie, R. A. & Zussman, J. 1997. *Disilicates and Ring Silicates. Rock-forming Minerals*. London: Geological Society.

Ding, Y. 1995. *AFM minerals in the Halifax Pluton*. Unpublished M.Sc. thesis. Dalhousie University, Halifax.

Douma, M. 1978. *Gravitational interpretation and modelling of the South Mountain Batholith, southern Nova Scotia*. Unpublished B.Sc. thesis. Dalhousie University, Halifax.

Edwards, B. R. & Russell, J. K. 1996. Influence of magmatic assimilation on mineral growth and zoning. *The Canadian Mineralogist* **34**, 1149–62.

Evensen, J. M. & London, D. 2003. Experimental partitioning of Be, Cs, and other trace elements between cordierite and felsic melt, and the chemical signature of S-type granite. *Contributions to Mineralogy and Petrology* **144**, 739–57.

Flood, R. H. & Shaw, S. E. 1975. A cordierite-bearing granite suite from the New England batholith, NSW, Australia. *Contributions to Mineralogy and Petrology* **52**, 157–64.

Ganguly, J. & Saxena, S. K. 1984 Mixing properties of alumosilicate garnets: constraints from natural and experimental data, and applications to geothermo-barometry. *American Mineralogist* **69**, 88–97.

Ginibre, C., Kronz, A. & Woerner, G. 2002. High-resolution quantitative imaging of plagioclase composition using accumulated backscattered electron images; new constraints on oscillatory zoning. *Contributions to Mineralogy and Petrology* **142**, 436–48.

Ham, L. J., Corey, M. C., Horne, R. J. & MacDonald, M. A. 1990. *Lithogeochemistry of the western portion of the South Mountain Batholith, Nova Scotia*. Nova Scotia Department of Mines and Energy, Open File Report 90–007.

Ham, L. J., Marsh, S. W., Corey, M. C., Horne, R. J. & MacDonald, M. A. 1989. *Lithogeochemistry of the eastern portion of the South Mountain Batholith, Nova Scotia*. Nova Scotia Department of Mines and Energy, Open File Report 89–001.

Ham, L. J. 1999. *Geological map of the Musquodoboit batholith*. Nova Scotia Department of Natural Resources Open File Report ME1999–2.

Harley, S. L., Thompson, P., Hensen, B. J. & Buick, I. S. 2002. Cordierite as a sensor of fluid conditions in high-grade metamorphism and crustal anatexis. *Journal of Metamorphic Geology* **20**, 71–86.

Hiroi, Y., Grew, E. S., Motoyoshi, Y., Peacor, D. R., Rouse, R. C., Matsubara, S., Yokoyama, K., Miyawaki, R., McGee, J. J., Su, S. C., Hokada, T., Furukawa, N. & Shibasaki, H. 2002. Ominelite, $(Fe,Mg)Al_3BSiO_9$ (Fe^{2+} analogue of grandidierite), a new mineral from porphyritic granite in Japan. *American Mineralogist* **87**, 160–70.

Holtz, F. & Johannes, W. 1991. Genesis of peraluminous granites. I. Experimental investigation of melt compositions at 3 and 5 kbar and various H_2O activities. *Journal of Petrology* **32**, 935–58.

Johannes, W. & Schreyer, W. 1981. Experimental introduction of CO_2 and H_2O into Mg-cordierite. *American Journal of Science* **281**, 299–317.

Kretz, R. 1983. Symbols for rock-forming minerals. *American Mineralogist* **68**, 277–9.

Lalonde, A. E. & Martin, R. F. 1983. The Baie-des-Moutons syenitic complex, La Tabatière, Québec. II. The ferromagnesian minerals. *The Canadian Mineralogist* **21**, 81–91.

L'Heureux, I. L. & Fowler, A. D. 1996. Isothermal constitutive undercooling as a model for oscillatory zoning in plagioclase. *The Canadian Mineralogist* **34**, 1137–47.

MacDonald, M. A. 2001. *Geology of the South Mountain Batholith, Southwestern Nova Scotia*. Nova Scotia Department of Natural Resources, Open File Report ME2001–2.

MacDonald, M. A., Corey, M. C., Ham, L. J. & Horne, R. J. 1992. An overview of recent bedrock mapping and follow-up petrological studies of the South Mountain Batholith, southwestern Nova Scotia, Canada. *Atlantic Geology* **28**, 7–28.

MacDonald, M. A. & Clarke, D. B. 1985. The petrology, geochemistry, and economic potential of the Musquodoboit batholith, Nova Scotia. *Canadian Journal of Earth Sciences* **22**, 1633–42.

MacDonald, M. A. & Horne, R. J. 1988. Petrology of the zoned, peraluminous Halifax Pluton, South-Central Nova Scotia. *Maritime Sediments and Atlantic Geology* **24**, 33–45.

Maillet, L. A. & Clarke, D. B. 1985. Cordierite in the peraluminous granites of the Meguma Zone, Nova Scotia. *Canadian Mineralogical Magazine* **49**, 695–702.

McKenzie, B. 1976. *Gravitational interpretation and modelling of the Meguma and intrusive granite in the Halifax area*. Unpublished B.Sc. thesis. Dalhousie University, Halifax.

McKenzie, B. & Clarke, D. B. 1975. Petrology of the South Mountain Batholith, Nova Scotia. *Canadian Journal of Earth Sciences* **12**, 1209–18.

O'Reilly, C. T. 1975. Gravitational interpretation and modelling of the South Mountain Batholith utilising two and three dimensional computer programming. Unpublished B.Sc. thesis, Dalhousie University, Halifax.

Perchuck, L. L., Aranovich, L. Y., Podlessskii, K. K., Lavrent'eva, I. V., Gerasimov, V. Y., Fed'kin V. V., Kitsul, V. I., Karasakov, L. P. & Berdnikov, N. V. 1985. Precambrian granulites of the Aldan shield, eastern Siberia, USSR. *Journal of Metamorphic Geology* **3**, 265–310.

Pereira, M. D. & Bea, F. 1994. Cordierite-producing reactions in the Peña Negra Complex, Avila Batholith, central Spain: the key role of cordierite in low pressure anatexis. *The Canadian Mineralogist* **32**, 763–80.

Phillips, G. N., Wall, W. J. & Clemens, J. D. 1981. Petrology of the Strathbogie batholith: a cordierite-bearing granite. *The Canadian Mineralogist* **19**, 47–63.

Schreyer, W. & Yoder, H. S. Jr 1964. The system Mg-cordierite-H_2O and related rocks. *Neues Jahrbuch für Mineralogie, Geologie und Paläontologie. Monatshefte* **101**, 271–342.

Shore, M. & Fowler, A. D. 1996. Oscillatory zoning in minerals: a common phenomenon. *The Canadian Mineralogist* **34**, 1111–26.

Singer, B. S., Dungan, M. A. & Layne, G. D. 1995. Textures and Sr, Ba, Mg, Fe, K, and Ti compositional profiles in volcanic plagioclase: clues to the dynamics of calc-alkaline magma chambers. *American Mineralogist* **80**, 776–98.

Tate, M. C. 1994. The nature and origin of enclaves in four peraluminous granitoid intrusions from the Meguma Zone, Nova Scotia. *Atlantic Geology* **30**, 205–15.

Tate, M. C. 1995. *The relationship between Late Devonian mafic intrusions and peraluminous granitoid generation in the Meguma Lithotectonic Zone, Nova Scotia, Canada*. Unpublished Ph.D. thesis. Dalhousie University, Halifax.

Villaseca, C. & Barbero, L. 1994. Chemical variability of Al-Ti-Fe-Mg minerals in peraluminous granitoid rocks from Central Spain. *European Journal of Mineralogy* **6**, 691–710.

SASKIA ERDMANN and D. BARRIE CLARKE, Department of Earth Sciences, Dalhousie University, Halifax, NS, Canada B3H 3J5.
e-mail: serdmann@dal.ca; clarke@dal.ca

MICHAEL A. MACDONALD, Nova Scotia Department of Natural Resources, P.O. Box 698, Halifax, NS, Canada B3J 2T9.
e-mail: mamacdon@gov.ns.ca

MS received 3 September 2003. Accepted for publication 9 June 2004.

Tourmaline and boron as indicators of the presence, segregation and extraction of melt in pelitic migmatites: examples from the Ryoke metamorphic belt, SW Japan

Tetsuo Kawakami

ABSTRACT: The mode of occurrence of borosilicates and the breakdown fronts of prograde tourmaline (tourmaline-out isograd) in three anatectic migmatite regions of the Ryoke metamorphic belt, SW Japan, are reported. The breakdown of tourmaline in the migmatite zones and release of boron into the melts, followed by the extraction of the boron-bearing melts from the migmatite zones occurred throughout the Ryoke metamorphic belt. Retrograde, magmatic tourmaline in interboudin partitions filled with leucosome is useful for calculating the degree of partial melting in the migmatites. Using boron contents in the leucosomes and pelitic schists, the degree of partial melting at the migmatite front of the Aoyama area is estimated to be 12 wt.%. Extraction of the boron-bearing melt is suggested by the boron-depleted nature of the migmatites. Connection of boudinage structures probably supplied the vertical pathways of the segregated melts, and major transport of the melts was accomplished by dyking. Irregularly shaped, amoeboid tourmaline locally occurs on the high-temperature side of the tourmaline-out isograds in the Yanai and Komagane areas, implying incomplete extraction of boron-bearing melts from those areas. Discriminating retrograde from prograde tourmaline enables correct recognition of the tourmaline-out isograd. The amount of retrograde tourmaline in migmatites can potentially be used as an indicator of the degree of melt extraction from them.

KEY WORDS: Borosilicate, breakdown, leucogranite, migmatite, partial melting.

Leucogranites formed by partial melting of pelitic rocks commonly contain tourmaline as a major constituent mineral (e.g. Nabelek *et al.* 1992; Scaillet *et al.* 1996). The formation mechanism of tourmaline-bearing leucogranites has been studied by high-temperature experiments (Scaillet *et al.* 1995; London *et al.* 1996; Wolf & London 1997), geochemical studies of the granites themselves (Nabelek *et al.* 1992; Nabelek & Bartlett 1998), and by petrological and geochemical studies of the likely metasedimentary source rocks (Kawakami 2001a, b; Wilke *et al.* 2002). Pelitic migmatite terranes are a possible source zone for peraluminous, boron-bearing melt, but there still is a discussion about whether the melt can be effectively extracted from them.

In the source region, previous studies have suggested that the trace amount of boron contained in muscovite alone is sufficient to produce tourmaline leucogranites via muscovite breakdown under low degrees of partial melting (Nabelek *et al.* 1992; Pereira & Shaw 1997). However, as shown by Sperlich *et al.* (1996) more than half of the whole-rock boron is contained in tourmaline in amphibolite-facies pelitic metamorphic rocks. During prograde muscovite breakdown, the boron in muscovite is considered to form, at least partly, overgrowths of tourmaline on pre-existing detrital or metamorphic tourmaline (Henry & Dutrow 1996; Kawakami & Ikeda 2003). As a result, almost all of the whole-rock boron is contained in tourmaline in the metapelites by the uppermost amphibolite-facies grade (e.g. Kawakami 2001b), although sillimanite and biotite can also be the potential reservoirs of boron (e.g. Grew & Hinthorne 1983; Henry & Dutrow 1996). In metapelites, tourmaline of the schorl-dravite series is common. Experimental studies have shown that tourmaline of such compositions ($\pm$ quartz) breaks down at $\sim$700–800°C at low pressures (e.g. Benard *et al.* 1985; von Goerne *et al.* 1999). Petrographic studies have also shown that tourmaline breaks down, or is generally absent, in upper amphibolite- to granulite-facies metapelites (e.g. Lonker 1988; Moran *et al.* 1992; Kawakami 2001a; Kawakami & Ikeda 2003), and that the boron released goes into the melt in most cases (Kawakami 2001a; Kawakami & Ikeda 2003). Other borosilicates, such as grandidierite, are formed in some cases as a product of tourmaline breakdown (Grew 1996). Once the boron resides in melt, it is compatible in the melt until it crystallises. Such boron-bearing melts crystallise tourmaline $\pm$ other borosilicates on cooling if the boron content is sufficiently high (e.g. London 1999). This kind of tourmaline, formed during the retrograde stage, is called 'retrograde tourmaline' (Kawakami & Ikeda 2003). From the distribution of retrograde borosilicates such as dumortierite, in the field, we can trace the movement of boron-bearing melts and fluids. In the present contribution, distributions of prograde and retrograde tourmaline in three areas of the high-temperature (T), low-pressure (P) type Ryoke metamorphic belt, SW Japan, are reported, with new data demonstrating the usefulness of retrograde tourmaline in analysing segregation and extraction processes of melt, particularly in determining the degree of partial melting of metapelites in the migmatite front. Mineral abbreviations follow that of Kretz (1983).

1. General geology of the Ryoke metamorphic belt

The Ryoke metamorphic belt, SW Japan, is a highly elongate high-T, low-P type metamorphic belt of approximately 800 km in length (Nakajima *et al.* 1990; Miyashiro 1965; Brown 1998;

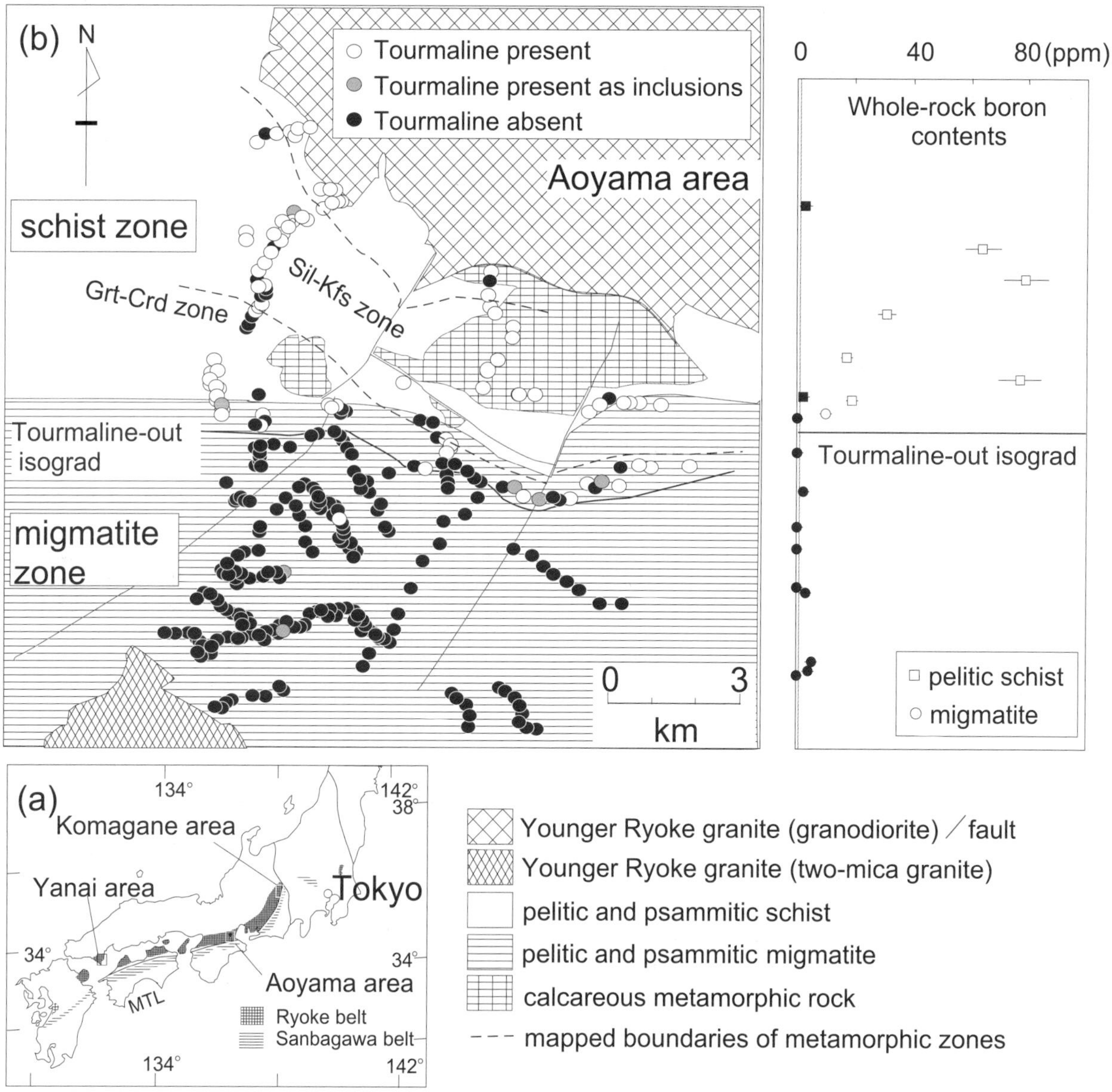

Figure 1 (a) Simplified geological map of SW Japan showing the locations of the Aoyama, Komagane and Yanai areas. (b) Geological map of the Aoyama area, modified from Yoshizawa *et al.* (1966), Hayama *et al.* (1982) and Yoshida *et al.* (1995). The mapped boundaries of the metamorphic zones (dashed, black lines), schist and migmatite zones, and the tourmaline-out isograd (bold, black line) are also shown (based on Kawakami 2001a). Whole-rock boron contents determined by prompt gamma neutron activation analysis are shown on the right-hand side of the figure. Solid circles and squares indicate that the rock is tourmaline-free. See Kawakami (2001b) for details.

Nakajima 1994; Suzuki & Adachi 1998). It is mainly composed of pelitic and psammitic metamorphic rocks and metacherts, and the peak metamorphism is considered to have reached granulite facies conditions (e.g. Ikeda 2002). The metamorphic belt grades into the unmetamorphosed Jurassic accretionary Mino-Tanba complex to the north (Wakita 1987). The low-T, high-P type Sanbagawa belt is located to the south of the Ryoke metamorphic belt, and the two belts are separated by a major strike-slip fault, the median tectonic line (MTL). Eastward, younging of the K-Ar and Rb-Sr ages of the Ryoke granites and metamorphic rocks has been ascribed to an eastward along-arc shift of granitic activity (Nakajima *et al.* 1990; Nakajima 1994). However, the chemical Th-U-total Pb isochron method (CHIME) of dating monazite from gneisses and granitoids gives ages between 102 and 98 Ma both in the east and west of the belt (Suzuki & Adachi 1998). Three field areas of the Ryoke metamorphic belt are dealt in the present paper: the Yanai, Aoyama and Komagane areas, from East to West (Fig. 1a). The detailed geology of each area is described below.

2. Geological setting and mode of occurrence of borosilicates

2.1. Aoyama area

In the Aoyama area (Fig. 1), the high-grade part of the Ryoke metamorphic rocks is exposed over a large area (Yoshizawa *et al.* 1966; Hayama *et al.* 1982; Takahashi & Nishioka 1994; Kawakami 2001a, b). The main rock types are metapelite and metapsammite, with subordinate amounts of metachert, metabasite and calcareous metasediments. The pelitic-psammitic rocks are schists in the northern half of the area, and metatexites to inhomogeneous diatexites (Brown 1973) in the southern half of the area. The schistosity in the northern part of the area generally strikes E–W to WSW–ENE, and dips toward either the N or S because of later upright fold with fold axis trending E–W to WSW–ENE. The foliation (migmatitic banding) in central and southern parts strikes NW–SE to WNW–ESE and dips, in most cases, toward NE to NNE. Intrafolial folds with axial planes parallel to the penetrative migmatitic layering are overprinted by upright folds with fold axes trending NW–SE to WNW–ESE (Takahashi & Nishioka 1994; Kawakami 2001a). The Aoyama area can be divided into three metamorphic zones: the Sil-Kfs zone where Ms+Qtz assemblage is unstable and Sil+Kfs+Bt is stable; the Grt-Crd zone where Grt+Crd+Bt ± Sil is stable; and the contact metamorphic zone recognised by the occurrence of randomly oriented andalusite porphyroblasts (Fig. 1b). All zones occur at grades above the second sillimanite isograd (Hayama *et al.* 1982). The garnet-cordierite isograd and the schist–migmatite boundary are almost coincident. The grade of regional metamorphism increases from N to S. The peak P–T conditions are estimated to be 3·0–4·0 kbar, 615–670 °C for the Sil-Kfs zone, and 4·5–6·0 kbar, 650–800 °C for the Grt-Crd zone (Kawakami 2001a, 2002). The introduction of spessartine content into garnet stabilised the Grt+Crd assemblage in the low-temperature part of the Grt-Crd zone, giving low-temperature estimates for the Grt+Crd assemblage. Sillimanite and biotite is included in cordierite in the melanosome, and the Grt+Crd assemblage is common in the Grt-Crd zone. Euhedral plagioclase that is indicative of the presence of melt is observed in the leucosome. These textures suggest that the dehydration melting reactions consuming biotite and sillimanite, such as

$$\text{Bt+Sil+Qtz=Grt+Crd} \pm \text{Kfs} \pm \text{Ilm+melt} \quad (1)$$

took place in the Grt-Crd zone and are responsible for the formation of migmatites (Kawakami 2001a). K-Ar dating of biotite and muscovite from the pelitic-psammitic metamorphic rocks gives ages of 71 and 61 Ma, respectively (Ishizaka 1966).

Besides the metamorphic rocks, massive granodiorite occurs at the northern edge of the area. This pluton cuts the schistosity of the metamorphic rocks, and its emplacement was accompanied by the development of a contact aureole (Takahashi & Nishioka 1994; Kawakami 2001a). The pluton contains xenoliths of the surrounding metamorphic rocks (Yoshida *et al.* 1995). Therefore, the intrusion is considered to post-date the regional metamorphism. K-Ar dating of biotite from the granodiorite gives ages of 67–69 Ma (Miyamura *et al.* 1981). Another pluton of massive two-mica granite occurs to the south of the metamorphic rocks. This pluton also cross-cuts the schistosity of the surrounding metamorphic rocks, but without evidence of contact metamorphism in the adjacent country rocks.

2.2. Mode of occurrence of tourmaline in the Aoyama area

Tourmaline is common (0·1 vol.%) in pelitic-psammitic rocks of the northern half of the area, but is almost absent in the southern half. It is schorl-dravite in composition (X_{Mg}=0·53–0·78) and occurs in the matrix as light-brown, euhedral grains with weak concentric core to rim chemical zoning. The exchange vectors $Al_2Mg_{-1}Si_{-1}$, $AlOMg_{-1}(OH)_{-1}$ and $CaMg_2{}^X\square_{-1}{}^YAl_{-2}$ or $CaMgO^X\square_{-1}{}^YAl_{-1}(OH)_{-1}$ are important in the tourmaline of this area (Kawakami 2001b). Kawakami (2001a, b) gave a detailed tourmaline composition and described analytical methods. These grains are aligned along the foliation of the rocks and are considered to be prograde. Some such tourmaline crystals include biotite. A line marking the disappearance of prograde tourmaline with increasing metamorphic grade can be mapped and is termed the 'tourmaline-out isograd'. In rocks near the isograd, irregularly shaped tourmaline surrounded by Sil+Crd intergrowth occurs locally, suggesting that tourmaline broke down to form sillimanite and cordierite at the isograd. Prograde tourmaline breakdown textures occur preferentially in patchily developed melanosomes in pelitic-psammitic migmatites. These rocks also contain leucosomes developed preferentially in inter-boudin partitions. Large euhedral crystals of tourmaline (X_{Mg}=0·49–0·55), up to 5 mm in diameter, occur in the leucosomes. These tourmalines rarely show chemical zoning. Plagioclase, apatite, andalusite and tourmaline are euhedral, suggesting their magmatic origin. Based on this textural evidence, Kawakami (2001a) considered that the tourmaline-out isograd was possibly formed through the breakdown of prograde tourmaline to produce Sil+Crd+boron-bearing melt according to the following reaction (modified from Kawakami 2001a):

$$\text{Tur} \pm \text{Ab} \pm \text{Kfs+Qtz=Sil+Crd+boron-bearing melts} \quad (2)$$

The boron-bearing melts probably segregated into the inter-boudin partitions, followed by the crystallisation of the leucosome melts which precipitated retrograde, magmatic tourmaline on cooling. Such interboudin partitions filled with tourmaline-bearing leucosome are common near the tourmaline-out isograd. Interboudin partitions are connected with each other across the structural layering where boudinage is intense (Fig. 2a, b).

It is significant that the tourmaline-out isograd lies in the northern margin of the migmatite zone close to the schist–metatexite boundary. Tourmaline is absent in nearly all the migmatites. Other borosilicates, such as dumortierite, were not observed. Whole-rock boron analyses showed that presence and absence of tourmaline almost completely controls the whole-rock boron contents and that the rocks on the high-temperature side of the tourmaline-out isograd are depleted in boron (Kawakami 2001b). The boron originally contained in the rocks now occurring on the high-temperature side of the tourmaline-out isograd probably went into the melts when the tourmaline broke down. It is considered that further segregation and extraction of boron-bearing melts and/or fluids from the migmatite zone by means of boudinage and dyking processes (Fig. 2a–d) probably caused the relative depletion of boron in this zone and that the boron is now observed in the tourmaline-bearing pegmatite veins to the N of the isograd (Kawakami 2001b). Tourmaline grains in pegmatite veins show Fe-richer compositions (X_{Mg}=0·27–0·40) than those in pelitic schists.

2.3. Yanai area

The Yanai area (Figs 1a & 3) consists of Jurassic sedimentary rocks of the Kuga Group and their metamorphosed equivalent, the Ryoke metamorphic rocks. They are intruded sequentially by three suites of Cretaceous granitic rocks: the Older Ryoke, the Younger Ryoke and the San-yo granitic rocks (Fig. 3) (Higashimoto *et al.* 1983). The metamorphic rocks are mainly pelitic and siliceous rocks. Basic, psammitic and calcareous rocks are rare (Higashimoto *et al.* 1983). Metamorphic zonal mapping utilising the dominant pelitic lithology has been undertaken by a number of authors (Higashimoto *et al.* 1983; Ikeda 1998a; Okudaira *et al.* 1993; Nakajima 1994). The present study follows Ikeda (1998a), who defined the following zones with ascending metamorphic grade: chlorite (Chl), chlorite-biotite (Chl-Bt), biotite (Bt), muscovite-cordierite (Ms-Crd), K-feldspar-cordierite (Kfs-Crd), sillimanite-K-feldspar (Sil-Kfs) and garnet-cordierite (Grt-Crd). The

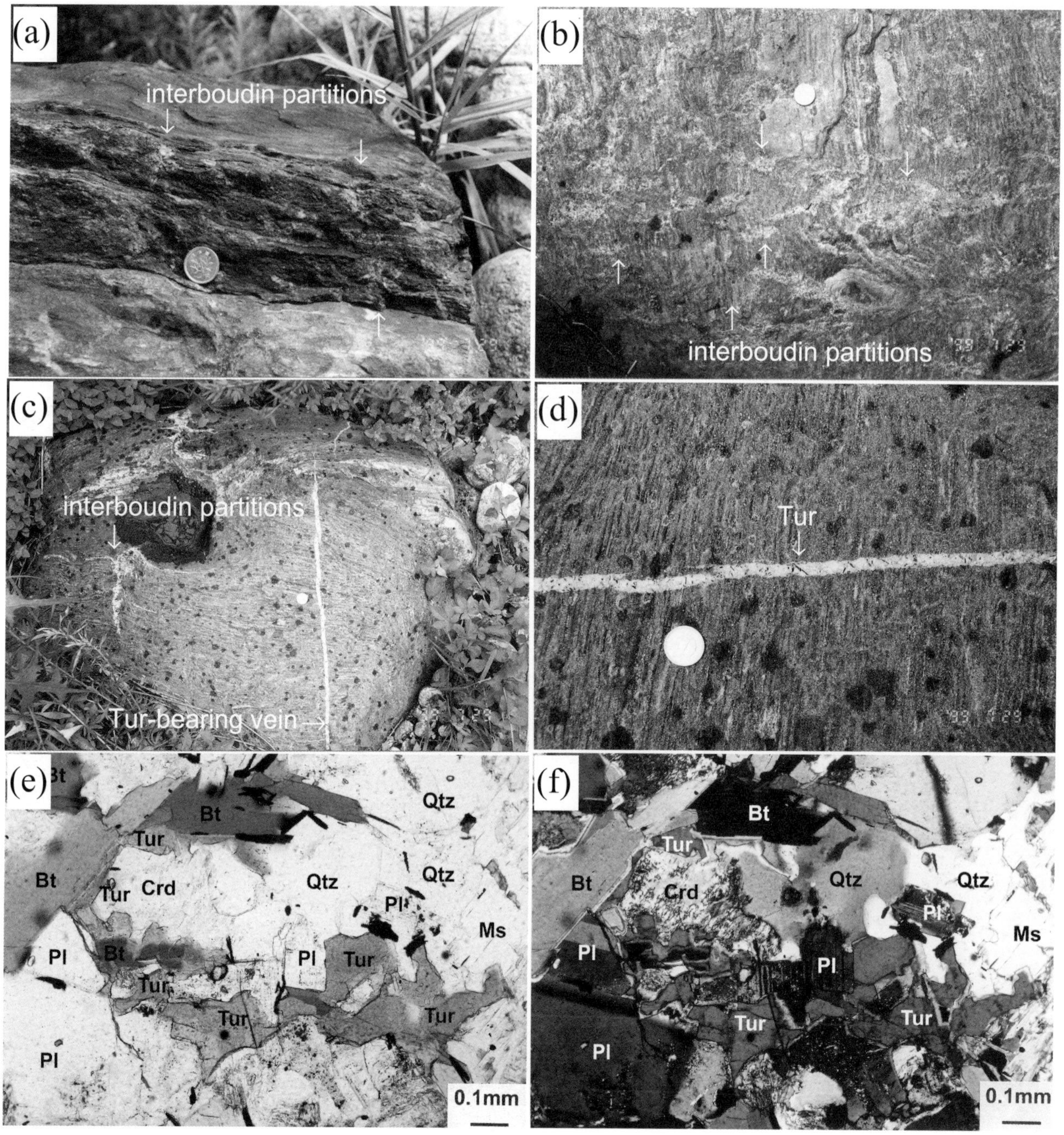

Figure 2 (a–f).

transition form andalusite to sillimanite takes place in the Kfs-Crd zone, and the beginnings of the Kfs-Crd and Sil-Kfs zones are both defined by the breakdown of the Ms+Qtz assemblage. The Grt-Crd zone yields orthopyroxene coexisting with hornblende in basic rocks, suggesting that it reached the granulite facies conditions (Ikeda 2002). For further detailed description on mineral assemblages, see Ikeda (1998a, b). The peak P–T estimates of each zone by Ikeda (2004) showed that the Sil-Kfs zone (4·57–6·05 kbar, 699–759 °C) records lower temperatures than, but almost the same pressure condition as the Grt-Crd zone (4·20–6·97 kbar, 782–873 °C). The Sil-Kfs zone is absent on the northern side of the Grt-Crd zone, giving a spatially asymmetric distribution of metamorphic zones. The metamorphic isograds of regional metamorphism mentioned above are generally discordant to the outline of weakly gneissose Younger Ryoke granites and massive San-yo granites which crosscut the metamorphic rocks (Higashimoto *et al.* 1983). These two groups of granite are associated with contact aureoles. On the other hand, gneissose Older Ryoke granites occur concordantly with the schistosity of high-grade metamorphic rocks and do not have contact aureoles (e.g. Nureki 1974; Higashimoto *et al.* 1983).

The pelitic rocks are schists in the northern half of the area up to the Ms-Crd zone, and are metatexites and inhomogeneous diatexites in the southern half of the area, in the Kfs-Crd, Sil-Kfs and Grt-Crd zones (Higashimoto *et al.* 1983; Ikeda 1998b; Brown 1998). The boundary between schists and migmatites is spatially close to the beginning of the Kfs-Crd zone. In migmatitic rocks of the Kfs-Crd and Grt-Crd zones, high modal amount of cordierite is found in the mesosome and sillimanite is only found as inclusion in cordierite. Based on these textures, Brown (1998) considered that the muscovite-consuming subsolidus reaction or dehydration melting reaction, such as

$$\mathrm{Ms+Pl+Kfs+Qtz=Sil+melt,}$$

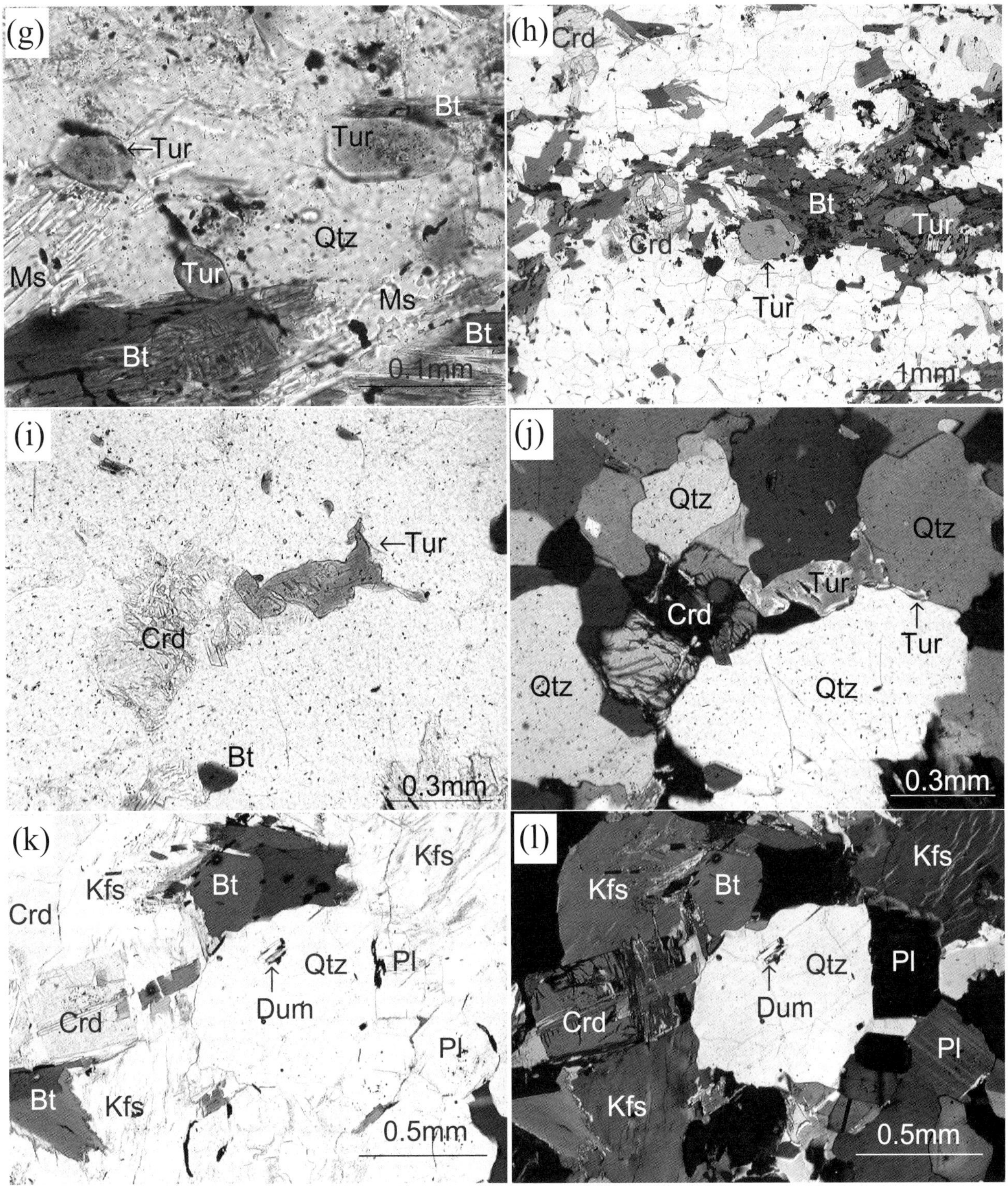

Figure 2 (g–l).

Figure 2 (a–d). Photographs of the mode of occurrence of interboudin partitions filled with tourmaline-bearing leucosome and the tourmaline-bearing leucogranite veins. These textures are commonly observed near the tourmaline-out isograd. The diameter of the coin is 2 cm. The location is the Aoyama area. (a) Interboudin partitions filled with tourmaline-bearing leucosome (perpendicular to the schistosity). (b) The same texture as (a), sub-parallel to the schistosity. (c) A tourmaline-bearing leucogranite vein discordantly cutting the foliation of pelitic schist and interboudin partitions filled with leucosome. The black spots in the rock matrix are caused by the rain drops. (d) Enlargement of (c): randomly oriented black tourmaline crystals can be observed in the vein. (e) Irregularly shaped retrograde tourmaline replacing biotite from the Yanai area. It is partly overgrown on cordierite, biotite and euhedral plagioclase. Plane-polarised light. (f) Cross-polarised light of (e). (g–l) Photomicrographs showing the mode of occurrence of tourmaline and dumortierite from the Komagane area: (g) Prograde tourmaline from the low-temperature side of the tourmaline-out isograd. Plane-polarised light; (h) Prograde tourmaline found near the tourmaline-out isograd. Plane-polarised light; (i) Retrograde tourmaline from the high-temperature side of the tourmaline-out isograd. Plane-polarised light; (j) Cross-polarised light of (i); (k) Dumortierite included in the interstitial quartz that is filling the grain boundaries of K-feldspar, biotite, euhedral plagioclase and subhedral cordierite. The sample is from the high-temperature side of the tourmaline-out isograd; (l) cross-polarised light of (k).

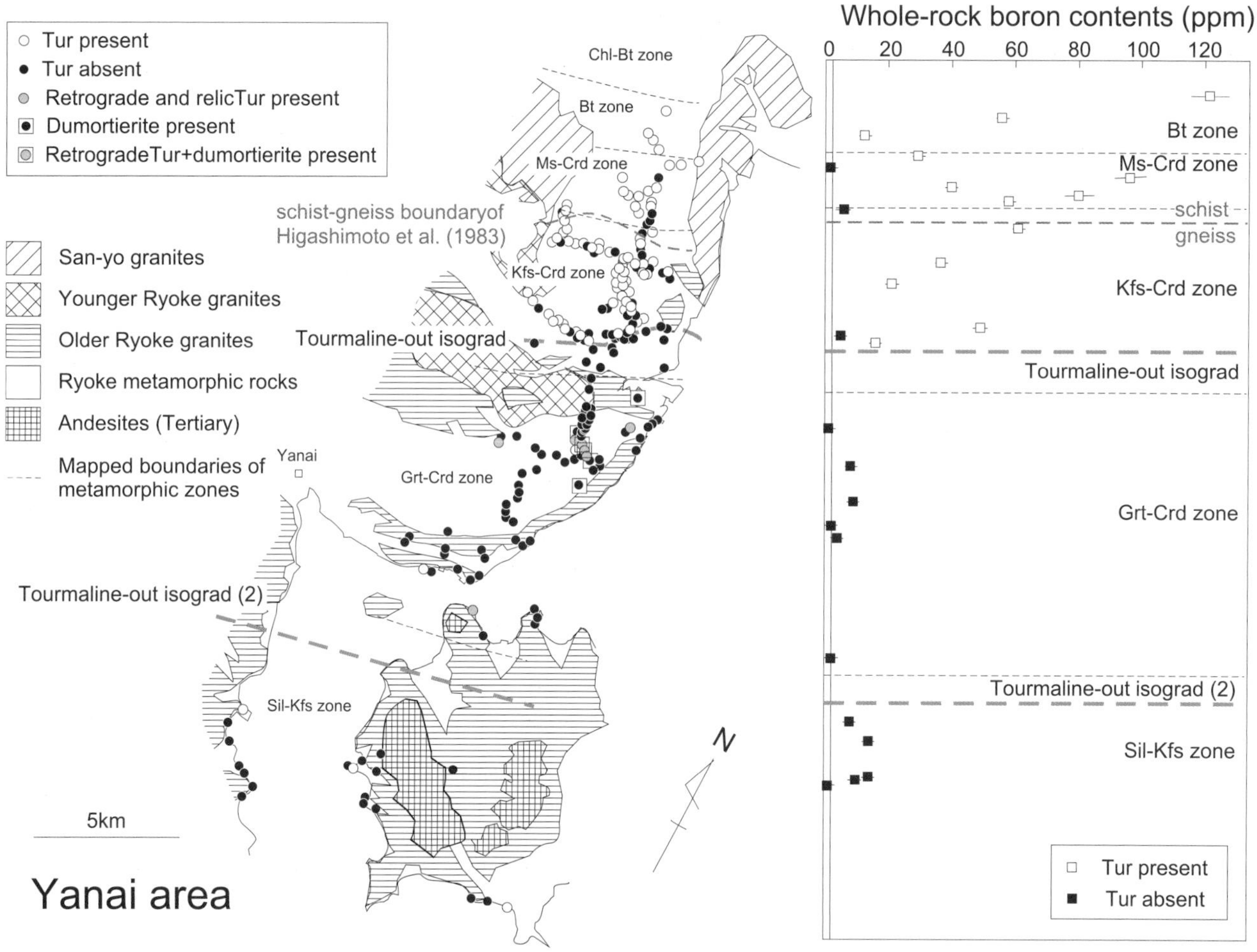

Figure 3 Geological map showing the field distribution of borosilicates in the Yanai area. The geological map is after Ikeda (1998a) and Okudaira *et al.* (1993). The mapped boundaries of the metamorphic zones (dashed, black lines; Ikeda 1998a), the schist–gneiss boundary (dashed grey line; Higashimoto *et al.* 1983) and the tourmaline-out isograd (bold, dashed grey line; Kawakami & Ikeda 2003 and present study) are also shown. Whole-rock boron contents determined by prompt gamma neutron activation analysis are shown on the right-hand side of the figure. Seven samples are newly analysed and added to the result of Kawakami & Ikeda (2003). See Kawakami & Ikeda (2003) for details.

took place at lower temperatures on a prograde evolution, followed by the biotite- and sillimanite-consuming, cordierite-forming dehydration melting reaction,

$$\mathrm{Bt+Sil+Qtz=Crd+Kfs+melt}, \qquad (3)$$

which was responsible for the formation of migmatites in the Kfs-Crd zone. The presence of K-feldspar and cordierite in the leucosome is consistent with this interpretation. Euhedral K-feldspar, plagioclase and subhedral cordierite against quartz in the leucosome, and the coexistence of garnet and cordierite in the Grt-Crd zone suggests that dehydration melting reaction 1, described above, also contributed to the formation of migmatites in this zone (Brown 1998).

The Yanai area can be divided into three structural domains on the basis of large-scale structures, such as the northern, central and southern domains (Okudaira *et al.* 1993; Okudaira 1996). The Grt-Crd zone belongs to the central domain and the Sil-Kfs zone belongs to the southern domain. The northern and southern domains are characterised by gentle upright folds. Fold axes gently plunge toward ESE in the northern domain, whereas they trend WNW–ESE to E–W in the southern domains. The geological structure of the central domain is characterised by overturned folds whose axial surfaces gently dipping toward NNE to NE (Okudaira *et al.* 1993). However, these folds are considered to postdate the main crystallisation of regional metamorphism, and the structures coeval to the regional metamorphism include distinct foliation (S1-foliation) that is formed parallel to the lithologic layering, and intrafolial folds with axial planes parallel to the S1-foliation. Okudaira *et al.* (1995) reported melt-filled fractures possibly formed by shearing along the foliation during the regional metamorphism. For further detailed descriptions on deformation structures, see Okudaira *et al.* (1993, 1995) and Okudaira (1996).

The time of peak metamorphism is considered to be Cretaceous. Rb-Sr dating of biotite from the metamorphic rocks gives ages ranging from 100 to 90 Ma (Banno & Nakajima 1992; Nakajima 1994), and CHIME dating of monazite from high-grade pelitic metamorphic rocks gives ages between 102 and 98 Ma (Suzuki & Adachi 1998). Three groups of granites give similar Rb-Sr and K-Ar biotite ages of 100 to 80 Ma (Nakajima *et al.* 1990; Nakajima 1994), while CHIME dating gives distinctly different ages; 95 Ma for the Older Ryoke, 90 Ma for the Younger Ryoke and 87–86 Ma for the San-yo granites (Suzuki *et al.* 1994, 1996; Suzuki & Adachi 1998).

2.4. Mode of occurrence of tourmaline in the Yanai area

Tourmaline is more common in pelitic rocks in the lower-grade areas up to the low-temperature part of the Kfs-Crd zone, whereas it is rare in the high-temperature part of the Kfs-Crd zone, and the Sil-Kfs and Grt-Crd zones (Fig. 2). Under plane-polarised light in thin section, tourmaline crystals are zoned in colour and commonly dark green (which is common in the cores of tourmaline from the Bt and Ms-Crd zones) to yellowish. Tourmaline crystals from the Bt and Ms-Crd zones show concentric, oscillatory compositional zoning consisting of dusty cores, inner and outer mantles, and rims. In contrast, tourmaline crystals in the Kfs-Crd and Sil-Kfs zones are rather homogeneous, having weak concentric chemical zoning consisting of cores and rims. Some of the tourmaline includes graphite and biotite, but in general, inclusion minerals are rare in tourmaline at higher grades. Tourmaline is rare in the Sil-Kfs zone, but is locally common in the melanosome layers which are concordant with the migmatitic banding. It shows no reaction texture with other minerals, and commonly coexists with sillimanite. From these features, tourmaline crystals in the Sil-Kfs zone rocks are probably prograde products and are stable at the peak P–T conditions of this zone. The rim composition of tourmaline becomes more aluminous with increasing metamorphic grade. Chemical zoning of tourmaline occurs via $^{X}\square AlNa_{-1}Mg_{-1}$ and $MgTi^{Y}Al_{-2}$ exchange vectors in the Bt and Ms-Crd zones where muscovite is stable, whereas it is governed by $Mg(OH)^{Y}Al_{-1}O_{-1}$, $CaMgO^{X}\square_{-1}{}^{Y}Al_{-1}(OH)_{-1}$ and $MgTi^{Y}Al_{-2}$ vectors in further higher-grade, muscovite-unstable zones such as the Kfs-Crd and Sil-Kfs zones (Kawakami & Ikeda 2003). Kawakami & Ikeda (2003) gave details of tourmaline compositions.

On the other hand, the rare tourmaline in the Grt-Crd zone can be divided into two types based on the textural relationships: The first type is surrounded by muscovite aggregates (themselves possibly after cordierite) and probably are relics remaining after a tourmaline breakdown reaction similar to reaction 2. The second type occurs as irregularly shaped tourmaline formed along the grain boundaries of the matrix minerals and is interpreted to be retrograde (Kawakami & Ikeda 2003). This tourmaline generation locally rims euhedral plagioclase crystals (Fig. 2e, f), suggesting that it formed during or after the crystallisation of melts. The colour zoning of this tourmaline varies from yellow to blue, reflecting the wide range of chemical composition, possibly controlled by the small equilibrium volume of the retrograde stage. In the Grt-Crd zone, minor dumortierite occurs locally in the pelitic migmatites. Dumortierite and retrograde tourmaline rarely occur in the same thin section, but they are not in contact with each other and no reaction texture is observed between them. Dumortierite commonly occurs along grain boundaries of matrix minerals, or replaces them, and is closely associated with fine-grained aggregates of muscovite of retrograde origin. These modes of occurrence of dumortierite are indicative of its retrograde formation. It is not the product of the breakdown reaction of tourmaline because the occurrence of the mineral is very local and no reaction texture between tourmaline is observed. Rather, the common coexistence with muscovite of retrograde origin suggests that the dumortierite precipitated from a boron-bearing aqueous fluid that possibly evolved from the boron-bearing melts formed in migmatites.

The occurrence of relic tourmaline and retrograde borosilicates makes the distribution of the prograde tourmaline unclear in the field. However, when relict and retrograde borosilicates are distinguished, the disappearance of prograde tourmaline is so abrupt that a line delimiting the distribution of prograde tourmaline in the pelitic rocks, the tourmaline-out isograd, can be defined at the southern part of the Kfs-Crd zone (Kawakami & Ikeda 2003). The occurrence of prograde tourmaline in the Sil-Kfs zone implies that another tourmaline-out isograd can be defined in addition to this 'ordinary' tourmaline-out isograd. This new isograd is termed the 'tourmaline-out isograd (2)' and occurs between the Sil-Kfs and Grt-Crd zones, as shown in Figure 3. Prograde tourmaline does not stably exist between these two 'tourmaline-out' isograds.

2.5. Komagane area

In the Komagane area of the Ryoke metamorphic belt, the following rock types occur: (1) medium- to high-grade pelitic and psammitic metamorphic rocks; (2) older and Younger Ryoke granites; and (3) Ikuta granite, which was emplaced between the Older and Younger Ryoke granites (Kuroda *et al.* 1982). Petrographic studies of the metamorphic rocks of the Komagane area (Figs 1a & 4) have been undertaken by many authors, including Hayama (1964), Ono (1969, 1977a, b, 1978), Morikiyo (1984, 1986, 1998), Hokada (1996, 1998) and Brown (1998). In this area, medium- to high-grade pelitic and psammitic metamorphic rocks are widely distributed and the grade of metamorphism increases toward the southeast (Hayama 1964; Morikiyo 1986; Hokada 1998). Metapelitic rocks are schistose in the north-western, lower-grade part of the area, whereas migmatites predominate in the south-eastern part of the area (Hokada 1998). The foliation of the metamorphic rocks strikes NE–SW and dips towards the NW. The areal distribution of euhedral plagioclase in migmatites can be mapped as shown in Figure 4, suggesting that partial melting of the pelitic rock prevailed to the high-temperature side of the euhedral plagioclase-in isograd. Cordierite also shows subhedral shape suggestive of growth in the melt (e.g. Ellis & Obata 1992). The main characteristic of this area is the occurrence of homogeneous diatexite (e.g. Morikiyo 1998) that mainly consists of euhedral K-feldspar and plagioclase, subhedral cordierite, biotite, and interstitial quartz. Clots of biotite (1–1·5 cm in diameter) are also present. Where plagioclase is subhedral, it also shows euhedral chemical zoning. The homogeneous diatexite contains blocks of metatexite of several tens of centimetres in diameter, implying that homogeneous diatexite was formed at a deeper level and intruded to the present level where metatexite prevails (Morikiyo 1998). However, the areal distribution of the homogeneous diatexite has not been mapped in detail.

Morikiyo (1984, 1986) divided the area into three metamorphic zones based on the formation of andalusite, the coexistence of andalusite+K-feldspar and the disappearance of andalusite. Hokada (1998) defined the cordierite-in isograd and explained the formation of cordierite in the region (Fig. 4). He also reported the first occurrence of orthopyroxene in a basic rock (Hokada 1996). The peak metamorphic conditions previously estimated are 650 ± 50 °C, 4·5–5·0 kbar around Takato city (Ono 1977b) and 700–750 °C, 3·0–4·0 kbar in the study area (Hokada 1998; based on his P–T path). On the other hand, Brown (1998) suggested from the similar textural features as in the Yanai area that dehydration melting reactions 1 and 3 are responsible for the migmatite formation in this region, and proposed a P–T path with a higher peak temperature condition (around 850 °C) than previous studies. The formation of migmatites in this area is probably caused by these reactions, although reaction 1 seems to be less important, judging from the rare occurrence of garnet in this area.

2.6. Mode of occurrence of tourmaline in the Komagane area

Ono (1977a) briefly described the disappearance of tourmaline and appearance of dumortierite in the high-grade rocks around Takato city. He assumed the reaction

$$\text{Tur (dravite)} + \text{Sil} + \text{Qtz} = \text{Ab} + \text{dumortierite} + \text{Crd}$$

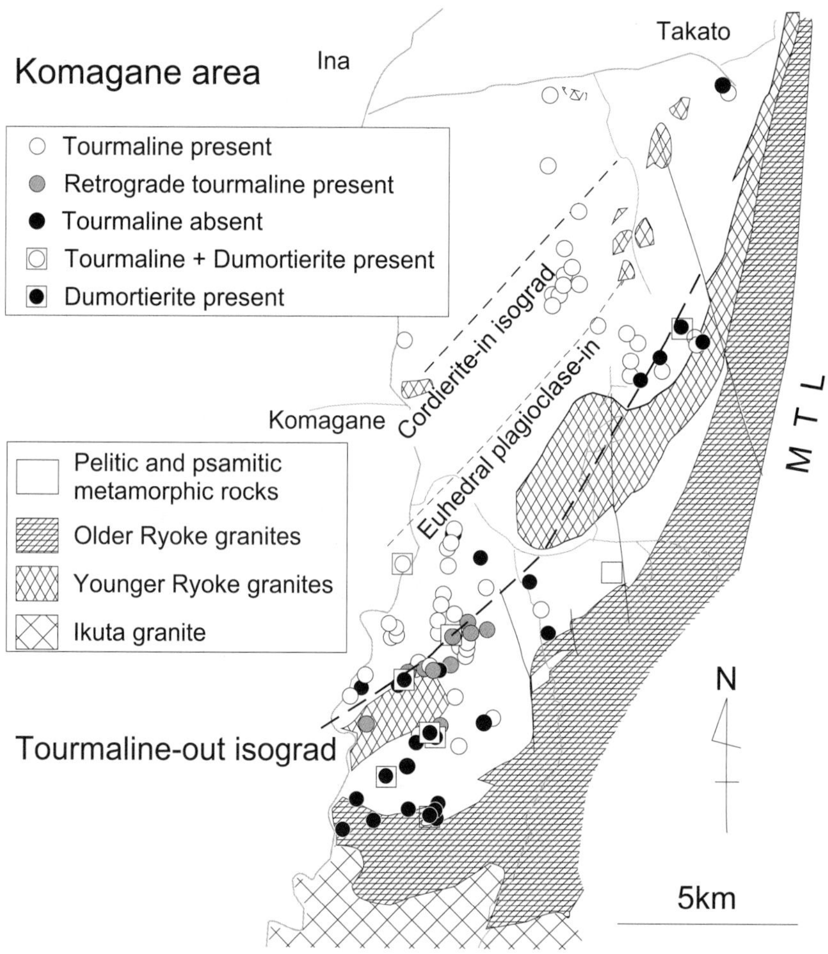

Figure 4 Geological map showing the field distribution of borosilicates in the Komagane area. The map is after Morikiyo (1986) and Hokada (1998). The mapped boundaries of the metamorphic zones (dashed, black line; Hokada 1998), the euhedral plagioclase-in isograd (dashed, fine black line; present study) and the tourmaline-out isograd (bold, dashed black line; present study) are also shown.

to have occurred, but concluded that the position of the isograd was difficult to define because of the rare occurrence of dumortierite. Takagi & Nagahama (1987) reported one dumortierite locality in the Komagane area. Other than these studies, descriptions of the occurrence of borosilicates are scarce in the Komagane area. In the north-western, low-grade part of the area, tourmaline grains are typically about 0·1 mm in diameter with greenish cores and yellow to brown rims (Fig. 2g). As the metamorphic grade increases, tourmaline becomes more coarsely grained up to several millimetres. It is yellow-coloured, having weak concentric chemical zoning consisting of the core and rim (Fig. 2h). However, chemical zoning of tourmaline has not been studied in detail in this area. These tourmaline grains are aligned parallel to the schistosity or migmatitic banding and are idiomorphic, indicating their formation during the prograde metamorphism. Some of them include biotite (Fig. 2h). The X_{Mg} value of the prograde tourmaline is 0·57–0·62. There is a tendency for pelitic rocks without prograde tourmaline to be common in the south-western, high-grade part of the area. Instead of prograde tourmaline, irregularly shaped tourmaline, formed along the grain boundaries of the matrix minerals (Fig. 2i, j), is common at these grades. These tourmaline grains are possibly retrograde products.

The occurrence of dumortierite is also limited to the high-grade area. Dumortierite commonly occurs along the grain boundaries of the matrix minerals such as plagioclase, cordierite, biotite and quartz. Dumortierite is locally found as an inclusion mineral of interstitial quartz that is interstitial to euhedral plagioclases and subhedral cordierites, textures of which are indicative of crystallisation in melts (Fig. 2k, l). It is also found to replace cordierite with muscovite aggregates of retrograde origin. These textures suggest that dumortierite is a retrograde product that possibly formed after or during the late stage of the crystallisation of the melts. Although breakdown textures involving tourmaline are not found, the common absence of prograde tourmaline and occurrence of retrograde borosilicates in the high-grade part of the area indicate that breakdown of tourmaline also took place in the Komagane area. Therefore, it is possible to define a tourmaline-out isograd at the boundary where prograde tourmaline becomes absent and retrograde borosilicates become relatively common (Fig. 4). It should be noted that this isograd lies on the high-grade side of the 'euhedral plagioclase-in' isograd.

Pegmatite veins containing tourmaline were reported from the highest-temperature part of the area (Ono 1978). Their distribution does not show any genetic relationship with nearby granitic plutons. The pegmatites occur approximately parallel to the axial plane of folded metamorphic rocks, suggesting that they intruded during the Ryoke metamorphism (Ono 1978). These pegmatites are potential candidates for the boron-bearing melt possibly extracted from the migmatite zone, as in the case of the Aoyama area.

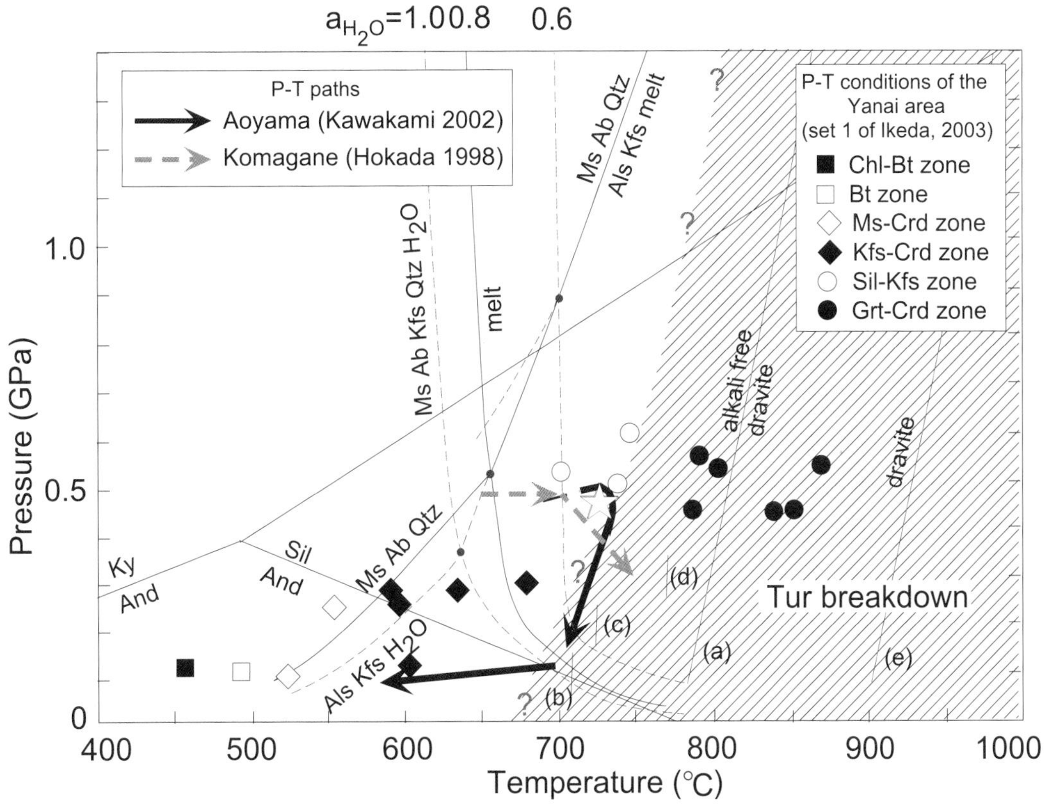

Figure 5 Pressure–temperature (P–T) paths of the low-temperature part of the Aoyama (Kawakami 2002; black arrows) and Komagane (Hokada 1998; grey arrows) areas, and the P–T conditions of the Yanai area (Ikeda 2003). The white star represents the provisional P–T location of the breakdown reaction of tourmaline in the Aoyama area (Kawakami 2001b). The melting curves, including muscovite, quartz, K-feldspar, aluminosilicate, albite and H_2O, are taken from Johannes & Holtz (1996). Some reaction curves are omitted for simplicity. The solid black lines represent the reaction curves for $a_{H_2O}=0{\cdot}8$ and the dashed grey lines represent those for $a_{H_2O}=0{\cdot}6$ and 1·0. Solid grey lines in the supra-solidus region are the selected experimental results of the high-temperature stability limit of tourmaline: (a) alkali free dravite+boron-rich fluid (Werding & Schreyer 1984); (b) alkali-free dravite without excess boron in the fluid (Werding & Schreyer 1984); (c) dravite+quartz with more than 4 wt.% B_2O_3 in the final fluid (von Goerne *et al.* 1999); (d) natural schorl+gneiss (Holtz & Johannes 1991); and (e) dravite (Schreyer & Werding 1997). The aluminosilicate phase diagram follows that of Holdaway (1971). The striped area is the hypothetical 'tourmaline unstable' P–T field for the metapelites of the Ryoke metamorphic belt, constrained by the field data and previous experimental studies. The low-temperature boundary of the tourmaline unstable zone with question ('?') marks represents the hypothetical P–T location of the breakdown reaction of tourmaline in the metapelites of the Ryoke metamorphic belt.

3. Discussion

3.1. Breakdown reaction of prograde tourmaline and its P–T conditions in the three studied areas

There are many migmatite and granulite terranes depleted in their whole-rock boron contents relative to contemporaneous or correlated lower-grade areas (e.g. Moran *et al.* 1992). Although studies on the areal distributions of tourmaline are not sufficiently available worldwide, the observations made here imply that most high-grade terranes are tourmaline-absent. Considering that the lower-grade equivalents of those high-temperature rocks are tourmaline-bearing, the low boron contents for the high-grade rocks probably reflect the instability of tourmaline at these temperatures and the extraction of boron-enriched melts. Kawakami (2001a) and Kawakami & Ikeda (2003) proposed that the breakdown of tourmaline occurred through a partial melting reaction. Alternatively, breakdown of tourmaline in the presence of the alkaline fluid (Morgan & London 1989) is sometimes favoured as an explanation of the low-temperature breakdown of tourmaline (e.g. Wilke *et al.* 2002). However, it should be noted that the infiltration of large amounts of alkaline fluid by percolation is required to explain the regional-scale, complete breakdown of tourmaline by this mechanism. Externally derived fluid may be difficult to infiltrate into every grain boundary to destabilise all the tourmaline grains because the alkaline fluid will react out with the rock quickly if the amount of the fluid is small. Internally derived alkaline fluid is another possible candidate to destabilise tourmaline, but as yet, we know little about the pH value of the fluid formed through the dehydration reactions. The similarity of the P–T conditions of the breakdown of tourmaline in the Aoyama, Yanai and Komagane areas (Fig. 5) implies that the breakdown of tourmaline in the Ryoke metamorphic belt is mainly controlled by the P–T conditions.

The provisional P–T conditions for the tourmaline-out isograd in the Aoyama area have been determined by Kawakami (2001a, b) and are shown as a star in Figure 5. That in the Yanai area can be constrained by the two 'tourmaline-out' isograds described above. First, because the tourmaline-out isograd is located within the high-temperature part of the Kfs-Crd zone, the tourmaline breakdown reaction line has to cross the P–T condition of the high-temperature part of the Kfs-Crd zone. Secondly, the tourmaline breakdown reaction should lie between P–T conditions of the Sil-Kfs and Grt-Crd zones from the constraints of the tourmaline-out isograd (2).

Therefore, provisional P–T location of the breakdown of tourmaline in the Yanai area should go through these two P–T conditions (Fig. 5). The P–T path of the Komagane area (Hokada 1998) is also shown in Figure 5. Considering that the provisional P–T location of the tourmaline-out isograd in the Komagane area should be on this P–T path, it seems that the P–T conditions for the tourmaline breakdown reaction are almost the same as those of the Aoyama area.

Although a decrease in whole-rock boron content is already observed on the low-temperature side of the tourmaline-out isograd in the Yanai area, the breakdown of tourmaline still controls the abundance of nearly half of the boron originally contained in the rock, if it is assumed that the original boron content in the pelitic rocks is ~120 p.p.m. Textural evidence shows that the breakdown of tourmaline in the three areas occurred in the region where partial melting of the pelitic rocks was taking place, which enabled the boron to be transferred into the melt directly when the tourmaline broke down.

3.2. Estimation of the degree of partial melting utilising boron contents of leucosome in interboudin partitions

Near the tourmaline-out isograd in the Aoyama area, tourmaline-bearing leucosomes are found in interboudin partitions with three-dimensionally closed shapes. This texture was interpreted by Kawakami (2001a) to be a frozen *in situ* melt, segregated from the surroundings on a limited length scale of less than 10 centimetres. Tourmaline with the breakdown texture was found in the surrounding pelitic schist, which lead Kawakami (2001a) to conclude that the boron in the leucosome was derived from the breakdown of tourmaline in the surroundings (closed-system behaviour of boron and melt within the sample scale). Using the whole-rock boron contents of the leucosome, surrounding host rock and pelitic schists (Kawakami 2001b), and assuming the closed-system behaviour of boron and melts within this sample, the degree of partial melting can be estimated. From a simple mass balance calculation, we have

$$C_o=C_1(1-F)+C_2F,$$

where C_o (p.p.m.) is the whole-rock boron content of the pelitic schist before the partial melting and breakdown of tourmaline, C_1 is that of the residue, C_2 is that of the leucosome melt, and F is the degree of partial melting. Adopting $C_o=48$ (average whole-rock boron contents of tourmaline-bearing pelitic schists at a lower grade than the tourmaline-out isograd), $C_1=17$ (boron contents of the surrounding matrix of the rock with leucosome-filling interboudin partitions) and $C_2=280$ (boron contents of the tourmaline-bearing leucosome in the interboudin partitions), the calculated degree of partial melting is 12 wt.%. This is equivalent to degree of partial melting of 12–14 vol.%, if we assume the density of leucosome melt to be 2·6–2·8 g cm^{-3} and that of residue to be 2·8–3·2 g cm^{-3}. The P–T conditions for the tourmaline-out isograd are estimated to be about 730°C, 4·5 kbar, and this estimate of degree of partial melting is consistent with thermodynamically calculated melt modes for 730°C, 5·0 kbar (e.g. White *et al.* 2001). However, it is difficult to compare this result with the experimentally obtained data on the degree of partial melting of pelites without extrapolation because the experiments are usually run at above 800°C (e.g. Patiño Douce & Johnston 1991; Montel & Vielzeuf 1997) and do not include boron. The degree of partial melting in the migmatite zone is probably much higher.

Boudinage probably played an important role in segregating melts into local low-pressure sites (Fig. 2a, b). Further development of the boudinage structures in the rock may have resulted in the connection of the interboudin partitions. Such connection would then provide vertical pathways for the extraction of the boron-bearing melt in the schists or layered migmatites. Tourmaline-bearing pegmatite veins found in the Aoyama and Komagane areas imply that the boron-bearing melts were extracted from the migmatite zone mainly by dyking.

3.3. Formation of retrograde tourmaline

The absence of retrograde tourmaline and the complete depletion of whole-rock boron in the migmatite zone of the Aoyama area suggest that extraction of the boron-bearing melts from the migmatite zone was almost complete there. However, the occurrence of irregularly shaped retrograde tourmaline and dumortierite in the Yanai and Komagane areas implies that the absence of retrograde borosilicates is rare, and that recognition of the retrograde tourmaline is of general importance. Barbey *et al.* (1996) reported the occurrence of amoeboidal tourmaline grown on biotite and antiperthite, or filling cracks in garnet from some granitic leucosomes of the Tibetan Slab migmatite. These crystals can be interpreted to be retrograde tourmaline.

There are two types of retrograde tourmaline: (1) irregularly shaped tourmaline, which occurs along the grain boundaries of other minerals (e.g. Barbey *et al.* 1996; Kawakami & Ikeda 2003) and is commonly observed in migmatites; and (2) euhedral, magmatic tourmaline, which is commonly observed in the leucosome pods and pegmatites (e.g. Michailidis & Kassoli-Fournaraki 1994; Kawakami 2001a, b). The textural differences between these two types are probably controlled by whether the tourmaline could grow in a free space or not. The former, irregularly shaped type possibly grew in crystal-rich, melt-poor environment. On the other hand, the latter, euhedral type possibly grew in a melt-rich environment so that it could grow its euhedral crystal faces well. These textural differences provide us with various kinds of information on the behaviour of boron-bearing melt, as discussed below.

3.3.1. Irregularly-shaped retrograde tourmaline. Irregularly shaped retrograde tourmaline is commonly observed in metatexites and inhomogeneous diatexites. These rocks are the potential 'source' of the peraluminous melt. What researchers would like to know is whether or not the melt was extracted from the source region. Irregularly shaped retrograde tourmaline can potentially be a tool to evaluate this. Although this type of retrograde tourmaline can be formed by the late crystallisation of tourmaline in leucosome pods, here the present author would like to concentrate on the irregularly shaped retrograde tourmaline found in the anatectic source rocks.

The boron-bearing melt crystallises tourmaline in the course of its retrograde, cooling P–T stage, provided that the melt has a sufficiently high concentration of boron (e.g. London 1999). The boron contents of the melts can be elevated through fractional crystallisation. Other elements necessary for tourmaline formation, such as Al and Mg, are probably supplied from restitic solid minerals, such as biotite and cordierite, because they are abundant in the source rocks. Therefore, in the anatectic rock rich in restite materials, such as migmatites and restite-rich S-type granites, back reaction between melt and restite (e.g. Kriegsman 2001) probably played an important role in the formation of the retrograde tourmaline, as suggested by the textural evidences of retrograde tourmaline replacing biotite (Fig. 2, f).

The characteristic of irregularly shaped retrograde tourmaline occurring at the grain boundaries of other minerals as amoeboidal crystals implies that only small amounts of evolved, boron-bearing melts that were trapped at grain boundaries could contribute to form the retrograde tourmaline in the migmatite. The presence of retrograde tourmaline in the source rock implies that boron-bearing melt was not completely extracted from the rock. Hence, the abundance of such retrograde tourmaline can tell us whether the extraction of the melt from the source rock was efficient or not. Relatively large amounts of retrograde tourmaline in the source rock may indicate that only small amounts of melt were extracted.

Melt-restite separation is probably more effective in the formation of metatexites than in diatexites. Since metatexites do not behave as a magma body, but instead, behave as a rock because of their low degree of partial melting, segregation of the melt is probably enhanced by deformation, such as shearing and boudinage, to segregate melt into low-pressure sites (Brown *et al.* 1995). On the other hand, because diatexites behave as a magma body, extraction of melt from them is possibly difficult even if aided by deformation, and therefore, the boron-bearing melt crystallises within the diatexite body, resulting in a substantial melt-restite back reaction that would favour the production of retrograde tourmaline. Therefore, it might be expected that metatexites and diatexites show distinctly different amounts of retrograde tourmaline formed by back-reaction.

3.3.2. Euhedral retrograde tourmaline. This type of retrograde tourmaline is easily discriminated from prograde tourmaline by its field occurrence because it commonly occurs in leucosome pods, interboudin partitions filled with leucosome and pegmatite veins. In such cases, it is compositionally different from prograde tourmaline (e.g. Kawakami 2001b). However, it is possible that euhedral retrograde tourmaline occurs in highly anatectic migmatites (diatexites). Discrimination of retrograde tourmaline from prograde tourmaline largely depends on textural evidence; therefore, euhedral retrograde tourmaline is difficult to distinguish from prograde tourmaline unless reaction textures with restite material are observed. Although this kind of euhedral tourmaline is absent or rare in the Aoyama and Yanai areas, relatively common ‘prograde tourmaline’ on the high-temperature side of the tourmaline-out isograd in the Komagane area (Fig. 4) could potentially be a retrograde product. Further studies are needed to solve this problem.

4. Acknowledgements

I am grateful to Dr T. Hokada for kindly permitting me to use his thin sections of the samples from the Komagane area. Thanks are also due to Professor Y. Hiroi and an anonymous reviewer for constructive reviews, Dr A. Christy and Professor D. Ellis for checking the English, and Miss V. M. Ingpen for editorial assistance. This study was financially supported by a Grant-in-Aid for JSPS Fellows (No. 05864) to T. Kawakami.

5. References

Banno, S. & Nakajima, T. 1992. Metamorphic belts of Japanese Islands. *Annual Review of the Earth and Planetary Sciences* **20**, 159–79.

Barbey, P., Brouand, M., Le Fort, P. & Pecher, A. 1996. Granite-migmatite genetic link: the example of the Manaslu granite and Tibetan Slab migmatites in central Nepal. *Lithos* **38**, 63–79.

Benard, F., Moutou, P. & Pichavant, M. 1985. Phase relations of tourmaline leucogranites and the significance of tourmaline in silicic magmas. *Journal of Geology* **93**, 271–91.

Brown, M. 1973. The definition of metatexis, diatexis and migmatite. *Proceedings of the Geologists' Association* **84**, 371–82.

Brown, M. 1998. Unpairing metamorphic belts: P–T paths and a tectonic model for the Ryoke Belt, southwest Japan. *Journal of Metamorphic Geology* **16**, 3–22.

Brown, M., Averkin, Y. A. & McLellan, E. L. 1995. Melt segregation in migmatites. *Journal of Geophysical Research*, **100**, 15655–79.

Ellis, D. J. & Obata, M. 1992. Migmatite and melt segregation at Cooma, New South Wales. *Transactions of the Royal Society of Edinburgh: Earth Sciences* **83**, 95–106.

Grew, E. S. 1996. Borosilicates (exclusive of tourmaline) and boron in rock-forming minerals in metamorphic environments. *In* Grew, E. S. & Anovitz, L. M. (eds) *Boron – Mineralogy, Petrology and Geochemistry. Reviews in Mineralogy* **33**, 387–502.

Grew, E. S. & Hinthorne, J. R. 1983. Boron in sillimanite. *Science* **221**, 547–9.

Hayama, Y. 1964. Progressive metamorphism of pelitic and psammitic rocks in the Komagane district, Nagano Pref., central Japan. *Journal of Faculty of Science, University of Tokyo, Section 2* **15**, 321–69.

Hayama, Y., Yamada, T., Ito, M., Kutsukake, T., Masaoka, K., Miyakawa, K., Mochizuki, Y., Nakai, Y., Tainosho, Y., Yoshida, M., Kawarabayashi, I. & Tsumura, Y. 1982. Geology of the Ryoke Belt in the eastern Kinki District, Japan – the phase-divisions and the mutual relations of the granitic rocks. *Journal of the Geological Society of Japan* **88**, 451–66. [In Japanese with English abstract.]

Henry, D. J. & Dutrow, B. L. 1996. Metamorphic tourmaline and its petrologic applications. *In* Grew, E. S. & Anovitz, L. M. (eds) *Boron – Mineralogy, Petrology and Geochemistry. Reviews in Mineralogy* **33**, 503–57.

Higashimoto, S., Nureki, T., Hara, I., Tsukuda, E. & Nakajima, T. 1983. Geology of the Iwakuni district. Quadrangle Series, Scale 1: 50,000. Geological Survey of Japan. [In Japanese with English abstract.]

Hokada, T. 1996. Orthopyroxene from a metabasite of the Ryoke belt in the Ina district, Nagano Prefecture, Central Japan: textural evidence of partial melting. *Journal of the Japanese Association of Mineralogists, Petrologists and Economic Geologists* **91**, 253–65.

Hokada, T. 1998. Phase relations and P–T conditions of the Ryoke pelitic and psammitic metamorphic rocks in the Ina district, Central Japan. *The Island Arc* **7**, 609–20.

Holdaway, M. J. 1971. Stability of andalusite and the aluminum silicate phase diagram. *American Journal of Science* **271**, 97–131.

Holtz, F. & Johannes W. 1991. Effect of tourmaline on melt fraction and composition of first melts in quartzofeldspathic gneiss. *European Journal of Mineralogy*, **3**, 527–36.

Ikeda, T. 1998a. Progressive sequence of reactions of the Ryoke metamorphism in the Yanai district, southwest Japan: the formation of cordierite. *Journal of Metamorphic Geology* **16**, 39–52.

Ikeda, T. 1998b. Phase equilibria and the pressure-temperature path of the highest-grade Ryoke metamorphic rocks in the Yanai district, SW Japan. *Contributions to Mineralogy and Petrology* **132**, 321–35.

Ikeda, T. 2002. Regional occurrence of orthopyroxene-bearing basic rocks in the Yanai district, SW Japan: evidence for granulite-facies Ryoke metamorphism. *The Island Arc* **11**, 185–92.

Ikeda, T. 2004. Pressure-temperature conditions of the Ryoke metamorphic rocks in Yanai district, SW Japan. *Contributions to Mineralogy and Petrology* **146**, 577–89.

Ishizaka, K. 1966. A geochronological study of the Ryoke metamorphic terrain in the Kinki district, Japan. *Memoirs of the College of Science, University of Kyoto, Series B* **33**, 69–102.

Johannes, W. & Holtz, W. 1996. *Petrogenesis and Experimental Petrology of Granitic Rocks*. Berlin: Springer-Verlag.

Kawakami, T. 2001a. Tourmaline breakdown in the migmatite zone of the Ryoke metamorphic belt, SW Japan. *Journal of Metamorphic Geology* **19**, 61–75.

Kawakami, T. 2001b. Boron depletion controlled by the breakdown of tourmaline in the migmatite zone of the Aoyama area, Ryoke metamorphic belt, SW Japan. *Canadian Mineralogist* **39**, 1529–46.

Kawakami, T. 2002. Magmatic andalusite from the migmatite zone of the Aoyama area, Ryoke metamorphic belt, SW Japan, and its importance in constructing the P–T path. *Journal of Mineralogical and Petrological Sciences*, **97**, 241–53.

Kawakami, T. & Ikeda, T. 2003. Depletion of whole-rock boron controlled by the breakdown of tourmaline and retrograde formation of borosilicates in the Yanai area, Ryoke metamorphic belt, SW Japan. *Contributions to Mineralogy and Petrology*, **145**, 131–50.

Kretz, R. 1983. Symbols for rock forming minerals. *American Mineralogist* **68**, 277–9.

Kriegsman, L. M. 2001. Partial melting, partial melt extraction and partial back reaction in anatectic migmatites. *Lithos* **56**, 75–96.

Kuroda, Y., Yamada, T., Fujimoto, K., O'hashi, H., Takahashi, K., Okano, O., Mochizuki, Y., Yamana, S. & Ohtomo, Y. 1982. A proposal for classification of the granitic rocks of the Ryoke Belt in the Chubu and Kinki districts in view of hydrogen isotope study. *Journal of the Geological Society of Japan* **88**, 541–54. [In Japanese with English abstract.]

Lonker, S. W. 1988. An occurrence of grandidierite, kornerupine, and tourmaline in southeastern Ontario, Canada. *Contributions to Mineralogy and Petrology* **98**, 502–16.

London, D. 1999. Stability of tourmaline in peraluminous granite systems: the boron cycle from anatexis to hydrothermal aureoles. *European Journal of Mineralogy* **11**, 253–62.

London, D., Morgan, G. B. VI & Wolf, M. B. 1996. Boron in granitic rocks and their contact aureoles. *In* Grew, E. S. & Anovitz, L. M. (eds) *Boron – Mineralogy, Petrology and Geochemistry. Reviews in Mineralogy* **33**, 299–330.

Michailidis, K. & Kassoli-Fournaraki, A. 1994. Tourmaline concentrations in migmatitic metasedimentary rocks of the Riziana and Kolchiko areas in Maceodnia, Northern Greece. *European Journal of Mineralogy* **6**, 557–69.

Miyamura, M., Yoshida, F., Yamada, N., Sato, T. & Sangawa, A. 1981. Geology of the Kameyama district. Quadrangle Series, scale 1: 50,000. Geological Survey of Japan. [In Japanese with English abstract.]

Miyashiro, A. 1965. *Metamorphic Rocks and Metamorphic Belts*. Tokyo: Iwanami Shoten Publishers. [In Japanese.]

Montel, J. M. & Vielzeuf, D. 1997. Partial melting of metagreywackes, Part II. Compositions of minerals and melts. *Contributions to Mineralogy and Petrology* **128**, 176–96.

Moran, A. E., Sisson, V. B. & Leeman, W. P. 1992. Boron depletion during progressive metamorphism: implications for subduction processes. *Earth and Planetary Science Letters* **111**, 331–49.

Morgan, G. B., VI & London, D. 1989. Experimental reactions of amphibolite with boron-bearing aqueous fluids at 200 Mpa: implications for tourmaline stability and partial melting in mafic rocks. *Contributions to Mineralogy and Petrology* **102**, 281–97.

Morikiyo, T. 1984. Carbon isotopic study on coexisting calcite and graphite in the Ryoke metamorphic rocks, northern Kiso district, central Japan. *Contributions to Mineralogy and Petrology* **87**, 251–9.

Morikiyo, T. 1986. Hydrogen and carbon isotope studies on the graphite-bearing metapelites in the northern Kiso district of central Japan. *Contributions to Mineralogy and Petrology* **94**, 165–77.

Morikiyo, T. 1998. Mode of occurrence of migmatites and metabasites in the Ryoke Metamorphic Belt. *In Excursion Guidebook for the 105th Annual Meeting of the Geological Society of Japan*, 205–30. [In Japanese.]

Nabelek, P. I., Russ-Nabelek, C. & Denison, J. R. 1992. The generation and crystallization conditions of the Proterozoic Harney Peak Leucogranite, Black Hills, South Dakota, USA: petrologic and geochemical constraints. *Contributions to Mineralogy and Petrology* **110**, 173–91.

Nabelek, P. I. & Bartlett, C. D. 1998. Petrologic and geochemical links between the post-collisional Proterozoic Harney Peak leucogranite, South Dakota, USA, and its source rocks. *Lithos* **45**, 71–85.

Nakajima, T. 1994. The Ryoke plutonometamorphic belt: crustal section of the Cretaceous Eurasian continental margin. *Lithos* **33**, 51–66.

Nakajima, T., Shirahase, T. & Shibata, K. 1990. Along-arc lateral variation of Rb-Sr and K-Ar ages of Cretaceous granitic rocks in Southwest Japan. *Contributions to Mineralogy and Petrology* **104**, 381–9.

Nureki, T. 1974. Contact metamorphism in the So-o district, Yamaguchi Prefecture, Japan – with special reference to the occurrence of sillimanite. *Memoirs of the Geological Society of Japan* **11**, 251–81.

Okudaira, T. 1996. Thermal evolution of the Ryoke metamorphic belt, southwestern Japan: tectonic and numerical modeling. *The Island Arc* **5**, 373–85.

Okudaira, T., Hara, I., Sakurai, Y. & Hayasaka, Y. 1993. Tectonometamorphic processes of the Ryoke belt in the Iwakuni-Yanai district, southwest Japan. *Memoirs of the Geological Society of Japan* **42**, 91–120.

Okudaira, T., Hara, I. & Takeshita, T. 1995. Emplacement mechanism of the Older Ryoke granites in the Yanai district, southwest Japan, with special reference to extensional deformation in the Ryoke metamorphic belt. *Journal of Science, Hiroshima University, Series C* **10**, 357–66.

Ono, A. 1969. Zoning of the metamorphic rocks in the Takato-Shioziri area, Nagano Prefecture. *Journal of the Geological Society of Japan*, **75**, 521–36. [In Japanese with English abstract.]

Ono, A. 1977a. Petrologic study of the Ryoke metamorphic rocks in the Takato Shiojiri Area, Central Japan. *Journal of the Japanese Association of Mineralogist, Petrologists and Economic Geologists* **72**, 453–68. [In Japanese with English abstract.]

Ono, A. 1977b. Temperature and pressure of the Ryoke Gneiss estimated by Garnet-Cordierite Geothermometer. *Journal of the Japanese Association of Mineralogists, Petrologists and Economic Geologists* **72**, 114–17.

Ono, A. 1978. Petrologic study of sillimanite-garnet-cordierite-bearing pegmatite intrusives from Takato, Central Japan. *Journal of the Japanese Association of Mineralogists, Petrologists and Economic Geologists* **73**, 185–90.

Patiño Douce, A. E. & Johnston, A. D. 1991. Phase equiribria and melt productivity in the pelitic system: implications for the origin of peraluminous granitoids and aluminous granulites. *Contributions to Mineralogy and Petrology* **107**, 202–18.

Pereira, M. D. & Shaw, D. M. 1997. Behaviour of boron in the generation of an anatectic complex: the Pena Negra complex, central Spain. *Lithos* **40**, 179–88.

Scaillet, B., Pichavant, M. & Roux, J. 1995. Experimental crystallization of leucogranite magmas. *Journal of Petrology* **36**, 663–705.

Scaillet, B., Holtz, F., Pichavant, M. & Schmidt, M. 1996. Viscosity of Himalayan leucogranites: implications for mechanisms of granitic magma ascent. *Journal of Geophysical Research* **101**, 27691–9.

Schreyer, W. & Werding, G. 1997. High-pressure behaviour of selected boron minerals and the question of boron distribution between fluids and rocks. *Lithos* **41**, 251–66.

Sperlich, R., Giere, R. & Frey, M. 1996. Evolution of compositional polarity and zoning in tourmaline during prograde metamorphism of sedimentary rocks in the Swiss Central Alps. *American Mineralogist* **81**, 1222–36.

Suzuki, K., Adachi, M. & Kajizuka, I. 1994. Electron microprobe observations of Pb diffusion in metamorphosed detrital monazites. *Earth and Planetary Science Letters* **128**, 391–405.

Suzuki, K., Adachi, M. & Nureki, T. 1996. CHIME age dating of monazites from metamorphic rocks and granitic rocks of the Ryoke belt in the Iwakuni area, Southwest Japan. *The Island Arc* **5**, 43–55.

Suzuki, K. & Adachi, M. 1998. Denudation history of the high T/P Ryoke metamorphic belt, southwest Japan: constraints from CHIME monazite ages of gneisses and granitoids. *Journal of Metamorphic Geology* **16**, 23–37.

Takagi, H. & Nagahama, H. 1987. The Ryoke Belt in the Hiki Hills, northeastern marginal area of the Kanto Mountains. *Journal of the Geological Society of Japan* **93**, 201–15. [In Japanese with English abstract.]

Takahashi, Y. & Nishioka, Y. 1994. Mode of plagioclase twinnings in Ryoke metamorphic rocks in the western area of Tsu City, Mie Prefecture. *Journal of the Japanese Association of Mineralogists, Petrologists and Economic Geologists* **89**, 261–8. [In Japanese with English abstract.]

von Goerne, G., Franz, G. & Robert, J.-L. 1999. Upper thermal stability of tourmaline + quartz in the system MgO-Al_2O_3-SiO_2-B_2O_3-H_2O and Na_2O-MgO-Al_2O_3-SiO_2-B_2O_3-H_2O-HCl in hydrothermal solutions and siliceous melts. *Canadian Mineralogist* **37**, 1025–39.

Wakita, K. 1987. The occurrence of Latest Jurassic – Earliest Cretaceous radiolarians at the Hida-Kanayama area in the Mino terrane, central Japan. *Journal of the Geological Society of Japan* **93**, 441–3.

Werding, G. & Schreyer, W. 1984. Alkali-free tourmaline in the system MgO-Al_2O_3-B_2O_3-SiO_2-H_2O. *Geochimica et Cosmochimica Acta* **48**, 1331–44.

White, R. W., Powell, R. & Holland, T. J. B. 2001. Calculation of partial melting equilibria in the system Na_2O–CaO–K_2O–FeO–Al_2O_3–SiO_2–H_2O (NCKFMASH). *Journal of Metamorphic Geology* **19**, 139–53.

Wilke, M., Nabelek, P. I. & Glascock, M. D. 2002. B and Li in Proterozoic metapelites from the Black Hills, U.S.A.: implications for the origin of leucogranitic magmas. *American Mineralogist* **87**, 491–500.

Wolf, M. B. & London, D. 1997. Boron in granitic magmas: stability of tourmaline in equilibrium with biotite and cordierite. *Contributions to Mineralogy and Petrology* **130**, 12–30.

Yoshida, F., Takahashi, Y. & Nishioka, Y. 1995. Geology of the Tsu-Seibu district. With geological sheet map at 1:50,000. Geological Survey of Japan. [In Japanese with English abstract.]

Yoshizawa, H., Nakajima, W. & Ishizaka, K. 1966. The Ryoke metamorphic zone of the Kinki district, Southwest Japan: accomplishment of a regional geological map. *Memoirs of the College of Science, University of Kyoto, Series B* **32**, 437–53.

TETSUO KAWAKAMI, Department of Earth Sciences, Faculty of Education, Okayama University, 700–8530, Okayama, Japan.
e-mail: tetsuo@cc.okayama-u.ac.jp

Present address: Department of Earth and Marine Sciences, Faculty of Science, The Australian National University, Canberra, ACT 0200, Australia.
e-mail: tetsuo.kawakami@anu.edu.au

MS received 3 September 2003. Accepted for publication 6 September 2004.

Transactions of the Royal Society of Edinburgh: Earth Sciences, **95**, 125–140, 2004

Low- and high-temperature granites

B. W. Chappell, A. J. R. White, I. S. Williams and D. Wyborn

ABSTRACT: I-type granites can be assigned to low- and high-temperature groups. The distinction between those groups is formally based on the presence or absence of inherited zircon in relatively mafic rocks of a suite containing less than about 68% SiO_2, and shown in many cases by distinctive patterns of compositional variation. Granites of the low-temperature group formed at relatively low magmatic temperatures by the partial melting dominantly of the haplogranite components *Qz*, *Ab* and *Or* in H_2O-bearing crustal source rocks. More mafic granites of this type have that character because they contain restite minerals, often including inherited zircon, which were entrained in a more felsic melt. In common with other elements, Zr contents correlate linearly with SiO_2, except sometimes in very felsic rocks, and Zr generally decreases as the rocks become more felsic. All S-type granites are apparently low-temperature in origin. After most or all of the restite has been removed from the magma, these granites may evolve further by fractional crystallisation. High-temperature granites formed from a magma that was completely or largely molten, in which zircon crystals were not initially present because the melt was not saturated in that mineral. High-temperature suites commonly evolved compositionally through fractional crystallisation and they may extend to much more mafic compositions through the production of cumulate rocks. However, it is probable that, in some cases, the compositional differences within high-temperature suites arose from varying degrees of partial melting of similar source rocks. Volcanic equivalents of both groups exist and show analogous differences. There are petrographic differences between the two groups and significant mineralisation is much more likely to be associated with the high-temperature granites. The different features of the two groups relate to distinctive source rock compositions. Low-temperature granites were derived from source rocks in which the haplogranite components were present throughout partial melting, whereas the source materials of the high-temperature granites were deficient in one of those components, which therefore, became depleted during the melting, causing the temperatures of melting to rise.

KEY WORDS: Cumulate, fractional crystallisation, mineralisation, restite, zircon cores, zircon saturation temperature, zirconium.

Granites outcrop over an area of 63 000 km^2 in the Lachlan Fold Belt (LFB) of southeastern Australia, and their distribution is shown on the map of Chappell *et al.* (1991). Those rocks cover about 20% of the area of exposed Palaeozoic rocks in the LFB. Related volcanic rocks are also an important component of the LFB and many of the statements that the present authors make in this paper also apply to those extrusive rocks, when the volcanic rocks are not specifically mentioned. These igneous rocks have been subdivided into I- and S-types (Chappell & White 1974, 1992, 2001) and the S-types comprise a little more than half of the total area of these rocks. Most of the other granites are I-type, with A-type granites (King *et al.* 2001) comprising only 0·6% of the total area of granite in this region.

It has been recognised for some time (e.g. Wyborn *et al.* 1987; Chappell *et al.* 1987) that there are two fundamentally distinct varieties of I-type granite in the LFB. These two groups differ, for example, in some petrographic features, in the patterns of chemical variation within comagmatic or cogenetic suites, in the relative proportions of felsic, intermediate and mafic rocks, and in the presence or absence of older inherited zircons. These differences reflect differences in the temperatures of formation of the granite magmas with consequent differences in the physical properties of the magmas and the way in which they evolved compositionally. In an earlier paper (Chappell *et al.* 1998), the authors distinguished these two groups, and referred to them as low- and high-temperature I-type granites. It is the purpose of the present contribution to further examine this subdivision, which is fundamental in terms of the origins of the granites and their prospectivity for related mineralisation. The present authors will also examine some criticisms of this subdivision that have been made since the 1998 paper. The terms low- and high-temperature will be used throughout this paper, although, in a sense, their validity will not be confirmed until the arguments for the existence of two such groups have been presented.

S-type granites, at least from all observations made to this time, are apparently always of low-temperature origin, and close analogues in that sense to the low-temperature I-type granites. Although there is no known group of high-temperature S-type granites with which those low-temperature S-type granites can be compared, the authors will examine the evidence for those rocks having formed at the lower magmatic temperatures.

In this paper, reference is made to the 'linear' compositional variations that are a distinctive feature of the low-temperature granite suites of the LFB. Such variations could, in principle, have resulted from processes of magma mixing or mingling. However, there are other compositional features of the granites of the LFB that preclude such processes having played any role in generating the larger scale variations within these suites of granites (Chappell 1996a; see section 3 of Chappell 2004). Those observations are probably more generally applicable.

The distinctive compositional differences between the two groups of I-type granite of the LFB are summarised in Table 1.

Table 1 Contrasting compositional characteristics of low- and high-temperature I-type granites of the Lachlan Fold Belt

Low-temperature	High-temperature
Rocks with less than 60% SiO_2 uncommon, with lowest SiO_2 contents near 55%	Mafic rocks common, with compositions ranging down to around 50% SiO_2
Compositions have a single SiO_2 mode near 68%, with a weak one near 76%	Compositions are strongly bimodal, with modes at about 60% and 75% SiO_2
Linear or 'simple' element variations	Curved or 'intricate' element variations
Compositional inflexions sometimes seen when fractional crystallisation occurred at felsic compositions	Compositional inflexions at compositions near 68% SiO_2 for elements such as Zr and Ba are characteristic
Volcanic rocks match the plutonic rocks closely in compositions	Volcanic rocks may be fractionated melt compositions and more felsic than the cumulate plutonic rocks

Of all of those features, it is the behaviour of Zr that provides the key to confirming the origins of those groups and that will now be discussed.

1. Contrasting patterns of Zr variation

There are two distinct patterns of Zr variation within the I-type granite suites of southeastern Australia. In one case, the most mafic rocks of a suite generally have SiO_2 contents below 55%, and from a low level of Zr at those compositions, the values rise with increasing SiO_2, with an inflexion in abundances if the SiO_2 contents pass above about 68% SiO_2. Such variations are illustrated by two of the diagrams for high-temperature granites on the left-hand side of Figure 1. For the low-temperature granites on the right-hand side of that figure, mafic rocks are less abundant, and for only one analysed sample is the SiO_2 content of such a rock less than 55% (54·4%). Linear variations against SiO_2 are characteristic of all elements, and the Zr contents generally decrease with increasing SiO_2. In a few cases, the Zr abundances remain fairly constant, and in rare instances, they increase with increasing SiO_2. Our understanding of this contrasting behaviour of Zr abundances has been greatly facilitated by zircon solubility experiments (Watson & Harrison 1983) and by small-scale U-Pb dating of zircon crystals using the Sensitive High Resolution Ion Microprobe at the Australian National University (SHRIMP) ion microprobe (Williams 1995), both discussed below.

The uppermost two suites in Figure 1, Boggy Plain (36 km^2) and Inlet (13 km^2) are both potassic and comprise single concentrically zoned plutons that range from gabbro through quartz monzodiorite to monzogranite in the case of Boggy Plain (Wyborn *et al.* 2001), and from mafic monzodiorite through monzonite to quartz monzonite for Inlet (Chappell 1966). The other four suites shown in Figure 1 are less potassic and more calcic in composition, and vary from quartz diorite or tonalite to granodiorite or monzogranite. Marulan (151 km^2; Carr *et al.* 1992), Cobargo (208 km^2) and Jindabyne (120 km^2; Hine *et al.* 1978) all contain several plutons (eight, four and nine respectively). Towgon Grange is a body of relatively mafic Cordilleran-like tonalite to granodiorite compositions with an area ~65 km^2 (Bryant *et al.* 1997). Together with the Inlet pluton, Towgon Grange is located in the younger New England Fold Belt to the northeast of the LFB, which is where all of the other suites referred to here are located.

The Boggy Plain pluton (Fig. 1) exemplifies one of the patterns of Zr variation. After Zr in that pluton has increased in amount by about a factor of five from the most mafic rocks, the compositions show an inflexion at close to 66% SiO_2. Because of their high Cr and low Zr contents, and the small positive Eu anomaly of one sample, at least the more mafic rocks of this body must represent cumulate rather than melt compositions (see also section 3.2), and this is thought to be the case for almost all of the rocks of the Boggy Plain pluton. The fact that the pluton is progressively zoned (with one major break) from 50·1% to 74·8% SiO_2, implies that all except perhaps the most felsic rocks formed as cumulates. The increase in Zr abundances with increasing SiO_2 in the early cumulates shows that zircon was not saturated in the melts from which those rocks formed, although some zircon eventually crystallised from trapped intercumulus melt. As the Zr concentrations in the main body of melt increased, the amounts of Zr in the more evolved trapped melts and, hence, in the cumulate rocks, also rose. Separated zircon grains from these more mafic rocks have irregular shapes that were apparently imposed by the earlier-formed minerals between which they crystallised, confirming their late crystallisation from trapped melt. Zircon became saturated in the melt at a composition corresponding to the inflexion in Zr abundances in the cumulate rocks. That happened both because of falling temperature and the changing composition of the melt, i.e. its bulk composition, but particularly, the increasing Zr content. Beyond that point, the melts remained saturated in zircon, and Zr decreased in abundance as that mineral separated from the melts and became incorporated in the cumulus minerals. There are several other zoned plutons in the BPS, but Boggy Plain is the one that has been studied in detail (Wyborn 1983). The other zoned plutons of the BPS may vary in the details of their compositional evolution, but all have in common an origin as cumulate rocks formed by sidewall crystallisation (Wyborn *et al.* 2001).

The pattern of Zr variation for the Marulan Suite is very similar to that of Boggy Plain, but with fewer felsic rocks. In contrast to Boggy Plain, the plutons of the Marulan Suite are relatively homogeneous and most of the variation within the suite results from the differences between the eight plutons. The present authors ascribe most of the variation in these rocks either to fractional crystallisation at depth, or to different degrees of partial melting of similar source rocks. Similar patterns of variation in Zr contents have been generated in a magma chamber at shallow levels in the crust (Boggy Plain) and at deeper levels (Marulan), and the mechanisms that generated those variations may have been different. The precise mechanism for the Marulan Suite is not immediately relevant to this discussion and will not be examined further here. The significant observation here is that the trend of increasing Zr with increasing SiO_2 for Marulan again corresponds to zircon undersaturation, and that of decreasing Zr with increasing SiO_2 to zircon saturation. These interpretations have been confirmed using SHRIMP. For sample NB6, 1550 grains were imaged by cathodoluminescence, no possible

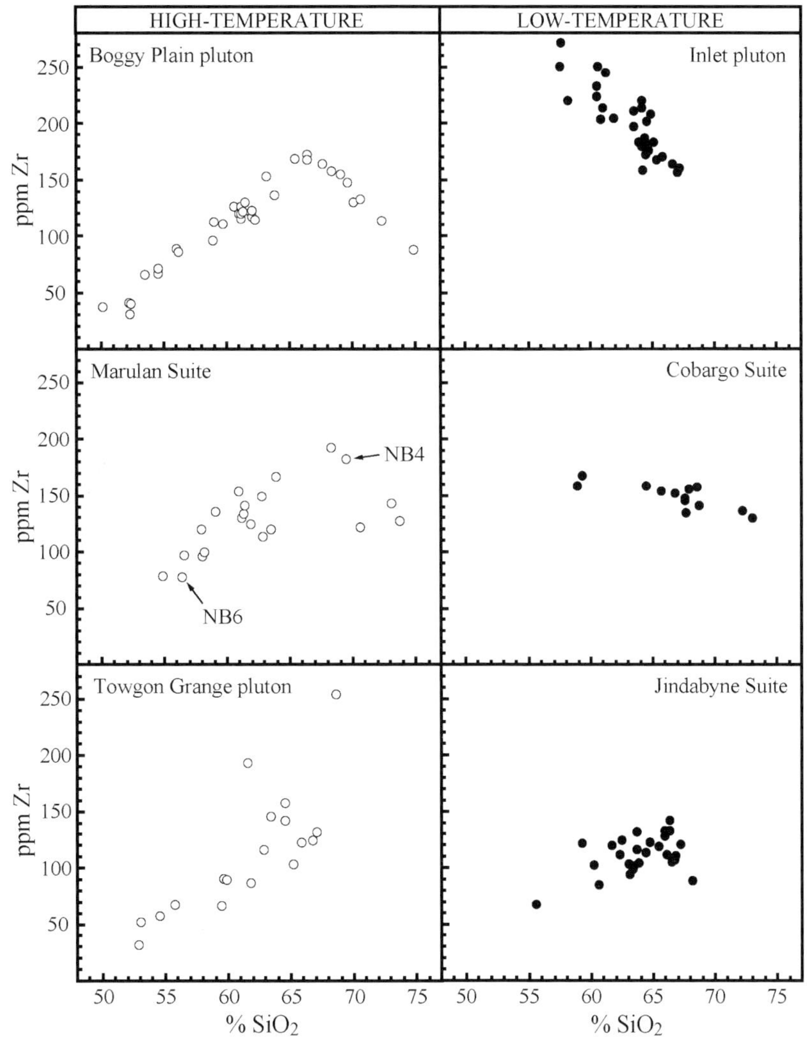

Figure 1 Variation diagrams for Zr in six granite suites of southeastern Australia. The suites on the left-hand side were derived from high-temperature magmas which that initially completely or largely molten. Variation of the other suites resulted from the separation of restite from a more felsic melt at lower temperatures. Age inheritance has been checked for the samples of Marulan labelled NB6 and NB4 using the SHRIMP ion probe (see text).

cores were identified and the 12 grains that were dated all gave Early Devonian crystallisation ages. For NB4, 2040 grains were imaged and 22 possible cores were identified and analysed, of which six proved to be older than Early Devonian (535, 1040, 1080, 1140, 2540 and 2770 Ma).

The critical observation for Marulan at this time is that there is a sequence of compositions that were formed from melts that were in part undersaturated in zircon, and in part saturated. The low-K Towgon Grange pluton shows similar variation, but no rocks are sufficiently felsic for the Zr contents to show an inflexion. The variation in this body has the same possible origins as that in the Marulan Suite.

Like Boggy Plain, the Inlet pluton is also near-concentrically zoned, from 57·5% to 67·2% SiO_2, but in this case the variation is considered to have been dominated by the accretion of restite crystals from the magma (Chappell *et al.* 1987) to the margins of the pluton, so that the rocks become more restite-free and, hence, more felsic towards the centre of the body.

Variation within the Cobargo Suite, like the Inlet pluton, was caused by varying degrees of separation of restite crystals from melt, as was generally the case for suites of the large I-type Bega Batholith (8940 km^2). The small Murrabrine pluton (1·4 km^2) of the Cobargo Suite is the most mafic (~59% SiO_2) intrusion in that batholith. Zr variations in the Inlet and Cobargo suites are distinctly different in character from those of the suites discussed earlier, with Zr having its maximum abundance in the most mafic rocks, decreasing progressively towards more felsic compositions. Such variation is consistent with zircon crystals having been present in the melts involved in the production of those granites which formed under conditions in which the melt was always saturated in zircon. This is confirmed by the presence of rare inherited Precambrian cores in the zircons from the Cobargo Suite. Whether or not the granites of the Inlet Suite contain inherited zircon has yet to be tested.

The Jindabyne Suite illustrates the much less common case in which, based on the presence of inherited zircon, the magma was apparently saturated in zircon (see section 4), but Zr abundances increase slightly with higher SiO_2 contents. The present authors interpret this as resulting from the partial melt being less felsic (~68% SiO_2), so that there was a higher concentration of Zr in such a melt at higher temperatures than in the bulk restite. Hence, the mass fraction of Zr retained in the restite was reduced to a level such that the content of Zr in the bulk restite was less than in the melt. Variation is again ascribed to varying degrees of separation of restite, but from a less felsic melt. It is difficult to see how the variation could have resulted from fractional crystallisation, given that the melt was saturated in zircon and Zr contents rise with increasing SiO_2.

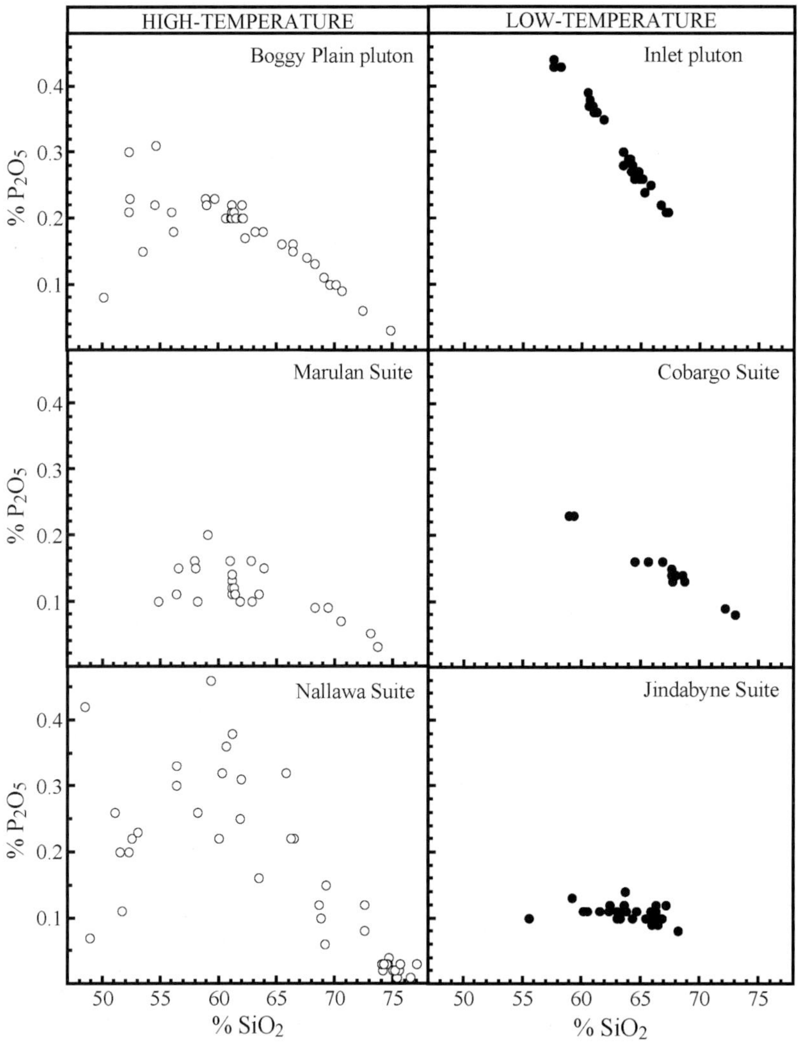

Figure 2 Variation diagrams for P_2O_5 in six granite suites of southeastern Australia. The three suites on the left-hand side are high-temperature in origin and those on the right are low-temperature. However, the pattern of variation for P_2O_5 in the Jindabyne Suite is consistent with a slightly higher temperature of formation than for the other two low-temperature suites.

While the Towgon Grange and Jindabyne suites do show similar Zr-SiO_2 systematics (Fig. 2), these two suites are regarded as high- and low-temperature in origin, respectively. This shows that, while Zr variations may be distinctive for the two groups, as shown by the other four plots in Figure 1, that is not necessarily the case when the Zr abundances increase with increasing SiO_2 contents in more mafic rocks and felsic rocks are not seen. Apart from Zr, there are significant compositional differences between these two suites. Towgon Grange has compositions that are more evenly distributed throughout their total range, and Jindabyne does not include any analysed rocks with SiO_2 contents of less than 55%. However, while those differences might possibly reflect the nature of the exposures of Jindabyne that are available for sampling, there are other elements for which that is very unlikely. Towgon Grange has a significantly higher ratio of Na/K, and substantially lower abundances, i.e. about half, of the elements Rb, Nb, Pb, Th, U and the light rare-earth elements (REE). In cases where discrimination using Zr abundances is ambiguous, the distinction of high- and low-temperature granites can generally be made using other criteria, as will now be discussed below.

Plagioclase crystals probably provide the clearest petrographic distinction between the two groups (see section 7; White & Chappell 2004). Except for the most felsic rocks, the plagioclase crystals in low-temperature granites commonly have corroded cores, with fairly uniform calcic compositions, surrounded by complexly zoned plagioclase. Such crystals are a feature of the Jindabyne Suite. Plagioclase crystals in the high-temperature granites are generally better-shaped and are uniform in composition. The corroded cores that are a feature of the low-temperature granites are not present, but in some cases, including the sample examined by White & Chappell (2004) from Towgon Grange, the plagioclase crystals have large uniform cores surrounded by relatively narrow more sodic rims. Such crystals are distinct in appearance from those with broader rims that surround corroded cores in the low-temperature granites, and are thought to indicate rapid cooling of the granites before they had fully crystallised. The presence or absence of inherited zircon is taken to be diagnostic of the low- and high-temperature granite suites (see 5.1). Such older zircons occur in the Jindabyne Suite, which is therefore certainly low-temperature. Their possible occurrence has not been checked in Towgon Grange, a member of the Cordilleran tonalite-like Clarence River Supersuite of Bryant *et al.* (1997); inherited zircon crystals are typically absent from such rocks.

Variations in Zr contents and conclusions that can be drawn about zircon saturation or undersaturation have the potential to provide information about the temperatures at which at least some of these rocks formed, using zircon saturation temperatures calculated from the information of Watson & Harrison (1983). However, before doing that, objections that have been made to the use of such temperatures for these rocks must be examined (see sections 3 and 4).

2. Contrasting patterns of variation for other elements

Elements other than Zr also show characteristic differences in their patterns of variation between the low- and high-temperature granite groups. Of these, the most instructive are P and Ba.

2.1. Phosphorus variations

Variations in P_2O_5 for six granite suites of southeastern Australia are shown in Figure 2. Cobargo and Inlet have patterns that are very characteristic of low-temperature I-type suites, with P falling sharply and in a very regular fashion from the most mafic to the most felsic compositions. Such variation results from the low solubility of P in metaluminous to weakly peraluminous felsic low-temperature granite melts (Harrison & Watson 1984; Mysen *et al.* 1999). This is consistent with the present authors' view that the magmas which produced all the rocks of those two suites comprised felsic melt and more mafic restite, so that the higher P contents of the less felsic rocks result from greater proportions of restite. Variation for the Jindabyne Suite is similar, but with a much lower gradient. That case probably corresponds to higher temperature felsic melt that had a higher P content that was closer to that of the associated restite.

P_2O_5 abundances in the most mafic rocks of the Boggy Plain and Marulan suites are scattered, but at more felsic compositions the contents fall very systematically with increasing SiO_2. For those more felsic compositions, the variation is analogous to that in the low-temperature granites. P for the Nallawa Suite is similar to the other two high-temperature suites, but with much more scatter. The Nallawa Suite is potassic, and grouped with Boggy Plain within the Boggy Plain Supersuite (BPS) (Wyborn *et al.* 2001). Rocks of this suite cover an area of about 130 km^2 in the Yeoval Batholith in the northern part of the LFB, ranging from diorite through to very felsic one-feldspar hypersolvus granites. The more felsic rocks are separated from the diorites by an approximately horizontal contact, which Wyborn *et al.* (2001) interpreted as the boundary between cumulate rocks in a zoned pluton and overlying felsic granites formed from the fractionating melt that escaped to the top of the magma chamber.

2.2. Barium variations

Ba variations in low- and high-temperature granites has been discussed in some detail by Chappell *et al.* (1998), and are illustrated for selected granite suites in Figure 3. This element behaves in a similar way to Zr in the high-temperature suites, but the variation is not related simply to saturation or undersaturation in a single mineral. During the initial stages of development of progressive cumulate compositions in the Boggy Plain pluton, plagioclase was the only precipitating mineral that contained Ba, and the bulk partition coefficient of the cumulus minerals was much less than one, so that the Ba contents of the remaining melt progressively increased (Fig. 3). The later appearance of biotite as a significant liquidus mineral led to both the melt and the rocks that formed from it being subsequently depleted in Ba, at rock compositions near 67% SiO_2. Variation in Ba in the Nallawa Suite is similar (Fig. 3), except that the inflexion in Ba contents occurs at the significantly higher SiO_2 content of about 74%. For Nallawa, the decrease in Ba abundances could have resulted from the appearance of K-feldspar as a liquidus mineral, rather than biotite, when the fractionating melt reached a haplogranite composition. The three Ba data that lie well above the main trend for the Nallawa Suite perhaps represent cumulative rocks formed by fractionation from that haplogranite melt, although that suggestion is not supported by higher Sr abundances. These observations for Boggy Plain and Nallawa are consistent with the formation of cumulate rocks by fractional crystallisation. That could also be so for the Marulan Suite, but in that case it could also have resulted from different degrees of partial melting, with biotite and K-feldspar, and therefore Ba, being largely removed from the source rocks during the earlier stages of that anatexis.

Ba variation in the low-temperature granite suites is always linear with increasing SiO_2 contents, sometimes increasing (e.g. Cobargo and Jindabyne) and sometimes decreasing (e.g. Inlet). When the Ba contents increase, some Ba was present in restite plagioclase, but that element was more abundant in the melt. When Ba abundances decrease with increasing SiO_2, a phase or phases that contain significant amounts of Ba, i.e. K-feldspar and/or biotite, was present in the restite.

The effects of later fractional crystallisation are sometimes observed in the most felsic rocks of those suites as rapidly decreasing Ba contents, but an example of such variation is not shown in Figure 3. The consistent linear variations at less-evolved compositions are distinct from the variations in the high-temperature suites and are difficult to reconcile with fractional crystallisation, as discussed in some detail for accumulative rocks by Chappell (1996b).

3. Accumulative granites in the Lachlan Fold Belt

Clemens (2001), in discussing S-type granites, stated that 'it is likely that the more mafic granites are cumulates, rather than former liquids'. He then correctly pointed out that zircon saturation temperatures calculated for such compositions would be meaningless. This is a legitimate concern which must be addressed. The present authors were aware of this possible problem when they proposed the low- and high-temperature groups in 1998. However, they also knew of strong arguments against any of the low-temperature granites of the LFB, both S- and I-type, being cumulates, which they did not enumerate at that time. The authors will examine these after here considering the terminology for such rocks and summarising the evidence that many of the high-temperature granites of the LFB did indeed have such a cumulate origin.

3.1. Cumulate, cumulative and accumulative rocks

The terminology for what are referred to here as cumulate, accumulate and accumulative rocks has recently been discussed by Chappell & Wyborn (2004). The use here of the term cumulate follows that of Irvine (1982) who stated that, 'A cumulate is defined as an igneous rock characterised by a cumulus framework of touching mineral crystals or grains that were evidently formed and concentrated primarily through fractional crystallisation.' The space between the cumulus minerals is occupied by material that crystallised from intercumulus liquid. Since some granites formed in precisely that way, it is desirable to use the term cumulate in exactly the way proposed by Irvine (1982).

Chappell & Wyborn (2004) have proposed that the term 'cumulative' be used for granites with compositions that resulted in part from the concentration of crystals within a melt, where those crystals did not form a cumulus framework. In general, such rocks result from any process that fractionates

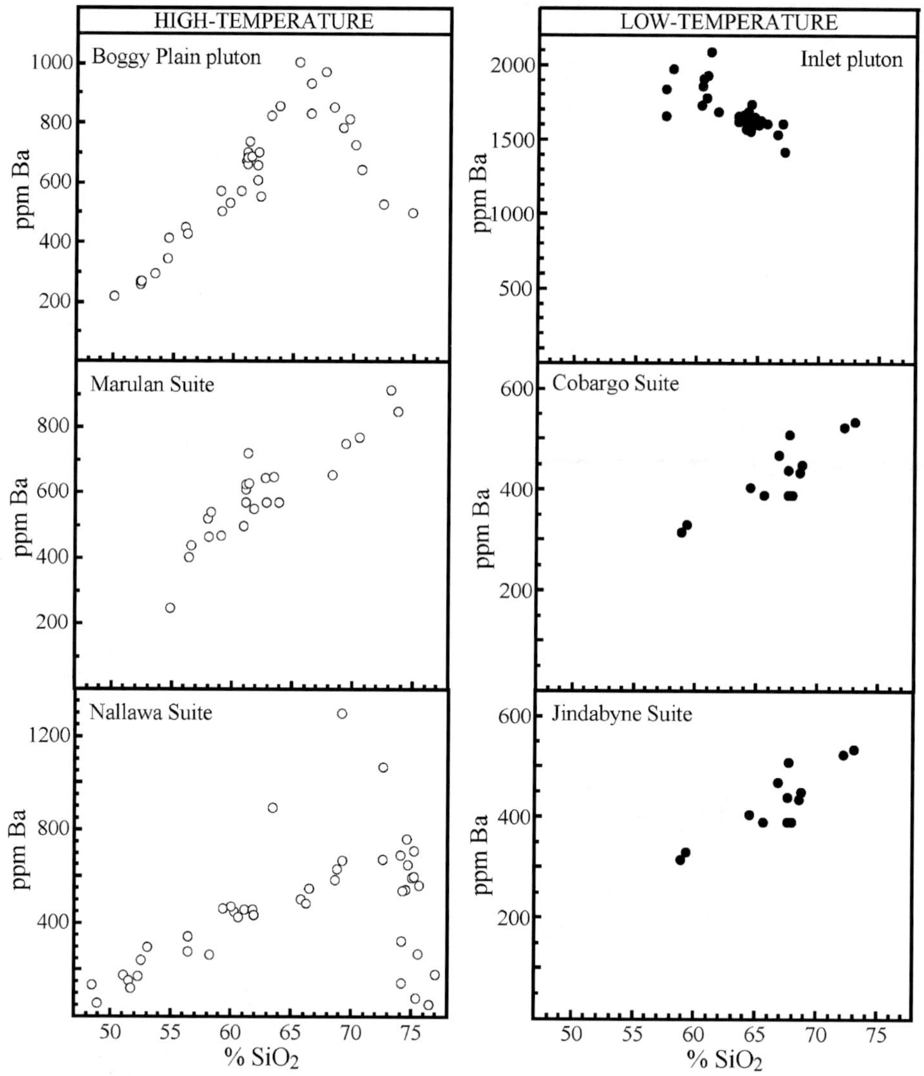

Figure 3 Variation diagrams for Ba in six granite suites of southeastern Australia. The three suites on the left-hand side are high-temperature in origin and those on the right are low-temperature.

melt from a mixture of melt and crystals, such as filter differentiation (Propach 1976) and sorting during flow. In some rocks, there is compositional evidence for the concentration of crystals by their movement relative to the supernatant melt in which they crystallised, but the cause of that movement is not known in that instance. Examples are positive Eu anomalies, or high abundances of Th and other elements resulting from the concentrating of monazite. Following Chappell & Wyborn (2004), the term 'accumulative' will be used here for such rocks.

Clemens (2001) regarded the mafic granites of the LFB as cumulates (accumulative in the present usage). McCarthy & Groves (1979) had previously proposed that the granites of the Blue Tier Batholith of Tasmania are cumulate rocks, and pointed out that the alternative scenario in which the granites represent melt compositions, would imply that 'each pluton was formed by a vast number of small separate intrusions, analogous to separate lava flows, each one different from the others'. Furthermore, they noted that this 'demands a second magma chamber at depth where fractional crystallisation occurred', an argument that could be applied more broadly to the other batholiths of the LFB. McCarthy & Groves (1979) were of the view that variations within the granites of the Blue Tier Batholith resulted from fractional crystallisation, so that the granites had to be either accumulative or represent crystallised melts. They did not recognise the other possibility that those granites formed from restite-bearing magmas that underwent crystal fractionation of restite from melt. There are some extremely evolved felsic granites in the Blue Tier Batholith that clearly represent melts formed by fractional crystallisation. But the less felsic granites are low-temperature in origin, and as the present paper shows, must therefore have evolved compositionally by the fractionation of restite. Apart from the minor occurrences of garnet-bearing cumulative rocks in the Blue Tier Batholith (see section 3.3), those granites have no compositional features which would suggest that they are accumulative.

3.2. Cumulate rocks of the Boggy Plain Supersuite

The rocks of the BPS show compositional variations that, in the present authors' view, unquestionably resulted from the production of cumulate rocks by fractional crystallisation. The observations, mainly compositional, which lead to that conclusion will now be summarised, so that the BPS can then be compared to and contrasted with other granite suites of the LFB.

The BPS comprises more than 25 plutonic complexes with a total area of approximately 2300 km^2, together with related volcanic rocks. These rocks are the most compositionally diverse among the granites and associated rocks of the LFB, ranging from 45·3% to 77·4% SiO_2 (Fig. 4). Two-pyroxene geothermometry on the cumulate rocks of the Boggy Plain

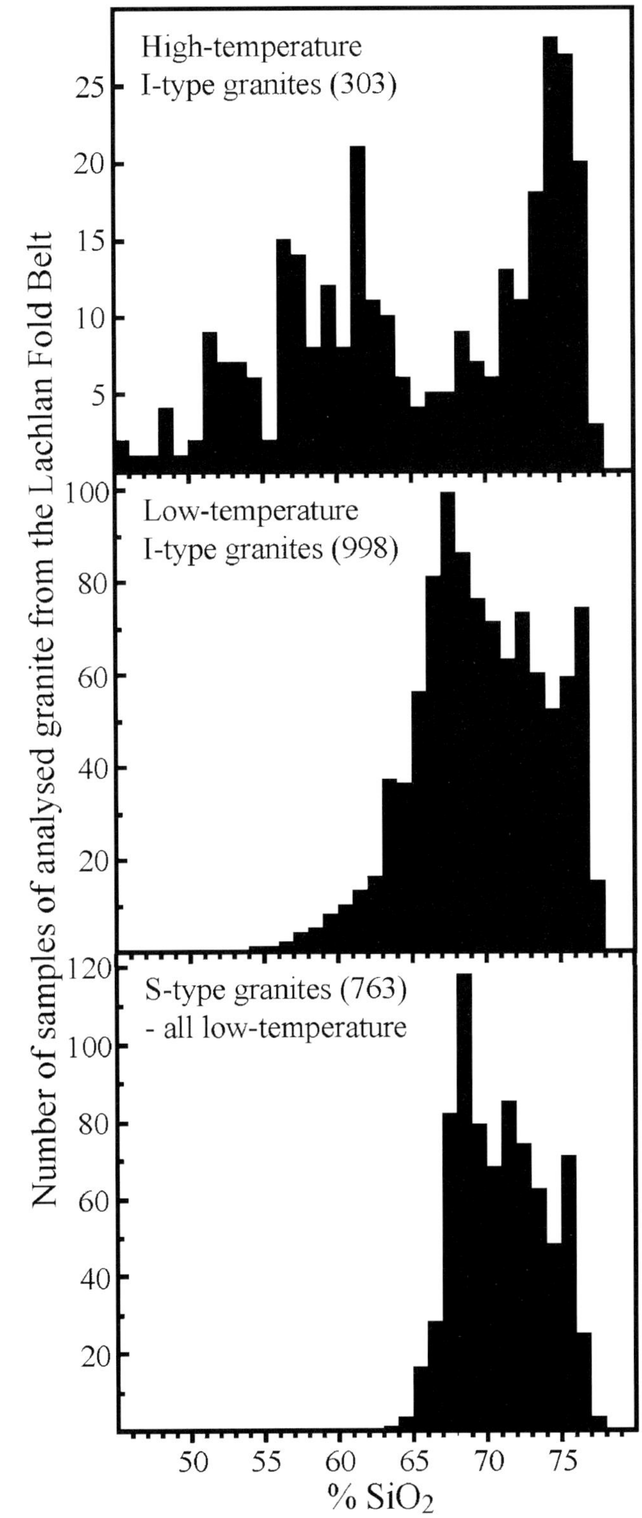

Figure 4 Histograms showing SiO_2 abundances in analysed high- and low-temperature I-type and S-type granites of the Lachlan Fold Belt.

pluton and magnetite-ilmenite geothermometry on volcanic rocks of the BPS indicate high temperatures for both, commonly in excess of 1000°C (Wyborn & Chappell 1986). That the more mafic rocks of the BPS represent cumulate rather than melt compositions is confirmed by extreme values for several elements, with a highest Cr content of 850 ppm, and lowest values for K_2O (0·06%), P_2O_5 (0·01%), Rb (0·8 ppm), Ba (11 ppm) and Zr (8 ppm) in the gabbros of the supersuite. None of these values could be those of mafic melts, and they suggest that the rocks are cumulates in the sense of Irvine (1982), with small amounts of intercumulus melt. That is confirmed by the continuous cryptic zoning in plutons such as Boggy Plain where, despite the extreme compositional range, there is only one discontinuity in the zoning at an internal contact (Wyborn *et al.* 2001). More generally, compositional trends within the BPS clearly indicate that the rocks formed progressively either as cumulates from an originally liquid or largely liquid magma, or else by direct crystallisation of the fractionated and more felsic melt. Many elements show trends that have inflexions, or are curved. W-Nb-Ta-Bi mineralisation is present in the most felsic rocks, which have melt compositions, as a result of those elements becoming concentrated in the melt during prolonged removal of crystals during fractional crystallisation. The volcanic rocks of the BPS are crystal-poor and more felsic, and they correspond to fractionated liquid lost during formation of the plutonic rocks (Wyborn & Chappell 1986).

The Boggy Plain pluton and the BPS in general show that fractional crystallisation can produce a wide range of compositions for several elements. Because granites of the BPS are so distinctly different in their mode of variation from others in the LFB, apart from the Marulan Suite, any suggestion that the variation within granite suites of the LFB arose generally by fractional crystallisation, and that those other suites contain cumulate rocks, or more generally accumulative rocks, must be examined very critically, as will now be done.

3.3. Arguments for the low-temperature granites not being accumulative

We have seen that mafic rocks of high-temperature granite suites, which are always I-type, may be cumulates, and that is always the case for the BPS. But that does not seem to be the case for the much more abundant low-temperature I-type suites in the LFB, and for the S-type granites, for four main reasons.

3.3.1. Rarity of low-silica rocks in low-temperature suites. Low-silica rocks that are cogenetic with granites, apart from those of the BPS and the Marulan Batholith, are very rare in the LFB (Fig. 4). About 26% of the analysed rocks of the BPS contain less than 59% SiO_2. Excluding the high-temperature granites of the Marulan Batholith, no other I-type rocks in the LFB contain less than 54·4% SiO_2 and only 0·8% of the total samples have less than 58% SiO_2. There is only one analysed S-type granite that contains less than 63% SiO_2. That sample with 57·6% SiO_2 has a concentration of garnet (Chappell & Wyborn 2004) and is quite distinct from other more mafic S-type granites of the LFB. In none of the other analysed granites did mafic minerals such as biotite (SiO_2 ~35%), hornblende (SiO_2 ~47%), pyroxenes (SiO_2 ~50%) or cordierite (SiO_2 ~50%) concentrate to the exclusion of minerals such as plagioclase (SiO_2=66·3% at An_{50}) and quartz. Rocks of the BPS are bimodal in SiO_2 content (Fig. 4), with mafic cumulate rocks and the complementary fractionated melt compositions being the most abundant, and only 10·4% of analysed rocks have compositions between 65% and 70% SiO_2. In contrast, for the other I-type and the S-type granites, 39·8% and 42·3% of the analyses, respectively, have SiO_2 contents in that interval.

3.3.2. Compositional equivalence of plutonic and volcanic rocks. Comagmatic plutonic and volcanic rocks of the LFB exhibit two distinct relationships (Wyborn & Chappell 1986). In the first case, the more mafic plutonic rocks represent crystal cumulates, and the volcanic rocks and the exclusively felsic plutons are the complementary fractionated melt. This situation is illustrated by the BPS, which includes rhyolitic lavas, the Mountain Creek Volcanics, which are much more felsic than the comagmatic or cogenetic cumulate plutonic rocks. In the second case, the plutonic and volcanic rocks can be matched fairly closely, and the compositional differences that distinguish between plutonic suites are also found among the volcanic suites. These comprise the low-temperature suites, and Wyborn & Chappell (1986) argued that those rocks must represent true magma (i.e. melt ± restite) compositions and cannot be cumulative rocks produced during fractional crystallisation. The volcanic rocks are not more felsic with a greater proportion of melts from which early-formed crystals had been removed, as would be expected if the more mafic granites are accumulative, and which is seen for the BPS. The equivalence in chemical compositions between plutonic and volcanic rocks of the low-temperature granite suites of the LFB strongly suggests that none of the former is accumulative.

3.3.3. Compositional variations within granite suites. Apart from the granites of the BPS and the Marulan Batholith, the variations for many elements within granite suites of the LFB are not consistent with those suites having formed progressively as a sequence of accumulative rocks. In no I-type suites that are regarded as low-temperature do elements such as P, Zr and Ba, that may appear as an important component of new liquidus phases during continuing fractional crystallisation, show inflexions on Harker diagrams. The low-temperature granite suites are characterised by linear trends on such diagrams. This implies that during fractionation, within a suite, the same minerals are generally being removed and in fixed relative proportions, as would generally be expected for restite minerals. If the separating minerals were being removed after progressive crystallisation, then the bulk compositional trends of the cumulative rocks that formed would reflect changes in mineral proportions and compositions. Chappell (1996a) modelled the variations of some trace elements in I-type suites of the Bega Batholith and the S-type Bullenbalong Suite, and showed that those variations cannot have resulted from the rocks forming as cumulates by fractional crystallisation.

3.3.4. Granites with trace element compositions that complement those of granites lying on a liquid line of descent are rare or absent. In the present authors' view, the less-felsic, low-temperature granites of the LFB evolved in composition through the progressive separation of entrained restite. However, there are some cases in which the more felsic magmas evolved further by fractional crystallisation after most or all of the restite had separated (e.g. Chappell & White 1998). In the latter cases, most of the compositional variation is seen among the trace elements since the major element compositions at that stage are generally close to those of minimum-temperature melts. Such trace element variations were discussed by Chappell (1999), and that paper basically assumed that these represented a sequence of melt compositions. That these changes in trace element abundances are those of melts rather than of a sequence of accumulative rocks is seen more easily for the fractionated S-type granites, in which the P abundances increase as the rock compositions become more evolved. These rocks have strongly peraluminous bulk compositions, with up to about 4% normative corundum. The data of Wolf & London (1994) show that P would be soluble in melts of such compositions, and therefore, behave incompatibly, which is what is observed in these rocks. Other distinctive features of these rocks are the low CaO contents, which decrease to about 0·4%, and the decreasing abundances of Th and the light REE apparently caused by monazite fractionation from the evolving melt (Chappell 1999). If these distinctive compositions are the result of fractional crystallisation, as they seem to be, then accumulative rocks with lower P, and higher Ca, Th and light REE must have also been produced. None of the approximately 770 S-type granites which the present authors have analysed from the LFB show those features, so it must again be concluded that accumulative S-type granites are not present in that area at the level to which the granites are exposed. But they are present in other regions; for example in the Variscan-age Sameiro granite near Braga in northern Portugal, where La abundances can be as much as 500 times chondrites (Dias *et al.* 2000).

Chappell & White (1998) reported on a sequence of fractionated melt compositions in the S-type Koetong Suite of the LFB. In the same area, there are granites of similar bulk composition, but with distinctly higher Sr contents, which the above authors regarded as cumulate rocks of the same suite. Sr isotopic data obtained subsequently show that cannot be the case since the 'cumulate' granites have less radiogenic initial Sr isotopic compositions than the Koetong Suite. They are a separate suite, apparently derived from sedimentary source rocks with higher Sr contents.

Strongly fractionated I-type melt compositions of the LFB are best characterised by their high abundances of Y and the heavy REE, combined, of course, with other broader features, such as high Rb/Sr ratios and low Ca contents. Once again, the complementary accumulate compositions to these are not seen among our approximately 1120 analyses of the dominant Devonian-age I-type granites of the LFB, but may be present in the Carboniferous granites in the northeastern part of the belt, which comprise about 5% of the total area of granite exposures of that region (Chappell & Wyborn 2004).

3.4. Implications of the lack of accumulative rocks among the low-temperature granites

We have seen that there is abundant evidence confirming that the more mafic rocks of what are here termed the low-temperature granite suites of the LFB are not accumulative. Hence, if these rocks are the products of fractional crystallisation, they are melt compositions to which it should be possible to apply zircon saturation thermometry. Before doing that, it is necessary to address the question of equilibrium between zircon crystals and melt, because Clemens (2001, 2003), for example, has argued that equilibrium is unlikely to be attained with respect to zircon and other minerals that are hosts to the trace elements in the rock.

4. The question of equilibrium between zircon crystals and melt

Watson & Harrison (1983) developed an equation relating temperature and the solubility of zircon in silicate melts of known bulk composition. Therefore, the temperature below which zircon would be saturated in a melt, a 'zircon saturation temperature', can be calculated. Such temperatures have previously been used to discuss the origins of granites in the LFB (Chappell *et al.* 1998, 2000).

The role of kinetics. The use of calculated zircon saturation temperatures in this way has been questioned by Clemens (2003). He stated that zircon dissolution into a silicate melt seems to be governed by kinetics rather than equilibrium phase relations and that equilibrium experiments, such as those of Watson & Harrison (1983), may have little bearing on actuality in petrogenesis. It is necessary to consider this further before using data derived from those experiments in the arguments being developed in this paper.

Watson (1996) has considered zircon dissolution and growth quantitatively on the basis of experiments on zircon solubility and Zr diffusivity in granitic melts. He stated unequivocally that dissolution and growth of zircon in nature is not limited by interface kinetics, but by Zr diffusion in the contacting melt. He concluded that only the largest protolith zircons (>120 μm) are likely to survive magmatic events which exceed a temperature of 850 °C. The old zircon cores in the LFB granites are more typically <50 μm in diameter. Also, if these suites evolved by fractional crystallisation, temperatures much higher than 850 °C would have been involved in production of the more mafic rocks, and associated, more felsic granites would have been derived from melts that had passed through that high temperature stage. All of this implies that isolated crystals of zircon would not survive immersion in a melt at temperatures above the zircon saturation temperature during the period involved in partial melting above that temperature, transport of magma, fractionation and cooling.

For the I-type granites, it could be argued that the crystals were once much larger and attained their present size through incomplete dissolution. For the S-type granites, that was apparently not the case. Williams (2001) noted that abrasion textures are preserved on the surfaces of zircon cores in the S-type Cooma Granodiorite and that no cores have the deep embayments normally indicative of partial dissolution. This implies that Zr in the melt phase was derived from fine-grained zircon or from other minerals, and that the larger zircon grains were preserved because that mineral was always present in excess of the amount required for dissolution in the melt, i.e. zircon was saturated in the melt. The Cooma Granodiorite is an uncommon kind of S-type granite in the LFB in that it is situated within a wide, high-temperature metamorphic aureole and it also is compositionally distinct from most of the other S-type granites (White & Chappell 1988). But the comments and conclusions of Williams (2001) about the surfaces of zircon cores also hold for the Jillamatong Granodiorite, part of the voluminous mafic S-type granites of the Bullenbalong Suite, clearly showing again that the melt phase involved in the production of those granites was saturated in zircon. At Cooma, Williams (2001) showed that the new zircon growth on detrital grains in the progressively metamorphosed metasediments starts when the metamorphic grade becomes high enough to generate a partial melt. This apparently involves a transfer of Zr, via the melt, from fine zircon grains and any decomposing Zr-bearing minerals, to the larger zircon grains. That is clearly a scenario in which zircon is saturated in the melt throughout the partial melting and later evolution of that granite. The observations of Jillamatong indicate that this can be applied more widely to other S-type granites.

Recently, Hanchar & Watson (2003), in discussing the experimental results of Watson & Harrison (1983), have stated that 'An implication of these results is that zircon dissolution in an undersaturated (in zircon) water-bearing granitic melt occurs instantaneously on geologic timescales with little possibility that such pre-existing zircon would survive a crustal melting event.'

No positive zirconium anomalies. Clemens (2003) accepted that many zircon crystals in granites are restitic and argued that zircons are located within mafic minerals, mainly biotite, in the source rocks. He stated that 'during partial melting, the biotite in the protolith decomposes, liberating many zircons that because of their small size, can be carried away as the melt rapidly segregates and ascends, leaving behind the coarser-grained restite phases'. Older zircon cores may be rare in the more felsic S-type granites of the LFB (e.g. Elburg 1996). However, for the more mafic S-type granites of that region, almost every crystal of zircon has an older core and those cores together comprise a large fraction (about 45%) of the total zircon content of the rock. But none of these rocks show anomalously high Zr contents, or a 'positive zirconium anomaly', that might be expected in some cases if they formed from a melt in which zircon crystals had been preferentially incorporated as they left the source rocks.

Granites on a liquid line of descent. Clemens (2003) also pointed out that many zircons in granitic rocks are found included in crystals of biotite, and once armoured by magmatic overgrowths, would no longer be vulnerable to attack by the melt. Certainly, biotite is the principal zircon host in many S-type granites, but in I-type granites this is commonly not the case (Williams *et al.* 1983). To examine this argument of Clemens (2003), it is probably best to look at some compositional evidence from the rocks themselves. If the low-temperature granite suites of the LFB evolved by fractional crystallisation, then they represent a sequence of melt compositions and not a sequence of accumulative rocks (see section 3), i.e. the rocks would lie on a liquid line of descent in the sense of Bowen (1928). This means that the more evolved or more felsic rocks would have formed by the equilibrium crystallisation of melt with the same composition as the granite itself, apart from volatile components. That melt would have been derived by the removal of precipitated crystals from less-evolved melts that had compositions corresponding to earlier positions on the line of descent. Individual zircon crystals might not have been removed because of their small size, but larger crystals of biotite that were armouring old zircons crystals that could, therefore, not react with the melt, would have separated. But all of these granites contain older zircons! The implication is that they were not armoured and were in a melt that was at a temperature less than that of zircon saturation.

4.1. Implications of zirconium variations within suites

Four patterns of Zr variation can be recognised within granite suites of the LFB, each with specific implications for zircon saturation or undersaturation, and for the use of zircon saturation temperatures. The first of these is the case of the high-temperature granites; the other three relate to the low-temperature granites. For the latter three cases, the compositional variations are ascribed, except in the most felsic rocks, to the fractionation of restite crystals. The implications of the different patterns of variations for the alternative fractional crystallisation model, will now be considered.

Zr variations with an inflexion. This is characteristic of the high-temperature granites (see section 1) and is illustrated by the diagrams on the left-hand side of Figure 1. Of course, the inflexion is only seen if the rocks as sampled reached sufficiently felsic compositions, which Towgon Grange in Figure 1 did not. In these cases, increasing Zr with increasing SiO_2 corresponds to undersaturation of Zr in the melt, conversely for decreasing Zr. This was confirmed by SHRIMP analysis of possible zircon cores in samples NB6 and NB4 from the Marulan Suite (see section 1). These relationships hold irrespective of whether the compositional variation resulted from fractional crystallisation (Boggy Plain), or from different degrees of partial melting of the source rocks (possibly Marulan and Towgon Grange). For Boggy Plain, all of the rocks are cumulates, except perhaps at the most felsic compositions. This implies the loss of Zr-rich melts from the system, probably as volcanic rocks (see Wyborn *et al.* 2001, fig. 4).

Decreasing Zr contents with increasing SiO_2. This is the most common pattern of variation in the low-temperature suites, both I- and S-type, illustrated by the I-type Inlet and Cobargo suites in Figure 1. If, hypothetically, the compositional changes in these suites were the result of fractional crystallisation, then this pattern of variation would confirm that the melt from which zircon crystals were precipitating was saturated in zircon. This is consistent with the observation that all such suites so far examined with SHRIMP contain inherited zircon. It would likewise show that the concerns raised by Clemens (2003) regarding the use of zircon saturation are not valid. It

Table 2 Average compositions of some low- and high-temperature mafic granites

	Low-temperature		High-temperature
	Mafic Inlet Suite G428 G42 G440*	Mafic Cobargo Suite AB126 AB127†	Mean of 17 PRB analyses (58–60% SiO_2)‡
SiO_2	57·74	59·10	59·03
TiO_2	0·84	0·86	0·86
Al_2O_3	15·87	17·19	16·68
FeO_{total}	5·72	5·40	6·38
MnO	0·12	0·11	0·11
MgO	4·42	3·43	3·56
CaO	6·08	6·44	6·67
Na_2O	3·19	3·78	3·33
K_2O	3·55	1·42	1·56
P_2O_5	0·43	0·23	0·16

*G428, G432 and G440 are the three most mafic analysed rocks (quartz monzodiorites) from the Inlet Suite of the New England Batholith.
†AB126 and AB127 are the two most mafic analysed rocks (quartz diorites) from the Cobargo Suite of the Bega Batholith in the Lachlan Fold Belt.
‡The PRB composition is the mean composition of all analysed granites between 58% and 60% SiO_2 from the Peninsular Ranges Batholith among the data of L. T. Silver and B. W. Chappell.

also means that zircon saturation temperatures calculated for such suites can be used to assess whether or not the rock compositions are those of melts (see section 5).

No significant changes in Zr contents with increasing SiO_2. In a few I-type low-temperature granite suites of the LFB (e.g. Glenbog; see figure in Chen *et al.* 1990), there is no significant change in Zr contents from mafic to felsic compositions. These are here interpreted as cases in which, fortuitously, the temperature of partial melting was such that the Zr content of the melt was close to that of the bulk restite. The compositional variations within the suite resulted from different degrees of separation of those two end-members, and the Zr contents remained unchanged. Any scenario that seeks to produce this type of constancy in Zr by fractional crystallisation would be quite complex, since it would imply not only a balance between the removal of Zr in precipitating minerals and the melt from which they were formed, but also the maintenance of that balance as both the temperature and bulk composition of the melt changed, both factors that affect the solubility of Zr.

Increasing Zr contents with increasing SiO_2. This is illustrated by the Jindabyne Suite in Figure 1. Braidwood is the only other suite in the LFB to show this pattern of variation so distinctly. In the absence of other information, increasing Zr contents could be interpreted to support an origin for the development of variation in these suites by fractional crystallisation, from a melt undersaturated in zircon. But it could also result from fractionation of restite that contains less Zr than the melt, and that is confirmed by the presence of inherited cores in some zircon crystals, implying saturation in zircon. One interesting point that can be made from this pattern of variation is that, if zircon crystals are armoured by crystals of biotite (Clemens 2003) or other mafic minerals, then those other minerals could not have been involved in a fractional crystallisation process to any extent, because the Zr contents would then have decreased. Therefore, if we accept the conclusions of Watson (1996) regarding the instability of crystals of zircon in melts in which they are undersaturated, fractional crystallisation can be ruled out as a mechanism for generating the variation in these suites. That conforms to the present authors' views that were based on other criteria.

5. Zircon saturation temperatures

As noted above, despite their distinctive 'linear' nature, the compositional variations within the low-temperature granite suites of the LFB cannot be the result of magma mixing or mingling (Chappell 1996a). Neither can they be a series of cumulative compositions produced by fractional crystallisation (see section 3). The possibility that these rocks represent sequences of melts resulting from that process, will now be addressed. As shown in the previous section, it is valid to do that using the data on zircon saturation furnished by Watson & Harrison (1983).

5.1. Zircon saturation temperatures of I-type granites

Calculated zircon saturation temperatures (Watson & Harrison 1983) for the two most mafic granites of the Cobargo Suite and the three most mafic rocks of the Inlet pluton are all in the interval from 740 to 762°C. Those temperatures represent the maximum values at which crystals of zircon could have been present in melts having the compositions of those rocks, under equilibrium conditions. However, rocks with bulk compositions close to the most mafic rocks of the Cobargo and Inlet suites (analyses in Table 2) could only have been complete melts at temperatures much higher than ~750°C. This shows that, given that the rocks cannot be accumulative (section 3), the more mafic rocks of the Cobargo and Inlet suites cannot represent melt compositions, and therefore, that variations within those suites certainly cannot have resulted from fractional crystallisation. The present authors' view is that melts were involved in the formation of the Cobargo and Inlet suites, but they were much more felsic than the mafic bulk rock compositions. This confirms that the differences in the chemical trends between the two groups of granites in Figure 1 relate to differences in temperature. Thus Boggy Plain and the other suites on the left-hand side of Figure 1 formed from high-temperature molten or largely molten magmas, in which zircon was soluble, except at the more felsic compositions of Boggy Plain and Marulan. In contrast, the other suites in Figure 1 were derived from partially molten magmas, existing at substantially lower temperatures, which consisted of a felsic melt containing both zircon and other restite crystals. All of the compositional features of the six suites are consistent with those observations. This subdivision into two groups of rocks which formed at relatively high or low temperatures is more generally applicable to the I-type granite suites of south-eastern Australia, and leads to their assignment as high- or low-temperature (see section 5.3).

It must be emphasised that the zircon saturation temperatures that are calculated do not tell us the temperatures at which these granites formed. That can only be done if the rock compositions are those of complete melts in which zircon is just saturated, and the first of those requirements does not apply here. These temperatures represent the maximum values at which those rock compositions could have formed as complete melts and still preserved inherited zircon, and through obtaining unrealistically low temperatures, it is argued that the rocks could not have been melts.

5.2. Zircon saturation temperatures of S-type granites

The more felsic S-type granites of the LFB may be free of older zircon (Elburg 1996), and in such cases, any older zircons would have been removed from melts by crystal fractionation. As far as is known, the more mafic S-type granites in the LFB invariably contain large amounts of inherited zircon, and virtually every zircon crystal in these rocks has an older core. The zircon saturation temperatures calculated for the most mafic rocks of the Bullenbalong Suite in the LFB are 810 ± 20°C. An estimate that 45% of the zircon in those rocks is inherited has been used earlier (see section 4), and on the basis of correspondingly lower Zr contents, the calculated zircon saturation temperatures would be less than 750°C. All of these temperature estimates are far too low for these mafic but quartz-rich rocks to have been completely molten. Once again, these data show that the melt component of the Bullenbalong Suite was quite felsic and the mafic rocks have that character because of the presence of abundant entrained restite. Also again, these calculated temperatures are limiting values and do not tell us the actual temperature of any melt.

5.3. The low- and high-temperature granite subdivision

Following Chappell *et al.* (1998), the present authors recognise a fundamental subdivision of granites into those of low- and high-temperature origin, based on the presence or absence of inherited zircon in relatively mafic rocks of a suite, and shown in many cases by distinctive patterns of compositional variation (see sections 1 and 2) and some petrographic features (see section 7). While this subdivision is based on studies in eastern Australia, it is much more widely applicable. I-type granites may belong to either group, but in the authors' experience, all S-type granites are low-temperature, based on the ubiquitous presence of older zircon, even in the less felsic rocks. A-type granites, as recognised in the LFB, and peralkaline rocks are regarded as high-temperature, but they are distinguished as such using different criteria (see section 5.4). Zircon saturation is a function of both the bulk composition of the melt and temperature, so that the distinction between all low- and high-temperature granites would not be at a specific temperature. However, temperature rather than composition is the dominant parameter in determining whether or not zircon would be saturated in melts of various metaluminous granitic compositions.

Overall, low-temperature granites appear to dominate over high-temperature granites, but the relative proportions of the two in various regions does vary widely. About 95% of the plutonic rocks of the LFB are low-temperature, and the figure for the extensive areas of granite in all of eastern Australia would be similar. Likewise, they dominate among the Caledonian granites of Scotland and Ireland, and the largest granite complex of that region, the Leinster granite, is S-type. That is also the case in regions in which S-type granites are common, such as the European Hercynian and Southeast Asia. On the other hand, high-temperature granites dominate throughout the Cordillera, and among the plutonic rocks of island arcs of varying maturity, such as New Britain and Japan.

The more felsic rocks of high-temperature suites may crystallise from melts that developed at high temperatures. An example is provided by the felsic rhyolites of the Mountain Creek Volcanics of the BPS, which have SiO_2 contents up to 73·5% and were erupted at temperatures above 1000°C (see section 6). Many of the felsic granites of the BPS would have formed from melts that were initially produced by fractional crystallisation at similar temperatures. But felsic granites that formed from melts produced by the fractionation of high-temperature melts to progressively lower temperatures are regarded as part of a high-temperature suite. Such felsic rocks are petrogenetically distinct, and should, in detail, be compositionally different from those that crystallised from melts produced by partial melting of older source rocks at low magmatic temperatures or by the fractional crystallisation of such partial melts.

Granites and related volcanic rocks form from magmas which can exist over a wide range of temperatures. Whitney (1988), for example, has estimated the temperature of partial melting involving the breakdown of different hydrous minerals. Tuttle & Bowen (1958) showed that, if free water is available, H_2O-saturated granitic melts may form by partial melting of quartzofeldspathic rocks in the middle and lower crust at temperatures as low as 630°C (Johannes & Holtz 1996). Rather felsic peraluminous melts may form, often in relatively small quantities, as a result of partial melting involving the breakdown of muscovite at temperatures below about 750°C. Production of more voluminous granites, such as the batholithic S-type granites of the LFB (White & Chappell 1988), corresponds to the breakdown of biotite, and melts would have had temperatures from about 750 to 850°C. The production of some I-type granites at lower temperatures could likewise be dominated by the destruction of biotite, and in this context, it should be noted that the most felsic I-type granites are almost universally weakly peraluminous (Chappell & White 1974). More mafic I-type granite magmas that have formed at these lower temperatures are metaluminous because of the presence of entrained restite. The production of I-type granite melts could continue to higher temperatures through the dehydration melting of hornblende. At some point, zircon would be soluble in the more mafic melt compositions and high-temperature granite melts are then being produced. Geothermometry on both the plutonic and volcanic rocks of the high-temperature BPS in the LFB indicates that at least some of those rocks formed at temperatures higher than 1000°C (see section 3.2). Hence, a range of temperatures of close to 400°C is indicated for the formation of those rocks broadly regarded as granitic.

Given that granite forms from magmas that exist over such a wide range of temperatures, it is reasonable to ask why they should be subdivided on the basis of temperature, given that all gradations of temperature within that range must have existed. The present authors propose this subdivision because the differences in temperatures of formation of the two groups correspond to significant diversity in other physical features of the magmas, in the compositions of the melt phase and in the mechanisms by which the granites evolve compositionally. At least in the early stages of their evolution, the high-temperature magmas are completely or largely molten, whereas the low-temperature magmas not only contain crystals of inherited zircon, but also of other restite minerals, immersed in a relatively felsic melt. In the latter case, the magmas change in composition, at least initially, by the fractionation of restite, in complete contrast to the high-temperature magmas that may evolve by fractional crystallisation. Consequently, the high-temperature granites are more prospective for associated mineralisation. Also, as Wyborn & Chappell (1986) pointed out, the granites that we now assign to these two groups have distinctly different compositional relationships with any associated volcanic rocks (see section 6). For these reasons, the low- and high-temperature division has a much broader significance than simply differences in temperature, and therefore, it is worthwhile making this distinction.

The temperature at which the transition from low- to high-temperature granites, defined on the basis of zircon saturation, occurs is not precisely known, for many reasons. The slow cooling of plutonic rocks means that possible geothermometers are reset to lower temperatures. The understanding of the relationship between volcanic and plutonic rocks has not yet reached the stage where the temperatures of volcanic rocks can be widely used. The primary compositions of high-temperature magmas are not known if that magma fractionated into cumulate rocks and more felsic melts. In cases where the mafic high-temperature magmas did not fractionate, they are not saturated in zircon, so the zircon saturation temperature cannot be used. Neither is their precise H_2O content known, which is needed for the liquidus temperature to be estimated.

Low-temperature granite magmas containing a minimum-temperature melt that is generally not saturated in H_2O are produced up to temperatures of 800–850°C, and correspond to the breakdown of muscovite, if present, and biotite. The non-minimum-temperature magmas are formed at higher temperatures (see section 9.1) as the incongruent melting of hornblende progressively supplies H_2O to the melt, in which zircon is still saturated. In the LFB, temperatures in excess of 1000°C have been measured for both plutonic and volcanic rocks of the BPS (see section 3.2) would certainly apply to many other high-temperature granites, but the lowest temperature at which such rocks could form is not known at present.

5.4. A-type granites

The least-fractionated A-type granites of the LFB have very high Zr contents (~500 ppm) with calculated zircon saturation temperatures ~900°C. Therefore, they are regarded as high-temperature, although is possible that those granites have high Zr contents because they crystallised from relatively dry melts and zircon saturation is dependent on water content (P. L. King, pers. comm.). Because these rocks do not generally extend to compositions much less felsic than 70% SiO_2, they are not distinguished as high-temperature using the criteria given in the previous section. The metaluminous to weakly peraluminous A-type granites have been referred to as the 'high-temperature felsic granites' by King *et al.* (2001). The A-type granites can be distinguished, when unfractionated, as evidenced by high Ba contents, by greater abundances of high-field-strength elements, such as Zr. In the LFB, the most Zr-rich unfractionated A-type granites are those of the Gabo Island Suite (Collins *et al.* 1982) and the least felsic rocks of the Danswell Creek Suite (King *et al.* 2001). The Gabo Island rocks are relatively uniform in composition, with an average Zr content of 490 ppm at 73% SiO_2, and calculated zircon saturation temperatures in the range 893 ± 10°C. The least felsic sample of the Danswell Creek Suite contains 535 ppm Zr at 70·5% SiO_2, corresponding to a calculated zircon saturation temperature of 897°C. It is likely that most A-type granites crystallised from melts that were only just saturated in zircon. As with more mafic high-temperature I-type granites, older zircon is probably not present in these rocks.

Peralkaline rocks are classified as A-type in some literature, as indeed they were when the term A-type was originally used by Loiselle & Wones (1979). Such a grouping is not favoured here (King *et al.* 1997) because peralkaline rocks themselves have distinctive and unique compositional features that are fundamentally different from those of the peraluminous to metaluminous granites that are regarded here as A-type. Peralkaline rocks typically have very high abundances of Zr, and are known to form at high temperatures, based on the geothermometry of volcanic rocks. However, it is not possible to examine Zr content relative to zircon-saturation in these rocks because the details of that relationship are unknown.

5.5. Origins of the low- and high-temperature granites

The origins of the low- and high-temperature granite groups, and the reasons for their existence, are examined in some detail by Chappell (2004) and will only be considered briefly here. The high-temperature granites, other than the A-types, are the more compositionally primitive group. They result from partial melting in the deep crust, or perhaps in modified mantle, of mafic rocks that are, in general terms, basaltic or dioritic to monzonitic in composition. They represent a relatively juvenile addition to the upper parts of the crust. Partial melting of lower-K source rocks produces the dominantly tonalitic to low-K granodioritic Cordilleran-like rocks that Pitcher (1982) referred to as Cordilleran I-types and Chappell & Stephens (1988) termed tonalitic I-types. The Clarence River Supersuite of Bryant *et al.* (1997), including the Towgon Grange pluton, is an example. While such rocks are characteristic of Cordilleran-like igneous activity, they are also found less commonly in more mature continental regions; for example, the Marulan Suite of the LFB. In that environment, they could be associated with granites with higher K contents derived from more K-rich source rocks, such as those of the Boggy Plain Supersuite (Wyborn *et al.* 2001). Chappell (2004) ascribed the origin of the high-temperature granites to depletion during partial melting of a haplogranite component, i.e. *Qz* (tonalites and quartz diorites), *Or* (monzonites and monzodiorites) or H_2O (A-type granites).

Low-temperature I-type granites are the products of partial melting of quartzofeldspathic source rocks in which all haplogranite components are available throughout melting. These sources may themselves have been of high-temperature origin. They are the Caledonian I-types of Pitcher (1982) or the granodioritic I-types of Chappell & Stephens (1988). Production of these granites represents a recycling of older components of the crust, and they may be spatially closely associated with S-type granites as they are in parts of the LFB (Chappell *et al.* 2000).

5.6. 'Hot' and 'cold' granites

Miller *et al.* (2003) have proposed 'hot' and 'cold' groups of granites, but these are not equivalent to the two groups that are being discussed here. Those two groups are probably both within this paper's low-temperature variety, and the data set of Miller *et al.* (2003) does not seem to include representatives of the high-temperature granites. In cases where Miller *et al.* (2003) had ion-microprobe data and images of internal zoning, the boundary between their two groups was set at ~10% of grains with cores, whereas here, except for the Zr-saturated members of fractionated high-temperature suites, a single older core would be taken as evidence for zircon saturation and the rock would be assigned to the low-temperature group.

Table 3 Petrographic characteristics of low- and high-temperature granites containing ~60% SiO_2

Low-temperature	High-temperature
Heterogeneous–mafic aggregates or clots	Even distribution of mafic minerals
Some rocks have large crystals of quartz	Quartz is commonly interstitial
Plagioclase crystals have corroded calcic cores (up to An_{80}) with rather uniform compositions	Calcic cores in plagioclase absent or rare. Zoning normally throughout crystals
Most rocks have randomly oriented crystals of plagioclase	Rare and weak parallel orientation of tabular plagioclase crystals
Mostly poorly shaped mafic minerals with rare and small relict clinopyroxene crystals in hornblende	Large tabular hornblende crystals dominated by relicts of partly reacted clinopyroxene

In addition to data on zircons, the present authors also make use of patterns of Zr variations in rock suites as indicators of zircon saturation. The detailed relationship between the present subdivision and that of Miller *et al.* (2003) remains to be elucidated.

6. Volcanic rocks

The granites of the LFB are closely associated in time and space with volcanic rocks. Two types of volcanic rocks can be recognised: those associated with granites that evolved dominantly by restite fractionation; and those related to granites which evolved by fractional crystallisation. These are the low- and high-temperature volcanic rocks, and most of the comments that have been made earlier about the two groups of granites also apply to these rocks.

The chemical compositions of some I- and S-type granites and relatively fresh volcanic rocks of the LFB can be matched very closely. This is seen for low-temperature suites, for which the associated volcanic and plutonic rocks have essentially the same range of compositions (Wyborn & Chappell 1986). For example, S-type volcanic rocks of low-temperature origin are widespread around Canberra. The range of aluminum saturation seen in the S-type granites is mirrored in the volcanic rocks. Strongly peraluminous volcanic rocks with ubiquitous phenocrysts of cordierite can be matched chemically with granites of the Bullenbalong Supersuite. Other weakly peraluminous volcanic rocks do not contain cordierite, and can be matched in composition with adjacent weakly peraluminous S-type granites of the Shannons Flat and Dalgety suites (Wyborn *et al.* 1981). To the east of Canberra, I-type volcanic rocks which are porphyritic in both ortho- and clinopyroxene can be matched closely in composition with I-type granites of the Bega Batholith (Wyborn & Owen 1986). The volcanic rocks of these suites, both I- and S-type, are invariably porphyritic, with the bulk of the phenocrysts (30–60%) being interpreted by the present authors as crystals of restite (Wyborn & Chappell 1986). Phenocrysts of plagioclase have compositions that match those of the relatively uniform plagioclase cores in granites of the same suite, with few outer zones seen in the volcanic phenocrysts.

The high-temperature volcanic rocks of the BPS are sparsely phyric (<10% phenocrysts) rhyolites with minor associated andesites and basaltic andesites. These compositions are exclusively I-type, and the rhyolites complement the more mafic cumulate granites of that supersuite. The rhyolites were derived from the felsic tops of magma chambers that evolved through convective fractionation (Wyborn *et al.* 2001). They erupted at very high temperatures, with temperatures determined by magnetite-ilmenite thermometry, for example, commonly in excess of 1000°C for the rhyolites of the Mountain Creek Volcanics of the BPS (Wyborn & Chappell 1986).

7. Petrographic differences between low- and high-temperature granites

There are distinctive petrographic differences between the low- and high-temperature granites. These have been discussed by White & Chappell (2004) and are summarised for rocks with SiO_2 contents close to 60% in Table 3. High-temperature granites have a more uniform distribution of mafic minerals, which is apparent in both hand sample and thin-section. This feature extends to granites of higher SiO_2 contents. The mafic minerals in low-temperature granites occur in aggregates and clots, which represent recrystallised crystals of primary restite (Chappell *et al.* 1987). At compositions around 60% SiO_2, quartz occurs as larger crystals in the low-temperature granites, whereas it is interstitial in the high-temperature rocks.

When new plagioclase is produced by melt-forming reactions in the source rocks, this is likely to be in equilibrium with the melt because of increased reaction rates with rising temperatures. In low-temperature granites, plagioclase crystals are likely to be retained in the magma and will appear in a granite as uniform cores within strongly zoned outer parts, or as uniform phenocrysts in volcanic rocks. These cores may be quite calcic, near An_{80}, because of the flattening of the plagioclase loop in the presence of quartz (White & Chappell 2004). The first plagioclase to crystallise from the melt will have a similar composition. With slow cooling, early-formed plagioclase will be progressively made over to more sodic plagioclase. However, local disequilibrium is more likely with falling temperatures, leading to the preservation of corroded early plagioclase and complex zoning. This explanation is consistent with experimental findings showing that plagioclase is a particularly difficult mineral to re-equilibrate with a melt at temperatures near the lower end of the range for granite magmas (Baschek & Johannes 1995). Disequilibrium relationships in plagioclase are always seen in low-temperature granites, but are much less common in the high-temperature granites where continuous zoning from the crystal centre to rim is typical.

8. Implications for mineralisation

In general, high-temperature granites are much more likely to be related to mineralisation, for the following reasons:

(1) During the partial melting process at higher temperatures, the solubility of trace metals in the melt will be greater than at lower temperatures.

(2) Granite suites that evolve by fractional crystallisation concentrate incompatible trace metals at that magmatic stage, enhancing the probability of the later significant mineralisation (Blevin & Chappell 1992). High-temperature suites, particularly those that are monzonitic rather than tonalitic, are more likely to undergo

fractional crystallisation, which may increase the abundances of some rare elements in the evolving melt, but will also cause the abundances of others to decrease.

(3) H_2O contents of the melts will also increase with such fractionation, so that mineralising fluids are more likely to be released (cf. the low-temperature suites as discussed by White 2001).

(4) High-temperature magmas are also largely molten magmas, and because of both the higher temperatures and that larger fraction of melt, will have a higher heat capacity and would introduce more heat into the country rocks during cooling.

This situation is illustrated by the granites of the LFB in which the Silurian to Carboniferous granites, which are dominantly low-temperature, are generally not associated with significant mineralisation, except for Sn-W in Tasmania and less important Sn associated with the Wagga Batholith (Blevin & Chappell 1992). In contrast, Ordovician-age high-temperature intrusive rocks in the northern LFB, typically high-K in character, are associated with major Cu-Au porphyry type mineralisation. For S-type granites, this is illustrated by Sn mineralisation associated with the Bullenbalong and Koetong suites of the LFB (Chappell 1996b). The Koetong Suite formed at slightly higher temperatures, and probably for that reason, the primary melt underwent fractional crystallisation after restite had been removed (Chappell & White 1998), so that Sn contents rose to a maximum measured level of 47 ppm. The more mafic granites of the Bullenbalong Suite are similar to Koetong in composition, but only one analysed sample shows slight effects of fractional crystillisation at felsic compositions with a Sn content of 10 ppm. These differences correlate with the reported amounts of Sn obtained from the areas in which those two groups of rocks are found, <10 t for Bullenbalong and ~5 kt for Koetong.

The better understanding of the low- and high-temperature granite groups that we are now developing, along with more criteria to recognise them, should have important implications for mineral exploration.

9. The future: Possible subdivisions of the low- and high-temperature granites?

The present authors have been aware for some time that it will probably be desirable at some stage to make a distinction between two groups within what are here termed low-temperature granites. In assembling and assessing information for this paper, it has likewise become apparent that two distinct groups of high-temperature granites are likely to emerge from further consideration of the data. Such subdivisions and their significance will be a fruitful area for future research and they will now be considered briefly.

9.1. Possible subdivision of the low-temperature granites

The present authors have long known of two distinct groups of granites that are both low-temperature in origin. No attempt has been made to formally subdivide these granites in this paper because, at this stage, the best criteria for distinguishing them are uncertain and have yet to be determined. These are what have previously been called minimum-temperature melt and non-minimum-temperature melt granites (Hine *et al.* 1978). All of these are low-temperature, they contain older crystals of zircon, and they are characterised by 'linear' compositional variations and evolve chiefly through the fractionation of restite. I- and S-type granites occur in both of these groups. The composition of the primary melt for the non-minimum temperature granites is less felsic and may evolve through fractional crystallisation after restite has been removed as, for example, in the S-type Koetong Suite (Chappell & White 1998). Among I-type granites, the Jindabyne Suite, and probably the Inlet Suite, in Figures 1, 2 and 3 are of this type. When satisfactory criteria to distinguish these two groups are developed, those formed at higher temperatures could be termed 'intermediate-temperature', or perhaps, 'transitional' granites.

9.2. Possible subdivision of the high-temperature granites

It is becoming apparent that there are also two distinct groups within the high-temperature granites, in this case resulting from compositional differences rather than temperatures of origin. It is likely that these compositional differences, which reflect corresponding features of the source rocks, cause these two groups of granites to evolve in quite different ways. The present work has been very much influenced by the presence of the BPS in the LFB, and features of that supersuite have been discussed at some length in this paper. Fractional crystallisation was the dominant process that produced a wide range of compositions, with distinct differences between the compositions of plutonic and volcanic rocks. In contrast, Silver (2003) reported that the volcanic rocks of the western Peninsular Ranges batholith, which are most commonly basaltic andesite, low-K andesite and sodic rhyolites, have relative abundances that are comparable to the most common plutonic rocks of that area. L. T. Silver (pers. comm.) also points out that the plutons of that batholith are compositionally uniform (in contrast again to the BPS), and has proposed that the compositional differences within the rocks of that batholith are a consequence of different degrees of partial melting, rather than fractional crystallisation. Observations such as these may eventually form a basis for subdividing the high-temperature granites into high-K and low-K or tonalitic groups, which evolve compositionally in different ways.

10. Acknowledgements

This paper has been significantly improved following detailed comments by Penny King and Jon Blundy. The first author has benefited greatly from conversations with Lee Silver about tonalite genesis. As this paper goes to press, we learn with great sadness of the death of Professor Wally Picher. He was a friend and mentor who greatly stimulated all of us in our study of granites. The recognition of the Caledonian and Cordilleran I-type granite groups in 1982 was a significant observation that foreshadowed some of the ideas developed in this paper. This is publication 382 of the Australian Research Council Key Centre for the Geochemical Evolution and Metallogenesis of the Continents (GEMOC).

11. References

Baschek, G. & Johannes, W. 1995. The estimation of NaSi-CaAl interdiffusion rates in peristerite by homogenization experiments. *European Journal of Mineralogy* **7**, 295–307.

Blevin, P. L. & Chappell, B. W. 1992. The role of magma sources, oxidation states and fractionation in determining the granite metallogeny of eastern Australia. *Transactions of the Royal Society of Edinburgh: Earth Sciences* **83**, 305–16.

Bowen, N. L. 1928. *The Evolution of the Igneous Rocks*. Princeton, NJ: Princeton University Press.

Bryant, C. J., Arculus, R. J. & Chappell, B. W. 1997. Clarence River Supersuite: 250 Ma Cordilleran tonalitic I-type intrusions in eastern Australia. *Journal of Petrology* **38**, 975–1001.

Carr, P. F., Chappell, B. W. & Jones, B. G. 1992. Geochemical contrasts between intersecting Devonian and Carboniferous magmatic belts in the eastern Lachlan Fold Belt [Abstract]. *Transactions of the Royal Society of Edinburgh: Earth Sciences* **83**, 487.

Chappell, B. W. 1966. *Petrogenesis of the granites at Moonbi, New South Wales*. Unpublished Ph.D. thesis. Canberra: Australian National University.

Chappell, B. W. 1996a. Magma mixing and the production of compositional variation within granite suites: evidence from the granites of southeastern Australia. *Journal of Petrology* **37**, 449–70.

Chappell, B. W. 1996b. Compositional variation within granite suites of the Lachlan Fold Belt: its causes and implications for the physical state of granite magma. *Transactions of the Royal Society of Edinburgh: Earth Sciences* **87**, 159–70.

Chappell, B. W. 1999. Aluminium saturation in I- and S-type granites and the characterization of fractionated haplogranites. *Lithos* **46**, 535–51.

Chappell, B. W. 2004. Towards a unified model for granite genesis. *Transactions of the Royal Society of Edinburgh: Earth Sciences* **95**, 1–10.

Chappell, B. W., White, A. J. R. & Wyborn, D. 1987. The importance of residual source material (restite) in granite petrogenesis. *Journal of Petrology* **28**, 1111–38.

Chappell, B. W., English, P. M., King, P. L., White, A. J. R. & Wyborn, D. 1991. Granites and related rocks of the Lachlan Fold Belt (1:1 250 000 scale map). Bureau of Mineral Resources, Geology and Geophysics, Australia.

Chappell, B. W., Bryant, C. J., Wyborn, D., White, A. J. R. & Williams, I. S. 1998. High- and low-temperature I-type granites. *Resource Geology* **48**, 225–36.

Chappell, B. W., White, A. J. R., Williams, I. S., Wyborn, D. & Wyborn, L. A. I. 2000. Lachlan Fold Belt granites revisited: high- and low-temperature granites and their implications. *Australian Journal of Earth Sciences* **47**, 123–38.

Chappell, B. W. & Stephens, W. E. 1988. Origin of Infracrustal (I-type) granite magmas. *Transactions of the Royal Society of Edinburgh: Earth Sciences* **79**, 71–86.

Chappell, B. W. & White, A. J. R. 1974. Two contrasting granite types. *Pacific Geology* **8**, 173–4.

Chappell, B. W. & White, A. J. R. 1992. I- and S-type granites in the Lachlan Fold Belt. *Transactions of the Royal Society of Edinburgh: Earth Sciences* **83**, 1–26.

Chappell, B. W. & White, A. J. R. 1998. Development of P-rich granites by sequential restite fractionation and fractional crystallisation: the Koetong Suite in the Lachlan Fold Belt. *Acta Universitatis Carolinae Geologica* **42**, 23–7.

Chappell, B. W. & White, A. J. R. 2001. Two contrasting granite types: 25 years later. *Australian Journal of Earth Sciences* **48**, 489–99.

Chappell, B. W. & Wyborn, D. 2004. Cumulate and cumulative granites and associated rocks. *Resource Geology* **54**, 227–40.

Chen, Y. D., Price, R. C., White, A. J. R. & Chappell, B. W. 1990. Mafic inclusions from the Glenbog and Blue Gum granite suites, southeastern Australia. *Journal of Geophysical Research* **95**, 17757–85.

Clemens, J. D. 2001. S-type granites – models and evidence. *In* Chappell, B. W. & Fleming, P. D. (eds). *S-type Granites and Related Rocks. Abstract volume. Australian Geological Survey Organisation. Record* **2001/02**, 31–2.

Clemens, J. D. 2003. S-type granitic magmas – petrogenetic issues, models and evidence. *Earth Science Reviews* **61**, 1–18.

Collins, W. J., Beams, S. D., White, A. J. R. & Chappell, B. W. 1982. Nature and origin of A-type granites with particular reference to south-eastern Australia. *Contributions to Mineralogy and Petrology* **80**, 189–200.

Dias, G., Noronha, F. & Ferreira, N. 2000. *Variscan Plutonism in the Central Iberian Zone, Northern Portugal, Eurogranites 2000 Field Meeting*. Braga: Universidade do Minho.

Elburg, M. A. 1996. U-Pb ages and morphologies of zircon in microgranitoid enclaves and peraluminous host granite: evidence for magma mingling. *Contributions to Mineralogy and Petrology* **123**, 177–89.

Hanchar, J. M. & Watson, E. B. 2003. Zircon saturation thermometry. *In* Hanchar, J. M. & Hoskin, P. W. O. (eds) *Zircon. Reviews in Mineralogy and Geochemistry* **53**, 89–112.

Harrison, T. M. & Watson, E. B. 1984. The behaviour of apatite during crustal anatexis: equilibrium and kinetic considerations. *Geochimica et Cosmochimica Acta* **48**, 1467–77.

Hine, R., Williams, I. S., Chappell, B. W. & White, A. J. R. 1978. Contrasts between I- and S-type granitoids of the Kosciusko Batholith. *Journal of the Geological Society of Australia* **25**, 219–34.

Irvine, T. N. 1982. Terminology for layered intrusions. *Journal of Petrology* **23**, 127–62.

Johannes, W. & Holtz, F. 1996. *Petrogenesis and Experimental Petrology of Granitic Rocks*. Berlin: Springer-Verlag.

King, P. L., White, A. J. R., Chappell, B. W. & Allen, C. M. 1997. Characterization and origin of the aluminous A-type granites of the Lachlan Fold Belt, southeastern Australia. *Journal of Petrology* **38**, 371–91.

King, P. L., Chappell, B. W., Allen, C. M. & White, A. J. R. 2001. Are A-type granites the high temperature felsic granites? Evidence from fractionated granites of the Wangrah Suite. *Australian Journal of Earth Sciences* **48**, 501–14.

Loiselle, M. C. & Wones, D. R. 1979. Characteristics and origin of anorogenic granites. *Geological Society of America, Abstracts with Programs* **11**, 468.

McCarthy, T. S. & Groves, D. 1979. The Blue Tier Batholith, northeastern Tasmania. *Contributions to Mineralogy and Petrology* **71**, 193–209.

Miller, C. F., McDowell, S. M. & Mapes, R. W. 2003. Hot and cold granites? Implications of zircon saturation temperatures and preservation of inheritance. *Geology* **31**, 529–32.

Mysen, B. O., Holtz, F., Pichavant, M., Beny, J. M. & Montel, J.-M. 1999. The effect of temperature and bulk composition on the solution mechanism of phosphorus in peraluminous haplogranitic magma. *American Mineralogist* **84**, 1336–45.

Pitcher, W. S. 1982. Granite type and tectonic environment. *In* Hsu, K. J. (ed.) *Mountain Building Processes*, 19–40. London: Academic Press.

Propach, G. 1976. Models of filter differentiation. *Lithos* **9**, 203–9.

Silver, L. T. 2003. The Cretaceous volcano-plutonic complex in the western Peninsular Ranges batholith, USA and Mexico. *Geological Survey of Japan, Interim Report 29 (Hutton Symposium V Abstracts)*, 139.

Tuttle, O. F. & Bowen, N. L. 1958. Origin of granite in the light of experimental studies in the system $NaAlSi_3O_8$-$KAlSi_3O_8$-SiO_2H_2O. *Geological Society of America Memoir* **74**.

Watson, E. B. 1996. Dissolution, growth and survival of zircons during crustal fusion: kinetic principles, geological models and implications for isotopic inheritance. *Transactions of the Royal Society of Edinburgh: Earth Sciences* **87**, 43–56.

Watson, E. B. & Harrison, T. M. 1983. Zircon saturation revisited: temperature and composition effects in a variety of crustal magma types. *Earth and Planetary Science Letters* **64**, 295–304.

White, A. J. R. 2001. Water, restite and granite mineralisation. *Australian Journal of Earth Sciences* **48**, 551–5.

White, A. J. R. & Chappell, B. W. 1988. Some supracrustal (S-type) granites of the Lachlan Fold Belt. *Transactions of the Royal Society of Edinburgh: Earth Sciences* **79**, 169–81.

White, A. J. R. & Chappell, B. W. 2004. Petrographic discrimination of low- and high-temperature I-type granites. *Resource Geology* **54**, 215–26.

Whitney, J. 1988. The origin of granite: the role and source of water in the evolution of granitic magmas. *Geological Society of America Bulletin* **100**, 1886–97.

Williams, I. S. 1995. Zircon analysis by ion microprobe: the case of the eastern Australian granites. *In Leon T. Silver 70th Birthday Symposium and Celebration*, 27–31. Pasadena, CA: The California Institute of Technology.

Williams, I. S. 2001. Response of detrital zircon and monazite, and their U-Pb isotopic systems, to regional metamorphism and host-rock partial melting, Cooma Complex, southeastern Australia. *Australian Journal of Earth Sciences* **48**, 557–80.

Williams, I. S., Compston, W. & Chappell, B. W. 1983. Zircon and monazite U-Pb systems and the histories of I-type magmas, Berridale Batholith, Australia. *Journal of Petrology* **24**, 76–97.

Wolf, M. B. & London, D. 1994. Apatite dissolution into peraluminous haplogranitic melts: an experimental study of solubilities and mechanisms. *Geochimica et Cosmochimica Acta* **56**, 4127–45.

Wyborn, D. 1983. *Fractionation processes in the Boggy Plain zoned pluton*. Unpublished Ph.D. thesis. Canberra: Australian National University.

Wyborn, D., Chappell, B. W. & Johnston, R. M. 1981. Three S-type volcanic suites from the Lachlan Fold Belt, southeast Australia. *Journal of Geophysical Research* **86**, 10335–48.

Wyborn, D., Turner, B. S. & Chappell, B. W. 1987. The Boggy Plain Supersuite: a distinctive belt of I-type igneous rocks of potential economic significance in the Lachlan Fold Belt. *Australian Journal of Earth Sciences* **34**, 21–43.

Wyborn, D., Chappell, B. W. & James, M. 2001. Examples of convective fractionation in high temperature granites from the Lachlan Fold Belt. *Australian Journal of Earth Sciences* **48**, 531–41.

Wyborn, D. & Chappell, B. W. 1986. The petrogenetic significance of chemically related plutonic and volcanic rock units. *Geological Magazine* **123**, 619–28.

Wyborn, D. & Owen, M. 1986. *Araluen, New South Wales – 1:100 000 map commentary*. Canberra: Bureau of Mineral Resources, Geology and Geophysics, Australia.

B. W. CHAPPELL, ARC National Key Centre for Geochemical Evolution and Metallogeny of Continents, Department of Earth and Planetary Sciences, Macquarie University, Sydney, NSW 2109, Australia.

A. J. R. WHITE, School of Earth Sciences, University of Melbourne, Melbourne, Victoria 3010, Australia.

I. S. WILLIAMS, Research School of Earth Sciences, The Australian National University, Canberra, ACT 0200, Australia.

D. WYBORN, Geodynamics Limited, PO Box 2046, Milton, Queensland 4064, Australia.

MS received 12 December 2003. Accepted for publication 9 September 2004.

Transactions of the Royal Society of Edinburgh: Earth Sciences, **95**, 141–159, 2004

Deciphering the petrogenesis of deeply buried granites: whole-rock geochemical constraints on the origin of largely undepleted felsic granulites from the Moldanubian Zone of the Bohemian Massif

Vojtěch Janoušek, Fritz Finger, Malcolm Roberts, Jiří Frýda, Christian Pin and David Dolejš

ABSTRACT: The prominent felsic granulites in the southern part of the Bohemian Massif (Gföhl Unit, Moldanubian Zone), with the Variscan ($\sim$340 Ma) high-pressure and high-temperature assemblage garnet+quartz+hypersolvus feldspar ± kyanite, correspond geochemically to slightly peraluminous, fractionated granitic rocks. Compared to the average upper crust and most granites, the U, Th and Cs concentrations are strongly depleted, probably because of the fluid and/or slight melt loss during the high-grade metamorphism (900–1050 °C, 1·5–2·0 GPa). However, the rest of the trace-element contents and variation trends, such as decreasing Sr, Ba, Eu, LREE and Zr with increasing SiO_2 and Rb, can be explained by fractional crystallisation of a granitic magma. Low Zr and LREE contents yield $\sim$750 °C zircon and monazite saturation temperatures and suggest relatively low-temperature crystallisation. The granulites contain radiogenic Sr ($^{87}Sr/^{86}Sr_{340}$= 0·7106–0·7706) and unradiogenic Nd ($\varepsilon^{340}_{Nd} = -4{\cdot}2$ to $-7{\cdot}5$), indicating derivation from an old crustal source. The whole-rock Rb–Sr isotopic system preserves the memory of an earlier, probably Ordovician, isotopic equilibrium.

Contrary to previous studies, the bulk of felsic Moldanubian granulites do not appear to represent separated, syn-metamorphic Variscan HP–HT melts. Instead, they are interpreted as metamorphosed (partly anatectic) equivalents of older, probably high-level granites subducted to continental roots during the Variscan collision. Protolith formation may have occurred within an Early Palaeozoic rift setting, which is documented throughout the Variscan Zone in Europe.

KEY WORDS: Austria, Czech Republic, high-grade metamorphism, Nd isotopes, partial melting, Sr isotopes, Variscan orogeny.

Felsic garnet ± kyanite-bearing high-pressure and high-temperature (HP–HT) granulites with broadly granitic compositions are prominent constituents in the central part of the European Variscides (e.g. Pin & Vielzeuf 1983; O'Brien & Rötzler 2003). These light-coloured, SiO_2-rich rocks are known from many places of the Bohemian Massif, namely from the Moldanubian (Lower Austria and southern Czech Republic; Fig. 1) and the Saxothuringian zones (the Saxonian Granulite Massif and Erzgebirge; Romer & Rötzler 2001; Rötzler & Romer 2001; O'Brien & Rötzler 2003). Since the Moldanubian granulites may provide insights into processes related to deep subduction and subsequent exhumation of the continental crust, research over the past 40 years has focused mainly on constraining their precise pressure–temperature–time (P–T–t) paths (for reviews, see O'Brien 2000; O'Brien & Rötzler 2003). However, their pre-metamorphic history has been largely neglected.

Concerning the protolith, the felsic granulites were considered as former rhyolites or granites which acquired their high-grade metamorphic character during the course of the Variscan collision (Fiala *et al.* 1987a, b; Vellmer 1992). Based on existing P–T estimates and fluid-absent melting models for crustal rocks, Roberts & Finger (1997) proposed that they probably experienced a certain degree of partial melting and melt persisted during much of their metamorphic history. Other authors (Vrána & Jakeš 1982; Vrána 1989; Jakeš 1997; Kotková & Harley 1999) went further and suggested that the granulites actually represent high-pressure granitic magmas which formed during the Variscan metamorphic cycle.

Clearly, the problem of granulite genesis cannot be resolved without involving additional aspects of granite petrology and geochemistry, constraining the protolith formation and metamorphic reworking. The present paper characterises the compositional variation of the felsic Moldanubian granulites and provides their petrogenetic interpretation. The authors demonstrate that most, if not all, of their whole-rock geochemical signature may be related to a pre-existing Early Palaeozoic igneous-rock suite, whereas the Variscan high-pressure metamorphism could have been nearly isochemical.

1. Geological setting and petrology

1.1. Moldanubian Zone

The Moldanubian Zone represents remnants of the crystalline core of the Variscan orogen in central Europe (e.g. Dallmeyer *et al.* 1995). In the southern part of the Bohemian Massif, shared by the Czech Republic and Austria, the Moldanubian Zone consists of several crustal segments with contrasting ages and complex polyphase deformational histories, intruded by numerous, mainly post-tectonic granitic masses (Dallmeyer *et al.* 1995; O'Brien 2000 and references therein). It has been

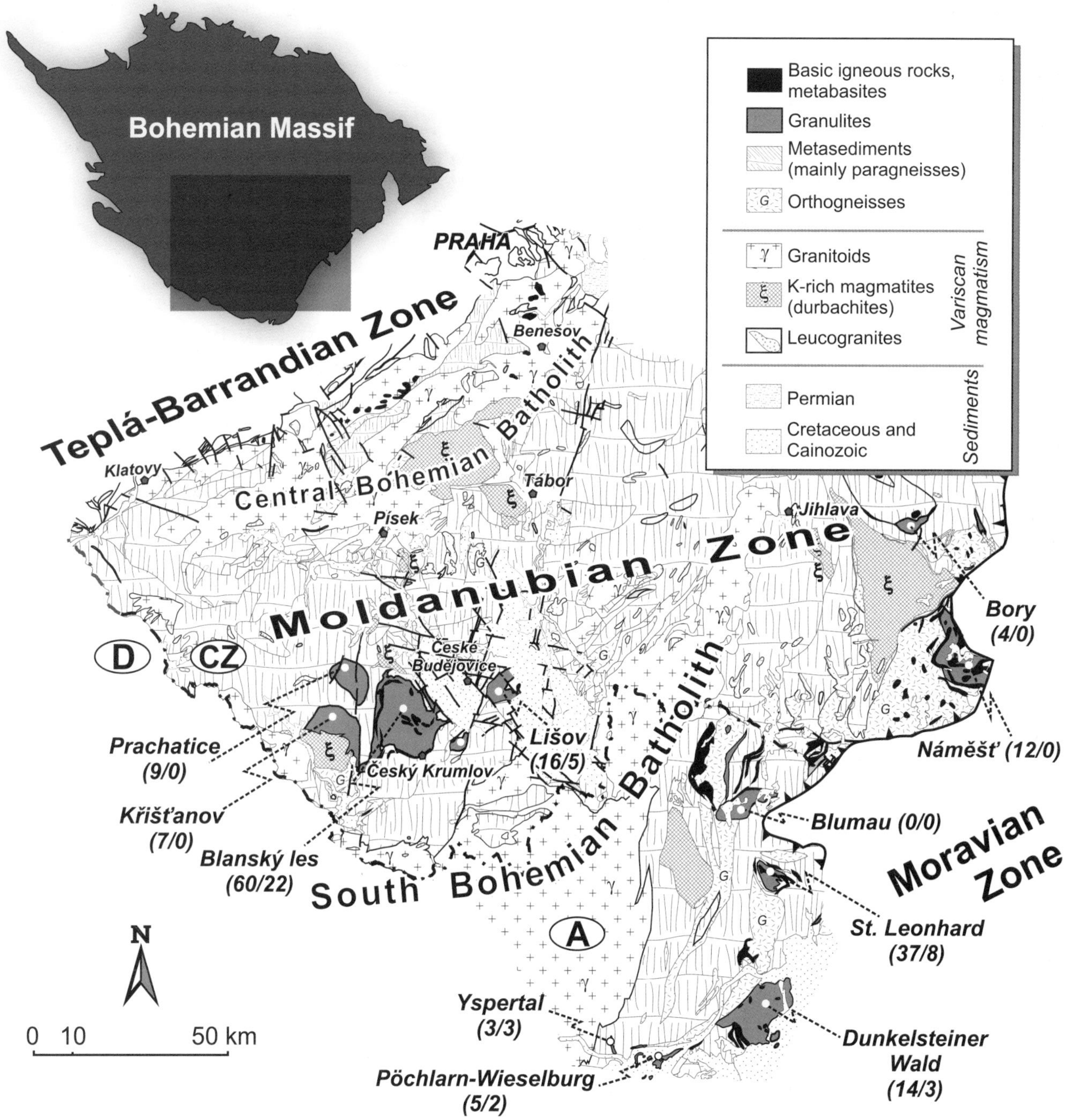

Figure 1 Geological sketch of the Moldanubian Zone in southern Bohemia, south-western Moravia and Lower Austria (simplified from the Czech Geological Survey map 1:500 000). The numbers of felsic granulite samples from larger massifs (total/new determinations) are shown.

subdivided into a structurally lower, mainly metasedimentary, amphibolite-facies Drosendorf Assemblage and an overlying, higher-grade, allochtonous Gföhl Assemblage (e.g. Matte *et al.* 1990; Fiala *et al.* 1995; Franke 2000).

The lower part of the Drosendorf Assemblage consists of the Monotonous Unit (Ostrong Unit) made up mainly of cordierite–biotite–sillimanite paragneisses, which are partly migmatitic and contain intercalations of orthogneisses and minor amphibolites. The upper part, termed the Varied (or Variegated) Unit, consists of paragneisses, orthogneisses, amphibolites, carbonate rocks, calc-silicate gneisses, quartzites and graphite schists.

The lower parts of the Gföhl Assemblage include mainly anatectic orthogneisses (Gföhl gneiss). The higher are dominated by granulites, associated with minor bodies of garnet and spinel peridotites, pyroxenites, dunites and eclogites (Fuchs & Matura 1976; Fiala *et al.* 1995).

1.2. Granulite bodies

Several granulite massifs occur in the Waldviertel region of Lower Austria. These are interpreted as the isolated klippen of an original regionally extensive nappe (Tollmann 1995). From SW to NE, these are the Yspertal, Pöchlarn–Wieselburg, Dunkelsteiner Wald, St. Leonhard and Blumau massifs (Fuchs & Matura 1976, Fig. 1). In the southern Czech Republic, the granulite bodies tend to be roughly oval-shaped and dome-like. The largest is the Blanský les Massif (24 × 4 km), followed by Křišťanov, Prachatice, Lišov, Krasejovka and other numerous small occurrences (Fig. 1). In southwest Moravia, the most prominent are the Náměšť and Bory bodies (Fiala *et al.* 1987a, b).

1.3. Petrology

The individual massifs consist of a heterogeneous blend of high-pressure crustal and upper mantle rocks assembled and exhumed in course of the Variscan orogeny. Felsic granulites with granitic compositions are by far the most abundant and occur in all the main massifs, accounting for more than 75–80 vol.% (Fiala *et al.* 1987a; Vellmer 1992). In comparison, mafic–intermediate pyroxene-bearing and sedimentary-derived types are rather rare.

The typical felsic granulites are characterised by the high-pressure assemblage of garnet, kyanite, quartz and hypersolvus feldspar. Rutile, zircon, apatite, ilmenite ± monazite are the common accessories (Fiala *et al.* 1987a; Carswell 1991; Vellmer 1992; O'Brien & Rötzler 2003). The estimated peak metamorphic conditions are 900–1050°C and 1·5–2·0 GPa (O'Brien & Rötzler 2003 and references therein).

Following the high-pressure metamorphic peak, the rocks were exhumed along a steep decompression path to mid-crustal levels and subjected to a medium-pressure overprint (0·8–1·2 GPa; 800–900°C: Cooke 2000). Further evolution was characterised by near isobaric cooling (0·8–0·5 GPa), variable retrogression of the original assemblages and intense deformation, reflecting the effects of stacking and subsequent exhumation of the Moldanubian nappe pile (Carswell & O'Brien 1993; Cooke 2000). These processes resulted in the breakdown of some garnet to a secondary biotite, kyanite to sillimanite or garnet with kyanite to hercynite–plagioclase coronas (Owen & Dostal 1996; O'Brien & Rötzler 2003). Most of the granulites were mylonitised and recrystallised, with pronounced foliations of platy quartz and fine K-feldspar–plagioclase mosaics. Regional differences in the metamorphic style do exist; in southern Bohemia (e.g. Prachatice and Blanský les massifs), the effects of a low-P, high-T overprint that led to the crystallisation of cordierite and the formation of partial melt patches can be observed (O'Brien 2000).

1.4. Geochronology

1.4.1. Carboniferous granulite-facies metamorphism, subsequent cooling and uplift. The numerous U–Pb zircon and monazite ages clustering around 340 Ma are interpreted by most authors as reflecting the granulite-facies metamorphic climax (van Breemen *et al.* 1982; Schenk & Todt 1983; Kröner *et al.* 1988; Aftalion *et al.* 1989; Wendt *et al.* 1994; Kröner *et al.* 2000; Friedl *et al.* 2003; Kotková *et al.* 2003). In contrast, Roberts & Finger (1997) proposed that much of the zircon could have crystallised only on a retrogression path. Svojtka *et al.* (2002) concluded, based on Sm–Nd ages for garnet cores with prograde zoning, that the HP–HT metamorphic peak may have been even c. 10–15 Ma older than the retrograde event.

However, it is worth noting that many of the dated zircons were enclosed in high-pressure phases (garnet and former hypersolvus feldspar; Aftalion *et al.* 1989; Kröner *et al.* 2000). The metamorphic peak was probably short-lived, as demonstrated by diffusion modelling of garnet zoning (Medaris *et al.* 1990, 2003). The exhumation rate was high since the granulite pebbles already occur in Upper Visean conglomerates (several mm year^{-1}: Vrána & Novák 2000; Kotková *et al.* 2003). Rapid uplift and cooling is compatible with the fact that zircons enclosed in late cordierite-bearing, decompression melt pockets also gave ages of c. 340 Ma (Kröner *et al.* 2000). Taken together, the age of the high-grade metamorphism is not likely to differ substantially from 340 Ma.

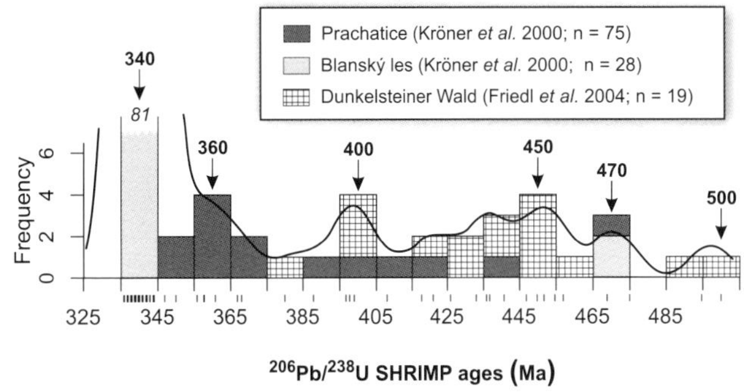

Figure 2 Histogram of frequencies for the $^{206}Pb/^{238}U$ SHRIMP zircon ages from the Moldanubian granulites. Data taken from Kröner *et al.* (2000; five samples from Prachatice and two from Blanský les massifs) and Friedl *et al.* (2004; sample *St*, Dunkelsteiner Wald).

1.4.2. Pre-Carboniferous ages. Devonian ages (c. 360–370 Ma) were obtained from some prismatic zircons as well as cores of multifaceted metamorphic grains (Wendt *et al.* 1994; Kröner *et al.* 2000). However, the SHRIMP dating of Kröner *et al.* (2000) and Friedl *et al.* (2003, 2004) revealed abundant oscillatory zoned zircon cores with concordant Ordovician (Fig. 2), as well as less prominent older ages. Friedl *et al.* (2003, 2004) assumed a c. 450-Ma-old igneous protolith for their sample from the Dunkelsteiner Wald.

The early Rb–Sr whole-rock dating of Austrian felsic granulites also yielded Ordovician ages (486 ± 11 and 450 ± 30 Ma; Arnold & Scharbert 1973, recalculated with the accepted decay constants of Steiger & Jäger 1977). Similar whole-rock Rb–Sr ages (~470 Ma) were obtained for felsic granulites from Blanský les (present work and A. Arnold, unpublished data, quoted by Kodym *et al.* 1978). Rb–Sr thin slab dating of the St. Leonhard Massif yielded a somewhat younger age of 428 ± 16 Ma (Frank *et al.* 1990). The significance of these pre-Variscan ages will be addressed below.

2. Whole-rock geochemistry

2.1. The database

The present authors have assembled a database in which new analyses supplement critically selected literature data in order to assess the overall whole-rock geochemical variability of the Moldanubian granulites. Data management, recalculation, plotting and statistical evaluation were facilitated using *GCDkit* (Janoušek *et al.* 2003a).

In this database, over 20 new samples complemented 93 previously published Austrian whole-rock analyses (Vellmer 1992; Becker 1997; Becker *et al.* 1999; Cooke 1999). Several granulite quarries show a striking alternation of darker and lighter bands, sometimes tightly folded (Fig. 3a, b). The banding could have formed by the Variscan metamorphic segregation or migmatisation. Alternatively, it may represent older, pre-Variscan heterogeneities (enclaves, dykes?) flattened and obscured by the strong deformation. In order to verify the presence of isotopic (dis-) equilibria, sample pairs were collected from adjacent darker (denoted by suffix *D*) and lighter (*L*) bands (*Popp-D/L*, *Ost-D/L* and *Ysp-D/L*). Petrographic sample descriptions are given in Appendix 1.

The database includes 50 new and ~180 old Czech whole-rock analyses (Kodym *et al.* 1978; Slabý 1983; Čadková *et al.* 1985; Strejček 1986; Fiala *et al.* 1987a; Gürtlerová *et al.* 1997). Peculiar K-feldspar–garnet perpotassic granulites, documented from a single small body within the Blanský les Massif (Vrána 1989), were not considered. Many of the new samples came

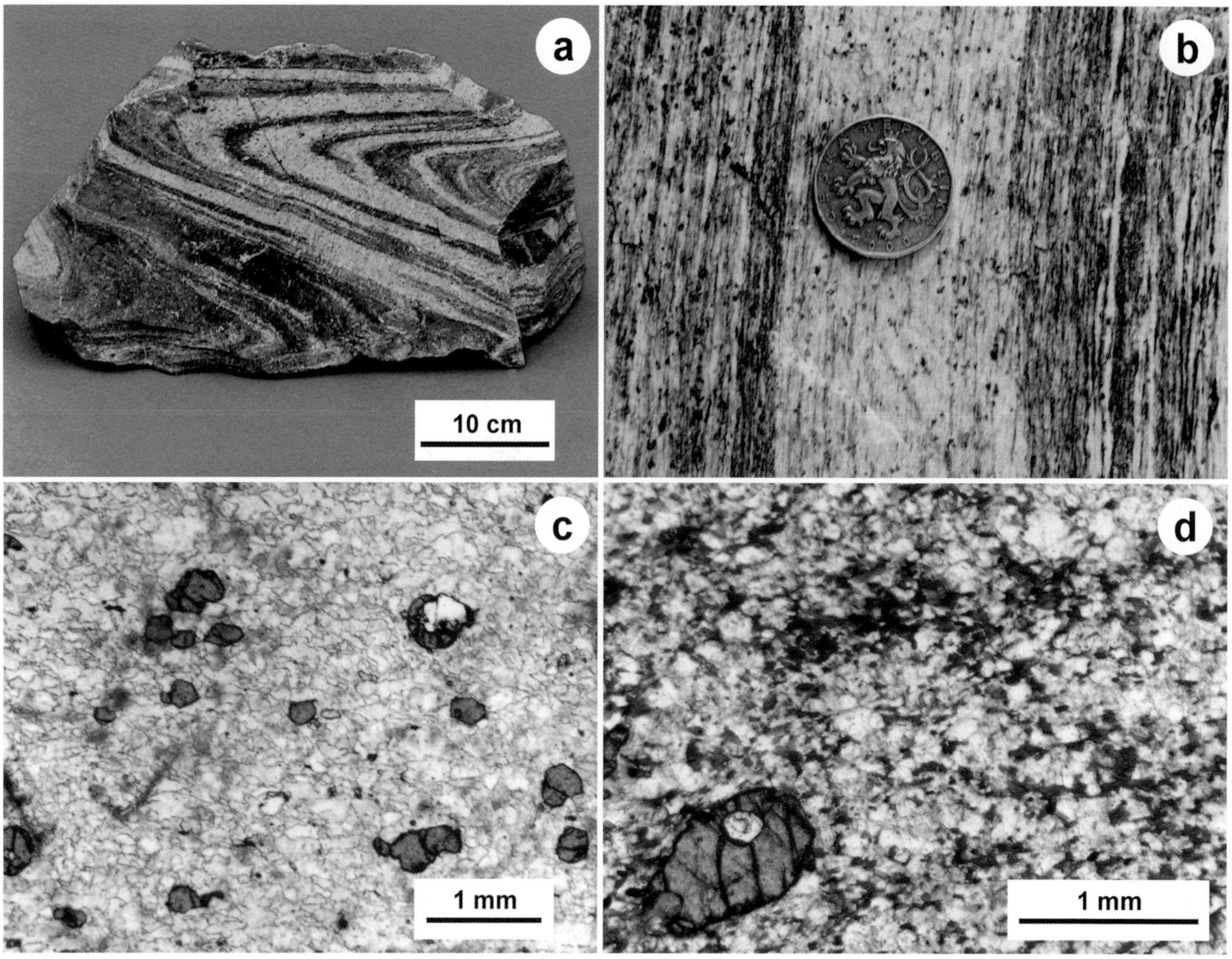

Figure 3 (a) Folded distinctly banded felsic granulite, Yspertal quarry. (b) Detail of the hand specimen of a felsic granulite with dark bands rich in biotite and strong mylonitic fabric, Plešovice quarry, Blanský les Massif. (c) Photomicrograph of felsic granulite *Popp-L* (Krug, St. Leonhard Massif). Garnet porphyroclasts are set in a fine-grained feldspar–quartz matrix. Plain polarised light. (d) Photomicrograph of dark facies of the sample from Osterberg, Pöchlarn–Wieselburg (*Ost-D*). A large garnet porphyroclast encloses a subhedral grain of mesoperthite. The strongly recrystallised feldspar–quartz matrix contains streaks of biotite. Plain polarised light.

from the borehole H-1 drilled in the Blanský les Massif (Appendix 1).

As noted above, caution should be taken to separate individual granulite groups with potentially distinct origin (Fiala *et al.* 1987a). For instance, the major-element binary plots do not follow simple patterns, but are characterised by inflections or discontinuities at SiO_2 ~60 and/or SiO_2 ~70 wt.% (Vellmer 1992). The present paper focuses on a rather coherent group of most common felsic garnet ± kyanite granulites, defined on the basis of geochemical variation diagrams as having SiO_2 >70 wt.%. In addition, they needed to be devoid of orthopyroxene, since the rare charnockitic types are suspected of forming a distinct geochemical and petrogenetic suite. In total, some 200 analyses of the felsic granulites were evaluated (Fig. 1).

2.2. Major elements

All Moldanubian granulites span a wide range of felsic (granitic) to mafic (gabbroic) compositions, as shown by the multicationic diagram P–Q (Debon & Le Fort 1983; Fig. 4). It is worth stressing that the sampling is strongly biased towards the less-common basic types. All the rocks are subalkaline and define a calc-alkaline trend in the Na_2O+K_2O–FeOt–MgO ternary diagram (AFM; Irvine & Baragar 1971; Fig. 5a). As noted already by Fiala *et al.* (1987b), the granulites can be characterised as medium- (mafic and transitional) to high-K (felsic) calc-alkaline rocks on the basis of the SiO_2–K_2O plot of Peccerillo & Taylor (1976) (Fig. 5b).

The major-element composition of the felsic granulites is very silicic (SiO_2=74·4 ± 2·3 wt.%, median ± 1σ; Table 1), slightly peraluminous (A/CNK=1·11 ± 0·08), K-rich (K_2O/Na_2O=1·54 ± 0·37) and with Fe prevailing strongly over Mg (FeOt=1·94 ± 0·71 wt.%, MgO=0·47 ± 0·35 wt.%, mg# ~30). At given SiO_2 contents, CaO (1·03 ± 0·48 wt.%) is rather high while Al_2O_3 is low (13·28 ± 0·91 wt.%). The Harker variation diagrams (Fig. 6) show strong negative correlations of SiO_2 with TiO_2, Al_2O_3, FeOt, MgO, CaO and mg#. In contrast, alkalis and the alumina saturation index (A/CNK, not shown) are scattered, potentially implying some metamorphic mobility.

2.3. Trace elements

2.3.1. LILE. All the Moldanubian granulites, including the mafic types, have typical crustal K/Rb ratios (i.e. between c. 120 and 500; Shaw 1968; Rudnick *et al.* 1985) (Fig. 7a; Table 1), and thus, are not significantly depleted in Rb (see also Fiala *et al.* 1987a, b; Vellmer 1992). The felsic types show wide range of Rb contents (28–362 p.p.m., 164 ± 59 p.p.m.) and Rb/Sr ratios (0·5–16·5, median=2·9). The SiO_2–Rb plot is characterised by a scattered positive trend (Fig. 6). The Rb/Cs

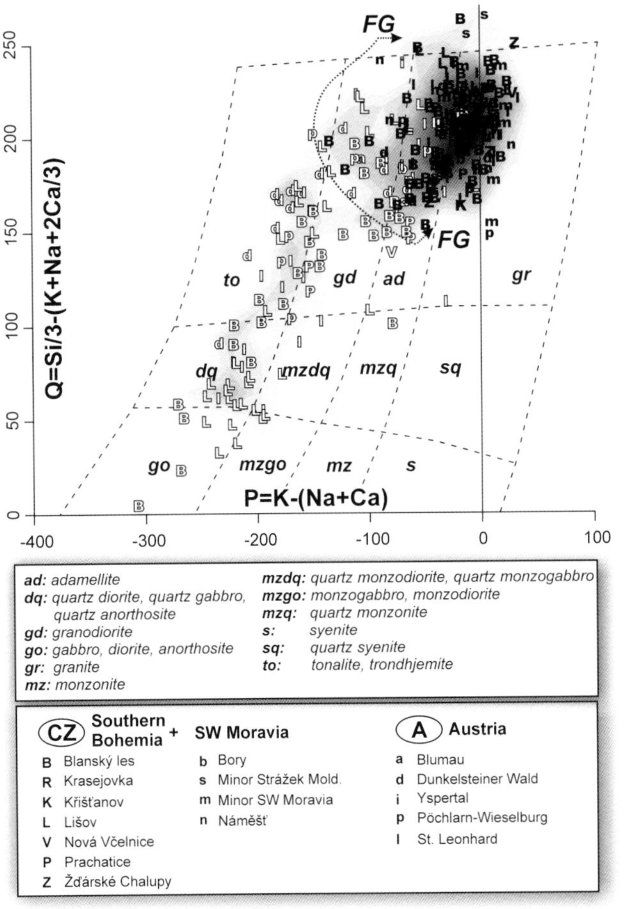

Figure 4 Frequency plots of Moldanubian granulites in the multi-cationic diagram P–Q (Debon & Le Fort 1983). *P* represents the proportion of K-feldspar to plagioclase and *Q* the quartz content. Plotting symbols represent individual granulite massifs. Filled letters stand for felsic granulite (*FG*) samples dealt with in detail in the text. The frequency of all available analyses is expressed by various shades of grey.

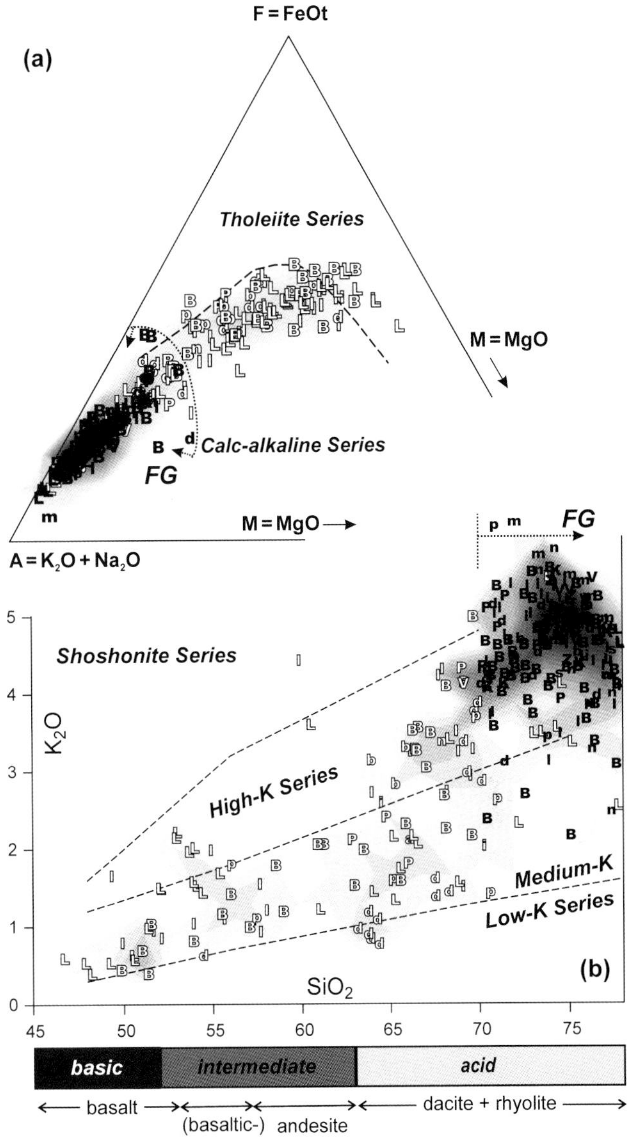

Figure 5 (a) AFM diagram ($A=Na_2O+K_2O$, F=FeOt, M=MgO: Irvine & Baragar 1971) illustrating the calc-alkaline trend shown by the Moldanubian granulites. (b) SiO_2–K_2O plot with the discrimination boundaries between the tholeiitic, calc-alkaline, high-K calc-alkaline and shoshonitic rocks after Peccerillo & Taylor (1976). Symbols as in Figure 4.

ratios for the felsic granulites (11–1100, median=122) are higher than the average continental crust (Rb/Cs=30; Taylor & McLennan 1985; or even 23; Wedepohl 1995). Strontium, and in particular Ba, vary over a wide range. Both show good covariation and sharp decrease with increasing silica (Figs 6 & 7 b–c).

2.3.2. U and Th. While many of the felsic Moldanubian granulites have Th/U (0·2–39, median=3·1), corresponding to normal igneous rocks (Rudnick & Presper 1990), a significant overall depletion of both elements exists (Fiala *et al.* 1987a), down to less than 0·05 times the concentration of average upper crust (Taylor & McLennan 1985).

2.3.3. REE. Chondrite-normalised REE plots resemble typical patterns of fractionated granitic rocks, showing moderate to weak enrichment in LREE (Ce_N/Yb_N=1·5–7·2, Ce_N/Sm_N=1·3–3·4; Fig. 8) and negative Eu anomalies (Eu/Eu*=0·14–0·62, $Eu^*=\sqrt{Sm_N Gd_N}$), the magnitude of which increases with increasing silica (see also Vellmer 1992; Jakeš 1997). There is a sharp decrease in LREE and MREE contents with increasing SiO_2, while the HREE contents remain relatively constant (Figs 6 & 8). Such variation is a common feature in felsic granitic magmas which have undergone fractional crystallisation, particularly at monazite saturation (Miller & Mittlefehldt 1982, 1984; Ayres & Harris 1997). Given the siliceous and slightly peraluminous nature of the studied rocks, the presence of monazite as a main LREE carrier is likely (Watt & Harley 1993; Broska *et al.* 2000). If one assumes that the whole-rock compositions correspond to those of monazite-saturated melt, the calculated crystallisation temperatures (Montel 1993) would be close to 750°C (Fig. 9a).

2.3.4. Zr. Zirconium concentrations in the felsic granulites are generally low (86 ± 50 p.p.m.). These decrease with increasing SiO_2 to less than 50 p.p.m. (Fig. 6), which is compatible with zircon fractionation. The zircon saturation temperatures (Watson & Harrison 1983) of 740 ± 48°C are comparable to those obtained from the monazite saturation model (Fig. 9b).

2.3.5. Spiderplots. Spiderplots for the felsic granulites (Fig. 10a) normalised to the average upper crust are characterised by conspicuous troughs in Nb, Sr, Hf, Zr and Ti. A similar feature is also seen for U and Th in most cases. Some elements tend to be variously depleted, starting with apparent crustal values (e.g. Ba, P and Nd). Rubidium and K seem to be in the range of the upper crust (if not somewhat higher) together with HREE and Y. In contrast, the LREE tend to be slightly lower than upper crustal values. With notable exception of U and Th, the patterns resemble those of felsic granites, variably modified by fractional crystallisation.

Table 1 Selected whole-rock geochemical data for the (mostly felsic) Moldanubian granulites

	Popp-D	Popp-L	Ost-D®	Ost-L	St	Du©	Ysp-D	Ysp-L	1059	1062	1064	1074	1076	1078
Locality*	1	1	2	2	3	4	5	5	1146	1178	1312	1372	1381	1405 m
Massif†	StL	StL	PW	PW	DW	DW	Yt	Yt	BL	BL	BL	BL	BL	BL
SiO_2	71·69	74·82	63·90	77·55	76·94	70·14	76·03	75·84	70·43	73·87	75·54	76·54	76·02	70·78
TiO_2	0·47	0·21	0·78	0·09	0·12	0·50	0·20	0·15	0·35	0·27	0·12	0·07	0·08	0·32
Al_2O_3	14·53	12·95	17·89	11·90	12·30	13·97	12·66	12·36	14·36	13·53	12·58	12·10	12·16	13·92
FeOt	2·95	1·72	3·66	1·52	1·66	4·05	1·84	1·64	4·01	2·22	1·72	1·56	1·54	2·79
MnO	0·05	0·04	0·03	0·04	0·04	0·07	0·05	0·03	0·05	0·03	0·02	0·03	0·02	0·04
MgO	0·83	0·51	2·82	0·17	0·21	0·94	0·40	0·24	1·54	0·55	0·58	0·28	0·31	0·93
CaO	1·63	0·84	3·02	0·69	0·73	3·22	0·82	0·67	2·50	1·69	1·41	0·78	0·95	1·68
Na_2O	3·12	2·37	5·20	3·49	3·20	3·69	2·95	3·16	3·94	3·64	4·21	2·66	2·78	3·34
K_2O	4·65	5·40	2·74	3·82	4·32	2·85	4·45	4·98	2·36	4·06	3·13	4·85	4·94	4·31
P_2O_5	0·18	0·16	0·21	0·10	0·15	0·12	0·16	0·14	0·08	0·07	0·03	0·13	0·15	0·18
H_2O^+/LOI	0·57‡	0·20‡	0·51‡	0·22‡	0·44‡	0·27‡	0·49‡	0·30‡	0·27	0·08	0·27	0·12	0·14	0·64
F	–	–	–	–	–	–	–	–	0·05	0·03	0·06	0·05	0·04	0·03
Total	100·67	99·22	100·76	99·59	100·11	99·82	100·05	99·51	99·94	100·04	99·67	99·17	99·13	98·96
mg#	33·4	34·6	57·9	16·6	18·4	29·3	27·9	20·7	40·7	30·7	37·6	24·3	26·4	37·3
A/CNK	1·11	1·15	1·05	1·07	1·09	0·93	1·13	1·05	1·06	1·01	0·98	1·10	1·04	1·05
K_2O/Na_2O	1·49	2·28	0·53	1·09	1·35	0·77	1·51	1·58	0·60	1·12	0·74	1·82	1·78	1·29
(p.p.m.)														
Rb	145	168	201	331	288	48	220	180	66	126	92	241	216	153
Sr	96	73	158	35	34	71	61	58	79	70	73	35	46	108
Ba	612	398	402	58	139	565	371	357	496	601	476	114	190	586
Nb	7·5	4·9	11·1	b.d.	4·8	4·3	6·3	4·1	7·4	5·1	2·9	b.d.	3·3	8·9
Zr	115	105	120	16	38	108	102	74	135	112	29	50	66	106
Hf	3·8	1·5	2·4	b.d.	2·1	2·1	2·7	1·0	2·1	1·7	0·9	b.d.	0·6	1·7
Ga	19	18	26	20	18	18	18	17	19	17	18	17	16	19
Th	2·1	2·8	b.d.	2·6	3·7	b.d.	3·2	b.d.	5·0	3·8	b.d.	3·4	2·4	5·6
U	2·2	2·9	b.d.	2·3	2·7	b.d.	5·2	3·0	3·8	1·6	1·0	3·0	2·2	2·7
Cr	41	61	120	11	11	75	22	18	–	–	–	–	–	–
Co	5·9	2·4	7·1	1·2	2·2	13·5	4·6	2·7	10·7	4·8	5·4	3·4	3·4	6·8
Pb	1·3	6·7	b.d.	b.d.	4·1	b.d.	5·5	4·7	b.d.	0·9	4·4	4·3	7·2	4·3
Zn	48	23	24	15	17	59	37	26	59·8	27·6	28·2	22·5	23·4	43·3
La	32·2	22·0	22·0	9·0	11·6	37·0	17·4	14·1	21·1	13·5	–	9·4	12·1	24·1
Ce	69·3	50·4	48·6	19·3	27·0	74·6	40·0	32·1	47·0	39·4	–	23·6	30·5	56·0
Pr	–	–	–	–	–	–	–	–	6·02	4·86	–	2·38	3·82	6·67
Nd	28·9	19·2	22·7	6·3	11·5	29·2	17·4	13·3	24·2	20·8	–	10·6	14·2	26·8
Sm	6·67	4·42	5·51	1·61	3·05	5·35	4·16	3·48	5·05	4·72	–	2·67	3·41	5·67
Eu	0·91	0·42	0·93	0·09	0·17	1·00	0·34	0·23	0·89	0·54	–	0·25	0·38	0·89
Gd	7·30	4·72	5·35	2·28	3·58	4·88	4·30	3·70	5·43	5·93	–	3·15	3·92	6·38
Tb	–	–	–	–	–	–	–	–	1·2	1·2	–	0·83	0·96	1·24
Dy	7·93	6·07	5·68	4·38	5·76	3·71	6·06	6·00	5·5	7·84	–	5·47	6·73	7·51
Ho	–	–	–	–	–	–	–	–	0·99	1·73	–	1·15	1·50	1·45
Er	4·43	3·92	3·47	3·25	3·51	1·98	3·8	3·81	3·53	5·54	–	4·07	4·92	4·35
Tm	–	–	–	–	–	–	–	–	0·33	0·55	–	0·52	0·60	0·49
Yb	4·21	4·06	3·94	3·82	3·78	1·89	3·97	4·12	3·25	4·84	–	4·11	4·57	3·31
Lu	0·62	0·57	0·60	0·52	0·49	0·28	0·52	0·56	0·5	0·7	–	0·58	0·65	0·44
Y	45·8	38·7	34·7	30·3	35·6	19·9	36·8	38·0	34·1	50·2	11·2	36·2	44·9	48·6
Eu/Eu*	0·40	0·28	0·52	0·14	0·16	0·60	0·25	0·20	0·52	0·31	–	0·26	0·32	0·45
Ce_N/Yb_N	4·3	3·2	3·2	1·3	1·9	10·2	2·6	2·0	3·7	2·1	–	1·5	1·7	4·4
Ce_N/Sm_N	2·5	2·8	2·1	2·9	2·1	3·4	2·3	2·2	2·2	2·0	–	2·1	2·2	2·4
K/Rb	265·5	266·5	113·3	95·9	124·4	490·8	167·6	229·3	299·1	266·9	284·0	167·0	189·9	233·4
Rb/Sr	1·5	2·3	1·3	9·5	3·6	8·5	3·1	0·7	0·8	1·8	1·3	7·0	4·7	1·4

*Localities: (1) Naturstein Popp, Krug: (2) Osterberg: (3) Steinaweg: (4) Lindwurmkreuz: (5) Gleisen. The rest: Holubov, borehole H-1 (with depth indicated). ®Mafic sample. ©Charnockite.

†Massifs (Fig. 1): (StL) St. Leonhard: (PW) Pöchlarn–Wieselburg: (DW) Dunkelsteiner Wald: (Yt) Yspertal: (BL) Blanský les. ‡ Loss on ignition: (b.d.) below detection limit; (–) not determined.

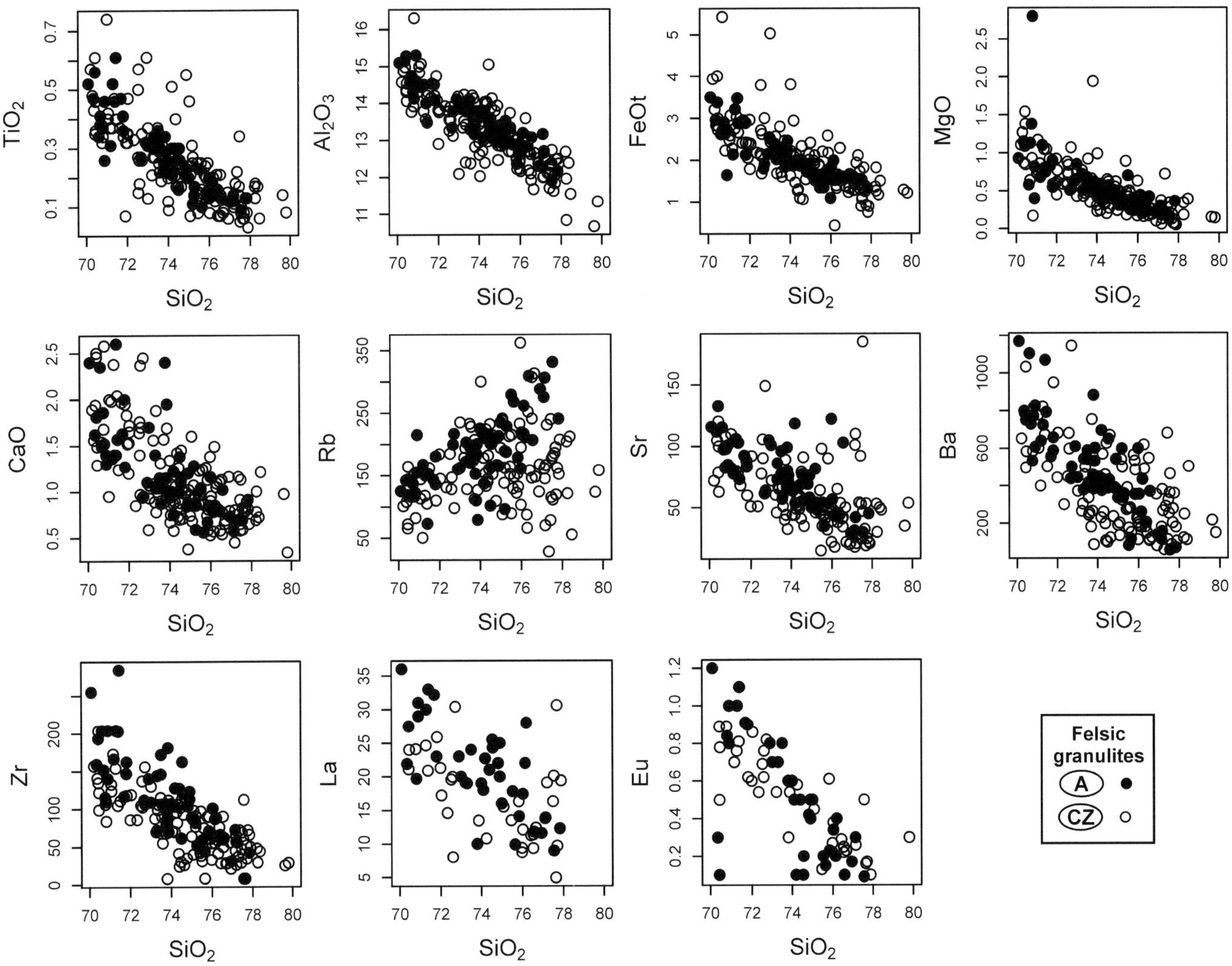

Figure 6 Binary plots of SiO_2 versus major-element oxides (wt.%), mg# and selected trace elements (p.p.m.) for felsic Moldanubian granulites from Austria (filled symbols) and the Czech Republic (empty symbols).

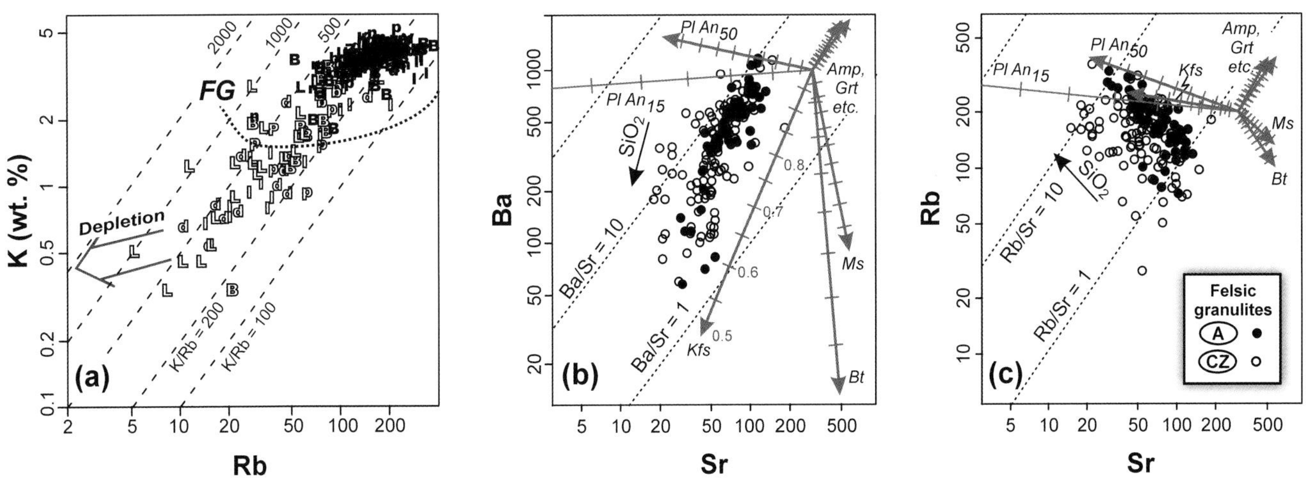

Figure 7 (a) Rb (p.p.m.) versus K (wt.%) diagram for all the Moldanubian granulites. Symbols as in Figure 4. *FG* felsic granulites. Depletion trend for granulites with K/Rb > 500 is taken from Jahn (1990). (b) Sr versus Ba and (c) Sr versus Rb plots (p.p.m.) for the felsic granulites from Austria (filled symbols) and Czech Republic (empty symbols). Labelled vectors correspond to up to 50% fractional crystallisation of the main rock-forming minerals. Trace-element distribution coefficients are from Hanson (1978; Kfs, Amp, Rb in Pl), Icenhower & London (1995; Bt and Ms) and Blundy & Shimizu (1991; Sr in Pl, assuming 750°C, An_{15} and An_{50}).

2.4. Radiogenic isotopes

2.4.1. Neodymium The ε_{Nd}^{340} values for the felsic granulites are distinctly negative (−4·2 to −7·5, median = −6·1; Table 2 & Fig. 11a). This is reflected by rather high mean crustal residence ages (T_{Nd}^{DM}: two-stage depleted-mantle Nd model ages; Liew & Hofmann 1988) of 1·39–1·64 Ga (Fig. 11a), which corresponds to derivation from mature continental crust. The systematic shift in Nd isotopic compositions

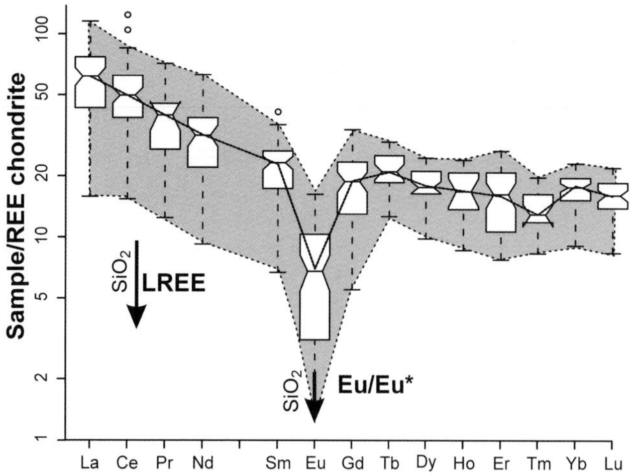

Figure 8 Chondrite-normalised plot for REE in felsic granulites. This new diagram ('spider boxplot') represents a combination of the more common chondrite-normalised REE plot with box and whiskers plots (boxplots). For each element, the box represents 50% of the population (being delimited by two quartiles); the horizontal line inside is the median and the 'whiskers' show the total range without outliers (denoted by small circles). The solid black arrows indicate trends with rising silica, if applicable. Normalisation values are from Boynton (1984).

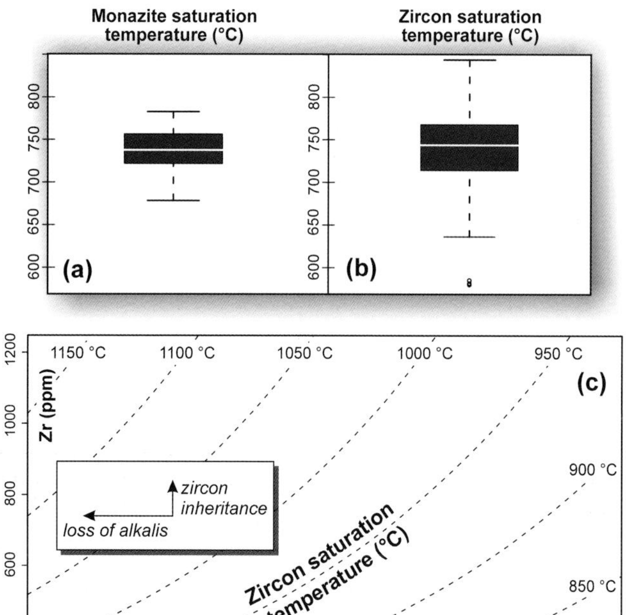

Figure 9 Monazite and zircon saturation calculations for felsic Moldanubian granulites. (a) Boxplot of monazite saturation temperatures (Montel 1993). (b) Boxplot of zircon saturation temperatures (Watson & Harrison 1983). (c) Binary plot of M = 100 (Na + K + 2Ca)/Al*Si versus Zr (p.p.m.) for the felsic granulites (see Fig. 4 for legend). Boxplots at the axes portray the statistical distribution of both parameters. The isotherms of Zr saturation levels were calculated using the model of Watson & Harrison (1983). The arrows indicate the effects of increasing zircon inheritance and (hypothetical) alkali loss, both of which would produce slight temperature overestimates.

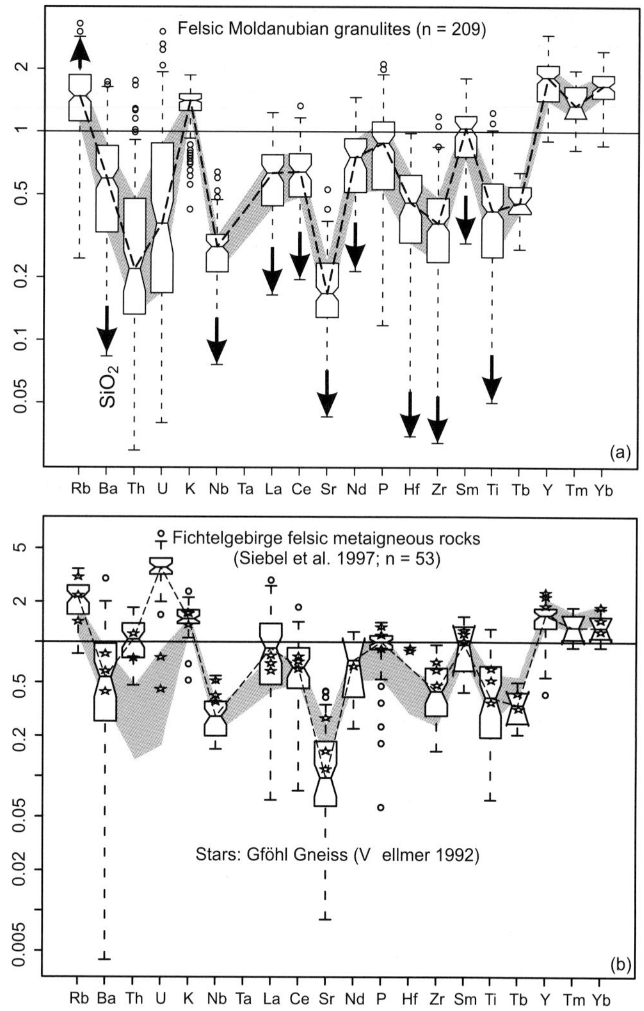

Figure 10 Average upper crust normalised (Taylor & McLennan 1985) 'spider boxplots' for felsic granulites (a) and acid metaigneous rocks from Fichtelgebirge (SiO_2 >70%; Siebel *et al.* 1997) (b). In the latter plot, stars denote three samples of Gföhl gneiss given by Vellmer (1992). An identical field corresponding to 50% of the felsic granulites is outlined in both diagrams. The solid black arrows indicate trends with increasing SiO_2.

between adjacent lighter (more negative ε_{Nd}^{340}) and darker facies (less negative ε_{Nd}^{340}) of the samples *Popp* (− 7·0/− 6·1) and *Ost* (− 7·5/− 5·7) rules out closed-system differentiation during the Variscan metamorphism. Small-scale isotopic disequilibrium was also evidenced by previous Sm–Nd (and U–Pb) garnet–whole-rock ages which show considerable scatter (470 to <330 Ma: Janoušek *et al.* 1996; Becker 1997; Chen *et al.* 1998; Prince *et al.* 2000). Conversely, granulites *Ysp-D* and *Ysp-L* have nearly the same ε_{Nd}^{340} values (− 6·2/− 6·4), in accordance with their mutually comparable major- and trace-element compositions.

2.4.2. Strontium. The newly measured Sr isotopic ratios show pronounced variations ($^{87}Sr/^{86}Sr_{340}$ ~0·711–0·770; Fig. 11b). The ratios demonstrate strong isotopic disequilibrium between adjacent lighter and darker samples *Popp* (0·730/0·719), *Ost* (0·753/0·715) and *Ysp* (0·771/0·738) at 340 Ma (Fig. 11b).

If one assumes an initial equilibrium state, the individual pairs would all give Early Palaeozoic ages in line with the results of the previous Rb–Sr dating (Arnold & Scharbert 1973; Frank *et al.* 1990). Similarly, the Blanský les granulites define an Ordovician isochron (469 ± 8 Ma; 2σ; $^{87}Sr/^{86}Sr_i$ ~0·706; Fig. 12a). Sample 1078, which is not included in the regression, could have been derived, along with 1076, from a

Table 2 Selected whole-rock Sr–Nd isotopic data for the (mostly felsic) Moldanubian granulites

Laboratory†	Popp-D CNRS	Popp-L CNRS	Ost-D® CNRS	Ost-L CNRS	St CNRS	Du© CNRS	Ysp-D CNRS	Ysp-L CNRS
Rb (p.p.m.)	162	209	212	288	260	72	249	284
Sr (p.p.m.)	97	65	137	37	34	99	43	30
^{87}Rb/^{86}Sr	4·85	9·36	4·49	22·9	22·5	2·11	16·9	27·9
^{87}Sr/^{86}Sr	0·74240	0·77558	0·73679	0·86389	0·86499	0·72075	0·81994	0·90574
2 s.e.	3	3	3	3	3	3	3	3
^{87}Sr/^{86}Sr$_{340}$*	0·71893	0·73026	0·71506	0·75322	0·75625	0·71055	0·73796	0·77059
Sm (p.p.m.)	7·64	5·53	5·6	1·88	3·29	6·34	4·59	3·53
Nd (p.p.m.)	34·0	23·8	24·8	7·03	12·1	35·2	18·5	13·6
^{147}Sm/^{144}Nd	0·1358	0·1402	0·1363	0·1617	0·1640	0·1090	0·1501	0·1571
^{143}Nd/^{144}Nd	0·512142	0·512200	0·512119	0·512266	0·512291	0·512073	0·512218	0·512224
2 s.e.	7	7	11	8	9	8	8	11
^{143}Nd/^{144}Nd$_{340}$*	0·511840	0·511888	0·511816	0·511906	0·511926	0·511830	0·511884	0·511874
ε_{Nd}^{340*}	− 7·0	− 6·1	− 7·5	− 5·7	− 5·4	− 7·2	− 6·2	− 6·4
T_{DM}^{Nd} (Ga)	1·60	1·53	1·64	1·50	1·47	1·62	1·54	1·55
^{87}Sr/^{86}Sr$_{470}$*	0·70993	0·71287	0·70672	0·71076	0·71453	0·70664	0·70651	0·71875
ε_{Nd}^{470*}	− 6·0	− 5·2	− 6·5	− 5·2	− 4·8	− 5·8	− 5·4	− 5·7

Laboratory†	1059 CGS	1062 CGS	1064 CGS	1074 CGS	1076 CGS	1078 CGS
Rb (p.p.m.)	72	134	99	254	230	158
Sr (p.p.m.)	72	63	66	26	39	102
^{87}Rb/^{86}Sr	2·65	5·84	4·04	26·4	16·1	4·30
^{87}Sr/^{86}Sr	0·72398	0·74516	0·73349	0·8784	0·81717	0·73802
2 s.e.	20	5	8	11	8	8
^{87}Sr/^{86}Sr$_{340}$*	0·71117	0·71691	0·713932	0·75045	0·73926	0·71722
Sm (p.p.m.)	6·7	5·1	0·7	2·8	4·7	4·3
Nd (p.p.m.)	33·0	20·9	2·20	10·0	20·8	19·1
^{147}Sm/^{144}Nd	0·1227	0·1475	0·19235	0·1693	0·1366	0·1361
^{143}Nd/^{144}Nd	0·512087	0·512206	0·512335	0·512301	0·512277	0·512199
2 s.e.	21	14	40	19	15	19
^{143}Nd/^{144}Nd$_{340}$*	0·511814	0·511878	0·511907	0·511924	0·511973	0·511896
ε_{Nd}^{340*}	− 7·5	− 6·3	− 5·7	− 5·4	− 4·4	− 5·9
T_{DM}^{Nd} (Ga)	1·64	1·55	1·50	1·48	1·4	1·52
^{87}Sr/^{86}Sr$_{470}$*	0·70626	0·70607	0·70643	0·70137	0·70936	0·70924
ε_{Nd}^{470*}	− 6·3	− 5·5	− 5·7	− 4·9	− 3·4	− 4·9

*Subscripts indicate the age to which were isotopic ratios corrected: 340 Ma (granulite-facies metamorphism: Kröner *et al.* 2000 and references therein) and 470 Ma (estimated protolith age): ε_{Nd}^{i} values were calculated using the Bulk Earth parameters given by Jacobsen & Wasserburg (1980); T_{DM}^{Nd} are two-stage Nd model ages based on Liew & Hofmann (1988). ®Mafic sample (SiO_2=63·90 wt.%; see Table 1). ©Charnockite.
†CNRS: Université Blaise Pascal, Clermont-Ferrand; analytical techniques described by Pin *et al.* (1990) and Roberts *et al.* (2000). CGS: Czech Geological Survey, Prague–Barrandov; Sm and Nd concentrations determined by isotope dilution, Rb and Sr by XRF (Harvey & Atkin 1981). Isotopic ratios were measured in static mode on a Finnigan MAT 262 spectrometer in the CGS. In both laboratories, the ^{143}Nd/^{144}Nd ratios were corrected for mass fractionation by normalisation to ^{146}Nd/^{144}Nd=0·7219, while ^{87}Sr/^{86}Sr ratios were normalised to ^{86}Sr/^{88}Sr=0·1194.

domain with a somewhat higher ^{87}Sr/^{86}Sr$_i$ (∼0·709; Fig. 12a). Alternatively, it may have been partly influenced by the nearby Variscan granitic dykes (Appendix 1). Nevertheless, the bootstrap spectrum for all the samples seems to confirm the Ordovician age (Fig. 12b).

Five thin-slab and one whole-rock Rb–Sr isotopic analyses from the St. Leonhard Massif were re-evaluated for comparison (Frank *et al.* 1990, without feldspar fraction and sample 1, rejected by the original authors because of substantial alteration; M. Thöni, pers. comm., 2004). The age obtained (425 ± 19 Ma, MSWD=5·8) seems to be distinctly younger than that for the Blanský les set. Again, the bootstrap spectrum (Fig. 12c), despite skewness, shows no ages lower than 400 Ma and the ∼430 Ma maximum remains rather well defined.

Furthermore, on the ^{87}Sr/^{86}Sr$_{340}$–ε_{Nd}^{340} plot (Fig. 11c), the granulites from the Czech Republic and Austria form broadly subparallel, partly overlapping fields. They run diagonally from less to more radiogenic Sr–Nd isotopic compositions, reaching ^{87}Sr/^{86}Sr$_{340}$ values >0·77, which are clearly above the initial ratios typical of common granitic rocks. The easiest explanation is that the data are not sufficiently age-corrected and that isotopic equilibrium during the Variscan was not attained. This agrees with the fact that the ε_{Nd}^{340} values become less negative with increasing silica along with the Sm/Nd ratios.

The significance of the individual Rb–Sr whole-rock ages should not be overrated since the system is known to be prone to resetting by metamorphism. Nonetheless, it is clear that most of the Sr isotopic ratios of the felsic granulites when corrected to Early Palaeozoic ages (see also Fig. 2) would yield ^{87}Sr/^{86}Sr$_i$ isotopic ratios expected for normal, slightly-peraluminous, crustally-derived granites (Fig. 11b; Table 2).

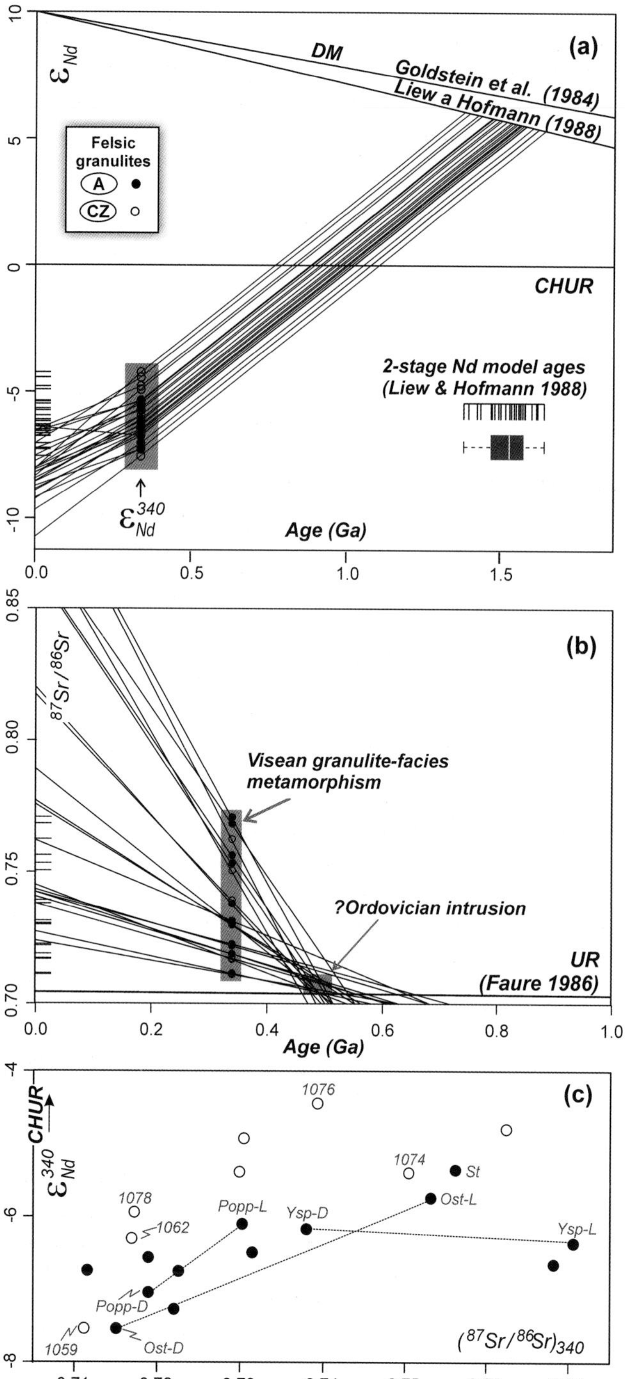

Figure 11 (a) Two-stage Nd development diagram for the felsic (except for *Ost-D*) Moldanubian granulites. DM = Depleted Mantle evolution lines after Goldstein *et al.* (1984) and Liew & Hofmann (1988). (b) Strontium isotopic plot. UR = Uniform (chondritic) Reservoir evolution line after Faure (1986). (c) Diagram of $(^{87}Sr/^{86}Sr)_{340}$ versus ε^{340}_{Nd} for felsic granulites. Tie lines join adjacent sample pairs from banded granulites (L = lighter; D = darker). The plot also includes data from Vellmer (1992), Valbracht *et al.* (1994) and Becker *et al.* (1999); new analyses are labelled by sample names.

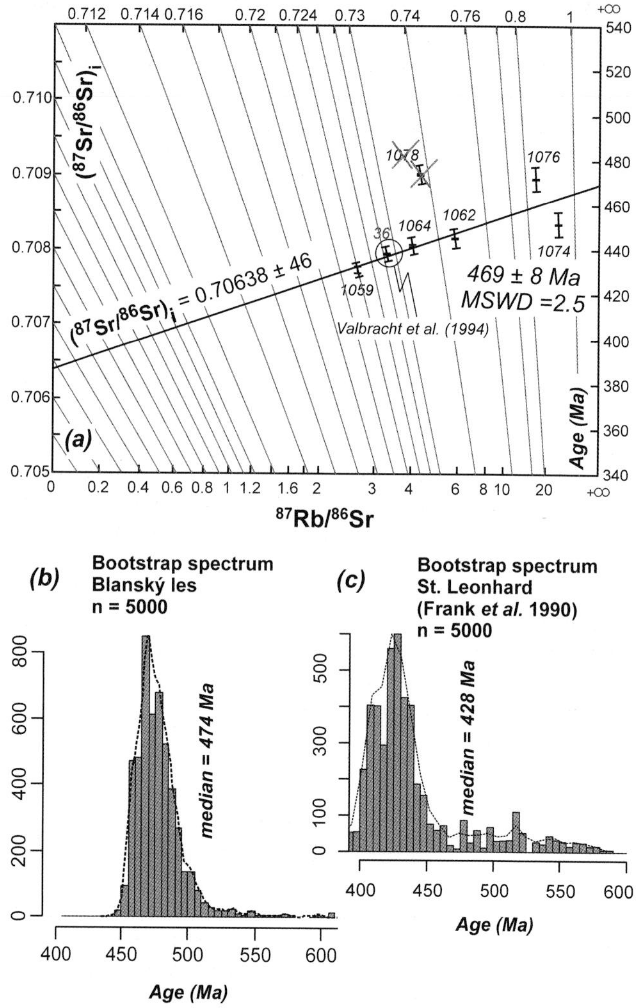

Figure 12 (a) Improved isochron diagram (Provost 1990) for Rb–Sr isotopic compositions of felsic granulites from the Blanský les Massif (this work and sample 36 of Valbracht *et al.* 1994). (b) Bootstrap spectrum (Diaconis & Efron 1983; Kalsbeek & Hansen 1989) obtained on the basis of 5000 randomly chosen data sets for seven points from Figure 12a (i.e. including 1078). (c) Bootstrap spectrum for six Rb–Sr isotopic analyses from the St. Leonhard Massif (Frank *et al.* 1990; feldspar fraction and sample 1 were omitted). The individual ages in both cases were calculated using the algorithm of York (1969) and the unpublished programme *BTRAP*.

3. Discussion

3.1. Pre-Variscan metagranites or Variscan high-P melts?

The CIPW-normative compositions of the felsic granulites fall close to a low-pressure minimum in the Ab–Qz–Or ternary diagram (Fiala *et al.* 1987a; Vrána 1989; Kotková & Harley 1999), and this demonstrates that they represent magma compositions with little cumulate or restite component. Indeed, most of the authors agreed that they were granitic or rhyolitic magmas at some stage of their evolution (Fiala 1987a, b; Carswell 1991; Vellmer 1992; Jakeš 1997; Kotková & Harley 1999). Even though there is near consensus on igneous heritage of the felsic Moldanubian granulites, it is important to discuss whether: (a) the bulk formed and segregated as high-pressure melts during Variscan subduction (Kotková & Harley 1999); or (b) they are merely Early Palaeozoic granites which were nearly isochemically metamorphosed during the Variscan collision, undergoing no or a negligible degree of partial melting.

In the present authors' view, there are several reasons why the latter model is more realistic:

3.1.1. Preservation of the pre-Carboniferous Rb–Sr whole-rock ages. As shown by the variation in the $^{87}Sr/^{86}Sr_{340}$ ratios and the preserved Ordovician/Silurian isochron ages, the whole-rock Rb–Sr system failed to re-equilibrate, even on a local scale, during the Variscan high-grade metamorphism. This leaves little scope for models which interpret the rocks as segregated Variscan granites and indicates that any granulite-facies partial melting did not lead to larger-scale chemical equilibrium. Given the correspondence of the Rb–Sr systematics with the SHRIMP ages for many inherited zircon cores, the Rb–Sr isotopes seem to have largely maintained memory of an older igneous protolith. The preservation of pre-metamorphic Sr isotopic signatures in high-grade meta-igneous rocks is by no means rare (e.g. Hradetzky & Lippolt 1993; Kühn *et al.* 2000). It could have been facilitated by the rarity of hydrous fluids, or melt immobility and isolation during the HP–HT stage, coupled with a presumably short duration of the metamorphic peak.

3.1.2. Zircon and monazite saturation temperatures. If the felsic granulites represent Variscan hot, restite-poor, high-PT melts, then they should have acquired Zr contents several times higher than observed (Fig. 9c). To invoke dramatic Zr-undersaturation (Kotková & Harley 1999) seems unrealistic because all felsic Moldanubian granulites contain abundant pre-Variscan, mainly Ordovician, zircon relics (Pöschl-Otrel 1995; Kröner *et al.* 2000; Friedl *et al.* 2003, 2004). For such low Zr concentrations, the saturation temperatures tend to be robust to changes in both the bulk magma composition and errors in the Zr determinations (Fig. 9c; see also Miller *et al.* 2003).

The fact that older zircon grains are usually overgrown by Variscan shells, interpreted as having crystallised from melt under granulite-facies conditions (Roberts & Finger 1997; Friedl *et al.* 2003), rules out the possibility that the inheritance is merely a result of shielding by restitic minerals. Thus, any partial melting during the Variscan high-grade metamorphism must have been volumetrically limited since considerably more restitic zircons should have dissolved otherwise.

Conversely, the observed strong negative SiO_2–Zr correlation (Fig. 6) seems to be indicative of zircon saturation in course of the melt fractionation (Chappell *et al.* 1998; Hoskin *et al.* 2000). As shown by the zircon saturation calculations, this must have been at temperatures much lower than 900–1000°C, and thus, most likely a feature of the protolith stage. As the volume proportion of pre-Ordovician cores is rather limited (Kröner *et al.* 2000; Friedl *et al.* 2003, 2004), the equivalent zircon and monazite saturation temperatures (~750°C) can be interpreted as a rough estimate of the crystallisation temperatures of the Early Palaeozoic protolith.

3.1.3. The HREE and Y signature. If the felsic granulites were essentially restite-depleted HP melts, then restite would inevitably be rich in garnet and kyanite. In such case, a significant HREE+Y depletion should be observed in the granulite samples, but this does not seem to be the case (Figs 8 & 10). The data of Kotková & Harley (1999), indicating that the granulite garnets typically show a decrease of garnet-compatible elements (HREE+Y) from core to rim, are indeed compatible with a Rayleigh-type fractionation during the crystal growth. However, this observation does not constrain the proportion of the melt involved (Otamendi *et al.* 2002).

3.2. Nature and origin of the igneous protolith to the Moldanubian granulites

3.2.1. Granite typology. The protolith to the Moldanubian granulites has probably corresponded to low-temperature granites (Miller *et al.* 2003). The geochemical characteristics resemble fractionated I-type granite suites, as defined in the Lachlan Fold Belt in eastern Australia. The similarities include high concentrations of Y and HREE at low-to-moderate Zr, a strongly negative Eu anomaly increasing with SiO_2 and generally low P_2O_5 contents (Chappell & White 1992; Chappell 1999). Compared to A-types, the Fe:Mg ratios and HFSE contents (Zr+Nb+Ce+Y ~ 200 ± 60 p.p.m.) of the granulites are too low (cf. Whalen *et al.* 1987; Eby 1990). The low Nb contents (mostly <10) of the granulites are a typical feature for granites derived from rather primitive, arc-type crustal sources (Pearce *et al.* 1984).

3.2.2. Possible source. With the assumption of granulite samples at SiO_2 ~70 wt.% to approximate the least differentiated granitic melts, this indicates relative paucity in Rb ($Rb^{[70]}$ ~100 p.p.m., superscript = approximate silica contents by weight) and Sr (~100 p.p.m., $Rb/Sr^{[70]}$ ~1), together with a high Ba content (600–800 p.p.m.).

The whole-rock geochemical character (A/CNK ~1·1, K_2O/Na_2O ~1·5, high Ba, low Rb, strongly negative ε^{470}_{Nd} at relatively unevolved $^{87}Sr/^{86}Sr_{470}$) argues for a partial melting of a quartzo–feldspathic source (orthogneiss or greywacke; Barbarin 1996 and references therein) fairly immature in terms of its major- and trace-element composition and with relatively low time-integrated Sm/Nd and Rb/Sr ratios. In addition, metapelitic parentage may be ruled on the basis of the high $CaO/Na_2O^{[70]}$ ratios (~0·6–0·8), fairly low $Al_2O_3/TiO_2^{[70]}$ (~25–40), $Rb/Ba^{[70]}$ (~0·1–0·3) and $Rb/Sr^{[70]}$ (~1) (Sylvester 1998). Apart from low Rb/Sr ratios, any significant role for muscovite dehydration melting can be discounted based on low $Sr/Ba^{[70]}$ = 0·1–0·2 (Harris & Inger 1992).

The distinction between melts produced from meta-psammitic parents and derived by anatexis of K-rich orthogneisses, respectively, remains fairly difficult using geochemical characteristics (e.g. Barbarin 1996; Sylvester 1998).

3.2.3. Magmatic evolution. Increasing Rb and decreasing Ba and Sr with increasing SiO_2 of the Moldanubian granulites (Figs 7b, c) are best explained by fractional crystallisation and are inconsistent with partial-melting models. Mass-balance calculations (Janoušek *et al.* 1997) indicate that unreasonably high degrees of partial melting (≫80%) would be required to yield a comparably broad compositional range of partial melts.

An origin by large-scale fractionation of more mafic (dioritic–tonalitic) parents is unlikely given the generally crustal-like Sr and Nd isotopic compositions of the felsic granulites. Indeed, the much rarer more mafic types have noticeably more primitive isotope compositions (Becker 1997; Janoušek *et al.* 2003b). In addition, mafic granulite types are fairly subordinate at the present surface level.

The trends seem to be compatible with fractional crystallisation involving K-feldspar>plagioclase from rather siliceous (SiO_2 ~70 wt.%) magma. Some role for biotite ± accessory oxide minerals (ilmenite?) is apparent in view of falling MgO, FeOt (Fig. 6), Nb and Zn contents (not shown). The drop in Zr and LREE can be accounted for by zircon and monazite fractionation, respectively (Fig. 6). The negative correlation of Sr, Ba, Zr, LREE and Eu/Eu* with the SiO_2, along with enrichment in Rb, are considered to be characteristic of evolved acid igneous rocks, crystallising feldspar, zircon and monazite dominated assemblages (Miller & Mittlefehldt 1984; Chappell & White 1992; Hoskin *et al.* 2000).

3.3. Geological correlation

It follows from the above that a specific, felsic igneous suite of Early Palaeozoic age should have been an important constituent of the subducted Gföhl terrane. In support of this, the associated Gföhl orthogneisses (e.g. Dudek *et al.* 1974; Cooke & O'Brien 2001) also show Ordovician protolith ages (Friedl *et al.* 2004) and whole-rock geochemical characteristics strongly resembling the felsic granulites (Fig. 10b; Vellmer 1992).

Geochemically similar felsic meta-igneous rocks can be found in the Saxothuringian Zone; for example, orthogneisses in the Orlica–Śnieżnik dome in Poland (Franke & Żelaźniewicz 2000; Turniak *et al.* 2000; Bröcker *et al.* 2003), or metagranites and metarhyolites in the Fichtelgebirge in Germany (Siebel *et al.* 1997; Wiegand 1997). Regrettably, detailed comparisons are hampered by a lack of reliable and sufficiently large whole-rock geochemical datasets.

Of the available data, the best characterised seem to be the meta-igneous rocks from the Fichtelgebirge (Siebel *et al.* 1997; Wiegand 1997). As follows from Figure 10b, their analyses with SiO_2 >70% match well the overall compositional spectrum for the felsic granulites in most of the elements save U and Th. Unfortunately, there is a dearth of data on Cs contents: the median of four samples analysed by Wiegand (1997) was 4·8 p.p.m., i.e. slightly lower than the upper crustal average of 5·8 p.p.m. (Wedepohl 1995). The most reliable data from felsic granulites (Vellmer 1992; Becker *et al.* 1999) seem to show that these are markedly Cs-impoverished (0·1–2·1 p.p.m., median=0·9).

The meta-igneous rocks from Fichtelgebirge were dated at 455–480 Ma (Siebel *et al.* 1997; Wiegand 1997 and references therein), which falls within the spectrum of inherited ages observed in the granulites. Despite considerable disturbance of the Rb–Sr isotopic system in many of the Fichtelgebirge samples, some still yield Ordovician ages and initial ratios (Siebel *et al.* 1997) comparing favourably with the mean of the felsic granulites at that time (~0·709). Moreover, the ε^{480}_{Nd} values (orthogneisses: $-3{\cdot}5 \pm 0{\cdot}4$, metarhyolites: $-4{\cdot}8 \pm 1{\cdot}1$) also match the mean ε^{480}_{Nd} ($-4{\cdot}9 \pm 1{\cdot}9$) for the felsic granulites and overlap with the data for the Gföhl gneiss ($-4{\cdot}4$ to $-6{\cdot}1$; Vellmer 1992).

3.4. Geochemical changes during the granulite-facies metamorphism

Most of the felsic Moldanubian granulites show undepleted LILE contents. Unlike in many other granulite terrains around the world (Rudnick *et al.* 1985; Rudnick & Presper 1990), prograde metamorphism appears to have occurred largely isochemically. This implies that high-pressure melting in felsic granulites was widely suppressed and small-fraction partial melts were (mostly?) incapable of segregating from their source.

Nevertheless, a chemical feature that needs explanation is the distinct U, Th and Cs depletion compared to the average upper crust or the otherwise geochemically similar Saxothuringian meta-igneous rocks. As proposed already by Fiala *et al.* (1987a, b) and Vellmer (1992), these elements could have been stripped out by fluid or small-scale melt during the high-pressure metamorphism. The strong Cs impoverishment (Vellmer 1992; Becker *et al.* 1999) is a well-known feature of many granulite-facies terrains (Rudnick & Taylor 1987; Rudnick & Presper 1990). This effect has been ascribed to the fractionation between K-feldspar and melt/fluid (Rudnick & Presper 1990; Hart & Reid 1991; Morgan & London 2003).

Minor fluid or melt loss is compatible with the preservation of the nearly pristine high-pressure mineralogy (garnet+kyanite+rutile) in many granulites. Otherwise, as argued by Powell & Downes (1990), Clemens & Droop (1998), Brown (2002) and White & Powell (2002), the retention of the aqueous fluids (or hydrous melt) should have, upon retrograde cooling, led to the back-reaction of alumosilicates and garnet to biotite and muscovite.

3.4.1. Role of fluids. On the basis of the experimental work by Keppler & Wyllie (1990), a major role for Cl- and CO_2–rich fluids can be probably discounted, since U and Th are almost equally depleted but undecoupled (Fig. 10a). Interestingly, rare glimmerite veins, occurring in Austrian peridotite bodies adjacent to granulite massifs, show high concentrations of LILE, U and Th, accompanied by low Rb/Cs and K/Rb ratios (Becker *et al.* 1999). These contain Sr and Nd with isotopic compositions which overlap with the granulites at c. 340 Ma. According to Becker *et al.* (1999), high Zn and F, low Sr, negative Eu anomaly, as well as low HFSE, all point to origin of glimmerites through crystallisation from a F-bearing aqueous-carbonic fluid released during the high-temperature, high-pressure devolatilisation in the granulite massifs.

3.4.2. Melt loss. A melt would be two orders of magnitude more efficient in partitioning U and Th than a pure H_2O-rich fluid (Keppler & Wyllie 1990). The present authors have constrained potential metamorphic assemblages, partial melting and volatile regimes by pressure–temperature and temperature–X(H_2O) pseudosections (Powell *et al.* 1998), calculated using the internally consistent thermodynamic dataset (Holland & Powell 1998) and the associated melt model (Holland & Powell 2001; White *et al.* 2001). Solid-solution models were taken from the following sources: (feldspar) Fuhrman & Lindsley (1988) and Wen & Nekvasil (1994); (biotite) Powell & Holland (1999); (garnet and phengite) Coggon & Holland (2002) and (clinopyroxene) Holland & Powell (1996); hydrous cordierite was considered ideal. Computations were performed with the Perplex computer software (Connolly & Petrini 2002) for average granulite composition of 81·03 mol.% SiO_2, 0·209 mol.% TiO_2, 8·54 mol.% Al_2O_3, 1·89 mol.% FeO_t, 0·868 mol.% MgO, 1·113 mol.% CaO, 3·17 mol.% Na_2O and 3·19 mol.% K_2O, at variable H_2O concentrations.

The representative felsic granulite has a relatively monotonous and modally insensitive pre-peak mineral assemblage: biotite (4·5–5·5 vol.%), white mica (5·4–6·4 vol.%), plagioclase (27·7–28·3 vol.%), K-feldspar (22·2–23·0 vol.%) and quartz (37·7–38·1 vol.%) with accessory garnet (0·2–0·3 vol.%), ilmenite or rutile (0·1–0·3 vol.%). The bulk-rock H_2O budget, prior to the mica dehydration reactions, is 0·49–0·51 wt.%. During prograde burial, breakdown of micas leads to the assemblage feldspar+garnet+kyanite+quartz+rutile+granitic melt. However, the fertility of the protolith is severely restricted by the low modal abundances of micas. The calculated isopleths of melt fraction, produced by dehydration melting, are illustrated in the pressure–temperature space together with the available P–T estimates for the Moldanubian granulites (Fig. 13).

The overall *dP/dT* slope of the melt-fraction isopleths is approximately parallel to the fluid-absent haplogranitic melting curve (Holtz *et al.* 2001) and shifts to progressively higher temperatures with increasing pressure. This explains both the infertility of the granulites at the presumed metamorphic peak conditions and the irregular distribution of the melt-fraction isopleths (Fig. 13). Initial melting, limited by mica availability, does not exceed 10 vol.% for any of the granulite decompression paths. Rapid increase of the melt fraction would occur only at about 1100 °C and 1 GPa and 1200 °C and 2 GPa because of the closely approaching fluid-absent haplogranitic solidus.

These results are in a very good agreement with estimated melt production at the peak P–T conditions (1000 °C and 1·6 GPa) of 5–15 vol.% (Roberts & Finger 1997). Such low degrees of partial melting suggest that granulites have undergone very limited anatexis. It can be argued that the partial melt would start to escape only following the breakdown of the framework of residual minerals, after reaching the Melt Escape Threshold (>20–25 vol.% melt) (Vigneresse *et al.* 1996; Sawyer 1999). However, some authors have suggested that as little as 5 vol.% of melt can be efficiently separated if deformation is involved (Rutter & Neumann 1995). On this basis, some degree of melt loss from the felsic granulites cannot be ruled out.

The present authors have estimated the effect of extraction of a small-degree partial melt on the LILE budget of an average felsic granulite. To their knowledge, no suitable K_ds exist for hypersolvus feldspar, and hence, they have employed

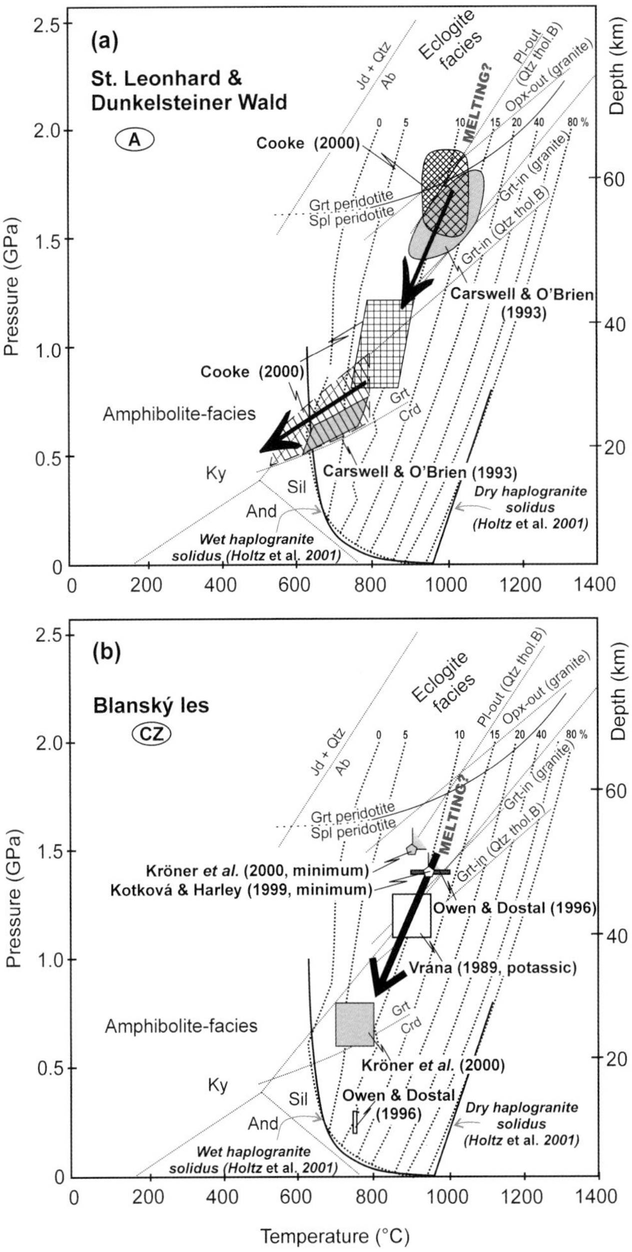

Figure 13 Pressure–temperature diagram for the Moldanubian granulites (including both felsic and mafic types). The water-saturated and dry haplogranite solidi are from Holtz *et al.* (2001), and the petrogenetic grid from Cooke (2000). Data sources: (a) Lower Austria (Carswell & O'Brien 1993; Cooke 2000); and (b) Blanský les Massif, Southern Bohemia (Vrána 1989; Owen & Dostal 1996; Kotková & Harley 1999; Kröner *et al.* 2000). Isopleths represent the volume fraction of partial melt produced by mica dehydration melting (see text for calculation procedure).

those for K-feldspar/rhyolitic melt (Fig. 14). Taking into account the considerable scatter in the K_d values, some 0·4–12·5% batch melting would be sufficient to deplete Cs from 4·8 p.p.m. (an average Fichtelgebirge orthogneiss, see preceding section) to 0·9 (felsic granulite). On the other hand, Sr and Ba concentrations would be nearly unaffected by comparably small degrees of partial melting (Fig. 14). Even for Rb, published K_ds (Leeman & Phelps 1981; Mahood & Hildreth 1983; Stix & Gorton 1990; Ewart & Griffin 1994) suggest moderately incompatible behaviour. Thus, Rb depletion would be imperceptible at lower degrees of melting. In this respect, it is worth noting that several samples seem to have slight excess radiogenic Sr (Fig. 11b) compatible with a minor Rb loss in course of the Variscan high-grade metamorphism. However,

Element	K_d (Kfs/rhyolite)	References*	Depletion factor† (5 % melting)	Depletion factor† (10 % melting)
Cs	0.002-0.061	1-3	0.02-0.36	0.01-0.22
Rb	0.11-0.74	1-4	0.51-0.91	0.34-0.83
Sr	2-22.1	3-4	0.99-1.05	0.99-1.10
Ba	1-44	1-4	0.94-1.05	0.89-1.10

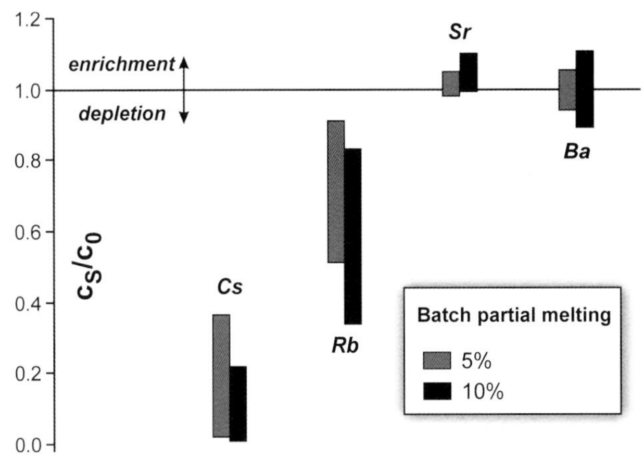

Figure 14 Modelling effects of slight melt loss on LILE budget (c_S/c_0=change in element concentration of the residue relative to the original source composition). Batch melting, estimated residual mineralogy 45% alkali feldspar+45% Qtz+10% Grt. Distribution coefficients K-feldspar/rhyolitic melt are from: (1) Leeman & Phelps (1981), (2) Mahood & Hildreth (1983), (3) Stix & Gorton (1990) and (4) Ewart & Griffin (1994); garnet and quartz were assumed to contain no appreciable concentrations of the modelled elements. †The concentration in the residue=concentration in the source × depletion factor.

the real significance of this observation might be doubtful since these are the most felsic samples, with $^{87}Rb/^{86}Sr$ ratios sensitive to small errors in Sr determinations (Jahn *et al.* 2000).

If a protolith composition similar to Fichtelgebirge meta-igneous rocks (Siebel *et al.* 1997) is assumed, depletion in a similar fashion to Cs would need to be invoked for Th (11→2·1 p.p.m.) and U (9→0·95 p.p.m.) as well. Rudnick & Presper (1990) demonstrated that both elements can strongly partition into melt/fluid, especially if the role of U–Th bearing accessory phases was limited.

In summary, additional work is needed to shed more light on the mechanism of the striking U, Th and Cs depletion in the Moldanubian granulites. Given the results of the thermodynamic modelling, only a small melt fraction was likely to be present at the high-grade metamorphic peak and any melt loss would be rather unlikely. This still would have little effect on the whole-rock geochemical signatures, including the LIL elements other than Cs.

3.5. Geotectonic setting

Taken together, the Ordovician meta-igneous crust with a composition resembling that preserved in the Fichtelgebirge could have yielded rocks similar to felsic Moldanubian granulites upon subduction to depths of at least 50–60 km (O'Brien 2000). This process was accompanied by surprisingly negligible geochemical modification.

The most appropriate geotectonic setting for the Ordovician igneous activity widespread throughout Variscan Europe (e.g. Pin & Marini, 1993) seems to be an intracontinental or back-arc rift, where crustal thinning and heating by upwelling mantle could have led to melting of various crustal lithologies (Siebel *et al.* 1997; Linnemann *et al.* 2000; Franke 2000). The high Nd model ages testify that the source of the Ordovician igneous rocks parental to the felsic granulites could have been a pre-existing crust, possibly Cadomian I-type granodiorites or geochemically undistinguishable immature psammitic (meta-) sediments. Such material was readily available at the active Gondwana margin, from which the Moldanubian and Saxothuringian crust were derived (e.g. Nance & Murphy 1996; Linnemann *et al.* 2000; Winchester 2002).

4. Conclusions

1. The majority of the felsic Moldanubian granulites do not seem to represent separated, high-pressure Variscan syn-collisional melts. The main arguments include:
   - concentrations of Zr far too low compared to calculated saturation levels at $T \geq 900°C$, coupled with presence of significant, mainly Ordovician, zircon inheritance;
   - consistently low zircon and monazite saturation temperatures;
   - preservation of Ordovician/Silurian whole-rock and thin slab Rb–Sr ages; and
   - high HREE+Y, ruling out the presence of larger amounts of garnet in the residue.
2. Instead, the felsic granulites are interpreted as metamorphosed (partly anatectic) equivalents of an older igneous complex.
3. Protoliths for the felsic granulites and Gföhl gneiss were compositionally similar to some meta-igneous rocks preserved in the Saxothuringian Zone; for instance, Ordovician–Silurian metarhyolites–orthogneisses from Fichtelgebirge. These were probably rift-related rocks which could have originated by remelting of Cadomian crust, such as I-type granodiorites or immature psammitic (meta-) sediments.
4. Trends of rapidly falling Sr, Ba, Zr, LREE and P with slightly increasing Rb and SiO_2 characteristic of evolved granites crystallising K-feldspar>plagioclase+biotite> zircon+monazite seem to have been inherited from the igneous stage.
5. During the Variscan orogeny, the Early Paleozoic acid meta-igneous crust was subducted to depths of at least 50–60 km. A moderately small melt fraction was presumably present during much of the metamorphic history. Most of the geochemical signature was conserved despite the HP–HT metamorphism, which led only to the loss of H_2O, U, Th and Cs.
6. The ultimate mechanism of the U, Th and Cs depletion is uncertain. It could have been caused by fluid loss or by removal of a small-degree (<10%) partial melt. While loss of such a small-degree partial melt would strip off most of the Cs, and presumably also U and Th, other LILE (in particular Rb, Ba and Sr) would be little affected.
7. In favourable cases (e.g. the absence of fluids and the short duration of the event), the Rb–Sr whole-rock system might be rather resistant to resetting during granulite-facies metamorphism.
8. Deep burial and HP–HT metamorphism of infertile crustal lithologies, such as felsic meta-igneous rocks, does not need to cause extensive partial melting. The resulting granulites would have largely undepleted geochemical signature and preserve much of geochemical information from their protolith.

5. Acknowledgements

This work originated during the research stay of V J at the Institute of Mineralogy, University of Salzburg, in the framework of the FWF Project 15133–GEO (FF). This is a continuation of previous FWF projects (M 287 and P 11674), which funded the research stay of M R in Salzburg from 1995 to 1998. Further support from the Czech Grant Agency and the Austrian–French cooperation programme Amadeus (2/96 and 4/97b) is gratefully acknowledged. The paper benefited from constructive and detailed reviews by P. Nabelek and an anonymous reviewer, as well as editorial handling by S. Harley. G. Friedl kindly provided her so-far-unpublished SHRIMP data and M. Thöni a table of Rb–Sr isotopic analyses plotted by Frank *et al.* (1990).

We would like to thank J. Trnková, along with V. Kopecký, J. Zeman and C. Bosq, for technical assistance in the isotope laboratories of Prague and Clermont–Ferrand, V. Sixta for major-element and REE chemical analyses, and C. Bertoldi for help with preparation of thin sections.

Thanks are also due to W. Siebel, B. Wiegand, J. Konopásek and Č. Tomek for advice, S. Vrána for his unbeatable knowledge of South Bohemian granulites, inspiring discussions and providing some samples, as well as P. O'Brien for organising a thought-provoking workshop on the (U)HP metamorphic rocks in 2002.

6. Appendix 1. Petrography of the studied samples

Popp-D/L: *Quarry of Natursteine Popp E of Ramsau, St. Leonhard Massif (GR 6892 3894)*
Ost-D/L: *Roadside exposure at Osterberg, on the Pöchlarn–Zelking road, Pöchlarn–Wieselburg Massif (GR 6873 3399)*

Sample pairs *Ost-L/D* and *Popp-L/D* represent adjacent lighter and darker granulite bands. Characteristic for the darker facies is the presence of considerable amounts of small biotite flakes (3–10%), which often define a weak foliation within the fine-grained (<0·1 mm) quartz–orthoclase–plagioclase matrix. Additionally, there is always garnet (<4 mm) and often some kyanite (<1 mm) in the darker bands.

The lighter varieties are typical felsic, massive granulites with generally very little biotite. They contain c. 2–5% garnet (0·2–4 mm across), the bigger of which often enclose inclusions of relict mesoperthite and sometimes show fine, exsolved rutile needles. The fine-grained (<0·1 mm) matrix consists of roughly 30–40% quartz, 25–30% K-feldspar and 25–30% sodic plagioclase. A few single, elongate kyanite crystals up to 4 mm are present, as well as accessory brown euhedral rutile (<1 mm).

Ysp-D/L: *Roadside quarry 'an der Gleisen', Yspertal (GR 6530 3422)*

The granulite at this locality is well banded and associated with a serpentinite body. Although sample *Ysp-D* is of clearly darker appearance than *Ysp-L*, there is no large compositional difference between the two. The former contains a small percentage of biotite (small, evenly distributed flakes) and has a finer grained, mylonitic groundmass. On the other hand, *Ysp-L* is a typical felsic granulite with rare garnets set in a quartzo–feldspathic matrix. Both contain some kyanite grains.

Du: *Small abandoned quarry west of Lindwurmkreuz, Dunkelsteiner Wald (GR 6903 3549)*

Sample *Du* is a dark-coloured granulite with a biotite-poor, charnockitic mineral paragenesis (Opx+Cpx+Grt) coming from a megaboudin c. 1–1·5 km long and up to 9 m thick. The sample contains large amounts of plagioclase (40%) and quartz (30%) and some K-feldspar (10%). All these minerals form a fine-grained mosaic.

St: *Road cut near Steinaweg, Dunkelsteiner Wald (GR 6953 3593)*

Leucocratic, fine- to medium-grained, quartz-rich felsic granulite with alternating Grt-free and Grt-bearing layers. Within the weakly folded matrix, numerous dark pink/red garnets are visible, up to 5 mm in diameter. Particularly striking features are the widespread pale blue kyanite grains up to 4 mm in length, often aligned to the foliation defined by quartz ribbons and sparse biotite. Compared to most of the acidic rocks found in the Austrian granulite occurrences, the peak metamorphic assemblage of Grt + Ky + hypersolvus feldspar + Qtz is remarkably preserved in this sample, a reason why it was chosen for a detailed geochronological study (Friedl *et al.* 2003, 2004). Within the matrix, abundant mesoperthitic K-feldspar porphyroclasts up to 3 mm across are visible with only minor recrystallisation around their margins to granoblastic fine-grained K-feldspar and subordinate plagioclase. Accessory phases are rutile, apatite, zircon and ilmenite.

1059–1078: *Borehole H-1 at Holubov, 8 km N of Český Krumlov, Blanský les Massif*

The borehole drilled in 1965 penetrated mainly granulitic rocks, with leucogranulites prevailing over more mafic types. The rest was accounted for by several tectonic slices of ultrabasic rocks and the whole assemblage was cut by infrequent dykes of Variscan aplites–pegmatites. While systematic petrographic account (including 50 modal analyses of stained granulite samples) was published by Kodym *et al.* (1978), Fediuková (1978) focused solely on mineralogy and chemistry of mafic phases.

The felsic granulites with peak metamorphic assemblage hypersolvus feldspar + quartz + garnet + rutile ± kyanite show effects of variable retrogression and mylonitic deformation. These processes resulted mainly in breakdown of garnet to a secondary biotite, as well as recrystallisation of the mesoperthite into a K-feldspar and plagioclase mosaic. The present samples coming from near the bottom of the borehole (c. 1100–1400 m; Fig. 15) were selected so that they would be as little affected as possible, all having inequigranular granoblastic texture with only indistinct plane-parallel fabric. However, differences do exist. Some (1074 and 1076) contain practically no biotite, still preserving large mesoperthite augen. The others were nearly completely recrystallised to an equilibrated mosaic of two feldspars and quartz (1062), and contain a large proportion of secondary biotite, sometimes altered to chlorite (1064 and 1078). The replacement of garnet porphyroclasts by lath-shaped biotite started at the margins and proceeded along cracks (1064). Rarely tiny flakes concentrated into thin stripes within the matrix, defining a weak foliation (1062).

In all cases, the mineral assemblage is dominated by quartz, K-feldspar (mesoperthitic microcline) and plagioclase ($\sim An_{10-20}$, in less felsic types ranging to $\sim An_{30}$). These minerals occur in about equal proportions in more felsic types, but the plagioclase content increases at the expense of Kfs to c. 40 vol.% in the less felsic ones (1059). Garnet porphyroclasts (up to 7 vol.%), which are often poikilitic, are conspicuous enclosing mainly subhedral perthite grains. The contents of garnet and rutile drop sharply from less to more silica-rich samples. The rutile forms both euhedral prismatic crystals in the matrix as well as, in less felsic samples (1059), a plethora of unmixed oriented needles in the garnet porphyroclasts (e.g. Zhang *et al.* 2003). In the two most felsic granulites, there were rather abundant kyanite porphyroclasts (1074 and 1078: up to 1 vol.%), some elongate and subhedral, others lobate, enclosing perthite grains. On the other hand, the least felsic sample, 1059, contained kyanite with hercynite coronas, the host mineral having been sometimes completely consumed. In all cases, common accessories were ilmenite, zircon and apatite.

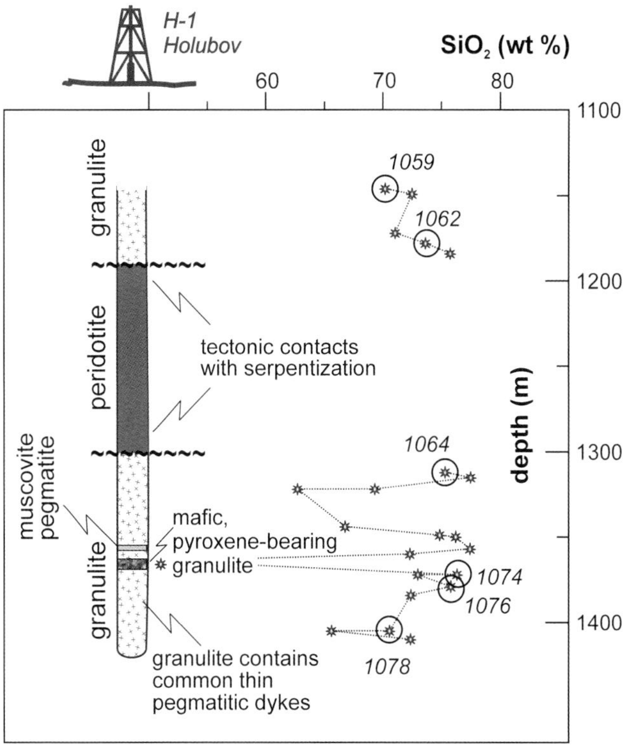

Figure 15 Simplified petrographic profile for the sampled part of borehole H-1, Holubov, Blanský les Massif. The location of the samples mentioned in the text as well as SiO_2 variations with depth are shown. In part after Kodym *et al.* (1978).

7. References

Aftalion, M., Bowes, D. R. & Vrána, S. 1989. Early Carboniferous U–Pb zircon age of garnetiferous, perpotassic granulites, Blanský les massif, Czechoslovakia. *Neues Jahrbuch für Mineralogie, Monatshefte* **4**, 145–52.

Arnold, A. & Scharbert, H. G. 1973. Rb–Sr Altersbestimmungen an Granuliten der südlichen Böhmischen Masse in Österreich. *Schweizerische mineralogische und petrografische Mitteilungen* **53**, 61–78.

Ayres, M. & Harris, N. 1997. REE fractionation and Nd-isotope disequilibrium during crustal anatexis: constraints from Himalayan leucogranites. *Chemical Geology* **139**, 249–69.

Barbarin, B. 1996. Genesis of the two main types of peraluminous granitoids. *Geology* **24**, 295–8.

Becker, H. 1997. Sm–Nd garnet ages and cooling history of high-temperature garnet peridotite massifs and high-pressure granulites from Lower Austria. *Contributions to Mineralogy and Petrology* **127**, 224–36.

Becker, H., Wenzel, T. & Volker, F. 1999. Geochemistry of glimmerite veins in peridotites from Lower Austria – implications for the origin of K-rich magmas in collision zones. *Journal of Petrology* **40**, 315–38.

Blundy, J. D. & Shimizu, N. 1991. Trace element evidence for plagioclase recycling in calc–alkaline magmas. *Earth and Planetary Science Letters* **102**, 178–97.

Boynton, W. V. 1984. Cosmochemistry of the rare earth elements: meteorite studies. *In* Henderson, P. (ed.) *Rare Earth Element Geochemistry*, 63–114. Amsterdam: Elsevier.

Bröcker, M., Lange, U., Mezger, K. & Żelaźniewicz, A. 2003. The orthogneisses of the Orlica–Śnieżnik dome (West Sudetes, Poland): Sr–Nd isotope characteristics and Rb–Sr geochronology. *Journal of the Czech Geological Society* **48**, 25–6.

Broska, I., Petrík, I. & Williams, C. T. 2000. Coexisting monazite and allanite in peraluminous granitoids of the Tribeč Mountains, Western Carpathians. *American Mineralogist* **85**, 22–32.

Brown, M. 2002. Retrograde processes in migmatites and granulites revisited. *Journal of Metamorphic Geology* **20**, 25–40.

Carswell, D. A. 1991. Variscan high P–T metamorphism and uplift history in the Moldanubian Zone of the Bohemian Massif in Lower Austria. *European Journal of Mineralogy* **3**, 323–42.

Carswell, D. A. & O'Brien, P. J. 1993. Thermobarometry and geotectonic significance of high-pressure granulites: examples from the Moldanubian zone of the Bohemian Massif in Lower Austria. *Journal of Petrology* **34**, 427–59.

Chappell, B. W. 1999. Aluminium saturation in I- and S-type granites and the characterization of fractionated haplogranites. *Lithos* **46**, 535–51.

Chappell, B. W., Bryant, C. J., Wyborn, D., White, A. J. R. & Williams, I. S. 1998. High- and low-temperature I-type granites. *Resource Geology* **48**, 225–35.

Chappell, B. W. & White, A. J. R. 1992. I- and S-type granites in the Lachlan Fold Belt. *Transactions of the Royal Society of Edinburgh: Earth Sciences* **83**, 1–26.

Chen, F., Hegner, E. & Todt, W. 1998. Garnet dating of granulites from the Variscan foldbelt in Central Europe; evidence for early and pre-Variscan high grade metamorphism. *Acta Universitatis Carolinae, Geologica* **42**, 221–2.

Clemens, J. D. & Droop, G. T. R. 1998. Fluids, P–T paths and the fates of anatectic melts in the Earth's crust. *Lithos* **44**, 21–36.

Coggon, R. & Holland T. J. B. 2002. Mixing properties of phengitic micas and revised garnet–phengite geothermometers. *Journal of Metamorphic Geology* **20**, 683–96.

Connolly, J. A. D. & Petrini, K. 2002. An automated strategy for calculation of phase diagram sections and retrieval of rock properties as a function of physical conditions. *Journal of Metamorphic Geology* **20**, 697–708.

Cooke, R. A. 1999. *The petrology of high-pressure granulites from the St. Leonhard Granulite Massif, Gföhl Unit, Lower Austria.* Unpublished Ph.D. thesis. University of Sheffield.

Cooke, R. A. 2000. High-pressure/temperature metamorphism in the St. Leonhard Granulite Massif, Austria: evidence from intermediate pyroxene-bearing granulites. *International Journal of Earth Sciences* **89**, 631–51.

Cooke, R. A. & O'Brien, P. J. 2001. Resolving the relationship between high P–T rocks and gneisses in collisional terranes: an example from the Gföhl gneiss–granulite association in the Moldanubian Zone, Austria. *Lithos* **58**, 33–54.

Čadková, Z., Jakeš, P., Haková, M. & Mrázek, P. 1985. Geochemical catalogue of the basic network. *In Lithogeochemical Database of the Czech Geological Survey, Prague.*

Dallmeyer, R. D., Franke, W. & Weber, K. (eds) 1995. *Pre-Permian Geology of Central and Eastern Europe.* Berlin: Springer-Verlag.

Debon, F. & Le Fort, P. 1983. A chemical–mineralogical classification of common plutonic rocks and associations. *Transactions of the Royal Society of Edinburgh: Earth Sciences* **73** (for 1982), 135–49.

Diaconis, P. & Efron, B. 1983. Computer-intensive methods in statistics. *Scientific American* **248**, 96–108.

Dudek, A., Matějovská, O. & Suk, M. 1974. Gföhl orthogneiss in the Moldanubicum of Bohemia and Moravia. *Krystalinikum* **10**, 67–78.

Eby, G. N. 1990. The A-type granitoids: a review of their occurrence and chemical characteristics and speculations on their petrogenesis. *Lithos* **26**, 115–34.

Ewart, A. & Griffin, W. L. 1994. Application of proton-microprobe data to trace-element partitioning in volcanic rocks. *Chemical Geology* **117**, 251–84.

Faure, G. 1986. *Principles of Isotope Geology.* Chichester: John Wiley & Sons.

Fediuková, E. 1978. Mafic minerals from granulites of the borehole Holubov (South Bohemian Moldanubicum). *Sborník geologických věd, Ložisková geologie–mineralogie* **19**, 169–98.

Fiala, J., Matějovská, O. & Vaňková, V. 1987a. Moldanubian granulites and related rocks: petrology, geochemistry, and radioactivity. *Rozpravy Československé akademie věd, řada matematických a přírodních věd* **97**, 1–102.

Fiala, J., Matějovská, O. & Vaňková, V. 1987b. Moldanubian granulites: source material and petrogenetic considerations. *Neues Jahrbuch für Mineralogie, Abhandlungen* **157**, 133–65.

Fiala, J., Fuchs, G. & Wendt, J. I. 1995. Moldanubian Zone – stratigraphy. *In* Dallmeyer, R. D., Franke, W. & Weber, K. (eds) *Pre-Permian Geology of Central and Eastern Europe*, 417–428. Berlin: Springer.

Frank, W., Hammer, S., Popp, F., Scharbert, S. & Thöni, M. 1990. Isotopengeologische Neuergebnisse zur Entwicklungsgeschichte der Böhmischen Masse. *Österreichische Beiträge zu Meteorologie und Geophysik* **3**, 185–228.

Franke, W. 2000. The mid-European segment of the Variscides: tectonostratigraphic units, terrane boundaries and plate tectonic evolution. *In* Franke, W., Haak, V., Oncken, O. & Tanner, D. (eds) *Orogenic Processes: Quantification and Modelling in the Variscan Belt*, 35–61. *Geological Society, London, Special Publication* **179**. London: Geological Society.

Franke, W. & Żelaźniewicz, A. 2000. The eastern termination of the Variscides; terrane correlation and kinematic evolution. *In* Franke, W., Haak, V., Oncken, O. & Tanner, D. (eds) *Orogenic Processes: Quantification and Modelling in the Variscan Belt*, 63–86. *Geological Society, London, Special Publication* **179**. London: Geological Society.

Friedl, G., Cooke, R., Finger, F., McNaughton, N. J. & Fletcher, A. 2003. U–Pb SHRIMP dating and trace element investigations on multiple zoned zircons from a South-Bohemian granulite. *Journal of the Czech Geological Society* **48**, 51–2.

Friedl, G., Finger, F., Paquette, J. L., von Quadt, A., McNaughton, N. J., & Fletcher, I. R. 2004. Pre-Variscan geological events in the Austrian part of the Bohemian Massif deduced from U/Pb zircon ages. *International Journal of Earth Sciences* **93**, 802–23.

Fuchs, G. & Matura, A. 1976. Zur Geologie des Kristallins der südlichen Böhmischen Masse. *Jahrbuch der Geologischen Bundesanstalt* **119**, 1–43.

Fuhrman, M. L. & Lindsley D. H. (1988) Ternary-feldspar modeling and thermometry. *American Mineralogist* **73**, 201–15.

Goldstein, S. L., O'Nions, R. K. & Hamilton, P. J. 1984. A Sm–Nd isotopic study of atmospheric dusts and particulates from major river systems. *Earth and Planetary Science Letters* **70**, 221–36.

Gürtlerová P., Čadek J., Čadková Z., Mrázek P., Adamová M. & Dušek P. 1997. Database of analytical determinations on maps of rock geochemical reactivity 1:50 000. *In Lithogeochemical Database of the Czech Geological Survey, Prague.*

Hanson, G. N. 1978. The application of trace elements to the petrogenesis of igneous rocks of granitic composition. *Earth and Planetary Science Letters* **38**, 26–43.

Harris, N. B. W. & Inger, S. 1992. Trace element modelling of pelite-derived granites. *Contributions to Mineralogy and Petrology* **110**, 46–56.

Hart, S. R. & Reid, M. R. 1991. Rb/Cs fractionation: a link between granulite metamorphism and the S-process. *Geochimica et Cosmochimica Acta* **55**, 2379–83.

Harvey, P. K. & Atkin, B. P. 1981. The rapid determination of Rb, Sr and their ratios in geological materials by X-ray fluorescence spectrometry using a rhodium X-ray tube. *Chemical Geology* **32**, 155–65.

Holland, T. & Powell, R. 1996. Thermodynamics of order-disorder in minerals: II. Symmetric formalism applied to solid solutions. *American Mineralogist* **81**, 1425–37.

Holland, T. & Powell, R. 1998. An internally consistent thermodynamic data set for phases of petrological interest. *Journal of Metamorphic Geology* **16**, 309–43.

Holland, T. & Powell, R. 2001. Calculation of phase relations involving haplogranitic melts using an internally consistent thermodynamic dataset. *Journal of Petrology* **42**, 673–83.

Holtz, F., Becker, A., Freise, M. & Johannes, W. 2001. The water-undersaturated and dry Qz–Ab–Or system revisited. Experimental results at very low water activities and geological implications. *Contributions to Mineralogy and Petrology* **141**, 347–57.

Hoskin, P. W. O., Kinny, P. D., Wyborn, D. & Chappell, B. W. 2000. Identifying accessory mineral saturation during differentiation in granitoid magmas: an integral approach. *Journal of Petrology* **41**, 1365–96.

Hradetzky, H. & Lippolt, H. J. 1993. Generation and distortion of Rb–Sr whole-rock isochrons – effects of metamorphism and alteration. *European Journal of Mineralogy* **5**, 1175–93.

Icenhower, J. & London, D. 1995. An experimental study of element partitioning among biotite, muscovite and coexisting peraluminous silicic melt at 200 MPa (H_2O). *American Mineralogist* **80**, 1229–51.

Irvine, T. N. & Baragar, W. R. A. 1971. A guide to the chemical classification of the common volcanic rocks. *Canadian Journal of Earth Sciences* **8**, 523–48.

Jacobsen, S. B. & Wasserburg, G. J. 1980. Sm–Nd evolution of chondrites. *Earth and Planetary Science Letters* **50**, 139–55.

Jahn, B. M. 1990. Origin of granulites: geochemical constraints from Archean granulite facies rocks of the Sino–Korean Craton, China. *In* Vielzeuf, D. & Vidal, P. (eds) *Granulites and Crustal Evolution*, 471–92. Dordrecht: Kluwer.

Jahn, B. M., Wu, F. & Chen, B. 2000. Granitoids of the Central Asian Orogenic Belt and continental growth in the Phanerozoic. *Transactions of the Royal Society of Edinburgh, Earth Sciences* **91**, 181–93.

Jakeš, P. 1997. Melting in high-P region – case of Bohemian granulites. *Acta Universitatis Carolinae, Geologica* **41**, 113–25.

Janoušek, V., Frýda, J. & Vokurka, K. 1996. Polyphase development of the Moldanubian granulites: evidence from the Rb–Sr and Sm–Nd isotopic study of the Blanský les granulite massif (Czech Republic). *In* Treloar, P. & O'Brien, P. J. (eds) *What Drives Metamorphism and Metamorphic Reactions: Heat Production, Heat Transfer, Deformation and Kinetics?* 43–4. Kingston upon Thames: Kingston University.

Janoušek, V., Rogers, G., Bowes, D. R. & Vaňková, V. 1997. Cryptic trace-element variation as an indicator of reverse zoning in a granitic pluton: the Říčany granite, Czech Republic. *Journal of the Geological Society (London)* **154**, 807–15.

Janoušek, V., Farrow, C. M. & Erban, V. 2003a. GCDkit: new PC software for interpretation of whole-rock geochemical data from igneous rocks. *Geochimica et Cosmochimica Acta* **67**, A186.

Janoušek, V., Vrána, S. & Finger, F. 2003b. Whole-rock geochemical and mineralogical constraints on the genesis of the metaigneous Variscan lower crust in Central Europe – case study of the Lišov granulite Massif, Southern Bohemia. *Journal of the Czech Geological Society* **48**, 72–3.

Kalsbeek, F. & Hansen, M. 1989. Statistical analysis of Rb–Sr isotope data by the 'bootstrap' method. *Chemical Geology (Isotope Geoscience Section)* **73**, 289–97.

Keppler, H. & Wyllie, P. J. 1990. Role of fluids in transport and fractionation of uranium and thorium in magmatic processes. *Nature* **348**, 531–3.

Kodym, O., Jakeš, P. & Schovánek, P. 1978. Granulite und ultramafische Gesteine aus der Strukturbohrung Holubov. *Sborník geologických věd, Geologie* **32**, 7–47.

Kotková, J., Gerdes, A., Parrish, R. R. & Novák, M. 2003. Pressure–temperature–time evolution of granulite clasts from Lower Carboniferous conglomerates – evidence for rapid exhumation at the eastern margin of the Variscan Bohemian Massif. *Geophysical Research Abstracts* **5**, 434.

Kotková, J. & Harley, S. L. 1999. Formation and evolution of high-pressure leucogranulites: experimental constraints and unresolved issues. *Physics and Chemistry of the Earth, Part A: Solid Earth and Geodesy* **24**, 299–304.

Kröner, A., Wendt, I., Liew, T. C., Compston, W., Todt, W., Fiala, J., Vaňková, V. & Vaněk, J. 1988. U–Pb zircon and Sm–Nd model ages of high-grade Moldanubian metasediments, Bohemian Massif, Czechoslovakia. *Contributions to Mineralogy and Petrology* **99**, 257–66.

Kröner, A., O'Brien, P. J., Nemchin, A. A. & Pidgeon, R. T. 2000. Zircon ages for high pressure granulites from South Bohemia, Czech Republic, and their connection to Carboniferous high temperature processes. *Contributions to Mineralogy and Petrology* **138**, 127–42.

Kühn, A., Glodny, J., Iden, K. & Austrheim, H. 2000. Retention of Precambrian Rb/Sr phlogopite ages through Caledonian eclogite facies metamorphism, Bergen Arc Complex, W-Norway. *Lithos* **51**, 305–30.

Leeman, W. & Phelps, D. 1981. Partitioning of rare earths and other trace elements between sanidine and coexisting volcanic glass. *Journal of Geophysical Research, B* **86** 10 193–9.

Liew, T. C. & Hofmann, A. W. 1988. Precambrian crustal components, plutonic associations, plate environment of the Hercynian Fold Belt of Central Europe: indications from a Nd and Sr isotopic study. *Contributions to Mineralogy and Petrology* **98**, 129–38.

Linnemann, U., Gehmlich, M., Tichomirowa, M., Buschmann, B., Nasdala, L., Jonas, P., Lützner, H. & Bombach, K. 2000. The mid-European segment of the Variscides: tectonostratigraphic units, terrane boundaries and plate tectonic evolution. *In* Franke, W., Haak, V., Oncken, O. & Tanner, D. (eds) *Orogenic Processes: Quantification and Modelling in the Variscan Belt*, 131–153. *Geological Society, London, Special Publication* **179**. London: Geological Society.

Mahood, G. A. & Hildreth, E. W. 1983. Large partition coefficients for trace elements in high-silica rhyolites. *Geochimica et Cosmochimica Acta* **47**, 11–30.

Matte, P., Maluski, H., Rajlich, P. & Franke, W. 1990. Terrane boundaries in the Bohemian Massif: result of large-scale Variscan shearing. *Tectonophysics* **177**, 151–70.

Medaris, L. G., Jr, Wang, H. F., Mísař, Z. & Jelínek, E. 1990. Thermobarometry, diffusion modelling and cooling rates of crustal peridotites: two examples from the Moldanubian Zone of the Bohemian Massif. *Lithos* **25**, 189–202.

Medaris, L. G., Jr. Wang, H. F., Jelínek, E. & Jakeš, P. 2003. Garnet peridotite in the Moldanubian Zone in the Czech Republic – a heat source for Variscan metamorphism? *Journal of the Czech Geological Society* **48**, 92–3.

Miller, C. F., McDowell, S. M. & Mapes, R. W. 2003. Hot and cold granites? Implications of zircon saturation temperatures and preservation of inheritance. *Geology* **31**, 529–32.

Miller, C. F. & Mittlefehldt, D. W. 1982. Depletion of light rare-earth elements in felsic magmas. *Geology* **10**, 129–33.

Miller, C. F. & Mittlefehldt, D. W. 1984. Extreme fractionation in felsic magma chambers; a product of liquid-state diffusion or fractional crystallization? *Earth and Planetary Science Letters* **68**, 151–8.

Montel, J. M. 1993. A model for monazite/melt equilibrium and application to the generation of granitic magmas. *Chemical Geology* **110**, 127–46.

Morgan, G. B., VI. & London, D. 2003. Trace-element partitioning at conditions far from equilibrium; Ba and Cs distributions between alkali feldspar and undercooled hydrous granitic liquid at 200 MPa. *Contributions to Mineralogy and Petrology* **144**, 722–38.

Nance, R. D. & Murphy, J. B. 1996. Basement isotopic signatures and Neoproterozoic paleogeography of Avalonian–Cadomian and related terranes in the circum-North Atlantic. *In* Nance, R. D. & Thompson, M. D. (eds) *Avalonian and Related Peri-Gondwanan Terranes of the Circum-North Atlantic*, 333–46. *Geological Society of America Special Paper* **304**. Boulder, Colorado and Lawrence, Kansas: Geological Society of America and University of Kansas Press.

O'Brien, P. J. 2000. The fundamental Variscan problem: high-temperature metamorphism at different depths and high-pressure metamorphism at different temperatures. *In* Franke, W., Haak, V., Oncken, O. & Tanner, D. (eds) *Orogenic Processes: Quantification and Modelling in the Variscan Belt*, 369–86. *Geological Society, London, Special Publication* **179**. London: Geological Society.

O'Brien, P. J. & Rötzler, J. 2003. High-pressure granulites: formation, recovery of peak conditions and implications for tectonics. *Journal of Metamorphic Geology* **21**, 3–20.

Otamendi, J. E., De la Rosa, J. D., Patiño Douce, A. E. & Castro, A. 2002. Rayleigh fractionation of heavy rare earths and yttrium during metamorphic garnet growth. *Geology* **30**, 159–62.

Owen, J. V. & Dostal, J. 1996. Contrasting corona structures in mafic granulite from the Blanský Les complex, Bohemian Massif, Czech Republic. *Canadian Mineralogist* **34**, 959–66.

Pearce, J. A., Harris, N. W. & Tindle, A. G. 1984. Trace element discrimination diagrams for the tectonic interpretation of granitic rocks. *Journal of Petrology* **25**, 956–83.

Peccerillo, R. & Taylor, S. R. 1976. Geochemistry of Eocene calc–alkaline volcanic rocks from the Kastamonu area, Northern Turkey. *Contributions to Mineralogy and Petrology* **58**, 63–81.

Pin, C., Binon, M., Belin, J. M., Barbarin, B. & Clemens, J. D. 1990. Origin of microgranular enclaves in granitoids – equivocal Sr–Nd evidence from Hercynian rocks in the Massif Central (France). *Journal of Geophysical Research, B* **95**, 17 821–8.

Pin, C. & Marini, F. 1993. Early Ordovician continental break-up in Variscan Europe: Nd–Sr and trace element evidence from bimodal igneous associations of the southern Massif Central, France. *Lithos* **29**, 177–96.

Pin, C. & Vielzeuf, D. 1983. Granulites and related rocks in Variscan Median Europe: a dualistic interpretation. *Tectonophysics* **93**, 47–74.

Pöschl-Otrel, K. 1995. *Untersuchungen an Zirkonen aus Granuliten der niederösterreichischen Waldviertels*. Unpublished M.Sc. thesis. Universität Salzburg, Austria.

Powell, R., Holland, T. & Worley, B. 1998. Calculating phase diagrams involving solid solutions via non-linear equations, with examples using Thermocalc. *Journal of Metamorphic Geology* **16**, 577–88.

Powell, R. & Downes, J. 1990. Garnet porphyroblast-bearing leucosomes in metapelites: mechanisms, phase diagrams, and an example from Broken Hill, Australia. *In* Ashworth, J. R. & Brown, M. (eds) *High-Temperature Metamorphism and Crustal Anatexis*, 105–23. London: Unwin Hyman.

Powell, R. & Holland, T. 1999. Relating formulations of the thermodynamics of mineral solid solutions: activity modeling of pyroxenes, amphiboles and micas. *American Mineralogist* **84**, 1–14.

Prince, C. I., Košler, J., Vance, D. & Günther, D. 2000. Comparison of laser ablation ICP-MS and isotope dilution REE analyses – implications for Sm–Nd garnet geochronology. *Chemical Geology* **168**, 255–74.

Provost, A. 1990. An improved diagram for isochron data. *Chemical Geology (Isotope Geoscience Section)* **80**, 85–99.

Roberts, M. P., Pin, C., Clemens, J. D. & Paquette, J. L. 2000. Petrogenesis of mafic to felsic plutonic rock associations: the calc–alkaline Querigut Complex, French Pyrénées. *Journal of Petrology* **41**, 809–44.

Roberts, M. P. & Finger, F. 1997. Do U–Pb zircon ages from granulites reflect peak metamorphic conditions? *Geology* **25**, 319–22.

Romer, R. L. & Rötzler, J. 2001. P–T–t evolution of ultrahigh-temperature granulites from the Saxon Granulite Massif, Germany. Part II: Geochronology. *Journal of Petrology* **42**, 2015–32.

Rötzler, J. & Romer, R. L. 2001. P–T–t evolution of ultrahigh-temperature granulites from the Saxon Granulite Massif, Germany. Part I: Petrology. *Journal of Petrology* **42**, 1995–2013.

Rudnick, R. L., McLennan, S. M. & Taylor, S. R. 1985. Large ion lithophile elements in rocks from high-pressure granulite facies terrains. *Geochimica et Cosmochimica Acta* **49**, 1645–55.

Rudnick, R. L. & Presper, T. 1990. Geochemistry of intermediate-to high-pressure granulites. *In* Vielzeuf, D. & Vidal, P. (eds) *Granulites and Crustal Evolution*, 523–50. Dordrecht: Kluwer.

Rudnick, R. L. & Taylor, S. R. 1987. The composition and petrogenesis of the lower crust; a xenolith study. *Journal of Geophysical Research, B* **92**, 13981–4006.

Rutter, E. H. & Neumann, D. H. K. 1995. Experimental deformation of partially molten Westerly Granite under fluid-absent conditions, with implications for the extraction of granitic magmas. *Journal of Geophysical Research, B* **100**, 697–715.

Sawyer, E. W. 1999. Criteria for the recognition of partial melting. *Physics and Chemistry of the Earth, Part A: Solid Earth and Geodesy* **24**, 269–79.

Schenk, V. & Todt, W. 1983. U–Pb Datierungen an Zirkon und Monazit der Granulite im Moldanubikum Niederösterreichs (Waldviertel). *Fortschritte der Mineralogie* **61**, 190–1.

Shaw, D. M. 1968. A review of K–Rb fractionation trends by covariance analysis. *Geochimica et Cosmochimica Acta* **32**, 573–601.

Siebel, W., Raschka, H., Irber, W., Kreuzer, H., Lenz, K. L., Höhndorf, A. & Wendt, I. 1997. Early Palaeozoic acid magmatism in the Saxothuringian belt: new insights from a geochemical and isotopic study of orthogneisses and metavolcanic rocks from the Fichtelgebirge, SE: Germany. *Journal of Petrology* **38**, 203–30.

Slabý, J. 1983. Modal composition and petrochemistry of granulites of the Lišov and Blanský les massifs (South Bohemia). *Časopis pro mineralogii a geologii* **28**, 41–60. [in Czech with English summary].

Steiger, R. H. & Jäger, E. 1977. Subcommission on Geochronology; convention on the use of decay constants in geo- and cosmochronology. *Earth and Planetary Science Letters* **36**, 359–62.

Stix, J. & Gorton, M.P. 1990. Variations in trace element partition coefficients in sanidine in the Cerro Toledo Rhyolite, Jemez Mountains, New Mexico: effects of composition, temperature and volatiles. *Geochimica et Cosmochimica Acta* **54**, 2 697–708.

Strejček, M. 1986. *Petrology of granulites in the Plešovice quarry near Český Krumlov*. Unpublished M.Sc. thesis. Charles University, Prague. [in Czech].

Svojtka, M., Košler, J. & Venera, Z. 2002. Dating granulite-facies structures and the exhumation of lower crust in the Moldanubian Zone of the Bohemian Massif. *International Journal of Earth Sciences* **91**, 373–85.

Sylvester, P. J. 1998. Post-collisional strongly peraluminous granites. *Lithos* **45**, 29–44.

Taylor, S. R. & McLennan, S. M. 1985. *The Continental Crust: Its Composition and Evolution*. Oxford: Blackwell.

Tollmann, A. 1995. Das Ausmass des variszischer Deckbaues im Moldanubikum. *Krystalinikum* **18**, 117–32.

Turniak, K., Mazur, S. & Wysoczanski, R. 2000. SHRIMP zircon geochronology and geochemistry of the Orlica–Śnieżnik gneisses (Variscan belt of Central Europe) and their tectonic implications. *Geodinamica Acta* **13**, 293–312.

Valbracht, P. J., Vrána, S., Beetsma, J. J., Fiala, J. & Matějka, D. 1994. Sr and Nd isotopic determinations in three Moldanubian granulite massifs in southern Bohemia. *Journal of the Czech Geological Society* **39**, 114.

van Breemen, O., Aftalion, M., Bowes, D. R., Dudek, A., Mísař, Z., Povondra, P. & Vrána, S. 1982. Geochronological studies of the Bohemian Massif, Czechoslovakia, and their significance in the evolution of Central Europe. *Transactions of the Royal Society of Edinburgh, Earth Sciences* **73** (for 1982), 89–108.

Vellmer, C. 1992. *Stoffbestand und Petrogenese von Granuliten und granitischen Gesteinen der südlichen Böhmischen Masse in Niederösterreich*. Unpublished Ph.D. thesis. Georg–August–Universität Göttingen.

Vigneresse, J.-L., Barbey, P. & Cuney, M. 1996. Rheological transitions during partial melting and crystallization with application to felsic magma segregation and transfer. *Journal of Petrology* **37**, 1579–600.

Vrána, S. 1989. Perpotassic granulites from southern Bohemia: a new rock-type derived from partial melting of crustal rocks under upper mantle conditions. *Contributions to Mineralogy and Petrology* **103**, 510–22.

Vrána, S. & Jakeš, P. 1982. Orthopyroxene and two-pyroxene granulites from a segment of charnockitic crust in southern Bohemia. *Věstník Českého geologického ústavu* **57**, 129–43.

Vrána, S. & Novák, M. 2000. Petrology and geochemistry of granulite clasts in the Visean Luleč conglomerate, Kulm in central Moravia, Czech Republic. *Bulletin of the Czech Geological Survey* **75**, 405–13.

Watson, E. B. & Harrison, M. 1983. Zircon saturation revisited: temperature and composition effects in a variety of crustal magma types. *Earth and Planetary Science Letters* **64**, 295–304.

Watt, G. R. & Harley, S. L. 1993. Accessory phase controls on the geochemistry of crustal melts and restites produced during water-undersaturated partial melting. *Contributions to Mineralogy and Petrology* **114**, 550–66.

Wedepohl, K. H. 1995. The composition of the continental crust. *Geochimica et Cosmochimica Acta* **59**, 1217–32.

Wen, S. & Nekvasil, H. 1994. SOLVCALC: an interactive graphics program package for calculating ternary feldspar solvus and for two-feldspar thermometry. *Computers and Geosciences* **20**, 1025–1040.

Wendt, J. I., Kröner, A., Fiala, J. & Todt, W. 1994. U–Pb zircon and Sm–Nd dating of Moldanubian HP/HT granulites from south Bohemia, Czech Republic. *Journal of the Geological Society (London)* **151**, 83–90.

Whalen, J. B., Currie, K. L. & Chappell, B. W. 1987. A-type granites: geochemical characteristics, discrimination and petrogenesis. *Contributions to Mineralogy and Petrology* **95**, 407–19.

White, R. W., Powell R. & Holland, T. 2001. Calculation of partial melting equilibria in the system Na_2O–CaO–K_2O–FeO–MgO–Al_2O_3–SiO_2–H_2O (NCKFMASH). *Journal of Metamorphic Geology* **19**, 139–53.

White, R. W. & Powell, R. 2002. Melt loss and the preservation of granulite facies mineral assemblages. *Journal of Metamorphic Geology* **20**, 621–32.

Wiegand, B. 1997. *Isotopengeologische und geochemische Untersuchungen zur prävariszischen magmatischen und sedimentären Entwicklung im saxothuringisch-moldanubischen Übergangsbereich (Grenzgebiet BRD/CR). Geotektonische Forschungen* **88**. Stuttgart: E. Schweizerbart'sche Verlagsbuchhandlung.

Winchester, J. A. 2002. Palaeozoic amalgamation of Central Europe: new results from recent geological and geophysical investigations. *Tectonophysics* **360**, 5–21.

York, D. 1969. Least-squares fitting of a straight line with correlated errors. *Earth and Planetary Science Letters* **5**, 320–4.

Zhang, R. Y., Zhai, S. M., Fei, Y. W. & Liou, J. G. 2003. Titanium solubility in coexisting garnet and clinopyroxene at very high pressure: the significance of exsolved rutile in garnet. *Earth and Planetary Science Letters* **216**, 591–601.

VOJTĚCH JANOUŠEK, FRITZ FINGER and MALCOLM P. ROBERTS, Institut für Mineralogie, Universität Salzburg, Hellbrunnerstraße 34, A-5020 Salzburg, Austria.
e-mail: vojtech.janousek@sbg.ac.at

MALCOLM P. ROBERTS, The Council for Geoscience, PO Box 5347, Walmer, Port Elizabeth 6065, South Africa. (Present address)

JIŘÍ FRÝDA, Czech Geological Survey, Klárov 3/131, 118 21 Prague 1, Czech Republic.

CHRISTIAN PIN, Département de Géologie, CNRS, Université Blaise Pascal, 5, rue Kessler, F-63038 Clermont-Ferrand, France.

DAVID DOLEJŠ, Department of Earth and Planetary Sciences, McGill University, 3450 rue University, Montreal, Quebec, Canada H3A 2A7.

MS received 10 November 2003. Accepted for publication 4 May 2004.

Transactions of the Royal Society of Edinburgh: Earth Sciences, **95**, 161–179, 2004

Variety and genesis of the pyroxene-bearing S- and I-type granitoids from the Hidaka Metamorphic Belt, Hokkaido, northern Japan

Toshiaki Shimura, Masaaki Owada, Yasuhito Osanai, Masayuki Komatsu and Hiroo Kagami

ABSTRACT: The high-*dT/dP*-type Hidaka Metamorphic Belt in Hokkaido, northern Japan, represents a tilted crustal section of a magmatic arc of Tertiary age. The highest metamorphic grades reached are granulite facies, and the syn-metamorphic granitic rocks are widely distributed in this metamorphic terrane. The granitic rocks are mainly tonalitic and granodioritic in composition, and are classified into peraluminous (S-type) and metaluminous (I-type) granitoids. A large amount of pyroxene-bearing S-type tonalites (garnet-orthopyroxene tonalite) is distributed in the Niikappu river region in the northern part of the Hidaka Metamorphic Belt. Pyroxene-bearing I-type tonalite (two-pyroxene hornblende tonalite) bodies are also distributed in this area.

The pyroxene-bearing tonalites are classified into several sub-types on the basis of their field occurrence, texture, mineral assemblage and geochemical features. Homogeneous I_H- and S_H-type tonalite are thought to represent original magmas, i.e. those which have been generated by partial melting of mafic metamorphic rocks and pelitic-psammitic metamorphic rocks, respectively. Model calculations assuming batch partial melting indicate that possible restites are garnet-two-pyroxene mafic granulite for I_H-type and garnet-orthopyroxene aluminous granulite for S_H-type.

The unexposed lowermost crust of the 'Hidaka crust' is thought to be composed of garnet-two-pyroxene mafic granulite, garnet-orthopyroxene aluminous granulite and metagabbros.

KEY WORDS: Crustal section, granite, granulite, lower crust, metaluminous, partial melting, peraluminous, Tertiary.

One of the important origins of the granitic magma is thought to be through crustal anatexis, producing the so-called S-type and I-type granites (Chappell & White 1974; White & Chappell 1977). S-type granitoids usually contain biotite, muscovite, garnet and cordierite, and I-type granitoids contain biotite and hornblende as Mg-Fe phases. However, some of these granitic rocks contain pyroxene as higher-temperature phases (e.g. Maeda *et al.* 1991; Shimura *et al.* 1992).

Under granulite facies conditions, pyroxene is also produced by the incongruent melting of crustal rocks (e.g. Vielzeuf & Holloway 1988; Vielzeuf & Montel 1994; Vielzeuf & Schmidt 2001). In metasedimentary rocks, pyroxene is produced by the dehydration melting reaction of biotite at temperatures of ca. 800–850°C (Vielzeuf & Holloway 1988; Vielzeuf & Montel 1994). Whilst in mafic metamorphic rocks, pyroxenes are produced by the dehydration melting reaction of hornblende at temperatures of ca. 850–900°C (Thompson 1990; Vielzeuf & Schmidt 2001). Thus, the pyroxene-bearing granitic rocks are important because they reveal the origin and the crystallisation of granitic magma.

The present contribution describes the mode of occurrence and chemical characteristics of the pyroxene-bearing S- and I-type tonalites from the Hidaka Metamorphic Belt (HMB), Hokkaido, northern Japan. Based on the results, the present authors discuss the crustal succession of the 'Hidaka crust' at the syn-metamorphic stage. In this paper, mineral abbreviations follow Kretz (1983).

1. The Hidaka Metamorphic Belt

1.1. Geological outline

The Hidaka Metamorphic Belt (HMB) is a high-*dT/dP* type metamorphic belt of Tertiary age, in Hokkaido, northern Japan (e.g. Komatsu *et al.* 1989, 1994; Osanai *et al.* 1991; Owada *et al.* 2003) (Fig. 1). The HMB consists of steeply eastward-dipping thrust sheets which are composed of mainly metamorphic (pelitic-psammitic rocks, intermediate rocks and mafic rocks), igneous (layered gabbros, massive gabbro-diorites and granitoids), and alpine-type ultramafic rocks. The base of the HMB is the Hidaka Main Thrust (HMT), where the metamorphic and igneous rocks are mylonitised with dextral shear senses (Toyoshima *et al.* 1994). Almost all igneous rocks are intruded into the metamorphic layers as syn-metamorphic suites. The HMB is thought to represent an eastward dipping island-arc type crustal section (Komatsu *et al.* 1989, 1994; Owada *et al.* 2003). A schematic succession of the 'Hidaka crust' is shown in Figure 2. The crustal section shallower than the HMT is directly observable. Although the deeper levels below the HMT are not exposed, a lower crust of this composition is presumed to have existed. Recent vivroseismic study has revealed a duplex structure for the HMB, interpreted to result from delamination of the lower-most crust of the eastern part of Hokkaido under the wedge of the western part (Arita *et al.* 1998; Tsumura *et al.* 1999; Ito 2000).

The HMB is divided into four metamorphic zones (Zones I–IV), from the top (east) to the bottom (west) of the thrust sheet, on the basis of the metamorphic mineral paragenesis (Osanai *et al.* 1986, 1991). Zones I, II, III and IV correspond to the prehnite-pumpellyite facies, greenschist/epidote-amphibolite facies, amphibolite facies and granulite facies, respectively.

1.2. Granitic rocks

The granitic rocks are widely distributed in the HMB (Fig. 1). They are mainly tonalitic or granodioritic in composition. In general, peraluminous [ASI >1; ASI: Aluminum Saturation

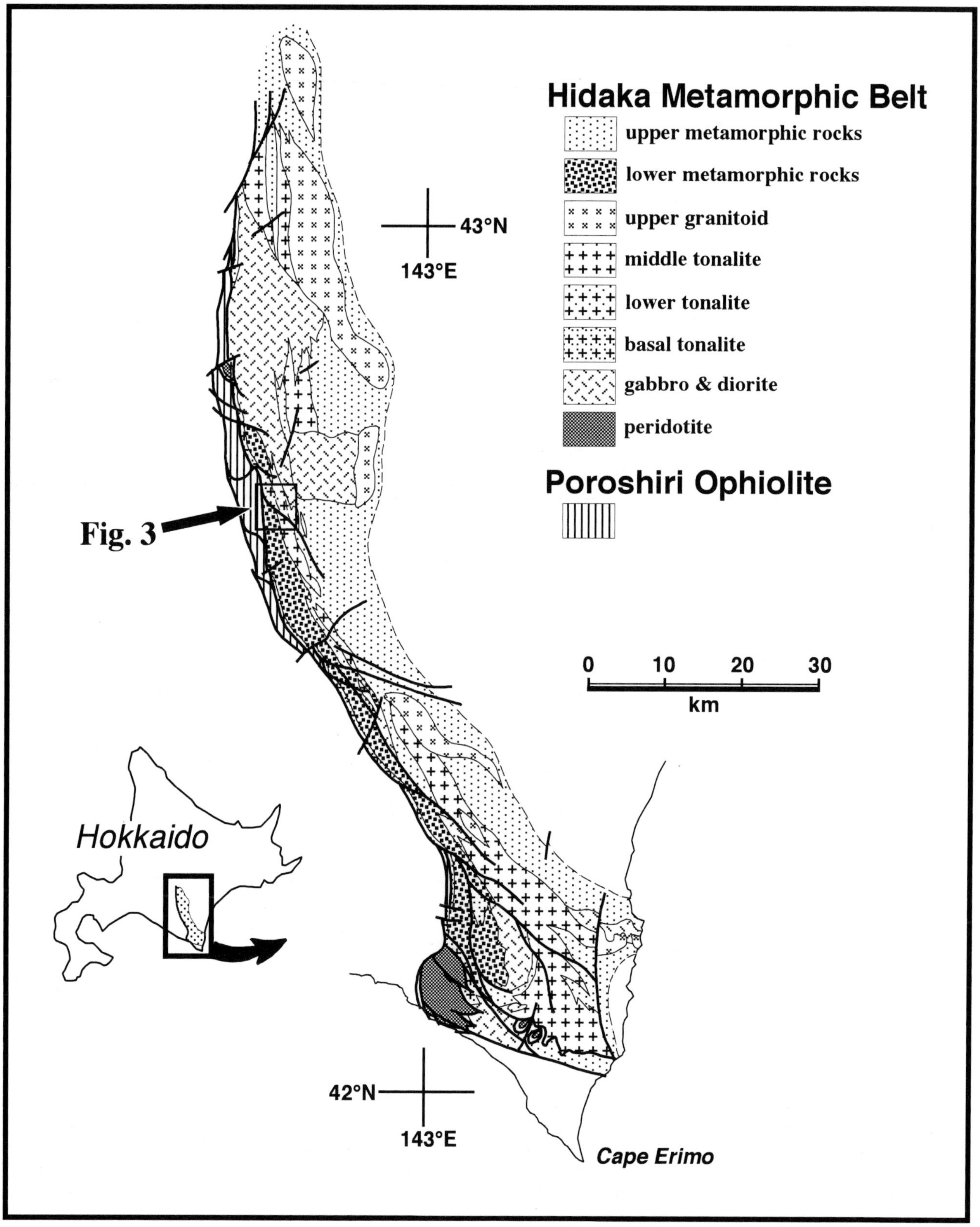

Figure 1 Simplified geological map of the Hidaka Metamorphic Belt, modified after Komatsu *et al.* (1986).

Index=mol $Al_2O_3/(CaO+Na_2O+K_2O)$] granitoids predominate over metaluminous (ASI <1) ones. The former and the latter groups correspond to the S- and I-type granitic rocks (Shimura *et al.* 1992; Shimura 1999), respectively. Since both groups contain ilmenite with absence of magnetite, they are classified into the Ilmenite-series, according to the classification of Ishihara (1977).

Since these granitic rocks have been emplaced into the various metamorphic layers at the syn-metamorphic stage (Shimura 1992; Shimura *et al.* 1997), the metamorphic grade of the wall rocks corresponds to the intrusive depth. The granitic rocks can be classified into four depth types: upper granite-granodiorite, middle granodiorite-tonalite, lower tonalite and basal tonalite (Shimura *et al.* 1992). Therefore, the granitic

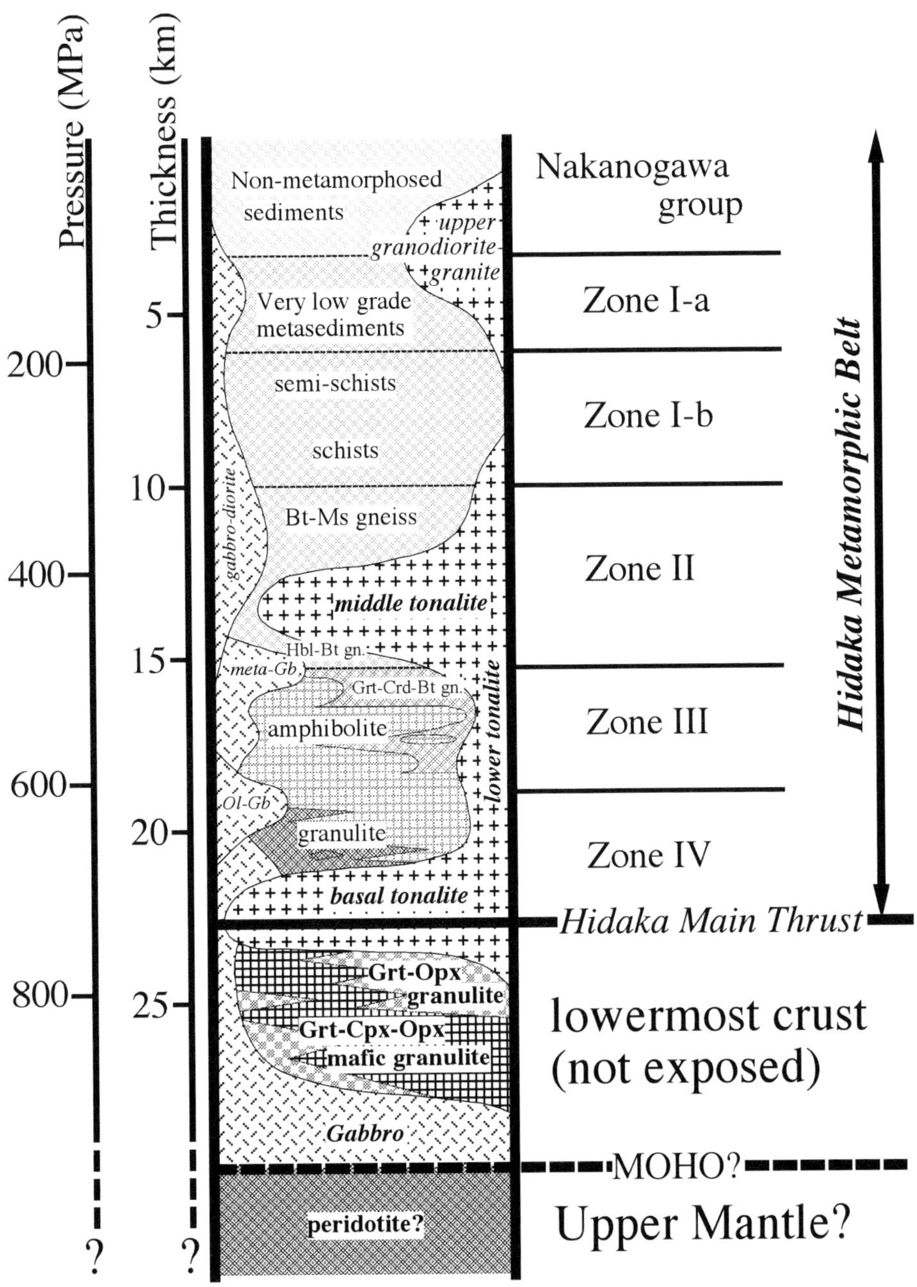

Figure 2 Idealised crustal succession of the Hidaka Metamorphic Belt, modified after Komatsu *et al.* (1986). Deeper levels below the Hidaka Main Thrust (lowermost crust) are not exposed, and are inferred from petrological study.

rocks of the HMB can be classified according to the intrusive levels and to their chemical characteristics (i.e. peraluminous/metaluminous types) (Table 1). Each type has characteristic mineral assemblages which reflect the depth of intrusion and rock composition. Pyroxenes occur only in the deeper-level tonalites.

2. Occurrence of pyroxene-bearing tonalite in the Niikappu river region

2.1. Geology and occurrences

Tonalitic rocks are widely distributed in the Niikappu River region, and the tonalitic suite has intruded into the granulites and lower amphibolite facies gneiss (Figs 1 & 3). A detailed account of the geology of this area is described in Shimura (1992). The Niikappu tonalite suite can be subdivided into the following three types based on depth: basal, lower and middle from west to east. The western edge of the Niikappu suite is mylonitised along the HMT, whereas the remaining part towards the east is weakly deformed and shows original igneous texture. According to mineral assemblages and chemical compositions, the Niikappu suite can be further subdivided into peraluminous and metaluminous groups. Both groups occur in association throughout the mass, although the former is more common.

The pyroxene-bearing S-type tonalite occurs within the basal tonalite (Table 1). The basal S-type tonalite is distributed in the western-most part of the Niikappu suite (Fig. 3). The exposed width is about 1200 m, including a 700m mylonitised zone. The boundary between the basal and the lower S-type tonalites can be marked by a sudden decrease in modal garnet, and the disappearance of coarse garnet grains (0·5–3 cm in diameter, Fig. 4B) around pelitic granulite enclaves. The

Table 1 Characteristic mineral assemblages of granitic rocks in the HMB, classified in terms of the S-type/I-type groups and intrusive levels (modified after Shimura *et al.* 1992). Mineral abbreviations are after Kretz (1983). I, II, III, and IV refer to metamorphic zones of the HMB (after Osanai *et al.* 1991).

Depth type	Rock type	Intrusive level	Characteristic mineral assemblage	
			Metaluminous (I-type)	Peraluminous (S-type)
Upper	Granite Granodiorite	Chl-Ms metasediment unit	Bt+Hbl+Pl+Qtz+Kfs	Bt+Pl+Qtz+Kfs
Middle	Granodiorite Tonalite	Bt-Ms gneiss and schist unit	Bt ± Hbl+Pl+Qtz ± Kfs	Bt ± Ms ± Crd+Pl+Qtz ± Kfs
Lower	Tonalite	Amphibolite unit	Bt+Hbl+Opx ± Cpx ± Cum+Pl+Qtz	Bt+Opx ± Crd+Pl+Qtz ± Kfs Bt+Grt ± Crd+Pl+Qtz ± Kfs
Basal	Tonalite	Granulite unit		Bt+Opx+Grt ± Crd+Pl+Qtz ± Kfs

disappearance of coexisting garnet-orthopyroxene is also observed.

The pyroxene-bearing I-type tonalites occur in the basal and the lower tonalites (Table 1, Fig. 3). The I-type tonalites are observed as narrow intrusive bodies less than ca. 100 m in width (Fig. 3). They usually have sharp contacts with S-type rocks. However, in some domains, I-type tonalites have irregularly shaped, vague boundaries with S-type tonalites, presumably because of mingling and mixing.

These tonalites, especially the S-type, enclose a large number of metamorphic enclaves. These enclaves are pelitic-psammitic granulite (garnet-orthopyroxene-biotite granulite and garnet-orthopyroxene-cordierite granulite), mafic granulite (orthopyroxene mafic granulite, two-pyroxene granulite and gabbroic gneiss), and metamorphosed ultramafic rocks (meta-harzburgite) (Fig. 4A). Through magmatic reactions with tonalite melts, unusual mineral assemblages have been formed around these enclaves (Shimura 2004). For example, 1–3-cm-thick leucocratic tonalite rims consisting of orthopyroxene, plagioclase and quartz occur around some mafic granulite enclaves. Similar reaction rims around ultramafic enclaves in the S-type tonalite consist of bronzite, phlogopite and highly magnesian cordierite with quartz and plagioclase. The reaction zones around pelitic-psammitic enclaves in the S-type tonalite are very wide, sometimes up to several metres, producing a heterogeneous tonalite that contains clots consisting of large garnet grains (0·5–3 cm in diameter), orthopyroxene and biotite (Fig. 4).

2.2. Type classification

The pyroxene-bearing tonalites, particularly the S-type tonalite, are heterogeneous lithofacies because of the abundance of metamorphic enclaves (Fig. 4A). These heterogeneous tonalites are classified into several types on the basis of their field occurrence, texture, mineral assemblage and geochemical features.

The first criterion for classification is the presence or absence of hornblende. Hornblende and peraluminous minerals (except for biotite) never coexist, and therefore, the pyroxene-bearing tonalities can be subdivided on this basis into metaluminous and peraluminous groups, corresponding to I- and S-type rocks.

The second criterion is based on the occurrence of enclaves. Rocks completely free from metamorphic enclaves or melanocratic clots show homogeneous lithofacies. This type is termed type-H, representing its homogeneous nature. On the other hand, the marginal facies of visible pelitic-psammitic enclaves are called type-P (pelitic-psammitic) (Fig. 4B). The marginal facies around mafic or ultramafic enclaves are then labelled, for example, type-M and type-U, but these are not observed in abundance. Other S-type rocks enclose heterogeneously distributed small melanocratic clots consisting of orthopyroxene, biotite, plagioclase, quartz and garnet. For this type, the present authors use the label type-C (clot) (Fig. 4C). Therefore, homogeneous metaluminous tonalite matrix and peraluminous tonalite matrixes correspond to I_H-type and S_H-type, respectively. Peraluminous tonalite around the pelitic metamorphic enclaves corresponds to S_P-type. S-type rocks enclosing heterogeneously distributed small melanocratic clots are called S_C-type.

Main types of pyroxene bearing tonalite can be classified on this basis into the two I-types (I_H and I_M) and three S-types (S_H, S_P and S_C).

3. Petrography

The petrographic observations and corresponding mineral chemical features of main five tonalite sub-types (I_H, I_M, S_H, S_P and S_C) and major enclaves (mafic granulite and pelitic-psammitic granulite) are described below. Mineral compositions were analysed by wavelength dispersive X-ray spectroscopy (JEOL JXA-8600MX) at the Graduate School of Science and Technology, Niigata University, Niigata, Japan. The oxide ZAF correction procedures were used and the analysed data were transmitted to a personal computer using the system of Shimura (1995). X_{Mg}=Mg/(Fe+Mg), An=100 × Ca/(Ca+Na+K) and Or=100 × K/(Ca+Na+K) in mol ratio.

3.1. I_H-type tonalite

This type has fewer metamorphic enclaves (especially the pelitic-psammitic metamorphic rocks) than S-type rocks. It has a hypautomorphic-granular texture and is composed mainly of orthopyroxene (X_{Mg}=0·52–0·59), clinopyroxene (X_{Mg}=0·74–0·75), brown-hornblende (X_{Mg}=0·61–0·68), cummingtonite (X_{Mg}=0·55–0·59), biotite (X_{Mg}=0·54–0·63), plagioclase (An=37–50) and quartz (Fig. 5A). Clinopyroxene occurs within the mafic I_H-type. Cummingtonite occurs around orthopyroxene, or between orthopyroxene and brown-hornblende. Accessory minerals are mainly ilmenite, apatite, zircon and titanite.

Orthopyroxene is medium grained (0·5–1·5 mm) and is mainly anhedral, but rarely subhedral in shape. Most of the orthopyroxene is enclosed by brown hornblende, and cummingtonite occurs between the orthopyroxene and hornblende (Fig. 5A). Where orthopyroxene is enclosed in biotite crystals,

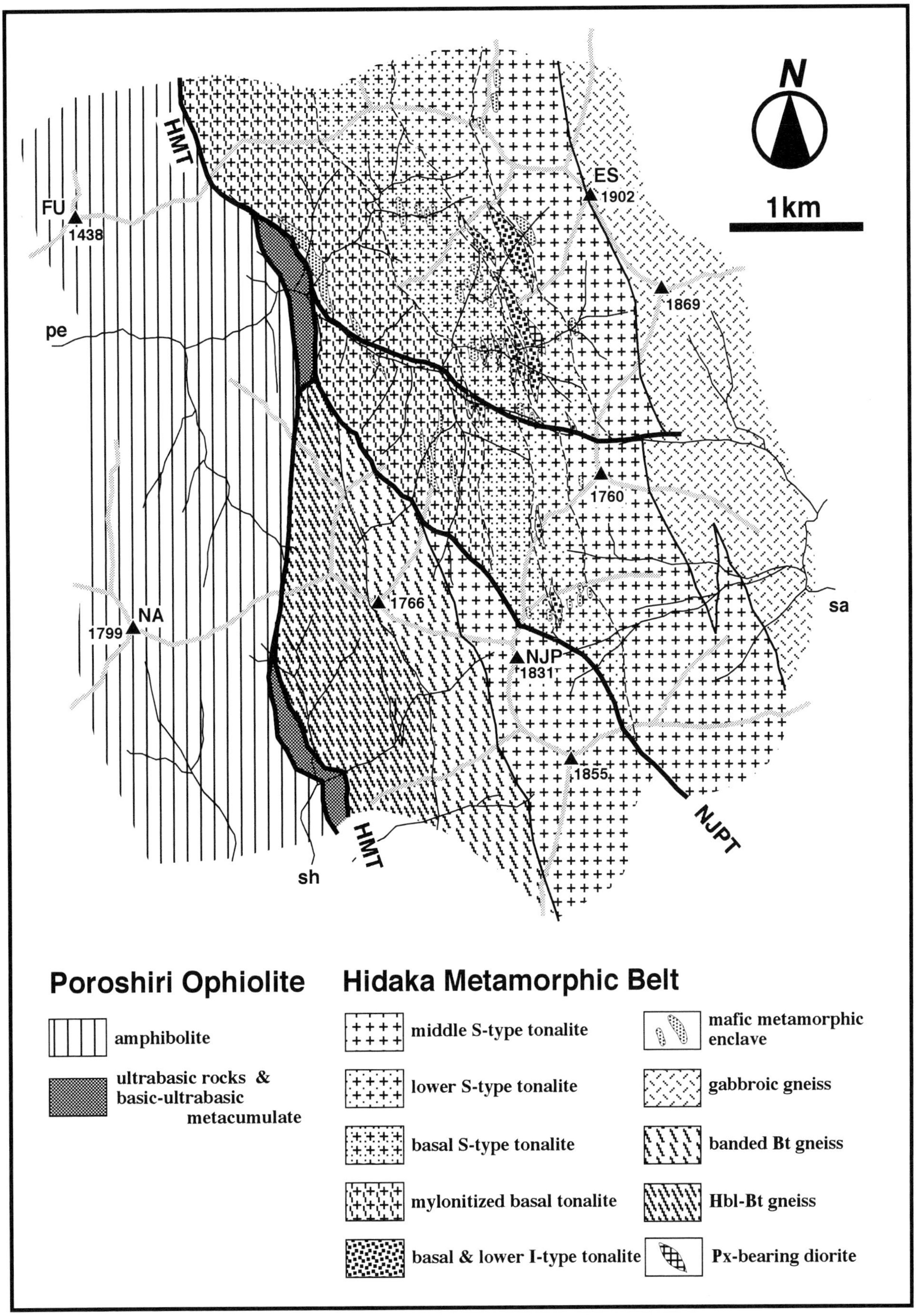

Figure 3 Geological map of the Hidaka Metamorphic Belt in the Mt. Esaoman-tottabetsu-dake region. (HMT) Hidaka Main Thrust; (NJPT) Namewakka Junction Peak Thrust. Triangles and numbers indicate mountain peaks and the altitudes. (FU) Mt. Futamata-yama; (ES) Mt. Esaoman-tottabetsu-dake; (NJP) Mt. Namewakka Junction peak; (NA) Mt. Namewakka-dake; (pe) Peppirigai-zawa River (upper-stream of Niikappu-gawa River); (sa) Satsunai-gawa River; (sh) Shunbetsu-gawa River. Modified from Shimura (1999).

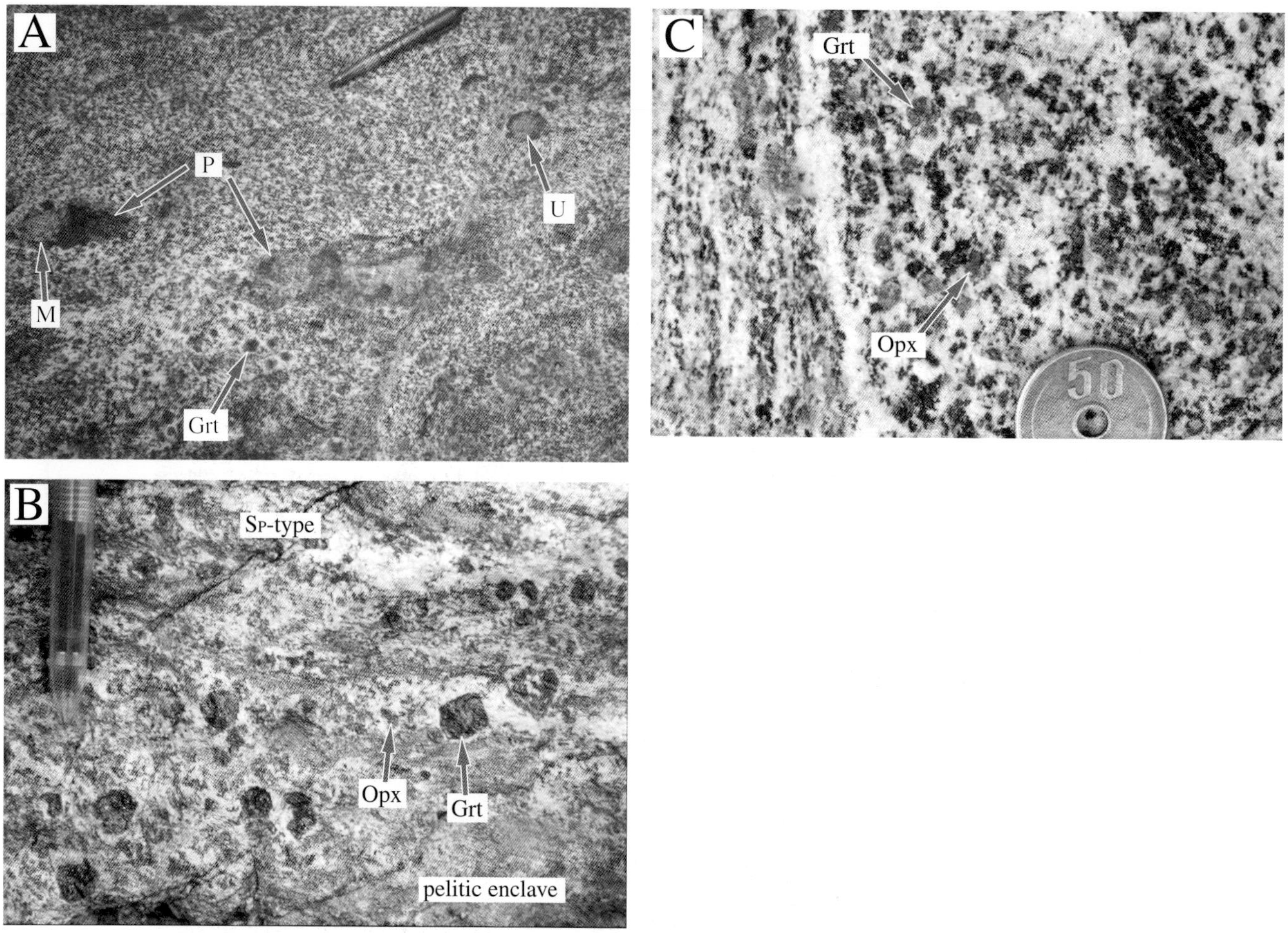

Figure 4 Mode of field occurrence of pyroxene-bearing tonalite: (A) The tonalite body shows very heterogeneous lithofacies, owing to various types of metamorphic enclaves such as mafic metamorphic enclaves (M), meta-ultramafic enclaves (U), and pelitic-psammitic granulite enclaves (P); (B) Close-up of S_P-type tonalite around the pelitic granulite enclaves. Note large garnet crystals and the leucocratic zone. The thickness of ballpoint pen for scale is 1 cm; (C) Close-up of S_C-type tonalite. Note melanocratic clots, which consist of garnet, orthopyroxene, and calcic plagioclase. The diameter of the coin is 2·1 cm.

cummingtonite rims do not occur around the orthopyroxene crystals. Orthopyroxene in the I_H-type has a low-ASI composition (Fig. 6).

Medium grained (0·5–1 mm) clinopyroxene is mainly anhedral, but rarely subhedral in shape. Plagioclase is poikilitically enclosed within clinopyroxene and orthopyroxene. Most of the clinopyroxene grains are rimmed by brown-hornblende.

It is rare that plagioclase encloses other minerals. Plagioclases of I_H-type show oscillatory-zoning in cores, and normal zoning at their rims.

3.2. I_M-type tonalite

I_M-type tonalite is composed mainly of orthopyroxene, brown-hornblende, cummingtonite, biotite, plagioclase and quartz. Cummingtonite normally forms a partial rim around orthopyroxene. Some I_M-type tonalite that surrounds two-pyroxene mafic granulite contains clinopyroxene coexisting with orthopyroxene. This clinopyroxene may have been incorporated from the two-pyroxene mafic granulite. Plagioclase has calcic cores (An=50–54), is compositionally similar to the plagioclase of the mafic metamorphic enclaves, and sometimes shows reverse zoning from An=70 to 84. In the latter case, the I_M-type tonalite has large pools of quartz, which enclose reverse zoned and corroded plagioclase crystals, and have myrmekite.

3.3. S_H-type tonalite

S_H-type tonalite is partly exposed in the basal tonalite body. This rock is free from metamorphic enclaves or melanocratic clots in hand specimens. It is commonly light grey in colour, medium-grained and normally massive, but it is sometimes weakly foliated. It encloses metamorphic enclaves (pelitic-psammitic granulite, mafic granulite and meta-harzburgite) in the outcrop, and forms S_P-type tonalites around pelitic-psammitic enclaves. Since S_C-type tonalite and S_H-type tonalite are intergradational, it is difficult to clearly discriminate one from the other in the field. The minerals in S_H type tonalite show compositional zoning consistent with a cooling history (Shimura *et al.* 1992, 1999).

S_H-type tonalite shows a granitic texture and is composed mainly of garnet (X_{Mg}=0·14–0·30), orthopyroxene (X_{Mg}= 0·42–0·51), cordierite (X_{Mg}=0·64–0·70), biotite (X_{Mg}=0·45–0·54), plagioclase (An=28–35), quartz and minor amounts of K-feldspar (Or=83–87) (Fig. 5B). Cummingtonite is occasionally observed around orthopyroxene. Accessory minerals such as ilmenite, graphite, apatite and zircon are present. Garnet grains in this type do not enclose other minerals, except for very small amount of apatite and ilmenite.

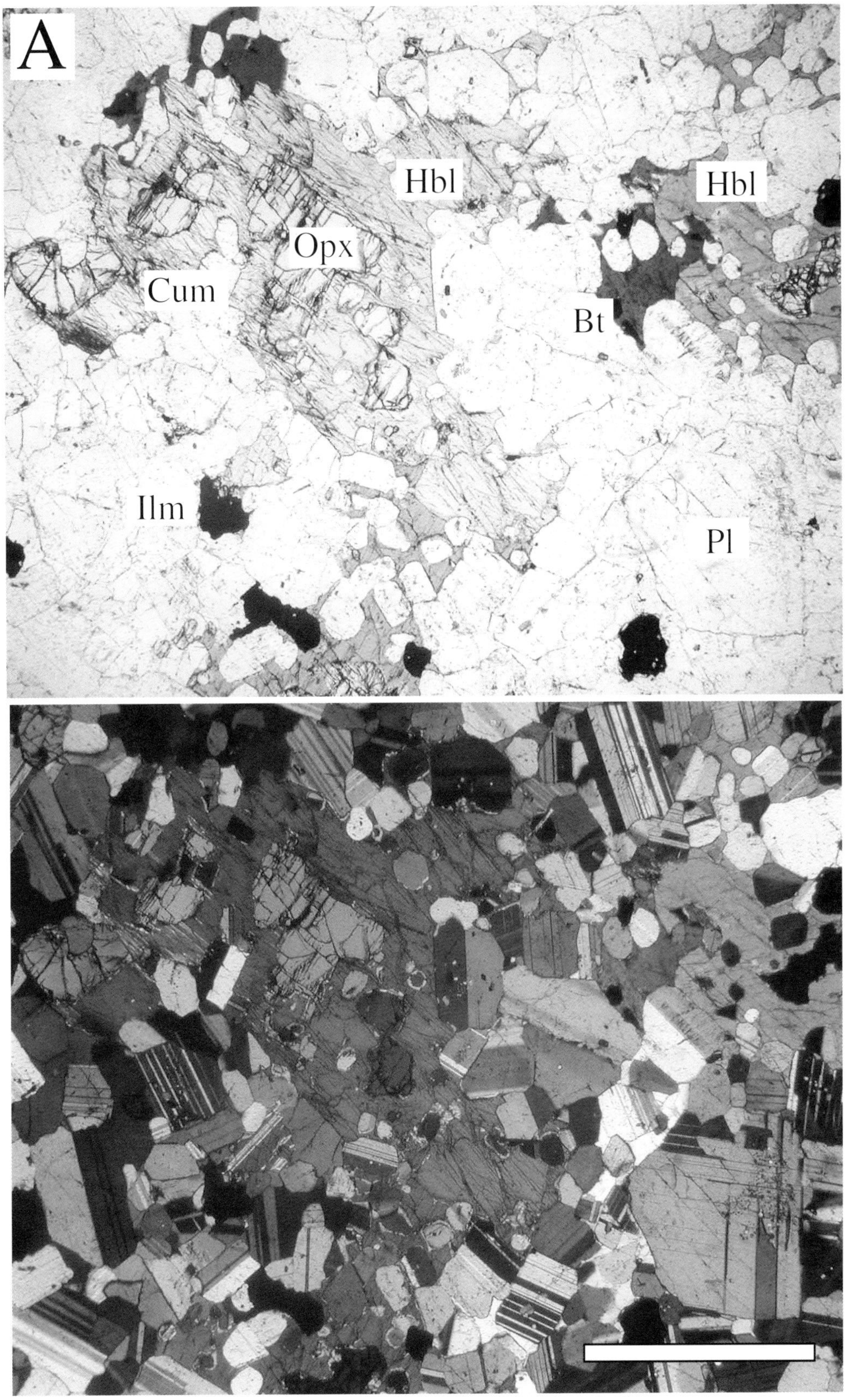

Figure 5 (A) A thin section of I_H-type tonalite, plane-polarised light (top) and crossed polars (bottom). Pyroxenes are enclosed by brown-hornblende. Cummingtonite occurs around orthopyroxene or between pyroxenes and brown-hornblende. Scale bar = 2 mm.

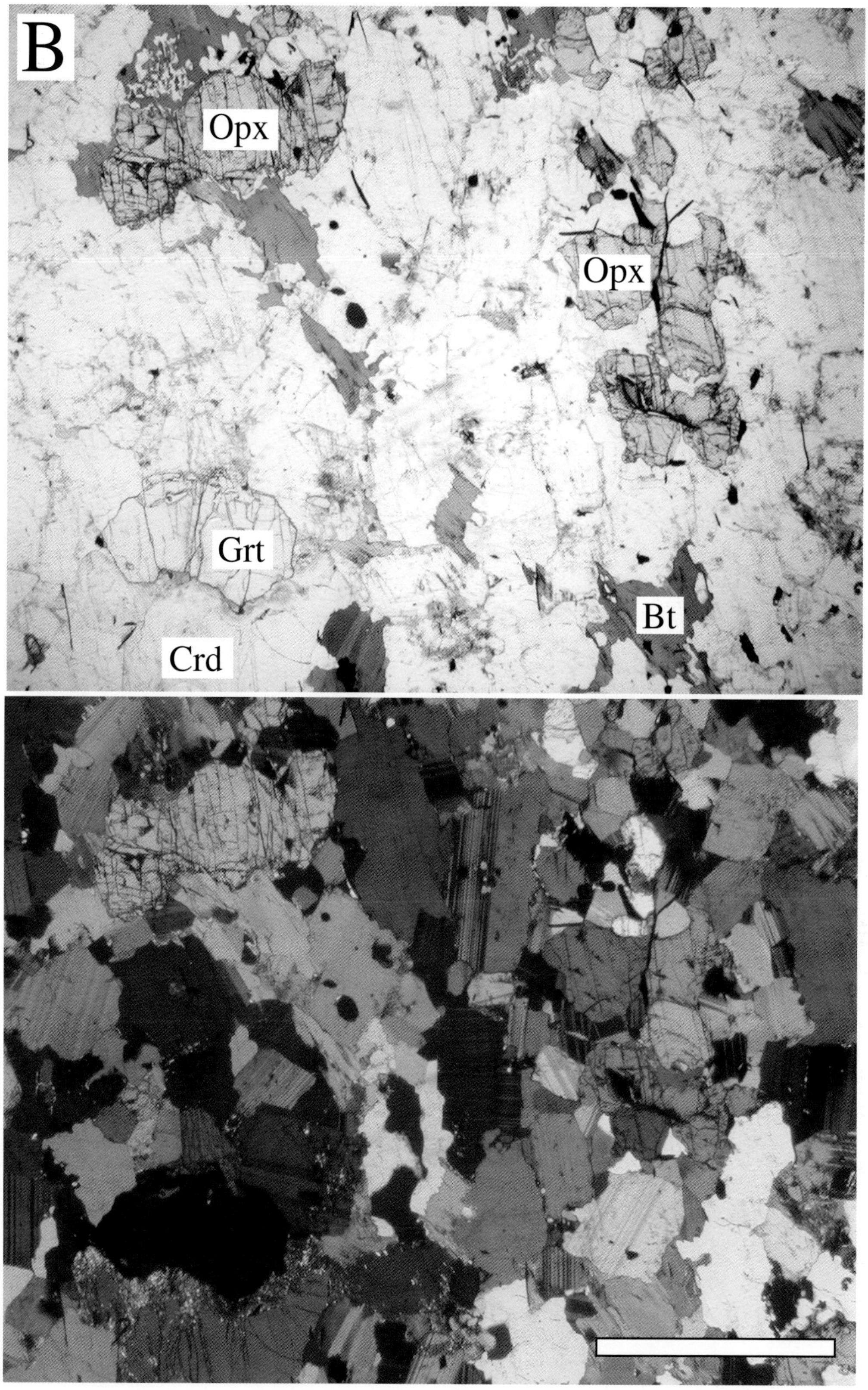

Figure 5 (B) A thin section of S_H-type tonalite, plane-polarised light (top) and crossed polars (bottom). Scale bar = 2 mm.

Figure 5 (C) Photomicrographs of a thin section of a mafic metamorphic enclave, plane-polarised light (top) and crossed polars (bottom). Scale bar = 2 mm.

Figure 5 (D) Photomicrographs of garnet-orthopyroxene-biotite granulite enclaves, plane-polarised light (top) and crossed polars (bottom). Scale bar = 2 mm.

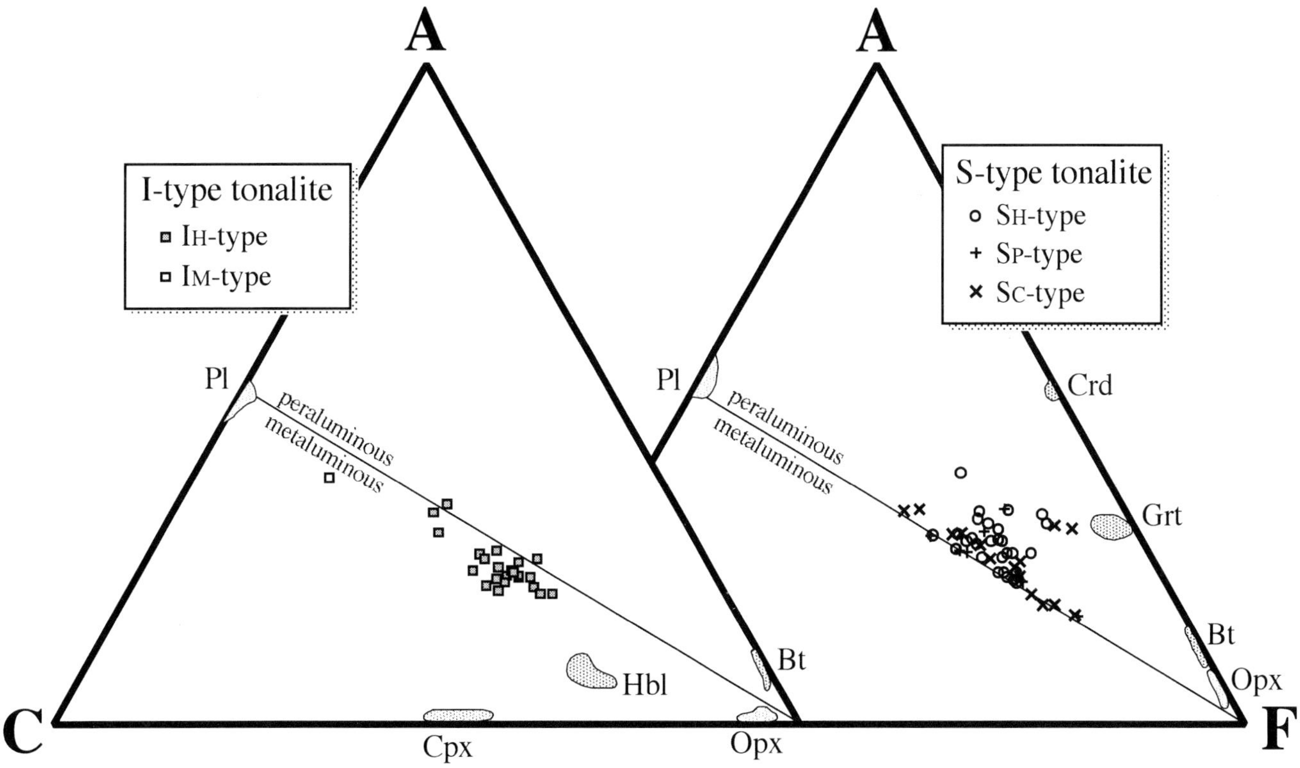

Figure 6 Bulk chemical composition of pyroxene bearing I- and S- type tonalites on ACF ($A=Al_2O_3-Na_2O-K_2O$, $C=CaO$, $F=FeO+MgO$ in mole ratio) diagram. Compositional ranges of minerals in the rocks are also shown. Note that orthopyroxenes from I-type and S-type tonalites have metaluminous and peraluminous composition, respectively.

3.4. S_P-type tonalite

S_P-type tonalite occurs around pelitic-psammitic metamorphic enclaves, or as veins in the enclaves (Fig. 4A, B). This rock is coarse-grained and leucocratic. It consists of garnet (0·5–3 cm in diameter;X_{Mg}=0·14–0·34), orthopyroxene (X_{Mg}=0·47–0·61), biotite (X_{Mg}=0·48–0·67), plagioclase (An=33–45), quartz and K-feldspar (Or=80–87). Cordierite occurs very rarely in the S_P-type tonalite. Accessory minerals are ilmenite, graphite, apatite and zircon. The S_P-type tonalite is a very heterogeneous lithofacies owing to the presence of melanocratic clots and leucocratic patches. The melanocratic clots are composed of medium-grained garnet, orthopyroxene, biotite, plagioclase and quartz. On the other hand, leucocratic patches are composed mainly of coarse- to medium-grained K-feldspar pools, plagioclase and quartz. The enclosing relations of the minerals in S_P-type are different to those of S_H-type tonalite; for example, garnet grains enclose orthopyroxene crystals.

3.5. S_C-type tonalite

S_C-type tonalite is intergradational with S_H- and S_P-type tonalites. The S_C-type tonalite is composed mainly of garnet (X_{Mg}=0·16–0·36), orthopyroxene (X_{Mg}=0·47–0·67), biotite (X_{Mg}=0·50–0·58), plagioclase (An=30–41) and quartz. Cordierite and K-feldspar are rarely observed in the S_C-type tonalite. Sometimes cummingtonite occurs around, or along cleavages in orthopyroxene. Accessory minerals are ilmenite, graphite, apatite and zircon. Some S_C-type tonalite shows orthocumulate or mesocumulate textures. In this case, plagioclase, garnet and some orthopyroxene are cumulus phases, while quartz and biotite are inter-cumulus phases.

The S_C-type tonalite contains small melanocratic clots which consist of orthopyroxene and/or garnet. These minerals occur not only in the melanocratic clots, but also as separate phases in the matrix. The enclosing relationships and the zoning pattern of the minerals in S_C-type tonalite are more irregular that those of S_H-type.

3.6. Mafic granulite enclaves

Mafic enclaves are gabbro, brown-hornblende amphibolite and mafic granulite (orthopyroxene granulite and two-pyroxene granulite). Brown-hornblende amphibolite and mafic granulite are distinguished by the occurrence of pyroxene. These are dark grey in colour and have schistosity.

Mafic granulites and the amphibolites from the metamorphic layers have MORB-like chemical characteristics (Osanai *et al.* 1989). They are thought to represent altered fragments from ophiolite blocks in an accretion complex (see Osanai *et al.* 1989; Mikoshiba 1999). The Niikappu tonalite suite intrudes into the mafic metamorphic layer and gabbroic suite at its northern edge. The tonalite suite encloses the mafic metamorphic rocks as blocks (Shimura 1992, 1999). The mineral assemblage of the mafic granulite enclaves is mainly orthopyroxene, clinopyroxene, brown-hornblende, cummingtonite and plagioclase (Fig. 5C). Accessory minerals are ilmenite, apatite, titanite and zircon. Mafic granulites are usually surrounded and veined by leucocratic tonalite. Modal orthopyroxene in mafic granulite increases in contrast with decreasing hornblende toward the tonalite matrix. The boundaries between the reaction rims (I_M-type tonalite) and tonalite matrix are relatively sharp, and the reaction rims are narrow compared with those of pelitic-psammitic enclaves (S_P-type).

3.7. Grt-Opx-Bt granulite enclaves

A large number of pelitic-psammitic enclaves occur in the basal tonalite (Fig. 4). Most of these enclaves are medium-grained, garnet-orthopyroxene-biotite granulite, and are less

Table 2 Representative whole rock chemical compositions of pyroxene-bearing I- and S- type tonalites. Representative mafic metamorphic rocks and pelitic-psammitic metamorphic rocks from the northern to central part of HMB are also shown. Sample 71017 is the starting material of experiments of Shimura *et al.* (1992). 71017, 71017AM, and 71017AL are different portion from the same outcrop.

Sample:	8PP21	8PP17	PP13	71017	71017AM	71017AL	8PP13	80405	92008	72707	94SS02	KE19	4HB29	3ME01	SS09	9090101
Rock:	Diorite	Tonal	Tonal	Tonal	Tonal	Tonal	Tonal	Tonal	Mafic	Mafic	Mafic	Mafic	Pelitic	Pelitic	Pelitic	Psammi
Type:		I_H	I_H	S_H	S_H	S_H	S_H									
(wt.%)																
SiO_2	52·95	59·82	57·02	64·63			61·82	62·01	49·51	49·70	48·84	52·15	68·36	63·89	62·88	61·29
TiO_2	1·22	1·09	1·19	0·71			0·93	0·89	1·25	1·70	1·42	1·04	0·75	0·93	0·92	1·02
Al_2O_3	16·28	17·35	16·95	16·07			16·80	17·30	14·72	15·72	15·47	15·96	14·81	16·10	17·59	16·20
Fe_2O_3	0·78	0·42	0·94	0·56			2·73	0·24	0·82	0·35	11·18	9·49	6·36	8·50	8·74	11·25
FeO	6·14	5·72	6·58	4·83			4·96	7·04	8·12	9·61						
MnO	0·14	0·10	0·11	0·08			0·03	0·09	0·19	0·15	0·18	0·18	0·06	0·11	0·10	0·08
MgO	6·98	3·60	4·76	2·41			3·10	2·80	11·68	7·45	8·63	6·60	2·99	2·85	3·01	3·07
CaO	9·37	6·40	6·06	3·28			2·57	1·95	11·94	11·26	11·34	9·73	2·16	2·52	2·35	3·17
Na_2O	2·94	4·01	3·71	3·86			3·41	3·09	1·66	3·14	2·42	3·52	2·17	3·64	3·86	4·20
K_2O	1·47	0·63	1·34	2·50			1·75	2·49	0·54	0·46	0·08	0·11	0·97	1·05	1·02	0·00
P_2O_5	0·17	0·19	0·23	0·10			0·08	0·09	0·03	0·16	0·10	0·10	0·04	0·16	0·06	0·17
$H_2O(+)$	0·70	0·02	0·02	0·36			0·81	0·72	0·00	0·07						
$H_2O(-)$	0·44	0·06	0·05	0·11			0·10	0·16	0·06	0·04						
Total	99·58	99·41	98·96	99·50			99·09	98·88	100·52	99·81	99·66	98·86	98·66	99·72	100·52	100·45
(p.p.m.)																
V	172	119	142	166	184	135	165	166	213	291	255	205	121	155	140	176
Cr	251	72	139	89	56	36	74	80	409	239	288	226	50	64	69	74
Co	52	81	60	44	90	131	109	97	71	66	54	63	106	127	133	112
Ni	28	21	40	34	56	31	33	35	212	69	96	83	29	27	38	34
Zn	75	75	79	74	93	75	109	84	69	62	50	104	85	79	71	–
Ga	16	19	18		20	18	20	18	13	17	15	16	15	17	18	14
Ge	1·6	1·5	1·5		1·5	1·8	1·4	1·6	2·2	1·8	1·7	1·4	1·3	1·4	1·3	0·9
Rb	51	10	36	47	42	49	45	56	6	3	–	–	29	26	29	–
Sr	275	357	328	432	460	441	272	253	279	181	108	159	204	252	291	259
Y	41·0	15·6	15·7	15·2	12·5	13·6	8·7	38·9	29·0	39·4	34·0	31·1	27·5	29·1	31·1	31·1
Zr	94	180	224	182	211	217	206	199	99	128	87	73	137	189	181	167
Nb	6·5	5·1	5·7	9·2	8·4	7·5	9·0	8·6	1·7	3·6	1·8	2·8	4·7	7·8	7·1	8·3
Ba	434	287	554	1147	891	1430	634	1220	358	19	17	68	498	220	619	29
La	9·30	11·04	10·34		25·44	22·19	24·03	25·23	2·56	4·09	2·08	5·40	22·37	21·88	25·99	24·00
Ce	25·85	23·66	22·49		49·99	43·80	51·37	54·59	8·81	12·87	7·67	15·67	46·97	48·61	57·38	54·17
Pr	3·31	2·49	2·39		4·74	4·17	5·01	5·32	1·40	1·84	1·24	2·03	4·90	4·94	5·44	5·42
Nd	18·22	11·69	11·48		19·64	17·34	21·38	23·83	8·68	11·09	7·77	10·94	21·49	21·74	23·20	24·10
Sm	5·90	2·72	2·75		3·74	3·31	4·58	5·10	3·11	4·08	3·11	3·52	4·49	4·63	4·73	5·01
Eu	1·09	1·32	1·16		1·56	1·61	1·38	1·34	1·06	1·51	1·12	1·16	1·08	1·14	1·44	0·98
Gd	6·29	2·72	2·67		2·86	2·63	3·47	4·49	4·03	5·30	4·30	4·18	3·89	4·28	4·05	4·64
Tb	1·21	0·45	0·45		0·41	0·40	0·45	0·87	0·79	1·04	0·87	0·83	0·69	0·75	0·78	0·84
Dy	7·08	2·46	2·57		2·20	2·25	1·89	5·66	4·73	6·35	5·41	5·03	4·31	4·51	5·03	5·06
Ho	1·40	0·50	0·52		0·41	0·47	0·29	1·32	0·98	1·31	1·13	1·03	0·90	0·93	0·98	1·06
Er	4·20	1·62	1·69		1·27	1·58	0·81	4·37	3·08	4·21	3·58	3·33	2·99	3·02	3·14	3·34
Tm	0·585	0·238	0·245		0·191	0·252	0·113	0·679	0·456	0·608	0·534	0·497	0·442	0·467	0·456	0·516
Yb	3·34	1·54	1·62		1·39	1·83	0·79	4·33	2·73	3·85	3·24	3·09	2·78	2·95	2·93	3·23
Lu	0·486	0·245	0·264		0·235	0·312	0·124	0·689	0·412	0·566	0·499	0·475	0·432	0·461	0·448	0·491
Hf	2·0	4·1	5·0		4·9	5·1	4·6	4·7	2·4	3·1	2·3	1·9	3·3	4·5	4·4	4·3
Ta	0·33	0·39	0·38		0·55	0·65	0·64	0·66	0·19	0·27	0·09	0·19	0·37	0·60	0·45	0·57
Tl	0·29	0·07	0·23		0·23	0·33	0·25	0·28	–	–	–	–	0·16	0·15	0·16	–
Pb	5	6	6	27	11	19	12	13	–	–	–	–	–	–	–	–
Th	1·35	1·04	1·34	4·2	4·04	3·24	7·47	8·53	0·06	0·13	0·06	0·48	7·93	7·48	8·77	8·83
U	0·55	0·47	0·61		1·07	1·28	1·19	1·33	0·03	0·09	0·04	0·23	1·05	1·39	0·72	1·26

than 5 cm (maximum=30 cm) in size. The enclaves mainly consist of garnet (X_{Mg}=0·16–0·34), orthopyroxene (X_{Mg}=0·55–0·61), biotite (X_{Mg}=0·68–0·72), plagioclase (*An*=35–45) and quartz, with accessory graphite, ilmenite and apatite as mineral assemblage. No K-feldspar is identified in these rocks. The garnet grains of these enclaves are 0·1–0·5 mm in grain-size diameter and euhedral, while orthopyroxene is 0·1–0·5 mm in size and anhedral (Fig. 5D).

3.8. Whole-rock geochemistry and sample selections

Whole rock major and trace element analyses were performed by means of X-ray fluorescence analyser at Niigata University.

Table 3 Sr and Rb isotope data of pyroxene-bearing I- and S-type tonalites.

Rock type	Sample no.	Rb (p.p.m.)	Sr (p.p.m.)	$^{87}Sr/^{86}Sr$	2σ ($\times 10^{-6}$)	$^{87}Rb/^{86}Sr$	Sr*I* (55 Ma)
I_H	8PP17	10·3	371	0·704593	12	0·0799	0·704530
I_H	PP13	37·2	352	0·704900	12	0·3058	0·704661
I_H	90804	46·0	457	0·704655	11	0·2909	0·704428
S_H	71017	46·5	432	0·705628	9	0·3113	0·705385
S_H	71017AL	48·6	441	0·705302	14	0·3185	0·705053
S_H	71017AM6	42·2	483	0·705212	14	0·2529	0·705014
S_H	71017AM7	43·1	478	0·705222	11	0·2610	0·705018
S_H	8PP13	46·1	287	0·706630	10	0·4647	0·706267
S_P	8ZA03	22·7	302	0·706046	12	0·2174	0·705876
S_P	PP07	34·3	311	0·706664	12	0·3189	0·706415
S_P	2ZA03–2	4·1	224	0·706572	10	0·0529	0·706531
S_P	2ZA03–3	7·7	213	0·706553	9	0·1045	0·706471
S_P	2ZA03–4	21·4	194	0·706352	12	0·3189	0·706103
S_P	2ZA02–M1	22·6	201	0·706562	15	0·3264	0·706307
S_C	80405	55·7	258	0·706683	12	0·6232	0·706196
S_C	8ZA04	25·4	205	0·705760	6	0·3590	0·705479

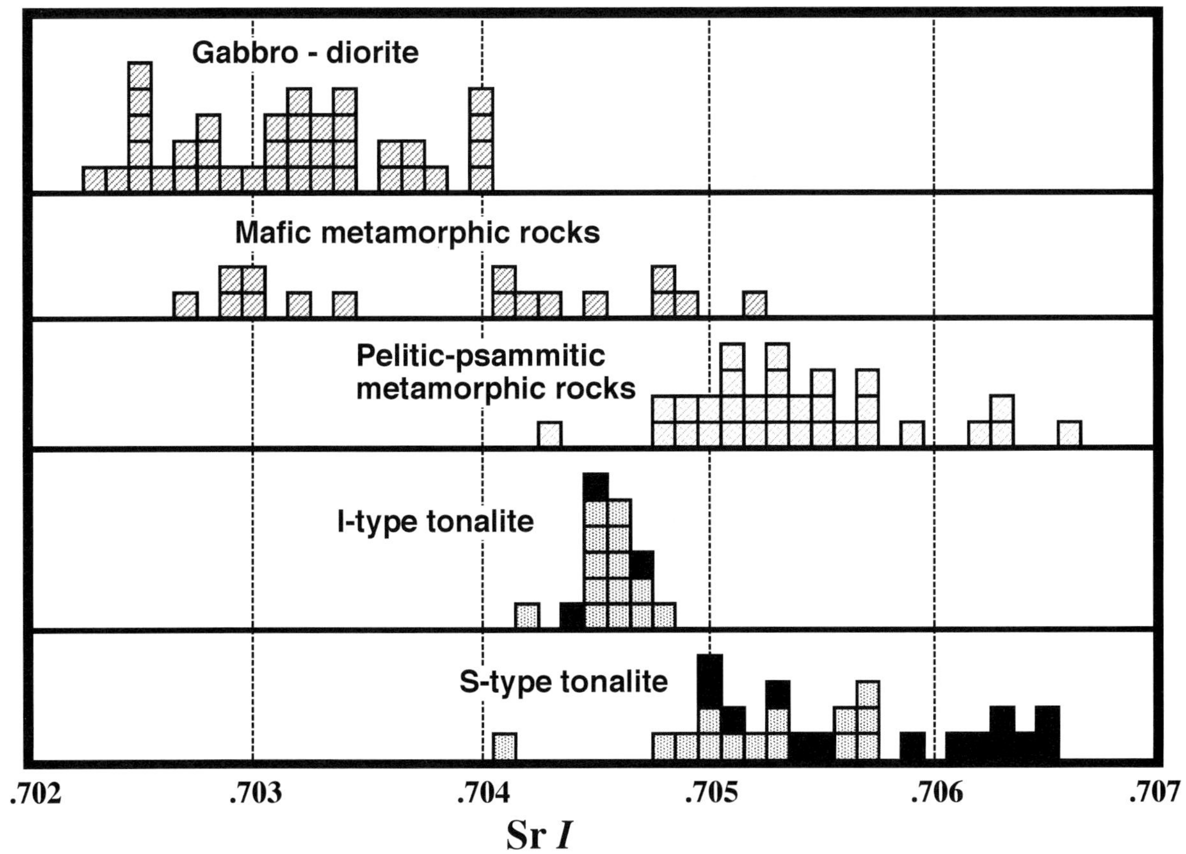

Figure 7 Histograms of initial $^{87}Sr/^{86}Sr$ ratios (Sr*I*) at 55 Ma for igneous rocks and metamorphic rocks in the HMB (modified after Shimura, 1999 and the compiled data therein). Solid squares indicate the pyroxene bearing tonalites of this study. The ranges of Sr*I* of I- and S-type tonalites are nearly identical to those of mafic metamorphic rocks and pelitic-psammitic metamorphic rocks, respectively.

The analytical methods are after Takahashi & Shuto (1997). FeO was determined by the $KMnO_4$ titration method. Major elements of the S- and I-type tonalites have already been reported in Shimura *et al.* (1992) and Shimura (1999), respectively. From those samples, a few typical samples of each rock type were chosen for rare-earth element (REE) and Sr isotope analysis. Matabasites and pelitic-psammitic metamorphic rocks have been sampled from enclaves in this tonalite body and from metamorphic layers of the HMB in adjacent areas. The REEs were analysed by ICP-MS by Activation Laboratories Ltd, Canada.

Isotope compositions were determined using the mass spectrometer (MAT-261) at the Graduate School of Science and Technology, Niigata University. All of the Rb and Sr

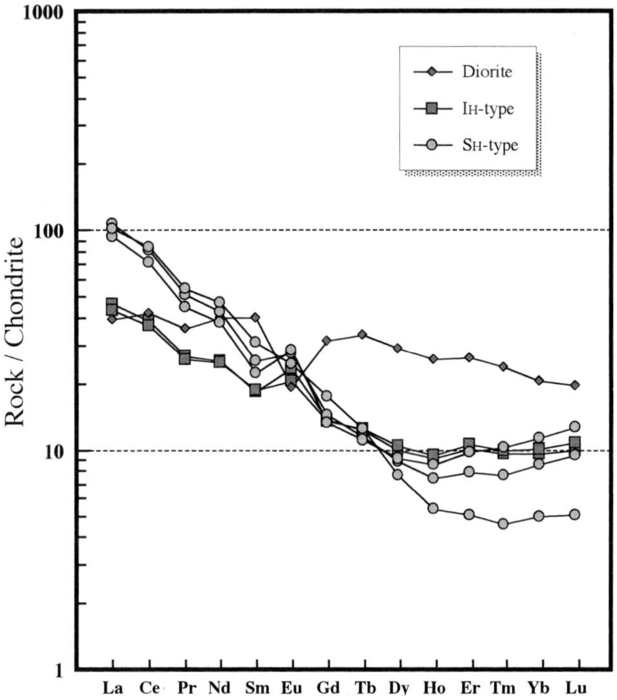

Figure 8 Chondorite-normalised REE pattern of pyroxene bearing I_H- and S_H- type tonalites. A sample of pyroxene-bearing diorite is also shown.

concentrations for dating were measured by the isotope dilution method. The 'Krogh-type' vessel and jacket (Krogh 1973) were employed to dissolve the samples. The analytical procedures are described in Kagami *et al.* (1987, 1989). The $^{87}Sr/^{86}Sr$ ratio for NBS987 was measured repeatedly three times during the present study, with a mean value of 0·710220 ± 0·000008 (2σ).

Selected analyses are given in Tables 2 and 3.

4. Origin of sub-types tonalite

4.1. Relationship between I_H- and I_M-type tonalites

As described above, based on petrographic results, it is rare that plagioclase encloses other minerals in I_H-type tonalite, and the plagioclase of I_H-type has oscillatory zoning in the core and normal zoning in the rim. This evidence indicates that the magma was crystallised through a normal cooling history. The crystallising sequence of minerals from the I_H-type tonalite magma is inferred from the enclosing and chemical relations of the minerals; for example, hornblende enclosing pyroxene and high *An* plagioclase (Fig. 5A). Such relationships indicate that the crystallisation sequence is (clinopyroxene, orthopyroxene, plagioclase) → (quartz, cummingtonite, hornblende) → biotite (Shimura *et al.* 1999). On the other hand, I_M-type tonalite has reverse zoned and corroded plagioclase, or sometimes has pyroxene xenocrysts derived from the mafic enclaves. Therefore, I_M-type tonalite is thought to be a mixture of fragments of disaggregated mafic granulite and homogeneous tonalite magma.

4.2. Relationship of S_H-, S_P-, and S_C-type tonalites

Because S_H-, S_P- and S_C-type tonalites are intergradational at the outcrop scale, it is more difficult to discuss their origins than those of I-type tonalite.

The crystallising sequence of minerals from the S_H-type tonalite magma is inferred from the enclosing relations of the

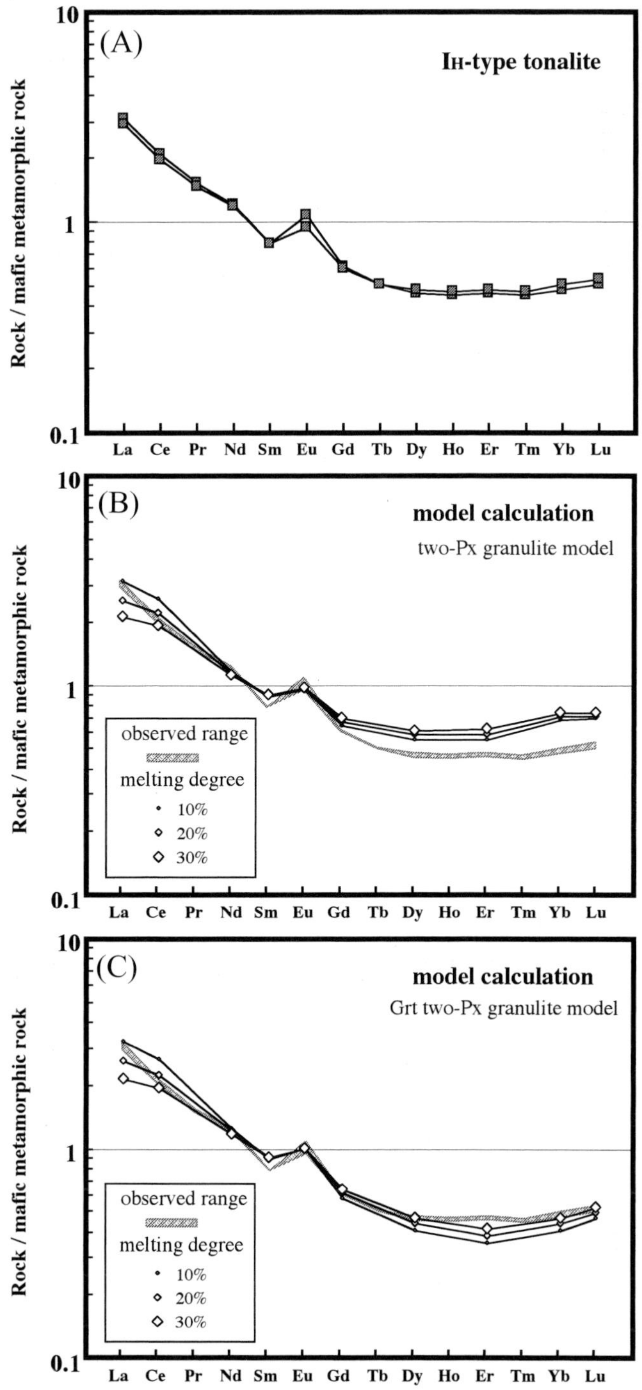

Figure 9 REE spidergram of pyroxene-bearing I_H-type tonalites normalised by average mafic metamorphic rocks (A). The calculated results of batch partial melting model, for I_H-type as two-pyroxene granulite restite model (B) and as garnet-two-pyroxene restite model (C).

minerals: high X_{Mg} garnet grains are enclosed in plagioclase and high *An* plagioclase grains are enclosed in orthopyroxene with poikilitic texture. This evidence suggests that the crystallisation sequence is garnet → plagioclase → orthopyroxene → biotite → quartz → cordierite → K-feldspar (Shimura *et al.* 1992).

The textural relationships of the minerals in S_P- and S_C-type tonalites are different to those of the S_H-type; for example, garnet encloses orthopyroxene. Patterns of chemical zoning in minerals of the S_P- and S_C-types are also different from those of S_H-type mineral grains. Garnet in S_P-type tonalite and in garnet-orthopyroxene granulite enclaves has normal zoning

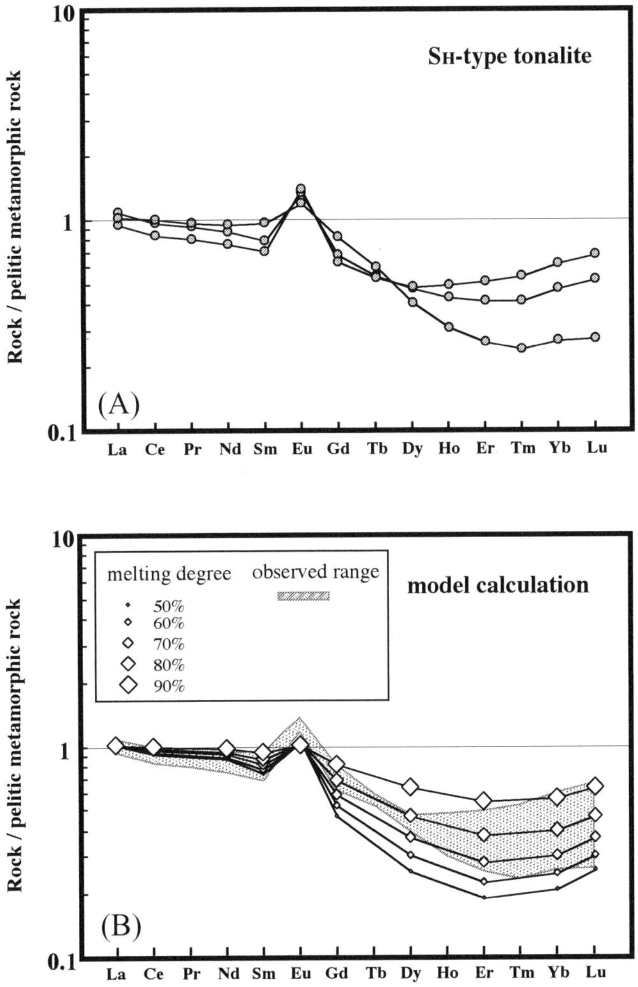

Figure 10 REE spidergrams of (A) pyroxene S_H-type tonalites, normalised by average pelitic-psammitic metamorphic rocks and (B) the calculated results of the batch partial melting model, for S_H-type tonalites.

from the core to the mantle, and has reverse zoning in the rim (Shimura *et al.* 1999). Normally zoned parts of garnet in S_P-type tonalite formed as *in situ* incongruent melting reaction products of the enclave, probably as a result of heating by the host tonalite magma (Shimura *et al.* 1999). S_P-type tonalite is a mixture of original magma, xenocryst and *in situ* melting products, and is thought to be of assimilation origin. On the other hand, the normally zoned parts of garnet in S_C-type tonalite may represent original restites derived from the lowermost parts of the crust. Since the garnet in S_H-type tonalite shows a simple cooling pattern of zoning, the present authors believe that the S_H-type tonalite represents the original magma.

4.3. Evolution of I_H- and S_H-type tonalites

As described in section 2.1, the western edge of the pyroxene-bearing tonalite area is mylonitised along with the HMT. The tectonic evolution of the HMB is subdivided into three stages: (1) formation of magmatic arc; (2) exhumation of the crust; and (3) brittle deformation (Toyoshima *et al.* 1994). Owada *et al.* (2003) reviewed the ages of the metamorphic and igneous events of the HMB. The above authors implied that the tectonothermal activity in the HMB mainly occurred at 56–50 and 23–18 Ma. Since the igneous rocks of the latter stage cut the foliation of mylonitised tonalites (Owada *et al.* 2003), they concluded that the tonalite activity is part of the earlier stage at 56–50 Ma. The initial $^{87}Sr/^{86}Sr$ ratios (Sr*I*) corrected to 55 Ma for those igneous and metamorphic rocks are summarised in Figure 7 (after Shimura 1999). This shows values of 0·7023–0·7040 for gabbros and diorites, 0·7027–0·7052 for mafic metamorphic rocks (N-MORB composition metabasites), 0·7043–0·7066 for pelitic-psammitic metamorphic rocks, 0·7042–0·7048 for I-type granitoids, and 0·7041–0·7065 for S-type granitoids (Table 3 & Fig. 7). The range of Sr*I* of the pyroxene-bearing I-type tonalite is almost the same as that of other pyroxene-free I-type tonalites. The range indicates that the I_H-type tonalite cannot be genetically related to the gabbroic magma, but may be related to mafic metamorphic rocks. On the other hand, pyroxene-bearing S_H-type tonalite is isotopically identical to the pelitic-psammitic granulite.

Figure 8 shows REE patterns of I_H- and S_H-type tonalites. Note that I_H-type is lower in LREE contents than S_H-type tonalite. The REE pattern of the I_H-type tonalite is distinct from that of diorite, which is considered to be a differentiation product from gabbroic magma (Fig. 8).

Melting reactions of mafic metamorphic rocks in the granulite facies have been considered by several works (Ellis & Thompson 1986; Clements 1990; Vielzeuf *et al.* 1990; Rushmer 1991); for example:

Hbl + Qtz = Cpx + Opx ± Grt + Pl + melt
Hbl + Bt + Qtz = Cpx + Opx + Pl + Kfs + melt
Hbl + Pl + Qtz = Cpx + more calcic Pl ± Ilm + melt
Hbl + Pl = Cpx + Opx + more calcic Pl + Ilm + Ttn ± Spl + melt

These melting reactions involve the incongruent breakdown of hornblende, which occurs at about 750–900 °C (Thompson 1990), and produce metaluminous granitic melts.

On the other hand, possible melting reactions for pelitic-psammitic metamorphic rocks in the granulite facies are biotite dehydration melting reactions described as (Green 1976; Vielzeuf & Holloway 1988; Le Breton & Thompson 1988; Patiño Douce & Johnston 1991):

Bt + Qtz = Opx + Kfs + melt
Bt + Qtz = Grt + Opx + Kfs + melt
Bt + Pl + Qtz = Opx + Kfs + melt

These reactions occur at about 850–900 °C and produce peraluminous granitic melt.

It is possible that the I- and S-type tonalite magmas are generated by partial melting of metabasite and pelitic-psammitic metamorphic rocks, respectively. If so, then we can assume that the melting reactions are the origin of the I- and S-type tonalites, respectively:

Metabasite → (Grt-) two-Pxgranulite (restite) + I-type magma

Metasedimentary rock → Grt-Opx granulite (restite) + S-type magma

To illustrate the melt/original rock relationships clearly, spidergrams of I_H-type tonalite normalised against mafic metamorphic rocks and S_H-type tonalite normalised against metasedimentary rocks are shown in Figures 9A and 10A, respectively. These have positive Eu anomalies. Tagiri *et al.* (1989) reported REEs of cordierite tonalite (pyroxene-absent middle S-type tonalite) in the southern part of the HMB. Their REE pattern of enriched cordierite tonalite had a negative Eu anomaly, but their depleted type has a similar pattern to that of the pyroxene-bearing S_H-type tonalites in the present study. This may indicate that crustal-scale fractionation has occurred.

A model calculation was performed on the REE patterns, assuming equilibrium batch partial melting, using the equation of Shaw (1970). The proportions of restite minerals for I_H-type tonalite are assumed from natural two-pyroxene mafic granulite enclaves in this tonalite body (e.g. Fig. 5C: 20% Opx,

Table 4 Mineral-melt partition coefficients for the model calculations. F1984: Fujikami *et al.* (1984); A1976: Arth (1976); IF1978: Irving & Frey (1978); B1994: Bea *et al.* (1994); NC1985: Nash & Crecraft (1985); DW1975: Drake & Weill (1975); GP1983: Green & Person (1985).

	Opx	Cpx	Hbl	Grt	Bt	Pl		Kfs	Ilm	Spn	Zrn	Mnz	Ap
						I-type	S-type						
La	0·031	0·047	0·500	0·076	0·06	0·228	0·302	1·01	7·1	2·0	1·30	49219	456
Ce	0·028	0·084	0·899	0·350	0·05	0·136	0·221	0·86	7·8		2·04	63871	569
Nd	0·028	0·183	2·890	0·530	0·08	0·115	0·149	0·51	7·6		3·35	76387	855
Sm	0·028	0·377	3·990	2·660	0·06	0·077	0·102	0·42	6·9	10·0	3·79	75289	1105
Eu	0·028	0·800	3·440	1·500	0·05	0·079	1·214	2·32	2·5		0·45	548	23·8
Gd	0·039	0·583	5·480	10·500	0·10	0·056	0·067	0·60			9·21	95815	2133
Dy	0·076	0·774	6·200	28·600	0·17	0·045	0·050	0·77	4·9		38·8	48888	3257
Er	0·153	0·708	5·940	42·800	0·22	0·040	0·045	1·14			165	29299	4231
Yb	0·254	0·633	4·890	39·900	0·12		0·041	0·64	4·1		278	6485	2216
Lu	0·323	0·665	4·530	29·600	0·20	0·046	0·039	0·96	3·6	6·0	923	7216	2981
Reference:	F1984	F1984	GP1983 A1976	IF1978 A1976	B1994	DW1975	F1984	B1994	NC1985	GP1983	B1994	B1994	B1994

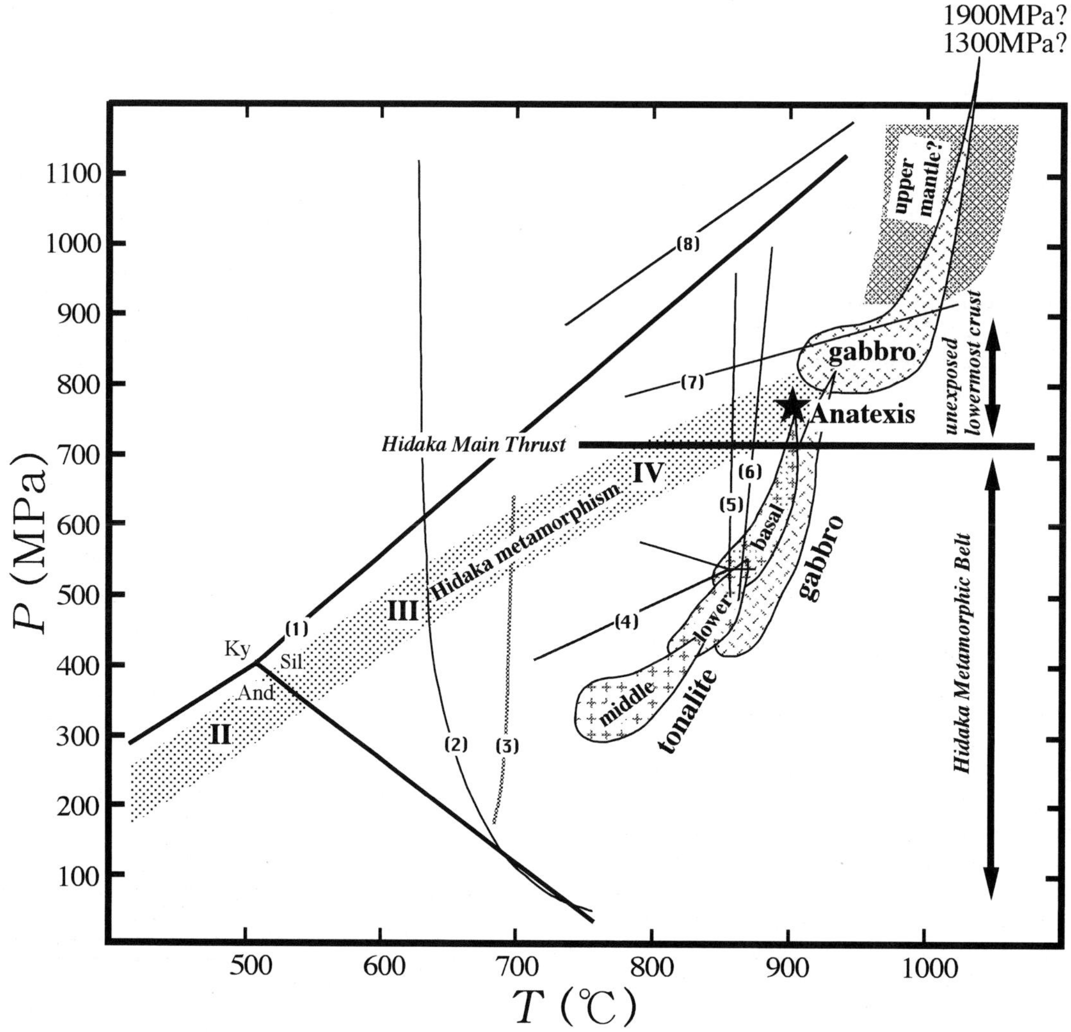

Figure 11 *P–T* condition of metamorphic rocks and S-type tonalite magmas in the Hidaka Metamorphic Belt. II, III, IV and a shadow belt are metamorphic grade and metamorphic geothermal gradient of the HMB in the peak metamorphism (after Osanai *et al.* 1991). Diapirs show intrusion of buoyant magma (after Shimura *et al.* 1992). *P–T* condition of gabbro source after Maeda and Saito (1997). X_{Mg} of the petrogenetic grid close to 0·6. Reactions: (1) reaction boundaries of Al_2SiO_5 polymorphs are after Holdaway & Makhopadhyay (1993); (2) Ms+Qtz+V=Als+Kfs+L and (4) Bt+Als+Qtz=Crd+Kfs+L in X_{H_2O}=1·0, Holdaway & Lee (1977); (3) S-type tonalite solidus, Shimura *et al.* (1992); (5) Bt+Sil+Qtz=Grt+Kfs+L, Vielzeuf & Holloway (1988); (6) Bt+Pl+Qtz=Grt+Opx+Kfs+L, Vielzeuf & Montel (1994); (7) and (8) are Grt-in reaction curve in high-*P* side on olivine tholeiite composition (Green & Ringwood, 1967) and quartz tholeiite composition (Ito & Kennedy, 1971), respectively.

20% Cpx, 25% Hbl, 35% Pl's and 0·01% accessory Spn and Ap). The proportions used in the case of the S_H-type tonalite are assumed from natural pelitic-psammitic granulite enclaves in this tonalite body (e.g. Fig. 5D): 30% Opx, 20% Grt, 37% Pl, 3% Kfs, 5% Ilm, and 0·01% accessory Zrn, Mnz and Ap. The mineral-melt partition coefficients of each element used in these calculation are shown in Table 4.

The results are shown in Figures 9B and 10B, respectively. The REE pattern of S_H-type tonalite can be successfully reproduced by the partial melting of pelitic-psammitic gneiss (Fig. 10B). On the other hand, for I_H-type tonalite, calculated melts contain high HREE in comparison to the actual range (Fig. 9B). Therefore, the present authors believe that the restite of the I-type tonalite is a garnet-two-pyroxene granulite. If 3% garnet is added to the restite, keeping the proportion of other phases the same, better agreement is obtained, as in Fig. 9C. Although this revised model has been produced simply by a trial-and-error method, the result of this model calculation is in better agreement with the observed range than the two-pyroxene granulite model. Therefore, the restite of the I-type tonalite is thought to be garnet-two-pyroxene mafic granulite. Garnet-two-pyroxene mafic granulite has not been observed in the HMB, although two-pyroxene mafic granulite and garnet-orthopyroxene mafic granulite have been reported (Owada *et al.* 2003).

The maximum *P–T* condition of the HMB is up to 730 MPa and 870°C (Osanai *et al.* 1991). At this condition, metamorphic rock can partially melt and the granulite unit of the HMB locally includes melt patches, but in small amounts (Owada *et al.* 2003). It is difficult to produce the large amounts of the granitic masses of the HMB from the thin granulite sequence (Fig. 2). Therefore, the magma generation level is thought to be deeper in the crust, in a deeper part of the HMB. This deeper unit is missing because of the up-thrust movement. The conditions of melting to produce these magmas are thought to be ca. 800 MPa and greater than 900°C (Osanai *et al.* 1997; Owada *et al.* 2003) (Fig. 11).

5. Conclusions

Pyroxene-bearing I- and S-type tonalites are abundant, and are intimately associated with each other in the Niikappu tonalite suite of the HMB. On the basis of the lithological evidence, they are mainly classified into I_H-, I_M- and S_H-, S_P- and S_C-type tonalites. Petrographical features indicate that the I_M- and S_P-type tonalites are likely to be the results of assimilation. S_C-type tonalite is considered to be the cumulate facies, which contains restite minerals. I_H- and S_H-type tonalites are thought to represent original magmas, i.e. those which have been generated by partial melting of mafic metamorphic rocks and pelitic-psammitic metamorphic rocks, respectively. Model calculations assuming batch partial melting indicate that possible restites are garnet-two-pyroxene mafic granulite for I_H-type tonalite and garnet-orthopyroxene aluminous granulite for the S_H-type.

The lowermost crust of 'the Hidaka crust' is thought to be composed of garnet-two-pyroxene mafic granulite, garnet-orthopyroxene aluminous granulite and gabbro. Mafic metamorphic rocks and gabbros of the HMB show MORB-like composition (Osanai *et al.* 1989; Mikoshiba 1999; Maeda & Kagami 1994, 1996). When these rocks cool isobarically, the lower most crust will be converted to garnet granulite or eclogite facies rock through reaction 7 (Fig. 11), which will potentially be subjected to delamination.

6. Acknowledgements

The authors wish to thank Dr Tony Kemp and Professor Masaaki Obata for their improvements to the manuscript. We also indebted to Dr K. Sajeev for his revision of the English. This work was supported by a Grant-in Aid for Scientific Research (No. 14540420, to T. Shimura) and a Grant for Promotion of Niigata University Research Projects.

7. References

Arita, K., Ikawa, T., Ito, T., Yamamoto, A., Saito, M., Kimura, G., Watanabe, T., Ikawa, T. & Kuroda, T. 1998. Crustal structure and tectonics of the Hidaka Collision Zone, Hokkaido (Japan), revealed by vibroseis seismic reflection and gravity surveys. *Tectonophysics* **290**, 197–210.

Arth, J. G. 1976. Behavior of trace elements during magmatic processes – a summary of theoretical models and their applications. *US Geological Survey Journal of Research* **4**, 41–7.

Bea, F., Pereira, M. D. & Stroh, A. 1994. Mineral/leucosome trace-element partitioning in peraluminous migmatite (a laser ablation-ICP-MS study). *Chemical Geology* **117**, 291–312.

Chappell, B. W. & White, A. J. R. 1974. Two contrasting granite types. *Pacific Geology* **8**, 173–4.

Clements, J. D. 1990. The granulite-granite connection. *In* Vielzeuf, D. & Vidal, Ph. (eds) *Granulites and Crustal Evolution*, 25–36. Dordrecht: Kluwer Academic Publishers.

Drake, M. J. & Weill, D. F. 1975. Partition of Sr, Ba, Ca, Y, Eu^{2+}, Eu^{3+} and other REE between plagioclase feldspar and magmatic liquid: an experimental study. *Geochimica et Cosmochimica Acta* **39**, 689–712.

Ellis, D. J. & Thompson, A. B. 1986. Subsolidus and partial melting reactions in the quartz-excess $CaO+MgO+Al_2O_3+SiO_2+H_2O$ system under water-excess and water-deficient conditions to 10 kb: some implications for the origin of peraluminous melts from mafic rocks. *Journal of Petrology* **27**, 91–121.

Fujikami, H., Tatsumoto, M. & Aoki, K. 1984. Partition coefficients of Hf, Zr and REE between phenocrysts and groundmasses. Proceedings of the fourteenth lunar and planetary science conference, Part 2. *Journal of Geophysical Research* **89**, B665–72.

Green, T. H. 1976. Experimental generation of cordierite- or garnet-bearing granitic liquids from a pelitic composition. *Geology* **4**, 85–8.

Green, T. H. & Person, N. J. 1985. Experimental determination of REE partition coefficients between amphibole and basaltic liquids at high pressure. *Geochimica et Cosmochimica Acta* **49**, 1465–8.

Green, T. H. & Ringwood, A. E. 1967. An experimental investigation of the gabbro-to-eclogite transformation and its petrological applications. *Geochimica et Cosmochimica Acta* **31**, 767–833.

Holdaway, M. J. & Lee, S. M. 1977. Fe-Mg cordierite stability in high-grade pelitic rocks based on experimental, theoretical, and natural observations. *Contributions to Mineralogy and Petrology* **63**, 175–98.

Holdaway, M. J. & Mukhopadyay, B. 1993. A reevaluation of the stability relations andalusite: thermochemical data and phase diagram for the aluminum silicates. *American Mineralogist* **78**, 298–315.

Irving, A. J. & Frey, F. A. 1978. Distribution of trace elements between garnet megacrysts and host volcanic liquids of kimberitic to rhyolitic composition. *Geochimica et Cosmochimica Acta* **42**, 771–87.

Ishihara, S. 1977. The magnetite-series and ilmenite-series granitic rocks. *Mining Geology* **27**, 293–305.

Ito, T. 2000. Crustal structure of the Hidaka collision zone and its foreland fold-thrust-belt, Hokkaido, Japan. *Journal of the Japanese Association for Petroleum Technology* **65**, 103–9.

Ito, K. & Kennedy, C. 1971. An experimental study of the basalt-garnet granulite-eclogite transition. *In* Heacosk, J. G. (ed.) *The structural and physical properties of the Earth's crust. Geophysical Monograph Series* **14**, 303–14. Washington, DC: American Geophysical Union.

Kagami, H., Iwata, M., Sanno, S. & Homma, H. 1987. Sr and Nd isotopic compositions and Rb, Sr, Sm and Nd concentrations of standard samples. *Technical Report, Institute for Study of Earth's Interior, Okayama University Series* **4**, 1–16.

Kagami, H., Yokose, H. & Homma, H. 1989. $^{87}Sr/^{86}Sr$ and 143Nd/^{144}Nd ratios of GSJ rock reference samples; JB-1a, JA-1 and JG-1a. *Geochemical Journal* **23**, 209–14.

Komatsu, M., Miyashita, S. & Arita, K. 1986. Composition and structure of the Hidaka metamorphic belt, Hokkaido – historical review and present status. *Monograph of the Association for the Geological Collaboration in Japan* **31**, 189–203.

Komatsu, M., Osanai, Y., Toyoshima, T. & Miyashita, S. 1989. Evolution of the Hidaka metamorphic belt, northern Japan. *In* Daly, J. S., Cliff, R. A. & Yardley, B. W. D. (eds) *Evolution of Metamorphic Belts. Geological Society, London, Special Publication* **43**, 487–93.

Komatsu, M., Toyoshima, T., Osanai, Y. & Arai, M. 1994. Prograde and anatectic reactions in the deep arc crust exposed in the Hidaka metamorphic belt, Hokkaido, Japan. *Lithos* **33**, 31–49.

Kretz, R. 1983. Symbols for rock-forming minerals. *American Mineralogist* **68**, 277–9.

Krogh, T. E. 1973. A low-contamination method for hydrothermal decomposition of zircon and extraction of U and Pb for isotopic age determinations. *Geochimica et Cosmochimica Acta* **37**, 485–94.

Le Breton, N. & Thompson, A. B. 1988. Fluid-absent (dehydration) melting of biotite in metapelites in the early stages of crustal anatexis. *Contributions to Mineralogy and Petrology* **99**, 226–37.

Maeda, J., Shimura, T., Arita, K, Osanai, Y., Murata, M., Bamba, M. & Suetake, S. 1991. Chemical features of orthopyroxene in peraluminous igneous rocks. *American Mineralogist* **76**, 1676–84.

Maeda, J. & Kagami, H. 1994. Mafic igneous rocks derived from N-MORB source mantle, Hidaka magmatic zone, central Hokkaido: Sr and Nd isotopic evidence. *Journal of the Geological Society of Japan* **100**, 185–8.

Maeda, J. & Kagami, H. 1996. Interaction of a spreading ridge and an accretionary prism: implications from MORB magmatism in the Hidaka magmatic zone, Hokkaido, Japan. *Geology* **24**, 31–4.

Maeda, J. & Saito, K. 1997 Role of mantle-derived primitive magma for generation of continental crust: inferences from the Hidaka metamorphic belt, central Hokkaido. *Memoirs of the Geological Society of Japan* **47**, 75–85.

Mikoshiba, M. 1999. Chemical characteristics of basic rocks from the Main Zone of the Hidaka metamorphic belt, Hokkaido: an affinity to mid-ocean ridge basalts. *Bulletin of the Geological Survey of Japan.* **50**, 613–34.

Nash, W. P. & Crecraft, H. R. 1985. Partition coefficients for trace elements in silisic magmas. *Geochimica et Cosmochimica Acta* **49**, 2309–22.

Osanai, Y., Arita, K. & Bamba, M. 1986. *P–T* conditions of granulite-facies rocks from the Hidaka metamorphic belt, Hokkaido, Japan. *Journal of the Geological Society of Japan* **92**, 793–808.

Osanai, Y., Owada, M. & Takasu, I. 1989. Original rock constitution of the Main Zone of the Hidaka metamorphic belt, Hokkaido, Japan. *Bulletin of the Fukuoka University of Education, Series III* **38**, 71–91.

Osanai, Y., Komatsu, M. & Owada, M. 1991. Metamorphism and granite genesis in the Hidaka Metamorphic Belt, Hokkaido, Japan. *Journal of Metamophoric Geology* **9**, 111–24.

Osanai, Y., Owada, M., Shimura, T., Kawasaki, T. & Hensen, B. J. 1997. Crustal anatexis and related acidic magma genesis in the Hidaka metamorphic belt, Hokkaido, northern Japan. *Memoirs of the Geological Society of Japan* **47**, 29–42.

Owada, M., Osanai, Y., Shimura, T., Toyoshima, T. & Katsui, Y. 2003. Crustal section and anatexis of lower crust due to mantle flux in the Hidaka metamorphic belt, Hokkaido, Japan. *Geological Survey of Japan, Interim Report* **28**, 81–102.

Patiño Douce, A. E. & Johnston, A. D. 1991. Phase equilibria and melt productivity in the pelitic system: implications for the origin of peraluminous granitoids and aluminous granulites. *Contributions to Mineralogy and Petrology* **107**, 202–18.

Rushmer, T. 1991. Partial melting of two amphibolites: contrasting experimental results under fluid-absent conditions. *Contributions to Mineralogy and Petrology* **107**, 41–59.

Shaw, D. M. 1970. Trace element fractionation during anatexis. *Geochimica et Cosmochimica Acta* **34**, 237–43.

Shimura, T. 1992. Intrusion of granitic magma and uplift tectonics of the Hidaka metamorphic belt, Hokkaido. *Journal of the Geological Society of Japan* **98**, 1–20.

Shimura, T. 1995. Data transmission system from EPMA to the file of spread sheet programs. *Geoinformatics* **6**, 31–40.

Shimura, T. 1999. Genesis of the pyroxene-bearing I-type tonalite and melting degree of the source rock, in the Hidaka Metamorphic Belt, northern Japan. *Journal of the Geological Society of Japan* **105**, 536–51.

Shimura, T. 2004. Variety of S-type granitoids in the Niikappu River area, Hidaka Metamorphic Belt, northern Japan. *Chishitsu News* **593**, 65–70.

Shimura, T., Komatsu, M. & Iiyama, J. T. 1992. Genesis of the lower crustal garnet-orthopyroxene tonalites (S-type) of the Hidaka Metamorphic Belt, northern Japan. *Transactions of the Royal Society of Edinburgh: Earth Sciences* **83**, 259–68.

Shimura, T., Komatsu, M., Tsutai, T., Owada, M. & Takahashi, Y. 1997. Thermal and chemical interaction between granitic magmas and wall rocks in the Hidaka metamorphic belt, northern Japan. *Memoirs of the Geological Society of Japan* **47**, 1–12.

Shimura, T., Komatsu, M., Iiyama, J. T. & Osanai, Y. 1999. Grt-Opx granulite enclaves in S-type tonalite of the Hidaka metamorphic belt, northern Japan. *Memoirs of the Geological Society of Japan* **53**, 21–34.

Tagiri, M. Shiba, M. & Onuki, H. 1989. Anatexis and chemical evolution of pelitic rocks during metamorphism and migmatization in the Hidaka metamorphic belt, Hokkaido. *Geochemical Journal* **23**, 321–37.

Takahashi, Y. & Shuto, K. 1997. Major and trace element analyses of silicate rock using X-ray fluorence spectrometer RIX3000. *The Rigaku-Denki Journal* **28**, 28–37.

Thompson, A. B. 1990. Heat, fluids, and melting in the granulite facies. *In* Vielzeuf, D. & Vidal, Ph. (eds) *Granulites and Crustal Evolution*, 37–57. Dordrecht: Kluwer Academic Publishers.

Toyoshima, T., Komatsu, M. & Shimura, T. 1994. Tectonic evolution of lower crustal rocks in relations to magmatic activities in the Hidaka metamorphic belt, Hokkaido, northern Japan. *The Island Arc* **3**, 182–98.

Tsumura, N., Ikawa, H., Ikawa, T., Shinohara, M., Ito, T., Arita, K., Morita, T., Kimura, G. & Ikawa, T. 1999. Delamination-wedge structure beneath the Hidaka Collision Zone, Central Hokkaido, Japan: inferred from seismic reflection profiling. *Geophysical Research Letters* **26**, 1057–60.

Vielzeuf, D., Clements, J. D., Pin, C. & Moinet, E. 1990. Granites, granulites, and crustal differentiation. *In* Vielzeuf, D. & Vidal, Ph. (eds) *Granulites and Crustal Evolution*, 59–85. Dordrecht: Kluwer Academic Publishers.

Vielzeuf, D. & Holloway, R. 1988. Experimental determination of the fluid-absent melting relations in the pelitic system. Consequences for crustal differentiation. *Contributions to Mineralogy and Petrology* **98**, 257–76.

Vielzeuf, D. & Montel, J. M. 1994. Partial melting of metagreywackes. Part I. Fluid-absent experiments and phase relationships. *Contributions to Mineralogy and Petrology* **117**, 375–93.

Vielzeuf, D. & Schmidt, M. W. 2001. Melting relations in hydrous systems revisited: application to metapelites, metagreywackes and metabasalts. *Contributions to Mineralogy and Petrology* **141**, 251–67.

White, A. J. R. & Chappell, B. W. 1977. Ultrametamorphism and granitoid genesis. *Tectonophysics* **43**, 7–22.

TOSHIAKI SHIMURA, Department of Geology, Faculty of Science, Niigata University, Ikarashi 2–8050, Niigata 950–2181, Japan.
e-mail: smr@gs.niigata-u.ac.jp

MASAAKI OWADA, Department of Earth Sciences, Yamaguchi University, Yoshida 1677–1, Yamaguchi 753–8512, Japan.

YASUHITO OSANAI, Division of Evolution of Earth Environments, Graduate School of Social and Cultural Studies, Kyushu University, Ropponmatsu 4–2–1, Fukuoka, 810–8560, Japan.

MASAYUKI KOMATSU, Division of Earth Sciences, Department of Biosphere-Geosphere Science, Faculty of Science, Ehime University, Bunkyo-cho 2–5, Matsuyama 790–8577, Japan.

HIROO KAGAMI, Graduate School of Science and Technology, Niigata University, Ikarashi 2–8050, Niigata 950–2181, Japan.

MS received 25 September 2003. Accepted for publication 9 September 2004.

Transactions of the Royal Society of Edinburgh: Earth Sciences, **95**, 181–198, 2004

Possible source rocks of Mesozoic granites in South Korea: implications for crustal evolution in NE Asia

Yong-Joo Jwa

ABSTRACT: The exposed Precambrian cratonic crust in South Korea is divided into two massifs – the Gyeonggi massif to the north and the Yeongnam massif to the south. Mesozoic granites intruded into both massifs and are mostly I-types. The Jurassic granites form extensive deep-seated batholiths, the Triassic granites are deep-seated stocks and the Cretaceous granites occur as volcanic-plutonic complexes.

The systematic variation of ε_{Nd} and SrI in the Korean Mesozoic granites could result from the mixing of two components in different proportions to produce the source of the granites. Although most Mesozoic I-type granites were apparently derived from more juvenile crust, the old evolved crustal components seem to have been incorporated in the magmas in various proportions. Mantle–crust mixing can account for the generation of the source of the Triassic and Cretaceous granites in the Gyeongsang basin. On the other hand, crust–crust mixing can feasibly produce the source of the Triassic and Jurassic granites in the Yeongnam massif, the Jurassic granites in the Gyeonggi massif, and the Cretaceous granites in the Yeongdong–Gwangju basin and the Okcheon belt. However, some Jurassic granites in the Yeongnam massif and Cretaceous granites in the Yeongdong–Gwangju basin can be also explained by the mantle–crust mixing. Combined geochemical and isotopic signatures indicate that a simple binary mixing model is inadequate to explain both the geochemical and isotopic data. The chemistry of the granites is considered likely to reflect the composition of the igneous protolith that derived from depleted mantle, which explains why most Mesozoic granites in South Korea are represented by I-types, regardless of their temporal and spatial position. Nd-Sr isotopic signatures of the Mesozoic granites and basement rocks indicate that the continental crust beneath the Korean peninsular is vertically structured by the successive underplating of mantle-derived materials. It is postulated that the crust is vertically stratified from the surface to the lowermost crust with late Archean to early Proterozoic, early to middle Proterozoic (ca. 1·9 Ga), middle Proterozoic (ca. 1·5 Ga), and late Proterozioc (younger than 1·5 Ga) components.

KEY WORDS: Igneous protolith, I-type granites, Nd-Sr isotopic signature, two-component mixing, vertically stratified crust.

The Korean Peninsula is located at the NE Eurasian continental margin and forms part of the amalgamated Sino-Korean and Yangtze cratons. Mesozoic granites in South Korea have been important in understanding temporal and spatial changes in crustal evolution and magma production at the continental margin (Jwa 1999; Cheong *et al.* 2002; Sagong *et al.* 2003). In the early Mesozoic, the China mainland had been formed by collision of the Sino-Korean craton with the Yangtze craton (Ames *et al.* 1993, 1996; Li *et al.* 1993; Chavagnac & Jahn 1996). This collision modified thermal and structural regimes, and eventually produced voluminous post-orogenic granites in South Korea during the middle Mesozoic (Jwa 1998). In the late Mesozoic, the active orthogonal subduction of the oceanic Kula plate beneath the continental margin resulted in the volcanic-plutonic complex of the SE part of South Korea (Jwa 1999; Sagong *et al.* 2003).

It has long been argued that the Mesozoic Korean granites were formed either by partial melting of the old crustal material during the Jurassic, or by fractional crystallisation of mafic magma that originated from younger igneous protoliths during the Cretaceous (Shimazaki & Lee 1981; Tsusue *et al.* 1986; Hong 1987). However, this was not supported by the geochemical nature of source rocks for the granites. Since the 1990s, Nd-Sr isotopic studies of the granites have required that crustal contamination was necessary to modify magma compositions (Kim *et al.* 1996; Cheong & Chang 1997; Cheong *et al.* 2002). Granite can be easily modified by the interaction of granitic magma with country rocks or with other magmas (DePaolo 1981; Collins 1996). On the contrary granite compositions can directly reflect the source composition, even mixed sources (Chappell *et al.* 1987, 2000). Nevertheless, whether the Korean Mesozoic granites directly reflect their source rocks has not been discussed.

Recently, Kwon & Sagong (2002) proposed that regional Pb-Nd-Sr isotopic differences of Mesozoic granites closely relate to their basement structure in terms of crustal thickness, regardless of their intrusion ages. They also suggested that crustal thickness may correlate with crustal formation age. However, the current contrast of the crustal thickness in South Korea is not so great ($\Delta d \leq 3$ km; Park *et al.* 2003), and it has recently been reported that Nd model ages of basement rocks do not show a positive correlation with crustal thickness (Lee *et al.* 2003). Thus, regional variation of the isotopic signatures of the granites is considered unlikely to be controlled by crustal thickness, but rather, may indicate temporal and spatial heterogeneity of the source rocks for the granites.

In this paper, the present author examines the geochemical and isotopic characteristics of the Mesozoic granites in South Korea, and then considers possible source rocks of the granites with implication for crustal evolution since the Proterozoic.

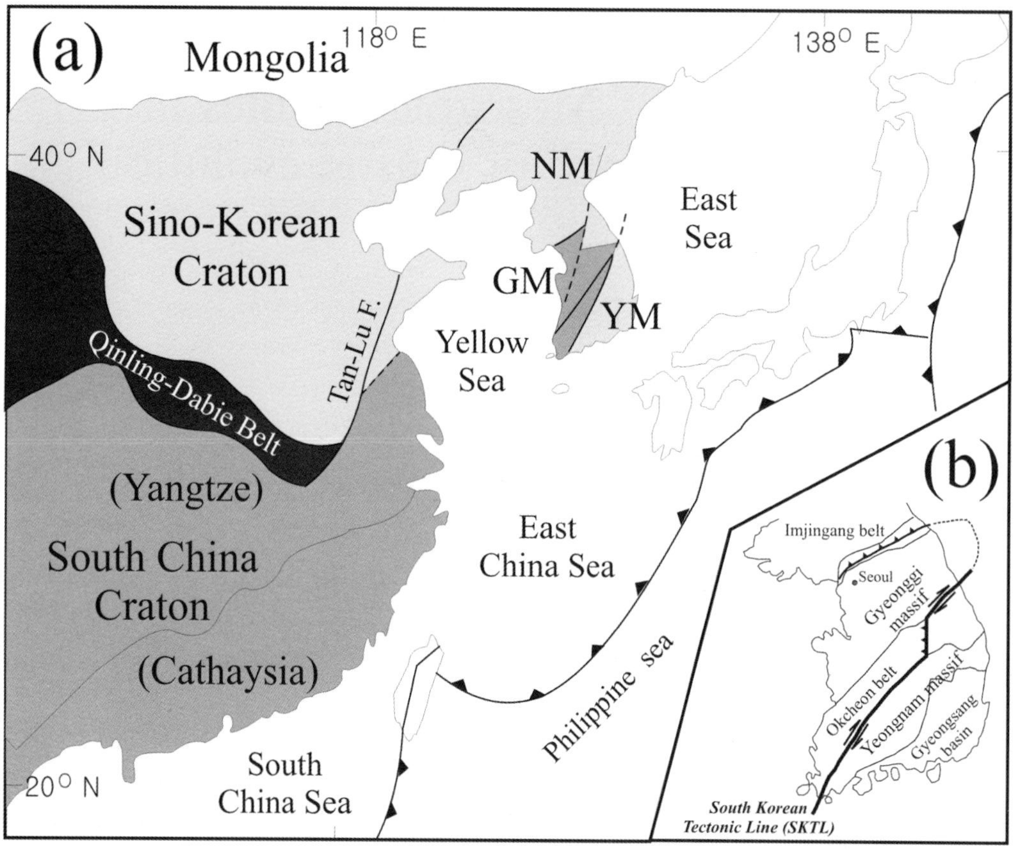

Figure 1 Tectonic province map of (a) NE Asia and (b) South Korea: (NM) Nangrim massif; (GM) Gyeonggi massif; (YM) Yeongnam massif. Modified from Chough *et al.* (2000), Ree *et al.* (2001) and Lee *et al.* (2003).

1. Tectonic setting in South Korea

Tectonically, South Korea can be divided into five provinces based on the distribution of sedimentary basins upon Precambrian basements (Figs 1 & 2). The pre-Cretaceous basement comprises two Precambrian massifs: the Gyeonggi massif in the north and the Yeongnam massif in the south. The two massifs exhibit Sinian tectonic trends (SW–NE), and are made up of Precambrian gneiss complexes with some Proterozoic to Jurassic metasedimentary cover (Lee 1987). The Okcheon belt between the two massifs consists of a NE non- to weakly metamorphosed Palaeozoic–Mesozoic sequence which grades into the Okcheon metamorphic belt to the SW. In the central region of the belt late Proterozoic amphibolites and metavolcanics also occur. The Cretaceous sedimentary basins in South Korea were formed by transtension driving the movement of sinistral strike-slip faults (Lee 1999). The sedimentary strata comprise non-marine sedimentary and andesitic to rhyolitic volcaniclastic rocks. The basins overlie the pre-Cretaceous sequences with a marked unconformity. The products of Cretaceous granitic magmatism are quite evident in two large Cretaceous basins: the Gyeongsang basin and the Yeongdong–Gwangju basin.

From a comprehensive review of the tectonic evolution of the Korean Peninsula, Chough *et al.* (2000) recently proposed a new tectonic boundary between the Gyeonggi massif and the Yeongnam massif on the basis of structural analysis of the early Palaeozoic basin in the NE part of South Korea. They named the new boundary the 'South Korean Tectonic Line' (SKTL). This SKTL may be an extension of the Triassic suture in China with its northern equivalent, the Imjingang belt (Fig. 1b), and it penetrates diagonally through the Okcheon belt (Fig. 2). Although the older basement overlain by the Okcheon belt is ambiguous, it is suggested from the direction of the SKTL that the NE part of the belt may belong to the Yeongnam massif and the SW part to the Gyeonggi massif. Thus, most of the early Palaeozoic metasediments, and late Proterozoic amphibolites and metavolcanics, in the Okcheon belt seem to have been underlain by older basement, as with the Gyeonggi massif (Ree *et al.* 2001).

There are two models to explain a possible eastward extension of the Triassic suture to South Korea: one is an indentation model by Yin & Nie (1993) and the other is a crustal detachment model by Li (1994). Yin & Nie's (1993) model explains the indented collision of the Yangtze craton with Korea without a significant vertical change of the crustal thickness. On the other hand, Li postulated a crustal detachment of E China in which he suggested that, during the Triassic collision between the Sino-Korean and Yangtze cratons, the upper crustal part of the Yangtze was detached from the lower crust and thrust over Korea. Although the tectonic evolution of Korea is still a matter of debate, the two models commonly assume that the Gyeonggi massif has a crustal affinity to the Yangtze craton. However, this matching of the Gyeonggi massif to the Yangtze should be carefully re-examined because of the different isotopic signatures recorded in both crusts (Cheong *et al.* 2000; Lee *et al.* 2003).

2. Three groups of Mesozoic granites in South Korea

The Mesozoic granites of South Korea have been divided into two groups based on intrusion age: the older granites of Jurassic age and the younger granites of the late Cretaceous to early Tertiary ages (Lee 1974; Kim 1975; Hong 1987). This two-fold discrimination is in accord with the conventional idea that the older granites are genetically associated with the

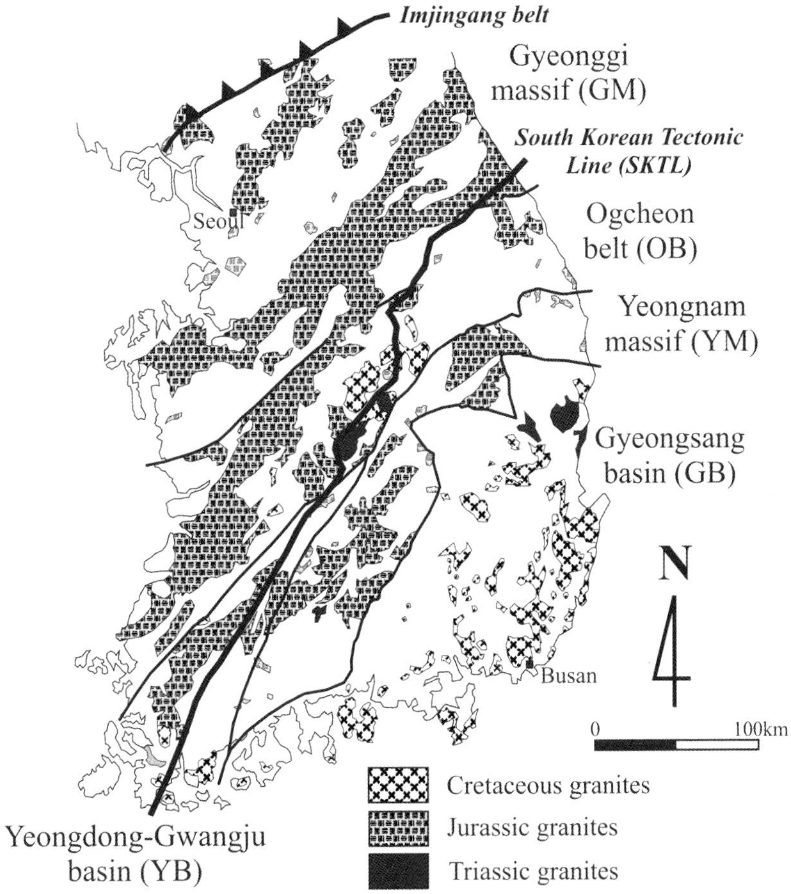

Figure 2 Distribution of Mesozoic granites in South Korea. The locations of the Imjingang belt and the NE South Korean Tectonic Line (SKTL) were modified after Chough *et al.* (2000). The SW SKTL is designed on the basis of the distribution of the Palaeozoic strata, which can be regarded as the lithologic boundary between the Gyeonggi massif and the Yeongnam massif.

Jurassic Daebo orogeny and the younger granites with the Cretaceous Bulguksa disturbance in South Korea (Lee 1987). The Jurassic granites form a large batholith and occur in the Gyeonggi massif (JG–GM), the Yeongnam massif (JG–YM) and the Okcheon belt (JG–OB) (Fig. 2). They are generally medium- to coarse-grained porphyritic biotite granites, but include various petrographic variations such as migmatitic granites, two mica granites and K-feldspar megacryst-bearing granites. The modal compositions of the granites show a broad spectrum from tonalite to syenogranite (Fig. 3b). The Cretaceous granites occur as shallow stocks and normally form volcano-plutonic complexes. They are distributed in the Okcheon belt (CG–OB), the Cretaceous Gyeongsang basin (CG–GB) and the Yeongdong–Gwangju basin (CG–YB) (Fig. 2). The granites are characterised by medium- to coarse-grained biotite granites and granite porphyry. They have modal compositions of granodiorite, monzogranite, syenogranite and alkali-feldspar granite (Fig. 3c).

New radiometric ages reportedly indicate that there are some Triassic granites in the Jurassic granites group (Kim & Turek 1996; Cheong & Chang 1997; Sagong 2000; Cheong *et al.* 2002). The Triassic granites crop out in a restricted area of the Okcheon belt, the Yeongnam massif (TG–YM) and the Gyeongsang basin (TG–GB), and occur as deep-seated stocks. They are mostly medium- to coarse-grained equigranular hornblende biotite granites, with subordinate amounts of K-feldspar megacryst-bearing granites. They have a general composition of granodiorite (Fig. 3a). The Triassic granites in the Okcheon belt are here regarded as the TG–YM according to the tectonic boundary of the SKTL.

From a comprehensive study of the radiometric ages of the Mesozoic granites in South Korea, Sagong *et al.* (2003) argued that the crystallisation ages of the granites clearly show three episodes of Mesozoic magmatism in South Korea: Triassic (248–210 Ma), Jurassic (197–158 Ma) and late Cretaceous–early Tertiary (110–50 Ma), with a late Jurassic–early Cretaceous hiatus. They suggested that the age structure of the granites was closely related to temporal changes in relative plate motion of the oceanic Izanagi plate to the eastern Eurasian continent during the Mesozoic.

3. Major and trace element geochemistry of Mesozoic granites

Some geochemical elements can be compared for the three groups of Mesozoic granites (Fig. 4). Most of the data were obtained from Jwa (1998), but a new data set for the TG–GB was obtained from Cheong *et al.* (2002). Representative geochemical data of the granites are presented in Table 1 according to the temporal and spatial distribution of the granite bodies. The chemical compositions shown in Table 1 are inferred as parental compositions of the granites which form well-defined plutonic bodies. The inferred parental compositions do not indicate the most mafic facies in the bodies, but are starting compositions which produced progressive derivatives with differentiation. It is recognised that each temporal group of the granites is geochemically distinct from the other groups.

The alumina saturation index (ASI) or A/CNK, calculated as the molar ratio of $Al_2O_3/(CaO+Na_2O+K_2O)$, is useful

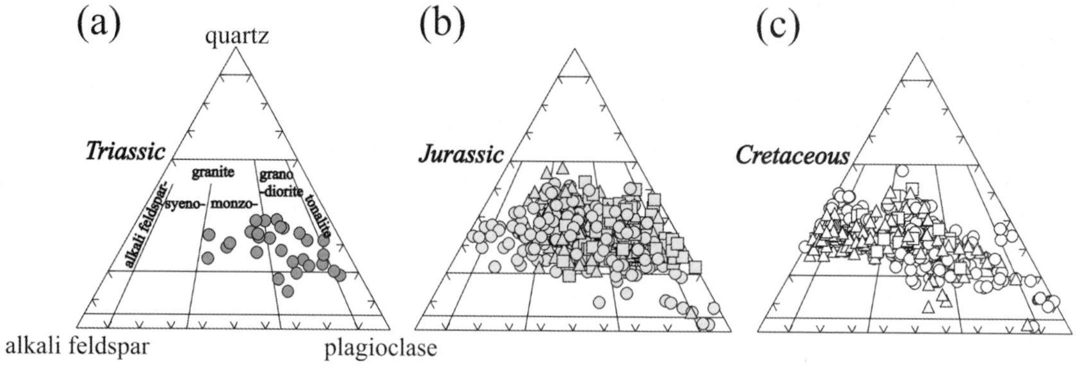

Figure 3 Modal compositions (quartz-alkali feldspar-plagioclase) of the Mesozoic granites in South Korea: (a) Triassic granites are generally represented by granodiorite and monzogranite. Modal data are exclusively from the granites in the Yeongnam massif (TG–YM); (b) Jurassic granites show a broad spectrum of tonalite to syenogranite: (triangles) granites in the Gyeonggi massif (JG–GM); (squares) granites in the Yeongnam massif (JG–YM); and (circles) granites in the Okcheon belt (JG–OB); (c) Cretaceous granites are mostly granodiorite, monzogranites, syenogranites and alkali feldspar granite: (triangles) granites in the Okcheon belt (CG–OB); (squares) granites in the Yeongdong–Gwangju basin (CG–YB); and (circles) granites in the Gyeongsang basin (CG–GB).

petrogenetically because the degree of Al-saturation often directly reflects the source rocks of the granites (Chappell 1999). The ASI is plotted against SiO_2 content in Fig. 4a. Triassic granites generally have an ASI lower than 1·1 and are metaluminous to weakly peraluminous. The ASI of the TG–YM is lower than that of the TG–GB, suggesting that the TG–YM magma is rich in a metaluminous mineral such as hornblende. The TG–YM actually contains up to 20 vol.% of hornblende, and sample J2204 in Table 1 includes 18 vol.% of hornblende (Sagong & Jwa 1997). The ASI of the Jurassic granites has a rough tendency to increase with SiO_2. It is apparent that, although the Jurassic granites show both metaluminous and peraluminous natures, the peraluminous granites are much more abundant. In particular, the peraluminosity becomes to be higher over 63% SiO_2, and the granites of high peraluminosity (ASI >1·1) occasionally contain small amounts of muscovite and/or garnet. Garnet-bearing granites rarely occur in the northern part of the Gyeonggi massif (JG–GM) and they mostly have an ASI of 1·1–1·2 (sample S24 in Table 1). It is noticeable that very high peraluminosity (ASI >1·2) is observed at high silica content (>70%) and the strongly peraluminous granites are characterised by fractionated two-mica granites in the JG–GM and the JG–OB (Jwa 1997). Thus, high peraluminosity observed in the Jurassic granites seems to have been a result of feldspar fractionation and the existence of a peraluminous mineral such as muscovite. The ASI of the Cretaceous granites varies with SiO_2 in a similar way to the Jurassic ones, but the granites of intermediate composition (around 65% SiO_2) do not show the same highly peraluminous character. High peraluminosity in the Cretaceous granites is also seen at high silica range of above 70%. Unlike the Jurassic granites, the Cretaceous granites do not contain any strongly peraluminous minerals such as muscovite and garnet. Thus, the high peraluminosity is unlikely to be caused by the presence of peraluminous minerals. Considering that the granites of high peraluminosity are strongly fractionated granites and that they are often related to intensive hydrothermal ore deposition, the dramatic increase in peraluminosity in the Cretaceous granites must have been caused either by the removal of feldspar from the granitic melt (Zen 1986; Chappell 1999), or by selective partitioning of mol CNK into fluids (Clarke & MacDonald 2003).

The ASI has been widely used to discriminate I- and S-type granites (Chappell & White 1974, 1992). Hine *et al.* (1978) suggested a value of 1·1 for the ASI to distinguish between I- and S-types in the Kosciusko Batholith, and Chappell (1999) also proposed an ASI=1·1 as a separating value between fractionated I- and S-type granites in the Lachlan Fold belt. Accepting this value of the ASI, all of the Triassic granites in South Korea are I-types (Fig. 4). The Jurassic granites are mainly I-types below 63% SiO_2, but include some S-types over 63%. The affinity to S-types is generally observed in highly differentiated facies like two-mica granite, which occupies a restricted area within a large intermediate to felsic I-type granitic complex (Jwa 1997). Except for rare examples of S-type garnet-bearing granites there are few, if any, Jurassic granites which are entirely S-types. Most Jurassic granites should be typologically assigned to felsic I-types. The Cretaceous granites are mostly I-types below 70% SiO_2, but some S-types appear in the higher silica ranges. However, this typological division using the ASI is not always consistent with the mineral assemblages in the granites. Since the Cretaceous granites do not normally contain any peraluminous mineral except biotite, their high peraluminosity at high silica values comes from the felsic derivatives through the removal of feldspar. Furthermore, melt-fluid separation that is ubiquitously found at the late stage of the Cretaceous magmatic evolution could lead the granite composition to higher peraluminosity. Therefore, it is considered unlikely that the high peraluminous Cretaceous granites are characterised by S-types. Instead, they are possibly strongly fractionated I-type granites.

CaO contents systematically decrease with SiO_2 contents (Fig. 4b), suggesting that Ca-rich phases such as hornblende and Ca-plagioclase progressively formed and were then removed from the granitic melt. Variations of Ca with silica content are almost the same in the three temporal groups of the granites.

P_2O_5 contents generally decrease with SiO_2 contents (Fig. 4c). Decreasing P with silica is a feature of I-type granites. P dissolves more in peraluminous melt because apatite is soluble in that melt (Montel *et al.* 1988; Pichavant *et al.* 1992; Wolf & London 1994). Thus, P is expected to increase with increasing peraluminosity. Although the ASI of the Jurassic and Cretaceous granites generally becomes greater at high silica contents, the P_2O_5 contents decrease over the same silica range. The reason for this negative correlation between P and ASI is not clear, but a possible explanation comes from a counterbalancing effect of silica content. Harrison & Watson (1984) predicted that apatite solubility decreases with fractionation and increasing SiO_2 content of the melt. For the felsic Jurassic and Cretaceous granites, the fractionation led to increased peraluminosity at high silica contents and low apatite.

The Rb contents of the Triassic granites are very low and do not vary with SiO_2 contents (Fig. 4d), indicating that the granites originated from Rb-depleted source and that the fractionation of feldspar was not an effective process during the magmatic evolution. It is clear that the TG–YM has higher Rb content than the TG–GB. The Jurassic granites show a progressive increase in Rb contents with SiO_2. In particular, a sharp increase in Rb over 75% SiO_2 is evident in the JG–GM and the JG–OB, and the feature is thought to be the result of feldspar fractionation in felsic melts. Variation of Rb with SiO_2 in the Cretaceous granites is similar to that in the Jurassic granites, but the relative contents of Rb are lower in the Cretaceous granites. The rapid increase in Rb as a result of feldspar fractionation can be also seen in the high silica Cretaceous granites, especially in the CG–OB and the CG–GB.

The Sr contents of the Triassic granites show very complicated variation trends (Fig. 4e). There are at least three different trends, two from the TG–GB and one from the TG–YM (also see samples YD15, YD06 and J2204 in Table 1). This difference may be indicative of different source-rock compositions. On the whole, the Sr contents progressively decrease with SiO_2 contents. Variation of Sr with silica in the Jurassic granites is also complicated and similar to that in the Triassic granites over the same compositional range. On average, the Sr content is highest in the JG–YM (see sample AD1412 in Table 1), intermediate in the JG–OB and lowest in the JG–GM. Thus, it can be said that the source rocks for the Triassic and Jurassic granites were heterogeneous in Sr contents. On the other hand, the Sr contents of the Cretaceous granites are lower than those of the Triassic and Jurassic granites. Therefore, it is apparent that the source of the Cretaceous granites was distinct from the Triassic and Jurassic granites.

The Zr contents of the Triassic granites are low and nearly constant regardless of their silica contents (Fig. 4f). This trend suggests a slight undersaturation of Zr in the Triassic melts. Over 60% SiO_2, the Jurassic granites show a systematic decline in Zr with silica, suggesting that the granitic melts were nearly saturated in Zr. The Zr contents in the Cretaceous granites decrease sharply with SiO_2, indicating that the granitic melts were obviously saturated in Zr. From the variation in the Zr contents, it can be said that the Triassic granites are possibly high-temperature granites, but the Jurassic and Cretaceous granites are characterised by low-temperature granites (Chappell *et al.* 1998, 2000).

4. Isotopic signatures of Mesozoic granites

Epsilon Nd values [ε_{Nd}(T)] and initial Sr isotopic ratios (SrI) were calculated at a reference age of 230 Ma for the Triassic granites, 170 Ma for the Jurassic granites and 70 Ma for the Cretaceous granites (Fig. 5a, b, c). The three reference ages represent the most intensive temporal peaks of granitic intrusion during the Mesozoic (Jwa 1998; Sagong *et al.* 2003). The Jurassic granites in the Okcheon belt (JG–OB) were allocated either to the JG–GM or to the JG–YM according to their relative distribution from the tectonic boundary of the SKTL (Fig. 2).

Sr and Nd isotopic ratios demonstrate two contrasting groups within the Triassic granites (Fig. 5a). The TG–GB is isotopically depleted (3·5 to −4·2 for ε_{Nd} and 0·7043–0·7069 for SrI), whereas the TG–YM is relatively enriched (−12·3 to −17·0 for ε_{Nd} and 0·7095–0·7105 for SrI). These two Triassic granites form a well-defined array in relative isotopic composition. This kind of array is consistent with the generation of the granites by the mixing of a depleted mantle component with an evolved crustal component (McCulloch & Chappell 1982). Applying the same model, it becomes apparent that the TG–GB granites are dominated by the depleted mantle component and the TG–YM granites by the crustal component in their source rocks. In particular, the isotopic compositions of the TG–GB most likely represent materials derived from the melting of subducted slab or island arc sources, partially contaminated with a crustal component.

The Jurassic granites are more evolved in radiogenic isotopic ratios compared with the Triassic granites (Fig. 5b), which suggests that they were derived from more-evolved crustal materials. More dispersed isotopic ratios as shown in the JG–YM would suggest more heterogeneous crustal components. Except for six data points showing higher SrI, both the JG–GM (−8·5 to −20·8 for ε_{Nd} and 0·7091–0·7188 for SrI) and the JG–YM (−8·5 to −20·4 for ε_{Nd} and 0·7080–0·7216 for SrI) generally share common Nd-Sr isotopic properties. Six anomalous data points selected from an A-type granite pluton in the SW Yeongnam massif require the involvement of more enriched Sr isotopic components in the source rocks. In the JG–GM, the enriched data are partly ascribed to the garnet-bearing granites (see sample S-24 in Table 1).

Three Cretaceous granites have quite distinctive isotopic signatures with each other: the most depleted in the CG–GB (2·5 to −4·4 for ε_{Nd} and 0·7024–0·7069 for SrI), the most enriched in the CG–OB (−13·4 to −17·6 for ε_{Nd} and 0·7122–0·7226 for SrI) and the intermediate in the CG–YB (−6·3 to −15·2 for ε_{Nd} and 0·7083–0·7129 for SrI). This kind of variation implies either different source materials or different degrees of contamination. For the genesis of the CG–GB, the contribution of crustal materials would have been minimal, but higher mantle input may appear to have operated. On the other hand, the large contribution of crust seems to have been required for the isotopic composition of the CG–OB. It is striking that the ε_{Nd} variation in the CG–OB is relatively restricted compared with the SrI. This variation may be ascribed to varying degrees of Rb fractionation in the source rocks.

5. Isotopic signatures of the exposed basement rocks

The exposed basement rocks, which are overlain by Palaeozoic–Mesozoic sedimentary rocks and intruded by the Mesozoic granites, can be found in both the Gyeonggi massif and the Yeongnam massif (Fig. 2). They consist of gneissic and metasedimentary rocks; the gneisses are classified petrographically as granite gneiss, porphyroblastic gneiss, biotite gneiss, banded gneiss and amphibolite, whereas the metasedimentary rocks are schist, quartzite and metapelite. The ASI values of the Precambrian rocks are mostly greater than 1·0, indicating a peraluminous character (Fig. 6). In particular, some paragneissic rocks show very strongly peraluminous characteristics (ASI $\geq$2·0).

Recent age determinations for the Precambrian basement rocks reportedly argue that the rocks have a broad spectrum of metamorphic and/or formation ages, mostly between 0·8 and 3·0 Ga (Cho 1999; Cho *et al.* 1999; Kim *et al.* 1999; Lee *et al.* 2001; Song *et al.* 2001; Park *et al.* 2000, 2001; Cho *et al.* 2002a, b). Figure 7a is a compilation of reported radiometric ages of the Precambrian basement rocks in terms of U-Pb mineral age, Pb-Pb whole rock age, Nd-Sm whole rock and mineral ages, and chemical isochron (CHIME) mineral age. The figure shows at least four peaks which indicate metamorphic events

Table 1 Representative analytical data of Mesozoic granites in South Korea

	Triassic				Jurassic						Cretaceous					
	Yeongnam massif		Gyeongsang basin		Gyeonggi massif		Okcheon belt		Yeongnam massif		Okcheon belt		Yeongdong–Gwangju basin		Gyeongsang basin	
Body:	Baegrok	Jeomchon	Youngdeog	Younghae	Hongcheon	Seoul	Boeun	Wolaksan	Namwon	Andong	Wolaksan	Sokrisan	Haenam	Gangjin	Gosung	Onjeongri
Sample no:	J2204	J2401CA	YD15	YD06	0406	S24	J1718	1501	N-12	AD1412	3008	BE48	MG-16	M4	PG3-1	OJ06
ASI	0·86	0·83	0·95	0·90	0·99	1·12	0·96	0·94	0·97	0·93	1·08	1·08	1·10	1·07	0·81	1·02
SiO_2 (wt.%)	57·16	66·26	62·91	57·87	66·41	73·62	66·50	65·32	65·49	62·56	72·28	74·93	72·74	69·57	60·49	66·53
TiO_2	0·72	0·40	0·86	0·79	0·57	0·08	0·55	0·80	0·61	0·80	0·32	0·06	0·24	0·32	0·86	0·50
Al_2O_3	17·88	14·49	16·99	16·34	16·45	14·26	15·60	16·27	15·85	18·19	14·73	13·23	14·45	16·04	15·81	15·44
Fe_2O_3 (T)	6·53	3·55	4·24	7·35	4·12	1·95	3·46	4·23	4·15	4·04	2·51	1·04	2·23	3·05	7·24	4·36
MnO	0·10	0·06	0·06	0·12	0·06	0·09	0·05	0·07	0·06	0·05	0·06	0·03	0·10	0·07	0·12	0·10
MgO	3·94	1·97	1·35	3·38	1·61	0·24	1·00	1·77	2·11	1·45	0·67	0·04	0·36	0·60	3·21	1·64
CaO	6·38	3·33	3·69	5·68	3·73	1·07	2·83	4·28	4·33	5·13	1·97	0·49	1·14	1·93	5·60	3·34
Na_2O	4·12	4·53	4·84	3·45	4·11	3·85	3·97	3·86	3·80	4·98	3·77	3·37	4·01	4·45	3·73	3·42
K_2O	2·13	3·64	2·97	2·06	2·79	4·12	4·15	2·86	2·09	1·82	3·59	5·32	4·15	3·77	2·86	3·22
P_2O_5	0·26	0·11	0·31	0·26	0·14	0·01	0·17	0·20	0·14	0·25	0·10	0·02	0·08	0·12	0·24	0·14
Ba (ppm)	981	934	1384	482	537	303	822	863	467	788	680	218	941	870	475	567
Rb	58	90	54	52	181	215	129	88	92	34	138	187	125	140	104	88
Sr	985	592	1173	668	404	80	504	577	527	1012	188	34	198	300	439	387
Th	4·7	7·0	4·5	10·2	18·5	29·0	8·8	9·2	11·0	7·6	18·5	17·0			13·2	13·9
Y	16·5	11·1	6·0	22·0	21·5	70·0	8·8	13·1	16·0	13·0	21·9	11·0	21·0	30·0	26·2	17·0
Zr	88	109			156	111	138	163	163	202	153	72			196	
La	29·5	27·4	29·8	23·0	36·6	35·0	21·4	45·0	21·1	41·9	50·0	23·3	48·9	59·0	28·9	28·1
Ce	56·4	50·1	59·5	47·5	76·6	60·0	40·5	72·0	44·0	82·5	78·0	54·8	90·1	102·0	62·2	51·3
Pr			7·1	6·2					9·3			5·3	9·7		7·4	5·9
Nd	29·5	20·4	24·7	20·6	36·8	29·0	16·2	35·0	19·6	38·7	30·0	20·2	34·7	41·0	29·3	21·0

Table 1 *Continued*

	Triassic				Jurassic						Cretaceous					
	Yeongnam massif		Gyeongsang basin		Gyeonggi massif		Okcheon belt		Yeongnam massif		Okcheon belt		Yeongdong–Gwangju basin		Gyeongsang basin	
Body:	Baegrok	Jeomchon	Youngdeog	Younghae	Hongcheon	Seoul	Boeun	Wolaksan	Namwon	Andong	Wolaksan	Sokrisan	Haenam	Gangjin	Gosung	Onjeongri
Sample no:	J2204	J2401CA	YD15	YD06	0406	S24	J1718	1501	N-12	AD1412	3008	BE48	MG-16	M4	PG3-1	OJ06
Sm	4·0	3·6	4·1	4·4	6·7	6·6	3·8	4·4	3·3	5·7	5·5	3·7	5·6	7·1	5·8	3·9
Eu	1·36	0·95	1·56	1·46	1·32		0·62	1·14	0·91	1·40	0·89	0·32	1·21	1·20	1·28	0·98
Gd	4·7	2·8	3·9	4·9	5·9		3·3		2·5			2·8	4·3		5·4	3·9
Tb			0·4	0·8				0·5	0·3		0·7	0·4			0·8	0·5
Dy	2·9	1·9	1·8	4·3	4·4		1·0			2·0		2·0	3·7	3·6	4·8	3·2
Ho			0·3	0·8					0·3			0·4	0·7		0·9	0·6
Er			0·6	2·4					0·8			1·3	2·1		2·7	1·9
Tm			0·08	0·36					0·09			0·20			0·39	0·29
Yb	1·35	0·92	0·42	2·30	1·54	5·40	0·39	0·90	0·50	0·90	0·35	1·35	2·18	1·90	2·70	1·94
Lu			0·06	0·36				0·14			4·30	0·19	0·36		0·39	0·30
$(La/Yb)_{CN}$	15·7	21·4	50·8	7·2	17·0	4·6	39·1	35·9	30·2	33·4	102·5	12·4	16·1	22·3	7·7	10·4
$^{87}Rb/^{86}Sr$	0·171	0·441	0·133	0·225	1·297	6·790	0·695	0·440	0·511	0·092	2·115	16·000	1·650	1·351	0·691	0·633
$^{87}Sr/^{86}Sr$	0·7099	0·7110	0·7053	0·7057	0·7157	0·7327	0·7141	0·7138	0·7145	0·7100	0·7170	0·7322	0·7100	0·7143	0·7056	0·7064
$^{147}Sm/^{144}Nd$	0·0826	0·1067	0·1004	0·1289	0·1097	0·1365	0·0998	0·0958	0·1010	0·0890	0·1050	0·1094	0·1079	0·1047	0·1246	0·1119
$^{143}Nd/^{144}Nd$	0·5118	0·5117	0·5126	0·5127	0·5117	0·5117	0·5117	0·5118	0·5117	0·5120	0·5118	0·5119	0·5123	0·5119	0·5126	0·5126
SrI (T)*	0·7093	0·7096	0·7049	0·7050	0·7125	0·7163	0·7124	0·7128	0·7133	0·7098	0·7149	0·7163	0·7084	0·7129	0·7049	0·7057
ε_{Nd} (T)*	−12·5	−14·5	3·0	4·8	−16·8	−17·1	−15·3	−13·7	−15·4	−9·8	−15·4	−14·3	−6·6	−13·8	0·4	−0·9
T_{DM} (Ga)	1·57	2·03	0·73	0·76	2·14	2·86	1·94	1·77	1·96	1·43	1·90	1·87	1·28	1·76	0·90	0·89
Reference**	1,17	2,17	3	3	4,17	5,6	7,17	8	9	10	8	11	12	13	14,15,17	16

*T is 230 Ma for Triassic granites, 170 Ma for Jurassic granites and 70 Ma for Cretaceous granites.
**Reference: (1) Sagong & Jwa (1997); (2) Choe & Jwa (1998); (3) Cheong *et al.* (2002); (4) Jwa (1988); (5) Kwon *et al.* (1994); (6) Kwon *et al.* (1999); (7) Jwa (1996); (8) Lee *et al.* (1995); (9) Na (1994); (10) Lee *et al.* (1999); (11) Cheong & Chang (1997); (12) Shin *et al.* (1997); (13) H.-N. Kim *et al.* (1997); (14) Jwa & Park (1996); (15) Jwa (1998); (16) Cheong *et al.* (1998); and (17) Y.-J. Jwa (unpublished results).

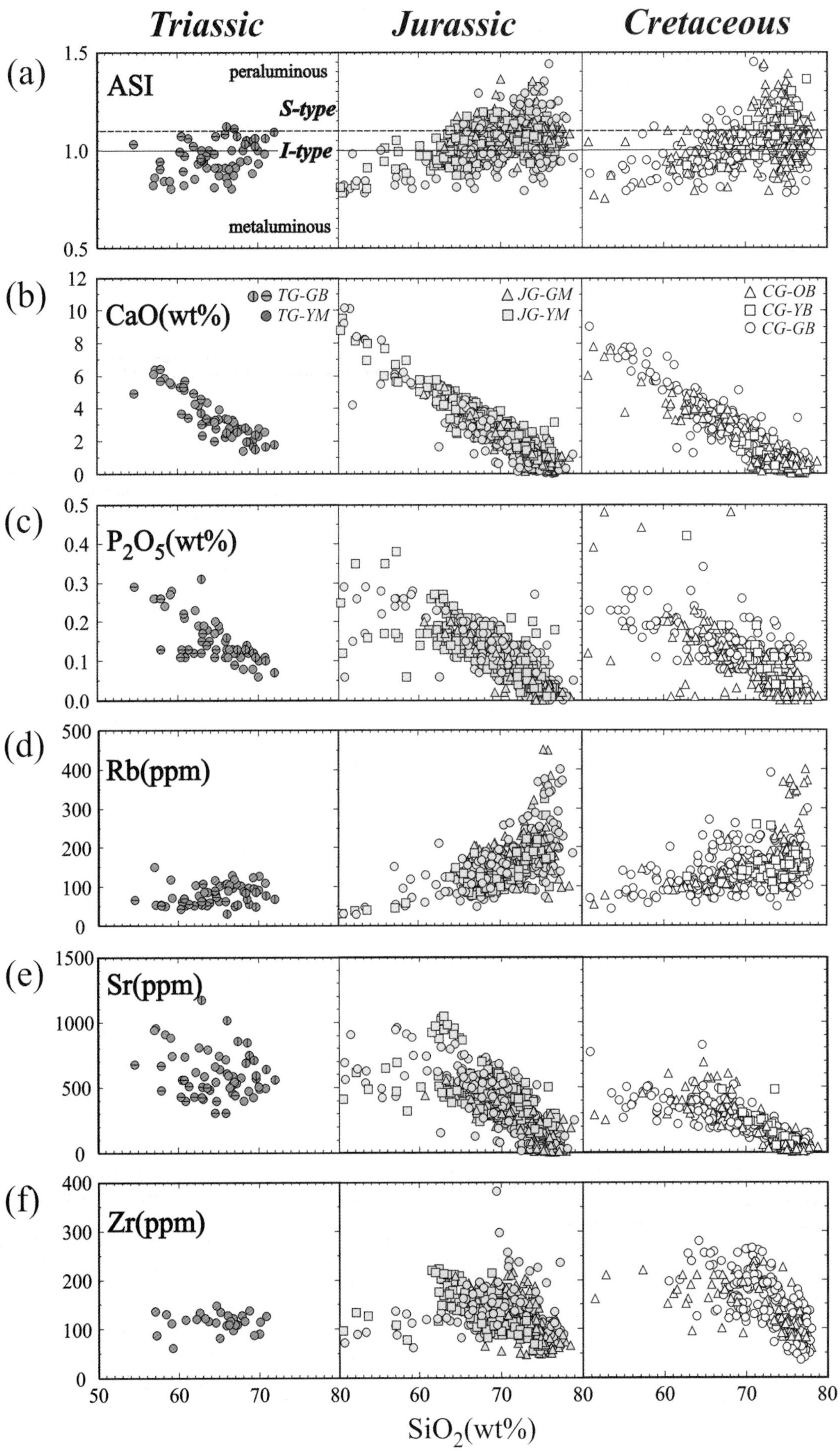

Figure 4 Harker diagrams for (a) alumina saturation index (ASI) (mol ratio), (b) CaO (wt. %), (c) P_2O_5 (wt.%), (d) Rb (ppm), (e) Sr (ppm) and (f) Zr (ppm). The geochemical variation in each temporal granite is distinct from the other granites. The circle with a bar indicates the Triassic granites in the Gyeongsang basin (TG–GB), and the other symbols are the same as those in Figure 3.

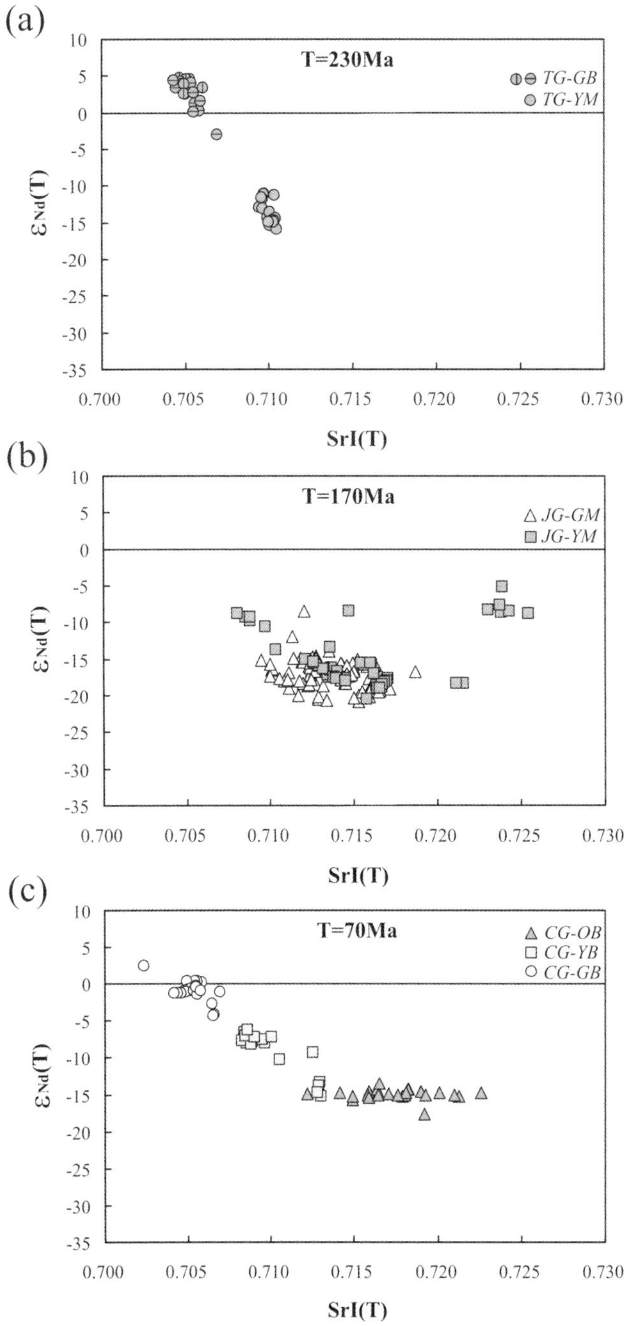

Figure 5 Nd-Sr isotopic relationship in the Mesozoic granites. The ε_{Nd}(T) and SrI(T) of the Triassic, Jurassic and Cretaceous granites were calculated at three reference ages: (a) 230 Ma, (b) 170 Ma and (c) 70 Ma. Abbreviations: (TG) Triassic granites; (JG) Jurassic granites; (CG) Cretaceous granites; (GM) Gyeonggi massif; (YM) Yeongnam massif; (OB) Okcheon belt; (YB) Yeongdong–Gwangju basin; and (GB) Gyeongsang basin. Data sources: (TG–GB) Cheong *et al.* (1998); (TG–YM) Cheong & Chang (1997) and Y.-J. Jwa (unpublished results); (JG–GM) Na (1994), Kim *et al.* (1996), Cheong & Chang (1997), Kwon *et al.* (1999), Y.-J. Jwa (unpublished results); (JG–YM) Na (1994), Kim *et al.* (1996), Cheong & Chang (1997) and J. I. Lee *et al.* (1999); (CG–OB) Na (1994), Lee *et al.* (1995), Kim *et al.* (1996) and Cheong & Chang (1997); (CG–YB) H.-N. Kim *et al.* (1997) and Shin *et al.* (1997); and (CG–GB) Kim *et al.* (1996), Shin *et al.* (1997), Cheong *et al.* (1998) and Y.-J. Jwa (unpublished results).

during the intervals of 0·8–1·0, 1·4–1·6, 1·7–2·0 and 2·1–2·5 Ga. This age distribution also indicates that both cratonic crusts – the Gyeonggi massif and the Yeongnam massif – had experienced similar orogenic events since the early Proterozoic. It is noticeable that the most intense metamorphic event that was recorded in the basement rocks occurred in the middle Proterozoic (ca. 1·9 Ga).

Figure 7b shows the relationship between the ε_{Nd} values calculated at a reference age of 170 Ma and the depleted mantle model age (T_{DM}, single-stage) for the basement rocks from two massifs. Palaeozoic–late Proterozoic metasediments and meta-basaltic rocks (amphibolites) in the Okcheon belt are also shown. To avoid the effect of rare-earth element (REE) fractionation on the T_{DM} age calculation, a limited range of Sm/Nd fractionation can be taken into account. The range of $-0{\cdot}2$ to $-0{\cdot}6$ of the $f_{(Sm/Nd)}$ value, as suggested by Jahn *et al.* (2000) and Wu *et al.* (2003), was taken to calculate the model ages in Figures 7b and 9. Considering that intense crustal fractionation had possibly occurred during the middle Proterozoic time (ca. 1·9 Ga), a two-stage depleted mantle model age can be obtained (DePaolo 1988). However, the contrast between the single- and two-stage model ages is relatively small: less than 0·4 Ga for ages older than 3·0 Ga, less than 0·2 Ga for ages of 2·5–3·0 Ga, and less than 0·1 Ga for ages of 2·0–2·5 Ga. Since there is no critical difference with age, the single-stage model age is exclusively used in the present paper.

The basement rocks in the Gyeonggi massif have old model ages ($\geq$1·8 Ga) and strongly negative ε_{Nd} values (-12 to -31). Most Yeongnam massif rocks have older model ages ($\geq$2·2 Ga) and strongly negative ε_{Nd} values (-17 to -27), but some have younger model ages (around ca. 1·5 Ga) and moderate negative ε_{Nd} values (-6 to -14). It is striking that the T_{DM} ages for the basement rocks are slightly older than the metamorphic episodes. Thus, the radiometric ages for the basement rocks seem unlikely to be of the formation ages of the exposed cratonic crusts. The metasediments of the Okcheon belt have model ages of 1·8–2·4 Ga, and ε_{Nd} values range from -12 to -15, whereas the amphibolites in the belt have model ages of 1·0–1·5 Ga and ε_{Nd} of $+2$ to -4. The cratonic crust beneath the Okcheon belt is still uncertain, but Sagong & Kwon (1998) argued that the basement of the Okcheon belt is an extension of the Gyeonggi massif from the fact that the Pb isotope data of the basement rocks beneath the Okcheon belt are consistent with a general trend of the Gyeonggi massif. The tectonic division according to the SKTL seems to support this crustal relationship between the Okcheon belt and the Gyeonggi massif (Fig. 2).

It is noted that some rocks of the Gyeonggi massif and the Okcheon belt cluster around 1·9 Ga of T_{DM} age. In particular, these rocks mostly have less-evolved ε_{Nd} values than most basement rocks from the Gyeonggi and Yeongnam massifs. This feature would imply that both the protolith of the metasediments in the Okcheon belt and some basement rocks in the Gyeonggi massif were derived from materials which were extracted from the mantle during the middle Proterozoic, i.e. the period of the most intense metamorphic event recorded in the basement rocks.

Although the T_{DM} ages for the exposed basement rocks largely overlap between 2·5 and 3·0 Ga, the ε_{Nd} values can be divided into four distinct isotopic groups (Fig. 7b). The Yeongnam massif is divided into a more-evolved group, YM(I), and a less-evolved group, YM(II). The YM(II) group is represented by granitic gneiss and occurs in the central part of the Yeongnam massif. The Gyeonggi massif is divided into a more-evolved group, GM(I), and a less-evolved group, GM(II). The GM(II) group includes the metasediments of the Okcheon belt. The amphibolites in the Okcheon belt are characterised by distinctive isotopic signatures and are designated as OB-v in Figure 7b. The GM(II) and the YM(II) groups occur in restricted areas within the massifs. The widespread basement rocks in the Gyeonggi massif [GM(I) group] and the Yeongnam massif [YM(I) group] have similar T_{DM} values, but slightly different ε_{Nd} values; the YM(I) group has higher ε_{Nd} values than the GM(I) group.

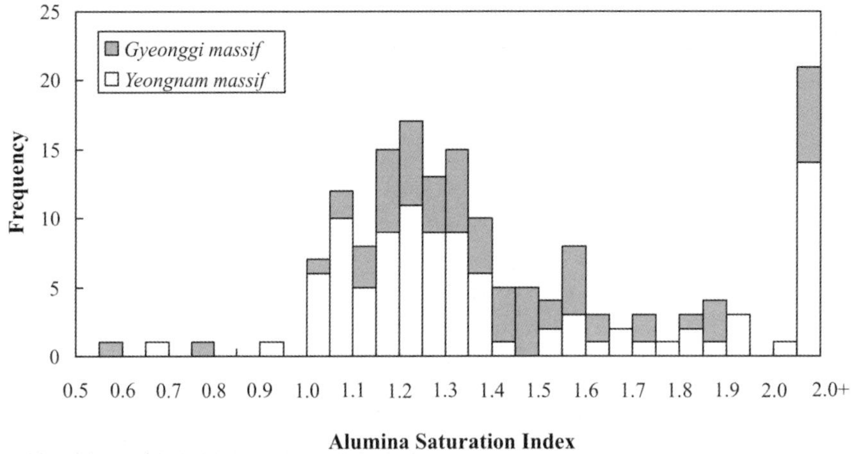

Figure 6 Alumina saturation index (ASI) for the Precambrian basement rocks in the Gyeonggi massif and the Yeongnam massif. The ASI values are mostly greater than 1·0, i.e. peraluminous. In particular some paragneissic rocks show very strongly peraluminous characteristics (ASI $\geq$2·0). Data sources: (Gyeonggi massif) Jwa (1988), Lee *et al.* (1990), Cho & Nam (1993), Lee & Choi (1994) and Cheong & Chang (1997); and (Yeongnam massif) Yun & Lee (1986), Chang *et al.* (1988), Song (1989), Cho (1994), Cho (2000) and Lee *et al.* (2001).

6. Discussion

6.1. Two-component mixing model

The systematic variation of ε_{Nd} and SrI in the Korean Mesozoic granites could result from the mixing of two components in different proportions to produce the source of the granites (Fig. 5). The possible depleted mantle component (DM) and crustal components [GM(I), GM(II), YM(I) and YM(II)] are shown in Figure 8, together with mixing lines. The DM component was taken to have 136 ppm of Sr, 8 ppm of Nd, +10 of $\varepsilon_{Nd}(0)$ and 0·7026 of SrI(0) (Saunders & Tarney 1984; Liew & McCulloch 1985; Taylor & McLennan 1985), and the isotopic ratios were recalculated for three reference ages: 230 Ma, 170 Ma, and 70 Ma. The crustal components for the four distinctive basement rocks were averaged from the raw data (see Fig. 7b), and their isotopic ratios were also recalculated for three reference ages.

For the Triassic granites, a possible crustal end-member is represented by the widespread Precambrian basement rocks in the Yeongnam massif. Although the TG–GB occurs in the Gyeongsang basin, the basement of the basin is often considered to be like the crust of the Yeongnam massif, given the tectonic relationship between them since the Triassic. Thus, two mixing lines were deduced from DM–YM(I) and DM–YM(II) pairs for the Triassic granites (Fig. 8a). The curvature of the mixing line is controlled by the relative Sr/Nd ratio, $K=(Sr/Nd)_{DM}/(Sr/Nd)_{CC}$, between the depleted mantle and the crustal material (CC) (DePaolo & Wasserburg 1979). The K-value for the DM–YM(I) is 3·3, and that for the DM–YM(II) is 0·5. In Figure 8a, it is clear that the TG–GB needs a small proportion of the crustal component to fit its isotopic composition. If the YM(I) was the crustal component that mixed with the mantle component, the proportion is constrained to less than 10%. However, the isotopic composition of the TG–GB deviates from the mixing line of the DM–YM(I), but it varies along the line of the DM–YM(II). Thus, it is more likely that the isotopic composition of the TG–GB was a result of the mixing of the DM with the YM(II). The mixing line for various proportions of the DM and the YM(II) shows that the amount of the YM(II) in the source rocks of the TG–GB must be 10–30% to explain the observed isotopic character of those granites. On the other hand, the isotopic composition of the TG–YM does not lie along the mixing line of the DM–YM(II). A more evolved composition than the YM(II) is required to produce the source of the TG–YM. Furthermore, the TG–YM deviates so far from the line of the DM–YM(I) that the isotopic composition of the TG–YM is barely explained by the two-component mixing of the DM and the basement rocks in the Yeongnam massif. An alternative explanation for the isotopic composition of the TG–YM could be through the partial melting of a mixed crustal source. If the YM(I) and the YM(II) are taken as two end-members of the mixed source, the TG–YM can be produced by mixing the YM(I) up to 30% with the YM(II). The K-value $[(Sr/Nd)_{YM(II)}/(Sr/Nd)_{YM(I)}]$ for the mixing line connecting the YM(I) and the YM(II) is 7·2.

Three mantle–crust mixing lines for the Jurassic granites are added in Figure 8b. The DM–GM(I) and the DM–GM(II) pairs have K-values of 4·0. A mixing line between the DM and the amphibolites in the Okcheon belt (OB-v) is also shown, and the K-value for the DM–OB-v is 1·3. Most of the JG–YM group have a more evolved isotopic signature than the YM(II), so the isotopic composition of the JG–YM cannot result from the two-component mixing of the DM–YM(II) (Fig. 8b). However, some granites lie along the DM–YM(II) mixing line at a crustal proportion of 60–80%. Mixing between the DM and the YM(I) cannot account for the JG–YM composition because the JG–YM deviates considerably from the DM–YM(I) line. Although some JG–YM granites with higher SrI are close to the curve of the DM–YM(I) at a crustal proportion of about 20%, an additional enriched Sr isotopic component is required for these granites. If mixing between the mantle and crust is not the case for the isotopic composition of the granites, then the partial melting of the mixed source can be taken into account. For the JG–YM, the crustal end-members consisting of the mixed source can be taken from the YM(I) and the YM(II). From the mixing line between two basement rocks shown in Figure 8b, it is assumed that the source rocks of the JG–YM could have included up to 70% of the YM(I) component.

The isotopic composition of the JG–GM is also not well explained by two-component mixing between the DM and the basement rocks in the Gyeonggi massif [GM(I) and GM(II)] (Fig. 8b). It is apparent that the source rocks of the JG–GM were isotopically distinct from the exposed basement rocks. Like the JG–YM, the JG–GM can be assumed to have been formed by mixing of the crustal sources. Since the mixture between the GM(II) and the GM(I) cannot make any composition similar to the JG–GM, the only feasible way to produce the JG–GM composition is to mix the component of the

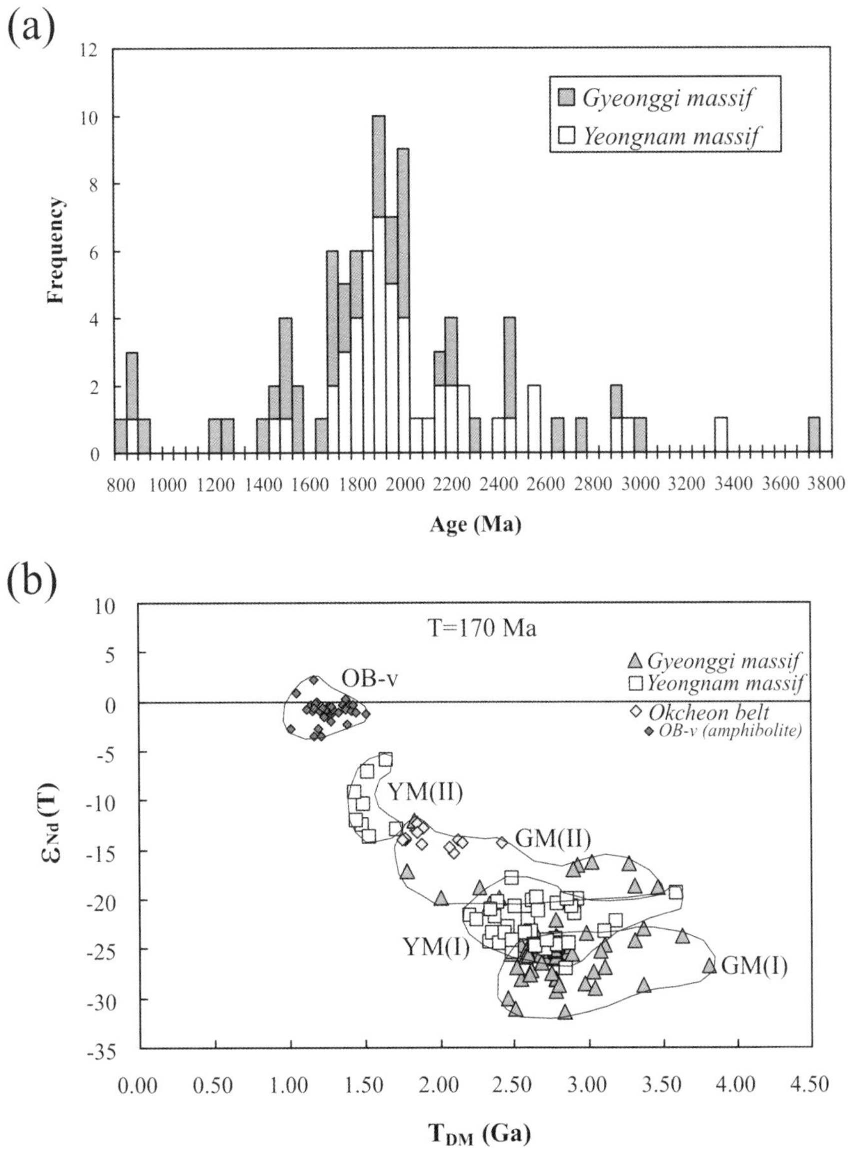

Figure 7 Radiometric ages (a) of the Precambrian basement rocks and (b) Nd isotopic signatures. Data sources for the radiometric ages: (Gyeonggi massif) Gaudette & Hurley (1973), Park & Cheong (1993), Lee *et al.* (1994b), Kwon *et al.* (1995a), Ree *et al.* (1996), Turek & Kim (1996), S. R. Lee *et al.* (1997), Sagong & Kwon (1998), Cho (1999), Cho *et al.* (1999), Kim *et al.* (1999), Lee *et al.* (2000), Cho *et al.* (2001) and Song *et al.* (2001); and (Yeongnam massif) Kwon & Jeong (1990), Park *et al.* (1993), Lee *et al.* (1994a), Suzuki & Adachi (1994), Kwon *et al.* (1995b), Park (1996), Turek & Kim (1996), B.-C Kim *et al.* (1997), Lee *et al.* (1997), Cho (1999), Lee, H. S. *et al.* (1999), Lee, S. G. *et al.* (1999), Cheong *et al.* (2000), Kwon & Kim (2000), Oh *et al.* (2000), Park *et al.* (2000), Lee *et al.* (2001), Park *et al.* (2001), Cho *et al.* (2002a), Cho *et al.* (2002b) and Kim *et al.* (2002). Data sources for the Nd isotopic signature: (Gyeonggi massif) Lee *et al.* (1992), Lee *et al.* (1994b), Lan *et al.* (1995), Cheong & Chang (1997), J.-H. Kim *et al.* (1997) and S. R. Lee *et al.* (1997); (Yeongnam massif) Na (1994), Lan *et al.* (1995), Cheong *et al.* (2000), Park *et al.* (2000), Lee *et al.* (2001); and (Okcheon belt) Kwon & Lan (1991), Na (1994), Lan *et al.* (1995), Lee & Chang (1997) and Lee *et al.* (1998).

GM(I) with that of the YM(II). However, it is difficult to evaluate the possible role of the YM(II) as a suitable protolith for the JG–GM in the crust–crust mixing because no basement rock that has a similar composition to the YM(II) has been found in the Gyeonggi massif (Fig. 7b). Nevertheless, the Gyeonggi massif has a record of the thermal event at 1·4–1·5 Ga, the same age as the formation of the YM(II). From the CHIME monazite ages for the gneissic rocks in the NE Gyeonggi massif, Cho (1999) argued that the massif had experienced regional metamorphism during 1·4–1·5 Ga. If this thermal event was closely related to the crustal formation at that time, the YM(II)-like crust is considered likely to have been underplated beneath the Gyeonggi massif. For such a case, the YM(II) can be taken as an end-member for the crust–crust mixing. The mixing line of the YM(II)–GM(I) demonstrates that the source of the JG–GM could have included up to 50% of the GM(I) component [$K=(Sr/Nd)_{YM(II)}/(Sr/Nd)_{GM(I)}=8{\cdot}9$].

The isotopic composition of the CG–OB does not lie on any mantle–crust mixing line. However, it varies along the crust–crust mixing line of the YM(II)–GM(II) (Fig. 8c), and the CG–OB is estimated to have included up to 80% of the GM(II) components in the source [$K=(Sr/Nd)_{YM(II)}/(Sr/Nd)_{GM(II)}=8{\cdot}9$]. The CG–YB can be split into two according to isotopic signature: the less-evolved granites have mixed isotopic compositions between the DM and the YM(II), and the more-evolved granites have intermediate compositions between the YM(II) and the CG–OB. The less-evolved CG–YB is represented by a proportion of 40–60% of the YM(II) components in the source. The more-evolved CG–YB

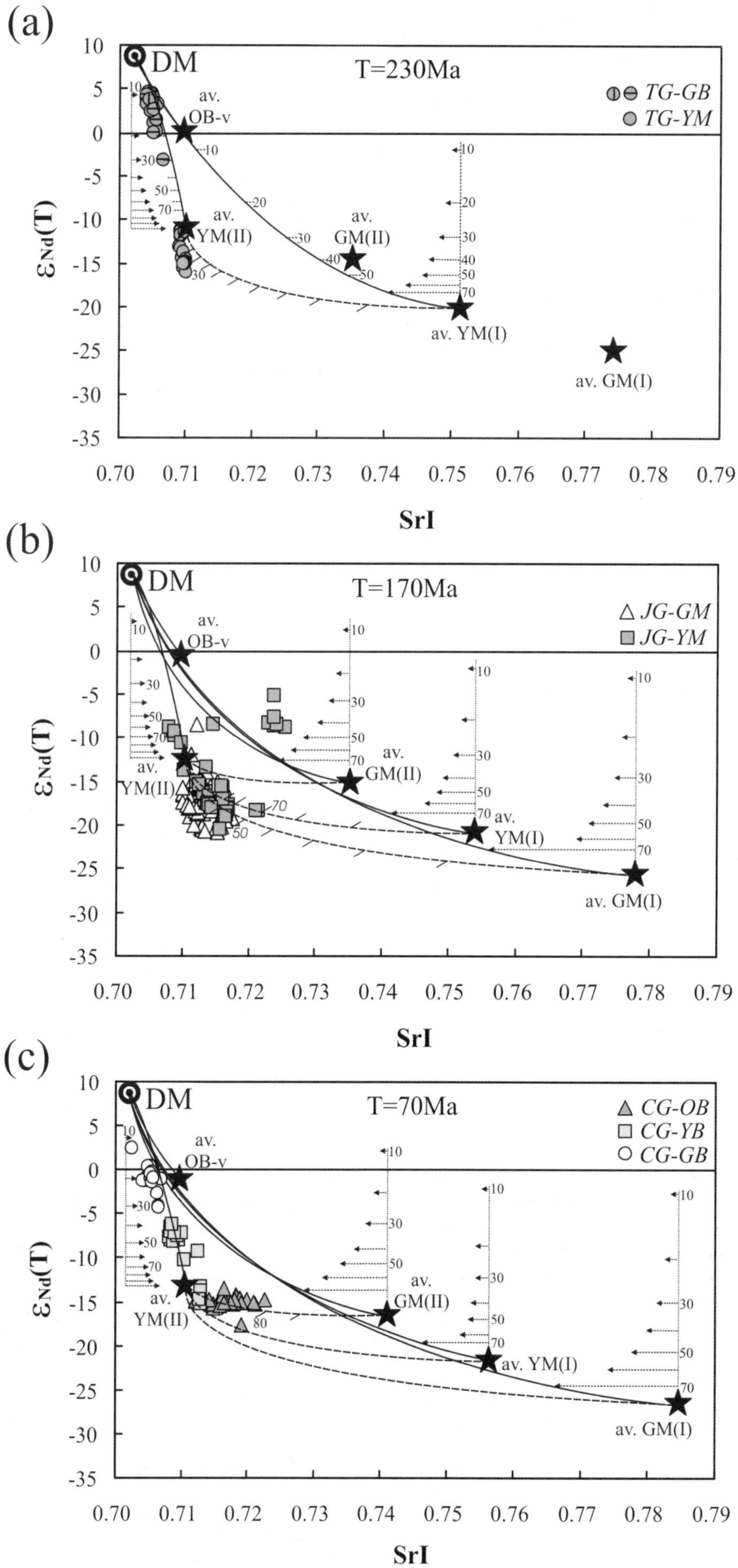

(a)
DM
T=230Ma
TG-GB
TG-YM
av. OB-v
av. YM(II)
av. GM(II)
av. YM(I)
av. GM(I)
εNd(T)
SrI
(b)
DM
T=170Ma
JG-GM
JG-YM
av. OB-v
av. YM(II)
av. GM(II)
av. YM(I)
av. GM(I)
εNd(T)
SrI
(c)
DM
T=70Ma
CG-OB
CG-YB
CG-GB
av. OB-v
av. YM(II)
av. GM(II)
av. YM(I)
av. GM(I)
εNd(T)
SrI

cannot be produced by the mixing of the DM and the YM(II), but possibly derived from the crustal mixture of the YM(II)–GM(II) pair with about 40% of the GM(II) component. The isotopic composition of the CG–GB lies on the mixing line of the DM–YM(II) at a crustal proportion of 10–30%. Thus, the CG–GB is likely to have been derived from the mixed source of the depleted mantle and the YM(II).

It is apparent from the above that no local exposed basement rocks could have been the sole source rocks for the Mesozoic granites in South Korea. Two-component mixing would appear to be the most feasible process in producing the source of the granites, but the end-members and degree of mixing are different for each group of the granites. Mantle–crust mixing can account for the generation of the source for the TG–GB, the JG–YM, the CG–GB and the CG–YB, whereas crust–crust mixing is a feasible process to produce the source of the TG–YM, the JG–YM, the JG–GM, the CG–YB and the CG–OB.

6.2. Geochemical nature versus isotopic signature

The Mesozoic granites in South Korea are characterised by I-type granites except for rare S-type, garnet-bearing JG–GM (Fig. 4a). In two-component mixing models, some granites are estimated to have been produced from the crustal sources and they have a more evolved crustal component varying in proportion from 30 to 80% (Fig. 8). Even the granites which include the depleted mantle component have varying proportions (10–80%) of crustal component. This is not consistent with their chemical compositions. Given that the basement rocks are strongly peraluminous (Fig. 6), the large amount of crustal component incorporated in the magma would increase the peraluminosity of the granites, and eventually, the granites could be typologically designated as S-types. Nevertheless, the high peraluminosity in the granites appears to be restricted to the high silica content, and it seems not to have been related to crustal contamination but probably to have been caused by the feldspar fractionation and/or the melt-fluid separation. Even though the TG–YM was possibly derived from mixed crustal components, the granites strongly demonstrate the metaluminous characteristics of I-types. Thus, it is speculated that the bulk geochemical nature of the granites was not greatly influenced by the chemistry of the basement rocks incorporated in the source. On the other hand, the chemistry of the granites is considered likely to reflect the composition of the igneous protolith like the DM and the YM(II) (for the igneous nature of the YM(II), see section 6·4). This would be why most Mesozoic granites in South Korea are represented by I-types, regardless of their temporal and spatial position. They have at least one igneous component in their source – the DM, the YM(II) or both. However, it is also recognised that there exists chemical heterogeneity in source rock composition. In particular, the TG–GB shows an isotopically uniform signature (Fig. 5), but demonstrates a dispersed variation pattern in P_2O_5 and Sr contents (Fig. 4). This would indicate that the granites were derived from a source that was isotopically uniform, but had highly variable chemistry.

6.3. Nd model ages of Mesozoic granites

The ε_{Nd} values and the depleted mantle model ages (T_{DM}) for the Mesozoic granites are plotted together in Figure 9 at the reference ages of 230, 170 and 70 Ma. As shown by two-component mixing models, the Mesozoic granites have mixed isotopic compositions either of depleted mantle plus crust, or of crust plus crust. Thus, the depleted mantle model ages of the granites are far from the crustal extraction age of each end-member, but possibly indicate the mixed model ages between two end-members (Arndt & Goldstein 1987). If the granites were produced by mixtures of an old crust with a young DM, the model age will reflect this mixture and give a minimum age of the crustal component. Therefore, it is expected that the T_{DM} ages of the granites will vary with the relative quantity of the mixed crustal components. In a sense, the granites can have a large range in the T_{DM} ages because of the various degrees of mixing between the end-members (McCulloch & Chappell 1982).

Apart from one data point, the TG–GB has a T_{DM} age of 0·6–1·0 Ga (Fig. 9a). In addition to the contrast of the ε_{Nd} values between the TG–GB and the TG–YM, the younger model age for the TG–GB is distinct from the age of the TG–YM (1·6–2·2 Ga). The Jurassic granites generally have more evolved ε_{Nd} values and older T_{DM} ages than the Triassic granites (Fig. 9b). The JG–YM can be divided into two groups on the basis of the ε_{Nd} values: less-evolved granites and more-evolved granites. The former has a T_{DM} age of 1·4–1·8 Ga, whereas the latter has a T_{DM} age of 1·7–3·3 Ga. The large range in the T_{DM} of the more-evolved JG–YM seems to be caused by the wide extent ($\sim$70%) of mixing between the YM(II) and the YM(I) (Fig. 8b). The JG–GM has a T_{DM} age of 1·6–3·1 Ga, and the broad range of the T_{DM} is likely to be a result of the various degrees ($\sim$50%) of mixing.

The Cretaceous granites show three different isotopic signatures (Fig. 9c). The CG–OB has a T_{DM} age of 1·9–2·8 Ga, and the range in the model age is thought to be attributable to the degree ($\sim$80%) of mixing between the YM(II) and the GM(II) components. The CG–YB has a T_{DM} age of 1·2–1·9 Ga, but the granites can be divided into two groups on the basis of the ε_{Nd} values. The less-evolved CG–YB has ages of 1·2–1·7 Ga, whereas the more-evolved group has model ages of 1·8–1·9 Ga. The CG–GB with depleted signatures has a T_{DM} age of 0·8–1·5 Ga.

From the T_{DM} distribution three groups of model ages are identified with a small degree of overlap: (1) 0·6–1·5 Ga, (2) 1·2–1·8 Ga and (3) $\geq$1·6 Ga. The youngest group consists of the TG–GB and the CG–GB, and is characterised by the high amount of mantle component in the mixed source (Fig. 8a, c). The middle group comprises the less-evolved granites from the JG–YM and the CG–YB. This group is represented by a small amount of mantle component in the mixed source, and has a similar range of ε_{Nd} value and T_{DM} age to the YM(II). The oldest group, having various proportions of the evolved crustal component such as the GM(I), GM(II) and YM(I), includes the TG–YM, the JG–GM, the CG–OB and the more-evolved granites from the JG–YM and the CG–YB. Since these three T_{DM} age groups are based on mixed model ages, each group has a minimum or maximum age constraint for the formation of the source rocks. For example, the youngest group is represented by the large amount of mantle-derived protolith and the extraction age of the mantle material is younger than 1·5 Ga. The middle group is characterised by a small amount of mantle-derived materials, and the extraction age of the

Figure 8 Mixing lines describing ε_{Nd}(T) and SrI(T) in mixtures of two components. The compositions for the depleted mantle (DM) are 136 ppm of Sr, 8 ppm of Nd, +10 of ε_{Nd}(0) and 0·7026 of SrI(0), and the isotopic compositions were recalculated with respect to the references ages of (a) 230 Ma, (b) 170 Ma and (c) 70 Ma. The average compositions of five crustal components [GM(I), GM(II), YM(I), YM(II) and OB-v] are shown. Solid lines indicate the DM–crust mixing, whereas broken lines represent the crust–crust mixing. The relative weight fraction (%) of the crustal component in the mixture is marked with a number. For abbreviations see caption to Figure 5. It is striking from the two-component mixing test that most of the data commonly include the YM(II) component as a crustal end-member.

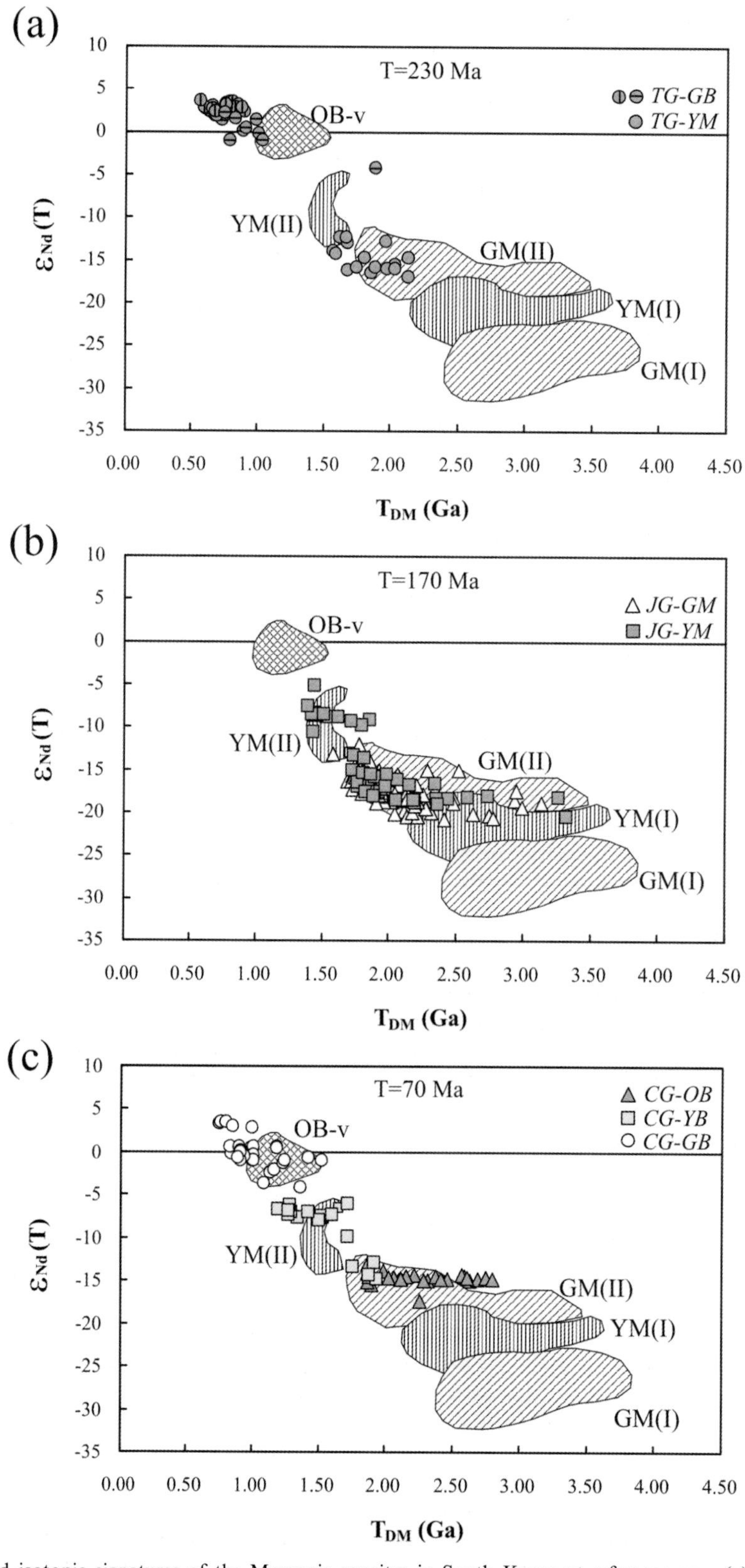

Figure 9 Nd isotopic signatures of the Mesozoic granites in South Korea at reference ages (a) 230 Ma, (b) 170 Ma and (c) 70 Ma. Each field for the basement rocks is taken from Figure 7b. The depleted mantle model ages (T_{DM}) for the granites are likely to indicate mixed model ages between the sources. From the T_{DM} distribution, the mixed source ages can be grouped into 0·6–1·5 Ga, 1·2–1·8 Ga and ≥1·6 Ga. Most Mesozoic granites have a T_{DM} age of 1·7–2·5 Ga, which is temporally consistent with an intense metamorphic event recorded in the basement rocks (Fig. 7a).

crustal source rocks would be slightly older than the range of 1·2–1·8 Ga. On the other hand, the oldest group originated from the crust–crust mixture would have a mean T_{DM} age between the two end-members. For example, the JG–GM can be produced by mixing of the YM(II) and the GM(I), so the mixed model age would ideally lie between the T_{DM} ages of the YM(II) and the GM(I). In Figure 9b, it is certain that the model ages of the JG–GM are older than the T_{DM} of the YM(II), but it is hard to compare the model ages of the granites with the T_{DM} of the GM(I) because both ages have a broad spectrum. Nevertheless, it can be said that a large proportion of the model ages for the JG–GM are not older than the minimum T_{DM} of the GM(I). Similarly, most mixed-formation ages of the sources for the more-evolved JG–YM would be younger than the YM(I), and those for the CG–OB and the more-evolved CG–YB are younger than that for the GM(I). In other words, although the mixed model ages indicate the ages intermediate between two formation ages of the crustal components, they are generally younger than the T_{DM} of the widely exposed basement rocks, the YM(I) and the GM(I).

6.4. Crustal evolution in South Korea

The crust–crust mixed sources have model ages older than 1·6 Ga, and in particular, most ages range from 1·7 to 2·5 Ga. Interestingly, this temporal range is very consistent with the age of intense metamorphic events recorded in the basement rocks (Fig. 7a). By compiling newly obtained Nd isotopic data for the Precambrian metamorphic rocks in the Gyeonggi massif, Lee *et al.* (2003) argued that crustal materials in the massif were largely extracted from the mantle during the late Archean (ca. 2·9–2·5 Ga). The above authors also suggested three groups of early Proterozoic tectonothermal events on the basis of U-Pb zircon ages: ca. 2·5–2·4 Ga, 2·3–2·2 Ga and 1·9–1·7 Ga. They thought that the three tectonothermal events would indicate the timing of the underplating of the juvenile crust beneath the late Archean crust. Among the three events, the youngest age (ca. 1·9 Ga) is well-recorded in the metamorphic rocks as a granulite-facies metamorphic event, and this middle Proterozoic event is also ubiquitous in the basement of the Yeongnam massif (Fig. 7a). Although a T_{DM} age of around 1·9 Ga is rarely observed in the basement rocks of the Gyeonggi and Yeongnam massifs, it is evident in the metasediments of the Okcheon belt (Fig. 7b). In addition, some inherited zircons in the Mesozoic granites mostly show upper intercept U-Pb concordia ages of 1·9–2·0 Ga (Turek & Kim 1995, 1996; Kim *et al.* 1999). Therefore, it is speculated that the middle Proterozoic crust which underplated beneath the early Proterzoic–late Archean crust would have been a possible protolith for the Okcheon metasediments and have acted as a crustal component to produce the Mesozoic granites (the CG–OB and the evolved CG–YB) in South Korea.

In Figure 8, mixing lines which incorporate most of the data commonly include the YM(II) component as an end-member. In addition, some granites have very similar T_{DM} ages to those of the YM(II) (Fig. 9). The YM(II) is mostly characterised by granitic gneiss in the central Yeongnam massif. Lee *et al.* (2001) obtained a Nd–Sm whole-rock age of 1·5 Ga for the granitic gneiss and ε_{Nd} values of +7. They interpreted that 1·5 Ga is the age of crustal formation of the igneous suite that was extracted from a highly depleted mantle (MORB-like) source. Therefore, the YM(II) is likely to have been formed 1·5 Ga ago by underplating of mantle-derived material to the old crust. It is apparent from the above and the record of metamorphic events (Fig. 7a) that the middle Proterozoic (ca. 1·9 Ga) crustal growth episode was followed by another underplating of depleted mantle-derived material at 1·5 Ga ago. This YM(II)-like crust seems likely to be the most feasible igneous protolith that predominantly contributed to the Mesozoic granites in South Korea (Fig. 8).

Late Proterozoic crustal growth following the underplating of the YM(II)-like crust is represented by amphibolites in the Gyeonggi massif and the Okcheon belt (Kwon *et al.* 1995a; Ree *et al.* 1996; Lee & Chang 1997; Lee *et al.* 1998). The isotopic signature of these amphibolites indicates that they were derived from the depleted mantle material and that the juvenile crust was underplated 1·0–1·5 Ga ago (Fig. 7b). Since the magmatic age of the amphibolites is ca. 760 Ma (Lee *et al.* 1998), it is considered likely that they were derived from source rocks separated shortly from the mantle. However, this late Proterozoic crust seems unlikely to have participated in the production of the Mesozoic granites in South Korea.

The mixed model ages for the TG–GB and the CG–GB suggest that more juvenile material which derived from the mantle would have been accreted to the lower crust during the late Proterozoic. The isotopic signatures of these granites in the Gyeongsang basin can be compared with those of the granites in the Central Asian Orogenic Belt (CAOB). Most Phanerozoic granites in the CAOB are characterised by low SrI (ca. 0·705), positive ε_{Nd} values and a young T_{DM} age ($\leq$1·5 Ga) (Jahn 2003), and the proportion of juvenile or mantle components for generation of the granites was estimated to vary from 70% to 100% (Jahn *et al*·2000). The isotopic signatures of the TG–GB and the CG–GB are very similar to those of the granites in the CAOB. In particular, the assumed mantle proportion added to the granite composition is 70–90% for the TG–GB and the CG–GB, which is also consistent with observation of the granites in the CAOB. The Phanerozoic granitic magmatism in the CAOB could have significantly contributed to crustal growth of massive juvenile continental crust in Central Asia (Jahn *et al.* 2000; Jahn 2003). Crustal growth in the Phanerozoic is not certain in E Asia, but the Mesozoic granites in the Gyeongsang basin may be a candidate for juvenile crustal accretion in E Asia. However, although the Phanerozoic granites in the Gyeongsang basin and in the CAOB have common isotopic signatures, the tectonic relationship between the two areas is still uncertain.

The isotopic systematics for the Mesozoic granites and basement rocks imply that the crust is likely to be vertically stratified beneath South Korea. The early–middle Proterozoic crust (T_{DM}=1·7–2·5) is possibly located beneath the exposed old cratonic crust, and under this early–middle Proterozoic crust is a YM(II)-like crust (T_{DM}=ca. 1·5 Ga). The lowest crust is characterised by the juvenile material and younger T_{DM} age of less than 1·5 Ga. The T_{DM} age for the OB-v supports a model of crustal growth during the 1·0–1·5 Ga. Such basaltic magmatism is often observed in the Gyeonggi massif, and their extrusion age ranges from 0·7 to 0·8 Ga (Kwon *et al.* 1995a; Ree *et al.* 1996; Cho *et al.* 2001). These radiometric ages for the amphibolites could also be evidence for crustal growth during 0·6–1·0 Ga.

7. Conclusions

The three Mesozoic granites in South Korea have distinctive geochemical compositions. Triassic granites are similar to high-temperature I-type granites in the sense of Chappell *et al.* (1998), and are characterised by low ASI, Rb and Zr, but high Sr. Except for rare S-type, garnet-bearing granites, the Jurassic granites are mostly similar to low-temperature I-type granites. High ASI and Rb at higher silica compositions indicate that the Jurassic granites experienced feldspar fractionation during the magmatic stage. Cretaceous granites are of low-temperature I-types and show strong fractionation at high silica contents. They are characterised by low Sr and rapid decrease of Zr with silica.

The systematic variation of ε_{Nd} and SrI in the Korean Mesozoic granites could have resulted from the mixing of two components in different proportions to produce the protolith of the granites. Mantle–crust mixing could account for the generation of the source of the Triassic and Cretaceous granites in the Gyeongsang basin, the Jurassic granites (less enriched) in the Yeongnam massif and Cretaceous granites (less enriched) in the Yeongdong–Gwangju basin. On the other hand, crust–crust mixing could feasibly have produced the source of the Triassic and Jurassic granites (more enriched) in the Yeongnam massif, the Jurassic granites in the Gyeonggi massif, and the Cretaceous granites (more enriched) in the Yeongdong–Gwangju basin and the Okcheon belt.

Combined geochemical and isotopic signatures indicate that a simple binary mixing model is not adequate to explain both geochemical and isotopic data. The chemistry of the granites is considered likely to reflect the composition of the igneous protolith that derived from the depleted mantle. This determines that most Mesozoic granites in South Korea are represented by I-types, regardless of temporal and spatial constraints.

The Nd-Sr isotopic signatures of the Mesozoic granites and basement rocks indicate that the continental crust beneath the Korean peninsular is vertically structured by the successive underplating of mantle-derived materials. The crust is modelled to be stratified, in sequence from surface to deepest crust as late Archean–early Proterozoic, early–middle Proterozoic (ca. 1·9 Ga), middle Proterozoic (ca. 1·5 Ga) and late Proterozoic (younger than 1·5 Ga).

8. Acknowledgements

This work was supported by the Korea Research Foundation Grant (KRF-2002–013-D00085), which made it possible for the author to stay for a year at GEMOC, Macquarie University, Australia. This paper has benefited from reviews by Masaaki Owada and an anonymous reviewer.

9. References

Ames, L., Tilton, G. R. & Zhou, G. 1993. Timing of collision of the Sino-Korean and Yangtze cratons: U-Pb zircon dating of coesite-bearing eclogite. *Geology* **21**, 339–42.

Ames, L., Zhou, G. Z. & Xiong, B. C. 1996. Geochronology and isotopic character of ultrahigh-pressure metamorphism with implications for collision of the Sino-Korean and Yangtze cratons, central China. *Tectonics* **15**, 472–89.

Arndt, N. T. & Goldstein, S. L. 1987. Use and abuse of crust-formation ages. *Geology* **15**, 893–5.

Chappell, B. W. 1999. Aluminium saturation in I- and S-type granites and the characterization of fractionated haplogranites. *Lithos* **46**, 535–51.

Chappell, B. W., White, A. J. R. & Wyborn, D. 1987. The importance of residual source material (restite) in granite petrogenesis. *Journal of Petrology* **28**, 1111–38.

Chappell, B. W., Bryant, C. J., Wyborn, D., White, A. J. R. & Williams, I. S. 1998. High- and low-temperature I-type granites. *Resource Geology* **48**, 225–35.

Chappell, B. W., White, A. J. R., Williams, I. S., Wyborn, D. & Wyborn, L. A. I. 2000. Lachlan Fold belt granites revisited: high- and low-temperature granites and their implications. *Australian Journal of Earth Sciences* **47**, 123–38.

Chappell, B. W. & White, A. J. R. 1974. Two contrasting granite types. *Pacific Geology* **8**, 173–4.

Chappell, B. W. & White, A. J. R. 1992. I- and S-type granites in the Lachlan Fold belt. *Transactions of the Royal Society of Edinburgh: Earth Sciences* **83**, 1–26.

Chang, H. W., Lee, M. S. & Lee, J. M. 1988. The origin of geochemical variations in the Hongjesa granite gneiss. *Journal of the Geological Society of Korea* **24**, 500–10.

Chavagnac, V. & Jahn, B. M. 1996. Coesite-bearing eclogite from the Bixiling Complex, Dabie Mountains, China: Sm-Nd ages, geochemical characteristics, and tectonic implications. *Chemical Geology* **133**, 29–51.

Cheong, C.-S., Kwon, S.-T., Kim, J.-M. & Chang, B.-U. 1998. Geochemical and isotopic study of the Onjeongri granite in the northern Gyeongsang basin, Korea: comparison with Cretaceous to Tertiary granitic rocks in the other part of the Gyeongsang basin and the inner zone of southwest Japan. *Journal of the Petrological Society of Korea* **7**, 77–97. [In Korean with English abstract.]

Cheong, C.-S., Kwon, S.-T. & Park, K.-H. 2000. Pb and Nd isotopic constraints on Paleoproterozoic crustal evolution of the northeastern Yeongnam massif, South Korea. *Precambrian Research* **102**, 207–20.

Cheong, C.-S., Kwon, S.-T. & Sagong, H. 2002. Geochemical and Sr-Nd-Pb isotopic investigation of Triassic granitoids and basement rocks in the northern Gyeongsang basin, Korea: implications for the young basement in the East Asian continental margin. *The Island Arc* **11**, 25–44.

Cheong, C.-S. & Chang, H.-W. 1997. Sr, Nd, and Pb isotope systematics of granitic rocks in the central Ogcheon belt, Korea. *Geochemical Journal* **31**, 17–36.

Cho, D.-L. 1999. CHIME dating of the early- to middle Proterozoic para- and orthogneisses in the Gosung-Gansung area, Gyeonggi massif, South Korea: their tectonic implications. *In* Cheong, D. K. *et al.* (eds) *Proceedings, the 6th Sino-Korean Joint Symposium on Crustal Evolution in Northeast Asia*, 47–53. Chuncheon: The Research Institute for Development of Earth Resources, Kangwon National University.

Cho, D.-L., Kwon, S.-T., Jeon, E.-Y. & Armstrong, R. 2001. SHRIMP U-Pb zircon geochronology of an amphibolite and a paragenesis from the Samgot unit, Yeoncheon complex in the Imjingang belt, Korea: tectonic implication. *Abstract with Program, the 56th Annual Meeting of the Geological Society of Korea*, 89. Daegu: Kyungpook National University, Geological Society of Korea.

Cho, D.-L., Kim, Y.-J. & Armstrong, R. 2002a. SHRIMP U-Pb zircon geochronology of the metasedimentary rocks and leucosome in the Jirisan area, Yeongnam massif: deposition of early Proterozoic (2002 Ma–1846 Ma) and regional metamorphism. *Abstract with Program, the 35th Annual Meeting of the Korean Society of Economic and Environmental Geology*, 283–5. Seoul: Hanyang University, Korea Society of Economic and Environmental Geology. [In Korean.]

Cho, D.-L., Kim, Y.-J. & Armstrong, R. 2002b. SHRIMP U-Pb zircon age of the Dongbok granite gneiss in the Jirisan area, Yeongnam massif and its tectonic implication. *Proceedings of the annual joint conference, Petrological Society of Korea and Mineralogical Society of Korea*, 31–3. Chonju: Chonbuk National University, Petrological Society of Korea. [In Korean.]

Cho, K. S. 1994. Petrochemical studies on the gneisses in the Gurye-Hadong area. *Journal of the Korean Earth Science Society* **15**, 286–94. [In Korean with English abstract.]

Cho, K. S. 2000. Petrochemical study on the granitic gneiss in the Gyrye-Suncheon, Korea. *Economic and Environmental Geology* **33**, 173–80. [In Korean with English abstract.]

Cho, K. S. & Nam, K. S. 1993. Geochemical characteristics and origin of the gneisses in the Jincheon-Eumseong area, Chungcheong-bukdo, Korea. *Journal of the Korean Earth Science Society* **14**, 274–80. [In Korean with English abstract.]

Cho, M., Lee, S. R. & Yi, K. 1999. Tectono-metamorphic evolution of the Hwacheon granulite complex, South Korea. *In* Cheong, D. K. *et al.* (eds) *Proceedings of the 6th Sino-Korean Joint Symposium on Crustal Evolution in Northeast Asia*, 3–4. Chuncheon: The Research Institute for Development of Earth Resourses, Kangwon National University.

Choe, W. H. & Jwa, Y.-J. 1998. Petrochemical characteristics of the granites in the Jeomchon area. *Journal of the Petrological Society of Korea* **7**, 37–52. [In Korean with English abstract.]

Chough, S. K., Kwon, S.-T., Ree, J.-H. & Choi, D. K. 2000. Tectonic and sedimentary evolution of the Korean peninsula: a review and new view. *Earth-Science Reviews* **52**, 175–235.

Clarke, D. B. & MacDonald, M. A. 2003. Chemical variation in Al_2O_3-CaO-Na_2O-K_2O space: controls on the peraluminosity of the South Mountain Batholith. *In* Arima, M. *et al.* (eds) *Abstract with Program, Hutton Symposium V, The Origin of Granites and Related Rocks*: *Geological Survey of Japan, Interim-Report* **29**, 24.

Collins, W. J. 1996. Lachlan Fold Belt granitoids: products of three-component mixing. *Transactions of the Royal Society of Edinburgh: Earth Sciences* **87**, 171–9.

DePaolo, D. J. 1981. Trace element and isotopic effects of combined wallrock assimilation and fractional crystallization. *Earth and Planetary Science Letters* **53**, 189–202.

DePaolo, D. J. 1988. *Neodymium Isotope Geochemistry: An introduction*. Berlin: Springer-Verlag.

DePaolo, D. J. & Wasserburg, G. J. 1979. Petrogenetic mixing model and Nd-Sr isotopic patterns. *Geochimica et Cosmochimica Acta* **43**, 615–27.

Gaudette, H. E. & Hurley, P. M. 1973. U-Pb zircon age of Precambrian basement gneisses of South Korea. *Geological Society of America Bulletin* **84**, 2305–6.

Harrison, T. M. & Watson, E. B. 1984. The behaviour of apatite during crustal anatexis: equilibrium and kinetic consideration. *Geochimica et Cosmochimica Acta* **48**, 1468–77.

Hine, R., Williams, I. S., Chappell, B. W. & White, A. J. R. 1978. Contrasts between I- and S-type granitoids of the Kosciusko Batholith. *Journal of the Geological Society of Australia* **25**, 219–34.

Hong, Y. K. 1987. Geochemical characteristics of Precambrian, Jurassic and Cretaceous granite in Korea. *Journal of the Korean Institute of Mining Geology* **20**, 35–60.

Jahn, B.-m. 2003. Generation of the juvenile crust in Central Asia. *In* Arima, M. *et al.* (eds) *Abstract with Program, Hutton Symposium V, The Origin of Granites and Related Rocks: Geological Survey of Japan, Interim-Report* **29**, 56.

Jahn, B.-m., Wu, F. & Chen, B. 2000. Massive granitoid generation in Central Asia: Nd isotope evidence and implication for continental growth in the Phanerozoic. *Episodes* **23**, 82–92.

Jwa, Y.-J. 1988. *Petrology, geochemistry and geochronology of the granitic rocks in the Inje-Hongcheon district, South Korea.* Unpublished Ph.D. thesis. University of Tokyo.

Jwa, Y.-J. 1996. Petrochemistry and magma process of Jurassic Boeun granodiorite in the central Ogcheon belt. *Journal of the Petrological Society of Korea* **5**, 188–99. [In Korean with English abstract.]

Jwa, Y.-J. 1997. Petrological characteristics of two-mica granites: examples from Cheongsan, Inje-Hongcheon, Yeongju and Namwon areas. *Journal of the Petrological Society of Korea* **6**, 210–25. [In Korean with English abstract.]

Jwa, Y.-J. 1998. Temporal, spatial and geochemical discriminations of granitoids in South Korea. *Resource Geology* **48**, 273–84.

Jwa, Y.-J. 1999. Cretaceous granitic magmatism in South Korea and its implication to the Mesozoic tectonic evolution of the eastern Eurasian continental margin. *Abstract with Program, Hutton Symposium IV, The Origin of Granites and Related Rocks: Clermont-Ferrand, France*, 153.

Jwa, Y.-J. & Park, J.-M. 1996. Petrology of the igneous rocks in the Goseong area: I. major element geochemistry and K-Ar radiometric age. *Economic and Environmental Geology* **29**, 561–73. [In Korean with English abstract.]

Kim, B-C., Choi, S.-J., Suzuki, K., Adachi, M., Obayashi, T. & Yu, K.-M. 1997. Provenance of Cretaceous sandstones in the southeastern Yongdong Basin, Korea: CHIME geochronology of detrital monazites. *Geoscience Journal* **1**, 37–49.

Kim, C.-B., Turek, A., Chang, H.-W., Park, Y.-S. & Ahn, K.-S. 1999. U-Pb zircon ages for Precambrian and Mesozoic plutonic rocks in the Seoul-Cheongju-Choongcheon area, Gyeonggi massif, Korea. *Geochemical Journal* **33**, 379–97.

Kim, C.-B. & Turek, A. 1996. Advances in U-Pb zircon geochronology of Mesozoic plutonism in the southwestern part of Ryeongnam massif, Korea. *Geochemical Journal* **30**, 323–38.

Kim, H.-N., Shin, I.-H., Ahn, K.-S., Park, W.-S. & Lee, C.-S. 1997. Petrochemistry and Sr, Nd isotopic compositions of granitoids in the Gangjin-Jangheung areas, Korea. *Journal of the Korean Earth Science Society* **18**, 433–42. [In Korean with English abstract.]

Kim, J. H., Song, Y. C. & Koh, H. J. 1997. Geology and Sr, Nd and Pb isotopic compositions of Precambrian granitoids in the Pyeongchang area, Korea. *Journal of the Geological Society of Korea* **33**, 27–35. [In Korean with English abstract.]

Kim, K. H., Park, S. S. & Na, C. K. 1996. Nd and Sr isotopic signatures of Mesozoic granitoids in South Korea. *Resource Geology* **46**, 215–26.

Kim, O. J. 1975. Study on the intrusion epochs of younger granites and their bearing to orogenesis in South Korea. *Journal of the Korean Institute of Mining Geology* **8**, 223–30.

Kim, Y.-B., Lim, S.-B., Choi, H.-I., Lee, C.-B., Kim, B,-C., Chun, H.-Y., Park, S.-H., Cho, D.-L., Koh, H.-J., Lee, S. R., Park, H.-S., Kim. J.-C. & Lee, Y.-S. 2002. New stratigraphy in the western part of Kangjin. *Abstract with Program, the 18th Annual Meeting of the Paleontological Society of Korea*, 42. Daejeon: Korea Institue of Geoscience and Mineral Resources, Paleontological Society of Korea. [In Korean.]

Kwon, S.-T., Cho, D.-L., Lan, C.-Y., Shin, K.-B., Lee, T. & Mertzman, S. A. 1994. Petrology and geochemistry of the Seoul granitic batholith. *Journal of the Petrological Society of Korea* **3**, 109–27.

Kwon, S.-T., Cho, M., Jeon, E.-Y., Lee, S. R. & Nakamura, E. 1995a. Sm-Nd and Rb-Sr isotopic study on Chuncheon amphibolites. *Abstract with Program, the 4th Annual Meeting of the Petrological Society of Korea*, 25. Busan: Busan National University, Petrological Society of Korea. [In Korean.]

Kwon, S.-T., Ree, J.-H., Park, K.-H. & Jeon, E.-Y. 1995b. Nature of contact between the Ogcheon belt and Yeongnam massif and the Pb-Pb age of granitic gneiss in Cheondong-ri, Danyang. *Journal of the Petrological Society of Korea* **4**, 144–52. [In Korean with English abstract.]

Kwon, S.-T., Lan, C.-Y. & Lee, T. 1999. Rb-Sr and Sm-Nd isotopic study of the Seoul granitic batholith in middle Korea. *Geoscience Journal* **3**, 107–14.

Kwon, S.-T. & Jeong, J. G. 1990. Preliminary Sr-Nd isotope study of the Hadong-Sanchung anorthositic rocks in Korea: implication for their origin and for the Precambrian tectonics. *Journal of the Geological Society of Korea* **26**, 341–9.

Kwon, S.-T. & Lan, C.-Y. 1991. Sm-Nd isotopic study of the Ogcheon amphibolite, Korea: preliminary report. *Journal of the Korean Institute of Mining Geology* **24**, 277–85. [In Korean with English abstract.]

Kwon, S.-T. & Sagong, H. 2002. Regional Pb-Nd-Sr isotopic differences of the Mesozoic granitoids in South Korea: implications for the basement structure. *Goldschmidt Conference Abstracts*, A426. Davos: Cambridge Publication.

Kwon, Y.-W. & Kim, H.-S. 2000. Geochronological study on the Punggi area, central part of the Sobaeksan massif. *Abstract with Program, the 55th Annual Meeting of the Geological Society of Korea*, 95. Daejeon: Chungnam National University, Geological Society of Korea. [In Korean.]

Lan, C.-Y., Lee, T., Zhou, X. H. & Kwon, S.-T. 1995. Nd isotopic study of Precambrian basement of South Korea: evidence for Early Archean crust? *Geology* **23**, 249–52.

Lee, D. S. (ed.) 1987. *Geology of Korea*. Geological Society of Korea. Seoul: Kyohak-sa.

Lee, D.-W. 1999. Strike-slip fault tectonics and basin formation during the Cretaceous in the Korean Peninsula. *The Island Arc* **8**, 218–31.

Lee, H.-S., Song, Y.-S., Park, K.-H. & Lee, S.-H. 1999. 2950Ma CHIME age of Precambrian gneiss in the central Sbaeksan massif. *Abstract with Program, the 54th Annual Meeting of the Geological Society of Korea*, 77–8. Dangjin: Korea Agricultural & Rural Infrastructure Corporation, Geological Society of Korea. [In Korean.]

Lee, I. H., Tsusue, A. & Honma, H. 1995. Mesozoic 'I-like' granitic rocks of the Wolagsan mass in the Ogcheon zone of South Korea. *Resource Geology* **45**, 53–65. [In Japanese with English abstract.]

Lee, J. I., Jwa, Y.-J., Park, C.-H., Lee, M. J. Moutte, J. & Kagami, H. 1999. Sr and Nd isotopic compositions of late Paleozoic Youngju and Andong granites in the northeastern Yeongnam massif, Korea. *Geochemical Journal* **33**, 153–65.

Lee, K.-S., Cheong, C.-S., Park, K.-H. & Chang, H.-W. 1997. Geochemical and Sm-Nd isotopic study of amphibolite from the Muju area, Korea. *Economic and Environmental Geology* **30**, 313–20. [In Korean with English abstract.]

Lee, K.-S., Chang, H.-W. & Park, K.-H. 1998. Neoproterozoic bimodal volcanism in the central Ogcheon belt, Korea: age and tectonic implication. *Precambrian Research* **89**, 47–57.

Lee, K.-S. & Chang, H.-W. 1997. Geochemistry and Sr-Nd-Pb isotopic systematics of the Ogcheon amphibolites from the central Ogcheon belt, Korea: implications for the source heterogeneity. *Geochemical Journal* **31**, 223–43.

Lee, S.-G., Shimizu, H., Masuda, A. & Song, Y. S. 1992. Crustal evolution of the Precambrian basement in the Korean peninsula. *Journal of the Petrological Society of Korea* **1**, 124–31.

Lee, S.-G., Masuda, A. & Kim, H. S. 1994a. An early Proterozoic leuco-granitic gneiss with the REE tetrad phenomenon. *Chemical Geology* **114**, 59–67.

Lee, S.-G., Song, Y. & Masuda, A. 1994b. 1. 2 Ga mineral isochron of Changhowon garnet gneiss. *Journal of the Geological Society of Korea* **30**, 62–8.

Lee, S.-G., Shin, S. C., Jin, M.-S., Chung, K.-S. & Ogasawara, M. 1999. Geochemistry of Precambrian granite from Taebaek area, Northeastern part of South Korea. *In* Cheong, D. K. *et al.* (eds) *Proceedings of the 6th Sino-Korean Joint Symposium on Crustal Evolution in Northeast Asia*, 122–3. Chuncheon: The Research Institute for Development of Earth Resources, Kangwon National University.

Lee, S.-G., Masuda, A., Shimizu, H. & Song, Y.-S. 2001. Crustal evolution history of Korean Peninsula in East Asia: the significance of Nd, Ce isotopic and REE data from the Korean Precambrian gneisses. *Geochemical Journal* **35**, 175–87.

Lee, S. H. & Choi, G. J. 1994. Geochemistry and chemical equilibria of coexisting minerals in the gneiss around the Daeheung talc deposits, Korea. *Journal of the Petrological Society of Korea* **3**, 138–55. [In Korean with English abstract.]

Lee, S. M. 1974. The tectonic setting of Korea, with relation to plate tectonics. *Journal of the Geological Society of Korea* **10**, 25–36.

Lee, S. M, Kim, H. S., Hong, S. T. & Park, C. S. 1990. Petrological studies on the metamorphic rocks in Wonju-Pyongchang area. *Journal of the Geological Society of Korea* **26**, 32–52.

Lee, S. R., Cho, M., Cheong, C.-S. & Park, K.-H. 1997. An early Proterozoic Sm-Nd age of mafic granulite from the Hwacheon area, South Korea. *Geoscience Journal* **1**, 136–42.

Lee, S. R., Cho, M., Yi, K.-W. & Stern, R. 2000. Early Proterozoic granulites in central Korea: tectonic correlation with Chinese cratons. *Journal of Geology* **108**, 729–38.

Lee, S. R., Cho, M., Hwang, J. H., Lee, B.-J., Kim, Y.-B. & Kim, J. C. 2003. Crustal evolution of the Gyeonggi massif, South Korea: Nd isotopic evidence and implications for continental growths of East Asia. *Precambrian Research* **121**, 25–34.

Li, S. G., Xiao, Y. L., Liou, D. L., Chen, Y. Z., Ge, N. J., Zhang, Z. Q., Sun, S. S., Cong, B. L., Zhang, R. Y., Hart, S. R. & Wang, S. S. 1993. Collision of the North China – Yangtze block and formation of coesite-bearing eclogite: timing and processes. *Chemical Geology* **109**, 89–111.

Li, Z. X. 1994. Collision between the North and South China blocks: a crustal detachment model for suturing in the region east of the Tanlu fault. *Geology* **22**, 739–42.

Liew, T. C. & McCulloch, M. T. 1985. Genesis of granitoid batholiths of Peninsular Malaysia and implications for models of crustal evolution: evidence from a Nd-Sr isotopic and U-Pb zircon study. *Geochimica et Cosmochimica Acta* **49**, 587–600.

McCulloch, M. T. & Chappell, B. W. 1982. Nd isotopic characteristics of S- and I-type granites. *Earth and Planetary Science Letters* **58**, 51–64.

Montel, J. M., Mouchel, R. & Pichavant, M. 1988. High apatite solubility in peraluminous melts. *Terra Cognita* **8**, 71.

Na, C.-K. 1994. *Genesis of granitoid batholiths of Okchon zone, Korea and its implications for crustal evolution.* Unplublished Ph.D. thesis. University of Tsukuba.

Oh, C. W., Jeon, E. Y., Park, B. Y., Ahn, K.-S. & Lee, J. H. 2000. Metamorphic evolution of granitic and porphyroblatic gneisses in the Seungju-Suncheon area, the southwestern part of the Soback-san Massif. *Journal of the Petrological Society of Korea* **9**, 121–41. [In Korean with English abstract.]

Park, J. C., Ham, I. K., Kim, W. & Choi, K. S. 2003. A study on crustal thickness of the southern Korean Peninsula using travel times of PmP phase. *Journal of the Geological Society of Korea* **39**, 225–34. [In Korean with English abstract.]

Park, K.-H. 1996. Pb-Pb age of marble from Muju, Korea. *Journal of the Petrological Society of Korea* **5**, 84–8. [In Korean with English abstract.]

Park, K.-H., Cheong, C.-S, Lee, K.-S. & Chang, H.-W. 1993. Isotopic composition of lead in Precambrian granitic rocks of the Taebaeg area. *Journal of the Geological Society of Korea* **29**, 387–95. [In Korean with English abstract.]

Park, K.-H., Song, Y.-S., Park, M.-E., Lee, S.-G. & Ryu, H.-J. 2000. Petrological, geochemical and geochronological studies of Precambrian basement in Northeast Asia region: 1. Age of the metamorphism of Jirisan area. *Journal of the Petrological Society of Korea* **9**, 29–39. [In Korean with English abstract.]

Park, K.-H., Kim, D.-Y. & Song, Y.-S. 2001. Sm-Nd mineral ages of charnockites and ilmenite-bearing anorthositic rocks of Jirisan area and their genetic relationship. *Journal of the Petrological Society of Korea* **10**, 27–35. [In Korean with English abstract.]

Park, K.-H. & Cheong, C.-S. 1993. Preliminary study for the dating of marbles from Precambrian Kyonggi metamorphic complex. *Journal of the Petrological Society of Korea* **2**, 130–8. [In Korean with English abstract.]

Pichavant, M., Montel, J. M. & Richard, L. R. 1992. Apatite solubility in peraluminous liquids: experimental data and an extension of the model of Harrison and Watson (1984). *Geochimica et Cosmochimica Acta* **56**, 3855–61.

Ree, J.-H., Cho, M., Kwon, S.-T. & Nakamura, E. 1996. Possible eastward extension of Chinese collision belt in South Korea: the Imjingang belt. *Geology* **24**, 1071–4.

Ree, J.-H., Kwon, S.-H., Park, Y., Kwon, S.-T. & Park, S.-H. 2001. Pretectonic and posttectonic emplacement of the granitoids in the south central Okcheon belt, South Korea: implications for the timing of strike-slip shearing and thrusting. *Tectonics* **20**, 850–67.

Sagong, H. 2000. *U-Pb sphene ages and feldspar Pb isotope ratios of the Phanerozoic granitoids in South Korea: Implications for the crustal evolution of Korean Peninsula.* Unpublished Ph.D. thesis. Yonsei University. [In Korean with English abstract.]

Sagong, H., Kwon, S. T. & Ree, J.-H. 2003. Origin of the 160–110 Ma hiatus in the Mesozoic magmatism of South Korea. *In* Arima, M. *et al.* (eds) *Abstract with Program, Hutton Symposium V, The Origin of Granites and Related Rocks: Geological Survey of Japan, Interim-Report* **29**, 125.

Sagong, H. & Jwa, Y.-J. 1997. Mineral chemistry and major element geochemistry of the granitic rocks in the Cheongsan area. *Journal of the Petrological Society of Korea* **6**, 185–209. [In Korean with English abstract.]

Sagong, H. & Kwon, S.-T. 1998. Pb-Pb age and uplift history of the Busan gneiss complex in the Okchon belt, Korea: a comparison with the Bagdalryeong gneiss complex in the Kyongki massif. *Geoscience Journal* **2**, 99–106.

Saunders, A. D. & Tarney, J. 1984. Geochemical characteristics of basaltic volcanism within back-arc basins. *In* Kokelaar, B. P. & Howells, M. F. (eds) *Marginal Basin Geology. Geological Society, London, Special Publication* **16**, 59–76.

Shimazaki, H. & Lee, M. S. 1981. Reconnaissance on I- and S-type granitoids in Southern Korea. *Journal of Geological Society of Korea* **17**, 189–93.

Shin, I.-H., Kim, H.-N. & Ahn, K.-S. 1997. Origin of granitoids in the southwestern part of Korea based on Sr, Nd isotopic ratio. *Journal of the Korean Earth Science Society* **18**, 99–104. [In Korean with English abstract.]

Song, Y.-S. 1989. Geochemistry of the Precambrian metamorphic rocks from the central Sobaegsan massif, Korea. *Journal of the Korean Institute of Mining Geology* **22**, 293–300. [In Korean with English abstract.]

Song, Y.-S., Park, K.-H., Park, M.-E., Cao, L., Jin, W., Zhang, X. & Ryu, H.-J. 2001. Petrological, geochemical and geochronological studies of Precambrian basement in Northeast Asia region: 2. Zircon ages of some metamorphic rocks from Gyeonggi massif. *Journal of the Petrological Society of Korea* **10**, 95–105. [In Korean with English abstract.]

Suzuki, K. & Adachi, M. 1994. Middle Precambrian detrital monazite and zircon from the Hida gneiss on Oki-Dogo Island, Japan: their origin and implications for the correlation of basement gneiss of Southwest Japan and Korea. *Tectonophysics* **235**, 277–92.

Taylor, S. R. & McLennan, S. M. 1985. *The Continental Crust: its Composition and Evolution.* Oxford: Blackwell.

Tsusue, A., Mizuta, T., Tamai, T., Kim, S. W. & Ohyoshi, A. 1986. Korean granitic rocks: trace element evidence regarding their differentiation: 1. Ba, Sr, and Rb relationships. *Mining Geology* **36**, 339–50.

Turek, A. & Kim, C.-B. 1995. U-Pb zircon ages of Mesozoic plutons in the Damyang-Geochang area, Ryeongnam massif, Korea. *Geochemical Journal* **29**, 243–58.

Turek, A. & Kim, C.-B. 1996. U-Pb zircon ages for Precambrian rocks in the southwestern Gyeonggi massif, Korea. *Geochemical Journal* **30**, 231–49.

Wolf, M. B. & London, D. 1994. Apatite dissolution into peraluminous haplogranitic melts: an experimental study of solubilities and mechanisms. *Geochimica et Cosmochimica Acta* **58**, 4127–45.

Wu, F.-y., Jahn, B.-m., Wilde, S. A., Lo, C.-H., Yui, T.-F., Lin, Q., Ge, W.-c. & Sun, D.-y. 2003. Highly fractionated I-type granites in NE China (II): isotopic geochemistry and implications for crustal growth in the Phanerozoic. *Lithos* **67**, 191–204.

Yin, A. & Nie, S. Y. 1993. An indentation model for the North and South China collision and the development of the Tan-Lu and Honam fault systems, eastern Asia. *Tectonics* **12**, 801–13.

Yun, H. S. & Lee, D. S. 1986. Petrochemical study on the Precambrian granitic rocks in the basement area of Hambaeg basin. *Journal of the Korean Institute of Mining Geology* **19**, 35–55. [In Korean with English abstract.]

Zen, E. 1986. Aluminium enrichment in silicate melts by fractional crystallization: some mineralogical and petrographic constraints. *Journal of Petrology* **27**, 1095–117.

YONG-JOO JWA, Department of Earth and Environmental Sciences, Gyeongsang National University, Jinju 660–701, Korea.
e-mail: jwayj@gsnu.ac.kr

MS received 28 November 2003. Accepted for publication 24 August 2004.

Transactions of the Royal Society of Edinburgh: Earth Sciences, **95**, 199–214, 2004

Lithium isotopic compositions of the New England Batholith: correlations with inferred source rock compositions

Colleen J. Bryant, Bruce W. Chappell, Victoria C. Bennett and Malcolm T. McCulloch

ABSTRACT: A strong correlation exists between the Li isotopic compositions of Carboniferous-Triassic granites from the New England Batholith, and the previously inferred involvement of sedimentary and mantle/infracrustal source components. Isotopically (Nd and Sr) juvenile, low-K, Cordilleran I-type granites of the Clarence River supersuite have δ^7Li = +2·2 to +8‰ similar to those of arc magmas, the inferred source of these granites (Bryant *et al.* 1997). Isotopic variability within this supersuite probably arises from heterogeneity within primary mantle-derived magmas, combined with subsequent modifications through interactions with crustal materials. Oxidised, high-K granites of the Moonbi Supersuite have more homogenous and slightly lighter Li isotopic compositions (δ^7Li = +1·9 to +4·2‰). The observed range of values lies within the range of arc magmas, and is consistent with partial melting of arc shoshonites within the crust (cf. Chappell 1978) or the involvement of high-K mantle-derived magmas (cf. Shaw & Flood 1981; Landenberger & Collins 1998). S-type granites of the Bundarra (δ^7Li = −0·1 to +2·1‰; average = +1·3‰; n = 6) and Hillgrove supersuites (δ^7Li = +0·4 to +1·7‰; average = +0·8‰) define a narrow range of isotopic compositions which are, overall, lower than those observed in NEB I-type granites or generally observed in primary arc magmas. Their isotopic compositions are equivalent to those typically observed in shales (primarily δ^7Li = −3·2 to +2·0‰; Moriguti & Nakamura 1998; Teng *et al.* 2004). No difference is evident in the isotopic compositions of the two S-type supersuites despite inferred differences in the degree of weathering experienced by the sedimentary protolith, or differences in mineralogy of the granites. Granites of the Uralla Supersuite, which have been have formed from mixtures of local meta-igneous and meta-sedimentary components, span a broad range of values (δ^7Li = −1·3 to +3·9‰) which overlap with both the sediment-poor New England Batholith I-type intrusions of the Clarence River and Moonbi supersuites, and the S-type granites of the Bundarra and Hillgrove supersuites. Lower δ^7Li values primarily occur in lower-K plutons from the northern portion of the Uralla Supersuite.

Overall, anatexis and magma differentiation do not appear to contribute to significant fractionation of Li isotopes relative to the inferred source components. However, subtly lower δ^7Li values, evident in the three leucogranites analysed herein, imply that subtle Li isotopic fractionation may occur in association with the exsolution of an aqueous fluid. Like most isotopic systems, the Li isotopic composition of rocks is not a definitive guide to source rock compositions, but given the results herein, the present authors suggest that it may play a very useful role in understanding crustal processes.

KEY WORDS: Crust, granite, lithium isotope, sediment.

Lithium (Li) is a rather enigmatic element whose isotopic variations have great potential to aid our attempts to constrain the nature of, and mechanisms involved in, the geochemical evolution of the Earth. Nevertheless, despite the recent dramatic increase in Li isotopic studies, we have little knowledge about the nature and behaviour of these isotopes in crustal systems. In this paper, the present authors examine the Li isotopic variability within the diverse I- and S-type granites of the New England Batholith (NEB), of eastern Australia, to ascertain if, and to what extent, the Li isotopic composition of these granites can be correlated with their previously inferred igneous and/or sedimentary source components.

The rationale for this project is as follows. Li is moderately incompatible during partial melting and fractional crystallisation, and hence, is enriched in the crust relative to the mantle. In addition, there is no apparent isotopic fractionation of Li during fractional crystallisation of mantle-derived magmas at high temperatures (>1050°C; Tomascak *et al.* 1999a). If this holds for all mantle-derived materials, then the net flux from the mantle to the crust must lie within the spectrum of mantle-derived magmas. Fresh MORB typically have δ^7Li = +3·4 to +5·2‰ (Chan *et al.* 1992; Moriguti & Nakamura 1998; Tomascak & Langmuir 1999; Elliott *et al.* 2002), similar to those observed in representative upper mantle peridotites (δ^7Li = +4·4 to +5·2‰; Brooker *et al.* 2000; Chan *et al.* 2002a). The Li isotopic signatures of both ocean island basalts and arc lavas are more variable. For example, Hawaiian lavas have δ^7Li = +2·9 to +5·7‰ (Tomascak *et al.* 1999a; Chan & Frey 2003; Pistiner & Henderson 2003), whereas lavas with a HIMU component have δ^7Li = +5 to +8‰ (Tomascak & Langmuir 1999; Jeffcoate & Elliott 2003; Nishio *et al.* 2003). Arc lavas have δ^7Li = +1 to +11‰, although the majority have isotopic compositions similar to other mantle-derived magmas (average arc lava = +4; n = 119; Moriguti & Nakamura 1998; Tomascak *et al.* 2000 2002; Chan *et al.* 2002a; Bouman *et al.* 2004).

Overall, the crust is isotopically lighter than the mantle-derived components from which it is ultimately derived because of Li isotopic fractionation within the crust (e.g. Teng *et al.* 2003, 2004). Processes likely to contribute to Li isotopic fractionation in the crust include weathering (Huh *et al.* 1998, 2001; Pistiner & Henderson 2003), sediment diagenesis (cf. Zhang *et al.* 1998; Chan & Kastner 2000), and dehydration reactions during metamorphism (e.g. Teng *et al.* 2003). During weathering, ^{6}Li is preferentially retained within the clay fraction, with the heavier isotope being carried by groundwaters into the rivers (δ^7Li=+6 to +32‰; Huh *et al.* 1998), and subsequently, the world's oceans (δ^7Li=~+31‰; Chan & Edmond 1988; You & Chan 1996; Huh *et al.* 1998; Moriguti & Nakamura 1998a; James & Palmer 2000; Nishio & Nakai 2002). Consequently, shales typically have relatively high Li concentrations, and tend to have lighter Li isotopic compositions (δ^7Li= − 3·1 to +3·4‰; average= − 1·2; n=13; Moriguti & Nakamura 1998; Teng *et al.* 2004) than mantle-derived magmas, although there is some overlap in their isotopic compositions.

The question considered in the present study is, if shales are typically isotopically lighter than mantle-derived rocks, are these differences still evident in S- and I-types granites which are known to be derived from the partial melting of sedimentary and igneous (mantle and/or crustal) components, respectively, and is there any blurring between the Li isotopic signatures in I-type granites which have incorporated significant sedimentary materials? To address these questions, the authors have analysed the Li isotopic compositions of a diverse group of granites from the NEB, which range from S-types derived from weathered materials, to granites with transitional I–S-type characteristics, to high-K granites and juvenile, Cordilleran-type I-type intrusions.

1. New England Batholith

The NEB (Fig. 1) comprises all of the granites of the southern New England Orogen (NEO) in eastern Australia. These granites intrude a collage of terranes which were generated in response to Devonian-Triassic intraoceanic-continental marginal arc activity (Flood & Aitchison 1988). The granites range in age from Carboniferous to Triassic, with almost all being ascribed to one of five major supersuites based on distinctive petrological, geochemical and isotopic characteristics (Shaw & Flood 1981). The latter ultimately reflect differences in the source components and processes contributing to the petrogenesis of individual intrusions.

1.1. Hillgrove Supersuite (S-type)

The granites of the Hillgrove Supersuite are the oldest in the NEB (303 Ma; Kent 1994; Landenberger *et al.* 1995). They consist of blue-grey, moderately to intensely deformed granodiorites and monzogranites which in several cases, are associated with well-developed regional aureoles and migmatites. As is typical of S-type granites, these rocks are strongly reduced, containing white K-feldspar, ilmenite and locally graphite, and have elevated δ^{18}O (>10‰; O'Neil *et al.* 1977). They also have lower CaO, Na_2O and Sr than the NEB I-type granites, consistent with the destruction of feldspar during surficial weathering of their source components. However, many other features typical of S-types are poorly developed and not strongly peraluminous. Excess alumina is primarily hosted within aluminous russet red biotite, and very rarely in almandine-rich garnet. Owing to their lower peraluminosity, there is no enrichment in P_2O_5 at higher silica contents (cf. Chappell 1999), such that the granites define 'I-type trends' on P_2O_5 Harker diagrams (Fig. 2). Given these characteristics, and the relatively juvenile Sr and Nd isotopic characteristics (typically initial ^{87}Sr/^{86}Sr=0·705–0·7065; Hensel *et al.* 1985; Fig. 3), the Hillgrove Supersuite has been interpreted to be derived through partial melting of immature sediments (greywackes) shed from the neighbouring volcanic arc (e.g. Landenberger & Collins 1995).

1.2. Bundarra Supersuite (S-type)

The Bundarra Supersuite incorporates an almost continuous, arcuate mass of biotite-bearing, typically coarse-grained and porphyritic (K-feldspar) S-type granites along the western portion of the batholith. These granites are compositionally restricted, high in silica (typically >71 wt.% SiO_2), reduced (white K-feldspar; ilmenite; δ^{18}O >12‰; O'Neil *et al.* 1977) and moderately peraluminous. Excess alumina is reflected in the occurrence of cordierite, muscovite and, less commonly, garnet. The higher peraluminosity of the melt is consistent with elevated P_2O_5 contents at high silica (Fig. 2). These petrological and geochemical features suggest that the sedimentary components involved in the petrogenesis of the Bundarra Supersuite underwent greater degrees of weathering relative to their Hillgrove Supersuite counterparts. The comparably low initial ^{87}Sr/^{86}Sr ratios (0·7048–0·7062; Flood & Shaw 1977; Hensel *et al.* 1985) and high initial ε_{Nd} values (− 0·2 to − 0·9; Hensel *et al.* 1985) together with low Nb concentrations indicate that these sedimentary materials were themselves derived through rapid reworking of the early, arc crust (e.g. Shaw & Flood 1981).

1.3. Clarence River Supersuite (I-type)

The Clarence River Supersuite incorporates a series of late Carboniferous-Permian, amphibole-biotite ± clinopyroxene ± orthopyroxene-bearing, calcic and sodic, predominantly tonalitic, granodioritic and dioritic intrusions which are characterised by low abundances of large ion lithophile elements, Th, U, Pb and high-field-strength elements. These granites are isotopically juvenile, having typical initial ^{87}Sr/^{86}Sr=0·703 to 0·704 and ε_{Nd}=+6 to +2 (Hensel *et al.* 1985; Bryant *et al.* 1997). These granites are most similar to those occupying the somewhat younger Cordilleran belts of North and South America (e.g. the western Peninsular Ranges Batholith and the Coastal Batholith). Intrusions of the Clarence River supersuites are probably formed as a direct product of arc magmatism and/or by partial melting of juvenile arc-derived materials (Bryant *et al.* 1997).

1.4. Moonbi Supersuite (I-type)

The Moonbi Supersuite (MSS) granites are strongly oxidised, hornblende-biotite (± augite)-bearing quartz monzonites, monzogranites and leucogranites which occur in the northern and southern extensions of the NEB. These intrusions are characterised by pink K-feldspar megacrysts, titanite and magnetite, and generally have elevated K_2O, Rb, Sr, Ba, Th, U, P_2O_5 and Pb, although some regional variations in granite geochemistry are evident within the supersuite. The granites analysed herein derive from the southern portion of the batholith, where the distinctive features outlined above are best developed. These granites have low Y and HREE and steep REE patterns, indicating the involvement of garnet at some point during their petrogenesis (Chappell 1978). Chappell (1978) suggested that the southern MSS near Tamworth was derived by partial melting of high-K source rocks (e.g. shoshonites) within the crust. Alternatively, it has been proposed that the high-K characteristics of these magmas derive from a potassic mantle-derived magma (e.g. shoshonite

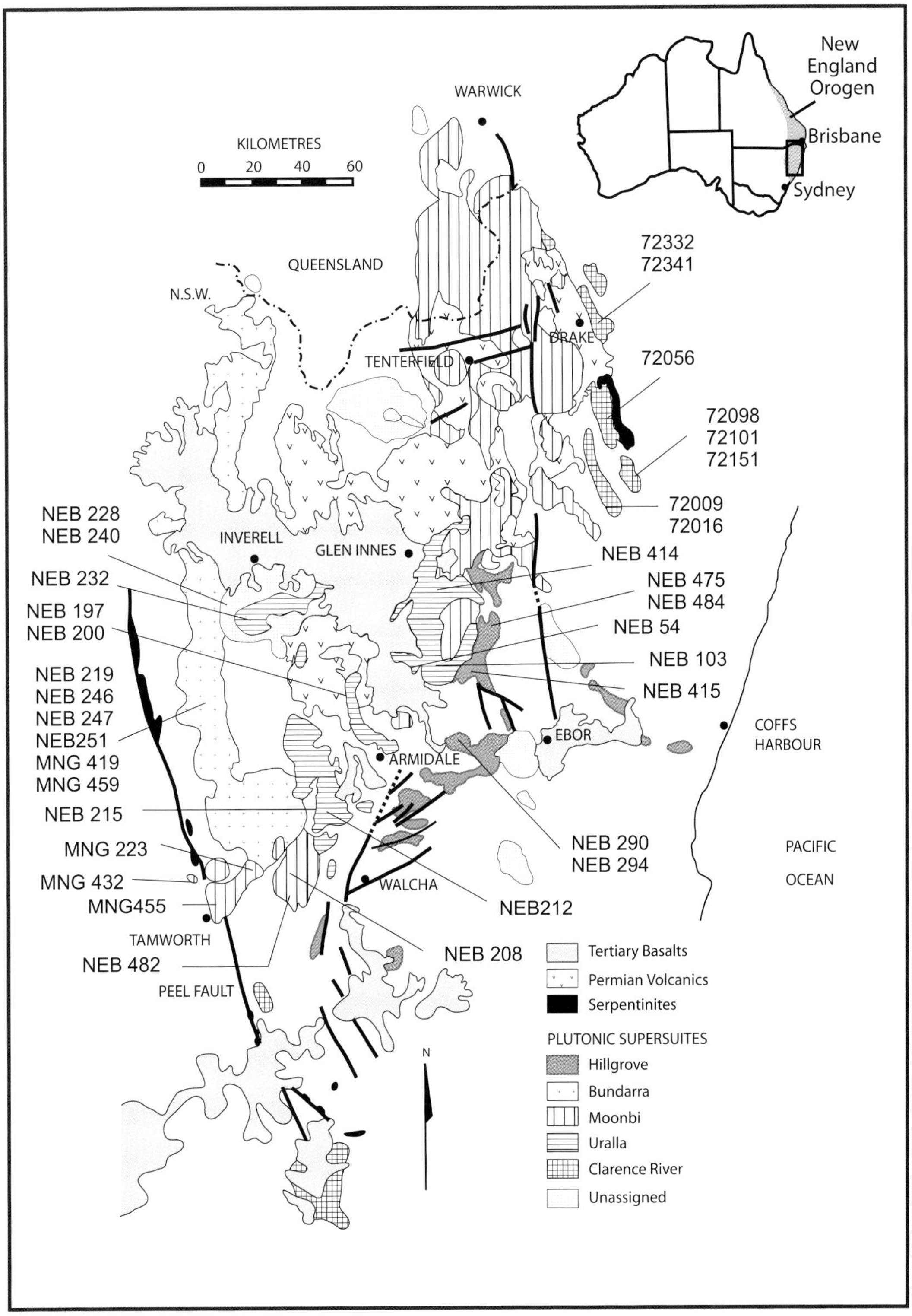

Figure 1 Map of the New England Batholith in eastern Australia (modified after Shaw & Flood 1981), illustrating the location of samples analysed for Li isotopic analysis. The labels indicate general location only. Refer to Table 1 for exact locations.

or lamprophyre) (Shaw & Flood 1981; Landenberger & Collins 1998). Irrespective of which of these models is selected, the sedimentary contribution is inferred to be low.

1.5 Uralla Supersuite (I-type)

The Uralla Supersuite consists of a series of compositionally diverse, hornblende-biotite ± clinopyroxene ± orthopyroxene-bearing I-type intrusions. Overall, these intrusions have many major and trace element characteristics which are intermediate between those of the high-K Moonbi and low-K Clarence River supersuites. The distinguishing features of this supersuite include lower redox states (lower Fe_2O_3/FeO; $\delta^{18}O$ typically >9; O'Neil *et al.* 1977), lower normative diopside and higher initial $^{87}Sr/^{86}Sr$ ratios (Shaw & Flood 1981). Shaw & Flood (1981) argued that the Uralla Supersuite was derived from both meta-igneous and meta-sedimentary components, plausibly representing combinations of Moonbi, Clarence River and S-type components.

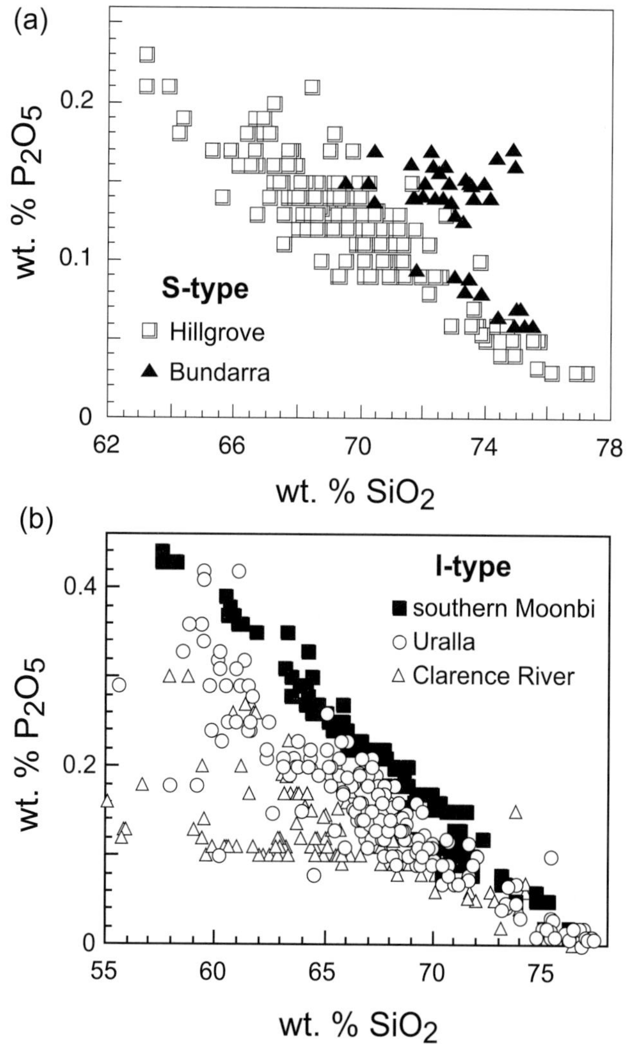

Figure 2 Scatterplots of P_2O_5 against SiO_2 in S- and I-type supersuites of the New England Batholith. The linear decrease in P_2O_5 with increasing silica in the Hillgrove Supersuite is similar to that observed in the I-type granites, and reflects their lower peraluminosity relative to those of the Bundarra Supersuite. The more peraluminous content of the Bundarra Supersuite probably results from the more extensive weathering of the source components.

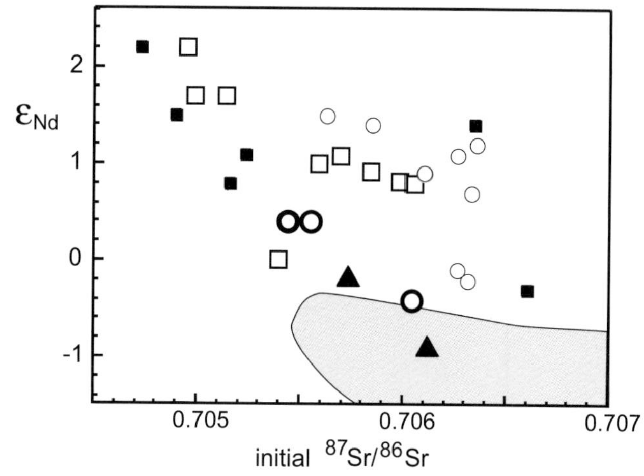

Figure 3 Scatterplot of initial ε_{Nd} versus initial $^{87}Sr/^{86}Sr$ for granites of the Hillgrove, Bundarra, Moonbi and Uralla supersuites. Data from Hensel *et al.* (1985), recalculated using the ages of Dirks *et al.* (1993), Shaw & Flood (1993) and Landenberger *et al.* (1995). The field is defined by typical sedimentary components from east of the Peel Fault (Zone B sediments of Hensel *et al.* 1985) recalculated to the composition at 260 Ma, the principal age of I-type granites in the New England Batholith.

1.6. Granite Summary

In a broad sense, the granites of the NEB may be viewed as a continuum of granite types ranging from juvenile, mantle-derived/infracrustal I-type granites in which there was minimal sediment input through to genuine S-type granites, which are likely to have been derived solely from reworked sedimentary material. Overall, the supracrustal contributions, or character, of the granites increases in the following order: Clarence River≤Moonbi<Uralla<Hillgrove≈Bundarra (e.g. Shaw & Flood 1981). This continuum does not imply a petrogenetic relationship between the granite supersuites, or endorse a specific petrogenetic model. Sedimentary components may be incorporated via a number of different mechanisms: (1) granites may be derived from a mixed source region; (2) the signature may be pre-protolithic (e.g. the parental arc magma that formed the granite protolith may have incorporated sedimentary components); (3) mixing of magmas derived from mantle and crustal source components during granite magma genesis; or (4) through contamination of granite magmas during differentiation within the crust. It should be noted that, even if one were to advocate a magma-mixing model in the petrogenesis of the I-type granites, a different mantle-derived magma would be required in the petrogenesis of each of the three supersuites (Landenberger & Collins 1998) and, in detail, within the supersuites.

The diversity of granite types evident within the NEB provides an excellent opportunity to assess the correlation between Li isotopic compositions and previously inferred supracrustal inputs.

2. Analytical methods and isotopic data

Fresh interior granite samples were broken into small chips between tungsten carbide plates. Powders were produced using a tungsten carbide puck and ring mill. A ~50-mg powder aliquot was then dissolved using standard HF and HNO_3 dissolution methods. Lithium was then isolated using cation-resin-based chromatographic techniques, as outlined in Bryant *et al.* (2003) for sea water. Samples were then diluted in 2% nitric acid, to a concentration of 100 ng g^{-1}.

The Li isotopic analyses were conducted using the ThermoFinnigan Neptune, multi-collector ICP-MS, located at the Research School of Earth Sciences of the Australian National University. Samples were introduced via a Glass Expansion (GE) Micromist 50 quartz nebuliser attached to a modified Scott's double-pass spray chamber. ^6Li and ^7Li were measured simultaneously in static mode in the outmost faraday cups (L4 and H4). The plasma was run at 800 W (warm plasma conditions) to minimise background interferences (Bryant *et al.* 2003). An on-peak measurement of 2% nitric acid was conducted prior to each analysis as a measure of the blank. The blank corrections at 800 W are negligible and, in general, do not alter the measured composition. Each sample analysis was bracketed (before and after) by a measurement of a 100 ng g^{-1} solution of the NIST L-SVEC Li_2CO_3 standard ($^7Li/^6Li = 12{\cdot}1163 \pm 0{\cdot}0098$; Moriguti & Nakamura 1998b) to correct the effects of machine mass bias.

Individual analyses of both the L-SVEC solution and the unknowns consisted of two blocks of 20 cycles, where each cycle contains 20 1-s integrations. Samples were run in blocks during which no adjustment was made to the tuning. Where possible, a purified sea water sample was run as an unknown

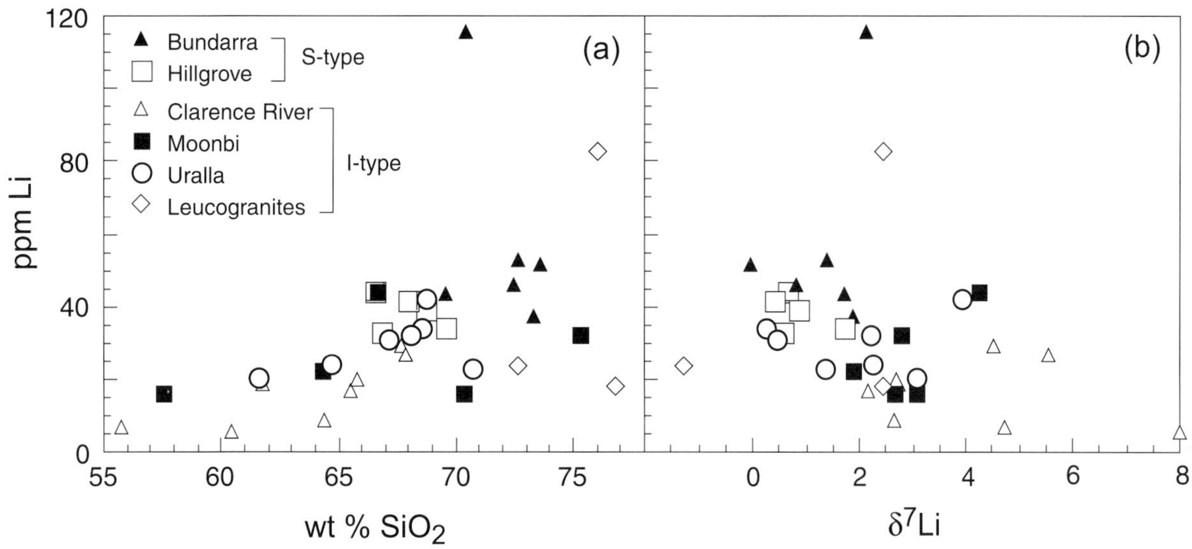

Figure 4 (a) Variation in Li concentrations as a function of SiO_2 in NEB I- and S-type granites, and (b) variation in Li isotopic composition as a function of Li concentrations.

at the beginning and end of each block. Between one and three repeat analyses (same solution) were conducted on the majority of samples over a 6-month interval.

Alterations to the instrumental setup, primarily the gas-flow parameters, generated some variability in measured sea water and associated sample analyses [$\delta^7Li_{seawater} = +30{\cdot}8 \pm 1{\cdot}5$‰ ($2\sigma$ SD); $n=24$ analyses on ~10 replicate sea water separations]. To ensure compatibility between the different analytical sessions, a second-order correction was applied using a linear correlation between sea water analyses obtained within each run and L-SVEC. The δ^7Li value for sea water was set at $\delta^7Li=30{\cdot}4$‰ (value from Bryant *et al.* 2003), which was marginally lower than the average value obtained in the present study. The overall correction was minor, with the mean Li isotopic compositions of the corrected and non-corrected results being very similar. Both sets of data are presented in Appendix 1. The final corrected results are reported relative to the NIST L-SVEC standard [i.e. $\delta^7Li = (^7Li/^6Li_{sample}/^7Li/^6Li_{standard} - 1) \times 1000$], and are plotted in Figures 4 and 5.

The present authors have plotted the 1σ errors for each of samples analysed in Figure 5, based on an average of between two and four analyses of the same solution. Samples for which only one analysis was conducted are represented as a point in Figure 5. The error on the samples can be assumed to be ~1·0‰. International rock standards BHVO-1, BHVO-2 and G-2 were run as unknowns to independently monitor the accuracy of the data.

The Li and Cs concentrations were determined by quadrupole ICP-MS (Agilent 7500), following a standard $HF–HNO_3–HCl$ digestion ~50 mg rock powder. The concentrations were determined using a bracketing technique, relative to BCR-2. Three replicate analyses of BHVO-2 gave values of $5{\cdot}3 \pm 0{\cdot}1$ (2σ) ppm Li. The 2σ standard deviations of the measured Li concentrations in the granite samples were <5%.

Major and trace element data for granites analysed for Li isotopic compositions are presented in Table 1. Major element data were obtained by X-ray fluorescence (XRF) spectrometry using fused discs (modified technique of Norrish & Hutton 1969). The trace elements (excluding Cs and Li) were measured on pressed powder pellets using the XRF methods described by Chappell (1991).

3. Results

3.1. Lithium concentrations

Li concentrations in NEB granites vary over almost two orders of magnitude (6–116 ppm; Fig. 4). Overall, NEB S-type granites contain higher Li (average=49 ppm; $n=11$) than their I-type counterparts (average=24 ppm; $n=20$; excludes leucogranites). With the exception of a single analysis from the Bundarra Supersuite, the range of Li concentrations in the two S-type supersuites is relatively small (33–53 ppm), lying within the range of post-Archaean shales (30–110 ppm; e.g. Moriguti & Nakamura 1998; Teng *et al.* 2004).

Although some overlap exists, the average Li concentrations of the NEB I-type supersuites decreases in the order Uralla>Moonbi>Clarence River. The Li concentrations in the most mafic Clarence River samples are equivalent to those observed in arc lavas from active subduction zones, whereas the abundances in many Uralla granites overlap with those of the S-type Hillgrove Supersuite. Overall, concentrations of Li within individual I-type supersuites increase with SiO_2 (Fig. 4), similar to that observed within volcanic systems (e.g. Trompetter *et al.* 1999; Bouman 2004).

Noticeably higher Li abundances (>80 ppm) are observed in two NEB samples. Extreme enrichment in Li concentrations may arise through extended fractional crystallisation of quartz and feldspar at minimum melt compositions (i.e. following the cessation of biotite, amphibole, muscovite and cordierite crystallisation). For example, Teng *et al.* (2004) observed four-fold enrichment in Li (from 44 to 187 ppm Li) with extended fractional crystallisation of the S-type Koetong Suite. Based on both Rb concentrations and Rb/Sr ratios, this is a probable explanation for the higher Li concentrations in NEB240, a leucogranite from Gilgai. The origin of Li enrichment in sample NEB246 is less obvious. This sample is petrographically distinct from other Bundarra samples analysed herein, consisting of large (0·5–4 cm) K-feldspar, biotite and quartz phenocrysts in a fine-grained matrix of quartz-K-feldspar-plagioclase and biotite. Unlike the leucogranite discussed above, this sample does not possess the extreme enrichment in Rb that is characteristic of highly fractionated, minimum melt granites. Nor does it contain an excessive abundance of restitic or magmatic biotite, muscovite or cordierite, which are likely to be the primary hosts for Li in these granites. An alternative

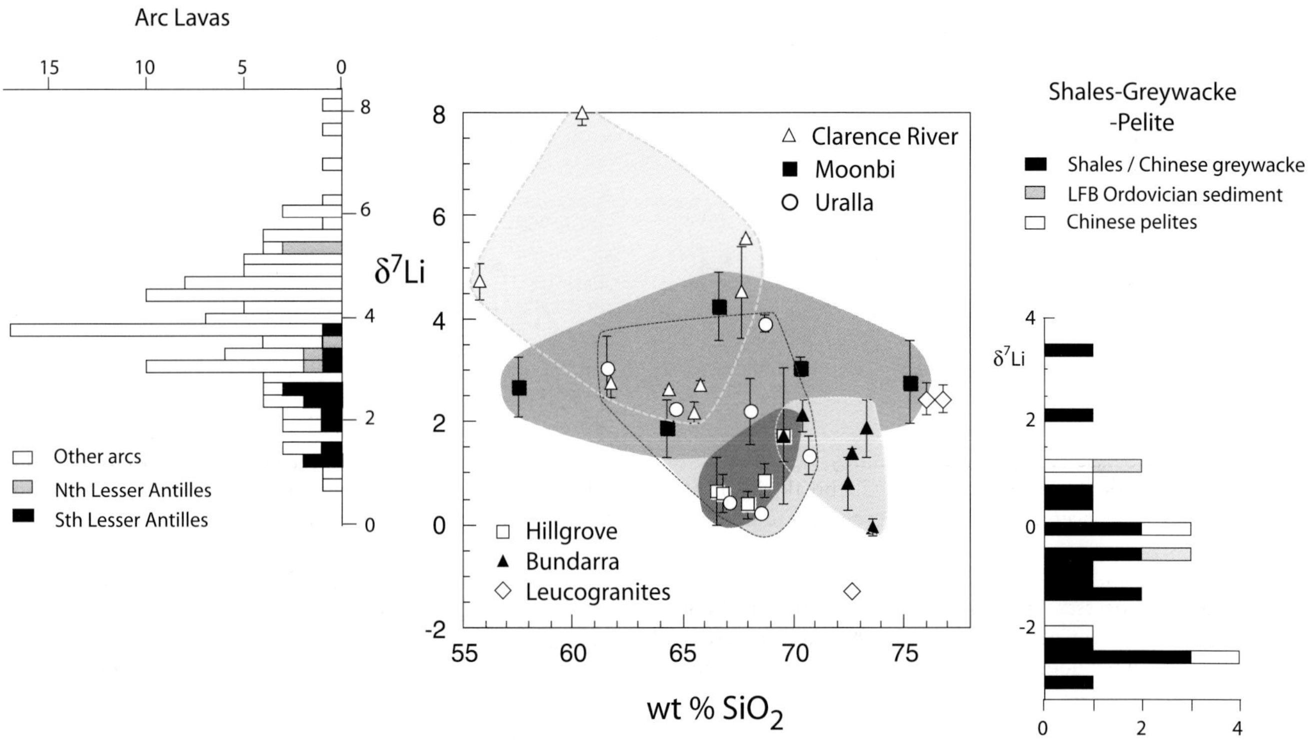

Figure 5 Variation in Li isotopic composition as a function of SiO_2. Error bars represent the 1σ standard deviation of multiple analyses of the same sample replicate. Fields for individual supersuites are defined by the 1σ error (excluding one sample from the Hillgrove Supersuite, which has a very large error). Plotted on the left is the frequency distribution of δ^7Li values in arc lavas derived from Moriguti & Nakamura (1998a), Tomascak *et al.* (2000, 2002), Chan *et al.* (2002a) and Bouman (2004). Lavas from the northern (Saba to Nevis) and southern Lesser Antilles (Monttserrat to Grenada) are plotted as a separate subsets. Plotted on the right is the frequency distribution of shales (Moriguti & Nakamura 1998a; Teng *et al.* 2004), as well as composites of greywackes and pelitic from China (Teng *et al.* 2004). Ordovician sediments from the Lachlan Fold and the Chinese pelitic composites are plotted separately.

explanation may include the involvement of a Li-rich fluid. However, there is no evidence of hydrothermal alteration within this sample or petrographic features indicative of volatile phase exsolution (e.g. veins or miarolitic cavities). The latter are evident in sample NEB240, the leucogranite from Gilgai.

3.2. Lithium isotopic composition

The NEB granites are extremely diverse in terms of their Li isotopic compositions, ranging from δ^7Li = − 1·3 to +8‰ (Figs 4 & 5). Although there is no direct correlation between the Li concentration and Li isotopic composition of NEB granites, the supersuites which have the highest Li concentrations also, on average, have the lowest δ^7Li values and *vice versa* (Figs 4 & 5).

The Hillgrove Supersuite has δ^7Li ranging from +0·4 to +1·7‰ (average = +0·8‰; n=5). Similar compositions are evident within the Bundarra Supersuite (δ^7Li = − 0·1 to +2·1‰; average = +1·3‰; n=6), despite inferred differences in the source rock compositions. The average Li isotopic composition of NEB S-type granites is δ^7Li = 1·2 ± 1·1(2σ 26‰).

In contrast, I-type granites (including the I-type leucogranites from Gilgai and Yellow Gap) have δ^7Li = − 1·3 to +8‰. The greatest heterogeneity is evident within the Clarence River Supersuite (δ^7Li = +2·2 to +8·0‰; average δ^7Li = +4·1‰, n=8), and more specifically, within the Jenny Lind intrusion (δ^7Li = +2·7 and +8·0‰, n=2). The Towgon Grange intrusion, which is chemically most similar to Jenny Lind, has δ^7Li = 2·7 to 4·7‰ (mean = 3·4‰, n=3). The two analyses of the Kaloe intrusion are both relatively high, and within error (δ^7Li = +4·5 to +5·6‰). A single analysis of the high-K Dumbudgery Creek Granodiorite, which on the basis of geochemical, and Nd and Sr isotopic data can be demonstrated to have assimilated more evolved upper crustal components (Bryant *et al.* 1997; see below) has relatively low δ^7Li (2·2‰). The overall range and average of the Clarence River Supersuite are similar to those reported for arc lavas (δ^7Li = +1 to +11‰; average = +4; n=119; Moriguti & Nakamura 1998; Tomascak *et al.* 2000, 2002; Chan *et al.* 2002a; Bouman 2004).

Granites of the MSS also overlap with arc lavas, although with values of δ^7Li = +1·9 to +4·2‰, they fall within the lower end of the spectrum. On average, these granites have slightly lower δ^7Li (mean = 2·8‰; n=5) and are more homogeneous than those of the Clarence River Supersuite.

Uralla Supersuite granites define a complete spectrum between the values observed in the MSS through to those observed in the S-type granites (δ^7Li = − 1·3 to +3·9‰; average = +1·5‰; n=7). Granites of the Wards Mistake (δ^7Li = 0·2‰) and Llangothlin (δ^7Li = 1·4‰) groups, which are characterised by lower K_2O, Rb and Th, for example, also have generally lower δ^7Li than those of the Mafic Uralla (δ^7Li = 3·1‰) and Mount Duval Groups (δ^7Li = +2·2 and +3·9‰). The δ^7Li values for the latter groups overlap with the heaviest values observed for the MSS.

4. Discussion

4.1. S-type granites

The overall higher Li concentrations in NEB S-type granites relative to their I-type counterparts are consistent with the preferential enrichment of Li during weathering of sedimentary source components. Virtually all post-Archaean shales analysed to date have Li concentrations in excess of the value calculated for bulk upper crust (~30 ppm; see Teng *et al.* 2004).

Li is concentrated in the clay fraction in soils and shales. At higher temperatures and pressures (i.e. in metamorphosed rocks), Li is primarily incorporated into micas (e.g. biotite, muscovite and chlorite), cordierite, garnet ± tourmaline (e.g. Bea *et al.* 1994; Visser *et al.* 1994; Icenhower & London 1995; Pereira & Shaw 1996). Studies of coexisting migmatites, leucogranites and restite from the low-pressure Peña Negra complex in Spain (Bea *et al.* 1994) indicate that, under small degrees of melting, a significant portion of the Li in the rocks resided in the restitic fraction, with Li concentrations in minerals decreasing in the order cordierite (454–465 ppm)> biotite (110–166 ppm)>garnet (12–46 ppm)>K-feldspar (5–8 ppm)>plagioclase (1–4 ppm). Significantly, with increasing grade, the major repositories of Li will break down at different times. Li concentrations of cordierites in the Bamble sector, Norway, decrease significantly across the amphibolite–granulite transition, reflecting progressive dehydration and decarbonation processes, possibly related to partial melting (Visser *et al.* 1994). Icenhower & London (1995) indicated that muscovite may be the dominant source of Li at lower temperatures. The release of Li from biotite possibly only occurs at higher temperatures. In S-type granites, Li is primarily hosted in micas, with the $D(Li)^{biotite}$ exceeding the $D(Li)^{muscovite}$ by as much as 4:1 (Carron & Lagache 1972; Papike *et al.* 1984; Kretz *et al.* 1989; Icenhower & London 1995). Experiments conducted at 600–750 °C at 200 MPa (H_2O), indicate the partition coefficients of Li between biotite and melt increase from 1·0 to 1·7 with decreasing temperature (Icenhower & London 1995).

The implication of the above is that, if the partition coefficients of ^{7}Li and ^{6}Li differ between minerals relative to melts ± fluid, there is ample opportunity to fractionate the Li isotopic composition of the granites relative to the source material. Whilst Li isotopic fractionation between minerals and fluids has been observed at temperatures >500 °C (e.g. Lynton *et al.* 1999; Teng *et al.* 2003; Zack *et al.* 2003), the possibility of inter-mineral and mineral-melt Li isotopic fractionation is unexplored. Nevertheless, the anecdotal evidence suggests that the overall degree of fractionation during anatexis and subsequent differentiation is small [i.e. equivalent to or less than the analytical uncertainty of ± 1‰] (2σ; Teng *et al.* 2004). Notably, Teng *et al.* (2004) observed identical compositions in mafic and felsic end-member compositions of the Koetong Suite from the Lachlan Fold Belt despite a four-fold increase in Li concentrations (44–187 ppm). Moreover, these granites have identical Li isotopic compositions to associated sedimentary rocks (Teng *et al.* 2004). The present authors similarly observed a close correspondence between the Li isotopic composition of NEB S-type granites and typical shale compositions (δ^7Li= − 3·2 to +3·4‰, with 12 of the 13 samples having δ^7Li= − 3·2 to +2·0‰; Moriguti & Nakamura 1998; Teng *et al.* (2004) (Fig. 5). Furthermore, the Li isotopic composition of sample NEB246 (δ^7Li_{NEB246}= +2·1‰), which contains 116 ppm Li, is equivalent to that observed in other Bundarra samples with half or even one-third the amount of Li (Fig. 4b).

There is no detectable difference between the Li isotopic composition of the Hillgrove and Bundarra supersuites, despite the observed differences in mineralogy, geochemistry, and O and Nd isotopic compositions (Fig. 3; O'Neil *et al.* 1977; Hensel *et al.* 1985), or the inferred differences in the degree of weathering experienced by the sedimentary protoliths.

4.2. I-type granites: Clarence River Supersuite

The origin of the Clarence River Supersuite is intimately related to arc magmatism, with the granites representing either the differentiates of primary arc magmas and/or partial melts of an infracrustal source of that composition (Bryant *et al.* 1997). The Li isotopic values of Clarence River granites (δ^7Li= +2·2 to +8·0‰; average= +4·1‰; n=8) clearly lie within the broad spectrum of compositions observed in arc systems (δ^7Li= +1 to +11‰; average= +4; n=119) (Fig. 5). Therefore, it is plausible that the entire Li isotopic heterogeneity evident within the Clarence River Supersuite is derived from the arc magmas themselves. What is the origin of this variability in arc lavas?

This is a difficult question to answer since the behaviour of Li in arc systems is not particularly well understood. The selective enrichment in Li/Y in arc lavas relative to MORB and OIB provide conclusive evidence that Li is readily mobilised in slab-derived fluids (e.g. Moriguti & Nakamura 1998a). Given that the incorporation of sea water (δ^7Li= +31‰) during low-temperature alteration of oceanic crust at mid-ocean ridges leads to heavier isotopic compositions (δ^7Li ≈ +5 to +20‰; e.g. Chan *et al.* 2002b) relative to 'mantle' values, Li isotopes were initially expected to provide a monitor of the fluid-derived component within arc magmas, and potentially, further constrain the relative contributions from altered oceanic crust and sediments. Following this rationale, Moriguti & Nakamura (1998) argued that the significant decrease in both δ^7Li (+7·6 to +1·1‰) and Li/Y across the Izu arc with increasing depth to the Wadati-Benioff zone (150–210 km) reflected binary mixing between a fluid derived primarily from the altered oceanic crust, and to a lesser extent, overlying sediments (δ^7Li_{fluid}=9·7‰) and the mantle wedge (δ^7Li ≈ 0‰). However, given that the majority of MORB and ocean island basalts are in the range of δ^7Li= +3·4 to +5·2‰ (Chan *et al.* 1992; Moriguti & Nakamura 1998a; Tomascak & Langmuir 1999; Tomascak *et al.* 1999a; Elliott *et al.* 2002), respectively, the latter values are almost certainly too low for unmodified mantle wedge. Based on comparative studies of the south Sandwich (δ^7Li= +1·6 to +7·6‰) and southwestern Washington Cascades arc (δ^7Li= +2·5 to +4·0‰), Chan *et al.* (2001) concluded that the subduction component itself might have a large range of compositions, depending on the thermal state of the system. Tomascak *et al.* (2000) similarly attributed differences between the 10 Ma (δ^7Li= +3·9 to +11·2‰) and younger adakitic (δ^7Li= +1·4 to +4·6‰) arc lavas from Panama to differing thermal states of the subducting lithosphere.

Despite the large isotopic variability reported in arc lavas, both the mode and mean are similar to MORB, i.e. although most arc lavas are enriched in Li, many do not show any isotopic anomalies. In the case of the Cascades arc, Chan *et al.* (2001) proposed that the presence of young and hot subducting lithosphere may lead to a weak slab-derived Li isotopic signal, and hence, MORB-like values in the resultant arc magmas. Based on a study of serpentinite seamounts in the Mariana forearc, which had δ^7Li= − 0·4 to +9·8‰, Benton & Tera (2000) argued that ^{7}Li is preferentially lost from the shallow portions of the subducted slab during the initial phases of subduction, leading to the incorporation of an isotopically light subduction component within the source of Mariana arc magmas. Zack *et al.* (2003) similarly argued that the extremely light Li isotopic compositions evident in eclogites (δ^7Li down to − 10‰) result from isotopic fractionation during the initial stages of metamorphism. Tomascak *et al.* (2002) argued that the convergence of Li isotopic ratios with MORB values results from the sequestration of slab-derived Li in the subarc mantle before it reaches the zone of melting. Bouman (2004)

Table 1 Major and trace element data for samples analysed for Li isotopic compositions*

	Bundarra Supersuite						Clarence River Supersuite								Hillgrove Supersuite		
Unit:	Banalasta		Copeton			Roumalla	Dumb Ck	Jenny Lind		Kaloe		Towgon Grange			Hillgrove		Kooka.
No:	MNG419	MNG459	NEB246	NEB247	NEB251	NEB219	R72056	R72332	R72341	R72009	R72016	R72101	R72098	R72151	NEB290	NEB294	NEB475
Lat.:	30·81°S	30·8°7S	29·88°S	29·90°S	30·11°S	30·61°S	29·34°S	28·72°S	28·81°S	29·50°S	29·45°S	29·41°S	29·40°S	29·43°S	30·54°S	30·54°S	30·06°S
Long.:	151·13°E	151·15°E	150·89°E	150·91°E	151·01°E	151·23°E	152·58°E	152·46°E	152·53°E	152·55°E	152·50°E	152·63°E	152·62°E	152·61°E	151·89°E	151·87°E	152·07°E
SiO_2	69·52	72·46	70·42	72·66	73·58	73·35	65·52	60·48	64·36	67·63	67·85	55·71	61·74	65·84	68·71	69·51	67·95
TiO_2	0·55	0·35	0·46	0·38	0·28	0·33	0·51	0·61	0·6	0·54	0·46	0·64	0·65	0·51	0·65	0·61	0·67
Al_2O_3	14·45	13·92	13·75	13·57	13·35	13·18	15·98	16·31	15·02	16·27	16·07	17·53	15·71	15·67	14·34	14·06	15·33
Fe_2O_3	0·45	2·47	2·81	2·44	2·09	1·86	2·26	2·64	2·34	1·35	1·09	2·77	2·21	1·98	4·14	3·72	4·44
FeO	3·01						1·81	3·48	2·65	1·63	1·55	4·8	3·98	2·23			
MnO	0·08	0·05	0·06	0·05	0·05	0·03	0·08	0·13	0·08	0·08	0·07	0·16	0·14	0·11	0·08	0·07	0·08
MgO	1·2	0·6	0·9	0·63	0·44	0·43	1·83	3·09	2·58	1·5	1·38	4·4	3·02	1·68	1·37	1·24	1·40
CaO	2	1·69	2·09	1·72	1·38	1·25	4·2	6·31	4·97	3·83	3·53	8·62	6·29	4·77	2·23	2·25	2·43
Na_2O	3·25	3·14	3·12	3·13	3·09	2·95	3·71	3·2	3·12	4·33	4·50	3·07	3·23	4·06	3·05	2·97	3·03
K_2O	3·98	4·24	4·19	4·32	4·37	5·01	2·87	1·85	2·69	2·05	2·12	0·89	1·83	1·74	4·03	4·25	3·41
P_2O_5	0·15	0·18	0·14	0·14	0·15	0·08	0·1	0·11	0·11	0·11	0·10	0·12	0·11	0·1	0·13	0·12	0·18
Li	43·8	46·0	116·0	52·7	51·5	37·1	16·7	5·9	8·7	29·1	26·8	6·7	18·7	19·8	39·0	34·5	41·9
Cs	13·5	13·4	25·4	16·1	15·4	11·2	3·3	1·6	3·0	3·8	11·2	1·1	2·8	3·1	15·4	11·0	10·3
Ba	600	540	570	495	420	505	410	320	330	370	370	148	270	285	625	635	619
Rb	157	185	197	193	204	187	109	53	81	65	80	20	51	47·5	167	172	128
Sr	186	158	117	111	101	105	295	300	281	343	317	366	251	295	145	135	154
Pb	22	21	14	21	19	19	10	6	4	10	11	5	6	5	21	21	21
Th	15·5	12·5	18·6	15·8	14	19	18·8	7	10	6·6	6·6	2·4	6·2	5·6	17·4	18	13
U	4	2·4	4·4	5·2	6·6	2·8	5·2	1·4	1·8	2·4	2·6	0·6	1·2	1·2	3·4	3·6	1
Zr	181	131	236	180	125	156	132	116	173	135	129	68	87	123	200	202	199
Nb	7·5	6·5	8·5	8	6·5	5	5·5	2	3·5	5·5	5·5	1·5	2·5	3	8	8	8
Y	33	36	38	36	33	27	18	18	24	16	17	16	21	19	32	32	38
La	25	21	29	22	17	24	20	13	18	14	13	9	12	14	25	25	27
Ce	58	45	66	54	42	62	44	29	44	33	27	19	28	29	62	63	54
Sc	13		10	8	6	6	13	26	20	8	7	33	26	14	16	13	16
V	51		30	22	16	18	85	143	108	40	35	193	146	70	59	52	74
Cr	19		23	11	6	5	8	27	34	13	10	38	21	8	30	25	39
Ni	9	4	5	4	3	3	6	9	14	9	6	19	12	4	13	9	10
Cu	14	7	7	7	5	7	10	21	78	3	4	57	84	9	18	14	14
Zn	67	43	75	45	43	32	38	59	34	50	45	70	60	42	71	61	68
Ga	18·4	18·2	16·2	16·2	16·4	15·6	16·8	15·6	15	16·4	16·6	17	14·6	14·8	17·2	16·4	17

Table 1 (Continued.)

	Hillgrove Supersuite			Moonbi Supersuite				Uralla Supersuite							Leucogranite		
Unit:	Kooka.	Tobermory	Bendem.	Congi Ck	Inlet	Moonbi	Walcha Rd	Glenore	Khatoun	Llangoth.	Mt Duval	Tingha	Uralla	Ward. Mist.	Yell. Gap	Gilgai	
No:	NEB484	NEB415	MNG223	NEB208	MNG432	MNG455	NEB482	NEB200	NEB215	NEB54	NEB197	NEB232	NEB212	NEB414	NEB103	NEB240	NEB228
Lat.:		30·21°S	30·88°S	30·91°S	30·94°S	30·96°S	30·99°S	30·21°S	30·64°S	30·12°S	30·33°S	29·95°S	30·54°S	30·14°S	30·15°S	29·85°S	29·94S
Long.:		152·01°E	151·22°E	151·29°E	150·91°E	151·11°E	151·39°E	151·55°E	151·42°E	151·7°E	151·53°E	151·21°E	151·43°E	151·89°E	151·82°E	151·12°E	151·03E
SiO_2	66·87	66·53	75·3	70·3	57·52	66·64	64·33	67·07	61·59	70·64	68·01	68·74	64·63	68·5	72·67	76·07	76·79
TiO_2	0·68	0·68	0·15	0·4	0·86	0·54	0·72	0·52	0·74	0·35	0·56	0·45	0·65	0·45	0·23	0·06	0·08
Al_2O_3	15·82	14·91	13·08	13·97	16·03	14·67	14·06	15·04	14·67	14·37	14·14	14·32	14·65	14·94	13·88	12·44	11·72
Fe_2O_3	4·67	0·91	0·28	2·53	2·06	3·79	4·67	0·87	5·87	0·56	1·04	2·96	4·67	0·87	0·8	0·9	1·11
FeO		3·51	1·06		3·91			2·14		2·02	2·65			2·38	0·82		
MnO	0·08	0·09	0·06	0·06	0·11	0·08	0·08	0·06	0·12	0·04	0·11	0·06	0·1	0·08	0·04	0·03	0·02
MgO	1·56	1·65	0·25	1·38	4·37	1·94	3·3	2·03	4·92	0·66	1·81	1·26	3·18	1·39	0·56	0·06	0·05
CaO	2·28	2·47	1·1	2·55	6·2	3·38	3·88	3·23	4·72	2·12	2·51	2·48	3·78	2·99	1·94	0·48	0·42
Na_2O	3·23	3·17	3·45	3·26	3·19	3·16	3·01	3·6	2·88	3·64	3·05	3·3	3·01	3·35	3·37	3·88	3·45
K_2O	3·4	3·64	4·75	4·33	3·26	4·59	4·64	3·63	3·03	4·15	3·86	4·62	3·63	4·22	4·56	4·6	4·72
P_2O_5	0·18	0·16	0·05	0·16	0·43	0·25	0·35	0·2	0·25	0·08	0·16	0·14	0·19	0·13	0·06	0·01	0·01
Li	33	43·9	32·3	16·3	16·4	44·3	22·6	30·9	20·3	22	32·3	42·1	24·6	34	23·5	82·9	18·2
Cs	8·7	8·8	13·9	23·3	10·4	23·2	14	19·8	8·3	12·3	8·8	32·2	9·4	8·3	13·7	34·8	12
Ba	551	580	420	565	1840	940	1050	720	810	770	565	685	650	730	605	18	6
Rb	145	143	288	210	184	245	209	154	115	160	159	235	143	184	169	476	330
Sr	178	181	160	339	909	412	572	645	345	192	188	220	262	233	173	8·5	5
Pb	20	20	49	46	27	50	42	15	20	29	23	26	20	20	28	43	40
Th	13	14·8	37	39·5	34	37·1	43·5	22	15·2	21·5	18	34·5	18	24	27	57	43
U	2	3·2	13·4	10·4	5·6	12·2	9·9	3·4	3·4	3·2	4·2	9·6	3·8	7	6·2	11·4	12
Zr	190	191	126	135	251	170	231	183	180	217	159	176	174	165	114	123	138
Nb	9	8·5	14	9	9·5	11·7	14	7·5	8·5	8·5	7·5	10	9	7	6·5	21	15
Y	33	33	26	13	15	17	24	22	23	35	28	30	22	38	19	231	89
La	22	27	19	28	39	30	37	37	26	33	24	30	27	31	28	49	37
Ce	52	85	41	57	82	60	83	81	63	74	59	71	62	93	62	132	99
Sc	17	17	2	8	22			11	22	12	14	10	17	12	5	2	13
V	79	71	11	43	140	89	101	65	105	24	60	42	77	58	21	–	–
Cr	49	34	2	37	127	64	129	47	276	5	53	35	141	21	6	–	–
Ni	11	13	1·5	14	47	17	36	26	68	2	15	10	37	4	2	–	2
Cu	18	18	1·5	47	17	6	11	12	24	0	13	5	19	2	2	–	–
Zn	75	81	35	42	48	44	54	42	79	45	78	42	69	54	37	17	52
Ga	18	18·2	16·2	15·8	17·4	16·7	17·9	18·8	16·8	15	16·6	16·2	17	16·2	13·8	20·6	20·6

*Abbreviations: (Dumb Ck) Dumbudgery Creek; (Kooka.) Kookabookra; (Bendem.) Bendemeer; (Congi Ck.) Congi Creek; (Ward. Mist.) Wards Mistake; (Langoth.) Langolthlin; (Yell. Gap) Yellow Gap.

similarly proposed a zone-refining process, where the slab-derived Li becomes equilibrated with the mantle wedge. Significantly, there does not appear to be a correlation between Li isotopes and geochemical vectors used to monitor the incorporation of sedimentary components via the subduction pathway (e.g. Tomascak *et al.* 2002; Bouman 2004).

Whilst the dynamics of Li in the subduction environment are complex, there are several observations worthy of note. Individual arc systems are potentially, although not necessarily, characterised by a large variation in Li isotopic compositions. Given that the Clarence River Supersuite falls within this range, it is possible that the entire variability within this supersuite derives from variability within the arc magmas themselves. Nevertheless, δ^7Li values $\leq$2‰ are relatively uncommon. For example, of the 119 analyses available in the literature, only nine have δ^7Li <2‰. Of these, three are from Izu-Bonin backarc volcanoes (Moriguti & Nakamura 1998a). The low values in these magmas may arise from isotopic fractionation of Li during the dehydration of subducted slabs (cf. Benton & Tera 2000). Given that backarc magmas are volumetrically insignificant compared to those generated at the volcanic front, these compositions cannot be regarded as being typical of arc magmas. A further three of the nine samples are from the Lesser Antilles, where there is some evidence that the lighter δ^7Li values are related to the involvement of terrigenous sedimentary components. Notably, the Lesser Antilles arc is characterised by marked along-arc variations in arc chemistry. The high Ba/La and low Th/Nb ratios in the Northern Lesser Antilles are indicative of an aqueous fluid-dominated subduction component, in which the contributions from sedimentary components are comparatively low (e.g. van Soest 2000). In contrast, southern Lesser Antilles lavas have low Ba/La, high Th/Nb (van Soest 2000), and highly evolved Nd, Sr and Pb isotopic characteristics (e.g. Davidson 1987; Thirlwall *et al.* 1996), indicative of higher sedimentary contributions. Whilst it is possible that these signatures were added via a subduction pathway, there is convincing evidence to suggest that the arc magmas interacted with sedimentary components in the arc crust (e.g. Davidson 1987; Smith *et al.* 1996; Thirlwall *et al.* 1996; van Soest *et al.* 2002). Significantly, southern Lesser Antilles lavas are isotopically lighter (δ^7Li=+1·3 to 3·7‰; average=2·4‰; n=13; Bouman 2004) than those from the northern Lesser Antilles (δ^7Li=+3·1 to +5·4‰; average=4·4‰; n=6; Bouman 2004; Fig. 5), and are biased toward lower values relative to other arc systems where sedimentary contributions are generally low (e.g. South Sandwich, δ^7Li=+3·0 to 6·2‰; average=4·34‰; n=10; Bouman 2004; Marianas, δ^7Li=+3·0 to 5·6‰; average=4·2‰; n=15; Bouman 2004; Central America, δ^7Li=+4·1 to 6·4‰; average=4·34‰; n=12; Chan *et al.* 2002a). Despite, the shift towards lower δ^7Li values, the southern Lesser Antilles lavas still lie within the spectrum of what are regarded as typical arc values (Fig. 5).

Following this logic, it is plausible that some Clarence River granites have also incorporated small volumes of isotopically lighter crustal components, and this may account for some of the lower δ^7Li values observed in this supersuite. There is independent evidence for this assertion. Based on higher concentrations of alkalis, Th, U and Pb, and more evolved Sr and Nd isotopic compositions, Bryant *et al.* (1997) argued that the Dumbudgery Creek Granodiorite underwent significant shallow-level crustal contamination. This pluton has the lowest δ^7Li value in the Clarence River Supersuite. Isotopic modelling of the Towgon Grange pluton indicates that the lowest δ^7Li value (2·7‰) can be generated by mixing 15% sediment (400 ppm Li, δ^7Li=+0·5‰) with a basaltic component containing 5 ppm Li and having δ^7Li=+6‰. Obviously, the identification of specific end-members is difficult. However, this simple calculation illustrates that some of the lighter Clarence River samples may have incorporated small amounts (<15%) of sediment, yet still lie within the range of observed arc values (as observed for the Lesser Antilles). The case for some level of crustal input is supported by the observation that Clarence River granites are more evolved in terms of their Sr and Nd characteristics than primitive Permian basalts in the southern New England Orogen (cf. Caprarelli & Leitch 1998). They also tend to have lower ε_{Nd} and are more reduced than intraoceanic arc lavas.

It is likely that isotopic heterogeneity within the Clarence River granites reflects both isotopic heterogeneity within primary arc magmas as well as subsequent modification through crustal processes.

4.3. I-type granites: Moonbi Supersuite

The MSS granites (δ^7Li=+1·9 to +4·2‰) also lie within the range of arc magmas (δ^7Li=+1 to +11‰; average=+4) and other mantle-derived magmas (Fig. 5). This is consistent with the previously proposed models of Chappell (1978; partial melting of a shoshonitic source) and Landenberger & Collins (1998; lamprophyric source component), which infer that the extreme enrichments in LILE and Th, for example, primarily arise from enrichment processes within the mantle, and that the overall contribution from sedimentary components to the MSS is comparatively low (e.g. Shaw & Flood 1981).

On the basis of Hf isotopes, Flood & Shaw (2003) argued that two felsic magmas were involved in the generation of the Walcha Road pluton, one derived from the differentiation of a mantle-derived magma, and the other from the partial melting of the crust. However, the Li isotopic compositions of Moonbi granites analysed herein are relatively homogeneous over the entire spectrum of compositions analysed (~20 wt.% SiO_2) (Fig. 5). This suggests that, if two components were involved in the petrogenesis of the MSS, the Li isotopic compositions (± Li concentrations) of those components were similar, the magmas were sufficiently homogenised prior to emplacement, or the amount of any isotopically lighter component was comparatively small. It is noteworthy that the isotopic shift resulting from the late-stage mixing of a felsic, differentiated mantle-derived magma and an isotopically lighter felsic crustal magma (model of Flood & Shaw 2003) would be less significant than if the mantle-derived magma were basaltic in composition owing to the higher Li concentrations in more differentiated magmas.

4.4. I-type granites: Uralla Supersuite

The lower redox states (lower Fe_2O_3/FeO), higher δ^{18}O (typically >9‰; O'Neil *et al.* 1977), lower normative diopside and higher initial ^{87}Sr/^{86}Sr ratios in the Uralla Supersuite are proposed to result from partial melting of a mixture of meta-sedimentary source components similar to that which gave rise to the NEB S-type granites, and the meta-igneous materials which gave rise to the Clarence River and Moonbi granites (Shaw & Flood 1981). The Li isotopic compositions in the Uralla Supersuite range between those observed in both the Clarence River and Moonbi granites down to the values observed in NEB S-type intrusions. Given that both the Clarence River and Moonbi I-type supersuites, which appear to have minor or non-existent sedimentary inputs, typically have higher δ^7Li values than NEB S-type granites, it is entirely plausible that mixtures between these source components will yield a diverse range of Li isotopic compositions that extends to lower δ^7Li values. Nevertheless, the Uralla Supersuite is internally very complex owing to regional variability in source components. Thus, it is necessary to examine each of the geochemical subgroups within the Uralla Supersuite individually.

Elevated K_2O, Rb, U and Th, and generally higher oxidation states in granites of the Mt Duval Group (e.g. Tingha, Mt Duval and Gwydir River) with respect to other Uralla granites are consistent with the involvement of Moonbi-like source components in these rocks and lower contributions from sedimentary components (cf. Shaw & Flood 1981). The range of Li isotopic compositions observed in these granites (δ^7Li = +2·2 to +3·9‰) are similar to those of the southern MSS (δ^7Li = +1·9 to +4·2‰), implying that, if additional sedimentary components were incorporated into the Mt Duval Group plutons relative to the southern MSS, this is not evident within either of the two plutons analysed herein.

The Khartoun (Mafic Group) and Uralla Granodiorite are both more mafic and significantly more reduced than the Mt Duval Group intrusions. Based on the data of Hensel *et al.* (1985), Mafic Group granites appear to incorporate a component that is more enriched in initial $^{87}Sr/^{86}Sr$ relative to MSS granites with similar ε_{Nd}, with the majority occurring within the field occupied by Hillgrove S-type granites (Fig. 3) (Hensel *et al.* 1985). The petrogenesis of this group is not clearly defined, but it is unlikely to result from simple mixtures of known source I- and S-type granite source components. For example, to generate the Nd-Sr characteristics of the Mafic Group granites would require ~50% mixing between a Moonbi or Clarence River mafic component, and the highest initial $^{87}Sr/^{86}Sr$ Hillgrove sedimentary component. However, this is not resolvable with the major and trace element compositions of the Mafic Group granites (<6 wt.% MgO, <300 ppm, <60 ppm Ni and <500 ppm Sr), given that the most mafic Hillgrove granites with ~63 wt.% SiO_2 contain ~2 wt.% MgO, 50 ppm Cr and 10 ppm Ni. Furthermore, although these granites are more reduced than the southern Moonbi and Mt Duval Group granites, the $\delta^{18}O$ values are equivalent [Mafic Group, $\delta^{18}O$ = 8·6–9·9‰; southern MSS, $\delta^{18}O$ = 9·1‰ and 8·6‰ (Congi and Walcha Road, respectively); Mt Duval Group, $\delta^{18}O$ = 9·9‰ (Gwydir River); O'Neil *et al.* 1977). It is possible that the higher $\delta^{18}O$, initial $^{87}Sr/^{86}Sr$ and possibly δ^7Li in these rocks derive from igneous and sedimentary components which had themselves interacted with sea water. Notably, oceanic crust that has experienced low-T alteration by sea water has δ^7Li values of +3·4 to +20·8‰ (Chan *et al.* 1992, 2002a). Marine sediments are variable in composition, but have δ^7Li values up to +14·5‰ (Chan & Kastner 2000; Bouman 2004). Based on the high MgO, Cr and Ni (for granites), it is unlikely that Mafic Group granites derive solely from the partial melting of very immature sedimentary components.

Granites of the Llangothlin (Llangothlin, Aberfoyle) and Wards Mistake groups (Wards Mistake, Cottesbrook) are located along the northern portion of the region occupied by granites of the Uralla Supersuite. These granites are geochemically distinct from the southern Uralla granites, requiring different source components. These granites have low Fe_2O_3/FeO ratios, white K-feldspar, and initial $^{87}Sr/^{86}Sr$ and ε_{Nd} compositions which are very similar to those observed in the Bundarra Supersuite. This is in stark contrast to the felsic granites of the northern MSS, located to the north, which are oxidised and have initial $^{87}Sr/^{86}Sr$ ~0·7045 and ε_{Nd} ~+2 (Sivell 2002), similar to those of the southern MSS. Consequently, it is argued that the Wards Mistake and possibly the Llangothlin groups have incorporated greater quantities of sedimentary components during their petrogenesis than the neighbouring northern MSS granites. The Li isotopic composition of the Wards Mistake intrusion is similar to those observed in the NEB S-type granites, and detectably lower than those of the southern Uralla granites. The combination of very low ε_{Nd}, relatively high initial $^{87}Sr/^{86}Sr$, and reduced oxidation states are consistent with the lower δ^7Li value being derived from a sedimentary component similar to that involved in the petrogenesis of the Bundarra Supersuite. Interestingly, low δ^7Li values (+0·5‰) are also evident within the Glenore Pluton, which occurs within the northernmost portion of the Mt Duval intrusion, and which is geographically the closest to the Wards Mistake granites.

The question remains, is sediment addition the only explanation for lower values in the Uralla Supersuite? This issue is addressed further in the following section.

4.5. Isotopic fractionation during anatexis and differentiation of I-type magmas?

There is very little concrete evidence on which to make an assessment of the degree of Li isotopic fractionation that may occur prior to, or during, the anatexis, or during the subsequent differentiation of I-type magmas. Whilst there is no evidence for isotopic fractionation between S-type granites and shales, or during subsequent differentiation of those magmas (Teng *et al.* 2004; refer to the discussion above), does this mean that isotopic fractionation is insignificant during the partial melting of infracrustal protoliths and subsequent differentiation of I-type magmas?

In basaltic and andesitic melts, the compatibility of Li decreases in the order olivine > clinopyroxene ≈ orthopyroxene > plagioclase (Jeffries *et al.* 1995; Bindeman *et al.* 1998; Trompetter *et al.* 1999; Seitz & Woodland 2000). In lherzolitic xenoliths, the concentrations of Li in pyroxene are greater than those of coexisting amphibole (Seitz & Woodland 2000). The mineral-melt partition coefficients in basaltic to andesitic compositions, including micas, are typically <1 (Jeffries *et al.* 1995; Bindeman *et al.* 1998; Schmidt *et al.* 1999; Trompetter *et al.* 1999), and in magmas of this composition, Li primarily resides in the melt (Trompetter *et al.* 1999). The strong correlation between the Li isotopic composition of MORB and upper mantle peridotites indicates that the degree of isotopic fractionation between olivine+pyroxene and melt during partial melting is insignificant. Similarly, no isotopic fractionation is observed during the fractional crystallisation of either OIB (Tomascak *et al.* 1999a) or arc lavas (Bouman 2004). On this basis, it can be reasonably argued that the partitioning of Li between olivine, pyroxene and plagioclase, and melt does not contribute to significant isotopic fractionation. This is of direct relevance to the granites of the Clarence River supersuites, owing to the overall dry nature of the parental magmas (Eggins & Hensen 1987; Bryant *et al.* 1997). In most cases, amphibole is late-stage, having formed through reaction of early-formed pyroxene with the melt and/or through sub-solidus alteration of clinopyroxene to actinolitic amphibole (Bryant *et al.* 1997). Hence, the degree of isotopic fractionation induced by differential partitioning of Li between minerals and melt during the petrogenesis of these intrusions is likely to be very small, i.e. the Li isotopic variations are related to the source components.

The partition coefficient of biotite (Icenhower & London 1995), and probably of amphibole, increases with decreasing temperature, such that, in rhyolitic/granitic compositions, much of the Li may be housed within these two minerals (e.g. Carron & Lagache 1971; Trompetter *et al.* 1999). Therefore, if the two Li isotopes are fractionated between minerals and melt, amphibole and biotite are the most likely candidates. Under conditions of small degrees of partial melting at lower temperatures, hornblende and biotite may be left in the residuum following melt segregation. Similarly, the separation of amphibole and biotite through restite unmixing or fractional crystallisation can contribute to the differential partitioning of Li between minerals and melt. However, given that the Li concentration of the magmas increases with SiO_2, the amount of Li removed in these minerals may, in many cases, be less than that partitioned into the melt.

Is there any evidence that fractionation of hornblende and biotite leads to significant Li isotopic fractionation either during partial melting or subsequent fractional crystallisation? Amphibole and biotite are the dominant ferromagnesian minerals in all southern MSS intrusions. These minerals are variably argued to be restitic (Chappell 1978) or precipitates of the melt (Shaw & Flood 1981). The uniformity of Li isotopic compositions of southern MSS granites over an almost 20 wt.% variation in SiO_2 (Fig. 5), together with their similarity with arc magmas world-wide, implies that the segregation of amphibole and biotite, either within the source or during subsequent differentiation did not contribute to detectable isotopic fractionation within these granites.

The fact that the Li isotopic compositions of the Clarence River and Moonbi granites are identical to those observed in arc lavas, and that both NEB and Lachlan Fold Belt S-type granites have δ^7Li values equivalent to shales, suggests that the overall processes involved in the genesis of these magmas did not contribute to significant fractionation of Li isotopic compositions in these granites. In these cases, there appears to be a direct correspondence between the observed Li isotopic composition of the source components and the resultant granites.

This brings us to origin of the diversity of Li isotopic compositions observed in the Uralla Supersuite, and why granites from the northern portion of this supersuite have lower δ^7Li values. Based on observations from the MSS, this is unlikely to be simply related to the fractionation of biotite and amphibole during magma differentiation, and therefore, is in some way attributable to variability in the isotopic composition of the source components.

4.6. Isotopically light crustal source components

The work of Teng *et al.* (2003) demonstrated that lower crustal granulitic xenoliths are isotopically light, having δ^7Li$= -13$ to $+7$‰, with an average of 0‰ (Teng *et al.* 2004). One of the difficulties with applying these results to the study of granites is that we do not know if these depleted compositions derive from the previous loss of granite melts or from dehydration reactions. Essentially all documented cases of Li isotope fractionation relate to the partitioning of Li between minerals and an aqueous fluid; for example weathering (Huh *et al.* 1998, 2001; Pistiner & Henderson 2003); sediment diagenesis (Zhang *et al.* 1998; Chan & Kastner 2000), sea floor alteration (e.g. Chan *et al.* 2002b), and muscovite-fluid partitioning experiments (Lynton *et al.* 1999). Zack *et al.* (2003) documented that Li isotopes undergo extreme isotopic fractionation during dehydration of the subducting slab. Similar processes may occur in the lower crust. Alternatively, the low σ^7Li values may result from the loss of granitic components during ultrametamorphism. Significantly, if the loss of granitic components leads to lower δ^7Li values in the lower crust, the close correspondence between the Li isotopic composition of S-type granites and shales would require that the granite magmas were actually derived from a source with even lower δ^7Li relative to the NEB I-types. In general, lower crustal granulites, or in the case of I-type granites, the eclogitic subducted slab, are not a suitable source for large-scale granitic magmatism since they have lost the critical water component required for anatexis. However, Li isotopes may be fractionated during pre-anatectic dehydration reactions, i.e. subsolidus metamorphic conditions. For example, Teng *et al.* (2004) reported a pelite composition with δ^7Li$= -5{\cdot}2$‰. Nevertheless, the strong correspondence between Lachlan Fold Belt S-type granites and local shales implies that net fractionation during anatexis is small (Teng *et al.* 2004). Interestingly, this could be combined with the observation of extremely light materials in the lower crust if ^{7}Li is preferentially lost during pre-anatectic metamorphic reactions relative to the shales, but subsequently enriched during the melting process. This would not only leave S-type granites with a similar isotopic composition as shales, but also the residual lower crust with substantially lower δ^7Li values than with which it started. At present, this is conjecture, and a great deal of work needs to be conducted regarding Li isotopic fractionation during crustal-related processes.

Can I-type granites be derived from an infracrustal (i.e. no sedimentary component) source with low δ^7Li values, and is this an adequate explanation for the northern Uralla Supersuite granites? Teng *et al.* (2004) recorded a Neoproterozoic diorite composite from China with δ^7Li$= -3{\cdot}4$‰, demonstrating that it is possible to have infracrustal source components with very low δ^7Li values. The origin of these low values is unknown. Given the age and crustal complexity of the crust, it entirely possible that these light Li isotopic compositions were acquired through the interaction of the Neoproterozoic magmas with an older, isotopically light granulitic component in the lower crust, similar to the scenario proposed by Teng *et al.* (2003) to explain the extremely low values in mafic cumulate xenoliths from the Chudleigh Suite in northeastern Australia. These xenoliths have Li isotopic compositions ranging from a mantle end-member component through to Proterozoic lower crust values (δ^7Li ~ -9‰). Given the juvenile nature of the New England Orogen, the existence of a significantly older, highly isotopically fractionated lower crustal basement is unlikely. Whilst the present authors do not rule out an origin by partial melting of an infracrustal source component with low δ^7Li for the northern Uralla Supersuite, the reduced character of these granites, combined with strongly evolved (for New England) Sr and Nd isotopic characteristics is entirely consistent with the previous interpretation of Shaw & Flood (1981) that these granites represent mixtures of infracrustal and supracrustal source components. Hence, the Li isotopic compositions of these granites are reasonably explained by variable involvement of those two source components.

4.7. Leucogranites: interpretation of Li isotopic compositions

One of the fundamental difficulties in granite studies is determining the origins of highly fractionated granites because of the convergence of geochemical trends as the minimum melt composition is approached. Highly fractionated I- and S-type granites may be distinguished on the basis of P_2O_5, Th and Y concentrations owing to the increased solubility of apatite in peraluminous melts and, hence, the delayed saturation of phosphate-bearing minerals (Chappell 1999). However, distinguishing between unfractionated felsic I-type granites generated from distinctly different source rocks can be very difficult.

Three leucogranite samples have been analysed, two from the Gilgai Granite and one from the Yellow Gap Micromonzogranite. Both are spatially associated with members of the Uralla Supersuite. The Yellow Gap Pluton appears to represent the highly differentiated carapace of the Wards Mistake Pluton (Gilligan *et al.* 1992), an intrusion that is characterised by low δ^7Li (0·2‰). The Gilgai Granite forms a rim around, and is locally interspersed with, the Tingha Monzogranite (δ^7Li$= +3{\cdot}9$‰), although a direct genetic link between the two has not been established. The Gilgai samples show clear evidence of volatile saturation (e.g. miarolitic cavities). Such features are also evident within the Yellow Gap Pluton, although they are not well developed within the analysed sample. The Gilgai granite has δ^7Li ~+2·4‰, whereas the Yellow Gap intrusion has δ^7Li= − 1·3‰. In both cases, the Li isotopic compositions of the leucogranites appear to be slightly lower than the more mafic granites with which they are associated. Nevertheless, the differences which are evident between the δ^7Li values of the more mafic granites appear to be maintained within the associated leucogranites. The possibility that the Li isotopic composition of leucogranites may in some way be tied to the more mafic compositions has significant implications for the classification and interpretation of strongly fractionated granite suites, for which there are few other useful chemical and isotopic indicators. However, further studies are required to evaluate if the observed differences between the more mafic and felsic suites are real (given an uncertainty ~1‰ on the analysis, and that the plutons are not necessarily cogenetic), and if so, to what extent do they relate to isotopic fractionation within the magma. Of particular relevance is the potential for isotopic fractionation during volatile phase exsolution (cf. Lynton *et al.* 1999). Water is critical to magma genesis, and given sufficient periods of fractional crystallisation at shallow crustal levels, a high proportion of crustal granites will evolve a separate aqueous fluid. Obviously, the timing of volatile saturation is dependent on a number of factors, but the probability of volatile saturation in felsic melts is much higher since, at this point, there are insufficient hydrous phases precipitating from the melt to prevent the exsolution of volatiles.

5. Conclusions

Significant variations are evident in the Li isotopic compositions of granites from the NEB (δ^7Li= − 1·3 to +8‰). The isotopic compositions within individual supersuites can be correlated with those of likely source components (based on previous, independent studies). The available evidence suggests that the degree of Li isotope fractionation experienced during anatexis and magma differentiation is low. The possible exceptions to this are the leucogranite samples, which may have experienced subtle isotopic fractionation during volatile exsolution. The results obtained herein, combined with the evidence from the Lesser Antilles arc (Bouman 2004), indicate that the incorporation of sedimentary components into isotopically heavier, mantle-derived or infracrustal magmas can contribute to lower δ^7Li values in the resultant magmas. Hence, Li may be a useful tool in assessing the involvement of sediment or other isotopically light (e.g. lower crustal) source components during crustal magmatic processes, particularly in regions of juvenile crust, where there is limited spread in radiogenic isotopic compositions. However, Li isotopic signatures are not unique indicators in that not all arc lavas have a uniformly high δ^7Li value, and not all sediments are isotopically light. Similarly, crustal processes other than weathering (e.g. lower crustal dehydration) may contribute to Li isotopic fractionation in the crust. Clearly, future studies addressing the behaviour of Li isotopes during metamorphism and anatexis are required. Such studies would have benefits not only to granite petrogenesis, but also, depending on the principal factors generating Li isotopic fractionation within the lower crust, to understanding the nature and mechanisms of crustal differentiation.

6. Appendix 1. Li isotopic data

Table 2 presents the Li isotopic data utilised herein. The final isotopic composition represents an average of isotopic analyses of a single sample solution. The column labelled 'number' refers to the number of analyses conducted on an individual solution. The average measured Li isotopic composition calculated relative to bracketing L-SVEC samples, and the associated 2σ standard deviation, are listed as δ^7Li and 2σ, respectively. The unknowns were run in blocks of approximately six to 10, in which the first and last unknowns were purified sea water Li.

A second-order drift correction was then applied to the granite analyses using a linear interpolation between the bracketing sea water values. The sea water values were normalised to δ^7Li=30·4‰, as obtained by Bryant *et al.* (2003). The latter value is marginally lower than the average of the measured sea water values obtained in the present study (δ^7Li=30·8‰), but lies well within the range of the reported values for sea water (δ^7Li=28·8–32·4‰; summarised in Bryant *et al.*, 2003; Jeffcoate *et al.*, 2004). The interpolation between L-SVEC and sea water was assumed to be linear. The averages of the corrected Li isotopic compositions are listed in Table 2 as δ^7Li_{SW} and the associated 2σ standard deviations as $2\sigma_{SW}$ The 2σ standard deviation for a single analyses can assumed to be ~1‰.

The values reported in the text and utilised in the various figures are the δ^7Li_{SW} values, except for samples indicated with an asterisk (*) in Table 2. In some cases, it was impossible to complete a sequence because of machine failure. In this case,

Table 2 Lithium concentration and isotopic composition of NEB granites

Sample no.	Unit	Number	δ^7Li_{SW}#	$2\sigma_{SW}$#	δ^7Li	2σ	Li (ppm)
Bundarra Supersuite							
MNG459	Banalasta	2	0·8	1·0	2·4	0·7	46·0
MNG419	Banalasta	2	1·7	1·0	2·4	0·6	43·8
NEB247	Copeton	3	1·4	0·2	1·5	1·0	52·7
NEB251*	Copeton	2			− 0·1	0·3	51·5
NEB246	Copeton	2	2·1	0·6	2·8	1·0	116·0
NEB219	Roumalla	2	1·9	1·1	2·7	1·9	37·1
Hillgrove Supersuite							
NEB290	Hillgrove	2	0·9	0·6	1·2	0·1	39·0
NEB294	Hillgrove	2	1·7	2·6	1·9	2·0	34·5
NEB484	Kookabookra	2	0·6	0·8	0·9	1·1	33·0
NEB475	Kookabookra	2	0·4	0·6	0·5	0·2	41·8 (42·0)
NEB 415	Tobermory	2	0·6	1·3	0·9	0·8	43·9

Table 2 Continued

Sample no.	Unit	Number	δ^7Li_{SW}#	$2\sigma_{SW}$#	δ^7Li	2σ	Li (ppm)
Uralla Supersuite							
NEB215	Khatoun	3	3·1	1·2	3·6	2·1	20·7 (19·8)
NEB212	Uralla	2	2·2	0·2	2·0	0·1	24·6
NEB197	Mount Duval	3	2·2	1·3	2·5	0·9	32·3
NEB232	Tingha	2	3·9	0·3	4·0	0·4	42·1
NEB 414*	Wards Mistake	1			0·2		34·7 (33·3)
NEB54	Llangothlin	3	1·4	0·8	1·8	1·2	22·2 (21·8)
NEB200*	Glenore	1			0·5		30·9
Moonbi Supersuite							
MNG432	Inlet	4	2·7	1·2	2·9	1·0	16·4
MNG455	Moonbi	3	4·2	1·3	4·3	0·7	44·3
MNG223*	Bendemeer	2	1·4	1·4	2·8	1·6	32·3
NEB482	Walcha Road	3	1·9	1·1	2·3	1·7	22·6
NEB208	Congi Creek	3	3·1	0·4			16·3
Clarence River Supersuite							
72056	Dumbudgery Ck	2	2·2	0·4	1·9	0·2	16·7
72332	Jenny Lind	3	8·0	0·5	8·1	0·1	5·9
72341	Jenny Lind	1	2·7		3·3		8·7
72016	Kaloe	1	5·6		6·2		27·3 (26·3)
72009	Kaloe	3	4·5	1·8	4·5	1·8	29·1
72098	Towgon Grange	3	2·8	0·5	2·7	1·5	18·7
72101	Towgon Grange	3	4·7	0·7	4·8	0·7	6·7
72151	Towgon Grange	2	2·7	0·2	3·1	0·7	19·8
Leucogranites							
EB228	Gilgai	2	2·4	0·5	2·8	0·2	18·2
NEB240*	Gilgai	3	1·4	0·8	2·44	0·7	82·9
NEB103	Yellow Gap	2	− 1·3	0·1	− 0·2	1·0	23·5
Standards							
BHVO-1					+6·1		
BHVO-2					+4·4		
G-2					− 1·7		
Sea water		24			+30·8	1·5	

the present authors have plotted the measured δ^7Li values. The difference between the uncorrected and 'sea water' corrected results are typically minimal, with the values the same within error.

7. Acknowledgements

Les Kinsley, Charlotte Allen, Mike Shelley and Graham Mortimer are all greatly thanked for their assistance with various analytical aspects of this study. CJB was funded by an Australian Research Council Postdoctoral Fellowship. We thank also thank Paul Tomascak and Simone Kaseman for their constructive and helpful reviews, and Simon Harley for his editorial efforts.

8. References

Bea, F., Pereira, M. D. & Stroh, A. 1994. Mineral/leucosome trace-element partitioning in a peraluminous migmatite (a laser ablation-ICP-MS study). *Chemical Geology* **117**, 291–312.

Benton, L. D. & Tera, F. 2000. Lithium isotope systematics of the Marianas revisited. *Goldschmidt Conference 2000, Journal of Conference Abstracts* **5**, 210. Cambridge, UK: Cambridge Publications.

Bindeman, I. N., Davis, A. M. & Drake, M. J. 1998. Ion microprobe study of plagioclase-basalt partition experiments at natural concentration levels of trace elements. *Geochimica et Cosmochimica Acta* **62**, 1175–93.

Bouman, C. 2004. *Lithium isotope systematics at subduction zones*. PhD thesis. Amsterdam, The Netherlands: Vrije Universiteit Amsterdam.

Bouman, C., Elliott, T. & Vroon, P. Z. 2004. Lithium inputs to subduction zones. *Chemical Geology* **212**, 59–79.

Brooker, R., Blundy, J. & James, R. 2000. Subduction-related mantle pyroxenites from Zabargad Island, Red Sea. *Goldschmidt Conference 2000, Journal of Conference Abstracts* **5**, 249.

Bryant, C. J., Arculus, R. J. & Chappell, B. W. 1997. Clarence River Supersuite: 250 Ma Cordilleran tonalitic I-type intrusions in eastern Australia. *Journal of Petrology* **38**, 975–1001.

Bryant, C. J., McCulloch, M. T. & Bennett, V. C. 2003. Impact of matrix effects on the accurate measurement of Li isotope ratios by inductively coupled plasma mass spectrometry (MC-ICP-MS) under 'cold; plasma conditions. *Journal of Analytical and Atomic Spectrometry* **18**, 734–7.

Caprarelli, G. & Leitch, E. C. 1998. Magmatic changes during the stabilisation of a Cordilleran fold belt: the Carboniferous–Triassic igneous history of eastern New South Wales, Australia. *Lithos* **45**, 413–30.

Carron, J. P. & Lagache, M. 1971. La distribution des elements alcalins Li, Na, K, Rb dans les mineraux essentiels des granites et granodiorites du Sud de la Corse. *Bulletin de la Societe Francaise de Mineralogie et de Cristallographie* **94**, 70–80

Carron, J. P. & Lagache, M. 1972. Etude du partage des elements alcalins, Na, K, Li, Rb, Cs, entre les mineraux de quelques roches granitiques de France. *International Geological Congress, Abstracts – Congres Geologique Internationale, Resumes* **24**, 297.

Chan, L. H., Edmond, J. M., Thompson, G. & Gillis, K. 1992. Lithium isotopic composition of submarine basalts – implications for the lithium cycle in the oceans. *Earth and Planetary Science Letters* **108**, 151–60.

Chan, L. H., Alt, J. C. & Teagle, D. A. H. 2002a. Lithium and lithium isotope profiles through the upper oceanic crust: a study of seawater-basalt exchange at ODP sites 504B and 896A. *Earth and Planetary Science Letters* **201**, 187–201.

Chan, L. H., Leeman, W. P. & You, C. F. 2002b. Lithium isotopic composition of Central American Volcanic Arc lavas: implications for modification of subarc mantle by slab-derived fluids, correction. *Chemical Geology* **182**, 293–300.

Chan, L. H., Leeman, W. P. & Tomarini, S. 2001. Lithium isotope compositions of South Sandwich Arc and Southwest Washington Cascades: a comparative study of arc processes. *EOS Transactions of the American Geophysical Union* **82**, 1293.

Chan, L. H. & Edmond, J. M. 1988. Variations of lithium isotope composition in the marine environment: a preliminary report. *Geochimica et Cosmochimica Acta* **52**, 1711–17.

Chan, L. H & Frey, F. A. 2003. Lithium isotope geochemistry of the Hawaiian plume: results from the Hawaii Scientific Drilling Project and Koolau Volcano. *Geochemistry, Geophysics and Geosystems* **4**(3), 8707, article No. 2002GC000365.

Chan, L. H. & Kastner, M. 2000. Lithium isotopic compositions of pore fluids and sediments in the Costa Rica subduction zone: implications for fluid processes and sediment contribution to the arc volcanoes. *Earth and Planetary Science Letters* **183**, 275–90.

Chappell, B. W. 1978. Granitoids from the Moonbi district, New England Batholith, eastern Australia. *Journal of the Geological Society of Australia* **25**, 267–83.

Chappell, B. W. 1991. Trace element analysis of rocks by X-ray spectrometry. *Advances in X-ray Analysis* **34**, 263–76.

Chappell, B. W. 1999. Aluminium saturation in I- and S-type granites and the characterization of fractionated haplogranites. *Lithos* **46, 535–51.**

Davidson, J. P. 1987. Crustal contamination versus subduction zone enrichment; examples from the Lesser Antilles and implications for mantle source compositions of island arc volcanic rocks. *Geochimica et Cosmochimica Acta* **51**, 2185–98.

Dirks, P. H. G. M., Offler, R. & Collins, W. J. 1993. Timing of emplacement and deformation on the Tia Granodiorite, southern New England fold belt, NSW; implications for the metamorphic history. *Australian Journal of Earth Sciences* **40**, 103–8.

Eggins, S. & Hensen, B. J. 1987. Evolution of mantle-derived, augite-hypersthene granodiorites by crystal-liquid fractionation; Barrington Tops Batholith, eastern Australia. *Lithos* **20**, 295–310.

Elliott, T., Thomas, A. L., Jeffcoate, A. B. & Niu, Y. 2002. Li isotope variations in the upper mantle. *Geochimica et Cosmochimica Acta* **66**, A214.

Flood, P. G. & Aitchison, J. C. 1988. Tectonostratigraphic terranes of the southern part of the New England Orogen. *In* Kleeman, J. D. (ed.) *New England Orogen Tectonics and Metallogenesis*, 7–10. Armidale, NSW: University of New England.

Flood, R. H. & Shaw, S. E. 1977. Two 'S-type' granite suites with low initial $^{87}Sr/^{86}Sr$ ratios from the New England Batholith, Australia. *Contributions to Mineralogy and Petrology* **61**, 163–73.

Flood, R. H. & Shaw, S. E. 2003. A reconnaissance $^{176}Hf/^{177}Hf$ investigation of the New England Batholith Eastern Australia. *Hutton Symposium V, Abstract with Program*, 31.

Gilligan, L. B., Brownlow, J. W., Cameron, R. G. & Henley, H. F. 1992. Dorrigo-Coffs Harbour 1:250 000 Metallogenic Map. Sydney: NSW Geological Survey.

Hensel, H. D., McCulloch, M. T. & Chappell, B. W. 1985. The New England batholith: constraints on its derivation from Nd and Sr isotopic studies of granitoids and country rocks. *Geochimica et Cosmochimica Acta* **49**, 369–84.

Huh, H., Chan, L. H., Zhang, L. & Edmond, J. M. 1998. Lithium and its isotopes in major world rivers: implications for weathering and the oceanic budget. *Geochimica et Cosmochimica Acta* **62**, 2039–51.

Huh, H., Chan, L. H. & Edmond, J. M. 2001. Lithium isotopes as a probe of weathering processes: Orinoco river. *Earth and Planetary Science Letters* **194**, 189–99.

Icenhower, J. & London, D. 1995. An experimental study of element partitioning among biotite, muscovite, and coexisting peraluminous silicic melt at 200 MPa (H_2O). *American Mineralogist* **80**, 1229–51.

James, R. H. & Palmer, M. R. 2000. The lithium isotope composition of international rock standards and solutions. *Chemical Geology* **166**, 319–26.

Jeffcoate, A. B., Elliott, T., Thomas, A. & Bouman, C. 2004. Precise, small sample size determinations of lithium isotopic compositions of geological reference materials and modern seawater by MC-ICP-MS. *Geostandards and Geoanalytical Research* **28**, 161–72.

Jeffcoate, A. & Elliott, T. 2003. Tracing recycled Li in the mantle: insights into mantle heterogeneities. *EOS Transactions of the American Geophysical Union* **84**(46), Fall Meeting Supplement, Abstract V52A–0416.

Jeffries, T. E., Perkins, W. T. & Pearce, N. J. G. 1995. Measurements of trace elements in basalts and their phenocrysts by laser probe microanalysis inductively coupled plasma mass spectrometry (LPMA-ICP-MS). *Chemical Geology* **121**, 131–44.

Kent, A. J. R. 1994. Geochronology and geochemistry of Palaeozoic intrusive rocks in the Rockvale region, southern New England Orogen, New South Wales. *Australian Journal of Earth Sciences* **41**, 365–79.

Kretz, R., Loop, J. & Hartree, R. 1989. Petrology and Li-Be-B geochemistry of muscovite-biotite granite and associated pegmatite near Yellowknife, Canada. *Contributions to Mineralogy and Petrology* **102**, 174–90.

Landenberger, B., Farrell, T. R., Offler, R., Collins, W. J. & Whitford, D. J. 1995. Tectonic implications of Rb-Sr biotite ages for the Hillgrove Plutonic Suite, New England fold belt, N.S.W., Australia. *Precambrian Research* **71**, 251–63.

Landenberger, B. & Collins, W. J. 1995. S-type granites of the Hillgrove plutonic suite, New England Batholith, eastern Australia; products of partial melting of an intermediate greywacke source. *In* Brown, M. & Piccoli, P. M. (eds) *The Origin of Granites and Related Rocks. U.S. Geological Survey Circular*, 86.

Landenberger, B. & Collins, W. J. 1998. Evidence for mantle control on the character of I-type granitoids; examples from the New England Batholith. *Geological Society of Australia; Abstracts* **49**, 259.

Lynton, S. J., Walker, R. J. & Candela, P. A. 1999. An experimental study of lithium isotopes and geochemistry in felsic systems. *EOS Transactions of the American Geophysical Union* **80**, Fall Meeting Supplement, Abstract V32D–05.

Moriguti, T. & Nakamura, E. 1998a. Across-arc variation of Li isotopes in lavas and implications for crust/mantle recycling at subduction zones. *Earth and Planetary Science Letters* **163**, 167–74.

Moriguti, T. & Nakamura, E. 1998b. High-yield lithium separation and the precise isotopic analysis for natural rock and aqueous samples. *Chemical Geology* **145**, 91–104.

Nishio, Y., Nakai, S., Kogiso, T. & Barsczus, H. 2003. Lithium isotopic composition of HIMU oceanic basalts: implications for the origin of HIMU component. *XXIII General Assembly of the International Union of Geodesy and Geophysics*, 178. Sapporo, Japan: Science Council of Japan.

Nishio, Y. & Nakai, S. I. 2002. Accurate and precise lithium isotopic determinations of igneous rock samples using multi-collector inductively coupled plasma mass spectrometry. *Analytica Chimica Acta* **456**, 270–81.

Norrish, K. & Hutton, J. T. 1969. An accurate X-ray spectrographic method for the analysis of a wide range of geological samples. *Geochimica et Cosmochimica Acta* **33**, 431–53.

O'Neil, J. R., Shaw, S. E. & Flood, R. H. 1977. Oxygen and hydrogen isotope compositions as indicators of granite genesis in the New England Batholith, Australia. *Contributions to Mineralogy and Petrology* **62**, 313–28.

Papike, J. J., Shearer, C. K., Simon, S. B. & Laul, J. C. 1984. Crystal chemical aspects of Li-Rb, and Cs partitioning between coexisting muscovite and biotite. *International Geological Congress, Abstracts* **27**, 135. Moscow.

Pereira, M. D. & Shaw D. M. 1996. B and Li distribution in the Peña Negra Complex; an alpha-track study. *American Mineralogist* **81**, 141–5.

Pistiner, J. S. & Henderson, G. M. 2003 Lithium-isotope fractionation during continental weathering processes. *Earth and Planetary Science Letters* **214**, 327–39.

Schmidt, K. H., Bottazzi, P., Vannucci, R. & Mengel, K. 1999. Trace element partitioning between phlogopite, clinopyroxene and leucite lamproite melt. *Earth and Planetary Science Letters* **168**, 287–99.

Seitz, H. M. & Woodland, A. B. 2000. The distribution of lithium in peridotitic and pyroxenitic mantle lithologies; an indicator of magmatic and metasomatic processes. *Chemical Geology* **166**, 47–64.

Shaw, S. E. & Flood, R. H. 1981. The New England Batholith, Eastern Australia: geochemical variations in space and time. *Journal of Geophysical Research* **86**, 10530–44.

Shaw, S. E. & Flood, R. H. 1993. A compilation of Late Permian and Triassic biotite Rb-Sr data from New England Batholith and areas to the southeast. *In Centre for Isotope Studies, Research Report. North Ryde, N.S.W.*, 151–5. Centre for Isotope Studies, CSIRO Mineral Research Laboratories.

Sivell, W. J. 2002. Sr-Nd isotopic constraints on origin of Stanthorpe Granite Group; role of a depleted mantle-derived component. *Geological Society of Australia, Abstracts* **67**, 216.

Smith, T. E., Thirlwall, M. F. & Macpherson, C. G. 1996. Trace element and isotope geochemistry of the volcanic rocks of Bequia, Grenadine islands, Lesser Antilles arc: a study of subduction enrichment and intra-crustal contamination. *Journal of Petrology* **37**, 117–43.

Teng, F., McDonough, W. F., Rudnick, R. L., Tomascak, P. B. & Saal, A. E. 2003. Lithium isotopic composition of the lower continental crust: a xenolith perspective. *EOS Transactions of the American Geophysical Union* **84**(46), Fall Meeting Supplement, Abstract VS1A–02.

Teng, F.-Z., McDonough, W. F., Rudnick, R. L., Dalpé, C., Tomascak, P. B., Chappell, B. W. & Gao, S. 2004. Lithium isotopic composition and concentration of the upper continental crust. *Geochimica et Cosmochimica Acta* **68**, 4167–78.

Thirlwall, M. F., Graham, A. M., Arculus, R. J., Harmon, R. S. & Macpherson, C. G. 1996. Resolution of the effects of crustal assimilation, sediment subduction, and fluid transport in island arc magmas: Pb-Sr-Nd-O isotope geochemistry of Grenada, Lesser Antilles. *Geochimica et Cosmochimica Acta* **60**, 4785–810.

Tomascak, P. B., Carlson, R. W. & Shirey, S. B. 1999a. Accurate and precise determination of Li isotopic compositions by multicollector sector ICP-MS. *Chemical Geology* **158**, 145–54.

Tomascak, P. B., Tera, F., Helz, R. T. & Walker, R. J. 1999b. The absence of lithium isotope fractionation during basalt differentiation; new measurements by multicollector sector ICP-MS. *Geochimica et Cosmochimica Acta* **63**, 907–10.

Tomascak, P. B., Ryan, J. G. & Defant, M. J. 2000. Lithium isotope evidence for light element decoupling in the Panama subarc mantle. *Geology* **28**, 507–10.

Tomascak, P. B., Widom, E., Benton, L. D., Goldstein, S. L. & Ryan, J. G. 2002. The control of lithium budgets in island arcs. *Earth and Planetary Science Letters* **196**, 227–38.

Tomascak, P. B. & Langmuir, C. H. 1999. Lithium isotope variability in MORB. *EOS Transactions of the American Geophysical Union* **80**, Fall Meeting Supplement, Abstract V11E–10.

Trompetter, W. J., Reyes, A. G., Vickridge, I. C. & Markwitz, A. 1999. Lithium and boron distributions in geological samples. *Nuclear Instruments and Methods in Physics Research, Section B: Beam Interactions with Materials and Atoms* **158**, 568–74.

van Soest, M. C. 2000. *Sediment subduction and crustal contamination in the Lesser Antilles: the geochemical and isotopic imprints on recent lavas and geothermal fluids.* Ph.D. thesis. Amsterdam: Vrike Universiteit.

van Soest, M. C., Hilton, D. R., Macpherson, C. G. & Mattey, D. P. 2002. Resolving sediment subduction and crustal contamination in the Lesser Antilles island arc; a combined He-O-Sr isotope approach. *Journal of Petrology* **43**, 143–70.

Visser, D., Kloprogge, J. T. & Maijer, C. 1994. An infrared spectroscopic (IR) and light element (Li, Be, Na) study of cordierites from the Bamble Sector, South Norway. *Lithos* **32**, 95–107.

You, C. F. & Chan, L. H. 1996. Precise determination of lithium isotopic composition in low concentration natural samples. *Geochimica et Cosmochimica Acta* **60**, 909–15.

Zack, T., Tomascak, P. B., Rudnick, R. L., Dalpé, C. & McDonough, W. F. 2003. Extremely light Li in orogenic eclogites: The role of isotope fractionation during dehydration in subducted oceanic crust. *Earth and Planetary Science Letters* **208**, 279–90.

Zhang, L., Chan, L.H. & Gieskes, J. M. 1998. Lithium isotope geochemistry of pore waters from Ocean Drilling Program Sites 918/919, Irminger Basin, *Geochimica et Cosmochimica Acta* **62**, 2437–50.

COLLEEN J. BRYANT, VICTORIA C. BENNETT and MALCOLM T. McCULLOCH, Research School of Earth Sciences, The Australian National University, Canberra, ACT 0200, Australia.
e-mail: Colleen.Bryant@anu.edu.au

BRUCE W. CHAPPELL, ARC National Key Centre for Geochemical Evolution and Metallogeny of Continents, Department of Earth and Planetary Sciences, Macquarie University, Sydney, NSW 2109, Australia.

MS received 3 September 2003. Accepted for publication 6 September 2004.

Transactions of the Royal Society of Edinburgh: Earth Sciences, **95**, 215–225, 2004

Timing of Early Proterozoic magmatism along the Southern margin of the Siberian Craton (Kitoy area)

Ulrike Poller, D. P. Gladkochub, T. V. Donskaya, A. M. Mazukabzov, E. V. Sklyarov and W. Todt

ABSTRACT: Several granites and basement gneisses along the southern margin of the Siberian Craton were sampled to obtain precise age constrains and compositional isotopic data. The analysed granite plutons are interpreted to have been emplaced between 1880 and 1850 Ma, and are related to Palaeoproterozoic collisional and post-collisional events. Pb-Pb whole data of a granulite (1884 ± 26 Ma) and a two mica granite (1821 ± 29 Ma) constrain these U-Pb single zircon ages. Sr, Nd and Pb-Pb isotope data reveal the crustal origin of the investigated rocks and the reworking of Archaean material. Nevertheless, a minor influence of a mantle component is still visible in the Pb isotopes. Geodynamically, the magmatic and metamorphic ages in the Kitoy area are linked to several Early Proterozoic events along the southern margin of the Siberian Craton. Since these events are all older than the assembly of Rodinia (about 1·4 Ga ago) the collisional processes are linked to the consolidation of the Siberian Craton itself in Early Proterozoic times.

KEY WORDS: Cathodoluminescence, geochronology, radiogenic isotopes, Rodinia, zircon.

Along the southern margin of the Siberian Craton, basement rocks such as granites, migmatites, granulites, carbonate and mafic suites appear frequently. They are overlain by Vendian-Cambrian to Cenozoic sedimentary sequences, for which very little isotopic and geochronological data exist.

The region sampled for this study is situated along the Kitoy River, 150 km west of Irkutsk in Central Siberia (see Fig. 1A). The river area is dominated by basement rocks. Additionally, Neoproterozoic dolerite dykes outcrop along the river (Fig. 1B). Sklyarov *et al.* (2003) dated the dykes by Sm-Nd and found an age of 743 ± 47 Ma. Ar-Ar dating in the same study resulted in plateau ages of 758 ± 4 Ma. These dykes cross-cut the granites and the basement rocks.

The studied area is situated within an Upper Archaean collisional zone enclosed by the Archaean Irkut and Oka blocks belonging to the Sharizhalgai terrane (see Fig. 1A). This Archaean collisional event is marked in the Kitoy area by collisional granites dated to be 2·53 Ga in age by the zircon U-Pb method (Sklyarov 2001) and is probably related to the assemblage of the Irkut and Oka blocks as part of the Sharizhalgai terrane (Sklyarov 2001, and references therein).

In general, the crystalline basement along the southern margin of the Siberian Craton consists of Archaean and Proterozoic rocks which suffered metamorphism under amphibolite to granulite facies conditions. One main magmatic stage is found in the Early Proterozoic, the next younger known event is the intrusion of Neoproterozoic dykes, which might reflect the break-up stage of the Rodinia supercontinent.

The goal of the present investigation was to present new, precise U-Pb single zircon age data to determine the emplacement of granites and granulites in the Kitoy area. This age determination will clarify if granite emplacement ages younger than 1·8 Ga exist, pointing to the assembly of Rodinia 1·4 Ga ago. In addition, the authors present new isotopic data for Pb, Sr and Nd for the main basement rocks of this region to describe the source material of the investigated rocks and to enable further comparison with terranes of the same age. For this study, they sampled two granites, a quartz-syenite to monzonite (according to De La Roche *et al.* 1980), a two-mica granite, two granulites to granulitic gneisses, a metabasite, an amphibolite and a garnet-bearing gneiss.

1. Analytical methods

The present authors selected about 15–20 kg of fresh material for each geochronological sample and prepared this by crushing, grinding and sieving (<500 μm). Heavy minerals were separated using a Wilfley table. Sieved concentrates of heavy minerals were subsequently separated with a Frantz magnetic separator, and heavy liquids refined the final (non-magnetic) zircon fraction. The authors prepared zircon mounts for each sample for cathodoluminescence (CL) imaging. The CL imaging was performed on a LEO 1530 field emission scanning electron microscope at the cosmochemistry department of the Max-Planck-Institut für Chemie, Mainz, Germany.

U-Pb measurements were carried out on single zircon grains (vapour digestion method; Wendt & Todt 1991), and for one granite sample, also on half zircon crystals, which were recovered from the CL mounts (CLC method; Poller 2000).

Single zircons were placed in a special teflon bomb with small holes for each grain (Wendt & Todt 1991). A ^{205}Pb-^{233}U mixed spike and 28N HF were added to each hole, and the bomb was twice placed in an oven at 200°C for about 5 days. After complete dissolution, the samples were dried down and 6N HCl was added to the holes, followed by one day in the oven at 200°C. At this point, zircons were completely dissolved, homogenised with the spike and ready for analysis.

After drying down, the samples were loaded on Re single filaments with a mixture of silica gel and H_3PO_4. The authors performed isotopic measurements on a MAT 261 mass spectrometer (Max-Planck-Institut für Chemie) in peak-jumping mode using a secondary electron multiplier. The total Pb blank was about 3 pg. For blank Pb corrections, the following ratios were used (analysed together with the samples): ^{206}Pb/^{204}Pb=21·12, ^{207}Pb/^{204}Pb=15·67. For the common Pb correction, the authors used the isotopic composition of the Stacey &

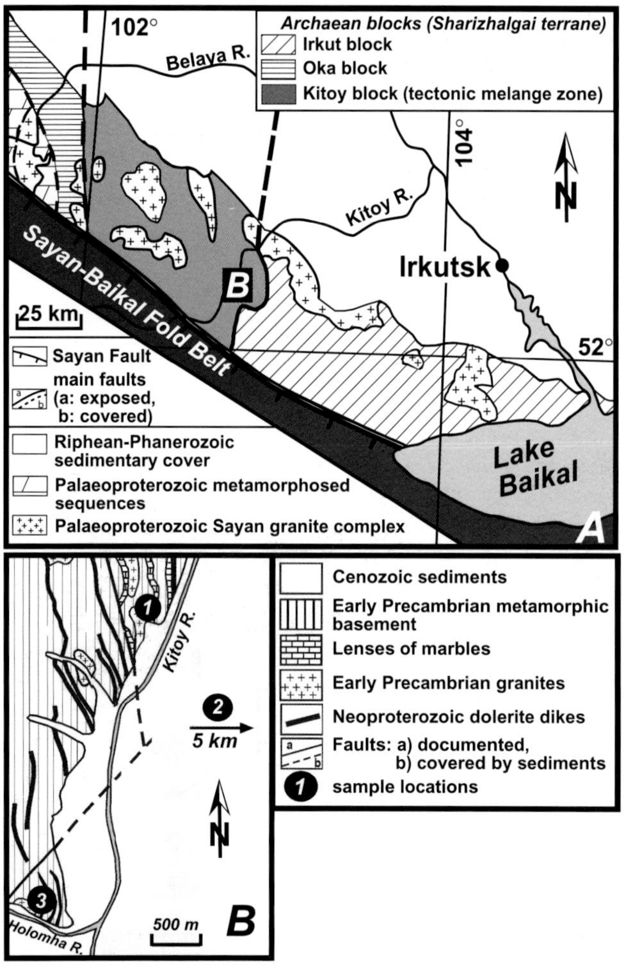

Figure 1 (A) Simplified geological map of the investigated southern margin of the Siberian Craton. (B) Geological map of the Kitoy area (modified after Sklyarov 2001).

Kramers (1975) Pb evolution line ($2{\cdot}0 \pm 0{\cdot}1$ Ga). All Pb ratios were corrected for fractionation using the NBS 981 and NBS 982 reference standards (Todt *et al.* 1996). U was corrected using a 'U nat' standard solution. The analyses were corrected with parallel-determined fractionation values scattering for Pb between 2·9‰ and 3·5‰ per Δ amu for the period of the measurements (Loveridge 1986). U-Pb age calculations are based on ISOPLOT by K. Ludwig (1992), Version 2.01 of May 1999. All errors refer to the 2σ deviation of the weighted mean from two to six blocks (see Table 4).

Fine-grained, whole-rock powders were used for geochemical and Sr-Nd isotope analyses. Rock chips were leached for common Pb whole-rock analyses. The results are listed in Table 3. Preparation and measurement of Sr, Nd and Pb followed the method described in Poller *et al.* (2001b).

Nd and Sr isotopic data in this work are presented as either present day values (0) or are recalculated to the emplacement ages determined by zircon geochronology (see Table 2).

Pb-Pb ages were calculated using regression lines for the $^{206}Pb/^{204}Pb$ and $^{207}Pb/^{204}Pb$ ratios. The errors are calculated based on the correlation of the regression lines.

The major and trace elements were analysed by X-ray influorescence (XRF) at the University of Mainz. Sm and Nd were measured on an Element 2 inductively coupled plasma mass spectrometry (ICP-MS) at the Max-Planck-Institut für Chemie using external calibration standards. All whole rock data are presented in Table 1.

2. Sample characterisation

In the present study, nine samples were collected from three outcrops in the Precambrian Kitoy block (Fig. 1). Five of these samples were used for zircon geochronology. In addition to these five, four other samples were analysed for their geochemical and isotopical characteristics.

The granite UP 3001 and the granulite UP 3014 were sampled from outcrop number 1 (see Fig. 1B) on the west shore of the Kitoy river (52°15.30′N, 102° 48.75′E). This outcrop covers about 2 km along the river, and exhibits migmatites, granulites and granulitic gneisses, intercalated lenses of marbles, and even Neoproterozoic doleritic dykes. This complex basement sequence was intruded by different granitic bodies. The granite UP 3005, the Qz-syenite UP 3002, the granulite to granulitic gneiss UP 3008 and the metabasite UP 3006 were sampled from outcrop number 2 located on the east side of the Kitoy River (52°07.01′N, 102°56.50′E). In this outcrop, the garnet-bearing and garnet-free granulites, metabasites and granites are represented.

The two-mica granite UP 3023, the amphibolite UP 3009 and the biotite-garnet-bearing gneiss UP 3011 occur within outcrop number 3, which spreads along the north shore of the Holomha River (52°11.74′N, 102°44.95′E).

2.1. Granites

The present authors analysed all sampled five granites for their geochemical and isotopic features prior to the U-Pb zircon dating. Sample UP 3001 is a typical example of the greyish, fine-grained granites in this area. It contains predominantly quartz and plagioclase, and a minor amount of colourless K-feldspar. The main mafic mineral is biotite and no amphiboles were found in this rock. Using the multi-element classification (Fig. 2A) of De La Roche *et al.* (1980), granite UP 3001 plots at high R1 values at the right-hand side of the syenogranite field. In contrast to this greyish granite, sample UP 3005 is dominated by red K-feldspar and plagioclase is less frequent. UP 3005 can be classified as syenogranite according to De La Roche *et al.* (1980). Sample UP 3002 is greyish because of the high content of quartz. Plagioclase is the second most abundant mineral, whereas K-feldspar is scarce. UP 3002 shows smooth layers defined by concentrations of biotites. As a result of major element distribution, it is classified as a monzonite to quartz-syenite (Fig. 2A).

In addition to these granites, the present authors also dated a two-mica granite from the Holomha River area (UP 3023). The primary petrographical difference between the Kitoy and the Holomha granites is the occurrence of muscovite together with biotite. Similar to UP 3005, the K-feldspars of the Holomha two mica granite are of red colour. Applying the De La Roche *et al.* (1980) classification, the Holomha sample is defined to be a syenogranite (similar to UP 3005; Fig. 2A).

2.2. Other basement rocks

To get an overview of the geochemical and isotopic composition of additional Kitoy area basement rocks, the authors sampled two granulites to granulitic gneisses, a biotite-garnet-bearing gneiss, an amphibolite and a metabasite. The biotite gneiss has the characteristics of a typical upper crustal rock and a high alumina saturation index (1.19), whereas the granulites, the amphibolite and the meta-basite have a more mafic composition with SiO_2 contents between 45 and 60 wt.%. The granulite UP 3014 contains ortho- and clinopyroxene, plagioclase, and high concentrations of accessory minerals such as titanite, apatite and zircon. The granulite UP 3008 contains garnet, orthopyroxene, plagioclase and quartz. The

Table 1 Major and trace element concentrations of the Kitoy rocks (analysed by XRF)

Sample	Kitoy granites		Kitoy Qz-syenite	Holomha 2-micra gr.	Kitoy gneiss	Kitoy granulites		Kitoy meta-bas.	Kitoy amphibolite
	3001	3005	3002	3023	3011	3014	3008	3006	3009
SiO_2	77·29	72·80	60·54	71·17	76·82	45·89	47·65	49·01	59·68
TiO_2	0·15	0·11	0·70	0·32	0·31	0·55	2·99	1·25	0·89
Al_2O_3	12·08	13·21	19·83	12·80	11·17	15·17	13·36	7·32	13·37
Fe_2O_3	0·79	1·44	3·32	3·05	4·32	9·63	19·78	15·96	10·22
MnO	0·01	0·01	0·03	0·03	0·10	0·66	0·23	0·25	0·14
MgO	0·13	0·32	1·22	0·27	0·27	3·85	5·93	12·85	4·24
CaO	1·18	0·90	3·92	0·76	1·65	20·50	9·50	9·45	6·77
Na_2O	2·50	2·51	4·82	2·97	2·27	0·41	0·30	1·23	3·35
K_2O	4·96	6·39	4·22	5·27	2·48	0·72	0·06	2·02	1·27
P_2O_5	0·03	0·06	0·33	0·06	0·04	n.d.	0·54	0·12	0·10
l.ign.†	1·70	0·65	0·80	2·84	0·62	2·0	n.d.	n.d.	0·26
Total	100·82	98·40	99·73	99·54	100·05	99·38	100·34	99·46	100·29
Ba	300	1679	981	457	769	297	39	649	329
Co	5	3	11	3	n.d.	24	54	74	31
Cr	21	11	10	8	15	169	93	1303	110
Cu	5	7	21	4	11	9	160	33	75
Ga	14	15	27	18	16	20	23	14	20
Nb	12	4	12	14	9	7	22	10	8
Ni	4	11	18	9	5	79	70	557	54
Pb	64	29	24	26	18	9	5	n.d.	7
Rb	150	209	173	286	80	53	3	102	32
Sc	3	1	3	n.d.	6	29	39	23	19
Sr	71	401	505	69	75	185	147	215	156
Th	36·6	5·3	36·2	29·2	20·7	9·0	3·4	0·5	4·5
U	2·7	n.d.	1·5	n.d.	2·3	2·7	1·5	n.d.	n.d.
V	9	9	46	18	6	133	459	218	175
Y	27	5	46	20	65	23	61	30	22
Zn	12	15	47	43	50	150	193	228	86
Zr	113	64	494	267	386	99	306	125	108
ASI	1·04	1·04	1·01	1·07	1·19	0·39	0·75	0·34	0·70
Mg no.	0·14	0·18	0·27	0·08	0·06	0·28	0·23	0·45	0·31

*n.d. = not detected (or below detection limit).
†l. ign. = loss of ignition.

samples UP 3006 and UP 3009 are pyroxene- and/or amphibole-bearing. The meta-basite UP 3006 contains very fresh pyroxenes. All of these samples were analysed using whole-rock geochemical and isotopic techniques; the granulite UP 3014 was dated using U-Pb zircon geochronology.

3. Results

3.1. Geochemistry and Sr-, Nd- and Pb-isotopes

The chemical and isotopic composition of the granites from Kitoy are of special interest because establishment of their geodynamic setting is crucial for the correct interpretation of the geochronological data. Using the alumina saturation index (Chappell & White 1974), all investigated granites have values between 1.01 and 1.07 (see Table 1). This implies that they have I-type as well as S-type characteristics, and a final classification by the ASI is not possible (Chappell 1999). In general, all four granites have alkalic composition according to the total alkali versus silica (TAS) classification of Cox *et al.* (1979) and Wilson (1989). With Na_2O+K_2O between 7·46 and 9·04, and SiO_2 contents above 60·5 wt.%, the Kitoy granites are falling into the syenite, granite and alkali-granite fields in this plot.

Mafic rocks from the Kitoy river have SiO_2 and Fe_2O_3 contents between 49·01 (UP 3006) and 59·7 wt.% (UP 3009), and between 10·22 and 19·78 wt.%, respectively. Using the TAS classification after Le Maitre *et al.* (1989), they fall into the basalt (UP 3006) and andesite (UP 3009) fields, and are all clearly subalkalic. The Mg-numbers (see Table 1, calculated from wt.% of MgO and Fe_2O_3) vary between 0·31 and 0·45.

Based on the trace element concentrations of Nb, Y, Zr and Rb, the granite UP 3001 and Qz-syenite UP 3002 can be classified as volcanic-arc granites following the discriminations of Pearce *et al.* (1984) and Harris *et al.* (1986), or as post-collisional granites according to Pearce (1996). The sample UP 3005, one of the Kitoy granites, falls into the collisional or into the common field of collisional and volcanic-arc granites. The two-mica granite UP 3023 plots into the collisional and post-collisional fields on different discrimination diagrams of Pearce *et al.* (1984) and Pearce (1996). However, it has been shown by Mukasa & Henry (1990), for example, that these classifications can be ambiguous. To illustrate the geodynamic implications, Figure 2B shows the less problematic diagram of Pearce *et al.* (1984) using Nb and Y concentrations in order to distinguish WPG, ORG and VAG together with COLG. All four Kitoy granites fall into the COLG+VAG field.

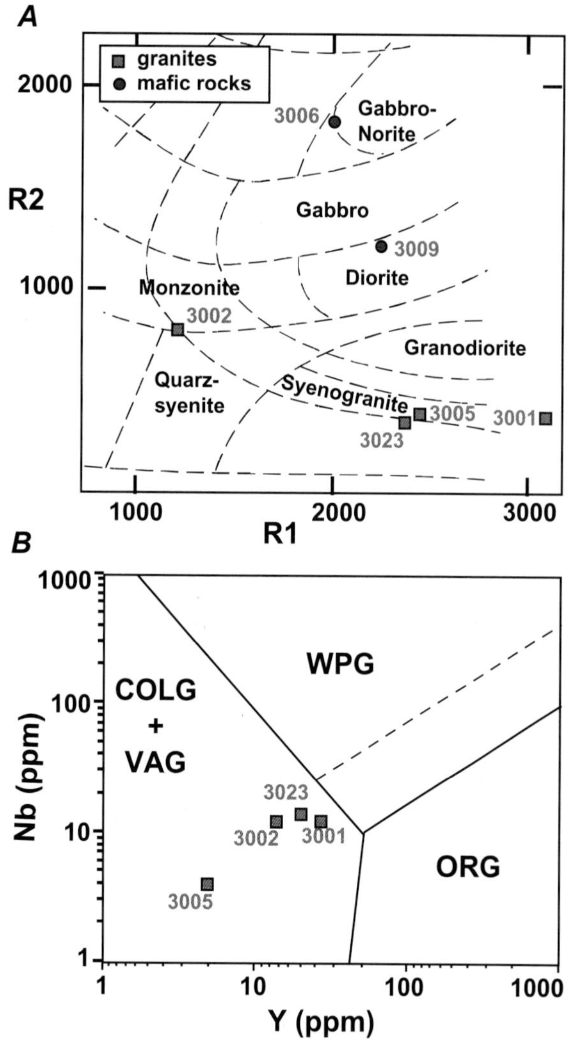

Figure 2 (A) Multi-element parameter (R1–R2) classification after De La Roche *et al.* (1980) applied for the Kitoy rocks. (B) Y versus Nb discrimination diagram after Pearce *et al.* (1984) showing the geodynamic characteristics of the Kitoy granites: (VAG) volcanic-arc granites; (COLG) collisional granites; (WPG) within-plate granites; and (ORG) ocean-ridge granites.

The Sr isotopes are given together with the Nd isotope data in Table 2. The Sr initial ratios (recalculated to 1850 Ma) vary from 0·731 to 0·867 for the granites, the mafic rocks have Sr $\text{initials}_{(1850)}$ of 0·718 and 0·730, and the granulite has an Sr $\text{initial}_{(1850)}$ of 0·706. These values are (as are Nd initial isotope ratios) inside the range McCulloch & Chappell (1982) have reported for I-type granites in Australia and Allegre & Ben Othman (1980) published for average Precambrian granitoids from the upper crust.

Recalculated $\varepsilon\,Nd_{(1850)}$ values vary between −6·8 for UP 3001 to −10·7 for UP 3005. Such ε Nd values are quite common for Proterozoic crustal derived rocks of Siberia (e.g. Jahn *et al.* 1998). However, the $^{147}Sm/^{144}Nd$ ratios of the Kitoy granites (0·077–0·116) are lower than the average continental crust value (0·12; White 1998). The Pb-Pb analyses of the whole rock splits are shown in Table 3 and in Figure 3. In the $^{206}Pb/^{204}Pb$ versus $^{207}Pb/^{204}Pb$ plot (Fig. 3A), the mafic rock UP 3006 is situated near the mantle evolution line after Doe & Zartman (1979). Sample UP 3008 has very radiogenic values and plots rather distant from the evolution lines. The Pb composition of the granites show a relative large variation (Table 3 and Fig. 3A). The granites UP 3001 and UP 3005 lie between the Pb evolution lines of the old upper crust and the Stacey & Kramers (1975) model, whereas the other two granites (UP 3002 and UP 3023) have a strong affinity towards the upper mantle evolution line.

For two samples, the two-mica granite UP 3023 and the granulite UP 3014, the Pb-Pb system could be used for the construction of Pb-Pb regression lines (Fig. 3B, C). Figure 3B shows the result for the Holomha two-mica granite. Six different whole rock splits (see Table 3) result in a regression line with correlation of 0·984 and an age of 1821 ± 29 Ma. The five whole rock splits of granulite sample UP 3014 also fit on a regression line (Fig. 3C and Table 3), resulting in an age of 1884 ± 26 Ma (r=0·986).

3.2. U-Pb zircon dating

The U-Pb zircon data for the granites are presented in Table 4 and Figure 5. In the case of the Kitoy granite, UP 3001, the present authors used both the vapour digestion and CLC method (Poller & Todt 2000). All other samples were dated only by the vapour digestion method (Wendt & Todt 1991).

Figure 4 shows several CL images of zircons from the Kitoy samples. For UP 3001, the authors show all five dated zircons. For the other dated samples, only two examples of typical CL images are given. The zircons of UP 3001 are all magmatic-zoned crystals without large core components. Only zircon 3001–18 shows a possible small, brightly luminescent core. In addition to homogeneous, magmatic-zoned grains, zircons with lines of bright luminescence within dark crystals or diffuse, quietly luminescent grains are present (e.g. 3001–24 or 3002–22; Fig. 4). Low luminescence results from high U concentrations within these zones, which typically decreases the CL signal because of radioactive destruction of the crystal lattice (Poller *et al.* 2001a). Zircons from other Kitoy granites contain dominantly magmatic crystals with few inherited cores (e.g. 3023–2). In sample UP 3023, core components are more widespread than in the other granites, but nevertheless, magmatic structures also occur (e.g. 3023–20). The granulite UP 3014 contains zircons dominated by magmatic growth (see Fig. 4).

The age determination of sample UP 3001 (see Fig. 5A) resulted in a discordia line through five grains (zircons UP 3001–17 and 3001–18 were not used for calculation because of their position parallel to the discordia line, and consequently, are only given in Table 4). The upper intercept yields an age of 1880 ± 11 Ma, but the lower intercept at 311 ± 57 Ma is not well defined.

The results for sample UP 3005 are similar. Figure 5B shows the discordia line through an age of 1875 ± 18 Ma. Again, the lower intercept is with 383 ± 100 Ma Palaeozoic in age.

The granulite UP 3014 (Fig. 5C) has a discordia line with an upper intercept of 1888 ± 61 Ma, which corresponds well with the Pb-Pb age (1884 ± 26 Ma; Fig. 3C). Like the Kitoy granites, the lower intercept is not well defined because of the small spread of the results, but appears Palaeozoic in age (225 ± 300 Ma).

The two-mica granite UP 3023 from the Holomha River presented in Fig. 5D has an upper intercept age of 1844 ± 14 Ma, and therefore, belongs to the 'younger' granitoid rocks of this region. This zircon age of 1844 ± 14 Ma is in good correspondence with the Pb-Pb whole rock data (1821 ± 29 Ma; Fig. 3B). Both the Pb-Pb results and the discordia through zero demonstrate that the Pb systematics of these granites have remained undisturbed since the Early Proterozoic.

The Qz-syenite UP 3002 is shown in Figure 5E and the nine measured zircons define an upper intercept age of 1853 ± 5 Ma. The lower intercept of the discordia is zero, reflecting recent Pb loss.

Table 2 Sm-Nd and Rb-Sr isotope data for the Kitoy rocks: Sr and Nd isotope ratios were analysed by TIMS; Sm-Nd and Rb-Sr ratios were calculated using ICP-MS and XRF data*

	$^{147}Sm/^{144}Nd$	$^{143}Nd/^{144}Nd$ today	± 2 σ m	ε Nd (1·85 Ga)	T DM 1 step	T DM 2 step (1·85 Ga)	$^{87}Rb/^{86}Sr$	$^{87}Sr/^{86}Sr$ (today) ± 2 sm	± 2 σ m	$^{87}Sr/^{86}Sr$ (1·85 Ga)
Kitoy granitoids and gneiss										
UP 3001	0·1158	0·511308	17	−6·8	2·71	2·74	0·5896	0·883026	28	0·867333
UP 3002	0·0768	0·510622	7	−10·4	2·66	3·03	0·0956	0·733618	28	0·731073
UP 3005	0·0864	0·510750	41	−10·7	2·75	3·05	0·1454	0·744967	28	0·741096
UP 3011	0·1128	0·510883	26	−14·4	3·23	3·33	0·2977	0·822732	47	0·814809
Kitoy meta-basite, granulite and amphibolite										
UP 3006	0·1387	0·511453	24	−9·4	3·20	2·95	0·1324	0·721726	15	0·718202
UP 3008	0·1343	0·511447	20	−8·5	3·05	2·87	0·0057	0·706100	34	0·705948
UP 3009	0·1130	0·510536	12	−21·2	3·73	3·86	0·0572	0·731484	31	0·729960

*The element ratio errors are estimated to be 0·5% (2σ level).

4. Discussion

The petrographic features, i.e. the major and trace element concentrations of the granites, suggest that they have affinities to I-type as well as to S-type rocks. When plotted in discrimination diagrams based on major and trace element abundances, variable geodynamic settings are suggested, with some dependence on the discrimination diagram used. Applying the granite classification features of Castro *et al.* (1991), the major elements and ASI indicate a possible H-type origin; however, the $^{87}Sr/^{86}Sr_{(ini)}$ values are rather high. Age-corrected $^{87}Sr/^{86}Sr_{(ini)}$ values (all calculated for an age of 1850 Ma) are very similar to present day ratios (see Table 2) and reflect a high influence from the upper crust. These upper crustal signatures of Sr isotopic ratios (0·73 to 0·88) indicate that older crustal material has been reworked and subsequently incorporated into the Kitoy granites.

The influence of reworked crust is also visible in the Nd model ages of the investigated rocks. The relatively low Sm/Nd ratios (0·0768 to 0·1158 for the granites) have a significant influence on Nd model ages. The ratios produce a scatter between 2·66 and 2·75 Ga in one-step calculations and 2·74 to 3·05 Ga in two-step calculations using the equation of Liew & Hofmann (1987; see Table 2). For mafic rocks and the biotite gneiss, the result is similar, with the oldest model ages being 3·73 Ga (one-step) and 3·86 Ga (two-step) for sample 3009. In addition to the Sm/Nd ratios, ages used to constrain the two-step evolution are also important. U-Pb results indicate that the dated granites have upper intercept ages between 1880 and 1853 Ma. Therefore, it is reasonable to assume the last reset of the Nd isotope system happened during the Early Proterozoic time period. Consequently, the present authors used an age of 1850 Ma to constrain the two-step evolution shown in Table 2. With regard to recently dated granulites which are Archaean in age (Poller *et al.* 2004), several outcrops along the southern margin of the Siberian Craton contain old Archaean source material. Therefore, two-step model ages of 2·74 to 3·85 Ga appear to be reasonable.

The Sr, Nd and Pb isotope systematics constrain a source material composed of primarily crustal components for the analysed granites. However, common Pb results demonstrate that a small amount of additional material from a mantle source has been incorporated.

At first glance, the Pb-Pb data (Fig. 3A) might imply a significant mantle influence for the Kitoy samples. However, evolution of the U-Pb system is actually responsible for the low $^{207}Pb/^{204}Pb$ ratios and the present day position of the granite data in Figure 3A. These samples probably lost uranium at the beginning of the Proterozoic. Between the Archaean and Proterozoic, the amount of available oxygen increased dramatically (Kump *et al.* 2001; Sleep 2001). These changed conditions allowed uranium to oxidise and to react to UO_2. In this state, uranium is highly soluble in water and is easily lost from U-Pb system. This seems to have happened to the source materials of UP 3001, UP 3002 and UP 3005. After uranium loss evolution in the $^{207}Pb/^{204}Pb$ versus $^{206}Pb/^{204}Pb$ diagram proceeded primarily along the x-axis ($^{206}Pb/^{204}Pb$), while the $^{207}Pb/^{204}Pb$ ratios have changed very little.

In general, Pb isotopes and geochemistry indicate that the source material for the analysed Kitoy granites is dominated by the crust, with only a minor influence of mantle-derived material.

The geochronological and CL zircon data indicate two different Early Proterozoic age groups at 1·88 and 1·85 Ga.

All measured zircons are discordant; however, the CL images show no evidence of metamorphic overgrowth or resorbed areas which indicate a specific Pb-loss event responsible for the discordance (see Fig. 4). Therefore, Pb loss seems to have been continuous for all samples and recent for the young granites.

Taking this into account, the upper intercept ages for the Kitoy granites UP 3001 and UP 3005 of 1880 ± 11 Ma and 1875 ± 18 Ma, respectively, reflect the intrusion of these granites. The lower intercepts of 357 ± 11 Ma and 383 ± 100 Ma are both Palaeozoic, and as mentioned above, are not well constrained because of the small scatter of the data. A metamorphic overprint can generally excluded because overprinting structures were not found in CL images. Additionally, there is no petrographic evidence for metamorphism after emplacement in any of the granites thin sections. Since similarly low intercept ages are known for this part of Siberia from earlier studies (e.g. Donskaya *et al.* 2002; Levitskii *et al.* 2002), a regional process may be responsible for these ages. Because of the lack of structural evidence, a low-temperature mobilisation of fluids, in combination with uplift, could explain the lower intercept ages. Metamorphism of this type has been reported during the Ordovician (480–460 Ma; Salnikova *et al.* 1998; Donskaya *et al.* 2001) for other parts of the Baikal area (Khamar Daban, Ol'khon). Fluid circulation could explain Pb mobilisation and Pb loss, and consequently, the lower intercept ages of the Kitoy granites.

Similar age results were found for the Kitoy granulite UP 3014. The upper intercept of 1888 ± 61 Ma is interpreted as the peak of granulite metamorphism because CL images of that sample show growth features similar to magmatic zoning.

Table 3 Pb-Pb whole rock isotope data for the Kitoy rocks*

Sample	$^{206}Pb/^{204}Pb$	$^{207}Pb/^{204}Pb$	$^{208}Pb/^{204}Pb$
3001-1	18·6154	15·7030	41·8376
3001-2	17·5493	15·6273	40·1284
3001-3	17·3130	15·5884	38·4339
3001-4	18·7949	15·7172	41·6477
3001-5	17·8882	15·6016	39·6470
3002-1	17·7160	15·3507	60·5995
3002-2	15·9073	15·2409	42·9143
3002-3	15·9036	15·2470	40·0658
3002-4	16·0359	15·2599	46·7169
3002-5	15·5152	15·1995	38·7599
3005-1	15·1140	15·1875	36·4086
3005-2	14·4125	15·2672	35·8913
3005-3	14·8575	15·1194	35·5516
3005-4	14·7922	15·1018	35·5708
3005-5	14·8575	15·0978	35·6368
3023-1	16·8235	15·2953	42·7375
3023-2	18·0257	15·4047	42·4104
3023-3	16·1044	15·1747	41·7984
3023-4	15·9020	15·1777	39·5031
3023-5	16·8563	15·2857	40·0758
3023-6	16·3654	15·2438	42·9585
3014-1	26·6197	16·5708	45·8378
3014-2	28·0333	16·7448	49·9306
3014-4	28·2122	16·7800	49·6560
3014-5	27·2003	16·6117	48·3103
3014-6	29·4680	16·8814	48·5534
3006-1	17·5423	15·4520	36·4172
3006-2	16·9265	15·4137	36·0727
3006-3	17·0794	15·4001	36·0870
3006-4	17·1887	15·4433	36·1264
3006-5	17·5774	15·5090	36·2963
3008-1	19·8209	16·0575	39·5383
3008-2	20·3847	16·0783	40·9945
3008-3	20·6976	16·1311	39·9983
3008-4	20·9582	16·1421	42·4978
3008-5	20·8388	16·1172	39·5598

*Pb data are corrected for fractionation.
The 2σ external reproducibility is <200 ppm.

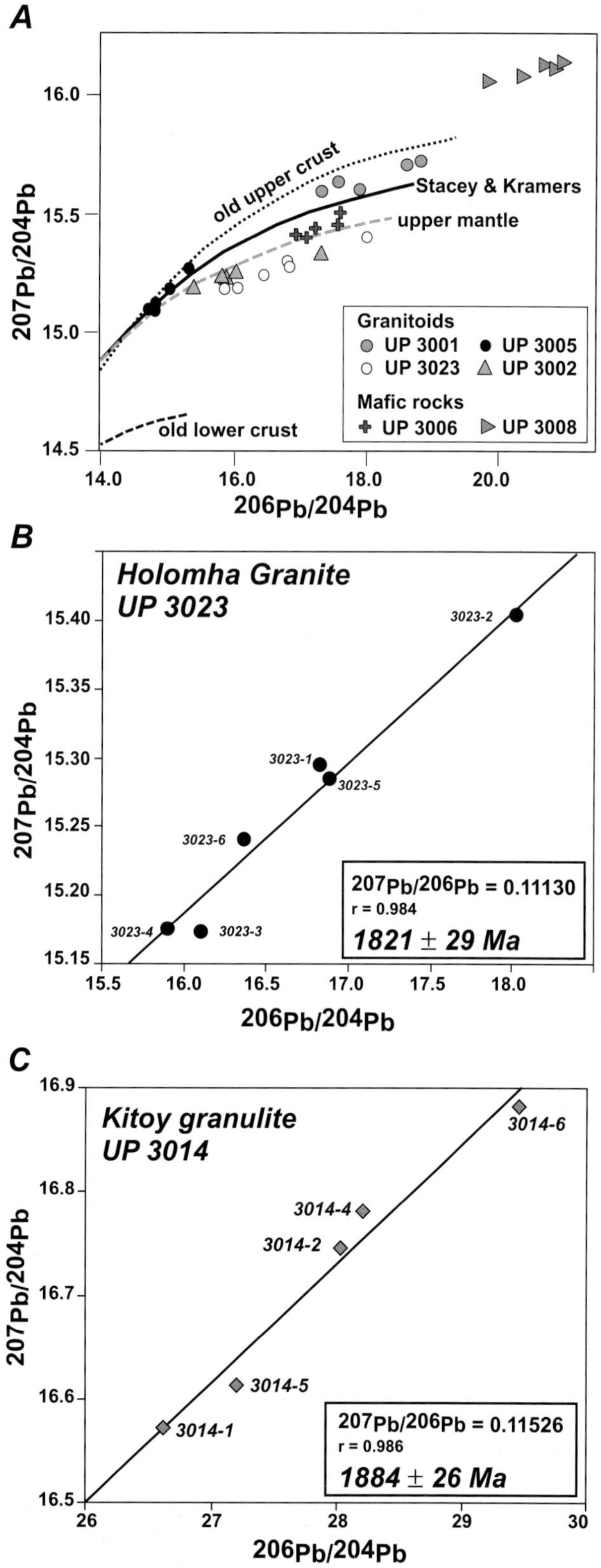

Figure 3 (A) $^{207}Pb/^{204}Pb$ versus $^{206}Pb/^{204}Pb$ diagram with Pb evolution lines after Doe & Zartman (1979) and Stacey & Kramers (1975) showing the Pb isotopic composition of several Kitoy rocks. (B) $^{207}Pb/^{204}Pb$ versus $^{206}Pb/^{204}Pb$ whole-rock isochron diagram for the Holomha granite UP 3023. (C) $^{207}Pb/^{204}Pb$ versus $^{206}Pb/^{204}Pb$ whole-rock isochron diagram for the Kitoy granulite UP 3014.

Internal structures like these are known to be found in high-grade migmatites or granulites (e.g. Hoskin & Black 2000; Poller & Todt 2000; Corfu *et al.* 2003). Newly grown zircons on the peak of, for example, granulite facies metamorphism often have 'magmatic' CL structures like these. Since the upper intercept age is not well defined, the granulite metamorphism cannot definitely be linked with either the 1·88 Ga or the 1·85 Ga granite intrusion. Recent SHRIMP analyses of the Kitoy granulite UP 3014 (Poller *et al.* 2004) resulted in concordant ages around 1855 Ma and suggest that the granulite metamorphism is connected to the second event in the Early Proterozoic.

The discordia of UP 3014 given in Figure 5C shows a lower intercept at 225 ± 300 Ma. This imprecise age cannot be linked to the granulite metamorphism because the neighbouring granites did not suffer similarly high metamorphic conditions. We cannot be certain of a Palaeozoic Pb loss like that of the old Kitoy granites because of the large error. Because the Pb isochron corresponds with the discordia age and documents the undisturbed nature of the Pb system since ≈ 1·8 Ga, recent Pb loss seems to be much more probable for this granulite. In summary, the only geologically meaningful explanation for the

Table 4 U-Pb zircon data for the Kitoy rocks: (VD) vapour digestion single zircon datings; and (CLC) cathodoluminescence controlled dating

No. sample methods	Measured atomic ratios*							Ages (Ma)†					
	Utot/Pb	$^{206}Pb/^{204}Pb$	2σ m‡	$^{207}Pb/^{206}Pb$	2σ m‡	$^{208}Pb/^{206}Pb$	2σ m‡	$^{206}Pb/^{238}U$	2σ m‡	$^{207}Pb/^{235}U$	2σ m‡	$^{207}Pb/^{206}Pb$	2σ m‡
2UP 3001, Kitoy granite													
1 3001-d VD	3·09	878	21	0·12679	29	0·16597	46	1504·6	12·6	1643·0	13·4	1824·8	15·4
2 3001-e VD	3·10	1047	13	0·12480	21	0·15620	58	1502·6	7·8	1644·7	7·7	1831·5	8·3
3 3001-6 CLC	3·37	936	15	0·12458	21	0·16799	55	1385·8	7·5	1560·4	8·6	1805·4	9·1
4 3001-12 CLC	3·48	967	7	0·12424	19	0·14936	37	1365·7	5·7	1548·3	6·3	1807·3	6·2
5 3001-17 CLC	3·38	1333	41	0·11926	18	0·15123	57	1393·1	10·4	1556·2	10·5	1785·1	10·8
6 3001-18 CLC	3·43	2010	35	0·11424	16	0·18551	48	1334·1	5·2	1506·4	5·6	1757·9	5·9
7 3001-24 CLC	3·01	3782	48	0·11655	16	0·06830	20	1618·3	8·4	1721·1	6·5	1848·6	3·7
UP 3002, Kitoy Qz-syenite													
8 3002-c VD	2·45	513	29	0·14037	121	0·37294	292	1633·1	35·9	1738·0	41·6	1866·7	60·0
9 3002-d VD	2·52	274	7	0·15773	158	0·46487	513	1556·0	54·2	1650·0	57·1	1771·8	81·4
10 3002-e VD	2·55	12694	1746	0·11423	24	0·25840	899	1636·4	18·9	1733·1	14·1	1852·0	6·6
11 3002-f1 VD	2·52	965	112	0·12418	31	0·20313	80	1754·7	22·2	1774·7	27·9	1798·4	39·2
12 3002-f2 VD	2·44	1264	34	0·12369	41	0·20400	98	1785·1	20·8	1814·9	16·9	1849·4	15·4
13 3002-g VD	2·55	344	5	0·15233	62	0·33241	80	1651·5	33·5	1744·2	29·9	1857·1	30·5
14 3002-h VD	2·57	797	7	0·12983	32	0·30108	114	1615·9	10·0	1722·3	10·5	1854·1	9·6
15 3002-i VD	2·42	741	14	0·13127	24	0·30024	101	1707·4	8·7	1774·2	9·9	1853·7	11·9
16 3002-j VD	2·65	3840	76	0·11671	18	0·18876	51	1662·0	5·9	1748·6	5·5	1853·8	3·9
UP 3005, Kitoy granite													
17 3005-a VD	4·50	1929	19	0·11081	24	0·09905	32	1114·1	7·2	1328·2	7·6	1692·5	6·9
18 3005-b VD	2·64	850	16	0·12898	23	0·23277	63	1652·2	20·3	1741·7	16·1	1850·9	15·3
19 3005-c VD	2·44	1059	23	0·12576	32	0·27078	76	1714·7	11·4	1776·4	11·2	1849·8	15·6
20 3005-d VD	2·65	902	9	0·12796	25	0·23978	69	1635·8	10·9	1732·5	10·0	1851·3	9·1
21 3005-e VD	2·39	1938	70	0·12003	21	0·22563	91	1786·5	26·1	1817·1	19·4	1852·4	11·6
22 3005-f VD	2·53	1357	49	0·12283	27	0·21577	108	1716·1	14·5	1777·1	13·8	1849·6	12·8
UP 3014, Kitoy granulite													
23 3014-a VD	3·37	2294	70	0·11875	36	0·11287	54	1420·9	7·1	1601·4	8·6	1847·7	9·6
24 3014-c VD	3·36	2822	169	0·11693	26	0·06389	26	1477·7	48·9	1631·8	34·1	1836·5	9·1
25 3014-d VD	2·88	2469	203	0·11983	28	0·09382	31	1654·5	8·6	1752·8	10·5	1872·1	12·1
26 3014-e VD	3·38	2028	72	0·11879	33	0·08801	27	1447·1	7·2	1613·2	8·8	1837·2	9·5
27 3014-f VD	3·51	406	3	0·14610	20	0·15445	58	1399·6	7·3	1589·4	8·9	1851·2	13·2
UP 3023, Kitoy two-mica granite													
28 3023-a VD	2·28	222	5	0·17480	32	0·45352	208	1732·9	22·2	1799·2	27·0	1876·9	36·8
29 3023-b VD	3·12	354	16	0·14997	59	0·39692	125	1315·2	8·6	1528·1	24·2	1837·0	45·3
30 3023-d VD	2·46	340	9	0·15179	25	0·39765	129	1631·2	9·9	1726·3	17·3	1843·6	27·0
31 3023-e VD	2·24	229	3	0·17079	22	0·47812	188	1727·9	11·2	1780·8	16·1	1843·5	25·2
32 3023-f VD	3·41	308	2	0·15603	29	0·39174	206	1230·2	7·7	1741·9	10·3	1840·2	16·0

*Corrected for fractionation.
†Corrected for blank, spike and common Pb.
‡2σ m errors refer to the 2σ deviation of the weighted mean of two to six blocks (22–66 ratios).

age data is that the granulite facies metamorphism happened in Early Proterozoic times and that this granulite has suffered only a recent Pb loss.

Figure 5D and E shows two analysed granites which are 30 Ma younger than the 'old' Kitoy granites. The Holomha two-mica granite and the Qz-syenite UP 3002 have crystallisation ages of 1844 ± 14 Ma and 1853 ± 5 Ma. These younger granites have discordia lines passing through zero, representing recent Pb loss. This is in contrast to the older granites discussed above. The younger granites seem unaffected by the Pb mobilisation in Palaeozoic times. This is also reflected in the Pb isochron of the Holomha granite (Fig. 3B), which corresponds to the zircon dating age.

5. Conclusions and summary

In general, the Early Proterozoic (about 2·0–1·9 Ga) was a time period with both high crust production and high continental growth rates (Condie 1998, 2000).

The granite emplacements and the granulite metamorphism in the Kitoy region happened between 1·88 and 1·85 Ga, and therefore, are older than the assumed assemblage of Rodinia (around 1·4 Ga).

Archaean Nd model ages and Pb isotopes demonstrate that old crustal material was reworked during the intrusion of the investigated granites. The data base of reliable geochemical and isotopic data as well as age determinations for the Siberian Craton is not yet large enough for direct comparisons with analogous counterparts during a collisional scenario. Nevertheless, the magmatic activity in Siberia during Early Proterozoic time is not limited to the Kitoy area. For other blocks of the Siberian Craton, Early Proterozoic activity was marked by high-grade metamorphic complexes with ages around 2·0–1·9 Ga (e.g. Aftalion *et al.* 1991; Bibikova *et al.* 1993), collisional granites about 2·0 Ga (e.g. Neymark *et al.* 1998) and 1·88–1·84 Ga post-collisional granites (Donskaya *et al.* 2002, 2003; Levitskii *et al.* 2002; Poller *et al.* 2004, in press).

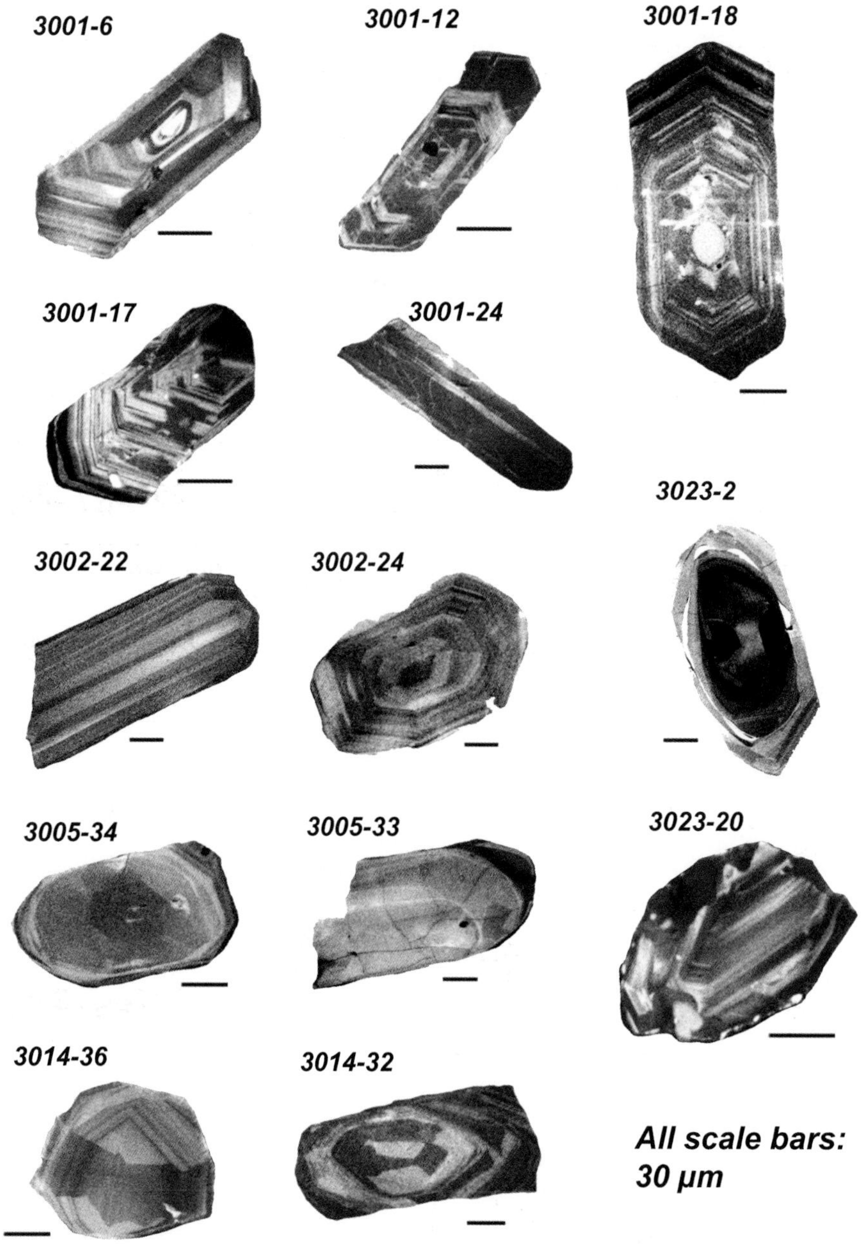

Figure 4 (A) Selected cathodoluminescence images of zircons from the Kitoy samples. The shown zircons of sample UP 3001 were used for the CLC dating (see Table 4 and Fig. 5A for results).

Consequently, these Early Proterozoic ages occur along the entire southern margin of the Siberian Craton. Two interpretations are possible: either this magmatic and metamorphic activity is linked to the consolidation of the Siberian Craton itself, or it is linked to the amalgamation of the Siberian Craton to an Early Proterozoic supercontinent (Condie 2002). In both cases, the magmatism is linked to continental collision with reworking of old crustal material.

Donskaya *et al.* (2002) and Levitskii *et al.* (2002) interpreted 1·86–1·84 Ga old post-collisional granite complexes in South Siberia as reflecting a late stage of the amalgamation of the Siberian Craton. In this context, the intrusion of the younger (1·85 Ga) Kitoy granites possibly documents the completion of the main collisional event in this part of the craton. Later (Palaeozoic) uplift movements are possibly reflected in the lower intercept ages of some of the Kitoy granites.

To summarise the results of the present study, the following points can be stated:

1. The Pb-Pb system remained closed since Early Proterozoic and constrains the crustal character of the source material with a small addition from a mantle source.
2. The Nd isotopes indicate the crustal origin of the granites. The Nd model ages reflect that the source material was separated from the mantle reservoir in Archaean times ($T_{DM(t)}$>2·7 Ga).
3. The first Early Proterozoic granites at the Kitoy area were emplaced 1880 Ma ago. Only 30 Ma later, around 1850 Ma ago, the Holomha two-mica granite and the Qz-syenite at the Kitoy area intruded.
4. The magmatic and metamorphic activities responsible for the Kitoy granites are linked to the consolidation of the Siberian Craton in Early Proterozoic.

6. Acknowledgements

We are grateful to J. Huth for his help at the scanning electron microscope and the cathodoluminescence, and to B. Stoll, M. Willbold and A. Gaab for their help during the ICP-MS

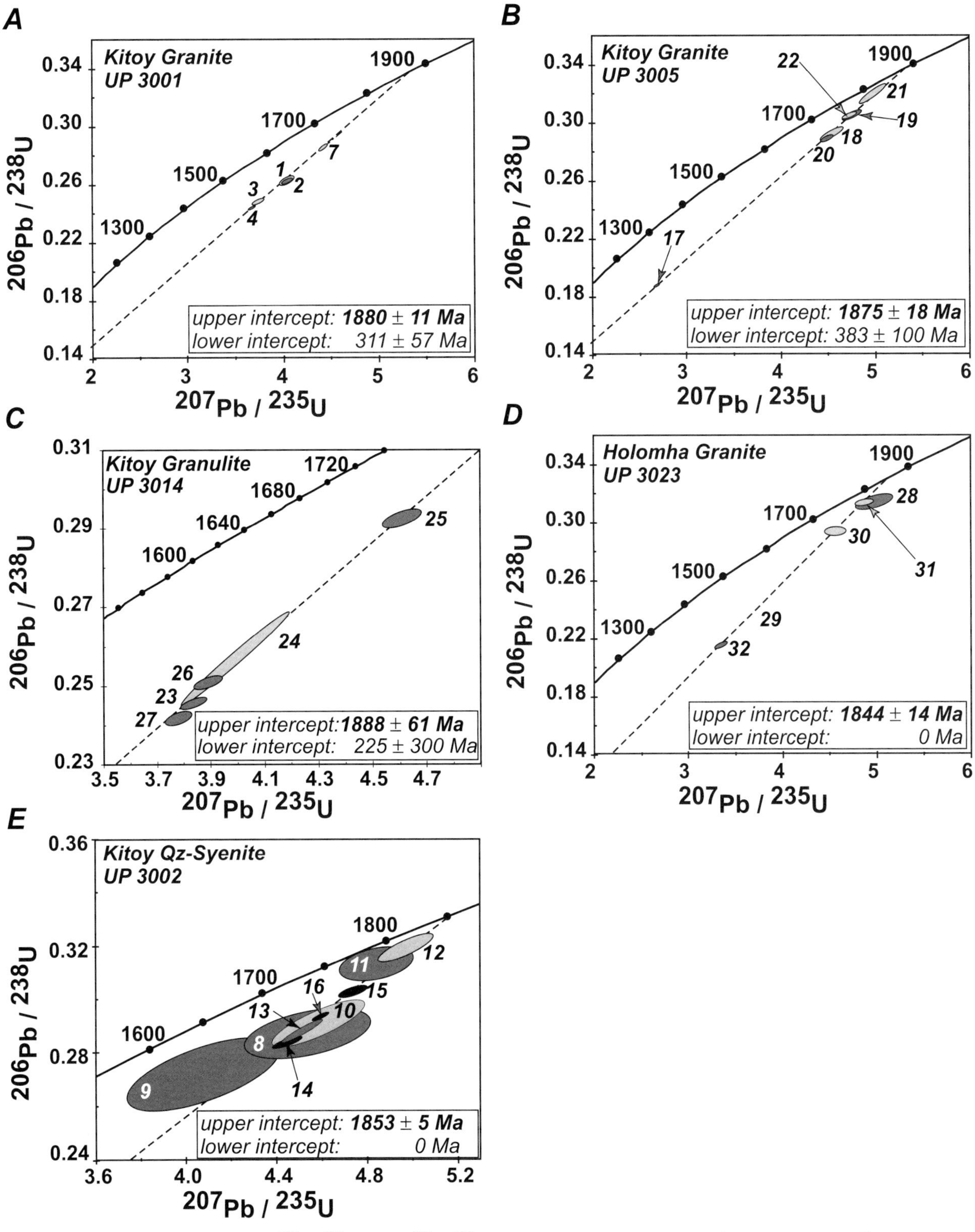

Figure 5 (A–E) Several $^{206}Pb/^{238}U$ versus $^{207}Pb/^{235}U$ discordia plots for the dated Kitoy granites, granulite gneiss and Qz-Syenite.

analyses. Many thanks also to D. Neuhäuser for her work in the clean laboratory. We thank A. Hofmann for the opportunity to perform this study at MPI. This work was supported by the DFG (PO 608/4-1 and PO 608/3-2), the Russian Foundation for Basic Researches (#04–05–64412) and the Russian Science Support Foundation. The reviews of C. Buchan and R. Hinton helped to improve this manuscript. J. Friedrich is greatly acknowledged for the correction of the style and the language.

7. References

Aftalion, M., Bibikova, E. V., Bowes, D. R., Hopwood, A. M. & Perchuk L. L. 1991. Timing of Early Proterozoic collisional and extensional events in the granulite-gneiss-charnokite-granite complex, lake Baikal, USSR: A U-Pb, Rb-Sr and Sm-Nd isotopic study. *Journal of Geology* **99**, 851–61.

Allegre, C. J. & Ben Othman, D. 1980. Nd-Sr isotopic relationship in granitoid rocks and continental crust development: a chemical approach to orogenesis. *Nature* **286**, 335–42.

Bibikova, E. V., Gracheva, T. V., Makarov, V. A. & Nozhkin, A. D. 1993. Age boundaries in the Early Precambrian geologic evolution of the Yenisei Range. *Stratigrafiya. Geologicheskaya Korrelyatsiya* **1**, 35–40.

Castro, A., Moreno-Ventas, I. & de la Rosa, J. D. 1991. H-type (hybridic) granitoids: a proposed revision of the granite-type classification and nomenclature. *Earth Science Reviews* **31**, 237–53

Chappell, B. W. 1999. Aluminium saturation in I- and S-type granites and the characterization of fractionated haplogranites. *Lithos* **46**, 535–51.

Chappell, B. W. & White, A. J. R. 1974. Two contrasting granite types. *Pacific Geology* **8**, 173–4.

Condie, K. C. 1998. Episodic continental growth and supercontinents: a mantle avalanche connection? *Earth and Planetary Science Letters* **163**, 97–108.

Condie, K. C. 2000. Episodic continental growth models: afterthoughts and extensions. *Tectonophysics* **322**, 153–62.

Condie, K. C. 2002. Break-up of a Paleoproterozoic supercontinent. *Gondwana Research* **5**, 41–3.

Corfu, F., Hanchar, J. M., Hoskin, P. W. O. & Kinny, P. 2003. Atlas of zircon textures. *In* Hanchar, J. M. & Hoskin, P. W. O. (eds) *Zircon. Reviews in Mineralogy and Geochemistry* **53**, 469–500.

Cox, K. G., Bell, J. D. & Pankhurst, R. J. 1979. *The Interpretation of Igneous Rocks*. London: George Allen and Unwin.

De La Roche, H., Letherrier, J., Grandclaude, P. & Marchal, M. 1980. A classification of volcanic and plutonic rocks using R1–R2 diagrams and major-element analyses – its relationship with current nomenclature. *Chemical Geology* **29**, 183–210.

Doe, B. R. & Zartman, R. E. 1979. Plumbotectonics, the Phanerozoic. *In* Barnes, H. L.(ed.) *Geochemistry of Hydrothermal Ore Deposits*, 22–70. New York, NY: Wiley Interscience.

Donskaya, T. V., Sklyarov, E. V., Gladkochub, D. P. & Mazukabzov, A. M. 2001. Early Paleozoic collisional events along southern margin of the Siberian craton (Northern segment of the Central Asian foldbelt). *Gondwana Research* **4**, 610–11.

Donskaya, T. V., Salnikova, E. B., Sklyarov, E. V., Gladkochub, D. P., Mazukabzov, A. M., Kovach, V. P., Yakovleva, S. Z. & Berezhnaya, N. G. 2002. Early Proterozoic postcollisional magmatism at the southern flank of the Siberian Craton: new geochronological data and geodynamic implications. *Doklady Earth Sciences* **383/2**, 125–8.

Harris, N. B. W., Pearce, J. A. & Tindle, A. G. 1986. Geochemical characteristics of collision-zone magmatism. *In* Coward, M. P. & Ries, A. C. (eds) *Collisional Tectonics. Geological Society, London, Special Publication* **19**, 67–81.

Hoskin, P. W. O. & Black, L. P. 2000. Metamorphic zircon formation by solid-state recrystallization of protolith igneous zircon. *Journal of Metamorphic Geology* **18**, 423–39.

Jahn, B.-M., Gruau, G., Caodevila, R., Cornichet, J., Nemchin, A., Pidgeon, R. & Rudnik, V. A. 1998. Archean crustal evolution of the Aldan Shield, Siberia: geochemical and isotopic constraints. *Precambrian Research* **91**, 333–63.

Kump, L. R., Kasting, J. F. & Barley, M. E. 2001. Rise of atmospheric oxygen and the 'upside-down' Archean mantle. *Geochemistry, Geophysics, Geosystems* **2** , paper number 2000GC000114.

Le Maitre, R. W., Bateman, P., Dudek, A., Keller, J., Lameyre Le Bas, M. J., Sabine, P. A., Schmid, R., Sorensen, H., Streckeisen, A., Woolley, A. R. & Zanettin, B. 1989. *A Classification of Igneous Rocks and Glossary Terms*. Oxford: Blackwell.

Levitskii, V. I., Mel'nikov, A. I., Reznitskii, L. S., Bibikova, E. V., Kirnozova, T. I., Kozakov, I. K., Makarov, V. A. . & Plotkina, Yu. V. 2002. Early Proterozoic postcollisional granites in southwestern Siberian Platform. *Russian Geology and Geophysics (Geologia i Geofizika)* **43**, 679–92.

Liew, T. C. & Hofmann, A. W. 1987. Precambrian crustal components, plutonic associations, plate environment of the Hercynian Fold Belt of central Europe: indications from a Nd and Sr isotopic study. *Contributions to Mineralogy and Petrology* **98**, 129–38.

Loveridge, W. D. 1986. Measurement of biases in the electron multiplier ion detection system of a Finnigan MAT Model 261 mass spectrometer. *International Journal of Mass Spectrometry* **74**, 197–206.

Ludwig, K. 1992. ISOPLOT a plotting and regression program for radiogenic isotope data, version 2.57. *U.S. Geological Survey Open File Report* **91–445**. Berkeley, CA: Geochronology Center.

McCulloch, M. T. & Chappell, B. W. 1982. Nd isotopic characteristics of S- and I-type granites. *Earth and Planetary Science Letters* **58**, 51–64.

Mukasa, S. B. & Henry, D. S. 1990. The San Nicolas batholith of coastal Peru: early Paleozoic continental arc or continental rift magmatism? *Journal of the Geological Society of London* **147**, 27–39.

Neymark, L. A., Larin, A. M., Nemchin, A. A., Ovchinnikova, G. V. & Rytsk, E. Yu. 1998. Geochemical, geochronological (U-Pb), and isotope (Pb, Nd) evidence for the anorogenic type of magmatism in the North Baikal volcanoplutonic belt. *Petrologiya* **6**, 139–64.

Pearce, J. A. 1996. Sources and settings of granitic rocks. *Episodes* **19**, 120–5.

Pearce, J. A., Harris, N. B. W. & Tindle, A. G. 1984. Trace element discrimination diagrams for the tectonic interpretation of granitic rocks. *Journal of Petrology* **25**, 956–83.

Poller, U. 2000. A combination of single zircon dating by TIMS and cathodoluminescence investigations on the same grain: the CLC-method – U-Pb geochronology for metamorphic rocks. *In* Pagel, M., Barbin, V., Blanc, Ph. & Ohnenstetter, D. (eds) *Cathodoluminescence in Geosciences*, 401–14. Berlin: Springer-Verlag.

Poller, U., Huth, J., Hoppe, P. & Williams, I. S. 2001a. REE, U, Th and Hf distribution in zircon from Western Carpathian granites: A combined cathodoluminescence and ion microprobe study. *American Journal of Science* **301**, 858–76.

Poller, U., Kohut, M., Todt, W. & Janak, M. 2001b. Nd, Sr, Pb isotope study of the Western Carpathians: implications for Paleozoic evolution. *Schweizerische Mineralogisch-Petrographische Mitteilungen* **81**, 159–74.

Poller, U., Gladkochub, D., Donskaya, T., Mazukabzov, A., Sklyarov, E. & Todt, W. 2004. New Archean to Proterozoic SHRIMP ages on granulites from the Siberian Craton. *Geochimica et Casmochimica Acta* **68**, Issue 11, Suppl. 1, 664.

Poller, U., Gladkochub, D., Donskaya, T., Mazukabzov, A., Sklyarov, E. & Todt, W. in press. Multistage magmatic and metamorphic evolution of the Siberian Craton: Archean and Proterozoic zircons ages revealed by SHRIMP and TIMS. *Precambrian Research.*

Poller, U. & Todt, W. 2000. U-Pb single zircon data of granitoids from the High Tatra Mountains (Slovakia): implications for the geodynamic evolution. *Transactions of Royal Society of Edinburgh, Earth Sciences* **91**, 235–43.

Salnikova, E. B., Sergeev, S. A., Kotov, A. B., Yakovleva, S. Z., Reznitskiy, L. Z. & Vasilyev, E. P. 1998. Dating of granulite metamorphism in the Sludyanskiy complex, Eastern Siberia. *Gondwana Research* **1**, 195–205.

Sklyarov, E. V. (Ed), 2001. Assembly and break-up of Rodinia supercontinent: evidence from South Siberia. *Guidebook and Abstract Volume, Workshop IGCP-40*. Irkutsk: Institute of the Earth's crust, Siberian Division, Russian Academy of Sciences.

Sklyarov, E. V., Gladkochub, D. P., Mazukabzov, A. M., Menshagin, Yu. V., Watanabe, T. & Pisarevsky, S. A. 2003. Neoproterozoic mafic dike swarms of the Sharyzhalgai metamorphic massif, southern Siberian craton. *Precambrian Research* **122**, 359–76.

Sleep, N. H. 2001. Oxygenating the atmosphere. *Nature* **410**, 317–19.

Stacey, J. S. & Kramers, J. D. 1975. Approximation of terrestrial lead isotope evolution by two stage model. *Earth and Planetary Sciences Letters* **26**, 207–21.

Todt, W., Cliff, R. A., Hanser, A. & Hofmann, A. W. 1996. Evaluation of a ^{202}Pb-^{205}Pb double spike for high-precision lead isotope analysis. *Geophysica Monograph* **95**, 429–37.

Wendt, J. I. & Todt, W. 1991. A vapour digestion method for dating single zircons by direct measurements of U and Pb without chemical separation. *Terra Abstracts* **3**, 507–8.

White, W. M. 1998. *Geochemistry. An on-line textbook*. [www document] URL http://www.geo.cornell.edo/geology/classes/chapters.html

Wilson, M. 1989. *Igneous Petrogenesis*. London: Unwin Hyman.

ULRIKE POLLER and WOLFGANG TODT, Max-Planck-Institut für Chemie, Abt. Geochemie, Postfach 3060, D-55020 Mainz, Germany.
e-mail: poller@mpch-mainz.mpg.de
e-mail: todt@mpch-mainz.mpg.de

DIMITRI P. GLADKOCHUB, TATIANA V. DONSKAYA, ANATOLI M. MAZUKABZOV and EUGENE V. SKLYAROV, Institute of the Earth's crust, Siberian Branch, Russian Academy of Sciences, Lermontov Street 128, 664033 Irkutsk, Russia.
e-mail: dima@gpg.crust.irk.ru (D. P. Gladkochub)
e-mail: tanlen@gpg.crust.irk.ru (T. V.Donskaya and A. M. Mazukabzov)
e-mail: skl@gpg.crust.irk.ru (E. V. Sklyarov)

MS received 2 September 2003. Accepted for publication 31 May 2004.

Transactions of the Royal Society of Edinburgh: Earth Sciences, **95**, 227–236, 2004

Two subgroups of A-type granites in the coastal area of Zhejiang and Fujian Provinces, SE China: age and geochemical constraints on their petrogenesis

Jian-Sheng Qiu, De-Zi Wang, Brent I. A. McInnes, Shao-Yong Jiang, Ru-Cheng Wang and Satoshi Kanisawa

ABSTRACT: Late Cretaceous (90–100 Ma) A-type granites are widespread in the coastal area of the Zhejiang and Fujian Provinces, SE China. According to mineralogical and geochemical characteristics, the A-type granites in this belt can be further divided into aluminous and peralkaline subgroups. The aluminous subgroup often contains aluminous-rich minerals (e.g. spessartine and Mn-rich muscovite), while the peralkaline subgroup usually contains riebeckite, arfvedsonite and aegirine. Geochemically, the aluminous A-type granites show lower Nb, Zr, Ga, Y and REE abundances, and lower FeO*/MgO and Ga/Al than the peralkaline subgroup. When they occur in the same area, the two subgroups of A-type granites display quite similar initial Nd isotopic compositions, which are indicative of mixing of ancient basement crustal rocks with variable amounts of mantle materials. Integrated geological and geochemical investigations indicate that both the aluminous and the peralkaline magmas are highly evolved and reflect the residual liquids left after high degrees of fractional crystallisation in a deep magma chamber. The present authors suggest that the mineralogical and geochemical differences between the aluminous and peralkaline subgroups are likely to have been generated via different differentiation paths controlled by varying fluorine contents of the parent magmas.

KEY WORDS: Aluminous A-type granites, geochemistry, Late Mesozoic, peralkaline A-type granites.

A-type granites have attracted a great deal of attention in the past few decades because of their unusual mineralogy and chemistry, and their potential value in reconstructing the tectonic setting of ancient terranes. It has been recognised that A-type granites can arise in a variety of within-plate and late- to post-orogenic settings (Whalen *et al.* 1987; Sylvester 1989; Eby 1992). A number of petrogenetic models have been proposed for the origin of A-type granites, including fractional crystallisation of mantle-derived mafic magmas, with or without crustal assimilation (Loiselle & Wones 1979; Turner *et al.* 1992; Kerr & Fryer 1993; Wickham *et al.* 1995; Smith *et al.* 1999; Volkert *et al.* 2000; Litvinovsky *et al.* 2002; Kebede & Koeberl 2003), and partial melting of crustal materials (Collins *et al.* 1982; Clemens *et al.* 1986; Creaser *et al.* 1991; Patino Douce 1997; Jung *et al.* 2000; Wu *et al.* 2002). It appears that no single model is suitable for all A-type granites from diverse geological settings, and their petrogenesis is still a subject of considerable controversy.

Late Cretaceous (90–100 Ma) A-type granites are widespread along an 800-km-long and 60–80-km-wide NNE-trending belt in the coastal area of Zhejiang and Fujian Provinces, SE China (Fig. 1). This belt may extend across the East China Sea to South Korea, with a total length of 1800 km (Hong *et al.* 1997). Although numerous studies have focused on determining geochemical features, source characteristics, petrogenesis and tectonic significance (Tu *et al.* 1984; Hong *et al.* 1987; Zhou & Wu 1994; Martin *et al.* 1994; Qiu *et al.* 1996, 1999, 2000; Wei *et al.* 2000), many uncertainties and controversies still remain as to the origin and formation process for these granites. Moreover, most of the previous studies are published in Chinese, and thus are not easily accessible to the international community. In this paper, the authors attempt to synthesise the available geochemical data, including that already published in the literature and their new data, to further elucidate the genesis of these A-type granites. The emphasis will be on the systematic comparison of two subgroups of A-type granites (aluminous and peralkaline, or AA and PA) in this district, particularly in terms of sources and petrogenesis.

1. Geological setting

Southeast China is composed of the Yangtze Block in the north-west and the Cathaysia Block in the southeast (Fig. 1, lower right-hand inset). It is generally believed that these two blocks were amalgamated along the Jiangshan-Shaoxing fault during the Neoproterozoic era, forming a unified South China Plate (Xing *et al.* 1992; Zhou & Zhu 1993; Li & McCulloch 1996; Chen & Jahn 1998). The Cathaysia Block can be further divided into a vast 'Southeast China Caledonian fold belt' and a much smaller tectonic unit ('Southeast maritime fold belt') along the southeastern coastal area (Huang 1978). These two tectonic units are separated by the Zhenghe-Dapu fault. The basement of the Cathaysia Block, which is composed of metamorphic rocks of greenschist and amphibolite facies, was formed during the Palaeo- to Mesoproterozoic. For example, the isotopic ages of the Mayuan, Badu and Longquan Groups, which crop out under the Mesozoic volcanic rocks in the southern Zhejiang and northwestern Fujian provinces, range from 1813 to 2080 Ma (Sm-Nd and zircon U-Pb techniques; Hu *et al.* 1991; Xu 1987). The entire Cathaysia Block has been

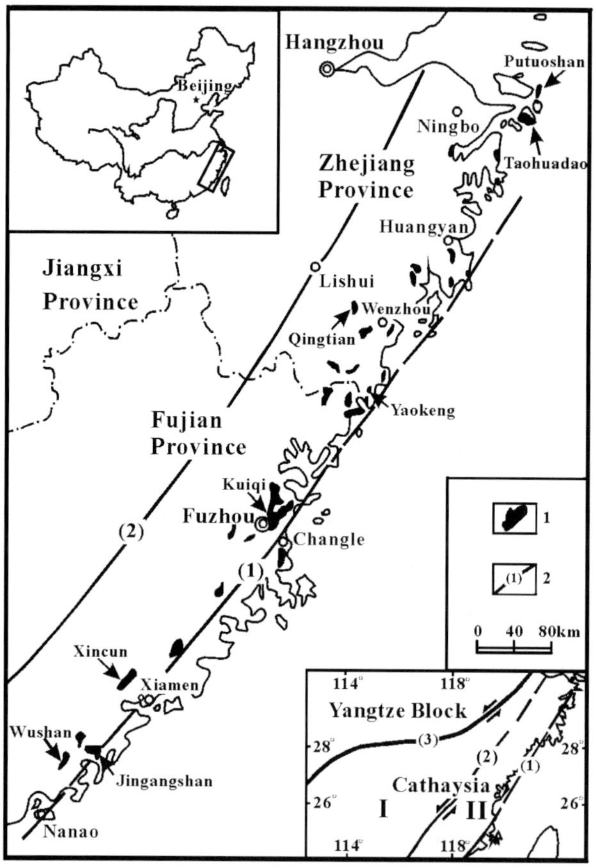

Figure 1 Distribution of A-type granites in the coastal area of the Fujian and Zhejiang provinces, SE China (modified after Tu *et al.* 1984). Tectonic divisions are shown in the lower right-hand inset: (I), Southeast China Caledonian fold belt; (II), Southeast maritime fold belt; (1) A-type granite (the plutons labelled with names are discussed in the present paper); and (2) fault [(1) Changle-Nan'ao fault; (2) Zhenghe-Dapu fault; and (3) Jiangshan-Shaoxing fault].

intruded by multi-stage granitic magmas since the Neoproterozoic and the magmatism shows a clear younging trend toward the coast. Geochemical and isotopic data reveal that the granitoids of Cathaysia Block have multiple origins and can be divided into three types: I-type, S-type and A-type (Wang & Zhou 2002). The spatial distribution of the different type granitoids is noteworthy in that, generally, the S-type granites occur predominantly inland to the west of the Zhenghe–Dapu fault, whereas the I-type granitoids are distributed mainly to the east of the fault. Therefore, the Zhenghe–Dapu fault roughly marks the boundary between I-type and S-type granites (Shu & Zhou 2002).

A-type granites occur mainly in the coastal area of the Zhejiang and Fujian provinces, in a belt subparallel to the Changle-Nan'ao fault (Fig. 1). Based on their mineralogical and geochemical characteristics, they can be further divided into two subgroups: aluminous and peralkaline A-types (AA and PA types). The majority of the A-type granites in this belt are alkali-feldspar granites of the AA subgroup, and PA granites are more rare. In the present paper, eight A-type granitic plutons have been selected for study (Fig. 1). Among them, four plutons (i.e. Taohuadao, Qingtian, Yaokeng and Kuiqi) are of PA character, while four plutons (i.e. Putuoshan, Xincun, Jingangshan and Wushan) belong to the AA subgroup.

2. Petrography

A-type granites in this belt are leucocratic, ash-grey to pink in colour, with coarse- to medium-grained granitic texture. They are usually homogeneous and rarely contain schlieren and mafic enclaves. Miarolitic cavities are fairly common in most coarse-grained varieties and are often filled by the distinctive minerals of each A-type subgroup. The vugs in PA granites are usually infilled with drusy quartz, alkali-feldspar, and occasionally, with alkaline mafic minerals and fluorite. Those in the AA subgroup are often filled by aggregates of pegmatitic quartz and alkali-feldspar, and sometimes, by aluminous-rich minerals, such as spessartine, muscovite and even corundum. The main rock-forming minerals for both subgroups are quartz and perthite. Plagioclase (albite-oligoclase) is common in the AA subgroup, but scarce in PA granites. Mafic minerals in the AA subgroup consist mainly of biotite, but in the PA subgroup, often include Na-rich amphibole and pyroxene. The Na-rich mafic minerals in PA granites usually occur as interstitial grains, indicating that they were crystallised at a late stage. Oxide minerals for both AA and PA subgroups are mainly magnetite with lesser ilmenite, implying that they were formed under a relatively oxidised condition. This contrasts with the general assumption that A-type granites evolved under relatively reduced conditions, with fO_2 less than the nickel-nickel oxide buffer (fO_2<NNO; Frost & Frost 1997; King *et al.* 1997). However, relatively oxidised A-type granites are also reported elsewhere, such as in the Amazonian Craton of Brazil (Dall'Agnol *et al.* 1999) and in the south-western USA (Anderson & Bender 1989). Aluminous-rich minerals (e.g. spessartine and corundum) also occur as accessory minerals in AA subgroup granites. Micrographic intergrowths of

Table 1 Zircon U-Pb isotopic data from selected A-type granites in the coastal area of Zhejiang Province, SE China.

Sample number	Spot number	Atomic ratios								Ages (Ma)		
		$^{206}Pb/^{204}Pb$	$^{208}Pb/^{206}Pb$	$^{206}Pb/^{238}U$	$2\sigma_m$	$^{207}Pb/^{235}U$	$2\sigma_m$	$^{207}Pb/^{206}Pb$	$2\sigma_m$	$^{206}Pb/^{238}U$	$^{207}Pb/^{235}U$	$^{207}Pb/^{206}Pb$
TZ-1 (Taohuadao)	1	187	2·2770	0·01467	12	0·09678	212	0·04786	92	93·9	93·8	92·1
	2	30	0·5349	0·01431	15	0·13810	148	0·06999	688	91·6	131·3	928·0
	3	63	0·5212	0·01414	40	0·15060	462	0·07721	2161	90·5	142·4	1127·0
QT-1												
(Qingtian)	1	221	0·1841	0·01585	60	0·10540	63	0·04823	205	101·4	101·8	110·6
	2	224	0·1877	0·01575	53	0·10490	56	0·04831	183	100·7	101·3	114·3
	3	34	0·1505	0·01588	66	0·10630	116	0·04852	445	101·6	102·6	124·8
P-104												
(Putuoshan)	1	2078	0·2889	0·01457	9	0·09617	494	0·04786	231	93·3	93·2	92·5
	2	1315	0·2539	0·01467	8	0·09694	595	0·04794	281	93·9	94·0	96·5

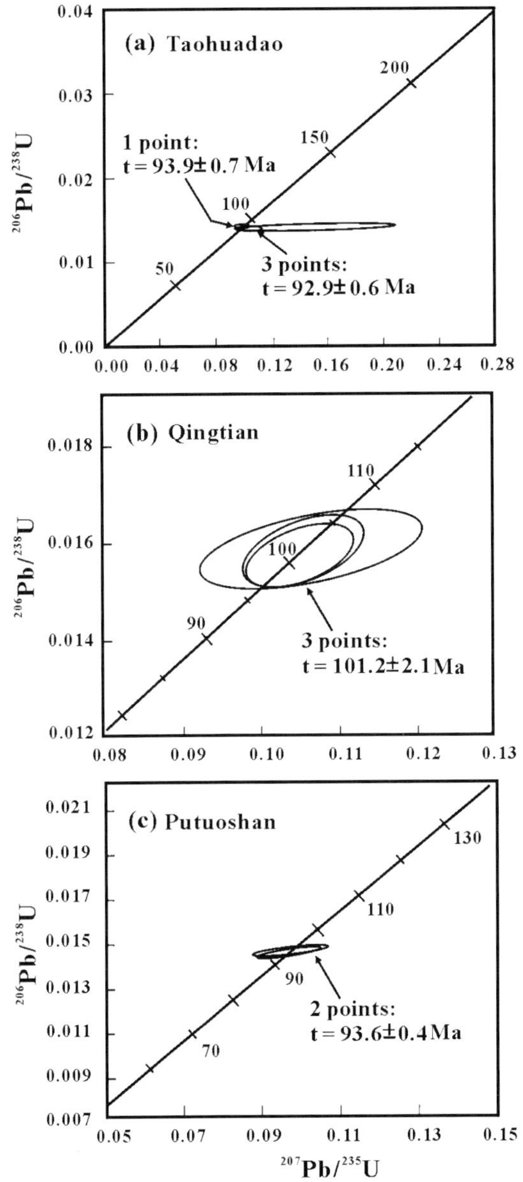

Figure 2 Results of single-grain zircon U-Pb isotopic analyses on A-type granites from Zhejiang Province, SE China.

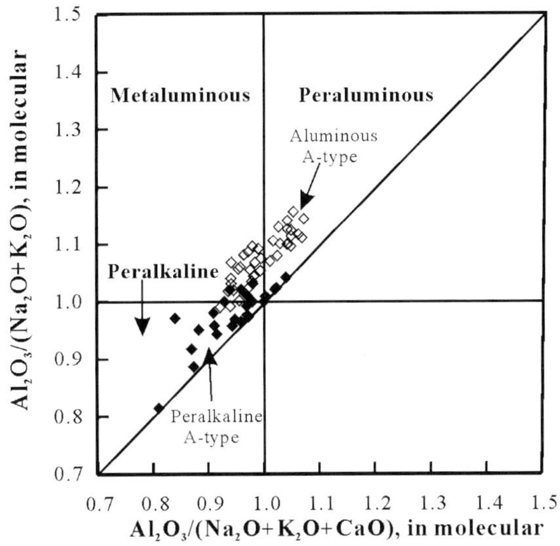

Figure 3 A/NK versus A/CNK plot for A-type granites in the coastal area of the Fujian and Zhejiang provinces, SE China. Solid diamonds represent the peralkaline subgroup and open diamonds represent the aluminous subgroup. Besides the data from this study, those of Hong *et al.* (1987), Martin *et al.* (1994) and Zhou & Wu (1994) are also included in this plot.

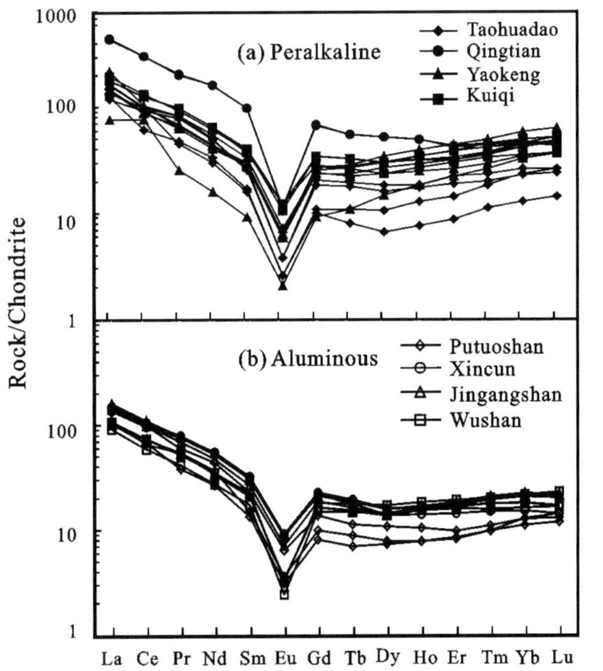

Figure 4 Chondrite-normalised rare-earth element patterns for A-type granites in the coastal area of the Fujian and Zhejiang provinces, SE China.

quartz and alkali feldspar (granophyric texture), interpreted as the result of rapid cooling of a high-level intrusion (Tsusue & Liu 1987), are always well developed. Hong *et al.* (1987) further suggested that the emplacement depth of the A-type granites in this belt is less than 3 km, which implies that they are subvolcanic granites.

3. Analytical methods

Major elements were measured by wet chemical methods and rare-earth elements (REEs) by inductively coupled plasma emission spectrometry (ICP-AES) at the State Key Laboratory for Mineral Deposits Research of Nanjing University, Nanjing, China. The other trace elements were analysed in fused lithium-tetraborate glass beads using standard X-ray fluorescence (XRF) techniques at the CSIRO Division of Exploration and Mining, North Ryde, New South Wales, Australia (Kuiqi, Xincun and Jingangshan intrusions), and at the Material Analysis Centre of Nanjing University (all other plutons). The analytical precision is estimated to be within 2–5% for major elements, and better than 5–10% for all trace elements. Fluorine content was determined using a selective ion-electrode method at the Institute of Mineralogy, Petrology and Economic Geology, Tohoku University, Sendai, Japan. Analytical errors are estimated to be less than 3% for all samples.

Zircon U-Pb dating was conducted at the Tianjin Institute of Geology and Mineral Resources, China Geological Survey, Tianjin, China, using a VG-354 mass spectrometer with a Daly collector. The blanks were 0·03–0·05 ng for Pb and 0·002–0·004 ng for U. Neodymium isotope analyses were carried out at the Isotopic Laboratory of the Geological Institute, Chinese Academy of Science, Beijing, China, using a VG-354 mass spectrometer. $^{143}Nd/^{144}Nd$ ratios were normalised against the value of $^{146}Nd/^{144}Nd=0{\cdot}7219$. During the period of sample

Table 2 Whole rock major (in wt.%) and trace element (in ppm) compositions of representative A-type granites in the coastal areas of the Zhejiang and Fujian provinces, SE China.*

Type:	Peralkaline														
Pluton:	Taohuadao				Qingtian		Yaokeng						Kuiqi		
Sample no.:	T-1	T-3	T-9	T-10	QT-2	QT-4	YK-2	YK-4	YK-5	YK-6	YK-7	YK-8	Fz-1	Fz-7	Fz-22
SiO_2	77·80	76·48	77·00	77·00	75·02	74·96	77·60	77·76	77·68	77·50	77·68	77·82	76·48	77·55	77·68
TiO_2	0·09	0·12	0·11	0·09	0·17	0·17	0·08	0·08	0·10	0·08	0·08	0·10	0·15	0·15	0·15
Al_2O_3	11·38	11·49	11·58	11·53	12·07	12·42	11·47	11·38	11·35	11·47	11·25	11·01	11·49	11·49	11·27
Fe_2O_3	1·03	1·19	1·15	1·13	1·11	1·19	1·25	1·24	1·16	1·17	1·54	1·34	1·38	1·39	1·32
FeO	0·58	0·44	0·46	0·59	1·19	1·05	0·06	0·07	0·23	0·25	0·07	0·16	0·22	0·13	0·20
MnO	0·11	0·12	0·14	0·13	0·08	0·07	0·01	0·03	0·08	0·09	0·14	0·11	0·08	0·05	0·05
MgO	0·08	0·07	0·08	0·01	0·06	0·03	0·01	0·02	0·02	0·02	0·02	0·18	0·40	0·05	0·05
CaO	0·49	0·52	0·49	0·35	0·39	0·31					0·14	0·24	0·96	0·51	0·33
Na_2O	4·06	4·26	3·80	4·74	3·94	4·01	3·77	3·64	3·90	3·90	4·05	3·98	4·32	4·53	4·49
K_2O	4·32	4·37	4·73	4·43	4·94	5·08	4·44	4·56	4·58	4·47	4·23	4·06	4·40	4·28	4·04
P_2O_5	0·01	0·01	0·01	0·01	0·01	0·02	0·05	0·05	0·04	0·02	0·03	0·04	0·02	0·01	0·01
ALK	8·38	8·63	8·53	9·17	8·88	9·09	8·21	8·20	8·48	8·37	8·28	8·04	8·72	8·81	8·53
AI	1·00	1·02	0·98	1·09	0·98	0·97	0·96	0·96	1·00	0·98	1·00	0·99	1·03	1·05	1·04
ASI	0·93	0·91	0·94	0·87	0·96	0·98	1·04	1·04	1·00	1·02	0·98	0·97	0·84	0·88	0·91
F	313	2395	1529	1180	1786	1638	306	867	1180	1522	1412	1382	976	1149	1282
Rb	208	241	196	187	240	240	290	290	313	314	333	351	243	264	232
Sr	4·9	15·0	5·2	8·0	17·0	18·0	9·0	5·0	8·0	9·0	14·0	6·0	34·0	7·0	5·0
Ba	72	48	42	34	92	79	73	70	64	57	58	78	16	2	<21
U	4·6	1·1	3·5	3·4	8·0	7·0	6·0	4·0	8·0	11·0	9·0	6·0	5·0	6·6	12·1
Th	12·2	18·9	14·9	16·0	21·0	22·0	30·0	30·0	43·0	19·0	38·0	33·0	28·4	22·0	64·5
Ga	20·8	22·6	21·2	21·2	22·9	23·7	24·0	22·0	22·0	22·0	21·0	22·0	21·0	22·0	23·0
Y	13·0	30·6	38·5	33·0	58·0	58·0	47·6	30·1	56·3	58·3	75·8	58·1	44·0	68·0	66·0
Nb	37	32	32	33	40	41	48	55	58	71	54	56	31	34	45
Zr	245	244	251	276	390	415	246	303	338	374	316	401	281	260	496
$10^4 \times$ Ga/Al	3·45	3·72	3·46	3·47	3·58	3·60	3·95	3·65	3·66	3·62	3·53	3·77	3·45	3·62	3·85
La	29·58	38·98	28·62	31·98	106·20	103·80	52·51	17·90	38·16	34·23	39·30	35·52	43·00	43·09	46·38
Ce	37·62	59·86	56·88	60·90	184·50	185·20	63·27	46·30	61·47	54·36	62·29	54·58	77·91	76·86	81·13
Pr	4·51	4·31	7·56	7·96	19·61	19·10	7·99	2·41	6·39	6·25	6·58	5·98	9·21	9·41	8·80
Nd	15·48	13·80	22·80	23·70	76·44	75·16	25·60	7·35	20·42	19·99	20·81	19·01	30·10	30·44	28·49
Sm	2·54	2·45	4·10	3·86	14·98	14·83	5·06	1·40	4·26	4·45	4·53	4·29	5·66	6·27	6·01
Eu	0·15	0·15	0·22	0·22	0·65	0·66	0·41	0·12	0·34	0·37	0·36	0·34	0·69	0·69	0·60
Gd	2·04	2·20	4·14	3·71	13·79	13·94	4·98	1·86	4·82	5·57	5·58	5·28	5·83	7·10	7·03
Tb	0·30	0·40	0·71	0·66	2·08	2·06	0·82	0·40	0·87	1·03	1·07	0·99	0·97	1·21	1·20
Dy	1·67	2·67	4·62	4·08	13·14	13·17	5·98	3·74	6·80	7·72	8·75	7·56	5·92	7·60	7·54
Ho	0·43	0·71	1·04	0·97	2·78	2·78	1·39	1·02	1·63	1·78	2·21	1·77	1·51	1·93	1·92
Er	1·42	2·32	3·61	3·11	7·16	6·95	4·34	3·65	5·20	5·26	7·26	5·49	4·98	6·36	6·31
Tm	0·28	0·47	0·60	0·50	1·15	1·12	0·75	0·67	0·92	0·90	1·24	0·96	0·86	1·06	1·14
Yb	2·19	4·01	4·39	3·90	7·95	7·61	5·89	5·55	7·42	7·08	9·90	7·73	6·08	7·39	8·41
Lu	0·36	0·67	0·67	0·61	1·11	1·09	0·94	0·92	1·23	1·18	1·60	1·31	0·93	1·11	1·32
ΣREE	98·57	133·00	139·96	146·16	451·54	447·47	179·93	93·29	159·93	150·17	171·48	150·81	193·65	200·52	206·28
$(La/Yb)_N$	9·69	6·97	4·68	5·88	9·58	9·78	6·39	2·31	3·69	3·47	2·85	3·30	5·07	4·18	3·96
δEu	0·19	0·19	0·16	0·18	0·14	0·14	0·25	0·23	0·23	0·23	0·22	0·22	0·36	0·32	0·28

analysis, the international rock standard BCR-1 gave $^{143}Nd/^{144}Nd=0{\cdot}512615 \pm 8$ (2σ). The maximum external error on $^{143}Nd/^{144}Nd$ ratios is $\pm 0{\cdot}000010$, and the corresponding maximum external error on the ε_{Nd} values is about $\pm 0{\cdot}2$.

4. Geochronology

Previous high-quality age data are available only for the A-type granites in Fujian province, and these were all obtained using whole-rock Rb-Sr isochron dating techniques. For example, Martin *et al.* (1994) defined an isochron age of 93 ± 1 (2σ) Ma (MSWD=2·8) and $I_{Sr}=0{\cdot}7069 \pm 0{\cdot}0007$ (2σ) for the Kuiqi pluton. Hong *et al.* (1987) reported a similar result from a 16-point whole-rock isochron of $91{\cdot}3 \pm 0{\cdot}7$ Ma with $I_{Sr}=0{\cdot}7059 \pm 0{\cdot}0030$ for the same intrusion. The whole-rock Rb-Sr isochron ages for Xincun and Jingangshan are $97{\cdot}12 \pm 2{\cdot}32$ Ma and 92 ± 3 Ma, respectively (Zhou & Wu 1994; Hong *et al.* 1987). There is no age difference between the AA and PA subgroup granites.

Table 1 gives the single zircon U-Pb results for the A-type granites in Zhejiang province. Young zircons usually contain higher abundances of ^{206}Pb and ^{238}U, and are commonly less

Table 2 *Continued*

Type:	Aluminous											
Pluton:	Putuoshan			Xincun			Jingangshan				Wushan	
Sample no.:	P-1	P-3	P-7	XC-3	XC-5	XC-8	Jg-3	Jg-5	Jg-6	Jg-9	WS-2	WS-3
SiO_2	77·10	77·39	78·09	77·11	77·46	76·83	77·47	77·13	77·21	77·34	78·04	78·79
TiO_2	0·06	0·12	0·08	0·12	0·12	0·13	0·12	0·12	0·13	0·13	0·09	0·07
Al_2O_3	12·25	12·64	12·41	12·58	12·26	12·62	12·37	12·26	12·26	12·15	12·06	11·56
Fe_2O_3	0·73	0·60	0·50	0·40	0·48	0·53	0·68	0·63	0·71	0·69	0·66	0·52
FeO	0·43	0·20	0·14	0·27	0·26	0·23	0·10	0·17	0·12	0·15	0·07	0·10
MnO	0·10	0·08	0·03	0·05	0·06	0·04	0·06	0·07	0·05	0·06	0·08	0·06
MgO	0·08	0·08	0·07	0·05	0·05	0·10	0·05	0·05	0·05	0·10	0·16	0·14
CaO	0·54	0·32	0·28	0·86	0·75	0·75	0·68	0·68	0·68	0·61	0·42	0·32
Na_2O	3·84	3·84	3·85	4·17	4·04	4·04	4·32	4·39	4·24	4·39	4·04	3·84
K_2O	4·76	4·62	4·61	4·55	4·34	4·59	4·22	4·22	4·28	4·22	3·98	3·88
P_2O_5	0·01	0·01	0·02								0·03	0·03
ALK	8·60	8·46	8·46	8·72	8·38	8·63	8·54	8·61	8·52	8·61	8·02	7·72
AI	0·94	0·90	0·91	0·94	0·93	0·92	0·94	0·96	0·95	0·97	0·91	0·91
ASI	0·98	1·06	1·05	0·94	0·96	0·97	0·96	0·94	0·95	0·94	1·03	1·04
F	130	138	416	1026	1327	913	1102	1134	1296	1263	1472	1567
Rb	222	177	238	241	244	243	269	227	250	240	224	229
Sr	31·2	34·0	32·0	1·0	3·0	1·0	2·0	6·0	6·0	3·0	5·0	5·0
Ba	129	186	66						7		74	74
U	2·8	4·0	3·6	12·8	9·8	10·5	10·6	10·4	10·4	9·1	6·0	6·0
Th	25·0	21·0	29·0	36·4	35·3	35·8	36·7	35·2	36·9	35·6	26·0	25·0
Ga	15·3	16·7	17·4	16·0	18·0	17·0	21·0	18·0	21·0	19·0	17·0	17·0
Y	15·5	16·8	17·0	26·0	28·0	24·0	31·0	32·0	31·0	31·0	32·0	33·0
Nb	20	20	21	24	21	17	32	26	29	22	25	25
Zr	84	98	86	66	70	72	88	93	82	85	90	78
$10^4 \times$ Ga/Al	2·36	2·50	2·65	2·40	2·77	2·54	3·21	2·77	3·24	2·95	2·66	2·78
La	26·18	34·32	24·44	32·38	33·92	35·70	33·76	37·93	24·36	23·94	25·34	21·78
Ce	39·92	60·23	41·24	58·96	62·32	64·92	60·39	67·41	45·19	44·73	45·30	35·88
Pr	5·12	5·66	3·59	7·19	7·18	7·50	6·57	7·34	5·07	5·07	4·91	3·93
Nd	16·97	20·30	12·41	24·18	25·56	25·47	21·94	25·20	15·75	15·78	16·18	12·71
Sm	2·30	3·49	2·05	4·65	4·92	4·92	4·06	4·57	3·18	3·23	3·43	2·65
Eu	0·21	0·37	0·18	0·50	0·53	0·52	0·41	0·51	0·20	0·20	0·19	0·14
Gd	2·03	2·81	1·66	4·53	4·66	4·40	3·82	4·32	3·36	3·37	3·68	3·07
Tb	0·33	0·42	0·26	0·71	0·72	0·66	0·61	0·69	0·57	0·56	0·66	0·55
Dy	1·99	2·73	1·85	4·01	4·05	3·49	3·52	3·97	3·60	3·45	4·34	3·67
Ho	0·44	0·59	0·44	0·92	0·92	0·79	0·88	0·95	0·92	0·87	1·03	0·86
Er	1·40	1·61	1·36	2·75	2·74	2·36	2·82	2·99	3·01	2·83	3·16	2·62
Tm	0·25	0·28	0·25	0·45	0·45	0·38	0·50	0·51	0·53	0·49	0·51	0·40
Yb	1·88	2·15	2·17	3·05	3·05	2·56	3·60	3·58	3·77	3·50	3·63	2·74
Lu	0·30	0·33	0·36	0·42	0·43	0·36	0·53	0·53	0·55	0·51	0·58	0·42
ΣREE	99·32	135·29	92·26	144·70	151·45	154·03	143·41	160·50	110·06	108·53	112·94	91·42
$(La/Yb)_N$	9·99	11·45	8·08	7·62	7·98	10·00	6·73	7·60	4·63	4·91	5·01	5·70
δEu	0·29	0·35	0·29	0·33	0·33	0·33	0·31	0·35	0·19	0·18	0·16	0·15

*ALK = K_2O+Na_2O; AI = $(K_2O+Na_2O)/Al_2O_3$, in molecular; ASI = $Al_2O_3/(Na_2O+K_2O+CaO)$, in molecular.

affected by experimental blanks and initial common lead contents. Therefore, the present authors report the weighted mean $^{206}Pb/^{238}U$ ages for the measured samples.

4.1. Taohuadao pluton

Of the three dated zircons, one is a transparent, long, colourless, euhedral prismatic grain; the other two are short, euhedral bipyramidal grains which are brown yellow in colour. The colourless grain plots on the concordia; however, the other two do not (Fig. 2a). The three grains yield a weighted mean $^{206}Pb/^{238}U$ age of $92·9 \pm 0·6$ Ma. If only the colourless grain is considered, an age of $93·9 \pm 0·7$ Ma is obtained. These two ages agree within analytical error, and hence, the authors interpret the age of $92·9 \pm 0·6$ Ma as the emplacement age of Taohuadao pluton. The much older $^{207}Pb/^{206}Pb$ apparent ages of 928–1127 Ma may represent the effect of inherited zircon in the analysed fraction, and may also imply that ancient crustal materials were involved in the genesis, as indicated by the highly negative $\varepsilon_{Nd}(t)$ values ($= -9·1$ to $-10·1$) of this intrusion.

4.2. Qingtian pluton

The three dated zircons are transparent, short euhedral prismatic and light yellow in colour. The data points are clustered on the concordia (Fig. 2b), yielding a weighted mean $^{206}Pb/^{238}U$ age of 101·2 ± 2·1 Ma. The authors consider this age to be the emplacement age of the Qingtian pluton.

4.3. Putuoshan pluton

Of the two dated zircons, one is transparent, short, euhedral, prismatic and light yellow in colour, and the other is a long, colourless, euhedral prismatic grain. Both data points are tightly clustered on the concordia, yielding a $^{206}Pb/^{238}U$ weighted mean age of 93·6 ± 0·4 Ma (Fig. 2c). This age is considered to be the time of emplacement of this pluton.

In summary, the new obtained ages of A-type granites in Zhejiang are similar to those obtained for plutons in Fujian. Their ages range predominantly from 100 to 90 Ma, indicating that A-type granites in this belt were predominantly generated in the Late Cretaceous.

5. Geochemistry

5.1. Major and trace elements

Major and trace element compositions for representative samples are listed in Table 2. The rocks are highly siliceous, and have a narrow silica range with SiO_2 contents of 74·96–78·79%. All samples are characterised by high contents of total alkalis, with K_2O+Na_2O values ranging from 8·04% to 9·17% in the PA subgroup and from 7·72% to 8·72% in the AA subgroup. They also have low abundances of Fe_2O_3 (0·40–1·54%), FeO (0·07–1·19%), MgO (0·01–0·40%) and CaO (<0·01–0·96%). Although many chemical compositions for both AA and PA rocks significantly overlap, there are still obvious differences between them. Compared with the PA subgroup, the AA subgroup granites are relatively rich in aluminum, with Al_2O_3 contents generally >12% and alumina saturation index (ASI) values commonly >0·95; however, their agpacity index (AI) values are usually <1·0. In a plot of A/NK versus A/CNK, the AA subgroup rocks fall mainly into the areas of metaluminous and weakly peraluminous, whereas those of the PA subgroup mostly plot in the peralkaline field (Fig. 3). In addition, the Fe_2O_3 content of the PA rocks are generally higher than that of the AA subgroup, indicating that the PA granites likely formed in a more oxidising environment.

Chondrite-normalised REE patterns of the studied A-type granites (Fig. 4) are moderately fractionated in light rare-earth elements (LREE), but roughly flat to slightly enriched in heavy rare-earth elements (HREE), and are quite different from the highly fractionated REE patterns of the within-plate A-type granites from Nigeria (cf. Bowden *et al.* 1987). The drop in LREE contents is coincident with the segregation of LREE-rich accessory phases such as monazite, allanite or titanite. All samples show pronounced negative Eu anomalies with δEu values varying from 0·14 to 0·36 (Table 2), consistent with fractional crystallisation dominated by a significant quantity of plagioclase. Total REE (ΣREE) contents generally overlap but are quite high in some PA granites (e.g. the Qingtian pluton). The high ΣREE content of the Qingtian pluton is likely a result of its higher titanite, apatite and allanite contents (Peng & Deng 1991). The fractionation of the HREE is more pronounced in the AA subgroup than in the PA subgroup, with average $(Gd/Yb)_N$ ratios of 0·98 and 0·75, respectively, suggesting that more HREE-bearing minerals (e.g. hornblende and pyroxene) might have been fractionated during the formation of the former.

Both subgroups display similar trace element patterns, with significant depletion in Sr, Ba, P and Ti, and enrichment in Rb, Th and U (Fig. 5). The Sr, Ba, P, Ti depletion could be related to fractionation of plagioclase, apatite and ilmenite. Ga/Al ratios are particularly diagnostic for A-type granites (Fig. 6) where most of the studied granites fall in the A-type granite field, as defined by Whalen *et al.* (1987). Compared to the PA rocks, the AA subgroup contains less Nb, Zr and Ga (especially Zr; Fig. 6b). The halogens, and in particular F, are

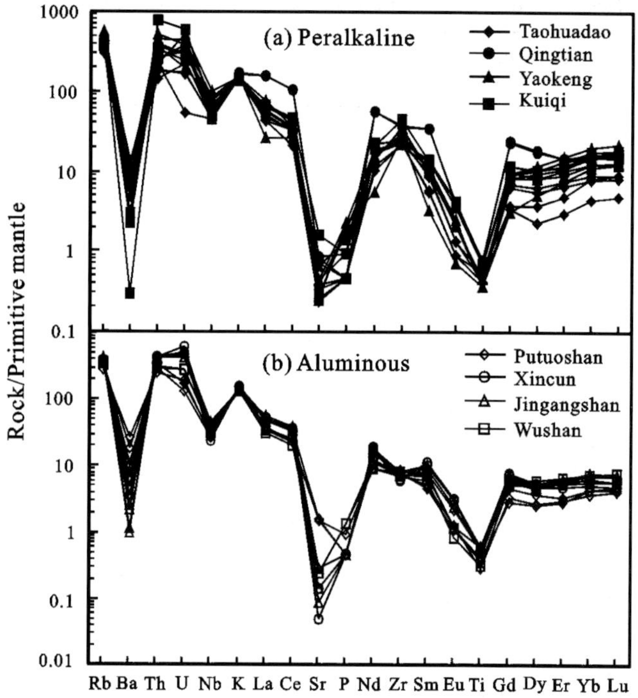

Figure 5 Primitive-mantle (PM) normalised trace element spidergrams for A-type granites from the coastal area of the Fujian and Zhejiang provinces, SE China. The PM values are from Sun & McDonough (1989).

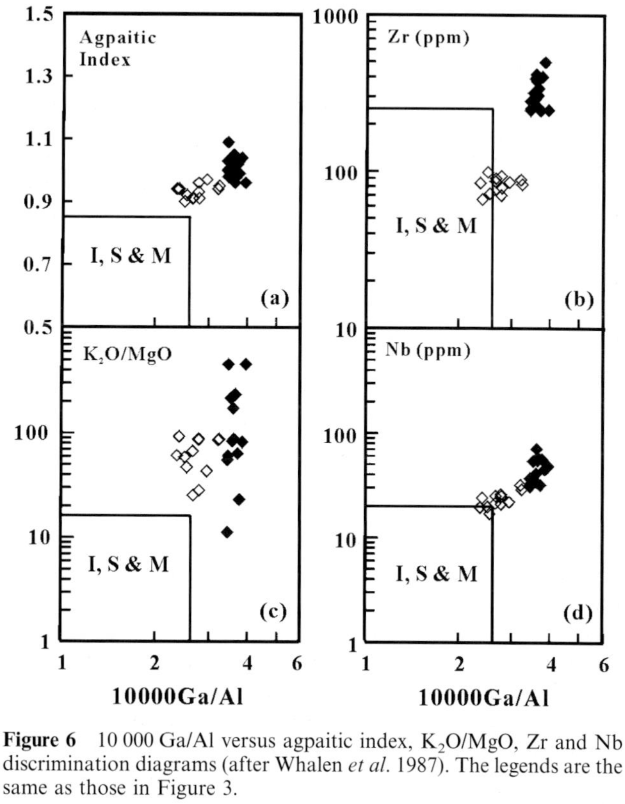

Figure 6 10 000 Ga/Al versus agpaitic index, K_2O/MgO, Zr and Nb discrimination diagrams (after Whalen *et al.* 1987). The legends are the same as those in Figure 3.

Table 3 Nd isotopic compositions of A-type granites in the coastal areas of the Zhejiang and Fujian provinces, SE China.*

Sample number	Pluton	Chemical type	t (Ma)	Sm (ppm)	Nd (ppm)	Sm/Nd	$^{147}Sm/^{144}Nd$	$^{143}Nd/^{144}Nd$	2σ	$\varepsilon_{Nd}(t)$	T_{DM2} (Ga)	M(%)
A-type granites in Zhejiang Province												
T-1	Taohuadao	PA	92·9	2·20	13·57	0·16	0·0980	0·512114	13	-9·1	1·63	33·7
T-3	Taohuadao	PA	92·9	2·45	20·13	0·12	0·0738	0·512045	18	-10·1	1·72	27·3
T-9	Taohuadao	PA	92·9	4·11	19·90	0·21	0·1248	0·512119	13	-9·3	1·64	32·4
T-10	Taohuadao	PA	92·9	4·07	20·24	0·20	0·1215	0·512125	21	-9·1	1·63	33·3
P-1	Putuoshan	AA	93·6	2·24	14·09	0·16	0·0962	0·512098	20	-9·3	1·65	32·0
Q-1	Qingtian	PA	101·2	10·38	50·09	0·21	0·1253	0·512173	9	-8·2	1·56	38·8
A-type granites in Fujian Province												
Fz-22	Kuiqi	PA	93·0	6·23	22·71	0·27	0·1659	0·512352	8	-5·2	1·31	53·5
Xkq-2	Kuiqi	PA	93·0	3·07	15·17	0·20	0·1222	0·512416	9	-3·5	1·17	61·1
Jg-5	Jingangshan	AA	92·0	3·09	14·83	0·21	0·1260	0·512372	8	-4·4	1·25	57·2
Jg-9	Jingangshan	AA	92·0	3·45	16·57	0·21	0·1258	0·512376	8	-4·3	1·24	57·6
Ws-3	Wushan	AA	95·0	2·73	12·59	0·22	0·1310	0·512342	7	-5·0	1·29	54·5

*$\varepsilon_{Nd}(t)=[(^{143}Nd/^{144}Nd)_i^S/(^{143}Nd/^{144}Nd)_i^{CHUR}-1]\times10^4$, where t=emplacement age, S=sample, $(^{143}Nd/^{144}Nd)_{CHUR}=0{\cdot}512638$, $(^{147}Sm/^{144}Nd)_{CHUR}=0{\cdot}1967$, $\lambda_{Sm}=0{\cdot}00654\ Ga^{-1}$. T_{DM2} are two-stage Nd model ages, see Wu *et al.* (2002) for details of calculation. M represents the degree of mantle components involved in the formation of A-type granites (see text for details).

considered to play a key role in concentrating the high-field-strength (HFS) elements (e.g. Nb and Zr) and Ga in A-type magmas through complex ion formation (Collins *et al.* 1982). Many workers have proposed that A-type granites have anomalously high F contents (Collins *et al.* 1982; Whalen *et al.* 1987); however, the fluorine contents of both subgroups studied here are highly variable and range from 313 to 2395 p.p.m. (PA subgroup), and from 130 to 1567 p.p.m. (AA subgroup; Table 2). The Nb and Zr contents show no correlation with increasing F. This suggests that the measured fluorine contents in A-type granites cannot be indicative of original magmatic contents. The presence of quartz-, albite-, microcline- and fluorite-rich miarolitic cavities and chloritised biotites suggests the presence of late circulating waters (Collins *et al.* 1982; Poitrasson *et al.* 1994). Perhaps the fluorine content of the A-type granite was modified during anion exchange reactions with these waters.

5.2. Nd isotopes

Nd isotopic compositions for representative samples are reported in Table 3. They fall into two distinct groups, one corresponding to the A-type granites in Zhejiang [$\varepsilon_{Nd}(t)=-8{\cdot}2$ to $-10{\cdot}1$], and the other to those in Fujian [$\varepsilon_{Nd}(t)=-3{\cdot}5$ to $-5{\cdot}2$). However, the $\varepsilon_{Nd}(t)$ values for the AA and the PA granites found in neighbouring district are quite similar. For example, the $\varepsilon_{Nd}(t)$ values of the Putuoshan (AA subgroup) and the Taohuadao (PA subgroup) in the northernmost part of this belt are $-9{\cdot}3$ and $-9{\cdot}1$ to $-10{\cdot}1$, respectively, which agree within analytical error. A similar case is also seen in Fujian, with $\varepsilon_{Nd}(t)$ values of $-4{\cdot}3$ to $-5{\cdot}0$ for the AA subgroup (Jingangshan and Wushan), and of $-3{\cdot}5$ to $-5{\cdot}2$ for the PA subgroup (Kuiqi). The good matches of $\varepsilon_{Nd}(t)$ values between the AA and PA subgroups argue strongly that both originated from a similar source.

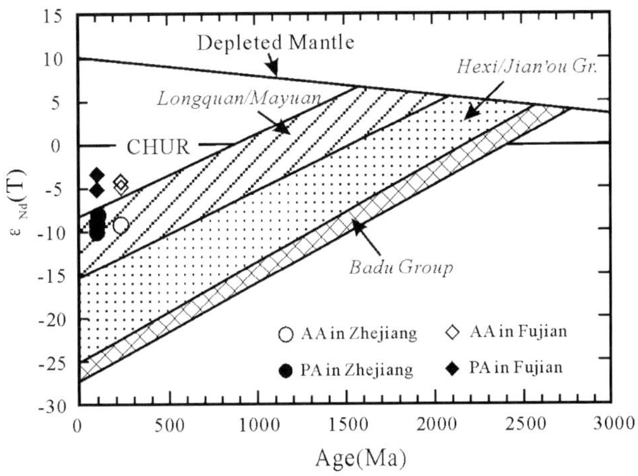

Figure 7 $\varepsilon_{Nd}(t)$ versus age plot for A-type granites in the coastal areas of the Zhejiang and Fujian provinces, SE China. Data points of the AA- and PA- subgroups are separated for clarity and no accurate age differences are implied. The crustal evolutionary trends are after Chen & Jahn (1998).

6. Discussion

6.1. Origin of the A-type magma: crust- or mantle-derived?

The genesis of A-type granites is an actively debated issue, and several sources and magmatic processes have been proposed (Collins *et al.* 1982; Clemens *et al.* 1986; Creaser *et al.* 1991; Poitrasson *et al.* 1994; Frost & Frost 1997). Of primary concern is the role of mantle materials in generation of A-type granites. Because Nd isotopes are scarcely remobilised by hydrothermal fluids, they can be used as a powerful tool for studying the genesis of A-type granites (Poitrasson *et al.* 1995). The A-type granites described in the present study possess moderately to highly negative $\varepsilon_{Nd}(t)$ values, and therefore, it is unlikely that they derive directly from the mantle. Chen & Jahn (1998) proposed that two crust-forming events took place in the Cathaysia Block. The first is represented by the formation of the protoliths of the Badu and Hexi metamorphic rocks during the Late Archaean to Palaeoproterozoic period (probably between 2·4 and 2·8 Ga), and the second by the formation of the protoliths of the Mayuan and Longquan Groups during Palaeo- to Mesoproterozoic times (2·5–1·6 Ga). On a $\varepsilon_{Nd}(t)$ versus T diagram (Fig. 7), all the samples plot either within or above the evolutionary trend defined by the second group of metamorphic rocks, indicating that re-melting of ancient basement rocks with variable amounts of mantle input is the most likely process of magma genesis. The two-stage Nd model ages (T_{DM2}) of A-type granites in Fujian (1·17–1·31 Ga) are younger than those for granites from Zhejiang (1·56–1·72 Ga), suggesting that the A-type granites in Fujian contain more of a mantle component or that the Fujian crust was more juvenile. According to the parameters proposed by Liu *et al.* (1990), the average Nd concentration of the crust in South China is 28 p.p.m. and $\varepsilon_{Nd}(t=95\ Ma) = -13{\cdot}8$. A simple binary mixing calculation shows that the degree of depleted-mantle-derived magma [Nd=14 p.p.m., $\varepsilon_{Nd}(t=95\ Ma)=9{\cdot}8$; Faure 1986] involved in the formation of the A-type granites ranged from 53·5% to 61·1% in Fujian, and from 27·3% to 38·8% in Zhejiang (Table 3). It follows that, during the Late Cretaceous, the depth of extension-related fracturing in Fujian was greater than that in Zhejiang, and increasing depth led to more mantle materials involved in the generation of A-type granites (Qiu *et al.* 1999).

The studied A-type granites are highly siliceous, have extremely high Rb/Sr ratios, and display distinct Sr, Ba, P, Ti and Eu depletions (Figs 4 & 5). These features clearly indicate that they were highly evolved, and thus, large amounts of fractional crystallisation are necessary to account for both major- and trace-element compositions. The mineralogical and chemical compositions of both subgroup granites are homogeneous, indicative of near-solidus crystallisation (Martin *et al.* 1994). This also characterises high degrees of fractional crystallisation, and consequently, the A-type granitic magmas can be considered as residual liquids. The fractional crystallisation process could have occurred either during magma emplacement or earlier in the magma chamber. Because of their high temperature and high halogen contents, A-type melts have low viscosities (Clemens 1986). This allows them to rapidly rise from their deep crustal sources to the subvolcanic level. Therefore, the fractional crystallisation probably occurred in a deep magma chamber rather than during magma ascent.

6.2. Relationship between the peralkaline and aluminous A-type granites

Loiselle & Wones (1979) first proposed the term 'A-type granite', and included both the peralkaline and the metaluminous to weakly peraluminous rocks in their A-type group. King *et al.* (1997) argued that rocks with PA associations are very different from AA subtype rocks, and that the two subgroups should be considered separately. The differences between PA and AA rocks have been extensively discussed by previous workers. Poitrasson *et al.* (1994, 1995) suggested that the aluminous subsolvus trend of A-type granites might be generated through the partial melting of metamorphosed, amphibole-bearing and K_2O-rich mafic cumulates which had not undergone a previous partial melting event, whereas the PA granites were probably formed by mantle-derived magmas which had interacted with the deep crust. King *et al.* (1997) indicated that the AA granites were derived by small degrees of partial melting of a felsic, infracrustal source region that was probably relatively reduced, with H_2O and F contents similar to those of the I-type source regions. In contrast, the above authors suggested that the PA rocks were likely to be generated by fractionation from an anhydrous mafic source. The above studies all advocate that the PA and AA rocks originate from contrasting materials and have different petrogenetic schemes. However, geochemical investigations demonstrate that the AA and PA granites of the present study have similar sources and correspond to the residual liquids left after high degrees of fractional crystallisation in a deep magma chamber. Therefore, the authors favour the model in which the differences between the PA and AA rocks are most likely to result from different differentiation trends rather than from different source materials. The halogens, and F in particular, are considered to be critical in determining fractionation trends in A-type magmas (Collins *et al.* 1982). In fluorine-rich magmas, the Ca-amphiboles are early crystalline phases, resulting in the magma becoming not only enriched in the alkali feldspar component, but ultimately peraluminous with progressive crystallisation. However, if the fluorine content is relatively low in the melt, then early amphibole crystallisation may not occur and anorthite-rich plagioclase could be the dominant crystallising phase, thus leading to a peralkaline trend (Collins *et al.* 1982). This result is consistent with the mineral associations and chemistry observed in the studied PA and AA granites. Therefore, the present authors suggest that the disparate differentiation paths, a consequence of variation in F contents in the parent magmas, might be the main reason for the mineral and chemical differences between the PA and AA granites.

7. Conclusions

1. Rb-Sr and U-Pb geochronology results demonstrate that A-type granites in the coastal area of the Zhejiang and Fujian provinces of SE China were generated mainly during the Late Cretaceous (100–90 Ma).

2. Two subgroups of A-type granites (AA and PA) are distinguished in this belt. The AA subgroup often contains spessartine and manganese-rich muscovite, whereas the PA rocks usually have riebeckite, arfvedsonite and aegirine. Geochemically, the AA granites show higher Al_2O_3 contents, lower Nb, Zr, Ga, Y and REE abundances, and lower FeO*/MgO and Ga/Al ratios than the PA subgroup.

3. The initial Nd isotopic compositions of the two subgroups of A-type granites occurring in neighbouring district are quite similar, and are all characterised by mixing of ancient basement rocks with variable amounts of mantle materials.

4. Combined geological and geochemical investigations indicate that both the AA and the PA magmas are highly evolved and reflect the residual liquids left after high degrees of fractional crystallisation in a deep magma chamber. The different differentiation paths, resulting from variations in the fluorine contents in the parent magmas, are inferred to be the most crucial factor in explaining the mineral and geochemical differences between the two subgroups of A-type granites.

8. Acknowledgements

The first author expresses his sincere thanks to Professor S. Ishihara for inviting the submission of this paper to this proceedings volume. Special thanks are also due to Professor X. M. Zhou for his guidance and concern during this research work. Dr Noreen Evans is gratefully acknowledged for her careful correction of the English. Constructive reviews by B. Collins and T. Ramo have substantially improved the final draft. This study was financially supported by grants from the National Natural Science Foundation of China (Grant Numbers 40221301, 40132010 and 40272036).

9. References

Anderson, J. L. & Bender, E. E. 1989. Nature and origin of Proterozoic A-type granitic magmatism in the southwestern United States of America. *Lithos* **23**, 19–52.

Bowden, P., Black, R., Martin, R. F., Ike, E. C., Kinnaird, J. A. & Batchelor, R. A. 1987. Niger-Nigerian alkaline ring complexes: a classic example of African Phanerozoic anorogenic mid-plate magmatism. *In* Fitton, J. G. & Upton, B. G. J. (eds) *Alkaline Igneous Rocks. Geological Society of America, Special Publication* **30**, 357–79.

Chen, J. F. & Jahn, B. M. 1998. Crustal evolution of southeastern China: Nd and Sr isotopic evidence. *Tectonophysics* **284**, 101–33.

Clemens, J. D., Holloway, J. R. & White, A. J. R. 1986. Origin of an A-type granite: experimental constraints. *American Mineralogist* **71**, 317–24.

Collins, W. J., Beams, S. D., White, A. J. R. & Chappell, B. W. 1982. Nature and origin of A-type granites with particular reference to Southeastern Australia. *Contributions to Mineralogy and Petrology* **80**, 189–200.

Creaser, R. A., Price, R. C. & Wormald, R. J. 1991. A-type granites revisited: assessment of a residual-source model. *Geology* **19**, 163–6.

Dall'Agnol, R., Scaillet, B. & Pichavant, M. 1999. An experimental study of a lower Proterozoic A-type granite from the eastern Amazonian craton, Brazil. *Journal of Petrology* **40**, 1673–98.

Eby, G. N. 1992. Chemical subdivision of the A-type granitoids: petrogenetic and tectonic implications. *Geology* **20**, 641–44.

Faure G. 1986. *Principles of isotope geology*. Chichester: John Wiley & Sons.

Frost, C. D. & Frost, B. R. 1997. Reduced rapakivi-type granites: the tholeiite connection. *Geology* **25**, 647–50.

Hong, D. W., Guo, W. J., Li, G. J., Kang, W. & Xu, H. M. 1987. *Petrology of the miarolitic granite belt in the southeast coast of Fujian province and their petrogenesis*. Beijing: Science and Technology Press of Beijing. [In Chinese.]

Hu, X. J., Xu, J. K., Kang, H. N., Tong, C. X., Chen, C. H. & Ye, G. G. 1991. Geological features of the Lower Proterozoic Badu Group of southwestern Zhejiang and its implications. *Regional Geology of China* **3**, 234–40. [In Chinese with English abstract.]

Huang, T. K. 1978. An outline of the tectonic characteristics of China. *Eclogae Geologicae Helvetiae* **71**, 611–35.

Jung, S., Hoernes, S. & Mezger, K. 2000. Geochronology and petrogenesis of Pan-African, syn-tectonic, S-type and post-tectonic A-type granite (Namibia): products of melting of crustal sources, fractional crystallization and wall rock entrainment. *Lithos* **50**, 259–87.

Kebede, T. & Koeberl, C. 2003. Petrogenesis of A-type granitoids from the Wallagga area, western Ethiopia: constraints from mineralogy, bulk-rock chemistry, Nd and Sr isotopic compositions. *Precambrian Research* **121**, 1–24.

Kerr, A. & Fryer, B. J. 1993. Nd isotope evidence for crustal-mantle interaction in the generation of A-type granitoid suites in Labrador, Canada. *Chemical Geology* **104**, 39–60.

King, P. L., White, A. J. R., Chappell, B. W. & Allen, M. 1997. Characterization and origin of aluminous A-type granites from the Lachlan Fold belt, Southeastern Australia. *Journal of Petrology* **38**, 371–91.

Li, X. H. & McCulloch, M. T. 1996. Secular variation in the Nd isotopic composition of Neoproterozoic sediments from the southern margin of the Yangtze block: evidence for a Proterozoic continental collision in the southeast China. *Precambrian Research* **76**, 67–76.

Litvinovsky, B. A., Jahn, B. M., Zanvilevich, A. N., Saunders, A., Poulain, S., Kuzmin, D. V., Reichow, M. K. & Titov, A. V. 2002. Petrogenesis of syenite-granite suites from the Bryansky complex (Transbaikalia, Russia): implications for the origin of A-type granitoid magmas. *Chemical Geology* **189**, 105–33.

Liu, C. S., Zhu, J C., Shen, W. Z. & Xu, S. J. 1990. Evolution of continental crust and material source (genesis) of granitoids in south China. *Geotectonica et Metallogenia* **14**, 125–38. [In Chinese with English abstract.]

Loiselle, M. C. & Wones, D. R. 1979. Characteristics and origin of anorogenic granites. *Geological Society of America, Abstracts with Program* **11**, 468.

Martin, H., Bonin, B., Capdevila, R., Jahn, B. M., Lameyre, J. & Wang, Y. 1994. The Kuiqi peralkaline granitic complex (SE China): petrology and geochemistry. *Journal of Petrology* **35**, 983–1015.

Patino Douce, A. E. P. 1997. Generation of metaluminous A-type granites by low-temperature melting of calc-alkaline granitoids. *Geology* **25**, 743–46.

Peng, Y. M. & Deng, C. W. 1991. A study on the Qingtian alkaline granite, Zhejiang province. *Journal of Nanjing University (Earth Sciences)***3**, 138–47. [In Chinese with English abstract.]

Poitrasson, F., Pin, C., Duthou, J. L. & Platevoet, B. 1994. Aluminous subsolvus anorogenic granite genesis in the light of Nd isotopic heterogeneity. *Chemical Geology* **112**, 199–219.

Poitrasson, F., Duthou, J. L. & Pin, C. 1995. The relationship between petrology and Nd isotopes as evidences for contrasting anorogenic granite genesis: example of the Corsican Province (SE France). *Journal of Petrology* **36**, 1251–74.

Qiu, J. S., Wang, D. Z., Peng, Y. M. & Zhou, J. C. 1996. Petrology, geochemistry and genesis of the Taohuadfao alkali granite in Zhoushan, Zhejiang province. *Journal of Nanjing University (Natural Sciences)* **32**, 80–9. [In Chinese with English abstract.]

Qiu, J. S., Wang, D. Z. & McInnes, B. I. A. 1999. Geochemistry and petrogenesis of the I- and A-type composite granite masses in the coastal area of Zhejiang and Fujian Provinces. *Acta Petrologica Sinica* **15**, 237–46. [In Chinese with English abstract.]

Qiu, J. S., Wang, D. Z., Kanisawa, S. & McInnes, B. I. A. 2000. Geochemistry and petrogenesis of aluminous A-type granites in the coastal area of Fujian province. *Geochimica* **29**, 313–21. [In Chinese with English abstract.]

Shu, L. S. & Zhou, X. M. 2002. Late Mesozoic tectonic framework of southeastern China. *Geological Review* **48**, 249–60. [In Chinese with English abstract.]

Smith, D. R., Noblett, J., Wobus, R. A., Unruh, D., Douglass, J., Beane, R., Davis, C., Goldman, S., Kay, G., Gustavson, B., Saltoun, B. & Stewart, J. 1999. Petrology and geochemistry of late-stage intrusions of the A-type, mid-Proterozoic Pikes Peak batholith (Central Colorado, USA): implications for petrogenetic models. *Precambrian Research* **98**, 271–305.

Sun, S. S. & McDonough, W. F. 1989. Chemical and isotopic systematics of oceanic basalts: implications for mantle composition and process. *In* Saunder, A. D. & Norry, M. J. (eds) *Magmatism in the ocean basin. Geological Society, London, Special Publication* **42**, 313–45. London: Geological Society.

Sylvester, P. J. 1989. Post-collisional alkaline granites. *The Journal of Geology* **97**, 261–80.

Tsusue, A. & Liu, J. 1987. A preliminarily consideration of the genesis of the Kuiqi mass near Fuzhou, southeastern China, as an example of A-type granite. *In*: Tu, G. C., Xu, K. W. & Qiu, Y. Z. (eds) *Petrogenesis and mineralization of granitoids*, 365–68. Beijing: Science Press.

Tu, G. C., Zhang, Y. Q. & Zhao, Z. H. 1984. Preliminary studies on two alkali-rich intrusive belts in South China. *In* Xu, K. Q. & Tu, G. C. (eds) *Geology of granites and their metallogenetic relations*, 21–37. Nanjing: Jiangsu Science and Techonology Press. [In Chinese.]

Turner, S. P., Foden, J. D. & Morrison, R. S. 1992. Derivation of some A-type magmas by fractionation of basaltic magma: an example from the Padthaway Ridge, South Australia. *Lithos* **28**, 151–79.

Volkert, R. A., Feigenson, M. D., Patino, L. C., Delaney, J. S. & Drake, A. A., Jr. 2000. Sr and Nd isotopic compositions, age and petrogenesis of A-type granitoids of the Vernon Supersuite, New Jersey Highlands, USA. *Lithos* **50**, 325–47.

Wang, D. Z. & Zhou, X. M. 2002. *The late Mesozoic granitic volcanic-intrusive complexes in southeastern China: constraints on petrogenesis and crustal evolution*. Beijing: Science Press. [In Chinese.]

Wei, C. S., Zheng, Y. F. & Zhao, Z. F. 2000. Hydrogen and oxygen isotope of A-type granites in the continental margins of eastern China. *Tectonophysics* **328**, 205–27.

Whalen, J. B., Currie, K. L. & Chappell, B. W. 1987. A-type granites: geochemical characteristics, discrimination and petrogenesis. *Contributions to Mineralogy and Petrology* **95**, 407–19.

Wickham, S. M., Litvinovsky, B. A., Zanvilevich, A. N. & Bindeaman, D. N. 1995. Geochemical evolution of Phanerozoic magmatism in Transbaikalia, East Asia: a key constraint on the origin of K-rich silicic magmas and the process of cratonization. *Journal of Geophysical Research* **100**, 15641–54.

Wu, F. Y., Sun, D. Y., Li, H. M., Jahn, B. M. & Wilde, S. 2002. A-type granites in northeastern China: age and geochemical constraints on their petrogenesis. *Chemical Geology* **187**, 143–73.

Xing, F. M., Xu, X., Chen, J. F., Zhou, T. X. & Foland, K. A. 1992. The Late Proterozoic continental accretionary history of the southeastern margin of Jiangnanian. *Acta Geologica Sinica* **5**, 283–99. [In Chinese with English abstract.]

Xu, B. T. 1987. Isotopic geochronology of paleo-basement metamorphic rocks in southern Zhejiang province. *Geological Review* **33**, 468–74. [In Chinese with English abstract.]

Zhou, X. R. & Wu, K. L. 1994. *Zhangzhou Composite Granite*. Beijing: Science Press. [In Chinese with English abstract.]

Zhou, X. M. & Zhu, Y. H. 1993. Late Proterozoic collisional orogen and geosuture in southeastern China: petrological evidence. *Chinese Journal of Geochemistry* **12**, 239–51.

JIAN-SHENG QIU, DE-ZI WANG, SHAO-YONG JIANG and RU-CHENG WANG, State Key Laboratory for Mineral Deposit Research, Department of Earth Sciences, Nanjing University, Nanjing 210093, China.
e-mails: jsqiuwyh@public1.ptt.js.cn; editors@public1.ptt.js.cn; shyjiang@public1.ptt.js.cn; rcwang@nju.edu.cn

BRENT I. A. MCINNES, CSIRO Exploration & Mining, Australian Resources Research Centre, Kensington, WA 6151, Australia.
e-mail: brent.mcinnes@csiro.au

SATOSHI KANISAWA, Institute of Mineralogy, Petrology and Economic Geology, Graduate School of Science, Tohoku University, Sendai 980–8578, Japan.
e-mail: kanisawa@mail.tains.tohoku.ac.jp

MS received 2 September 2003. Accepted for publication 9 September 2004.

Transactions of the Royal Society of Edinburgh: Earth Sciences, **95**, 237–248, 2004

Geochemical, Sr and Nd isotopic characteristics and tectonic implications for three stages of igneous rock in the Late Yanshanian (Cretaceous) orogeny, SE China

Cheng-Hong Chen, Wayne Lin, Ching-Ying Lan and Chi-Yu Lee

ABSTRACT: Late Yanshanian magmatism in SE China includes three stages of thermal event induced by the interaction between the continental margin of Eurasia and the paleo-Pacific plate during the Cretaceous period. Products of syn-orogenic magmatism (130–110 Ma) include high-Al gabbros (HAG), and gneissic tonalite, trondhjemite and granodiorite (TTG), which intruded into the deep basement (18–24 km). Rocks of the post- and an-orogenic magmatism are shallow-level (6–8 km) I-type granitoids (110–99 Ma), and miarolitic A-type granites plus rhyolite-dominate bimodal volcanics (94–81 Ma), respectively. Geochemically, HAG and TTG belong to the medium-K calc-alkaline affinity with high Sr/Y, whereas other granitoids are mainly high-K calc-alkaline to shoshonitic rocks with low Sr/Y. Sr and Nd isotope compositions suggest different sources of HAG and TTG from other rocks. Progressive depletions of Ba, Sr, Eu and P from I- to A-type granites reflect partial melting of felsic granulites from hydrous to dry conditions, whereas high Sr/Y in HAG and TTG are compatible with dehydration melting of amphibolites. Tectonic models which accommodate HAG and TTG may involve thickening of the lithosphere to convert the pre-existing lower-crust basic rocks into amphibolites. It was followed by basaltic underplating which is attributed to delamination of the thickened lithosphere and led to triggering of crust melting under extension.

KEY WORDS: Amphibolite, basaltic underplating, dehydration melting, HAG and TTG.

Granitoids in S China are distributed across a vast area ($\sim$250,000 km^2), and thus, play an important role in our understanding of the late Mesozoic geohistory of crustal accretion in the Cathaysia Block – one of the major granitic provinces in the world. It is generally believed that the Yanshanian orogeny outlined the present-day geological configuration that has developed since the Jurassic (Li 2000), by which the Early Yanshanian (EY, late Jurassic) dominated the Cathaysia interior, whereas the Late Yanshanian (LY, early Cretaceous) dominated the southeast coastal area.

Influenced by subduction of the paleo-Pacific plate, LY magmatism has been attributed to the change of plate convergence rate (Northrup *et al.* 1995). Basically, products of the LY magmatism are more restricted in a NE–SW trending zone – the Southeast Coast Magmatic Belt (SCMB) – that is separated from the Cathaysia interior to the west and the Changle-Nanao Metamorphic Belt (CNMB) to the east by two deep faults (Fig. 1). Previous investigations have revealed that rock assemblages are mainly shallow-level, calc-alkaline, I-type granitoids ranging from granodiorite to alkali feldspar granite, and succeeding A-type granites and basalt-rhyolite bimodal volcanics (Chen *et al.* 2000). The CNMB, although intruded by undeformed granitoids as well, is a shear zone characterised by ductile deformation and greenschist to amphibolite facies metamorphism, leading to the development of strong schistosity and gneissosity in the pre-existing and simultaneous igneous rocks (Tong & Tobisch 1996). For the presence of such tectonic uplift towards the subduction zone, the CNMB has been proposed to represent the metamorphic core complex of the SCMB, or the LY syn-orogenic products (Chen *et al.* 2000). With a number of ^{40}Ar/^{39}Ar mineral ages reported for these metamorphic rocks (Yang *et al.* 1997; Wang & Lu 2000), the regional exhumation rates of this metamorphic core complex, as calculated from cooling rates, were estimated to vary from 1·3 to 0·8 Km/Myr in different terranes during 130–110 Ma (Chen *et al.* 2002).

In the CNMB, mafic intrusive rocks are distributed sporadically from Matsu southward to Dongshan, including Lienhuashan, Daiqianshan, Qinglanshan, Taohuashan and Dongyue (Fig. 1). The associated rocks of these mafic intrusives are mostly granitic gneiss (e.g. tonalite, trondhjemite and granodiorite) and migmatite. Their contact relationships are rather complicated. The outcrop at Qinglanshan is reflected by a mylonitic ductile shear zone, indicating a tectonic contact. Mafic rocks in Daiqianshan show an obvious cumulate layering structure, characterised by a chilled gabbro base that grades into medium grained gabbro, or diorite. Those in Lienhuashan occur mainly as stocks and dykes, or as enclaves within the granodiorite host (Wang 2002). In the SCMB, gabbros (e.g. Kuanshan and Shanfang) are precursors of the large I-type granitic complexes (Pitcher 1997).

Although the structure features and dynamics of this ductile shear zone have been extensively studied (Tong & Tobisch 1996; Wang & Lu 1997, 2000), the association of mafic and gneissic rocks has seldom been discussed on the basis of magmatic processes contemporaneous with the ductile deformation and mylonisation. Here, in addition to two ^{40}Ar/^{39}Ar hornblende ages of mafic rocks to supplement previous dating works in the CNMB, the present authors describe new major and trace element contents, and Sr and Nd isotope ratios for both the mafic and associated gneissic rocks. These geochemical and isotopic characteristics can reveal the genesis of

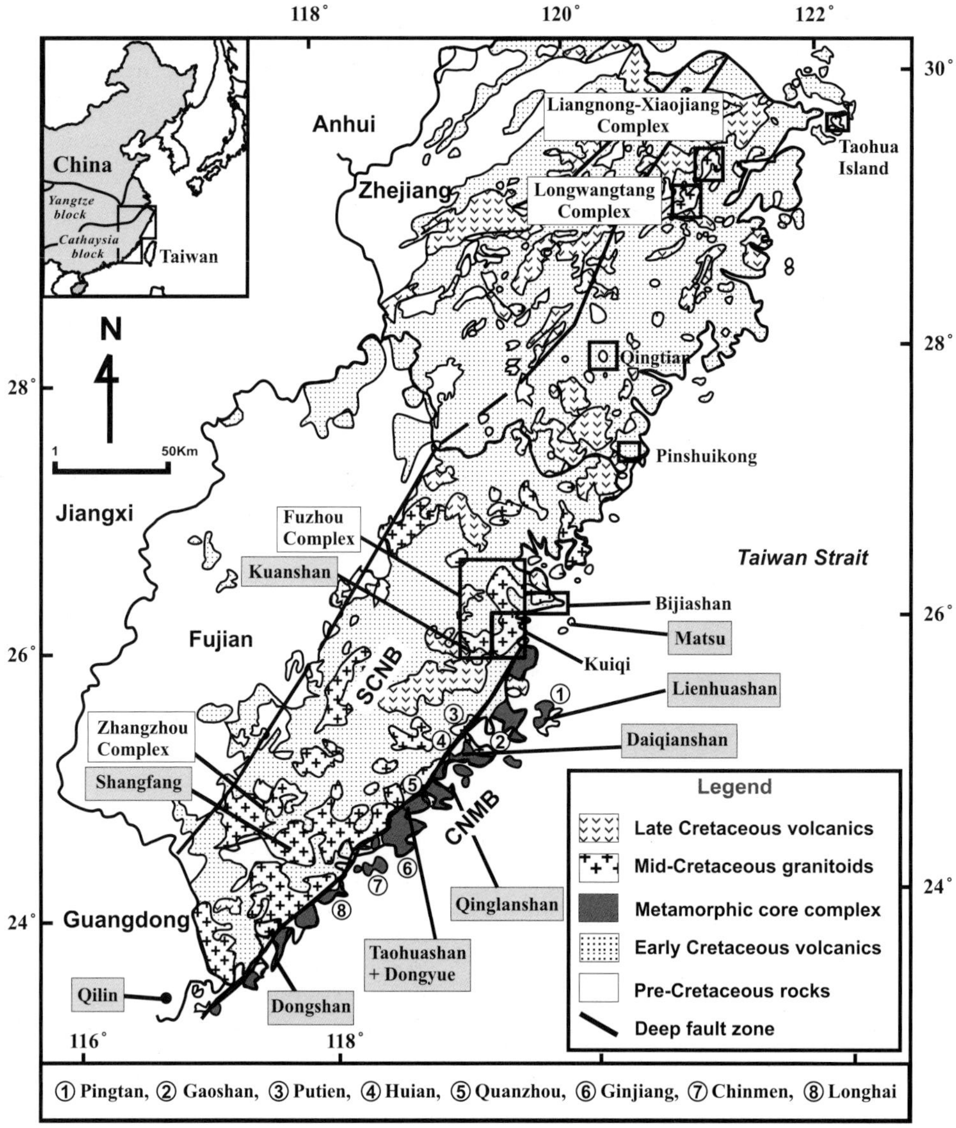

Figure 1 Localities of gabbros (shaded rectangles) and gneissic rocks (numbered circles), and distribution of igneous complex (open rectangles) and volcanic rocks in the Southeast Coast Magmatic Belt (SCMB) and the Changle-Nanao Metamorphic Belt (CNMB), which are two tectonic units in the Cathaysia block (inset). Other localities in boxes represent A-type granites.

such deep-seated rocks in a collisional tectonic setting, which can further be used to differentiate the post-orogenic, I-type granitoids and an-orogenic, A-type granites (Chen *et al.* 2000), as well as the pre-existing late Jurassic granitoids in S China (Chen & Jahn 1998). On this basis, the tectonic evolution in the SE China continental margin may be deciphered.

1. Analytical methods

Mineral compositions were analysed with electron probe X-ray microanalysers including a JEOL JXA-8900R (Institute of Earth Sciences, Academia Sinica, Taipei, Taiwan) and a Shimadzu-ARL EMX-SM7 (Department of Geosciences, National Taiwan University, Taipei, Taiwan). Data reductions for the atomic number effect, absorption and fluorescence were calculated with the ZAF correction computer program (Chen & Tung 1984). Amphibole analysis is particularly useful for $^{40}Ar/^{39}Ar$ age dating and the application of the Al in the hornblende geobarometer (Table 1).

Fresh amphiboles were dated using the step-heating $^{40}Ar/^{39}Ar$ age spectrum technique described by Lo & Lee (1994). Here, the present authors only describe $^{40}Ar/^{39}Ar$ age plateau, and Cl/K and Ca/K spectra for two mafic rocks which had not been firmly dated before (i.e. Daiqianshan and Shanfang). Because the inverse slope correlation ages are similar to the plateau ages, they are not shown. Those of the gneissic rocks are based on previous works (Yang *et al.* 1997; Chen *et al.* 2002).

In the present study, 56 mafic and gneissic rocks were measured for their major and trace element abundances, and 16 samples among them were measured for Sr and Nd isotope compositions using the same methods as described in Chen *et al.* (2000). The analytical uncertainties are generally better than 5% for all elements. $^{87}Sr/^{86}Sr$ and $^{143}Nd/^{144}Nd$ ratios were corrected for mass fractionation by normalising to $^{86}Sr/^{88}Sr=0{\cdot}1194$ and $^{146}Nd/^{144}Nd=0{\cdot}7219$, respectively. The standards used were SRM987 for Sr ($^{87}Sr/^{86}Sr=0{\cdot}710226$), with a long-term reproducibility of 0·000038 (95% confidence level), and the La Jolla (UCSD) standard for Nd ($^{143}Nd/^{144}Nd=0{\cdot}511845$), with a long-term reproducibility of 0·000020. Whole-rock chemical analyses, and Sr and Nd isotope ratios for representative samples are presented in Table 2.

Table 1 Representative amphibole compositions of various rock types in SE China: (1) Xiaojiang; (2) Liangnong; (3) Fuzhou; (4) Zhangzhou; (5) Putien; (6) Huian; (7 & 8) Pingtan; (9) Gaoshan; (10) Ginjiang. (Mgr) monzogranite; (Gdr) granodiorite; (Mig) migmatite; (Gab) gabbro; (Gn) gneiss; (tFeO) total Fe as FeO; and (n.d.) not determined.

Lithology	1 (Mgr)	2 (Mgr)	3 (Gdr)	4 (Mgr)	5 (Mig)	6 (Mig)	7 (Gab)	8 (Gn)	9 (Gn)	10 (Gn)
wt.%										
SiO_{2reset}	50·06	52·27	46·66	50·23	45·42	45·75	43·71	42·75	41·90	43·91
TiO_2	1·06	0·57	1·04	0·55	1·64	1·52	1·01	0·87	0·87	0·84
Al_2O_3	4·38	3·54	6·08	5·65	7·57	7·54	12·01	10·82	10·83	11·68
tFeO	12·92	10·87	14·19	12·71	16·93	16·97	11·56	19·92	20·73	19·20
MnO	0·64	0·53	0·93	1·44	0·81	0·81	0·28	0·62	0·75	0·35
MgO	15·73	16·92	14·09	14·87	11·65	11·82	14·45	8·82	8·35	7·78
CaO	11·11	12·34	12·12	11·82	11·65	11·60	11·67	11·35	11·78	11·86
Na_2O	1·32	0·75	1·34	1·30	1·49	1·46	1·75	1·35	1·28	1·25
K_2O	0·48	0·30	0·67	0·45	0·88	0·88	0·62	1·41	1·38	1·40
Cr_2O_3	0·28	0·24	0·04	0·03	n.d.	n.d.	0·07	n.d.	n.d.	n.d.
Total	97·98	98·33	97·48	99·04	98·04	98·35	97·13	97·91	97·87	98·27
Molecular proportion based on 23 oxygen										
Si	7·280	7·475	6·951	7·241	6·807	6·832	6·429	6·506	6·454	6·537
Al	0·752	0·597	1·069	0·962	1·338	1·328	2·082	1·941	1·959	2·149
Na	0·371	0·209	0·492	0·358	0·432	0·422	0·499	0·397	0·376	0·371
K	0·089	0·055	0·128	0·082	0·167	0·168	0·117	0·293	0·275	0·279
P* (Kb)	0·6	<0·6	2·1	1·6	3·4	3·3		6·2	6·3	7·2

*P (± 0·6 Kb)= − 3·01+4·76 Al (Schmidt 1992).

2. Results

2.1. Mineral composition and geobarometric data

Mafic rocks generally consist of the plagioclase, amphibole, pyroxene and spinel groups of minerals, with minor amounts of olivine, biotite and quartz in a few samples. Highly calcic plagioclases are present in the cumulated rocks; for example, some Lienhuashan (An_{97-85}), Daiqianshan (An_{93-86}) and Taohuashan (An_{88-74}) samples. On the other hand, less calcic plagioclases are found in other Lienhuashan (An_{48-34}) gabbros, indicating a large degree of crystal fractionation. The precursory gabbros of Kuanshan (An_{93-83}) show cumulate textures, but those of Shanfang (An_{82-67}) are less obvious.

Olivine is only found in the mafic rocks of Kuanshan (Fo_{79-77}) and Lienhuashan (Fo_{68}). Both orthopyroxene and clinopyroxene are present in the mafic rocks, except those from Dongyue, where only clinopyroxene is found. Orthopyroxenes are mainly enstatite (En_{81-80} in Kuanshan and En_{72-65} in other areas), while clinopyroxenes are mainly diopside ($Wo_{42}En_{43}Fs_{15}$–$Wo_{48}En_{40}Fs_{12}$) and minor augite ($Wo_{30}En_{53}$ Fs_{17} in Kuanshan). The amphiboles are the calcic varieties; those with high Al_{IV} content (1·4–1·9) have $(Na+K)_A \geq 0{\cdot}5$, while others with low Al_{IV} content (0·8–1·3) have $(Na+K)_A$ <0·5, corresponding to pargasitic and edenitic hornblendes, respectively (Fig. 2a). Brown micas are found in a few mafic rocks in Lienhuashan, Taohuashan and Shanfang, ranging from phlogopite [$Fe^{+2}/(Mg+Fe^{+2})$, or Fe′ = 17·4] in Lienhuashan to biotite (Fe′ = 31·8 to 38·7) in other two areas. Magnetites contain some amounts of Cr_2O_3 (2·8–13·6 wt.%) and Al_2O_3 (2·8–4·3 wt.%) which replace ferric iron. Following the method of Powell & Powell (1977), coexisting magnetite and exsolved ilmenite in Shanfang gabbro give rise to equilibration temperatures of 465–430°C, which is indicative of subsolidus oxidation.

Gneissic rocks are generally grey in colour and consist mainly of quartz, sphene, K-feldspar, plagioclase, biotite and hornblende. In a few localities (e.g. Chinmen), K-feldspars are pinkish and associated with white micas. Quartz may be myrmekitically included in plagioclase and/or graphically intergrown with K-feldspar. Plagioclase and K-feldspar compositions vary with rock types: An_{36-33} and Or_{91-90} for tonalite, An_{20} and Or_{92-89} for trondhjemite, and An_{40-28} and Or_{96-90} for granodiorite (Lan *et al.* 1997). Biotites are separated into two groups: Fe′ = 51–57 in tonalite and trondhjemite, and Fe′ = 69–76 in granodiorite. Calcic amphiboles are commonly present in the gneissic rocks and construct a similar trend with the mafic rocks, which is clearly distinguished from the other obtained from the granitoids of the major igneous complexes in the SCMB (Fig. 2a).

The alumina contents of amphibole in the granitic rocks can be used as pressure indicators, but are restricted to those in presence of quartz, plagioclase, K-feldspar, biotite, sphene and Fe-Ti oxides (Hammerstrom & Zen 1986). Using the geobarometer of Schmidt (1992), a few examples show that the pressure condition differs in rock types and in tectonic units (Fig. 2b). The CNMB hornblendes are characterised by higher total Al contents, by which the host granitic gneisses are estimated to have formed under a pressure of 6·2–7·2 kb in Pingtan, Gaoshan and Ginjiang, while migmatites were formed at a lower pressure of ~3·3 kb in Huian and Putien. On the other hand, those estimated for the plutons of major igneous complexes in the SCMB are much lower (Table 1).

2.2. Time of mafic emplacement and metamorphism

To constrain the time of thermal event, available U-Pb zircon and $^{40}Ar/^{39}Ar$ hornblende and biotite ages are adopted (Table 3). Because the closing temperatures of hornblende and biotite have been suggested to be 550 ± 20°C and 300 ± 30°C (Rollinson 1993), the similarity of these two mineral ages in one intrusive body would indicate a rapid cooling, and thus, can be used to represent the time close to the event. However, mafic rocks are based on the hornblende age alone for the lack of biotite in the dated samples. The $^{40}Ar/^{39}Ar$ age spectra of amphiboles separated from mafic rocks of Daiqianshan (125·1 ± 2·3 Ma) and Shanfang (105·2 ± 1·9 Ma) are shown in Figure 3. Such results are in accord with those available for

Table 2 Chemical compositions of representative gabbros and gneissic rocks in SE China (sample numbers in brackets): (1) Daiqianshan; (2 & 7) Pingtan; (3) Taohuashan; (4) Shanfang; (5, 9 & 10) Chinmen; (6) Dongshan; (8) Huian. (Gab) gabbro; (Tonal) tonalite; (Trond) trondhjemite; (Gdr) granodiorite; (c) cumulate; (n) gneiss; and (tFe_2O_3) total Fe as Fe_2O_3

Lithology	1 (FJ920826) [Gab (c)]	2 (FJ9405) [Gab]	3 (FJ9416) [Gab (c)]	4 (TSFO1) [Gab]	5 (20HSHO3) [Gab (n)]	6 (FJ9337) [Gab (n)]	7 (FJ9501) [Tonal (n)]	8 (FJ9413) [Tonal (n)]	9 (KMB-2) [Tonal (n)]	10 (SKM7-1) [Gdr (n)]
wt.%										
SiO_2	37·41	51·67	41·88	47·29	53·74	49·20	60·73	66·48	70·98	70·60
TiO_2	2·19	1·15	1·47	0·78	1·22	1·09	0·73	0·52	0·33	0·29
Al_2O_3	17·95	18·82	16·38	19·28	19·77	16·71	18·73	16·09	14·74	15·35
tFe_2O_3	16·50	10·26	17·65	10·61	9·30	10·53	6·30	4·36	3·49	2·79
MnO	0·13	0·18	0·17	0·18	0·09	0·45	0·10	0·10	0·07	0·09
MgO	7·30	3·79	7·39	7·42	2·58	7·10	1·29	1·10	0·56	0·54
CaO	13·94	8·85	12·14	11·22	7·04	8·54	4·67	3·85	2·46	2·39
Na_2O	1·34	2·82	0·77	1·96	4·26	3·44	3·72	3·56	4·03	3·60
K_2O	0·21	0·71	0·29	0·22	1·70	1·81	1·48	2·36	2·00	3·58
P_2O_5	1·09	0·36	0·05	0·22	0·32	0·35	0·26	0·14	0·10	0·08
LOI	0·68	0·65	0·32			0·39	1·35	0·39	1·35	0·72
Total	98·73	99·26	98·52	99·18	100·02	99·59	99·36	98·95	100·60	100·03
ppm										
Sc	58	26·5	42·5			36	11	12·5		6
V	407	177	745	246	210	214	19	53		10
Cr	0·7	4·1	8·0	173	4·5	190	5	3	28	18
Co	37·9	20·9	60·9	38·0	13·6	31·7	8·1	7·4	3·31	1
Ni	3·4	5·4	10·4	70		62·3	2·7	4·8	12·8	6
Ga	21	20	21	22	24	19	22·5	17	17	15
Rb	2	19	13	6	160	99	63	61	139	104
Sr	829	677	630	797	728	410	419	425	281	369
Y	20·2	15·4	9·7	11·3	25·9	29·0	17·3	21·9	21·6	24
Zr	114	114	57	12	171	219	366	181	211	160
Nb	1·5	4·1	1·5	0·5	7·8	8·3	6·3	7·5	11·6	9·2
Cs	0·1	0·9	1·2	1·2	4·7	2·5	1·3	1·2	2·5	
Ba	74	297	73	91	238	156	621	1081	874	967
La	4·16	12·01	5·94	7·58	24·78	27·32	32·59	21·94	49·6	42·7
Ce	12·79	26·14	12·55	14·80	53·00	54·84	66·50	45·48	89·5	76·1
Pr	2·47	3·66	1·76	1·89	7·17	7·50	7·77	5·50	8·86	
Nd	14·25	15·78	7·63	8·91	27·93	30·48	27·81	19·92	31·20	33·95
Sm	4·18	3·25	1·76	2·15	5·85	6·08	5·02	3·78	5·23	5·76
Eu	1·45	1·16	0·59	0·99	1·53	1·83	1·20	1·06	1·07	0·98
Gd	4·65	3·01	1·73	2·23	5·45	5·33	4·06	3·90	4·72	
Tb	0·72	0·47	0·29	0·35	0·81	0·86	0·63	0·64	0·63	1·15
Dy	4·03	2·63	1·67	1·90	4·78	4·79	3·34	3·70	3·38	
Ho	0·76	0·53	0·34	0·37	0·94	0·94	0·64	0·77	0·66	
Er	1·85	1·41	0·92	1·02	2·67	2·63	1·76	2·16	1·86	
Tm	0·25	0·21	0·14	0·14	0·38	0·42	0·26	0·33	0·28	
Yb	1·41	1·32	0·91	0·90	2·53	2·67	1·69	2·27	1·81	2·87
Lu	0·21	0·22	0·15	0·14	0·38	0·44	0·27	0·36	0·29	0·47
Hf	3·02	2·71	1·40	0·44	4·63	5·14	8·92	5·07	5·01	5·37
Ta	0·08	0·25	0·15	0·03	0·52	0·48	0·75	0·55	0·44	0·95
Th	0·18	1·60	2·01	0·77	5·92	2·24	11·7	5·9	17·3	15·2
U	0·04	0·27	0·58	0·26	1·53	1·61	1·41	1·29	2·34	3·7
I_{Sr}*	0·7059	0·7056	0·7056	0·7057	0·7061	0·7056			0·7065	0·7051
$eNd_{(T)}$*	− 3·1	− 2·4	− 4·4	− 3·1	− 3·0	− 2·3			− 2·4	− 2·9

*In deriving I_{Sr} and $\varepsilon Nd_{(T)}$, the age of 105 Ma is adopted for the gneissic rocks which have no dating data.

gabbroic rocks in the CNMB (Table 3), and support the present authors' previous proposition that basic magmas were generally emplaced between 130 and 110 Ma (Chen *et al.* 2000). It is also noted that the U-Pb zircon age of gabbro (115·2 ± 1·2 Ma) in Pingtan is similar to the $^{40}Ar/^{39}Ar$ hornblende (114·8 ± 1·3 Ma) and biotite (115·0 ± 1·4 Ma) ages of the associated gneiss (Table 3), which is indicative of nearly simultaneous syn-orogenic magmatic activities and metamorphism.

A number of paired $^{40}Ar/^{39}Ar$ hornblende and biotite ages for some gneissic rocks and migmatites in the CNMB show insignificant or only slight differences (Table 3). Accordingly, gneissic rocks of Pingtan and Gaoshan were postulated to have formed at 115 and 110 Ma, respectively. Combined with two

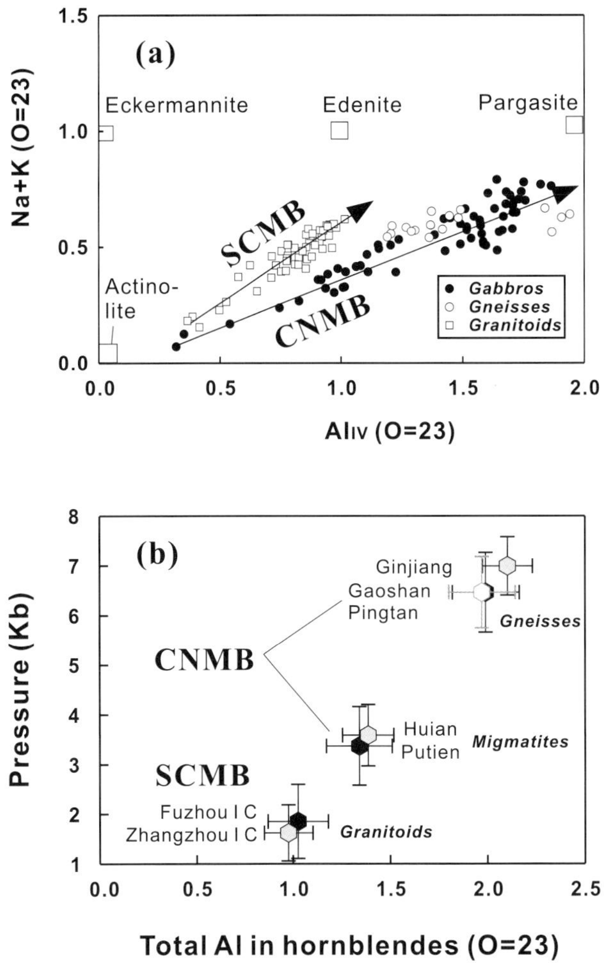

Figure 2 (a) Amphibole distribution trends for gabbroic and gneissic rocks in the Changle-Nanao Metamorphic Belt (CNMB), and I-type granitoids in the Southeast Coast Magmatic Belt (SCMB). (b) The pressure of rock suites estimated from the Al-in-amphibole geobarometer (Table 1). The bars denote the range of hornblende compositions in the sample.

U/Pb zircon age data on gneissic granodiorites near Huian (~130 Ma; Li *et al.* 2003) and one datum from Dongshan (121 Ma; Tong & Tobisch 1996), it can be concluded that gneissic rocks were formed at the same time as the mafic intrusions. The $^{40}Ar/^{39}Ar$ hornblende age of migmatites in Putien (108·6 ± 2·7 Ma; Chen *et al.* 2002) is similar to that of mylonites in Huian (109·2 ± 2·2 Ma; Wang & Lu 2000). Other migmatites show younger $^{40}Ar/^{39}Ar$ hornblende and biotite ages of ~102 Ma (Huian) and ~92 Ma (Longhai). Therefore, migmatisation and mylonisation could be related to the contact aureoles of shallower intrusions (Fig. 2b) in later stages (e.g. Lin *et al.* 1997).

2.3. Major and trace elements, and Sr and Nd isotopes

There is a wide range of silica contents for mafic rocks (SiO_2=37·4–53·7 wt.%) and associated gneisses (SiO_2=60·7–71·0 wt.%). In contrast with other late Mesozoic igneous rocks in this area, they generally belong to the medium-K calc-alkaline affinity (Fig. 4). A few other felsic gneisses which fall within the high-K affinity may represent pre-existing granitiods affected by thermal metamorphism (see below). The high alumina and low magnesium contents of the mafic rocks (Al_2O_3=15–26 wt.% and MgO=6–8 wt.%) reveal them to be the more differentiated high-Al gabbros (HAGs) and cumulates. The CIPW normative compositions were calculated in order to reveal the relationship between HAGs and associated gneisses in the An-Ab-Or diagram (Fig. 5), and hence, these gneisses belong to the assembly of peraluminous tonalite-trondhjemite-granodiorite (TTG) that can be extended from the high-Al basalts of Wolf & Wyllie (1994).

The chondrite-normalised, multi-element distribution patterns for the TTG display different patterns from the granitoids which appeared during other stages of the LY orogeny, and even the EY orogeny (Fig. 6). A systematic decrease in Ta and Nb depletion, and an increase in Ba, Sr, P, Eu and Ti depletion from the syn- to an-orogenic stages (Fig. 6b–d) is a sign of less involvement by subduction-related calc-alkaline magmatism, or in other words, more involvement by the crustal component. The pre-orogenic rocks with moderate depletion in Ba, Nb, Sr, P and Ti (Fig. 6a) are comparable with the LY post-orogenic rocks. Similarly, the decrease in Sr/Y ratios with time is also significant in differentiating the rock associations in the temporal distribution of the LY orogeny, with the pre-orogenic rocks mostly corresponding to those which appeared during the post-orogenic stage (Fig. 7).

The age-corrected Sr and Nd isotopic ratios for gabbroic rocks and associated gneissic rocks are presented in Figure 8. The $\varepsilon Nd_{(T)}$ falls in the −1·3 to −4·4 range, the I_{Sr} constant is 0·706 for gabbros, and the associated gneissic rocks share similar values [$\varepsilon Nd_{(T)}$= −2·3 to −4·9; I_{Sr}=0·705–0·707]. As a whole, they are distributed in a small field, overlapping with mafic dykes which are more depleted than the younger shallow intrusives and volcanics in the SCMB, as well as the late Jurassic felsic granites in the Cathaysia interior (Fig. 8). Therefore, the Sr and Nd isotope data imply that gabbro and gneissic rocks are most probably derived from a common source distinguished from the products of both the earlier and later thermal events.

3. Discussion

3.1. Magma differentiation between gabbroic and gneissic rocks

Based on the chondrite-normalised rare-earth element (REE) patterns (Fig. 9a), mafic rocks can be grouped into two types, the gabbro and the cumulate. The latter is particularly characterised with an Eu(+) anomaly including the precursory gabbro of Kuanshan (KUS01). With various degrees of crystal fractionation, some gabbros become more fractionated and so possess higher total REE abundances. Despite the great majority of mafic rocks displaying light REE (LREE)-enriched patterns, the Daiqianshan samples (e.g. FJ920826) are depleted in LREE. Their whole-rock compositions are high in Ti and P, but low in Si, Ni and Cr content (Table 2), reflecting the highly fractionated type of cumulates with a minor role for feldspar. As for gneissic rocks, they show similar REE patterns to the gabbroic rocks (Fig. 9b). Crystal fractionation must involve significant amounts of feldspar to account for samples which present Eu(−) and Eu(+) anomalies.

Figure 9c shows the REE patterns for gabbroic and associated gneissic rocks in the Pingtan Island. These rocks, which exhibit parallel slopes slightly enriched with LREE, have SiO_2=51·7–60·7 wt.% clustering in the centre. More basic rocks (e.g. FJ9406, SiO_2=46·5 wt.%) contain lower REE abundances with an Eu(+) anomaly whereas more fractionated rocks (e.g. FJ9401, SiO_2=68·5 wt.%) display higher REE abundances with Eu(−) anomaly. This can be explained by magmas derived with different degrees of partial melting from a common source and followed by a large degree of crystal fractionation, in which FJ9401 represents the fractionate and FJ9406 the cumulate. Therefore, the present case clearly

Table 3 $^{40}Ar/^{39}Ar$ hornblende and biotite ages for gabbroic and associated gneissic rocks in SE China compared with available U-Pb zircon ages: (Hb) hornblende; (Qz) quartz; (Zr) zircon; (Bi) biotite and (Mn) monazite

Sample number	Locality	Lithology	Size (km²)	Age (Ma)	Method	Reference
Pre-orogenic						
KM8-3	Chinmen	Gneiss	20	139·4 ± 0·4	U-Pb (Zr)	Yui *et al.* (1996)
Syn-orogenic						
FJ920826	Daiqianshan	Hb gabbro	3·5	125·1 ± 2·3	Ar/Ar (Hb)	Present study
TSF01	Shanfang	Gabbro	1	105·2 ± 1·9	Ar/Ar (Hb)	Present study
FJ9407	Lienhuashan	Hb gabbro		131·2 ± 0·6	Ar/Ar (Hb)	Yang *et al.* (1997)
ZK302-60	Taohuashan	Qz gabbro	4	106·5 ± 0·2	U-Pb (Zr)	Xu *et al.* (1999a)
Pin4-5	Pingtan	Hb gabbro	<10	115·2 ± 0·2	U-Pb (Zr)	Dong *et al.* (1997)
FJ9501	Pingtan	Gneiss		114·8 ± 1·3	Ar/Ar (Hb)	Yang *et al.* (1997)
				115·0 ± 1·4	Ar/Ar (Bi)	
Pu1-4	Huian	Gneiss	~1·5	131·0 ± 0·5	U-Pb (Zr)	Li *et al.* (2003)
Pu2-1	Huian	Gneiss		129·9 ± 0·8	U-Pb (Zr)	Li *et al.* (2003)
DS-25	Dongshan	Gneiss		121·5 ± 2·8	U-Pb (Zr)	Tong & Tobisch (1996)
FJ05	Gaoshan	Gneiss		109·6 ± 2·6	Ar/Ar (Hb)	Chen *et al.* (2002)
				105·3 ± 2·1	Ar/Ar (Bi)	
FJ07	Putien	Migmatite		108·6 ± 2·7	Ar/Ar (Hb)	Chen *et al.* (2002)
				106·7 ± 2·6	Ar/Ar (Bi)	
D58-7	Huian	Mylonite		109·2 ± 2·2	Ar/Ar (Hb)	Wang & Lu (2000)
Post-orogenic						
KM10-3	Chinmen	Granite	1	100·9 ± 0·5	U-Pb (Mn)	Yui *et al.* (1996)
KM3	Chinmen	Gneiss	20	100·2 ± 0·9	Ar/Ar (Hb)	Lo *et al.* (1993)
				97·4 ± 0·9	Ar/Ar (Bi)	
FJ08	Huian	Migmatite		102·4 ± 2·5	Ar/Ar (Hb)	Chen *et al.* (2002)
				101·4 ± 2·5	Ar/Ar (Bi)	
FJ17	Longhai	Migmatite		91·6 ± 1·9	Ar/Ar (Hb)	Chen *et al.* (2002)
				86·9 ± 1·8	Ar/Ar (Bi)	

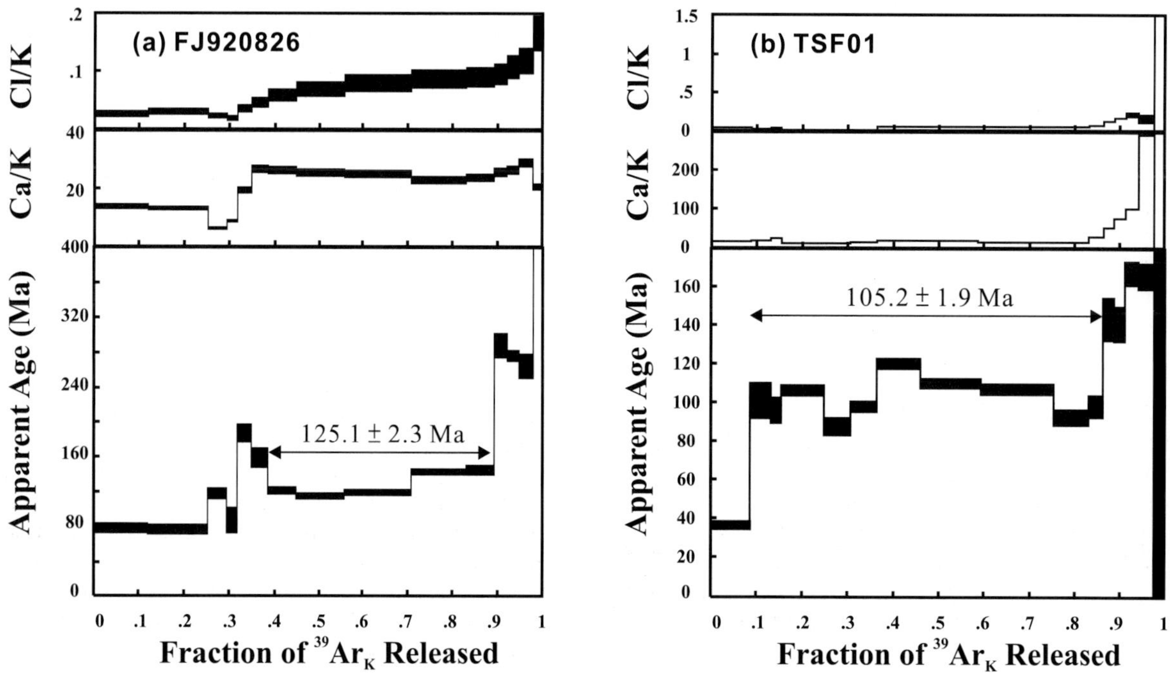

Figure 3 Cl/K and Ca/K spectra, and $^{40}Ar/^{39}Ar$ hornblende plateau ages for two gabbroic rocks in SE China, including (a) the Daiqianshan cumulate and (b) the Shanfang gabbro.

demonstrates that intrusions of basic magma in a single location could lead to the formation of a wide range of rock association. Large-scale mixing between gabbroic and granitic magmas in Pingtan is not consistent with geochronological constraints (Xu *et al.* 1999a). Those more felsic rocks which display intermediate compositions could be derived from gabbros, by small amounts of assimilation (Griffin *et al.* 2002) during magma differentiation.

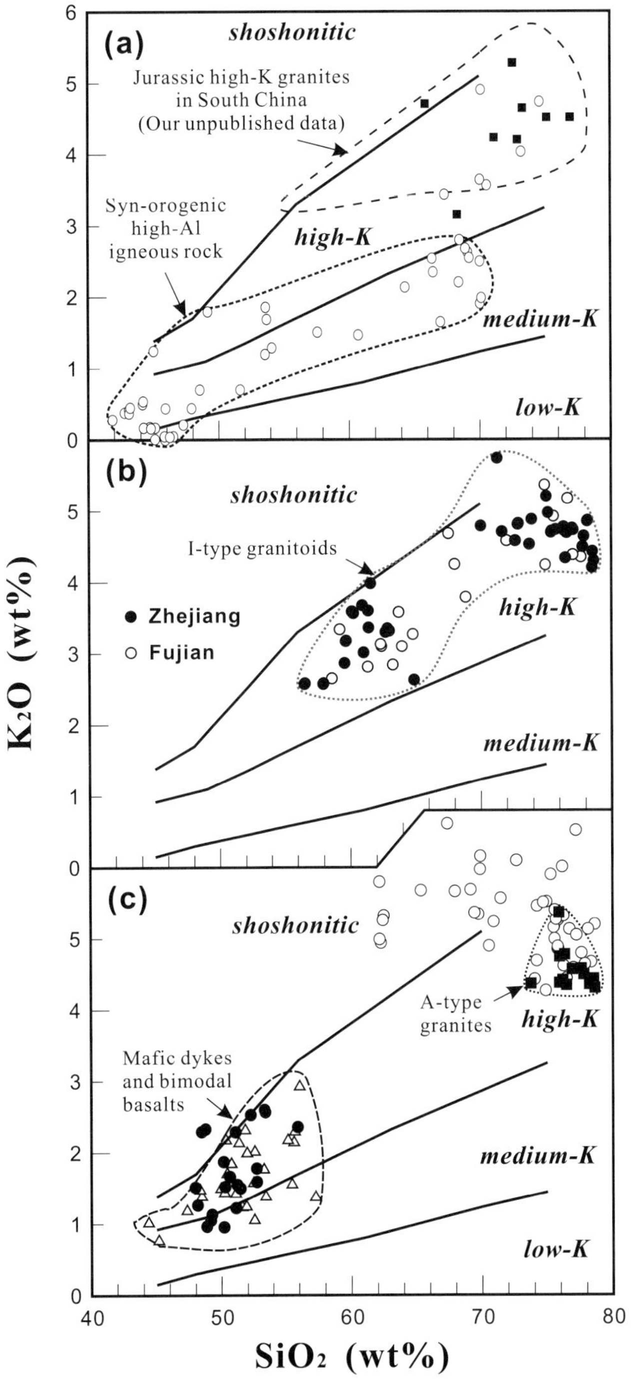

Figure 4 K_2O versus SiO_2 plots for (a) pre- and syn-orogenic, (b) post-orogenic, and (c) an-orogenic igneous rocks in SE China. Some deform rocks which are distributed in the field of Jurassic high-K granites in (a) are suggested to be derivatives of Early Yanshanian (late Jurassic) magmatism. The subdivision of calc-alkaline rocks follows the scheme of Le Maitre (1989). Symbols in (a): (solid squares) granite gneisses >130 Ma (pre-orogenic); and (open circles) gabbros and associated gneissic rocks (syn-orogenic). Symbols in (c): (solid circles) bimodal basalts; (open triangles) mafic dykes; (solid squares) A-type granites; and (open circles) bimodal rhyolites. In addition to the present study, data sources also include unpublished data by the author and co-workers.

3.2. Geochemical and isotopic characteristics of igneous rocks

Geochemical and isotopic characterisations are apparently distinguishable in rocks which occurred at three different stages in the LY orogeny. As a whole, the early-stage gabbros and the associated gneissic rocks are characterised by possessing lower K, higher Al and Sr/Y, and more depleted Sr and Nd isotope compositions. They can be separated into HAGs and TTG, respectively. Rocks from the later stages become more potassic and even shoshonitic, contain less Ba, Sr, P, Eu, Ti and Sr/Y, and show more enriched Sr and Nd isotopic compositions. Furthermore, the HAGs and TTG are geochemically and isotopically distinguished from the EY granitoids (i.e. the pre-orogenic rocks) which are widely distributed in S China (Figs 4, 6, 7 & 8).

In order to assess the influence of subduction on the source of basic rocks, late Jurassic mafic dykes (140 Ma) in the Cathaysia interior and late Cretaceous basic rocks (94–76 Ma) in the SE China were compared with the HAGs by using petrogenetic indicators such as Th/Nb and Zr/Nb ratios (Fig. 10). Since the late Jurassic and late Cretaceous samples are indistinguishable, and both originate in the subcontinental lithospheric mantle (Li & McCulloch 1998; Chen *et al.* 2000), whereas the HAGs are apparently lower in Th and Zr contents than these rocks, but relatively unaffected by Nb (the subduction-related indicator), the HAGs are suggested to have formed in a source region other than the metasomatised mantle wedge or the subcontinental lithospheric mantle. Using the partition coefficients for Th and Zr in amphibole and plagioclase (Rollinson 1993), they would be compatible elements in an amphibolite source. Hence, simple batch melting of amphibolitic basic rocks with chemical compositions similar to the 140-Ma mafic dykes would be expected to produce lower values of Th and Zr (Fig. 10).

Although TTG in the CNMB show weak signatures of arc magmas (e.g. Ta-Nb-Ti depletions in Fig. 6b), the high Sr/Y ratios are characteristic of TTG formed under thickened crust (e.g. in the Andes; Atherton & Petford 1993). In fact, the late Jurassic lithosphere has already been influenced by ancient subduction, as shown by the presence of an arc signature of the 140-Ma mafic dykes in SE China (Li & McCulloch 1998). The unique geochemical characters in HAGs and TTG relative to the earlier and later igneous rocks in the continental margin of SE China indicate that the onset of the LY magmatism, or the syn-orogenic magmatism, could have been accompanied by an increase in lithosphere thickness during transpression (Brown & Solar 1999).

3.3. Magma generation in various orogenic stages

Magmas generated in the syn-orogenic stage are complicated: their possible genesis may have included the intrusion of mantle-derived melts, partial melting of mafic residues, migmatisation between mafic magmas and the country rocks, metasomatism caused by ductile deformation in the shear zone, and anatexis of the deep crust (Pitcher 1997). Recent studies have revealed that heat transferred from basaltic underplating in the lower crust plays an important role to the generation of granitic magmas (Bonin *et al.* 1998; Petford *et al.* 2000). As basic intrusions in the CNMB emplaced at a depth of 18–24 km, with a temperature >900°C (Huppert & Sparks 1988), they would cause partial melting of adjacent crust materials, forming felsic granites, or lead to the thermal metamorphism or in-site melting of wall rocks, forming migmatites. Basically, they are rather small in dimension.

The appearance of HAG and TTG in the CNMB is significant. In terms of time sequence, they predate the major felsic magmatism in the LY orogeny and occur mainly along the strike-slip fault. Geochemically, they are distinguishable from common arc magmas in having a higher Sr/Y ratio, but not one that is high enough for adakites (Martin 1999). Syn-orogenic HAG and tonalite-trondhjemite-granodiorite have been found in the Cordillera Blanca of the Peruvian Andes and the Klamath of North America (Atherton & Petford 1993; Barnes *et al.* 1996). They all indicate the

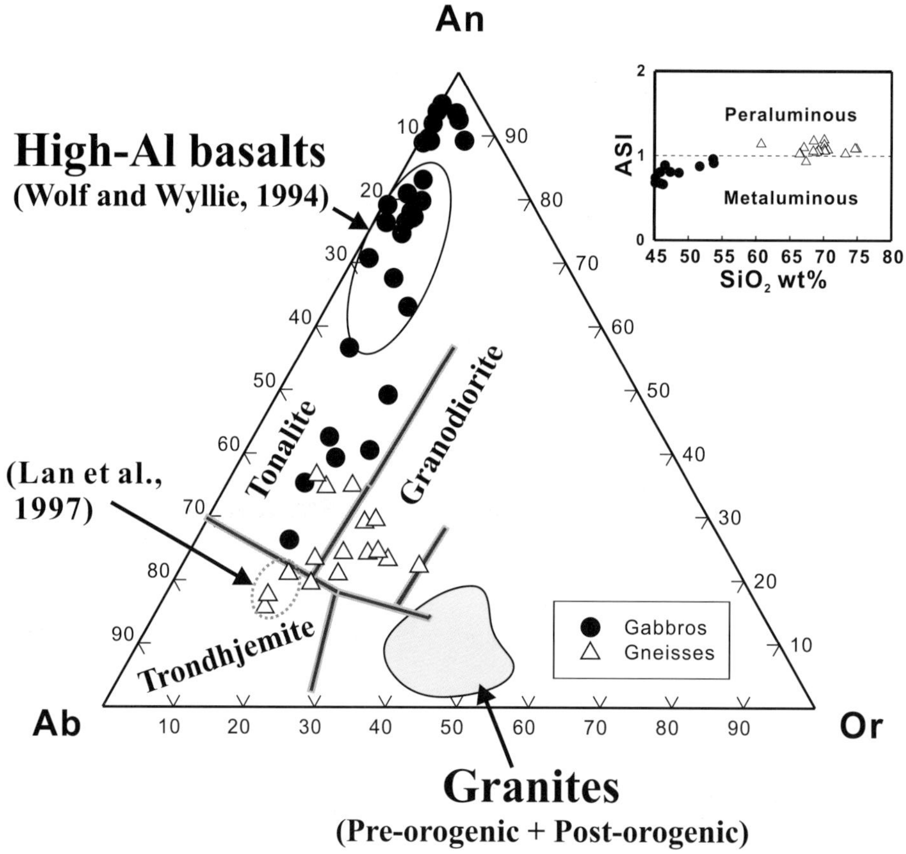

Figure 5 Normative An-Ab-Or triangular plots for gabbros and the associated gneissic rocks as compared with the granites which include both deform (pre-orogenic) and undeform (post-orogenic) samples. The gneissic rocks constitute a peraluminous tonalite-trondhjemite-granodiorite assembly when taking the alumina saturation index (ASI) versus SiO_2 plots (upper right) into consideration.

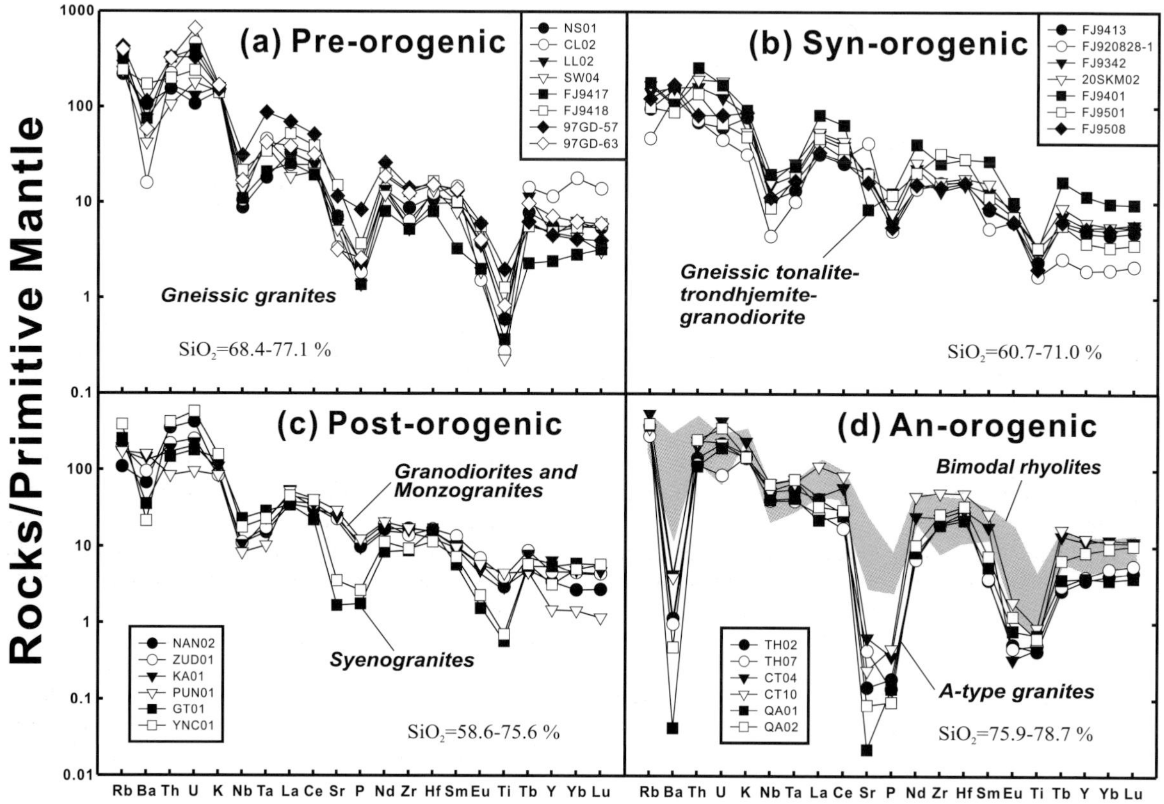

Figure 6 Primitive mantle-normalised, multi-element distribution patterns for pre-, syn-, post- and an-orogenic igneous rocks in SE China. Two typical late Jurassic granites in S China (97GD-57 and -63) are incorporated in (a) for comparison with the pre-orogenic gneisses in the Changle Nanao Metamorphic Belt. Normalising values are from McDonough & Sun (1995).

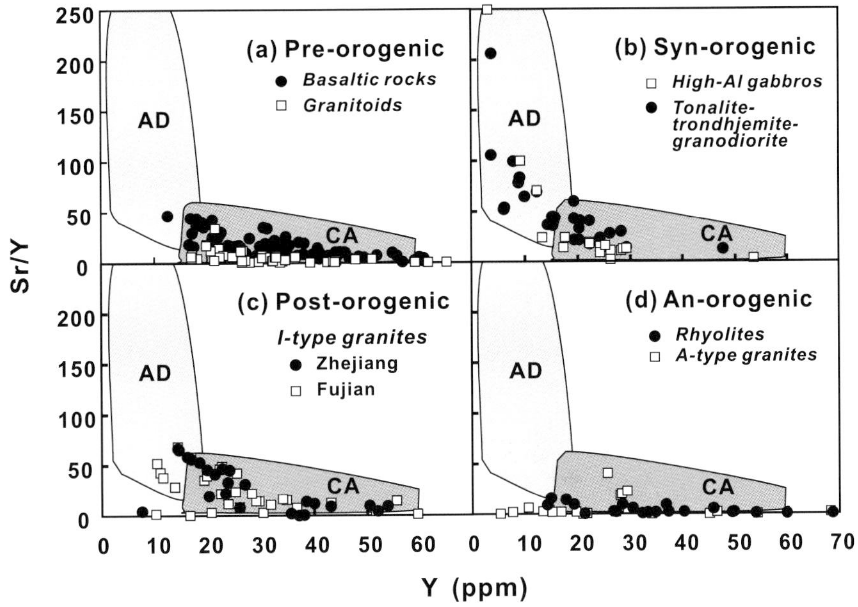

Figure 7 Sr/Y versus Y plots for pre-, syn-, post- and an-orogenic igneous rocks in SE China. Fields for adakite (AD) and classical arc (CA) are taken from Martin (1999).

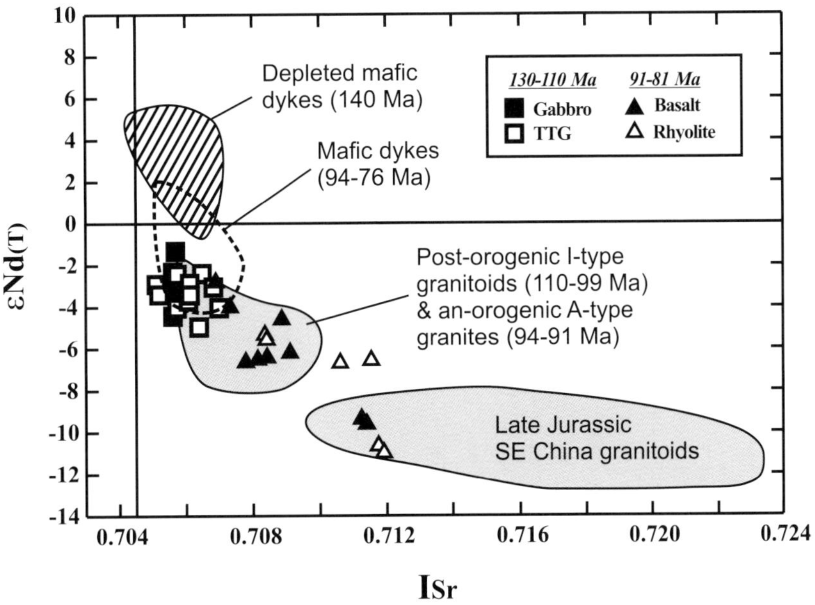

Figure 8 $\varepsilon Nd_{(T)}$ versus I_{Sr} plots for gabbros and tonalite-trondhjemite-granodiorite (TTG), as compared with the late Cretaceous shallow-level I-type granitoids, A-type granites and bimodal volcanics in the Southeast Coast Magmatic Belt (Chen *et al.* 2000). The fields of the 140-Ma and 94–76-Ma mafic dykes (Li & McCulloch 1998; Chen *et al.* 2000) and the late Jurassic granitoids in S China (Chen & Jahn 1998) are used to constrain the mantle and crust sources.

evolution of arc magmas towards a more sodic trend in the early stage of orogeny – typical of the Cordillera orogenesis away from the subduction zone.

Two different processes for the generation of high-Al and Na-trending arc magmas have been suggested: high-pressure partial melting of basic rocks with garnet in the residue (Johnson *et al.* 1997); and partial melting of subducted slab under garnet amphibolite to eclogite conditions (Drummond *et al.* 1996). The present authors believe that the syn-orogenic HAG and TTG in the CNMB represent partial melts of amphibolite under pressure, and that TTG was formed at P/T conditions of 3–5 kb and 650–800 °C because of dehydration of amphibole, whereas HAG are large degree partial melts from the same materials formed at 5–10 kb and 900 °C, based on the experimental works of Wolf & Wyllie (1994). When garnet is involved in the amphibolite residues, melts have a higher Sr/Y ratio because of the retention of Y and HREE in the garnet. On the other hand, slab-melting needs to be under conditions of high shear stresses, incipient subduction and/or near volcanic front to account for the more adakite-like rocks.

This model requires the pre-existence of basic residues in the lower crust to serve as the source for melting. It is suggested that generation of the widely distributed late Jurassic high-K, calc-alkaline granitoids in S China was under the continental rifting environment in conjunction with the formation of extensional basins (Li 2000). Such a tectonic environment could facilitate decompressional melting of the upwelled mantle, and result in a large amount of basic magma emplacing at the base of the lower crust to trigger crustal melting for the generation of the extensional type of the granitic magmas (Bonin *et al.* 1998). Therefore, voluminous basic magmas are postulated to have emplaced in the Cathaysia crust prior to the

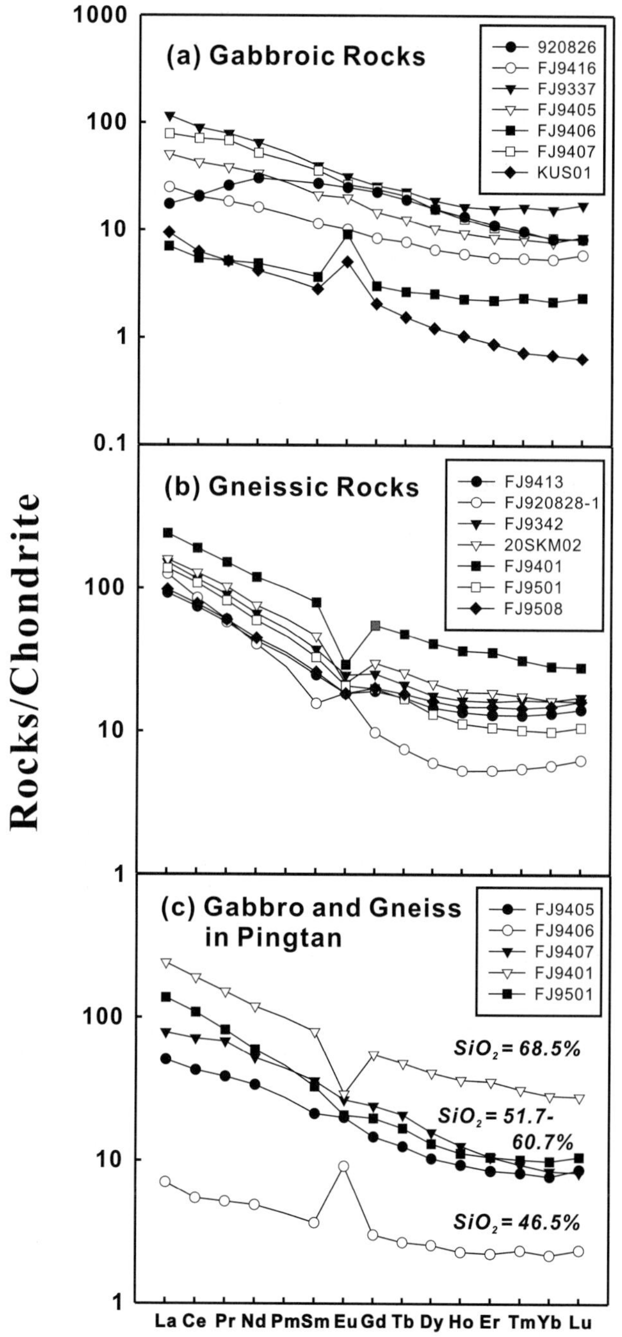

Figure 9 Chondrite-normalised rare earth element distribution patterns for (a) gabbroic rocks and (b) gneissic rocks in the Southeast Coast Magmatic Belt, and (c) both gabbroic and gneissic rocks in Pingtan. Normalising values are from McDonough & Sun (1995).

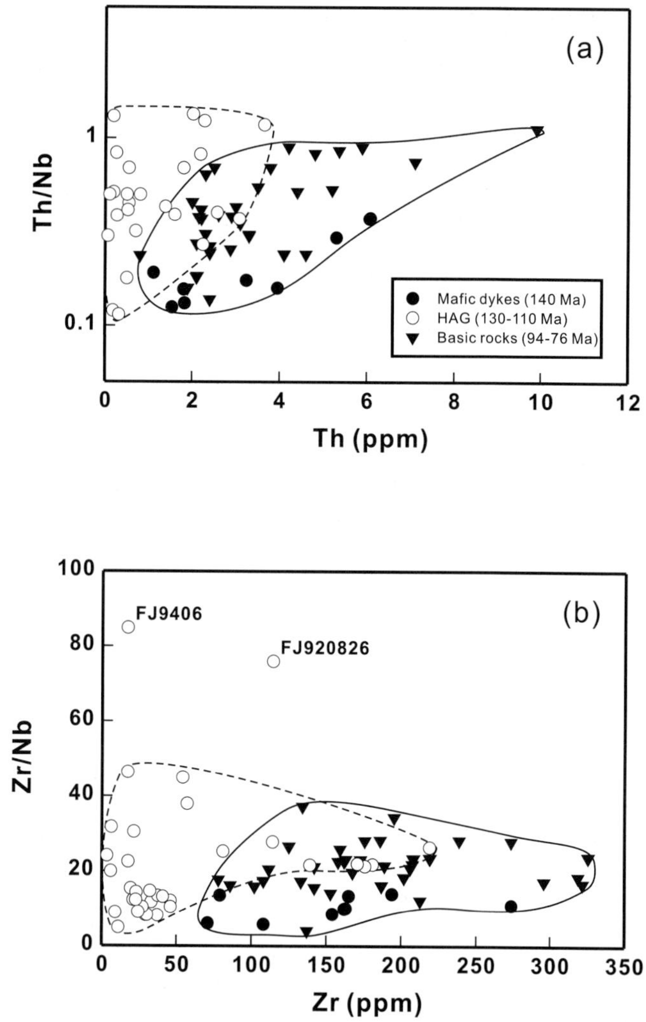

Figure 10 Distinction of high-Al gabbros from pre- and post-orogenic basic rocks in SE China using (a) Th/Nb versus Th and (b) Zr/Nb versus Zr plots. Note that a few cumulates which have high Zr/Nb ratios may reflect discordant Zr concentrations in the sample.

onset of the LY orogeny. Parts of these basic layers were then metamorphosed into amphibolites in a thickened crust during syn-orogenic collision. The direct evidence to support the existence of a thickened crust is the finding of gabbroic granulite xenoliths captured by the Neogene basaltic pipes in Qilin (Fig. 1). Such xenoliths, dated as 112 ± 18 Ma on a Sm-Nd isochron (Xu *et al.* 1999b), were estimated to more probably equilibrate under an extraordinarily high pressure of 15–17 kb (Y. Xu *et al.* 1999).

In the post-orogenic stage, rocks are dominantly high-K calc-alkaline I-type shallow intrusives which constitute the major igneous complexes in the SCMB. Magmatism occurred at this stage, lasting for ~ 12 Mys; it was intensive, and is thought to relate to a combined lithospheric extension and basaltic underplating (Chen *et al.* 2000). The succeeding emplacement of near-surface A-type granite and eruptions of A- and I-type rhyolite intercalating basalts reflect further extension. In sum, I- and A-type magma generation was not only related to the extent of dryness in the source region, but was also controlled by the degree of lithospheric extension in different areas (Chen *et al.* 2000).

3.4. Tectonic evolution

To account for the geochemical and isotopic characters of the staged igneous rocks, the following tectonic evolution in SE China is suggested. In the late Jurassic period, the Cathaysia block was dominated with high-K, calc-alkaline granitoids. The extensional environment (Li 2000) caused the emplacement of underplated basalts. Thickening and downwarping of the lithosphere because of plate convergence converted the preceded underplating basalts into amphibolites. The sizes of gabbro and TTG plutons are small (Table 3), which is typical of those formed under a compression environment (Vigneresse 1995). Dehydration melting of hydrous minerals in the lower-crust amphibolites initiated the LY high-Al magmatism, and partial melting of amphibolites in deep shear zones was responsible for those with a high Sr/Y ratio. The HAG and TTG formed at this stage were metamorphosed into gneisses along the major shear zone caused by the plate subduction.

As a consequence of the thickened lithosphere, detachment or delamination of the lithosphere root could have taken place

(Gvirtzman & Nur 1999). Moreover, this could have resulted in a lithospheric extension and new-phase underplating of basaltic magmas (Turner *et al.* 1992). Therefore, heat transferred from these magmas could have triggered the partial melting of the lower and middle crust necessary for producing I-type magmas. The isotopic analogy between I- and A-type granites is the evidence needed to indicate a common source. On this basis, dry melting of the residue after extraction of I-type magmas heated by further basaltic underplating was the main mechanism by which A-type magmas were formed.

The final phase of LY magmatic evolution can be explained by the increasing degree of lithospheric extension, which led to regional emplacements and eruptions of A-type magmas and associated basaltic melts to form miarolitic granites, mafic dykes and bimodal volcanism. I-type magmas could also be simultaneously formed where a lesser degree of lithospheric stretching had happened. The entire LY magmatism terminated when the basaltic magmas were no longer underplated.

4. Conclusions

The LY magmatism in SE China provides a good model for unravelling cyclic magmatic evolution in a subduction-initiated orogenesis. The amphibole geobarometer and age dating revealed that the LY magmatism can be separated into deep and shallow intrusions in the early and late stages, reflecting compressive and extensional tectonism in the LY orogeny. Magmatism of compressive tectonism (130–110 Ma) produced HAG and TTG suites. These magmas intruded into the deep basement along the ductile shear zone and thermally metamorphosed pre-existing igneous rocks into migmatites. Geochemically, these high-Al rocks are also marked by a high Sr/Y ratio but a low K_2O content, resembling of materials derived from dehydration melting of amphibolites at depth. The more depleted Sr and Nd isotope compositions in the later felsic rocks support their derivation from pre-existing basic layers in the lower crust.

At the beginning of extensional tectonism (110–99 Ma), rocks became dominated by high-K, calc-alkaline I-type granitoids. Following a short non-magmatic period, miarolitic A-type granites (94–91 Ma) and bimodal volcanics (91–81 Ma) were the essence of magmatism. Based on the general similarity of the Sr and Nd isotope compositions among these rocks, the progressive melting of felsic granulites from wet to dry conditions is responsible for the generation of I- and A-type magmas as a result of constant basaltic underplating. The common existence of simultaneous mafic dykes (94–76 Ma), which differ from HAG in geochemical and isotopic composition, and represent derivatives from the lithospheric mantle, is evidence of basaltic underplating.

To account for the geochemical and isotopic variations of LY magmatism and the heat budget under the lithospheric extension, lithosphere detachment or delamination is suggested. Basically, subduction of paleo-Pacific plate during the early Cretaceous resulted in thickened-crust, pre-existing basic layers in the lower crust which most probably emplaced during Jurassic EY extensional magmatism and were metamorphosed into amphibolites. As a consequence of crustal thickening, lithospheric delamination or detachment took place. The upwelled mantle brought up extra heat, causing an immense melting of the crustal material through basaltic underplating. Magmas thus generated were emplaced at a higher level than extensional environments. Until lithospheric extension became less effective, orogenic magmatism terminated.

5. Acknowledgements

Professor W. S. Chen (National Taiwan University) provided some of the study samples. Drs X. Li (Guangzhou Institute of Geochemistry) and C. H. Chen (Academia Sinica now at Shih Chien University Kaohsiung Campus) are thanked for arranging the ICP–MS and electron microprobe analysis. Comments from two anonymous reviewers greatly helped to improve the manuscript. We acknowledge the research grants from the National Science Council (NSC-89, 90, 91–2116–M002–047, 012, 032/Chen) which supported our investigations in S China.

6. References

Atherton, M. P. & Petford, N. 1993. Generation of sodium-rich magmas from newly underplated basaltic crust. *Nature* **362**, 144–6.

Barnes, C. G., Petersen, S. W., Kistler, R. W., Murray, R. & Kays, M. A. 1996. Source and tectonic implications of tonalite-trondhjemite magmatism in the Klamath Mountains. *Contributions to Mineralogy and Petrology* **123**, 40–60.

Bonin, B., Azzouni-Sekkal, A., Bussy, F. & Ferrag, S. 1998. Alkali-calcic and alkaline post-orogenic granite magmatism: petrologic constraints and geodynamic setting. *Lithos* **45**, 45–70.

Brown, M. & Solar, G. S. 1999. The mechanism of ascent and emplacement of granite magma during transpression: a syntectonic granite paradigm. *Tectonophysics* **312**, 1–33.

Chen, C-H., Lin, W., Lu, H., Lee, C. Y., Tien, J. L. & Lai, Y. H. 2000. Cretaceous fractionated I-type granitoids and metaluminous A-type granites in SE China: the Late Yanshanian post-orogenic magmatism. *Transactions of the Royal Society of Edinburgh: Earth Sciences* **91**, 195–205.

Chen, C.-H. & Tung, T. C. 1984. On-line data reduction for electron microprobe analysis. *Acta Geologica Taiwanica* **22**, 196–200.

Chen, J. F. & Jahn, B. 1998. Crustal evolution of southeastern China: Nd and Sr isotopic evidence. *Tectonophysics* **284**, 101–33

Chen, W. S., Yang, H. C., Wang, X. & Huang, H. 2002. Tectonic setting and exhumation history of the Pingtan-Dongshan metamorphic belt along the coastal area, Fujian Province, southeast China. *Journal of Asian Earth Sciences* **20**, 829–40.

Dong, C., Zhou, X., Li, H., Ren, S & Zhou, X. 1997. Late Mesozoic crust mantle interaction in southeastern Fujian – Isotopic evidence from the Pingtan igneous complex. *Chinese Science Bulletin* **42**, 495–8.

Drummond, M. S., Defant, M. J. & Kepezhinskas, P. K. 1996. Petrogenesis of slab-derived trondhjemite-tonalite-dacite/adakite magmas. *Transactions of the Royal Society of Edinburgh: Earth Sciences* **87**, 205–15.

Griffin, W. L., Wang, X., Jackson, S. E., Pearson, N. J., O'Reilly, S. Y., Xu, X. & Zhou, X. 2002. Zircon chemistry and magma mixing, SE China: in-situ analysis of Hf isotopes, Tonglu and Pingtan igneous complexes. *Lithos* **61**, 237–69.

Gvirtzman, Z. & Nur, A. 1999. Plate detachment, asthenosphere upwelling, and topography across subduction zones. *Geology* **27**, 563–6.

Hammerstrom, J. M. & Zen, E. 1986. Aluminum in hornblende: an empirical geobarometer. *American Mineralogists* **71**, 1297–313.

Huppert, H. E. & Sparks, R. S. J. 1988. The generation of granitic magmas by intrusion of basalt into continental crust. *Journal of Petrology* **29**, 599–624.

Johnson, K., Barnes, C. G. & Miller, C. A. 1997. Petrology, geochemistry and genesis of high-Al tonalite and trondhjemites of the Cornucopia stock, Blue Mountains, northeastern Oregon. *Journal of Petrology* **38**, 1585–611.

Lan, C. Y., Chung, S. L. & Mertzman, S. A. 1997. Mineralogy and geochemistry of granitic rocks from Chinmen, Liehyu and Dadan Islands, Fujian. *Journal of the Geological Society of China* **40**, 527–58.

Le Maitre, R. W. 1989. *A Classification of Igneous Rocks and Glossary of Terms: Recommendation of the IOU Subcommission on the Systematics of Igneous Rocks*: Oxford: Blackwell.

Li, W., Zhou, X. & Li, X. 2003. U-Pb and $^{40}Ar/^{39}Ar$ dating of deformed igneous rocks from the Changle-Nanao fault. *Chinese Journal of Geology* **38**, 22–30. [In Chinese with English abstract.]

Li, X. H. 2000. Cretaceous magmatism and lithospheric extension in southeast China. *Journal of Asian Earth Sciences* **18**, 293–305.

Li, X. H. & McCulloch, M. T. 1998. Geochemical characteristics of Cretaceous mafic dykes from northern Guangdong, SE China: age, origin and tectonic significance. *In* Flower, M. F. J., Chung, S. L., Lo, C. H. & Lee, T. Y. (eds), *Mantle Dynamics and Plate Interactions in East Asia. Geodynamics Series* **27**, 405–19. Washington, DC: American Geophysical Union.

Lin, W., Chen, C.-H. & Lee, C. Y. 1997. Origin of REE-depleted leucogranites in Chinmen Island, SE Fujian. *Journal of the Geological Society of China* **40**, 587–606.

Lo, C. H., Onstott, T. C. & Wang Lee, C. M. 1993. $^{40}Ar/^{39}Ar$ dating of plutonic/metamorphic rocks from Chimen island off southeast China and its tectonic implications. *Journal of the Geological. Society of China* **36**, 35–55.

Lo, C. H. & Lee, C. Y. 1994. $^{40}Ar/^{39}Ar$ method of K-Ar age determination of geological samples using Tsing-Hua Open-pool Reactor (THOR). *Journal of the Geological Society of China* **37**, 143–64.

McDonough, W. F. & Sun, S. S. 1995. The composition of the earth, *Chemical Geology* **120**, 223–53.

Martin, H. 1999. Adakitic magmas: modern analogues of Archaean granitoids. *Lithos* **46**, 411–29.

Northrup, C. J., Royden, L. H. & Burchfiel, B. C. 1995. Motion of the Pacific plate relative to Eurasia and its potential relation to Cenozoic extension along the eastern margin of Eurasia. *Geology* **23**, 719–22.

Petford, N., Cruden, A. R. McCaffrey, K. J. W. & Vigneresse, J.-L. 2000. Granite magma formation, transport and emplacement in the Earth's crust. *Nature* **408**, 669–73.

Pitcher, W. S. 1997. *The Nature and Origin of Granite*, 2nd edn. London: Chapman & Hall.

Powell, R. & Powell, M. 1977. Geothermometry and oxygen barometry using coexisting iron-titanium oxides: a reappraisal. *Mineralogical Magazine* **41**, 257–63.

Rollinson, H. R. 1993. *Using Geochemical Data: Evaluation, Presentation, Interpretation*. Singapore: Longman Science & Technology.

Schmidt, M. W. 1992. Amphibole composition in tonalite as a function of pressure: an experimental calibration of the Al-in-hornblende barometer. *Contributions to Mineralogy and Petrology* **110**, 304–10.

Tong, W. X. & Tobisch, O. T. 1996. Deformation of granitoid plutons in the Dongshan area, southeast China: constraints on the physical conditions and timing of movement along the Changle-Nanao shear zone. *Tectonophysics* **267**, 303–16.

Turner, S., Sandiford, M. & Foden, J. 1992. Some geodynamic and compositional constraints on 'postorogenic' magmatism. *Geology* **20**, 931–4.

Vigneresse, J. L. 1995. Control of granite emplacement by regional deformation. *Tectonophysics* **249**, 173–86.

Wang, Z. 2002. The origin of the Cretaceous gabbros in the Fujian coastal region of SE China: implications for deformation-accompanied magmatism. *Contributions to Mineralogy and Petrology* **144**, 230–40.

Wang, Z. H. & Lu, H. F. 1997. Evidence and dynamics for the change of strike-slip direction of the Changle-Nanao ductile shear zone, southeastern China. *Journal of Asian Earth Sciences* **15**, 507–15.

Wang, Z. H. & Lu, H. F. 2000. Ductile deformation and $^{40}Ar/^{39}Ar$ dating of the Changle-Nanao ductile shear zone, southeastern China. *Journal of Structural Geology* **22**, 561–70.

Wolf, M. B. & Wyllie, P. J. 1994. Dehydration-melting of amphibolite at 10 kbar: the effects of temperature and time. *Contributions to Mineralogy and Petrology* **115**, 369–83.

Xu, X., Dong, C., Li, W. & Zhou X. 1999a. Late Mesozoic intrusive complexes in the coastal area of Fujian, SE China: the significance of the gabbro-diorite-granite association. *Lithos* **46**, 299–315.

Xu, X., Zhou, X., O'Reilly, S. Y. & Tang, H. 1999b. Exploration for the lower crustal materials and granite genesis in southeast China. *Acta Petrologica Sinica* **15**, 217–23. [In Chinese with English abstract.]

Xu, Y., Lin, C. & Shi, L. 1999. The geotherm of the lithosphere beneath Qilin, SE China: a re-appraisal and implications for P-T estimation of Fe-rich pyroxenites. *Lithos* **47**, 181–93.

Yang, H. C., Chen, W. S., Lo, C. H., Chen, C. H., Huang, H., Wang, H. & Wang Lee, C. 1997. $^{40}Ar/^{39}Ar$ thermochronology of granitoids from the Pingtan-Dongshan Metamorphic Belt and its tectonic implication. *Journal of the Geological Society of China* **40**, 559–85.

Yui, T. F., Heaman, L. & Lan, C. Y. 1996. U-Pb and Sr isotopic studies on granitoids from Taiwan and Chinmen-Leiyu and their tectonic implications. *Tectonophysics* **263**, 61–7.

CHENG-HONG CHEN and CHI-YU LEE, Department of Geosciences, National Taiwan University, Taipei 106, Taiwan.
e-mail: chench@ntu.edu.tw
WAYNE LIN, Energy and Resources Laboratories, Industrial Technology Research Institute, Hsinchu 310, Taiwan.
CHING-YING LAN, Institute of Earth Sciences, Academia Sinica, Taipei 115, Taiwan.

MS received 2 September 2003. Accepted for publication 4 August 2004.

Mafic rocks from the Ryoke Belt, southwest Japan: implications for Cretaceous Ryoke/San-yo granitic magma genesis

Takashi Nakajima, Hiroyuki Kamiyama, Ian S. Williams and Kenichiro Tani

ABSTRACT: Mafic rocks in the Ryoke belt, the Cretaceous granitic province in Southwest Japan, occur in two modes: (1) as mafic dykes and pillow-shaped enclaves, and (2) as isolated kilometre-sized bodies of gabbroic cumulate. The dykes and pillows have fine-grained textures with thin radiating plagioclase laths, indicative of quenching. The gabbroic cumulates are predominantly coarse-grained and commonly lithologically layered.

SHRIMP zircon U-Pb ages of both types of mafic rocks are in the range 71–86 Ma, late Cretaceous. The mafic rocks become younger eastwards, matching the along-arc age trend of the associated Cretaceous granites (Nakajima *et al.* 1990). Both types of mafic rocks were apparently generated during the same magmatic event that produced the Ryoke/San-yo granites. The mafic dykes and pillows are aphyric basaltic-andesites to andesites (SiO_2 54–60 wt.%), with microphenocrysts of biotite and hornblende. They have a composition which is similar to mafic rocks from the northern Sierra Nevada, and also to medium-K calc-alkaline rocks from present-day arc volcanics. The gabbroic cumulates are mostly pyroxene-hornblende gabbros (SiO_2 43–52 wt.%). Their bulk-rock chemical compositions are mostly unlike any magma compositions.

Both types of mafic rocks from the Ryoke belt have relatively high $^{87}Sr/^{86}Sr$ initial ratios (SrI), 0·7071–0·7097, which are similar to those of the associated granites. The granites were formed either by fractional crystallisation of the mafic magmas, or by partial melting of newly formed mafic rocks at depth. The high SrI indicates that the mafic magmas were derived from enriched mantle or mixed with enriched crustal materials. Even if the mixing occurred between primitive basaltic magma and metasedimentary rocks, then the basaltic andesite–andesite magmas must have contained more than 60% mantle-derived components. The Cretaceous magmatism in Southwest Japan represents a major episode of crustal growth by additions from the upper mantle in an arc setting.

KEY WORDS: continental arc, crustal evolution, granites, magma source, SHRIMP U-Pb age, Sr isotope.

The late Mesozoic Circum-Pacific granitic provinces comprise the largest Phanerozoic granitic province on earth. The granitic arcs occur at continental margins, having formed by Cordilleran-type orogeny related to the subduction of oceanic plates rather than by continental collision. Cordilleran-type orogeny is one of the major processes by which Phanerozoic continental crust was produced, thus the understanding of the generation of subduction-related granitic magmas is essential to understanding global crustal growth.

The Cretaceous granitic province of southwest Japan is a part of the Circum-Pacific granitic province. It is of similar size to the Sierra Nevada and Peninsular Range Batholiths in the western United States. The province formed in a continental arc setting at the eastern Eurasian continental margin prior to separation of the Japanese Islands from the continent during the Miocene (Otofuji & Matsuda 1984). The granites form three arc-parallel zones; from back-arc to fore-arc the San-in, San-yo and Ryoke zones respectively (Fig. 1). It has been suggested that a transect from the San-yo to the Ryoke zone represents a crustal cross section through the eastern margin of the Eurasian continent down to mid-crustal depths (12–14 km; Nakajima 1994, 1996). The granites of the San-yo and Ryoke zones are mostly of I-type and ilmenite-series (Takahashi *et al.* 1980) and are classified as VAG (volcanic arc granite) by their Y, Nb and Rb contents (Nakajima 1996).

Large granitic provinces commonly contain some associated mafic rocks. These can provide important clues to the genesis of the granitic magmas and their tectonic setting (e.g. Sisson *et al.* 1996; Ducea & Saleeby 1998; Collins 2002). In the granitic province of SW Japan, mafic rocks are most abundant in the Ryoke belt, where they occur sporadically but ubiquitously throughout the whole area. This paper presents a geological, petrographical and geochemical description of the mafic rocks from the Ryoke belt, and discusses their petrogenesis in relation to the generation of the associated granitic magmas.

1. Mafic rocks from the Ryoke Belt

1.1. Occurrence of the mafic rocks

The exposed area of mafic rocks in the Ryoke belt is about 1–2% of the exposed area of granitic rocks (Fig. 1). The former occur either as fine-grained dykes and pillows closely associated with granitic rocks at the outcrop scale, or as discrete bodies of medium- to coarse-grained gabbroic cumulate.

1.2. Dykes and pillows

The mafic dykes range in thickness from ~0·3 to 3 m. Although some apparently intruded after complete solidification

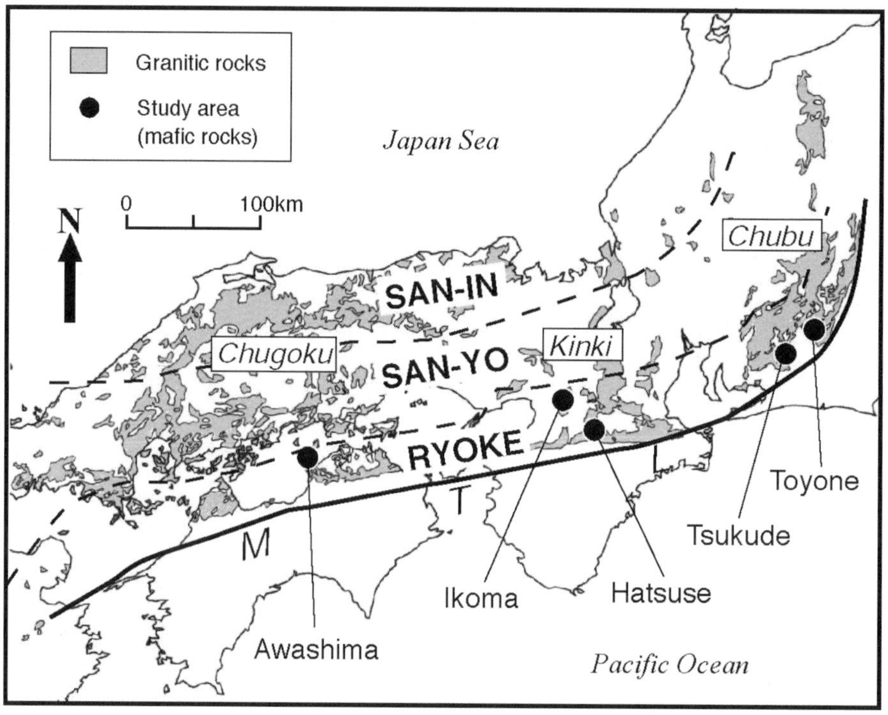

Figure 1 Cretaceous granitic province of SW Japan, showing the distribution of mafic rocks associated with the granitoids of the present study. MTL denotes the Median Tectonic Line, an arc-parallel active fault.

of the granitic host rocks, others are sharply cut by veins indistinguishable from the host granite, consistent with syn-plutonic dyke emplacement while the granite was partially molten (Fig. 2a). The mafic pillows are generally 0·1–0·5 m diameter, and commonly have crenulate chilled margins with convex-outward lobes separated by sharp, inward-pointing cusps. The volume of mafic pillows relative to granitic matrix within an outcrop varies. The layer of granite separating pillows can be as thin as a few millimetres. In such cases, some pillows are moulded against one another and appear to have been mutually deformed (Fig. 2b). These occurrences strongly indicate that both mafic and granitic magmas coexisted in a partially-molten state. Similar occurrences of mafic dykes and pillows have been reported from other bimodal magmatic provinces (e.g., Wiebe 1973; Mattson *et al.* 1986; Falkner *et al.* 1995; Snyder *et al.* 1997; Wiebe *et al.* 2001; Kapp *et al.* 2002).

The dykes and pillows have similar textures and, using the nomenclature of Streckeisen (1976), range in composition from diorite to quartz diorite. They are fine-grained, and characterised by thin radiating laths of plagioclase (Fig. 2c). Average grain size is commonly <0·1 mm. Most of the rocks are aphyric, but some are porphyritic, with large euhedral crystals of plagioclase (~2 mm) set in a fine-grained matrix. The mafic minerals are dominated by subhedral brown to greenish-brown hornblende, with subordinate biotite and opaque oxides. Some rocks contain small amounts of interstitial quartz.

1.3. Gabbroic cumulates

The gabbroic cumulates are medium- to coarse-grained, and occur as isolated bodies up to several kilometres in diameter. Although they are closely associated with granitic rocks on a regional scale, their intrusion relationships to the granites remain unclear. The gabbroic rocks are commonly layered, the relative proportions of plagioclase and brown hornblende varying regularly on a scale of tens of millimetres (Fig. 2d).

Hornblende gabbro (Fig. 2e), with approximately equal amounts of plagioclase and mafic minerals, is the dominant rock type, but anorthositic rocks are also common. Plagioclase, hornblende, and minor opaque oxides are ubiquitous. The plagioclase crystals mostly consist of euhedral homogeneous cores surrounded by thin, strongly zoned rims. Hornblende occurs either as small crystals interstitial to euhedral to subhedral plagioclase, or as large poikilitic crystals up to ~0·1 m. Both clinopyroxene and orthopyroxene are common, but they only occur in small amounts as corroded cores in large poikilitic crystals of hornblende. Olivine is rare and only occurs as small crystals included in hornblende.

2. Magmatic ages of the mafic rocks

2.1. Various ages and various stories

Before discussing the petrogenetic relationships between the mafic and granitic rocks, two facts need to be established: (1) that the mafic and granitic rocks are contemporaneous; and (2) that the chemical composition of each rock type, especially of the mafic rocks, could be the composition of a magma.

As mentioned above, there is strong field evidence that the mafic magmas that formed the pillows and at least some of the dykes were contemporaneous with the host granitic magma. However, field observations have not been able to determine the temporal relationships between the isolated bodies of gabbroic cumulates and the associated granitic rocks. Recently, several workers have measured pre-Cretaceous Sm-Nd mineral ages of 180–240 Ma on those gabbroic cumulates and concluded that they represent older basement to SW Japan (Kagami *et al.* 1995; Okano *et al.* 2000). However, Rb-Sr ages measured on the same samples are much younger (<120 Ma), thus the emplacement ages of the gabbros have remained a subject of debate. Zircon U-Pb age measurements are less ambiguous.

2.2. SHRIMP U-Pb ages

Zircon for SHRIMP U-Pb age measurement was separated from six Ryoke mafic rocks. The zircon occurs as fine elongate to needle-like grains in the fine-grained pillows and dykes, and

Figure 2 (a) Pillow-shaped mafic enclave swarm in granitic rocks, Toyone village, Chubu district. (b) Mafic synplutonic dyke with granitic back veins, Hatsuse village, Kinki district. Width of the dyke is approximately 30 cm. (c) Photomicrograph of a mafic enclave. (d) Layered gabbro, Awashima Island, Chugoku district. (e) Photomicrograph of hornblende gabbro. Abbreviations: (plg) plagioclase; (hbl) hornblende.

as coarse blocky to moderately elongate grains in the coarse-grained gabbros. In common with zircon from most other mafic rocks, the zircon zoning structures (as revealed by cathodoluminescence imaging) are relatively simple; broad prism-parallel zones in the elongate grains, broad concentric and sector zones in the blocky grains (Fig. 3). This contrasts with the fine oscillatory zoning common in zircon from granitic rocks. Also in contrast to granites, because mafic magmas are initially strongly zircon undersaturated, older inherited zircon is relatively rare.

Six mafic rocks (a pillow, a dyke and four cumulate gabbro-type rocks) were collected from a wide area over a distance of 500 km between the Chubu and Chugoku districts in SW Japan, and their Zircon U-Pb ages were measured using the SHRIMP II at the Australian National University. Some of the samples were collected from the same bodies on which pre-Cretaceous ($\sim$200 Ma) Sm-Nd mineral ages had previously been measured (Kagami *et al.* 1995; Okano *et al.* 2000). Sample preparation, analysis and data processing followed procedures based on those described by Williams *et al.* (1996).

Figure 3 A cathodoluminescence image of zircon from Ryoke hornblende gabbro 99120203 showing strongly elongated, needle-like shapes with the simple broad zonation characteristic of most zircon from mafic rocks.

Pb/U ratios were determined relative to Palaeozoic reference zircon Temora 1 (417 Ma; Black *et al.* 2003). Ages were calculated using the constants recommended by the IUGS Subcommission on Geochronology (Steiger & Jäger 1977). Uncertainties in the ages are cited as 95% confidence limits, and include allowance for uncertainty in the Pb/U calibration (0·30–0·34%).

The zircon analyses are listed in Table 1 and plotted on Tera-Wasserberg concordia diagrams in Figure 4. To minimise the effects of calibration errors, some of the samples were run together in the same analytical session. These sessions also included analyses of zircon from some Ryoke granites, which will be reported elsewhere. With a few exceptions, the analyses from each sample cluster around a mixing line between common Pb and a single Cretaceous radiogenic end member, consistent with most of the zircon in each sample being the same age and having suffered little or no radiogenic Pb loss. The zircon in all samples is characterised by moderate to high U contents (150–1870 ppm) and generally moderate to high Th/U (0·5–2·6), although the low Th in some grains results in a few Th/U values being very low (<0·1).

99120101A: Fine-grained basaltic andesite from a mafic pillow in a synplutonic composite dyke, Toyone village, Mikawa area, Chubu district. Eight zircons and 2 zircon cores analysed (Fig. 4a). Six of the zircons have equal radiogenic $^{206}Pb/^{238}U$ within analytical uncertainty, equivalent to a weighted mean age of 71·5 ± 1·1 Ma. One grain ~93 Ma, cores ~166 Ma and ~1·85 Ga.

00021701A: Fine-grained mafic synplutonic dyke, Hatsuse village, Kinki district. Ten zircons analysed (Fig. 4b). Nine zircons have almost equal radiogenic $^{206}Pb/^{238}U$ within analytical uncertainty, equivalent to a weighted mean age of 75·5 ± 0·8 Ma. One grain ~89 Ma.

99120203: Coarse-grained hornblende gabbro, Tsukude village, Mikawa area, Chubu district. Eleven zircons analysed (Fig. 4c). All have almost equal radiogenic $^{206}Pb/^{238}U$ within analytical uncertainty, equivalent to a weighted mean age of 72·4 ± 1·2 Ma.

01012103: Coarse-grained hornblende gabbro, Awashima Island, Setouchi area, eastern Chugoku district. Ten zircons analysed (Fig. 4d). All have equal radiogenic $^{206}Pb/^{238}U$ within analytical uncertainty, equivalent to a weighted mean age of 86·0 ± 1·2 Ma.

00021401A: Coarse-grained layered gabbro, Ikoma Mountain, Kinki district. Ten zircons analysed (Fig. 4e). All have equal radiogenic $^{206}Pb/^{238}U$ within analytical uncertainty, equivalent to a weighted mean age of 82·0 ± 0·9 Ma.

00021504: Coarse-grained anorthositic gabbro, Ikoma Mountain, Kinki district. Ten zircons analysed (Fig. 4f). The scatter is beyond analytical uncertainty, even if the analysis with lowest $^{206}Pb/^{238}U$ is omitted. The weighted mean radiogenic $^{206}Pb/^{238}U$ of the other nine analyses is equivalent to an age of 83·2 ± 1·3 Ma.

There is no systematic difference between the zircon ages of the pillow, dyke, and cumulate gabbros and all measured ages are in the range 71 to 86 Ma, Late Cretaceous. There is no zircon evidence to support the suggestion that some of the mafic rocks are Jurassic or older. The few grains and cores of older zircon in the dyke and pillow samples (~89 Ma, ~93 Ma, ~166 Ma, ~1·85 Ga) are probably inherited from the host granites, many of which contain a small amount of older zircon thought to be contributed by a sediment-derived component in the magmas. No inherited zircon was found in the cumulate gabbro samples.

2.3. Contemporaneous magmatism with the Ryoke/San-yo granitoids

The zircon U-Pb ages of the Ryoke mafic rocks seem to be independent of rock type, but related to sample locality. The rocks are older in the west (86 Ma in the Chugoku district), younger in the east (71–72 Ma in the Chubu district), and intermediate in between (75–83 Ma in the Kinki district) (Fig. 5). This eastward younging of the Ryoke mafic rocks seems to match the along-arc variation in granite ages documented by Nakajima *et al.* (1990) and Nakajima (1994) (Fig. 6). Namely,

Table 1 SHRIMP zircon U-Pb isotopic analyses of zircon from Ryoke mafic rocks

	Pb* ppm	U ppm	Th ppm	Th/U	% ^{206}Pb† comm.	^{207}Pb/ ^{206}Pbm	Å}	^{238}U/ ^{206}Pbm	Å}	^{208}Pb*/ ^{206}Pb	Å}	^{208}Pb*/ ^{232}Th	Å}	^{206}Pb*/ ^{238}U	Å}	Apparent ages (Ma)			
																208/232	Å}	206/238	Å}
Grain spot gabbros																			
99120203																			
1·1	12	806	1182	1·47	0·83	0·0551	13	85·02	86	0·4506	98	0·00358	9	0·01166	12	72·3	1·8	74·7	0·8
2·1	10	678	1027	1·51	0·68	0·0537	12	87·87	99	0·4839	83	0·00361	8	0·01130	13	72·9	1·5	72·5	0·8
3·1	9	672	804	1·20	1·19	0·0583	20	89·58	96	0·3837	93	0·00354	10	0·01103	12	71·4	1·9	70·7	0·8
5·1	8	623	664	1·07	0·67	0·0535	23	89·62	103	0·3442	100	0·00358	11	0·01108	13	72·2	2·3	71·1	0·8
7·1	9	611	976	1·60	0·45	0·0516	17	89·90	155	0·5164	139	0·00358	12	0·01107	19	72·2	2·4	71·0	1·2
10·1	6	487	456	0·94	1·09	0·0574	17	86·41	86	0·2762	93	0·00337	12	0·01145	12	68·1	2·4	73·4	0·7
9·1	6	429	419	0·98	1·24	0·0588	24	85·75	108	0·3033	121	0·00358	15	0·01152	15	72·2	3·0	73·8	0·9
11·1	3	325	6	0·02	1·63	0·0624	23	86·07	147	0·0082	80	0·00540	531	0·01143	20	108·9	106·9	73·3	1·3
4·1	3	242	172	0·71	2·04	0·0661	27	88·28	135	0·2466	143	0·00385	23	0·01110	17	77·7	4·7	71·1	1·1
6·1	2	210	145	0·69	3·19	0·0767	59	89·80	239	0·2136	268	0·00334	43	0·01078	30	67·4	8·7	69·1	1·9
8·1	3	207	141	0·68	1·78	0·0637	45	86·53	216	0·2104	202	0·00352	35	0·01135	29	71·0	7·1	72·8	1·8
00021401A																			
6·1	16	958	1301	1·36	0·11	0·0486	9	78·14	85	0·4399	63	0·00414	8	0·01278	14	83·5	1·6	81·9	0·9
5·1	10	633	703	1·11	0·49	0·0521	15	77·97	69	0·3520	72	0·00404	9	0·01276	12	81·5	1·9	81·8	0·7
1·1	9	544	726	1·34	0·79	0·0549	12	77·40	49	0·4295	78	0·00412	8	0·01282	8	83·2	1·7	82·1	0·5
2·1	9	529	837	1·58	0·56	0·0527	14	78·40	57	0·4991	95	0·00400	10	0·01268	9	80·7	1·9	81·3	0·6
3·1	9	436	1151	2·64	0·77	0·0547	15	78·39	72	0·8324	154	0·00399	8	0·01266	12	80·5	1·7	81·1	0·8
4·1	4	257	426	1·66	1·25	0·0591	20	77·83	124	0·5229	179	0·00399	15	0·01269	20	80·6	3·1	81·3	1·3
10·1	4	235	420	1·79	1·38	0·0603	21	75·26	77	0·5468	142	0·00401	13	0·01310	14	80·9	2·6	83·9	0·9
9·1	3	217	9	0·04	1·45	0·0609	22	77·07	138	0·0129	64	0·00393	195	0·01279	23	79·3	39·3	81·9	1·5
7·1	3	211	28	0·13	1·50	0·0614	22	74·89	140	0·0390	70	0·00384	70	0·01315	25	77·4	14·1	84·2	1·6
8·1	2	207	19	0·09	1·56	0·0619	38	75·49	130	0·0200	101	0·00280	142	0·01304	23	56·5	28·6	83·5	1·5
00021504																			
3·1	24	1724	1299	0·75	0·19	0·0494	14	78·08	64	0·2413	51	0·00409	9	0·01278	11	82·6	1·9	81·9	0·7
2·1	22	1507	1181	0·78	0·31	0·0505	13	76·66	72	0·2447	53	0·00406	10	0·01300	12	81·9	2·0	83·3	0·8
9·1	12	925	369	0·40	0·20	0·0495	9	75·81	77	0·1333	37	0·00440	13	0·01317	14	88·8	2·7	84·3	0·9
10·1	10	776	417	0·54	0·44	0·0516	12	79·34	78	0·1706	78	0·00398	19	0·01255	12	80·3	3·8	80·4	0·8
5·1	7	530	301	0·57	0·84	0·0553	14	78·08	109	0·1747	61	0·00391	15	0·01270	18	78·8	3·0	81·3	1·1
7·1	8	520	415	0·80	0·63	0·0535	14	75·77	100	0·2545	68	0·00418	13	0·01311	17	84·3	2·6	84·0	1·1
1·1	7	491	386	0·79	0·47	0·0520	14	74·44	87	0·2515	68	0·00428	13	0·01337	16	86·3	2·6	85·6	1·0
6·1	6	440	377	0·86	0·60	0·0531	15	78·23	79	0·2689	74	0·00399	12	0·01271	13	80·4	2·4	81·4	0·8
8·1	6	423	329	0·78	0·57	0·0529	18	74·89	88	0·2471	107	0·00422	19	0·01328	16	85·1	3·8	85·0	1·0
4·1	6	411	314	0·76	0·95	0·0563	21	76·31	93	0·2235	91	0·00380	16	0·01298	16	76·6	3·3	83·1	1·0
01012103																			
10·1	10	638	683	1·07	0·39	0·0513	13	74·46	72	0·3485	78	0·00436	11	0·01338	13	87·9	2·2	85·7	0·8
8·1	8	448	571	1·27	1·10	0·0578	23	71·80	117	0·3888	106	0·00421	14	0·01377	23	84·8	2·8	88·2	1·4
5·1	6	352	419	1·19	1·87	0·0648	29	73·91	87	0·3625	155	0·00405	18	0·01328	16	81·6	3·7	85·0	1·0
6·1	5	295	270	0·91	0·70	0·0541	26	74·29	117	0·3105	139	0·00454	22	0·01337	21	91·5	4·4	85·6	1·4
7·1	4	250	363	1·45	0·10	0·0486	30	73·65	213	0·4624	169	0·00432	21	0·01356	40	87·2	4·3	86·9	2·5
9·1	4	228	216	0·95	2·15	0·0674	33	70·83	117	0·2872	134	0·00419	21	0·01382	23	84·4	4·3	88·5	1·5
1·1	3	193	168	0·87	3·59	0·0806	73	72·49	190	0·2771	240	0·00422	39	0·01330	37	85·1	7·8	85·2	2·3
4·1	3	187	169	0·90	2·34	0·0691	38	73·15	128	0·2776	166	0·00410	26	0·01335	24	82·8	5·2	85·5	1·5
2·1	2	175	140	0·80	3·67	0·0812	36	76·54	156	0·2361	150	0·00371	25	0·01259	26	74·8	5·0	80·6	1·7
3·1	2	153	130	0·85	0·76	0·0546	22	74·37	152	0·2861	212	0·00447	35	0·01335	28	90·2	7·0	85·5	1·8
Syn-plutonic dykes																			
99120101A																			
7·1	23	1867	1396	0·75	0·25	0·0497	7	90·05	81	0·2370	35	0·00351	6	0·01108	10	70·8	1·2	71·0	0·6
2·1	18	1260	315	0·25	0·22	0·0498	11	68·72	59	0·0764	33	0·00444	20	0·01452	13	89·5	4·0	92·9	0·8
3·1	10	763	833	1·09	0·65	0·0534	12	88·98	97	0·3554	93	0·00363	10	0·01116	12	73·3	2·1	71·6	0·8
4·1	20	685	512	0·75	0·27	0·0518	8	38·32	35	0·2265	48	0·00789	19	0·02603	24	158·8	3·7	165·6	1·5
10·1	221	670	228	0·34	0·02	0·1132	11	3·16	4	0·0994	32	0·09233	329	0·31639	432	1785·0	60·9	1772·1	21·2
6·1	8	639	558	0·87	0·53	0·0522	13	94·01	104	0·2771	69	0·00335	10	0·01058	12	67·7	2·0	67·8	0·8
1·1	7	561	518	0·92	0·95	0·0562	13	88·65	118	0·2835	76	0·00343	11	0·01117	15	69·2	2·1	71·6	1·0
9·1	6	492	278	0·57	0·53	0·0523	16	88·10	114	0·1838	66	0·00367	14	0·01129	15	74·1	2·9	72·4	0·9
5·1	5	353	396	1·12	1·03	0·0568	18	90·17	149	0·3546	102	0·00347	12	0·01098	18	70·0	2·4	70·4	1·2
8·1	4	342	205	0·60	0·89	0·0556	17	88·19	120	0·1899	96	0·00355	19	0·01124	15	71·7	3·8	72·0	1·0

Continued on next page

Table 1 *Continued*

	Pb* ppm	U ppm	Th ppm	Th/U	% ^{206}Pb† comm.	^{207}Pb/^{206}Pbm	Å}	^{238}U/^{206}Pbm	Å}	^{208}Pb*/^{206}Pb	Å}	^{208}Pb*/^{232}Th	Å}	^{206}Pb*/^{238}U	Å}	Apparent ages (Ma) 208/232	Å}	206/238	Å}
00021701A																			
10·1	21	1465	1774	1·21	–	0·0474	7	85·17	31	0·3785	41	0·00367	6	0·01174	4	74·0	1·1	75·3	0·3
6·1	20	1211	2111	1·74	–	0·0470	8	84·88	55	0·5477	83	0·00370	6	0·01179	8	74·7	1·2	75·6	0·5
7·1	16	1155	1253	1·08	0·08	0·0483	11	85·97	59	0·3396	52	0·00364	6	0·01162	8	73·5	1·3	74·5	0·5
9·1	15	1108	1229	1·11	0·03	0·0478	8	86·40	69	0·3447	55	0·00360	7	0·01157	9	72·6	1·3	74·2	0·6
5·1	14	1031	979	0·95	0·24	0·0497	7	83·50	33	0·2969	41	0·00374	9	0·01195	5	75·4	1·7	76·6	0·3
1·1	14	1027	769	0·75	0·13	0·0487	16	83·41	83	0·2366	55	0·00378	10	0·01197	12	76·3	2·0	76·7	0·8
4·1	17	1024	1751	1·71	0·16	0·0490	9	84·82	52	0·5418	86	0·00373	6	0·01177	7	75·2	1·3	75·4	0·5
2·1	13	957	960	1·00	0·20	0·0494	9	84·77	82	0·3200	52	0·00376	7	0·01177	11	75·8	1·5	75·4	0·7
8·1	12	897	209	0·23	0·08	0·0486	10	71·61	52	0·0743	32	0·00445	19	0·01395	10	89·8	3·9	89·3	0·6
3·1	9	666	601	0·90	0·05	0·0480	14	84·54	78	0·2934	63	0·00384	9	0·01182	11	77·6	1·8	75·8	0·7

*Corrected for common Pb of Broken Hill galena composition using ^{207}Pb/^{206}Pb
Measured value not corrected for common Pb
†Percentage of total ^{206}Pb that is common ^{206}Pb.

at any given site in SW Japan, the granitic and mafic magmatism appears to have been synchronous, implying that the two types of magmatism are genetically related.

3. Magmatic composition of the mafic rocks

3.1. Cumulate gabbros versus dykes and pillows

Field observations and petrography show that the dykes and pillows are quenched liquids, whereas the cumulate gabbros are accumulations of minerals precipitated from magma. The dyke and pillow chemical compositions probably correspond to magma compositions, but the gabbro compositions do not. Representative whole rock chemical analyses of the Ryoke mafic rocks and some Ryoke granites and metasediments are listed in Table 2.

The dykes and pillows analysed so far are mostly basaltic andesites to andesites (53–60% SiO_2), whereas the gabbroic cumulates are more mafic (43–52% SiO_2). Some of the latter contain very high Al_2O_3 (25–29%) and could not have been magmas. Plotted on Harker variation diagrams, the Al_2O_3, MgO and CaO contents of the gabbroic cumulates deviate markedly from the magmatic trend defined by the dykes, pillows and granitic rocks (Fig. 7a, b, c). Co-linearity on Harker diagrams does not necessarily prove that rocks are comagmatic, so it is not yet possible to establish whether the Ryoke granitic rocks, dykes and pillows constitute a single magma series. The diagrams do indicate, however, that the compositions of the gabbroic cumulates cannot be used as magma compositions in any petrogenetic discussion.

3.2. Arc affinities of the mafic magma compositions

In contrast to the large range in chemical composition of the gabbroic cumulates, the dykes and pillows are all basaltic andesites and andesites (SiO_2 53–60%; K_2O 1·3–1·8%; TiO_2 0·8–1·1%; Rb 36–58 ppm; Sr 370–500 ppm; Nb 6–10 ppm and FeO*/MgO ratio 2·0–2·5: Table 1), similar to primitive mafic rocks from the northern Sierra Nevada (Tani 2002), and to medium-K series calc-alkaline rocks from modern volcanic arcs (Fig. 8a). The dykes and pillows have compositions near the boundary between Miyashiro's tholeiite and calc-alkaline fields on a SiO_2-FeO*/MgO diagram, a common feature of arc volcanics, whereas the cumulate gabbros mostly plot deep in the tholeiite field (Fig. 8b). Recent claims that the Ryoke mafic rocks are mostly tholeiitic continental flood basalt (Kagami *et al.* 2000) are possibly based on a misinterpretation of gabbro compositions as magma compositions, as the supporting analyses are mostly of gabbros and granulites. Our analyses (Table 2) show that the Ryoke mafic magmas were characterised by high Sr and Zr and moderately low Nb contents, features of arc rocks, especially the volcanic rocks from continental arcs (Table 3).

The basaltic andesite–andesite of the dykes and pillows has low Ni and Cr contents and a relatively high FeO*/MgO ratio, consistent with moderate fractionation of a primary magma composition.

3.3. Sr isotopes

There is commonly a strong negative correlation between εNd and SrI in arc magmas. The Cretaceous granites from SW Japan are a typical example of this behaviour (Okudaira *et al.* 2003, fig. 6). In such a case, SrI alone (without ε_{Nd}) is a useful petrogenetic indicator. Furthermore, the isotopic evolution line on a SrI-ε_{Nd} diagram is concave upward, making SrI a more sensitive parameter than ε_{Nd} in the study of isotopically evolved magmas. Here we have measured SrI as a tracer of the processes of mafic and granitic magma generation.

In the Chugoku district, in western SW Japan, the SrI of the Cretaceous granites increases across the arc, being higher in the Ryoke zone than in the San-yo zone (Nakajima 1996). It remains unclear whether the variation is continuous, or if there are several isotopically discrete crustal segments as proposed by Kagami *et al.* (1992). The same across-arc SrI gradient is not found everywhere in SW Japan. In eastern SW Japan, for example, SrI is lower in the Ryoke zone than in the San-yo zone (Nakajima 1996, fig. 9). The changes in the across-arc SrI trends are principally due to along-arc changes in the SrI of the San-yo granites, the SrI of the Ryoke granites along-arc being on average almost constant (Nakajima 1996).

All the Ryoke mafic rocks – dykes, pillows and cumulate gabbros – have relatively high SrI, 0·7071–0·7097, higher than most Quaternary subduction-related basaltic rocks. The SrI of most of the granitic rocks is in the same range (0·707–0·709), although some are even higher, up to 0·712 (Fig. 9). In any given region, the SrI of the mafic and granitic rocks is similar. The SrI of the Ryoke metasediments is much higher, 0·714–0·721.

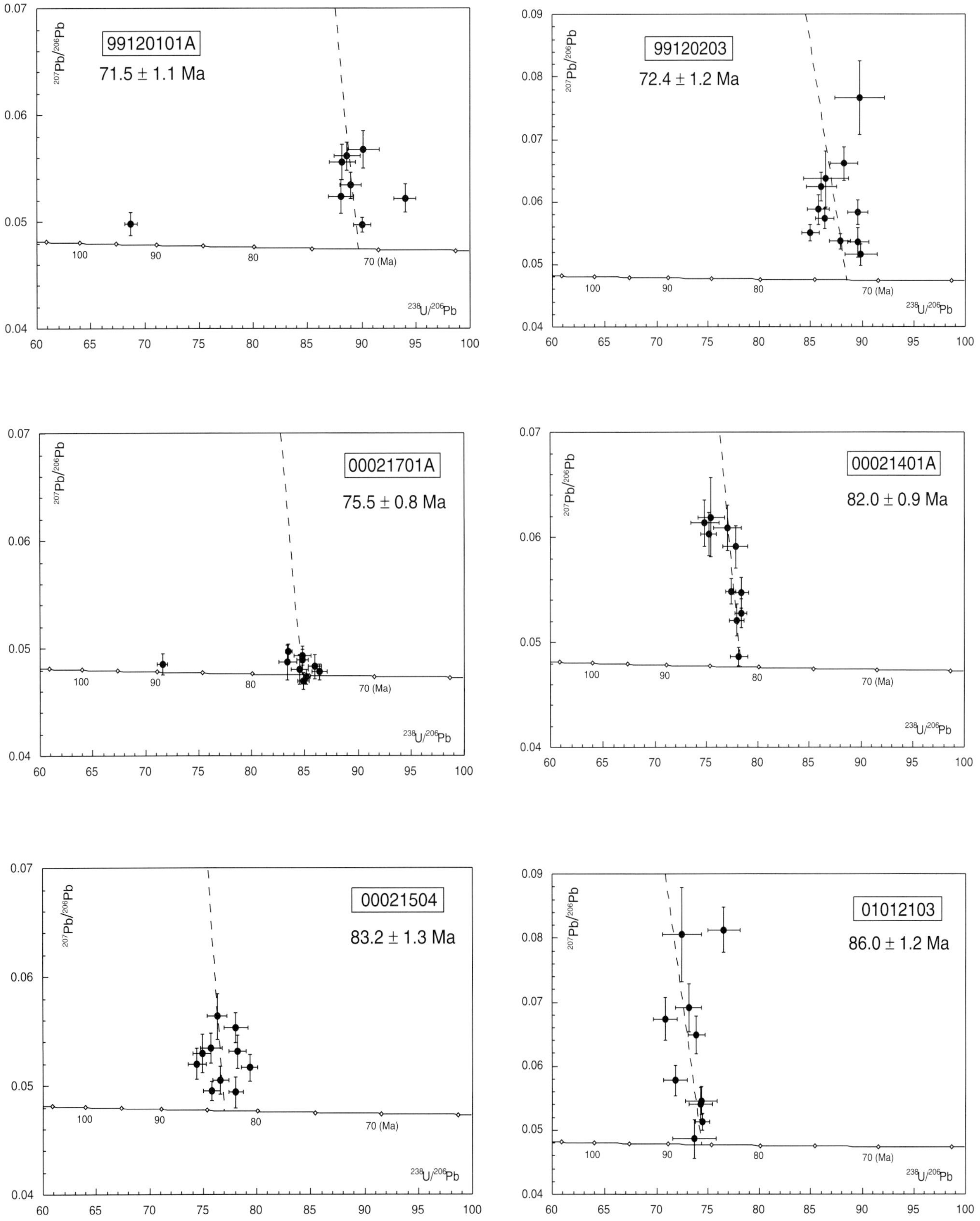

Figure 4 Concordia plots of SHRIMP U-Pb isotopic analyses of zircon from Ryoke mafic rocks before correction for common Pb. Dashed lines show the locus of mixing with common Pb. Analytical uncertainties one standard error, age uncertainties 95% confidence limits.

4. Discussion

4.1. Magma sources

4.1.1. Mantle-derived component versus crustal component. The most striking aspect of the geochemical data for the Cretaceous plutonic rocks in Southwest Japan is the rather small variation in initial Sr isotopic composition over the entire range of rock types. This suggests that all the plutonic rocks were derived from similar source materials. The presence of mafic rocks in close spatial and temporal association with the granitic rocks implies an involvement of mantle-derived juvenile material in the latter's petrogenesis.

The mafic dykes and pillows, which approximate melt compositions, have high FeO*/MgO ratios and low Ni and Cr

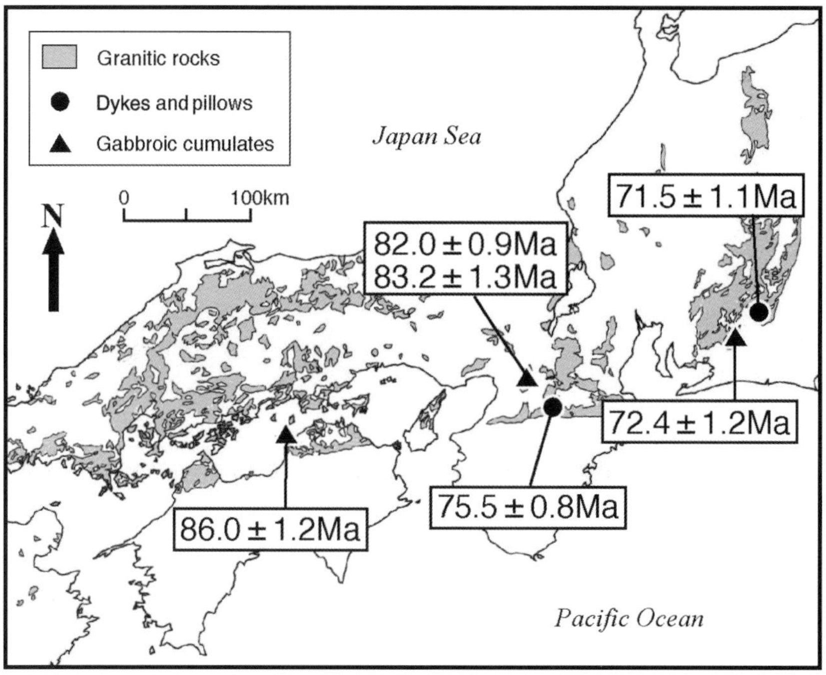

Figure 5 Sample localities and SHRIMP zircon U-Pb ages of the analysed Ryoke mafic rocks.

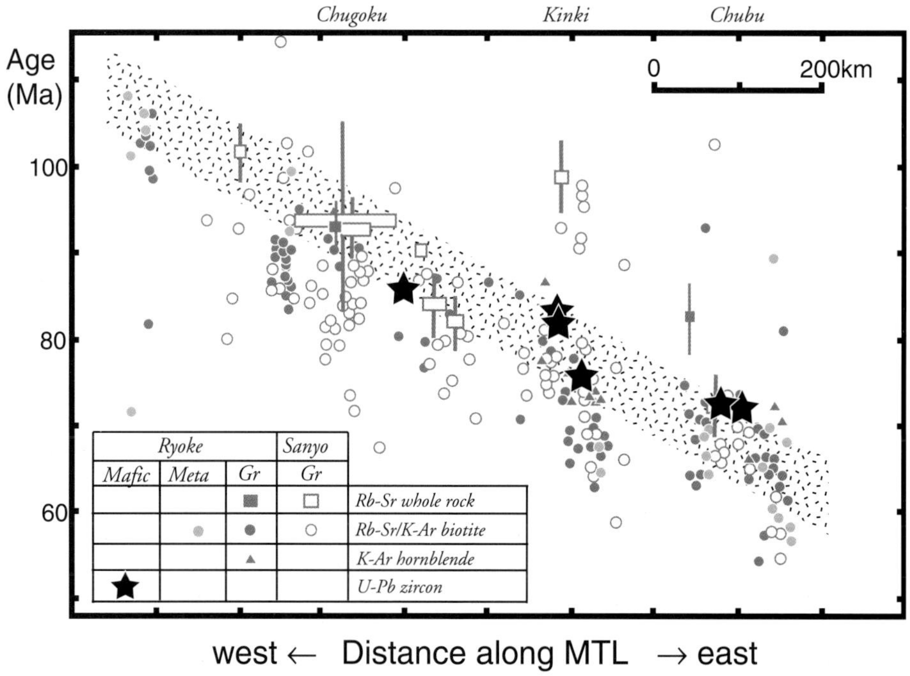

Figure 6 Along-arc variation of SHRIMP zircon ages of mafic rocks plotted with the ages of granitoids and metamorphic rocks from SW Japan. The along-arc position of each point is defined as the normal projection from the sample locality onto the Median Tectonic Line. The stippled band is a trace of the along-arc variation trend of the Rb-Sr whole rock isochron ages of the granitoids after Nakajima *et al.* (1990).

contents. Such evolved compositions argue against equilibration of the melts with mantle peridotite. The mafic magmas must have evolved from mantle-derived primitive magmas in the deep crust.

Another important fact is that none of the mafic rocks in the Ryoke belt have isotope signatures of depleted mantle. The high SrI could be interpreted in two different ways. Either the mafic rocks were produced from primitive basaltic magma with a high SrI (~0·7075) derived directly from enriched mantle, or the SrI of the magmas has been elevated by the incorporation of a crustal component. Although the mantle- and crust-derived end members cannot be directly identified, validity of crustal contamination as an explanation for the enriched isotopic signature of the Ryoke mafic rocks can be evaluated for possible ranges of mantle- and crust-derived end member compositions. In the following discussions, the SiO_2 and Sr contents and the SrI are used as key parameters.

4.1.2. Characterisation of the mantle-derived component. The outcome of any calculation assuming mixing between crust- and mantle-derived components depends heavily on the SrI, Sr and SiO_2 assumed for the mantle-derived basaltic magma. If, for example, the SrI of the basaltic magma is 0·708, no crustal component need be involved. The modern arc basalts show considerable chemical variation. We will give

Table 2 Representative analyses of the Ryoke mafic and granitic rocks, and metasediments

	95101702 Gr	91030804 Gr	99112902A Gr	99112902B D/P	99120101A D/P	99120203 G.C.	00021402 G.C.	01012103 G.C.	99120201 Gn	95101602B Gn	95101801B Gn
%											
SiO_2	71·76	71·19	69·85	54·95	59·92	50·81	46·31	48·71	65·50	67·04	68·40
TiO_2	0·34	0·38	0·43	1·13	0·82	1·08	0·85	0·25	0·74	0·74	0·66
Al_2O_3	14·36	14·72	15·22	17·68	17·74	17·86	18·78	24·71	17·46	17·25	15·98
Fe_2O_3	2·56	3·13	4·11	9·04	6·42	10·37	10·69	4·73	5·57	5·53	5·19
MnO	0·04	0·04	0·08	0·16	0·11	0·17	0·19	0·08	0·09	0·07	0·09
MgO	0·72	0·61	1·28	4·91	2·98	6·33	8·53	5·14	2·07	1·99	1·88
CaO	1·77	2·41	3·49	7·56	6·48	10·36	13·24	14·69	1·33	1·44	1·25
Na_2O	2·77	3·23	3·93	3·00	3·60	2·23	1·22	1·41	2·47	2·21	2·00
K_2O	5·55	4·17	1·46	1·40	1·77	0·68	0·17	0·23	4·59	3·61	4·24
P_2O_5	0·14	0·11	0·13	0·16	0·17	0·12	0·03	0·05	0·16	0·12	0·10
Total	100·01	100·00	99·99	100·00	100·00	100·01	100·00	100·00	99·99	100·00	100·00
ppm											
Zr	168	251	179	82	103	84	15	19	182	186	158
Y	21	23	19	21	15	27	11	6	30	24	18
Sr	125	218	343	368	471	351	511	442	189	177	140
Rb	258	128	51	40	58	17	<1	6	168	143	165
Ba	460	814	374	275	499	192	65	71	858	587	779
Zn	55	42	71	102	92	107	80	38	100	89	94
Cu	3	3	3	43	7	26	11	10	24	16	35
Ni	5	<2	5	14	6	15	15	18	24	18	19
Cr	13	6	11	89	34	61	87	156	60	73	64
V	29	14	25	214	124	206	288	150	99	127	85
Nb	18	13	10	7	8	7	2	4	14	13	16
SrI	0·70892	0·70813	0·70807	0·70806	0·70734	0·70951	0·70751	0·70726	0·71775	0·71616	0·71565

(Gr) granitic rocks; (D/P) dykes and pillows type mafic rocks; (G.C.) Gabbroic cumulates; (Gn) gneisses.

some constraints on these chemical parameters and then evaluate the mantle/crust contributions with the various values in possible range of the chemistry.

First, the SiO_2 content of mantle-derived primitive basalt is controlled by phase equilibria. The SiO_2 content of a basaltic melt in equilibrium with mantle peridotite (ol+opx ± cpx ± spinel) is relatively insensitive to the melting temperature, but can be changed significantly by varying pressure and the amount of dissolved water (e.g. Hirose and Kushiro 1993; Hirose and Kawamoto 1995; Gaetani and Grove 1998). Melts generated in the presence of H_2O have significantly higher SiO_2 than those produced under dry conditions. Decreasing pressure increases the SiO_2 content of the melt. The final equilibration site of ascending basaltic melts and mantle peridotite is near the asthenosphere/lithosphere boundary, where ascent of the basaltic melts changes from porous flow to crack-dominated melt flow (Tatsumi and Eggins 1995, pp. 134–138). Consequently, if a very thick lithosphere and dry conditions for partial melting are assumed, a lower limit for the SiO_2 content of the mantle-derived primitive basalt can be obtained. Melting experiments on dry peridotites by Hirose and Kushiro (1993) have shown that the SiO_2 content of a dry partial melt decreases with increasing pressure from ~51 wt.% at 10 kbar to ~46 wt.% at 30 kbar. A pressure of 30 kbar corresponds to the base of a thick lithosphere. It is therefore reasonable to assume that the primitive arc basalts, which normally contain substantial amounts of dissolved water, would have SiO_2 contents of 46–51 wt.% or higher.

On the other hand, Sr content and SrI are much harder to constrain. Actually, they have remarkably wide ranges in modern arc basalts. It is nevertheless possible to make an informed evaluation of the alternative 'enriched mantle model' and 'crustal contamination model' for the production of the Ryoke mafic rocks.

4.1.3. Enriched mantle model. Present-day arc basalts have a variety of isotopic compositions. Most intraoceanic arc basalts have a relatively low SrI of ~0·7035 (e.g. Dixon & Batiza 1979; Tatsumi *et al.* 1991; Taylor & Nesbitt 1995; Ewart *et al.* 1998), while continental arc basalts commonly are slightly more enriched, SrI >0·704 (e.g., Kimura *et al.* 2002). Some parts of the mantle beneath continental arcs are especially enriched. Mukhopadhyay & Manton (1994) reported mantle xenoliths from beneath the Sierra Nevada batholith with SrI ranging from 0·7044 to 0·7082, concluding that those xenoliths represent old subcontinental lithospheric mantle. Many mafic rocks from continental arcs also have enriched isotopic signatures, consistent with their derivation, at least in part, from enriched mantle. This might be either old subcontinental lithospheric mantle (e.g. Young *et al.* 1992; Sisson *et al.* 1996; Ratajeski *et al.* 2001; Lucassen *et al.* 2002) or less enriched mantle contaminated by recycling of old continental material via a subduction zone (Tommasini *et al.* 1995).

If the source of magmas beneath SW Japan during the Cretaceous is assumed to be highly enriched mantle (SrI=0·7073–0·7083), then the petrogenetic model is simple (Fig. 10a). The magmas which formed the dykes and pillows could be derived from mantle-derived primitive basalt (SiO_2=48 wt.%) by moderate fractionation. The gabbroic cumulates and Ryoke granites could have been formed from those magmas in turn by crystal accumulation and further fractionation or partial melting, respectively (Fig. 10a). No crustal component need be involved. Although there is no direct evidence for such enriched mantle peridotite beneath SW Japan during the Cretaceous, namely beneath the Eurasian

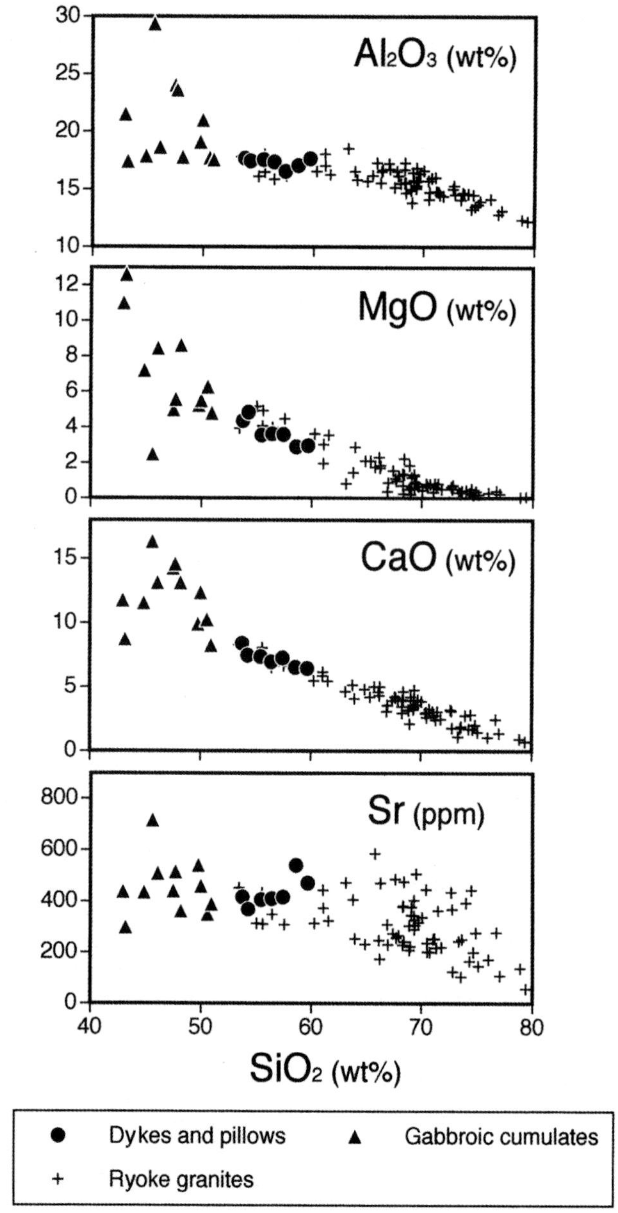

Figure 7 Harker variation diagrams for the Ryoke mafic and granitic rocks: (a) SiO_2-Al_2O_3; (b) SiO_2-MgO; (c) SiO_2-CaO; (d) SiO_2-Sr.

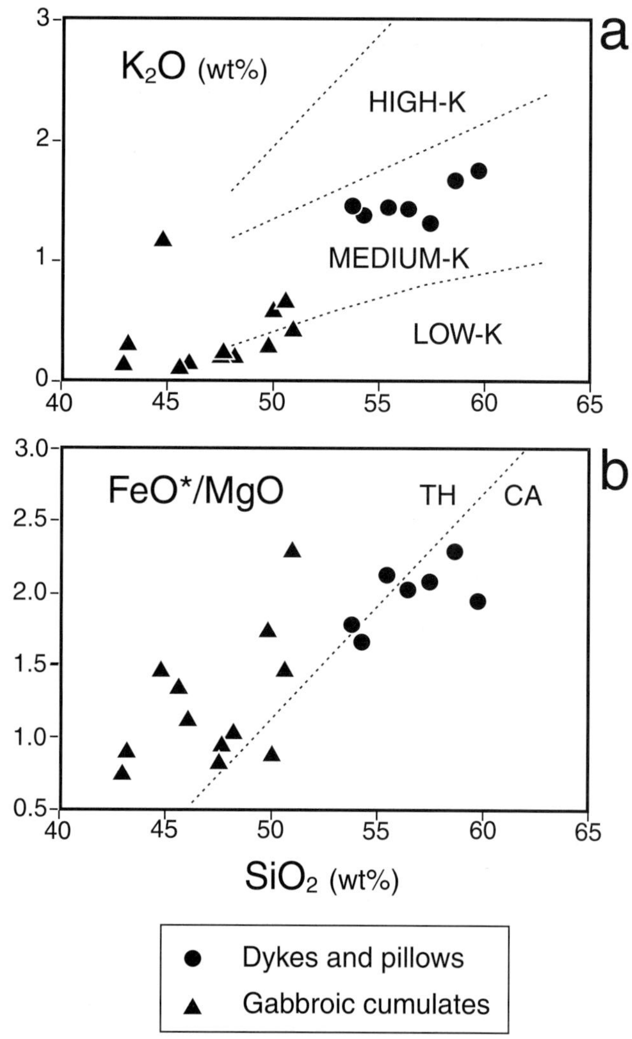

Figure 8 (a) SiO_2-K_2O diagram. Field boundaries are from Tatsumi and Eggins (1995). Ryoke mafic dykes and pillows belong to the medium-K series. (b) SiO_2-FeO*/MgO diagram. Ryoke mafic dykes and pillows plot close to the tholeiite/calc-alkali boundary of Miyashiro (1974), whereas the gabbro cumulates plot mostly in the tholeiite field.

continental margin, in some respects this model accords with reality.

The proposed enriched mantle beneath the Eurasian continental margin might be old subcontinental lithospheric mantle related to the South China craton, along the margin of which the Cretaceous accretional orogen of SW Japan developed. However, involvement of isotopically enriched sedimentary components sinking with the subducting oceanic plate could also raise the SrI of the overlying mantle wedge. If the eastward-younging trend of the Cretaceous granitic rocks in SW Japan (Fig. 6) was caused by the eastward migration of the RTT (ridge–trench–trench) junction as Nakajima *et al.* (1990) suggested, then the subducting plate would be young and hot, causing the overlying sediments to melt at some depth, isotopically enriching the Sr-poor peridotite of the mantle wedge above.

4.1.4. Crustal contamination model. Most of the mafic rocks and mantle xenoliths from continental arcs have much lower SrI than the Ryoke mafic rocks. The few that do not have mostly been affected by crustal contamination. Furthermore, the high $\delta^{18}O$ of the Ryoke mafic rocks (8·5–9·9‰; Ishihara & Matsuhisa 2002) can also be interpreted to indicate involvement of sedimentary components in the deep crust, rather than in the mantle source region.

If the mantle-derived primitive basaltic magmas were less enriched than the Ryoke mafic rocks, then they must have incorporated some enriched crustal material in order to form the hybrid parental magma (SrI = ~0·7075). The 'dykes and pillows' magma and the gabbroic cumulates would have been generated from this parental magma by fractionation, so the crustal component must have been incorporated, and the magma homogenised at deep in the lower crust.

A possible enriched crustal component in the lower crust is restitic metasediment. This is the metamorphosed equivalent of the Middle to Late Jurassic accretionary complex of the Mino–Tanba–Kuga belt in Southwest Japan. They are mainly pelitic and psammitic rocks, originated from the mountain ranges formed by the collision of the North China and South China blocks, so have isotopic compositions much more enriched (SrI = 0·714–0·721) than the Ryoke mafic rocks.

Some possible SrI–SiO_2 mixing relationships between metasediment and mantle-derived primitive basalt to form hybrid parental magmas are illustrated in Fig. 10b. The primitive basalt is assumed to have 46 wt.% SiO_2 (the minimum likely value given by phase equlibria), SrI = 0·705 and

Table 3 Compositional ranges for selected elements from mafic rocks in various tectonic settings

	SiO_2 (%)	Nb (ppm)	Sr (ppm)	Zr (ppm)
Southwest Japan				
MME	54–60	6·3–7·9	370–480	82–103
Gabbro	45–55	2·1–2·4	300–700	6–7
Continental arc				
Andes (Quat)	51–54	2·3–12	480–650	50–170
Sierra Nevada	52–57	8–9	350–750	169–700
New England FB	54–58	13–22	450–630	270–380
Oceanic island arc				
Sunda	50–55	2–4	400–460	82–134
Izu	50–55	<1	170–420	40–80
Continental(?) island arc				
NE Japan				
Nekoma	55–60	4·0–5·9	300–360	84–130
Bandai	60–61	5·6–5·9	200–320	120–130
Continental flood basalt				
Decan, Snake River	46–51	15–37	220–370	170–400
N-MORB	48–51	2–3	90–125	70–85

Data from Thompson *et al.* (1983), Kimura *et al.* (2002), Taylor & Nesbitt (1998), Foden (1983), Thorpe *et al.* (1984), Marriner & Milward (1984), Hickey *et al.* (1986), Schilling *et al.* (1983), Feeley & Davidson (1994), Tani (2002) and Sisson *et al.* (1996).

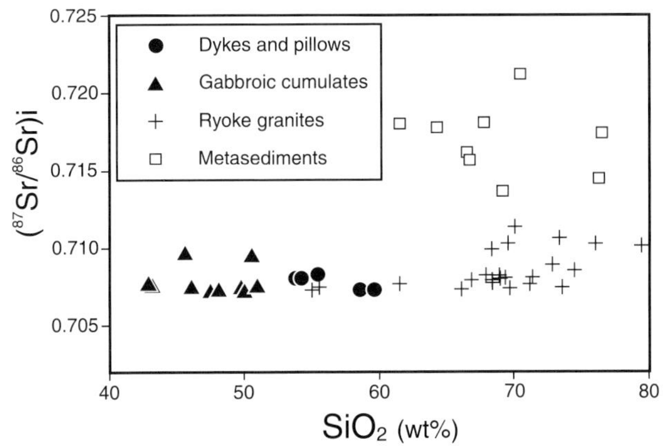

Figure 9 SiO_2-SrI systematics of the two types of Ryoke mafic rocks, Ryoke granites and metasediments.

Sr = 350 ppm (as is common in continental arc basalts). Two model metasediments are considered, the first with 67 wt.% SiO_2, SrI = 0·717 and Sr = 170 ppm (based on the average of the Ryoke high-grade pelitic gneisses) (Table 2), and the second with 60 wt.% SiO_2 such as the melt-depleted lower crustal restitic metasediments from the Ivrea–Verbano zone (Sinigoi *et al.* 1995, 1996). The SiO_2 contents of the lower crustal materials involved in producing the hybrid magmas are likely to be within this range. It is shown that the proportion of crustal material required to produce the SrI observed in the Ryoke mafic magmas is 30–40%, producing a hybrid magma with 51 and 53 wt.% SiO_2, respectively, which can fractionate to form the 'dykes and pillows' magma.

However, the crustal contamination model is only possible for a limited composition range of the assumed mixing end members. Figure 11 shows the SiO_2 versus Sr mixing relationships for the same crustal contamination model system as illustrated in Figure 10b. The assumed SiO_2 content of the crustal metasedimentary component is 67 wt.% in Figure 11a, 60 wt.% in Figure 11b. The bold lines trace the composition of hybrid magmas with SrI = 0·7075 (similar to the dykes and pillows), formed by mixing with mantle-derived primitive basalts with different Sr contents.

The Sr content of the parental basalt magma cannot be as low as 100–200 ppm because no realistic olivine-pyroxene fractionation of the resultant hybrid could produce the high Sr content of the dykes and pillows. An Sr-content of 350–400 ppm in the parental magma is much more likely. The SiO_2 content of the primitive basalt is constrained to be as low as 46 wt.%, because if it is much higher, the hybrid magma becomes too SiO_2-rich to produce the 'dykes and pillows' magma by fractionation (Fig. 10b). Similarly, higher SiO_2 contents of the crustal metasediment component would make the hybrid magma more Si-rich.

This modelling shows that it would be possible to produce the Ryoke mafic rocks by crustal contamination of primitive basaltic magmas without involving a highly-enriched mantle source. However, only a limited range of end member compositions is permitted if the model is to be realistic. Even assuming the basaltic end member to be derived from primitive mantle (SrI = 0·705), the maximum likely crustal contribution to the hybrid magma is ~40% (Fig. 10b). It is concluded, therefore, that the major component of the Ryoke mafic magmas is mantle-derived.

4.2. Granitic magmatism and crustal evolution

4.2.1. Generation of granitic magma in the Ryoke belt. The broad isotopic similarity between the mafic and granitic rocks in the Ryoke belt suggests that the two rock types are genetically related. One possibility, which is consistent with the synchroneity of the mafic and granitic magmatism, is that the granitic magmas were produced by fractional crystallisation of the successive batches of mafic magma represented by the dykes and pillows. If so, then the isolated bodies of gabbroic cumulate might be some of the solid residue complementary to the granitic magmas (Fig. 11). Another possibility is that the granitic magmas were produced by partial melting of the mafic rocks in the lower crust. Because the granitic rocks are closely associated with the mafic rocks in space and time, it is conceivable that the intrusion of successive batches of mafic magma induced partial melting of mafic rocks formed just beforehand. In either case, the mafic magmatism has played both a thermal and a material role in the extensive Cretaceous granitic magmatism in SW Japan.

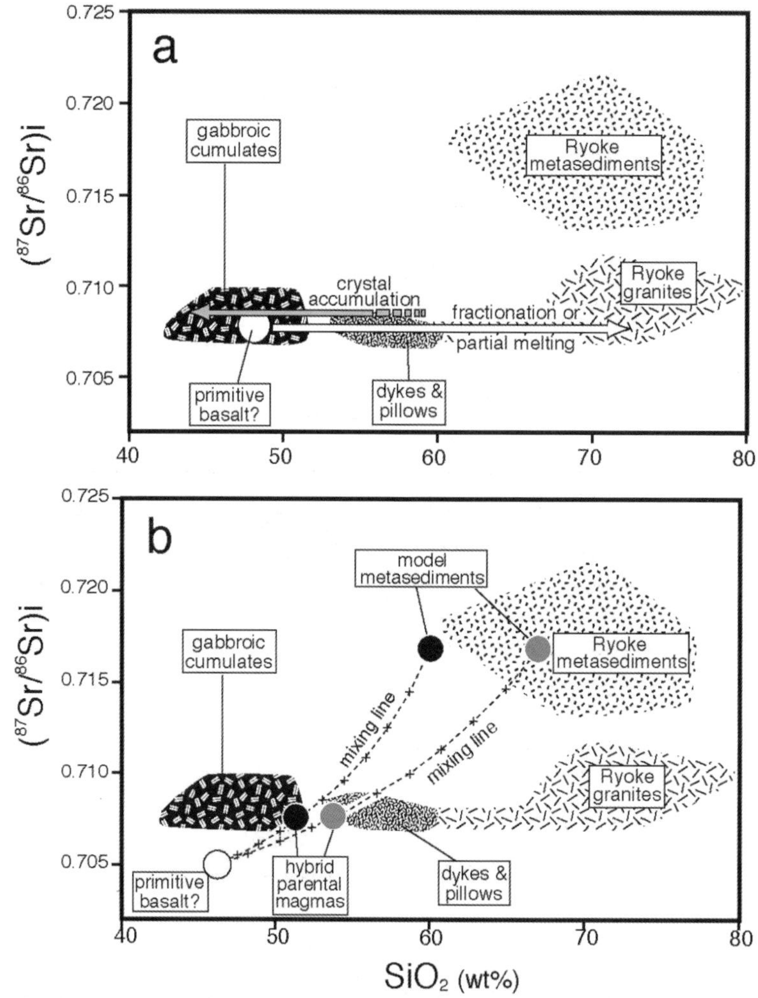

Figure 10 Generation models for the Ryoke mafic and granitic rocks on a SiO_2-SrI system: (a) Enriched mantle model assuming the source mantle of SrI = ~0·7075, which generates a mantle-derived enriched primitive basaltic magma, which generates the 'dykes and pillows' magma only with fractionation; (b) crust contamination model assuming less enriched source mantle and mixing with the crustal components to form hybrid parental magmas. Two cases of mixing involving high-grade Ryoke metapelite (SiO_2 = 67 wt.%) and more SiO_2-depleted model metasediments (SiO_2 = 60 wt.%), respectively, are shown.

It is argued above that the mafic rocks in the Ryoke belt are derived predominantly from the mantle. This in turn means that the majority of the granitic rocks are also composed mostly of mantle-derived components. Only a few of the more felsic granites have SrI slightly higher than the mafic rocks, which can be ascribed to minor assimilation of additional metasediment. S-type granites are generated mostly from recycled upper crust, so do not contribute to the growth of continental crust. Most of the Ryoke granites are I-type, although they include some two mica- or garnet-bearing varieties (Nakajima 1996). The predominance of I-type granites accords with a large proportion of juvenile material being derived from mantle. The Cretaceous granitic rocks in SW Japan therefore provide a record of crustal growth by addition of predominantly mantle-derived material during arc magmatism.

4.2.2. The Cretaceous lower crust of SW Japan. As discussed above, the granitic rocks of SW Japan were probably formed from mafic rocks by either fractional crystallisation or partial melting. Because mafic rock produces only small amounts of granitic magma by either process, the amount of mafic to ultramafic cumulate or residue remaining must have been an order of magnitude greater than the volume of the resultant granites. However, the volume of mafic rock exposed in SW Japan is negligibly small, compared to the huge volumes of granite. This suggests that the mafic lower crust was thicker by an order of magnitude than the newly-formed granitic upper crust. If the newly-formed granitic upper crust were 5 km thick, then a mafic lower crust 50 km thick would be required. There is no evidence now for such a thick mafic layer under SW Japan, so much of it must have been recycled back into the mantle by delamination (Rudnick 1995). However, in the Cretaceous, judging from the inferred P–T conditions of the Ryoke metamorphism, the geothermal gradient beneath SW Japan was high, 45–60°C/km (Nakajima 1996). Although this high geotherm is estimated from metamorphic petrology on the middle to upper crustal rocks, a high geotherm is presumed to also have prevailed in the lower crust, because it would have originated from the active underplating of basaltic magma from mantle to the lowermost crust. Such a high gradient would shift the garnet stability field to higher pressure, increasing the depth at which the crust would be sufficiently dense to delaminate, thereby possibly allowing a relatively thick mafic underplate to develop.

5. Conclusions

Although mafic rocks are volumetrically a minor component of the Ryoke belt, they provide important clues to the granitic

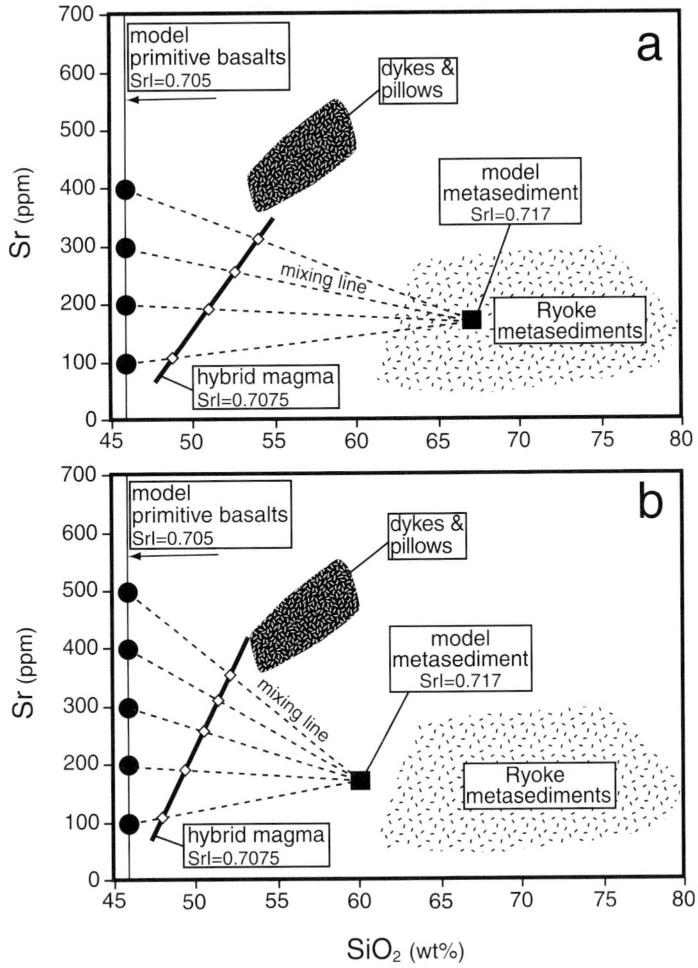

Figure 11 SiO_2-Sr systematics showing the mixing relationships of the crust contamination model. Mixing between possible mantle-derived primitive basalts with 46 wt.% SiO_2 and SrI=0·705 and (a) representative high-grade Ryoke metasediments (SiO_2=67 wt.%, Sr=170 ppm and SrI=0·717) and (b) model lower crustal metasediments (SiO_2=60 wt.%, Sr=170 ppm and SrI=0·717). Each of the broken lines represents a mixing line joining the metasediments and the primitive basalt with different Sr content. Bold lines show the compositions of the hybrid magma with SrI=0·7075, produced by mixing between the metasediments and the primitive basalts with varying Sr content. In both cases, the primitive basaltic magma with limit ranges of Sr content can generate the 'dykes and pillows' magma through olivine-pyroxene fractionation. See text for discussion.

magmatism and crustal evolution in Cretaceous SW Japan. The mafic rocks occur either as dykes and pillows set in granitic matrix or as separate bodies of gabbroic cumulate. The gabbroic cumulates do not have the composition of magmas, whereas the dykes and pillows have the composition of arc-related basaltic andesite or andesite. SHRIMP zircon U-Pb ages confirm that throughout the Ryoke belt, the mafic rocks formed contemporaneously with the associated granitic rocks, becoming younger eastward along the arc. The broad isotopic similarity between the mafic and granitic rocks suggests that the granitic rocks were produced either by fractional crystallisation of mafic magmas or by partial melting of newly-formed mafic rocks at depth. The close spatial and temporal association of the mafic and granitic rocks, together with their Sr isotopic similarity, points to thermal and material roles for the mafic magmatism in the extensive granitic magmatism.

The enriched Sr isotopic signature of the Ryoke mafic rocks could indicate either of two possible petrogenesis: (1) Derivation of the magmas from a highly-enriched mantle source that might be either subcontinental lithospheric mantle or less enriched mantle contaminated by addition of melt derived from subducting sediment; or (2) Deep crustal sedimentary contamination of primitive basaltic magma derived from a less enriched mantle source. Regardless of the mechanism of enrichment, it is concluded that the Cretaceous magmatism in Southwest Japan represents a major episode of crustal growth by additions from the upper mantle during arc magmatism.

6. Acknowledgements

We thank Professor S. Yoshikura and Dr. S. Atsuta for guiding us (TN, HK and KT) to their beautiful mafic rock field in Shodo-Shima island, which inspired this work. We are particularly grateful for the kindness and generosity of Professors Y. Tainosho and T. Kutsukake in providing us with helpful field information and encouragement, while knowing that our interpretation was entirely different from theirs. One of us (TN) is grateful to Professor T. Ireland for allowing and encouraging our SHRIMP work at RSES, even though his group already had some unpublished SHRIMP age measurements on Ryoke mafic rocks. We thank Professor C. F. Miller and Dr. S. Ishihara for their insightful reviews. This work was undertaken as part of the project 'Formation and growth of the continental crust' supported by the Ministry of International Trade and Industry, Japan.

7. References

Black, L. P., Kamo, S. L., Allen, C. M., Aleinikoff, J. N., Davis, D. W., Korsch, R. J. & Foudoulis, C. 2003. TEMORA 1: a new zircon standard for Phanerozoic U-Pb geochronology. *Chemical Geology* **200**, 155–70.

Collins, W. J. 2002. Nature of extensional accretionary orogens. *Tectonics* **21** (4), 10.1029/2000TC001272.

Dixon, T. H. & Batiza, R. 1979. Petrology and chemistry of recent lavas in the northern Marianas: implications for the origin of island arc basalts. *Contributions to Mineralogy and Petrology* **70**, 167–81.

Ducea, M. N. & Saleeby, J. B. 1998. The age and origin of a thick mafic-ultramafic keel from beneath the Sierra Nevada batholith. *Contributions to Mineralogy and Petrology* **133**, 169–85.

Ewart, A., Collerson, K. D., Regelous, M. Wendt, J. I. & Niu, Y. 1998, Geochemical evolution within the Tonga-Kermadec-Lau arc-back-arc systems: the role of varying mantle wedge composition in space and time. *Journal of Petrology* **39**, 331–68.

Falkner, C. M., Miller, C. F., Wooden, J. L. & Heizler, M. T. 1995. Petrogenesis and tectonic significance of the calc-alkaline, bimodal Aztec Wash pluton, Eldorado Mountains, Colorado River extensional corridor. *Journal of Geophysical Research* **100**, 10453–76.

Feeley, T. C. & Davidson, J. P. 1994, Petrology of calc-alkaline lavas at Volcan Ollague and the origin of compositional diversity at Central Andean stratovolcanoes. *Journal of Petrology* **35**, 1295–340.

Foden, J. D. 1983. The petrology of the calkalkaline lavas of Rindjani volcano, East Sunda arc: a model for island arc petrogenesis. *Journal of Petrology* **24**, 98–130.

Gaetani, G. A. & Grove, T. L. 1998. The influence of water on melting of mantle peridotite. *Contributions to Mineralogy and Petrology* **131**, 323–46.

Hickey, R. L., Frey, F. A. & Gerlach, D. C. 1986. Multiple sources for basaltic arc rocks from the southern volcanic zone of Andes (34–41S): trace element and isotopic evidence for contributions from subducted oceanic crust, mantle and continental crust. *Journal of Geophysical Research* **91**, 5963–83.

Hirose, K. & Kawamoto, T. 1995. Hydrous partial melting of lherzolite at 1GPa: the effect of H_2O on the genesis of basaltic magmas. *Earth and Planetary Science Letters* **133**, 463–73.

Hirose, K. & Kushiro, I. 1993. Partial melting of dry peridotites at high pressures: determination of compositions of melts segregated from peridotite using aggregates of diamond. *Earth and Planetary Science Letters* **114**, 477–89.

Ishihara, S. & Matsuhisa, Y. 2002. Oxygen isotopic constraints on the geneses of the Cretaceous-Paleogene granitoids in the Inner Zone of Southwest Japan. *Bulletin of the Geological Survey of Japan* **53**, 421–38.

Kagami, H., Iizumi, S., Tainosho, Y. & Owada, M. 1992. Spatial variations of Sr and Nd isotope ratios of Cretaceous–Paleogene granitoid rocks, Southwest Japan Arc. *Contributions to Mineralogy and Petrology* **112**, 165–77.

Kagami, H., Yuhara, M., Tainosho, Y., Iizumi, S., Owada, M. & Hayama, Y. 1995. Sm-Nd isochron ages of mafic igneous rocks from the Ryoke Belt, Southwest Japan: remains of Jurassic igneous activity in a late Cretaceous granitic terrane. *Geochemical Journal* **29**, 123–35.

Kagami, H., Yuhara, M., Iizumi, S., Tainosho, Y., Owada, M., Ikeda, Y., Okano, O., Ochi, S., Hayama, Y. & Nureki, T. 2000. Continental basalts in the accretionary complexes of the South-west Japan Arc: constraints from geochemical and Sr and Nd isotopic data of metadiabase. *Island Arc* **9**, 3–20.

Kapp, J. D., Miller, C. F. & Miller, J. S. 2002. Ireteba pluton, Eldorado Mountains, Nevada: late, deep source, peraluminous magmatism in the Cordilleran Interior. *Journal of Geology* **110**, 649–69.

Kimura, J.-I., Yoshida, T. & Iizumi, S. 2002. Origin of low-K intermediate lavas at Nekoma volcano, NE Honshu arc, Japan: Geochemical constraints for lower-crustal melts. *Journal of Petrology* **43**, 631–61.

Lucassen, F., Escayola, M., Romer, R. L., Viramonte, J., Koch, K. & Franz, G. 2002. Isotopic composition of Late Mesozoic basic and ultrabasic rocks from the Andes (23–32S) – implications for the Andean mantle. *Contributions to Mineralogy and Petrology* **143**, 336–49.

Marriner, G. F, & Milward, D. 1984. The petrology and geochemistry of Cretaceous to recent volcanism in Columbia: the magmatic history of an accretionary plate margin. *Journal of the Geological Society, London* **141**, 473–86.

Mattson, S. R., Vogel, T. A. & Wilband, J. T. 1986. Petrochemistry of the silicic-mafic complexes at Vesturhorn and Austurhorn, Iceland: evidence for zoned/stratified magma. *Journal of Volcanology and Geothermal Research* **28**, 197–223.

Miyashiro, A. 1974. Volcanic rock series in island arcs and active continental margins. *American Journal of Science* **274**, 321–55.

Mukhopadhyay, B. & Manton, W. I. 1994. Upper-mantle fragments from beneath the Sierra Nevada batholith: Partial fusion, fractional crystallization, and metasomatism in a subduction-related ancient lithosphere. *Journal of Petrology* **35**, 1417–50.

Nakajima, T. 1994. The Ryoke plutonometamorphic belt: crustal section of the Cretaceous Eurasian continental margin. *Lithos* **33**, 51–66.

Nakajima, T. 1996. Cretaceous granitoids in SW Japan and their bearing on the crust-forming process in the eastern Eurasian margin. *Transactions of the Royal Society of Edinburgh: Earth Sciences* **87**, 183–91.

Nakajima, T., Shirahase, T. & Shibata, K. 1990. Along-arc lateral variation of Rb-Sr and K-Ar ages of Cretaceous granitic rocks in Southwest Japan. *Contributions to Mineralogy and Petrology* **104**, 381–9.

Okano, O., Sato, T. & Kagami, H. 2000. Rb-Sr and Sm-Nd isotopic studies of mafic igneous rocks from the Ryoke plutono-metamorphic belt in the Setouchi area, Southwest Japan: implications to their genesis and thermal history. *Island Arc* **9**, 21–36.

Okudaira, T., Yuhara, M., Ikeda, T. & Nakajima, T. 2003. Mid-Cretaceous plutono-metamorphic complex of the Ryoke and San-yo zones in the Yanai district, SW Japan: *In* Shimura, T. & Ishihara, S. (eds) *Hutton Symposium V, Field Guidebook. Geological Survey of Japan, Interim-Report* **28**, 23–40.

Otofuji, Y. & Matsuda, T. 1984. Timing of rotational motion of Southwest Japan inferred from paleomagnetism. *Earth and Planetary Science Letters* **70**, 373–82.

Ratajeski, K., Glazner, A. F. & Miller, B. V. 2001. Geology and geochemistry of mafic to felsic plutonic rocks in the Cretaceous intrusive suite of Yosemite Valley, California. *Geological Society of America Bulletin* **113**, 1486–502.

Rudnick, R. L. 1995. Making continental crust. *Nature* **378**, 571–8.

Schilling, J.-G., Zajac, M., Evans, R., Johnston, T., White, W., Devine, J. D. & Kingsley, R. 1983. Petrologic and geochemical variations along the Mid-Atlantic Ridge from 27N to 73N. *American Journal of Science* **283**, 510–86.

Sinigoi, S., Quick, J. E., Mayer, A. & Demarchi, G. 1995. Density controlled assimilation of underplated crust, Ivrea Zone, Italy. *Earth and Planetary Science Letters* **129**, 183–91.

Sinigoi, S., Quick, J. E., Mayer, A. & Budahn, J. 1996. Influence and stretching and density contrasts on the chemical evolution of continental magmas: an example from the Ivrea-Verbano Zone. *Contributions to Mineralogy and Petrology* **123**, 238–50.

Sisson, T. W., Grove, T. L. & Coleman, D. S. 1996. Hornblende gabbro sill complex at Onion Valley, California, and a mixing origin for the Sierra Nevada batholith. *Contributions to Mineralogy and Petrology* **126**, 81–108.

Snyder, D., Crambes, C. Tait, S. & Wiebe, R. A. 1997. Magma mingling in dykes and sills. *Journal of Geology* **105**, 75–86.

Steiger, R. H. & Jäger, E. 1977. Subcommission on geochronology: convention on the use of decay constants in geo- and cosmochronology. *Earth and Planetary Science Letters,* **36**, 359–62.

Streckeisen, A. 1976. To each plutonic rock its proper name. *Earth Science Review* **12**, 1–33.

Takahashi, M., Ishihara, S. & Aramaki, S. 1980. Magnetite-series/ilmenite-series vs. I-type/S-type granitoids. *Mining Geology, Special Issue* **88**, 13–28.

Tani, K. 2002. *Evolution of granitic magma chamber in the Donner Pass region granodiorite, Sierra Nevada, U.S.A.* Postgraduate Master's thesis. University of Tokyo.

Tatsumi, Y., Murasaki, M., Arsadi, E. M. & Nohda, S. 1991. Geochemistry of Quaternary lavas from NE Sulawesi: transfer of subduction components into the mantle wedge. *Contributions to Mineralogy and Petrology* **107**, 137–49.

Tatsumi, Y. & Eggins, S. 1995. *Subduction Zone Magmatism*. Boston: Blackwell Scientific.

Taylor, R. N. & Nesbitt, R. W. 1995. Arc volcanism in an extensional regime at the initiation of subduction: a geochemical study of Hahajima, Bonin Islands, Japan: *In* Smellie, J. L. (ed.) *Volcanism Associated with Extension at Consuming Plate Margins. Geological Society, London, Special Publication* **81**, 115–34. London: Geological Society.

Taylor, R. N. & Nesbitt, R. W. 1998. Isotopic characteristics of subduction fluids in an intra-oceanic setting, Izu-Bonin Arc, Japan. *Earth and Planetary Science Letters* **164**, 79–98.

Thompson, R. N., Morrison, M. A., Dickin, A. P. & Hendry, G. L. 1983. Continental flood basalts ... arachnid rule OK? *In* Hawksworth, C. J. & Norry, M. J. (eds) *Continental Basalts and Mantle Xenoliths*, 158–85. Nantwich: Shiva.

Thorpe, R. S., Francis, P. W. & O'Callaghan, L. 1984. Relative roles of source composition, fractional crystallization and crustal contamination in the petrogenesis of Andean volcanic rocks. *Philosophical Transactions of the Royal Society of London* **A310**, 675–92.

Tommasini, S., Poli, G. & Halliday, A. N. 1995. The role of sediment subduction and crustal growth in Hercynian plutonism: isotopic and trace element evidence from the Sardinia–Corsica batholith. *Journal of Petrology* **36**, 1305–32.

Wiebe, R. A. 1973. Relations between coexisting basaltic and granitic magmas in a composite dike. *American Journal of Science* **273**, 130–51.

Wiebe, R. A., Frey, H. & Hawkins, D. P. 2001. Basaltic pillow mounds in the Vinalhaven intrusion, Maine. *Journal of Volcanology and Geothermal Research* **107**, 171–84.

Williams, I. S., Buick, I. S. & Cartwright, I. 1996. An extended episode of early Mesoproterozoic metamorphic fluid flow in the Reynolds Range, central Australia. *Journal of Metamorphic Geology* **14**, 29–47.

Young, E. D., Wooden, J. L., Shieh, Y.-N. & Farber, D. 1992. Geochemical evolution of Jurassic diorites from the Bristol Lake region, California, USA, and the role of assimilation. *Contributions to Mineralogy and Petrology* **110**, 68–86.

TAKASHI NAKAJIMA, Institute of Geoscience, Geological Survey of Japan, AIST, Higashi, Tsukuba 305–8567, Japan.
e-mail: tngeoch.nakajima@aist.go.jp

HIROYUKI KAMIYAMA, Institute of Seismology and Volcanology, Graduate School of Science, Hokkaido University, Sapporo, Japan.

IAN S. WILLIAMS, Research School of Earth Sciences, The Australian National University, Canberra, ACT 0200, Australia.

KENICHIRO TANI, IFREE, Japan Marine Science and Technology Center, Natsushima, Yokosuka 237–0061, Japan.

MS received 13 October 2003. Accepted for publication 6 September 2004.

Transactions of the Royal Society of Edinburgh: Earth Sciences, **95**, 265–276, 2004

Petrochemical characteristics of felsic veins in mantle xenoliths from Tallante (SE Spain): an insight into activity of silicic melt within the mantle wedge

Yohei Shimizu, Shoji Arai, Tomoaki Morishita, Hisayoshi Yurimoto and Fernando Gervilla

ABSTRACT: Felsic and related veins within mantle-derived peridotite xenoliths from Tallante, Spain, were examined in order to understand the mantle-wedge processes related to the behaviour of Si-rich melt. The thickest part of the vein has a quartz diorite lithology, and is composed mainly of quartz and plagioclase, with pyroxenes, hydrous mineral, apatite, zircon and rutile present as minor phases. The thinner parts are free of quartz and predominantly composed of plagioclase. Orthopyroxene always intervenes between the internal part (plagioclase ± quartz) and host peridotite, indicating that it is a product of interaction between silica-oversaturated melt and olivine. This indicates that a sufficiently high melt/wall rock ratio enabled the melt to retain its silica-oversaturated character.

The quartz diorite part has adakite-like geochemical signatures, except for negative Ba, Rb Eu and Sr anomalies, and positive Th and U anomalies. These negative anomalies indicate that fractionation of plagioclase and hydrous minerals was achieved between the upper most mantle and the slab melting zone. The shape of the rare-earth element (REE) pattern of clinopyroxene in quartz diorite is strikingly similar to that of clinopyroxene phenocrysts from Aleutian adakites. However, the former has one order higher REE contents than the latter, except for Eu which shows a prominent negative spike. This feature was caused by the precipitation of large amounts of plagioclase and small amounts of clinopyroxene from a fractionated adakitic melt before and during solidification. This adakitic melt was produced by partial melting of a detached and sinking slab beneath the Betic area in the Tertiary.

KEY WORDS: Adakite, orthopyroxene, peridotite xenolith, quartz diorite.

The origin of the continental crust is an important problem for our understanding of the Earth's history. It has been controversial, and many workers have discussed the highly specific structure and composition of the continental crust (e.g. Taylor & McLennan 1985; Rudnick & Fountain 1995). Taylor & McLennan (1985) suggested an andesitic composition for the continental crust. Kelemen (1995) more recently pointed out that the continental crust is similar in composition to high-Mg andesites, and suggested that the high-Mg andesitic character has been achieved by melt/rock reaction in the upper mantle based on experimental constraints and phase equilibra calculations. Tatsumi (2000a, b) examined the roles of slab melting and delamination of the lower crust by means of geochemical modelling, and concluded that the delamination process has been important in the formation of the continental crust.

Defant & Drummond (1990) described the adakites as volcanic rocks derived from partially molten oceanic slab. Adakite has high Al_2O_3 wt.%, positive Sr anomaly, and high Sr/Y and La/Yb ratios compared with normal arc magma. The adakitic melt occurs in a specific setting where hot and/or young oceanic plate is subducted. Experimental studies have demonstrated that the melting of downgoing oceanic slab can produce melts similar to adakite in composition (Rapp *et al.* 1991; Sen & Dunn 1994; Rapp & Watson 1995; López & Castro 2001). Defant & Drummond (1990) also pointed out the similarities between adakite and high-Al tonalite, trondhjemite and granodiorite (TTG). The high-Al TTG is a major component of Archaean crust and is possibly formed by slab melting as a result of a higher geothermal gradient during the Archaean. This hypothesis is a plausible explanation for the origin of the Archaean continental crust. However, there are also compositional differences between them: the adakites generally have a higher MgO content than the experimentally duplicated slab melt and high-Al TTG (Rapp *et al.* 1999; Martin 1999; Smithies 2000). This is because of the assimilation and fractional crystallisation (mantle AFC) of slab melts during its ascent through the mantle, as suggested both by natural (Kay *et al.* 1993) and experimental adakite studies (Rapp *et al.* 1999).

There is little information related to the slab-melt activity within the mantle from mantle peridotite studies because of the scarcity of mantle xenoliths derived from the mantle wedge. Recently, Arai *et al.* (2003) reported quartz diorite veining in a mantle peridotite xenolith from Tallante, SE Spain. The relationship between the peridotite and such a felsic vein provides an indication of how the slab melt ascends through the mantle wedge.

In this paper, the present authors describe detailed petrochemical characteristics of the quartz diorite and related veinlets from mantle xenoliths of Tallante, SE Spain, in order to explore the behaviour of slab melts within the mantle wedge.

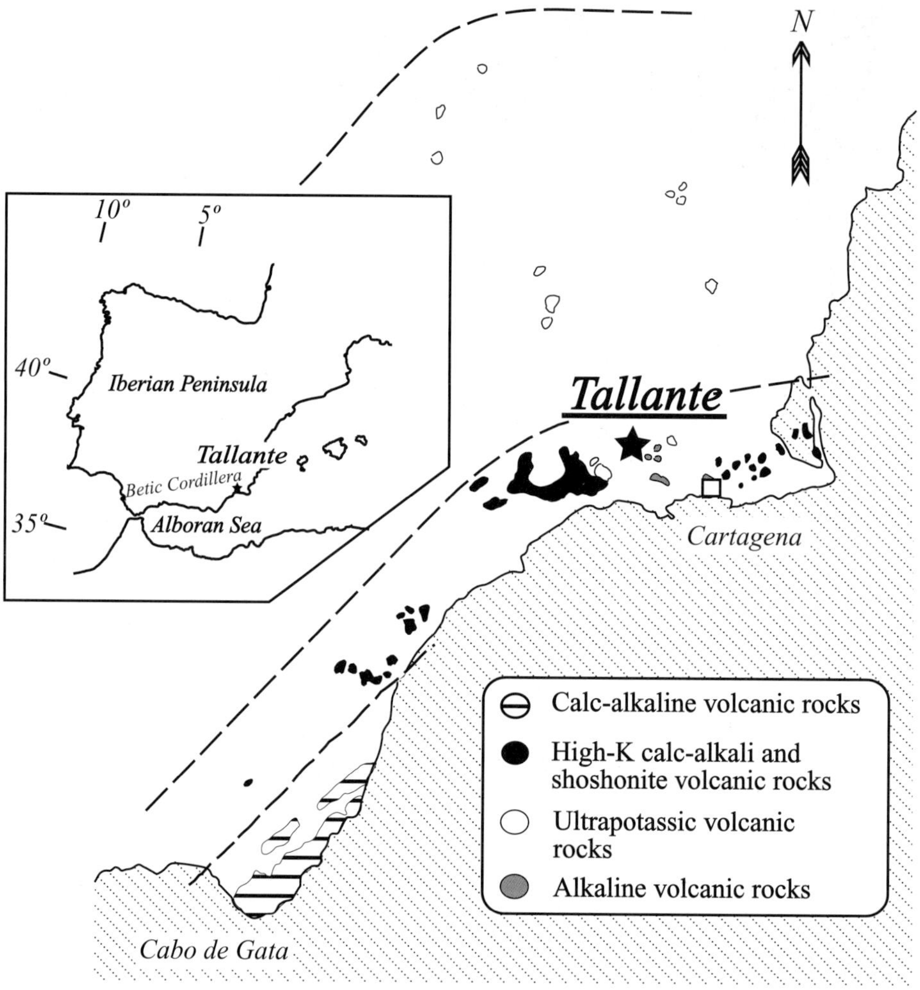

Figure 1 Locality map of Tallante, SE Spain. The distribution of Tertiary volcanic rocks around Tallante is after Ancochea & Nixon (1987). The K content of the volcanic rocks systematically increases northward.

1. Geological background and petrography of peridotite xenoliths

The Neogene volcanic rocks in SE Spain (Fig. 1) consist of calc-alkaline, high-K calc-alkaline, shoshonite, ultrapotassic rocks (lamproite) (Venturelli *et al.* 1988) and alkali basalts (López-Ruiz & Rodoríguez-Badiola 1980). The composition and eruptive age of the calc-alkaline rocks regularly change northward; their K_2O content increases considerably northward (e.g. Ancochea & Nixon 1987). This relationship has been interpreted as a result of northward subduction of slab in the Alboran Sea at the end of Oligocene (Araña & Vegas 1974). The seismic tomography indicates that there possibly is a detachment slab beneath the region (Blanco & Spakman 1993). Turner *et al.* (1999) ascribed the calc-alkaline and related magmas to anatexis of lithosphere beneath the Betic region.

The alkali basalt in Tallante (NW Cartagena, Province of Murcia) includes abundant mantle-derived ultramafic xenoliths (e.g. Dupuy *et al.* 1986; Ancochea & Nixon 1987; Capedori *et al.* 1989). These ultramafic xenoliths occur as rounded or elliptically shaped ejecta within host alkali-basaltic pyroclastics. The ultramafic xenoliths are less than 25 cm in diameter, and they are predominantly lherzolite and harzburgite and occasionally pyroxenites. The ultramafic xenoliths are classified into two types, Groups I and II in the sense of Frey & Prinz (1978). Group I comprises lherzolite, harzburgite and pyroxenites, while Group II is hornblende-bearing clinopyroxenite. Rocks of both groups sometimes form composite xenoliths (Irving 1980). Group I rocks can further be divided into two types; namely, spinel peridotite and plagioclase peridotite in mineral assemblage. Group I peridotites have equigranular texture and, in a rare case, porphyroclastic texture with olivine porphyroclast (cf. Mercier & Nicolas 1975). Olivine and pyroxenes often show kink bands and pyroxenes contain exsolution lamellae. Pyroxenes show more prominent exsolution lamellae in plagioclase peridotite than in spinel peridotite. Plagioclase, which shows slight chemical zoning, is observed as an interstitial phase, especially surrounding grains of spinel. Hydrous minerals (amphibole and mica) are rarely found as interstitial grains, although substantial amounts of hydrous minerals (amphiboles and rarely mica) with plagioclase occur along the contact between Group I and II rocks of the composite xenoliths. Group I rocks occasionally contain lenticular fine-grained aggregates of pyroxenes and spinel (Fig. 2). Group II rocks have igneous textures, and their mineral assemblage is clinopyroxene+amphibole+magnetite, with or without phlogopite, olivine and orthopyroxene.

2. Petrography of the quartz diorite and related veins

Peridotites from Tallante are characterised by secondary, white, network-like veins (Fig. 2; Arai *et al.* 2003), which are less than 8 mm in thickness, and cut or replace primary minerals. Each vein is mainly composed of plagioclase, and modal quartz can be found only at the thick (>1 mm) parts of

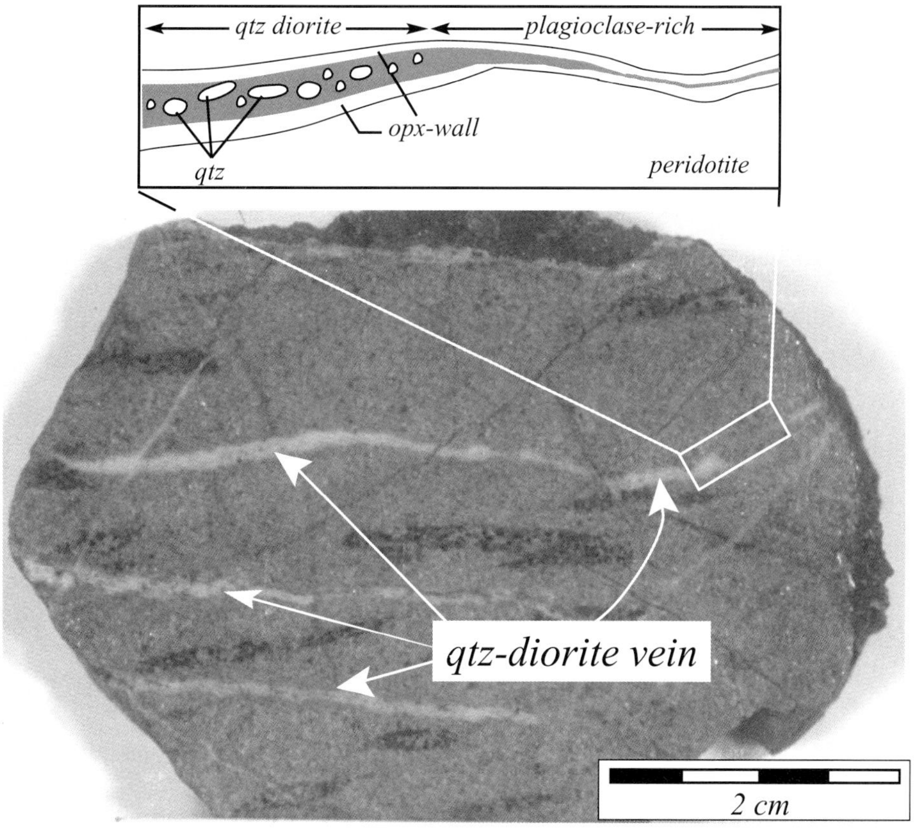

Figure 2 Sawed surface of the peridotite xenolith penetrated by a network of white veins, which are mainly composed of plagioclase. Quartz is found only in thick (>1-mm) parts, which have quartz diorite lithology. Foliation is defined by dark lenticular aggregates of spinel and pyroxenes: (qtz) quartz; and (opx) orthopyroxene.

the vein, which is quartz diorite, as described below (Fig. 2). Orthopyroxene occurs between olivine of the host peridotite and the internal part (plagioclase ± quartz) (Fig. 3b). In very thin (<0·1-mm) veins, orthopyroxene is sporadically present between plagioclase and olivine (Fig. 3a). Clinopyroxene and orthopyroxene in the host peridotite tend to have more prominent exsolution lamellae when the rock contains plagioclase-rich veins.

The thick part of the veins (=quartz-bearing) reveals important information. It contains small amounts of orthopyroxene, clinopyroxene, amphibole, phlogopite, apatite, rutile, zircon and glass in addition to plagioclase and quartz. The presence of quartz is specially remarkable. In the thickest part, quartz is anhedral, and is usually 100–200 μm, reaching up to 750 μm in diameter. It occurs only at the centre of the vein, and is surrounded by plagioclase and/or glass. The plagioclase is less than 1·5–2 mm in diameter and tends to be coarser toward the centre of the vein. Plagioclase is slightly zoned, and there is no K-feldspar. Apatite, rutile and zircon are observed as inclusions in plagioclase, quartz and/or glass. Rutile and zircon are 10–100 μm across, and rutile is more abundant than zircon. Apatite often occurs as needles in the felsic part of the vein. Glass occurs along the grain boundaries of quartz, plagioclase and orthopyroxene, and never occurs as inclusions. Clinopyroxene is rare, but is very coarse (2 mm across) in the quartz-bearing part of the vein. The coarse clinopyroxene grain is also in contact with quartz and glass, and is partly enclosed by plagioclase. Amphibole and phlogopite are interstitial and rarely show a linear arrangement.

We obtained the modal composition of the quartz-bearing part of the vein by measuring the area of all phases on photomicrographs using image data processing software (NIH image and Adobe Photoshop) (Table 1). Orthopyroxene lining the olivine wall was excluded from this estimate. Quartz and plagioclase occupy more than 90 vol.% of the vein by volume, and are classified as 'quartz diorite' according to the nomenclature of Streckeisen (1976); we adopt the term quartz diorite vein hereafter (Fig. 2).

3. Analytical methods

Minerals and glass were analysed for major elements by a JEOL JXA-8800 electron probe X-ray microanalyser (WDS) at the Centre for Cooperative Research of Kanazawa University, Kanazawa, Japan. Microprobe analysis was conducted with an acceleration voltage of 20 keV and sample currents of 20 nA with a 3-μm probe diameter. Oxides, and natural and synthetic minerals were used as standards. ZAF correction procedure was applied to all raw data. Secondary standards, including olivine, clinopyroxene, spinel and alkali feldspar, were analysed to ensure accuracy for the major elements. The cation fractions of Mg, Fe^{2+}, Al, Cr and Fe^{3+} in spinel were calculated assuming spinel stoichiometry after allotting all of Ti to the ulvöspinel molecule. Selected microprobe analyses are listed in Table 2.

Selected clinopyroxenes were analysed *in situ* for rare-earth elements (REEs), and Ti, Sr, Zr and Y using a Cameca IMS 3f ion-microprobe (SIMS) at the Tokyo Institute of Technology, Tokyo, Japan. The analytical procedure used on the ion microprobe was after Yurimoto *et al.* (1989) and Wang & Yurimoto (1993). Ion microprobe analyses are listed in Table 3.

The present authors tried to determine the trace element composition of the bulk quartz diorite vein. The rock sample with the vein was sliced into thin slabs with a diamond saw. Peridotitic portions around the quartz diorite veins were eliminated first by grinding the cut surface. After that, the

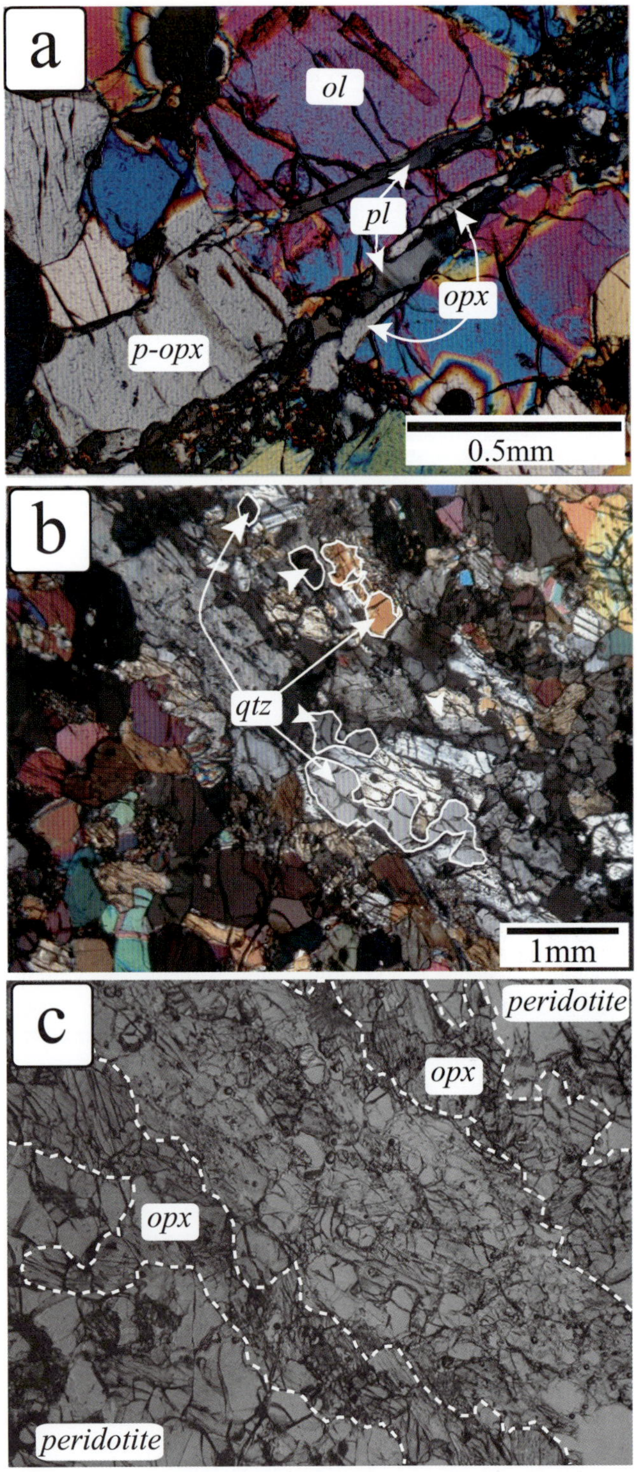

Figure 3 Photomicrographs of secondary veinlets from Tallante, SE Spain: (a) Lherzolite with a thin plagioclase vein associated with orthopyroxene. Primary olivine (ol) and orthopyroxene (p-opx) are cut and replaced by the vein. Note the orthopyroxene (opx) forms between the olivine and plagioclase (pl). Crossed-polarised light. (b) The quartz diorite vein within the peridotite. Note that the quartz grains (qtz) are at the centre of the vein, and thin orthopyroxenite [see (c)] is between the felsic vein and host olivine. Crossed-polarised light. (c) The same as (b), with plane-polarised light. Orthopyroxenite (opx) is highlighted by white broken lines.

remaining fragments were ground to a powder using an agate mortar. They eventually obtained 4 g of powder sample of the vein. The trace-element composition of the vein was determined on fused glass (prepared by mixing the rock powder with one third $LiBO_2$) by laser ablation (193 nm ArF excimer) inductively coupled plasma mass spectrometry (Agilent 7500S)

Table 1 Modal composition of the quartz-diorite vein determined by the area measured for each phase on photomicrogphs

Modal composition	Quartz diorite vein (vol.%)
Plagioclase	73·46
Quartz	16·91
Orthopyroxene	2·90
Clinopyroxene	2·69
Glass	3·67
Apatite	0·33
Rutile	0·04
Total	100·00

at the Research School of Earth Sciences, Australian National University, Canberra, Australia. Analytical data are listed in Table 4.

4. Geochemistry

4.1. Olivine and spinel

Peridotites with veins tend to show wider ranges of Fo contents of olivine (=89·3–91·4) and Cr# [=Cr/(Cr+Al) atomic ratio] of spinel (=0·11–0·33) than vein-free peridotites (Fig. 4). However, there is no difference in TiO_2 content and the $Fe^{3+}/(Fe^{3+}+Al^{3+}+Cr^{3+})$ ratio of spinel. The peridotite with the thickest, i.e. quartz diorite, vein has a low Fo content of olivine (=89·9–89·3) and Cr# of spinel (=0·114–0·156), and differs markedly from the peridotites with thinner veins in their olivine-spinel chemical variations (Fig. 4). The Cr# of spinel in the orthopyroxene wall of quartz diorite vein is slightly high (=0·13–0·17) relative to that in the host peridotite (=0·12–0·16), while the $Fe^{3+}/(Al+Cr+Fe^{3+})$ ratio and TiO_2 wt.% are the same.

4.2. Clinopyroxene

The clinopyroxene core in the peridotite with the veins has a relatively homogeneous composition En_{47} Fs_4 Wo_{48} with Al_2O_3 and Cr_2O_3 contents of 5·52–6·32 wt.% and 1·07–1·23 wt.%, respectively. However, it commonly shows chemical zoning; for example, Al_2O_3 wt.% decreases from core (=6·32 wt.%) to rim (=4·65 wt.%), while Cr_2O_3 wt.% slightly increases from core (=0·74 wt.%) to rim (=0·91 wt.%), with a constant Mg# [=100*Mg/(Mg+Fe) atomic ratio] of 92·4–92·7. The TiO_2 content is 0·30–0·74 wt.%. The Na_2O content of clinopyroxene is clearly lower (<0·5 wt.%) in the peridotite with the quartz diorite vein than in that with thin veins (0·5–1wt.%).

The clinopyroxene in the quartz diorite vein is apparently different in composition from the primary clinopyroxenes in peridotites (Fig. 5). It shows an extremely low content of TiO_2 (<0·2 wt.%), Cr_2O_3, Al_2O_3 (=0·71–1·0wt.%) and Na_2O (=0·14–0·20wt.%), and a relatively high CaO content (=22·7–24·1 wt.%) compared with the peridotite clinopyroxenes (Fig. 5; Table 4). A coarse clinopyroxene grain within the quartz diorite vein shows chemical zoning from the interior to the exterior of the vein (Fig. 5). Both Mg# and Al_2O_3 increase from the inside rim (89·9 and 0·74 wt.%, respectively) toward the outside rim (93·2 and 1·04 wt.%, respectively) (Fig. 5).

4.3. Orthopyroxene

Orthopyroxene associated with the veins is apparently lower in Al_2O_3, Cr_2O_3 and CaO contents than ordinary peridotite

Table 2 Selected electron microprobe analyses of constituent minerals and glass in Tallante xenoliths. A glass composition of slab-derived melt from a peridotite xenolith from Batan Island, the Philippines (Schiano *et al.* 1995), is also listed for comparison: (Mg#) Mg/(Mg+Fe) atomic ratio except for spinel [Mg# for spinel is $Mg/(Mg+Fe^{2+})$ atomic ratio, for which Fe^{2+} was calculated assuming spinel stoichiometry]; (Cr#) Cr/(Cr +Al) atomic ratio for spinel; and (An) Ca/(Ca+Na) atomic ratio for plagioclase.

	Olivine						
Rock:	Spinel peridotite	Plagioclase peridotite	Sp peridotite with vein	Pl peridotite with vein	Peridotite with quartz diorite vein	Composite xenolith	Group II
Sample no.	(990430–8)	(990430–7)	(990806–5)	(TAL-17)	(9901015–8)	(990806–21)	(990414–20)
SiO_2	40·84	40·54	41·18	40·99	41·23	40·15	38·83
TiO_2	0·01	0·01	0·01	0·02	0·00	0·03	0·00
Al_2O_3	0·00	0·01	0·00	0·00	0·04	0·00	0·01
Cr_2O_3	0·00	0·00	0·01	0·00	0·00	0·00	0·01
FeO	8·86	9·53	10·27	10·15	10·33	14·27	22·68
MnO	0·09	0·11	0·16	0·14	0·10	0·20	0·36
MgO	49·65	49·38	48·92	48·27	48·29	45·52	38·02
CaO	0·03	0·04	0·04	0·03	0·02	0·06	0·07
Na_2O	0·00	0·01	0·00	0·00	0·01	0·02	0·02
K_2O	0·02	0·01	0·00	0·00	0·00	0·00	0·02
NiO	0·37	0·36	0·38	0·38	0·46	0·31	0·11
Total	99·86	100·00	100·97	99·98	100·49	100·56	100·13
Mg#	0·910	0·90	0·894	0·894	0·892	0·850	0·749
An (Cr#)							

	Clinopyroxene								
Rock:	Spinel peridotite	Plagioclase peridotite	Spinel peridotite with vein	Plagioclase peridotite with vein (core)	Plagioclase peridotite with vein (rim)	Peridotite with quartz diorite vein	Quartz diorite vein) Inside	Quartz diorite vein) Outside	Composite xenolith	Group II
Sample no.	(990430–8)	(990430–7)	(990806–5)	(990414–15)	(990414–16)	(991015–8)	(991015–9)	(991015–10)	(990806–21)	(990414–20)
SiO_2	51·59	51·56	53·61	52·70	51·42	51·77	54·92	55·58	51·56	52·92
TiO_2	0·45	0·63	0·30	0·50	0·48	0·61	0·16	0·10	0·77	0·49
Al_2O_3	6·16	5·65	4·10	4·65	6·35	4·46	0·74	1·04	5·23	3·27
Cr_2O_3	1·35	0·95	1·08	1·05	1·04	0·63	0·01	0·05	0·96	0·33
FeO	1·99	2·19	2·19	2·22	2·22	2·10	3·51	2·40	2·38	6·18
MnO	0·07	0·04	0·09	0·10	0·06	0·15	0·12	0·03	0·13	0·20
MgO	15·51	15·40	16·33	15·78	15·09	16·23	17·44	18·45	15·29	14·38
CaO	21·28	22·16	22·57	23·29	22·82	22·63	23·86	24·08	22·31	20·89
Na_2O	1·01	0·79	0·66	0·71	0·82	0·46	0·17	0·18	0·77	1·00
K_2O	0·01	0·02	0·01	0·01	0·03	0·03	0·01	0·02	0·00	0·03
NiO										
Total	99·42	99·39	100·94	101·01	100·33	99·07	100·94	101·93	99·40	98·76
Mg#	0·932	0·926	0·930	0·926	0·923	0·933	0·898	0·932	0·919	0·805
An (Cr#)										

Table 2 *Continued*.

Rock:	Prthopyroxene											Spinel					
	Spinel peridotite	Plagioclase peridotite	Spinel peridotite with vein	Spinel peridotite with vein (vein)	Plagioclase peridotite with vein	Plagiolcase peridotite with vein (vein)	Peridotite with quartz diorite vein	Quartz diorite vein Core	Quartz diorite vein Rim	Composite xenolith	Group II	Spinel peridotite	Plagioclase peridotite	Spinel peridotite with vein	Plagioclase peridotite with vein	Peridotite with quartz diorite vein	Composite xenolith
Sample no.	(990430–8)	(990430–7)	(990806–3)	(990806–3)	(TAL17)	990806–26)	(9901015–8)	(991015–8)	(991015–8)	(990806–21)	(990806–16)	(990430–8)	(990430–7)	(990806–5)	(TAL17)	(quartz)	(990806–20)
SiO_2	55·94	55·82	56·26	57·78	56·63	56·66	55·56	57·79	57·41	55·14	55·23	0·01	0·02	0·00	0·03	0·02	0·04
TiO_2	0·08	0·11	0·15	0·07	0·23	0·07	0·18	0·10	0·10	0·12	0·06	0·05	0·08	0·11	0·10	0·07	0·09
Al_2O_3	3·31	3·48	3·92	2·09	2·93	2·62	4·69	1·43	1·93	3·46	1·98	49·30	50·88	46·68	40·50	57·21	52·13
Cr_2O_3	0·41	0·50	0·68	0·26	0·60	0·31	0·49	0·04	0·00	0·38	0·00	18·75	16·98	22·95	27·98	11·47	13·78
FeO	5·66	5·97	5·72	5·70	6·47	6·09	6·34	6·93	6·66	7·71	13·87	9·99	10·78	10·33	13·48	10·28	14·26
MnO	0·11	0·09	0·11	0·13	0·15	0·13	0·13	0·10	0·13	0·19	0·35	0·11	0·11	0·12	0·19	0·14	0·18
MgO	33·93	33·06	33·02	34·48	33·16	33·90	32·08	34·66	34·21	32·34	27·94	20·05	20·13	19·88	17·23	21·05	18·69
CaO	0·59	0·70	0·94	0·70	1·05	0·71	1·02	0·28	0·35	0·57	0·91	0·01	0·01	0·00	0·01	0·00	0·01
Na_2O	0·04	0·06	0·03	0·02	0·03	0·03	0·01	0·00	0·00	0·01	0·06	0·00	0·00	0·01	0·00	0·02	0·00
K_2O	0·01	0·01	0·01	0·00	0·02	0·02	0·01	0·04	0·00	0·01	0·02	0·01	0·01	0·03	0·01	0·00	0·00
NiO												0·26	0·31	0·27	0·20	0·39	0·27
Total	100·08	99·80	100·84	101·23	101·27	100·54	100·51	101·37	100·78	99·93	99·34	98·53	99·31	100·38	99·73	100·65	99·44
Mg#	0·914	0·908	0·911	0·915	0·901	0·908	0·900	0·899	0·901	0·882	0·782	0·805	0·799	0·795	0·719	0·807	0·740
An (Cr#)												0·203	0·182	0·248	0·316	0·118	0·150

Rock:	Plagioclase					Amphibole			Phlogopite			Glass		
	Plagioclase peridotite	Plagioclase peridotite with vein	Plagioclase peridotite with vein (vein)	Quartz diorite vein Core	Quartz diorite vein Rim	Plagioclase-rich vein	Composite xenolith	Group II	Plagioclase-rich vein	Composite xenolith	Group II	Quartz diorite vein Inside	Quartz diorite vein Outside	Slab-derived melt (Schiano *et al.* 1995)
Sample no.	(990430–7)	(990414–1)	(990414–1)	(991015–8)	(991015–8)	(990806–5)	(990414–19)	(990414–20)	(990414–15)	(990806–21)	(990414–20)	(991015–8)	(991015–8)	(IVI)
SiO_2	55·43	54·40	53·55	54·87	59·22	44·01	42·01	42·00	38·99	38·07	38·52	72·37	67·81	59·69
TiO_2	0·03	0·01	0·01	0·00	0·01	1·57	4·36	4·00	1·00	3·72	4·44	0·36	0·84	0·12
Al_2O_3	29·14	29·19	29·12	28·94	25·86	14·04	13·31	12·77	16·69	16·60	14·65	13·37	12·85	18·47
Cr_2O_3	0·00	0·04	0·00	0·00	0·00	1·76	1·26	0·02	0·28	0·34	0·06	0·04	0·03	
FeO	0·11	0·07	0·17	0·08	0·00	3·10	6·12	9·18	3·22	5·22	9·21	1·22	1·20	2·71
MnO	0·00	0·00	0·07	0·05	0·00	0·05	0·08	0·11	0·00	0·03	0·05	0·05	0·02	0·11
MgO	0·03	0·02	0·11	0·03	0·05	18·17	15·46	14·04	22·87	20·89	18·04	2·07	1·81	1·11
CaO	11·83	11·44	12·24	11·60	9·12	11·94	10·90	10·91	0·01	0·02	0·03	5·13	5·30	4·10
Na_2O	4·80	4·64	5·35	4·80	5·55	2·98	3·10	2·89	0·82	1·01	0·97	2·16	0·95	5·07
K_2O	0·04	0·30	0·21	0·36	0·43	0·02	0·86	1·19	9·93	8·39	8·38	0·96	1·11	2·47
NiO														
Total	101·40	100·11	100·83	100·73	100·25	97·64	97·46	97·11	93·81	94·29	94·36	97·73	91·92	93·85
Mg#						0·912	0·818	0·731	0·926	0·877	0·777	0·751	0·729	0·422
An (Cr#)	57·6	57·6	55·8	57·1	47·6									

Table 3 Rare-earth element (REE), Ti, Sr, Zr and Y abundances (p.p.m.) of representative clinopyroxenes in Tallante peridotite xenoliths determined by SIMS.

Rock:	Peridotite with the vein			Quartz diorite vein	Spinel peridotite	Plagioclase peridotite	Group II
Sample:	99041432	02	99080614	(991015–8)	(99080607)	(99080624)	Group II-26)
La	5·95	9·84	9·49	35·93	1·53	1·89	13·60
Ce	9·63	17·84	27·65	152·45	4·63	8·37	36·59
Pr	1·19	1·62	3·38	24·24	0·74	1·51	5·13
Nd	5·22	6·52	12·27	99·13	3·78	8·66	21·23
Sm	3·35	3·22	3·45	32·84	1·44	3·57	5·97
Eu	1·18	1·12	0·92	0·45	0·62	0·91	2·33
Tb	0·98	0·56	0·84	7·13	0·36	0·83	1·17
Gd	4·42	2·92	4·32	36·41	1·76	4·31	6·17
Dy	8·53	4·53	6·23	40·79	3·15	6·46	6·23
Ho	1·71	1·12	1·18	5·38	0·56	1·17	0·96
Er	4·24	2·98	3·38	10·76	1·78	3·50	2·91
Tm	0·58	0·45	0·46	1·18	0·25	0·51	0·38
Yb	3·19	2·57	2·11	7·65	1·14	2·19	1·60
Lu	0·53	0·37	0·35	1·54	0·18	0·37	0·27
Ti	2827	2935	4498	1802	2194	4220	5056
Sr	46·9	41·9	6·19	4·84	54·5	5·79	9·77
Zr	34·9	43·7	57·1	35·5	22·9	53·5	48·8
Y	40·1	24·4	29·1	198·3	15·2	27·1	36·4
La/Yb	1·869	3·831	4·713	4·698	1·350	0·866	8·484
Ti/Zr	80·934	67·219	78·884	50·747	95·865	78·904	103·712
Sr/Y	1·167	1·720	0·209	0·124	3·591	0·215	0·268

orthopyroxenes (Fig. 6). Orthopyroxene around the quartz diorite vein has extremely low values of Al_2O_3 (0·84–2·48 wt.%), Cr_2O_3 (0–0·26 wt.%), CaO (0·28–0·64 wt.%) and Mg# (85·7–90·2). In the thickest wall orthopyroxenite, one can recognise a chemical gradient from the contact with the interior quartz diorite (Mg#=86·3) to the outer contact (Mg#=90·3) with host peridotite, where the Mg# is the same as that of host peridotite olivine. Individual orthopyroxene grains in wall orthopyroxenite often show chemical zoning; for example; the Al_2O_3 content increases from the core (=2·48 wt.%) to the rim (=0·84 wt.%).

4.4. Plagioclase

Plagioclase in the thin, quartz-free veins is relatively constant in composition, showing An_{49-57} [An=100*Ca/(Ca+Na) atomic ratio] and 0·08–0·22 wt.% of K_2O. Plagioclase in the quartz diorite vein has a relatively wide compositional range, from An_{37} to An_{58}. It often shows remarkable zoning when it is in contact with glass (Fig. 7); namely, An content decreases from the core to the mantle, but increases again towards the rim (=contact with glass).

4.5. Hydrous minerals (amphibole and phlogopite)

Amphibole is pargasite in the veins, and is kaersutite in composite xenoliths and Group II. The values of Mg# for pargasite and phlogopite in the veins are 89·0–91·7 and 87·0–91·0, respectively. The hydrous minerals in the veins are higher in Mg# than those of Group II (77·4–84·7 and 66·2–78·9 wt.%, respectively) and related composite xenoliths (71·6–88·8 for pargasite and 66·2–91·3 for phlogopite). Hydrous minerals clearly have higher contents of TiO_2 in Group II and related composite xenoliths than in the veins (=0·81–1·96 wt.%).

4.6. Glass

It is noteworthy that the glass (Table 2) is high in MgO, SiO_2 and Mg#, as compared with slab-derived melts in mantle xenoliths (Schiano *et al.* 1995; Kilian & Stern 2002), typical adakites (Martin 1999), TTG and Phanerozoic Na-rich granitoids (Smithies 2000), and plots off the 'mantle AFC' trend of slab melts on the MgO- SiO_2 diagram (e.g. Smithies 2000). The glass also has low Al_2O_3 and Na_2O contents relative to slab-derived melts (Schiano *et al.* 1995; Kilian & Stern 2002).

4.7. Bulk composition of the quartz diorite vein

The present authors calculated the bulk chemical composition of the quartz diorite vein based on the modal composition (Table 1) and the averaged composition of each phase (Table 2). The quartz diorite vein is rich in Al_2O_3 and CaO, and depleted in Na_2O and K_2O relative to the average adakite and high Al-TTG (Table 4).

The trace element composition of the quartz diorite vein is listed in Table 4. The La/Yb ratio of the quartz diorite vein is about 37·9, and is higher than that of the average adakite, and lower than those of high-Al TTG (Drummond *et al.* 1996) and Phanerozoic Na-rich granitoids (Smithies 2000). The Sr/Y ratio is apparently lower in the vein than in adakite, TTG and Phanerozoic Na-rich granitoids (Table 4).

The quartz diorite vein is characterised by a strongly fractionated REE pattern with a negative Eu anomaly (Fig. 8a). This pattern is similar to those of adakites and TTG, except for the negative Eu spike of the former (Fig. 8a). The quartz diorite vein normalised by primitive mantle abundances shows weak negative anomalies of high-field-strength (HFS) elements (Ti, Nb, Zr and Hf), Eu, Sr, Rb and Ba, and enrichment of Th and U (Fig. 8b). It is similar in incompatible element distribution pattern to adakite and TTG, except for Sr: the quartz diorite exhibits a prominent negative Sr anomaly while adakite shows a positive Sr anomaly, and TTG does not show any Sr anomaly (Fig. 8b).

Table 4 The major-element bulk composition calculated by modal and chemical compositions of minerals, and the trace-element composition determined by ICP-MS analysis of a quartz diorite vein, compared with the average adakite and trondhjemite, tonalite and granodiorite (TTG) compositions (Drummond *et al.* 1996). Note that the Sr/Y ratio is lower in the quartz diorite vein than in adakite and TTG.

	Quartz diorite vein (present study)	Adakite (Drummond *et al.* 1996)	TTG (Drummond *et al.* 1996)
Major elements (wt.%)			
SiO_2	62·89	63·89	70·2
TiO_2	0·16	0·61	0·33
Al_2O_3	21·80	17·4	15·74
Cr_2O_3	0·11		
FeO	0·42	4·21	2·56
MnO	0·14	0·18	0·14
MgO	1·52	2·47	1·19
CaO	9·58	5·23	3·17
Na_2O	3·59	4·4	4·87
K_2O	0·30	1·52	1·88
P_2O_5	0·14	0·19	0·12
Total	100·36	100	100
Mg#	86·6	51·1	43·1
Trace elements (p.p.m.)			
Ti	778		
V	29	72	38
Cr	781	54	36
Ga	5		
Rb	2·94	30	50
Sr	66·6	869	495
Y	6·5	9·5	6·8
Zr	42	117	149
Nb	14	8·3	5·4
Cs	0·7	1·19	1·73
Ba	39	485	746
La	18·6	17·5	29·8
Ce	28·2	34·6	51·6
Nd	9·5	20·1	19·9
Sm	1·89	3·1	2·7
Eu	0·44	0·97	0·91
Gd	1·79	2·25	2·14
Tb		0·37	0·25
Dy	1·44	1·43	1·16
Er	0·56	0·76	0·59
Yb	0·5	0·91	0·46
Lu	0·162	0·15	0·19
Hf	1·13	3·5	3·4
Ta	1·9	0·53	0·55
Pb	0·181		
Th	31	3·52	5·98
U	5	0·99	1·13
Sr/Y	10·2	91·5	72·8
La/Yb	37·2	19·2	64·8

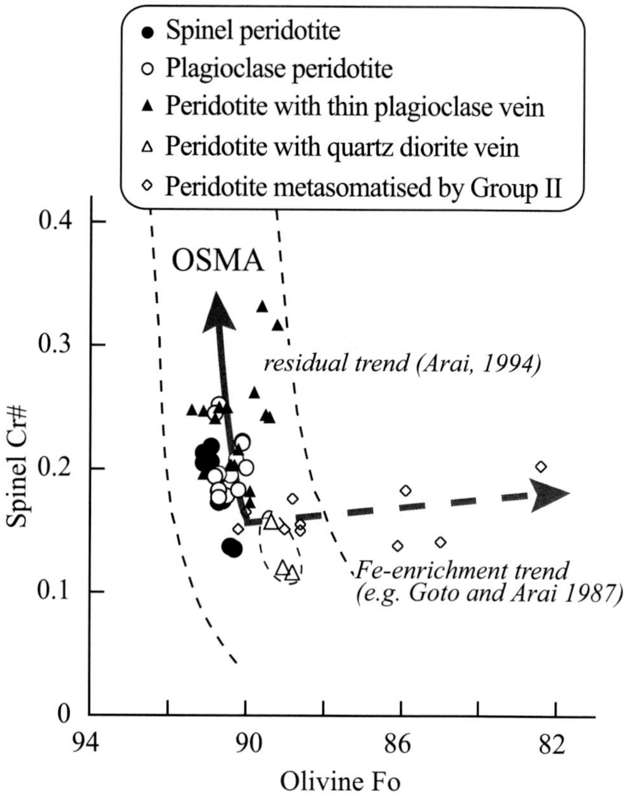

Figure 4 Relationships of the Fo content of olivine and the Cr# [5Cr/(Cr+Al)] atomic ratio of spinel in Tallante peridotite xenoliths. The olivine-spinel mantle array (OSMA), is a spinel peridotite restite trend in terms of the Fo-Cr# relation (Arai 1994). The Fe-enrichment trend shows the metasomatism by Fe-rich melt (Goto & Arai 1987).

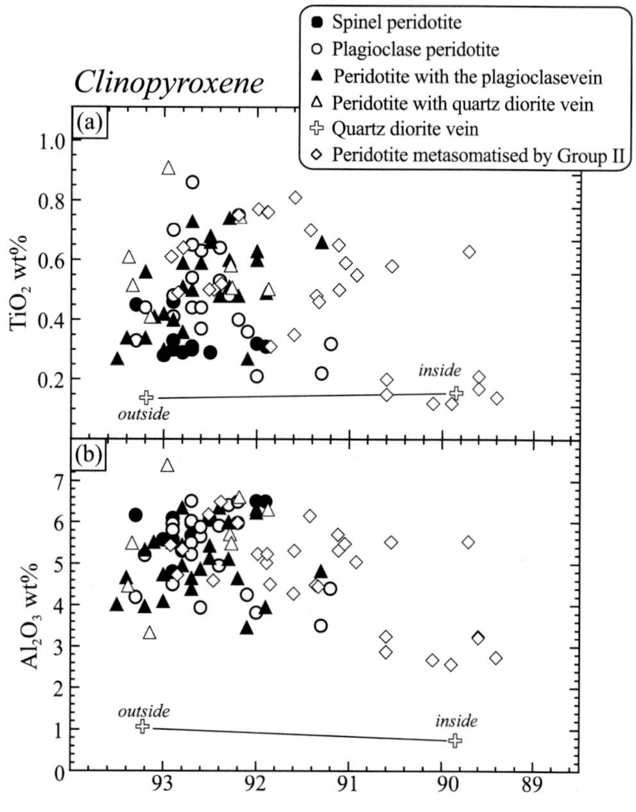

Figure 5 Relationships between TiO_2 and Al_2O_3 content and Mg# [=100*Mg/(Mg+Fe) atomic ratio] in clinopyroxenes in the Tallante peridotite xenoliths, SE Spain. Note that clinopyroxene in quartz diorite vein is distinctively low in Ti and shows a clear chemical zoning in Mg#.

4.8. Trace element compositions of clinopyroxenes

Selected clinopyroxenes from spinel and plagioclase peridotites, peridotites with the veins, quartz diorite vein, composite xenolith, and Group II were analysed *in situ* for REEs and Ti, Sr, Zr and Y by SIMS (Table 3; Fig. 9).

The Sr, Zr and Ti contents of clinopyroxene in the peridotites with the veins show compositional ranges similar to those

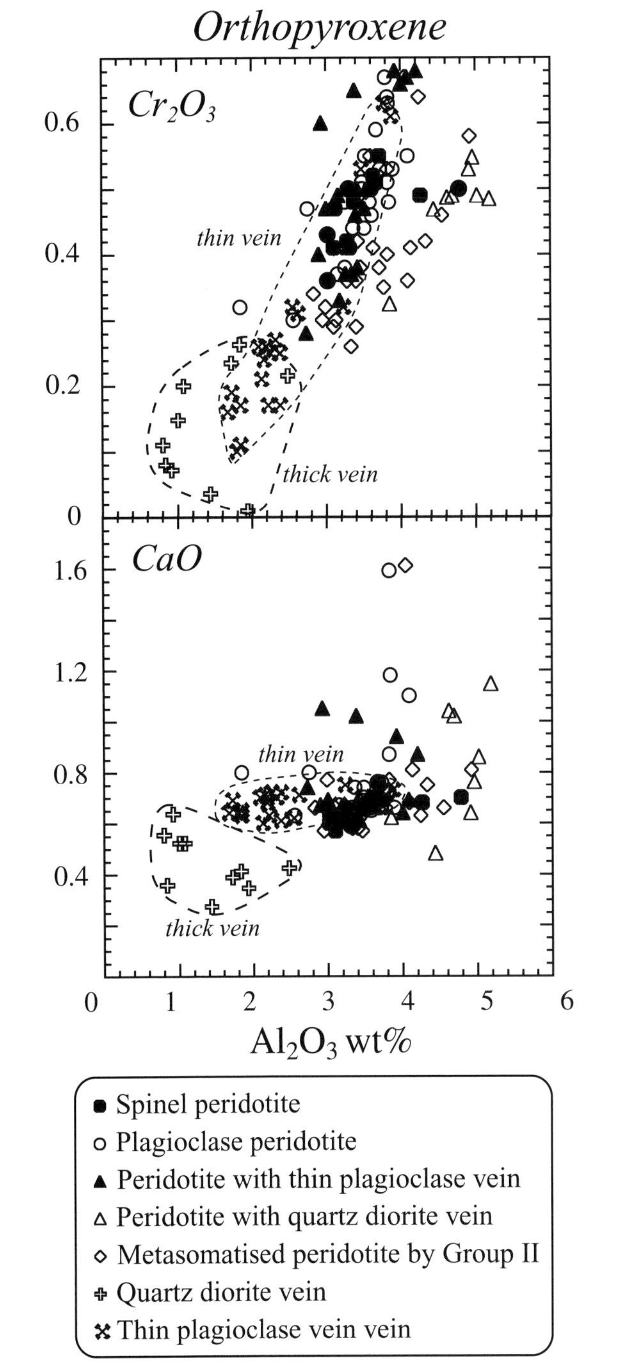

Figure 6 Relationships between the CaO, Cr_2O_3 and Al_2O_3 content of orthopyroxenes in Tallante peridotite xenoliths, SE Spain. Note that orthopyroxene of the thick vein (quartz diorite) is very low in these components and becomes enriched in them to approach mantle orthopyroxene in chemistry with a decrease in the size of the vein.

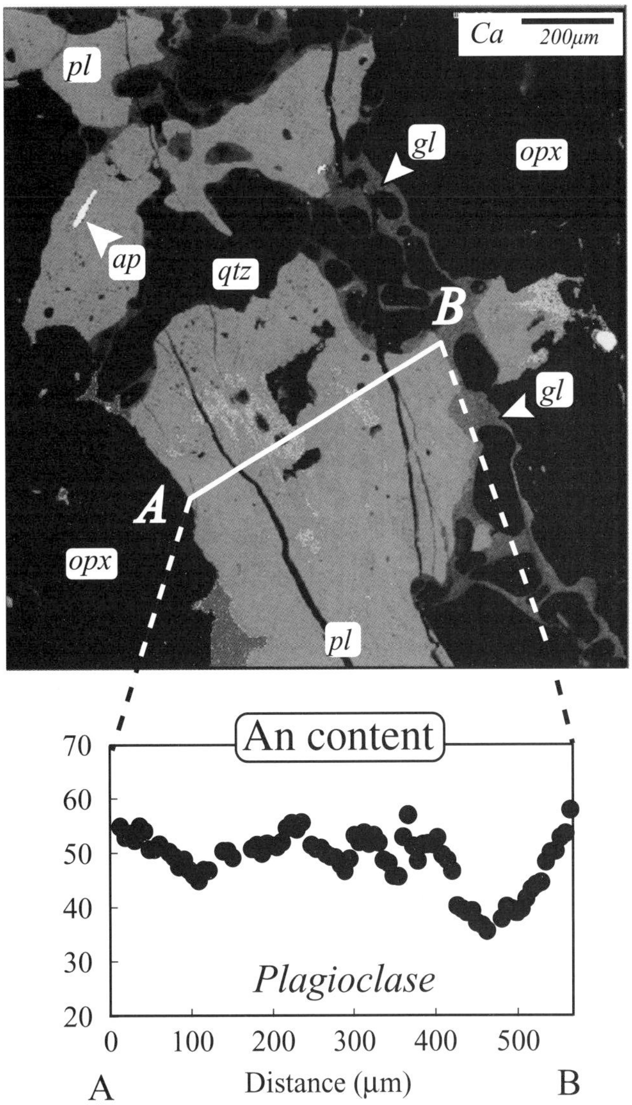

Figure 7 Compositional profile for a plagioclase grain in a quartz diorite vein (Ca compositional map by microprobe). Note that An content decreases from the core to the mantle, but increases again toward the rim (=contact with glass): (pl) plagioclase; (ap) apatite; (gl) glass; (opx) orthopyroxene; and (qtz) quartz.

in vein-free spinel and plagioclase peridotites. The La/Yb ratio of clinopyroxene is relatively higher (=1·9–4·7) in the peridotite with the vein than in spinel peridotite (=1·3–2·3) and plagioclase peridotite (=0·8–1·9). The REE contents of clinopyroxene are generally similar and the chondrite-normalised patterns are flat (Fig. 9). It is noticeable that clinopyroxene is remarkably enriched in LREE (especially, La and Ce) in peridotites with the veins (Fig. 9).

Clinopyroxene in the quartz diorite vein is apparently different from peridotite clinopyroxene in chemistry (Fig. 9). It is characterised by a high content of Y (=198 p.p.m.) and low contents of Ti (=1801 p.p.m.) and Sr (=4·8 p.p.m.) (Table 3). Thus, the Sr/Y ratio is extremely low (Sr/Y=0·024), although the La/Yb ratio is relatively high (=4·7) (Table 3). It has remarkably high REE contents except for Eu (Fig. 9; Table 3). The REE distribution pattern for the quartz diorite clinopyroxene demonstrates a prominent Eu negative spike (Fig. 9).

5. Discussion

The presence of modal quartz in the vein within mantle peridotite xenoliths from Tallante is direct evidence for the activity of silica-oversaturated melt within the upper mantle. The orthopyroxenite intervening between the quartz diorite and host peridotite indicates that it is a reaction product between a slilica-oversaturated melt and host olivine. This is consistent with orthopyroxene chemistry (Fig. 6): orthopyroxene as a reaction product of the vein approaches peridotite orthopyroxene composition with decreases in the size of vein (Fig. 6). When the veins are thin, the melt/wall rock ratio is low and the silica-oversaturated character of the melt is not preserved. The preservation of quartz only in the centre of the vein indicates that the reaction which consumes the SiO_2 component did not reach completion in a large melt conduit. It may also mean that the orthopyroxene, which can be in equilibrium with both olivine and silica-oversaturated melt, helps, to some extent, keep the silica-oversaturated within the

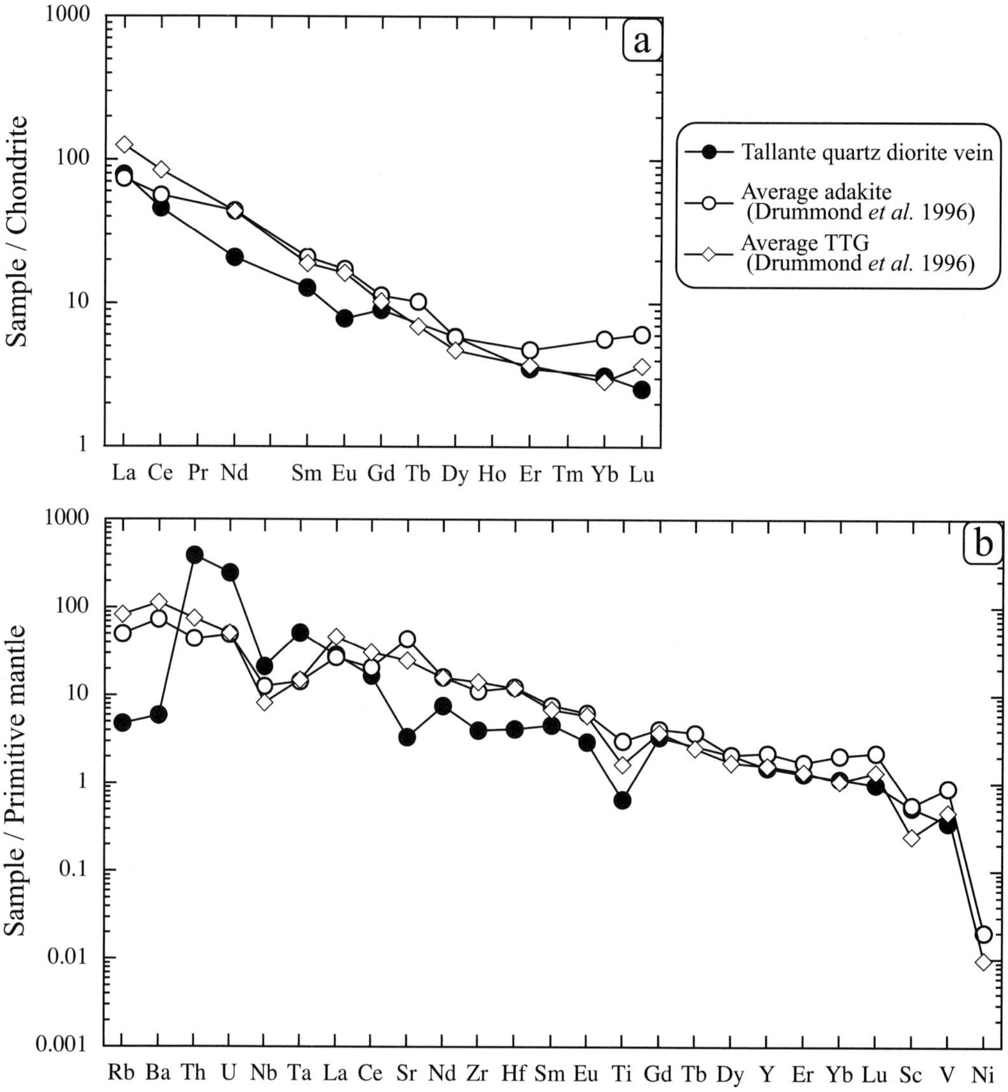

Figure 8 (a) Chondrite (McDonough & Sun 1995) normalised rare-earth element patterns and (b) primitive mantle (McDonough & Sun 1995) normalised trace element distribution patterns of the Tallante quartz diorite vein, adakite and trondhjemite, tonalite and granodiorite (TTG). The quartz diorite vein is similar to adakite and TTG (Drummond *et al.* 1996), except for its negative anomalies of Rb, Ba, Eu and Sr, and positive anomalies of Th and U.

olivine-rich environment. These observations and interpretations suggest that the silica-oversaturated melt can erupt on the Earth's surface only if the melt/wall rock ratio is high within the mantle wedge and/or the orthopyroxene armour is thick enough to work effectively. Kepezhinskas *et al.* (1995) also reported a fine-grained adakitic vein composed of quartz, feldspar, amphibole, Fe-Ti oxide and glass in a mantle peridotite xenolith from the north Kamchatka arc. However, the Kamchatka vein is very fine-grained and has no orthopyroxenite lining, indicating that it has been quenched and provides no information on the behaviour of silica-oversaturated melt within the mantle. The formation of orthopyroxene by the reaction between Si-rich melts and mantle olivine may contribute to Si-enrichment of the mantle wedge (e.g. Kelemen *et al.* 1998).

There are two possibilities for the origin of silica-rich melt within the mantle peridotite. One is the low degree of partial melting of mantle periodtite (e.g. Schiano & Clocchiatti 1994; Baker *et al.* 1995; Hirschmann *et al.* 1998; Schiano *et al.* 1998). However, low-degree partial melt is generally known to occur as inclusions of mantle minerals (e.g. Takahashi 1997; Schiano *et al.* 1998), and this is, therefore, not the case for the Tallante quartz diorite vein that cuts and replaces the host periodtite (Figs 2 & 3). The content of total alkalis in the quartz diorite vein is also very low, while the low-degree partial melt is high in total alkalis (Schiano & Clocchiatti 1994; Baker *et al.* 1995; Hirschmann *et al.* 1998). The other possibility is a derivation from partial melting of the oceanic slab (e.g. Schiano *et al.* 1995; Kepezhinskas *et al.* 1995; Kilian & Stern 2002). This is consistent with the petrochemical characteristics of Tallante quartz diorite vein. It is similar in REE and incompatible trace element abundances to adakite and to TTG (Fig. 8). However, it is noteworthy that the Tallante quartz diorite exhibits a negative Eu anomaly (Fig. 8), indicating precipitation of plagioclase at deeper levels. Plagioclase precipitation is plausible because the Tallante mantle peridotite was derived from the stability field of plagioclase of the shallowest upper mantle (<10 kbar; Kushiro & Yoder 1966). Subtraction of some phases including phlogopite, amphibole and rutile may be necessary because the quartz diorite is also poor in TiO_2, K_2O and Na_2O (Table 2), and is lower in Ba, Rb and Sr than adakite and TTG (Fig. 8b). The highly silicic glass in the quartz diorite vein (Table 2) has possibly formed by *in situ* partial melting of quartz diorite by heating of host alkali basalt because plagioclase in contact with glass shows a reverse zoning (Fig. 7).

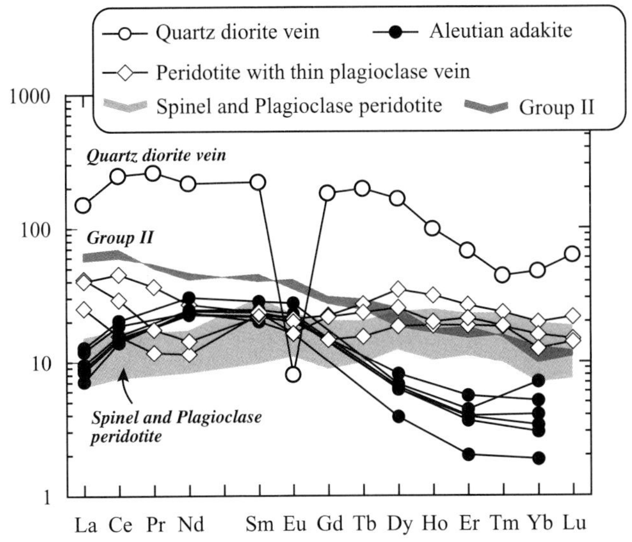

Figure 9 Chondrite-normalised rare-earth element (REE) patterns of selected clinopyrocenes from Tallante mantle xenoliths and clinopyroxene phenocrysts of an adakite from the Aleutian arc (Yogodzinski & Kelemen 1998). The chondrite values are from McDonough & Sun (1995). Note that the clinopyroxene of the quartz diorite vein is apparently higher in REE content than other clinopyroxenes from Tallante. The clinopyroxene of the quartz diorite is very similar in the shape of its REE pattern to that of adakite, except for a negative Eu anomaly. Clinopyroxene from peridotites with the veins shows clear enrichment of LREE compared with other peridotite clinopyroxenes.

The extremely high abundances of REEs of clinopyroxene in the quartz diorite relative to other clinopyroxenes from peridotites and Group II pyroxenites (Fig. 9) clearly indicate the exotic nature of the quartz diorite. The shape of REE pattern in the quartz-diorite clinopyroxene is strikingly similar to that in clinopyroxene phenocrysts in an adakite from the Aleutians (Yogodzinski & Kelemen 1998), except for the prominent negative Eu anomaly of the former (Fig. 9). This is consistent with precipitation of the Tallante clinopyroxene from an adakite melt highly fractionated by precipitation of large amount of plagioclase, which is poor in REEs except for Eu.

The adakitic melt of Tallante may have been derived from a detached slab that is sinking within the asthenospheric mantle beneath the Betic area (Blanco & Spakman 1993; Zeck 1996). Heating of the sinking slab by high-temperature asthenosphere thus produced the adakitic melt.

6. Conclusions

White veins have been commonly found in peridotite xenoliths from Tallante, SE Spain. The thick veins (>1 mm, up to 8 mm) have quartz diorite lithology, and the thinner examples consist only of plagioclase. Orthopyroxene intervenes between the vein (plagioclase ± quartz) and host olivine, indicating an interaction between a silica-oversaturated melt and mantle olivine. The silica-oversaturated character of slab-derived melts can be preserved if the melt/wall olivine ratio is sufficiently high and the orthopyroxene armour works effectively.

The quartz diorite vein from Tallante has adakitic geochemical signatures. Its clinopyroxene is remarkably similar to phenocryst clinopyroxene in adakite from the Aleutians in REE distribution pattern, except for one-order higher abundances of REEs and the strong negative spike of Sr of the former. This feature was obtained by coprecipitation of clinopyroxene and large amounts of plagioclase from an adakitic melt within a confined conduit within the upper mantle, and Tallante quartz diorite is a frozen melt of fractionated adakite affinity. The adakitic melt of Tallante was produced by partial melting of a detached and sinking slab beneath the Betic area in Tertiary times.

7. Acknowledgements

We are indebted to J. Uesugi, M. Kida and C. Tanaka for their assistance in the field. TM is greatly indebted to Steve Eggins, Mike Shelley and Charlotte Allen, Australian National University (ANU), for their guidance in LA-ICP-MS analysis at the ANU. We thank A. Ishiwatari and A. Toramaru for daily discussions. Critical reviews by M. Arima and A. Castro were helpful in revising the manuscript.

This study was partly financed by the 21st Century COE project 'Environmental Monitoring and Prediction of Long- and Short-Term Dynamics of Pan-Japan Sea Area: Construction of Monitoring Network and Assessment of Human Effects', led by K. Hayakawa, Kanazawa University, Kanazawa, Japan.

8. References

Ancochea, E. & Nixon, P. H. 1987. Xenoliths in the Iberian Peninsula. *In* Nixon, P. H. (ed.) *Mantle Xenoliths*, 119–25. Chichester: John Wiley & Sons.

Arai, S. 1994. Characterization of spinel peridotites by olivine-spinel compositional relationships: review and interpretation. *Chemical Geology* **111**, 191–204.

Arai, S., Shimizu, Y. & Gervilla, F. 2003. Quartz diorite veins in a peridotite xenolith from Tallante, Spain: implications for reaction and survival of slab-derived SiO_2-oversaturated melt in the upper mantle. *Proceedings of the Japan Academy, Series B* **79**, 145–50.

Araña, V. & Vegas, R. 1974. Plate tectonics and volcanism in the Gibraltar arc. *Tectonophysics* **24**, 197–212.

Baker, M. B., Hirschmann, M. M., Ghiorso, M. S. & Stolper, E. M. 1995. Compositions of near-solidus peridotite melts from experiments and thermodynamic calculations. *Nature* **375**, 308–11.

Blanco, M. J. & Spakman, W. 1993. The P-wave velocity structure of the mantle below the Iberian Peninsula: evidence for subducted lithosphere below southern Spain. *Tectonophysics* **221**, 13–34.

Capedori, S., Venturelli, G., Salvioli-Mariani, E., Crawford, A. J. & Barbieri, M. 1989. Upper-mantle xenoliths and megacrysts in an alkali basalt from Tallante, south-eastern Spain. *European Journal of Mineralogy* **1**, 685–99.

Defant, M. J. & Drummond, M. S. 1990. Derivation of some modern arc magmas by melting of young subducted lithosphere. *Nature* **347**, 662–5.

Drummond, M. S., Defant, M. J. & Kepezhinskas, P. K. 1996. Petrogenesis of slab-derived trondhjemite-tnalite-dacite/adakite magmas. *Transactions of the Royal Society of Edinburgh: Earth Sciences* **87**, 205–15.

Dupuy, C., Dostal, J. & Boivin, P. A. 1986. Geochemistry of ultramafic xenoliths and their host alkali basalts from Tallante, southern Spain. *Mineralogical Magazine* **50**, 231–9.

Frey, F. A. & Prinz, M. 1978. Ultramafic inclusions from San Carlos, Arizona: petrologic and geochemical data bearing on their petrogenesis. *Earth and Planetary Science Letters* **38**, 129–76.

Goto, K. & Arai, S. 1987. Petrology of peridotite xenoliths in lamprophyre from Shingu, southwestern Japan: implications for origin of Fe-rich mantle peridotites. *Mineralogy and Petrology* **37**, 137–55.

Hirschmann, M. M., Barker, M. B. & Stolper, E. M. 1998. The effect of alkalis on the silica content of mantle-derived melts. *Geochimica et Cosmochimica Acta* **62**, 883–902.

Irving, A. J. 1980. Petrology and geochemistry of composite ultramafic xenoliths in alkali basalts and implications for magmatic processes within the mantle. *American Journal of Science* **280–A**, 389–426.

Kay, S. M., Ramos, V. A. & Marques, M. 1993. Evidence in Cerro Pampa volcanic rocks for slab-melting prior to ridge collision in southern South America. *Journal of Geology* **101**, 703–14.

Kelemen, P. B. 1995. Genesis of high Mg# andesites and the continental crust. *Contributions to Mineralogy and Petrology* **120**, 1–9.

Kelemen, P. B., Hart, S. R. & Bernstein, S. 1998. Silica enrichment in the continental upper mantle via melt/rock reaction. *Earth and Planetary Science Letters* **164**, 387–406.

Kepezhinskas, P. E., Defant, M. J. & Drummond, M. S. 1995. Na metasomatism in the island-arc mantle by slab melt-mantle interaction: evidence from mantle xenoliths in the North Kamchatka. *Journal of Petrology* **36**, 1505–27.

Kilian, R. & Stern, C. R. 2002. Constraints on the interaction between slab melts and the mantle wedge from adakitic glass in peridotite xenoliths. *European Journal of Mineralogy* **14**, 25–36.

Kushiro, I. & Yoder, H. S. 1966. Anorthite-forsterite and anorthite-enstatite reactions and their bearing on the basalt-eclogite transformation. *Journal of Petrology* **7**, 337–62.

López, S. & Castro, A. 2001. Determination of the fluid-absent solidus and supersolidus phase relationships of MORB-derived amphibolites in the range 4–14 kbar. *American Mineralogist* **86**, 1396–403.

López-Ruiz, J. & Rodoríguez-Badiola, E. 1980. La región volcánica Neógena del sureste de Espãna. *Estudios Geológicos* **36**, 5–63.

Martin, H. 1999. Adakitic magmas: modern analogues of Archaean granitoids. *Lithos* **46**, 411–29.

McDonough, W. F. & Sun., S.-S. 1995. The composition of the Earth. *Chemical Geology* **120**, 223–53.

Mercier, J.-C. C. & Nicolas, A. 1975. Textures and fabrics of upper mantle peridotites as illustrated by xenoliths from basalts. *Journal of Petrology* **16**, 454–87.

Rapp, R. P., Watson, E. B. & Miller, C. F. 1991. Partial melting of amphibolite/eclogite and the origin of Archean trondhjemites and tonalites. *Precambrian Research* **51**, 1–25.

Rapp, R. P., Shimizu, N., Norman, M. D. & Applegate, G. S. 1999. Reaction between slab-derived melts and peridotite in the mantle wedge: experimental constrains at 3·8 GPa. *Chemical Geology* **160**, 335–56.

Rapp, R. P. & Watson, E. B. 1995. Dehydration melting of metabasalt at 8–32 kbar: implications for continental growth and crust-mantle recycling. *Journal of Petrology* **36**, 981–931.

Rudnick, R. L. & Fountain, D. M. 1995. Nature and composition of the continental crust: a lower crustal perspective. *Reviews of Geophysics* **33**, 267–309.

Schiano, P., Clocchiatti, R., Shimizu, N., Maury, R., Jochum, K. P. & Hofmann, A. W. 1995. Hydrous, silica-rich melts in the sub-arc mantle and their relationships with erupted arc lavas. *Nature* **377**, 595–600

Schiano, P., Buourdon, B., Clocchiatti, R., Massare, D., Varela, M. E. & Bottinga, Y. 1998. Low-degree partial melting trends recorded in upper mantle minerals. *Earth and Planetary Science Letters* **160**, 537–50.

Schiano, P. & Clocchiatti, R. 1994. Worldwide occurrence of silica-rich melts in sub-continental and sub-oceanic mantle minerals. *Nature* **368**, 621–4.

Sen, C. & Dunn, T. 1994. Experimental modal metasomatism of a spinel lherzolite and the production of amphibole-bearing peridotite. *Contributions to Mineralogy and Petrology* **119**, 422–32.

Smithies, R. H. 2000. The Archaean tonalote-trondhjemite-granodiorite (TTG) series is not an analogue of Cenozoic adakite. *Earth and Planetary Science Letters* **182**, 115–25.

Streckeisen, A. 1976. To each plutonic rock its proper name. *Earth Science Reviews* **12**, 1–33.

Takahashi, N. 1997. Incipient melting of mantle peridotites observed in the Horoman and Nikanbetsu peridotite complexes, Hokkaido, northern Japan. *Journal of Mineralogy, Petrology, and Economic Geology* **92**, 1–24.

Tatsumi, Y. 2000a. Continental crust formation by crustal delamination in subduction zones and complementary accumulation of the enriched mantle I component in the mantle. *Geochemistry Geophysics Geosystems* Paper number 2000GC000094.

Tatsumi, Y. 2000b. Slab melting: its role in continental crust formation and mantle evolution. *Geophysical Research Letters* **27**, 3941–4.

Taylor, S. R. & McLennan, S. M. 1985. *The continental crust: its composition and evolution*. Oxford: Blackwell.

Turner, S. P., Platt, J. P., George, R. M. M., Kelley, S. P., Pearson, D. G. & Nowell, G. M. 1999. Magmatism associated with orogenic collapse of the Betic-Alboran domain, SE Spain. *Journal of Petrology* **40**, 1011–36.

Venturelli, G., Mariani, E. S., Foley, S. F., Capedri, S. & Crawford, A. J. 1988. Petrogenesis and crystallization of Spanish lamproitic rocks. *Canadian Mineralogist* **26**, 67–79.

Wang, W. & Yurimoto, H. 1993. Analysis of rare earth elements in garnet by SIMS. *Annual Report of the Institute of Geoscience, the University of Tsukuba* **19**, 87–91.

Yogodzinski, G. M. & Kelemen, P. B. 1998. Slab-melting in the Aleutians: implications of an ion probe study of clinopyroxene in primitive adakite and basalt. *Earth and Planetary Science Letters* **158**, 53–65.

Yurimoto, H., Yamashita, A., Nishida, N. & Sueno, S. 1989. Quantitative SIMS analysis of GSJ rock reference samples. *Geochemical Journal* **23**, 215–36

Zeck, H. P. 1996. Betic-Rif orogeny: subduction of Mesozoic Tethys lithosphere under eastward drifting Iberia, slab detachment shortly before 22 Ma, and subsequent uplift and extensional tectonics. *Tectonophysics* **254**, 1–16.

Y. SHIMIZU, S. ARAI and T. MORISHITA, Graduate School of Natural Science and Technology, Kanazawa University, Kakuma-machi, Kanazawa 920-1192, Japan.

H. YURIMOTO, Department of Earth and Planetary Sciences, Faculty of Science, Tokyo Institute of Technology, 2-12-1 Ookayama, Meguro-ku, Tokyo 152–8550, Japan.

F. GERVILLA, Instituto Andaluz de Ciencias de la Tierra, University of Granada-C.S.I.C., Fuentenueva S/N, E-18002, Granada, Spain.

MS received 2 October 2003. Accepted for publication 7 September 2004.

Transactions of the Royal Society of Edinburgh: Earth Sciences, **95**, 277–295, 2004

Granites, dynamic magma chamber processes and pluton construction: the Aztec Wash pluton, Eldorado Mountains, Nevada, USA

Brian E. Harper, Calvin F. Miller, G. Christopher Koteas, Nicole L. Cates, Robert A. Wiebe, Daniel S. Lazzareschi and J. Warner Cribb

ABSTRACT: The mid-Miocene Aztec Wash pluton is divisible into a relatively homogeneous portion entirely comprising granites (the G zone, or GZ), and an extremely heterogeneous zone (HZ) that includes the products of the mingling, mixing and fractional crystallisation of mafic and felsic magmas. Though far less variable than the HZ, the GZ nonetheless records a dynamic history characterised by cyclic deposition of the solidifying products of the felsic portion of a recharging, open-system magma chamber.

Tilting has exposed a 5-km section through the GZ and adjacent portions of the HZ. A porphyry is interpreted as a remnant of a chilled roof zone that marks the first stage of felsic GZ intrusion. Subsequent recharging by felsic and mafic magma, reflected by repeated cycles of crystal accumulation and melt segregation in the GZ and emplacement of mafic flows in the HZ, rejuvenated and maintained the chamber. Kilometre-scale lobes of mafic HZ material were deposited as prograding tongues into the GZ during periods of increased mafic input. Thus, they are lateral equivalents of the cumulate GZ granites with which they interfinger. Conglomerate-like units comprising rounded, matrix-supported intermediate clasts in cumulate granite are located immediately above the lobes. These ‘conglomerates’ appear to represent debris flows shed from sloping upper surfaces of the lobes. Thus, the GZ can be viewed as comprising distal facies, remote from the site of mafic recharging in the HZ, and the HZ as comprising proximal facies.

Elemental chemistry suggests that the GZ cumulate granites represent a second-stage accumulation from an already evolved melt, and that coarse, more mafic, feldspar + biotite + accessory mineral ± hornblende rocks trapped between mafic sheets in the HZ are the initial cumulates. Fractionated melt accumulated roofward and laterally, and was the direct parent of the ‘evolved’ GZ cumulates. The most highly fractionated, fluid-rich melts accumulated at the roof.

KEY WORDS: Cumulate, deposition, mingling, open-system, transport.

Both direct new evidence and theoretical considerations bearing on the behaviour of felsic magma systems suggest that they can be long-lived and dynamic and complex in their mechanical behaviour (e.g. Wiebe 1974; Weinberg & Leitch 1998; Wiebe & Collins 1998; Bergantz 1999; Collins *et al.* 2000; Koyaguchi & Kaneko 2000; Reid & Coath 2000; Schmitt *et al.* 2002; Vazquez & Reid 2002). In contrast to volcanic sequences, which are series of snapshots of the portions of magma systems that successfully reached the Earth's surface, plutons are solidified, taphonomic remnants of transient magmas. In other words, they are the death assemblages of magmas and reflect physical processes that arranged and rearranged the materials that remained beneath the surface. Thus, the plutonic record is biased toward – but not limited to – the latter portions of the lifetime of a magmatic system. This record preserves invaluable documentation of facets of magma system histories that are largely destroyed in erupted products.

Much of the Aztec Wash pluton reveals spectacular evidence for multiple recharging, and repeated mingling and mixing of mafic and felsic magmas. Past efforts in studying this magma system have primarily focused on what is referred to as the heterogeneous zone (HZ), the portion of the pluton comprising rocks ranging from troctolites and quenched basalts to leucogranites that preserves clear evidence for an open-system history (Falkner *et al.* 1995; Patrick & Miller 1997; Coiner *et al.* 2003). The remainder of the pluton, about 30% of the exposed area, shows far less variability. Apart from a study of accessory mineral assemblages and textures, which demonstrated that it is not a simple, monotonous, closed-system sequence (Robinson & Miller 1999), the granite zone (GZ) has received little attention. In this paper, the present authors summarise subtle but well-preserved field, textural and geochemical evidence that documents a protracted history and diverse physical processes within the GZ. This record provides insights into: (1) the roles of upward growth by accumulation and downward propagation of a solidification front (e.g. Marsh 1995) in pluton construction; (2) mechanical instability and material transport of solidifying magmatic products; (3) recognisable depositional facies and preserved time horizons; and (4) the relationship between input magmas and preserved plutonic products, with implications for magma chambers as compositional filters.

1. Geological setting

1.1. Regional setting

The Aztec Wash pluton is exposed in the Eldorado Mountains, southernmost Nevada, USA, 80 km south of Las Vegas. The Eldorado Mountains lie within the northern Colorado River extensional corridor, a 50–100-km-wide extensional belt (e.g. Howard & John 1987; Faulds *et al.* 1990). East–west extension here occurred primarily between 16 and 13 Ma (Anderson

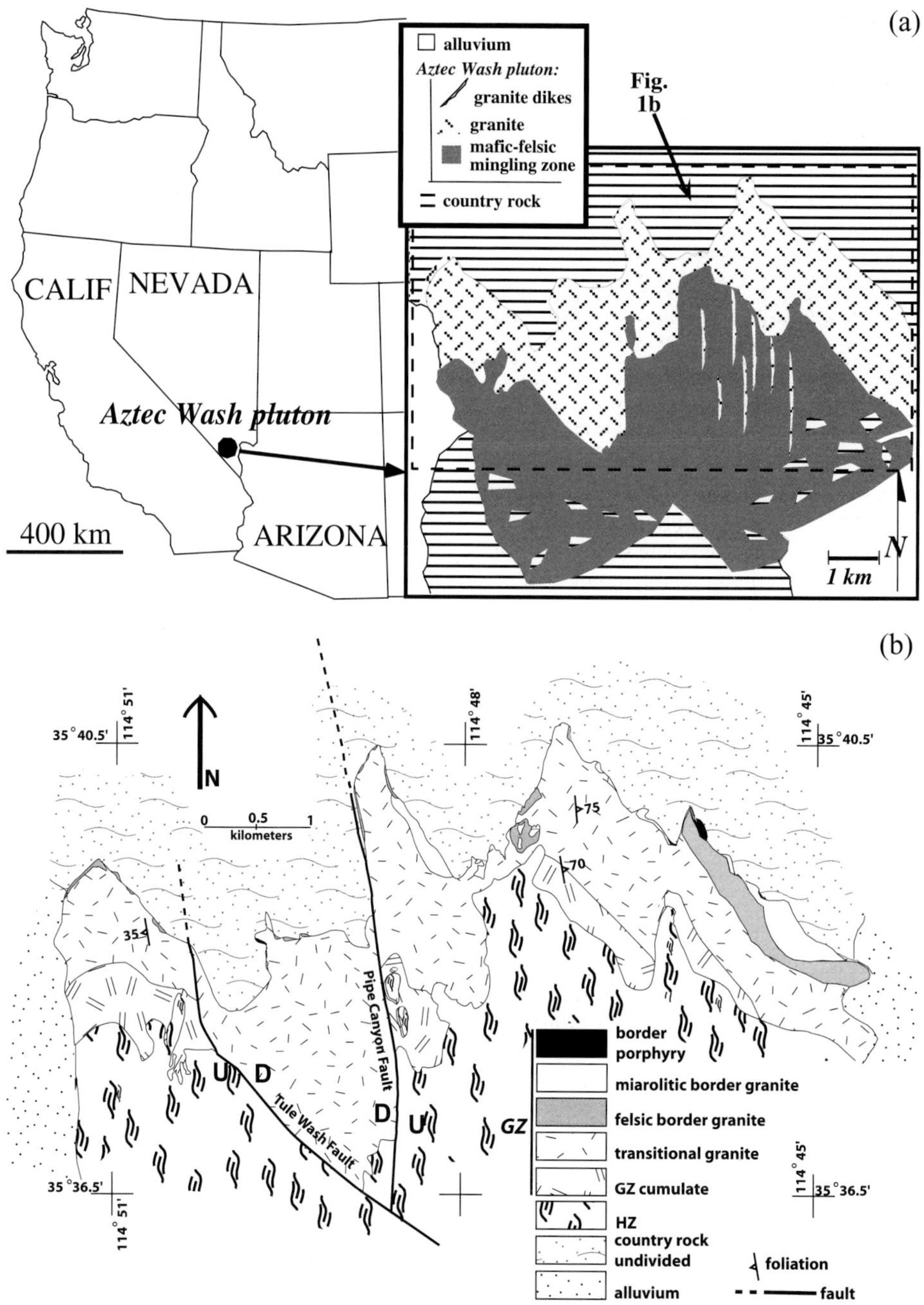

Figure 1 (a) Location and simplified geology of the Aztec Wash pluton (the area shown in Fig. 1b is indicated by a dashed box). (b) Detailed map of the granite zone.

et al. 1972; Faulds *et al.* 1992, 1995, 2001, 2002; Gans *et al.* 1992). Volcanism spanned a longer interval, beginning at ~20 Ma and continuing until ~8 Ma; initially voluminous intermediate lavas gave way to less abundant, essentially bimodal volcanism during or just after peak extension, and finally, to mafic volcanism as extension waned (Gans & Bohrson 1998; Feuerbach *et al.* 1999; Faulds *et al.* 2001). Plutons near the latitude of Aztec Wash are restricted to an interval spanning the onset of large-magnitude, regional extension (~15·5–17·5 Ma; Faulds *et al.* 1995, 2001; Metcalf *et al.* 1995; Howard *et al.* 1996; Bachl *et al.* 2001; Cates *et al.* 2003; Miller *et al.* 2003). Large crustal blocks are tilted up to 90° and expose crustal sections as much as 15 km thick, revealing both overlying volcanic sections and cross-sections through the plutons (e.g. Faulds *et al.* 1995; Bachl *et al.* 2001). Metcalf (2004) demonstrates the general coherence between the compositions of the intrusive and extrusive suites.

1.2. Aztec Wash pluton

The Aztec Wash pluton is exposed over an area of 50 km^2. It is broadly divisible into a relatively homogeneous GZ, discussed in detail in the present paper, and a larger HZ. The pluton straddles a NW-trending antiformal accommodation zone within the extensional corridor (e.g. Faulds *et al.* 2001) and is divided into three structural blocks by faults (Fig. 1). Magmatic structures within the pluton (see below) and paleomagnetic data indicate that the western block, on the WSW side of the Tule Wash fault, is tilted gently WNW; the eastern block, east of Pipe Canyon fault, is tilted steeply ENE (Patrick & Miller 1997; Coiner 2003; Faulds *et al.* 1992; J. E. Faulds, pers. comm.). Tilt markers are absent within the central, down-dropped block.

Rocks of the HZ display extraordinary diversity in composition and texture, on microscopic to map-scale (Falkner *et al.* 1995; Patrick & Miller 1997; Coiner *et al.* 2003). Rock types include troctolite, hornblende gabbro, quenched trachybasalt and trachybasaltic andesite, diverse fine- to medium-grained granites, and a host of texturally variable rocks of intermediate composition. Silica concentrations range from 42 to 78 wt.%. The dominant features within the HZ are alternating mafic and felsic sheets (Fig. 2a). Mafic sheets range in thickness from ~1 to 50 m and felsic sheets from <1 cm to ~30 m. Felsic sheets are mostly quartz monzodiorites and quartz monzonites characterised by abundant coarse, euhedral feldspars; these rocks are interpreted as cumulates deposited from granitic magma (Patrick & Miller 1997; see discussions in following sections). Mafic sheets are almost invariably asymmetric: they typically have chilled, crenulate margins against the felsic sheet on one side, indicating that cooler felsic melt was present in the adjacent sheet when the mafic sheet was emplaced. The mafic sheets coarsen toward their interior, then gradually become more felsic and lack chilling and crenulate margins against the felsic sheet on the opposite side. Common flame structures on the chilled margin side point toward the mafic sheet, and load casts are convex in the opposite direction, indicating that the chilled side is the base of the mafic sheet. These structures apparently formed as higher density mafic magma sank into the granitic cumulate mush. Similarly, common felsic pipes – tubes of granitic material that penetrate the mafic sheets – are interpreted to have formed as escape structures through which felsic melt ascended (Wiebe & Collins 1998); their orientations serve as effective indicators of the palaeovertical.

The mafic sheets are interpreted essentially as lava flows that spread on the effective base of the granitic magma chamber, quenching against the cooler substrate of accumulated crystals from the overlying granitic magma (Falkner *et al.* 1995; Patrick & Miller 1997; Coiner *et al.* 2003; cf. Wiebe 1994; Wiebe & Collins 1998). The upper, less mafic parts of the mafic sheets reflect both the ascent of fractionated magma and the progressive hybridisation with felsic magma. Hornblende gabbros and troctolites are cumulates that apparently formed from thicker, slower cooling sheets. The HZ contains well over 100 discrete mafic sheets, each one representing a mafic recharge event.

Orientations of sheets, load casts, flame structures and pipes yield a consistent pattern of tilting within the pluton (Patrick & Miller 1997; Coiner 2003). The eastern block, east of Pipe Canyon fault, is tilted 40–90° to the NE to E. The western block, SW of Tule Wash fault, is tilted more gently (~10–30°) to the NW. Restoring the larger eastern block to its pre-tilt configuration reveals a section through the pluton that is about 5 km thick.

The contact between the HZ and GZ is sharp but irregular on map-scale. In the large eastern block, three large NNW-trending lobes are evident. In pre-tilt reconstructions (cf. Fig. 3), these lobes can be seen as tongues extending out into the GZ with subhorizontal bases and N-sloping upper surfaces. Each lobe comprises many mafic and felsic sheets, as well as apparent zones of hybridisation. The upper surfaces of the lobes are marked by a thin but distinct horizon of mechanical mixing between HZ lithologies and overlying granite. The authors view these lobes as lateral equivalents of the granite to the north, with deposition in the HZ dominated by mafic input and simultaneous GZ accumulation by deposition of granite crystals.

$^{40}Ar/^{39}Ar$ and SHRIMP U-Pb zircon studies indicate that the age of the pluton is 15·7 Ma (Falkner *et al.* 1995; Cates 2003; Cates *et al.* 2003; M. T. Heizler, pers. comm.). Detailed SHRIMP analysis of nine samples suggests that the interval of solidification for most of the pluton was on the order of 0·2 My (Cates 2003).

After solidification, the pluton was intruded by swarms of very large felsic dykes; most are on the order of 5–10 m thick, but some are locally as thick as 100 m (Steinwinder *et al.* 2004). Most of these dykes are fine-grained granite and rhyolite which are very similar in composition to the granites of the pluton, but some carry abundant quenched mafic pillows in a felsic host. Limited $^{40}Ar/^{39}Ar$ and SHRIMP analysis suggest that these dykes are only slightly younger than the pluton, i.e. ~15·5 Ma (Falkner *et al.* 1995; Cates 2003).

2. Methods

The background for the present research was reconnaissance field and petrographic work and geochemistry (16 elemental analyses) in the GZ by Falkner (1993; Falkner *et al.* 1995) and Robinson (1997; Robinson & Miller 1999) and related work in the HZ by Patrick (1996; Patrick & Miller 1997). The present authors carried out detailed mapping and field investigations, added 46 new elemental analyses, and reinterpreted the history and processes recorded within the GZ. Field work emphasised observation of textures and fabrics, sampling, and detailed investigations of country rock contacts. The textures and modal mineralogy of more than 70 new thin sections were described. The bulk of this work is described by Harper (2003).

Thin slabs of samples were powdered for elemental analysis in an alumina ceramic shatterbox. Thirty-three samples were analysed for major elements, Rb, Sr, Ba and Zr by X-ray fluorescence (XRF) at Middle Tennessee State University (MTSU), Murfreesboro, TN, USA, using an Oxford MDX 1080^+ multi-dispersive XRF spectrometer. Trace element concentrations were determined by analysis of pressed powder pellets and major element concentrations by analysis of fused glass disks with a $Li_2B_4O_7$–$LiBO_2$ flux. A combination of USGS standards and high-SiO_2 samples analysed by Activation Laboratories (to extend the range of standards to encompass Aztec Wash compositions) were used as standards for XRF. Thirty samples (17 of them also analysed at MTSU) were analysed by Activation Laboratories, a commercial lab in Ancaster, Ontario, Canada. Activation Laboratories utilised inductively coupled plasma spectrometry (ICP), ICP-mass spectrometry and instrumental neutron activation analysis to determine the concentrations of 62 elements. The 17 duplicate Actlabs+MTSU analyses were effectively identical for elements analysed in both laboratories (SiO_2 differed by <0·5 wt.% in most cases, Al_2O_3 by <0·3 wt.%, other oxides by <0·1 wt.%, Rb, Sr and Zr by <10 ppm); the lone exception was Ba, which was systematically lower at MTSU. For samples analysed in both laboratories, the present authors use the Actlabs data, but the differences are invisible on the data figures (Ba values for samples analysed only at MTSU are omitted). Also included in the figures are 16 analyses from Falkner (1993), Patrick (1996) and Robinson (1997). Major element analyses are normalised to 100%, excluding ignition loss, in Table 1 and Figure 4. Initial totals, excluding ignition loss, are almost all 98–100%, so the adjustment is small, but normalisation facilitates comparison of analyses, especially of SiO_2. All analysed samples appeared to be homogeneous, entirely lacking enclaves of contrasting lithologies, except for some of the samples referred to as HZ cumulates. Some of these samples contained small amounts of microenclave material.

Figure 2 Field photographs: (a) Alternating mafic and felsic sheets in the HZ. The mafic sheets are interpreted as dense basaltic flows onto a substrate of accumulated granitic crystals deposited from overlying felsic magma. Note the crenulate contacts between the mafic sheets and the underlying felsic sheets, and the flame structures protruding from the felsic into the overlying mafic sheets. (b) Granite zone (GZ) cumulate, characterised by coarse, subhedral to euhedral touching feldspars. (c) Pods of fine-grained intermediate rock in a coarse GZ cumulate matrix, suggestive of a matrix-supported conglomerate (debris flow?). (d) Cumulate dominated by rapakivi-textured feldspar, typical of exposures at the edge of the GZ, adjacent to mafic rocks of the heterogeneous zone (HZ); field of view 15 × 10 cm. (e) Miarolitic granite from the roof of the pluton. (f) Aphanitic, felsic enclaves in GZ transitional granite (these are unusually large enclaves). (g) Sheet of felsic cumulate from the HZ, trapped between underlying and overlying mafic sheets.

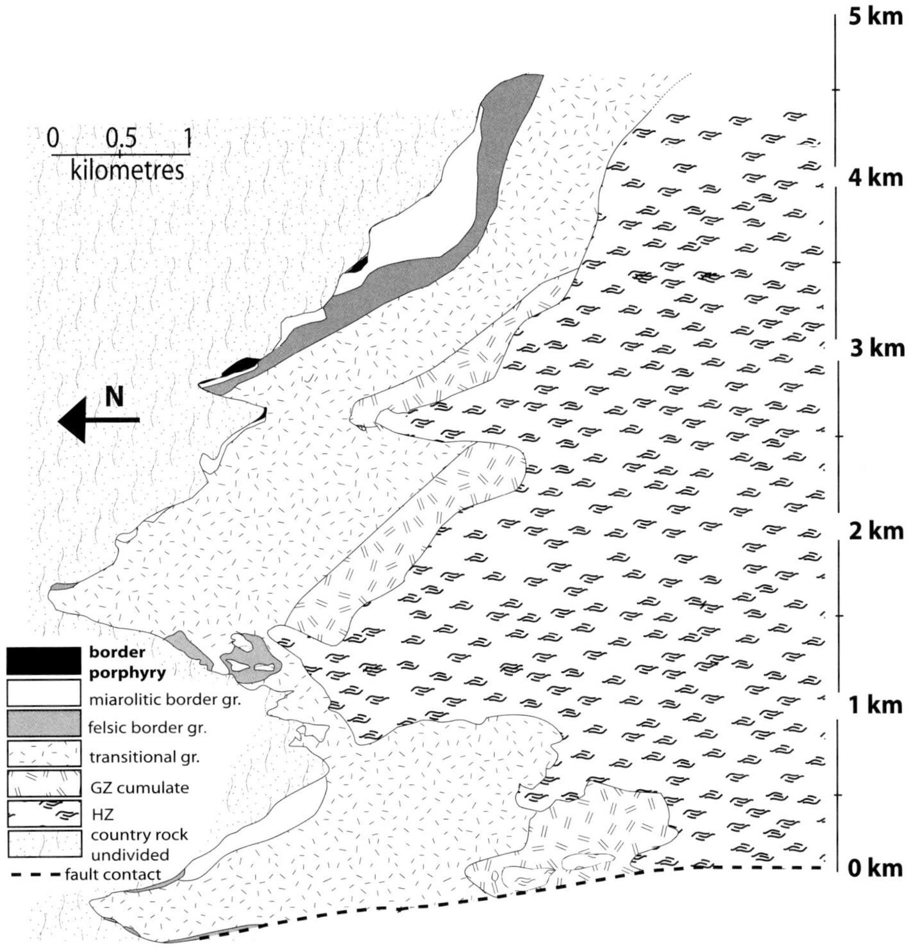

Figure 3 Cross-section of the granite zone (GZ), restored to its original orientation. Restoration assumes 62° of ENE tilting (toward N10E), based on average orientation of magmatic layering in GZ and of palaeohorizontal indicators in the adjacent heterogeneous zone. No vertical exaggeration.

3. Field and petrographic characteristics

3.1. Lithologic units

The GZ of the pluton is homogeneous insofar as it is entirely made up of granite. Nonetheless, it shows very distinct variability and can be subdivided into seven units, five of them mappable on the scale of Figure 1. The present authors also discuss a widespread suite of enclaves in the GZ and a felsic lithology in the HZ that is relevant to evolution of the GZ granites (HZ felsic cumulate). All of these rocks contain K-feldspar, plagioclase, quartz, biotite, sphene, magnetite, zircon and apatite, and most also have allanite, but in varying proportions and textural relations. Except in the HZ felsic cumulate, K-feldspar is the most abundant mineral. It is generally pale pink, but in the most felsic variants it is white. Hornblende and chevkinite are present in the GZ and HZ cumulates, and ilmenite occurs mostly as inclusions in sphene (Robinson & Miller 1999; Cates 2003).

3.1.1. Granite zone cumulate (coarse-grained granite). Relatively coarse-grained, pale pink to tan granites, characterised by abundant, subhedral to euhedral K-feldspar and plagioclase, are common in the GZ (Fig. 2b). The ~3–10-mm feldspars are closely packed and commonly touching, and constitute a largely feldspar grain-supported fabric. Quartz, though relatively abundant (15–20%), is interstitial and only evident with a hand lens or in thin section. The colour index (CI, sum of modal abundance of dark minerals) is ~4–8 (biotite+magnetite ± uncommon hornblende). Sphene is larger and more abundant than in other GZ rocks. Ovoid, fine-grained mafic enclaves <3 cm in diameter are sparsely distributed. The rocks richest in these small enclaves (up to ~3%) are close to HZ/GZ contacts, tend to be coarsest, and locally contain abundant, much larger (1–10 m), fine-grained, ovoid enclaves of intermediate composition (Fig. 2c). These enclaves lack chilled, crenulate margins, indicating that they were transported as clasts rather than representing locally quenched magma. Rapakivi-textured K-feldspars, rare elsewhere in the GZ, are also common in these areas adjacent to contacts with the HZ (Fig. 2d). In some areas, GZ cumulates have a faint to pronounced foliation defined by alignment of large feldspar grains. In the large block east of the Pipe Canyon fault, this foliation strikes consistently NNW (~350–360°); where measurable, dips are mostly 70–90°E.

The coarse, euhedral and touching feldspar framework suggests accumulation. The relative abundance of biotite and sphene (which also are part of the proposed cumulate assemblage), the scarcity of quartz (mostly grown from trapped melt), and the presence in some areas of clasts of mafic to intermediate rock support this interpretation.

3.1.2. Transitional granite (medium-grained granite). The GZ cumulates grade into medium-grained granite (~2–6-mm feldspars) that has a lower CI (~2–5) and more abundant and obvious quartz. Quartz is mostly interstitial, but in more felsic variants, it forms more prominent grains. Mafic enclaves are very rare or absent, and no foliation is evident.

3.1.3. Chilled border granite (fine-grained granite). A fine- and uniform-grained variant (not shown on Fig. 1) that differs from the more abundant granites primarily in grain size (<~1-mm feldspars, CI ~3–6) is present locally along the margin against country rock, grading within 10s of metres into

Table 1 Major and selected trace element concentrations of representative samples of the GZ and related rocks.

Sample unit	HAW-10 (trans)	HAW-21 (trans)	HAW-35 (trans)	HAW-56 (trans)	HAW-60 (trans)	HAW-64 (trans)	NAW-Z50 (trans)	NAW-Z52 (trans)	HAW-5 (GZ cum)	HAW-18 (GZ cum)	HAW-15 (fels bord)	HAW-36 (fels bord)	HAW-59 (fels bord)	NAW-Z26 (miaro)	HAW-38 (miaro)
SiO_2 (wt.%)	72·83	72·33	72·43	73·28	73·42	74·37	73·93	73·59	71·73	70·81	76·77	76·81	77·58	76·7	77·57
Al_2O_3	13·94	14·28	14·09	13·92	13·96	13·39	13·66	13·57	14·56	15·12	12·73	12·7	12·44	12·76	12·3
Fe_2O_3(tot)†	1·75	1·78	1·76	1·76	1·61	1·55	1·57	1·78	1·94	1·92	0·81	0·76	0·57	0·78	0·68
MnO	0·04	0·03	0·04	0·03	0·04	0·04	0·03	0·03	0·04	0·03	0·03	0·04	0·02	0·03	0·04
MgO	0·59	0·55	0·54	0·51	0·49	0·46	0·45	0·54	0·63	0·59	0·16	0·16	0·09	0·18	0·1
CaO	1·56	1·43	1·48	1·17	1·19	0·92	1·05	1·32	1·53	1·62	0·53	0·67	0·51	0·56	0·54
Na_2O	3·53	3·85	3·68	3·6	3·81	3·58	3·39	3·46	3·98	3·95	3·65	3·72	3·69	3·69	3·8
K_2O	5·37	5·34	5·58	5·32	5·11	5·33	5·57	5·34	5·11	5·47	5·17	4·99	4·98	5·14	4·83
TiO_2	0·3	0·32	0·32	0·32	0·29	0·28	0·27	0·32	0·35	0·37	0·13	0·14	0·1	0·14	0·11
P_2O_5	0·08	0·09	0·09	0·09	0·08	0·08	0·07	0·07	0·11	0·12	0·02	0·03	0·02	0·02	0·02
LOI‡	1·23	0·50	0·93	0·45	0·48	0·46	0·53	1·38	0·75	0·83	0·23	0·57	0·36	0·50	0·49
Total§	99·27	99·39	99·38	100·11	98·68	100·23	99·80	99·79	99·13	99·02	99·91	99·74	99·96	100·31	100·43
Rb (ppm)	167	176	174	180	178	175	206	171	149	134	197	206	223	206	190
Sr	209	270	228	213	221	170	180	198	300	354	61	37	21	45	59
Ba	604	692	694	603	662	524	540	569	830	1051	145	115	30	121	42
Y	18	17	18	17	16	14	13	17	18	17	13	15	6	12	11
Zr	181	223	205	231	216	176	177	207	211	214	105	124	99	125	84
Hf	5·7	6·4	6·0	6·6	6·2	5·6	5·7	6·5	6·4	6·0	4·2	5·0	4·0	5·0	3·8
Nb	23·5	25·2	25·4	23·1	24·4	19·9	26·2	23·2	20·4	20·2	27·6	36·4	19·6	26·5	26·0
Ta	1·76	1·79	1·81	1·92	1·61	2·07	2·67	1·75	2·15	1·84	2·04	2·90	1·89	2·21	2·56
La	60·9	62·7	60·2	69·8	59·9	68·1	61·3	61·1	72·2	66·8	39·3	41·1	35·1	41·6	37·2
Ce	107	113	113	117	105	106	96·3	100	121	121	67·1	71·5	45·6	63·8	57·1
Pr	10·4	11·1	11·1	11·5	10·2	10·4	9·49	10·3	12·7	11·9	5·69	6·12	3·23	5·66	4·70
Nd	33·7	36·4	35·5	37·6	32·4	30·7	28·5	30·8	40·4	40·7	15·1	16·7	6·96	14·7	10·9
Sm	4·95	5·26	5·31	4·98	4·61	4·29	4·05	5·06	6·00	6·01	2·06	2·35	0·70	1·96	1·33
Eu	0·774	0·912	0·858	0·824	0·765	0·701	0·672	0·803	1·03	1·12	0·219	0·229	0·073	0·214	0·098
Gd	3·38	3·72	3·73	3·74	3·29	3·20	2·88	3·22	4·27	3·94	1·47	1·69	0·44	1·34	1·05
Tb	0·50	0·53	0·53	0·53	0·47	0·43	0·43	0·51	0·61	0·59	0·26	0·31	0·09	0·25	0·19
Dy	2·86	3·01	3·02	2·75	2·67	2·43	2·40	2·89	3·32	3·24	1·69	2·03	0·59	1·57	1·31
Ho	0·55	0·57	0·58	0·55	0·51	0·49	0·47	0·54	0·66	0·61	0·38	0·45	0·16	0·35	0·32
Er	1·65	1·73	1·73	1·86	1·53	1·48	1·41	1·71	1·91	1·85	1·25	1·51	0·61	1·19	1·13
Tm	0·271	0·270	0·270	0·294	0·243	0·258	0·231	0·275	0·308	0·280	0·224	0·267	0·144	0·234	0·236
Yb	1·73	1·71	1·80	1·83	1·55	1·63	1·59	1·75	1·95	1·89	1·57	1·92	1·16	1·64	1·56
Lu	0·269	0·270	0·286	0·285	0·240	0·248	0·252	0·268	0·286	0·264	0·252	0·303	0·207	0·258	0·250
Th	27·1	28·4	31·6	31·2	26·8	31·0	33·9	33·9	28·6	24·6	38·8	42·0	41·8	35·8	36·5
U	5·07	5·26	5·02	5·88	4·35	5·99	6·20	4·83	4·20	3·51	6·32	9·35	8·17	8·01	7·24

Key: (trans) transitional granite; (GZ cum) GZ cumulate; (fels bord) felsic border granite; (miaro) miarolitic border granite; (felsite) felsite (dykes and pods); (chilled) chilled border granite; (brd porph) border porphyry; (enclave) felsic enclave; and (HZ cum) HZ cumulate.

†Fe_2O_3(tot): Total Fe reported as Fe_2O_3.

‡LOI: Loss on ignition. §Total: Total of all oxides+LOI, as reported; the oxides listed above are concentrations normalised to 100%.

Table 1 *Continued.*

Sample unit	HAW-83 (miaro)	HAW-20 (felsite)	NAW-Z53 (chilled)	HAW-80 (chilled)	HAW-84 (chilled)	CAW-Z106 (brd porph)	HAW-7 (enclave)	DAW-2 (enclave)	DAW-5 (enclave)	DAW-9 (enclave)	DAW-38 (enclave)	DAW-41 (enclave)	DAW-50 (enclave)	NAW-Z16 (HZ cum)	NAW-Z51 (HZ cum)
SiO_2 (wt.%)	76·72	77·83	73·22	74·07	72·7	71·85	71·2	71·28	71·18	70·66	71·37	71·38	71·79	60·03	65·37
Al_2O_3	12·57	12·28	14·14	13·73	14·42	14·55	14·99	14·67	14·41	15·44	14·95	15·26	15·18	19·07	17·03
Fe_2O_3(tot)†	0·86	0·42	1·74	1·63	1·79	1·94	2·13	2·41	2·18	1·68	1·84	2·02	1·33	5·18	3·74
MnO	0·03	0·01	0·04	0·03	0·04	0·04	0·05	0·06	0·05	0·04	0·04	0·05	0·02	0·06	0·03
MgO	0·21	0·07	0·54	0·57	0·55	0·76	0·98	0·81	0·67	0·35	0·42	0·54	0·24	1·79	1·38
CaO	0·58	0·47	1·43	0·51	1·61	1·41	1·84	1·98	1·47	1·22	1·29	1·78	1·12	4·33	3·05
Na_2O	3	3·4	3·37	3·35	3·51	3·51	4·39	4·17	3·7	4·39	4·16	4·05	4·26	5·34	3·83
K_2O	5·87	5·4	5·19	5·79	5·05	5·51	3·99	4·13	5·79	5·8	5·41	4·47	5·75	2·85	4·74
TiO_2	0·13	0·09	0·26	0·25	0·25	0·34	0·32	0·36	0·41	0·34	0·35	0·3	0·24	1·02	0·63
P_2O_5	0·03	0·02	0·07	0·07	0·08	0·09	0·12	0·12	0·13	0·06	0·17	0·15	0·06	0·32	0·2
LOI‡	0·51	0·22	0·96	0·87	0·65	1·87	1·39	0·64	0·68	0·42	0·47	0·76	0·40	0·76	0·90
Total§	99·21	99·82	99·42	99·93	100·10	99·47	99·45	100·41	99·54	100·15	100·35	100·29	100·07	99·40	99·91
Rb (ppm)	183	230	154	141	133	152	123	144	201	187	138	101	203	58	80
Sr	128	48	331	227	360	325	396	366	290	299	318	367	309	880	799
Ba	295	100	1321	1347	1376	854	1058	1092	886	867	919	1340	940	1146	2580
Y	5	5	15	17	15	16	10	12	18	18	16	14	11	59	13
Zr	91	76	169	187	173	248	141	207	219	291	250	241	189	700	394
Hf	3·3	3·8	5·2	4·9	4·9	6·9	4·4	5·8	6·6	9·0	7·9	6·8	6·1	16·8	9·2
Nb	17·1	20·6	15·1	16·1	14·6	26·2	15·9	16·4	27·8	29·5	24·6	21·5	22·6	28·1	13·0
Ta	1·23	2·18	1·10	1·21	1·52	1·76	0·99	0·81	2·38	2·32	2·06	1·50	1·53	2·78	1·05
La	36·5	48·6	53·3	57·5	53·3	53·9	49·7	55·1	64·1	79·4	80·5	66·3	66·1	124	108
Ce	55·8	58·8	89·5	105	93·2	98·6	75·9	95·4	126	143	144	116	108	304	164
Pr	4·23	3·78	8·91	9·90	9·07	8·04	7·50	7·34	10·7	11·3	11·6	9·32	7·97	33·6	16·7
Nd	9·98	7·63	28·6	33·4	31·4	29·0	24·6	29·4	42·6	43·3	44·1	35·5	28·3	125	51·1
Sm	1·08	0·66	4·57	5·13	4·62	4·68	3·85	4·41	6·66	6·31	6·56	5·12	4·00	23·6	6·13
Eu	0·215	0·093	0·890	0·894	0·907	0·849	0·983	0·937	1·10	1·19	1·17	1·17	0·720	3·58	1·61
Gd	0·70	0·34	3·19	3·50	3·23	3·38	2·89	2·74	4·01	3·78	3·90	3·22	2·12	18·3	4·06
Tb	0·10	0·07	0·49	0·50	0·50	0·47	0·38	0·43	0·62	0·59	0·61	0·50	0·38	2·86	0·49
Dy	0·60	0·47	2·61	2·97	2·76	2·56	1·97	2·30	3·33	3·36	3·35	2·74	2·12	14·2	2·42
Ho	0·15	0·12	0·51	0·58	0·53	0·48	0·37	0·41	0·59	0·58	0·59	0·49	0·37	2·46	0·45
Er	0·53	0·47	1·54	1·69	1·64	1·55	1·06	1·31	1·84	1·90	1·83	1·55	1·22	6·09	1·35
Tm	0·091	0·106	0·251	0·261	0·251	0·243	0·156	0·230	0·315	0·327	0·307	0·258	0·218	0·801	0·198
Yb	0·73	0·82	1·69	1·78	1·74	1·67	1·07	1·52	2·01	2·21	1·99	1·68	1·46	4·56	1·42
Lu	0·143	0·160	0·259	0·266	0·258	0·251	0·175	0·245	0·297	0·330	0·290	0·260	0·231	0·581	0·238
Th	33·6	53·1	15·5	14·8	14·8	30·8	19·4	22·8	30·9	36·0	33·2	15·7	31·7	15·9	16·2
U	4·43	6·08	1·83	1·80	2·09	5·39	3·78	5·18	5·22	6·31	4·92	1·44	4·51	2·08	1·69

Key: (trans) transitional granite; (GZ cum) GZ cumulate; (fels bord) felsic border granite; (miaro) miarolitic border granite; (felsite) felsite (dykes and pods); (chilled) chilled border granite; (brd porph) border porphyry; (enclave) felsic enclave; and (HZ cum) HZ cumulate.

†Fe_2O_3(tot): Total Fe reported as Fe_2O_3.

‡LOI: Loss on ignition. §Total: Total of all oxides+LOI, as reported; the oxides listed above are concentrations normalised to 100%.

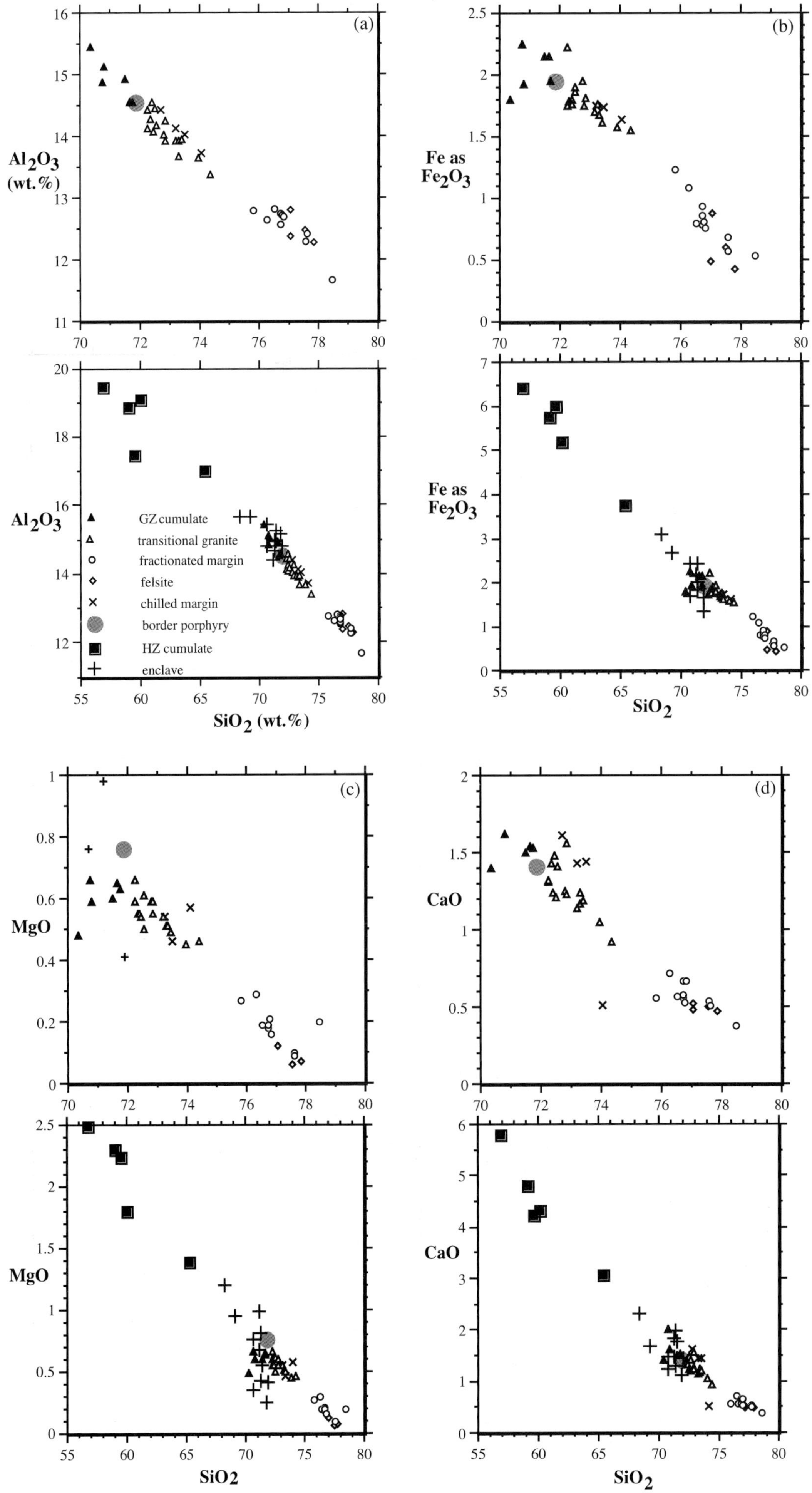
(a)
(b)
(c)
(d)
Al2O3 (wt.%)
Fe as Fe2O3
Al2O3
SiO2 (wt.%)
SiO2
MgO
CaO
SiO2
GZ cumulate
transitional granite
fractionated margin
felsite
chilled margin
border porphyry
HZ cumulate
enclave

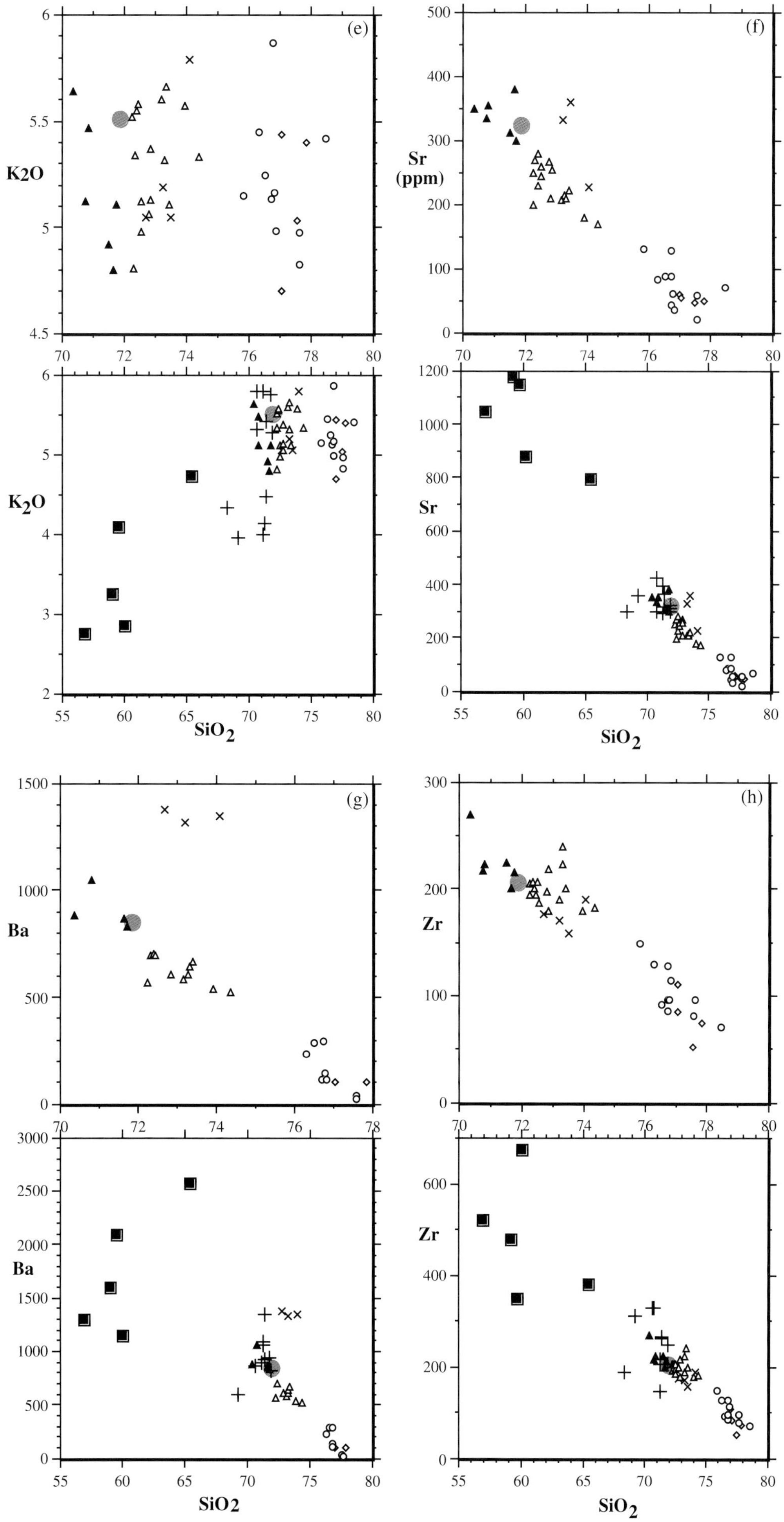
(e)
(f)
(g)
(h)
K2O
Sr (ppm)
Sr
Ba
Zr
SiO2

transitional granite. Feldspars are less prominent than in the previously described units and quartz is present as discrete grains as well as interstitially. Mafic enclaves and foliation are absent.

3.1.4. Border porphyry. Very locally, primarily on the east side, the rock in contact with wall rock is a light grey porphyry. The porphyry has an aphanitic to very fine-grained phaneritic (largely granophyric) groundmass and contains K-feldspar, plagioclase and biotite phenocrysts. Although it is similar to the chilled border granite in showing evidence for quenching, it differs in its much finer groundmass and the presence of prominent feldspar phenocrysts.

3.1.5. Felsic border granite (fine- to medium grained leucogranite). Along and near much of the margin, the medium-grained transitional granite (not shown on Fig. 1) grades into a white, generally finer and more felsic granite (feldspars ~1–4 mm, CI <2). Quartz occurs as distinct, subhedral grains, and is commonly as large and prominent as feldspar. Mafic enclaves and foliation are absent.

3.1.6. Miarolitic border granite (miarolitic leucogranite). The felsic border granite locally grades into a similarly felsic rock (CI <1), distinguished by abundant miarolitic cavities, that is exposed immediately adjacent to the country rock (Fig. 2e). Granophyric textures are abundant, and mafic enclaves and foliation are absent.

3.1.7. Felsite dykes and pods. Fine-grained phaneritic to aphanitic felsite (or aplite) dykes (not shown on Fig. 1) cut all GZ units except the chilled border granite and border porphyry. These dykes are small (mostly 1–10 cm, rarely up to 50 cm in width), extremely felsic and granophyric (not shown in Fig. 1). Felsic pods are similar in composition, texture and distribution.

3.1.8. Felsic enclaves. Light grey enclaves (not shown on Fig. 1) are widespread in the granites of the GZ (Fig. 2f). The enclaves are aphanitic to very fine-grained phaneritic; most are porphyritic, with phenocrysts of feldspar ± biotite, and, in general, slightly more mafic than the host granite. Most are small (~2–10 cm) and ovoid, but they range upward to several metres in maximum dimension, and some of the large ones are elongate, or tabular and angular. A large exposure (~0·1 km^2) in the western GZ comprises lithologies which appear to be identical to the enclaves.

3.1.9. Heterogeneous zone felsic cumulate (coarse-grained heterogeneous zone quartz monzonite-quartz monzodiorite). Although the HZ is not the focus of this study, the HZ felsic cumulate (not distinguished from the remainder of the HZ in Fig. 1) is relevant to interpretation of the GZ granites. The present authors refer to this unit as felsic to distinguish it from the mafic, clinopyroxene ± olivine-rich cumulates of the HZ. The felsic cumulate, the most widespread granitoid in the HZ, occurs as thin to thick sheets and lenses (<1 cm to ~30 m thick) between mafic sheets (Fig. 2g). It is as coarse as or coarser than the GZ cumulates, and considerably poorer in quartz (<10%) and richer in mafic minerals (CI typically ~15). Quartz is restricted to interstices between feldspars and is generally not visible in hand sample, and plagioclase is more abundant than K-feldspar. Accessory minerals are very abundant, and hornblende, though subordinate to biotite, is much more common than in GZ cumulates. Feldspar-defined foliation is absent in many sheets, but in places it is much more distinct than that in the GZ cumulate. This foliation is subparallel to the orientation of the cumulate sheet, which is, in turn, subparallel to paleohorizontal indicators in the surrounding, more mafic rocks. Small, fine-grained mafic enclaves are common.

The same lines of textural and mineralogic evidence support a cumulate origin for HZ cumulates as for GZ cumulates, but the characteristics are more extreme: less quartz, more biotite and early accessory minerals, and in some cases, a more pronounced feldspar-defined foliation. The accumulating crystals appear to have been more tightly packed (less trapped melt), probably because of compaction beneath the overlying mafic sheet (Patrick & Miller 1997; cf. Wiebe 1993, 1994).

3.2. Field relationships of the units

The spatial distribution of the GZ units and variants within units provides some of the strongest evidence for the processes that acted within the Aztec Wash magma chamber. Figure 1b shows the map pattern of units within the GZ. Figure 3 shows a reconstructed, true-scale cross-section of the large, structurally continuous block east of the Pipe Canyon fault. Patterns are similar in the smaller blocks to the west, but less extensive exposure and more gently dipping structures provide less complete sections. Important features include the following:

(1) The GZ cumulates contact only the HZ and the transitional granite, and do not reach the pluton margin. Their contacts with the HZ are invariably sharp; in contrast, they grade imperceptibly into the transitional granite. They are localised in the southern and western parts of the GZ.
(2) Attitudes of magmatic foliation in the GZ cumulates are parallel to palaeohorizontal markers in the adjacent part of the HZ (cf. Coiner 2003).
(3) The GZ cumulate rocks are coarsest on the NE sides of HZ lobes. Rounded enclaves of HZ material are abundant within 10–20 m of these NE (upper) contacts. Smaller clasts are mafic to intermediate, whereas the large clasts are uniformly intermediate. This narrow zone strongly resembles a matrix-supported conglomerate (Fig. 2b). The feldspar matrix here contains scattered, locally abundant, rapakivi-textured grains (Fig. 2c). Robinson & Miller (1999) referred to these areas as 'buffer zones' at the HZ–GZ contact, and interpreted them to represent a limited region of thermal, chemical and mechanical interaction between the two main zones of the pluton.
(4) Granites on the west sides of HZ lobes are generally transitional granites or finer-grained cumulates that are poorer in enclaves than their northeastern counterparts.
(5) The transitional granites are interspersed with, but lie generally to the north and east of, the GZ cumulate, and have gradational contacts with all of the other GZ units except the miarolitic and border porphyry units. They are also present along much of the E-facing portion of the wall rock contact.
(6) The chilled border granite is present only locally at the wall rock contact and grades in a short distance to transitional granite.
(7) The felsic border granite contacts all other units except the cumulate. It forms an almost continuous marginal

Figure 4 Elemental compositions of granite zone (GZ) units and heterogeneous zone (HZ) felsic cumulates. SiO_2 versus (a) Al_2O_3, (b) $Fe_2O_3^t$, (c) MgO, (d) CaO, (e) K_2O, (f) Sr, (g) Ba and (h) Zr. The upper panel for each part shows only GZ units; the lower panel expands the elemental range, and includes HZ cumulates and felsic enclaves. 'Fractionated border' combines felsic and miarolitic border granites. Major oxides in wt.%, and trace elements in ppm.

zone along W- and SW-facing parts of the margin. It is also present locally along other portions of the border.

(8) The miarolitic border granite is present only along the west–southwest-facing border and contacts only the felsic and miarolitic border granite and wall rock. Like the felsic border granite, it reaches its maximum thickness in the central part of the eastern contact.

(9) The felsic and miarolitic border granites reach their greatest combined thickness (0·6 km) on what is inferred to be the central part of the eastern contact (farther south, this contact is covered).

(10) The border porphyry unit is restricted to local exposures adjacent to wall rock along the NE contact and along what is probably a structurally analogous wall rock contact in the central block of the pluton. Felsic and miarolitic border granites contact the border porphyry unit on its interior (plutonward) side; these contacts are sharp.

(11) Felsic enclaves are present sporadically throughout the GZ, except for the felsic border on the east and equivalent portions of the margin in the western blocks.

4. Elemental chemistry

Each of the GZ granite units is restricted in elemental composition, and for the most part, the units are distinct from one another (Fig. 4). For example, SiO_2 and Sr range from 70 to 72 wt.% and 300 to 360 ppm, respectively, in GZ cumulates; 72 to 74 wt.% and 170 to 290 ppm in transitional granites; and 76 to 78 wt.% and 20 to 130 ppm in felsic and miarolitic border granites and felsite dykes and pods. Three of four chilled border samples have 73 wt.% SiO_2 and 330–360 ppm Sr; the fourth is somewhat more felsic (74 wt.% and 227 ppm). The single border porphyry sample is very similar to the chilled border samples (72 wt.% SiO_2, 325 ppm Sr). These principal granite units as a group define extremely coherent trends for all elements plotted versus SiO_2, except that the chilled border unit (but not the border porphyry) lies distinctly above the trends for Sr and especially for Ba, as defined by the other samples (Fig. 4). Rare earth element (REE) patterns of the GZ cumulates, transitional granites, border porphyry and chilled borders are tightly coherent (Fig. 5a, b). All show strong light REE enrichment, nearly flat heavy REE patterns and modest negative Eu anomalies. Subtle differences show up more clearly when REE concentrations are normalised to the average composition of the chilled border. Transitional granites have slightly more negative Eu anomalies and GZ cumulates are slightly richer in all REE than the chilled border samples (Fig. 5d, e). The more felsic rocks (felsic and miarolitic border granites and felsite dykes and pods) are poorer in all REE, with especially pronounced depletion in middle REE (Fig. 5b,d). All are strongly depleted in Eu relative to the other granites, but, because of their extreme middle REE depletion, some lack pronounced Eu anomalies.

The felsic enclaves and the HZ felsic cumulates, though showing more scatter, define similar elemental trends which appear to coincide with and extend to lower SiO_2 those of the granite units of the GZ (Fig. 4). Felsic enclaves are generally less felsic than the GZ granites (68–72 wt.% SiO_2), but rather variable and less geochemically coherent. The HZ felsic cumulates are much poorer in SiO_2 (57–65 wt.%) and richer in compatible elements (e.g. 800–1200 ppm Sr). The REE patterns of the enclaves are generally similar to those of the GZ cumulates, transitional granites, chilled border granites and border porphyry, but considerably more variable (Fig. 5c, e). The HZ cumulates are much richer in REE than any of the other units, especially the middle REE (Fig. 5c). They are Eu-rich, and two of three analysed samples lack Eu anomalies, resulting in positive anomalies when normalised to average chilled border (Fig. 5e).

5. Interpretation

5.1. Evolution of granitic magma in the Aztec Wash pluton: geochemical constraints

Field relations and geochemical and petrographic coherence strongly indicate that the principal units of the GZ (cumulate and transitional granites; felsic, miarolitic and chilled border granites; and felsite dykes) are closely related. The border porphyry, felsic enclaves and HZ cumulates may also be products of the same or closely related felsic magma, but field and geochemical evidence is equivocal. Below, the present authors discuss a simple model for relationships among the first six units listed in section 3.1 and address constraints upon their connections to the latter three.

As suggested previously, the textures of the rocks referred to as GZ cumulate granites indicate formation by crystal accumulation. The textures of felsite dykes and fine-grained rocks at wall rock contacts suggest quenching of crystal-poor magmas. Elemental abundance trends (Fig. 4) and textures are broadly consistent with a fractional crystallisation sequence in which the cumulate granites represent initial crystal accumulation, the transitional granites crystal-poorer material, and the most felsic units (felsite dykes, and felsic and miarolitic border granites) highly evolved, segregated crystal-poor magma.

Uncertainties in partition coefficients (D's) limit the reliability of quantitative trace element modelling. Even for the most diagnostic and well-characterised elements and minerals, published D's for felsic systems vary by factors of at least 2 or 3 (e.g. Arth 1976; Henderson 1982; Rollinson 1993; White *et al.* 2003). Nonetheless, useful semiquantitative approximations are possible. The simple trace element model presented in Figure 6 for the co-evolution of melts and extracted crystalline assemblage is based on typical published D's for Rb, Sr and Ba, the estimated modal proportions of the crystallising assemblage (from the modes of subhedral minerals in GZ cumulates), and the assumption that the chilled border granite represents input melt. Specific input parameters are listed in the figure legend. Modifying modal abundances within a reasonable range has a negligible effect for the modelled elements. Adjusting D's within published ranges has a small effect on trends, but a pronounced effect on lengths of the fractionation vectors.

The modelling suggests the following:

(1) Well-defined trends in the data that parallel predicted trends support the interpretation that the rocks represent the products of crystallisation of K-feldspar + plagioclase + biotite from the same or very similar initial magmas.

(2) The overall range of compositions is too great to be explained by varying degrees of equilibrium crystal-melt segregation. For example, Sr and Ba concentrations differ by factors of $\sim$20 and 30, respectively, from least to most evolved rocks, but bulk D's for both elements must have been considerably less than 10. In the present model, roughly 50% perfect (Rayleigh) fractional crystallisation is required to produce the observed range; the calculated fraction is strongly influenced by the D's used, but it is relatively insensitive (within reasonable limits) to assumed initial melt composition.

The REE and Zr data are consistent with the above constraints. Modelling REE trends or Zr concentrations using

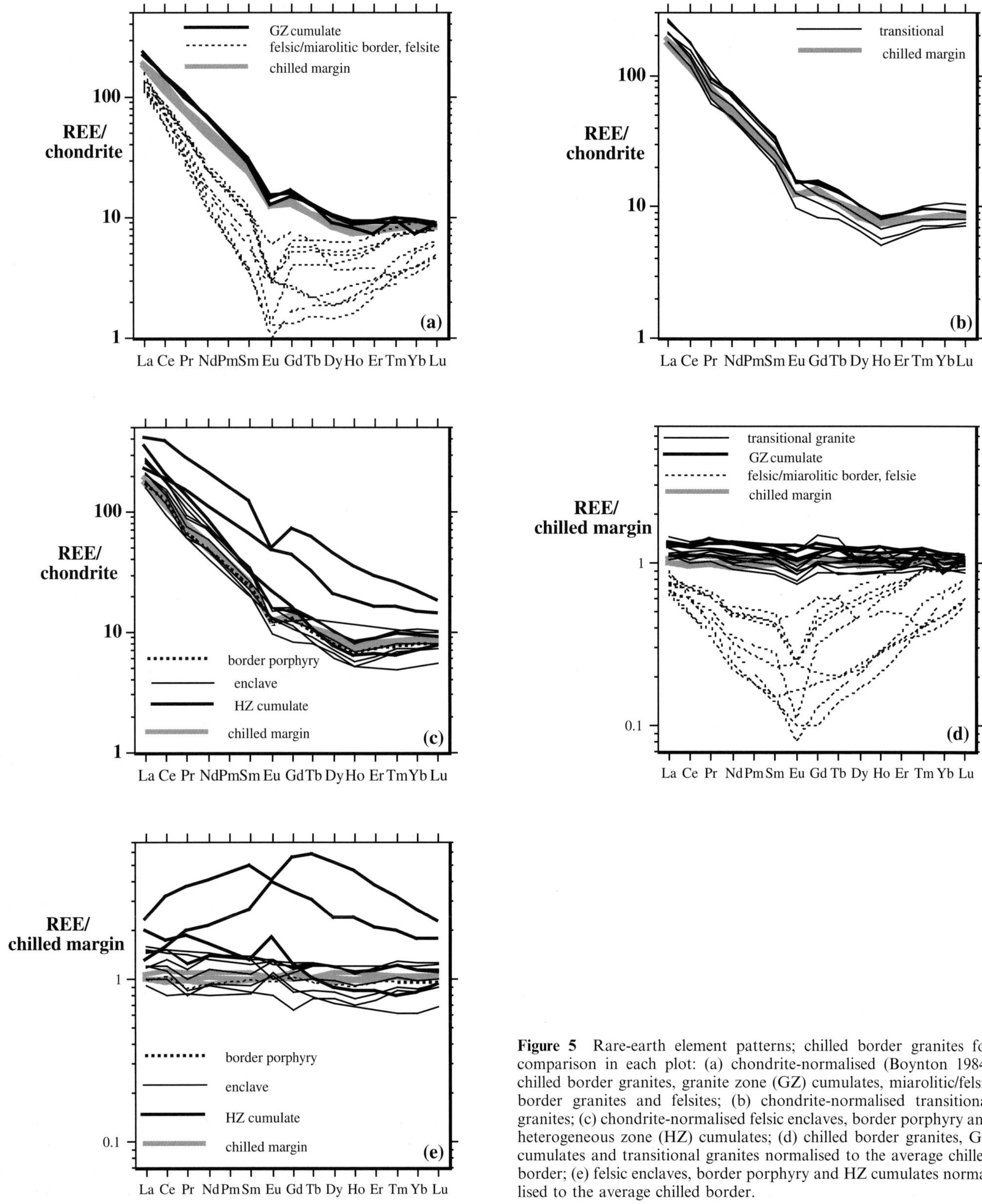

Figure 5 Rare-earth element patterns; chilled border granites for comparison in each plot: (a) chondrite-normalised (Boynton 1984) chilled border granites, granite zone (GZ) cumulates, miarolitic/felsic border granites and felsites; (b) chondrite-normalised transitional granites; (c) chondrite-normalised felsic enclaves, border porphyry and heterogeneous zone (HZ) cumulates; (d) chilled border granites, GZ cumulates and transitional granites normalised to the average chilled border; (e) felsic enclaves, border porphyry and HZ cumulates normalised to the average chilled border.

the Rayleigh equation is of little value because petrographic evidence suggests that melts were generally saturated in zircon and chevkinite and/or allanite (Robinson & Miller 1999; Cates 2003). Because Zr is an essential structural (stoichiometric) constituent (ESC) of zircon and REE are ESCs of chevkinite and allanite, these elements do not behave in dispersed (Rayleigh) fashion in melts that are saturated in these minerals. Nonetheless, Zr concentrations and varying REE patterns provide a useful test of the fractional crystallisation model. The systematic changes in REE patterns and Zr from the least to most felsic granites are entirely consistent with removal of the observed mineral assemblage of feldspars+biotite+accesories from evolving magma (Figs 4h & 5a, b, d). Diminishing light REE and Zr concentrations in progressively more felsic rocks reflect both the accumulation of early accessories and the lower levels of these elements required to maintain zircon and allanite/chevkinite saturation in the evolving melt as the temperature dropped. Zircon saturation temperatures (Watson & Harrison 1983) calculated for rocks that appear to have been melt-rich drop systematically from $800 \pm 10\,^{\circ}C$ in chilled border and transitional granites to $745 \pm 10\,^{\circ}C$ in fractionated rocks (felsic border and miarolitic granites and felsites). Extreme depletion in middle REE in the fractionated rocks was presumably a response to fractional crystallisation

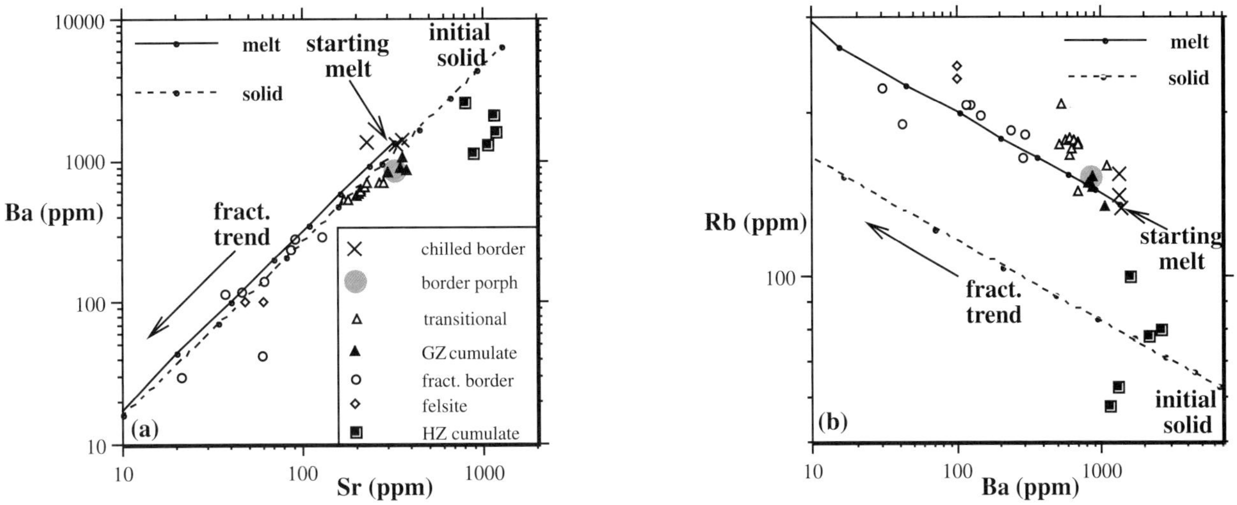

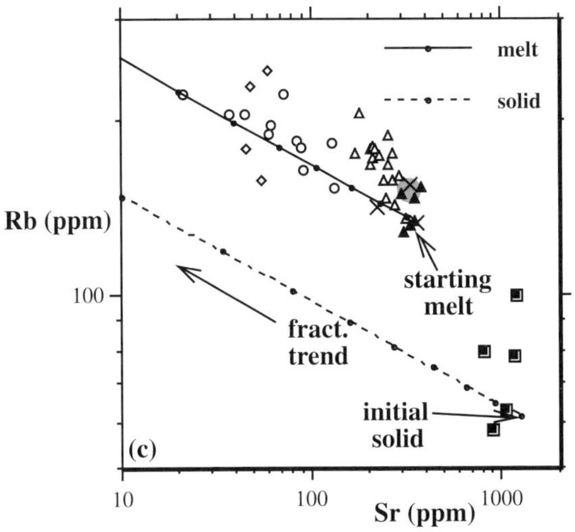

Figure 6 Simple fractional crystallisation model for the evolution of dispersed (trace) elements in melt and coexisting solid. The model applies the Rayleigh fractional crystallisation equation $C_l/C_o = F^{D-1}$ (C_l = concentration of element in melt, C_o = initial concentration of element in melt, F = fraction of melt remaining, $D = D^{s/l}$ = bulk partition coefficient of element for crystallising assemblage versus melt); instantaneous concentration of element in crystallising solid = D × C_l. Parameters: mode of crystallising assemblage (based on estimates of early assemblage from cumulates): 56% K-feldspar, 35% plagioclase, 6% biotite, 3% others; $D^{s/l}$'s (estimated from typical literature values – Arth 1976; Henderson 1982; Rollinson 1993; White *et al.* 2003): Sr: Ksp 4, plag 5, bio 0·2; Rb: Ksp 0·4, plag 0·1, bio 3·2; Ba: Ksp 7, plag 0·6, bio 10; initial melt composition: mean of four chilled margin granites. Tick marks on the melt and solid evolution lines represent decrements of 0·1 in F (remaining melt fraction). (a) Ba versus Sr; (b) Rb versus Sr; (c) Rb versus Ba.

of apatite and especially the abundant sphene, which have very high D's for these elements (e.g. Simmons & Hedge 1978; Rollinson 1993). All of these accessories, including zircon, probably contributed to the modest heavy REE depletion.

Despite the general consistency with a simple fractional crystallisation model, there is a discrepancy that indicates that the solidification history was more complex than the simple, monotonic model. The GZ cumulate granites are too evolved to have been derived directly from melt equivalent to the chilled border granites (Fig. 6). They are comparable in Sr and Rb and poorer in Ba than the chilled border granites. Given their very K-feldspar+plagioclase-rich compositions, their effective D's for Ba and Sr should have been at least 2 or 3, even if they contained abundant trapped melt, and therefore, concentrations in excess of 2000 and 600 ppm, respectively, would be expected if they were extracted from a chilled border-like parent; Rb D's must have been <1, yielding cumulates with concentrations appreciably lower than those observed (<140 ppm). There is no evidence of input into the pluton of magma with an appropriate composition for a direct parent of the GZ cumulates. Instead, the GZ cumulates probably formed from an evolved melt that had already been depleted in Ba and Sr and enriched in Rb compared to the input magma. Assuming that the GZ rocks represent a loosely packed aggregation of crystals with 50% trapped melt and applying the same D's for accumulated minerals as the authors used for the general model, the coexisting melt would have had about 190 ppm Rb, 135 ppm Sr, and 320 ppm Ba. Based on the textural evidence, this evolved melt was undersaturated or was just reaching saturation in quartz when accumulation took place, and, as a consequence, these rocks are slightly poorer in SiO_2 than the input magma, in spite of the fact that coexisting melt was presumably more siliceous than the input. Accumulation of biotite and accessory minerals accounts for the slightly elevated Fe, Mg, Ti, P, LREE and Eu of the HG cumulates.

In contrast to the GZ cumulates, the HZ felsic cumulates have appropriate Ba- and Sr-rich compositions for feldspar-rich accumulations from melts equivalent to the chilled border granites (Fig. 6). Barium contents are somewhat lower than predicted, which could be a consequence of the lower abundance of cumulate K-feldspar in most HZ cumulates than in the GZ cumulates from which the model mode was derived. Rare earth element patterns are also inconsistent with a direct relationship between putative input magmas and the GZ cumulate granites, but they support a link between the input magmas and HZ cumulates (Fig. 5c, d, e). The GZ cumulates have patterns that differ only subtly from those of the chilled border or transitional granites, rather than displaying the enrichment in light and, especially, middle REE and Eu expected for feldspar- and accessory mineral-rich cumulates. However, the HZ cumulates display the predicted patterns. Based on these lines of geochemical evidence, the authors consider it likely that the initial cumulates deposited from input granitic magmas are preserved within the HZ.

Their similar elemental compositions permit the border porphyry and the chilled border granite to represent the same evolutionary stage – probably quenched input melt. However, different texture, sharper contact with other units, and restriction to association with the felsic and miarolitic border granites suggest that the border porphyry was emplaced under different circumstances. The quenched textures of the felsic enclaves also are consistent with their being quenched input magmas, but they are mostly less felsic than the chilled border and the porphyry. Furthermore, the felsic enclaves are sufficiently variable that they are unlikely to have all had identical origins. They may represent input magmas which were mildly but variably contaminated by mafic magmas in or beneath the chamber, and subsequently quenched – possibly as dykes or at the margin – and disaggregated. Thus, in the authors' view, the chilled margin granite, border porphyry and felsic enclaves represent similar input magmas into the Aztec Wash chamber. The enclaves experienced variable contamination, probably in the HZ (Lazzareschi *et al.* 2004; Ericksen *et al.* 2004), and the aphanitic border porphyry may have quenched against a very cold wall early in the intrusive history. Furthermore, the large dyke swarm that cuts the pluton mostly comprises granites and rhyolites that have compositions indistinguishable from the chilled border granite (Coiner 2003; Steinwinder *et al.* 2004), consistent with this being the composition of major, continuing input into the system.

5.2. Punctuated history of the granite zone of the Aztec Wash magma chamber

Field relations in the GZ are grossly consistent with the simple evolutionary model outlined in section 5.1 (GZ cumulate to transitional granite to felsite dykes and felsic/miarolitic border granites, with chilled border granite and border porphyry representing parental input), but in detail, they suggest greater complexity. On the whole, compositions and textures indicate that cumulate rocks are more abundant at deeper levels, and the roof of the pluton was clearly a site of collection of highly felsic, fractionated melt. However, there is no simple, monotonic upward trend in either composition (Fig. 7) or texture – compositions and textures oscillate between more and less felsic, and between coarser and more cumulate-like and finer and less cumulate-like. Furthermore, small felsite dykes and pods with similarly fractionated compositions are present throughout the sequence. It is unlikely that all of these rocks mark the same segregation event – there is no evidence for continuous dyke networks, and it is implausible that trapped melts throughout a several-kilometre section all reached the same stage of fractionation simultaneously. A more likely interpretation is that fractionated melt was expelled into dykes and pods repeatedly as the granite accumulated.

The geochemical considerations cited in section 5.1 indicate that what appears to be the input magma is not the direct parent of cumulates currently exposed within the GZ – rather, these cumulates were deposited from melt that had already evolved from its initial composition. Furthermore, the border porphyry suggests an early stage of intrusion and solidification that preceded the crystal accumulation and melt segregation that produced most of the rocks of the GZ. Combining these interpretations with the evidence discussed in the preceding paragraph for punctuated, repeating episodes of crystal accumulation and melt segregation, a picture emerges of a felsic magma system in which events are generally, but *not* invariably, preserved in normal stratigraphic sequence (rocks higher in the pluton lithified later than those lower in the pluton). The system did not evolve in monotonic fashion.

5.3. Physical model for the evolution of the felsic portion of the Aztec Wash magma chamber

The relationship between the GZ and the HZ supports the notion of a complex, punctuated, potentially protracted history. The HZ records repeated recharging by mafic magmas, and physical, chemical and thermal interaction there between mafic and felsic magma (Falkner *et al.* 1995; Patrick & Miller 1997; Cheversia 2001; Coiner 2003). The prograding tongues of mafic and hybrid material that marked periods of copious mafic recharge interfingered laterally with accumulating felsic material to the north (see also Coiner 2003). These interactions must have induced thermal fluctuations in the adjacent GZ magmas and local mechanical perturbations at the HZ/GZ interface. Limited whole rock isotopic data (Falkner *et al.* 1995; Ericksen *et al.* 2004) and coherent elemental trends, as described in section 4, indicate that, for the most part, the GZ behaved as a chemically closed system (except immediately adjacent to its contact with the HZ). However, varying accessory mineral assemblages and Pb isotopic zoning in feldspars in the GZ (Robinson & Miller 1999; Cheversia *et al.* 2000, Cheversia 2001) suggest that the HZ may have subtly influenced the chemistry of the GZ.

The fluctuating physical influences of HZ processes provide a possible mechanism for inducing repeated cycles of crystal accumulation and melt segregation. The crystal fraction in the GZ is likely to have been sensitive to temperature fluctuations resulting from recharging of the HZ by mafic magmas. The dominant HZ influence on the GZ was probably thermal; cycles of crystal resorption (e.g. Robinson & Miller 1999), and of rapid growth and accumulation were probable consequences. Gravitational instability on sloping GZ/HZ interfaces also led to mechanical mixing of HZ clasts into GZ cumulates, perhaps as debris flows. The thickness of the lens of crystal-poor felsic magma within the GZ probably fluctuated widely, but it may generally have been thin and, at times, disappeared. Fractionated felsic melt segregated from the mush, locally concentrating in fractures and gashes to form the felsite dykes and pods. More generally, this fractionated melt ascended to the effective roof of the magma chamber – at times the lower edge of a coherent crystal mush zone (solidification front), and at times the country rock contact – eventually forming the felsic and miarolitic border zone rocks.

The border porphyry may be a remnant of an early upper solidification front, perhaps formed when the pluton was still small, thin and sill-like, and rapidly chilled. Felsic cumulates in the HZ appear to represent initial accumulations of crystals from input felsic magmas; the resulting moderately fractionated melts were the parents from which the GZ cumulates formed.

In summary, the present authors propose the following evolution of the Aztec Wash magma chamber/pluton (emphasising the granitic portion, now the GZ) (see Fig. 8). Except for steps 1, 8 and 9, these processes were repeated, so 2–7 do not represent a simple progression.

(1) Initial intrusion of felsic magma, which quickly solidified adjacent to the roof: Except for the very sparsely preserved border porphyry, there is no direct record of the evolution of the remainder of this magma. Most was presumably eroded by and disseminated into later, underlying felsic magma.
(2) Intrusion of mafic and felsic magmas: The chamber was recharged repeatedly by mafic magma throughout most of the remainder of its history to form the HZ. The evidence is less clear for felsic recharging, but it is unlikely that such a large mass of magma could have been emplaced in a single event and the repeated similar cycles of fractionation are more easily explained if there was felsic replenishment. The geometry of the stacks of mafic sheets that are the principal building blocks of the lobes, terminating to the north and thickening to the south, requires that mafic feeders were beneath the HZ to the south. Absence of

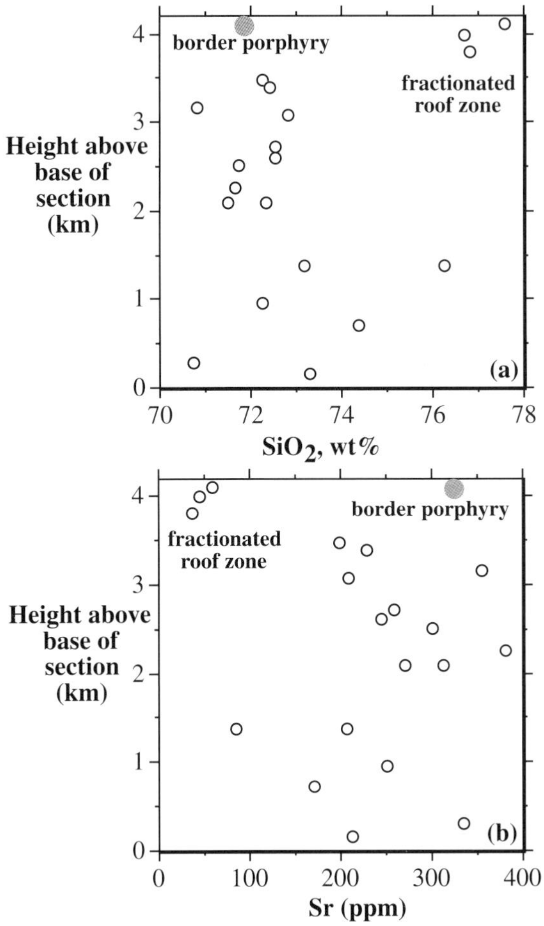

Figure 7 Height of the samples in reconstructed chamber versus elemental composition. Felsites, enclaves, heterogeneous zone cumulates, and distal samples from near the northern margin are excluded: (a) wt.% SiO_2 versus height; (b) ppm Sr versus height.

exposed felsic feeders in the GZ suggests that, if the chamber was recharged by felsic magma, it was probably injected into the HZ.

(3) Accumulation of crystals from the granitic magma and emplacement of overlying lava-flow-like mafic sheets, forming repeated sequences: HZ felsic cumulates were trapped between underlying older and overlying younger mafic flows. Fractionated melt segregated from the compacted felsic cumulates, accumulated roofward and spread to the distal, northern part of the chamber (GZ).

(4) Accumulation of crystals from the fractionated magma to form GZ cumulates, segregation of highly fractionated melt: The GZ cumulates formed from mildly fractionated melt from step 3; these cumulates graded into overlying transitional granites with smaller initial crystal fractions and more-evolved melt. As crystallisation progressed, highly fractionated melts injected fractures in the semisolid cumulate pile to form the felsite dykes and ascended to the roof as fluid-rich, highly fractionated melt, eventually forming the felsic border and miarolitic granites.

(5) Periodic increase in mafic input, resulting in expanding lobes of HZ mafic and hybrid material: The bases of the lobes mark progradation toward the north of the HZ as mafic sheets – essentially lava flows – were emplaced on top of GZ felsic cumulate mush.

(6) Debris flows from the tops of HZ lobes, mechanically mixing with GZ crystal mush.

(7) Waxing and waning of the melt-rich portion of the GZ, in response to varying thermal input from the HZ (and to felsic recharging?): This resulted in repeated cycles both of cumulate deposition, and probably of redissolution of crystals. Repeated development and destruction of an upper solidification front is likely.

(8) Final solidification.

(9) Continuing injection of magma similar to the initial granitic input after solidification of the pluton, emplaced as very large granite, rhyolite and mafic-felsic composite dykes (Falkner *et al.* 1995; Patrick & Miller 1997; Steinwinder *et al.* 2004).

6. Implications

6.1. Granite cumulates

Granite cumulates are widely considered unlikely or at least unusual, mainly because the viscosities of felsic melts are perceived to be too high to permit effective accumulation of crystals. However, recent viscosity measurements (Baker 1996; Scaillet *et al.* 1998) have indicated that crystal settling rates of more than a metre per year are likely in typically hydrous felsic magmas at temperatures of ~750–800°C, and field-based studies provide convincing evidence for downward crystal accumulation (e.g. Wiebe & Collins 1998; Bachl *et al.* 2001). Both the GZ and the HZ cumulates provide good examples of crystal accumulation, based on their fabrics and field relations. Elemental chemistry is equivocal in the case of the GZ cumulates – they are only slightly enriched in crystal-compatible elements compared with the presumed input magma; according to the present interpretation, this is because they precipitated from evolved melt. In contrast, the HZ felsic cumulates are strongly enriched in compatible elements, and based on both chemistry and texture, appear to have been especially crystal-rich, probably in part because of effective melt extraction through loading by dense overlying mafic flows. The height of the column of melt from which crystals accumulated remains uncertain – it could plausibly have been anywhere on a scale from metres to a several kilometres. In some cases, cumulates were probably temporarily deposited at one site and then remobilised as debris flows (see below).

6.2. Depositional processes in granites

The cumulate granites are part of the evidence for a protracted, cyclic depositional history in the Aztec Wash magma chamber (Fig. 8). The authors view this pluton as revealing a sedimentary sequence, both in a broad stratigraphic sense – for the most part, it was constructed from the bottom up – and in a more detailed way through its preservation of sedimentary structures that record processes remarkably similar to those that are observed at the Earth's surface. The most striking of these features, such as load casts and flame structures, are evident in the HZ (cf. Wiebe 1974; Patrick & Miller 1997; Wiebe & Collins 1998; Wiebe *et al.* 2002; Coiner 2003), but the GZ and the GZ/HZ contact also document the complex depositional system. Intertonguing of the HZ and GZ demonstrates prograding of the HZ complex onto laterally equivalent depositional mushes (GZ cumulates). The semisolid, sloping upper surfaces of these lobes were gravitationally unstable and shed debris into cumulate 'sediment' of the adjacent GZ. The general process is comparable to growth of a volcanic edifice within a depositional basin – mafic sheets are analogous to the flows that build the volcano, and cumulates are analogous to clastic sediment deposited within the basin as the volcano grows. Because magmas comprise solid particles in a fluid matrix, it is not surprising that they should behave in ways that are comparable to sediment. As with sediment, the solid particles may settle through the fluid, or if the particle fraction is sufficiently high, the material may move as a density current.

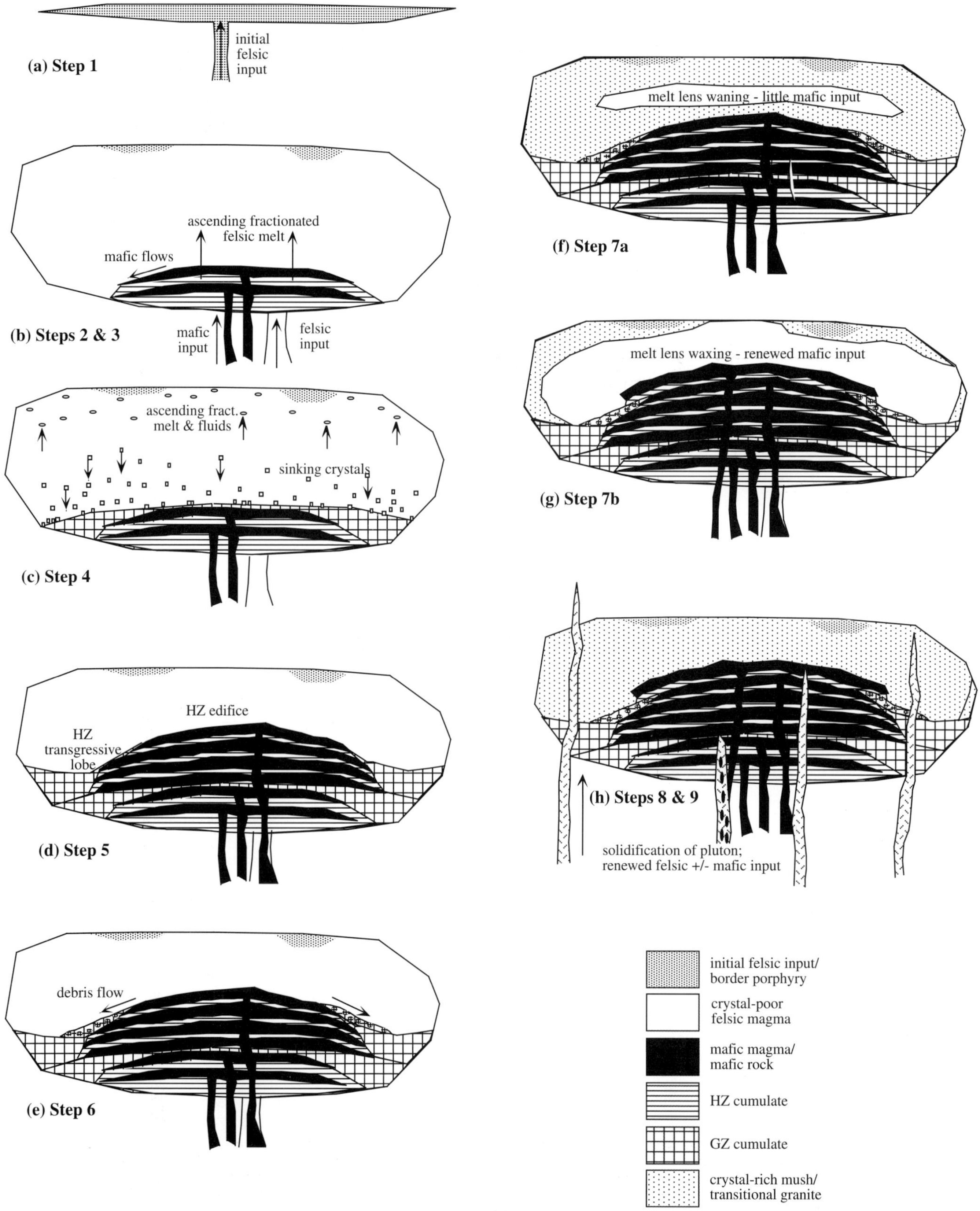

An important distinction is that the solid/fluid ratio varies not simply as a consequence of segregation processes, but also through precipitation and dissolution of crystals. Thus, gain or loss of heat or water can dramatically change material properties and behaviour of the 'sediment'. The permanence of the plutonic stratigraphic record can be compromised not only by mechanical erosion and mass wasting, but also by thermally or fluid-induced dissolution of deposits that diminishes the crystal fraction.

6.3. Plutons and the subsurface record of eruption and magma filtering

A major goal of understanding plutons is deciphering integrated intrusive–extrusive magmatic systems: How do the materials that remain permanently in magma chambers to form plutons relate to erupted products? What are the processes that trigger eruptions and how are they manifested in the plutonic record? There are thick volcanic sequences of

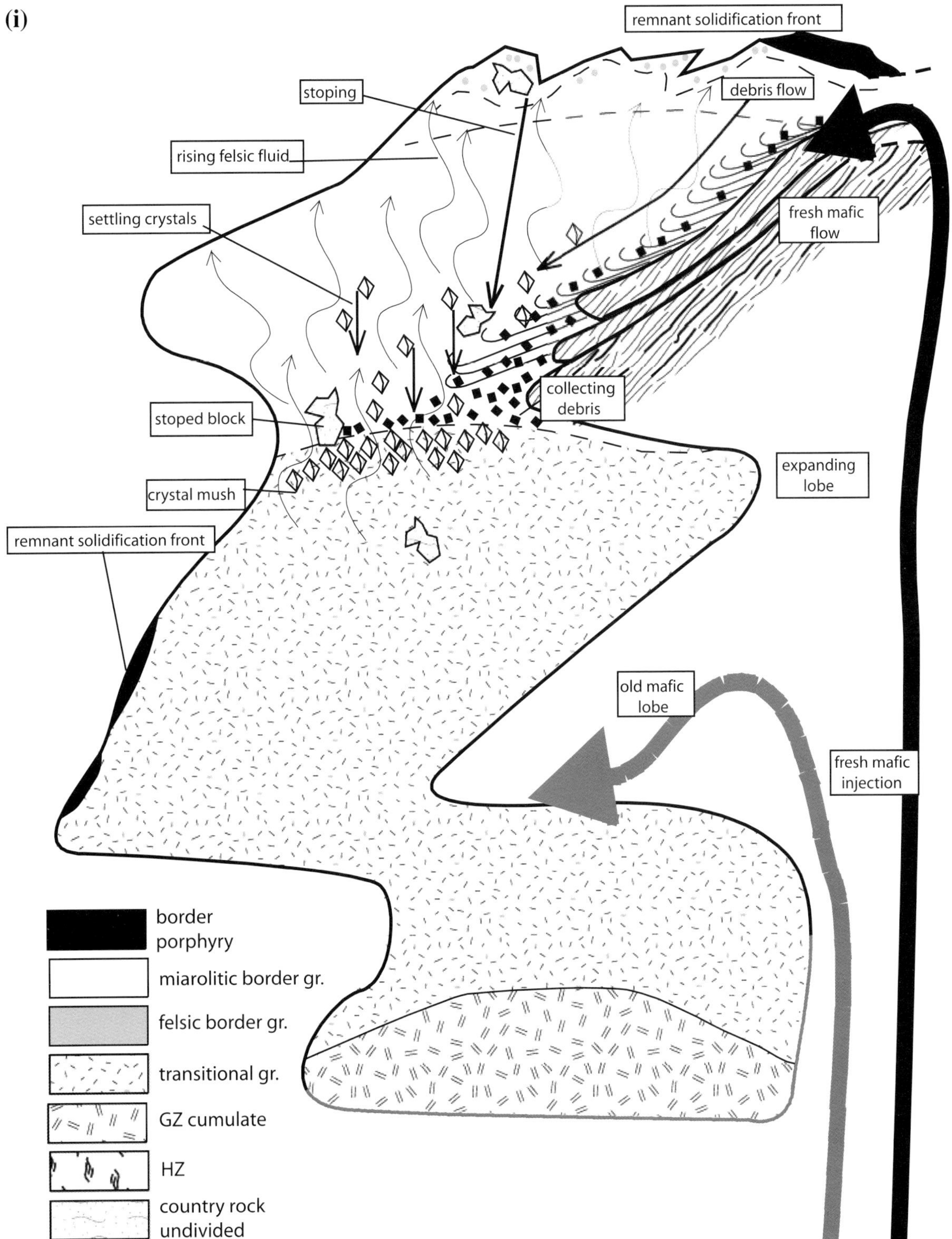

Figure 8 Conceptual view of active processes in and adjacent to the granite zone (GZ) portion of the Aztec Wash magma chamber. (a–h) Sequential (but in part repeating) processes which constructed and modified the pluton (see section 5.3 of the text): (a) step 1 – initial felsic input to form border porphyry; (b) steps 2 and 3 – mafic (lava flow-like) and felsic input, initial deposition of felsic cumulates, primarily in the heterogenous zone (HZ), and felsic melt segregation; (c) step 4 – deposition of GZ cumulates from evolved melt, ascent of highly felsic melt and water-rich fluids; (d) step 5 – construction of an HZ edifice with lobes prograding into the GZ; (e) step 6 – mass wasting of mechanically unstable materials from the slopes of the HZ edifice; (f) step 7a – periodic waning of the melt-rich lens in the GZ as thermal input diminishes; (g) step 7b – growth of the felsic melt lens with mafic (and/or felsic) replenishment and consequent introduction of heat; (h) steps 8 and 9 – pluton solidifies, continuing felsic and lesser mafic input forms large dykes. (After step 1, the pluton is shown as the same size for simplicity; presumably it was growing as magma was added and it may have periodically lost magma to eruption.) (i) A schematic view of the dynamics of the Aztec Wash magma chamber.

appropriate age that have erupted from the Aztec Wash chamber in nearby areas of southern Nevada and north-western Arizona (Gans *et al.* 1989, 1992; Faulds *et al.* 2001, 2002; Metcalf 2004). These sequences may contain the complementary record necessary to evaluate the eruptive history of the chamber and to compare erupted with unerupted products. As noted above, the authors interpret the GZ to include rocks that attained their compositions as both crystal-rich, uneruptible materials (cumulates) and highly mobile melt-rich materials (chilled border rocks, border porphyry, fractionated felsic/miarolitic border rocks and felsites, and felsic enclaves). The mobile magmas represented by the latter GZ rocks, which had the potential to erupt, are essentially bimodal. They include the equivalents of low-silica ($\sim$73 wt.% SiO_2) rhyolite, probably the input felsic magma, and fractionated high-silica rhyolite ($\sim$77 wt.% SiO_2). There is no obvious record of melt-rich magmas with compositions intermediate between these two extremes ever accumulating in appreciable quantities. Therefore, the authors suggest that it is these two compositions, rather than those of the main plutonic phases, that should be sought as the erupted manifestation of this felsic system (it is possible, but less likely, that more crystal-rich and less mobile magmas with up to $\sim$40% phenocrysts also erupted – perhaps like the GZ cumulates and some of the transitional granites; e.g. Bachmann *et al.* 2002). The record of eruptions and eruption triggers is most likely to be preserved in the HZ, which was the most dynamic part of the chamber, and the site of material and energy input (cf. Snyder & Tait 1996; Eichelberger *et al.* 2000; Hawkins & Wiebe 2000).

7. Acknowledgements

This work was supported by NSF grant EAR-0107094. Previous studies by Claudia Falkner, Dave Patrick and Delores Robinson formed the basis of the present investigation. Lorrie Coiner, Mary Beth Cheversia, Thomas Steinwinder, Heather Bleick, Shelly Ericksen, Jonathan Miller, Jim Faulds, Rick Hazlett and David Furbish all contributed many useful insights in the field and in continuing discussions. The paper was greatly improved as a consequence of very thorough and thoughtful reviews by Giray Ablay, Rod Metcalf and Roberto Weinberg.

8. References

Anderson, R. E., Longwell, C. R., Armstrong, R. L. & Marvin, R. F. 1972. Significance of K-Ar ages of Tertiary rocks from the Lake Mead region, Nevada-Arizona. *Geological Society of America Bulletin* **83**, 273–88.

Arth, J. G. 1976. Behavior of trace elements during magmatic processes; a summary of theoretical models and their applications. *Journal of Research of the U.S. Geological Survey* **4**, 41–7.

Bachl, C. A., Miller, C. F., Miller, J. S. & Faulds, J. E. 2001. Construction of a pluton: evidence from an exposed cross-section of the searchlight pluton, Eldorado Mountains, Nevada. *Geological Society of America Bulletin* **113**, 1213–28.

Bachmann, O., Dungan, M. A. &. Lipman, P. W. 2002. The Fish Canyon magma body, San Juan volcanic field, Colorado. Rejuvenation and eruption of an upper-crustal batholith. *Journal of Petrology* **43**, 1469–503.

Baker, D. R. 1996. Granitic melt viscosities. Empirical and configurational entropy models for their calculation. *American Mineralogist* **81**, 126–34.

Bergantz, G. W. 1999. On the dynamics of magma mixing by reintrusion; implications for pluton assembly processes. *Journal of Structural Geology* **22**, 1297–309.

Boynton, W. W. 1984. Cosmochemistry of the rare earth elements. *In* Henderson, P. (ed.) *Rare Earth Element Geochemistry*, 63–114. New York, NY: Elsevier.

Cates, N. L. 2003. *Perspectives on the development of an intermediate, open-system magma chamber, Aztec Wash pluton, Nevada.* Unpublished Master's thesis. Vanderbilt University, Nashville, Tennessee.

Cates, N. L., Miller, J. S., Miller, C. F., Wooden, J. L., Ericksen, S. & Means, M. 2003. Longevity of plutonic systems: SHRIMP evidence from Aztec Wash and Searchlight Plutons, Nevada. *Geological Society of America, Abstracts with Programs* **35**, 63.

Cheversia, M. B. 2001. *Investigating open system processes in magma chambers by in-situ Pb isotope analysis of feldspars, Aztec Wash and Mt. Perkins plutons, NV-AZ.* Unpublished Master's thesis. Vanderbilt University, Nashville, Tennessee.

Cheversia, M. B., Miller, C. F., Metcalf, R. V. & Coath, C. D. 2000. Documenting open-system growth of feldspars by in-situ Pb isotope analysis: Aztec Wash and Mt. Perkins plutons, Nevada-Arizona. *Geological Society of America Abstracts with Program* **32**, A-149.

Coiner, L. V. 2003. *Mafic-felsic interaction and cyclic stratigraphy in the Aztec Wash Pluton, Nevada.* Unpublished Master's thesis. Vanderbilt University, Nashville, Tennessee.

Coiner, L. V., Cates, N. L., Miller, C. F. & Harper, B. E. 2003. Cyclic stratigraphy in the Aztec Wash Pluton, Nevada. Vertical construction of a composite intrusion. *Geological Society of America, Abstracts with Programs* **35**, 63.

Collins, W. J., Richards, S. W., Healy, B. & Wiebe, R. A. 2000. Granite magma transfer, pluton construction, the role of coeval mafic magmas, and the metamorphic response. *Southeastern Lachlan Fold Belt. 15th Australian Geological Convention*, Field Guide FP3, 143. Sydney: Geological Society of Australia.

Eichelberger, J. C., Chertkoff, D. G., Dreher, S. T. & Nye, C. J. 2000. Magmas in collision: rethinking chemical zonation in silicic magmas. *Geology*, **28**, 603–6.

Ericksen, S. M., Miller, J. S., Miller, C. F., Harper, B. E. & Aggarwal, J. K. 2004. Isotopic constraints on the evolution of felsic magma in the Aztec Wash pluton, Eldorado Mountains, Nevada. *Geological Society of America, Abstracts with Programs* **36**, 8.

Falkner, C. M. 1993. *History and petrogenesis of Aztec Wash pluton, Eldorado Mountains, Nevada.* Unpublished Master's thesis. Vanderbilt University, Nashville, Tennessee.

Falkner, C. M., Miller, C. F., Wooden, J. L. & Heizler, M. T. 1995. Petrogenesis and tectonic significance of the calc-alkaline, bimodal Aztec Wash pluton, Eldorado Mountains, Colorado River extensional corridor. *Journal of Geophysical Research* **100**, 204–27.

Faulds, J. E., Geissman, J. W., Mawer, C. K. 1990. Structural development of a major extensional accommodation zone in the Basin and Range Province, northwestern Arizona and southern Nevada; implications for kinematic models of continental extension. *In* Wernicke, B. P. (ed.) *Basin and Range extensional tectonics near the latitude of Las Vegas, Nevada. Memoir – Geological Society of America* **176**, 37–76.

Faulds, J. E., Geissman, J. W. & Shafiqullah, M. 1992. Implications of paleomagnetic data on Miocene extension near a major accommodation zone in the Basin and Range Province, northwestern Arizona and southern Nevada. *Tectonics* **11**, 204–27.

Faulds, J. E., Feuerbach, D. L., Reagan, M. K., Metcalf, R. V., Gans, P. & Walker, J. D. 1995. The Mt. Perkins block, northwestern Arizona: an exposed cross section of an evolving, preextensional to synextensional magmatic system. *Journal of Geophysical Research* **100**, 15,249–66.

Faulds, J. E., Feuerbach, D. L., Miller, C. F. & Smith, E. I. 2001. Cenozoic evolution of the northern Colorado River extensional corridor, southern Nevada and northwest Arizona. *Utah Geological Association Publication 30 – Pacific Section American Association Petroleum Geologists Publication* **GB78**, 239–271.

Faulds, J. E., Olson E. L., Harlan, S. S. & McIntosh, W. C. 2002. Miocene extension and fault-related folding in the Highland Range, southern Nevada: a three-dimensional perspective. *Journal of Structural Geology* **24**, 861–86.

Feuerbach, D. L., Faulds, J. E. & Reagan, M. K. 1999. Interrelations between magmatism and extension in a major extensional accommodation zone, southern Nevada and northwest Arizona. *In* Faulds, J. E. (ed.) *Cenozoic Geology of the northern Colorado River Extensional Corridor, southern Nevada, northwestern Arizona: Economic implications of regional segmentation structures.* Nevada Petroleum Society 1999 Field Trip Guidebook, Reno, NV., 115–37.

Gans, P. B., Mahood, G. A. & Schermer, E. R. 1989. Synextensional magmatism in the Basin and Range Province: a case study from the eastern Great Basin. *Geological Society of America, Special Paper* **233**, 53.

Gans, P. B., Darvall, P. & Faulds, J. E. 1992. Miocene extension in the Colorado River extensional corridor; a comparison with slow-spreading ridges. *Geological Society of America, Abstracts with Programs* **24**, 25.

Gans, P. B. & Bohrson, W. A. 1998. Suppression of volcanism during rapid extension in the Basin and Range Province, United States. *Science* **279**, 66–8.

Harper, B. E. 2003. *Granite accumulation and fractionation in the "Homogeneous Zone" of the Aztec Wash Pluton, Eldorado Mountains, Nevada.* Unpublished Senior Honours thesis, Vanderbilt University, Nashville, Tennessee.

Harper, B. E., Miller, C. F., Wiebe, R. A., Cates, N. L. & Cribb, J. W. 2003. Granite accumulation and fractionation in a dynamic, open-system magma chamber, Aztec Wash Pluton, Eldorado Mtns, Nevada. *Geological Society of America, Abstracts with Programs* **35**, 63.

Hawkins, D. P. & Wiebe, R. A. 2000. Geological signatures of explosive eruptions in granitic plutons: examples from coastal Maine, USA. *Geological Society of Australia, Abstracts* **59**, 213.

Henderson, P. 1982. *Inorganic Geochemistry*. Oxford: Pergamon Press.

Howard, K. A., Wooden, J. L. & Simpson, R. W. 1996, Extension-related plutonism along the Colorado River extensional corridor. *Geological Society of America, Abstracts with Programs* **28**, A-450.

Howard, K. A. & John, B. E. 1987. Crustal extension along a rooted system of imbricate low angle faults: Colorado River Extensional Corridor, California, and Arizona. *In* Coward, M. P., Dewey, J. F. & Hancock, P. L. (eds) *Continental Extensional Tectonics. Geological Society, London, Special Publication* **28**, 299–311.

Koyaguchi, T. & Kaneko, K. 2000. Thermal evolution of silicic magma chambers after basalt replenishments. *Transactions of the Royal Society of Edinburgh: Earth Sciences* **91**, 46–60.

Lazzareschi, D., Miller, C. F., Koteas, G. C. Ericksen, S. M. & Hazlett, R. W. 2004. Fine grained felsic enclaves in a bimodal magma chamber, Aztec Wash pluton, Eldorado Mountains, Nevada: origins and implications. *Geological Society of America, Abstracts with Programs* **36**, 90.

Marsh, B. D. 1995. Solidification fronts and magmatic evolution. *Mineralogical Magazine* **60**, 5–40.

Metcalf, R. V. 2004. Volcanic-plutonic links, plutons as magma chambers and crust-mantle interaction: a lithospheric scale view of magma systems. *Transactions of the Royal Society of Edinburgh: Earth Sciences* **95**, 357–74.

Metcalf, R. V., Smith, E. I., Walker, J. D., Reed, R. C. & Gonzales, D. A. 1995. Isotopic disequilibrium among commingled hybrid magmas; evidence for a two-stage magma mixing-commingling process in the Mt. Perkins Pluton, Arizona. *Journal of Geology* **103**, 509–27.

Miller, J. S., Cates, N. L., Miller, C. F., Wooden, J., Means, M. A. & Ericksen, S. 2003. Magmatic construction and duration of solidification of Searchlight pluton, Eldorado Mountains, Nevada (USA). *EOS Transactions of the American Geophysical Union* **84**, F1630.

Patrick, D. W. 1996. *Implications of the style of intrusion into a composite, recharging magma chamber: the Aztec Wash pluton.* Unpublished Master's thesis. Vanderbilt University, Nashville, Tennessee.

Patrick, D. W. & Miller, C. F. 1997. Processes in a composite, recharging magma chamber; evidence from magmatic structures in the Aztec Wash Pluton, Nevada. *In* Li, Z., Qi J. & Zhang, Z. (eds) *Proceedings of the 30th international geological congress; igneous petrology. Proceedings of the 30th International Geological Congress* **15**, 121–35.

Reid, M. R. & Coath, C. D. 2000. In situ U-Pb ages of zircons from the Bishop Tuff; no evidence for long crystal residence times. *Geology* **28**, 443–6.

Robinson, D. M. 1997. *Investigating magma chamber dynamics through the examination of accessory minerals: the Aztec Wash pluton.* Unpublished Master's thesis. Vanderbilt University, Nashville, Tennessee.

Robinson, D. M. & Miller, C. F. 1999. Record of magma chamber processes preserved in accessory mineral assemblages. *American Mineralogist* **84**, 1346–53.

Rollinson, H. 1993. *Using geochemical data: evaluation, presentation, interpretation.* London and Chichester: Longman Scientific & Technical and John Wiley & Sons.

Scaillet, B., Holtz, F. & Pichavant, M. 1998. Phase equilibrium constraints on the viscosity of silicic magmas: 1, Volcanic-plutonic comparison. *Journal of Geophysical Research* **103**, 27,257–66.

Schmitt, A. K., Lindsay, J. M., de Silva, S. & Trumbull, R. B. 2002. U-Pb zircon chronostratigraphy of early-Pliocene ignimbrites from La Pacana, north Chile: implications for the formation of stratified magma chambers. *Journal of Volcanology and Geothermal Research* **120**, 43–53.

Simmons, E. C. & Hedge, C. E 1978. Minor-element and Sr-isotope geochemistry of Tertiary stocks, Colorado Mineral Belt. *Contributions to Mineralogy and Petrology* **67**, 379–96.

Snyder, D. & Tait, S. 1996. Magma mixing by convective entrainment. *Nature* **379**, 529–31.

Steinwinder, T. R., Miller, C. F., Faulds, J. E., Koteas, G. C. & Ericksen, S. M. 2004. Transition from plutonism to voluminous diking in the Eldorado mountains, northern Colorado River extensional corridor. *Geological Society of America, Abstracts with Programs* **36**, 8.

Vazquez, J. A. & Reid, M. R. 2002. Time scales of magma storage and differentiation of voluminous high-silica rhyolites at Yellowstone caldera, Wyoming. *Contributions to Mineralogy and Petrology* **144**, 274–85.

Watson, E. B. & Harrison, T. M. 1983. Zircon saturation revisited: temperature and composition effects in a variety of crustal magma types. *Earth and Planetary Science Letters* **64**, 295–304.

Weinberg, R. F. & Leitch, A. M. 1998. Mingling in mafic magma chambers replenished by light felsic inputs; fluid dynamical experiments. *Earth and Planetary Science Letters* **157**, 41–56.

White, J. C, Holt, G. S., Parker, D. F. & Ren, M. 2003. Trace-element partitioning between alkali feldspar and peralkalic quartz trachyte to rhyolite magma; Part I, Systematics of trace-element partitioning. *American Mineralogist* **88**, 316–29.

Wiebe, R. A. 1974. Coexisting intermediate and basic magmas, Ingonish, Nova Scotia. *Journal of Geology* **82**, 74–87.

Wiebe, R. A. 1993. The Pleasant Bay layered gabbro-diorite, Coastal Maine: ponding and crystallization of basaltic injections into a silicic magma chamber. *Journal of Petrology* **34**, 461–89.

Wiebe, R. A. 1994. Silicic magma chambers as traps for basaltic magmas; the Cadillac Mountain intrusive complex, Mount Desert Island, Maine. *Journal of Geology* **102**, 423–37.

Wiebe, R. A., Blair, K. D., Hawkins, D. P. & Sabine, C. P. 2002. Mafic injections, in situ hybridization, and crystal accumulation in the Pyramid Peak granite, California. *Geological Society of America Bulletin* **114**, 909–20.

Wiebe, R. A. & Collins, W. J. 1998. Depositional features and stratigraphic sections in granitic plutons: implications for the emplacement and crystallization of granitic magma. *Journal of Structural Geology* **20**, 1273–89.

BRIAN E. HARPER[1], CALVIN F. MILLER, G. CHRISTOPHER KOTEAS and NICOLE L. CATES[2],
Department of Geology, Vanderbilt University, Nashville, TN 37235, USA.

[1]Present address: Environmental Design International Inc., 200 South Michigan Avenue, Suite 700, Chicago, IL 60604, USA.
[2]Present address: Department of Geological Sciences, University of Colorado, Boulder, CO 80309–0399, USA.

ROBERT A. WIEBE, Department of Earth Sciences, Franklin and Marshall College, Lancaster, PA 17604–3003, USA.

DANIEL S. LAZZARESCHI, Department of Geology, Pomona College, Claremont, CA 91711, USA.

J. WARNER CRIBB, Department of Geosciences, Middle Tennessee State University, Murfreesboro, TN 37132, USA.

MS received 3 September 2003. Accepted for publication 4 August 2004.

Transactions of the Royal Society of Edinburgh: Earth Sciences, **95**, 297–307, 2004

Late Archaean Kenogamissi complex, Abitibi Subprovince, Ontario, Canada: doming, folding and deformation-assisted melt remobilisation during syntectonic batholith emplacement

Keith Benn

ABSTRACT: The Kenogamissi complex represents a large exposure of folded Late Archaean crystalline crust exposed within the Abitibi Subprovince, Ontario, Canada. It is composed of an heterogeneous amphibolite-grade orthogneiss unit, and several generations of batholiths and plutons of tonalite, granodiorite and granite composition. Together, the various units represent granitic magmatism during the period from 2740 Ma to 2660 Ma. Structural mapping and petrographic studies were focused on the orthogneiss unit (2723 Ma), on the newly defined Roblin tonalite-granodiorite batholith (ca. 2713 Ma) and on the highly strained metavolcanic rocks within the deformation aureole that surrounds the Kenogamissi complex. Structural analysis indicates that the Kenogamissi complex was emplaced into the greenstones as a dome that caused severe flattening and recumbent F2 refolding of earlier F1 folds in the deformation aureole. Doming is interpreted to be caused by the emplacement and inflation of tonalite-granodiorite batholiths, such as the Roblin Batholith, into the actively folding Swayze greenstone belt. Continued regional folding resulted in F3 refolding of F1 and F2 in the deformation aureole. Continued regional folding also deformed and folded the Kenogamissi complex and resulted in further uplift and emplacement of the complex into the greenstone belt. The early-formed magmatic foliation and compositional layering in the Roblin Batholith were folded by F3 while the batholith was still a crystal mush, and an F3 axial-surface magmatic foliation was locally formed. Folding of the partially molten Roblin Batholith also resulted in the remobilisation of fractionated liquids into shear zones which formed on the limbs of the F3 magmatic folds. Similar structures are present in the orthogneiss unit and are interpreted to represent remobilisation of melts which intruded the orthogneiss at the time of emplacement of the Roblin Batholith. The formation of the dykes on sheared fold limbs may be attributed to increased dilatancy during localised shearing of the crystal mush. Deformation-assisted remobilisation and extraction of fractionated liquids, and the possible transport of the fractionated liquids to higher levels in the crystallising Roblin Batholith, may have played a role in its magmatic differentiation.

KEY WORDS: Granite, melt extraction, pluton emplacement, Swayze greenstone belt.

The Late Archaean Abitibi Subprovince is one of the largest and most studied granite-greenstone terranes in the Superior Province of the Canadian Shield, and indeed, on Earth. It includes several greenstone belts, the largest of which are the mineral-rich Abitibi belt and the Swayze belt, which is situated in the SW part of the Abitibi Subprovince. In the Ontario (western) portion of the Abitibi Subprovince (Fig. 1), the supracrustal rocks have been lithologically and, to a lesser extent, structurally mapped in some detail. The mapping, together with petrological and geochronological studies have provided a fairly robust stratigraphic framework for the supracrustal rocks (Ayer *et al.* 2002a). Large batholiths within the Abitibi Subprovince have received less attention, especially from a structural point of view. The present paper describes new structural data from and interpretations of the Kenogamissi complex, a large crystalline complex in the western Abitibi Subprovince.

The Kenogamissi complex includes a suite of plutons of tonalite, granodiorite and granite compositions which represent magmatism spanning some 80 Ma, from 2740 Ma to 2660 Ma. Some of the plutonic rocks have been strongly deformed and locally transformed to orthogneiss of diorite and tonalite compositions, whereas other plutons preserve pristine igneous textures. In this paper, attention is focused on some key results of mapping and petrographic study of the orthogneiss, and a large, newly defined tonalite-granodiorite batholith, the Roblin Batholith, which together represent a large portion of the Kenogamissi complex.

In the course of mapping, it was demonstrated that the Kenogamissi complex represents a refolded dome emplaced into previously folded greenstones of the upper crust. The host greenstone belt and the Kenogamissi complex were then folded together. Emplacement of the Roblin Batholith occurred during doming and folding. The results and interpretations presented here suggest that the Kenogamissi dome was emplaced during the single long-lived, E–W upright folding event that deformed the entire Abitibi Subprovince, ca. 2715 Ma through 2700 Ma.

The Roblin Batholith was folded while it remained a crystallising mush, resulting in magmatic folds, in local overprinting of early-formed magmatic structures and in remobilisation of fractionated melts within the batholith. Magma remobilisation structures are documented within the orthogneiss and also within the Roblin Batholith, and it is suggested that magmatic differentiation of the batholith may have been enhanced by the regional deformation and folding.

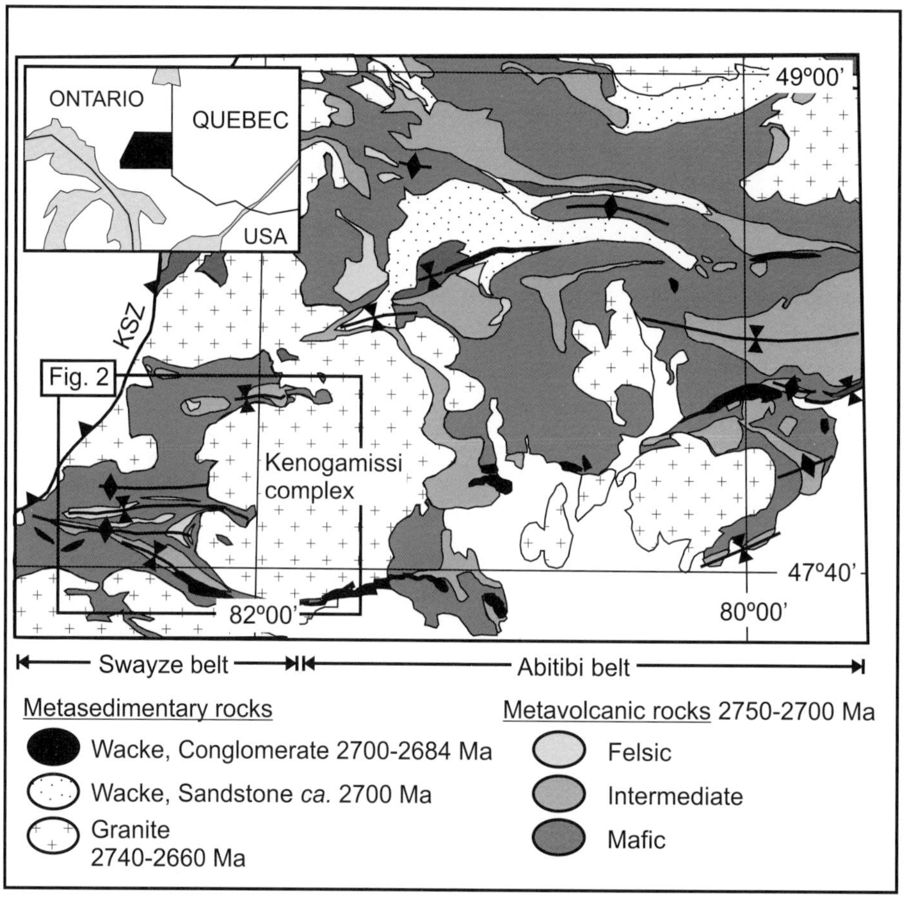

Figure 1 Map of the Abitibi Subprovince in Ontario, Canada, compiled from 1:100 000-scale maps published by the Ontario Geological Survey.

1. Geological setting

The Late Archaean Abitibi Subprovince is situated within the south-eastern portion of the Superior Province, in the Canadian Shield. In Ontario, the Abitibi Subprovince comprises several greenstone belts, of which the Abitibi greenstone belt (Fig. 1) is the largest and the most extensively studied. The smaller Swayze greenstone belt is situated in the western part of the Abitibi Subprovince (Figs 1 & 2), where its western limit corresponds to the Kapuskasing structure, a Proterozoic uplift of Archaean high-grade metamorphic rocks corresponding to a crustal-scale thrust or flower structure (Percival 1988; Nitescu & Halls 2002). The Kenogamissi complex, a large crystalline complex composed of multiple plutons of distinctly different ages, straddles the boundary between the Abitibi and Swayze greenstone belts (Fig. 1).

On its western and southern borders, the Kenogamissi complex crops out within greenschist-grade mafic to felsic metavolcanic rocks of the Swayze greenstone belt (Fig. 2). The metavolcanic units which host the Kenogamissi complex were deposited between approximately 2750 and 2705 Ma (Fig. 2; U-Pb zircon geochronological data reported in Heather 1999). The Swayze and Abitibi greenstone belts are deformed by one generation of upright, moderately doubly-plunging, E–W-trending regional-scale folds. In the region depicted in Figure 2, the older metavolcanic units, the Chester and Marion Groups, crop out in the south, whereas the younger metavolcanic units, the Trailbreaker Group, and the Newton and Brett Lake Formations, crop out in the north, suggesting that, viewed in profile, the regional-scale folds define an asymmetrical pattern with southward vergence.

2. Rock units and geochronology

The main rock types of the Kenogamissi complex are discussed with reference to Figure 2, which was compiled and modified from existing 1:50 000-scale maps (Heather 1999) with contacts between plutonic units added or modified according to field mapping by the present author. The plutonic units are presented in relative chronological age (oldest to youngest) based on cross-cutting relationships documented during mapping and on published single-zircon geochronological data (Heather 1999; Ayer *et al.* 2002b).

2.1. Amphibole-biotite tonalite

The amphibole-biotite tonalite has an apparent crystallisation age constrained by U-Pb zircon dates of $2741 \pm 2{\cdot}5$ to $2746 \pm 1{\cdot}3$ Ma which were obtained for samples from the south-western lobe of the Kenogamissi complex. The dates for the amphibole-biotite tonalite unit are comparable to the oldest exposed volcanic units of the Swayze and Abitibi greenstone belts. Tonalite plutons of similar ages elsewhere in the Abitibi Subprovince are interpreted to represent magma chambers associated with the older metavolcanic units (Sutcliffe *et al.* 1993). That interpretation is adopted here for the amphibole-biotite tonalite unit of the Kenogamissi complex. The unit is compositionally homogeneous and it has undergone partial dynamic recrystallisation under amphibolite-grade conditions, although igneous textures are locally preserved (Becker & Benn 2003). As a result, the rock is generally well foliated.

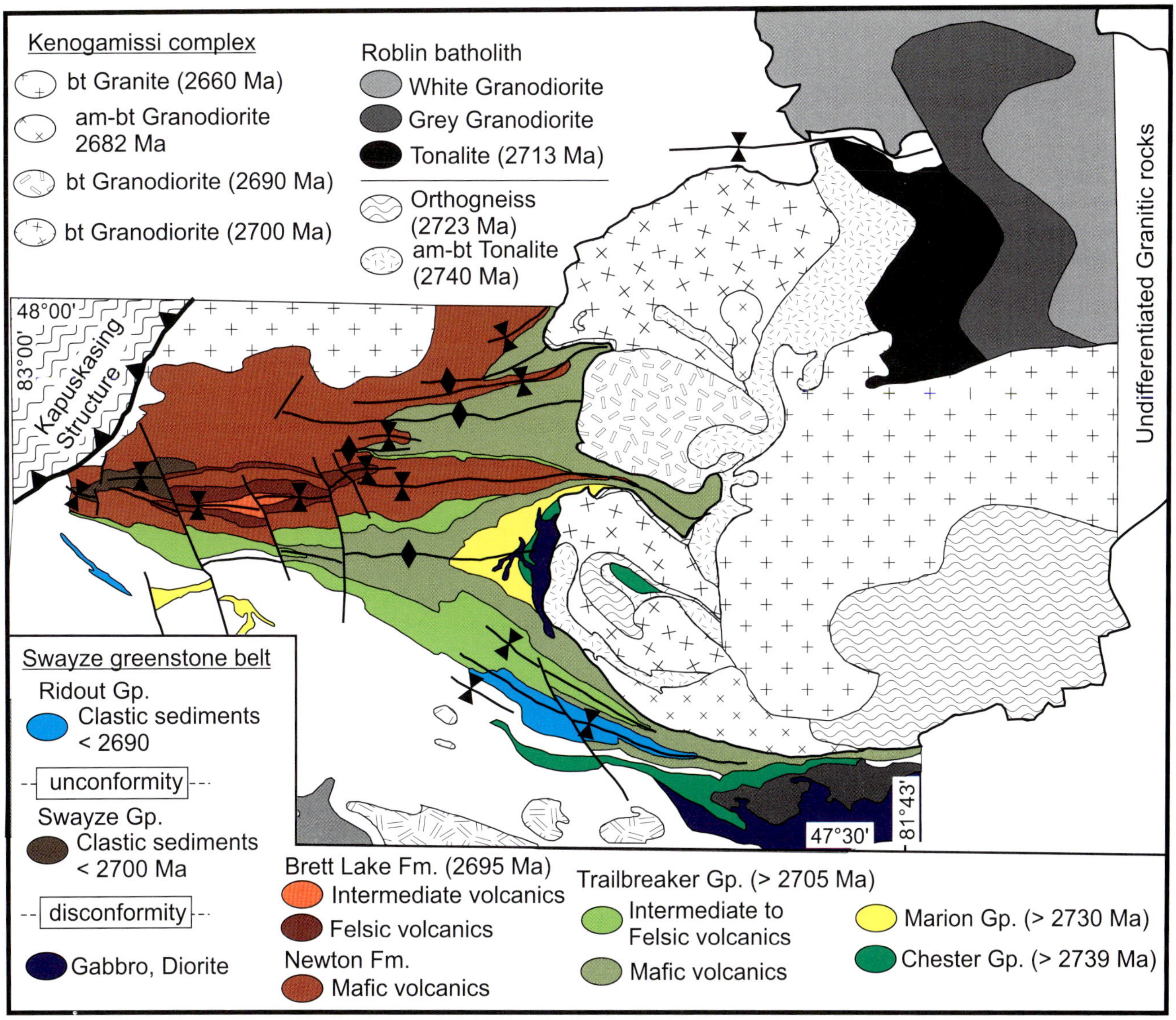

Figure 2 Geological map of the Swayze greenstone belt, and the central and western parts of the Kenogamissi complex (location indicated in Fig. 1). Compiled from 1:50 000-scale maps, and geochronological data published in Heather (1999) and Ayer *et al.* (2002b), with the internal contacts of the Kenogamissi complex modified according to mapping by the present author.

2.2. Orthogneiss

The orthogneiss unit is compositionally heterogeneous. It is mainly composed of medium-grained, mesocratic amphibole-biotite tonalite and finer grained biotite leucotonalite, with a lesser but significant proportion represented by fine- to medium-grained amphibolite, diorite, melanocratic tonalite and biotite granodiorite. Cross-cutting and xenolith relationships indicate the orthogneiss is composed of multiple generations of intrusions, most of which have been isoclinally folded (Fig. 3A).

Microprobe analyses (this author's unpublished data) revealed the amphiboles of the mesocratic tonalite to be hastingsite to hastingsitic ferro-hornblende in composition. The mesocratic tonalite component of the orthogneiss commonly contains mafic inclusions of melanotonalite to diorite composition (seen above the compass in Fig. 3B) which are interpreted to represent boudinaged xenoliths. The characteristic granoblastic microstructures and quartz ribbons (Fig. 4A) record dynamic recrystallisation and fabric development under the amphibolite-grade metamorphic conditions which accompanied isoclinal folding.

One U-Pb zircon date of 2723 Ma has been published for an outcrop of orthogneiss located within a few tens of metres of the one shown in Figure 3A. The orthogneiss is quite heterogeneous in composition, and may include components with different ages. Further geochronological study will be required to determine the ages of the different components.

2.3. The Roblin Batholith

Three easily distinguishable and mappable units are grouped together into a newly defined batholith, here named the Roblin Batholith (Fig. 2). The structurally lowest unit of the batholith is composed of tonalite that grades upward into grey and white granodiorite units.

The tonalite unit is typically strongly foliated and contains biotite ± dark green amphibole as ferrosilicate minerals. A U-Pb zircon date for a compositionally homogeneous outcrop of this unit yielded 2713 ± 3 Ma, which is interpreted to represent the crystallisation age.

The grey granodiorite unit overlies the tonalite. No intrusive relationships between this unit and the tonalite unit were documented in the field, and it is interpreted that the two units are gradational one to the other. Typically, amphibole is absent from this unit, which is characterised by large K-feldspar phenocrysts which, along with biotite crystals,

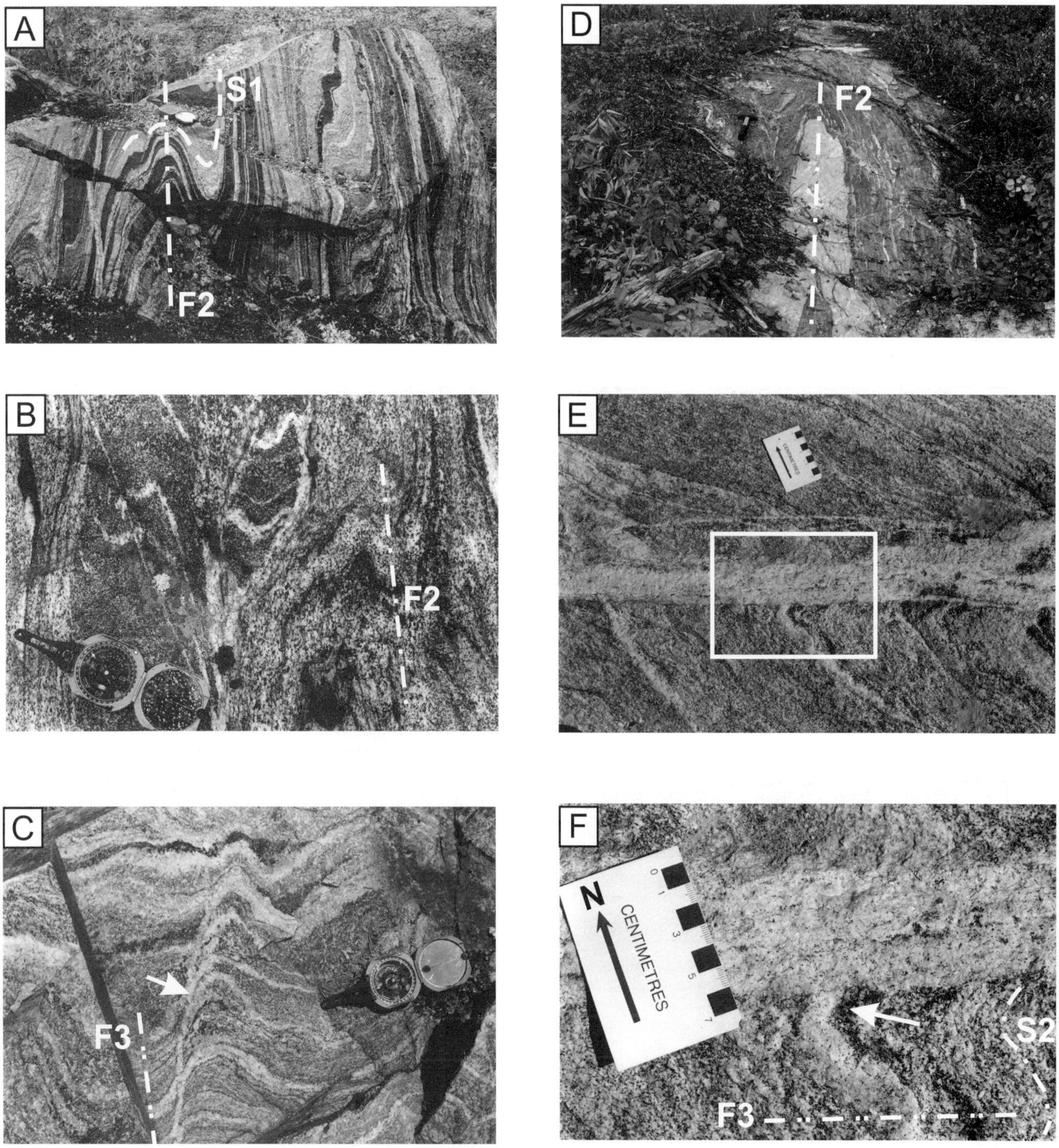

Figure 3 (A) F2 isoclinal folds of S1 and of the compositional layering in the orthogneiss unit. The F2 folds at that locality have been steepened by F3 folding. Compass for scale. (B) Boudins of melanocratic tonalite (darker) containing F2-folded leucocratic veins, hosted within a medium-grained tonalite facies that has also been deformed by F2. Compass for scale. (C) Compositional banding in the orthogneiss unit folded by F3. The arrow indicates a leucocratic dyke on the limb of an F3 fold that cross-cuts the banding and the S2 foliation, and that terminates in a banding-parallel sill. The outcrop surface is vertical. Compass for scale. (D) An F2-folded granodiorite dyke (light grey) that crops out within the deformation aureole of the Kenogamissi complex. The F2 fold has been steepened by F3 folding. Hammer for scale. (E) Leucocratic dyke emplaced on the sheared limb of an F3 fold in the tonalite unit of the Roblin Batholith. The rectangle shows the location of Figure 3F. (F) Detail of the dyke shown in Figure 3E, showing its contiguity with the S2 foliation-parallel leucocratic bands.

define the foliation. It is distinguished from the unit described next by the relative abundance of biotite (15–20% modal abundance), which is responsible for the grey colour of the rock in outcrop.

The white granodiorite (to granite) unit is distinguished from the previous one by its white colour in outcrop because of its more leucocratic composition. Biotite represents only about 5–8% modal percentage of the rock and it contains abundant K-feldspar phenocrysts. Blocks of the grey granodiorite unit are indentified within this unit. The magmatic foliation within the blocks is locally concordant to the magmatic foliation in the host white granodiorite, but in other cases, it is discordant. The blocks are interpreted to represent autoclaves of the earlier crystallised grey granodiorite components of the batholith.

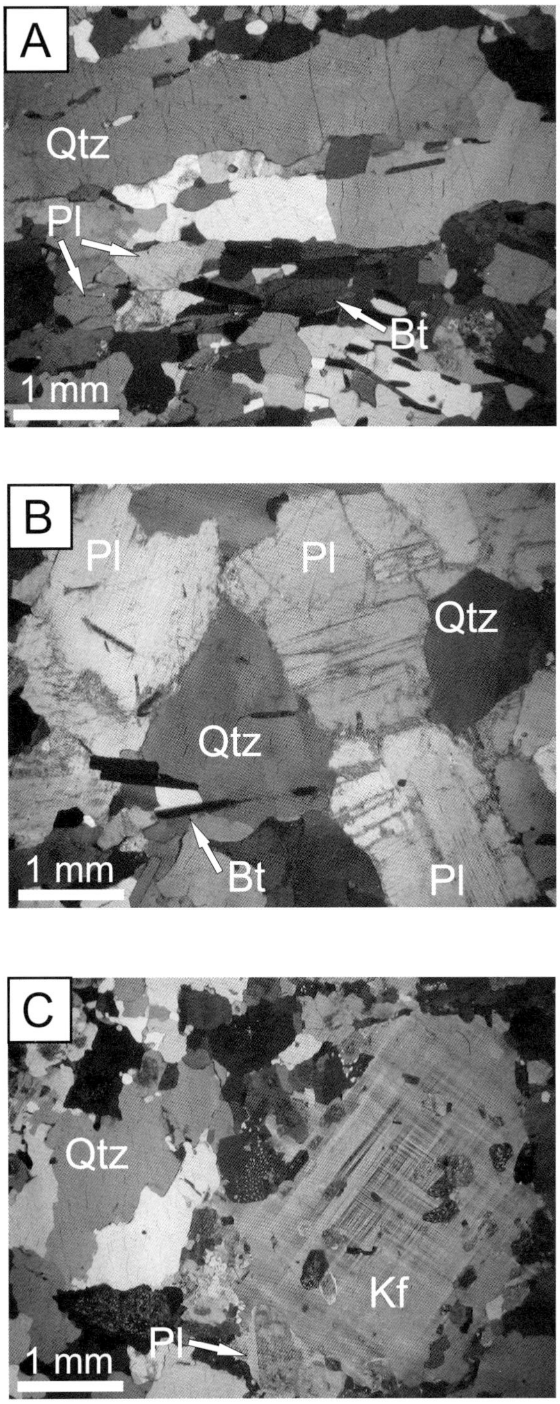

Figure 4 Photomicrographs of textures in three rock units: (A) Tonalite orthogneiss with a fine-grained dynamically recrystallised quartzo-feldspathic matrix and highly elongate quartz ribbons; (B) Tonalite unit of the Roblin Batholith. The sample has a well-preserved igneous texture with subhedral to euhedral plagioclase crystals and interstital quartz grains which have slightly undulose extinction; (C) White granodiorite unit of the Roblin Batholith. The igneous texture is indicated by the presence of subhedral to euhedral K-feldspar phenocrysts and plagioclase crystals, and coarse-grained interstitial quartz that shows evidence of minor subgrain development. Abbreviations: (Qtz) quartz; (Pl) plagioclase; (Bt) biotite; and (Kf) K-feldspar.

2.4. Biotite granodiorites

Two biotite granodiorite units are indentified within the Kenogamissi complex (Fig. 2) based on geochronological data and textural observations. Samples of the older unit yielded zircon dates within the range 2700 ± 2 Ma (Ayer *et al.* 2002b) from outcrops in the south-western lobe of the Kenogamissi complex. The younger of the biotite granodiorite units is considered to belong to a suite of plutons which yielded zircon dates of ca. 2690 Ma (Heather 1999). The two units may be distinguished in the field by the common presence of biotite-rich enclaves in the younger one (Heather 1999). Both biotite granodiorite units are texturally distinct from the granodiorites of the Roblin Batholith in that they are devoid of phenocrysts.

2.5. Amphibole-biotite granodiorite

This unit crops out mainly along the southern margin of the Kenogamissi complex. It contains abundant phenocrysts of K-feldspar and it is easily distinguished from the other granodiorite units by the ubiquitous presence of dark green amphibole, and also by the common presence of subangular xenoliths or enclaves of amphibolitic material which are a few centimetres in size. A published U-Pb zircon date for the unit is 2682 ± 3 Ma (Heather 1999).

2.6. Biotite granite

The youngest intrusive unit in the study area is a fine- to medium-grained biotite granite, commonly associated with abundant coeval pegmatite. The rock is easily identified by its distinctive pink colour in outcrop and by its common association with pegmatite. Observations of cross-cutting relationships in the field show that this unit intruded all other plutonic units in the study area. It forms abundant dykes which occur in most outcrops, and a large map unit is defined (Fig. 2). A U-Pb zircon date for this unit is 2660 Ma.

3. Structural geology

In this section, the present author summarises the essential elements of the structural geology of the study area. The focus is on structures in the marginal greenstones which form a narrow (10^2 to 10^3 m) deformation aureole around the Kenogamissi complex, on metamorphic structures in the orthogneiss unit and on magmatic structures in the Roblin Batholith. Melt remobilisation structures in the orthogneiss and in the Roblin Batholith are also documented.

Figure 6A is a map of the Kenogamissi complex that includes representative measurements of structures in the orthogneiss and the Roblin Batholith. Figure 6B is a profile through the batholith along the line A′–B′ in Figure 6A. Figure 7 presents equal-area nets of structural measurements.

3.1. Structures in the marginal greenstones

The marginal greentones of the deformation aureole surrounding the Kenogamissi complex have been deformed by three generations of folds. Two of them can be neatly summarised with reference to the fold interference pattern in Figure 5A, which is a photograph of an horizontal outcrop surface. The outcrop is located in Figure 6B.

Figure 5A shows a type 3 refold interference pattern (Ramsay 1967) involving F1 and F2 folds (see also the inset in Fig. 6B). Both generations of folds, F1 and F2, have an associated axial planar foliation (S1 and S2, respectively).

The F2 folds in Figure 5A are in an upright orientation (the axial surfaces dip steeply) because of rotation by later E–W- to NW–SE-trending regional F3 folds (Fig. 6A) which are upright and gently to moderately doubly plunging. The upright F3 folds record the regional horizontal bulk N–S shortening that deformed the Abitibi Subprovince. Prior to F3 folding, the F2 folds must have been shallowly dipping (recumbent) such that they were refolded during the F3 bulk horizontal shortening event.

Refold interference patterns of type 3 require that the two generations of folds, in this case F1 and F2, have axial surfaces which were perpendicular (or nearly so) and fold axes which

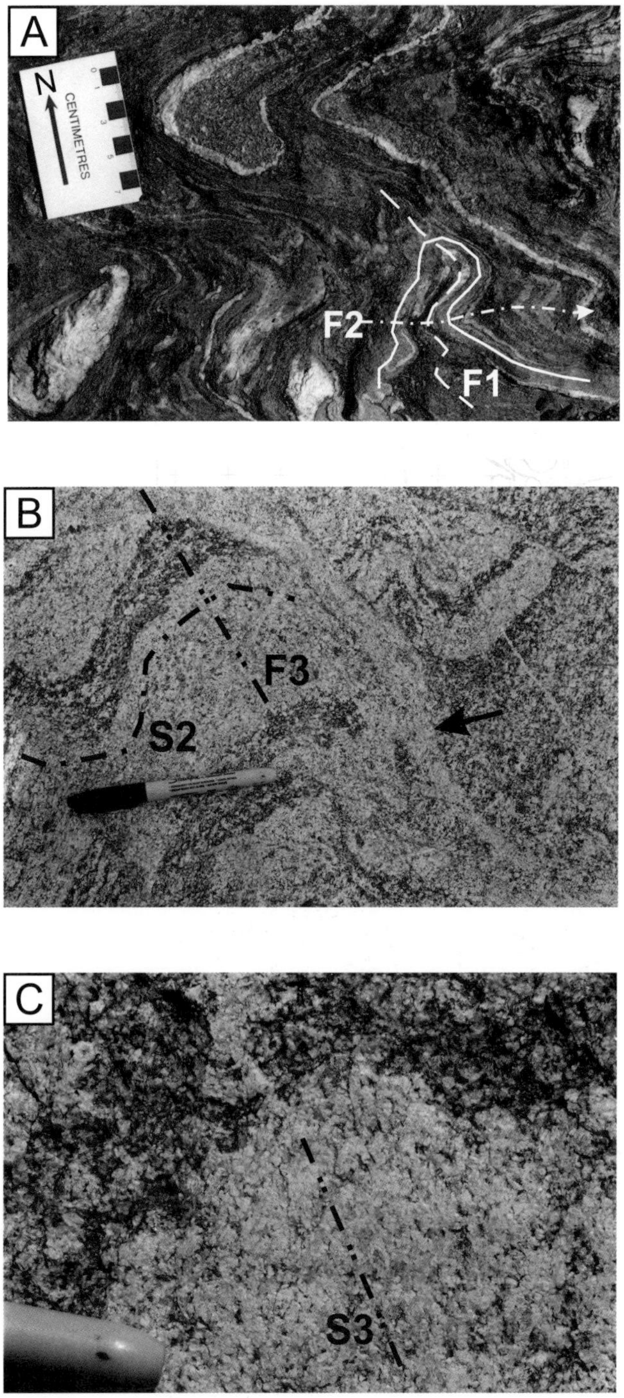

Figure 5 (A) Photograph of a horizontal greenstone outcrop surface in the deformation aureole of the Kenogamissi complex. See Figure 6B for the location. The type 3 fold interference pattern is caused by refolding of F1 folds by F2 folds. (B) F3 magmatic folds in an outcrop where the grey and white granodiorite units of the Roblin Batholith are interlayered. Note the leucocratic dyke (indicated by an arrow) within the sheared limb of an F3 fold. Pen for scale. (C) Detail of the contact between layers of grey and white granodiorite, in the hinge of an F3 magmatic fold in the granodiorites of the Roblin Batholith. The S3 foliation, defined by bt crystals, overprints the S2 foliation in the leucocratic layer. The pen tip provides the scale.

were parallel to oblique (Thiessen 1986). The type 3 refold pattern in Figure 5A requires that the axial surfaces of F1 folds were upright, perpendicular to originally recumbent axial surfaces of F2 folds. The type 3 pattern also requires that the F1 hinges were parallel to oblique to the F2 hinges. F2 folds are also documented in the orthogneiss unit (section 3.2; Fig. 7C).

3.2. Metamorphic structures in the orthogneiss

The orthogneiss unit is characterised by a penetrative gneissic banding and a parallel mineral foliation (Fig. 3A). The foliation is interpreted to represent an S1 fabric, as it is folded around the hinges of isoclinal F2 folds. It is interpreted to be correlative with the S1 foliation in the marginal greenstones. A mineral lineation, defined mainly by dark green prismatic amphibole, was measured in the S1 plane and it is interpreted to represent an L1 lineation.

The S1 foliation is folded into isoclinal F2 folds (Fig. 3A, B) which are commonly rootless. The constructed F2 fold axis, based on the distribution of S1 poles, suggests a NE–SW-trending and shallowly plunging fold axis (Fig. 7A). Dispersion of the S1 poles from the great circle in Figure 7A may be partly caused by later F3 folding. The L1 lineation orientations are quite scattered (Fig. 7B), presumably because of the subsequent F2 and F3 folding. The F2 fold hinges measured in the field define a NE–SW striking, vertical partial girdle caused by later F3 folding about a SE-trending axis (Fig. 7C). The distribution of the F2 fold hinges on a great circle (Fig. 7C) suggests that the axis of later F3 folding was roughly perpendicular to the F2 hinges.

In Figure 3 parts A and B, the axial surfaces of F2 folds have been tilted to an upright orientation by F3 folding. Many other outcrops have been mapped where the isoclinal F2 folds are in a recumbent orientation within hinge zones of the later F3 folds. The F2 folds in the orthogneiss unit are interpreted to have formed as recumbent folds which record intense flattening during the same event that caused F2 recumbent folding in the overlying marginal greenstones.

3.3. Magmatic structures in the Roblin Batholith

Igneous textures are well preserved in most outcrops of the three units of the Roblin Batholith. Petrographic study reveals the preservation of euhedral to subhedral biotite, plagioclase and K-feldspar crystals, and interstital quartz grains, the deformation of which is limited mainly to undulose extinction and minor subgrain development (Fig. 4B, C; cf. criteria for indentifying igneous fabrics in Bouchez *et al.* 1990; Paterson *et al.* 1989; Vernon 2000). Textural evidence for dynamic recrystallisation of quartz is observed in thin sections from some outcrops of the tonalite unit. The pervasive foliation that is present in all three units is interpreted to be of magmatic origin. It is defined by biotite and plagioclase crystals in the tonalite unit, and also by K-feldspar crystals in the grey and white granodiorite units. Near the mapped contact between the grey and white granodiorite units, compositional banding is also parallel to, and highlights, the foliation (Fig. 5B). This observation suggests an interlayering of the two units in a gradational transition zone between them.

The magmatic foliation in the Roblin Batholith is folded by one generation of upright folds, interpreted to be F3 (Fig. 5B, C), so the foliation must predate the F3 folds. Within the deformation aureole that surrounds the Kenogamissi complex, granodiorite dykes, thought to represent Roblin Batholith magmas, cross-cut the S1 foliation in the greenstones, are folded by F2 (Fig. 3D) and contain an S2 axial planar foliation, suggesting post-D1, pre- to syn-D2 emplacement of the Roblin Batholith. Based on these observations, the magmatic foliation in the Roblin Batholith is designated an S2 foliation.

The originally shallow dip angle of the S2 magmatic fabric is preserved in the hinge zones of the F3 folds, but in many outcrops, S2 has been rotated into steeper orientations (Fig. 7D). In the mapped area, measured F3 magmatic fold hinges trend to the southeast and have an average plunge angle of about 30° (Fig. 7E). The F3 folds of the magmatic fabric in

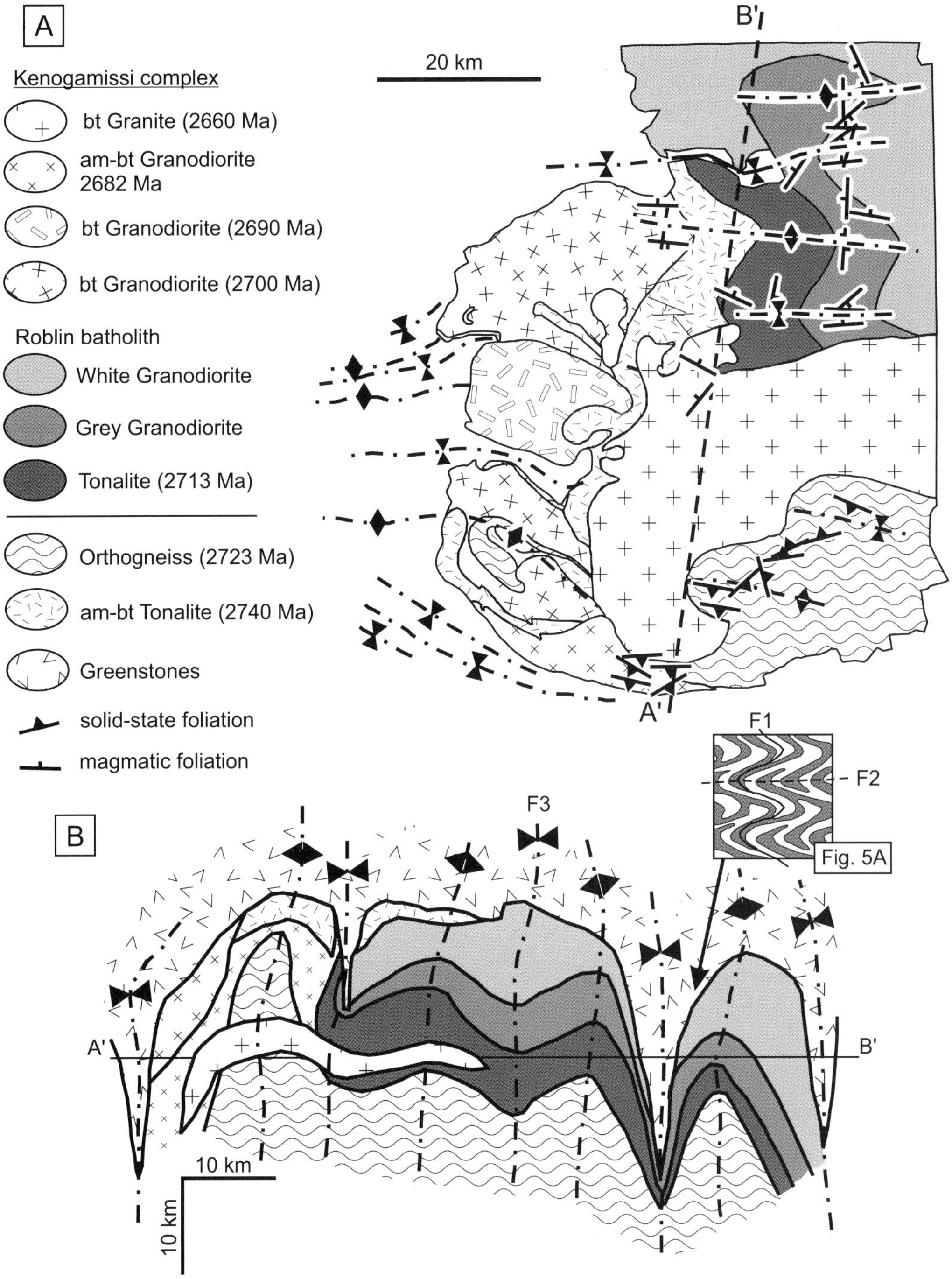

Figure 6 (A) Map of the Kenogamissi complex showing representative measurements of metamorphic foliation in the orthogneiss unit and in deformed rocks near the southern margin of the complex, and magmatic foliations in the Roblin Batholith. (B) Profile through the Kenogamissi complex based on mapping by the present author in the orthogneiss unit and in the Roblin Batholith, and on existing geological maps (Heather 1999). No vertical exaggeration is implied; however, the vertical dimensions of units are approximate. The inset shows the location of Figure 5A and an ideal F3 refold interference pattern. The profile line is shown in Figure 6A.

Figure 5B occurred while the granodiorites were still partially molten, as indicated by the preservation of igneous textures in the rock and also by the presence of the melt remobilisation structures discussed in the next section. Locally, F3 folding also resulted in the formation of an S3 axial planar foliation, recognised in the field by alignments of biotite crystals (Fig. 5C). Petrographic criteria indicate that the S3 foliation in the granodiorites is also a magmatic fabric.

Interpretations regarding the emplacement of the Roblin Batholith and its role in doming of the Kenogamissi complex must account for the observations and structural data in the previous paragraphs. First, the Roblin Batholith was emplaced during the D2 deformation event, which gave rise to the recumbent F2 folds in the orthogneiss and in the greenstones of the deformation aureole. Secondly, during or following emplacement, a horizontal magmatic foliation was developed throughout the batholith. Thirdly, the magmatic fabric was then folded by F3 while the Roblin Batholith was still a crystal mush.

3.4. Melt remobilisation structures

In the orthogneiss unit and in the three units of the Roblin Batholith, outcrop-scale evidence suggests deformation-assisted extraction and remobilisation of magmatic liquids during the F3 folding event. Examples of melt remobilisation structures are shown in Figures 3C, E, F and 5B.

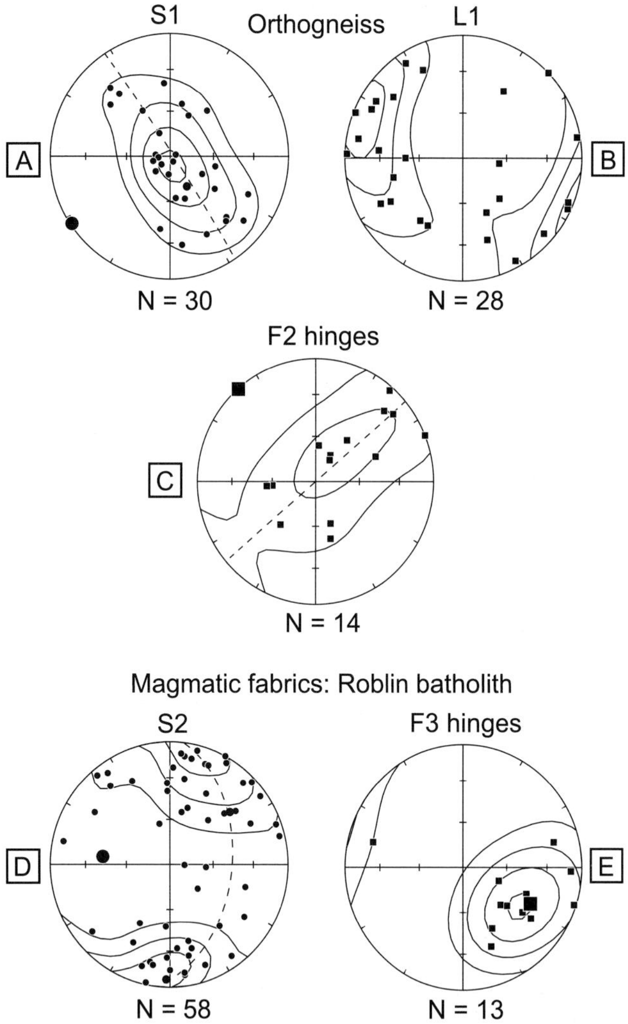

Figure 7 Lower-hemisphere equal-area projections of structural measurements: (A) S1 in the orthogneiss unit. The dashed great circle is the best-fit plane and the large dot indicates the constructed F2 fold axis; (B) L1 in the orthogneiss unit; (C) F2 fold hinges in the orthogneiss unit. The dashed great circle is the best fit plane and the large square indicates the constructed F3 fold axis; (D) S2 magmatic foliation in the Roblin Batholith. The dashed great circle is the best-fit plane and the large dot indicates the constructed F3 fold axis; (E) Field measurements of F3 fold hinges in the Roblin Batholith. The large square indicates the average orientation.

Figure 3C shows an example from the orthogneiss unit, where a leucocratic dyke a few centimetres wide (indicated by the arrow) cross-cuts the folded S2 foliation and the gneissic banding. The photograph is of a vertical outcrop surface. Near the top of the photograph the dyke terminates within one of the leucocratic bands, which is, in essence, a sill. The leucocratic dyke is parallel to, and it is located on, the limb of a mesocopic-scale F3 fold (Fig. 3C). The presence of syn-F3 melt remobilisation structures within the orthogneiss unit raises a question as to the origin of the melt. Detailed petrographic study of the orthogneiss revealed no mineral assemblage that could be attributed to melting reactions, and therefore, the present author rules out the possibility that the rocks could represent migmatites.

According to available geochronological data, the orthogneiss is 10 Ma older than the younger Roblin Batholith. It is unlikely that the vertical dyke in Figure 3C was preserved through the F1 and F2 folding events. It is more likely that the leucocratic dykes and sills in the orthogneiss represent magma that intruded the orthogneiss during emplacement of the tonalite-granodiorite magmas which formed the Roblin Batholith and possibly other, as-yet-unrecognised intrusions.

Figure 3 parts E, F are photographs of an horizontal outcrop surface and they show a similar melt remobilisation structure within the tonalite unit of the Roblin Batholith. There, a leucocratic dyke is in petrographic continuity with S2-parallel leucocratic bands of the same composition within the tonalite. The dyke is parallel to and situated on the limb of an F3 fold. The interpretation is that differentiated magmatic liquid formed narrow bands parallel to S3, and that the liquid was extracted and remobilised into dykes which formed on the sheared limbs of upright F3 folds, as in the outcrop depicted in Figure 3C.

Figure 5B corresponds to a dipping outcrop surface (45° dip angle) in the Roblin Batholith where the grey and white granodiorites are interlayered. The S2 foliation defined by K-feldspar phenocrysts is parallel to the compositional banding. A leucocratic dyke (indicated by the arrow) is parallel to and situated on the limb of an F3 fold of the S2 magmatic foliation and of the compositional banding. The compositional layers in the granodiorite do not match up across the dyke, indicating that the dyke was emplaced within a magmatic shear zone that corresponds to the sheared limb of the F3 fold.

4. Discussion

The present paper provides an overview of the structural geology and the structural evolution of the Kenogamissi complex, a large granitoid-gneiss complex that crops out within greenschist-grade supracrustal assemblages of the Late Archaean Abitibi Subprovince. The study has concentrated on the marginal greenstones in the deformation aureole that surrounds the Kenogamissi complex, on the orthogneiss unit that crops out in the southern part of the Kenogamissi complex and on the newly defined Roblin Batholith that makes up much of its northern part. Figure 8 summarises the interpreted history of deformation and batholith emplacement within the study area, from ca. 2720 through 2700 Ma.

From the profile in Figure 6B, the original prefolding dimensions of the Roblin Batholith are estimated to be at least 80 km in horizontal dimension and about 10 km in thickness. With regards to the vertical thickness of the batholith determined from the profile, it should be noted that the upper margin of the batholith was projected tens of kilometres along periclinal (doubly plunging) fold axes through some regions which require further detailed mapping. Therefore, the thickness of the batholith shown in the profile is estimated to be accurate to ± 2 km. Also, the thickness may well have been increased and decreased locally during syncrystallisation F3 folding if magma was squeezed upward into the cores of the F3 folds shown in Figure 6B.

Mapping results and interpretations of field and petrographic observations indicate that the orthogneiss and the Roblin Batholith underwent the regional folding event that deformed the host greenstone terrane. Field observations also indicate that F3 folding was ongoing while the Roblin Batholith was still partially molten, and that folding led to deformation-assisted melt extraction and remobilisation within the orthogneiss and the Roblin Batholith.

In this section, the present author discusses the emplacement of the Kenogamissi complex as a dome that was inflated by the emplacement of the Roblin Batholith, within the folding greenstone belt. The potential petrological significance of the evidence for deformation-assisted melt remobilisation within the Kenogamissi complex is also discussed.

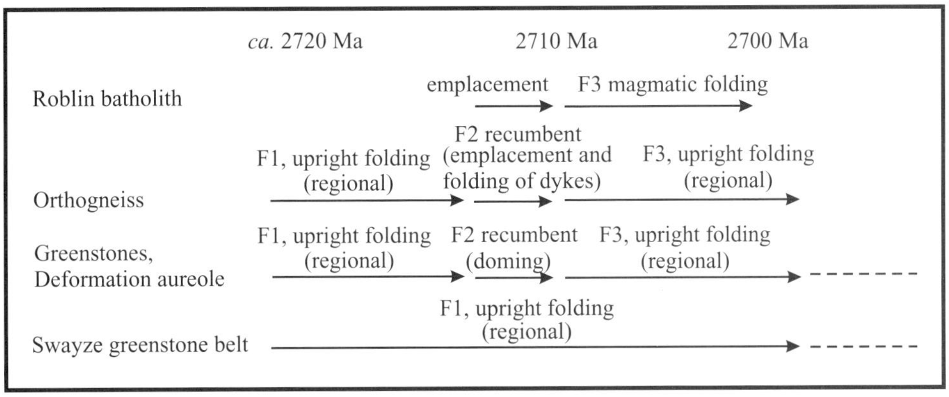

Figure 8 Summary of deformation and batholith emplacement in the study area.

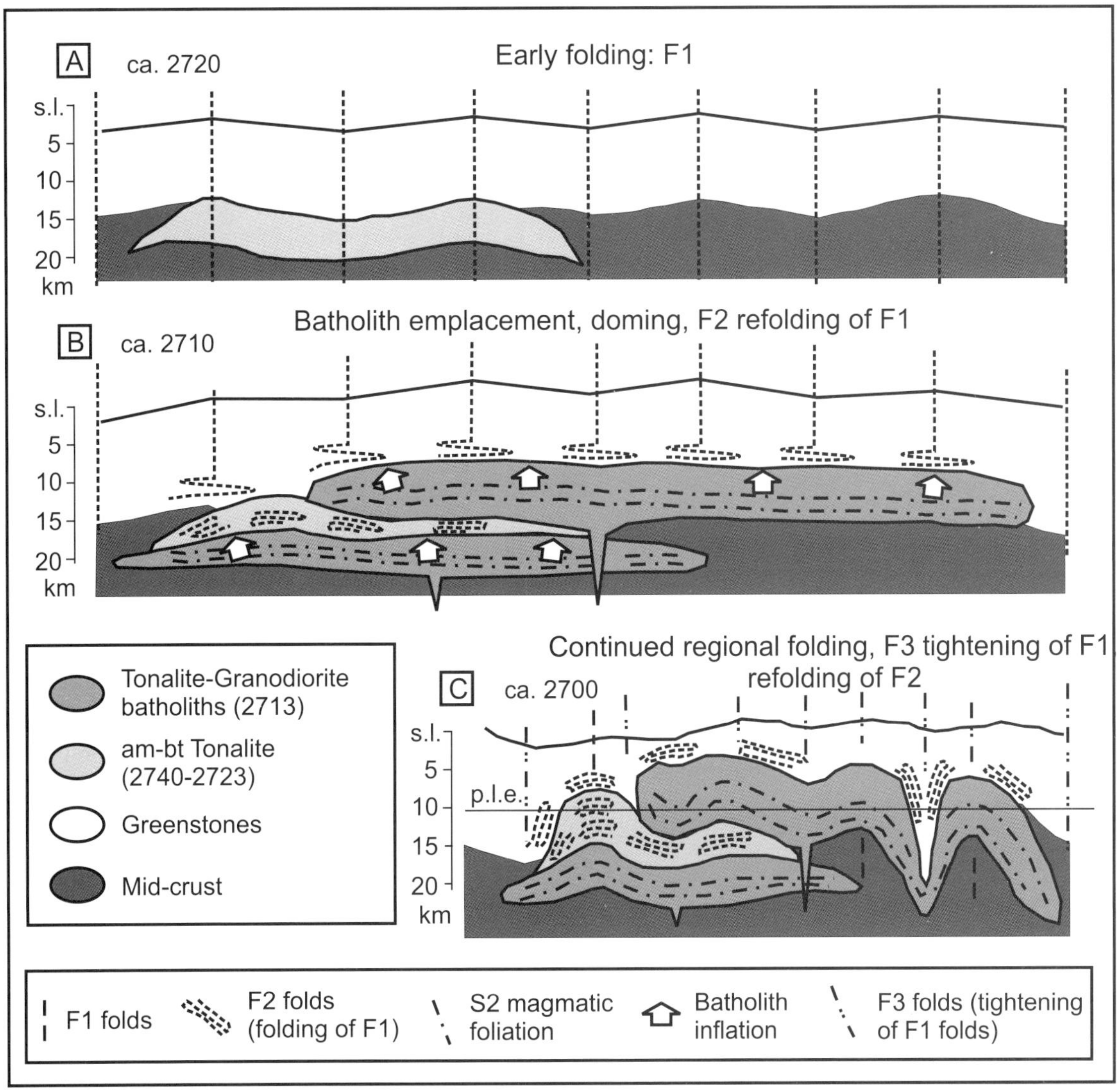

Figure 9 Schematic depiction of the emplacement of the Kenogamissi complex within the greenstones of the Swayze belt by folding, doming and batholith emplacement. See the text for explanations and discussion. Abbreviations: (s.l.) sea level; and (p.l.e.) present level of erosion. The schematic model shown here is approximately to scale.

4.1. Doming and emplacement of the Kenogamissi complex

Based on the mapping, structural analysis and petrographic study, a model is presented for the different stages of evolution of the Kenogamissi complex and the emplacement of the Roblin Batholith. The model, presented schematically in Figure 9, deals only with the evolution of the Kenogamissi complex until ca. 2700 Ma. Other plutonic units, shown in Figure 2, were emplaced later.

The tonalite plutons dated >2720 Ma are here considered to represent magma chambers associated with the metavolcanic units of similar age (Sutcliffe *et al.* 1993). As such, they are interpreted to be tabular intrusions emplaced at the base of the upper crust (Fig. 9A). Doming of the Kenogamissi complex is shown to have caused severe flattening and recumbent refolding of earlier upright F1 folds in the deformation aureole. Because only one generation of folding is recognised in the Swayze greenstone belt outside of the deformation aureole, the present author interprets that the refolded F1 folds in the deformation aureole represent those regional folds. Upright, regional F1 axial surfaces are shown to deform the greenstones and the early tonalite plutons in Figure 9A.

Emplacement of tonalite-granodiorite batholiths, such as the Roblin Batholith, is shown in Figure 9B. Emplacement and inflation of the batholiths resulted in doming of the Kenogamissi complex and uplift of the overlying greenstones. It also resulted in refolding of the F1 folds in the earlier tonalite plutons, and in the deformation aureole, by recumbent F2 folds (Fig. 9B), giving rise to the refold interference pattern seen in Figure 5A and to recumbent F2 folds in the orthogneiss. At this stage, magmas associated with emplacement of the tonalite-granodiorite batholiths intruded the orthogneiss as dykes and sills. Intrusions of those magmas during doming resulted in recumbent folding of tonalite and granodiorite dykes, giving rise to the heterogeneous compositional makeup of the isoclinally folded orthogneiss. Also at this stage, the magmatic S2 foliation and compositional layering were forming in the Roblin Batholith.

Figure 8C shows F3 folding of the Roblin Batholith, and refolding of F2 folds and of the S2 foliation in the Roblin Batholith. The present author interprets that F3 in the Kenogamissi complex and in its deformation aureole is caused by the continued regional N–S bulk shortening and regional folding which gave rise to the early folds shown in Figure 9A. This interpretation is consistent with F3 having affected the Roblin Batholith while it was still a magma mush, causing remobilisation of fractionated melts, as shown in Figure 5B. It is also consistent with the fact that only one generation of E–W to NW–SE-trending regional folds is documented in the overlying Swayze greenstone belt. Therefore, F3 folding in the Swayze belt in Figure 9C is seen as continued tightening of the regional F1 folds. In other words, F1 regional folding of the Swayze greenstone belt is correlated with both F1 and F3 in the deformation aureole surrounding the Kenogamissi complex. In the Kenogamissi complex and in the deformation aureole, the regional folding event was punctuated by recumbent F2 folding associated with doming.

4.2. Deformation-assisted melt remobilisation

It is widely recognised that deformation-enhanced melt segregation and extraction plays a fundamental role in the chemical differentiation of the continental crust and in the transport of granitic melts from deep zones of partial melting to magma emplacement sites at shallower crustal levels (Clemens & Mawer 1992; Brown 1994; Sawyer 1996; Brown & Solar 1999; Brown 2004). From the observations presented here, it seems that tectonically driven deformation, specifically folding, has played a potentially important role in remobilisation and transport of fractionated liquids within the orthogneiss unit and also within the partially solidified Roblin Batholith. An important interpretation that arises from those observations is that deformation-assisted remobilisation of melts may have been important in the formation of compositional layering (injection of sills, Fig. 3C) and also in the magmatic differentiation of the batholith.

In the examples shown in Figures 3C, E, F and 5B, the leucocratic dykes are located on, and are parallel to, the sheared limbs of F3 folds. In the cases of leucocratic dykes in the Roblin Batholith, the folds are synmagmatic. The present author interprets the melts to have been injected into the deforming orthogneiss unit from the Roblin Batholith or from an underlying, unexposed intrusion of the same age as the Roblin Batholith. At the time of F3 folding, the latest of the injected melts had only partially crystallised and fractionated, such that liquid was available to be remobilised during that folding event (Fig. 3C).

The documented leucocratic dykes are small-scale features, of the order of a few centimetres to 15 cm in thickness. However, their occurrences are widespread and common, suggesting that they may represent a significant remobilisation of magmatic liquid. Also, it cannot be ruled out that similar but larger dykes may remain as yet unidentified.

In the Kenogamissi complex, deformation-assisted remobilisation of magmatic liquids, which potentially contributed to the fractionation of the Roblin Batholith, was the result of tectonic folding of the partially molten Roblin Batholith and of the orthogneiss that contained injected melt. The present author is not aware of melt remobilisation structures of the type described here, located on the sheared limbs of upright magmatic folds, having been previously documented in the literature. Recently published work on the Archaean Woodstock monzogranite (Pilbara Craton) has documented emplacement of leucocratic granitic magmas into magmatic shear zones which formed within the folding, partially molten monzogranite (Pawley & Collins 2002; Pawley *et al.* 2002). In that case, the magmatic shear zones cross-cut magmatic fold limbs in the host pluton and they are interpreted to have formed in response to a bulk dextral shearing of the crystallising monzogranite. The work of Pawley and colleagues, together with the results presented here, indicate that deformation-assisted extraction and remobilisation of fractionated melts within crystallising intrusions may be an important, although previously unrecognised process in syntectonic plutons and batholiths. As seen in Figure 3C, it can lead to the formation of magmatic compositional layering (sills). It may also result in significant redistribution of mass and affect chemical differentiation histories of the deforming magma bodies.

Understanding the full significance of the deformation-enhanced melt remobilisation and transport within the Kenogamissi complex will require further, more detailed structural, microstructural and petrological studies. In particular, it will be important to discover whether the types of melt remobilisation structures documented here caused the redistribution of large volumes of melt, and at what length and time scale the melt redistribution was efficient. However, it is reasonable to speculate that deformation-assisted extraction and redistribution of melt within mushy plutons which are subjected to regional tectonic deformation, in particular folding, may represent a previously unrecognised though important mechanism for the magmatic differentiation of syntectonic plutons and batholiths.

It will also be important to better understand the significance of the localisation of melt remobilisation structures, such as the dykes documented here, on the limbs of folds. In magmatic shear zones it is expected that mechanical interactions of rigid solid phases will result in localised dilatation, as is the case in shear zones developed during flow of unconsolidated granular materials such as sand (Mandl *et al.* 1977; Du Bernard *et al.* 2002). It has also been suggested that shear localisation and increased dilatancy may occur in deforming, crystallising granitic magmas and lead to melt segregation (Vigneresse & Tikoff 1999). That is in agreement with the structures documented in the present paper. It is speculated that dilatation caused by shearing of crystal mush on the limbs of magmatic folds in the Roblin Batholith led to the migration of melts to the fold limbs and the formation of the leucocratic dykes.

5. Conclusions

Mapping, structural analysis and petrographic study of the orthogneiss unit and the three units of the Roblin Batholith provide insights into the history of the doming and emplacement of the Late Archaean Kenogamissi complex into the greenstones of the Swayze greenstone belt in the Abitibi Subprovince, Ontario. Doming resulted from the emplacement and inflation of tonalite-granodiorite batholiths, such as the Roblin Batholith, in the lower stratigraphic levels of the greenstone belt and possibly in the underlying crust. Doming led to the horizontal flattening of earlier tonalite plutons and of the greenstones within the deformation aureole that surrounds the Kenogamissi complex, and the formation of recumbent F2 folds. The recumbent, doming-related folds refolded earlier, regional E–W to NW–SE-trending F1 regional folds in the earlier plutons and in the deformation aureole. Injection of magmas related to the tonalite-granodiorite batholiths into the earlier plutons and concomitant flattening caused by doming resulted in the heterogeneous compositional layering in the isoclinally (F2) folded orthogneiss unit.

The doming and related recumbent folding punctuated one long-lived regional folding event that continued during and after doming, folding the orthogneiss and the Roblin Batholith while the latter remained a magma mush, forming F3 folds in the orthogneiss, the Roblin Batholith and in the deformation aureole. Effectively, F1 folds and F3 folds within the deformation aureole of the Kenogamissi complex both correlate with the one generation of regional folds in the overlying and surrounding greenstone belt.

The tectonic deformation and folding of the Kenogamissi complex resulted in syntectonic extraction and redistribution of fractionated melts within the deforming Roblin Batholith magma mush and within the orthogneiss. Magma was extracted and remobilised into dykes which formed on the sheared limbs of magmatic folds. The concentration of melt on the limbs of magmatic folds, which led to the formation of the dykes, was probably allowed by dilatancy caused by shearing of the crystal-liquid suspension (magma mush) on the fold limbs.

Further structural and petrological study will be required to evaluate the length and time scales over which deformation-assisted melt remobilisation was efficient within the Kenogamissi complex. However, it is suggested that deformation-assisted melt remobilisation and redistribution may prove to be an important process in the magmatic differentiation histories of syntectonic plutons and batholiths.

6. Acknowledgements

This research was funded by a Natural Sciences and Engineering Research Council of Canada Discovery Grant. Insightful comments and suggestions by journal reviewers R. Anma and R. Weinberg were very helpful in improving the final manuscript. I also wish to thank M. Brown for discussions which focused my ideas and improved my interpretations. The organisers of the Hutton V Symposium on Granites and Related Rocks in Toyohashi, Japan, and the editors of this special issue are also thanked for the opportunity to present this work at the symposium and to publish it in the journal.

7. References

Ayer, J. A., Amelin, Y., Corfu, F., Kamo, S. L., Ketchum, J., Kwok, K. & Trowell, N. 2002a. Evolution of the southern Abitibi greenstone belt based on U-Pb geochronology: authochthonous volcanic construction followed by plutonism, regional deformation and sedimentation. *Precambrian Research* **115**, 63–95.

Ayer, J. A., Ketchum, J. W. F. & Trowell, N. F. 2002b. New geochronological and neodymium isotopic results from the Abitibi greenstone belt, with emphasis on the timing and the tectonic implications of Neoarchean sedimentation and volcanism. Summary of field work and other activities. *Ontario Geological Survey, Open File Report 6100*, pp. 5.1–5.16.

Becker, J. K. & Benn, K. 2003. The Neoarchean Rice Lake batholith and its place in the tectonomagmatic evolution of the Swayze and Abitibi granite-greenstone belts, northeastern *Ontario. Ontario Geological Survey, Open File Report 6105*, p. 42.

Bouchez, J. L., Gleizes, G., Djouadi, T. & Rochette, P. 1990. Microstructure and magnetic susceptibility applied to emplacement kinematics of granites: the example of the Foix pluton (French Pyrenees). *Tectonophysics* **184**, 157–71.

Brown, M. 1994. The generation, segregation, ascent and emplacement of granite magma: the migmatite-to-crustally-derived granite connection in thickened orogens. *Earth-Science Reviews* **36**, 83–130.

Brown, M. 2004. The mechanism of melt extraction from lower continental crust of orogens. *Transactions of the Royal Society of Edinburgh: Earth Sciences* **95**, 35–48.

Brown, M. & Solar, G. S. 1999. The mechanism of ascent and emplacement of granite magma during transpression: a syntectonic granite paradigm. *Tectonophysics* **312**, 1–33.

Clemens, J. D. & Mawer, C. K. 1992. Granitic magma transport by fracture propogation. *Tectonophysics* **204**, 339–60.

Du Bernard, X., Eichhubl, P. & Aydin, A. 2002. Dilation bands: a new form of localized failure in granular media. *Geophysical Research Letters* **29**, doi:10.1029/2002GL015966.

Heather, K. B. 1999. *Geology of the Swayze greenstone belt, Ontario. Open File 3384*. Ottawa: Geological Survey of Canada.

Mandl, G., de Jong, L. N. J. & Maltha, A. 1977. Shear zones in granular material. *Rock Mechanics* **9**, 95–144.

Nitescu, B. & Halls, H. C. 2002. A gravity profile across southern Saganash Lake fault: implications for the origin of the Kapuskasing structural zone. *Canadian Journal of Earth Sciences* **39**, 469–80.

Paterson, S. R., Vernon, R. H. & Tobisch, O. T. 1989. A review of criteria for the identification of magmatic and tectonic foliations in granitoids. *Journal of Structural Geology* **11**, 349–63.

Pawley, M. J., Collins, W. J. & Van Kranendonk, M. J. 2002. Origin of fine-scale sheeted granites by incremental injection of magma into active shear zones: examples from the Pilbara Craton, NW Australia. *Lithos* **61**, 127–39.

Pawley, M. J. & Collins, W. J. 2002. The development of contrasting structures during the cooling and crystallisation of a synkinematic pluton. *Journal of Structural Geology* **24**, 469–83.

Percival, J. 1988. Deep geology out in the open. *Nature* **335**, 671.

Ramsay, J. G. 1967. *Folding and Fracturing of Rocks*. New York, NY: McGraw-Hill Book Company.

Sawyer, E. W. 1996. Melt segregation and magma flow in migmatites: implications for the generation of granite magmas. *Transactions of the Royal Society of Edinburgh: Earth Sciences* **87**, 85–94.

Sutcliffe, R. H., Barrie, C. T., Burrows, D. R. & Beakhouse, G. P. 1993. Plutonism in the southern Abitibi Subprovince: a tectonic and petrogenetic framework. *Economic Geology* **88**, 1359–75.

Thiessen, R. 1986. Two-dimensional refold interference patterns. *Journal of Structural Geology* **8**, 563–73.

Vernon, R. H. 2000. Review of microstructural evidence of magmatic and solid-state flow. *Electronic Geosciences* **5**. (www documents.) URL http://link.springer-ny.com/link/service/journals/10069/bibs/0005001/00050002.htm

Vigneresse, J. L. & Tikoff, B. 1999. Strain partitioning during partial melting and crystallizing felsic magmas. *Tectonophysics* **312**, 117–32.

KEITH BENN, Ottawa-Carleton Geoscience Centre and Department of Earth Sciences, University of Ottawa, Ottawa, Ontario K1N6N5, Canada.
e-mail: kbenn@uottawa.ca

MS received 20 October 2003. Accepted for publication 14 September 2004.

Transactions of the Royal Society of Edinburgh: Earth Sciences, **95**, 309–317, 2004

Growth of wedge-shaped plutons at the base of active half-grabens

S. W. Richards and W. J. Collins

ABSTRACT: Combined field and geophysical data show that plutons from the Bega Batholith are elongate, meridional, wedge-shaped bodies which intruded during a period of regional east–west extension in the Palaeozoic eastern Lachlan orogen, eastern Australia. Plutons within the core of the batholith have intruded coeval, syn-rift sediments and co-magmatic volcanics. The batholith is bound by high-temperature, dip-slip faults, and contains several major NE-trending transtensional faults which were active during batholith construction. In the central part of the batholith, the Kameruka pluton is an asymmetric, eastward-thickening, wedge-shaped body with the base exposed as the western contact, which is characterised by abundant, shallow-dipping schlieren migmatites which contain recumbent folds and extensional shear bands. A shallow (<30°), east-dipping, primary magmatic layering in the Kameruka pluton steepens progressively westward, where it becomes conformable to the east-dipping basal migmatites. The systematic steepening of the layering is comparable to sedimentary units formed during floor depression in syn-rift settings. The present authors suggest that the wedge-shaped plutons of the Bega Batholith are the deeper, plutonic expression of a hot, active rift. The batholith was fed and sustained by injection of magma through sub-vertical dykes. Displacement along syn-magmatic, NE-trending faults suggests up to 25 km of arc-perpendicular extension during batholith construction. The inferred tectonic setting for batholith emplacement is a continental back-arc, where modern half-extension rates of 20–40 mm yr^{-1} are not unusual, and are sufficient to emplace the entire batholith in ~1 Ma. This structural model provides a mechanism for the emplacement of some wedge-shaped plutons and is one solution to the 'room problem' of batholith emplacement.

KEY WORDS: Bega Batholith, extension, Lachlan Fold Belt, pluton emplacement.

The generation and emplacement of granitic magmas are fundamental aspects of crustal evolution, yet the tectonic control on these two processes is vigorously debated. Granitic magmas may be generated during lithospheric extension, an ideal environment for adiabatic decompression melting of the asthenosphere (e.g. McKenzie & Bickle 1988), which supplies advective heat to the crust via basaltic intrusion, leading to granite magma production (e.g. Thompson 1999; Petford & Gallagher 2001). Intuitively, extensional environments also seem capable of solving the enigmatic 'room problem' associated with the emplacement of large batholiths, but tectonic strain rates are considered to be generally too low to accommodate pluton growth (e.g. Paterson & Fowler 1993). Alternatively, magma emplacement may occur during orogenic contraction or transpression because syn-magmatic upright folds, sub-vertical foliations and S/C-type fabrics are both temporally and spatially associated with the plutons (e.g. Brown & Solar 1998; Paterson & Miller 1998).

Plutons emplaced during compression are generally thicker and preserve steeper margins with respect to their thin, tabular counterparts emplaced during extension (Vigneresse 1995; Benn *et al.* 2000). Many plutons are also asymmetric and wedge-shaped (e.g. Vigneresse 1995; Cruden 1998), with growth commonly explained by lifting of the roof (e.g. Dixon & Simpson 1987; Benn *et al.* 2000) or subsidence of the floor (Cruden 1998). Such models overcome the perceived accommodation problems associated with low strain rates, but imply minor tectonic control on generating space for batholith emplacement. However, such plutons can be generated in an extensional setting at recorded tectonic strain rates, as explained below.

The present authors offer field and geophysical evidence that describes the geometry, internal structure and contact relations of plutons in the Bega Batholith in the eastern Lachlan orogen, eastern Australia, which suggest that it formed in a graben. Internal, primary magmatic features suggest that these plutons can be described like 'sedimentary' deposits formed as a series of cumulate layers on an aggrading magma chamber floor, with the layers recording a progression of pluton growth (Weibe & Collins 1998). From the orientation of way-up structures, the present authors deduce that the plutons underwent progressive tilting to form wedge-shaped structures during construction and inflation at the base of an actively developing half-graben. This model implies that a major tectonic control on batholith construction and pluton growth rates can occur within a rift setting.

1. The Bega Batholith

The 420–405-Ma-old Bega Batholith (Fig. 1) is over 300 km long, but only 50 km wide, and comprises a series of N–S trending suites (Chappell 1996), which contain one or more co-genetic plutons. Plutons and suites of plutons exhibit a pronounced N–S alignment: the Glenbog suite extends over 150 km along the western margin of the batholith, but is <10 km wide. The suites exhibit remarkably coherent geochemical patterns along strike, but rapid chemical variation across strike (Chappell 1996). The ages of the suites decrease eastward (Chappell *et al.* 1991). The proportion of plutonic to coeval supracrustal rock decreases northwards in the batholith (Fig. 1). These supracrustal rocks are dominated by volcaniclastics and co-genetic volcanics (e.g. Wyborn & Chappell

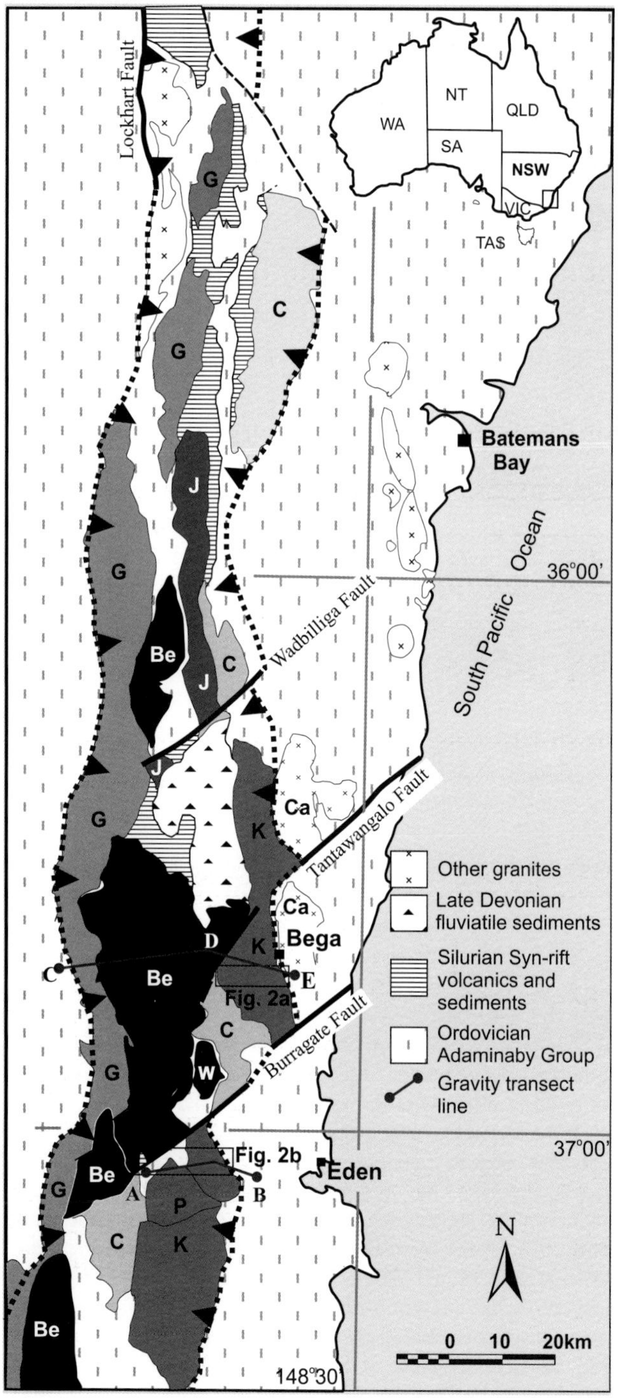

Figure 1 Map of the Bega Batholith, southeastern Australia, showing relevant suites of the batholith and syn-rift supracrustal rocks restricted to the northern extent, but within the bounds of the batholith (suites after Chappell 1996): (K) Kameruka; (P) Pericoe; (Be) Bemboka; (C) Candelo; (G) Glenbog; (J) Jillicambra; and (Ca) Cobargo. Dashed lines with barbs represent the inferred limit of graben based on gravity profiles and field relations. Note contrasting amounts of offset for different suites on the Burragate and Tantawangalo faults. Cover rocks beyond the batholith margin have been removed for clarity.

1986), and are contained within a narrow structural lithotectonic belt that is the northern extension of the Bega Batholith.

The highly elongate batholith is parallel to, and highlights, an intense, N–S trending structural grain developed within the poly-deformed, regional, low-grade turbiditic metasediments of Ordovician age. This relationship suggests a strong structural control on batholith construction. However, although the plutons intruded highly cleaved metasediments (Fig. 1), they are generally non-foliated. This suggests that they post-dated the major contractional deformation event.

The batholith is bound on the eastern and western sides by high-T, dip-slip shear zones, but they are largely obliterated by later intrusion. Most of the western edge of the batholith is occupied by the Glenbog suite, which has intruded a steep, N–S striking, east-dipping, high-temperature mylonite zone that is preserved at the northwest extremity of the batholith. This zone, preserved within the 30-km-long Lockhart complex, is characterised by tectonically interleaved lenses of gneissic granite and 'amphibolite'. The 'amphibolite' contains euhedral plagioclase and hornblende laths which define a steep inward dipping lineation (62→110). These laths are strongly aligned, but not strained, suggesting that the mafic body was cooling from magma during fault movement. These features imply magma emplacement into an active fault zone.

On the eastern side of the batholith, the eastern margin of the Kameruka pluton is sharp and abuts a narrow (<0·3-km-wide) cordierite-biotite-muscovite-quartz bearing hornfels. The hornfels is a high-strain, high-T aureole characterised by isoclinal folding and shearing. It contains a sub-vertical foliation with a steep lineation defined by elongate, 3-mm-long cordierite porphyroblasts, and P–T conditions of ~500° and 1–2 kbar are estimated. Enclaves within the adjacent granodiorite also define a steep sub-vertical lineation. These features suggest sub-vertical movement along ductile faults within the aureole and adjacent granite during emplacement of the Kameruka pluton, similar to that observed on the western side of the batholith.

The Cobargo suite of plutons lies to the east of the Kameruka pluton, beyond the high-T, high-strain zone at the eastern margin of the batholith (Fig. 1). These plutons have subrounded shapes and sharp discordant margins. The aureoles are narrow and characterised by hornfels containing small, randomly oriented cordierite grains, typical of static growth during emplacement. These characteristics differ considerably from other plutons within the Bega Batholith. Furthermore, the Cobargo suite appears to be younger (~395 Ma) than the main batholith (Chappell *et al.* 1991). For these structural and geochronological reasons, the present authors consider that the Cobargo suite formed well after the bounding shear zones associated with major batholith emplacement.

Age variations across the batholith highlight the variable timing of the emplacement of the various plutons and suites of plutons. The relative timing of pluton emplacement is well constrained based on cross-cutting and intrusive contact relations. The Glenbog suite and Kameruka pluton cross-cut the Candelo suite, and therefore, establish that the Candelo suite as one of the earliest emplaced suites in the Bega Batholith. Mixing between the Pericoe and Kameruka plutons suggests that the Pericoe pluton intruded late during construction of the Kameruka pluton. In turn, the Bemboka pluton, which intrudes the Kameruka pluton along its northwestern margin, has been determined as the last significant suite emplaced in the Bega Batholith. The order of emplacement tightly constrains the model of construction of the Bega Batholith.

Several NE-trending strike-slip faults (i.e. Tantawangalo, Burragate and Wadbilliga) cut obliquely across the batholith, and all show differential right-lateral offset for each suite (Fig. 1). For example, the western Glenbog suite shows no major offset along these faults, but the Kameruka suite is offset ~25 km and the Candelo suite is displaced ~31 km by the Burragate fault. The Bemboka suite is partly offset, but only on its eastern side. Farther north, the Wadbilliga fault offsets the Jillicambra pluton by ~12 km, but the Candelo suite is offset by only ~5 km. Strike-slip movement along the faults as well as the lack of any metamorphic grade change across the

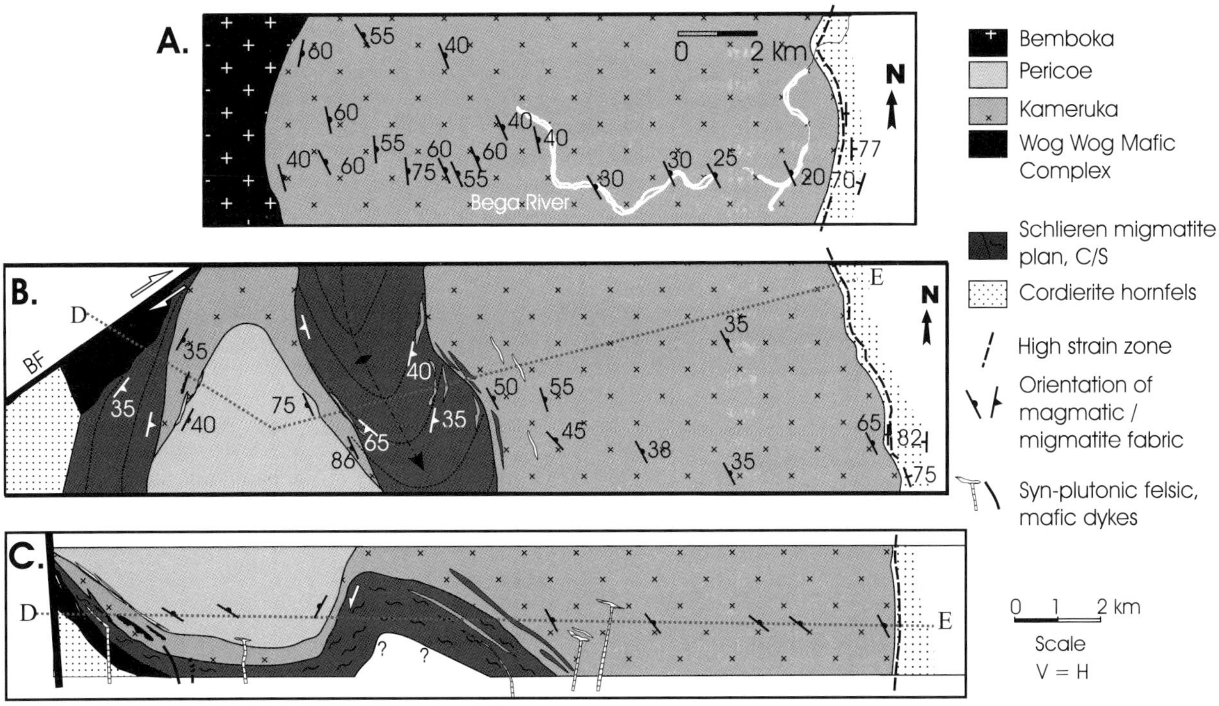

Figure 2 (a & b) Traverses across the Kameruka suite plutons (see Fig. 1 for location) showing systematic steepening of east-dipping primary magmatic features, which become concordant with structurally underlying migmatites and a sheeted gabbro complex at the western contact (b). (c) Cross-section of DŁ shows inferred pluton geometry, based on primary fabrics and the structure of basal migmatites from the southern transect.

faults indicate that these offsets are real values, therefore, this differential movement suggests that these oblique faults were also active during granite emplacement.

1.1. Contact relationships of the Bemboka and Kameruka suites

The Bemboka pluton is granodioritic, approximately 60 km long and <25 km wide, and is widest in the middle, tapering to the north and south (Fig. 1). The northern margin is overlain by shallow-dipping, fine-grained sediments which have been metamorphosed to cordierite hornfels at the contact. These sediments are the southern continuation of the supracrustal assemblages described from farther north (Fig. 1). In contrast, the southern margin either bounds the Burragate fault, or consists of intensely deformed diatexite migmatites which are characterised by an intense, shallow-dipping schlieren foliation. The migmatites occur at the base of plutons within the core of the batholith, forming a high-temperature zone which separates the plutons from underlying, upright-folded metasediments. Somewhat surprisingly, although much of the Bemboka pluton abuts the Burragate fault in the south, it is not deformed by it. Rather, the granites have intruded mylonitised granite and metasediment. This is further evidence for syn-faulting emplacement.

The Kameruka suite, which includes the Kameruka and Pericoe plutons, extends for >160 km, but is a maximum of 14 km wide. The Kameruka pluton is dominated by a coarse-grained megacrystic granodiorite containing pink K-feldspar grains of 3–10 cm in length. It preserves a variably developed, east-dipping primary magmatic fabric, which steepens progressively to the west (Fig. 2). The fabric is highlighted by internal layering, defined by variable grainsize and composition, and is associated with other features including rare alternating granite and mafic layers, flame-type injections and chilled lobate contacts along the lower margins of mafic layers, rare felsic 'pipes' in mafic layers (Fig. 3a), and channels of mafic enclaves. These features are taken as evidence for crystal deposition on a progressively aggrading floor of crystal mush in the pluton, but they are also indicators of 'way-up' (Weibe & Collins 1998). These geopetal indicators indicate top to the east, which requires that the western contact is the basal one.

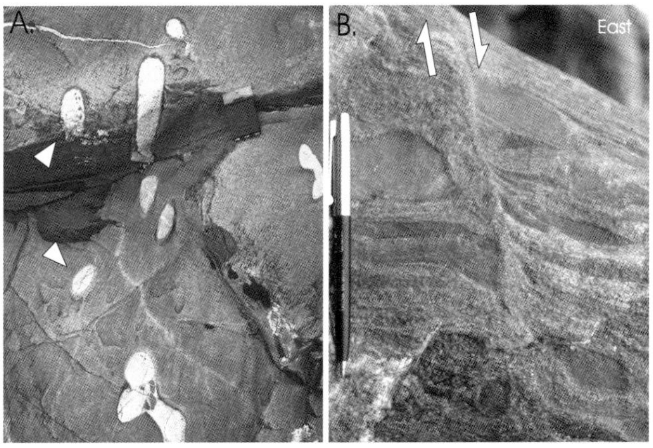

Figure 3 (a) Steeply plunging orientation of felsic granite 'pipes' intrusive into a mafic body near the base of the Kameruka pluton. Arrows indicate sections of the same pipe (base of photo ~50 cm). The 'pipes' are formed during upward, forceful injection of felsic material into the base of an overlying, viscous mafic layer. The process is primarily gravity-driven, and hence, the pipes consistently form in vertical orientations and can be used to determine the palaeo-horizontal (see Weibe and Collins 1998). (b) Well-developed, sub-horizontal schlieric fabric in the migmatites below the Kameruka pluton. The fabric is cut and deflected by leucosome-rich, sub-vertical shear zones with a normal sense of movement (pen for scale). The photo is representative of normal shears observed at the centimetre to decimetre scale.

Like the Bemboka pluton, the Kameruka has contrasting pluton–country-rock contact relationships. The hornfelsic eastern margin contrasts with the western margin, which is gradational from granite to migmatite. The western contact is characterised by a complex, 500-m-wide zone where migmatite, megacrystic granite and a myriad of hybrid phases are

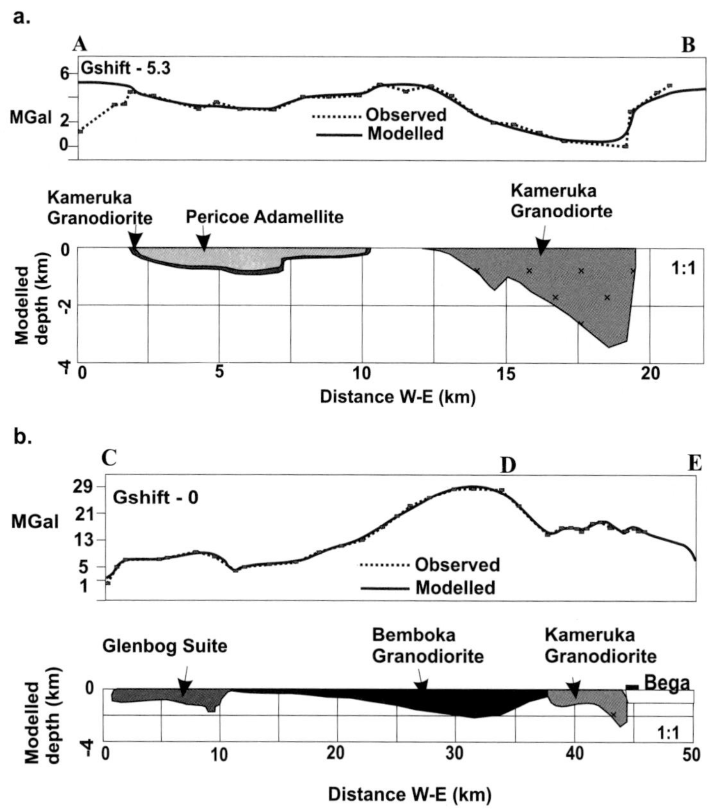

Figure 4 (a & b) Observed and modelled gravity profiles across the traverses in Figure 1 showing the modelled wedge-shaped geometry of the Kameruka and other Bega Batholith plutons. The model profile fits closely with the observed data with the addition of the polygons. The location and geometry is, in part, constrained by surface field measurements. In both transects, the eastern and western limit of the modelling was constrained by either major structures or rapid changes in rocktype. In (a), the dip in the observed gravity profile on the western margin reflects the presence of a pluton belonging to the Candelo super suite, which has been removed for this discussion. One major outlying point has been removed from transect C–E.

intimately inter-sheeted, indicating that the migmatites developed coevally with magma emplacement. Abundant cordierite-rich metapelitic blocks contain sillimanite and biotite hosted in a leucosome matrix, suggesting P–T conditions of >600 °C and 2–3 kbar. This reflects greater depths of equilibration compared with the eastern margin, and supports the way-up criteria that the western margin is the lower (basal) contact of the Kameruka pluton. Thus, the hornfelsed contacts of Bemboka and Kameruka appear to be the upper or roof contacts, whereas the migmatites appear to be the basal contacts of the plutons, and of the batholith.

The migmatites exhibit an intense, sinuous, sub-horizontal to moderately east-dipping schlieric foliation that anastomoses around remnant metapelitic blocks, and resemble σ-porphyroclasts within mylonites. They also contain local, syn-migmatitic recumbent folds outlined by relict bedding, schlieren and folded microgranite dykes. Steeply inclined, leucosome-rich normal shears cut the shallow-dipping migmatitic fabric (Fig. 3b). Significantly, the migmatites are extensively hybridised with the granites, with leucosomes commonly containing the characteristic K-feldspar megacrysts of the Kameruka granodiorite. Thus, the granite emplacement and migmatite formation occurred synchronously.

The migmatite zone, which extends for ~500 m below the Kameruka pluton (Fig. 2c), is concordantly underlain by a 600-m-thick, sheeted gabbro complex which, in turn, is discordantly underlain by highly cleaved, low-grade metasediments. The underlying gabbro complex is layered at the metre-scale with thicker units exhibiting way-up structures, including rhythmically graded layers with chilled, fine-grained bases and coarse-grained tops, which show top to the east. Metre-scale diapirs of coarse dioritic magma have intruded into overlying graded layers, and also indicate top to the east (Collins *et al.* 2000). The orientation of the sheets and dykes is sub-parallel to the fabric within the overlying migmatites and granite layers in the Kameruka pluton, but directly below, the structure abruptly reverts to upright folds, typical of the surrounding low-grade country rock (Fig. 2). The steep axial plane foliation in the underlying turbidites is overprinted by stellate clusters of muscovite and chlorite, indicating contact metamorphism by the gabbro. The metamorphic overprinting and structural discontinuity between the gabbro and underlying metasediments shows that the gabbro was emplaced after regional folding, as is the case for the plutons of the batholith.

The Kameruka pluton is also host to numerous felsic microgranitic dykes which cut through the underlying migmatites, but terminate as diffuse layers at various levels in the pluton. The field relations indicate that they intruded upward through the lower parts of the pluton before spreading laterally along discrete planar surfaces, concordant with primary magmatic layering. Because they emanate from below the pluton but are part of the geochemical system, Collins *et al.* (2000) interpreted these dykes as the feeder conduits to the Kameruka pluton.

2. Gravity data

Two-dimensional (2D) gravity modelling along two E–W transects across the Bega Batholith (Fig. 1) was used to constrain the overall pluton shape. Measurements were taken at 400–500-m spacings and 500- to 1-km spacings along two E–W trending transects across the southern and central Bega Batholith, respectively (Fig. 4). The elevation and position of gravity stations were obtained at an accuracy of 0·5 m and 0·01 m, respectively, using a Trimble TSC1 real-time differential Global Positioning System. Gravity data were measured using a SCINTREX autograv with a precision of ± 0·05 mGal. Figure 4 illustrates the final bouguer anomaly after corrections; free air, bouguer (2·6), latitude correction and regional E–W gradient. Direct-forward modelling (2D) using Model 2D (Roach 1990) allowed construction of multiple polygons with the density of each polygon (pluton) derived from laboratory measurements of the densities of 22 selected, un-weathered field samples of the four main rocktypes, which include the Pericoe Adamellite (av. $2{\cdot}70 \times 10^3$ kg m^{-3}), Kameruka Granodiorite (av. $2{\cdot}72 \times 10^3$ kg m^{-3}), schlieren migmatite (av. $2{\cdot}78 \times 10^3$ kg m^{-3}) and regional schists (av. $2{\cdot}58 \times 10^3$ kg m^{-3}). The densities of an additional 12 samples were measured from the Bemboka Granodiorite (av. $2{\cdot}77 \times 10^3$ kg m^{-3}) and Glenbog suite plutons, corresponding with the rocktypes encountered on the northern transect (Fig. 1). The density of each sample was determined according to the relationship:

Density=(Weight in air)/[(Weight in air)–(Weight in water)].

The average of each rocktype was used in the gravity modelling.

Recent seismic refraction profiles for the eastern Lachlan orogen indicate that Ordovician turbidites extend to ~15 km depth, below which mafic Cambrian and early Ordovician mafic rocks predominate (Finlayson *et al.* 2002). The plutons of the Bega Batholith have projected depths of $<$4 km based on the present authors projected surface structural data and gravity models, which are similar to those recognised elsewhere in the central and eastern Lachlan orogen, based on seismic-reflection profiling (Trzebski *et al.* 1999; Trzebski *et al.* 2001; Glen *et al.* 2002; Lennox *et al.* 2002). Therefore, the authors have modelled the plutons as enclosed within metasediments.

The northern transect exhibits a predominantly positive gravity anomaly corresponding with plutons with higher densities than the enclosing schists (av. $2{\cdot}58 \times 10^3$ kg m^{-3}). The final gravity model for the southern transect (Fig. 4b) shows a wedge-shaped profile that extends westward to include the shallow, slightly basin-shaped Pericoe pluton. The northern transect, including the Glenbog Granodiorite, the Bemboka Granodiorite and the Kameruka Granodiorite show a consistent deepening to the east profile with the Kameruka Granodiorite, similar to that modelled for the southern transect (Fig. 1). The Kameruka and Bemboka plutons display a consistent steeper, deeper eastern margin and a shallower, less-inclined base to the west. The consistent asymmetric shape of the Bega Batholith plutons suggests a fundamental structural control on emplacement and pluton growth.

3. The Bega Batholith as a graben

The presence of bounding, high-T, inward-dipping, dip-slip faults in the batholith suggests that it occupies a graben structure. Although the present authors have not determined sense of shear on the faults, the restriction of Silurian and Early Devonian supracrustal rocks to within the batholith, with older basement (Ordovician) rocks beyond the batholith margins (Fig. 1), indicates that it was a structural depression. Given that these coeval supracrustal rocks include bimodal volcanics, such as the Carrawang basalt (Glen 1995), it is apparent that the batholith was emplaced into an active graben. The lowermost syn-rift supracrustal rocks are regarded as the roof of the batholith, and the migmatites are the base of the graben.

The migmatites at the batholith base are considered to be part of an extensional detachment. They contain recumbent folds, shallow-dipping foliations and late melt-rich shear zones with normal sense of movement (Fig. 3b), which imply that the migmatite zone and overlying pluton formed during lithospheric extension. Furthermore, this kilometre-thick zone is underlain by non-foliated gabbros and upright folded metasediments (Fig. 2c). Therefore, the migmatites form a distinct, shallow-dipping zone of extensional shear, and this is best described as a detachment surface that developed synchronously with batholith emplacement.

The structural evidence for widespread E–W extension in the eastern Lachlan orogen throughout the Silurian–Devonian, during intrusion of the extensive batholiths, is the formation of north-trending rift basins, some with volcanic-hosted massive sulphide (VHMS) mineralisation (Glen 1992, 1995). The basins comprise typical syn- and post-rift sediments, which were inverted during later contraction (Glen 1995; Collins 2002b). An extensional model is also consistent with seismic reflection profiles of the central–eastern Lachlan orogen, which show Wyangala Batholith plutons localised within an inverted graben (Glen *et al.* 2002, fig. 12). The profiles indicate that steep, normal faults link at shallow depths ($<$5 km) with major listric faults which extend into the middle crust (10–15 km). These faults were reactivated during later compressive deformation. Equally, all the Silurian–Devonian rifts in the Lachlan Fold Belt have been structurally inverted (e.g. Glen 1992, 1995), and the Lockhart fault is considered to be the high-T manifestation of this structural inversion.

4. Growth of the Bega Batholith plutons in active half-grabens

Results from the geophysical survey support inferences from field studies that the Kameruka pluton is an asymmetric, wedge-shaped body, thickening eastward, with the base exposed as the western contact. The Bemboka pluton is also wedge-shaped, but much shallower (Fig. 4). Both plutons have basal contacts dominated by migmatites which exhibit a shallow-dipping, syn-migmatitic schlieren foliation that parallels the shallow dipping base of the pluton, but is highly oblique to the underlying upright metasediments. In the Kameruka pluton, primary depositional features have been interpreted to have formed by progressive vertical accumulation of crystal mush layers, similar to a vertically aggrading sedimentary sequence (Weibe & Collins 1998; Collins *et al.* 2004). Given that the depositional features define the palaeo-horizontal (e.g. Fig. 3a), the systematic E–W steepening of the structures (Fig. 2) appears to record progressive eastward tilting of an initially sub-horizontal pluton floor during growth. These features are consistent with the cantilever-style floor depression model proposed by Cruden (1998), but the gradual steepening of the internal primary features with depth closely resemble progressive syn-rift style deposition of sediments in developing half-grabens (Fig. 4).

Construction of the batholith was initiated by sub-horizontal emplacement of the sheeted gabbros in the uppermost crust, probably when the mafic magmas achieved hydrostatic equilibrium (Fig. 5a) and/or a level of crust containing sub-horizontal structural anisotropies, which facilitated sheeted emplacement of the magma. Migmatite formation initiated at this stage, as evidenced by fragments of migmatite within the gabbroic layers and leucosomes cross-cutting some gabbro sheets. Localised, lateral and vertical migration of leucosome caused preferential development of

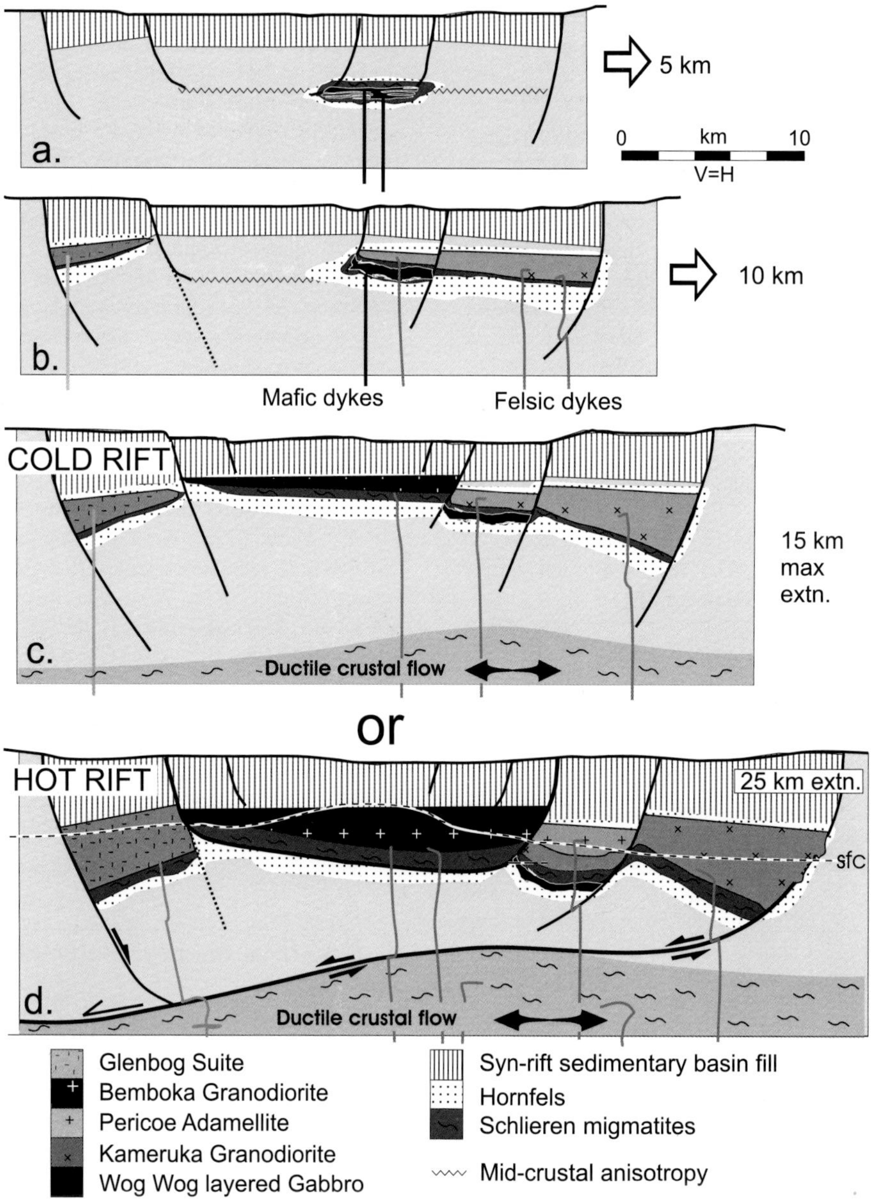

Figure 5 Model of batholith emplacement showing the stages of intrusion and inflation relative to half-graben development: (a) Sub-vertical normal faults propagate from the surface to link with the sheeted gabbro complex, which was probably emplaced at the level of hydrostatic equilibrium or at a structural discontinuity in the crust. (b) With ongoing extension, the faults propagate, allowing half-graben development and emplacement of the Kameruka pluton as sub-horizontal sheets above the gabbro complex. (c) Continued magma replenishment from dykes accompanies progressive tilting of the chamber floor at the base of individual half-grabens, allowing a maximum of 15 km eastward directed extension during 'cold' rifting. (d) Alternatively, stretching is focused at the batholith base, producing the shallow-dipping extensional fabric observed within the migmatites. Accelerated ductile stretching within this basal zone during 'hot' rifting can provide the additional 10 km of extension.

schlieren migmatites above the garbbroic body, which subsequently developed into a leucosome-rich shear zone. Heat and deformation were focused into the thermally softened detachment, and it became the locus for sheeted emplacement and construction of the overlying plutons (Fig. 5b, d).

However, much of the late extension is accommodated by the NE-trending strike-slip transfer faults. The Burragate, Tantawangalo and Wadbilliga faults record varying amounts of eastward displacement of the Kameruka and Candelo suites during emplacement of the Bemboka pluton (Fig. 1). Note that much of the eastward displacement of Kameruka and Candelo suite plutons, corresponding to widening of the batholith near Bega (Fig. 1), can be accommodated by removal of the Bemboka suite along the Burragate and Tantawangalo faults. To the north, the NE-striking Wadbilliga fault dextrally offsets the Jillicambra pluton by some 20 km, but the adjacent Glenbog suite is not offset. This fault also corresponds to a zone of widening in the batholith (Fig. 1). These relationships suggest that these NE-trending transfer faults facilitated emplacement of the Bemboka suite plutons.

5. How are plutons constructed in developing half-grabens?

Felsic and mafic dykes continuously fed magma to the pluton (Collins *et al.* 2000), ensuring the longevity of the magmatic system. The microgranitic dykes, which can be modelled as a parental liquid to the Kameruka suite granites (Healy 1999), migrated through the migmatites underlying the Kameruka pluton and intruded into the granite as syn-plutonic dykes. Additionally, migmatite fragments within many of the granite

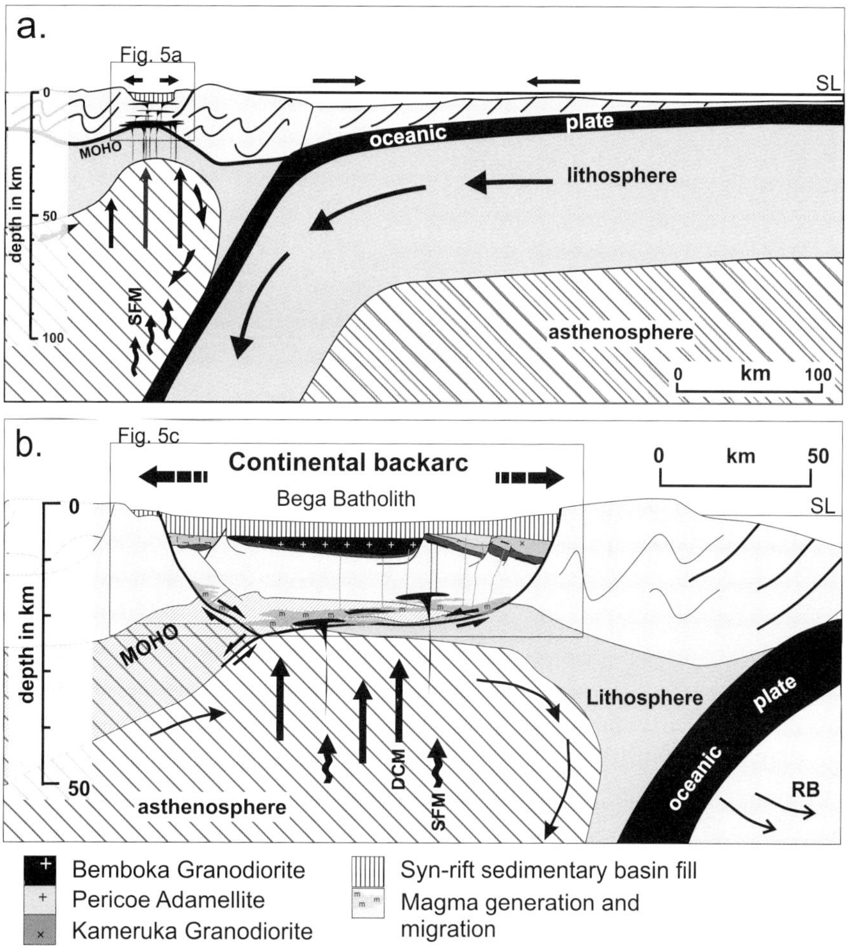

Figure 6 Tectonic setting of the Bega Batholith within a developing continental back-arc: (a) Extension is driven by the eastward retreat of the oceanic plate, resulting in lithospheric thinning, as well as initial generation and emplacement of mafic magmas associated with decompression and slab flux melting (Collins 2002a). (b) Construction of the Bega Batholith occurs in the developing back-arc. Eastward movement of the lower plate is accommodated by extension on the major west-dipping detachment that links to the zone of ductile stretching. The western margin of the system is bound by an antithetic normal fault that remained stationary during extension, causing preferential displacement on the eastern margin of Bega Batholith plutons. Key: (RB) slab rollback; (DCM) asthenospheric decompression melting; (SFM) slab flux melting; and (SL) sea level.

'feeder' dykes indicate that the dykes were able to intrude through actively deforming migmatitic rocks at the base of the pluton. The dykes intruded throughout the deformational and magmatic history of the migmatites and Kameruka pluton, as indicated by the abundant evidence for hybridisation between the dykes and migmatites, and by the presence of folded dykes in the migmatites. These syn-plutonic and syn-migmatitic dykes suggest magma ascent via fractures which penetrated through overlying active structures. Possibly the ascent rate (magma pressure) was sufficient for magmas to fracture through the underlying mid-crustal detachment, or that structure did not exist during much of the emplacement history. If the latter is correct, which the present authors consider more likely, the envisaged listric faults bounding the graben structure and enclosing the batholith did not link until the advanced stages of extension (Fig. 5c or d).

6. Tectonic control on magma emplacement rates

On the tectonic scale (Fig. 6a, b), the present authors consider that the batholith represents the base of a series of tilted half-grabens, represented by the N-trending plutons, developed during the eastward retreat of an arc–back-arc system, above a west-dipping subduction zone (cf. Collins 2002a). The measured but variable eastward displacement of the eastern side of the batholith (Fig. 1) indicates that this margin was the migrating boundary during extension. The lack of displacement along the western margin (Glenbog suite) suggests that this boundary is the major antithetic fault to the extensional system during batholith construction. This might indicate that the master detachment dipped eastward under the batholith, but this configuration would place the western (continent) side of the batholith as the lower retreating plate (terminology after Lister *et al.* 1986). However, the variable eastern margin of the batholith suggests that the major detachment dipped westwards, beneath the batholith. The lower plate migrated eastwards during the retreat of the subducting oceanic plate (Fig. 5d)

An eastern, retreating lower plate is consistent with the eastward younging of granite suites in the Bega Batholith, and with the general eastward migrating nature of plutonism in the eastern Lachlan Fold belt (e.g. Collins & Vernon 1992; Collins 2002a). It is consistent with slab rollback models for construction of the modern-day southwestern Pacific (e.g. Lister *et al.* 1986), and with uniformitarian tectonic models which invoke eastward (outboard) growth of the Palaeozoic eastern Lachlan, above a west-dipping subduction zone (e.g. Powell 1984; Collins & Vernon 1992; Foster & Gray 2000).

The composition of primitive mafic rocks in the batholith also favours extensional models. Primitive basaltic syn-plutonic dykes and gabbros from the batholith, including the sheeted gabbro complex, have compositions which closely resemble the active arc–back-arc basin system of the Lau Basin-Taupo Volcanic Zone in the southwest Pacific (Collins 2002b). These mafic rocks are being generated in regions where the lithosphere is <30 km thick and the tectonic regime is clearly extensional (e.g. Walcott 1987). They are forming during the eastward retreat of the west-subducting Pacific plate. A similar geodynamic environment can be envisaged for the Lachlan orogen during emplacement of the Bega Batholith.

The active rift model requires that some batholiths can grow at tectonic rates. Using the present authors' emplacement model for the Bega Batholith, the rift is 25–50 km wide and initially bound by normal faults which developed near the margin of the peripheral plutons (Figs 1 & 5). Based on approximate reconstructions of pre-rift geometry and experimental simulation of 'cold' rift basin development (Ellis & McClay 1988), the batholith could have formed by half-graben development during ~24% extension (Fig. 5a–c), which accounts for ~15 km eastward translation, some 10 km less than the 25 km measured movement.

The necessary additional 10 km of movement can be explained by ductile stretching at the base of plutons in a 'hot' rift environment. Post-Kameruka arc-perpendicular displacement is ~17 km, much of it recorded along the NE-trending faults during Bemboka suite emplacement (Fig. 1), but the cold rift model only accounts for ~3 km of this displacement. Thus, the present authors suggest that the additional stretching occurred by ductile flow along a shallow-dipping detachment focused in the migmatites observed directly below the Bemboka pluton (Fig. 5d). The detachment may have initiated as the western extension of the migmatites at the base of the Kameruka suite plutons (Fig. 5b). Extension below the migmatites may have been accommodated by pure shear above a master mid-crustal detachment, which is inferred to form by this stage (Fig. 5d). In plan view, this late extension was asymmetric and scissor-like, increasing in degrees of displacement southward and maximised along the NE-trending oblique transfer faults, where the batholith is widest (Fig. 1).

A southwest Pacific arc–back-arc basin system has already been used as a modern analogue for the eastern Lachlan in Silurian–Devonian times (Collins 2002b). Half-spreading rates for this modern system vary between 38 and 11 mm yr^{-1} (e. g. Malahoff *et al.* 1982), which imply that the maximum 25 km, arc-perpendicular extension on the Burragate fault could have developed in 1–2 Ma. Thus, it is possible that the entire batholith was emplaced in wedge-like, half-graben structures at inflation rates compatible to the 2-Ma period calculated by Petford *et al.* (2000), using the power-law inflation curve. However, Coleman *et al.*, (2004) have demonstrated that magma chambers which undergo incremental construction during repeated injection of felsic magma (Weibe & Collins 1998) may remain active for up to 10 Ma. This implies that construction of the Bega Batholith may have also been long-lived and that the rate of chamber or batholith construction may be, in part, controlled by the rate of extension. The data imply that magma inflation rates can be readily achieved at tectonic rates in extensional arc–back-arc environments. The present authors suggest that the plutons of the Bega Batholith underwent progressive chamber construction during repeated injection of felsic and mafic magma at the base of active half-grabens.

7. Acknowledgements

We thank Robin Offler for an early review of the manuscript, and Bob Wiebe, Jean-Louis Vigneresse, Tim Rolph and Paul Lennox for assistance in the field and many discussions. This study was partly funded by an Australian Postgraduate Award scholarship and ARC Large Grant No. A0001203. We would also like to thank Jean-Louis Vigneresse, Sandy Cruden, Scott Paterson, Keith Benn and Ryo Anma for constructive reviews of and comments on the paper.

8. References

Benn, K., Odonne, F., Lee, S. K. Y. & Darcovich, K. 2000. Analogue scale models of pluton emplacement during transpression in brittle and ductile crust. *Transactions of the Royal Society of Edinburgh: Earth Sciences* **91**, 111–21.

Brown, M. & Solar, G. S. 1998. Granite ascent and emplacement during contractional deformation in convergent orogens. *Journal of Structural Geology* **20**, 1365–93.

Chappell, B. W. 1996. Compositional variation within granite suites of the Lachlan Fold Belt: its causes and implications for the physical state of granite magma. *Transactions of the Royal Society of Edinburgh: Earth Sciences* **87**, 159–70.

Chappell, B. W., White, A. J. R. & Williams, I. S. 1991. A transverse section through granites of the Lachlan Fold Belt: Second Hutton Symposium Guide. Bureau of Mineral Resources, Geology and Geophysics Record 1992/22.

Coleman, D. S., Gray, W. & Glazner, A. F. 2004. Rethinking the emplacement and evolution of zoned plutons: geochronologic evidence for incremental assembly of the Tuolumne Intrusive Suite, California. *Geology* **52**, 433–6.

Collins, W. J. 2002a. Hot orogens, tectonic switching, and creation of continental crust. *Geology* **30**, 535–8.

Collins, W. J. 2002b. Nature of extensional accretionary orogens. *Tectonics*, **21**, 10. 1029/2000TC001272.

Collins, W. J., Richards, S. W., Healy, B. & Ellison, P. I. 2000. Origin of mafic enclaves by two-stage hybridisation in and below granitic magma chambers. *Transactions of the Royal Society of Edinburgh: Earth Sciences* **91**, 47–71.

Collins, W. J. & Vernon, R. H. 1992. Palaeozoic arc growth, deformation and migration across the Lachlan Fold Belt, southeastern Australia. *Tectonophysics* **214**, 381–400.

Cruden, A. R. 1998. On the emplacement of tabular granites. *Journal of the Geological Society, London* **155**, 853–62.

Dixon, J. M. & Simpson, D. G. 1987, Centrifuge modeling of laccolith intrusion. *Journal of Structural Geology* **9**, 87–103.

Ellis, P. G. & McClay, K. R. 1988, Listric extensional fault systems – results of analogue model experiments. *Basin Research* **1**, 55–70.

Finlayson, D. M., Johnstone, D. W., Korsch, R. J., Glen, R. A. & Leven, J. H. 2002. Seismic imaging and crustal architecture across the Lachan Transverse Zone, a possible early cross-cutting feature of Eastern Australia. *Australian Journal of Earth Sciences* **49**, 311–21.

Foster, D. A. & Gray, D. R. 2000. Evolution and structure of the Lachlan Fold Belt (orogen) of eastern Australia. *Annual Reviews in Earth and Planetary Science* **28**, 47–80.

Glen, R. A. 1992. Thrust, extensional and strike-slip tectonics in an evolving Palaeozoic orogen – a structural synthesis of the Lachlan Orogen of southeastern Australia. *Tectonophysics* **214**, 341–80.

Glen, R. A. 1995. Thrusts and thrust-associated mineralization in the Lachlan orogen. *Economic Geology* **90**, 1402–29.

Glen, R. A., Johnstone, D. W., Lawrie, K. C., Finlayson, D. M., Shaw, R. D., Korsch, R. J., Direen, N. G. & Jones, L. E. A. 2002. Crustal structure of the Ordovician Macquarie Arc, Eastern Lachlan Orogen, based on seismic-reflection profiling. *Australian Journal of Earth Sciences* **49**, 323–48.

Healy, B. 1999. *The nature and effects of wall-rock assimilation at the base of a silicic magma chamber, Australia*. Unpublished Ph.D. thesis. University of Newcastle, Australia.

Lennox, P. G., Trzebski, R., Johston, A. J. & Scott, M. M. 2002. Structural and gravity modelling along two transects of the southern Wyangala Batholith, Molong Zone, Eastern Lachlan Fold Belt, NSW. 16th Australian Geological Convention, *Geological Society of Australia, Abstracts* **67**, 131.

Lister, G. S., Etheridge, M. A. & Symonds, P. A. 1986. Detachment faulting and the evolution of passive continental margins. *Geology* **15**, 246–50.

Malahoff, A., Feden, R. H. & Fleming, H. S. 1982. Magnetic anomalies and tectonic fabric of marginal basins north of New Zealand. *Journal of Geophysical Research* **87**, 4109–25.

McKenzie, D. & Bickle, M. J. 1988. The volume and composition of melt generated by extension of the lithosphere. *Journal of Petrology* **29**, 625–79.

Paterson, S. R. & Fowler, T. K., Jr. 1993. Re-examining pluton emplacement processes. *Journal of Structural Geology* **15**, 191–206.

Paterson, S. R. & Miller, R. B. 1998. Magma emplacement during arc-perpendicular shortening: an example from the Cascades crystalline core, Washington. *Tectonics* **17**, 571–86.

Petford, N., Cruden, A. R., McCaffrey, K. J. W. & Vigneresse, J. L. 2000. Granite magma formation, transport and emplacement in the Earth's crust. *Nature* **408**, 669–73.

Petford, N. & Gallagher, K. 2001. Partial melting of mafic (ambphibolitic) lower crust by periodic influx of basaltic magma. *Earth and Planetary Science Letters* **193**, 483–99.

Powell, C. McA. 1984. Ordovician to earliest Silurian: marginal sea and island arc; Silurian to mid-Devonian dextral transtensional margin; Late Devonian and Early Carboniferous: continental magmatic arc along the eastern edge of the Lachlan Fold Belt. *In* Veevers, J. J. (ed.) *Phanerozoic Earth History of Australia*, 329–37. Oxford: Oxford University Press.

Roach, M. 1990. Interactive 2D Gravity and Magnetic Modeling For IBM PC Compatible Computers. CODES, University of Tasmania.

Thompson, A. B. 1999. Some time–space relationships for crustal melting and granitic intrusion at various depths. *In* Castro, A., Fernandez, C. & Vigneresse, J. L. (eds) *Understanding Granites: Integrating New and Classical Techniques. Geological Society, London, Special Publication* **168**, 7–25.

Trzebski, R., Lennox, P. G. & Palmer, D. 1999. Contrasts in morphogenesis and tectonic setting during contemporaneous emplacement of S- and I-type granitoids in the Eastern Lachlan Fold Belt, southeastern Australia. *In* Castro, A., Fernandez, C. & Vigneresse, J. L. (eds) *Understanding Granites: Integrating New and Classical Techniques. Geological Society, London, Special Publication* **168**, 123–40.

Trzebski, R., Lennox, P. G. & Palmer, D. 2001. A new gravity traverse across the northern Wyangala Batholith (Eastern Lachlan Fold Belt). *Geological Society of Australia Abstracts* **64**, 185–6.

Vigneresse, J. L. 1995. Control of granite emplacement by regional deformation. *Tectonophysics* **249**, 173–86.

Walcott, R. I. 1987. Geodetic strain and the deformational history of the North Island of New Zealand during the late Cainozoic. *Philosophical Transactions of the Royal Society of London* **321**, 162–81.

Wiebe, R. A. & Collins, W. J. 1998. Depositional features and stratigraphic sections in granitic plutons: implications for the emplacement and crystallization of granitic magma. *Journal of Structural Geology* **20**, 1273–89.

Wyborn, D. & Chappell, B. W. 1986. The petrogenetic significance of chemically related plutonic and volcanic rock units (Australia). *Geological Magazine* **123**, 619–28.

SIMON RICHARDS, Earth Materials: Structure and Tectonics, Research School of Earth Sciences, The Australian National University, Canberra, ACT 0200, Australia.
e-mail: simon.richards@anu.edu.au

BILL COLLINS*, Department of Geology, School of Environmental and Life Sciences, University of Newcastle, Callaghan, NSW 2308, Australia.

*Current address: School of Earth Sciences, James Cook University, Townsville, Queensland 4814, Australia.

MS received 16 December 2003. Accepted for publication 24 August 2004.

Transactions of the Royal Society of Edinburgh: Earth Sciences, **95**, 319–337, 2004

Crustal control on the redox state of granitoid magmas: tectonic implications from the granitoid and metallogenic provinces in the circum-Japan Sea Region

K. Sato, S. V. Kovalenko, N. P. Romanovsky, M. Nedachi, N. V. Berdnikov and T. Ishihara

ABSTRACT: Felsic magmatism has occurred over a large region of East Asia since Jurassic times and has provided important mineral resources such as tin, tungsten, base metals and gold. The circum-Japan Sea region preserves various geological records of active continental margins, including Jurassic to Early Tertiary magmatic arcs and subduction zones and pre-Jurassic continental basements, which were separated by the opening of the Japan Sea during the Miocene. The felsic magmatism in this region shows a wide variation in terms of redox state and related mineralisation, encompassing east–west contrasts around the Pacific Ocean. A review of granitoids and associated ore deposits in this region indicates that the character of the crust, sedimentary versus igneous, is an essential factor to control the redox state, and a tectonic setting may be an additional factor in some cases.

The reduced-type granitoids, characterised by tin mineralisation, were generated in carbon-bearing sedimentary crust which was composed mainly of accretionary complex material and not influenced by previous magmatism. Involvement of sedimentary materials is corroborated by oxygen, sulphur and strontium isotope data. The oxidised-type granitoids, characterised by gold or molybdenum mineralisation, were generated in igneous crust which was depleted in reducing agents as a result of previous magmatism. Granitoid magmatism in a given area tends to become more oxidised with time.

Jurassic accretionary complexes in East Asia are thought to have been largely displaced from the original place of accretion and stacked up against the northeastern margin in the Khingan and Sikhote–Alin Mountains. This region, dominated by sedimentary crust, was subsequently subjected to Cretaceous felsic magmatism and converted to a large province of reduced-type granitoids and tin–tungsten mineralisation. Diverse geodynamic processes, including the change of the arc-trench system, the creation and collapse of the back-arc basin and the collision of continents, may have prepared many favourable sites for the generation of reduced-type granitoids in northeast Asia. These processes may have resulted in a remarkable contrast with the Pacific margin of North America, where repeated arc magmatism during the Mesozoic formed granitoid batholiths of the oxidised-type.

The granitoid types may also be controlled by the tectonic setting and mode of magma emplacement. In the northern Kitakami area of Northeast Japan, Early Cretaceous episodic magmatism occurred in a Jurassic accretionary complex, and formed the oxidised-type granitoids accompanied by submarine bimodal volcanism associated with kuroko mineralisation. Granitoids of fissure-filling type emplaced under extensional environments may be oxidised, irrespective of basement geology, because of insignificant crustal input.

KEY WORDS: Accretionary complex, Cretaceous, granitoid type, Jurassic, Khingan, Korea, metallogeny, sedimentary crust, Sikhote–Alin, tin.

Granitoids and mineralisation around the Pacific Ocean show a remarkable contrast between the east and west sides. The Asian side is dominated by reduced-type granitoids and characterised by tin mineralisation. On the other hand, the Pacific margin of North America is dominated by oxidised-type granitoids and characterised by the lack of tin (e.g. Ishihara 1984). However, when examined closely East Asia is not homogeneous. For example, southeast China is very rich in tin, whilst northeast China and the Korean Peninsula are very poor, in fact practically free of Phanerozoic tin deposits (e.g. Sato 1982b; KIER 1983; Chinese Academy of Geological Sciences 1999). Far East Russia is very rich in tin, as the Sikhote–Alin, Khingan and Yana–Kolyma have been major tin-producing regions since the time of the Soviet Union (e.g. USSR Ministry of Geology 1987). The Japanese Islands include tin-rich and tin-barren provinces (e.g. Ishihara & Sasaki 1973).

Therefore, the circum-Japan Sea is a good area to increase the understanding of the genesis of granitoid types and related mineralisation (Sato 2000, 2003). In this paper, the authors review the characteristics of granitoids and mineralisation in the southern part of Far East Russia, with special emphasis on the redox state of granitoids and its correlation with basement geology, and compare them with the geological data for the Korean Peninsula and the Japanese Islands from the viewpoint of the tectonic history of East Asia.

The ilmenite- and magnetite-series classification of Ishihara (1977) is based essentially on the redox state of granitoid magmas. The two series granitoids are similar in major element composition except for their Fe^{3+}/Fe^{2+} ratio. Therefore, the two series can be named reduced-type and oxidised-type, respectively (Sato *et al.* 1992); these terms are used in this paper. The two types of granitoids are easily discriminated by measurement of magnetic susceptibility, even in the field, because ferric iron exists mainly in magnetite in the oxidised-type rocks. Magnetic susceptibility data for granitoids in Japan show a clear bimodal distribution (Sato *et al.* 1992), typically $<1 \times 10^{-3}$ SI for the reduced type and $10–50 \times 10^{-3}$ SI for the oxidised type. The boundary between the two types was placed at 100×10^{-6} emu/g (Ishihara *et al.* 1981), which approximately corresponds to 3×10^{-3} SI (Sato & Ishihara 1983). Accompanying mineralisation is quite contrasting. Tin is a representative ore element of the reduced-type granitoids, whilst gold and molybdenum mineralisations are related to the oxidised-type granitoids. These correlations, established in Japan, have also been confirmed in eastern Australia (e.g. Blevin & Chappell 1992) and in Far East Russia as described below.

Mesozoic to Cenozoic granitioids in the circum-Japan Sea region consists mainly of granite, granodiorite and tonalite. Most of them have chemical compositions of the I-type of Chappell & White (1974, 1992) (Takahashi *et al.* 1980; Jwa 1998; Jin *et al.* 2001; Sato & Chappell, unpublished data). The S-type or A-type granitoids do occur, but their occurrences are very limited. This paper focuses on a regional variation of the redox state of granitoids and its correlation with the metallogenic provinces from the viewpoint of the geodynamic history of East Asia. Individual petrochemical features including the I- and S-type classification will be reported elsewhere.

1. Khingan and Sikhote–Alin Mountains, Far East Russia

The Khingan and Sikhote–Alin Mountains (Fig. 1) are characterised by tin mineralisation related to Cretaceous felsic magmatism. Jurassic mineralisation is not known in this region of Far East Russia, although both Jurassic and Cretaceous mineralisations related to felsic magmatism are widely distributed in Korea and China.

The pre-Cretaceous basement of the Khingan and Sikhote–Alin Mountains is divided into two tectonic units: an older unit of Precambrian to Palaeozoic continental blocks on the continental side, and a younger unit of Jurassic to Cretaceous accretionary complexes and turbidite deposits on the oceanic side (Figs 2A, 5 & 11). The continental blocks, the Bureya massif in the north and Khanka massif in the south, consist mainly of Proterozoic to Palaeozoic igneous and metamorphic rocks, and subordinate Palaeozoic to Mesozoic sedimentary and volcanic sequences (e.g. Khanchuk *et al.* 1996; Sato *et al.* 2002, 2003b).

Cretaceous to Palaeogene felsic igneous activities occurred over a large region, including the above two units (Fig. 1). The igneous regions in southern Far East Russia are divided into two belts: the Khingan–Okhotsk Belt and Sikhote–Alin Belt (Fig. 2A). These two belts are considered to be juxtaposed arcs with differing geotectonic histories, showing an unusually broad distribution of igneous rocks in the Khingan and northern Sikhote–Alin Mountains, up to 700 km in an E–W direction.

1.1. The Khingan–Okhotsk Belt

The Khingan–Okhotsk Belt consists of many volcano–plutonic complexes, which are composed of granitoids and felsic volcanic rocks and associated with tin mineralisation (Sato *et al.* 2002). The Khingan (Kh), Badzhal (Bz) and Komsomolsk (Km) deposits are representative large tin deposits (Fig. 2A). The total length of this tin-associated volcano–plutonic belt exceeds 700 km, although both ends are not fully characterised. The northeastern end near the Okhotsk Sea appears to overlap with the Sikhote–Alin Belt. The southwestern extension beyond the Amur River, a national border with China, is somewhat enigmatic. Cretaceous volcanic rocks occur in the Xiao Hinggan Ling (Fig. 1), but no remarkable tin deposit is known in this region of northeast China (Chinese Academy of Geological Sciences 1999). Granitoids are widely exposed in this region, but they are mostly Palaeozoic age, as in the Bureya massif within Russia (Krasnyi & Peng 1996). The Khingan–Okhotsk tin belt is thought to be terminated near the Amur River.

Granitoids from the tin-associated volcano–plutonic complexes typically gave very low magnetic susceptibility values ($<1 \times 10^{-3}$ SI), and associated volcanic rocks are characterised by low Fe^{3+}/Fe^{2+} ratios of whole-rock chemical analyses (Mishin & Romanovsky 1994). This evidence indicates that the Khingan–Okhotsk Belt consists essentially of reduced-type rocks. This is consistent with the result of the aeromagnetic survey; the Khingan–Okhotsk Belt does not give rise to any significant magnetic anomalies (Fig. 2B). The northern part of this belt near the Okhotsk Sea is not well studied, but flat anomaly patterns suggest the paucity of magnetic minerals.

Regional variation of the magnetic anomalies correlates well with the tectonic provinces around the Khingan–Okhotsk Belt (Fig. 2A, B). It is remarkable that the flat pattern coincides with the regions of the accretionary complex in which the volcano–plutonic complexes of the Khingan–Okhotsk Belt are emplaced. The southern Siberian craton, Bureya massif and East Sikhote–Alin Volcano–Plutonic Belt (Fig. 2A) are characterised by a rugged magnetic anomaly pattern, indicating wide distribution of magnetite-bearing igneous rocks. Granitoids in the East Sikhote–Alin Volcano–Plutonic Belt have been corroborated to be the oxidised type, based on magnetic susceptibility data (Romanovsky *et al.* 1996). The complicated magnetic anomaly patterns in northern Sikhote–Alin may be biased by the presence of Neogene volcanic rocks.

Igneous activity of the Khingan–Okhotsk Belt has been thought to have begun in the Early Cretaceous (Barremian) and lasted throughout the entire Late Cretaceous (Faure & Natal'in 1992; Natal'in 1993). However, recent radiometric dating suggests that magmatism in this tin-rich belt was a much shorter episodic event in the middle Cretaceous (Sato *et al.* 2002). Radiometric age data for major tin deposits and the surrounding igneous rocks cluster at 95 Ma, although the deposits are separated by 100–200 km. This result suggests a short-lived tin-associated magmatic event in contrast with the long-lasted magmatism in Sikhote–Alin and the Japanese Islands (Sato *et al.* 1992, 2002). The Khingan–Okhotsk Belt is also clearly discriminated from the northern Sikhote–Alin Belt not only in age but also in metallogeny; the latter being characterised by gold mineralisation (Sato *et al.* 1996).

Cretaceous accretionary complexes are distributed to the southeast of the Khingan–Okhotsk Volcano–Plutonic Belt (Natal'in 1993; Sato *et al.* 2002). The Kiselevka–Manona complex (KM in Fig. 2A), the youngest accretionary complex in the Khingan and Sikhote–Alin Mountains, is distributed along the Amur suture, which is thought to be a tectonic boundary with Jurassic accretionary complexes in central Sikhote–Alin (SM in Fig. 2A). Detailed studies on radiolarian assemblages using specimens from good exposures near the Amur River (Zyabrev 1996; Markevich *et al.* 1997) indicate

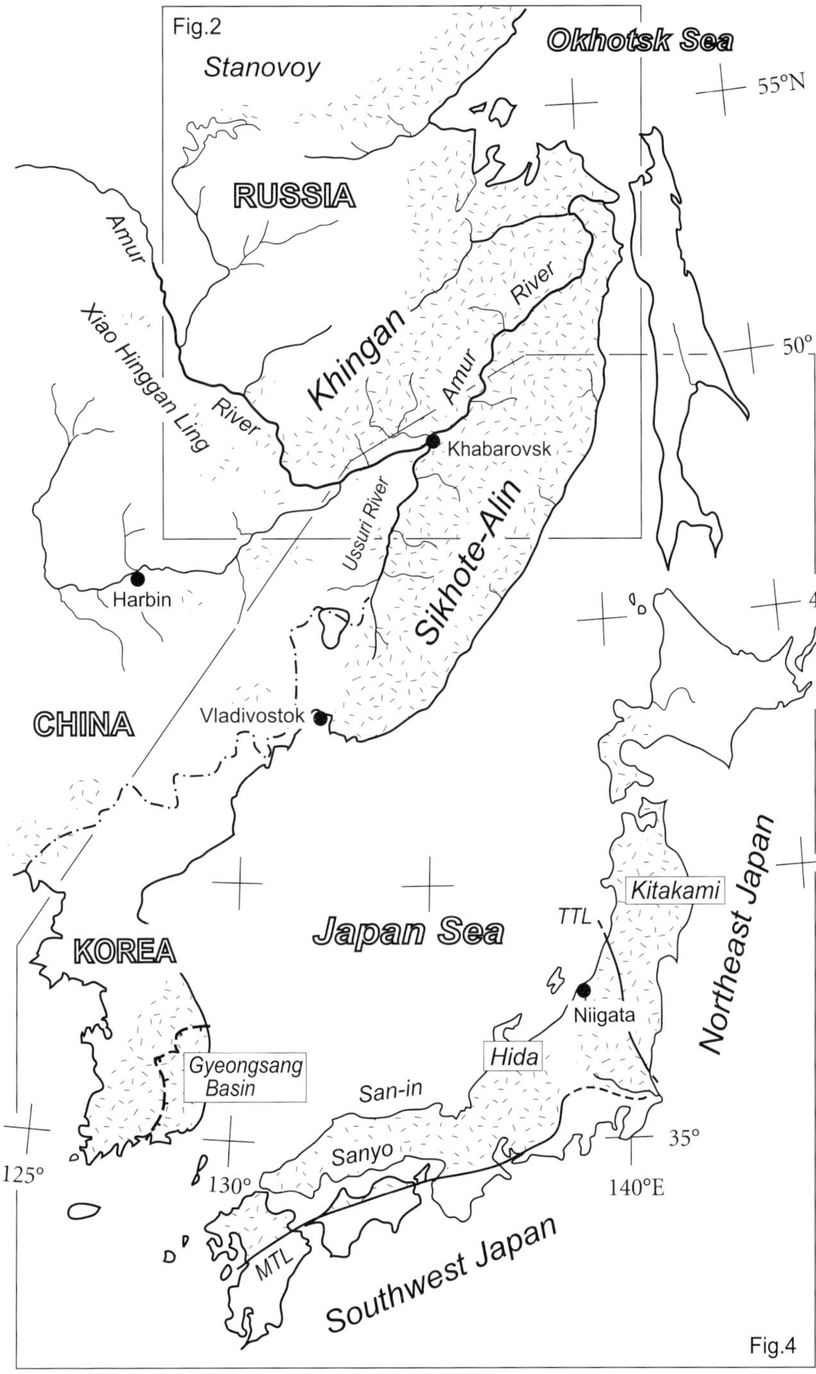

Figure 1 Map showing the distribution of Cretaceous–Palaeogene felsic magmatism in the circum-Japan Sea region, compiled from Sato *et al.* (1992), KIGAM (1995a) and Krasnyi & Peng (1996). (TTL) Tanakura Tectonic Line; (MTL) Median Tectonic Line. The national border between Russia and China coincides with the Amur and Ussuri Rivers. Xiao Hinggan Ling is also expressed as Lesser Khingan Mountains.

that the age of accretion was Albian–Cenomanian (ca. 110–90 Ma) in the middle Cretaceous. The apparent parallelism, in time and space, between the volcano–plutonic belt and the accretionary complex suggests that the Khingan–Okhotsk Belt and the Kiselevka–Manona complex represent a paired magmatic arc and subduction complex (Sato *et al.* 2002).

The Jurassic accretionary complex in central Sikhote–Alin occurs to the southeast of the Kiselevka–Manona complex, that is, the oceanic side is older than the continental side. These relationships between the accretionary complexes and igneous rocks within the Khingan–Okhotsk Belt and northern Sikhote–Alin Belt (Figs 2A, 3) suggest vigorous and complicated tectonic activity and magmatism over a relatively short period during the middle Cretaceous.

The Kiselevka–Manona complex includes Early Cretaceous bedded chert with basic tuffs and lavas below the middle Cretaceous clastic units (Nechaev *et al.* 1996; Zyabrev 1996; Markevich *et al.* 1997), indicating relatively young volcanism on an oceanic plate prior to subduction. Based on heavy mineral assemblages Nechaev *et al.* (1996) suggested the existence of an Early Cretaceous immature volcanic arc and back-arc basin in the present Lower Amur region. Sato *et al.* (2002) proposed a model of Sunda-style tectonics for the generation of the episodic reduced-type magmatism in the Khingan–Okhotsk Belt, assuming the existence of young back-arc basin such as the Andaman Basin (Fitch 1972; Hamilton 1979; Curray *et al.* 1979, 1982). That is, subduction of the young back-arc basin may have caused the episodic

(A)

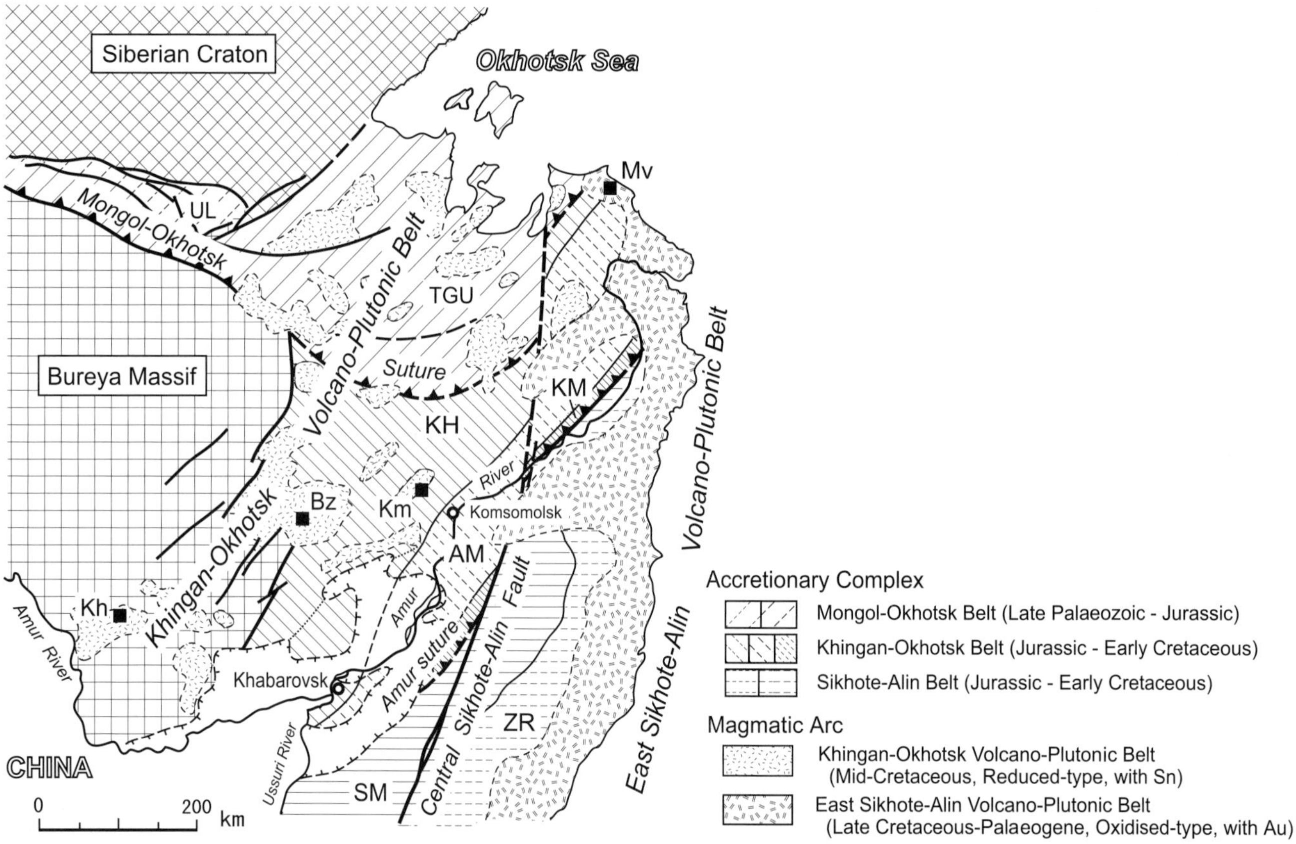

(B)

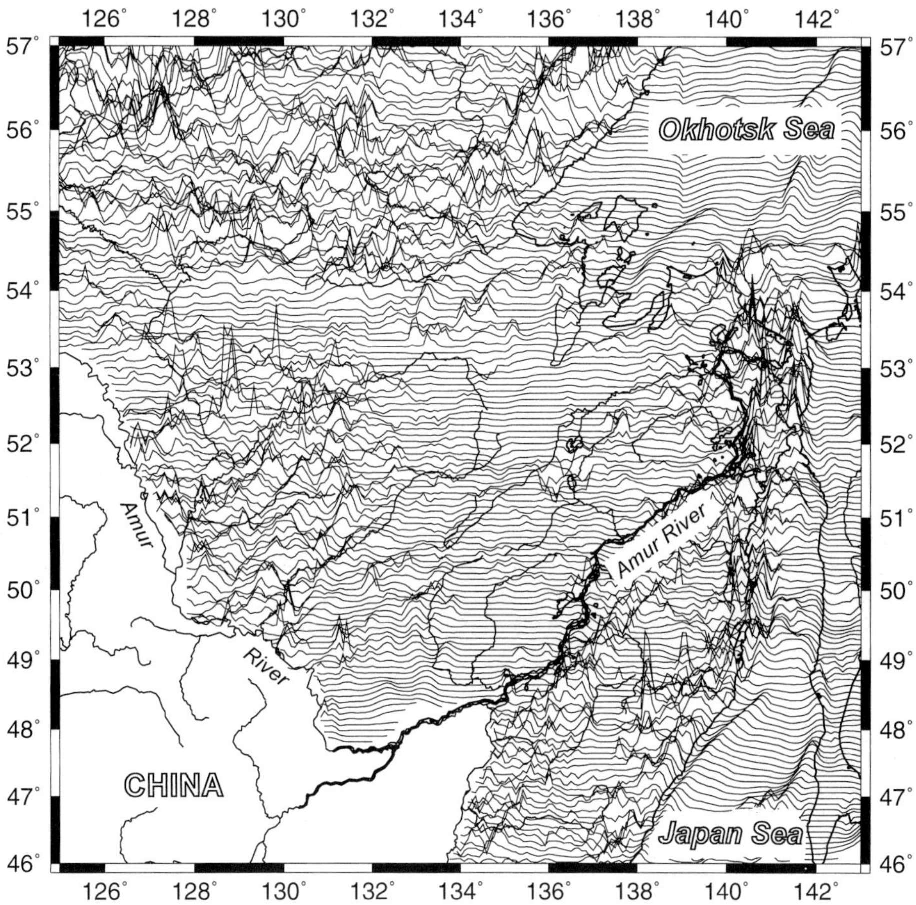

Figure 2 (A) and (B).

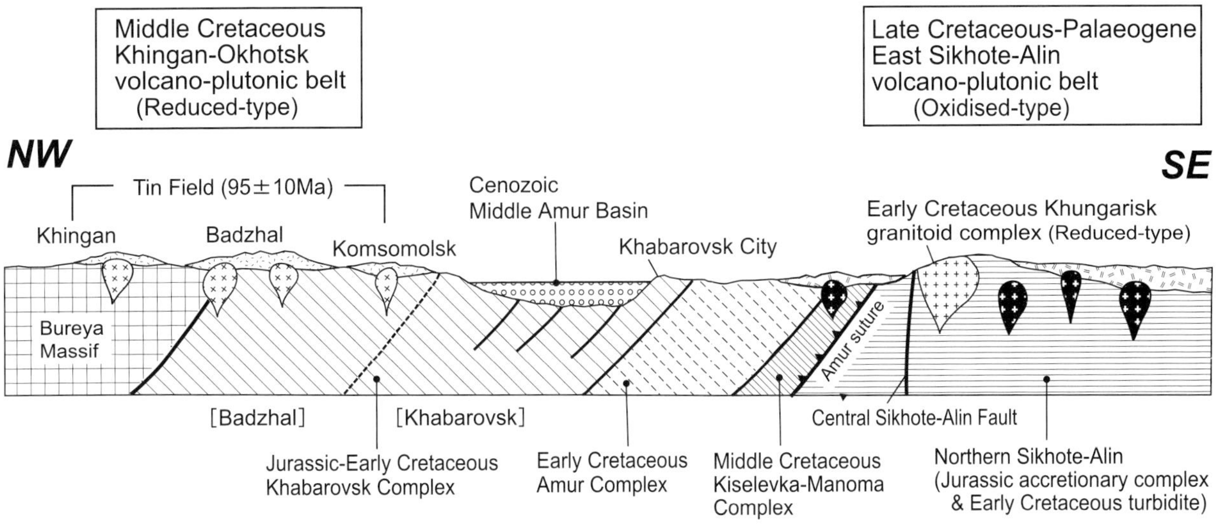

Figure 3 Schematic NW–SE cross section through the Khingan–Okhotsk active margin and northern Sikhote–Alin (Sato *et al.* 2002); modified from Faure & Natal'in (1992) and Natal'in (1993). Episodic magmatism and mineralisation in the Khingan–Okhotsk Tin Belt (95 ± 10 Ma) is coeval with the middle Cretaceous (Albian–Cenomanian) Kiselevka–Manoma complex (Voinova *et al.* 1995; Zyabrev 1996; Nechaev *et al.* 1996). The Khungarisk complex a conventional name for a group of Early Cretaceous granitoid plutons in north-central Sikhote–Alin.

magmatism in a similar manner to the Miocene tin-associated magmatism in the Outer Zone of Southwest Japan, which was caused by the subduction of the very young Shikoku Basin in relation to the opening of the Japan Sea (e.g. Kobayashi & Sato 1979; Furukawa & Tatsumi 1999).

1.2. The Sikhote–Alin Belt

The Sikhote–Alin Belt hosts Cretaceous–Palaeogene felsic magmatism which is associated with various mineral resources including tungsten, tin, base metals and gold (Fig. 4). The major part of the basement consists of Jurassic to Early Cretaceous accretionary complexes and turbidite deposits, but a Precambrian to Palaeozoic continental fragment of the Khanka massif occurs in southwestern Sikhote–Alin (Figs 5 & 11). Granitoid magmatism began at about 130 Ma in central Sikhote–Alin, and appears to have continued throughout the Cretaceous. Magmatism migrated eastwards to the present coastal regions, in the late Late Cretaceous to Palaeogene, resulting in the formation of the East Sikhote–Alin Volcano–Plutonic Belt. Detailed age variations in the granitoid magmatism within Sikhote–Alin require further study, but it is significant that magmatism lasted for more than 70 m.y. in contrast with the episodic Khingan–Okhotsk magmatism (Sato *et al.* 2002).

Compiled magnetic susceptibility data for most of granitoid plutons reveals regional distributions of the reduced- and oxidised-type granitoids in Sikhote–Alin (Romanovsky *et al.* 1996). In view of age and associated mineralisation as well as the magnetic susceptibility data, granitoid magmatism in Sikhote–Alin is divided into the following three provinces (Fig. 4): (1) the central Sikhote–Alin weakly magnetic province ($<1 \times 10^{-3}$ SI) composed essentially of the Cretaceous reduced-type granitoid plutons accompanied by tungsten and tin mineralisation; (2) the east Sikhote–Alin highly magnetic province (~ 3–30×10^{-3} SI), composed mainly of late Late Cretaceous to Palaeogene oxidised-type granitoid plutons and volcanic rocks (East Sikhote–Alin Volcano–Plutonic Belt) that are locally accompanied by gold mineralisation (Sato *et al.* 1996, fig. 1); and (3) the southwest Sikhote–Alin highly magnetic province (~ 5–25×10^{-3} SI), composed of Late Cretaceous oxidised-type granitoid plutons accompanied by gold mineralisation (Fig. 5). Province 1 constitutes a major part of Sikhote–Alin and includes many tungsten and tin deposits, such as the Vostok-2 (W skarn), Tigrinoe (Sn+W greisen) and Dubrovsk (Sn vein) deposits (Sato *et al.* 1993a; Khanchuk *et al.* 1996). Therefore, this province apparently represents the whole Sikhote–Alin Belt as a tungsten-tin metallogenic province. The magmatism and mineralisation in this province occurred over a long time interval (ca. 130–70 Ma), but their detailed variations within the province, particularly their relationships with those in province 2 require further reliable age data (Matsuda *et al.* 1998). Care has to be taken with dating because granitoids of province 1 near the coastal region are often influenced by the younger igneous activity of province 2. Provinces 1 and 3 correspond approximately to the region of Jurassic to Early Cretaceous sedimentary rocks and the Precambrian–Palaeozoic continental fragment, respectively (Figs 4 & 11). The relationship between these two provinces is well observed in the southernmost Sikhote–Alin as described below (Fig. 5), where the coast provides good exposures for the observation of across-arc variation of granitoids and basement rocks, compared with other areas in Sikhote–Alin.

Relationships between granitoid types, basement terranes and mineralisation in the southernmost Sikhote–Alin are

Figure 2 (A) Tectonic map showing the distribution of accretionary terranes and volcano–plutonic belts around the Khingan–Okhotsk Belt. Compiled from Natal'in & Zyabrev (1989), Faure & Natal'in(1992), Natal'in (1993), Krasnyi & Peng (1996), Romanovsky *et al.* (1996) and Sato *et al.* (2002). (TGU) Tukuringra–Dzhagdinsk, Galamsk and Ulbansk; (UL*) Uniya–Bomsk and Lansk; (KH) Khabarovsk; (AM*) Amur; (KM) Kiselevka–Manoma; (SM) Samarka; (ZR*) Zhuravlevka. (* = turbidite dominated). (Solid squares) large ore deposits; (Kh) Khingan (Sn); (Bz) Badzhal (Sn); (Km) Komsomolsk (Sn+W+Cu); and (Mv) Mnogovershinnoe (Au+Ag). (B) Magnetic anomaly map of the same region based on the data by USSR Ministry of Geology (1977). Profile interval is five minutes. Note that the flat pattern corresponds to the region of accretionary complex and igneous rocks of the Khingan–Okhotsk Volcano–Plutonic Belt which lacks significant magnetite, in contrast to the higher relief pattern for the southern Siberian craton, Bureya massif and East Sikhote–Alin Volcano–Plutonic Belt.

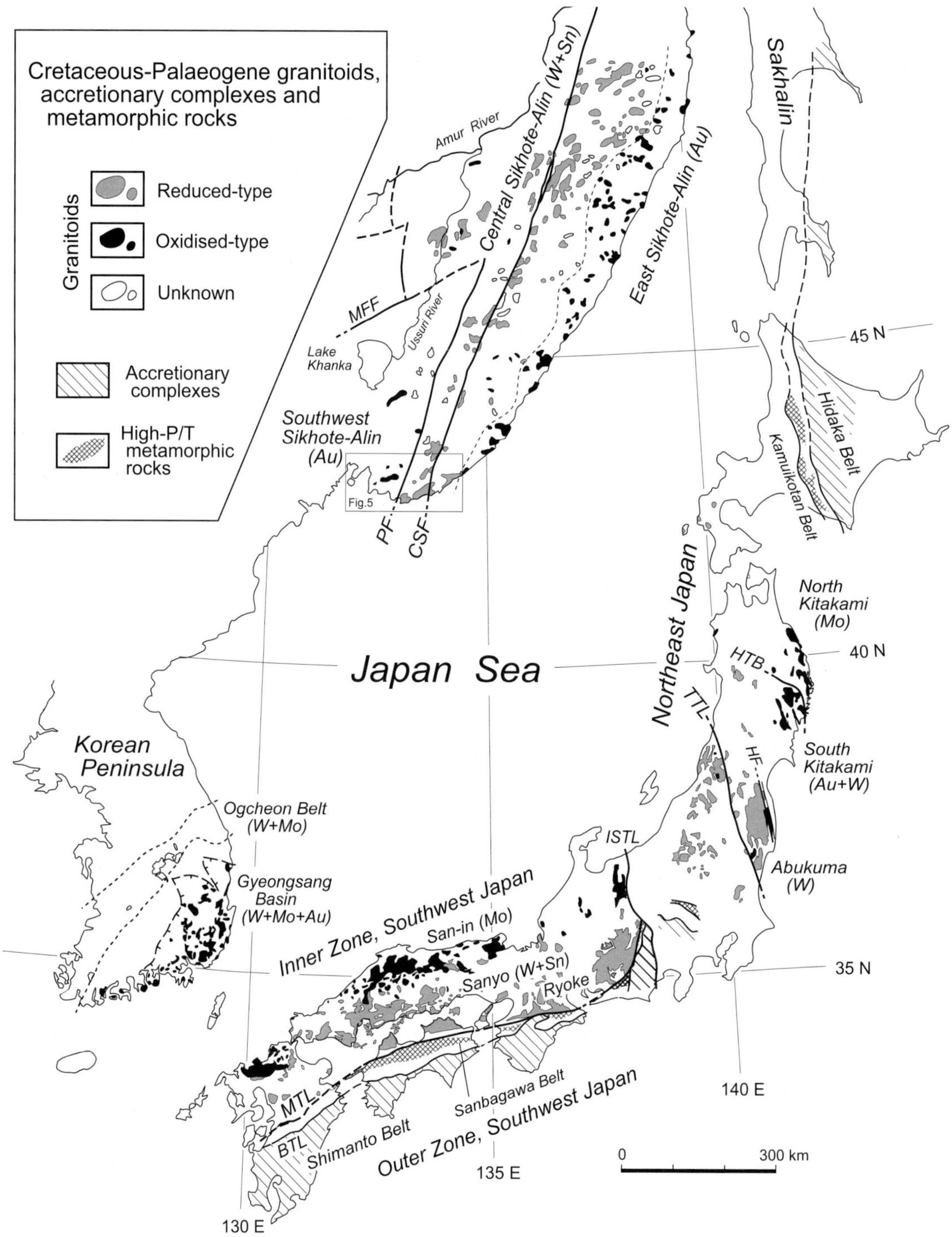

Figure 4 Distribution of Cretaceous–Palaeogene granitoids and metallogenic provinces in the circum-Japan Sea region. The distribution of reduced- and oxidised-type granitoids is based on Sato *et al.* (1992) for the Japanese Islands, Romanovsky *et al.* (1996) for Sikhote–Alin and Jin *et al.* (2001) for the Korean Peninsula. The metallogenic provinces were compiled from Ishihara *et al.* (1992) for the Japanese Islands, Sato *et al.* (1993a 1996) and Khanchuk *et al.* (1996) for Sikhote–Alin, and KIER (1983), KIGAM (1995b) and Jin *et al.* (2001) for the Korean Peninsula. Iron and base metals are excluded for simplicity because of their occurrences in both type granitoid provinces. Granitoid plutons in the central Ogcheon Belt in Korea mostly show low magnetic susceptibility values ($<1 \times 10^{-3}$ SI) corresponding to reduced-type, but this may be partly due to secondary alteration (see text). Both the Ogcheon Belt and the Gyeongsang Basin are characterised by tungsten and molybdenum mineralisation and the lack of tin. Accretionary complexes, high-P/T metamorphic rocks are from Geological Survey of Japan (1992), Kimura (1994) and Teraoka & Okumura (2003). See Figure 9 for ages of the geologic units. Major faults are: (BTL) Butsuzo Tectonic Line; (CSF) Central Sikhote–Alin Fault; (HF) Hatakawa Fault; (HTB) Hayachine Tectonic Belt; (ISTL) Itoigawa–Shizuoka Tectonic Line; (MFF) Mishan–Fushun Fault; (MTL) Median Tectonic Line; (PF) Partizansk Fault; and (TTL) Tanakura Tectonic Line. The MTL divides Southwest Japan into Inner Zone and Outer Zone.

shown in Figure 5. The Khanka massif, which is intruded by Late Cretaceous granitoid plutons, consists mainly of so-called 'Sergeevka gabbro'. This lithology consists of mafic to intermediate intrusive and metamorphic rocks probably of late Neoproterozoic to Cambrian age, with lesser amounts of Ordovician to Permian granitoids and Late Palaeozoic to Mesozoic volcanic and sedimentary rocks (Khanchuk *et al.* 1996). The eastern part of the massif occurs as a nappe on the

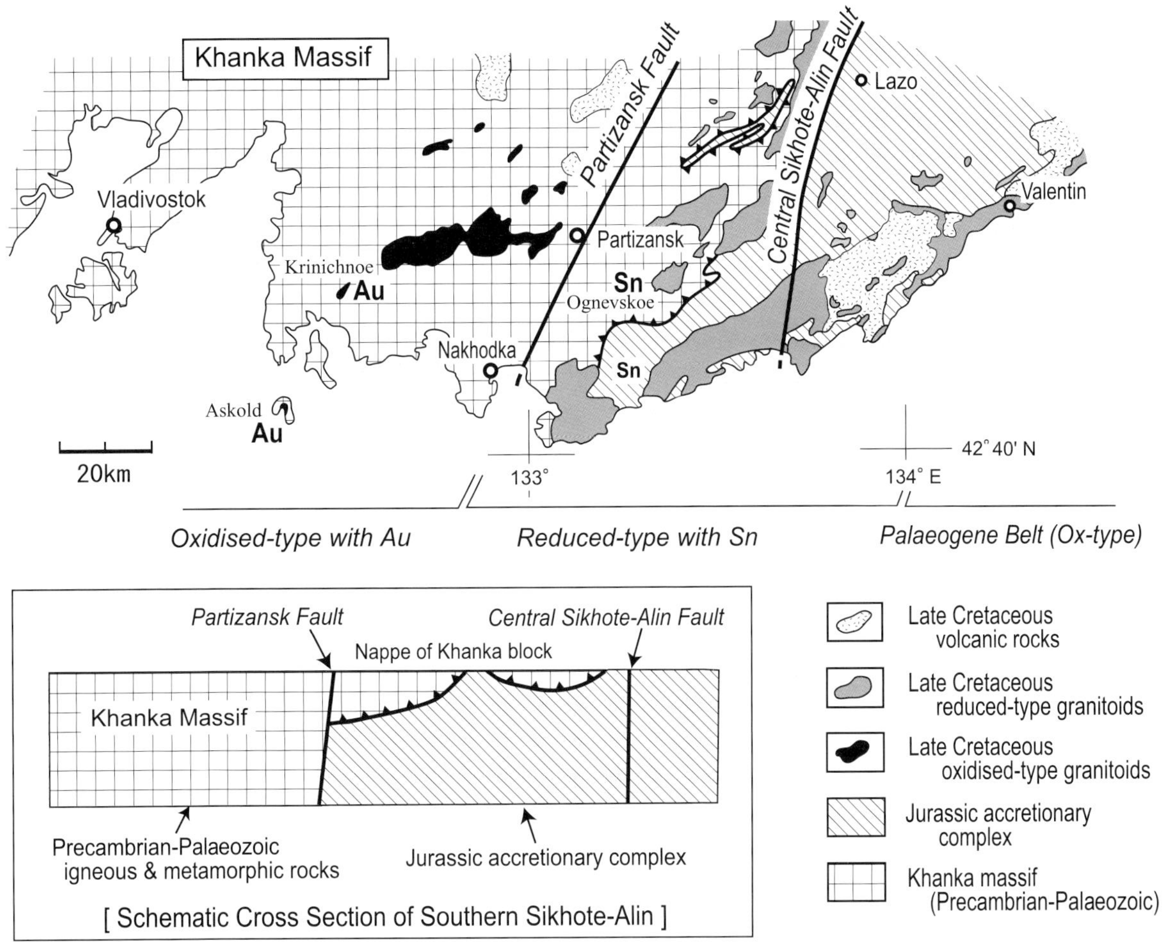

Figure 5 Simplified geologic map of southernmost Sikhote–Alin. The oxidised- and reduced-type plutons (ca. 80–90 Ma) are accompanied by Sn and Au mineralisation, respectively. Note that the distribution of the two type plutons correlates well with the Partizansk Fault. Rocks of the Khanka massif to the east of the Partizansk Fault occur as a nappe on the Jurassic accretionary complex (Sato *et al.* 1993b; Khanchuk *et al.* 1996) .

Jurassic accretionary complex, as shown in the cross section of Figure 5. The underlying Jurassic complex is exposed as a window in the nappe to the northeast of Partizansk City (Fig. 5).

The distribution of the reduced- and oxidised-type plutons correlates well with the Partizansk Fault; the oxidised-type on the west side and the reduced-type on the east side (Fig. 5). Plutons on both sides of the fault are nearly the same age (*ca.* 80–90 Ma), suggesting that the fault and nappe structures may have formed during the Early Cretaceous. Contrasting crust types are separated by the Partizansk Fault: continental crust of igneous affinity occurs to the west and accretionary wedge material of sedimentary affinity occurs to the east (Fig. 8). The Central Sikhote–Alin Fault occurs within the Jurassic accretionary complex and is not a significant boundary with respect to granitoid and metallogenic provinces, probably because the crust on both sides is similar.

The two types of granitoids separated by the Partizansk Fault have contrasting oxygen isotope compositions (Fig. 6). The reduced type has higher $\delta^{18}O$ values than the oxidised type, as in the Japanese Islands (e.g. Sato *et al.* 1992). The $\delta^{18}O$ values show a positive correlation with initial $^{87}Sr/^{86}Sr$ ratios (Fig. 7). These isotopic trends strongly suggest involvement of sedimentary materials in the generation of reduced-type granitoid magmas. Carbonaceous matter in the sedimentary rocks may have played an essential role in the formation of reducing conditions, as suggested by the isotopic studies on Japanese granitoids (Ishihara 1977; Sasaki & Ishihara 1979, 1980; Shibata & Ishihara 1979). The oxidised-type granitoids on the west side of the Partizansk Fault have low $\delta^{18}O$ values and initial $^{87}Sr/^{86}Sr$ ratios similar to the typical oxidised-type granitoids in Japan and Korea (Fig. 7). Involvement of sedimentary materials, if any, may have been insignificant for the genesis of the oxidised-type granitoids.

The authors consider that the contrasting features of granitoid types and mineralisation on each side of the Partizansk Fault basically reflect the remarkable difference in the dominant crust; sedimentary crust on the east and igneous crust on the west. The eastern side is essentially a zone of accretion, and had never been influenced by magmatism prior to the Late Cretaceous granitoid activity. On the other hand, the western side is a Precambrian–Palaeozoic continental block which had been subjected to magmatism prior to the Late Cretaceous granitoid activity (Fig. 8). This pre-Cretaceous repeated magmatism may have exhausted any reducing agents, leading to the formation of the oxidised-type granitoids in the younger magmatic event.

2. Korean Peninsula

Granitoids are widely distributed in the Korean Peninsula. Phanerozoic granitoids are divided by age into three major groups: Permian–Triassic, Jurassic and Cretaceous–Palaeogene (Fig. 8). The occurrence of Permian–Triassic granitoids is relatively minor except in the northeastern part of North Korea (KIGAM 1995a). Jurassic granitoid batholiths are widely distributed throughout the entire Korean Peninsula (Fig. 11), and are referred to as the Daebo granite. The Cretaceous–Palaeogene granitoids, the Bulguksa granite, are distributed

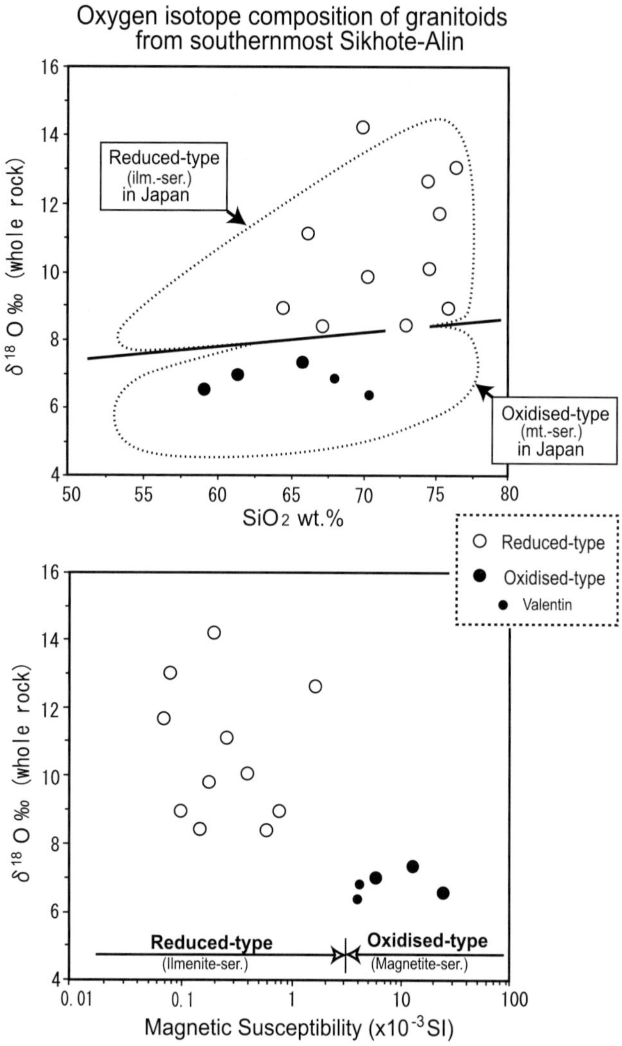

Figure 6 Contrasting oxygen isotope composition between the oxidised- and reduced-type granitoids from the Late Cretaceous plutons in southernmost Sikhote–Alin (see Fig. 5). Also included are data for a late Late Cretaceous weakly magnetic pluton near Valentin, which may be the western margin of the East Sikhote–Alin Volcano–Plutonic Belt (K.Sato, unpublished data). Data areas for the Japanese granitoids are based on Sato *et al.* (1992). The standard is SMOW.

mainly in the southern Korean Peninsula, particularly in the Gyeongsang Basin (Fig. 4). The Bulguksa granite tends to occur as relatively small plutons which often contain miarolitic cavities and are accompanied by volcanic equivalents, suggesting a shallower level of emplacement than that of the Daebo batholiths (e.g. Jin *et al.* 2001). Precambrian granitoids also occur in the Korean Peninsula; they are distributed mainly in North Korea, but are minor in South Korea (KIGAM 1995a).

A large number of radiometric ages have been reported for granitoids, mainly from South Korea (e.g. KIGAM 1995a). The age data are widely scattered from about 250 Ma to about 50 Ma without a clear gap between the two clusters around 170 Ma (Daebo) and 90 Ma (Bulguksa). The large age range is at least partly due to slow cooling of large batholiths, or secondary thermal events which are recorded in dated materials of low closure temperatures; such as biotite in a K-Ar system. Recent chronological studies, including U-Pb dating for zircon, suggest that the emplacement ages are more limited, to 200–160 Ma for the Daebo granite and 110–50 Ma for the Bulguksa granite, with a hiatus between the two (Sagong *et al.* 2003).

Relationships between granitoid types and mineralisation in the southern Korean Peninsula have been extensively studied since the 1980s. The Bulguksa granite consists mostly of the oxidised type, but a few plutons in the central Ogcheon Belt (Fig. 4), a Palaeozoic fold belt between the Gyeonggi massif in the north and Ryongnam massif in the south, are dominated by granitoids of low magnetic susceptibility corresponding to the reduced type (Ishihara *et al.* 1981). The low magnetic susceptibility values are at least partly due to secondary oxidation of magnetite to haematite (Jin *et al.* 2001); the granitoid types have not been fully discriminated. The Bulguksa magmatism in both the Gyeongsang Basin and the Ogcheon Belt is accompanied by tungsten, molybdenum, base metals and gold mineralisation. This magmatism has provided major mineral resources of South Korea (Shimazaki *et al.* 1981, 1987; Sato *et al.* 1981; KIER 1983).

The Daebo granite consists mainly of the reduced type, in contrast to the Bulguksa granite; the oxidised-type granitoids do occur but the genesis of their complicated distribution has not been resolved, although late stage pink granites tend to be magnetite-bearing (Jin *et al.* 2001; M.S. Jin 2003, pers. comm.). Gold-quartz veins occur in fault zones near the Daebo granite (e.g. KIGAM 1995b; Choi 2002), but their genetic relationship to granitoid magmatism is controversial because of age gaps between the granitoid emplacement and mineralisation (Jin *et al.* 2001). The apparent paucity of Jurassic mineralisation could be due to the relatively deep erosion levels of the Daebo batholiths. Permian–Triassic and Precambrian granitoids have very low magnetic susceptibility values ($<1 \times 10^{-3}$ SI) corresponding to the reduced type (Jin *et al.* 2001). The Permian–Triassic plutons are not accompanied by any significant mineralisation, but the Precambrian plutons in the northeast Ryongnam massif are accompanied by tin-bearing pegmatites.

The Bulguksa granite in the Gyeongsang Basin is characterised by low $\delta^{18}O$ values and initial $^{87}Sr/^{86}Sr$ ratios, which are similar to the typical oxidised-type granitoids in Japan and Sikhote–Alin, in contrast to the Daebo granite which is characterised by intermediate $\delta^{18}O$ values and high initial $^{87}Sr/^{86}Sr$ ratios (Fig. 7). The Bulguksa granite of low magnetic susceptibility in the central Ogcheon Belt (Fig. 4) shows crustal strontium and neodymium isotope signatures which are similar to the Daebo granite, rather than to the Bulguksa granite in the Gyeongsang Basin with its juvenile isotopic signatures (Kim *et al.* 1996), suggesting a similarity of source materials regardless of their emplacement ages. These granitoids with enhanced initial $^{87}Sr/^{86}Sr$ ratios are significantly different from the reduced-type granitoids in Japan and Sikhote–Alin (Fig. 7). The pre-Jurassic basement in the Korean Peninsula is dominated by Precambrian to Palaeozoic metasedimentary rocks (Lee 1988) which have very high initial $^{87}Sr/^{86}Sr$ ratios (Kim *et al.* 1996, Cheong & Chang 1997) and contain many graphite deposits (Gallagher 1963). The metasedimentary rocks, widely distributed in the Precambrian basement, may have been incorporated in the generation of the reduced-type granitoids with high initial $^{87}Sr/^{86}Sr$ ratios.

3. Japanese Islands

The relationships between granitoids and mineralisation in the Japanese Islands were extensively studied in 1970–80s, with special reference to their redox state (e.g. Tsusue & Ishihara 1974; Ishihara 1977, 1979; Czamanske *et al.* 1981). Compilation of the relevant metallogenic studies, combined with geochronological and isotopic studies (e.g. Sasaki & Ishihara 1979, 1980; Shibata & Ishihara 1979) established the granitoid and metallogenic provinces which are consistent with the major tectonic divisions of the Japanese Islands (Fig. 4) (Ishihara *et al.* 1992; Sato *et al.* 1992).

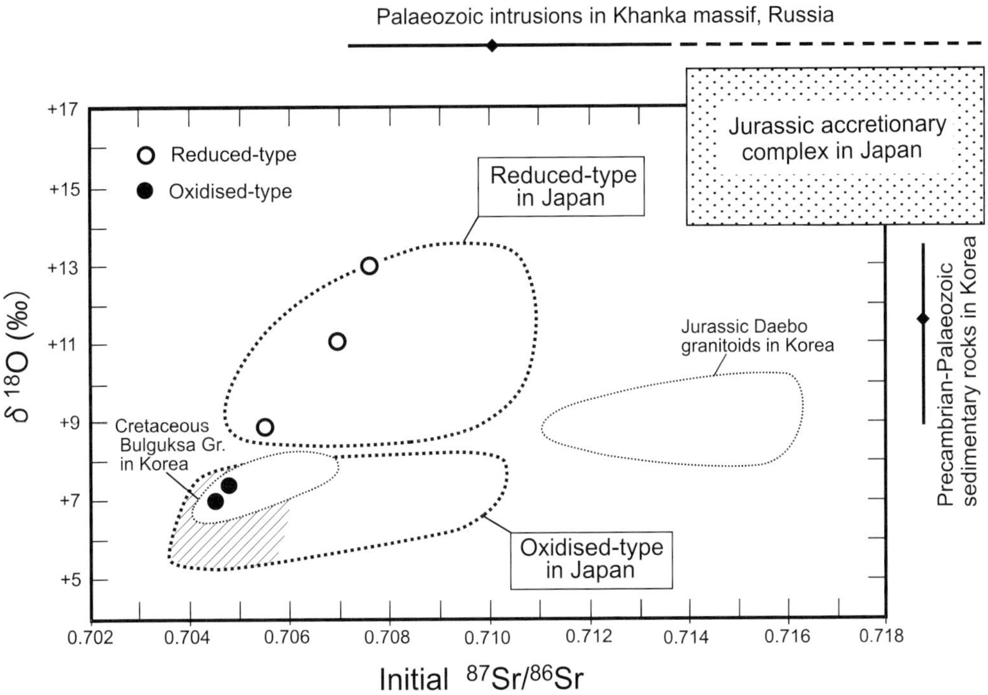

Figure 7 $\delta^{18}O$ (SMOW) versus initial $^{87}Sr/^{86}Sr$ (at 85 Ma) plot for the representative reduced- and oxidised-type granitoids in southernmost Sikhote–Alin (Fig. 5). Also shown are data areas for granitoids and the Jurassic accretionary complex (metasedimentary rocks from the Ryoke Belt) in Japan (Matsuhisa *et al.* 1982; Ishihara and Matsuhisa 2002). The oxidised-type granitoids in Japan generally have low initial $^{87}Sr/^{86}Sr$ ratios (hatched), but rarely have high values (~0·710) in the eastern San-in Belt within Southwest Japan (see Fig. 4). Data areas for the Korean granitoids are based on Kim *et al.* (1992, 1996); Jurassic Daebo (mainly reduced-type) and Cretaceous Bulguksa in the Gyeongsang Basin (oxidised-type), excluding the data for altered rocks. Available data for basement rocks are limited, but their ranges are shown by bars with diamonds of average values: Precambrian to Palaeozoic sedimentary rocks ($\delta^{18}O = +8{\cdot}9$ to $+13{\cdot}4$‰ ($n=3$)) in the southern Korean Peninsula (Kim *et al.* 1992, 1996) and Palaeozoic gabbroic to granitic intrusions (initial $^{87}Sr/^{86}Sr = 0{\cdot}707$ to $0{\cdot}720$ ($n=7$)) in the Voznesenka area within the Khanka massif (Ryazantseva *et al.* 1995).

Granitoids in Japan are divided by age into four groups: Palaeozoic, Jurassic, Cretaceous–Palaeogene and Neogene (e.g. Sato *et al.* 1992). Cretaceous–Palaeogene granitoids have the largest exposures and are most important with respect to metallogeny (e.g. Ishihara *et al.* 1992). The other groups have limited exposures and are less important or insignificant for metallogeny, although Middle Miocene granitoids emplaced in the Cretaceous–Palaeogene accretionary complex within the Outer Zone of Southwest Japan are characteristically accompanied by tin mineralisation (Fig. 9). Occurrences of Palaeozoic and Jurassic granitoids are very limited, and they are not accompanied by any significant mineralisation. The lack of Jurassic mineralisation of granitoid affinity in Japan is similar to that of the Khingan and Sikhote–Alin Mountains, and contrasts with the Korean Peninsula and eastern China.

The pre-Cretaceous basement of the Japanese Islands consists essentially of Jurassic accretionary complexes, as in the Khingan and Sikhote–Alin Mountains, with local occurrences of Palaeozoic continental fragments (e.g. South Kitakami, NE Japan), Palaeozoic to Early Mesozoic accretionary complexes (northwestern SW Japan) and late Palaeozoic–Jurassic metamorphic and igneous rocks (Hida, central Japan). The Jurassic Mino–Tanba terrane in Southwest Japan, for example, is thought to be a southern extension of the Jurassic Samarka terrane in Sikhote–Alin prior to the opening of the Japan Sea during the Miocene (Fig. 11) (Mizutani & Kojima 1992), suggesting a similar tectonic history between Japan and Sikhote–Alin in Jurassic to Palaeogene times (Figs 8 & 9).

The Cretaceous–Palaeogene granitoid magmatism in Japan changed its position and character with time (Figs 4 & 9). Magmatism was initiated in the Kitakami area of Northeast Japan at about 130 Ma, and ceased around 40 Ma in the western San-in Belt of Southwest Japan (e.g. Sato *et al.* 1992). The Early Cretaceous granitoids in Northeast Japan to the east of the Tanakura Tectonic Line (TTL in Figs 1 & 4) generally have less silicic compositions than the Japanese average and are characterised by low initial $^{87}Sr/^{86}Sr$ ratios (SrI <0·706) (Shibata & Ishihara 1979). They are divided into two provinces, Kitakami in the north and Abukuma in the south, with a boundary along the Hatakawa Fault (HF in Fig. 4). The granitoids in the Kitakami area (ca. 130–110 Ma) belong to the oxidised type, whilst those in the Abukuma area (ca. 120–100 Ma) are mainly of reduced type (Figs 4 & 9). The granitoids in Kitakami show the most juvenile isotopic features (SrI <0·705) regardless of basement geology: Palaeozoic continental fragment in the south and Jurassic accretionary complex in the north. The mineralisation is molybdenum in North Kitakami, gold with tungsten in South Kitakami and tungsten in Abukuma (Figs 4 & 9).

Granitoids in Southwest Japan were emplaced over a long time period from about 100 Ma to about 40 Ma (e.g. Sato *et al.* 1992; Suzuki & Adachi 1998). The granitoids in the Ryoke and Sanyo Belts within the Inner Zone to the north of the Median Tectonic Line (MTL in Figs 1 & 4) are relatively old (ca. 100–65 Ma) and mostly of reduced type (Figs 4 & 9) except for those at the western end in northern Kyushu near the Korean Peninsula. The granitoids in the Ryoke high-T/P metamorphic belt are generally barren, but those in the Sanyo Belt are associated with tungsten and tin mineralisation. The granitoids in the San-in belt are younger than in the Ryoke and Sanyo Belts (ca. 70–40 Ma) and of oxidised type with molybdenum mineralisation (Figs 4 & 9).

Isotopic compositions are variable within each belt of Southwest Japan, but the reduced-type granitoids have higher

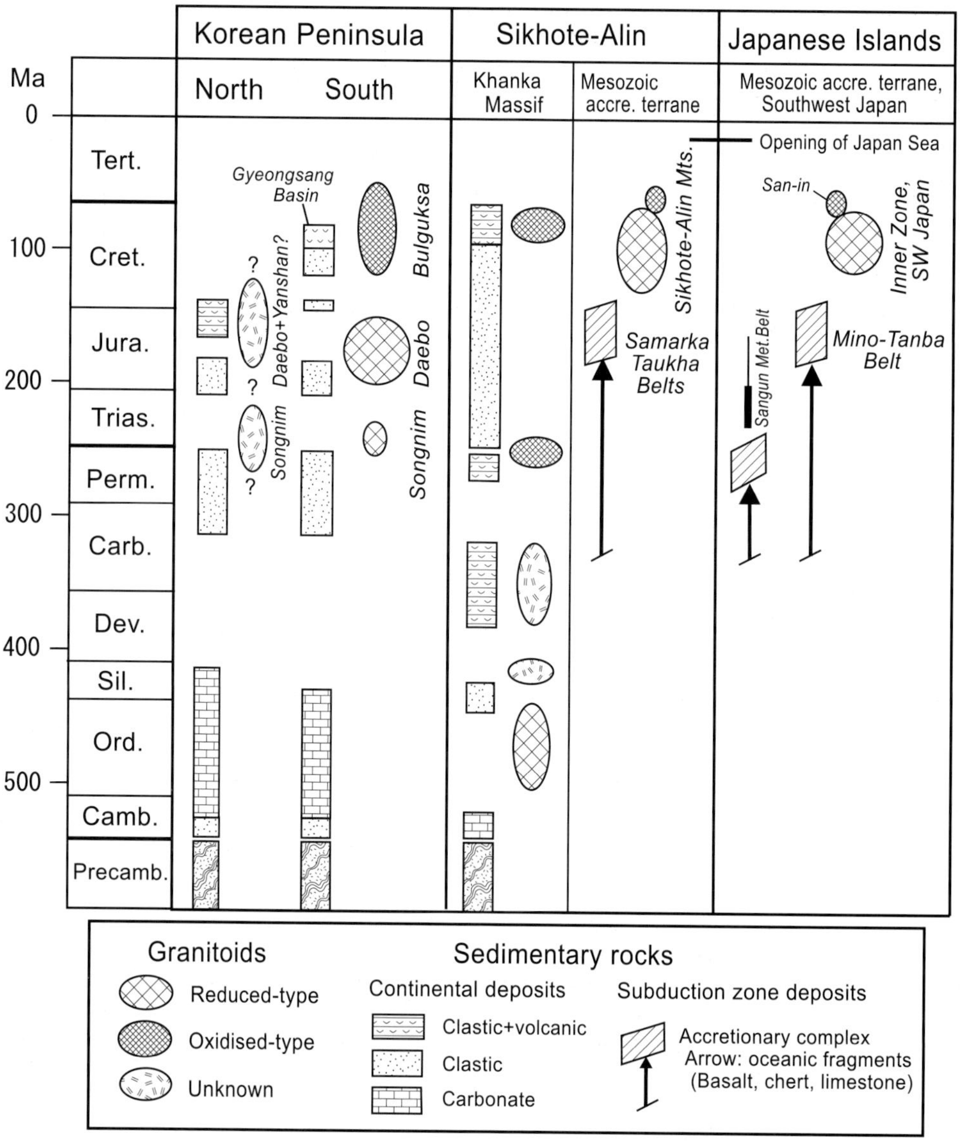

Figure 8 Diagram showing Phanerozoic granitoid magmatism in the circum-Japan Sea region. Figures 8 and 9 were compiled from following sources: for the Korean Peninsula: Ishihara *et al.* (1981), Lee (1988), KIGAM (1995a), Institute of Geology, State Academy of Sciences, DPR of Korea (1996) and Jin *et al.* (2001); for Sikhote–Alin: Mizutani & Kojima (1992), Sato *et al.* (1993a, b), Romanovsky *et al.* (1996), Khanchuk *et al.* (1996), Sato *et al.* (2003a) and this study; for the Japanese Islands: Isozaki & Itaya (1990), Mizutani & Kojima (1992), Geological Survey of Japan (1992), Ishihara *et al.* (1992) and Sato *et al.* (1992). Accretion age of the Jurassic accretionary complexes ranges from the latest Triassic to earliest Cretaceous, but mostly from Middle Jurassic to the earliest Cretaceous (Isozaki 1997; Matsuoka *et al.* 1998; Nakae 2000). Geologic ages are based on Gradstein *et al.* (1995). Little is known about either detailed age or types of granitoids in the northern Korean Peninsula. Note that the Palaeozoic history is remarkably different between the Khanka massif and Korean Peninsula, suggesting that they were situated in different continental blocks before their amalgamation in late Permian to Triassic times.

$\delta^{18}O$ values and initial $^{87}Sr/^{86}Sr$ ratios than the oxidised-type granitoids (Fig. 7). The oxidised-type granitoids in the San-in Belt generally have low initial $^{87}Sr/^{86}Sr$ ratios (<0·706), but occasionally show high values (~0·710) in its eastern part (Shibata & Ishihara 1979; Kagami *et al.* 1992; K. Sato, unpublished data).

The sulphur isotope composition of granitoids and related ore deposits shows a remarkable contrast between the reduced and oxidised types (Sasaki & Ishihara 1979, 1980). The $\delta^{34}S$ (CDT) values tend to be negative in the reduced-type granitoid belts (−14 to +2‰), and positive in the oxidised-type granitoid belts (−1 to +9‰). Gabbroids in the two granitoid belts show a consistent sulphur isotope trend, with $\delta^{34}S$ values of −4 to +3‰ in the reduced-type belt and +1 to +7‰ in the oxidised-type belt. These contrasts between the two granitoid belts are consistent with the results of oxygen and strontium isotope studies (Shibata & Ishihara 1979; Matsuhisa *et al.* 1982). The contrasts suggest that the reduced-type magmas admixed with the sediments of accretionary complexes rich in isotopically light biogenic sulphur, whilst the oxidised-type magmas were not influenced by the sedimentary sulphur and preserved signatures of the isotopically-heavy sea water sulphur which was introduced into magma systems through subduction of oceanic crust.

4. Discussion

4.1. Evolution of crust and granitoid types

4.1.1. Crustal control on redox state of granitoid magmas. A review of the granitoid and metallogenic provinces in the circum-Japan Sea region indicates that the nature of the crust,

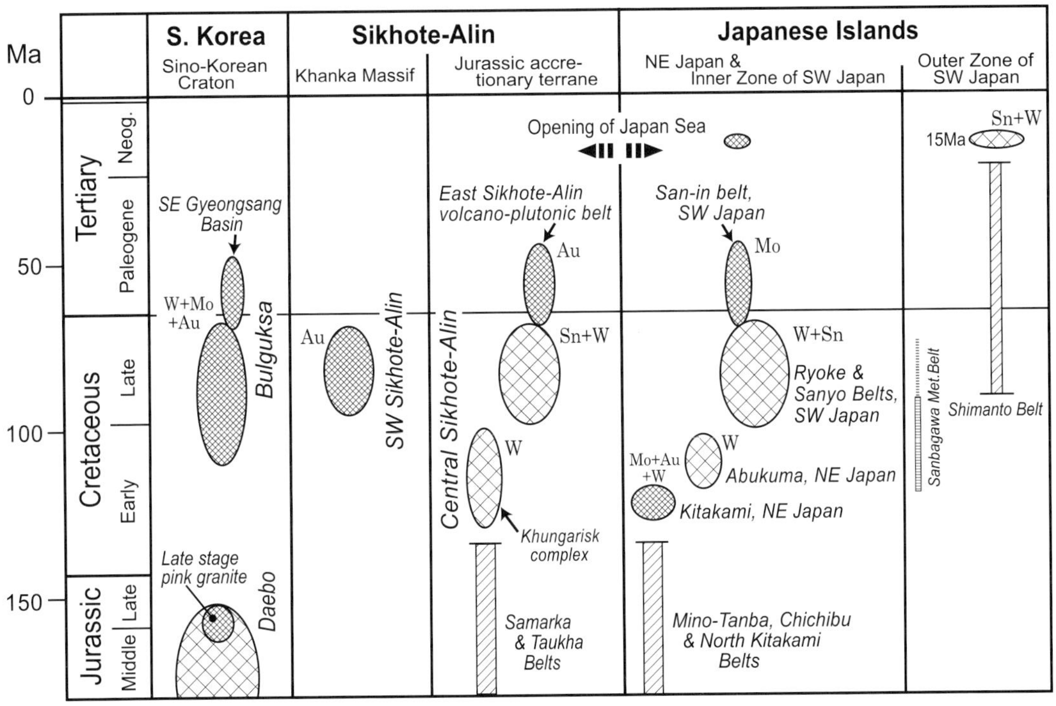

Figure 9 Detailed variation of granitoid types and related mineralisation in the circum-Japan Sea region since the Late Jurassic. Patterns are the same as in Figure 8. Note that younger magmatism tends to be of oxidised-type. See text for the unusual setting of the Kitakami area in Northeast Japan. Chemical symbols beside ellipsoids denote characteristic mineralisation. W mineralisation is related to both the reduced- and oxidised-type granitoids, but mineralogical features of W ores are different between the two types; skarn ores, for example, clearly reflect the different redox state of the two type granitoids (Sato 1980, 1982a).

whether sedimentary or igneous, is the most important control for the redox state of granitoid magmas. The reduced-type granitoids are generated in sedimentary crust which is composed mainly of accretionary complex material and has never been influenced by magmatism prior to the granitoid magmatism. On the other hand, the oxidised-type granitoids are generated in igneous crust which has been formed and/or modified by previous magmatism and depleted in reducing reagents such as sedimentary carbon. Therefore, the redox state of granitoid magmas in a given area tends to become more oxidised with time.

The distribution of the reduced-type and oxidised-type granitoids in the circum-Japan Sea region clearly reflects the tectonic histories of the crustal domains in which they are emplaced. Cretaceous reduced-type granitoids, which are most abundant in this region (Fig. 4), were formed in the terranes comprised of Jurassic accretionary complexes (Fig. 11). Cretaceous oxidised-type granitoids on the continental side were emplaced in older igneous crust; Precambrian to Palaeozoic continental block in the Khanka massif and Jurassic Daebo magmatic belt in the southern Korean Peninsula (Figs 4, 5 & 8). In the Kitakami area of Northeast Japan arc, however, Cretaceous granitoids occur in an unusual geologic setting (Fig. 9). They are of oxidised-type regardless of the nature of crust: Jurassic accretionary complex in the north compared with a Palaeozoic continental fragment in the south. This apparent anomaly will be discussed later.

The latest Cretaceous to Palaeogene granitoids of the oxidised type are distributed in coastal areas around the Japan Sea (Fig. 4): east Sikhote–Alin, San-in of Southwest Japan and eastern margin of the Gyeongsang Basin. The Palaeogene magmatism may have occurred in a relatively narrow zone within the Cretaceous magmatic belt, but this zone was fragmented and now occurs at the present coastal areas due to the opening of the Japan Sea during the Miocene. Miocene granitoids are minor and not described in this paper. It is noticed, however, that the plutons emplaced in the Shimanto accretionary complex are of reduced type, while those emplaced in the San-in Belt are of oxidised type (Fig. 9).

The relationship between the nature of the crust and the redox state of granitoids is distinct in the southernmost Sikhote–Alin, where Late Cretaceous granitoids were emplaced on both sides of the Partizansk Fault (Fig. 5), a deep fault between the Precambrian–Palaeozoic igneous crust of the Khanka massif and the Jurassic accretionary complex which had never been influenced by magmatism before the Late Cretaceous granitoid activity (Fig. 8). The accretionary complex is overlain by a nappe of the Khanka block (Fig. 5). The fault and nappe structure may have formed during the Early Cretaceous, and granitoid plutons were emplaced at about 80–90 Ma. Thickness of the nappe is estimated to be less than 8 km (S. V. Kovalenko, unpublished data), and the underlying Jurassic complex is locally exposed as a window. These relationships indicate that the nature of the crust of deep levels is more important than that of shallow levels for controlling the redox state of granitoid magmas, although the exact thickness of the nappe at the time of granitoid emplacement is not known.

Oxidised-type granitoid plutons are occasionally associated with marginal reduced facies along contacts with carbon-bearing sedimentary rocks (e.g. Sato 1991). The redox state of granitoid magmas could be locally changed in relatively shallow levels after intrusion, but it is difficult to explain the regional distribution of granitoid types by such a local process. Increasing Fe/(Fe+Mg) ratio for biotite and hornblende with increasing SiO_2 content of host granitoids from Sikhote–Alin suggests that oxidation during crystallisation near the level of intrusion (Czamanske *et al.* 1981) was not a general process (Sato *et al.* 1998). Occurrences of early formed magnetite in the oxidised-type granitoids also suggest that the granitoid types reflect the intrinsic conditions of granitoid magma genesis in the lower crustal environment. The granitoid provinces shown in Figure 4 may reflect the redox state of their lower crustal source regions, although they may be locally modified by secondary processes.

The geologic relationship between granitoid types and structure of crust in the southernmost Sikhote–Alin (Fig. 5), coupled with oxygen and strontium isotope data (Figs 6 & 7), demonstrates that involvement of sedimentary rocks containing carbonaceous matter resulted in the generation of reduced-type magmas. This result is consistent with the systematics established in the Japanese Islands, suggesting that mantle-derived magmas assimilated lower crustal sedimentary rocks and fractionated to form the reduced-type granitoids. The mantle-derived magmas may also have assimilated lower crust of the Khanka massif, but they did not encounter sedimentary materials at deep levels.

Examples from the Korean Peninsula are not as simple in character. The Jurassic Daebo granite is dominated by the reduced-type, and characterised by intermediate $\delta^{18}O$ values and very high initial $^{87}Sr/^{86}Sr$ ratios in contrast with other reduced-type granitoids in the circum-Japan Sea region (Fig. 7). These isotopic features and their peraluminous compositions (Jin *et al.* 2001) suggest involvement of sedimentary protoliths present in old continental crust. Occurrences of many graphite deposits in Precambrian–Palaeozoic metasedimentary rocks in the Korean Peninsula (Gallagher 1963) suggest reducing conditions of the crust at the time of Daebo magmatism. It is noticed, however, that the Daebo granite also includes a significant proportion of oxidised-type rocks, and their distribution and genesis are not well understood (Jin *et al.* 2001). Precambrian granitoids occur in the southern Korean Peninsula, although their distribution is limited (e.g. KIGAM 1995a). The basement for the Jurassic granitoid magmatism could be heterogeneous. Precambrian granitoids are widely distributed in the northern Korean Peninsula, implying that Jurassic granitoids in such areas are dominated by oxidised type, although little is known about them. The redox type of the Cretaceous granitoid plutons in the Ogcheon Belt is also a matter for further study (Fig. 4). Low magnetic susceptibility (Jin *et al.* 2001) and isotopic composition similar to the Daebo granite (Kim *et al.* 1996) apparently suggest reduced type, but accompanying mineralisation is similar to the oxidised-type Bulguksa granite in the Gyeongsang Basin (KIER 1983). These granitoids appear to be altered oxidised-type plutons which are isotopically similar to the Jurassic oxidised-type plutons.

4.1.2. Role of tectonic setting and mode of magma emplacement: implications from the Kitakami area. General relationships between granitoid types and the lithology of crust discussed above do not fit with the Kitakami area of Northeast Japan (Fig. 9). The Kitakami area is divided into two subdivisions: the Palaeozoic continental fragment in the south and the Jurassic accretionary complex in the north. Early Cretaceous granitoids were emplaced into these two sub-divisions at 120 ± 10 Ma (e.g. Sato *et al.* 1992), but almost all plutons are composed of the oxidised-type granitoids except for local occurrences of marginal reduced facies along the contact with host sedimentary rocks. The granitoid plutons are elongated in a N–S direction (Fig. 4) and locally accompanied by submarine bimodal volcanic sequences (Harachiyama Group) associated with kuroko mineralisation at the Taro mine (Ebiko 1965; Kanisawa 1974). These granitoids and volcanic equivalents are unconformably overlain by conglomerate and sandstone (Miyako Group) of late Aptian to early Albian age (ca. 110 Ma) (e.g. Mori *et al.* 1992). Radiometric ages of granitoids and volcanic rocks fall in a narrow range between 130 and 110 Ma, regardless of various closure temperatures (Kawano & Ueda 1965; Shibata *et al.* 1978; Sato *et al.* 1992).

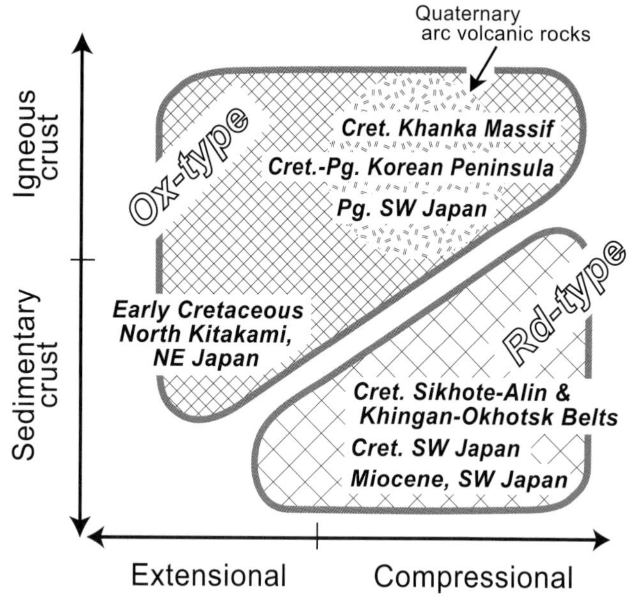

Figure 10 Schematic illustration of the relationship between granitoid types and the lithological and tectonic conditions of the crust. Quaternary arc volcanic rocks generally belong to the oxidised type. Abbreviations: (Cret) Cretaceous; (Pg) Palaeogene; (Ox) oxidised; (Rd) reduced.

These lines of evidence indicate that granitoids in the Kitakami area were emplaced at relatively shallow levels under an extensional environment within a short period and were quickly cooled and eroded. The elongated shape of the plutons and their low initial $^{87}Sr/^{86}Sr$ ratios ($<0{\cdot}705$) suggest a style of magma emplacement such as fissure-filling without significant interaction with crustal materials. The plutons have partly adakitic compositions, suggesting the genesis of granitoid magmas by partial melting of oceanic crust due to the subduction of a spreading ridge (Tsuchiya & Kanisawa 1994). The short-lived episodic magmatism in an extensional condition in the Kitakami area could be related to such an unusual tectonic setting.

The relationship between the Early Cretaceous Harachiyama Group and the underlying Jurassic accretionary complex is apparently conformable (e.g. Mori *et al* 1992). Radiolarian fossils in a clastic matrix of the accretionary complex near the Taro mine were identified to be of late Middle Jurassic age (*ca.* 160 Ma) (Matsuoka & Oji 1990), and sericite-quartz rocks from the Taro deposit were dated by the K-Ar method at 128 Ma (recalculated average age, Ueda *et al.* 1970). Therefore, the time gap between the beginning of Cretaceous felsic magmatism and the cessation of Jurassic subduction processes in the Taro area was shorter than 32 Ma. Felsic magmatism in the Kitakami area could have begun in a zone near a trench which was still in a marine environment, without significant uplift of the accretionary complex above sea level. It is likely that the temperature of the crust was relatively low, which promoted the brittle deformation favourable for the fissure-filling type emplacement of magmas and prevented the significant reaction of ascending magmas with host rocks. Stress and temperature conditions of the crust are considered to be additional important factors for the generation of granitoid types.

Figure 10 summarises the relationship between the granitoid types and their lithological and tectonic settings. If the intrusion of granitoid magmas are of fissure-filling type under extensional environments, they tend to be of oxidised type regardless of lithology of crust, as the crust is passive and does not contribute to the magma chemistry. A Miocene oxidised-type granitoid pluton occurs in Tsushima, an island between Southwest Japan and the Korean Peninsula (Figs 1 & 9). This pluton has very low initial $^{87}Sr/^{86}Sr$ ratios (0·7037) (Shibata & Ishihara 1979). It could be an example of such an oxidised magma emplaced in igneous crust under extensional conditions related to the opening of the Japan Sea.

4.2. Geodynamic background of large tin provinces in northeast Asia

4.2.1. Distribution of Jurassic igneous rocks and accretionary complex. Jurassic felsic magmatism occurred extensively along the continental margin of East Asia, particularly in southeast China and the Korean Peninsula. However, this magmatism did not extend towards the north; the Bureya and Khanka massifs are practically free of Jurassic igneous rocks, although Jurassic continental deposits are present (Fig. 11). Granitoids are widely distributed in these massifs and over a large region around the Songliao Basin, but they are mostly Palaeozoic to Triassic in age (e.g. Krasnyi & Peng 1996; Wu *et al.* 2000).

The Jurassic magmatism appears to have extended from the Sino-Korean Craton towards the northwest within northeast China (Fig. 11). Volcanic rocks are widely distributed in the Da Hinggan Ling (Great Khingan Mountains), and granitoids and volcanic rocks are sporadically distributed in the areas east of Chanchung and Harbin. These areas are far from the continental margin (400–1400 km). Magmatism in the Da Hinggan Ling continued to Early Cretaceous times (e.g. Krasnyi & Peng 1996) and it is characterised by alkalic and bimodal composition (Ministry of Geology and Mineral Resources of China 1990; Lin *et al.* 1998). Early Cretaceous A-type plutons have chemical features suggesting an anorogenic environment (Wu *et al.* 2002). It is more likely that the magmatism in the Da Hinggan Ling was rift-related, not subduction-related. The Songliao Basin, which may have begun to form in the Late Jurassic (Ma *et al.* 1989; Okada 1999), also suggests an extensional tectonic environment in the inner continental region of northeast China.

The Bureya and Khanka massifs and a region to the south of the Mongol–Okhotsk suture are characterised by the lack of significant magmatism in the Jurassic (Fig. 11). The paucity of igneous rocks cannot be explained by erosion because of the occurrences of Jurassic continental deposits. These continental margin areas may have been passive or transform margins in the Jurassic (Fig. 12). The Stanovoy Range to the north of the Mongol–Okhotsk suture is a region of Jurassic magmatism (Fig. 11), suggesting subduction of the oceanic plate of the Mongol–Okhotsk Ocean beneath the Siberian Craton during the Jurassic (Fig. 12).

4.2.2. Reorganisation of accretionary complex. The Jurassic accretionary complexes to the east of the Bureya and Khanka massifs are not accompanied by significant arc magmatism on the continental side (Fig. 11). These areas may have been a transform margin during the Jurassic (Fig. 12), and the accretionary complexes are thought to be allochthonous units. This insight is consistent with the idea of secondary translation of accretionary complex proposed by Mizutani & Yao (1991) based on the palaeomagnetic data; low palaeolatitude values for the Jurassic accretionary complexes in Southwest Japan and pre-Jurassic continental fragments within them suggest their remarkable northward drift (e.g. Hirooka 1990).

Jurassic to Early Cretaceous plant fossils in eastern Eurasia are divided into two types: Tetori-type in the north from northeast China to Siberia, and Ryoseki-type in the south from southeast China to the Malay Peninsula, with an EW-trending transitional zone between the two provinces (Kimura 1987; Ohana & Kimura 1995). Distribution of the Ryoseki-type floras extends towards the north to Japan (Outer Zone of SW Japan and NE Japan) and Sikhote–Alin, showing an unusually curved boundary with the Tetori-type floras. For example, the Tetori- and Ryoseki-type floras occur in the Hida and Kitakami areas, respectively. This boundary may be reorganised to a normal pattern if we assume northward displacement of the Ryoseki-type areas in Japan and Sikhote–Alin.

Otsuki & Ehiro (1992) suggested large-scale left-lateral displacements along major faults in Northeast Japan in the middle Cretaceous. Early to mid-Cretaceous time appears to be a unique stage of vigorous tectonic activity in the circum-Japan Sea region. Large scale left-lateral displacements have also been suggested for the Tan-Lu and other NE–SW trending faults in the continental margin of East Asia (e.g. Xu *et al.* 1993). Xu *et al.* (1993) and Chen (1993) suggested that major horizontal displacement occurred during the Early Cretaceous, although the fault activity has a long history up to the Recent as evidenced by large earthquakes along this fault. Sinistral tectonic movement may have prevailed over a wide region of the continental margin as well as in the accretionary terrane (Fig. 11), although details of the times and distances of left-lateral displacements require further study.

The authors consider that the accretionary complexes in the Khingan and Sikhote–Alin Mountains were largely displaced from the original place of accretion, and formed into an unusually wide zone through the amalgamation with autochthonous sediments which may have been derived from the Mongol–Okhotsk suture zone and deposited in a transform margin off the Bureya massif (Fig. 12). The original place of accretion is thought to be an active margin of the Korean Peninsula and southeast China where Jurassic igneous rocks are widely distributed. This idea is illustrated in Figure 12, showing the configuration of continents based on palaeomagnetics (Enkin *et al.* 1992) and representative velocity vectors of oceanic plates with respect to East Asia (Maruyama & Seno 1986) in the Jurassic and Cretaceous. The Khingan and Sikhote–Alin regions are thought to be a transform margin in the Jurassic, as mentioned above, but changed to a convergent margin in the Cretaceous; Cretaceous convergence may have been oblique in the Early Cretaceous and orthogonal in the Late Cretaceous. The Mongol–Okhotsk Ocean is thought to have closed from west to east and disappeared at its eastern end in Early Cretaceous times (e.g. Zonenshain *et al.* 1990). The collision of East Asia with Siberia, combined with the oblique convergence of the palaeo-Pacific plate in the Early Cretaceous, could have resulted in large left-lateral displacements in the eastern margin of East Asia. This may be a major cause for the displacement of the accretionary terranes and the slices of continental fragments within them. The Early Cretaceous Amur basin proposed in the Sunda-style tectonic model (Sato *et al.* 2002) may have been formed in this process.

4.2.3. Background of tin-associated reduced-type magmatism. The displaced and stacked accretionary terranes and adjacent continental margins (Fig. 11) were changed to a zone of extensive felsic magmatism in Cretaceous times (Figs 1 & 4). The Cretaceous vigorous magmatism could be related to the remarkably high velocity of convergence as well as subduction of a spreading ridge (e.g. Uyeda & Miyashiro 1974; Jahn 1974; Maruyama *et al.* 1997), although the real cause is not well understood. The magmatism might have been a response to such a global process as superplume activity, because unusually large production rates of oceanic crust in middle Cretaceous times (ca. 125–80 Ma, Larson 1991) are synchronous with the peak of granitoid magmatism not only in the circum-Japan Sea region but also in the western North America (e.g. Armstrong 1988).

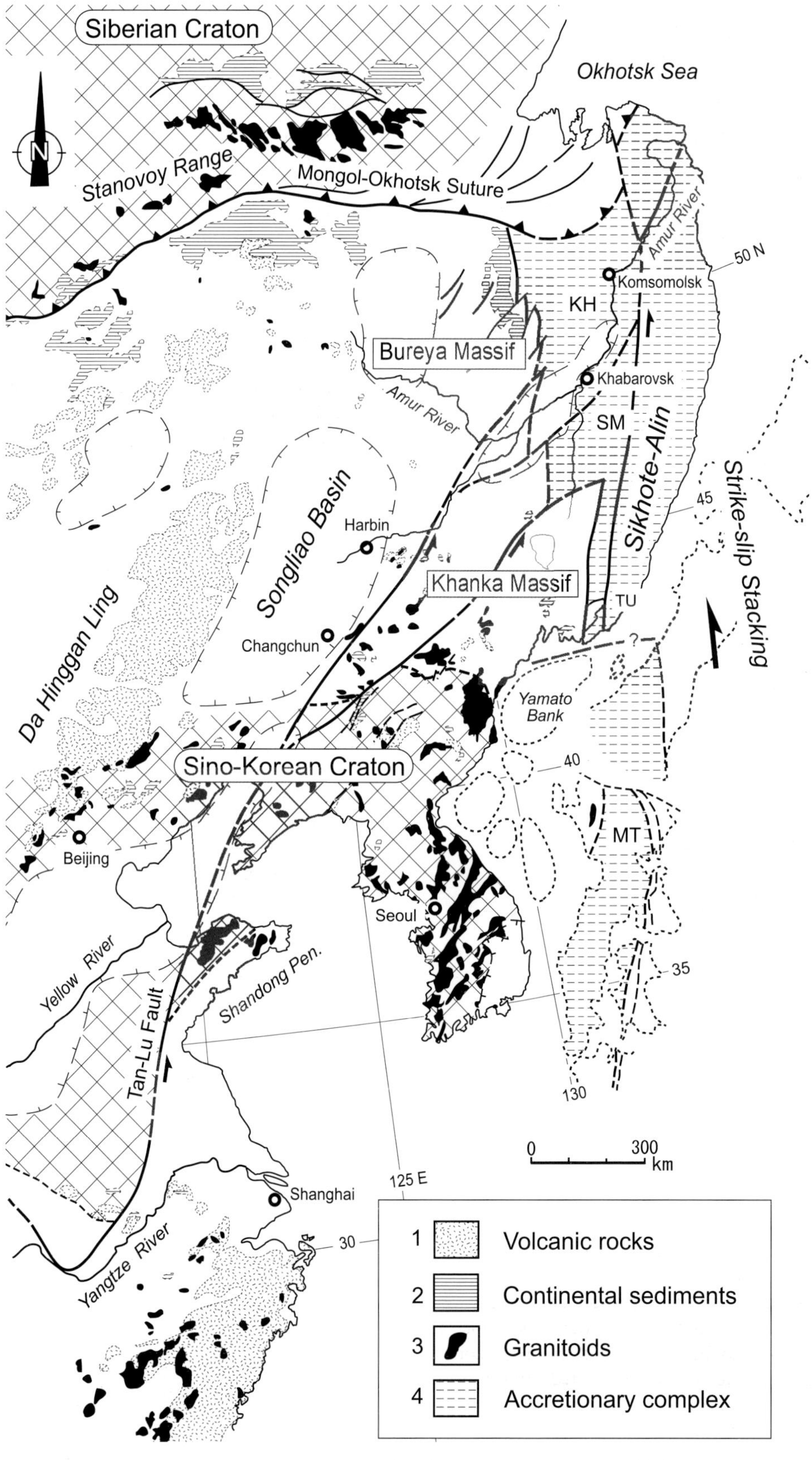

Figure 11 Distribution of Jurassic geologic units in East Asia. Modified from Sato (2000, 2003) with reference to Sato *et al.* (1992), Krasnyi *et al.* (1996) and Teraoka & Okumura (2003). Note that Jurassic igneous rocks are absent or scarce in the Bureya and Khanka massifs near Jurassic subduction complexes which are widely distributed in the Khingan and Sikhote–Alin Mountains. Jurassic magmatism occurred widely in east China and in the Korean Peninsula, but the magmatism extended to the Da Hinggan Ling (Great Khingan Mountains) in northeast China which is far away from the region of subduction complexes. Reconstruction of the Japanese Islands is based on palaeomagnetic studies (e.g. Otofuji *et al.* 1985) and bathymetric data of the Japan Sea. Accretionary complexes are: (KH) Khabarovsk; (MT) Mino–Tanba; (SM) Samarka; (TU) Taukha. Those in the Outer Zone of Southwest Japan (Chichibu) and Northeast Japan (North Kitakami) are omitted.

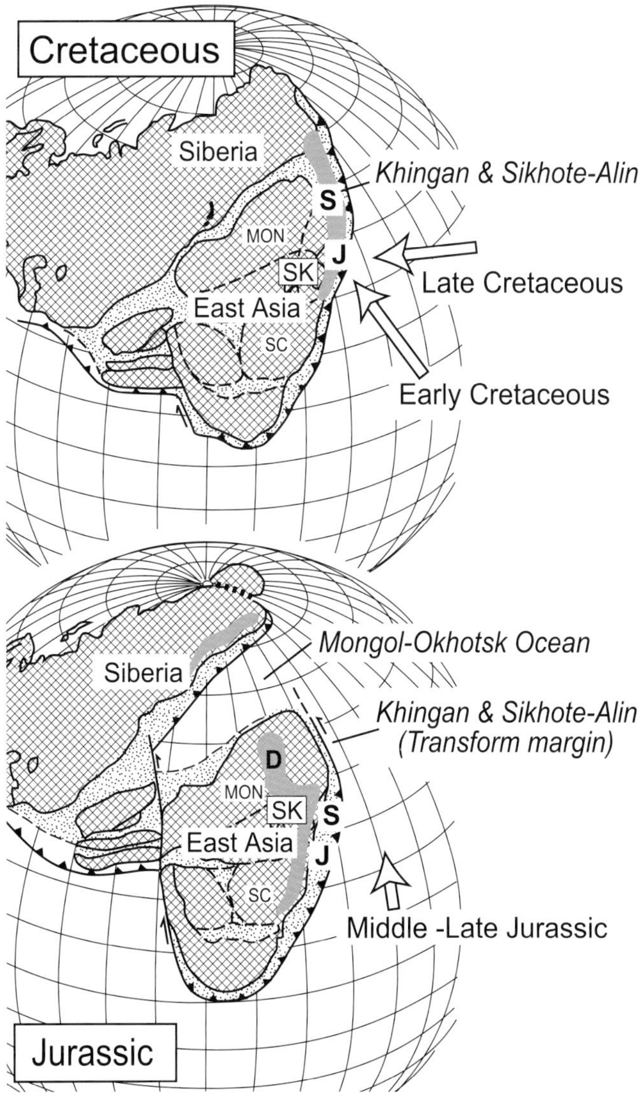

Figure 12 Distribution of continents and approximate velocity vectors of oceanic plate with respect to Asia in the Jurassic and Cretaceous, based on Enkin *et al.* (1992) and Maruyama & Seno (1986). Arrows are representative velocity vectors. Grey areas in the Cretaceous and Jurassic globes correspond to the areas of magmatism shown in Figures 1 and 11, respectively. Note that Jurassic magmatism in the Da Hinggan Ling (D) lasted to the Early Cretaceous, but is not shown in the Cretaceous globe. (J) Japanese Islands; (MON) Mongol Block; (S) Sikhote–Alin; (SC) South China Block; (SK) Sino-Korean Craton.

The Cretaceous felsic magmatism in the circum-Japan Sea region formed the reduced-type magmas accompanied by tin mineralisation in the regions of accretionary complexes, as exemplified in the Khingan–Okhotsk Belt, central Sikhote–Alin Belt and the Inner Zone of Southwest Japan (Fig. 9). Involvement of sedimentary materials in the generation of granitoid magmas may also be common in collision zones. Late Jurassic to Early Cretaceous tin-associated granitoids in the Yana–Kolyma region could be such examples (Parfenov 1991; Parfenov *et al.* 1999; Layer *et al.* 2001). Development of the sedimentary terrane before magmatism is the genetic background of the metallogenic provinces characterised by tin in northeast Asia.

4.3. East–west contrast around the Pacific

4.3.1. Tin-poor provinces in the circum-Japan Sea region. As discussed above, granitoid magmatism tends to become more oxidised with time, due to modification of the crust from sedimentary affinity to igneous affinity. The Cretaceous magmatism was superimposed on the Khanka massif and southern Korean Peninsula igneous crust, and resulted in the formation of the oxidised-type granitoids which are free of tin (Figs 4 & 9). The Palaeogene magmatism was localised to the present coastal areas of the Japan Sea, where extensive magmatism occurred in the Cretaceous (Fig. 4). The overprinted Palaeogene activity resulted in the formation of the oxidised-type granitoids, which are accompanied by molybdenum or gold, but not by tin (Fig. 9).

Granitoids are widely distributed in northeast China. Their ages range from Precambrian to Cretaceous, but granitoid types are not well known. However, the paucity of tin and occurrences of gold and molybdenum deposits in this region (Chinese Academy of Geological Sciences 1999) imply that the granitoids, at least those of later stages, are dominated by the oxidised type. Permian to Early Cretaceous A-type granitoids have been reported to be magnetite-bearing (Wu *et al.* 2002).

4.3.2. Pacific margin of North America. The Pacific margin of North America is a typical tin-poor region, as indicated at the beginning of this paper. Arc magmatism in the Pacific margin occurred for a long time during the Mesozoic (e.g. Dickinson 1976; Armstrong 1988; Burchfiel *et al.* 1992). The Cretaceous Sierra Nevada batholith is a representative and well-studied example of the oxidised-type granitoids (Bateman *et al.* 1991). The batholith was emplaced in a Jurassic magmatic arc (e.g. Dickinson 1981; Bateman 1983). The repeated magmatism in a convergent margin over a long time period may be a reason why the reduced-type granitoids and tin deposits are scarce in the Pacific margin of North America. One of the exceptional occurrences of the reduced-type granitoid belt is seen in the Cretaceous Peninsular Ranges Batholith (Gastil *et al.* 1990). The boundary between the oxidised- and reduced-type granitoid belts coincides approximately with the boundary of the Triassic–Jurassic volcanic arc to the west and the back-arc sedimentary basin to the east (Gastil 1993). This relationship, supported by higher $\delta^{18}O$ values (Silver *et al.* 1979, cited in Gastil *et al* 1990) for the reduced-type belt suggests generation of the reduced-type granitoids in sedimentary crust.

4.3.3. East–west contrast. The reduced-type granitoids are minor in the Pacific margin of North America, whilst they are dominant in East Asia. The oxidised-type granitoids occur in East Asia, as mentioned above, but their proportion within the whole of East Asia is relatively small. Generation of the reduced-type granitoid magmas requires a change from sedimentary terranes to magmatic belts which may be caused by the change of subduction system or collision of continents. This process was not common in the monotonous convergent margin of western North America. The remarkable contrast of granitoid types and mineralisation around the Pacific may be a consequence of different geodynamic histories between the east and west sides.

5. Concluding remarks

Mesozoic felsic magmatism provided major mineral resources in East Asia. The style of magmatism and mineralisation is highly variable within the circum-Japan Sea region, encompassing the contrast between the east and west side of the Pacific rim. The granitoid magmatism and related mineralisation in this region has been examined from the viewpoint of the redox state of granitoid magmas and its relation with geodynamic history of the area of magmatism. It is concluded that the reduced-type granitoids were formed in the sedimentary crust which is composed mainly of accretionary complexes and have never been influenced by previous magmatism, whilst the oxidised-type granitoids were formed in the igneous crust which is depleted in carbonaceous matter due to previously repeated or intense magmatism. Thus granitoid activity in a given area tends to be changed to oxidised-type with time.

A wide zone of sedimentary crust in the Khingan and Sikhote–Alin Mountains in Far East Russia is thought to have been formed by a strike-slip stacking of Jurassic accretionary complexes from the original place of accretion near the Korean Peninsula or southeast China. This zone provided favourable sites for the reduced-type felsic magmatism accompanied by tin mineralisation in the Cretaceous; the magmatism may have been caused by vigorous subduction of the palaeo-Pacific plate. The Early Cretaceous oblique subduction may have caused the left-lateral strike-slip displacement of the accretionary complexes, including sliced continental fragments towards the north, under a Sunda-style tectonic regime, and resulted in the formation of the Khingan–Okhotsk tin belt in mid-Cretaceous times (Sato *et al.* 2002). Early Cretaceous granitoids are of reduced-type in central Sikhote–Alin, but of oxidised-type in the Kitakami area (Figs 4 & 9), suggesting different tectonic environments; transpressional in central Sikhote–Alin, but transtensional in the Kitakami area. However, palinspastic reconstruction of the Early Cretaceous magmatism is not easy because of probable secondary displacement or modification of geologic units by Late Cretaceous–Palaeogene activities. The general scenario of Cretaceous tectonic activity and magmatism over the whole circum-Japan Sea region has great potential for more detailed study.

The change of sedimentary terranes to sites of magmatism is an essential factor for the generation of the reduced-type granitoid provinces and tin-rich metallogenic provinces. Such a situation may have occurred at various stages in the geodynamic history of northeast Asia, in contrast with western North America, where a monotonous subduction system operated for a long period in the Mesozoic. The Late Jurassic–Early Cretaceous tin province in the Yana–Kolyma region of northeast Russia is thought to have been formed by a collision activity that may be another suitable process for the formation of reduced-type granitoid magmas through the involvement of sedimentary materials. Eurasia is a collage of cratons and continental blocks cemented by Phanerozoic orogenic belts (e.g. Maruyama *et al.* 1989). Collision processes may have provided a large amount of sediments to surrounding oceans. Weathering and sedimentary processes in the biosphere may have prepared ^{18}O, ^{32}S and ^{87}Sr-enriched isotopic features as well as C-bearing reducing conditions in the sedimentary crust. This is a recycling process and, at the same time, a reducing process affecting continental crust. In northeast Asia, the recycled and reduced sedimentary crust was involved in new magmatism over large regions, resulting in the formation of the provinces of the reduced-type granitoids and tin mineralisation.

6. Acknowledgements

We thank I. N. Kigai, A. A. Brublevsky, S. M. Rodionov, S. V. Korenbaum, A. I. Khanchuk, V. G. Gonevchuk, G. A. Gonevchuk, V. P. Simanenko, V. P. Nechaev, S. V. Vysotskiy, and M. Enjoji for their help in the field work and administrative matters. Thanks are also due to Mrs T. Hirose, M. Obata and S. Hirano of the Consulate-General of Japan at Vladivostok and Khabarovsk, Mrs T. Nagai and Y. Yukimatsu of the Japanese Embassy in Moscow for their support to our activity in Russia. Helpful suggestions for age dating by Dr. K. Suzuki of Nagoya University, kind correction of the English by Dr. L. C. Hsu of Nevada University, and critical reviews of the manuscript by Drs. B. Jahn, C. Hart and E. Horikoshi are also greatly appreciated. This work was carried out as a part of an international cooperative research project, 'Metallogeny of the Circum-Japan Sea Region', authorised by the Committee on Science and Technology Cooperation between Japan and Russia, with financial support from the Science & Technology Agency and Ministry of Education of Japan and the Russian Academy of Sciences.

7. References

Armstrong, R. L. 1988. Mesozoic and early Cenozoic magmatic evolution of the Canadian Cordillera. *Geological Society of America, Special Paper* **218**, 55–91.

Bateman, P. C. 1983. A summary of critical relations in the central part of the Sierra Nevada batholith, California, U.S.A. *Geological Society of America, Memoir* **159**, 241–54.

Bateman, P. C., Dodge, F. C. W. & Kistler, R. W. 1991. Magnetic susceptibility and relation to initial 87Sr/86Sr for granitoids of the central Sierra Nevada, California. *Journal of Geophysical Research* **96**, 19,555–68.

Blevin, P. L. & Chappell, B. W. 1992. The role of magma sources, oxidation states and fractionation in determining the granite metallogeny of eastern Australia. *Transactions of the Royal Society Edinburgh: Earth Sciences* **83**, 305–16.

Burchfiel, B. C., Cwan, D. S. & Davis, G. A. 1992. Tectonic overview of the Cordilleran orogen in the western United States. *In* Burchifiel, B. C., Lipman, P. W. & Zoback, M. L. (eds) *The Geology of North America Volume G-3, The Cordilleran Orogen: Conterminous U.S.*, 407–79. Boulder, Colorado: Geological Society of America.

Chappell, B. W. & White, A. J. R. 1974. Two contrasting granite types. *Pacific Geology* **8**, 173–4.

Chappell, B. W. & White, A. J. R. 1992. I- and S-type granites in the Lachlan Fold Belt. *Transactions of the Royal Society Edinburgh: Earth Sciences* **83**, 1–26.

Chen, P. 1993. Timing of displacement along the Tancheng-Lujian fault zone and the migration of Late Mesozoic volcanism in eastern China. *In* Xu, J. (ed.) *The Tancheng-Lujian Wrench Fault System*, 51–74. Chichester: John Wiley & Sons.

Cheong, C. S. & Chang, H. W. 1997. Sr, Nd, and Pb isotope systematics of granitic rocks in the central Ogcheon Belt, Korea. *Geochemical Journal,* **31**, 17–36.

Chinese Academy of Geological Sciences 1999. *Minerogentic Series Map of Mineral Deposits of China, 1:5,000, 000*. Beijing: Geological Publishing House.

Choi, S. G. 2002. The genetic implications of Mesozoic gold-silver mineralization, South Korea. *Mesozoic Sedimentation, Igneous Activity and Mineralization in South Korea. Korea Institute of Geoscience and Mineral Resources, Special Publication* **1**, 216–26.

Curray, J. R. Moore, D. G., Lawver, L. A., Emmel, F. J., Raitt, R. W., Henry, M. & Kieckhefer, R. 1979. Tectonics of the Andaman Sea and Burma. *In* Watkins, J. S., Montadert, L. & Dickerson, P. W. (eds) *Geological and Geophysical Investigations of Continental Margins. AAPG Memoir* **29**, 189–98. Tulsa, Oklahoma: American Association of Petroleum Geologists (AAPG).

Curray, J. R., Emmel, F. J., Moore, D. G. & Raitt, R. W. 1982. Structure, tectonics and geological history of the Northeastern Indian Ocean. *In* Narin, A. E. M. & Stehli, F. G. (eds) *The Ocean Basins and Margins*, **6**, 399–45. New York, NY: Plenum.

Czamanske, G. K., Ishihara, S. & Atkin, S. A. 1981. Chemistry of rock-forming minerals of the Cretaceous–Paleocene batholith in southwestern Japan and implications for magma genesis. *Journal of Geophysical Research* **86**, 10431–69.

Dickinson, W. R. 1976. Sedimentary basins developed during evolution of Mesozoic-Cenozoic arc-trench system in western north America. *Canadian Journal of Earth Sciences.* **13**, 1268–87.

Dickinson, W. R. 1981. Plate tectonics and the continental margin of California. *In* Ernst, W. G. (ed.) *The Geotectonic Development of California*, 1–28. Englewood Cliffs, New Jersey: Prentice-Hall.

Ebiko, T. 1965. Geologic features and prospecting of the Taro mine, Iwate Prefecture, Japan. *Mining Geology* **15**, 109–13. [In Japanese, with English abstract.]

Enkin, R. J., Yang, Z., Chen, Y. & Courtillot, V. 1992. Paleomagnetic constraints on the geodynamic history of the major blocks of China from the Permian to the Present. *Journal of Geophysical Research* **97**, 13,953–89.

Faure, M. & Natal'in, B. 1992. The geodynamic evolution of the eastern Eurasian margin in Mesozoic times. *Tectonophysics* **208**, 397–411.

Fitch, T. J. 1972. Plate convergence, transcurrent fault, and deformation adjacent to Southeast Asia and the western Pacific. *Journal of Geophysical Research* **77**, 4432–60.

Furukawa, Y. & Tatsumi, Y. 1999. Melting of a subducting slab and production of high-Mg andesite magmas: unusual magmatism in SW Japan. *Geophysical Research. Letters* **26**, 2271–4.

Gallagher, D. 1963. *Mineral Resources of Korea, Volume 6B: Non-Metallics and Miscellaneous Metals*. USOM/Korea: Mining Branch, Industry & Mining Division.

Gastil, R. G. 1993. Prebatholithic history of peninsular California. *Geological Society of America, Special Paper* **279**, 145–56.

Gastil, R. G., Diamond, J., Knaack, C., Walewender, M., Marshall, M. Boyles, C., Chadwick, B. & Erskine, B. 1990. The problem of the magnetite/ilmenite boundary in southern and Baja California, California. *Geological Society of America, Memoir* **174**, 19–32.

Geological Survey of Japan 1992. *Geological Map of Japan, 1:1,000,000* (3rd edn). Tsukuba, Japan: Geological Survey of Japan.

Gradstein, F. M., Agterberg, F. P., Ogg, J. G., Hardenbol, J., Veen, P. V., Thierry, J. & Huang, Z. 1995. A Triassic, Jurassic and Cretaceous time scale. *In* Berggren, W. A., Kent, D. V., Aubry, M. P. & Hardenbol. J. (eds) *Geochronology Time Scales and Global Stratigraphic Correlation. SEPM Special Publication* **54**, 95–126. Tulsa, Olkahoma: Society for Sedimentary Geology (SEPM).

Hamilton, W. 1979. Tectonics of the Indonesian Region. *United States Geological Survey Professional Paper,* **1078**. Washington, DC: United States Government Printing Office.

Hirooka, K. 1990. Paleomagnetic studies of pre-Cretaceous rocks in Japan. *In* Ichikawa, K., Mizutani, S., Hara, I., Hada, S. & Yao, A. (eds) *Pre-Cretaceous Terranes of Japan, Publication of IGCP* **224**. *Pre-Jurassic Evolution of Eastern Asia*, 401–5. Osaka, Japan: International Geological Correlation Program (IGCP).

Institute of Geology, State Academy of Sciences, DPR of Korea 1996. *Geology of Korea*. Pyongyang: Foreing Languages Books Publishing House.

Ishihara, S. 1977. The magnetite-series and ilmenite-series granitic rocks. *Mining Geology* **27**, 293–305.

Ishihara, S. 1979. Lateral variation of magnetic susceptibility of the Japanese granitoids. *Journal of the Geological Society of Japan* **85**, 509–23.

Ishihara, S. 1984. Granitoid series and Mo/W-Sn mineralization in East Asia. *Geological Survey of Japan, Report* **263**, 173–208.

Ishihara, S., Lee, D. S. & Kim, S. Y. 1981. Comparative study of Mesozoic granitoids and related W-Mo mineralization in southern Korea and southwestern Japan. *Mining Geology* **31**, 311–20.

Ishihara, S., Sasaki, A., & Sato. K. 1992. *Metallogenic Map of Japan, 1:2,000,000 Map Series* **15-2, 3**. Tsukuba, Japan: Geological Survey of Japan.

Ishihara, S. & Matsuhisa, Y. 2002. Oxygen isotope constraints on the genesis of the Cretaceous-Paleogene granitoids in the Inner Zone of Southwest Japan. *Bulletin of the Geological Survey of Japan* **53**, 421–38.

Ishihara, S. & Sasaki, A. 1973. *Metallogenic Map of Japan, 1:2,000,000 Map Series* **15-1**. Tsukuba, Japan: Geological Survey of Japan.

Isozaki, Y. 1997. Jurassic accretionary tectonics of Japan. *The Island Arc* **6**, 25–51.

Isozaki, Y. & Itaya, T. 1990. Chronology of Sanbagawa metamorphism. *Journal of Metamorphic Geology* **8**, 401–11.

Jahn, B. 1974. Mesozoic thermal events in southeast China. *Nature* **248**, 480–3.

Jin, M. S., Lee, Y. S. & Ishihara, S. 2001. Granitoids and their magnetic susceptibility in South Korea. *Resource Geology* **51**, 189–203.

Jwa, Y. J. 1998. Temporal, spatial and geochemical discriminations of granitoids in South Korea. *Resource Geology* **48**, 273–84.

Kagami, H., Iizumi, S., Tainosho, Y. & Owada, M. 1992. Spatial variations of Sr and Nd isotope ratios of Cretaceous-Paleogene granitoid rocks, Southwest Japan Arc. *Contributions to Mineralogy and Petrology,* **112**, 165–77.

Kanisawa, S. 1974. Granitic rocks closely associated with the Lower Cretaceous volcanic rocks in the Kitakami Mountains, Northeast Japan. *Journal of the Geological Society of Japan* **80**, 355–67.

Kawano, Y. & Ueda, Y. 1965. K-Ar dating on the igneous rocks in Japan (2)-Granitic rocks in Kitakami massif. *Journal of the Japanese Association of. Mineralogy, Petrology and Economic Geology* **53**, 143–54. [In Japanese, with English abstract.]

Khanchuk, A. I., Ratkin, V. V., Ryazantseva, M. D., Golozubov, V. V. & Gonokhova, N. G. 1996. *Geology and Mineral Deposits of Primorsky Krai*. Vladivostok: Dalnauka.

Kim, K. H., Satake, H. & Mizutani, Y. 1992. Oxygen isotopic compositions of Mesozoic granitic rocks in South Korea. *Mining Geology* **42**, 311–22.

Kim, K. H., Park, S. S. & Na, C. K. 1996. Nd and Sr isotopic signatures of Mesozoic granitoids in South Korea. *Resource Geology* **46**, 215–26.

Kimura, G. 1994. The latest Cretaceous-early Paleogene rapid growth of accretionary complex and exhumation of high pressure series metamorphic rocks in northwestern Pacific margin. *Journal of Geophysical Research* **99**, 22147–64.

Kimura, T. 1987. Geographic distribution of Palaeozoic and Mesozoic plants in East and Southeast Asia. *In* Taira, A. & Tashiro, M. (eds) *Historical Biogeography and Plate Tectonic Evolution of Japan and Eastern Asia*, 135–200. Tokyo: Terrapub.

Kobayashi, K. & Sato, T. 1979. Tectonics and evolution of ocean-continent boundary zone (I). *Iwanami Earth Science Series* **11**, 175–252. [In Japanese.]

Korea Institute of Energy and Resources (KIER) 1983. *Metallogenic Map of Korea, 1:1,000,000 scale*. Seoul, Korea: KIER.

Korea Institute of Geology, Mining & Materials (KIGAM) 1995a. *Isotope Age Map of Plutonic Rocks in Korea., 1:1,000,000 scale*. Taejon, Korea: KIGAM.

Korea Institute of Geology, Mining & Materials (KIGAM) 1995b. *Isotope Age Map of Ore Deposits in Korea., 1:1,000,000 scale*. Taejon, Korea: KIGAM.

Krasnyi, L. I. & Peng, Y. (eds) 1996. *Geological Map of Amur Region and Adjacent Areas, scale 1:2,500,000*. A. P. Karpinsky All-Russian Geological Research Institute (VSEGEI), Amurgeolkom and Bureau of Geology and Mineral Resources of Heilongjiang Province. Harbin, Sankt-Petersburg & Blagoveshchensk.

Larson, R. L. 1991. Latest pulse of the Earth: evidence for a mid-Cetaceous superplume. *Geology* **19**, 547–50.

Layer, P. W., Newberry, R., Fujita, K., Parfenov, L., Trunilina, V. & Bakharev, A. 2001. Tectonic setting of the plutonic belts of Yakutia, northeast Russia, based on 40Ar/39Ar geochronology and trace element geochemistry. *Geology* **29 (2)**, 167–70.

Lee, D. S. (ed.) 1988. *Geology of Korea*. Seol, Korea: Geological Society of Korea, Kyohak-Sa, Publishing Co.

Lin, Q., Ge, W., Sun, D., Wu, F., Won, C. K., Min, K. D., Jin, M. S., Lee, M. W., Kwon, C. S. & Yun, S. H. 1998. Tectonic significance of Mesozoic volcanic rocks in northeastern China. *Scientia Geologica Sinica* **33**, 129–39. [In Chinese, with English abstract.]

Ma, L., Yang, J. & Ding, Z. 1989. Songliao Basin-An intracratonic continental sedimentary basin of combination type. *In* Zhu, X. (ed.) *Chinese Sedimentary Basins, Sedimentary Basins of the World* **1**, 77–87. Amsterdam: Elsevier.

Markevich, P. V., Zyabrev, S. V., Filippov, A. N. & Malinovskiy, A. I. 1997. East flank of the Kiselevka-Manoma terrane: An island arc fragment in the accretionary prism. *Geology of the Pacific Ocean* **13**, 295–338.

Maruyama, S., Liou, J. G. & Seno, T. 1989. Mesozoic and Cenozoic evolution of Asia. *In* Ben-Avraham, Z. (ed.) *The Evolution of the Pacific Ocean Margins. Oxford Monographs on Geology and Geophysics* **8**, 75–99. New York: Oxford University Press.

Maruyama, S., Isozaki, Y., Kimura, G. & Terabayashi, M. 1997. Paleogeographic maps of the Japanese Islands: Plate tectonic synthesis from 750 Ma to the present. *The Island Arc* **6**, 121–42.

Maruyama, S. & Seno, T. 1986. Orogeny and relative plate motions: Example of the Japanese Islands. *Tectonophysics* **127**, 305–29.

Matsuda, T., Sakiyama, T, Enami, R., Otofuji, Y., Sakhno, V. G., Matunin, A. P., Kulinich, R. G. & Zimin, P. S. 1998. Fission-track ages and magnetic susceptibility of Cretaceous to Paleogene volcanic rocks in southeastern Sikhote-Alin, Far East Russia. *Resource Geology* **48**, 285–90.

Matsuhisa, Y., Sasaki, A., Shibata, K. & Ishihara, S. 1982. Source diversity of the Japanese granitoids: An O, S and Sr isotope approach. *Abstracts of the 5th International Conference on Geochronology, Cosmochronology and Isotope Geology*, 239–40. Nikko, Japan.

Matsuoka, A., Yamakita, S., Sakakibara, M. & Hisada, K. 1998. Unit division for the Chichibu Composite Belt from a view point of accretionary tectonics and geology of western Shikoku, Japan. *Journal of the Geological Society of Japan* **104**, 634–53. [In Japanese.]

Matsuoka, A. & Oji, T. 1990. Middle Jurassic radiolarian fossils from the Magisawa Formation in the Taro Belt, North Kitakami Mountains. *Journal of the Geological Society of Japan* **96**, 239–41. [In Japanese.]

Ministry of Geology and Mineral Resources of China 1990. *Geologic Map of China, 1:500,0000*. Beijing: Geological Publishing House.

Mishin, L. F. & Romanovsky, N. P. 1994. Reduction-oxidation environments and metallogeny of igneous ore systems of the southern Far East. *Geology of the Pacific Ocean* **9**, 997–1011.

Mizutani, S. & Kojima, S. 1992. Mesozoic radiolarian biostratigraphy of Japan and collage tectonics along the eastern continental margin of Asia. *Palaeogeography, Palaeoclimatology, Palaeoecology* **96**, 3–22.

Mizutani, S. & Yao, A. 1991. Radiolarians and terranes: Mesozoic geology of Japan. *Episodes* **14**, 213–16.

Mori, K., Okami, K. & Ehiro, M. 1992. Paleozoic and Mesozoic sequences in the Kitakami Mountains (A05). *In* Adachi, M. & Suzuki, K. (eds). *29th International Geological Congress (IGC) Field Trip Guidebook* vol. 1, 81–114. Nagoya: Nagoya University.

Nakae, S. 2000. Regional correlation of the Jurassic accretionary complex in the Innner Zone of Southwest Japan. *Memoir of the Geological Society of Japan* **55**, 73–98. [In Japanese, with English abstract.]

Natal'in, B. A. 1993. History and modes of Mesozoic accretion in Southeastern Russia. *The Island Arc* **2**, 15–34.

Natal'in, B. A. & Zyabrev, S. V. 1989. Structure of Mesozoic Rocks of the Amur River Valley. *Field trip guide book for International Symposium on Tectonics, Energy and Mineral resources of Northwest Pacific, Khabarovsk, September, 1989*. Khabarovsk: Far East Branch of USSR Academy of Sciences. [In Russian and English.]

Nechaev, V. P., Markevich, P. V., Malinovsky, A. I., Philippov, A. N. & Vysotskiy, S. V. 1996. Tectonic setting of the Cretaceous sediments in the Lower Amur region, Russian Far East. *Journal of the Sedimentological Society of Japan* **43**, 69–81.

Ohana, T. & Kimura, T. 1995. Late Mesozoic phytogeography in eastern Eurasia with special reference to the origin of angiosperms in time and site. *Journal of the Geological Society of Japan* **101**, 54–69. [In Japanese, with English abstract.]

Okada, H. 1999. Plume-related sedimentary basins in East Asia during the Cretaceous. *Palaeogeography, Palaeoclimatology, Palaeoecology* **150**, 1–11.

Otofuji, Y., Matsuda, T. & Nohda, S. 1985. Opening mode of the Japan Sea inferred from the palaeomagnetism of the Japan Arc. *Nature* **317**, 603–4.

Otsuki, K. & Ehiro, M. 1992. Cretaceous left-lateral faulting in Northeast Japan and its bearing on the origin of geologic structure of Japan. *Journal of the Geological Society of Japan* **98**, 1097–112. [In Japanese, with English abstract.]

Parfenov, L. M. 1991. Tectonics of the Verkhoyansk-Kolyma Mesozoides in the context of plate tectonics. *Tectonophysics* **199**, 319–42.

Parfenov, L. M, Vetluzhskikh, V. G., Gamyanin, G. N., Davydov, Yu. V., Deikunenko, A. V., Kostin, A. V., Nikitin, V. M., Prokop'yev, A. V., Smelov, A. P., Supletsov, V. M., Timofeyev, V. F., Fridovskiy, V. Yu., Kholmogorov, A. I. & Yakovlev, Ya. V. 1999. Main metallogenic units of the Sakha Republic (Yakutia), Russia. *International Geology Review* **41**, 425–56.

Romanovsky, N. P., Gurovich, V. G. & Sato, K. 1996. Magnetic susceptibility and metallogeny of granitoids in the Circum-Japan Sea region. *Geology of the Pacific Ocean* **12**, 965–76.

Ryazantseva, M. D., Gerasimov, N. S. & Govorov, I. N. 1995. Rubidium-strontium isochrones and petrogenesis of igneous rocks in the Voznesensk ore-bearing area, Primorie. *Geology of the Pacific Ocean* **11**, 579–96.

Sagong, H., Kwon, S. T. & Ree, J. H. 2003. Origin of the 160-110 hiatus in the Mesozoic magmatism of South Korea. *5th Hutton Symposium, Abstract with Program,* 125. *Geological Survey of Japan, Interim Report* **29**.

Sasaki, A. & Ishihara, S. 1979. Sulfur isotopic composition of the magnetite-series and ilmenite-series granitoids in Japan. *Contributions to Mineralogy and Petrology* **68**, 107–15.

Sasaki, A. & Ishihara, S. 1980. Sulfur isotope characteristics of the granitoids and related mineral deposits in Japan. *Proceedings of the 5th International Association on the Genesis of Ore Deposits (IAGOD) Symposium*, 325–35. Stuttgart: E. Schweizerbart'sche Verlagsbuchhandlung.

Sato, K. 1980. Tungsten skarn deposit of the Fujigatani mine, Southwest Japan. *Economic Geology* **75**, 1066–82.

Sato, K. 1982a. Characteristics of tungsten skarns in Japan, two contrasting types. *In* Hepworth, J. V. & Yu, H. Z. (eds) *Proceedings of the Symposium on Tungsten Geology, Jiangxi, China, 1981*, 203–9. Bandung ESCAP/RMRDC (Economical and Social Commission for Asia and the Pacific/Regional Mineral Resources Development Centre) of the United Nations and Geological Publishing House in Beijing.

Sato, K. 1982b. Tungsten deposits in China. *Chishitsu(Geology) News*, 31–44. [In Japanese.]

Sato, K. 1991. Miocene granitoid magmatism at the island-arc junction, central Japan. *Modern Geology* **15**, 367–99.

Sato, K. 2000. Granitoids and mineralization in the Khingan-Okhotsk and Sikhote-Alin regions, Far East Russia: Crustal structure and granitoid series. *Gekkan-Chikyuu (Earth Monthly), Special Publication* **30**, 162–70. [In Japanese.]

Sato, K. 2003. Metallogeny in the circum Japan Sea region: crustal structure and granitoid series. *In* Society of Resource Geology (ed.) *Resource and Environmental Geology*, 61–70. Tokyo: Society of Resource Geology. [In Japanese.]

Sato, K., Shimazaki, H. & Chon, H. T. 1981. Sulfur isotopes of the ore deposits related to felsic magmatism in the southern Korean peninsula. *Mining Geology* **31**, 321–6.

Sato, K., Ishihara, S. & Shibata, K. 1992. Granitoid Map of Japan. *Geological Atlas of Japan* (2nd edn). Tsukuba: Geological Survey of Japan.

Sato, K., Lavrik, N. I. & Vrublevsky, A. A. 1993a. Geology and mineral deposits in Skhiote-Alin, Russia. *Chishitsu(Geology) News* **468**, 16–26. [In Japanese.]

Sato, K., Ishihara, T., Vrublevsky, A. A. & Ishihara, S. 1993b. Distribution of magnetic anomalies and igneous rocks in southern Sikhote-Alin, Far East Russia. *Chishitsu(Geology) News* **470**, 18–28. [In Japanese.]

Sato, K., Rodionov, S. M. & Enjoji, M. 1996. Mnogovershinnoe Au deposit in northern Sikhote-Alin, Far East Russia. *Chishitsu (Geology) News* **501**, 37–44. [In Japanese.]

Sato, K., Ishihara, T., Rodionov, S. M., Nedachi, M., Vrublevsky. A. A., Khanchuk, A. I., Romanovsky, N. P. & Enjoji, M. 1998. Mid-Cretaceous episodic magmatism and tin mineralization in the Khingan-Okhotsk volcano-plutonic belt, Far East Russia. *Abstract for Resource Geology 1998 Annual Meeting, P-2*. Tokyo: Society of Resource Geology.

Sato, K., Vruvlevsky, A. A., Rodionov, S. M., Rmanovsky, N. P. & Nedachi, M. 2002. Mid-Cretaceous epispdic magmatism and tin mineralization in Khingan-Okhotsk volcano-plutonic belt, Far East Russia. *Resource Geology* **52**, 1–14.

Sato, K., Suzuki, K., Nedachi, M., Terashima, S., Kovalenko, S. V., Ryazantseva, M. D, Vrublevsky. A. A. & Khanchuk, A. I. 2003a. Fluorite deposits at Vonesenka in the Khanka massif, Russia: Geology and age of mineralization. *Resource Geology* **53**, 193–211.

Sato, K., Nedachi, M., Suzuki, K., Ishihara, T., Vrublevsky. A. A., Rodionov, S. M., Berdnikov, N. A., Romanovsky, N. P., Kovalenko, S. V., Korenbaum, S. A. & Khanchuk, A. I. 2003b. Granitoids and mineralization in the Circum-Japan Sea region: Crustal structure and granitoid types. *5th Hutton Symposium, Abstract with Program*, 133. *Geological Survey of Japan, Interim Report* **29**.

Sato, K. & Ishihara, S. 1983. Chemical composition and magnetic susceptibility of the Kofu granitic complex. *Bulletin of the Geological Survey of. Japan* **34**, 413–27. [In Japanese, with English abstract.]

Shibata, K., Matsumoto, T., Yanagi, T. & Hamamoto, R. 1978. Isotopic ages and stratigraphic control of Mesozoic igneous rocks in Japan. *In The Geologic Time Scale, AAPG Studies in Geology* **6**, 143–64. Tulsa, Oklahoma: American Association of Petroleum Geologists (AAPG).

Shibata, K. & Ishihara, S. 1979. Initial 87Sr/86Sr ratios of plutonic rocks from Japan. *Contributions to Mineralogy and Petrology* **70**, 381–90.

Shimazaki, H., Sato, K. & Chon, H. T. 1981. Mineralization associated with Mesozoic felsic magmatism in Japan and Korea. *Mining Geology* **31**, 297–310.

Shimazaki, H., Shibara, K., Uchiumi, S., Lee, M. S. & Kaneda, H. 1987. K-Ar ages of some W-Mo deposits and their bearing on metallogeny of South Korea. *Mining Geology* **37**, 395–401.

Silver, L. T., Taylor, H. P. & Chappell, B. 1979. Some petrological, geochemical, and geochronological observations of the Peninsular Ranges batholith near the international border of the U. S. A. and Mexico. *In* Abbott, P. L. & Todd, V. R. (eds) *Mesozoic Crystalline Rocks: San Diego, California*, 83–110. San Diego State University: Department of Geological Sciences.

Suzuki, K. & Adachi, M. 1998. Denudation history of the high T/P Ryoke metamorphic belt, southwest Japan: constraints from CHIME monazite ages of gneisses and granitoids. *Journal of Metamorphic Geology* **16**, 23–37.

Takahashi, M., Aramaki, S. & Ishihara, S. 1980. Magnetite-series/ilmenite-series vs. I-type/S-type granitoids. *Mining Geology, Special Issue* **8**, 13–28.

Teraoka, Y. & Okumura, K. 2003. *Geological Map of East Asia, scale 1:3,000,000*. Tsukuba: Geological Survey of Japan.

Tsuchiya, N. & Kanisawa, S. 1994. Early Cretaceous Sr-rich silicic magmatism by slab melting in the Kitakami Mountains, northeast Japan. *Journal of Geophysical Research* **99**, 22,205–220.

Tsusue, A. & Ishihara, S. 1974. The iron-titanium oxides in the granitic rocks of Southwest Japan. *Mining Geology* **24**, 13–30. [In Japanese, with English abstract.]

USSR Ministry of Geology 1977. *Magnetic Anomaly Map of USSR, scale 1:2,500,000*. Leningrad: USSR Ministry of Geology. [In Russian.]

USSR Ministry of Geology 1987. *Map for useful minerals in the USSR., 1:5,000,000*. Leningrad: USSR Ministry of Geology. [In Russian.]

Ueda, Y., Yamaoka, K. & Onuki, H. 1970. K-Ar dating on the host rocks of the Taro mine, Iwate Prefecture-specially on the origin of ore deposits. *Journal of the Japanese Association of Mineralogy, Petrology and Economic Geology* **63**, 259–65. [In Japanese, with English abstract.].

Uyeda, S. & Miyashiro, A. 1974. Plate tectonics and the Japanese Islands: A synthesis. *Geological Society of America, Bulletin* **85**, 1159–70.

Voinova, I. P., Zyabrev, S. V. & Prihodko, V. S. 1995. Petrochemistry of Early Cretaceous intraplate oceanic basalts of the Kiselevka-Manoma terrane, northern Sikhote Alin. *Geology of the Pacific Ocean* **11**, 971–90.

Wu, F., Jahn, B., Wilde, S. & Sun, D. 2000. Phanerozoic crustal growth: U-Pb and Sr-Nd isotopic evidence from the granites in northeastern China. *Tectonophysics* **328**, 89–113.

Wu, F., Sun, D., Li, H., Jahn, B. & Wilde, S. 2002. A-type granites in northeastern China: age and geochemical constraints on their petrogenesis. *Chemical Geology* **187**, 143–73.

Xu, J., Ma G., Tong, W., Zhu, G. & Lin, S. 1993. Displacement of the Tancheng-Lujian wrench fault system and its geodynamic setting in the northwestern circum-Pacific. *In* Xu, J. (ed.) *The Tancheng-Lujian Wrench Fault System*, 51–74. Chichester: John Wiley & Sons.

Zonenshain, L. P., Kuzmin, M. I., & Natapov, L. M. 1990. Geology of the USSR: A Plate- Tectonic Synthesis. *American Geophysical Union, Geodynamic Series* **21**.

Zyabrev, S. V. 1996. Cretaceous radiolarian fauna from the Kiselyovsky subterrane, the youngest accretionary complex of the Russian continental far east: Paleotectonic and paleogeographic implications. *The Island Arc* **5**, 140–55.

K. SATO and T. ISHIHARA, Geological Survey of Japan, Higashi 1-1-1 Central 7, Tsukuba, 305-8567, Japan.
e-mail: sato.gsj@aist.go.jp

S. V. KOVALENKO, Committee on Geology and Mineral Resources of Primorie Province, Vladivostok, Russia.

N. P. ROMANOVSKY and N. V. BERDNIKOV, Institute of Tectonics and Geophysics, Russian Academy of Sciences, Khabarovsk, Russia.

M. NEDACHI, Faculty of Sciences, Kagoshima University, Kagoshima, Japan.

MS received 4 December 2003. Accepted for publication 12 September 2004.

Transactions of the Royal Society of Edinburgh: Earth Sciences, **95**, 339–356, 2004

Source and redox controls on metallogenic variations in intrusion-related ore systems, Tombstone-Tungsten Belt, Yukon Territory, Canada

Craig J. R. Hart, John L. Mair, Richard J. Goldfarb and David I. Groves

ABSTRACT: The Tombstone, Mayo and Tungsten plutonic suites of granitic intrusions, collectively termed the Tombstone-Tungsten Belt, form three geographically, mineralogically, geochemically and metallogenically distinct plutonic suites. The granites (*sensu lato*) intruded the ancient North American continental margin of the northern Canadian Cordillera as part of a single magmatic episode in the mid-Cretaceous (96–90 Ma). The Tombstone Suite is alkalic, variably fractionated, slightly oxidised, contains magnetite and titanite, and has primary, but no xenocrystic, zircon. The Mayo Suite is sub-alkalic, metaluminous to weakly peraluminous, fractionated, but with early felsic and late mafic phases, moderately reduced with titanite dominant, and has xenocrystic zircon. The Tungsten Suite is peraluminous, entirely felsic, more highly fractionated, reduced with ilmenite dominant, and has abundant xenocrystic zircon. Each suite has a distinctive petrogenesis. The Tombstone Suite was derived from an enriched, previously depleted lithospheric mantle, the Tungsten Suite is from the continental crust including, but not dominated by, carbonaceous pelitic rocks, and the Mayo Suite is from a similar sedimentary crustal source, but is mixed with a distinct mafic component from an enriched mantle source.

Each suite has a distinctive metallogeny that is related to the source and redox characteristics of the magma. The Tombstone Suite has a Au-Cu-Bi association that is characteristic of most oxidised and alkalic magmas, but also has associated, and enigmatic, U-Th-F mineralisation. The reduced Tungsten Suite intrusions are characterised by world-class tungsten skarn deposits with less significant Cu, Zn, Sn and Mo anomalies. The Mayo Suite intrusions are characteristically gold-enriched, with associated As, Bi, Te and W associations. All suites also have associated, but distal and lower temperature Ag-Pb- and Sb-rich mineral occurrences. Although processes such as fractionation, volatile enrichment and phase separation are ultimately required to produce economic concentrations of ore elements from crystallising magmas, the nature of the source materials and their redox state play an important role in determining which elements are effectively concentrated by magmatic processes .

KEY WORDS: Gold, granites, metallogeny, oxidation state, petrogenesis, tungsten, Yukon.

Regional-scale metallogenic and ore-element variations of intrusion-related ore systems have been ascribed to a variety of processes. The tectonic setting, magma composition, degree of fractional crystallisation and redox state all potentially play a role in determining the metallogeny of an intrusion-related mineral deposit district (e.g. Ishihara 1981; Titley 1982, 1991; Blevin & Chappell 1992; Barton 1996; Blevin *et al.* 1996; Lang & Titley 1998). Controversies have typically focused on the relative importance of either the source materials or magmatic-hydrothermal processes in enriching a system in a particular metal suite. Economic geologists commonly emphasise the role that magmatic-hydrothermal processes play in intrusion-related metal enrichments, whereas igneous petrologists emphasise the importance of the nature of the source materials.

The Tombstone-Tungsten Belt (TTB) in Yukon Territory, Canada, forms a 750-km-long belt of dominantly felsic intrusions which were emplaced across the ancient western North American continental margin from 96 to 90 Ma. The TTB is the innermost and youngest of numerous plutonic belts which intruded the continental margin in the mid-Cretaceous. Although magmatism occurred over a short time interval and across a single tectonic element, pluton compositions vary considerably along the belt, as do the associated mineral occurrence types. Several deposits and numerous occurrences of gold, molybdenum, silver, tin, copper and uranium, as well as world-class tungsten deposits, are associated with plutons of the TTB (Hart *et al.* 2000). This well-mineralised region represents the eastern part of the Tintina Gold Province (Hart *et al.* 2002), which notably hosts numerous intrusion-related gold systems across Yukon and interior Alaska (Thompson *et al.* 1999; Goldfarb *et al.* 2000; Lang *et al.* 2000). Based on pluton distribution, nature and composition, the belt can be divided into three main groups of intrusions. From west to east, they comprise the Tombstone, Mayo and Tungsten plutonic suites, with each having its own distinctive metallogenic signature.

This paper presents data on the geological, mineralogical and geochemical features which characterise each of the plutonic suites. In particular, petrology, and major, minor and trace element geochemical and isotope data are applied to characterise and compare the three suites. This is used to constrain the nature of the potential source materials which generated their magmas, and the processes which contributed to the diversity of lithologies. In addition, the metallogeny of each suite is evaluated in the context of these magmatic source

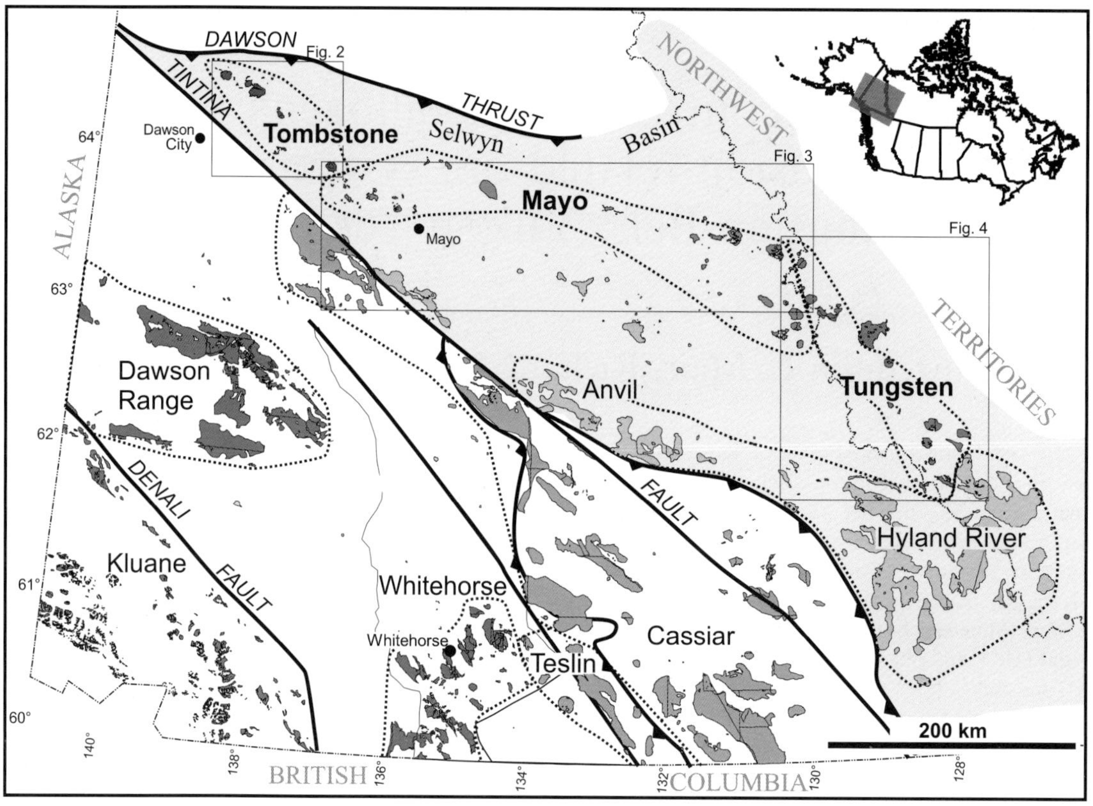

Figure 1 Regional tectonic elements of Yukon Territory and the distribution of mid-Cretaceous plutonic suites (shown in various shaded tones). The Tombstone-Tungsten Belt includes the distribution of the Tombstone, Mayo and Tungsten plutonic suites. The light shading shows the distribution of the Selwyn Basin. The dotted lines are the distribution limits of the plutonic suites. The dark lines are faults, and those with teeth are thrust faults. The light lines are tectonic boundaries.

variations, with particular emphasis on the role of oxidation state and fractionation.

1. Tectonic setting

The northern North American Cordillera is underlain by numerous tectonic elements which include outboard accreted island arc and oceanic terranes, accreted terranes with pericratonic affinities, and variably displaced or deformed portions of the ancient continental margin (Gabrielse *et al.* 1991). Whereas the mechanisms of terrane assembly and accretion remain contentious, most of the assembly and growth of the current continental margin occurred during ongoing convergent tectonism between Early Jurassic and mid-Cretaceous times (Gabrielse & Yorath 1991; Plafker & Berg 1994). Following the waning of collision and accretionary tectonics, extensive magmatism occurred across much of the newly assembled continental margin during mid-Cretaceous times (Armstrong 1988). Hundreds of batholiths and plutons of the mid-Cretaceous magmatic episode are distributed in broadly orogen-parallel belts and geographically restricted clusters. Plutons within each belt have similar characteristics and ages, and as such, comprise plutonic suites (Mortensen *et al.* 2000; Hart *et al.* 2004a; Fig. 1). There is a large variation in the characteristics between the plutonic suites: some are typical arc-related, metaluminous and calc-alkaline I-types (e.g. Whitehorse suite; Hart 1997), others comprise peraluminous granitoids which were mainly derived from crustal melts (e.g. Cassiar suite; Driver *et al.* 2000), and others consist of variably alkalic, locally silica-undersaturated plutons (e.g. Tombstone Suite; Anderson 1982).

Among the most interesting and economically prospective of the mid-Cretaceous plutonic suites are those which comprise the most inboard plutonic belt, the Tombstone, Mayo and Tungsten suites (Fig. 1). These plutons, which form the TTB, were emplaced into a thick package of variably calcareous, quartzose and argillaceous Neoproterozoic, and locally carbonaceous, Palaeozoic strata, which were deposited within the Selwyn Basin along the ancient North American margin. Although carbonate platforms comprise much of the ancient North American margin, the Selwyn Basin developed in response to Neoproterozoic and Palaeozoic rifting events which resulted in an attenuated crust, and the development of euxinic shale basins and associated Pb-Zn sedimentary exhalative deposits (e.g. Faro; Abbott *et al.* 1986). Selwyn Basin strata were structurally thickened and locally metamorphosed to lower greenschist facies as a result of outboard terrane accretion immediately prior to mid-Cretaceous plutonism (Murphy 1997). Numerous features associated with the plutons, such as east-trending dyke swarms and parallel sub-vertical extensional veins, suggest that the region underwent a period of post-collisional extension immediately following compression. As such, the TTB intrusions were emplaced during post-collision in an inboard part of the continental margin.

2. Characteristics of the Tombstone-Tungsten Belt

More than 100 plutons, as well as numerous dykes and sills, of the TTB form the most inboard plutonic belt of the mid-Cretaceous magmatic episode. The plutons intruded the northern Selwyn Basin, to the south of the long-lived and probably basin-controlling Dawson thrust fault (Fig. 2; Abbott 1995). Within the TTB, the Tombstone, Mayo and Tungsten plutonic suites are defined on the basis of their distributions and lithological similarities. The term 'plutonic suites' is used since it is common in the North American lexicon for a regional

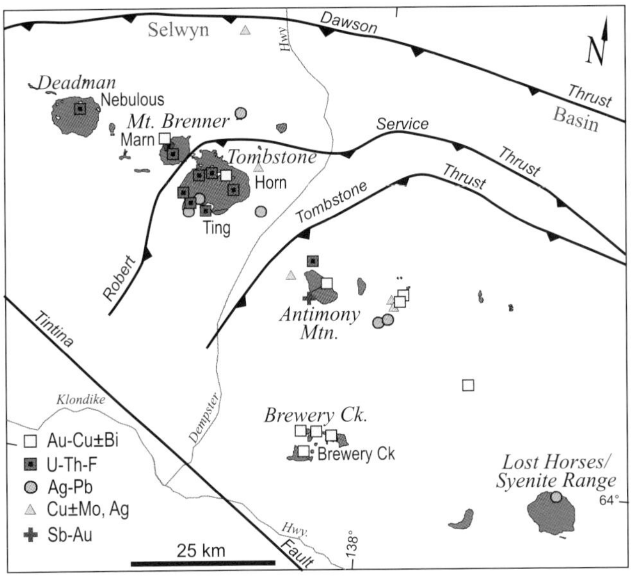

Figure 2 Distribution of the Tombstone Plutonic Suite and associated intrusion-related mineral occurrences. The pluton names are italicised. Significant mineral occurrences are named. Geological information is taken from Gordey & Makepeace (2003). Mineral deposit information is taken from Yukon Minfile (2003).

group of coeval and lithologically similar plutons. This approach differs somewhat from the use of 'suites' in the Australian context, which is more dependent upon the magma source according to geochemical and isotopic parameters, and typically refers to several lithologies within a single or numerous plutons (White 1995; Chappell 1996).

Each of these three plutonic suites in the TTB forms a linear belt, with associated dykes paralleling their trends. The Tombstone Suite is mostly located along a 120-km-long, southeast trend that parallels the Tintina Fault (Fig. 1). The plutons were discordantly emplaced into folded and faulted Neoproterozoic to Mesozoic sedimentary rocks in the western part of the Selwyn Basin (Fig. 2). Notably, the plutonic belt cuts the major Cretaceous structural features, such as the Robert Service and Tombstone thrust faults (Fig. 2), indicating a negligible role played by the faults in magma focusing. However, the trends of the Tombstone and Mayo suites are interpreted to reflect structural trends associated with Neoproterozoic rift events.

The Mayo Suite of intrusions forms an ESE-trending, 370-km-long belt from the Tintina Fault to the eastern border of the Yukon Territory (Fig. 3). Country rocks are dominantly folded and thrust-faulted quartzose and argillaceous strata of the Neoproterozoic Hyland Group, with carbonaceous shale and chert of the Devono-Mississippian Earn Group in the east.

The Tungsten Suite of intrusions forms an almost 200-km-long, SE-trending linear belt from Macmillan Pass to south of the townsite of Tungsten (Fig. 4). The plutons intrude carbonaceous and calcareous Palaeozoic strata, and lesser quartz-rich and pelitic Neoproterozoic strata of the easternmost Selwyn Basin.

In east-central Alaska, intrusions of equivalent age and lithology to the Mayo and Tombstone Suites form the Fairbanks-Salcha and Livengood suites, respectively, which crop out in the world-class Fairbanks gold district south of the Tintina Fault (Hart *et al.* 2004a). These intrusions are interpreted as the western parts of the Mayo and Tombstone Suites, which were offset by ~430 km of dextral transcurrent displacement on the Tintina Fault during the Late Cretaceous and Early Tertiary (Mortensen *et al.* 2000; Murphy & Mortensen 2003).

Plutons throughout the TTB have well-developed, resistant-weathering, contact-metamorphic aureoles which are about half the width of adjacent intrusions. None of the plutons have associated volcanic rocks. Equilibrium contact metamorphic assemblages and fluid inclusion data from syn-plutonic quartz-veins constrain pluton emplacement pressures to mostly 1·0–2·5 kbar (Baker & Lang, 2001; Marsh *et al.* 2003). Extensive geochronology (in Gordey & Anderson 1993; Murphy 1997; Mortensen *et al.* 2000; Coulson *et al.* 2002; Hart *et al.* 2004b; J. K. Mortensen, unpublished results) indicates that the TTB plutons were emplaced over a narrow time interval of 96–90 Ma with the Tombstone Suite of intrusions being emplaced at the younger end of the range. Essentially contemporaneous ^{40}Ar-^{39}Ar dates from magmatic biotite and hornblende with U-Pb zircon dates (Coulson *et al.* 2002; Hart *et al.* 2004b) indicate that the region has not been affected by post-intrusion thermal events.

3. Plutonic suites

Intrusions of the TTB commonly occur as isolated plutons, particularly within the Tombstone and Tungsten suites. However, the Mayo intrusions typically occur in clusters, commonly with numerous associated dykes, which collectively comprise intrusive complexes.

The Tombstone plutons are well-zoned, comprising several distinct lithological phases which form nested concentric zones or phases adjacent to each other. The Mayo intrusions most commonly consist of a main phase stock with gradational lithological variations potentially indicative of multiple intrusive phases. In addition, these plutons characteristically have a variety of associated dykes, including intermediate to mafic lamprophyres, porphyritic granitic dykes, and subordinate aplites and pegmatites. The Tungsten Suite intrusions have comparatively minor, within-pluton, lithological phase variations and lack associated intermediate to mafic phases. However, plutons of the Tungsten Suite commonly have associated pegmatites, aplites and greisens. Intrusions of all suites are wallrock inclusion-poor, with only sparse occurrences of microdiorite enclaves. The Tungsten and Mayo intrusions are weakly porphyritic, with white megacrystic alkali feldspar. The characteristics of the plutonic suites are summarised in Table 1.

3.1. Nomenclature

After first designating those plutons along the border between Yukon Territory and Northwest Territories (NWT) as the Selwyn Plutonic Suite (Anderson 1983), Anderson (1988) and Woodsworth *et al.* (1991) subsequently included all mid-Cretaceous plutons in the Selwyn Basin within this group. The plutons were subdivided into two-mica, transitional and hornblende-bearing phases. Anderson (1988) further encouraged the additional use of more regional names, such as Anvil Plutonic Suite (Fig. 1).

The Tombstone Plutonic Suite has traditionally referred to those geographically restricted plutons in the westernmost part of the TTB which have alkalic affinities (Tempelman-Kluit 1970; Woodsworth *et al.* 1991), with the Tombstone Mountain Batholith being the type locality. The Tombstone Plutonic Suite was subsequently used by Mortensen *et al.* (1995) and Murphy (1997) to refer to all plutons which intruded along the northern margin of the Selwyn Basin between Dawson City and the NWT border (Fig. 1). Furthermore, Mortensen *et al.* (1995, 2000) renamed the Selwyn plutonic suite intrusions near the NWT border, the Tungsten Plutonic Suite, but included only those smaller plutons which were *circa* 94 Ma, thus

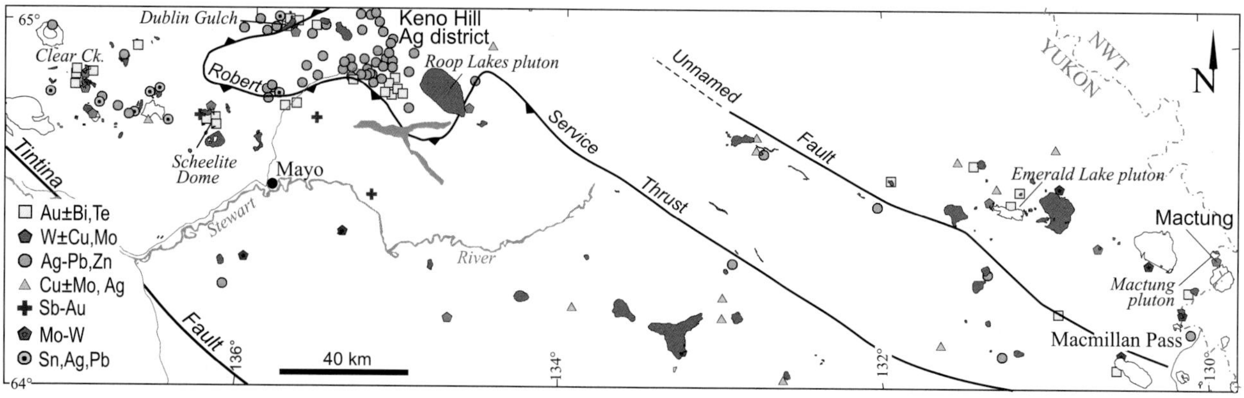

Figure 3 Distribution of the Mayo Plutonic Suite across central Yukon and associated intrusion-related mineral occurrences. Clear Creek, Scheelite Dome and Dublin Gulch also have significant placer gold. Pluton names are italicised. Significant mineral occurrences are named. Mineral occurrences often have the same name as the pluton. Unfilled plutons belong to other suites. Geological information is taken from Gordey & Makepeace (2003). Mineral deposit information is taken from Yukon Minfile (2003).

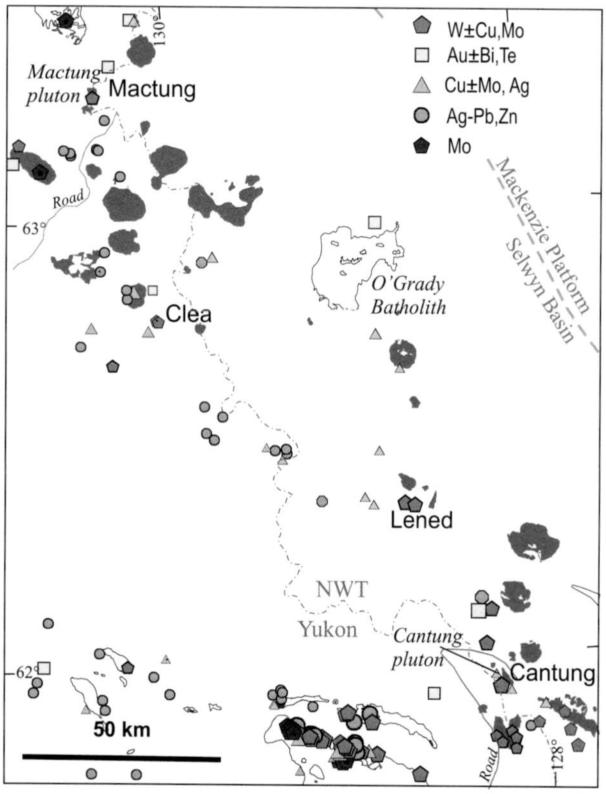

Figure 4 Distribution of the Tungsten Plutonic Suite in eastern Yukon and westernmost Northwest Territories (NWT) and associated intrusion-related mineral occurrences. Outlined, unshaded plutons do not belong to the Tungsten Plutonic Suite. The O'Grady Batholith is alkalic and a distal member of the Tombstone Suite. Pluton names are italicised. Significant mineral occurrences are named. Mineral occurrences often have the same name as the pluton. Geological information is taken from Gordey & Makepeace (2003). Mineral deposit information is taken from Yukon Minfile (2003).

excluding the batholiths which Anderson (1988) had included within the Selwyn Plutonic Suite.

The term Tombstone Plutonic Suite is herein retained as originally applied to the dominantly alkalic intrusive rocks near the Tintina Fault, but in contrast to Murphy (1997) and Mortensen *et al.* (1995), does not include the east-trending belt of intrusions in central Yukon Territory. These plutons, between the Tombstone and Tungsten suites, are designated the Mayo Plutonic Suite. The Tungsten Plutonic Suite, as applied to the small peraluminous intrusions near the Yukon Territory-NWT border by Mortensen *et al.* (2000), is retained. Together, these three suites form the TTB. The Tombstone Suite intrusions are characteristically alkalic and metaluminous; the Mayo intrusions are sub-alkalic and mostly metaluminous, and distinctive from the more felsic Tungsten Suite intrusions, which are dominantly peraluminous.

3.2. Tombstone Plutonic Suite

Tombstone Suite intrusions are characterised by their alkalic character, limited distribution and larger, well-zoned plutons (Smit *et al.* 1985; Anderson 1987, 1988; Abercrombie 1990; Gordey & Anderson 1993; Duncan *et al.* 1998). They include six main intrusions, as well as swarms of sills, dykes and small stocks along their trend (Fig. 2). Two outliers from the linear distribution are the isolated Emerald Lake Pluton (Fig. 3) and the O'Grady Batholith (Fig. 4), which occur several hundred kilometres east of the other intrusions in the suite.

Tombstone plutons are typically circular to sub-circular, relatively large (10–80 km^2 in plan), and typically composed of multiple concentric zones of lithologically distinct phases. The dominant rock types are coarse-grained alkali-feldspar syenite and quartz syenite. Many plutons have mafic marginal phases, such as pyroxenite and hornblende diorite, as well as central, more felsic, monzonitic or granitic phases which are nested in the syenite. Tombstone intrusions also locally contain distinctive alkalic, quartz-undersaturated phases such as tinguaite (feldspathoid syenite).

The dominant mafic minerals include augite, aegerine, hornblende, local biotite and lesser melanitic garnet. Hornblende (arfvedsonite) typically replaces earlier formed augite. Fe-Ti accessory phases are magnetite and titanite; other accessory minerals include zircon, fluorite, apatite and nepheline (Anderson 1988; Abercrombie 1990).

3.3. Mayo Plutonic Suite

Mayo Suite plutons are generally small in the west (1–5 km^2) and larger in the east (20–80 km^2), but the anomalously large Roop Lakes Pluton (125 km^2) is notable (Fig. 3). This suite has characteristic spatial and temporal associations with intrusion-related gold deposits. Mayo intrusions typically form isolated plutonic centres, or clusters of small felsic to mafic plutons. Intrusions are typically dominated by a main porphyritic quartz monzonite that characteristically includes less-evolved magmatic phases and locally forms weakly composite plutons (e.g. Dublin Gulch). Alternatively, some main phase plutons are cut by monzonite to diorite stocks and dykes (e.g. Scheelite Dome). Three small plutons in the Clear Creek area are dominated by a main-phase quartz monzonite, whereas more

Table 1 Summary of the characteristics of the plutonic suites*

Characteristic	Tombstone (92–90 Ma)	Mayo (95–92 Ma)	Tungsten (97–94 Ma)	References (1, 2, 9, 10)
Dominant lithologies	Alkali feldspar syenite to quartz syenite	Monzonite to granodiorite	Granite to monzogranite	
Pluton size	Moderate	Small	Small	
Plutons	Zoned, mafic margins, felsic cores	Simple, later mafic phases	Simple, textural variations, some fractionated phases	
Grain size	Coarse, cumulate	Medium to fine grained, locally porphyritic	Medium grained, weakly porphyritic	
Mafic phases	Pyroxene (aegerine-augite) >hornblende>biotite	Biotite>hornblende, clinopyroxene common	Biotite≫muscovite	7
Dominant Fe-Ti indicator mineral	Magnetite>titanite	Titanite	Ilmenite≫titanite	
Accessory minerals	Epidote, allanite, melanite, apatite, fluorite, zircon	Allanite, apatite, zircon	Garnet, apatite, monazite, tourmaline, allanite, zircon	4, 7
SiO_2 range	50–70%	55–75%	66–76%	3, Table 3
ASI	Metaluminous, except where highly fractionated (0·65–1·1)	Metaluminous to weakly peraluminous (0·6–1·15)	Weakly peraluminous (1·0–1·2)	
Alkalinity	Alkaline to peralkaline	Subalkaline	Subalkaline	
Fe_2O_3/FeO	0·2–1·1	0·15–0·45	0·1–0·3	Figure 11
Average magnetic susceptibility ($\times 10^{-3}$ SI)	1·79	0·11	0·16	Figure 10
Cr (p.p.m.)	Most <20, some 20–80	Most 20–100, some 100–600	Most <20	3, 10
Inherited zircons	None	Some	Considerable	
Initial Sr ratio	0·710–0·720	0·7115–0·7140	0·717–0·737	3, 4, 7, 8
ε_{Nd_T}	−7 to −9	−8 to −13	−13 to −15	3, 5, 10
Oxygen isotopes	9–11	11–14	9–13	6, 7, 10
ZST	820°C	780°C	750°C	Table 3
Associated mineralisation	Au-Cu-Bi, U-Th-F	Au-Bi-Te, W, As Ag-Pb	W, Cu-Zn-Mo	
Characterisation	Alkalic, slightly oxidised, metaluminous, radiogenic, syenite cumulates	Metaluminous, moderately reduced, radiogenic, biotite granodiorite	Weakly peraluminous, reduced, radiogenic, biotite granite	

*Abbreviations: (ASI) aluminum saturation index; (ZST) zircon saturation temperature; (NA) not available.
References: (1) Anderson (1993, in Gordey & Anderson 1993); (2) Murphy 1997; (3) Lang (2000), (4) Abercrombie (1990); (5) Farmer *et al.* (2000); (6) Marsh *et al.* (2003); (7) Anderson (1988); (8) Gareau (1986); (9) Coulson *et al.* (2002); and (10) C. Hart & J. Mair, unpublished data.

Table 2 Examples of mineralisation associated with the plutonic suites

Deposit/occurrence	Metals	Mineralisation style	Ore minerals	Gangue minerals	Reference
Tombstone Plutonic Suite					
Marn	Au-Cu-Bi	Skarn	Pyrrhotite, chalcopyrite, bismuthinite, electrum	Hedenbergite, almandine garnet	Brown & Nesbitt (1987)
Horn	Au-Cu-Bi	Skarn	Pyrrhotite>pyrite, chalcopyrite, gold, bismuthinite	Hedenbergite, garnet, fluorite	
Brewery Creek-Moosehead Zone	Au-As-Sb	Vein stockwork	Pyrite, arsenopyrite	Kaolinite	Lindsay *et al.* (2000)
Ting/Noting	U-Th-F	Disseminated, veins	Uraninite, molybdenite	Fluorite	Yukon Minfile (2003)
Mayo Plutonic Suite					
Dublin Gulch	Au-Bi-W	Sheeted veins	Po, bismuthinite, scheelite	Quartz, K-spar, sericite	Maloof *et al.* (2001)
Scheelite Dome	Au-As	Replacements, stockworks	Arsenopyrite, pyrite, pyrrhotite	Quartz, sericite, carbonate	Mair *et al.* (2000)
Clear Creek	Au-Bi-W	Sheeted veins	Pyrite, bismuthinite, scheelite	Quartz, K-spar	Marsh *et al.* (2003)
Ray Gulch	W	Skarn	Scheelite	Diopside	Brown *et al.* (2002)
Keno Hill	Ag-Pb	Veins	Galena	Quartz, siderite	Lynch *et al.* (1990)
Tungsten Plutonic Suite					
Mactung	W-Cu	Skarn	Scheelite, chalcopyrite, pyrrhotite	Diopsite, garnet, actinolite	Atkinson & Baker (1986)
Cantung	W-Cu	Skarn	Scheelite, chalcopyrite, pyrrhotite	Diopside, epidote, garnet	Bowman *et al.* (1985)
Clea	W-Cu	Skarn	Scheelite, chalcopyrite, pyrrhotite	Diopside, garnet, vesuvianite, biotite, fluorite	Yukon Minfile (2003)

mafic lithologies dominate three others in the same area. A variety of porphyritic, aplitic and pegmatitic felsic dykes, as well as calc-alkaline lamprophyre dykes, typically cut all plutons and their hornfelsed aureoles.

The early, mainly quartz monzonite plutons are weakly to moderately porphyritic, with white potassium feldspar phenocrysts which are locally >1 cm in length. The phenocrysts occur in a fine- to medium-grained matrix of predominantly quartz and feldspar, with lesser biotite. Biotite is more abundant than hornblende, and clinopyroxene, although typically rare, is common in some plutons. Accessory minerals include titanite, allanite, apatite and zircon. Intermediate intrusions are mineralogically similar to the earlier, more felsic main phase intrusions, but with a greater proportion of clinopyroxene, biotite, hornblende and titanite. These more mafic phases locally contain unusual textures, with two generations of clinopyroxene. Early clinopyroxene is typically Mg-rich augite (Mg# >80) and is poorly preserved. Finer euhedral diopside grains (Mg# ~60) occur amongst quartz, and are interpreted to have formed late in the crystallisation sequence. Lamprophyre dykes include both sub-alkalic amphibole-rich spessartites and alkalic phlogopite/biotite-rich minettes (cf. Rock *et al.* 1991). Deuteric alteration is most extensive in intermediate to mafic intrusive rocks, with pyroxenes altered to amphibole and biotite. Titanite is ubiquitous and magnetite is locally present only in late granitic dykes, whereas ilmenite is rare, occurring only as fine inclusions within mafic phenocrysts in intermediate to felsic intrusions.

3.4. Tungsten Plutonic Suite

Intrusions of the Tungsten Suite are distinguished by their felsic, more homogeneous and consistently peraluminous character, as well as their strong association with tungsten skarn mineralisation (Anderson 1982, 1983, in Gordey & Anderson 1993). Plutons are typically small, generally less than 15 km^2 in surface exposure.

Tungsten Suite plutons are dominated by medium-grained quartz monzonite to monzogranite. Notably, these plutons almost always lack both amphibole and clinopyroxene, which are common in quartz monzonite intrusions of the Mayo Suite. Plutons typically feature gradational zoning to more felsic or peraluminous phases. Late-stage aplite or pegmatite dykes are locally common, as are tourmaline veinlets and rusty-weathering greisen-like alteration on joint planes in pluton cupolas.

Biotite, the dominant mafic mineral, ranges from brown to reddish, and commonly contains black pleochroic halos from radiation damage. Although biotite is typically the sole mica, primary muscovite is locally present in fractionated phases. Orthoclase forms both irregular coarse ophitic grains which incorporate earlier-formed biotite and plagioclase, as well as euhedral to subhedral simple-twinned laths. Ilmenite, the dominant oxide phase, typically occurs as fine grains in biotite. Subhedral plagioclase grains commonly feature oscillatory zoning, with sericitised grain cores. Accessory phases include allanite, zircon, apatite and ilmenite, with trace amounts of garnet and tourmaline in more highly fractionated phases, aplites and pegmatites. Magnetite is not present. Secondary ilmenite is strongly anhedral and associated with areas of chloritised biotite. Weak to moderate deuteric alteration is characterised by chloritisation of biotite, sericitisation and carbonate alteration of plagioclase.

4. Intrusion-related metallogeny

Each of the three plutonic suites has its own distinctive metallogenic association (Table 2), although a wide range of factors, such as P-T of emplacement, nature of country rock, structure and water-rock conditions, result in the diverse range of gold and/or tungsten-rich mineralisation styles as described by Hart *et al.* (2000). However, all suites appear to have spatially associated Ag-Pb-Zn- and As-Sb-bearing veins which are peripheral to the plutons and are products of the final stages of intrusion-related hydrothermal activity.

The alkalic Tombstone plutons tend to be associated with mineral occurrences which are dominated by either a Au-Cu-Bi or U-Th-F association (Fig. 2). Enrichments of native gold (to tens of p.p.m.) are hosted in pyrrhotite-rich skarns which contain percentage-level copper and anomalous bismuth concentrations, such as characteristic of the Marn (Brown & Nesbitt 1987) and Horn deposits. An arsenic-rich epizonal gold deposit, Brewery Creek (0·3 Moz Au; Lindsay *et al.* 2000), is localised within, and adjacent to, monzonitic Tombstone Suite sills and stocks (Fig. 2). Four of the Tombstone Suite intrusions host U-Th-F mineralisation, with two of the occurrences being large-tonnage, low-grade deposits in disseminated and veinlet-hosted styles. Locally, highly-fractionated parts of the Tombstone intrusions contain Au-, Cu- and Bi-bearing quartz veins and vug fillings, such as in the Emerald Lake Pluton (Fig. 3), and Sn- and Ag-rich tourmaline greisens, such as in the Syenite Range Pluton (Fig. 2). Mineralisation associated with the Tombstone Suite is characterised by high halogen contents, as indicated by the presence of fluorite, scapolite, and locally, sodalite.

The Mayo intrusions are dominantly metaluminous, are most consistently associated with gold mineralisation and have the highest abundance of associated mineral occurrences (Fig. 3). Approximately 1 Moz of placer gold have been recovered from creeks which drain areas underlain by the intrusions of the Mayo Suite and surrounding hornfels. Gold occurrences include intrusion-hosted Au-Bi-Te-W quartz-alkali-feldspar sheeted vein arrays such as at Dublin Gulch (4 Moz Au; Hitchins & Orssich 1995; Maloof *et al.* 2001), Au ± W skarns and country-rock-hosted Au-As quartz vein arrays, such as those at Scheelite Dome (Mair *et al.* 2000). Solitary, shear- and fissure-hosted Au-As veins are also common features proximal to, or within, Mayo intrusions. Other types of mineralisation include tungsten skarns, Ag-Pb-Zn ± Sb veins and tin-bearing greisens. The area containing the Mayo Suite of intrusions also hosts significant silver-rich quartz veins, with >200 Moz Ag recovered from veins of the Keno Hill district (Lynch *et al.* 1990). However, the genetic relationship of these veins to the intrusions is unclear. The Fort Knox gold deposit in Alaska (7 Moz Au) is hosted in a Mayo Suite intrusion that has been offset by the Tintina Fault.

The peraluminous Tungsten Suite intrusions are associated with the largest tungsten deposits in North America: the Cantung Deposit (Bowman *et al.* 1985; Gerstner *et al.* 1989) hosts approximately 9 Mt of 1·4% WO_3 and the Mactung Deposit (Dick & Hodgson 1982; Atkinson & Baker 1986; Fig. 4) contains approximately 57 Mt of 0·95% WO_3. Both deposits are developed in Lower Cambrian carbonates adjacent or above Tungsten Suite intrusions. Some intrusions have highly fractionated phases which contain garnet, muscovite and tourmaline. Each deposit consists of several stratabound ore zones. The best ores (>1·5% WO_3 and 0·2% Cu) are characterised by pyrrhotite-rich (>15%) pyroxene skarn with scheelite. The molybdenum content of the scheelite is low. Garnet-bearing skarns are much less common, contain less pyrrhotite and are lower grade. The skarns locally have sub-economic grades of zinc and molybdenite. Elsewhere, Tungsten Suite intrusions generate small Sn, Mo and Au enrichments.

5. Geochemistry

The three plutonic suites are characterised by distinct major, minor and trace element geochemical signatures (Table 3). The Tombstone and Mayo intrusions have wide compositional variations, with SiO_2 contents ranging from 50 to 75 wt.%, whereas Tungsten intrusions have more restricted SiO_2 contents from 65 to 75 wt.% (Fig. 5A). The Tombstone Suite rocks are distinguished from those of the Mayo Suite by their higher alkalinity and lower MgO contents (and Mg#), particularly at ≤65% SiO_2 (Fig. 5B). The Tombstone intrusions which contain <58% SiO_2 are commonly nepheline normative, mainly because of their elevated alkali contents.

Aluminum saturation is also variable within the different suites (Fig. 5C). The Tombstone and Mayo suites display a wide range of aluminum saturation index (ASI) values, which is consistent with their broad range in SiO_2 (cf. Chappell & White 2001). Generally, rocks with <65% SiO_2 are metaluminous, with an ASI value <1, whereas those Mayo intrusions with SiO_2 values >70% have ASI values slightly >1, as do those few highly fractionated Tombstone Suite examples. Tungsten Suite intrusions are entirely weakly peraluminous, with all values slightly >1.

The use of Rb, Sr and Ba as fractionation indices is complicated by the relatively enriched nature of the intermediate to mafic intrusions of the Tombstone and Mayo suites (Fig. 6). Intrusions of the Tungsten Suite, which have a restricted SiO_2 range, feature a strong increase in rubidium content with increasing SiO_2. Similarly, intrusions of the Tombstone Suite show a clear increase in rubidium content with increased SiO_2 content, although this occurs over a greater SiO_2 range. In contrast, Mayo Suite intrusions are characterised by only a slight increase in rubidium with increased SiO_2. Strontium shows strong depletion with increasing SiO_2 content in both the Tombstone and Tungsten suites, whereas the Mayo Suite has no well-defined trends. Notably, the strontium contents of Mayo Suite intrusions are considerably higher than those of the Tungsten intrusions at a given SiO_2 content.

Interpretation of primitive mantle-normalised trace-element diagrams (Fig. 7) indicates that all groups are strongly enriched in large ion lithophile elements (LILEs) and light rare-earth elements (LREEs). In addition, all suites have distinct negative titanium and niobium anomalies (Fig. 7A). Most intrusions of the TTB also exhibit a negative phosphorous anomaly, but it is absent in the mafic phases of the Mayo intrusions. Chondrite-normalised REE plots indicate strong enrichments of LREE, with negative slopes for intrusions of all groups (Fig 7B). The Tombstone and Mayo suites generally have no obvious europium anomalies, whereas more fractionated intrusions of the Tungsten Suite have pronounced negative europium anomalies. On the trace-element discrimination diagrams of Pearce *et al.* (1984), each suite overlaps both the syn-collisional and volcanic-arc fields.

The variation in chromium concentration is pronounced (Fig. 8). Almost all Tungsten Suite intrusions have concentrations below the analytical detection level (e.g. <20 p.p.m. Cr). The Tombstone Suite intrusions similarly contain low chromium levels, although more primitive intrusions contain as much as 100 p.p.m. Cr. The Mayo Suite intrusions have elevated, but variable chromium levels; felsic intrusions mostly contain <50 p.p.m., intermediate phases show 50–200 p.p.m., and mafic phases contain as much as 700 p.p.m. Cr. Chromium content is concomitant with the presence of chromium diopside.

6. Isotopic data

Isotopic (Sr, Nd, and O) compositions can reflect the nature of magma source regions in the middle to lower crust or mantle, and/or the extent of contamination from upper crustal sources (DePaolo *et al.* 1992). Isotopic data for plutons from the TTB are currently limited (Lang 2000), with most available data from intrusions in the Syenite Range (Abercrombie 1990), Clear Creek (Farmer *et al.* 2000; Marsh *et al.* 2003), Emerald Lake (Smit *et al.* 1985) and Scheelite Dome (Mair 2004). Initial strontium isotope ratios for Tombstone Suite rocks are 0·710–713 and ε_{Nd} values range from − 7·6 to − 10·3 (Abercrombie 1990; Lang 2000). The present authors' unpublished whole-rock oxygen isotopic values range from 9–11 per mil. The Mayo intrusions have somewhat similar initial strontium isotope ratios from 0·712 to 0·714 (Kuran *et al.* 1982; Lang 2000),

Table 3 Representative geochemical data for the Tombstone (Deadman Pluton), Mayo (Dublin Gulch Pluton) and Tungsten (Mactung Pluton) plutonic suites*

Sample	Deadman 1	Deadman 2	Deadman 3	Deadman 4	Dublin main phase	Dublin dyke 1	Dublin Qtz-syenite	Dublin pegmatite dyke	Dublin dyke 2	Mactung main stock	Mactung tourm peg	Mactung dyke in stock	Mactung dyke
SiO_2	57·8	64·7	58·6	68·3	67·2	59·6	69·5	61·0	59·8	68·1	72·6	72·3	75·7
TiO_2	0·304	0·198	0·303	0·175	0·459	0·551	0·343	0·592	0·673	0·487	0·02	0·176	0·054
Al_2O_3	19·3	17·6	21·1	16·1	15·2	14·2	16·0	16·7	14·7	15·5	15·9	14·2	13·1
FeO (total)	2·85	1·92	2·53	1·8	3·07	5·12	1·9	1·64	5·58	3·53	0·34	1·89	1·01
Fe_2O_3	1·65	0·92	0·53	1·0	0·57	0·92	0·5	0·34	1·08	0·63	0·24	0·39	0·21
FeO	1·2	1·0	2·0	0·8	2·5	4·2	1·4	1·3	4·5	2·9	0·1	1·5	0·8
MnO	0·13	0·05	0·06	0·04	0·04	0·1	0·02	0·04	0·1	0·07	0·05	0·04	0·05
MgO	0·11	0·24	0·23	0·26	1·42	4·71	0·66	1·61	4·37	1·13	0·14	0·43	0·17
CaO	3·76	2·34	0·85	1·53	3·46	5·49	2·63	3·94	6·41	3·18	0·65	1·96	1·07
Na_2O	3·59	5·84	2·98	5·38	2·59	2·14	2·57	2·0	2·41	2·68	4·41	2·66	3·19
K_2O	8·6	5·43	12·1	4·9	4·5	5·44	4·37	9·81	4·28	4·45	5·32	5·17	4·31
P_2O_5	0·03	0·02	0·02	0·01	0·13	0·16	0·08	0·16	0·17	0·16	0·02	0·05	0·0066
Cr_2O_3	0·005	0·005	0·005	0·005	0·005	0·03	0·006	0·005	0·02	0·005	0·005	0·005	0·005
LOI	0·65	0·6	0·4	0·65	0·85	1·05	1·05	1·8	1·1	0·95	0·45	0·5	1·4
H_2O+	0·5	0·3	0·7	0·2	0·7	0·8	1·0	0·4	1·0	1·0	0·6	0·5	0·6
Total	98·8	99·4	99·4	99·4	99·2	98·9	99·5	99·8	99·8	100·4	99·9	99·5	100·1
ASI	0·90	0·95	0·88	0·98	1·13	0·90	1·28	0·78	0·98	1·17	1·05	1·06	1·08
Mg#	0·04	0·11	0·08	0·13	0·32	0·48	0·26	0·50	0·44	0·24	0·29	0·19	0·14
Cl	347	30	157	69	200	165	25	194	124	68	25	25	25
F	1050	479	1520	10	930	271	478	10	1790	2350	117	561	441
B	47	45	29	37	24	20	31	27	22	27	381	25	35
S (%)	0·005	0·005	0·005	0·005	0·005	0·005	0·005	0·005	0·03	0·005	0·005	0·005	0·005
CO_2 (%)	0·06	0·3	0·01	0·4	0·23	0·06	0·005	0·39	0·32	0·08	0·06	0·005	0·01
V	38	23	15	21	32	91	21	40	83	47	−5	−5	11
Cr	−20	−20	−20	−20	30	272	−20	37	178	57	−20	−20	−20
Co	1	1	2	1	7	18	2	3	16	5	−1	1	2
Ni	22	−20	−20	−20	21	60	−20	23	65	136	−20	−20	−20
Cu	−10	−10	−10	−10	−10	−10	−10	−10	12	−10	−10	−10	−10
Zn	87	−30	108	−30	59	85	124	−30	89	100	−30	−30	33
Ga	22	24	32	27	20	19	22	19	20	23	28	20	17
Ge	1·2	1·2	1·5	1·3	1·2	1·4	1·3	2·8	1·5	2·1	5·0	2·2	1·3
As	7	−5	−5	−5	−5	5	10	−5	15	−5	−5	−5	−5
Rb	197	178	705	211	198	238	213	302	221	275	468	355	206
Sr	3380	1110	599	637	465	467	476	640	438	365	70	89	262
Y	11·4	17·2	30·4	15·5	18·1	22·9	4·6	32·6	23·0	20·9	58·0	45·9	18·9
Zr	175	414	518	287	224	204	152	296	191	198	29	78	135
Nb	12·8	22·9	40·3	24·0	14·2	15·5	13·9	22·6	14·0	16·4	26·4	18·5	12·1
Mo	−2	−2	2	−2	−2	−2	−2	−2	−2	−2	−2	−2	−2

Table 3 *Continued*

Sample	Deadman 1	Deadman 2	Deadman 3	Deadman 4	Dublin main phase	Dublin dyke 1	Dublin Qtz-syenite	Dublin pegmatite dyke	Dublin dyke 2	Mactung main stock	Mactung tourm peg	Mactung dyke in stock	Mactung dyke
Ag	−0·5	−0·5	−0·5	−0·5	−0·5	−0·5	−0·5	−0·5	−0·5	−0·5	−0·5	−0·5	−0·5
In	−0·1	−0·1	−0·1	−0·1	−0·1	−0·1	−0·1	−0·1	−0·1	0·1	−0·1	−0·1	−0·1
Sn	2	2	3	2	1	3	4	8	4	12	7	11	5
Sb	0·6	0·5	0·7	0·2	0·3	0·6	3·8	0·2	0·8	0·5	0·2	−0·2	0·5
Cs	5·0	3·5	67·6	3·4	7·5	6·6	11·2	8·4	8·1	27·9	21·4	16·7	14·2
Ba	8590	2760	857	1660	1380	1620	2930	3780	1370	993	56	119	603
La	34·0	43·4	95·4	41·9	49·4	59·6	22·5	62·1	51·6	55·2	12·4	15·3	56·2
Ce	69·7	75·5	164	71·9	88·4	99·4	39·4	111	94·7	103	28·7	31·1	100
Pr	9·12	7·69	15·8	7·01	9·22	10·4	4·09	12·3	9·88	11·2	3·69	3·74	11·5
Nd	36·8	25·3	49·9	22·1	32·0	36·4	14·3	44·9	34·2	38·8	15·3	14·6	40·7
Sm	6·16	4·57	8·63	3·88	5·86	6·42	2·68	9·18	6·17	6·99	7·17	5·22	7·62
Eu	1·43	1·12	1·69	0·894	1·28	1·34	1·01	1·61	1·25	1·46	0·175	0·433	1·38
Gd	3·57	3·37	5·60	2·84	4·35	4·34	1·55	6·70	4·58	4·82	7·79	6·06	5·69
Tb	0·42	0·57	1·02	0·48	0·68	0·78	0·22	1·13	0·76	0·75	1·86	1·39	0·74
Dy	1·80	2·87	5·00	2·40	3·31	3·99	0·88	5·81	3·89	3·60	9·46	7·48	3·49
Ho	0·29	0·54	0·96	0·48	0·62	0·76	0·13	1·11	0·76	0·69	1·72	1·52	0·63
Er	0·85	1·53	2·64	1·30	1·59	2·01	0·32	2·94	2·07	1·88	4·64	4·09	1·79
Tm	0·096	0·207	0·366	0·190	0·203	0·276	0·040	0·427	0·284	0·239	0·830	0·668	0·230
Yb	0·78	1·38	2·24	1·14	1·27	1·71	0·25	2·75	1·83	1·57	5·55	3·81	1·59
Lu	0·111	0·217	0·312	0·173	0·184	0·249	0·032	0·446	0·279	0·226	0·843	0·572	0·231
Hf	4·3	9·8	10·8	7·9	6·2	5·6	4·5	8·4	5·3	5·5	4·4	4·1	4·3
Ta	0·74	1·15	2·40	1·15	1·20	1·24	1·10	1·53	1·15	1·48	18·4	4·09	1·73
W	0·6	0·6	3·2	0·8	2·2	0·5	1·7	377	2·1	0·8	2·8	0·9	0·6
Tl	1·29	0·63	2·51	0·73	0·90	1·27	1·56	0·79	1·12	2·11	2·63	1·97	1·19
Pb	55	13	73	13	16	22	30	8	20	38	51	44	26
Bi	0·1	−0·1	−0·1	−0·1	−0·1	−0·1	0·2	−0·1	−0·1	1·4	0·5	−0·1	−0·1
Th	10·6	29·7	68·5	27·4	25·3	31·6	13·2	29·1	25·7	27·4	13·2	21·2	25·8
U	3·93	6·44	11·0	10·7	5·66	4·65	7·79	5·21	6·58	7·30	15·0	20·9	15·4
Zircon saturation temperatures (calculated using equations of Watson & Harrison 1983)													
M	2·08	1·88	1·71	1·53	2·18	2·41	1·29	2·18	2·18	1·50	1·33	1·39	1·26
Temp K	1018	1108	1144	1102	1055	1008	1063	1055	1055	1070	933	1369	1055
Temp C	745	835	871	829	781	735	790	781	781	797	660	1096	781

*Whole-rock data by XRF at XRAL-SGS, Canada. Trace element data by ICP-MS at Actlabs, Canada. Rocks were milled in agate. Major oxides as percentage, others p.p.m. FeO determined by titration. Fe_2O_3 is calculated from FeO total and FeO. Sulphur by leuco, Cl and F by specific ion. Negative values are less than detection limit.

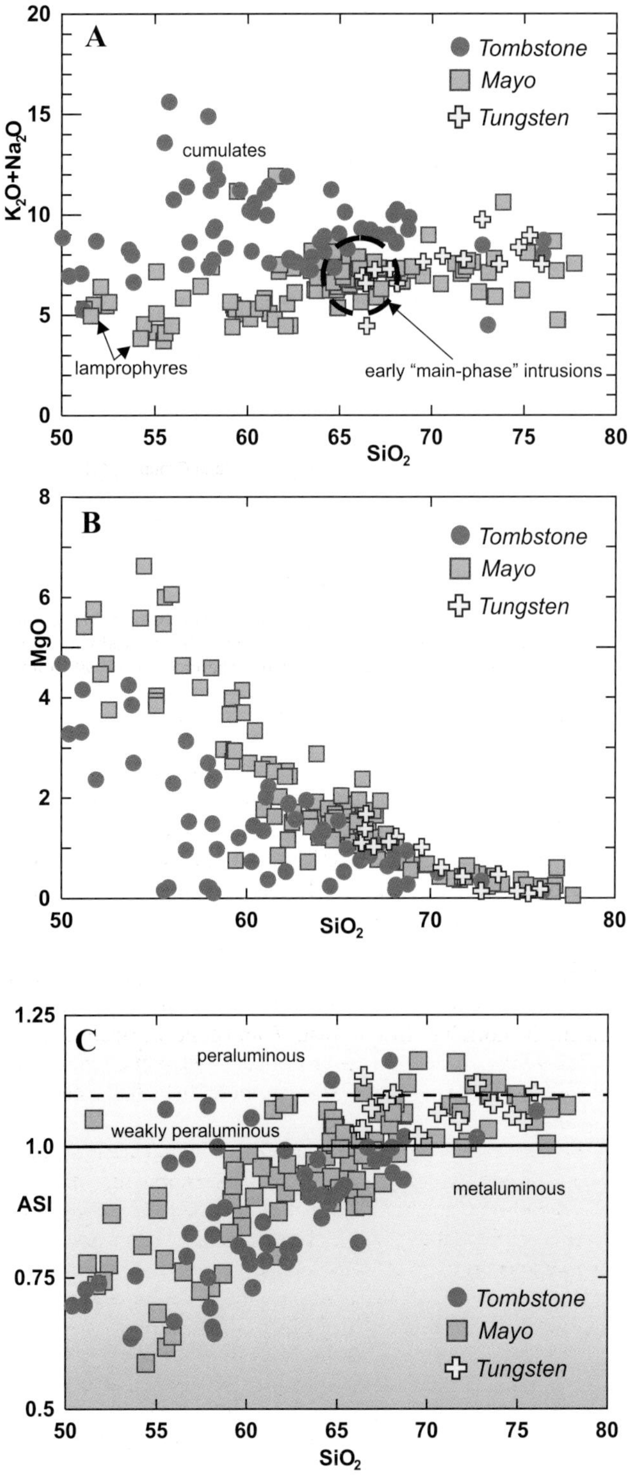

Figure 5 Geochemical plots for the Tombstone, Mayo and Tungsten plutonic suites. All geochemical plots were composed using selected data from C. Hart and J. Mair (unpublished results), Lang (2000) and Abercrombie (1990), and Table 3: (A) alkali-silica plot (ASI=molar Al/Ca+Na+K); (B) MgO-silica plot; and (C) ASI-silica plot.

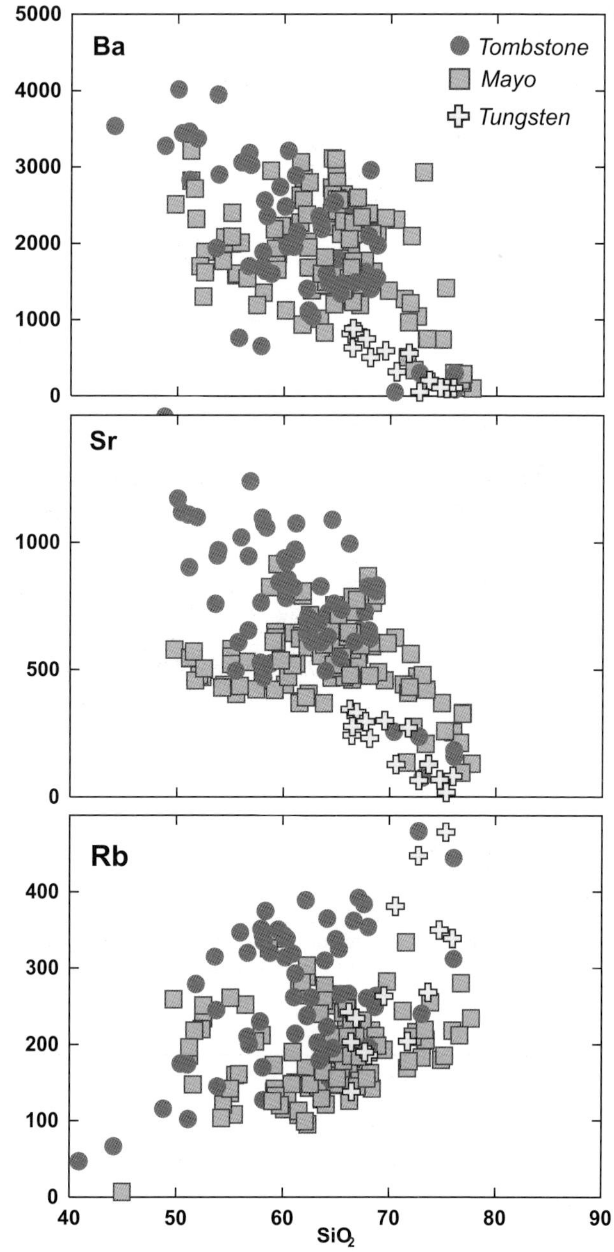

Figure 6 Variation diagrams of Ba, Sr and Rb versus silica for the Tombstone, Mayo and Tungsten plutonic suites emphasising varying source compositions and fractionation trends between the suites.

and ε_{Nd} values of −8·3 to −12·5 (Farmer *et al.* 2000; Lang 2000). Whole-rock oxygen isotope values have been determined only for Clear Creek intrusions and range from 9·5–13·9 per mil (Marsh *et al.* 2003; Mair 2004). Tungsten Suite intrusions have very high initial strontium isotope ratios of 0·712 to 0·748 (Godwin *et al.* 1980; Gareau 1986; Lang 2000), and two rocks have ε_{Nd} values of −13 to −15 (Lang 2000). Oxygen isotopes for Tungsten Suite rocks include values of 9–13 per mil for the Cantung Pluton (Bowman *et al.* 1985; Anderson 1988).

Both strontium and neodymium isotopic ratios are highly radiogenic in intrusions of all three plutonic suites. Notably, the more mafic and basic compositions, such as diorite, tinguaite and lamprophyres, also yield similar highly radiogenic values. However, Tombstone rocks may be slightly less radiogenic than Mayo Suite rocks, and Tungsten Suite rocks are significantly more radiogenic (Lang 2000). This suggests that the more easterly plutons of the Tungsten Suite have larger contributions from the middle to upper crust. Limited analyses of Neoproterozoic sedimentary rocks in the Selwyn Basin indicate initial strontium isotope ratios (at ~95 Ma) of 0·720–0·760 (Kuran *et al.* 1982) and ε_{Nd} values of ~ −16 to −25 (Garzione *et al.* 1997; Creaser and Erdmer 1997). Consequently, the parent magmas to the plutons appear to be composed of large proportions of highly radiogenic crustal melts, but the plutons are less radiogenic than the country rocks (Fig. 9).

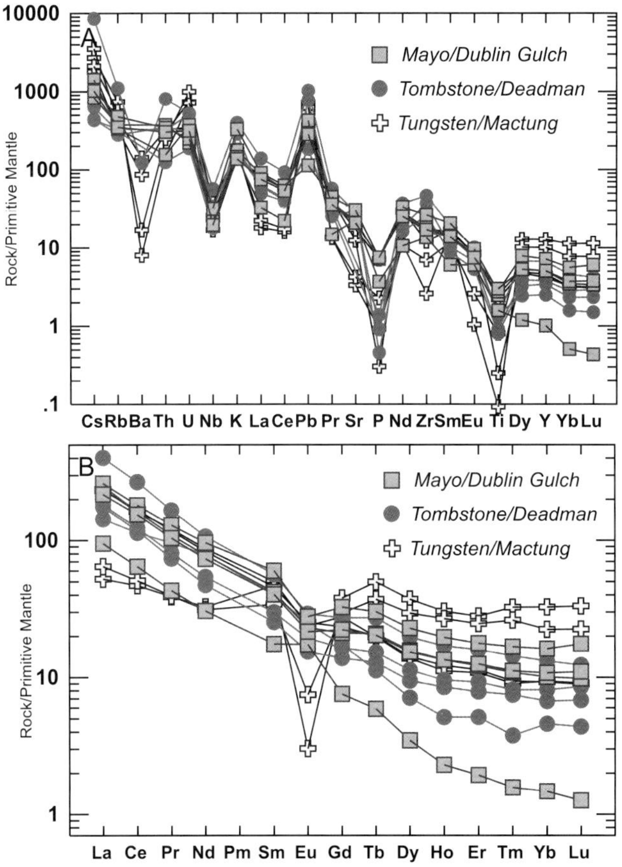

Figure 7 Multi-element variation diagrams for representative plutons of the Tombstone, Mayo and Tungsten plutonic suites. Data are normalised to the primitive mantle values of Sun & McDonough (1989): (A) LILE-data; and (B) rare-earth elements. Data from Table 3, and associated unpublished data.

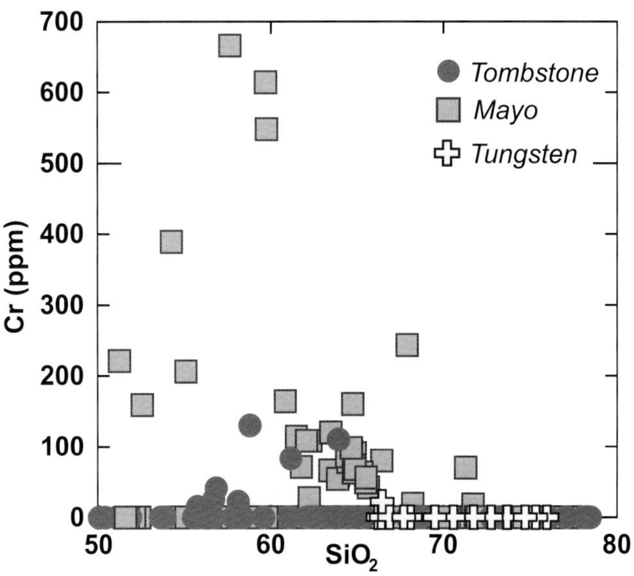

Figure 8 Plot of Cr (p.p.m.) versus SiO_2 for the Tombstone, Mayo and Tungsten plutonic suite granitoids emphasising the Cr-enrichment in the Mayo intrusions. High values are interpreted to come, at least in part, from xenocrystic Cr-diopside. Overlapping data points with low Cr are at or below the detection limit of 10 p.p.m.

7. Redox state

The oxygen fugacity of a granitic magma can reflect the redox state of the source region, as well as influence the processes of metal enrichment in the magma (Ishihara 1981; Candela 1989; Carmichael 1991; Blevin & Chappell 1992). Several features

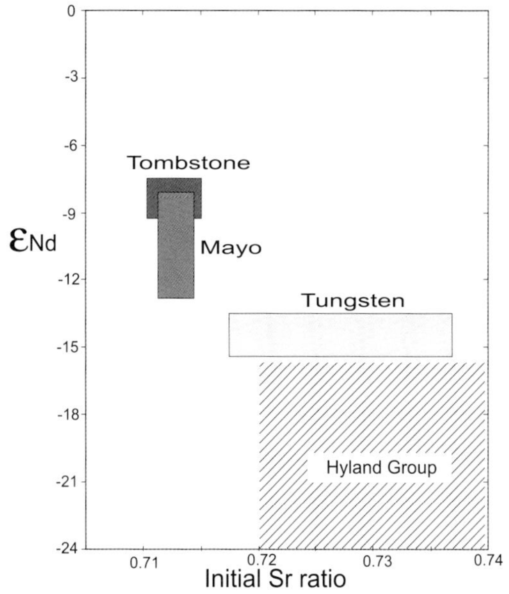

Figure 9 Plot of ε_{Nd} versus initial Sr ratio for Tombstone, Mayo and Tungsten plutonic suites using data compiled from Godwin *et al.* (1980), Kuran *et al.* (1982), Gareau (1986), Abercrombie (1990), Lang (2000) and Farmer *et al.* (2000). Hyland Group field taken from data of Garzione *et al.* (1997) and Creaser & Erdmer (1997) .

are indicative of the oxidation state of a pluton; these include its magnetic character, Fe-Ti mineralogy and Fe_2O_3/FeO ratio. The Mayo and Tungsten Suite intrusions have, for the most part, low and flat aeromagnetic signatures which are impossible to differentiate from the low background values of the sedimentary country-rock sequences (Hart *et al.* 2000). However, where these plutons intruded carbonaceous (i.e. reduced) sedimentary strata, they have contact metamorphic aureoles with elevated magnetic signatures, because of pyrrhotite development in hornfels. In contrast, Tombstone intrusions have an inherently zoned or heterogeneous magnetic character reflecting their varied lithologies, some of which do contain magnetite.

These aeromagnetic data are supported by magnetic susceptibility studies of the granitic rocks. Measurements using a KT-9 hand-held magnetic susceptibility meter on freshly broken rock surfaces indicate that the Tombstone Suite rocks have a higher average magnetic susceptibility ($1{\cdot}79 \times 10^{-3}$ SI) and a wider range (Fig. 10) than Mayo and Tungsten Suite rocks (average 0·11 and $0{\cdot}16 \times 10^{-3}$ SI, respectively).

The dominant Fe-Ti-bearing accessory mineral can also reflect the oxidation state of a magma and be characterised as magnetite or ilmenite-series (Ishihara 1977, 1981). Most of the intrusive rocks in the region are titanite-bearing, but some Tombstone rocks contain significant magnetite. In the Mayo intrusions, titanite is dominant and ilmenite is rare. Magnetite is generally restricted to late granite dykes and occurs as scattered grains in some mafic dykes. Ilmenite is common in the Tungsten Suite intrusions, whereas magnetite and titanite are absent.

Whole-rock ferric:ferrous ratios of fresh granitic rocks provide an approximation of the redox potential for rocks lacking subsolidus oxidation (Burkhard 1991). Generally, the intrusions have relatively low Fe_2O_3/FeO ratios, but a general trend of increased Fe_2O_3/FeO with increasing SiO_2 content suggests increased fO_2 with progressive magmatic differentiation for rocks of all three suites. The Tombstone intrusions have the highest Fe_2O_3/FeO ratios, whereas the Tungsten Suite intrusions have the lowest (Fig. 11). In the Tombstone intrusions, the higher proportion of Fe_2O_3/FeO is expressed mineralogically as Fe^{3+}-bearing silicates (andraditic garnet,

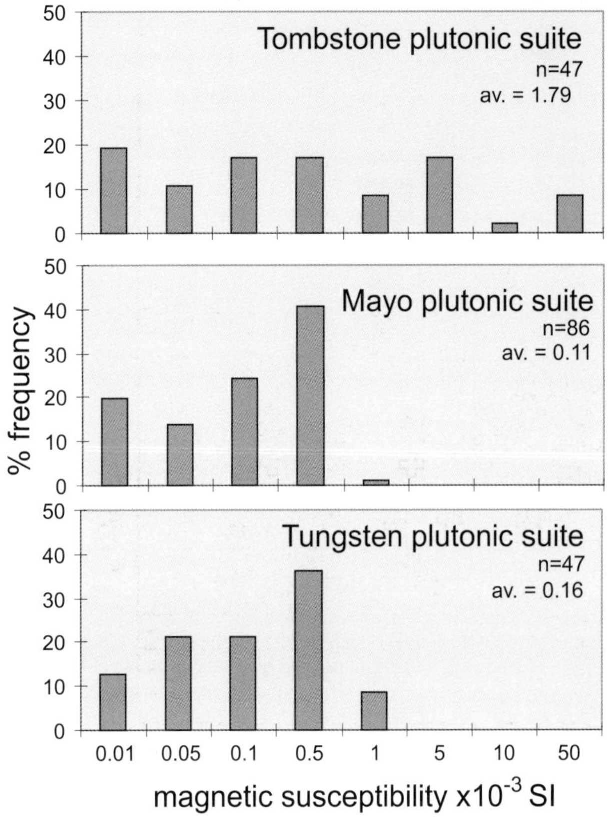

Figure 10 Histograms of magnetic susceptibilities for intrusive rocks of the Tombstone, Mayo and Tungsten plutonic suites. Measurements were made determining averages of between six and 10 readings of various, freshly broken rock surfaces of a single sample using a handheld KT-9 Kappameter with a pin spacer to account for surface roughness. Most plutons are represented by at least four or five samples.

hornblende), as well as local enrichments of magnetite. The least-felsic Tungsten Suite intrusions plot below the QFM buffer (Fig. 11), and pyrrhotite is locally present in some highly-fractionated Tungsten Suite aplitic phases. The Mayo intrusions are slightly reduced, with the greatest number of samples near the QFM buffer. Although many samples plot above the QFM buffer, most do not contain magnetite, which suggests that other parameters, in addition to oxygen fugacity, control the nature of the Fe-Ti minerals. Also notable is the dominance of titanite in the Mayo Suite rocks which are below the QFM buffer, which is uncommon in most igneous rocks.

8. Zircon character

Several authors have suggested that the presence and nature, or lack of, inherited zircon are useful guides for interpretations of the nature and thermal state of the source region for granitic melts (e.g. Watson & Harrison 1983; Chappell *et al.* 1987, 1998). Zircons in Tombstone Suite intrusions are large (to millimetre length), typically anhedral, and show only weak element zonation, with a notable absence of inherited cores (Fig. 12A). Zircons in rocks of the Mayo Suite are dominantly euhedral and contain variable proportions, as much as 10%, of ancient inherited cores which are overprinted with thick, oscillatory-zoned, euhedral magmatic zircon rims (Fig. 12B). Tungsten Suite intrusions contain zircons which are similar to those of the Mayo intrusions, but are smaller with thinner oscillatory-zoned rims, and contain a higher percentage of old inherited cores, many lacking overgrowths (Fig. 12C). Although the xenocrystic zircons may reflect the nature of the source material, their presence likely reflects low zircon saturation temperatures (Miller *et al.* 2003). Zircon saturation temperatures are modelled estimates of zircon solubility conditions in a melt that depends, in part, on melt composition and zirconium concentrations (Watson & Harrison 1983). It provides the best measurement of the temperature of a magma at its source (Miller *et al.* 2003). The Tombstone Suite intrusions have zircon saturation temperatures of about 820°C, which is about 40 and 70°C higher than average Mayo and Tungsten Suite temperatures, respectively.

9. Discussion

9.1. Source regions of magmas

Constraints on the source regions and redox states of each plutonic suite can be related to the contrasting element characteristics of their associated mineralisation.

The Tombstone intrusions have a wide range of silica content that is indicative of a protracted fractional crystallisation history. Variation diagrams of rubidium and strontium versus SiO_2 also support significant fractionation (Fig. 6). The lack of negative europium anomalies in the REE diagrams is interpreted to reflect the suppression of early plagioclase growth because of the high alkalinity or internal water pressure of the parental magma. Cumulate textures suggest that density separation of mineral phases has strongly influenced the compositional variations. All rocks are alkalic, with ultrapotassic compositions for those which host accumulations of potassium feldspar. Silica-undersaturated phases are feldspathoid-bearing. Non-cumulate mafic phases which contain approximately 50–55% SiO_2 are shoshonitic in composition. The strong enrichment of LILE, LREE and volatiles, and the high alkalinity, and high neodymium and strontium isotopic values, suggest that a parental mafic melt was derived from a lithospheric mantle source that had been enriched by metasomatism long before fusion and melt generation. However, the relatively low Mg#, and low chromium and nickel contents of these rock are best explained by early fractionation of olivine, pyroxene and spinels (Nelson *et al.* 1986).

The nature of the enriched components in intrusions of the Tombstone Suite suggests that metasomatism was subduction related, because the mafic rocks, as well as the more evolved intrusions, feature distinct negative Nb-Ti anomalies – a characteristic commonly attributed to melt generation in a subduction-related setting (Pearce 1982; Tatsumi & Eggins 1995). More-evolved phases probably resulted from fractionation, potentially with a degree of crustal assimilation. This is supported by U, Th and Zr contents which all increase significantly as SiO_2 increases from 50% to 60%. Rocks with >70% SiO_2 represent highly fractionated, quartz-rich intrusions, and at >75% SiO_2, are weakly peraluminous and locally contain tourmaline. Zircon saturation temperatures for the Deadman Pluton range from 750 to 850°C (Table 3), and the lack of inherited zircon suggests that any assimilation of crustal material occurred at temperatures >850°C. In addition, many of the zircons for the Deadman Pluton are anhedral, and are interpreted to have formed late in the crystallisation sequence, where they concentrated available uranium and thorium.

The Tungsten Suite intrusions have numerous characteristics to indicate that they are mainly derived from an ancient crustal source. The abundance of inherited zircon, the variable, although highly radiogenic, strontium and neodymium isotope values, the positive $\delta^{18}O$ values, the moderately to strongly reduced state of the rocks, a lack of associated intermediate and mafic phases, and the elevated ASI values are all consistent with a dominantly sedimentary-rock crustal source. However, the intrusions are only weakly peraluminous, are

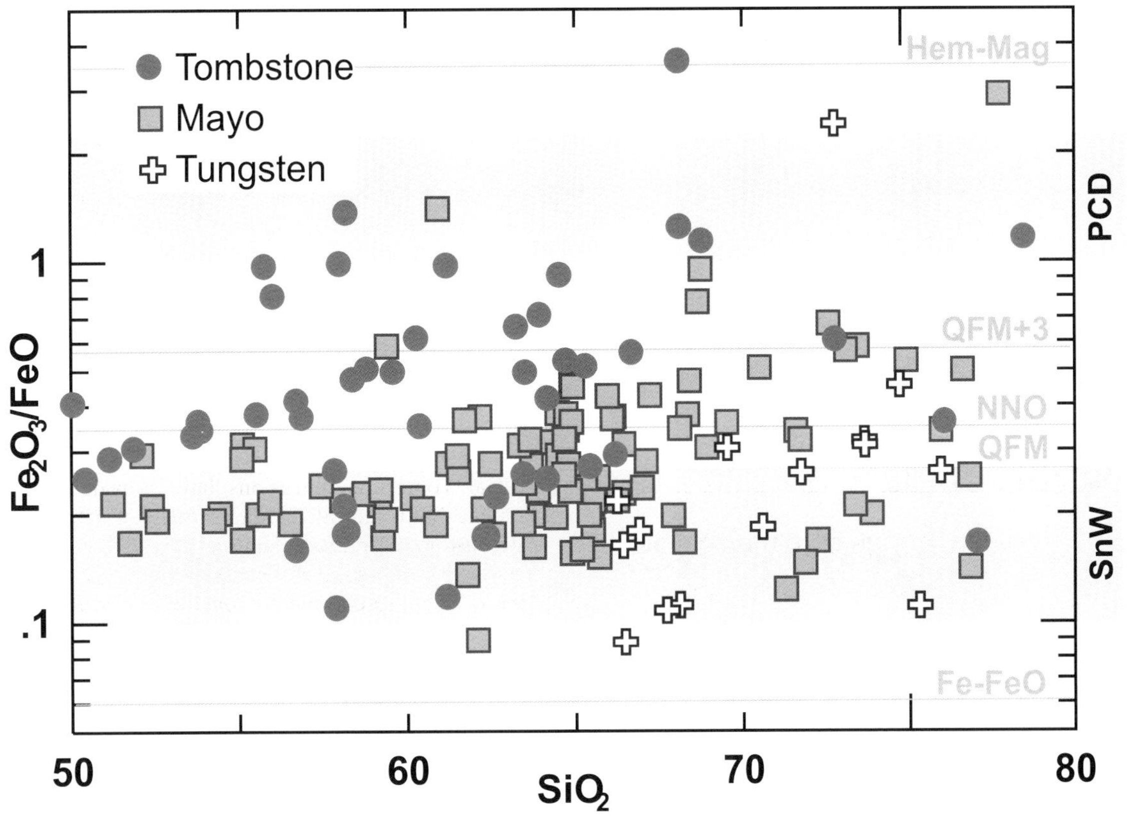

Figure 11 Changes in the oxidation state (Fe_2O_3/FeO ratio) of the Tombstone, Mayo and Tungsten intrusive rocks with fractionation, as represented by SiO_2. Note the increased oxidation state with fractionation for all three suites. The scale is in log units to give greater resolution at low levels. The anomalously elevated values probably reflect partial secondary oxidation from weathering. The approximate ranges for oxidation states of porphyry copper deposits (PCD) and tin-tungsten deposits (SnW) are indicated by grey shading. The calculated approximate positions of selected oxygen buffers are shown (NNO=nickel-nickel oxide; QFM=quartz-fayalite-magnetite; Hem-Mag=hematite-magnetite) and were derived using the MELTS (Ghiorso & Sack 1994) calculator for multi-component silicate liquids by inputting the absolute chemical compositions of all three suites, and constrained with T=850°C and P=2000 bars.

generally less peraluminous than typical S-type granites (cf. Chappell & White 2001), and are relatively rich in calcium and strontium for granitoids derived from metasedimentary rocks. These factors could be caused by either a mafic crustal component, or a significant volume of meta-carbonate (calc-silicate) rocks amongst the quartzo-feldspathic and pelitic strata in the source region. Deep seismic profiles of the region (Snyder *et al.* 2002) show that Proterozoic platform sedimentary rocks are imaged at the Moho boundary. It is likely that these strata were involved in melting to re-establish the Moho after depression through deep crustal isotherms in response to upper crustal thickening in the Late Jurassic to Early Cretaceous. Zircon saturation temperatures calculated for the main phases of the Mactung Pluton range from 770 to 800°C (Table 3). Inherited ancient zircon cores are abundant in the Mactung intrusion. The temperature probably corresponds to the point at which the inherited zircons were entrained, and does not necessarily rule out a more protracted petrogenesis.

Early, main-stage quartz monzonites to monzogranites of the Mayo Suite are geochemically similar to Tungsten Suite plutons (e.g. Fig. 5). They have positive $\delta^{18}O$ values, and evolved neodymium and strontium isotope characteristics. However, many early Mayo plutons contain at least a small proportion of hornblende, and some contain abundant clinopyroxene. This mineralogical difference, in combination with a lower proportion of inherited xenocrystic zircon, and an association with intermediate to mafic intrusions, indicates that either a mafic crustal component was included in the magma, or that early intrusions contained a component of both mantle- and crustally-derived material. In addition, Mayo Suite intrusions contain elevated chromium contents, and are generally enriched in strontium relative to the Tungsten Suite intrusions for a given silica content. Zircon saturation temperatures for felsic rocks of the Mayo Suite range from 770 to 790°C (Table 3), and the presence of inherited zircon suggests that they were incorporated during crustal melting close to, or slightly above, the temperatures estimated for zircon saturation. These zircon characteristics, along with the less pronounced fractionation trends in rubidium and strontium data, suggest that Mayo Suite intrusions had considerably more interaction with crustal rocks than did the bodies of the Tombstone Suite.

Mafic rocks of the Mayo Suite have similarities to some Tombstone Suite rocks in that they are alkalic in nature, volatile-rich and formed relatively late in the plutonic history of the region. However, the mafic rocks of the Mayo Suite typically have relatively higher Mg#s and contain higher chromium concentrations. These are related to the common occurrence of xenocrystic chromium-diopside, which indicates that these mafic Mayo phases are unlikely to have undergone considerable fractionation (Mair 2004). Intermediate lithological phases typically contain elevated chromium, and in some cases, have textures indicative of mixing mafic and felsic magmas. These phases probably result from the interaction between evolving felsic magmas in the crust and mantle-derived mafic magmas (Mair 2004).

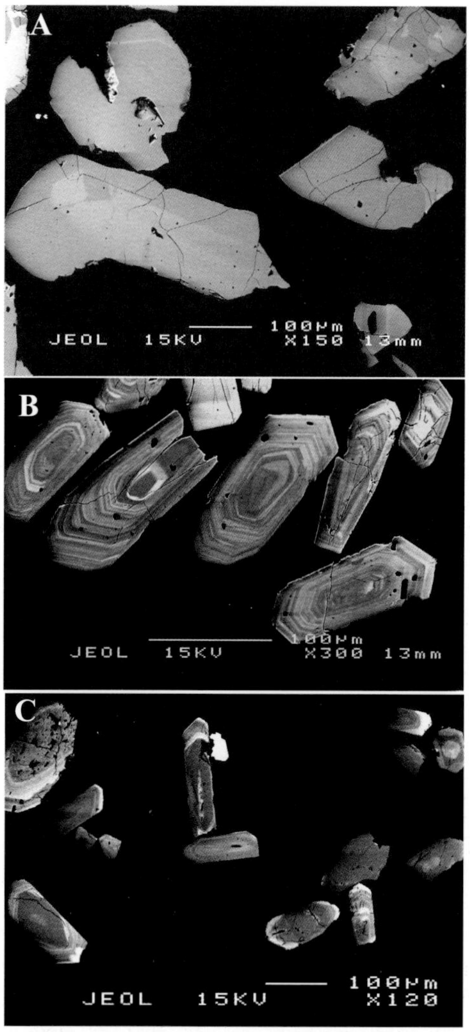

Figure 12 Backscatter scanning electron micrograph images of zircons: (A) Tombstone Suite zircons are large, mainly anhedral to subhedral and have faint internal variations; (B) Mayo Suite zircons have large, finely oscillatory rims on early cores, some of which are xenocrystic; and (C) Tungsten Suite zircons have thinner, oscillatory rims on a greater proportion of xenocrystic cores, but also include xenocrystic zircons lacking overgrowths.

Isotope data for rocks of the Mayo Plutonic Suite are intermediate between those for the Tombstone and Tungsten Suite plutons. The variations between the suites are interpreted to reflect the proportion of mantle-sourced material, with the Tombstone intrusions having the greatest component of material derived from a mantle source, followed by the Mayo Suite. The radiogenic nature of mantle-derived phases is attributed to ancient metasomatism of the lithospheric mantle, which also enriched LILE, LREE, volatiles and other incompatible phases (cf. Nelson *et al.* 1986; Carlson & Irving 1994). In addition, neodymium and strontium ratios may well have been shifted further to more-radiogenic values by increased crustal contributions. However, the values for mantle-derived rocks are considerably lower than ratios for both upper crustal strata of the Selwyn Basin and intrusions of the Tungsten Suite.

9.2. Redox characteristics of source materials

Intrusions of the TTB range from magnetite-dominant relatively oxidised, to ilmenite-dominant reduced bodies. These variations are interpreted to primarily reflect the varying proportion of magma contributed from different sources. The Selwyn Basin crustal rock package is dominated by marine siliciclastic rocks, some of which are basinal or off-shelf facies which contain graphite, and thus, can be considered as strong reductants (Ishihara 1977, 1981). In contrast, mantle-derived phases are typically relatively oxidised because of oxidising metasomatic agents which are introduced to the mantle wedge above subduction zones (cf. Carmichael *et al.* 1996). In mafic intrusions of the Mayo Suite, the presence of sparse magnetite, rare barite inclusions in mafic phenocrysts and high Mg# pyroxenes suggest that these rocks may have originally been oxidised. The Tombstone Suite rocks are the most alkalic, show the least evidence of crustal interactions, and are the most oxidised. Conversely, the isotope data suggest a greater component of middle to upper crustal material in the Tungsten Suite intrusions – these rocks will typically contain a higher proportion of reduced carbon, and therefore are the most reduced.

9.3. Relationships between the redox state of magmas and metallogeny

The associations between the oxidation state and the metallogenic expression of intrusion-related hydrothermal systems have been well documented (e.g. Ishihara 1977, 1981; Blevin & Chappell 1992). Ishihara (1977, 1981) indicated that the compatibility and partitioning behaviour of metals is strongly influenced by the oxidation state of a magma, which can be estimated based on the dominance of either magnetite or ilmenite. The chalcophile Cu ± Au association is generally associated with oxidised mantle-derived magmas generated in a subduction zone setting (Candela 1989). A high oxidation state is optimal for preventing early formation of sulphide globules in a magma, which can efficiently sequester chalcophile elements such as copper and gold (Burnham & Ohmoto 1980). Molybdenum enrichment is also favoured by oxidised conditions, but at a redox state lower than is typical of that associated with porphyry copper deposits (Ishihara 1981).

Numerous studies have indicated that the lithophile metal association of Mo, W and Sn is also strongly influenced by redox conditions (Ishihara 1981; Blevin & Chappell 1992; Candela 1995). Redox conditions influence the valency of some elements, which, in turn, determines the compatibility of such elements in accessory minerals. For example, tin enrichment in fractionating melts is favoured under reduced conditions, where it occurs mostly as Sn^{2+}, rather than as Sn^{4+}. In more oxidised magmas, Sn^{4+} is favoured, but readily substitutes for Fe^{3+} and Ti^{4+} in mafic silicate minerals and in Fe-Ti phases (such as titanite), thus preventing tin enrichment (Ishihara 1978). Tungsten enrichment is similarly favoured by reduced conditions, although associated intrusions span a broader range of redox environments (Ishihara 1981). At higher fO_2, tungsten has affinities with molybdenum and copper, being concentrated in minerals such as molydoscheelite (Sato 1980); under more reducing conditions, tungsten has affinities with tin (Ishihara 1981).

Other ore-associated elements which are variably enriched in the TTB systems are tellurium and bismuth. Bismuth is commonly strongly enriched in reduced gold and tungsten skarns, and by inference, the concentration of bismuth through magmatic processes appears to be favoured by an intermediate to reduced oxidation state (Meinert 1992). Conversely, tellurium is commonly enriched in epithermal mineralisation associated with oxidised alkalic intrusive complexes (Jensen & Barton 2000; Cooke & McPhail 2001), suggesting that its enrichment may be controlled by factors other than just redox effects. However, controls of bismuth and tellurium concentrations are equivocal since they are anomalous in a variety of different deposit styles which developed under contrasting redox conditions (Ciobanu *et al.* 2003).

The precise role that redox conditions play in generating gold-enriched magmas remains enigmatic. Although gold is a siderophile element, it has strong chalcophile tendencies, and the formation of early sulphide globules in the melt, particularly copper- and iron-bearing sulphides, could effectively reduce the gold concentration in the residual melt (Cygan & Candela 1995). As a consequence, magmas too oxidised to precipitate sulphide minerals may be more effective at concentrating gold during fractionation, and indeed, gold-enriched magmatic-hydrothermal systems are recognised as being associated with highly oxidised magmas (Candela 1989). However, other workers have suggested that gold in oxidised melts can be sequestered by magnetite; Tilling *et al.* (1973) reported values of >100 p.p.m. Au in magnetite. Supporting this concept are intrusions in central Alaska with Fe_2O_3/FeO ratios below 0·6, which typically lack magnetite and plot below the NNO buffer, and appear to be more prospective for gold mineralisation (Leveille *et al.* 1988; Newberry & Solie 1995; McCoy *et al.* 1997). Recently, however, sequestering of gold by magnetite has been shown experimentally to be only moderately efficient in felsic melts, even those that are gold-saturated (Cygan & Candela 1995; Simon *et al.* 2002, 2003). Despite these contradictory factors, gold is known to be highly incompatible in most silicate phases. Taken together, these factors indicate that redox state probably has a large influence on the partitioning of gold in the magmatic environment during fractionation.

The intrusions and associated mineral occurrences of the TTB show metal correlations which are mostly consistent with the redox associations outlined above. The relatively more-oxidised Tombstone Suite intrusions are associated with mineralisation dominated by Cu-Au ($\pm$ Bi) and U-Th signatures. Although the Tombstone intrusions are considerably less oxidised than those related to typical porphyry copper deposits, they are sufficiently oxidised to prevent early sulphide formation in the melt, into which copper and gold could partition (Cygan & Candela 1995). Uranium-Th mineralisation is interpreted to result from fluids generated during extended fractionation of undersaturated melts, which, at high temperatures, suppresses zircon crystallisation and uranium sequestering.

The Tungsten Suite intrusions are reduced and associated with large tungsten deposits which also have minor enrichments of Cu, Mo, Zn and Sn. Tungsten enrichment in the melts is favoured by the precipitation of ilmenite over titanite since W^{4+} may substitute for Ti^{4+} in titanite because of their common valency (Ishihara 1978). The lack of extensive tin mineralisation may be caused by insufficient fractionation of Tungsten Suite intrusions, stripping of Sn^{2+} from the melt by biotite crystallisation or a lack of tin in the melt source area. North America, as a whole, is anomalously lacking in significant tin deposits (Ishihara 1981; Taylor 1979).

The Mayo intrusions are generally characterised by an intermediate oxidation state. They are only slightly more oxidised than intrusions of the Tungsten Suite and slightly more reduced than those of the Tombstone Suite. Consequently, they have a metal association that is broadly intermediate between that of the other suites, such that many Mayo intrusions are predominantly associated with gold deposits, but also includes subordinate tungsten mineralisation. The Mayo Suite-related gold deposits are characteristically associated with enrichments of Bi, Te and As, and these ore systems are also generally copper-poor. In addition, tungsten mineralisation is widespread in the Mayo intrusions, but unlike deposits associated with Tungsten Suite intrusions, high tonnage resources are not recognised.

10. Conclusions

Among the numerous potential controls on the regional metallogeny of intrusion-related ore systems, the nature of the source rocks and the corresponding redox state of the magmas are among the most important. The TTB forms a single magmatic belt that was emplaced within the ancient North American continental margin in the mid-Cretaceous. Magmatism post-dates a period of terrane collision and crustal thickening. The nature of the plutons and their associated mineral deposits varies considerably across the TTB, becoming more reduced and radiogenic, and tungsten-rich/gold-poor to the southeast. On the basis of these variations, the belt is divisible into three plutonic suites, which, from west to east, are the Tombstone, Mayo and Tungsten suites. The plutons are compositionally variable, but comprise dominantly granites to monzonites. There is a general trend from more metaluminous intrusions in the north-west (Tombstone Suite) to peraluminous intrusions in the south-east (Tungsten Suite). All plutons have high initial strontium isotope ratios (0·710–0·730) and high whole-rock oxygen isotope values (+11 to +13 $\delta^{18}O$), and most plutons have low magnetic susceptibilities and low primary oxidation state (Fe_2O_3/FeO ratio ~0·2–0·3).

The Tombstone Suite intrusions are alkalic and metaluminous, and include silica-saturated and undersaturated phases which result from fractionation and the effects of density separation of minerals to form cumulates. These relatively oxidised plutons have accessory magnetite, and have associated Au-Cu-Bi and U-Th metallogeny.

Plutons of the Tungsten Suite are peraluminous and have many features of S-type granitoids, such as locally abundant muscovite, garnet and tourmaline. These plutons are reduced since they lack magnetite and titanite, but contain minor ilmenite and locally pyrrhotite. The plutons have significant xenocrystic zircon. Plutons of the Tungsten Suite are associated with world-class tungsten deposits, but lack association with significant gold occurrences.

Intrusions of the Mayo Suite have some characteristics which are difficult to reconcile within the generalisations of 'granite-series' models (e.g. Ishihara 1981; Blevin *et al.* 1996). They are characterised by titanite as the dominant iron- and titanium-bearing accessory mineral, whereas magnetite and ilmenite are typically too scarce to be diagnostic. As such, they are too reduced to be magnetite-series, but the titanite suggests the magmas are too oxidised to be considered ilmenite-series. They have characteristics which support neither an I-type nor an S-type designation, in the sense of Chappel & White (1992). They are prodigious mineralisers despite being weakly reduced, highly radiogenic and locally peraluminous, low-temperature granites. However, the plutons commonly contain amphibole and some contain clinopyroxene. Although felsic intrusions are volumetrically dominant, they are intimately associated with intermediate and mafic intrusions considered to represent hybrid phases developed through mixing with mafic magmas. The Mayo Suite intrusions host significant gold mineralisation, with elevated Bi, As, W and Te.

The distinctive lithological, geochemical and isotopic parameters which characterise each plutonic suite are interpreted to reflect the varying proportions of different source materials. The main sources are sialic crust, and a mantle source that is strongly enriched in LILE, LREE, volatiles and other incompatible phases, and locally contains depleted element enrichments. The influence of the crustal source gradually increases from north-west to south-east across central Yukon Territory, such that most granites in the south-east crystallised from melts which had a predominantly crustal source. The different source regions had a strong primary influence on the oxidation states of the parent magmas. Redox effects probably controlled the behaviour of metals during fractionation of the magmas, either promoting or inhibiting their residual concentration. Although the magma compositions, redox state and nature of associated mineralisation of TTB intrusions are relatively compatible with existing models of intrusion-related ore systems (e.g. Ishihara 1981; Blevin *et al.* 1996), it is remarkable for such variation in intrusion compositions and metallogeny to occur along a single contemporaneous magmatic belt.

11. Acknowledgements

Lara Lewis compiled the magnetic susceptibility data and assisted in the figure preparation. Continued support from the Yukon Geological Survey is appreciated by Craig Hart, and Rich Goldfarb recognises the support of the U.S. Geological Survey's Minerals Program. Previous Hutton Symposium volumes provided much of the basis for the senior author's granite acumen, and works by Shunso Ishihara and Phil Blevin established the foundations for integrating magmas and metallogenic processes. We thank K. Sato and an anonymous reviewer for their time and effort in reviewing this manuscript. The editorial assistance and patience of Vicki Ingpen is appreciated. This manuscript represents a portion of the senior author's Ph.D. dissertation at the Centre for Global Metallogeny at the University of Western Australia.

12. References

Abbott, J. G. 1995. Dawson Fault, a periodically reactivated Windemere-age rift transform. *Geological Society of America Cordilleran Division Abstracts with Program* **27**, 1.

Abbott, J. G., Gordey, S. P. & Tempelman-Kluit, D. J. (1986) Setting of sediment-hosted stratiform lead-zinc deposits in Yukon and northeastern British Columbia. *In* Morin, J. A. (ed.) *Mineral Deposits of the Northern Cordillera. Institute of Mining and Metallurgy, Special Volume* **37**, 1–18.

Abercrombie, S. M. 1990. *Petrology, geochronometry and conomic geology: the Zeta tin-silver prospect, Arsenic Ridge, west-central Yukon (115P/14 and 116A/03)*. M.Sc. thesis. Vancouver: The University of British Columbia.

Anderson, R. G. 1982. Geology of the Mactung pluton in Niddery Lake map area and some of the plutons in Nahanni map area, Yukon Territory and District of Mackenzie. *Current Research, Part A, Geological Survey of Canada Paper* **82-1A**, 299–304.

Anderson, R. G. 1983. Selwyn Plutonic suite and its relationship to tungsten skarn mineralization, southeastern Yukon and District of Mackenzie. *Current Research, Part B, Geological Survey of Canada, Paper* **83-1B**, 151–63.

Anderson, R. G. 1987. Plutonic rocks in the Dawson map area, Yukon Territory. *Current Research, Part A, Geological Survey of Canada, Paper* **1987–1A**, 689–97.

Anderson, R. G. 1988. An overview of some Mesozoic and Tertiary plutonic suites and their associated mineralization in the northern Canadian Cordillera. *In* Strong, D. F. & Taylor R. P. (eds) *Recent Advances in the Geology of Granite-related Mineral Deposits. Canadian Institute of Mining and Metallurgy, Special Volume* **39**, 96–113.

Armstrong, R. L. 1988. Mesozoic and early Cenozoic magmatic evolution of the Canadian Cordillera. *In* Clark, S. P., Jr, Burchfiel, B. C. & Suppe J. (eds) *Processes in Continental Lithospheric Deformation. Geological Society of America Special Paper* **218**, 55–92.

Atkinson, D. & Baker, D. J. 1986. Recent developments in the geologic understanding of MacTung. *In* Morin, J. A. (ed.) *Mineral Deposits of the Northern Cordillera. Canadian Institute of Mining and Metallurgy, Special Volume* **37**, 234–44.

Baker, T. & Lang J. R. 2001. Fluid inclusion characteristics of intrusion-related gold mineralization, Tombstone-Tungsten magmatic belt, Yukon Territory, Canada. *Mineralium Deposita* **36**, 563–82.

Barton, M. D. 1996. Granitic magmatism and metallogeny of southwestern North America. *Transactions of the Royal Society of Edinburgh: Earth Sciences* **87**, 261–80.

Blevin, P. L., Chappell, B. W. & Allen V. M. 1996. Intrusive metallogenic provinces in eastern Australia based on granite source and composition. *Transactions of the Royal Society of Edinburgh: Earth Sciences* **87**, 281–90.

Blevin, P. L. & Chappell, B. W. 1992. The role of magma sources, oxidation states and fractionation in determining the granite metallogeny of eastern Australia. *Transactions of the Royal Society of Edinburgh: Earth Sciences* **83**, 305–16.

Bowman, J. Covert, J. J., Clark, A. H. & Mathieson, G. A. 1985. The CanTung E zone scheelite skarn orebody, Tungsten, Northwest Territories: oxygen, hydrogen, and carbon isotope studies. *Economic Geology* **80**, 1872–95.

Brown, I. J. & Nesbitt, B. E. 1987. Gold-copper-bismuth mineralization in hedenbergitic skarn, Tombstone Mountains, Yukon. *Canadian Journal of Earth Sciences* **24**, 2362–72.

Brown, V. S., Baker, T. & Stephens, J. R. 2002. Ray Gulch tungsten skarn, Dublin Gulch, central Yukon: gold-tungsten relationships in intrusion-related ore systems and implications for gold exploration. *In* Emond, D. S., Weston, L. H. & Lewis, L. L. (eds) *Yukon Exploration and Geology 2001*, 259–68. Whitehouse, Yukon: Exploration and Geological Services Division, Indian and Northern Affairs Canada.

Burkhard, D. J. M. 1991. Temperature and redox path of biotite-bearing intrusives: a method applied to S- and I-type granites from Australia. *Earth and Planetary Science Letters* **104**, 89–98.

Burnham, C. W. & Ohmoto, H. 1980. Late-stage processes of felsic magmatism. *Mining Geology Special Issue* **8**, 1–12.

Candela, P. A. 1989. Felsic magmas, volatiles and metallogenesis. *In* Whitney, J. A. & Naldrett, A. J. (eds) *Ore Deposits Associated With Magmas. Reviews in Economic Geology* **4**, 223–33.

Candela, P. A. 1995. Felsic magmas, volatiles and metallogenesis. *In* Thompson, J. F. H. (ed.) *Magmas, Fluids and Ore Deposits. Mineralogical Association of Canada Short Course Series* **23**, 223–33.

Carlson, R. W. & Irving, A. J. 1994. Depletion and enrichment history of the subcontinental lithospheric mantle: An Os, Sr, Nd and Pb isotopic study of ultramafic xenoliths from the northwestern Wyoming Craton. *Earth and Planetary Science Letters* **126**, 457–72.

Carmichael, I. S. E. 1991. The redox state of basic and silicic magmas: a reflection of their source regions? *Contributions to Mineralogy and Petrology* **106**, 129–41.

Carmichael, I. S. E., Lange, R. A. & Luhr, J. F. 1996. Quaternary minettes and associated volcanic rocks of Mascota, western Mexico: a consequence of plate extension above a subduction modified mantle wedge. *Contributions to Mineralogy and Petrology* **124**, 302–33.

Chappell, B. W. 1996. Compositional variation within granite suites of the Lachlan fold Belt: its cause and implications of the physical state of granite magma. *Transactions of the Royal Society of Edinburgh: Earth Sciences* **88**, 159–70.

Chappell, B. W., White, A. J. R. & Wyborn, D. 1987. The importance of residual source material (restite) in granite petrogenesis. *Journal of Petrology* **28**, 1111–38.

Chappell, B. W., Bryant, C. J., Wyborn, D., White, A. J. R. & Williams, I. S. 1998. High- and low-temperature I-type granites. *Resource Geology* **48**, 225–35.

Chappell, B. W. & White, A. J. R. 1992. I- and S-type granites in the Lachlan Fold belt. *Transactions of the Royal Society of Edinburgh: Earth Sciences* **83**, 1–26.

Chappell, B. W. & White, A. J. R. 2001. Two contrasting granite types: 25 years later. *Australian Journal of Earth Sciences* **48**, 489–99.

Ciobanu, C. L., Cook, N. J., Bogdanov, K., Kiss, O. & Vucovic, B. 2003. Gold enrichment in deposits of the Banatitic magmatic and metallogenic belt, southeastern Europe. *In* Eliopoulus, D. G. *et al.* (eds) 7th Biennial SGA Meeting, Mineral Exploration and Sustainable Development, Athens, 1153–6. Rotterdam: Millpress Science Publishers.

Cooke, D. R. & McPhail, D. C. 2001. Epithermal Au-Ag-Te mineralisation, Acupan, Baguio district, Philippines: numerical simulations of mineral deposition. *Economic Geology* **96**, 109–31.

Coulson, I. M., Villeneuve, M., Dipple, G. M., Duncan, R., Russell, J. K. & Mortensen, J. K. 2002. Time-scales of assembly and thermal history of a composite felsic pluton: constraints from the Emerald Lake area, northern Canadian Cordillera, Yukon. *Journal of Volcanology and Geothermal Research* **114**, 331–56.

Creaser, R. A. & Erdmer, P. 1997. Mixed signals from the miogeocline: geochemical and Nd isotopic constraints from the Selwyn Basin. *In* Cook, F. & Erdmer, P. (eds) *Lithoprobe Snorcle Workshop, Lithoprobe Report* **56**, 74–5.

Cygan, G. L. & Candela, P. A. 1995. Preliminary study of gold partitioning among pyrrhotite, pyrite, magnetite, and chalcopyrite at 600 to 700°C, 140 MPa (1400 bars). *In* Thompson, J. F. H. (ed.) *Magmas, Fluids and Ore Deposits. Mineralogical Association of Canada Short Course* **23**, 129–37.

DePaolo, D. J., Perry, F. V. & Baldridge, W. S. 1992. Crustal versus mantle sources of granitic magmas: a two parameter model based on Nd isotopic studies. *Transactions of the Royal Society of Edinburgh: Earth Sciences* **83**, 439–46.

Dick, L. A. & Hodgson, C. J. 1982. The MacTung W-Cu-(Zn) contact metasomatic and related deposits of the northeastern Canadian Cordillera. *Economic Geology* **77**, 845–67.

Driver, L. A., Creaser, R. A., Chacko, T. & Erdmer, P. 2000. Petrogenesis of the Cretaceous Cassiar batholith, Yukon-British Columbia, Canada: implications for magmatism in the North American Cordilleran Interior. *Geological Society of America Bulletin* **112**, 1119–33.

Duncan, R. A., Russell J. K., Hastings N. L. & Anderson R. G. 1998. Relationships between chemical composition, physical properties and geology of the mineralized Emerald Lake pluton, Yukon Territory. *Geological Survey of Canada Current Research 1998-A* 1–11.

Farmer, G. L., Mueller, S., Marsh, E., Goldfarb, R. J. & Hart, C. J. R. 2000. Isotopic evidence on sources of Au-related mid-Cretaceous Tombstone Plutonic Suite granitic rocks, Clear Creek district, Yukon. *Geological Society of America, Cordilleran Section Abstracts with Programs* A-13.

Gabrielse, H., Monger, J. W. H., Wheeler, J. O. & Yorath, C. J. 1991. Morphogeological belts, tectonic assemblages, and terranes. *In* Gabrielse, H. & Yorath, C. J. (eds) *Geology of the Cordilleran Orogen in Canada. Geological Survey of Canada, Geology of Canada* **4**, 15–28.

Gabrielse, H. & Yorath, C. J. 1991. Tectonic synthesis. *In* Gabrielse, H. & Yorath, C. J. (eds) *Geology of the Cordilleran Orogen in Canada. Geological Survey of Canada. Geology of Canada* **4**, 97–124.

Gareau, S. A. 1986. *Petrology and geochronology of the Gun Claim Pluton and its aureole, eastern Selwyn Basin, Yukon Territory*. B.Sc thesis. Vancouver: The University of British Columbia.

Garzione, C. N., Patchett, P. J., Ross, G. & Nelson, J. 1997. Provenance of sedimentary rocks in the Canadian Cordilleran miogeocline: an Nd isotopic study. *Canadian Journal of Earth Sciences* **34**, 1603–18.

Gerstner, M. R, Bowman, J. R. & Pasteris, J. D. 1989. Skarn formation at the Macmillan Pass tungsten deposit (MacTung), Yukon and Northwest Territories, I. P-T-X-V characterization of the methane-bearing skarn-forming fluids. *Canadian Mineralogist* **27**, 545–63.

Ghiorso, M. S., & Sack, R. O. 1994. Chemical mass transfer in magmatic processes. IV. A revised and internally consistent thermodynamic model for the interpolation and extrapolation of liquid-solid equilibria in magmatic systems at elevated temperatures and pressures. *Contributions to Mineralogy and Petrology* **119**, 197–212.

Godwin, C. I., Armstrong, R. I. & Thompson, K. M. 1980. K-Ar and Rb-Sr dating and the genesis of tungsten at the Clea tungsten skarn property, Selwyn Mountains, Yukon Territory. *Canadian Institute of Mining and Metallurgy Bulletin* **73**, 90–3.

Goldfarb, R. J., Hart, C. J. R., Miller, L., Miller, M., & Groves, D. I. 2000, Tintina Gold Belt – a global perspective. *In The Tintina Gold Belt: Concepts, Exploration, and Discoveries. British Columbia and Yukon Chamber of Mines Special Volume* **2**, 5–34.

Gordey, S. P. & Anderson, R. J. 1993. Evolution of the Northern Cordilleran miogeocline, Nahanni map area (105I), Yukon and Northwest Territories. *Geological Survey of Canada, Memoir* **428**, 1–214.

Gordey, S. P. & Makepeace, A. 2003. *Yukon Digital Geology: Yukon Geological Survey Open File 2003–9D*. [Two CD-ROMS.]

Hart, C. J. R. 1997. A transect across Stikinia: Geology of the northern Whitehorse map area, southern Yukon Territory (105D13–16). *Exploration and Geological Services Division, Yukon, Indian and Northern Affairs Canada Bulletin* **8**, 1–112.

Hart, C. J. R., Baker, T. & Burke, M. J. 2000. New exploration concepts for country-rock-hosted intrusion-related gold systems: Tintina Gold Belt in Yukon. *In The Tintina Gold Belt: Concepts, Exploration, and Discoveries. British Columbia and Yukon Chamber of Mines Special Volume* **2**, 145–71.

Hart, C. J. R., McCoy, D. T., Goldfarb, R. J., Smith, M., Roberts, P., Hulstein, R., Bakke, A. A., & Bundtzen, T. K. 2002. Geology, exploration and discovery in the Tintina Gold Province. *In* Goldfarb R. J. & Neilson, R. (eds) *Geology, Exploration and Discovery in the Tintina Gold Province, Alaska and Yukon. Society of Economic Geologists Special Volume* **9**, 241–74.

Hart, C. J. R., Goldfarb, R. J., Lewis, L. L. & Mair, J. L. 2004a. The northern Cordilleran mid-Cretaceous Plutonic Province: ilmenite/magnetite-series granitoids and intrusion-related mineralization. *Resource Geology* **54**, 253–80.

Hart, C. J. R., Villeneuve, M. J., Mair, J. L., Goldfarb, R. J., Selby, D., Creaser., R. A. & Wijns, C. 2004b. The duration of magmatic-hydrothermal ore systems: comparative U-Pb SHRIMP & TIMS, Re-Os & Ar-Ar geochronology of mineralizing plutons in Yukon and Alaska. *In* Cassidy, K. F. *et al.* (eds) *Cutting-Edge Developments in Economic Geology, Abstract Volume*. Perth: Society of Economic Geologists.

Hitchins, A. C. & Orssich, C. N. 1995. The Eagle zone gold-tungsten sheeted vein porphyry deposit and related mineralization, Dublin Gulch, Yukon Territory. *In* Schroeter, T. A. (ed.) *Porphyry Deposits of the Northwestern Cordillera of North America. Canadian Institute of Mining and Metallurgy Special Volume* **46**, 803–10.

Ishihara, S. 1977. The magnetite-series and ilmenite-series granitic rock. *Mining Geology* **27**, 293–305.

Ishihara, S. 1978. Metallogenesis in the Japanese island-arc system. *Journal of the Geological Society* **135**, 389–406.

Ishihara, S. 1981. The granitoid series and mineralisation. *Economic Geology 75th Anniversary Volume* 458–84.

Jensen, E. P. & Barton, M. D. 2000. Gold deposits related to alkaline magmatism. *In* Hagemann, S. G. & Brown, P. E. (eds) *Gold in 2000. Reviews in Economic Geology* **13**, 279–314.

Kuran, V. M., Godwin, C. I. & Armstrong, R. L. 1982. Geology and geochronometry of the Scheelite Dome tungsten-bearing skarn property, Yukon Territory. *Canadian Institute of Mining and Metallurgy Bulletin* **75**, 137–42.

Lang, J. R. 2000. Regional and system-scale controls on the formation of copper and/or gold magmatic-hydrothermal mineralization. *Final Technical Report, Mineral Deposit Research Unit* 115 pp.

Lang, J. R., Baker, T., Hart, C. J. R. & Mortensen, J. K. 2000, An exploration model for intrusion-related gold systems. *Society of Economic Geology Newsletter* **40**, 6–15.

Lang, J. R. & Titley, S. R. 1998. Isotopic and geochemical characteristics of Laramide magmatic systems in Arizona and implications for the genesis of porphyry copper deposits. *Economic Geology* **93**, 138–70.

Leveille, R. C. A., Newberry, R. J. & Bull, K. F. 1988. An oxidation state-alkalinity diagram for discriminating some gold-favourable plutons: an empirical and phenomenological approach. *Geological Society of America Abstracts with Programs* **20**, A142.

Lindsay, M., Baker, T., Diment, R., Hart, C. J. R. & Oliver, N. 2000. The magmatic and structural setting of the Brewery Creek Gold mine, central Yukon. *In* Emond, D. & Weston, L. (eds) *Yukon Exploration and Geology 1999. Exploration and Geological Services Division, Indian and Northern Affairs Canada* 219–28.

Lynch, G. J. V., Longstaffe, F. J. & Nesbitt, B. E. 1990. Stable isotopic and fluid inclusion indications of large-scale hydrothermal paleoflow, boiling, and fluid mixing in the Keno Hill Ag-Pb-Zn district, Yukon Territory, Canada. *Geochimica et Cosmochimica Acta* **54**, 1045–59.

Maloof, T. L., Baker, T. & Thompson, J. F. H. 2001. The Dublin Gulch intrusion-hosted gold deposit, Tombstone plutonic suite, Yukon Territory, Canada. *Mineralium Deposita* **36**, 583–93.

Mair, J. L. 2004. Tectonic setting, magmatism and magnetic-hydrothermal systems at Scheelite Dome, Tombstone Gold Belt, Yukon: Critical constraints on intrusion-related gold systems. PhD thesis. Perth: The University of Western Australia.

Mair, J. L., Hart, C. J. R., Goldfarb, R. J., O'Dea, M. & Harris, S. 2000. Structural controls on mineralization at the Scheelite Dome gold prospect. *In* Emond, D. & Weston, L. (eds) *Yukon Exploration and Geology 1999. Exploration and Geological Services Division, Indian and Northern Affairs Canada* 165–76.

Marsh, E. E., Goldfarb, R. J., Hart, C. J. R. & Johnson, C. J. 2003. Geology and geochemistry of the Clear Creek intrusion-related gold occurrences, Tintina Gold Province, Yukon, Canada. *Canadian Journal of Earth Sciences* **40**, 681–99.

McCoy, D., Newberry, R. J., Layer, P., DiMarchi, J. J., Bakke, A., Masterman, J. S. & Minehane, D. L. 1997. Plutonic-related gold deposits of interior Alaska. *In* Goldfarb, R. J. & Miller, L. (eds) *Mineral Deposits of Alaska. Economic Geology Monograph* **9**, 191–241.

Meinert, L. D. 1992. Skarns and skarn deposits. *Geoscience Canada* **19**, 145–62.

Miller, C. F., Meschter McDowell, S. & Mapes, R. W. 2003. Hot and cold granites? Implications of zircon saturation temperatures and preservation of inheritance. *Geology* **31**, 529–32.

Mortensen, J. K., Murphy, D. C., Hart, C. J. R. & Anderson, R. G. 1995. Timing, tectonic setting, and metallogeny of Early and mid-Cretaceous magmatism in Yukon Territory. *Geological Society of America, Abstracts with Programs* **27**, 65.

Mortensen, J. K., Hart, C. J. R., Murphy, D. C. & Heffernan, S. 2000. Temporal evolution of early and mid-Cretaceous magmatism in the Tintina Gold Belt. *The Tintina Gold Belt: Concepts, Exploration, and Discoveries. British Columbia and Yukon Chamber of Mines Special Volume* **2**, 49–57.

Murphy, D. C. 1997. Geology of the McQuesten River Region, Northern McQuesten and Mayo Map Area, Yukon Territory, (115P/14, 15, 16; 105M/13, 14). *Exploration and Geological Services Division, Yukon, Indian and Northern Affairs Canada, Bulletin* **6**, 122 pp.

Murphy, D. C. & Mortensen, J. K. 2003. Late Paleozoic and Mesozoic features constrain displacement on Tintina Fault and limit large-scale orogen-parallel displacements in the northern Cordillera. *Geological Association of Canada Abstracts with Programs* **28**, 151.

Nelson, D. R., McCulloch, M. T. & Sun, S. S. 1986. The origin of ultrapotassic rocks as inferred from Sr, Nd and Pb isotopes. *Geochimica et Cosmochimica Acta* **50**, 231–45.

Newberry, R. J. & Solie, D. N. 1995. Data for plutonic rocks and associated gold deposits in Interior Alaska. *Alaska Division for Geological and Geophysical Surveys, Public Data File* **95–25**, 62 pp.

Pearce, J. A. 1982. Trace element characteristics of lavas from destructive plate boundaries. *In* Thorpe, R. S. (ed.) *Andesites: Orogenic Andesites and Related Rocks*. Chichester: John Wiley & Sons.

Pearce, J. A., Harris, N. B. W. & Tindle, A. G. 1984. Trace element discrimination diagrams for the tectonic interpretation of granitic rocks. *Journal of Petrology* **25**, 956–83.

Plafker, G. & Berg, H. C. 1994. Overview of the geology and tectonic evolution of Alaska. *In* Plafker, G. & Berg, H. C. (eds) *The Geology of Alaska. Geological Society of America, The Geology of North America* **G-1**, 998–1021.

Rock, N. M. S., Bowes, D. R. & Wright, A. E. 1991. *Lamprophyres*. Glasgow: Blackie & Son Ltd.

Sato, K. 1980. Tungsten skarn deposit of the Fujigatani mine, southwest Japan. *Economic Geology* **75**, 1066–82.

Simon, A. C., Candela, P. A., Piccoli, P. M., Pettke, T. & Heindrich, C. A. 2002. Gold solubility in magnetite. *Geological Society of America Abstracts with Programs* **34**, Paper 82–11.

Simon, A. C., Pettke, T., Candela, P.A., Piccoli, P. M. & Heindrich, C. A. 2003. Experimental determination of Au solubility in rhyolite melt and magnetite: Constraints on magmatic Au budgets. *American Mineralogist* **88**, 1644–51.

Smit, H., Armstrong, R. L. & van der Heyden, P. 1985. Petrology, chemistry, and radiogenic isotope (K-Ar, Rb-Sr, and U-Pb) study of the Emerald Lake pluton, eastern Yukon Territory. *Current Research, Part B, Geological Survey of Canada Paper* **85–1B**, 347–59.

Sun, S.-S. & McDonough, W. F. 1989. Chemical and isotopic systematics of oceanic basalts: implications for mantle composition and processes. *In* Saunders, A. D. & Norry, M. J. (eds) *Magmatism in the Ocean Basins. Geological Society, London, Special Publication* **42**, 313–46. London: Geological Society.

Snyder, D. B., Clowes, R. M, Cook, F. A., Erdmer, P., Evenchick, C. A., van der Velden, A. J. & Hall, K. W. 2002. Proterozoic prism arrests suspect terranes: insights into the ancient cordilleran margin from seismic reflection data. *GSA Today* **October 2002**, 4–10.

Tatsumi, Y. & Eggins, S. 1995. *Subduction Zone Magmatism*. Oxford: Blackwell Science.

Taylor, R. G. 1979. *Geology of Tin Deposits. Developments in Economic Geology II*. Amsterdam: Elsevier.

Tempelman-Kluit, D. J. 1970. Stratigraphy and structure of the 'Keno Hill Quartzite' in Tombstone River-Upper Klondike River map-areas, Yukon Territory (116B/7, B/8). *Geological Survey of Canada Bulletin* **180**, 101 pp.

Tilling, R. I., Gottfried, D. & Row, J. J. 1973. Gold abundance in igneous rocks: bearing on gold mineralization. *Economic Geology* **68**, 168–86.

Thompson, J. F. H, Sillitoe, R. H., Baker, T., Lang, J. R. & Mortensen, J. K. 1999. Intrusion-related gold deposits associated with tungsten-tin provinces. *Mineralium Deposita* **34**, 197–217.

Titley, S. R. 1982. Geological settings of porphyry copper deposits, southeast Arizona. *In* Titley, S. R. (ed.) *Advances in Geology of the Porphyry Copper Deposits of Southwestern North America*, 37–58. Tucson, AZ: University of Arizona Press.

Titley. S. R. 1991. Correspondence of ores of silver and gold with basement terranes in the American Southwest. *Mineralium Deposita* **26**, 66–71.

Watson, E. B. & Harrison, T. M. 1983. Zircon saturation revisited: temperature and compositional effects in a variety of crustal magma types. *Earth and Planetary Science Letters* **64**, 295–304.

White, A. J. R. 1995. Suite concept in igneous geology. *Proceedings of Leon T. Silver 70th Birthday Symposium and Celebration*. Pasadena, CA: California Institute of Technology, 113–16.

Woodsworth, G. J., Anderson, R. G., & Armstrong, R. L. 1991. Plutonic regimes. *In* Gabrielse, H. & Yorath, C. J. (eds) *Geology of the Cordilleran Orogen in Canada. Geological Survey of Canada, Geology of Canada* **4**, 491–531.

Yukon Minfile 2003. Yukon Minfile 2003 – A database of mineral occurrences in Yukon. Yukon Geological Survey.

CRAIG J. R. HART, Centre for Global Metallogeny, University of Western Australia, Nedlands, WA 6009, Australia, and Yukon Geological Survey, Box 2703 (K-10), Whitehorse, Yukon Y1A 2C6, Canada.
e-mail: craig.hart@gov.yk.ca

JOHN L. MAIR and DAVID I. GROVES, Centre for Global Metallogeny, University of Western Australia, Nedlands, WA 6009, Australia.

RICHARD J. GOLDFARB, United States Geological Survey, Box 25046 (MS964), Denver, Colorado 80225, USA.

MS received 5 January 2004. Accepted for publication 14 September 2004.

Transactions of the Royal Society of Edinburgh: Earth Sciences, **95**, 357–374, 2004

Volcanic–plutonic links, plutons as magma chambers and crust–mantle interaction: a lithospheric scale view of magma systems

Rodney V. Metcalf

ABSTRACT: The northern Colorado River extensional corridor (NCREC, USA) provides an excellent record of coeval volcanic and mid- to upper-crustal (<13 km) plutonic suites. The NCREC is a 50–100-km-wide zone that records late Tertiary lithospheric extension, volcanism, continental sedimentation and plutonism. Compilation of published studies of NCREC magmatic rocks permits an assessment of volcanic–plutonic links, magma sources and magmatic processes. The volcanic sections provide an excellent record of magma compositions (basalt, trachyandesite, trachyte and rhyolite) which span a 9-million-year period in the Miocene age (20–11 Ma).

Contemporaneous Miocene plutons span a similar compositional range (gabbro, diorite, quartz monzonite and granite) and were emplaced during a 4·5-million-year interval from 17 to 12·5 Ma. Geochemical and isotopic compositions and compositional trends allow direct correlation between plutonic and volcanic suites across the entire compositional range. Petrogenetic models demonstrate that intermediate magmas formed by a combination of magma mixing and fractional crystallisation involving mantle-derived mafic with crustal-derived felsic end-member magmas. Plutons exhibit a variety of features which suggest magma chamber processes, including (1) mafic cumulate sequences, (2) felsic cumulate sequences, and (3) magma mingling and advanced stages of magma mixing. Thus, the NCREC plutonic-volcanic record provides a link between magmatic processes recorded in pluton magma chambers and magmatic products in the form of extrusive igneous rocks. The NCREC plutons represent upper crustal magma chambers which connected volcanic eruptive centres to deeper-level magma chambers, and ultimately, to zones of mantle and crustal melting.

KEY WORDS: Basin and Range, crustal growth, magma mixing.

Volcanic rocks represent the physical end point of magma systems and are geologically instantaneous recorders of magmas transported from depth. Volcanologists call upon various magma chamber processes to explain a myriad of macroscopic to microscopic features in volcanic rocks, and observed compositional diversity in individual eruptions and larger volcanic suites. However, such attempts to understand magma chamber processes are, in reality, remote sensing techniques. Although it may seem intuitive that plutons and batholiths represent subvolcanic magma chambers, the petrology community has been slow to view plutons as recorders of magma chambers processes. Many petrologists view plutons as late-stage residual magmas which primarily record the thermal death of magma systems. However, an increasing number of studies have demonstrated that plutons have protracted histories and can provide important insight into magma chamber processes (Barnes *et al.* 1990; Larsen & Smith 1991; Wiebe 1993a, b, 1994; Falkner *et al.* 1995; Metcalf *et al.* 1995; Poli *et al.* 1996; Patrick & Miller 1997; Wiebe & Collins 1998; Collins *et al.* 2000; Bachl *et al.* 2001; Wiebe *et al.* 2002; Hawkins & Wiebe 2003). Such studies offer the potential for greater detail in our understanding of the underpinnings of volcanic systems.

Interaction of mantle-derived mafic and crustal-derived felsic magmas is increasingly considered a major process in magma systems above continental subduction zones and in continental rift zones (Reid *et al.* 1983; Frost & Mahood 1987; Holden *et al.* 1987; Poli *et al.* 1989; Coleman *et al.* 1995; Falkner *et al.* 1995; Metcalf *et al.* 1995; Collins 1996; Sisson *et al.* 1996; Ratajeski *et al.* 2001; Ernst *et al.* 2003). Crustal underplating of mafic mantle-derived magmas can promote the generation of primary crustal melts of felsic composition (Huppert & Sparks 1988; Bergantz 1989). Petrogenetic models which call upon the interaction of mafic mantle-derived and felsic crustal-derived magmas to generate intermediate magmas imply lithospheric scale magma systems. In such systems, plutons potentially represent cross-sections of magma conduits and magma chambers where magmatic processes produce the diverse suites of cogenetic magmas which reach the earth's surface as volcanic eruptions. However, the question remains, do plutons really provide a realistic portrait of magma chambers and magma conduits? Can we integrate petrogenetic studies from volcanic and plutonic systems to provide a greater understanding of magmatic processes? What are needed to address these questions are integrated studies of coeval suites of volcanic and plutonic rocks where genetic links can be established.

The northern Colorado River extensional corridor (NCREC, USA) provides an excellent record of coeval volcanic and mid- to upper-crustal (<13 km) plutonic suites. The present paper is a compilation of published geological, geochronological and geochemical data intended to provide a first-order integration of the Miocene-Pliocene volcanic–plutonic record in NCREC. This volcanic–plutonic province provides evidence of crust–mantle interaction on a lithospheric scale, insights into magma chamber processes and an opportunity to combine contemporaneous volcanic and plutonic histories. In many respects, the magmatic suites of the NCREC are similar to calc-alkaline magma systems typical of continental convergent margins.

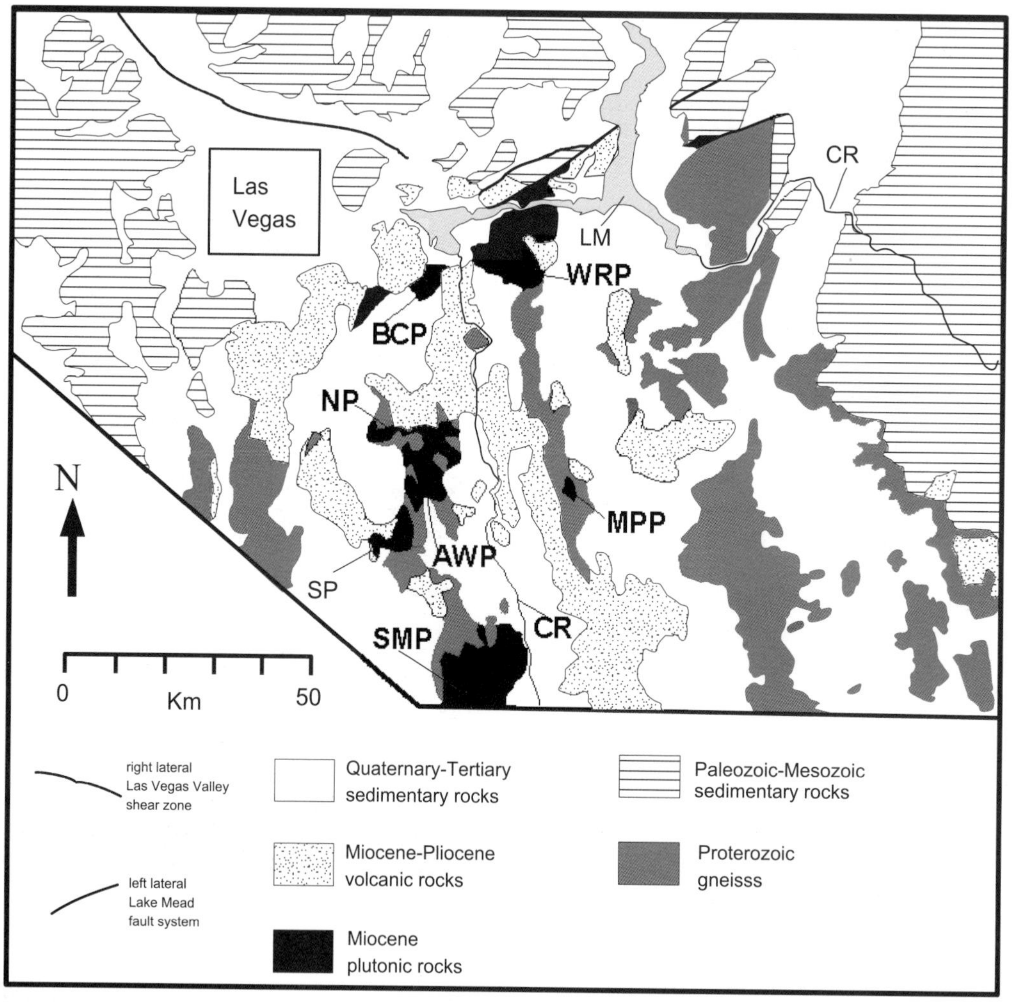

Figure 1 Generalised geological map of the northern Colorado River extensional corridor based on Stewart (1984) and Reynolds (1989). Abbreviations: (AWP) Aztec Wash pluton; (BCP) Boulder City pluton; (NP) Nelson pluton; (MPP) Mount Perkins pluton; (SMP) Spirit Mountain pluton; (SP) Searchlight pluton; (WRP) Wilson Ridge pluton; (CR) Colorado River; and (LM) Lead Mead.

1. Methods and data sources

In the NCREC, a combination of Late Cenozoic extensional faulting and entrenchment by the Colorado River has exposed a suite of coeval volcanic and plutonic rocks to palaeodepths of about 13 km. Detailed studies of the magmatic and tectonic history of the NCREC have spanned some 30 years, beginning with the pioneering work of Anderson (1971). A sufficient geological, geochronological and petrological framework now exists to permit the compilation of a fairly extensive data set. The geochemical data set used here contains 490 whole-rock geochemical analyses published in journals articles, guidebooks and theses. All of the geochemical data plots in the present paper (Figs 3 & 8–11) are based on data from the following sources: Mills (1984, 1994), Myers (1984), Feuerbach (1986), Naumann (1987), Schmidt (1987), Cole (1989), Smith *et al.* (1990), Bradshaw (1991), Bridwell (1991), Cascadden (1991), Larsen & Smith (1991), Daley & DePaolo (1992), Bradshaw *et al.* (1993), Morikawa (1993), Falkner *et al.* (1995), Faulds *et al.* (1995), McDaniel (1995), Metcalf *et al.* (1995), Boland (1996), Danielson (1998), Camin & Metcalf (1999) and Bachl *et al.* (2001).

Methods of correlating volcanic–plutonic records may differ depending on pluton exposure level and the degree of volcanic rock preservation, but in general, may include: (1) physical (mapped) connections between shallow plutons and volcanic centres; (2) geochronological correlations; (3) lithological, petrographic, geochemical and isotopic correlations; and (4) petrogenetic correlations. The compiled data set permits correlations based on absolute and relative ages, chemical and isotopic compositions, and petrogenetic links.

2. Geology of the northern Colorado River extensional corridor

The NCREC is a 50–100-km-wide area of medium to large magnitude crustal extension that is a part of the Basin and Range province (Howard & John 1987; Faulds *et al.* 1990). The NCREC is located between the relatively unextended Colorado Plateau to the east and the Spring Mountains region to the west (Fig. 1). The Lake Mead fault zone and the Las Vegas Valley shear zone bound the NCREC to the north. In the NCREC, a sequence of Miocene to Pliocene volcanic rocks (19–4·5 Ma) and intercalated continental sedimentary rocks rests on a basement of Proterozoic (1·7–1·8 Ga) granulite facies metasedimentary and metaigneous basement rocks, and minor Cretaceous plutonic rocks (Wooden & Miller 1990; Kapp *et al.* 2002).

During the Miocene, the NCREC was fragmented into a complex set of north-northwest trending fault blocks bounded by gently to steeply dipping normal faults. Volcanism and continental sedimentation was broadly coeval with the major episode of extension throughout the NCREC. The onset of both extensional faulting and volcanism migrated northward with time in the NCREC, with an average rate of about

3 cm year^{-1} (Faulds *et al.* 1994; Glazner & Bartley 1984; Smith & Faulds 1994). Volcanism began in the southern portion of the area shown in Figure 1 between 20 and 17 Ma ago, but did not began to the Lake Mead area in the north until 14·3–12 Ma ago (Faulds *et al.* 1994, 1995). At a given latitude, volcanism preceded extensional faulting by about 2 Ma (Faulds *et al.* 1994, 1995).

3. NCREC volcanic record

The volcanic rock record of the NCREC has been discussed in detail (Anderson *et al.* 1972; Smith *et al.* 1990; Fitton *et al.* 1991; Daley & DePaolo 1992; Bradshaw *et al.* 1993; Feuerbach *et al.* 1993; Faulds *et al.* 1995) and can be divided into three groups based on stratigraphic position, age relative to faulting and composition (Figs 2 & 3). These subdivisions are: (1) a pre- to syn-extensional group (20–11 Ma); (2) a late-extensional group (11–8 Ma); and (3) a post-extensional group (6–4 Ma).

3.1. Group 1 volcanic rocks (20–11 Ma)

Group 1 volcanic rocks consist of a compositionally expanded suite of alkali basalt, trachyandesite, trachydacite and rhyolite emplaced as lava flows and pyroclastic deposits. Group 1 volcanic rocks form what has been referred to as a trans-alkaline trend (Lameyre & Bonin 1991) on a traditional alkali-silica plot with mafic rocks largely falling in the alkali field and felsic rocks in the subalkaline field (Fig. 3a). On the AFM plot, Group 1 volcanic rocks display a typical calc-alkaline evolution trend (Fig. 3b). Numerous individual Group 1 eruptive centres have been recognised (Schmidt 1987; Weber & Smith 1987; Smith *et al.* 1990; Bridwell 1991; McDaniel 1993; Smith & Faulds 1994; Faulds *et al.* 1995) based primarily on the location of rhyolite dome complexes and caldera features. Where basal sections are exposed, Group 1 rocks rest unconformably on Precambrian basement. Group 1 volcanic rocks crop out as variably tilted ($\sim 90°$ to $\sim 10°$) and highly faulted stratigraphic sections which are intercalated with continent-derived clastic sedimentary rocks (Fig. 2c). In many sections, dips shallow up section with steeply dipping (45–90°) basal units overlain by younger units with progressive shallower dips (30–10°), suggesting development in growth-fault basins (Faulds *et al.* 1994, 1995; Mills 1994; McDaniel 1995).

Geochronologic control on Group 1 volcanic rocks is provided by published K-Ar and $^{40}Ar/^{39}Ar$ isotopic ages, including both mineral and whole-rock ages, and indicates volcanic activity for a period of 9 million years between 20 and 11 Ma. Two stratigraphic sections illustrate Group 1 age control. Faulds *et al.* (1995) published eight $^{40}Ar/^{39}Ar$ ages from a volcanic section preserved in a growth-fault basin west of the Mount Perkins pluton. Basalt at the base of the Mount Perkins section rests on Precambrian gneiss, dips 90° west and yielded a whole-rock 19·9 ± 0·5 Ma; a rhyolite up section (dip $\sim 55°$) yielded a biotite age of 15·7 ± 0·1 Ma; overlying basalts and intercalated felsic tuffs (dips $\sim 55°$ to 15°) yielded progressive younging of ages up-section from 14·5 ± 0·1 to 12·78 ± 0·02 Ma; the section is capped by nearly flat-lying basalt with a whole-rock age of 11·35 ± 0·1 Ma.

Mills (1994) compiled ages from a variety of sources (Anderson *et al.* 1972; Darvall *et al.* 1991; Feuerbach *et al.* 1991) for a section of Group 1 volcanic rocks exposed in Black Canyon, west of Hoover Dam. The Black Canyon section consists of intercalated flows and breccias of basalt, trachyandesite and dacite, dacitic tuff, and continental conglomerate. A basal trachyandesite sequence spans an age range from 18·5 to 14·19 Ma (based on both K-Ar and $^{40}Ar/^{39}Ar$ dates from Anderson *et al.* 1972; Darvall *et al.* 1991; Faulds & Gans, in Mills 1994). Overlying this basal sequence are intercalated dacite tuffs and flows, basaltic andesite flows, and conglomerate with ages ranging from 13·9 Ma at the base to 12·66 Ma at the top (Faulds & Gans, unpublished $^{40}Ar/^{39}Ar$ dates reported in Mills 1994). The Black Mountain conglomerate overlies the uppermost volcanic unit in the Black Canyon section; an ashfall tuff layer near the base of this conglomerate yielded a $^{40}Ar/^{39}Ar$ sanidine age of 11·72 ± 0·06 Ma (Williams 2003).

Figure 2 (A) Multiple lava flows of 6–4 Ma post-extensional Group 3 alkali basalt at Fortification Hill. (B) Tilted and faulted plateau of late-extensional Group 2 tholeiite basalt at Malpais Flattop. (C) Steeply dipping section of syn-extensional Group 1 andesites and tuffs west of Mount Perkins.

A variety of meso- and micro-scale features point to magma mingling and open-system behaviour prior to or during eruption of Group 1 volcanic rocks. Smith *et al.* (1990) reported Group 1 basaltic rocks with disequilibrium phenocryst assemblages; for example, euhedral olivine coexisting with embayed quartz and sodic plagioclase within intermediate composition

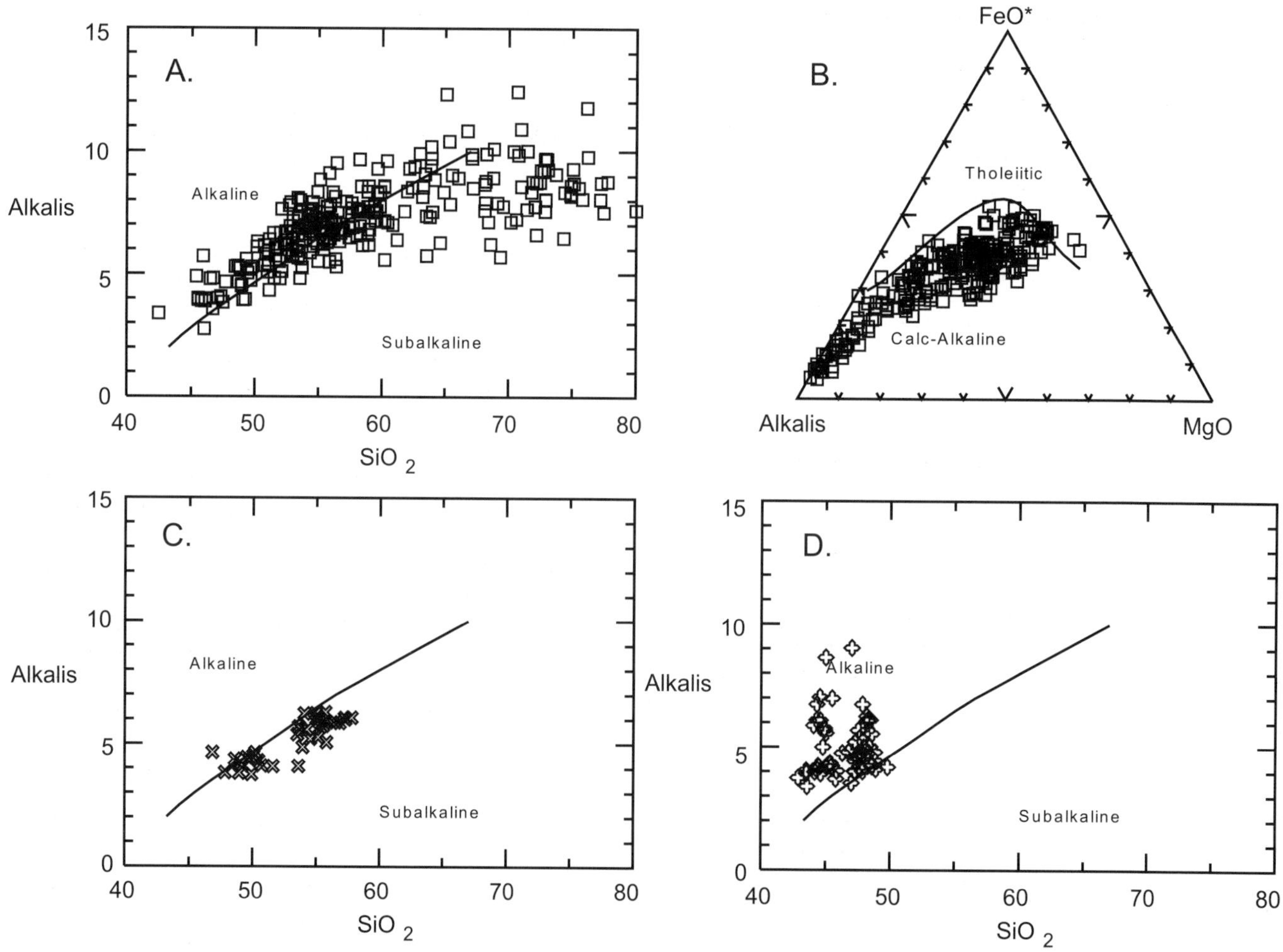

Figure 3 Compositional variability of three main volcanic groups in the northern Colorado River extensional corridor (A & B) Pre- to syn-extensional Group 1 alkali basalts-trachyandesite-trachyte-dacite-rhyolite erupted between 20 and 11 Ma. (C) Late-extensional (Group 2) tholeiite basalt and basaltic andesite erupted between 11 and 8 Ma. (D) Post-extensional (Group 3) alkali basalt erupted between 4 and 6 Ma.

rocks (see also Naumann 1987). These workers also reported rounded, embayed quartz surrounded by a reaction overgrowth of clinopyroxene in Group 1 trachyandesite (see Smith *et al.* 1990, fig. 5A, p. 176). This latter feature is similar to quartz ocelli reported from mafic enclaves formed during magma mingling in plutonic environments (Hibbard 1991). Also common in NCREC Group 1 volcanic rocks are plagioclases with embayed cores, oscillatory overgrowths and fretted textures consistent with magma mingling and mixing (Naumann 1987; Smith *et al.* 1990; McDaniel 1995).

Basaltic enclaves enclosed in dacitic flow rocks have been reported from three Group 1 locales: the Black Canyon dacite (~13 Ma) west of Hoover Dam (Mills 1994), in dacite (~14·2 Ma; Thompson 1985) from Boulder Wash on the north shore of Lake Mead (Naumann 1987; Smith *et al.* 1990) and in the dacite of Powerline Road (~12 Ma) in the River Mountains (Smith *et al.* 1990) east of Hoover Dam. At the Black Canyon locale, a biotite, plagioclase hornblende phyric dacite (~62 wt.% SiO_2) hosts vesiculated, 1–10-cm mafic enclaves (~48 wt.% SiO_2) exhibiting crenulate margins and containing rare plagioclase phenocrysts (Mills 1994). At the Boulder Wash locale, dacite hosts 2–50-cm basaltic enclaves which display crenulate margins. These enclaves are reported to contain olivine and plagioclase phenocrysts set in a matrix of glass and plagioclase microlites (Smith *et al.* 1990). At the River Mountains locale, dacite containing ~1-cm enclaves of pyroxene basalt and hornblende andesite makes up 20–40% of outcrops (Smith *et al.* 1990).

3.2. Group 2 volcanic rocks (11–8 Ma)

Group 2 volcanic rocks consist of generally sub-alkaline, tholeiitic basalt and basaltic andesite (Fig. 3c) emplaced as isolate lava flows of limited extent (Cascaddan 1991; Feuerbach *et al*, 1991, 1993; Daley & DePaolo 1992; Bradshaw *et al.* 1993; McDaniel 1995). Group 2 basalts form gently dipping plateau which typically cap sections of intercalated Group 1 volcanic rocks and continental conglomerates in growth fault basins (Fig. 2b). Although Group 2 basalts are cut by high-angle normal faults, offsets are typically minor, consistent with eruption late in the extensional episode. Age control on Group 2 basalts is provided by published K-Ar whole-rocks ages for basalts at Callville Mesa on the north shore of Lake Mead which range from $10{\cdot}46 \pm 0{\cdot}23$ to $8{\cdot}49 \pm 0{\cdot}2$ Ma (Feuerbach *et al.* 1991). Anderson *et al.* (1972) reported K-Ar ages of 10·5–8·5 Ma for Group 2 basalts at Malpais Flattop south of Hoover Dam. Feuerbach *et al.* (1993) quoted $^{40}Ar/^{39}Ar$ ages from Faulds & Gans (unpublished data) for Malpais Flattop basalts of $10{\cdot}6 \pm 0{\cdot}5$ and $9{\cdot}7 \pm 0{\cdot}5$ Ma. Basaltic andesites of Group 2 contain xenoliths and xenocrysts of felsic crustal material at Calville Mesa (Feuerbach *et al.* 1993). Such xenolithic material has been taken as evidence that Group 2 basaltic andesite evolved from Group 2 basalt by crustal assimilation (Feuerbach *et al.* 1993).

3.3. Group 3 volcanic rocks (6–4 Ma)

Group 3 volcanic rocks consist of numerous alkali basalts flows and minor cinder cones erupted from a series of vents extending some 50 km south of Lake Mead (Figs 2a & 3d). Group 3 basalts were erupted onto Precambrian basement, faulted and tilted Miocene conglomerates (overlying Group 1 volcanic rocks) and Miocene plutonic rocks exposed in fault blocks (Mills 1994). Vents for Group 3 lava flows appear to be localised along high-angle, extension-related faults which cut Group 1 volcanic rocks and intercalated conglomerate, Precambrian basement gneiss and Miocene plutonic rocks (Fig. 2a). That Group 1 basalts overlie, but are not cut by, such faults indicate that these lava flows were post-extensional (Fig. 2a). Age control on Group 3 basalts comes from whole K-Ar ages which vary from 5·8 to 4·3 Ma (Anderson *et al.* 1972; Feuerbach *et al.* 1991).

4. NCREC plutonic record: magma chambers and magma conduits

4.1. Emplacement history

Eight major Miocene plutons crop out in the NCREC (Fig. 1). Plutons were emplaced into Precambrian basement gneiss and/or contemporaneous Miocene volcanic rocks. Limited Al-in-hornblende barometry data (Metcalf & Smith 1991; Falkner *et al.* 1995; Metcalf *et al.* 1995; Patrick & Miller 1997; Bachl *et al.* 2001) indicates emplacement at shallow crustal levels (13–2 km). Plutons are predominantly granite to granodiorite, but significant mafic and intermediate rocks are present in the form of early gabbro-diorite intrusions, microgranitoid enclaves, syn-plutonic dykes and late aphanitic dykes. A general emplacement sequence for a NCREC pluton is: (1) gabbro-diorite, including olivine-clinopyroxene bearing cumulates; (2) granite pluton with associated mafic microgranitoid enclaves, synplutonic dykes and depositional sheets; and (3) aphanitic dykes of mafic to felsic compositions. The complete emplacement sequence is not apparent in all NCREC plutons.

Geochronologic control on plutonic rocks is provided by published K-Ar, $^{40}Ar/^{39}Ar$ and U-Pb isotopic ages for seven of the eight plutons and indicates plutonic activity for a period of 4·5 million years from 17 to 12·5 Ma. Age constraints for two plutons are based solely on less reliable K-Ar biotite ages, i.e. the Boulder City pluton at 13·8 ± 0·5 (Anderson *et al.* 1972) and the Nelson pluton at 16·93 ± 0·48 (Faulds *et al.* 1992). Age control for the Spirit Mountain pluton is provided by a single U-P zircon age of 17 Ma (Howard *et al.* 1996). A single $^{40}Ar/^{39}Ar$ biotite age of 15·96 ± 0·06 Ma for a granite from the Mount Perkins pluton was reported by Faulds *et al.* (1995). The age of the Wilson Ridge pluton is constrained by two K-Ar biotite ages (13·4 ± 0·4 and 13·5 ± 0·4 Ma; Larsen & Smith 1991) from the deeper southern portion of the pluton and a $^{40}Ar/^{39}Ar$ biotite age (12·5 ± 0·06 Ma; Anderson *et al.* 1994) from the northern hypabyssal portion of the pluton. Both $^{40}Ar/^{39}Ar$ biotite and U-Pb zircon ages are reported from the Searchlight pluton and are concordant at 16·4 ± 0·5 Ma (Bachl *et al.* 2001). Multiple ages have been reported for the Aztec Wash pluton including a K-Ar biotite age of 16·21 ± 0·45 Ma (Faulds *et al.* 1992), a $^{40}Ar/^{39}Ar$ biotite age of 15·7 ± 0·04 Ma (Falkner *et al.* 1995) and a U-Pb zircon (discordant, lower intercept) age of 15·14 ± 0·62 Ma (Falkner *et al.* 1995). More recently, Cates *et al.* (2003) reported U-Pb zircon SHRIMP ages from the Searchlight pluton ranging from 17 to 15·7 Ma (± 0·2–0·4 Ma) and from the Aztec Wash pluton ranging from 15·8 to 15·5 (± 0·1–0·15 Ma), suggesting protracted intervals pluton construction and crystallisation (see also Harper *et al.* 2004).

4.2. Mount Perkins pluton

The Mount Perkins pluton is a small (~6 km^2) composite intrusive body located in the central Black Mountains, Arizona (Figs 1 & 4). The pluton was emplaced into Precambrian gneissic country rock at a palaeodepth of 5–7 km (Metcalf *et al.* 1995) immediately prior to the onset of extensional faulting (Faulds *et al.* 1995). Field relationships indicate that the pluton was emplaced as a sequence of the four separate magma batches (phase 1 through phase 4; Fig. 4), each crystallising before emplacement of subsequent magma (Metcalf *et al.* 1995). Phase 1 consists of gabbro and diorite; phase 2 is dominated by mafic (diorite-qtz diorite) to intermediate (granodiorite) compositions which host mafic microgranitoid enclaves; phase 3 is composed of a bimodal suite comprising granite that hosts mafic microgranitoid enclaves; and phase 4 consists of a suite of planar aphanitic dykes of basaltic to rhyolitic composition. Mafic microgranitoid enclaves are clinopyroxene-bearing hornblende diorite and hornblende quartz diorite.

4.2.1. Cumulates and fractional crystallisation: phase 1 gabbro. Phase 1 consists primarily of isotropic coarse-grained hornblende gabbro with minor diorite and quartz diorite. Also present is a layered sequence of olivine hornblende gabbro and layered hornblende gabbro. A 10-m-thick layer of olivine hornblende gabbro with weak modal layering can be traced parallel to the pluton/gneiss margin. Inboard of the olivine hornblende gabbro layer is 15-m-thick sequence of layered hornblende gabbro with layering defined by centimetre-scale modal variations in hornblende and plagioclase, and by plagioclase and hornblende mineral alignment.

Mineralogical and chemical evidence suggests that phase 1 olivine hornblende gabbro samples are cumulates of olivine, clinopyroxene and magnetite which crystallised from a primitive basaltic magma sourced in the mantle. This evidence includes: (1) variably high modal percentages of olivine (≤35%), clinopyroxene (≤30%) and magnetite (≤45%); (2) variably high MgO (8–21 wt.%), FeO (7–16 wt.%), Ni (104–926 ppm) and Cr (521–1729 ppm); (3) variably low modal percentages of plagioclase (2–45%); and (4) variably Al_2O_3 (6·8–16 wt.%) extending to low values (Danielson 1998). Phase 1 cumulate rocks demonstrate that primitive mafic magmas reached shallow crustal levels and fractionated early in the pluton's history. Interestingly, phase 1 isotropic hornblende gabbro is not the daughter magma from this episode of cumulate fractionation. The isotropic hornblende gabbro instead represents a somewhat differentiated mafic magma that differed in isotopic (Nd-Sr-Pb) and trace element composition from the magmas which formed the cumulate sequence (Danielson 1998; Metcalf *et al.* 1999).

4.2.2. Magma mingling in a conduit: phases 2, 3 and 4. Mafic microgranitoid enclaves and enclosing host rocks of phases 2 and 3 in the Mount Perkins pluton record an incipient stage of magma commingling among a suite of compositionally contrasting magmas (Metcalf *et al.* 1995). Enclaves have ovoid, pillow-like shapes commonly bounded by fine-grained (chilled), crenulated margins, range in size from a few centimetres to several metres across, and make up 30–70% of outcrops which contain them (Fig. 5a, b). Outcrop-scale differences in chemical, and Nd and Sr isotopic compositions between enclaves and surrounding host rocks (differences of 6·8–10·6 ε_{Nd} units; differences of 0·0036–0·0046 $^{87}Sr/^{86}Sr$ units) record a state of compositional disequilibrium (Metcalf *et al.* 1995, fig. 7, p. 519). Significant *in situ* modification of magma compositions caused by either elemental or isotopic diffusion was limited to areas immediately adjacent to the enclave-host interface (Metcalf *et al.* 1995). Collectively, enclaves, their

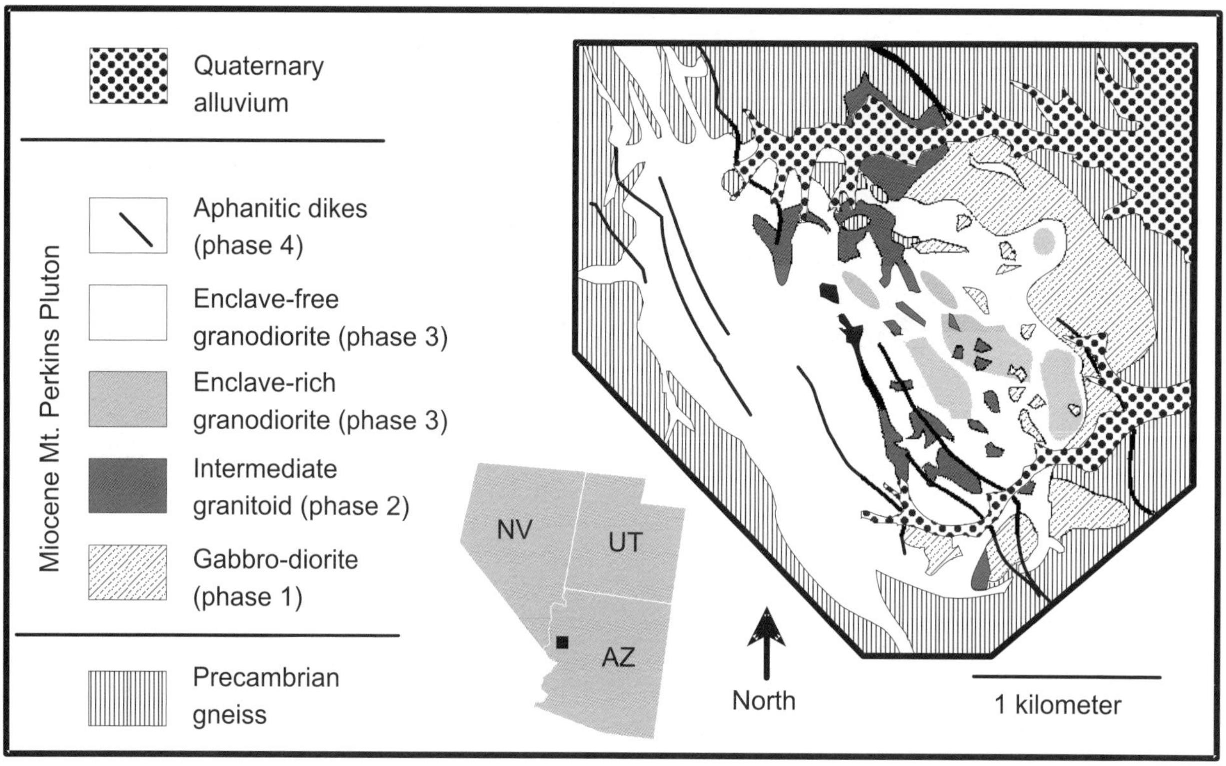

Figure 4 Generalised geological map of the Mount Perkins pluton after Metcalf *et al.* (1995).

enclosing host rocks and the enclave-free granitoid rocks define a continuous mafic to felsic compositional suite (SiO_2 range of 50–74 wt.%) and exhibit a covariance among major elements, trace elements, and Nd and Sr isotopes (Fig. 6). These data point to a single cogenetic suite. Given the isotopic disequilibrium discussed above, the process(es) that produced the suite must have operated at a deeper level and prior to commingling events recorded in the pluton. Thus, phases 2 and 3 of the Mount Perkins pluton are interpreted as a magma conduit or transfer zone.

Metcalf *et al.* (1995) modelled the petrogenetic evolution of the phase 2 and 3 suite as a combined process of magma mixing and fractional crystallisation (MFC), as depicted in Figure 6. Isotopic variation within the suite indicates the importance of two-component open-system processes, in this case magma mixing. Inflections in trace element data (LREE, Ba, Sr) on Harker diagrams (Fig. 6; see also Metcalf *et al.* 1995, fig. 5, p. 517) cannot be explained by simple binary mixing and require a component of fractional crystallisation in magma evolution; Metcalf *et al.* (1995) modelled this magma evolution as a two-stage process. Stage 1 involved combined fractional crystallisation of a mafic magma (represented by the most primitive enclave) with mixing of felsic magma (represented by phase 3 enclave-free granite) (Fig. 6). Stage 2 involved simple mixing of intermediate hybrid magmas from stage 1 with felsic magma (represented by phase 3 enclave-free granite) (MFC) (Fig; 6). As discussed in a later section, the petrogenetic evolution of these Mount Perkins magmas appears to provide insight into petrologic processes which operated on a much larger scale within the NCREC magma systems.

Phase 4 dykes appear to have a petrogenetic history similar to that described above for phases 2 and 3 (Camin & Metcalf 1999). However, post-magmatic hydrothermal alteration of the dykes has affected their whole-rock chemical and isotopic compositions, and complicates a straightforward petrogenetic interpretation.

4.3. Wilson Ridge pluton: a large-scale magma mixing chamber

The Wilson Ridge pluton is the largest exposed pluton in the NCREC (Figs 1 & 7). The pluton provides a north-south cross-sectional view of a major magma chamber where mafic and felsic magmas interacted to produce intermediate hybrid magmas (Feuerbach 1986; Smith *et al.* 1990; Larsen & Smith 1991; Metcalf *et al.* 1993). Two main rock suites are recognised in the pluton, the older Horsethief Canyon diorite (52–54 wt.% SiO_2), consisting of coarse-grained clinopyroxene-bearing hornblende diorite and pegmatitic diorite, and the younger Teakettle Pass suite (55–70 wt.% SiO_2), consisting of biotite quartz monzonite, hornblende quartz monzodiorite and hornblende monzodiorite (Larsen & Smith 1991). The east, west and northern margins are bounded by faults or are overlain by Miocene sedimentary deposits or Pliocene Group 3 basalt flows. The south margin of the pluton is a steep-walled intrusive contact with Precambrian basement. Grain size within the pluton varies from coarse-grained at the southern end to fine-grained and aphanitic at the northern end, suggesting that palaeo-up is to the north. Consistent with this interpretation are Al-in-hornblende geobarometry results for samples from the southern end of the pluton which yield pressures from 2·6 to 3·43 ± 0·23 kbar (total Al = 1·2–1·36 per 23 oxygen, all necessary phases present, calibration of Anderson & Smith 1994, T correction using plagioclase with X_{AN} between 0·28 and 0·42). These results suggest that the Wilson Ridge pluton represents a north-tilted body that extended from near surface hypabyssal levels to depths of 9·7–12·5 km.

Horsethief Canyon diorite is a pre-cursor mafic unit that crops out in a small area near the southwestern margin of the pluton and occurs primarily as angular xenoliths within the younger, more voluminous Teakettle Pass suite. The Teakettle Pass suite is dominated by biotite quartz monzonite that forms the upper (northern) two-thirds of the pluton. The Teakettle

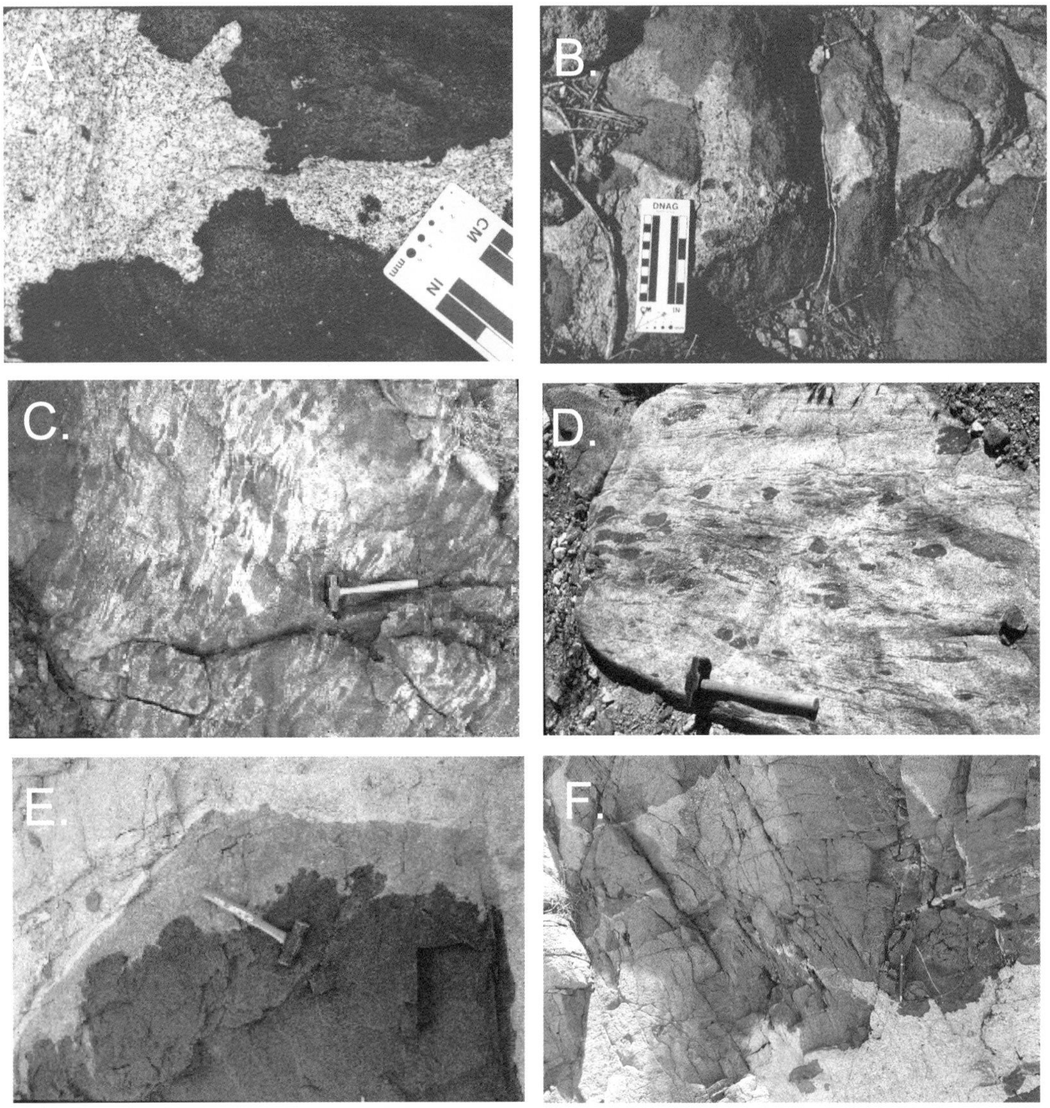

Figure 5 Field photos of the magma mingling and mixing features in northern Colorado River extensional corridor plutons: (A) Mafic enclave exhibiting a chilled crenulated margin enclosed in a granitic host; phase 3 unit in the Mount Perkins pluton. (B) Mafic enclaves enclosed in an intermediate host rock; phase 2 unit in the Mount Perkins pluton. (C) Mafic enclave swarm defining a disrupted syn-plutonic mafic dyke in the Teakettle Pass unit of the Wilson Ridge pluton. (D) Mafic schlieren in the Teakettle Pass unit of the Wilson Ridge pluton. (E) Double enclaves from the heterogeneous zone of the Aztec Wash pluton. F. Mafic sheets and pipes in the heterogeneous zone of the Aztec Wash pluton.

Pass suite in the southern portion of the pluton is extremely heterogeneous, composed of hornblende monzodiorite and hornblende quartz monzodiorite host rocks which contain up to 70% mafic microgranitoid enclaves (Fig. 5c). The enclaves are hornblende and/or biotite-bearing diorite and quartz diorite. A complete continuum exists from enclaves which are pillow-like with crenulate, chilled margins to ellipsoidal shapes (Fig. 5c) to mafic schlieren (Fig. 5d), and ultimately, to the mafic constituents (e.g. hornblende) in foliated quartz monzodiorite and monzodiorite (Larsen & Smith 1991; Metcalf *et al.* 1993). Magmatic foliations are apparent in host rocks and enclaves, and are defined by the parallel alignment of biotite, hornblende and plagioclase (Fig. 5d). Such fabrics are often parallel to elongate enclaves and schlieren, and a macroscopic foliation defined by alignment of enclaves (Fig. 5c, d).

Aphanitic dikes of mafic, intermediate and felsic compositions cut both the Horsethief Canyon diorite and the Teakettle Pass suite (Feuerbach 1986; Larsen & Smith 1991). Dykes are typically parallel to pluton-bounding faults, are planar in nature and were emplaced following crystallisation of the main pluton. However, some mafic dykes have crenulate margins and appear to be syn-plutonic. The margins of some mafic dykes change character along strike, with crenulate margins where cutting felsic host rocks and sharp, straight margins where cutting older mafic enclaves. Many mafic enclave swarms have dyke-like geometries, suggesting disruption of syn-plutonic dykes within the magma chamber (Fig. 5c). Collectively, these field observations are consistent with a protracted history of mafic magma injection into a silicic magma chamber and dynamic mixing to produce intermediate hybrid magmas.

Larsen & Smith (1991) used trace-element modelling to evaluate the whether or not intermediate rocks of the Wilson Ridge pluton were the products of magma mixing or combined MFC processes. Trace-element models were constrained using major element and petrographic data; the best-fit models were based on MFC calculations (equations from DePaolo 1981). An age-equivalent ($\sim$13 Ma) Group 1 basalt (near by River Mountains stratovolcano, Smith *et al.* 1990) was used as the mafic parent. A quartz monzonite from the Wilson Ridge pluton was used as a felsic end member in the MFC models. The MFC models of Larsen & Smith (1991) produced good

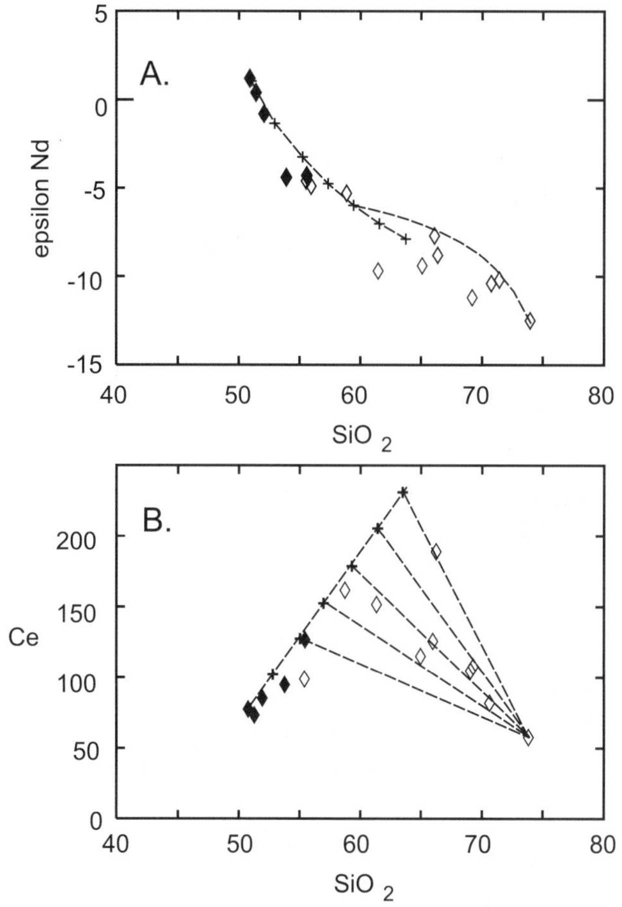

Figure 6 Geochemical data and petrogenetic models for phases 2 and 3 of the Mount Perkins pluton. The closed symbols are enclaves and the open symbols are granitoid host rocks. The models are after Metcalf *et al.* (1995): (a) stage 1 involves combined fractional crystallization of basaltic magma and mixing with felsic magma; (b) stage 2 involves mixing of stage 1 hybrid magmas with felsic magma.

agreement between calculated daughter magmas and observed compositions in individual samples of Wilson Ridge diorite, monzodiorite and quartz diorite. Metcalf *et al.* (1995) pointed to the Wilson Ridge pluton as one possible analogue for the type of mixing chamber that must have existed at depth beneath the Mount Perkins pluton.

4.4. Aztec Wash pluton: magma chamber recharge and mixing

The geology and petrogenesis of Aztec Wash pluton (Fig. 1) has been discussed in detail by Falkner *et al.* (1995) and Patrick & Miller (1997), and is the focus of continuing detailed investigations (Cates *et al.* 2003; Coiner *et al.* 2003; Miller *et al.* 2003; Harper *et al.* 2004). What is emerging from these studies is a protracted, dynamic history of magma recharge, magma mingling/mixing and fractional crystallisation involving formation of both mafic and felsic cumulates. What follows is a brief overview of the Aztec Wash pluton as it relates to the present study; the reader is directed to Harper *et al.* (2004; see also Falkner *et al.* 1995; Patrick & Miller 1997; Robinson & Miller 1999) for a detailed discussion of the pluton and its history.

The Aztec Wash pluton is a shallow (4–7 km in depth), ENE-tilted Miocene pluton ($\sim 50\ km^2$) emplaced into Proterozoic gneiss, Cretaceous granite and Miocene Nelson pluton (Falkner *et al.* 1995). The pluton is divided into two portions, a northern zone composed of granite, and a larger southern heterogeneous zone composed of mafic, intermediate and felsic rocks (Fig. 5e, f). The heterogeneous zone preserves sequences of mafic sheets and enclaves interlayered with felsic rocks (Fig. 5g). Features within the heterogeneous zone (asymmetric sheets with crenulated bases, felsic pipes penetrating mafic sheets and cumulate textures in felsic rocks) suggest repeated injection of mafic magma into the lower cumulate sections of a crystallising felsic magma chamber (Falkner *et al.* 1995; Patrick & Miller 1997; Miller *et al.* 2003; Harper *et al.* 2004). The mineralogy (hornblende+plagioclase ± clinopyroxene) and the whole-rock compositions (SiO_2 >50 wt.% and MgO <5 wt.%) of the chilled mafic sheets and enclaves suggest that mafic replenishments had evolved significantly from mantle-derived parental magmas prior to arrival (Falkner *et al.* 1995). However, some mafic sheets preserve rhythmic layering rich in clinopyroxene ± olivine and are interpreted as cumulates from a fairly primitive basaltic magma (Falkner *et al.* 1995, Patrick & Miller 1997).

The interaction of mafic and felsic magmas in the heterogeneous zone produced a broad, though volumetrically small, spectrum of intermediate hybrid magmas. Intermediate hybrid magmas have isotopic (Nd-Sr-Pb) compositions intermediate between mafic sheets and granite, indicating open-system processes. Whole-rock chemical data shows inflections for some elements on Harker diagrams (Falkner *et al.* 1995). As with the Mount Perkins magmas, intermediate magmas in the Aztec Wash pluton appear to record a combination of fractional crystallisation and magma mixing (Falkner *et al.* 1995). Also present within the pluton are felsic (K-feldspar) cumulates and the corresponding evolved high-silica rhyolite magmas. The latter are found in what is interpreted as the roof zone of the restored pluton (Miller *et al.* 2003). Cross-cutting the pluton are a series of north-trending late-stage dykes. The dyke suite is largely a bimodal mafic-felsic, but intermediate composition dykes are present. The chemical and isotopic compositions and trends of the dyke suite match that of the pluton suggesting a cogenetic relationship (Falkner *et al.* 1995).

4.5. Searchlight pluton: fractional crystallisation in a felsic magma chamber

The Miocene Searchlight pluton (Fig. 1) appears to record fractional crystallisation of felsic magma in an upper crustal silicic magma chamber (Bachl *et al.* 2001). The Searchlight pluton represents a 10-km-thick tilted cross-section through the magma chamber from its roof on the west to its floor on the east. Hornblende geobarometry suggests palaeodepths of 3 km (roof) to 13 km (floor) (Bachl *et al.* 2001). The pluton preserves a roof zone solidification front of quartz monzonite, and middle and lower sections which record complementary zones of differentiated felsic magma and felsic cumulates (K-feldspar, plagioclase, hornblende, biotite and accessory minerals). The differentiated rocks of the middle zone differ from parent magmas (roof zone quartz monzonite) with strong negative Eu anomalies and depletions in MREE relative to the roof (Bachl *et al.* 2001). Mafic components in the Searchlight pluton are restricted to mafic pods and mafic enclaves in the lower section of the pluton, and a larger (kilometre-scale) mafic pod in the middle granite zone (Bachl *et al.* 2001). These mafic rocks appear to record magma mingling, but little if any magma mixing (Bachl *et al.* 2001).

5. Correlation of volcanic and plutonic records

5.1. Age

Isotopic ages for NCREC plutonic rocks (~17–13 Ma) demonstrate that plutonism was coeval with Group 1

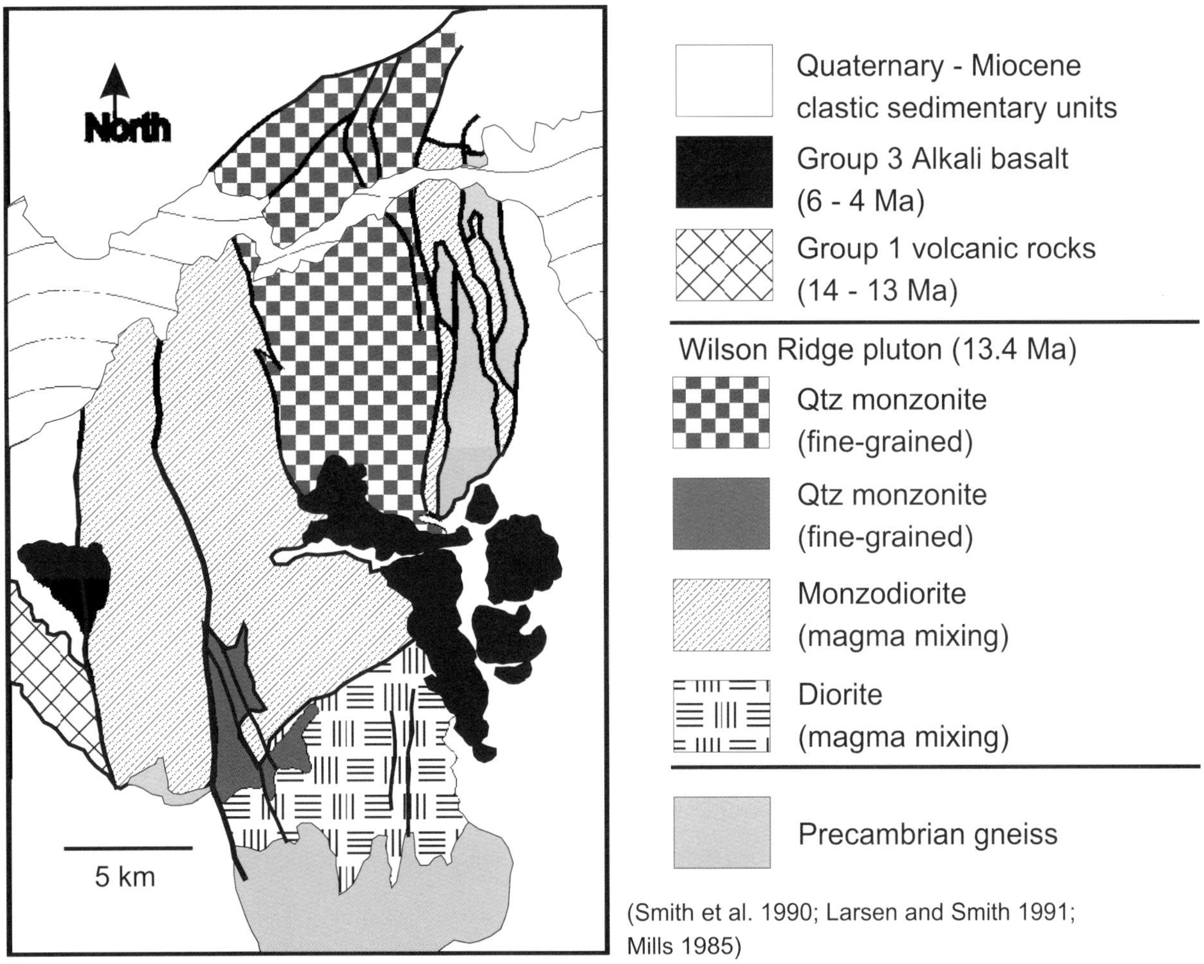

Figure 7 Generalised geological map of the Wilson Ridge pluton (modified from Smith *et al.* 1990; Larsen & Smith 1991).

volcanism (~20–12 Ma). Although plutonism represents a somewhat narrower time interval (<4·5 million years) than contemporaneous Group 1 volcanism (~8 Ma), this may be a consequence of a more limited geochronologic database for the plutonic record. The plutonic field relationships discussed above clearly point to a protracted history for the construction and solidification of NCREC plutons, but for some of these plutons, only a single isotopic age is available (e.g. Mount Perkins) so that the duration of plutonism is not known. For plutons with multiple reported ages, durations of perhaps a million years are indicated.

5.2. Whole-rock compositions

The NCREC plutonic rocks and Group 1 volcanic rocks display a notable overlap of elemental and isotopic compositional trends (Figs 8, 9, 10). Plutonic rocks show the same transalkaline trend on the alkali-silca plot and the same calc-alkaline evolution trend on the AFM plot as seen in Group 1 volcanic rocks (Fig. 8a, b). Plutonic and Group 1 volcanic rocks display overlapping linear trends on Harker diagrams for a number of major and trace elements (e.g. TiO_2, CaO, FeO and Rb) (Figs 8c, d, e, 9a). In contrast, a significant number of elements show inflections at SiO_2 values of 58–60 wt.% which range from fairly subtle (e.g. MgO, Fig. 8f) to rather profound (e.g. LREE, Sr, Ba and Zr; Fig. 9b–f). These inflections are apparent in both the volcanic and plutonic suites, and the data sets are coincident for all elements. Neodymium and Sr isotopic compositions vary systematically with SiO_2 contents (Fig. 10). As with elemental compositions, these isotopic trends are observed in both plutonic and volcanic data sets and the trends overlap. Whole-rock chemical and isotopic data trends provide a clear link between the volcanic and plutonic suites.

5.3. Petrogenetic correlations: magma evolution

The clear overlap of plutonic and volcanic compositional trends implies that the plutonic rocks are not only coeval with Group 1 volcanic rocks, but also co-genetic in origin. The term co-genetic (as opposed to co-magmatic) is used here to denote common magma sources and common petrogenetic processes. The correlation of Nd and Sr isotopic data with differentiation monitors such as SiO_2 point to the importance of open-system processes in the petrogenetic evolution of the volcanic–plutonic suite (Fig. 10). Inflections observed on Harker-style variation diagrams for a large number of elements are inconsistent with a simple mixing origin (Figs 8f, 9). As with the Mount Perkins pluton, these data require magma evolution via a two-stage open-system process. Field and petrographic observations in plutonic and volcanic rocks point to magma mixing as the dominant open system process (Fig. 5). In addition, cumulate assemblages in the plutonic suites (both mafic and felsic) provide evidence of fractional crystallisation. Collectively, field and whole-rock compositional evidence point to combined processes of fractional crystallisation and magma mixing in the evolution of the NCREC magma system.

The magma evolution models presented for the Mount Perkins pluton (Fig. 6) could be applied, in a general way, to the entire NCREC magma system. A general conclusion of such a model is that NCREC intermediate magmas are

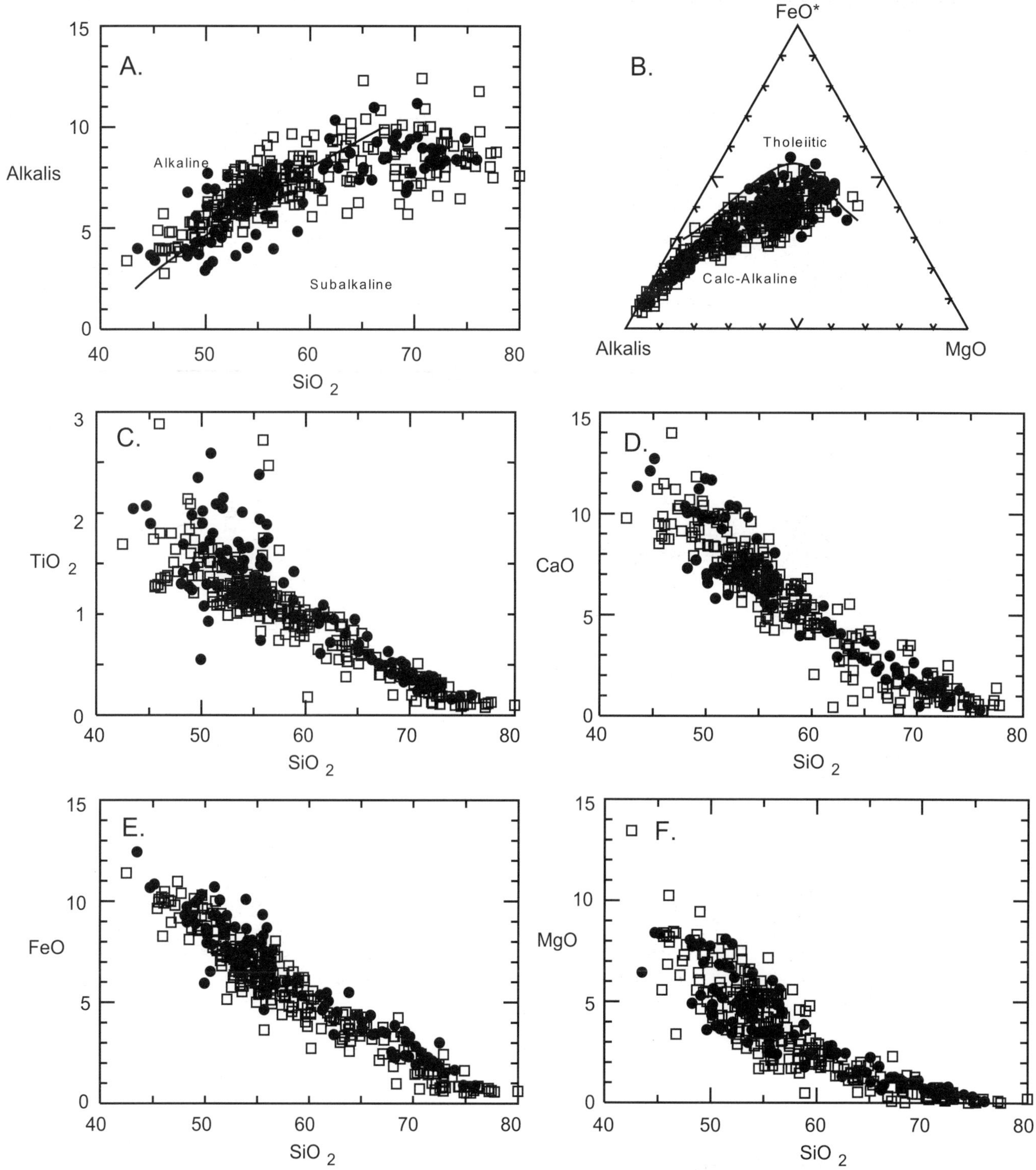

Figure 8 Correlation of major element whole-rock compositions of Group 1 volcanic rocks (open squares) with coeval plutonic rocks (filled circles).

hybrids formed by the interaction of mafic mantle-derived primary magmas and felsic crustal-derived primary magmas. The following sections address the nature and sources of mafic and felsic end-member magmas.

5.4. Mafic magma sources

5.4.1. Temporal variations in NCREC basaltic magma sources. The mafic character of Group 1, 2 and 3 basaltic rocks [e.g. high MgO (4–10 wt.%), high Cr (50–450 ppm) and the presence of olivine phenocrysts (Cascaddan 1991; McDaniel 1995; Mills 1994)] points to a peridotite source in the mantle. Much has been written regarding mantle sources for Cenozoic basalts of the Basin and Range province, including the NCREC. These discussions focused on data from trace element and isotopic studies (Perry *et al.* 1988; Farmer *et al.* 1989; Fitton *et al.* 1991; Kempton *et al.* 1991; Bradshaw *et al.* 1993; Feuerbach *et al.* 1993; Harry & Leeman 1995; Hawkesworth *et al.* 1995).

Group 1 basalts are fairly typical of extension-related basalts throughout the Basin and Range province. These basalts exhibit subduction-like trace element signatures typified by high LIL concentrations, enriched LREE and elevated LIL/HFSE ratios (e.g. Ba/Nb), low ε_{Nd} ($-2{\cdot}5$ to $-9{\cdot}5$) and high initial $^{87}Sr/^{86}Sr$ (Sri=0·7065–7095) (Fig. 11). Such

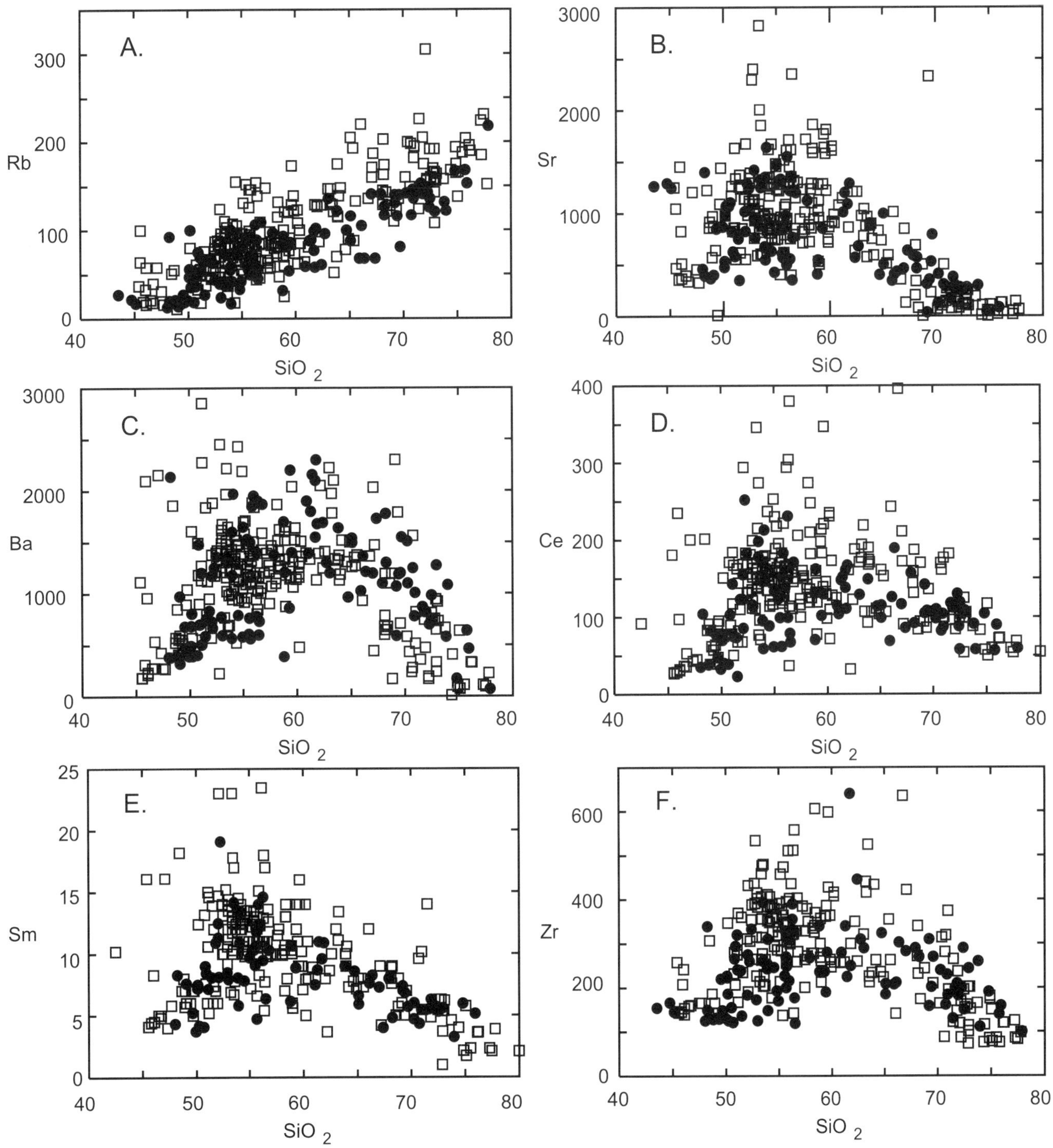

Figure 9 Correlation of trace element whole-rock compositions of Group 1 volcanic rocks (open squares) with coeval plutonic rocks (filled circles).

compositions must reflect either a source in ancient, enriched lithospheric mantle or a large measure of crustal contamination of asthenospheric-derived magmas. Owing to the lack of compositional or microstructural evidence of significant crustal contamination, the general consensus is that such Basin and Range basalts reflect melting of enriched subcontinental lithospheric mantle (Perry *et al.* 1988; Farmer *et al.* 1989; Fitton *et al.* 1991; Kempton *et al.* 1991; Bradshaw *et al.* 1993; Feuerbach *et al.* 1993; Leeman 1995; Hawkesworth *et al.* 1995; Metcalf & Smith 1995).

Subalkaline, tholeiitic basalts, such as those of Group 2, are rare in the Basin and Range province, and appear to be confined to the NCREC (Feuerbach *et al.* 1993; McDaniel 1995; Metcalf *et al.* 1997). Group 2 basalts are distinctive in exhibiting less enrichment in incompatible trace elements than Group 1 basalts, in particular HFSEs such as Nb, a feature commensurate with their subalkaline nature. Ratios of LIL/HFSE vary considerably, but range to fairly high values (e.g. Ba/Nb, Fig. 11). The isotopic compositions of Group 2 basalt are similar to Group 1 basalt with low ε_{Nd} (-3 to $-8{\cdot}5$) and high initial $^{87}Sr/^{86}Sr$ (Sri=0·705–0·708), suggesting a lithospheric mantle source. The subalkaline and less-enriched compositions of Group 2 basalts could simply reflect higher degrees of partial melting, as compared to Group 1 basalt. However, both Bradshaw *et al.* (1993) and McDaniel (1995), noted that Group 1 basalt has steeper HREE patterns and depleted Y-values while Group 2 basalt has flat HREE patterns and more enriched Y-values, suggesting that differences

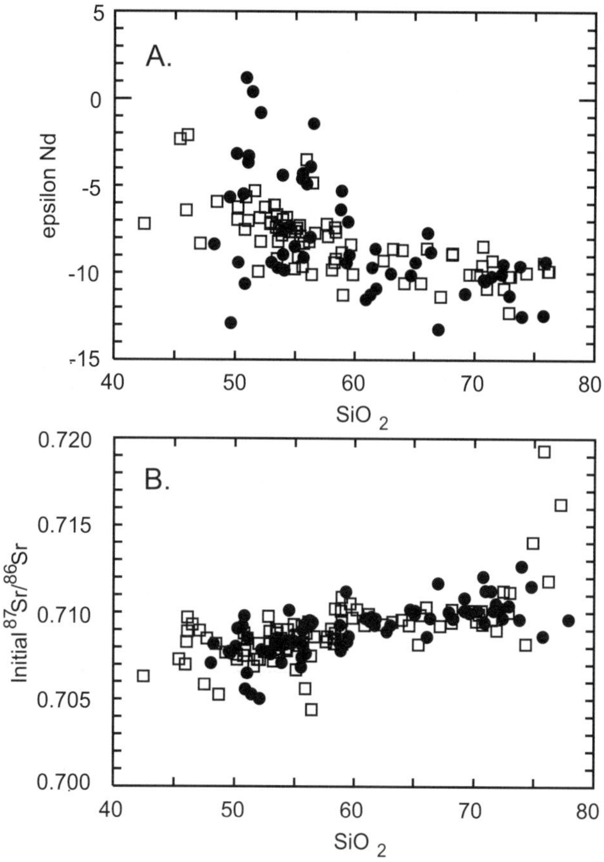

Figure 10 Correlation of isotopic whole-rock compositions of Group 1 volcanic rocks (open squares) with coeval plutonic rocks (filled circles).

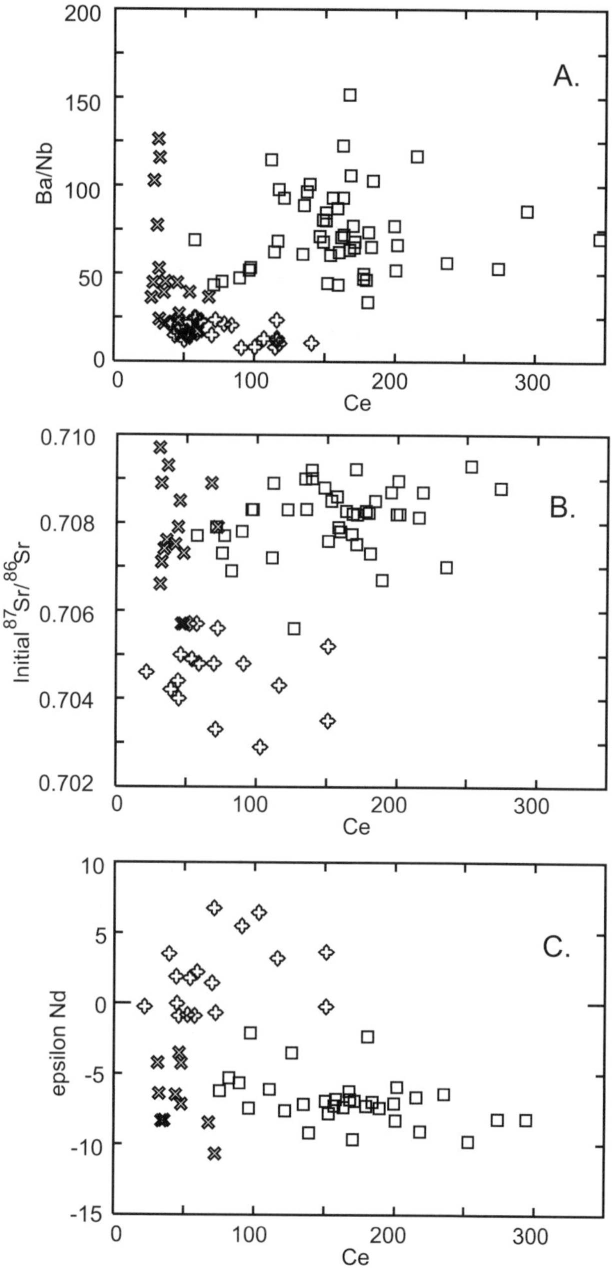

Figure 11 Discrimination among Group 1, 2 and 3 basalts based on trace element and isotopic compositions. The samples plotted have SiO_2 compositions between 46 and 54 wt.%.

may reflect partial melting of garnet-bearing peridotite source and spinel-bearing peridotite sources, respectively. Thus, it appears that Group 1 and Group 2 basalts record two distinct lithospheric mantle sources, with Group 2 basalts reflecting a less-enriched lithospheric mantle source at shallower depths.

Group 3 basalts have distinctive compositions which suggest derivation by partial melting of peridotite from the asthenosphere. Group 3 basalts display variable but depleted isotopic compositions (ε_{Nd} 1·5 to +7; Sri=0·7027–0·706) (Fig. 11) similar to ocean island basalt (OIB). In Basin and Range literature, it is common to reference basalt trace compositions to average OIB as a method for evaluating a potential asthenospheric source. Group 3 basalts have trace-element compositions which are very similar to average OIB (e.g. low Ba/Nb ratios, Fig. 11) (cf. Fitton *et al.* 1991; Feuerbach *et al.* 1993). The occurrence of OIB-like basalt is restricted to a few localities in the Basin and Range; all such basalts are young (<6 Ma) and represent the latest volcanic activity at each locale. The shift from basalt production in the lithospheric mantle to production in the asthenosphere is generally attributed to asthenospheric upwelling in response to extension-related lithospheric thinning (Fitton *et al.* 1991; Harry & Leeman 1995).

5.4.2. Correlation of NCREC basalts with mafic plutonic components. Mafic components in plutonic rocks are of three major types: (1) early gabbro to diorite intrusive units (e.g. phase 1 in the Mount Perkins pluton and Horsethief Canyon diorite in the Wilson Ridge pluton); (2) mafic microgranitoid enclaves and sheets hosted in generally more felsic rocks (recording mingling and mixing events among contrasting magmas); and (3) late dykes cutting plutonic units (either as mafic dykes or as enclaves in intermediate and felsic dykes). Many mafic samples have MgO (6·4–8·1 wt.%), Ni (65–438 ppm) and Cr (97–240 ppm) contents which point to a mantle source. Also consistent with a mantle origin are cumulates of olivine and clinopyroxene observed in both the Mount Perkins and Aztec Wash plutons. These cumulates indicate that, in some instances, fairly primitive magmas reached crustal levels represented by these plutons. In contrast, many of the mafic components in the plutonic suites have more evolved compositions compared to those expected for primary mantle-derived magmas. Some of these mafic components were affected by magmatic processes within exposed plutons (magma chambers), but others (chilled enclaves, late plutonic dykes and gabbros lacking co-magmatic felsic components) clearly evolved prior to arrival and must record magma chamber processes at deeper levels of the lithosphere. Despite these issues, it is possible to use the most mafic non-cumulate plutonic rocks to make correlations with mafic components in the volcanic record and inferences about mafic magma sources, as discussed below.

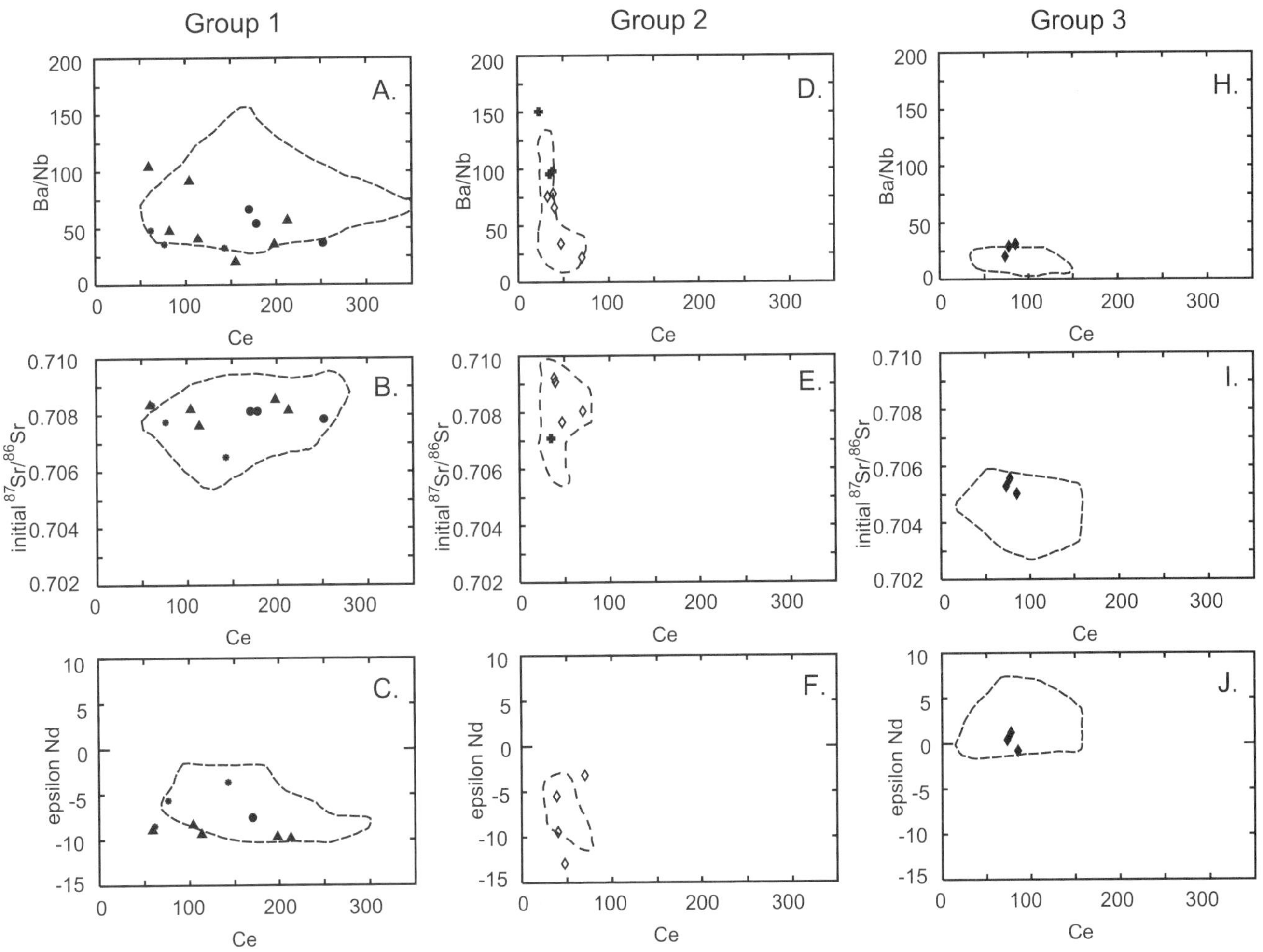

Figure 12 Correlation of mafic plutonic components with Group 1, 2 and 3 basalts: (A–C) Mafic enclaves from the Wilson Ridge (filled triangles), Aztec Wash (filled circles) and Searchlight (asterisk) plutons compared with fields defined by Group 1 basalts. The field outlines are based on the Group 1 samples shown in Figure 11. (D–F) The precursor isotropic gabbro from Mount Perkins (open diamonds) and late mafic dykes from the Spirit Mountain pluton (filled crosses) compared with fields defined by Group 2 basalts. The field outlines are based on the Group 2 samples shown in Figure 11. (G–I). Mafic enclaves from phases 2 and 3 of the Mount Perkins pluton (filled diamonds) compared with fields defined by Group 3 basalts. The field outlines are based on the Group 3 samples shown in Figure 11.

Given the geochronologic, compositional and petrogenetic correlations which are evident between NCREC plutonic and Group 1 volcanic suites, one would predict that mafic plutonic components (1) directly correlate with Group 1 basaltic rocks and (2) record an enriched, lithospheric mantle source. This is indeed the case for many, but not all, of the mafic components in the data set. Mafic microgranitoid enclaves from the Wilson Ridge (3–7·7 wt.% MgO), Aztec Wash (4·7–6·4 wt.% MgO) and Searchlight (5–9·7 wt.% MgO) plutons correlate fairly well with Group 1 basaltic rocks in terms of trace element, and Nd and Sr isotopic compositions (Fig. 12a, b, c). In contrast, mafic enclaves from phase 2 and 3 of the Mount Perkins pluton share trace element, and Nd and Sr isotopic affinities with Group 3 basalts (Fig. 12h, i, j). In addition, the phase 1 isotropic gabbro from the Mount Perkins pluton shares trace element, and Nd and Sr isotopic affinities with Group 2 basalts, as do late dykes cutting the Spirit Mountain pluton (Bradshaw 1991) (Fig. 12d, e, f). With the exception of the Spirit Mountain dykes, these mafic rocks can be shown by field relations to be same age (e.g. mingled and mixed enclaves) or older (pre-cursor gabbros) than the rocks used to date the plutons. Despite the restriction of plutonic ages to between 17 and 13 Ma, mafic components in the plutonic system can be correlated by composition to basaltic rocks which do not appear in the volcanic record until between 11 and 6 Ma. It would appear that, for the period prior to 11 Ma, the plutonic rocks provide a more complete record of magma types and magma sources than do contemporaneous volcanic rocks.

5.5. Felsic magma source

The preceding petrologic discussions have treated NCREC felsic rocks as a proxy end-member for MFC (mixing) processes. Implicit in those discussions is the assumption that such rocks approximate independent crustal melts rather than representing products of those same MFC processes. This section attempts to justify that assumption and discusses the nature of the crustal source. Rhyolites and felsic plutonic rocks in the NCREC are fairly uniform in major element, trace element and isotopic composition (Figs 8, 9, 10). Some variability does exist in rare-earth element (REE) data, with some samples showing negative Eu anomalies and depleted MREE (Morikawa 1994; McDaniel 1995; Falkner *et al.* 1995; Bachl *et al.* 2001). However, the latter features are likely the result of fractional crystallisation in shallow crustal magma chambers (e.g. Searchlight pluton; Bachl *et al.* 2001). A few rhyolite samples have elevated initial $^{87}Sr/^{86}Sr$ (Fig. 10), but these may simply reflect contamination in the upper crust. In general, though, there is more compositional uniformity (less scatter) at the felsic end of the data set than seen at either intermediate or mafic compositions (Figs 8, 9, 10).

Small fractions of inherited Proterozoic zircons have been reported from NCREC granites (Falkner *et al.* 1995; Metcalf *et al.* 1995), pointing to an ancient crustal component in their magmas. However, neodymium and Sr isotopic compositions for NCREC felsic rocks are intermediate between that of local Proterozoic basement ($\varepsilon_{Nd} \sim -19$ and $^{87}Sr/^{86}Sr \sim 0{\cdot}72$ in the late Miocene time; Bennett & DePaolo 1987) and typical mantle values represented, for example, by NCREC basaltic rocks (Fig. 10). An obvious explanation is that these felsic magmas represent crust–mantle hybrids. DePaolo *et al.* (1992) used a modelled Neodymium Crustal Index (NCI) to estimate the proportions of crust and mantle components in Tertiary granites of the western USA. A Miocene NCREC crust–mantle hybrid granite with $\varepsilon_{Nd} = -11$ would have a NCI of 0·2 (20% crust) using a mantle component with $\varepsilon_{Nd} = -9$ similar to Group 1 basalt (DePaolo *et al.* 1992). However, if the mantle component were Group 3 basalt (e.g. Mount Perkins pluton), and assuming interaction with the same Proterozoic crust, the NCI clearly would be significantly higher (e.g. above 0·8). It would be fortuitous indeed to find a varied suite of mantled-derived mafic magmas such as those in the NCREC, each mixing with a Proterozoic crustal component in just the right proportions, to produce felsic magmas of such similar isotopic composition. It seems far more likely that the most felsic of NCREC magmas ($\sim$72 wt.% SiO_2) approximate primary partial melts of a fairly homogeneous hybrid crust with little, if any, Miocene mantle component present. In discussing the felsic component for the Aztec Wash pluton, Falkner *et al.* (1995) suggested that crust–mantle interactions during Mesozoic magmatism, involving a fairly uniform mantle source and Proterozoic basement, could have produced a hybrid crust that, in turn, became the source for NCREC felsic magmas in Miocene time. Falkner *et al.* (1995) noted that surface exposures of Mesozoic hybrid granitoid rocks in the nearby eastern Mojave province have Nd and Sr isotopic compositions appropriate for such a crustal source. While a homogeneous crustal magma source the size of the NCREC seems fairly improbable, and calls upon a major crustal homogenisation event in the Mesozoic, it appears to be the most plausible explanation for the observed data.

6. Discussion

6.1. The NCREC volcanic–plutonic super-suite

Correlation of the NCREC volcanic and plutonic records provides a link between magmatic products represented by lava flows and evidence of magma chamber processes found in the pluton suites. The magmatic rocks in NCREC are probably best viewed as a volcanic–plutonic super-suite composed of numerous individual suites represented by individual plutons and volcanic centres. Geochronologic data clearly document the coeval nature of the Group 1 volcanic suite and exposed plutons. Whole-rock major element, trace element and isotopic correlations between plutonic rocks and Group 1 volcanic rocks are consistent with a cogenetic origin. Significant from a petrogenetic standpoint are inflections on trace-element Harker diagrams where volcanic trends are closely matched by plutonic trends (Fig. 10). Equally significant are the Nd and Sr isotopic variations which correlate with these inflections. Such trends must be explained by a two-stage petrogenetic history that involved open-system behaviour in combination with fractional crystallisation. Thus, the volcanic and plutonic records are not only linked by age and composition, but by common petrogenetic processes, and as such, these rocks should be viewed as a single volcanic–plutonic super-suite.

The NCREC plutonic rocks provide a direct record of the magma chamber processes and complement petrogenetic interpretations based on whole-rock chemical and isotopic data. Specifically, the plutonic rocks show that magma mixing (as opposed to assimilation) was the primary open-system process. Collectively, the Wilson Ridge and Aztec Wash plutons provide evidence for repeated injections of mafic magmas into upper crustal felsic magma chambers where magma interactions produced intermediate hybrid magmas. In particular, the Wilson Ridge pluton appears to provide an example of a large-scale magma chamber where magma mixing produced significant amounts of hybrid intermediate magmas. Both the Mount Perkins (phase 1) and Aztec Wash plutons provide evidence for limited fractional crystallisation of mafic magmas in these upper-crustal magma chambers. The Searchlight pluton provides evidence of upper crustal fractional crystallisation of felsic magmas. Thus, physical evidence is found in the plutonic record that supports the magma chamber processes implied by petrogenetic modelling.

Perhaps the most surprising result of the present study is the recognition in the mafic plutonic record of compositions which correlate not only to coeval Group 1 basalts, but the younger Group 2 and Group 3 basalts as well. What appears in the volcanic record to be a straightforward stratigraphic succession of three different basalt compositions is shown in the plutonic record to be coeval at about 16 Ma. Thus, basaltic magma compositions are apparent in the plutonic environment some 5–10 million years prior to their eruption as lava flows at the surface. It would seem that the plutonic record gives a more complete picture of the magma system during the syn-extensional period between 20 and 11 Ma. The presence of such a diverse suite of mafic magma compositions during the syn-extensional period may help explain the large amount of scatter seen in the chemical and isotopic data sets at mafic and intermediate compositions. This point is particularly apparent in the Nd isotopic data, where much of the scatter is attributable to the Mount Perkins data.

6.2. Lithospheric-scale magma systems

The NCREC volcanic–plutonic super-suite should be viewed as the upper-crustal portion of a lithospheric-scale magma system. Mafic components in the plutonic record clearly can be correlated with mafic lava flows found in the volcanic record. More importantly, these mafic components require a source in the mantle. Trace element and isotopic data indicate both lithospheric and asthenospheric mantle sources. An independent crustal source for NCREC felsic magmas is indicated by whole-rock chemical and isotopic data, combined with the presence of inherited Proterozoic zircons. Intermediate magmas in the NCREC can be linked to mantle-derived mafic end-members and crustal-derived felsic end-members via MFC processes. At one point or another in the history of NCREC magmatism, nearly every magma composition was erupted, from primitive basalts and rhyolites which approximate end-member magmas to a full spectrum of intermediate hybrid magmas to rhyolites fractionated in shallow magma chambers. The NCREC plutons can be linked genetically to both the eruptive products and the inferred primary magma sources. As such, the plutons provide links between magma source regions which are the physical starting points of magma systems and the volcanic centres which are the physical end points of magma systems. Therefore, the NCREC plutons should be viewed as two-dimensional slices through three-dimensional, lithospheric-scale magma systems.

Three lines of evidence suggest that exposed NCREC magma chambers (plutons) were underlain by still deeper magma chambers, where MFC processes modified primary mantle-derived magmas prior to arrival in the upper crust. First, most of the mafic magmas which appear as enclaves in NCREC plutons already had evolved significantly from mantle-derived parental magmas prior to mingling (Larsen & Smith 1991; Falkner *et al.* 1995; Metcalf *et al.* 1995; Patrick & Miller 1997), but corresponding mafic cumulate sequences are rare and of low volume. This implies that primary mantle-derived mafic magmas stalled in the lower crust and fractionated prior to reaching crustal levels represented by the NCREC plutons. Secondly, suites of quenched mafic enclaves in NCREC plutons define clear MFC magma evolution trends which require subjacent magma chambers (e.g. phases 2 and 3 in the Mount Perkins pluton). Because the enclaves are quenched, the MFC trends must predate mingling at the current exposure level. This situation is akin to chemically zoned eruptive units (e.g. ash flow tuffs) which require a chemically zoned subjacent magma chamber. Finally, the late-stage dykes which post-date pluton solidification include intermediate hybrid magmatic products (e.g. the Mount Perkins, Wilson Ridge and Aztec Wash plutons). These dykes are aphanitic and planar, and must reflect late-stage activity during the thermal decline of the upper crustal magma system. Because the dykes have compositions which can be related to the plutons they cut, including hybrid intermediate rocks, magma hybridisation must have continued at depth, either in magma chambers or in conduits during magma transport (e.g. Collins *et al.* 2000). Collectively, these observations suggest that upper-crustal magma chambers were themselves connected to deeper-level magma chambers and magma conduits. The Cretaceous Sierra Nevada batholith, with mid-crustal levels exposed in the north ($\sim$12 km, mingled and mixed magmas) and deeper levels in the south ($\sim$22 km, mafic cumulates) may provide an example of the deeper portions of such a magma system (Frost & Mahood 1987; Ague & Brimhall 1988; Sams & Saleeby 1988; Coleman *et al.* 1995; Sisson *et al.* 1996). The concept of crust–mantle hydridisation and differentiation as a multistage process occurring at various crustal levels and among variously hybridised magmas was discussed in some detail by Barbarin & Didier (1992).

7. Summary and conclusions

The NCREC provides an excellent record of coeval and cogenetic volcanic and plutonic activity between about 17 and 13 Ma. Chemical and isotopic data from the plutonic suites allows direct correlation with the contemporaneous volcanic rocks. Thus, to a first approximation, plutonic rocks can be equated to volcanic rocks which represent magma compositions. Petrogenetic interpretations based on chemical and isotopic data view intermediate rocks as the products of a combination of magma mixing and fractional crystallisation involving mantle-derived mafic and crustal-derived felsic parental magmas. The NCREC plutons provide a record of the inferred magma chamber processes, including mafic cumulate sequences, felsic cumulate sequences, magma mingling features and features indicative of more advanced stages of magma mixing.

The NCREC volcanic–plutonic super-suite appears to represent the upper-crustal (<13 km) portion of a lithospheric-scale magma system characterised by interaction of mantle and crustal melts. This upper-crustal volcanic–plutonic complex must have been connected to zones of mantle and crustal melting by conduits and deeper magma chambers where primary mantle melts stalled and were modified by MFC process prior to introduction into upper-crustal magma chambers.

Three mafic magma types are recognised in the NCREC and record partial melting of both lithospheric and asthenospheric mantle sources. All three mafic magmas can be identified as mafic components in the plutonic suite at or before about 16 Ma and appear to have served as end-members for MFC processes. Two of the mafic magma types, including the one derived from the asthenosphere, are not recognised in the volcanic suite until after the upper crustal plutons had solidified. Thus, the plutons appear to record asthenospheric melting some 10 million years before such magmas appear in the volcanic suites.

The felsic magmas appear to be crustal melts generated by partial melting of fairly homogeneous source. Because of mixing with a single felsic end-member, the compositions of NCREC intermediate hybrid magmas begin to converge at silica values above about 60 wt.%, regardless of the composition of the mafic end-member. This observation is significant when one considers that intermediate and felsic rocks are abundant in many batholiths, and mafic rocks are scarce, absent or overlooked. Many studies in granitoid batholiths view (or assume) quartz diorite and tonalite as primary crustal magmas, and use them to image source regions (e.g. Gromet & Silver 1988; Silver & Chappell 1988; Chappell 1996). Such an assumption applied to intermediate rocks of the NCREC would have led to some erroneous conclusions.

Plutonic studies have been advanced greatly in recent years by the recognition that plutons can be viewed as magma chambers with protracted histories (Barnes *et al.* 1990; Larsen & Smith 1991; Wiebe 1993a, b, 1994; Falkner *et al.* 1955; Metcalf *et al.* 1995; Poli *et al.* 1996; Patrick & Miller 1997; Collins 1998; Collins *et al.* 2000; Bachl *et al.* 2001; Wiebe *et al.* 2002; Hawkins & Wiebe 2003). The integration of volcanic and plutonic records has the potential to make additional advances and provide greater understanding of magma systems. Primarily, plutonic studies can provide physical evidence that permits evaluation of the magma chamber processes implied by petrogenetic modelling in volcanic rocks. Finally, we need to begin to view plutons as parts of larger magma systems, as two-dimensional slices through three-dimensional lithospheric-scale magma systems.

8. Acknowledgements

I am indebted to a number of key researchers in the NCREC, most notably Gene Smith, Calvin Miller and Jim Faulds, and their many students. The concepts discussed in this paper have been greatly influenced by discussions with, papers by and field trips shared with Bob Wiebe, Bill Collins, Calvin Miller, Cal Barnes, Ron Vernon, Bernard Barbarin and David Hawkins. Calvin Miller and Nancy McMillan provided thoughtful reviews of the manuscript.

9. References

Anderson, J. L. & Smith, D. R. 1994. The effect of temperature and fugacity on the Al-in-hornblende geobarometer. *American Mineralogist* **80**, 549–59.

Anderson, R. E. 1971. Thin skin distension in Tertiary rocks of southeastern Nevada. *Geological Society of America Bulletin* **82**, 42–58.

Anderson, R. E., Longwell, C. R., Armstrong, R. L. & Marvin, R. F. 1972. Significance of K-Ar ages of Tertiary rocks from the Lake Mead region, Nevada-Arizona. *Geological Society of America Bulletin* **83**, 273–88.

Anderson, R. E., Barnhard, T. P. & Snee, L. W. 1994. Roles of plutonism, mid-crustal flow, tectonic rafting and horizontal collapse in shaping the Miocene strain field of the Lake Mead area, Nevada and Arizona. *Tectonics* **13**, 1381–410.

Ague, J. J. & Brimhall, G. H. 1988. Regional variations on bulk chemistry, mineralogy, and the compositions of mafic and accessory minerals in the batholiths of California. *Geological Society of America Bulletin* **100**, 891–911.

Bachl, C. A., Miller, C. F., Miller J. S. & Faulds, J. E. 2001. Construction of a pluton: evidence from an exposed cross section of the Searchlight pluton, Eldorado Mountains, Nevada. *Geological Society of America Bulletin* **113**, 1213–28.

Barbarin, B. & Didier, J. 1992. Genesis and evolution of mafic microgranular enclaves through various types of interaction between coexisting felsic and mafic magmas. *Transactions of the Royal Society of Edinburgh: Earth Sciences* **83**, 145–54.

Barnes, C. G., Allen, C. M., Hoover, J. D. & Brigham, R. H. 1990. Magmatic components of a tilted plutonic system, Klamath Mountains, California. *In* Anderson, J. L. (ed.) *The Nature and Origin of Cordilleran Magmatism. Geological Society of America Memoir* **174**, 331–46.

Bennett, V. C. & DePaolo, D. J. 1987. Proterozoic crustal history of the western United States as determined by Neodymium isotopic mapping. *Geological Society of America Bulletin* **99**, 674–85.

Bergantz, G. W. 1989. Underplating and partial melting: implications for melt generation and extraction. *Science* **245**, 1093–5.

Boland, K. A. 1996. *The petrogenesis of andesites produced during regional extension: examples from the northern McCullough Range, NV and Xitle Volcano, Mexico*. Unpublished MSc thesis, University of Nevada, Las Vegas.

Bradshaw, T. K. 1991. *Tectonics and magmatism in the Basin and Range province of the western United States*. Unpublished PhD dissertation, Open University, Milton Keynes.

Bradshaw, T. K., Hawkesworth, C. J. & Gallagher, K. 1993. Basaltic volcanism in the southern Basin and Range: no role for mantle plume. *Earth and Planetary Science Letters* **116**, 45–62,

Bridwell, H. L. 1991. *The Sloan Sag: a mid-Miocene volcanotectonic depression, north-central McCullough Mountains, southern Nevada*. Unpublished MSc thesis, University of Nevada, Las Vegas.

Camin, M. C. & Metcalf, R. V. 1999. Petrogenetic evolution of the late stage dikes of the Miocene Mt. Perkins pluton, Arizona: Evidence of open system processes. *Geological Society of America Abstracts with Programs* **31–6**, A-43.

Cascadden, T. E. 1991. *Style of volcanism and extensional tectonics in the eastern Basin and Range Province, northern Mohave County, Arizona*. Unpublished MSc thesis, University of Nevada, Las Vegas.

Cates, N. L., Miller, J. S., Miller, C. F., Wooden, J. L., Erickson, S. & Means, M. 2003. Longevity of plutonic systems: SHRIMP evidence from Aztec Wash and Searchlight plutons. *Geological Society of America Abstracts with Programs* **25–2**, 63.

Chappell, B. W. 1996. Compositional variation within granite suites of the Lachlan Fold Belt: its causes and implications for the physical state of granite magma. *Transactions of the Royal Society of Edinburgh: Earth Sciences* **87**, 159–70.

Cole, E. D. 1989. *Petrogenesis of late Cenozoic alkalic basalt near the eastern boundary of the Basin-and-Range, upper Grand Wash Trough, Arizona and Gold Butte, Nevada*. Unpublished MSc thesis, University of Nevada, Las Vegas.

Coleman, D. S., Glazner, A. F., Miller, J. S., Bradford, K. J., Frost, T. P., Joye, J. L. & Bachl, C. A. 1995. Exposure of a Late Cretaceous layered mafic-felsic magma system in the central Sierra Nevada batholith, California. *Contributions to Mineralogy and Petrolology* **120**, 129–36.

Collins, W. J. 1996. Lachlan Fold Belt granitoids: products of three-component mixing. *Transactions of the Royal Society of Edinburgh: Earth Sciences* **87**, 171–81.

Collins, W. J., Richards, S. R., Healy, B. E. & Ellison, P. I. 2000. Origin of heterogeneous mafic enclaves by two-stage hybridization in magma conduits (dykes) below and in granite magma chambers. *Transactions of the Royal Society of Edinburgh: Earth Sciences* **91**, 27–45.

Coiner L. V., Cates, N. L., Miller, C. F. & Harper, B. E. 2003. Cyclic stratigraphy in the Aztec Wash pluton, Nevada: vertical construction of a composite pluton. *Geological Society of America Abstracts with Programs* **25–1**. 63.

Daley, E. E. & DePaolo, D. J. 1992. Isotopic evidence for lithospheric thinning during extension: Southern Great Basin. *Geology* **20**, 104–8.

Danielson, L. R. 1998. *Petrogenesis of the phase 1 gabbro-diorite of the Miocene Mt. Perkins pluton, northwest Arizona: implications for the role of mantle derived mafic magmas in calc-alkaline magma evolution*. Unpublished MSc thesis, University of Nevada, Las Vegas.

Darvall, P., Gans, P. B. & Lister, G. S. 1991. Normal faulting in the Eldorado Mountains, southeastern Nevada. *Geological Society of America Abstracts with Programs* **23–2**, 17.

DePaolo, D. J. 1981. Trace element and isotopic effects of combined wall-rock assimilation and fractional crystallization. *Earth and Planetary Science Letters* **53**, 189–202.

DePaolo, D. J., Perry, F. V. & Baldridge, W. S. 1992. Crustal versus mantle sources of granitic magmas: a two-parameter model based in Nd isotopic studies. *Transactions of the Royal Society of Edinburgh: Earth Sciences* **83**, 439–46.

Ernst, W. G., Coleman, D. S. & Van de Ven, C. M. 2003. Petrochemistry of granitic rocks in the Mount Bancroft area: implications for arc evolution, central White Mountains, easternmost California. *Geological Society of America Bulletin* **115**, 499–512.

Falkner, C. M., Miller, C. F., Wooden, J. L. & Heizler, M. T. 1995. Petrogenesis and tectonic significance of the calc-alkaline, bimodal Aztec Wash pluton, Eldorado Mountains, Colorado River extensional corridor. *Journal of Geophysical Research* **100**, 10,453–5,226.

Farmer, G. L., Perry, F. V., Semken, S., Crowe, B. Curtis, D. & DePaolo, D. J. 1989. Isotopic evidence of the structure of subcontinental lithospheric mantle in the southern Nevada. *Journal of Geophysical Research* **94**, 7885–98.

Faulds, J. E., Geissman, J. W. & Mawer, C. K. 1990. Structural development of a major extensional accommodation zone in the Basin and Range Province, northwestern Arizona and southern Nevada: implications for kinematic models of continental extension. *In* Wernicke, B. P. (ed.) *Basin and Range Extensional Tectonics Near the Latitude of Las Vegas, Nevada. Geological Society of America Memoir* **176**, 37–76.

Faulds, J. E., Gans, P. B. & Smith, E. I. 1994. Spatial and temporal patterns of extension in the northern Colorado River extensional corridor, northwestern Arizona and southern Nevada. *Geological Society of America Abstracts with Programs* **26–2**, 51.

Faulds, J. E., Feuerbach, D. L., Reagan, M. K., Metcalf, R. V., Gans, P. B. & Walker, J. D. 1995. The Mt. Perkins block, northwestern Arizona: an exposed cross section of an evolving, pre-extensional magma system. *Journal of Geophysical Research*. **100**, 15,249–66.

Feuerbach, D. L. 1986. *Geology of the Wilson Ridge pluton: a mid-Miocene quartz monzonite intrusion in the northern Black Mountains, Mohave County, Arizona and Clark County, Nevada*. Unpublished MSc thesis. University of Nevada, Las Vegas.

Feuerbach, D. L., Smith, E. I., Shafiqullah, M. & Damon, P. E. 1991. New K-Ar dates for late-Miocene to early Pliocene mafic volcanic rocks in the Lake Mead area. Nevada and Arizona. *Isochron West* **57**, 17–20.

Feuerbach, D. L., Smith E. I., Walker, J. D. & Tangeman, J. A. 1993. The role of the mantle during crustal extension: constraints from geochemistry of volcanic rocks in the Lake Mead area, Nevada and Arizona. *Geological Society of America Bulletin* **105**, 1561–75.

Fitton, J. G., James, D. & Leeman, W. P. 1991. Basic magmatism associated with Cenozoic extension in the western United States: compositional variations in space and time. *Journal of Geophysical Research* **96–B8**, 13,693–711.

Frost, T. P. & Mahood, G. A. 1987. Field, chemical and physical constraints on mafic-felsic magma interaction in the Lamark Granodiorite, Sierra Nevada, California. *Geological Society of America Bulletin* **99**, 272–91.

Glazner, A. F. & Bartley, J. M. 1984. Timing and tectonic setting of Tertiary low-angle normal faulting and associated magmatism in the southwestern United States. *Tectonics* **3**, 385–96.

Gromet, P. & Silver, L. T. 1988. REE variations across the Peninsular Ranges batholith: implications for batholithic petrogenesis and crustal growth in magmatic arcs. *Journal of Petrology* **28**, 75–125.

Harper, B. E., Miller, C. F., Koteas, G. C., Cates, N. L., Wiebe, R. A., Lazzareschi, D. S. & Cribb, J. W. 2004. Granites, dynamic magma chamber processes, and pluton construction: the Aztec Wash pluton, Eldorado Mountains, Nevada, USA. *Transactions of the Royal Society of Edinburgh: Earth Sciences* **95**, 277–95.

Harry, D. L. & Leeman, W. P. 1995. Partial melting of melt metasomatised subcontinental mantle and the magma source potential of the lower lithosphere. *Journal of Geophysical Research* **100**, 10,255–70.

Hawkesworth, C. J., Turner, S., Gallagher, K., Hunter, A., Bradshaw, T. & Rogers, N. 1995. Calc-alkaline magmatism, lithospheric thinning, and extension in the Basin and Range *Journal of Geophysical Research* **100**, 10,271–86.

Hawkins, D. P. & Wiebe, R. A. 2003. High-precision temporal constraints on the construction of a periodically replenished, subvolcanic magma chamber: the Silurian Vinalhaven intrusion, Maine, USA. *Fifth Hutton Symposium on the Origin of Granites and Related Rocks, Abstracts with Programs. Geological Survey of Japan Interim-Report* **29**, 43.

Hibbard, M. J. 1991. Textural anatomy of twelve magma-mixed granitoid systems. *In* Didier, J. & Barbarin, B. (eds) *Enclaves and granite petrology, Developments in Petrology* **13**, 431–44.

Holden, P., Halliday, A. N. & Stephens, W. E. 1987. Neodymium and strontium isotope content of microdiorite enclaves points to mantle input to granitoid production. *Nature* **330**, 53–6.

Hooper, P. R., Bailey, D. G. & Holder, G. A. M. 1995. Tertiary calc-alkaline magmatism associated with lithospheric extension in the Pacific Northwest. *Journal of Geophysical Research* **100**, 10,303–416.

Howard, K. A., Wooden, J. L., Simpson, R. W. & Pease, V. L. 1996. Extension-related plutonism along the Colorado River extensional corridor. *Geological Society of America Abstracts with Programs* **28–7**, 251.

Howard, K. A. & John, B. E. 1987. Crustal extension along a rooted system of imbricate low angle faults: Colorado River extensional corridor, California and Arizona. *In* Coward, M. P. (ed.) *Continental Extensional Tectonics. Geological Society, London, Special Publication* **28**, 299–311.

Huppert, H. E. & Sparks, R. S. J. 1988. The generation of granitic magmas by intrusion of basalt into continental crust. *Journal of Petrology* **29**, 599–624.

Kapp, J. D., Miller, C. F. & Miller, J. S. 2002. Ireteba pluton, Eldorado Mountains, Nevada: late, deep-source, peraluminous magmatism in the Cordilleran interior. *Journal of Geology* **110**, 649–69.

Kempton, P. D., Fitton, J. G., Hawkesworth, C .J. & Ormerod, D. S. 1991. Isotopic and trace element constraints on the composition and evolution of the lithosphere beneath the southwestern United States. *Journal of Geophysical Research* **96**, 13,713–35.

Lameyre, J. & Bonin, B. 1991. Granites in the main plutonic series. *In* Didier, J. & Barbarin, B. (eds) *Enclaves and Granite Petrology, Developments in Petrology* **13**, 3–17.

Larsen, L. L. & Smith, E. I. 1991. Mafic enclaves in the Wilson Ridge Pluton, northwestern Arizona: implications for the generation of a calc-alkaline intermediate pluton in an extensional environment. *Journal of Geophysical Research* **95**, 17,693–716.

McDaniel, S. M. 1995. *Geochemical evolution of a mid-Miocene synextensional volcanic complex: the Dolan Springs volcanic field, northwestern Arizona.* Unpublished MSc thesis. University of Nevada, Las Vegas.

Metcalf, R. V., Smith, E. I. & Mills, J. G. 1993. Magma mixing and commingling in the northern Colorado River extensional corridor: constraints on the production of intermediate magmas. *In* Larhen, M. M, Trexler, J. H., Jr. & Spinosa, S. (eds) *Crustal Evolution of the Great Basin and Sierra Nevada. Geological Society of America Guidebook*, 35–55.

Metcalf, R. V., Smith E. I., Walker, J. D., Reed, R. C. & Gonzales, D. A. 1995. Isotopic disequilibrium among commingled hybrid magmas: evidence of a two-stage magma mixing-commingling process in the Mt. Perkins Pluton, Arizona. *Journal of Geology* **103**, 509–27.

Metcalf, R. V., Danielson, L. R. & Camin, M. C. 1999. Role of mafic magmatism in the genesis of intermediate I-type granitoid suites: Miocene Mt. Perkins pluton, Arizona. *In* Barbarin, B. (ed.) *The Origin of Granites and Related Rocks IVth Hutton Symposium Abstracts, Documents du BRGH* **290**, 32.

Metcalf, R. V. & Smith, E. I. 1991. Hornblende geobarometry from two Miocene plutons: implications regarding uplift and block rotation during Basin and Range extension. *Geological Society of America Abstracts with Programs* **23–5**, A245.

Miller, C. F., Koteas, G. C., Erickson, S. M., Steinwinder, T. R., Miller, J. S., Wiebe, R. A., Wooden, J. L., Faulds, J. E., Cates, N. L., Coiner, L. V. & Harper, B. S. 2003. Depositional history of a bimodal, recharging magma chamber: construction of the Aztec Wash pluton, Nevada, USA. *Fifth Hutton Symposium on the Origin of Granites and Related Rocks, Abstracts with Programs, Geological Survey of Japan Interim-Report* **29**, 91.

Mills, J. G., Jr. 1985. *The geology and geochemistry of volcanic and plutonic rocks in the Hoover Dam 7 1/2 minute quadrangle, Clark County, Nevada and Mohave County, Arizona.* Unpublished MSc thesis. University of Nevada, Las Vegas.

Mills, J. G., Jr. 1994. Geologic map of the Hoover Dam Quadrangle, Arizona and Nevada. *Nevada Bureau of Mines and Geology* **Map 102**.

Morikawa, S. A. 1993. *The geology of the Tuff of Bridge Spring: southern Nevada and northwestern Arizona.* Unpublished MSc thesis. University of Nevada, Las Vegas.

Myers, I. A. 1984. *Geology and mineralization at the Cyclopic Mine, Mohave County, Arizona.* Unpublished MSc thesis. University of Nevada, Las Vegas.

Naumann, T. R. 1987. *Geology of the central Boulder Canyon quadrangle, Clark County, Nevada.* Unpublished MSc thesis. University of Nevada, Las Vegas.

Patrick, D. W. & Miller, C. F. 1997. Processes in a composite, recharging magma chamber: evidence from magmatic structures in the Aztec Wash pluton, Nevada. *Proceedings of the 30th International Geological Congress* **15**, 121–35.

Perry, F. V., Baldridge, W. S. & DePaolo, D. J. 1988. Chemical and isotopic evidence for lithospheric thinning beneath the Rio Grande rift. *Nature* **332**, 432–434.

Poli, G., Ghezzo, C. & Conticelli, S. 1989. Geochemistry of granitc rocks from the Hercynian Sardonia-Corsica batholith: implication for magma genesis. *Lithos* **23**, 247–66.

Poli, G., Tommasini, S. & Halliday, A. N. 1996. Trace element and isotopic exchange during acid-basic magma interaction processes. *Transactions of the Royal Society of Edinburgh: Earth Sciences* **87**, 225–32.

Ratajeski, K., Glazner, A. F. & Miller, B. V. 2001. Geology and geochemistry of mafic to felsic plutons in the Cretaceous intrusive suite of the Yosemite Valley, California. *Geological Society of America Bulletin* **113**, 1486–502.

Reid, J. B., Jr, Evans, O. C. & Fates, D. G. 1983. Magma mixing in granitic rocks of the central Sierra Nevada, California. *Earth and Planetary Science Letters* **66**, 243–61.

Reynolds, S. J. 1998. Geologic Map of Arizona, *Arizona Geological Survey, Report* **26**.

Robinson, D. M. & Miller, C. F. 1999. Record of magma chamber processes preserved in accessory minerals. *American Mineralogist* **84**, 1346–53.

Sams, D. B. & Saleeby, J. B. 1988. Geology and petrotectonic significance of crystalline rocks of the southernmost, Sierra Nevada, California. *In* Ernst, W.G. (ed.) *Metamorphism and Crustal Evolution of the Western United States. Ruby Volume* **VII**, 865–93. Englewood Cliffs, NJ: Prentice Hall.

Schmidt, C. S. 1987. *A mid-Miocene caldera in the central McCullough Mountains, Clark County, Nevada.* Unpublished MSc thesis. University of Nevada, Las Vegas.

Silver, L. T. & Chappell, B. W. 1988. The Peninsular Ranges batholith: an insight into the evolution of the Cordilleran batholiths of southwestern North America. *Transactions of the Royal Society of Edinburgh: Earth Sciences* **79**, 105–21.

Sisson, T. S., Grove, T. L. & Coleman, D. S. 1996. Hornblende gabbro sill complex at Onion Valley, California, and a mixing origin for the Sierra Nevada batholith. *Contributions to Mineralogy and Petrology* **126**, 81–108.

Smith E. I., Feuerbach, D. L., Naumann, T. R. & Mills, J. E. 1990. Mid-Miocene volcanic and plutonic rocks of the Lake Mead area of Nevada and Arizona: production of intermediate igneous rocks in an extensional environment. *In* Anderson, J. L. (ed.) *The Nature and Origin of Cordilleran Magmatism. Geological Society of America Memoir* **174**, 169–94.

Smith, E. I. & Faulds, J. E. 1994. Patterns of Miocene magmatism in the northern Colorado River extensional corridor, Nevada, Arizona and California. *Geological Society of America Abstracts with Programs* **26–2**, A-93.

Stewart, J. H. 1980. Geology of Nevada; a discussion to accompany the geologic map of Nevada. *Nevada Bureau of Mines and Geology, Special Publication* **4**, 132 pp.

Thompson, K. G. 1985. *Stratigraphy and petrology of the Hamblin–Cleopatra Volcano, Clark County, Nevada.* Unpublished MSc thesis, University of Texas at Austin.

Weber, M. E. & Smith, E. I. 1987. Structural and geochemical constraints on the reassembly of disrupted mid-Miocene volcanoes in the Lake Mead-Eldorado Valley area of southern Nevada. *Geology* **15**, 553–6.

Wiebe, R. A. 1993a. Basaltic injections into floored silicic magma chambers. *Eos Transactions of the American Geophysical Union* **74–1**, 3.

Wiebe, R. A. 1993b. The Pleasant Bay compositie layered intrusion, coastal Maine: ponding and crystallization of basaltic injections into a silicic magma chamber. *Journal of Petrology* **34**, 461–89.

Wiebe, R. A. 1994. Silicic magma chambers as traps for basaltic magmas: the Cadillac mountain intrusive complex, Mount Desert Island, Maine. *Journal of Geology* **102**, 423–37.

Wiebe, R. A., Blair, K. D., Hawkins, D. P. & Sabine, C. P. 2002. Mafic injections, in situ hybridization, and crystal accumulation in the Pyramid Peak granite, California *Geological Society of America Bulletin* **114**, 909–20.

Wiebe, R. A. & Collins, W. J. 1998. Depositional features and stratigraphy sections in granitic plutons: implication for the emplacement of granitic magma. *Journal of Structural Geology* **20**, 1273–89.

Williams, M. M. 2003. *Depositional history of the Black Mountains Conglomerate, Mohave County, Arizona: Sedimentary response to Miocene extension*. Unpubished MSc thesis, University of Nevada, Las Vegas.

Wooden, J. L. & Miller, D. M. 1990. Chronologic and isotopic framework for Early Proterozoic crustal evolution in the eastern Mojave Desert region, SE California. *Journal of Geophysical Research* **95**, 20,133–46.

RODNEY V. METCALF, Department of Geoscience, University of Nevada–Las Vegas, Las Vegas, NV 89154-4010, USA.
e-mail: metcalfr@unlv.nevada.edu

MS received 20 October 2003. Accepted for publication 24 August 2004.

Transactions of the Royal Society of Edinburgh: Earth Sciences, **95**, 375–390, 2004

Fragmentation, nucleation and migration of crystals and bubbles in the Bishop Tuff rhyolitic magma

Guilherme A. R. Gualda, David L. Cook, Rahul Chopra, Liping Qin, Alfred T. Anderson Jr and Mark Rivers

ABSTRACT: The Bishop Tuff (USA) is a large-volume, high-silica pyroclastic rhyolite. Five pumice clasts from three early stratigraphic units were studied. Size distributions were obtained using three approaches: (1) crushing, sieving and winnowing (reliable for crystals >100 µm); (2) microscopy of ~1 mm^3 fragments (preferable for crystals <100 µm); and (3) computerised X-ray microtomography of ~1 cm^3 pumice pieces.

Phenocryst fragments coated with glass are common, and the size distributions for all crystals are concave-upward, indicating that crystal fragmentation is an important magmatic process.

Three groups are recognised, characterised by: (1) high-density (0·759–0·902 g cm^{-3}), high-crystal content (14·4–15·3 wt.%) and abundant large crystals (>800 µm); concave-downward size distributions for whole crystals indicate late-stage growth with limited nucleation, compatible with the slow cooling of a large, gas-saturated, stably stratified magma body; (2) low-density (0·499 g cm^{-3}), low-crystal content (6·63 wt.%) and few large crystals; the approximately linear size distribution reveals that nucleation was locally important, perhaps close to the walls; and (3) intermediate characteristics in all respects.

The volumetric fraction of bubbles inversely correlates with the number of large crystals. This is incompatible with isobaric closed-system crystallisation, but can be explained by sinking of large crystals and rise of bubbles in the magma.

KEY WORDS: Crystal content, crystal size distribution, magmatic processes, pumice density, rhyolites.

Large-volume pyroclastic deposits are erupted snapshots of evolving plutons, and their stratigraphic variations help to stimulate and constrain ideas about the structure and evolutionary processes in buried bodies of magma (e.g. Smith & Bailey 1966; Hildreth 1977, 1979). Importantly, rhyolitic rocks, with varying crystal contents, allow a direct assessment of the evolution of silicic magma bodies in their initial stages of evolution.

In this work, the present authors describe a quantitative textural study of rhyolite pumice clasts from the Bishop Tuff, USA. Of particular interest to them is the apparent paucity or absence of euhedral phenocrysts of quartz smaller than about 200 µm in diameter. They assess the meaning of an apparent lack of small phenocrysts of quartz by relating the phenocryst size distribution to both the total crystal content and the density of pumice clasts (a convenient proxy for their pre-eruptive bubble content, magma density and depth in the pre-eruptive magma body; Wallace *et al.* 1999).

Moreover, Anderson *et al.* (2000), based on the composition of melt inclusions and zoning patterns in sanidine, suggested that crystal sinking and bubble rising were important magmatic processes. Such processes directly influence size distributions and other textural parameters, and the present authors assess whether their data are compatible with the relative movement of bubbles and quartz crystals within the Bishop Tuff magma.

1. Geology

The Bishop Tuff rhyolite raises some intriguing puzzles. Hildreth (1979) noted that the crystal content of Bishop Tuff pumice clasts varies from about 5 to about 25 wt.%. Normally, we expect crystallisation to increase as temperature decreases. For the Bishop magma, in general, the reverse was true. That the earliest erupted products derive mainly from the uppermost parts of a pre-eruptive body (Smith & Bailey 1966) has been broadly confirmed by Wallace *et al.* (1999) through documentation of melt inclusions in sequentially erupted Bishop Tuff rhyolite. The earliest erupted pumice clasts have, in general, the fewest crystals (Hildreth 1979), the most differentiated compositions and the most pre-eruptive gas (Wallace *et al.* 1999). These features, together with the reversed compositional zoning of later-erupted sanidine phenocrysts (Anderson *et al.* 2000), suggest that crystals sank and bubbles rose as the phenocrysts grew. Although bubbles and crystals apparently moved relative to each other, convective mixing of the Bishop magma appears negligible because phenocrysts within individual clasts of pumice have the same compositions and zoning patterns, and are not mixtures of crystals which occur in early- and late-erupted pumices (Hildreth 1979; Peppard *et al.* 2001). It is problematic to explain why early- and late-erupted phenocrysts failed to mix because big and little phenocrysts would sink at different rates, and this would cause mixing.

The present authors studied variously dense pumice clasts from the distal Chalfant Pumice Quarry (localities 94 and16/17 of Wilson & Hildreth 1997; Fig. 1), which is 42 km from the initial vent on the south rim of the Long Valley Caldera, the source of the Bishop Tuff. The sample selection emphasised large and rapidly cooled pumice clasts (Wallace *et al.* 2003), which would afford a range of studies, including those that benefit from analysing uncracked, bubble-free melt inclusions.

Figure 1 Field photograph taken at the Chalfant Quarry, locality 94 of Wilson & Hildreth (1997). Fall units F6, F7, F8, F9 and ash-flow unit Ig2Ea are indicated, as well as the approximate location of the contacts between them. Samples Ia-13 and Ia-15 were collected from unit Ig2Ea at this locality. Samples F7–12, F7–14 and F8–15 were collected from the indicated fall units at nearby localities at the Chalfant Quarry.

These pumice clasts are from stratigraphic plinian fall units F7, F8 and ignimbrite (pyroclastic flow) unit Ig2Ea of Wilson & Hildreth (1997). About 70 cm of fall unit F9 intervenes between F8 and Ig2Ea, and the present samples span a vertical range of about 2·5 m at the site. The Chalfant Pumice Quarry is near the SE distal termination of the remaining outcrops of the Bishop Tuff, where both the basal part of the pyroclastic flow (Ig2Ea) as well as the underlying plinian fall deposits contain glassy melt inclusions which cooled rapidly (Wallace *et al.* 2003). The samples came from above and below a thin layer of ash near the top of F8 that Wilson & Hildreth (1997) interpreted as resulting from a non-eruptive interlude when only fine ash continued to settle out, unaccompanied by pumice clasts. Above this stratigraphic position, rare pumice clasts containing pyroxene phenocrysts first appear, as do accidental fragments of Glass Mountain rhyolite, and these features are thought to reflect the opening of eastward-propagating vents which tapped different parts of the magma body as well as wall rock (Hildreth & Mahood 1986). Thus, the vents for Ig2Ea included newly tapped magma, but the F7 and F8 samples evidently derived from a more restricted vent near the southern rim of the Long Valley caldera. Newly opening vents plausibly erupted magma through the walls and/or roof of the pre-eruptive magma body, and this may enhance the chances of detecting variations in nucleation and crystallisation related to proximity to the margins of the magma body.

Two of the present authors' pumice clasts are from Wilson & Hildreth's (1997) fall unit F7 (present samples F7–12 and 14), one is from F8 (F8–15) and the other two are from Ig2Ea (Ia-13 and 15). The F7 pumice clasts are from the upper 30 cm of this unit as exposed on the southern face of the Chalfant Quarry. The F8 clast is from the upper 20 cm of this unit exposed on the northern face of the quarry (178–198 cm below the base of the overlying Ig2Ea pyroclastic flow). The Ig2Ea clasts are from a horizon of large clasts about 30 cm above the base of the pyroclastic flow (top of fall unit 9) that is exposed on the south wall of the quarry (Fig. 1). Within the individual sampling intervals, the present authors made no further stratigraphic distinction, and the relative stratigraphic positions of the two F7 and the two Ig2Ea clasts are not known.

2. Methods

The present authors aimed to establish the size distribution of phenocrysts in pumice clasts. Accordingly, they used a combination of three approaches: (1) crushing, sieving, winnowing and hand-picking using dissecting and petrographic microscopes and refractive index immersion oils; (2) optical microscopy of milligram-sized pumice fragments; and (3) X-ray tomography.

The authors also intended to determine the relative volumetric abundances of crystals, glass and vesicles. Their approach was to measure the mass fraction of crystals and the bulk density of pumice clasts. They assumed the average densities of glass and crystals to be 2·3 and 2·7 g cm^{-3}, respectively.

The data presented in this paper were obtained by six people, each one following basically the same procedures. However, some of the observations are subjective and may have differed from person to person. Therefore, the authors describe their methods in detail.

2.1. Pumice density

A total of 15 pumice clasts were originally selected. Each one was described noting its size, shape, and variations in colour and vesicles. The surface was cleaned using a stiff brush and a teasing needle under a dissecting microscope. The bulk density of each pumice clast was determined using the following:

$$\rho_{clast} = \frac{\rho_w m_{clast}}{m_{clast} - m_{submerged} + m_{putty}} \quad (1)$$

where ρ_w is the density of water; m_{clast} is the weight of the dry clast and $m_{submerged}$ is the weight (generally <0) of the clast submerged in water. To avoid infiltration of water, each pumice clast was wrapped using Silly Putty© before submersion, and m_{putty} is the weight of the putty used, submerged in water. All weights were obtained using a triple beam balance (± 0·02 g).

Five pumice clasts, representative of the variation in density observed, were selected for detailed crystal size distribution determination.

2.2. Physical separation of crystals

The first approach used to obtain crystal size distributions consisted of physically separating the crystals from the glassy matrix. The present authors followed Skirius (1990), crushing and sieving each clast, and separating crystals from pumice by winnowing in water. Significant material is lost during some of these processing procedures, as noted below.

2.2.1. Crushing. The main concern during crushing was artificial fragmentation of phenocrysts. Virtually all phenocrysts larger than about 0·5 mm in the pumice clasts have cracks (see below), generally intersecting with burst melt inclusions, and they fall apart when the supporting glass around them is crushed. To minimise fragmentation, the authors crushed the samples using a wooden pestle and cardboard mortar. Crushing was performed in several steps; after each step, the partly crushed material was sieved using a nylon sieve with 1·8-mm openings, so that smaller fragments were removed and not repeatedly crushed.

2.2.2. Sieving. The present authors sieved the crushed pumice using nylon sieves with openings of 1·8, 0·8, 0·5, 0·2 and 0·1 mm, as measured using a microscope stage micrometer. Sieve openings vary by about ± 10% of the average value. Each size fraction was weighed using a Mettler balance (± 0·00001 g).

The material was sieved in a fume hood with a moderate draft of 49 cm s^{-1}, which is sufficient to carry away quartz crystals smaller than about 50 µm in diameter. To avoid significant loss of material, the authors blocked the lower 8 cm of the hood inlet window and held the sieve within a few centimetres of the collecting paper on the bottom of the hood. Notwithstanding their precautions, they lost 4–20 wt.% of the pumice during crushing and sieving. The range of values for the mass fraction of crystals that was obtained (5–17 wt.%) is within that documented by Hildreth (1977) based on point counting for stratigraphically equivalent pumice clasts. However, since most of the crystal mass is in the larger sizes, it is likely that a significant proportion of quartz crystals smaller than about 30 µm was lost (20 µm for magnetite and 50 µm for pumice).

2.2.3. Winnowing. The winnowing procedure was designed to separate the crystals from the pumice fragments and is similar to that employed by Skirius (1990). It comprised adding tap water to the sieved size fractions and pouring it off into a larger beaker using a rocking motion to keep the low-density pieces of pumice suspended while the crystals settled. The winnowing procedure was repeated several times for each size category, and in some cases, acetone was used as the final liquid for winnowing. In contrast, Skirius (1990) did not sieve the crushed material prior to winnowing, with the consequence that a larger amount of fine material tended to be lost. Hence, the present authors consider that their method provides somewhat more reliable results for the total mass fraction than that of Skirius (1990).

The efficiency of the concentrating procedure was tested using heavy liquids. For crystals larger than 100 µm, roughly 80–90% are recovered, while only 50% of the crystals smaller than 100 µm are retrieved. In all cases, the particles smaller than 100 µm were captured into beakers and allowed to stand for hours to days while the liquid (usually a mixture of water and acetone) evaporated. All crystal and pumice concentrates were weighed after drying. The authors lost from 10 to 22 wt.% of the original pumice mass after crushing, sieving and winnowing.

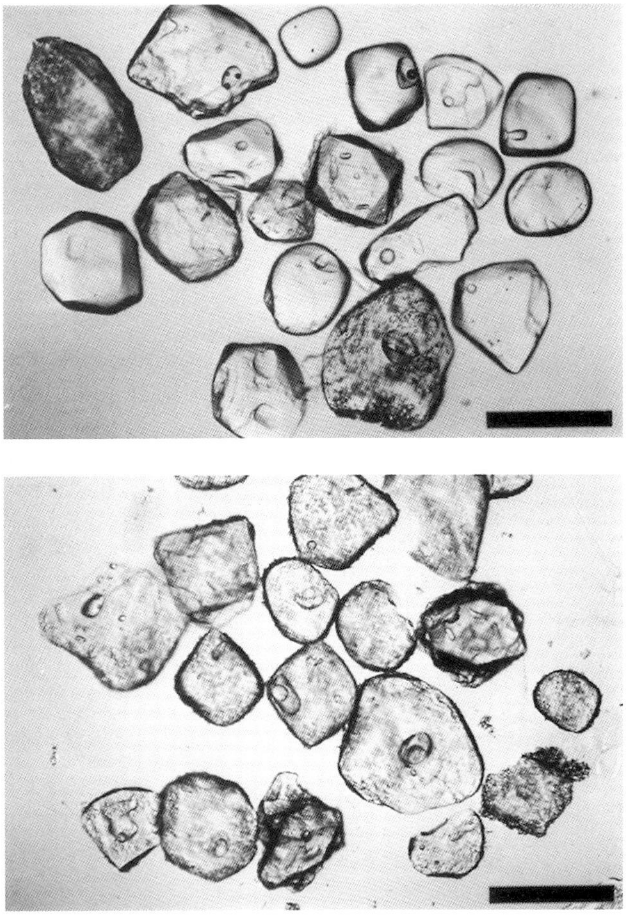

Figure 2 Micrograph showing a group of selected whole quartz crystals (size category=200–500 µm, sample F7–12) in two immersion liquids, with refraction indices of 1·492 (top) and 1·544 (bottom). The relative positions of crystals are distinct in the two photos, but corresponding grains can be identified using their morphology and melt inclusions. Note that the crystal boundaries are enhanced with the 1·492 refractive index liquid, whilst the glass coatings are clearly visible with refractive index 1·544. Plane-polarised light. The scale bar is approximately 500 µm long.

2.2.4. Petrography. The crystal concentrates were placed on watch glasses and immersed in refractive index oil so that petrographic observations could be made using dissecting and petrographic microscopes. The main advantage of this procedure over standard thin sectioning is that the crystals can be seen whole, without being sectioned, so that the three-dimensional (3D) morphology of the crystals can be observed. Moreover, this method allows the study of a very large number of crystals from each pumice clast, which would require an impractically large number of thin sections.

Most frequently, the present authors used an immersion oil with refractive index close to that of the pumice glass (n=1·492). This liquid tends to hide the glass and is best for observation of the morphology of the crystals (Fig. 2a). In other cases, an immersion oil with refractive index close to that of quartz (n=1·544) was used, which allows for the recognition of the extent of glass coating around the crystals and the properties of the glass (Fig. 2b).

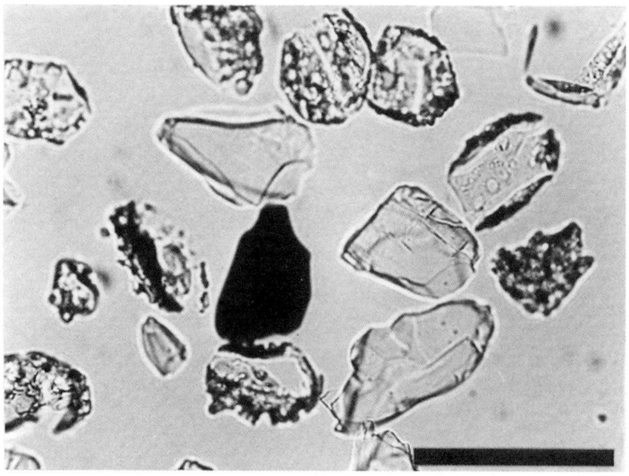

Figure 3 Micrograph showing a typical assortment of crystals in the 100–200 μm size category (sample F7–12). The immersion liquid has a refractive index of 1·544. The focal point above the crystals shows a diagnostic Becke line surrounding the crystals of sanidine in the centre of the frame; their cleavage traces and lack of attached glass are also diagnostic. Most quartz crystals are partially surrounded by attached, vesicular glass with thin walls between vesicles; uncoated boundaries (crystal in the top right corner) are invisible because of the matching refractive indices of the immersion liquid and quartz. Sodic plagioclase is almost invisible (grain to the left of top right corner) because of a refractive index slightly lower than that of the immersion liquid. In this view, four grains of sanidine, 11 of quartz, one of plagioclase and one of magnetite (or ilmenite) would be counted. Plane-polarised light. The scale bar is approximately 200 μm long.

2.2.5. Selection and counting. The same grain mounts were used for the counting procedure. Because of the present authors' interest in the fragmentation process, whole crystals were first separated from crystal fragments. By 'whole crystal', the authors mean a crystal or crystal fragment that appears to be at least half of an individual crystal. Regardless of whether the crystal is round or faceted, they selected crystals whose shapes suggested an approximation to a large portion of a subequant grain. When working with the 1·544 refractive index oil, the authors used the additional criterion that at least half of the crystal surface should be coated with glass. That would imply that at the time of quenching the crystal was immersed in silicate melt, rather than being fragmented during the crushing procedure. Examples of whole crystals and crystal fragments can be seen in Figure 3.

The total number of whole crystals and crystal fragments of all minerals was then estimated. Procedures for these operations varied with size category, sample size and person. In the larger size categories, because of the lower number of crystals, all whole crystals and fragments could be individually counted. For the smaller size categories, whole crystals were usually counted individually, while the number of fragments had to be estimated. Generally, the crystal fragments were dispersed as uniformly as possible, and they were then counted in a few representative microscopic fields of view; the ratio between the area counted and the total area was then used to obtain a final value for the whole area. For the smallest size categories (usually <100 μm), only a fraction of the concentrate was used because of the very large number of fragments. The fraction used was weighed so that the values could be adjusted for the total mass available in that size category.

Subjectivity enters in deciding what is a whole crystal. All of the authors used the same definition (see above), but evaluation of this is operator-dependent. They regard their counts to be accurate to within about 10% relative for crystals larger than 200 μm, and to within about 30% for crystals between 100 and 20 μm.

Finally, reference in the text to 'all crystals' means the sum of all whole crystals and crystal fragments of all minerals. 'All quartz' means all whole crystals and crystal fragments of quartz, and similarly for other minerals.

2.2.6. Further tests. In order to test the quality of the size distribution data, additional tests were performed using magnetite to monitor their processing procedure, and it was presumed that the collection efficiency for crystals of other minerals was similar for the <100-μm-diameter size fraction.

They tested their crushing and sieving procedure by dissolving 3·7 g of a 5·16-g fragment of a low-density pumice clast (sample BT01-BSH1b) in fluoboric acid (HBF_4). Another portion of the same clast was subject to the crushing procedure described here. The crystal yield by solution was 4·8 wt.% compared to 4·6 wt.% by crushing and sieving. The dissolved clast yielded 160 crystals of magnetite per gram, compared to a yield of 78 crystals per gram for the crushed portion. The smallest magnetite crystals obtained were 25 and 20 μm in diameter in the dissolved and crushed separates, respectively. The clast is internally variable in vesicle size and shape, and the mass of the crushed portion was more than 10 times that of the dissolved portion. The magnetite distribution also may be non-uniform, and differences in the amounts of magnetite recovered are regarded to be consistent with the loss of less than half of the crystals larger than about 20 μm. Most of the losses occurred during sieving and are mainly of pumiceous particles. Although 20-μm-diameter magnetite crystals have a Stokes settling speed of about 1 mm s^{-1} in water, and are expected to be retained during winnowing and wet transfers, the authors cannot rule out significant losses of crystals smaller than about 50 μm, as might occur due to surface tension and electrostatic effects. Bindeman (2003) reported similar results for zircon crystals separated chemically from Bishop pumice.

Another problem with the crushing procedure is that small crystals may remain embedded within larger pieces of pumice, and they may be hidden in the low-density fractions. However, magnetic separation using a strong horseshoe magnet can collect magnetite-bearing pumice fragments, if the magnetite is more than about 30% by weight of the particle. Using 0·98 g of the pumice concentrate in the size range 100–200 μm of sample F7–12, the authors recovered 74 magnetite-bearing pumice fragments. That corresponds to 1335 such crystals in the 17·68-g clast, and the magnetite crystals can be seen to be mostly 50–20 μm in diameter. When compared to the 512 magnetite crystals recovered by crushing and sieving in the <100-μm size category, it can be seen that about 28% of the <100-μm magnetite crystals were collected.

2.3. Microscopy of pumice fragments

Given the interest in having good estimates for the number of crystals <100 μm and the problems of loss of material in that size category during the crushing and sieving of the pumice clasts, an alternative approach was devised. It consisted of studying individual milligram-sized fragments of pumice.

The authors selected a few pumice fragments of 1–2 mm in diameter from each of the five samples. Each chip was weighed using a Mettler balance (± 0·00001 g) after being dried overnight at 150 °C. Most chips weighed between 0·0003 and 0·0011 g. An evacuable bell jar was used to pump the air out of vesicles and impregnate the chips with 1·492 refractive index oil. Vacuum was applied to pumice fragments immersed in oil so that bubbles escaping from the fragment would help scrub away loose crystal fragments from the surfaces of the pumice.

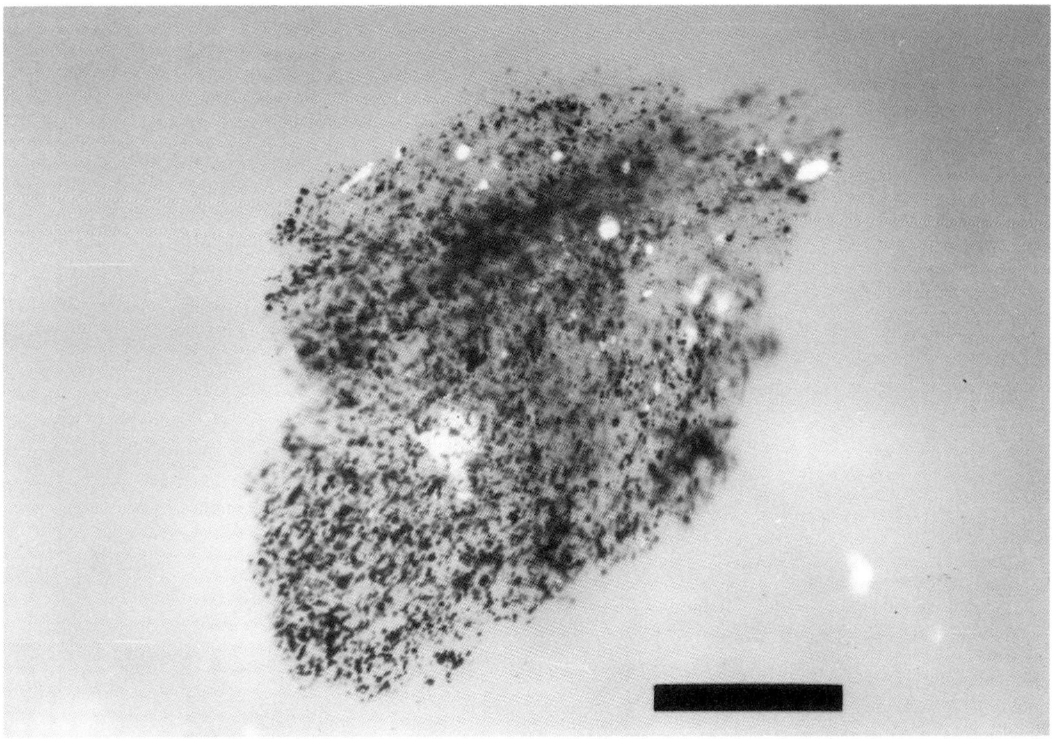

Figure 4 Micrograph of a pumice chip from sample F7–12, selected from the ash concentrate in the 800–1800 μm size category. The fragment was immersed in oil with a refractive index of 1·492 and subjected to low vacuum in an evacuable bell jar to pump the air out of vesicles and impregnate them with oil. The glass is invisible because of its refractive index matching that of the immersion liquid. Dark spots are non-impregnated vesicles which strongly scatter light. This chip yielded exceptionally good impregnation with oil, and crystal fragments are clearly visible under partially cross-polarised light, as illustrated by bright spots on the upper half of the chip. The scale bar is approximately 500 μm long.

After a few hours, the bubbles coalesced and broke; next, more vacuum was applied until negligible bubbles formed. Readmittance of air into the bell jar forced oil into the vesicles.

Ideally this procedure would render the glass invisible and leave an in-depth view of the crystals in the pumice fragment. Surprisingly – in view of the works of Whitham & Sparks (1986) and Sparks & Brazier (1982) suggesting that all vesicles are interconnected – it was found that, even with vacuum assisted impregnation for a week, many small vesicles remained unimpregnated and this rendered the thicker pumice fragments only translucent or even opaque because of the scattering of light by the low refractive index vesicles.

In two of the present samples, the pumice chips were sufficiently transparent after impregnation to see imbedded crystals down to a size of about 1 μm (Fig. 4). The three remaining chips had to be broken (using a teasing needle) into smaller fragments in order to improve the transparency. Using a low-power objective, the depth of field is adequate to sharply image crystals scattered throughout the chips using two or three focal levels. The authors examined the impregnated pumice fragments with partially crossed polarisers to enhance the visibility of crystals. They counted the crystals in size categories of 50–100 and 20–50 μm, in addition to those previously reported during crushing and sieving. Knowing the mass of the pumice chips allowed immediate computation of the number of crystals per gram in each size category.

2.4. X-ray tomography

The third method employed was to image small portions of the same pumice clasts using computerised X-ray tomography. X-ray tomography provides 3D images of samples in which the intensity of each volume element (voxel) reflects the linear X-ray attenuation coefficient. This is a function of the atomic number and density of the material (Rivers *et al.* 1999; Sutton *et al.* 2002), as well as of the X-ray energy used. Distinct phases can be recognised provided that there is sufficient contrast in atomic number and/or density.

The fundamental advantage of this technique is that no special treatment of the samples is required, so that the resulting images can be used to investigate the spatial distribution of crystals and crystal fragments in three dimensions. Additionally, the tomographic images can be used to independently yield size distributions, which can be compared to those obtained by the methods previously described.

2.4.1. Data collection and treatment. Data were collected on the bending magnet beamline of the GeoSoilEnvironCARS at the Advanced Photon Source (APS), Argonne National Laboratory (ANL), using a 22-keV monochromatic X-ray beam. The samples were rotated by 180° in 720 steps of 0·25° each. Transmitted X-rays were converted to visible light using a YAG scintillator, and the light was imaged onto a CCD camera using a 5 × microscope objective. For each rotational step, one camera image of 625 × 411 pixels was recorded. The images were reconstructed into a 3D volume using a Fast Fourier Transform algorithm. The reconstructed volumes consist of 625 × 625 × 411 voxels, with each voxel being 17·1 μm on each edge (Rivers *et al.* 1999; Sutton *et al.* 2002).

2.4.2. Image analysis. Inspection of the images showed that grains of quartz, alkali-feldspar, magnetite and zircon can be identified and easily distinguished from the vesiculated glassy matrix (Fig. 5); however, distinction of plagioclase and quartz proved unviable. For easier identification, images were displayed using a false colour scheme.

Ideally, the absorption within a particular mineral is constant and diagnostic of its density and mean atomic number. However, examination of large crystals revealed that they are

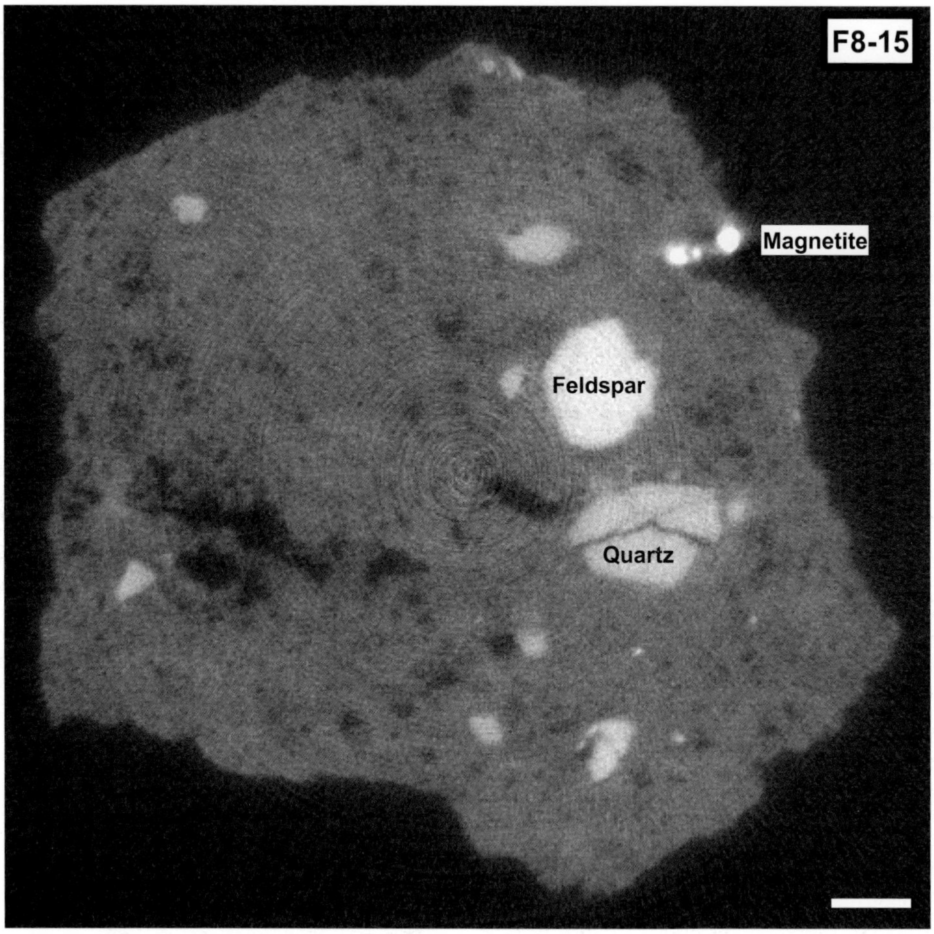

Figure 5 Tomographic image showing the distinction between quartz, alkali-feldspar, magnetite, glass and vesicles. The greyscale is proportional to X-ray attenuation by the sample. The dark area surrounding the pumice is air, the rhyolitic glass is dark grey and vesicles are clearly visible as dark areas. Attenuation by quartz, alkali-feldspar and magnetite increase in this order, and crystals of these minerals can be identified; plagioclase and quartz are indistinguishable because of similar mean atomic numbers. Note that crystal shapes are readily visible only for grains larger than 500 μm, with recognition becoming difficult for grains smaller than 200 μm. Concentric streaks are artefacts of the computerised reconstruction process and indicate the axis of rotation during data collection. The scale bar is approximately 1000 μm long.

characterised by rims about 3 voxels wide (about 48 μm), along which the recorded grey-level intensity gradually decreases to values similar to that of the background observed within the glassy matrix. Consequently, only crystals larger than ~50 μm in the smallest dimension can be positively identified in the images. Also, distinction between quartz and alkali-feldspar is problematic for crystals smaller than 200 μm. For simplicity, the authors assumed that all such crystals were quartz crystals, which introduces an error that is probably less than about 30% since quartz comprises about 70% of the phenocryst minerals by volume.

2.4.3. Petrography. The analysed volumes can be seen on a computer screen by displaying successive two-dimensional images in sequence along a selected direction. Three orthogonal directions (i.e. X, Y and Z) can be selected. Relevant petrographic observations were made by watching these 'movies' of the analysed volumes. During the petrographic studies, the present authors concentrated on (1) the shapes of the crystals; (2) the extent of fracturing of crystals; and (3) the presence or absence of aggregates.

2.4.4. Size distributions. For each one of the samples, the authors selected a box-shaped volume to be analysed for size distributions; the volume was arbitrarily located within the total analysed volume and was selected to be large enough so that approximately 100 crystals of any size and mineral were documented. Hence, the selected volume varied from one sample to another. Close to the boundaries, the selected volumes included several partial crystals; only the crystals whose centres-of-volume lie within the selected volume were counted.

All crystals within the selected volume were measured and classified in size categories. All measurements were made in a fixed arbitrary system of coordinates in which the volume was reconstructed. For non-equant crystals, this measurement procedure is problematic, but the authors focus their discussions on quartz, which is characteristically equant. Moreover, in order to compare the results with those of sieving and crushing, they classified the crystals in size categories using the intermediate dimension of the crystals, which is less sensitive to the orientation of the orthogonal set of coordinates.

The pumice samples are porous, and comparisons between tomographic measurements and the results of crushing and sieving assume that the small sample used for tomography and the larger crushed portion have the same bulk density. The error introduced by this assumption is probably not more than about 10% since the tomographic images revealed unexceptional distributions of bubbles. However, the statistical uncertainty in the number of tomographically imaged crystals of quartz larger than about 800 μm is near 100% for the present sample volumes.

3. Results

In this paper, the present authors focus on data for quartz crystals for three main reasons: (1) quartz is the most abundant mineral, corresponding to about 70% by volume of the phenocrystic assemblage; (2) given its abundance, the present size distribution data are more reliable; and (3) previous studies on melt inclusions (Wallace *et al.* 1999; Anderson *et al.* 2000) and zoning patterns (Peppard *et al.* 2001) have also focused on quartz and provided important information for the interpretation of the present data.

3.1. Petrography

Microscopic observations using grain mounts and tomographic images provide important information on magmatic processes in the Bishop magma and are fundamental for the interpretation of the size distribution data.

Whole quartz phenocrysts can be divided into two main categories (Figs 6 & 7): (1) euhedral crystals, which show typical bipyramidal shapes; and (2) round crystals, mostly subequant. In both cases, crystals are always glass-coated. Whole crystals as small as 200 μm are relatively common, but whole crystals smaller than 100 μm were not observed. The present authors conclude that virtually all quartz crystals smaller than 100 μm are fragments of larger crystals. However, it should be pointed out that neither optical resolution (effectively about 5 μm) nor tomographic resolution (about 17 μm) allowed unambiguous discernment of crystal faces on crystals smaller than about 50 μm.

Quartz fragments predominate in all size categories in the crushed and sieved material. The present observations show that virtually all phenocrysts larger than about 0·5 mm have cracks so that it may be hypothesised that extensive fragmentation took place during crushing. However, the authors observe that visible cracks generally intersect burst melt inclusions. Many of the fragmented crystals are coated with glass, showing that the fragments were immersed in melt or glass prior to eruption (Fig. 6). Furthermore, the tomographic images show several examples of broken crystals which have matching pieces close together (Fig. 7). Hence, they conclude that fragmentation is an important pre-eruptive process in the Bishop magma and that is in agreement with the size distribution data (see section 3.3.2 below).

Aggregates of crystals were not observed in the crushed and sieved material. In the tomographic images, aggregates were rarely recognised (Fig. 7), but could not be unambiguously distinguished from multiply fractured crystals. Thus, the authors' observations indicate that clustering was not an important process in erupted Bishop magma.

3.2. Pumice densities

A total of 52 pumice clast densities were obtained by Skirius (1990) and Wallace *et al.* (1999), and encompass values between 0·26 and 1·13 g cm^{-3} (Fig. 8). The present authors' density determinations are within that range (0·499–0·902 g cm^{-3}; Fig. 8), and this range includes more than 70% of all available values. Skirius (1990) has also determined the total mass fraction of crystals, and the results are compared to those obtained by the present authors in Figure 9. The distribution is characteristically bimodal, with a scarcity of crystal fraction values between 10 and 12 wt.%. It is important to point out that this 'gap' is observed in both fall and ash-flow units.

Despite the limited number of samples, the present data suggest that there is a positive correlation between the bulk density and the weight fraction of crystals in pumice clasts (Fig. 10); the same relationship is not evident in the data of Skirius (1990). However, it should be noted that this may be caused by the somewhat improved winnowing procedure employed by the authors. The origin of the correlation will be discussed below. Importantly, the authors conclude that, both in terms of bulk density and crystal contents, the pumice clasts analysed in this work are representative of pumice from stratigraphically similar parts of the Bishop Tuff.

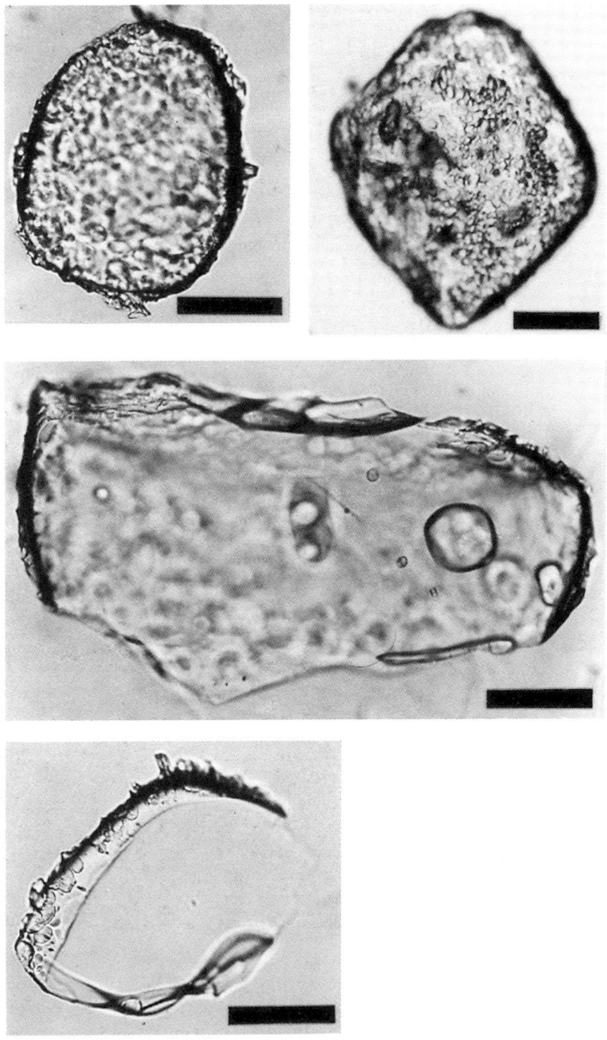

Figure 6 Micrographs showing the four main types of quartz identified in the present samples. The immersion liquid has a refractive index of 1·544. Top left: Round, glass-coated whole quartz crystal; although the crystal lacks facets, it is largely coated, indicating that, at the time of eruptive quenching, it was an isolated grain surrounded by melt. Top right: Partly faceted, glass-coated whole quartz crystal; the edges and corners are round, but the sides of the crystal are partly flat $\{1\,\bar{1}\,0\,1\}$ faces. Middle: Irregular, mostly glass-coated quartz crystal; the left and right terminations of the crystal are partly faceted and coated with finely vesicular glass; the top of the crystal is mostly conchoidal and covered with smooth or coarsely vesicular glass, showing pre-eruptive fracturing; a few melt inclusions are observed within this grain. Bottom: Partly glass-coated quartz crystal; most of the crystal is glass-free; glass-coating is restricted to the vesicular glass that coats the upper surface and the smooth glass coating the bottom surface; since less than half of the crystal is coated with glass, the crystal is not whole, and it probably broke during laboratory crushing. Plain-polarised light. The scale bars are 200 μm long, except for the top right one, which is 100 μm long.

3.3. Crystal size distributions

The major aims of the study were to: (1) document the minimum size of phenocrysts of quartz; and (2) evaluate crystal size evidence relevant to crystal sinking and mixing.

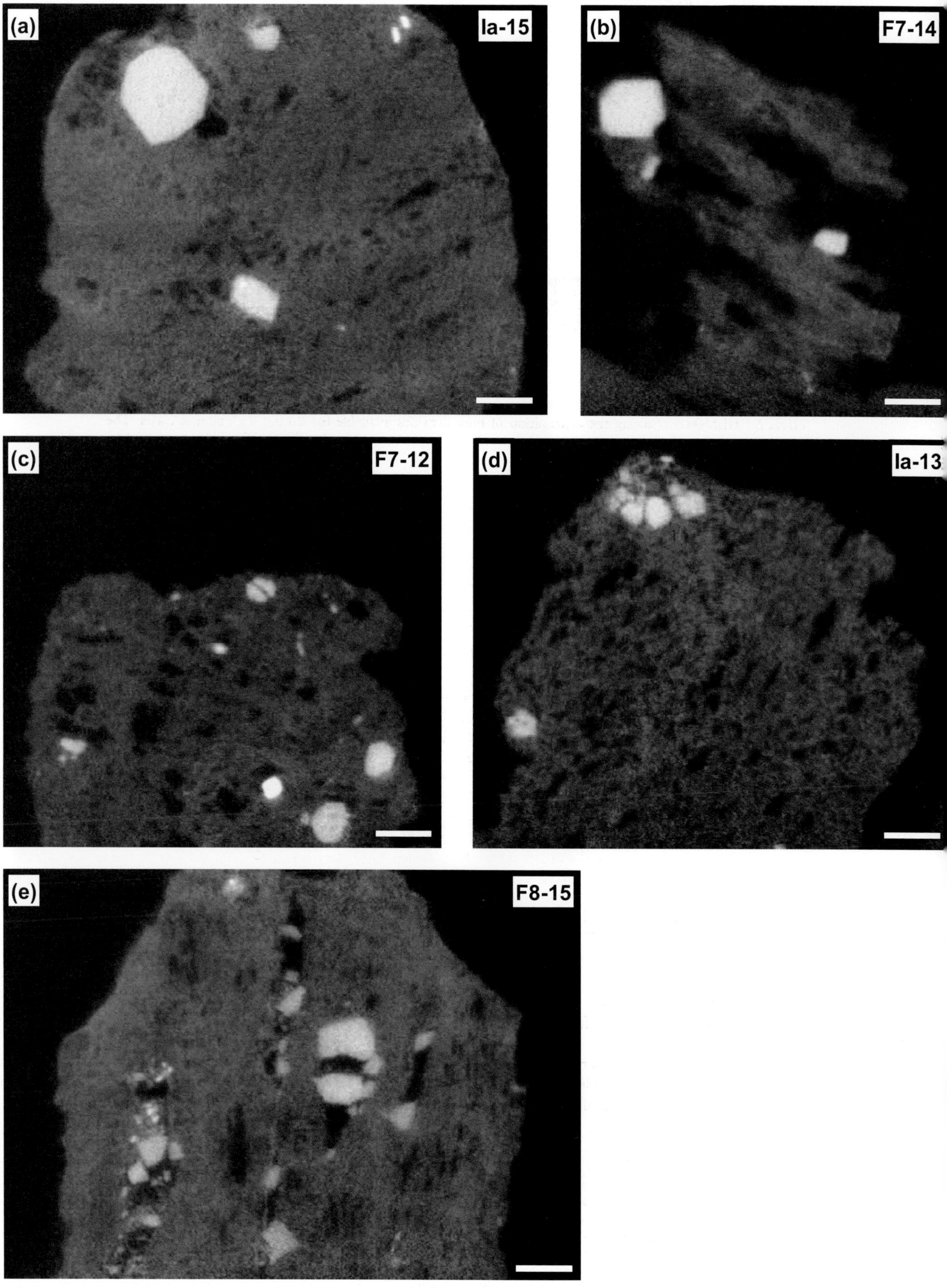
(a)
Ia-15
(b)
F7-14
(c)
F7-12
(d)
Ia-13
(e)
F8-15

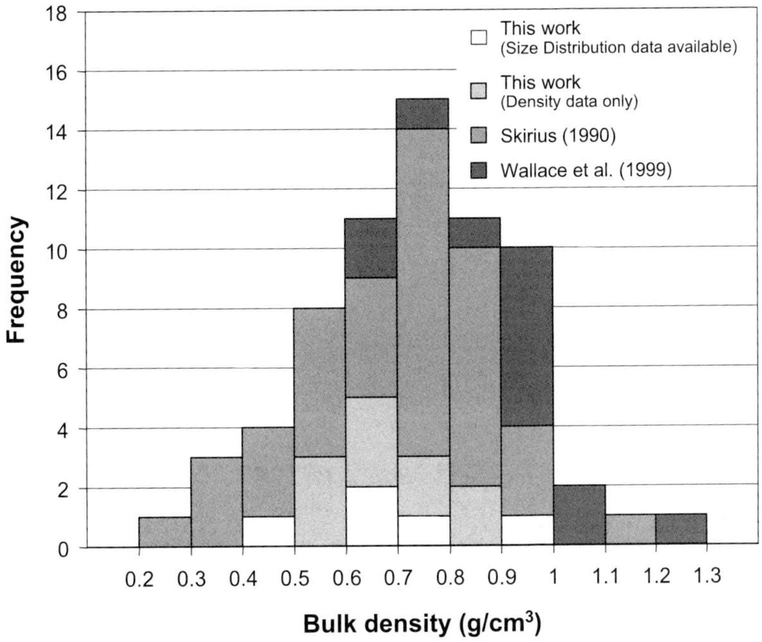

Figure 8 Histogram showing the distribution of bulk densities available for Bishop Tuff pumice clasts. The samples studied in the present work cover most of the variation previously observed by Skirius (1990) and Wallace *et al.* (1999).

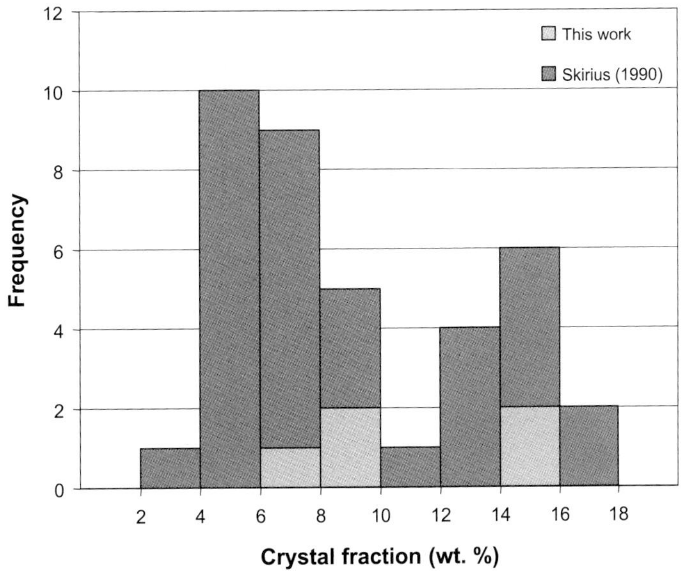

Figure 9 Histogram showing the distribution of the total crystal mass fraction of Bishop Tuff pumice clasts. The samples studied in the present work are within the range previously observed by Skirius (1990) and Wallace *et al.* (1999). Note the scarcity in crystal contents between 10 and 12 wt.% crystals. See the text for a discussion.

Documentation and interpretation of the size distribution of quartz phenocrysts toward these two goals faced two main obstacles.

First, the low number density of large crystals required crushing, sieving and winnowing in order to yield a significant number of millimetre-sized crystals, especially euhedral or whole phenocrysts. As discussed above, tiny crystals are partially lost in such processing. In addition, most of the phenocrysts smaller than about 100 μm remain trapped in incompletely crushed pumice fragments where they are poorly or not at all visible because of light scattering by vesicles in the pumice.

Figure 7 Selected tomographic images showing the shapes of quartz crystals: (a) Euhedral quartz crystal (top left) with distinct crystal faces in the high-density sample Ia-15; note the low abundance and small size of the vesicles; feldspar tablet is seen in the lower portion of the frame; magnetite and an additional quartz crystal are seen in the upper part. (b) Euhedral quartz crystal in sample F7–14, which has intermediate density; note the large sizes of the vesicles in this portion of the fragment. (c) Round quartz crystals (bottom right and top centre) in the low-density sample F7–12; note the widespread distribution of small vesicles; the bright crystal in the lower portion of the frame is magnetite; other small disseminated crystals can be observed. (d) A cluster of small quartz crystals in intermediate-density sample Ia-13; clusters are rare and cannot be unambiguously distinguished from fragments of broken crystals sitting close to each other; note the disseminated distribution of the small vesicles. (e) Broken crystals of quartz in high-density sample F8–15; the crystal in the centre is broken, but the two pieces are held in proximity, showing that fragmentation is at least partly pre-eruptive; fragments are observed in other large vesicles as well and presumably formed in a similar way; note that, while large vesicles are common, small vesicles have a more restricted distribution. The scale bar is approximately 1000 μm long in all images.

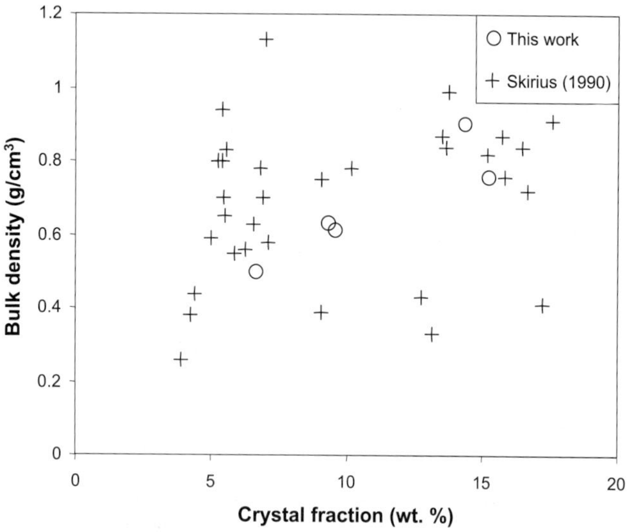

Figure 10 Diagram showing the relationship between total crystal mass fraction and bulk density of Bishop Tuff pumice clasts. A positive correlation is observed for the data presented in this work. The poor correlation for the data of Skirius (1990) may reflect the somewhat improved techniques used and a more restricted stratigraphic sampling interval used in this work. See the text for a discussion.

Secondly, many phenocrysts are broken into pieces. The tiny fragments of glass-coated quartz can behave as crystal nuclei, but if their origin is by fragmentation (rather than by classical nucleation), crystal size may reflect processes other than nucleation and growth.

In the discussion that follows, the present authors first compare the results of the distinct methods, which guided their choice of data for each size category. They then discuss the results in light of the objectives outlined above.

3.3.1. Comparison of the different methods. Each of the methods the authors used to obtain size distributions has inherent drawbacks. Based on the previous discussions, they expected the crushing, sieving and winnowing procedure to suffer from loss of small particles, at least for the size category <100 μm. On the other hand, optical microscopy of pumice fragments should yield good estimates for small crystals, particularly those smaller than 200 μm; nevertheless, the number of larger crystals contained in the sample volume is too small to be statistically meaningful. Finally, the tomographic procedure reveals that many of the pieces of fractured crystals are still close together and this yields an improved assessment of whole crystals in the size range between 200 and 500 μm; however, larger crystals are too few in the analysed volume (~ 1 cm^3), while crystals smaller than 50 μm cannot be positively identified.

In order to choose the best estimate for each size category, the three methods were compared taking into account the uncertainties caused by counting statistics. In Figure 11, the authors plot the results for all quartz crystals in each one of the five samples. It can be seen that the number densities are usually in agreement for crystals larger than 100 μm, especially when comparing the crushing, sieving and winnowing procedure with the tomographic data. It is also evident that the tomographic data yields much larger uncertainties, so the values obtained by crushing and sieving are preferred. On the other hand, the values for the size fraction 50–100 μm are discordant and systematically higher for the optical microscopy procedure. Hence, the authors consider that these latter values are preferred for crystals smaller than 100 μm. In all figures which follow, the values displayed are a combination of the optical microscopy method (crystals smaller than 100 μm) and the sieving and crushing method (crystals larger than 100 μm).

In Figures 12a and 14a, the size distribution data are presented on a cumulative basis with the sizes converted into the ϕ scale ($\phi = -\log_2$[diameter]), commonly used by sedimentologists. Given that the data were obtained by sieving of the crushed pumice clasts, the authors consider this to be the best representation. However, in most crystal size distribution studies (e.g. Cashman & Marsh 1988; Marsh 1988, 1998; Higgins 2000), data are usually presented as semi-log plots of number density versus size. In this kind of diagram, the canonical size distribution is represented by a straight line (cf. discussion in Marsh 1998), which corresponds to a binomial distribution of sizes (Eberl *et al.* 1998, 2002). Based on this assertion, size distributions are frequently interpreted according to their departure from this canonical binomial shape. From this perspective, it is advantageous to present the data in this fashion, as done in Figures 12b and 14b.

3.3.2. Pre-eruptive fragmentation. As discussed above, petrography provides strong evidence that fragmentation is an important magmatic process. Given that large crystals tend to be more susceptible to breakage than small crystals, size distributions may be affected by fragmentation; the present authors would predict enrichment in the number of small crystals and depletion in the number of large crystals.

The pattern observed for the distribution of sizes of all crystals is strikingly similar between the samples, approximately linear in Figure 12a and concave upward in Figure 12b. This pattern can be interpreted as a function of enrichments in either small or large crystals, or both. The main problem here is how to define the slope characteristic of each sample. For instance, if the slope is defined based on the size interval 200–800 μm, both the smallest and the largest size categories are enriched in crystals. However, if the slope is selected to correspond to that observed in the size category 800–1800 μm, then the enrichment is interpreted to be in the size categories with size <200 μm. In any case, it seems inevitable to conclude that the number density in the smallest size category is significantly larger than that expected considering the overall slope of the size distributions (cf. Fig. 12b).

In Figure 13, the authors plot the ratio between whole crystals and all crystals as a function of size fraction. The trends clearly show an increase in the number of fragments relative to whole crystals toward smaller crystal sizes, suggestive of preferred fragmentation of large crystals.

Given the petrographic evidence for fragmentation and the scarcity of small whole crystals in these size fractions (see below), the authors conclude that this enrichment is caused by preferred fragmentation of large crystals. Even though the processing of samples can partially account for this fragmentation, a similar pattern is observed using the tomographic images, so they conclude that significant fragmentation took place before or during the eruption of the Bishop magma. Accumulation of large crystals cannot be ruled out at this point, and this problem will be discussed in more detail in a subsequent section.

3.3.3. Nucleation. Models of crystal growth and nucleation commonly yield increasing numbers of crystals with decreasing size. While growth eliminates small crystals, nucleation is responsible for supporting their large numbers. Hence, the abundance of small crystals relative to large crystals can be used to assess the importance of nucleation during magmatic evolution.

The present authors assumed that the size distributions of whole crystals adequately reflect the size distributions prior to fragmentation. Even though this assumption could be in error, large deviations are only expected for the larger crystals

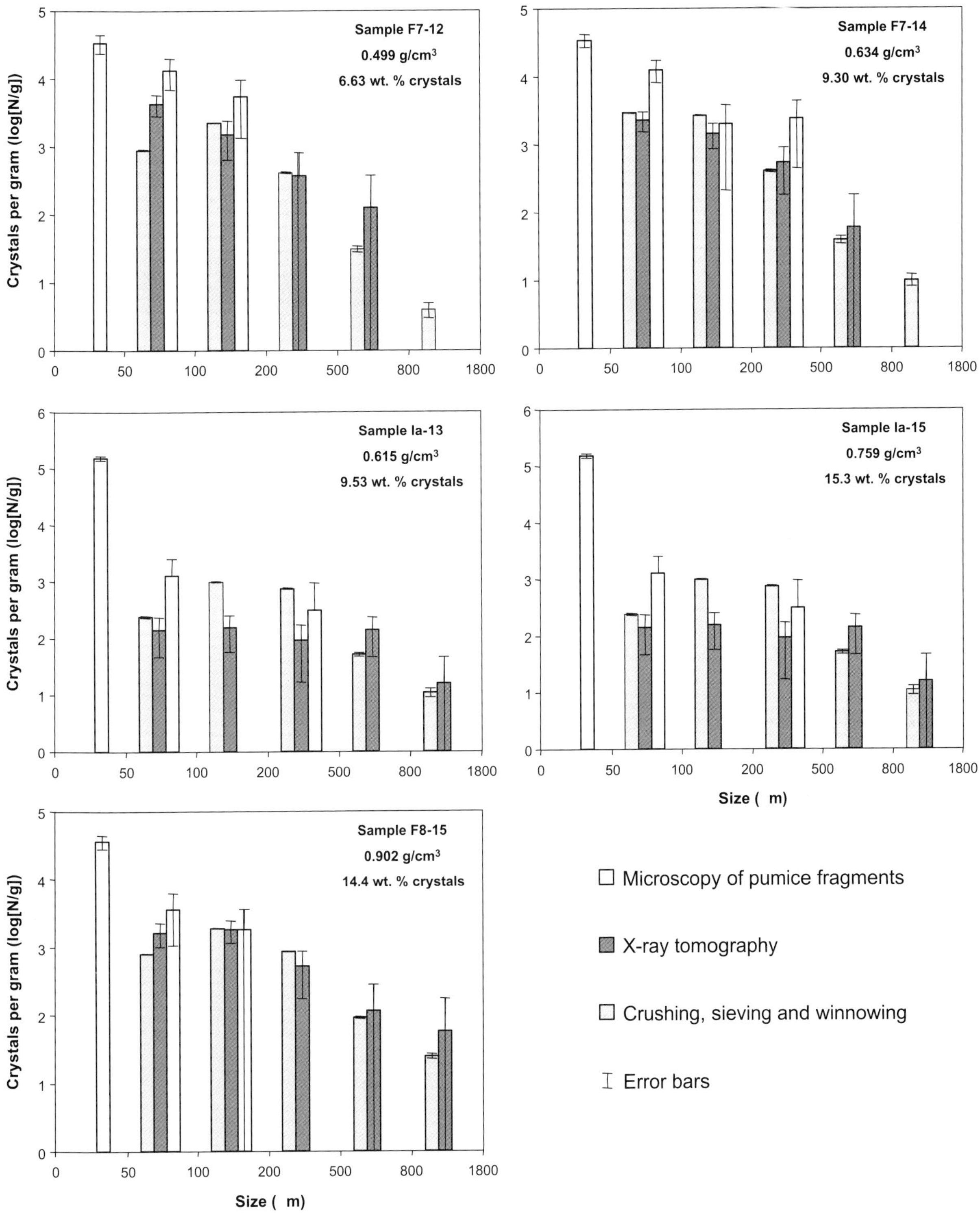

Figure 11 Semi-logarithmic histograms comparing the results for the three methods used to obtain size distributions for all samples. The data are for all crystals counted in each size category. Error bars reflect only uncertainties in counting statistics ($2\sqrt{N}$) and do not include possible loss during processing; for smaller size categories, some of the bars are too small to be visible. Counting statistics are best for the crushing and sieving method, but loss is evident in the 50–100 μm and <50 μm size categories, for which the microscopy of pumice fragments is superior. The tomography results are in agreement with the other methods.

because small whole crystals have a smaller chance of breakage. Given that their attention will be centred on the small crystals, the authors consider that the assumption can be regarded as reasonable.

The present qualitative observations suggest that small whole quartz crystals are scarce or lacking, and the size distributions largely support this conclusion (Fig. 14; see also Bindeman 2003). Three of the samples (i.e. F7–14, F8–15 and Ia-13) show an approximately linear pattern with a marked decrease in the amount of small crystals, first observable in the size category 100–200 μm and evident in the size category <100 μm. These patterns suggest that the magma represented by these pumice clasts was characterised by diminished nucleation, at least during its final stages of evolution. Importantly, this may reflect low supersaturation conditions, which would be suggestive of a large, conductively cooling body.

On the other hand, the least dense of the pumice clasts (F7–12, bulk density 0·499 g cm^{-3}) shows a contrasting

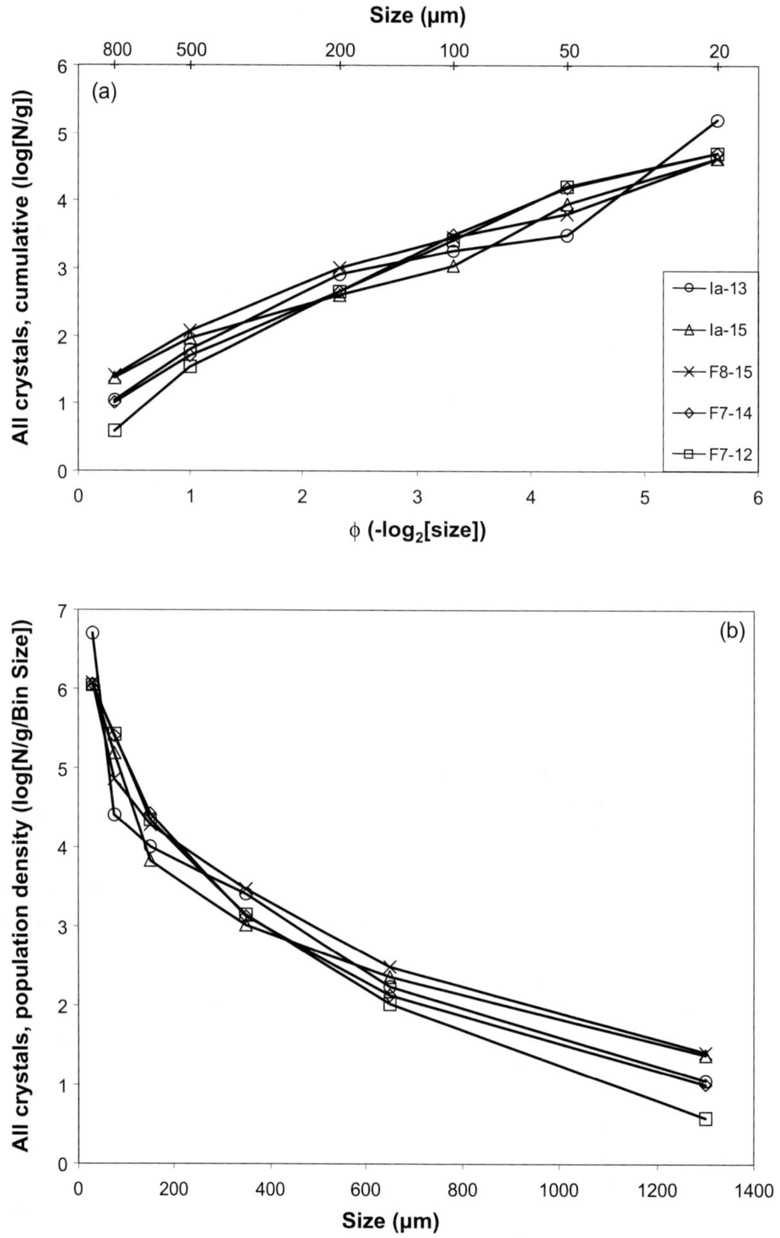

Figure 12 Size distributions for all crystals: (a) Log–log diagram showing the cumulative distribution of sizes; (b) Semi-logarithmic diagram showing the variation in population density with size. Note that all distributions are strikingly similar to each other, with a marked enrichment in crystals with a diameter <200 μm. The concentration of small crystals is likely to be caused by fragmentation of larger crystals, at least partly magmatic in origin.

pattern, more akin to the canonical binomial shape, with an increasing number density toward smaller crystals. A possible conclusion is that this sample derives from a distinct parcel of the Bishop magma, which was subject to greater nucleation rates.

Finally, in comparison to the previous groups, sample Ia-15 shows a less regular size distribution. This probably reflects a more complex crystallisation history.

3.3.4. Crystal sinking and bubble rising. Previous studies by Anderson and co-authors (Wallace *et al.* 1995; Anderson *et al.* 2000) have provided evidence for crystal sinking and bubble rise in the Bishop magma.

Sinking of quartz and sanidine phenocrysts in the Bishop magma is indicated by three features: (1) some late-erupted quartz phenocrysts have near-rim melt inclusions which are relatively rich in dissolved CO_2 and have relatively high gas-saturation pressures, suggesting that pressure increased as individual crystals grew (Anderson *et al.* 2000); (2) some phenocrysts of sanidine have Ba-rich rims, indicating that the melt from which the crystals grew became progressively richer in Ba with growth (Anderson *et al.* 2000); and (3) the near-rim, CO_2-rich melt inclusions in quartz phenocrysts mirror the zonation of the sanidine phenocrysts and have high Ba, which correlates positively with the dissolved CO_2 (Anderson *et al.* 2000). Whereas crystals could grow as pressure increased without sinking relative to their surroundings (the whole magma could descend and come under greater pressure), relative movement of crystals and melt is required to explain the rimward increases in Ba in sanidine as well as in near-rim, quartz-hosted, melt inclusions. Anderson *et al.* (2000) also argued that the relatively long time implied for extensive crystal sinking cannot be ruled out and is consistent with some radiometric studies (Christensen & Halliday 1996).

Using a closed-system model for gas-saturated crystallisation in the early- and mid-erupted parts of the Bishop Tuff magma, Wallace *et al.* (1995) inferred an upward increase in

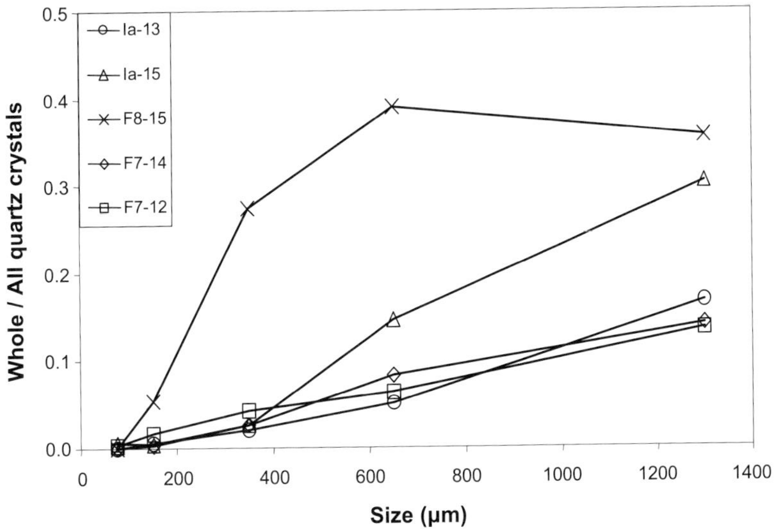

Figure 13 Diagram showing the relationship between the ratio whole crystals/all crystals as a function of size for each pumice clast. Note the relative increase of fragments toward smaller size fractions, suggestive of preferred fragmentation of large crystals.

the mass fraction of pre-eruptive exsolved gas from about 1 wt.% to as much as 6 wt.%. The inferred increase in exsolved gas is greater than can be explained by exsolution with decompression, and the total (dissolved+exsolved) H_2O increases from about 6 wt.% at about 200 MPa to more than about 10 wt.% at 150 MPa (Wallace *et al.* 1995, figs 2 & 4). Wallace *et al.* (1995) concluded that the gradient in exsolved gas may also have existed before crystallisation and entrapment of melt inclusions. The upward increase in inferred gas mass fraction suggests that gas had accumulated into the early-erupted magma, plausibly by the rise of bubbles.

In view of the above-mentioned evidence for sinking of quartz crystals and rise of bubbles, one of the present authors' goals was to evaluate crystal size evidence relevant to crystal sinking and mixing. Their thought was that big crystals should sink faster than little crystals; thus, if crystal accumulation occurred, it probably affected the large crystals more than the small ones. The prediction is that most of the variation in crystal mass fraction resides in the big crystals.

In Figure 15, the variation in mass fraction of all crystals per size category is shown. It can be seen that all samples have similar contents in the size category <200 μm, but they differ in the abundance of crystals larger than 500 μm. Two samples (i.e. F8–15 and Ia-15) are characterised by increasing contents in the size categories larger than 500 μm, while the other three samples (i.e. F7–12, F7–14 and Ia-13) display the opposite trend. Among the latter, sample F7–12 shows the lowest abundance of large crystals, while the other two are very similar to each other (also in bulk density). These three groups are also distinguished by the total crystal mass fraction, which correlates positively with the mass fraction of large crystals (Fig. 16). Interestingly, the group richer in crystals is separated from the others by the gap in total mass fraction of crystals (see Fig. 9), as illustrated in Figure 16.

As demonstrated by Wallace *et al.* (1999), the volume percentage of vesicles (porosity) of the pumice clasts strongly correlates with their pre-eruptive exsolved gas contents (as determined in melt inclusions; see Wallace *et al.* 1999, fig. 12). Hence, the present authors use the relationship between porosity and the mass fraction of crystals in the size category >800 μm to explore the possible migration of bubbles in the Bishop magma (Fig. 17). A negative correlation between the abundance of crystals in this category with porosity is evident. The addition of crystals to the size fraction >800 μm by accumulation would cause a large increase in the mass fraction in this size category, with a corresponding slight decrease in the porosity. Figure 17 indicates that a combination of crystal accumulation and bubble migration can explain the variation observed in the present samples.

In contrast, crystal growth would cause simultaneous variation in crystal mass fraction and bubble contents, and detailed treatment of this problem is beyond the scope of this paper. However, if the magma crystallised by cooling under gas-saturated conditions, both exsolved gas contents and mass fraction of big crystals (>800 μm) should increase simultaneously. Evidently, the expected increase in both porosity and mass fraction of big crystals is inconsistent with the authors' observations, lending support to a crystal sinking and bubble rise model.

Interpretation of the present data is non-unique, and other, more complex processes can probably explain the trends observed in Figure 17. However, a combination of crystal sinking and bubble rise is implied from the evidence presented by Wallace *et al.* (1995) and Anderson *et al.* (2000), and such a combination can explain the otherwise unexpected correlation.

4. Discussion

The present results yield some insights into the evolution of the Bishop magma. Based on the petrographic, density and size distribution data, the authors recognise three main groups.

The first group is characterised by the low-density sample, F7–12, which has a correspondingly low total mass fraction of crystals and low proportion of large crystals, but a high vesicle content. Moreover, the abundance of small whole crystals suggests that nucleation may have continued until a relatively late stage in the evolution of this sample.

At the other end of the spectrum, sample F8–15 has high density, high mass fraction, abundant big crystals and typically low vesicularity. The amount of whole crystals decreases below about 200 μm and indicates negligible nucleation, at least in the final stages of pre-eruptive evolution. The characteristics of F8–15 are in marked contrast with sample F7–12, suggesting that these samples represent distinct portions of the Bishop magma. Importantly, the two groups are separated by a gap given by the scarcity of samples with total crystal fraction between 10 and 12 wt.% (Skirius 1990 and present work); the

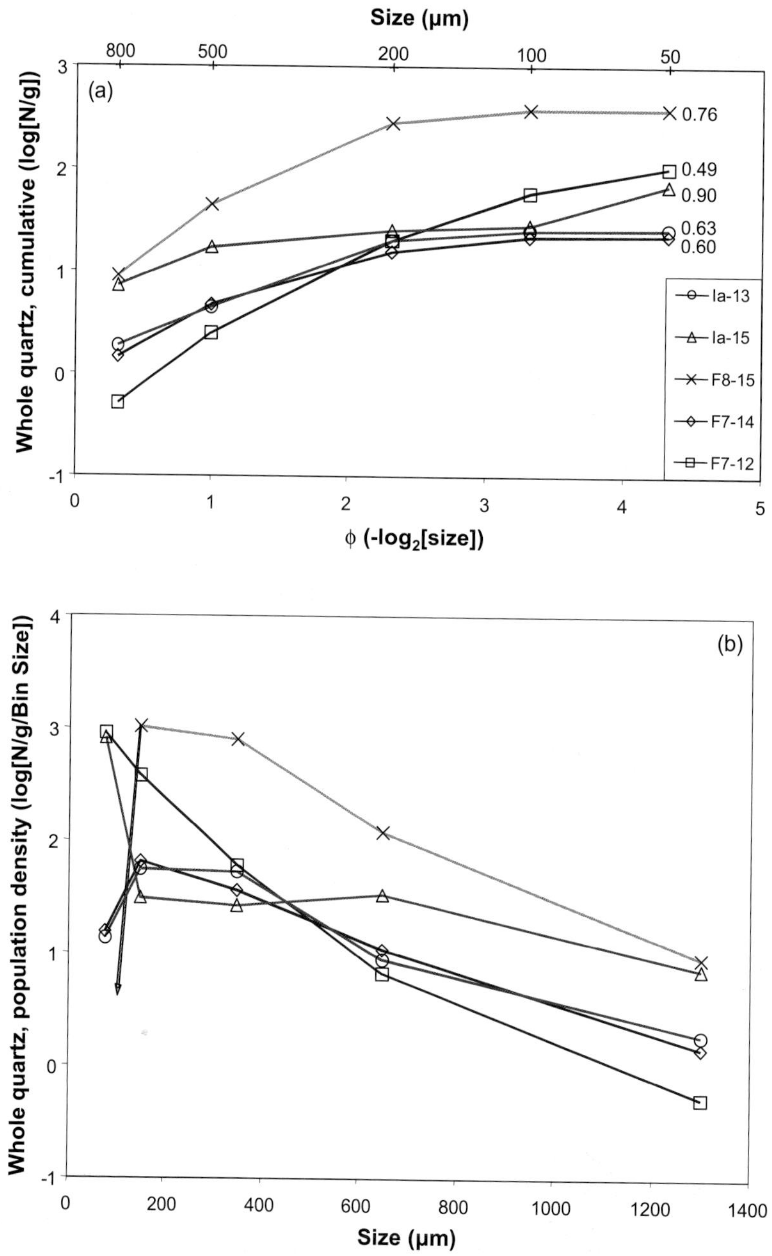

Figure 14 Size distributions for whole quartz crystals: (a) Log–log diagram showing the cumulative distribution of sizes; the numbers are bulk densities of each pumice clast. (b) Semi-logarithmic diagram showing the variation in population density with size. Three of the samples show a marked depletion in crystals with a diameter <200 μm; the scarcity of small crystals is interpreted as resulting from negligible nucleation during a late-stage pre-eruptive period. One sample approximates the canonical linear distribution of sizes. See the text for a discussion.

two portions of magma may be complementary. Sample Ia-15 is similar to F8–15, except for the size distribution of whole crystals, which reveals a more complex crystallisation history.

The two remaining samples (i.e. F7–14 and Ia-13) are strikingly similar to each other and intermediate, in all respects, between the previous groups. It is noteworthy that this intermediate character is also shared by the crystal size distributions, both in terms of mass fractions of total and whole crystal contents. The shared associations among the three groups and the observed gap in crystal mass fraction suggest that they may derive from distinct parcels of magma. The gap referred to above may reflect some zones in the magma being inaccessible to the eruptive vents which formed the Bishop Tuff.

5. Conclusions and suggestions for further work

The density, crystal abundance, and size and shape of pumice clasts from three stratigraphic units of the Bishop Tuff rhyolite varies in a correlated pattern. Including published data on Bishop pumice clasts, there is a bimodal distribution of crystal contents with a gap between about 10 and 12 wt.% of crystals. The gap may be an artifact of sampling, or it may reflect discrete eruptive segments of the pre-eruptive body of magma.

Three stages with distinct characteristics are recognised, and probably reflect separate but related parts of the pre-eruptive Bishop magma. The space-time relations between the stages are still incompletely documented, and a more detailed discussion awaits further work, especially on the composition

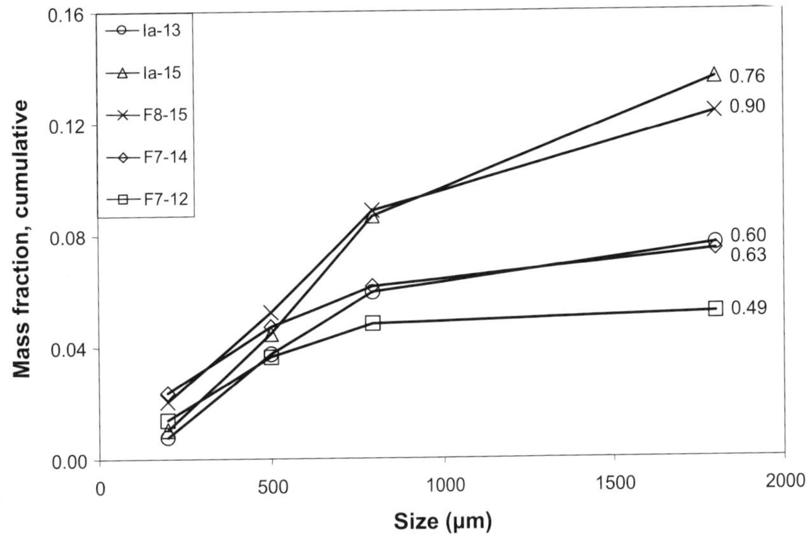

Figure 15 Diagram showing the cumulative distribution of total mass fraction as a function of size. The numbers are the bulk densities of each pumice clast. Note that three groups can be distinguished, especially based on the amount of large crystals. See the text for a discussion.

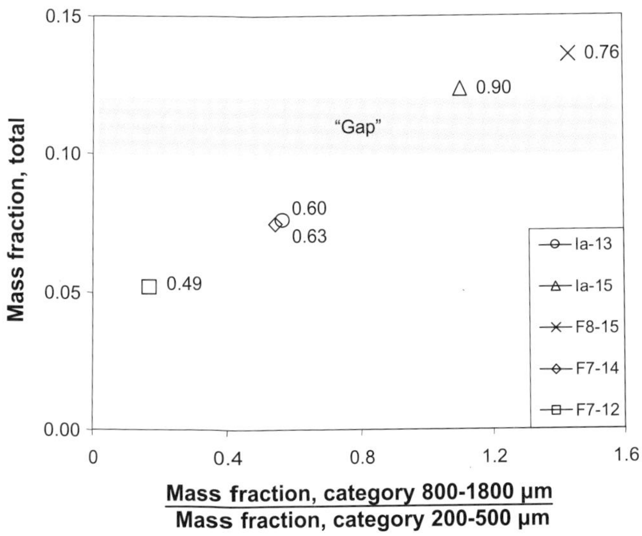

Figure 16 Binary diagram showing the relationship between the ratio of the mass fractions in the 800–1800 and 200–500 μm size categories, and the total crystal mass fraction for each pumice clast. Note the strong positive correlation, demonstrating that the distinction between different samples is in the amount of large crystals. The 'gap' in total crystal fraction is also indicated.

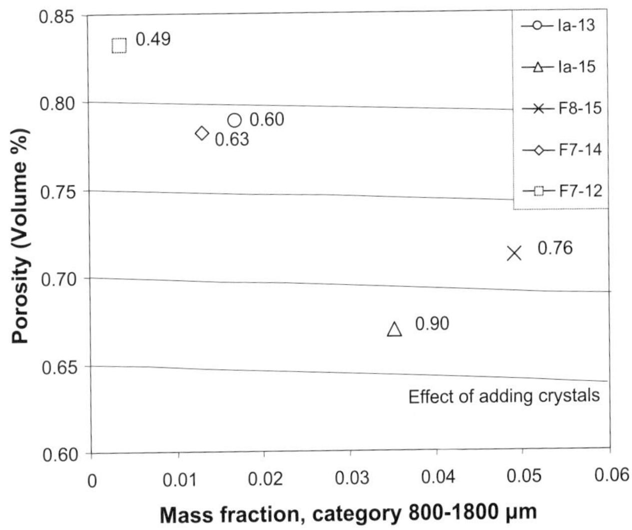

Figure 17 Binary diagram showing the relationship between the mass fraction in the 800–1800 μm size category and the porosity of the pumice clasts. The addition of large crystals without change in the glass or vesicle contents leads to a reduced change in the porosity as compared to that seen for the mass fraction of large crystals; this effect is shown in the figure for selected starting porosity values and no crystals originally present in the 800–1800 μm category. Adding bubbles promotes no change in the mass fraction, and therefore, would cause a vertical shift. Gas-saturated crystallisation would simultaneously increase the porosity and the mass fraction. The negative correlation observed can be satisfactorily explained by a combination of crystal sinking and bubble rising.

and volatile contents of melt inclusions and zoning of quartz and other minerals.

The present textural study suggests that crystal fragmentation, as well as nucleation and growth, affected the size distribution of crystals. The abundance of tiny anhedral crystals in all clasts suggests an interval of crystal fragmentation prior to eruption.

The paucity of whole phenocrysts of quartz smaller than about 200 μm in diameter in some clasts suggests an interval of growth without nucleation prior to eruption. This suggests both slow cooling and stability against convection, since slow cooling would minimise thermal supersaturation (supercooling), and for gas-saturated magmas at low pressure, a stable density gradient would form and limit both convection and supersaturation caused by decompression in ascending parcels of magma (Wallace *et al.* 1995, 1999). Slow cooling can only occur for a body of magma that is big, and convection can only be avoided if the magma is stably stratified, as with a gassy top. Consequently, it is reasonable to consider that a lack of small phenocrysts points to a body of magma that is big, gassy and stably stratified.

A negative correlation exists between the content of large crystals and the pumice density (Fig. 17), and is consistent with sinking of large crystals and rise of bubbles within the Bishop magma, as suggested by Anderson *et al.* (2000) and Wallace *et al.* (1995).

6. Acknowledgements

This work was initiated as a class project involving five graduate students under the supervision of A. T. Anderson. Four are co-authors of this paper, and we would like to acknowledge the participation of Julia Barzyk in the early stages of the project. We also thank Ian Steele for his help during the project. Wes Hildreth, Colin Wilson and Paul Wallace provided stimulating discussions. Reviews by Stephen Blake and an anonymous reviewer are appreciated. Financial support was provided by NSF grants to A. T. Anderson.

7. References

Anderson, A. T., Davis, A. M. & Lu, F. 2000. Evolution of Bishop Tuff rhyolitic magma based on melt and magnetite inclusions and zoned phenocrysts. *Journal of Petrology* **41**, 449–73.

Bindeman, I. N. 2003. Crystal sizes in evolving silicic magma chambers. *Geology* **31**, 367–70.

Cashman, K. V. & Marsh, B. D. 1988. Crystal size distributions (CSD) in rocks and the kinetics and dynamics of crystallization. II: Makaopuhi lava lake. *Contributions to Mineralogy and Petrology* **99**, 292–305.

Christensen, J. N. & Halliday, A. N. 1996. Rb-Sr ages and Nd isotopic compositions of melt inclusions from the Bishop Tuff and the generation of silicic magma. *Earth and Planetary Science Letters* **144**, 547–61.

Eberl, D. D., Drits, V. A. & Srodon, J. 1998. Deducing growth mechanisms for minerals from the shapes of crystal size distributions. *American Journal of Science* **298**, 499–533.

Eberl, D. D., Kile, D. E. & Drits, V. A. 2002. On geological interpretations of crystal size distributions: constant vs. proportionate growth. *American Mineralogist* **87**, 1235–41.

Higgins, M. D. 2000. Origin megacrysts in granitoids by textural coarsening: a crystal size distribution (CSD) study of microcline in the Cathedral Peak granodiorite, Sierra Nevada, California. *In* Castro, A., Fernández, C. & Vigneresse, J. L. (eds) *Understanding Granites: integrating New and Classical Techniques*, 207–20. London: Geological Society.

Hildreth, E. W. 1977. *The magma chamber of the Bishop Tuff: gradients in temperature, pressure and composition*. Ph.D. thesis. University of California, Berkeley.

Hildreth, W. 1979. The Bishop Tuff: evidence for the origin of compositional zonation in silicic magma chambers. *Geological Society of America, Special Paper* **180**, 43–75.

Hildreth, W. & Mahood, G. A. 1986. Ring-fracture eruption of the Bishop Tuff. *Geological Society of America Bulletin* **97**, 396–403.

Marsh, B. D. 1988. Crystal size distributions (CSD) in rocks and the kinetics and dynamics of crystallization. I: Theory. *Contributions to Mineralogy and Petrology* **99**, 277–91.

Marsh, B. D. 1998. On the interpretation of crystal size distributions in magmatic systems. *Journal of Petrology* **39**, 553–99.

Peppard, B. T., Steele, I. M., Davis, A. M., Wallace, P. J. & Anderson, A. T. 2001. Zoned quartz phenocrysts from the rhyolitic Bishop Tuff. *American Mineralogist* **86**, 1034–52.

Rivers, M. L., Sutton, S. R. & Eng, P. 1999. Geoscience applications of X-ray computed microtomography. *Proceedings of SPIE, Developments in X-Ray Tomography II* **3772**, 78–86.

Skirius, C. M. 1990. *Pre-eruptive H_2O and CO_2 content of the plinian and ash-flow Bishop magma*. Ph.D. thesis. University of Chicago, Illinois.

Smith, R. L. & Bailey, R. A. 1966. The Bandelier Tuff: a study of ash-flow eruption cycles from zoned magma chambers. *Bulletin Volcanologique* **29**, 83–104.

Sparks, R. S. J. & Brazier, S. 1982. New evidence for degassing processes during explosive eruptions. *Nature* **295**, 218–20.

Sutton, S. R., Bertsch, P. M., Newville, M., Rivers, M., Lanzirotti, A. & Eng, P. 2002. Microfluorescence and microtomography analyses of heterogeneous earth and environmental materials. *In* Fenter, P. A., Rivers, M. L., Sturchio, N. C. & Sutton, S. R. (eds) *Applications of Synchrotron Radiation in Low-Temperature and Environmental Science. Reviews in Mineralogy and Geochemistry* **49**, 429–483. Washington, DC: Mineralogical Society of America.

Wallace, P. J., Anderson, A. T. & Davis, A. M. 1995. Quantification of pre-eruptive exsolved gas contents in silicic magmas. *Nature* **377**, 612–16.

Wallace, P. J., Anderson, A. T. & Davis, A. M. 1999. Gradients in H_2O, CO_2, and exsolved gas in a large-volume silicic magma system: interpreting the record preserved in melt inclusions from the Bishop Tuff. *Journal of Geophysical Research – Solid Earth* **104**, 20097–122.

Wallace, P. J., Dufek, J., Anderson, A. T. & Zhang, Y. X. 2003. Cooling rates of Plinian-fall and pyroclastic-flow deposits in the Bishop Tuff: inferences from water speciation in quartz-hosted glass inclusions. *Bulletin of Volcanology* **65**, 105–23.

Whitham, A. G. & Sparks, R. S. J. 1986. Pumice. *Bulletin of Volcanology* **48**, 209–23.

Wilson, C. J. N. & Hildreth, W. 1997. The Bishop Tuff: new insights from eruptive stratigraphy. *Journal of Geology* **105**, 407–39.

GUILHERME A. R. GUALDA, DAVID L. COOK, RAHUL CHOPRA, LIPING QIN and ALFRED T. ANDERSON JR, Department of the Geophysical Sciences, The University of Chicago, 5734 S. Ellis Avenue, Chicago, IL 60637, USA.

MARK RIVERS, Consortium for Advanced Radiation Sources, The University of Chicago, Building 434A, 9700 South Cass, Argonne, IL 60439, USA.

MS received 2 September 2003. Accepted for publication 9 September 2004.

AUTHOR INDEX

Anderson, A. T. Jr, 375
Arai, S., 265
Arima, M., vii

Baumgartner, L. P., 49
Behrens, H., 59
Benn, K., 297
Bennett, V. C., 199
Berdnikov, N. V., 319
Brown, M., 35
Bryant, C. J., 199

Cates, N. L., 277
Chappell, B. W., 1, 125, 199
Chen, C-H., 237
Chopra, R., 375
Clarke, D. B., 99
Collins, W. J., 309
Cook, D. L., 375
Cribb, J. W., 277

Dolejš, D., 141
Donskaya, T. V., 215

Erdmann, S., 99

Finger, F., 141
Frýda, J., 141

Gervilla, F., 265
Gladkochub, D. P., 215
Goldfarb, R. J., 339
Groves, D. I., 339
Gualda, G. A. R., 375

Harley, S. L., v, 87
Harper, B. E., 277
Hart, C. J. R., 339
Holtz, F., 59

Ishihara, S., vii, 23
Ishihara, T., 319

Janoušek, V., 141
Jiang, S-Y., 227
Jwa, Y-J., 181

Kagami, H., 161
Kamiyama, H., 249
Kanisawa, S. W., 227
Kawakami, T., 111
Komatsu, M., 161
Koteas, G. C., 277
Kovalenko, S. V., 319

Lan, C-Y., 237
Lazzareschi, D. S., 277
Lee, C-Y., 237
Lin, W., 237
Liu, M., 73

McCulloch, M. T., 199
MacDonald, M. A., 99
McInnes, B. I. A., 227
Mair, J. L., 339
Marsh, B. D., 49
Mazukabzov, A. M., 215
Metcalf, R. V., 357
Miller, C. F., 277
Morishita, T., 265

Nabelek, P. I., 73
Nakajima, T., vii, 249
Nedachi, M., 319

Olsen. S. N., 49
Osani, Y., 161
Owada, M., 161

Pin, C., 141
Poller, U., 215

Qin, L., 375
Qiu, J-S., 227

Richards, S. W., 309
Richet, P., 59
Rivers, M., 375
Roberts, M., 141
Romanovsky, N. P., 319

Sato, K., 319
Scaillet, B., 59
Shimizu, Y., 265
Shimura, T., 161
Sklyarov, E. V., 215
Stephens, W. E., v

Tani, K., 249
Thompson, P., 87
Todt, W., 215

Vigneresse, J. L., 11

Wang, D-Z., 227
Wang, R-C., 227
White, A. J. R., 125
Whittington, A., 59
Wiebe, R. A., 277
Williams, I. S., 125, 249
Wyborn, D., 125

Yurimoto, H., 265